11/23 R3-

Cornerstones

Series Editors
Charles L. Epstein, *University of Pennsylvania, Philadelphia*
Steven G. Krantz, *University of Washington, St. Louis*

Advisory Board
Anthony W. Knapp, *State University of New York at Stony Brook, Emeritus*

Anthony W. Knapp

Basic Algebra

Along with a companion volume
Advanced Algebra

Birkhäuser
Boston • Basel • Berlin

Anthony W. Knapp
81 Upper Sheep Pasture Road
East Setauket, NY 11733-1729
U.S.A.
e-mail to: aknapp@math.sunysb.edu
http://www.math.sunysb.edu/~aknapp/books/b-alg.html

Cover design by Mary Burgess.

Mathematics Subject Classicification (2000): 15-01, 20-02, 13-01, 12-01, 16-01, 08-01, 18A05, 68P30

Library of Congress Control Number: 2006932456

ISBN-10 0-8176-3248-4 eISBN-10 0-8176-4529-2
ISBN-13 978-0-8176-3248-9 eISBN-13 978-0-8176-4529-8

Advanced Algebra ISBN 0-8176-4522-5
Basic Algebra and *Advanced Algebra* (Set) ISBN 0-8176-4533-0

Printed on acid-free paper.

©2006 Anthony W. Knapp

All rights reserved. This work may not be translated or copied in whole or in part without the written permission of the publisher (Birkhäuser Boston, c/o Springer Science+Business Media LLC, 233 Spring Street, New York, NY 10013, USA) and the author, except for brief excerpts in connection with reviews or scholarly analysis. Use in connection with any form of information storage and retrieval, electronic adaptation, computer software, or by similar or dissimilar methodology now known or hereafter developed is forbidden.

The use in this publication of trade names, trademarks, service marks and similar terms, even if they are not identified as such, is not to be taken as an expression of opinion as to whether or not they are subject to proprietary rights.

9 8 7 6 5 4 3 2 1

www.birkhauser.com (EB)

To Susan

and

To My Algebra Teachers:

Ralph Fox, John Fraleigh, Robert Gunning,
John Kemeny, Bertram Kostant, Robert Langlands,
Goro Shimura, Hale Trotter, Richard Williamson

CONTENTS

Contents of Advanced Algebra	x
List of Figures	xi
Preface	xiii
Dependence Among Chapters	xvii
Standard Notation	xviii
Guide for the Reader	xix

I. PRELIMINARIES ABOUT THE INTEGERS, POLYNOMIALS, AND MATRICES — 1

1. Division and Euclidean Algorithms — 1
2. Unique Factorization of Integers — 4
3. Unique Factorization of Polynomials — 9
4. Permutations and Their Signs — 15
5. Row Reduction — 19
6. Matrix Operations — 24
7. Problems — 30

II. VECTOR SPACES OVER $\mathbb{Q}$, $\mathbb{R}$, AND $\mathbb{C}$ — 33

1. Spanning, Linear Independence, and Bases — 33
2. Vector Spaces Defined by Matrices — 38
3. Linear Maps — 42
4. Dual Spaces — 50
5. Quotients of Vector Spaces — 54
6. Direct Sums and Direct Products of Vector Spaces — 58
7. Determinants — 65
8. Eigenvectors and Characteristic Polynomials — 73
9. Bases in the Infinite-Dimensional Case — 77
10. Problems — 82

III. INNER-PRODUCT SPACES — 88

1. Inner Products and Orthonormal Sets — 88
2. Adjoints — 98
3. Spectral Theorem — 104
4. Problems — 111

IV. GROUPS AND GROUP ACTIONS — 116

1. Groups and Subgroups — 117
2. Quotient Spaces and Homomorphisms — 128
3. Direct Products and Direct Sums — 134
4. Rings and Fields — 140
5. Polynomials and Vector Spaces — 147
6. Group Actions and Examples — 158
7. Semidirect Products — 166
8. Simple Groups and Composition Series — 170
9. Structure of Finitely Generated Abelian Groups — 174
10. Sylow Theorems — 183
11. Categories and Functors — 188
12. Problems — 198

V. THEORY OF A SINGLE LINEAR TRANSFORMATION — 209

1. Introduction — 209
2. Determinants over Commutative Rings with Identity — 212
3. Characteristic and Minimal Polynomials — 216
4. Projection Operators — 224
5. Primary Decomposition — 226
6. Jordan Canonical Form — 229
7. Computations with Jordan Form — 235
8. Problems — 239

VI. MULTILINEAR ALGEBRA — 245

1. Bilinear Forms and Matrices — 246
2. Symmetric Bilinear Forms — 250
3. Alternating Bilinear Forms — 253
4. Hermitian Forms — 255
5. Groups Leaving a Bilinear Form Invariant — 257
6. Tensor Product of Two Vector Spaces — 260
7. Tensor Algebra — 274
8. Symmetric Algebra — 280
9. Exterior Algebra — 288
10. Problems — 292

VII. ADVANCED GROUP THEORY — 303

1. Free Groups — 303
2. Subgroups of Free Groups — 314
3. Free Products — 319
4. Group Representations — 326

VII. ADVANCED GROUP THEORY (Continued)

5.	Burnside's Theorem	342
6.	Extensions of Groups	344
7.	Problems	357

VIII. COMMUTATIVE RINGS AND THEIR MODULES — 367

1.	Examples of Rings and Modules	367
2.	Integral Domains and Fields of Fractions	378
3.	Prime and Maximal Ideals	381
4.	Unique Factorization	384
5.	Gauss's Lemma	390
6.	Finitely Generated Modules	396
7.	Orientation for Algebraic Number Theory and Algebraic Geometry	408
8.	Noetherian Rings and the Hilbert Basis Theorem	414
9.	Integral Closure	417
10.	Localization and Local Rings	425
11.	Dedekind Domains	434
12.	Problems	439

IX. FIELDS AND GALOIS THEORY — 448

1.	Algebraic Elements	449
2.	Construction of Field Extensions	453
3.	Finite Fields	457
4.	Algebraic Closure	460
5.	Geometric Constructions by Straightedge and Compass	464
6.	Separable Extensions	469
7.	Normal Extensions	476
8.	Fundamental Theorem of Galois Theory	479
9.	Application to Constructibility of Regular Polygons	483
10.	Application to Proving the Fundamental Theorem of Algebra	486
11.	Application to Unsolvability of Polynomial Equations with Nonsolvable Galois Group	488
12.	Construction of Regular Polygons	493
13.	Solution of Certain Polynomial Equations with Solvable Galois Group	501
14.	Proof That π Is Transcendental	510
15.	Norm and Trace	514
16.	Splitting of Prime Ideals in Extensions	521
17.	Two Tools for Computing Galois Groups	527
18.	Problems	534

X. MODULES OVER NONCOMMUTATIVE RINGS — 544
1. Simple and Semisimple Modules — 544
2. Composition Series — 551
3. Chain Conditions — 556
4. Hom and End for Modules — 558
5. Tensor Product for Modules — 565
6. Exact Sequences — 574
7. Problems — 579

APPENDIX — 583
A1. Sets and Functions — 583
A2. Equivalence Relations — 589
A3. Real Numbers — 591
A4. Complex Numbers — 594
A5. Partial Orderings and Zorn's Lemma — 595
A6. Cardinality — 599

Hints for Solutions of Problems — 603
Selected References — 697
Index of Notation — 699
Index — 703

CONTENTS OF *ADVANCED ALGEBRA*

I. Transition to Modern Number Theory
II. Wedderburn–Artin Ring Theory
III. Brauer Group
IV. Homological Algebra
V. Three Theorems in Algebraic Number Theory
VI. Reinterpretation with Adeles and Ideles
VII. Infinite Field Extensions
VIII. Background for Algebraic Geometry
IX. The Number Theory of Algebraic Curves
X. Methods of Algebraic Geometry

LIST OF FIGURES

2.1.	The vector space of lines $v + U$ in $\mathbb{R}^2$ parallel to a given line U through the origin	55
2.2.	Factorization of linear maps via a quotient of vector spaces	56
2.3.	Three 1-dimensional vector subspaces of $\mathbb{R}^2$ such that each pair has intersection 0	62
2.4.	Universal mapping property of a direct product of vector spaces	64
2.5.	Universal mapping property of a direct sum of vector spaces	65
3.1.	Geometric interpretation of the parallelogram law	91
3.2.	Resolution of a vector into a parallel component and an orthogonal component	93
4.1.	Factorization of homomorphisms of groups via the quotient of a group by a normal subgroup	132
4.2.	Universal mapping property of an external direct product of groups	136
4.3.	Universal mapping property of a direct product of groups	136
4.4.	Universal mapping property of an external direct sum of abelian groups	138
4.5.	Universal mapping property of a direct sum of abelian groups	139
4.6.	Factorization of homomorphisms of rings via the quotient of a ring by an ideal	146
4.7.	Substitution homomorphism for polynomials in one indeterminate	150
4.8.	Substitution homomorphism for polynomials in n indeterminates	156
4.9.	A square diagram	193
4.10.	Diagrams obtained by applying a covariant functor and a contravariant functor	193
4.11.	Universal mapping property of a product in a category	195
4.12.	Universal mapping property of a coproduct in a category	197
5.1.	Example of a nilpotent matrix in Jordan form	231
5.2.	Powers of the nilpotent matrix in Figure 5.1	232
6.1.	Universal mapping property of a tensor product	261
6.2.	Diagrams for uniqueness of a tensor product	261

6.3.	Commutative diagram of a natural transformation $\{T_X\}$	265
6.4.	Commutative diagram of a triple tensor product	274
6.5.	University mapping property of a tensor algebra	279
7.1.	Universal mapping property of a free group	305
7.2.	Universal mapping property of a free product	320
7.3.	An intertwining operator for two representations	330
7.4.	Equivalent group extensions	349
8.1.	Universal mapping property of the integral group ring of G	371
8.2.	Universal mapping property of a free left R module	374
8.3.	Factorization of R homomorphisms via a quotient of R modules	376
8.4.	Universal mapping property of the group algebra RG	378
8.5.	Universal mapping property of the field of fractions of R	380
8.6.	Real points of the curve $y^2 = (x-1)x(x+1)$	409
8.7.	Universal mapping property of the localization of R at S	428
9.1.	Closure of positive constructible x coordinates under multiplication and division	465
9.2.	Closure of positive constructible x coordinates under square roots	466
9.3.	Construction of a regular pentagon	496
9.4.	Construction of a regular 17-gon	500
10.1.	Universal mapping property of a tensor product of a right R module and a left R module	566

PREFACE

Basic Algebra and its companion volume *Advanced Algebra* systematically develop concepts and tools in algebra that are vital to every mathematician, whether pure or applied, aspiring or established. These two books together aim to give the reader a global view of algebra, its use, and its role in mathematics as a whole. The idea is to explain what the young mathematician needs to know about algebra in order to communicate well with colleagues in all branches of mathematics.

The books are written as textbooks, and their primary audience is students who are learning the material for the first time and who are planning a career in which they will use advanced mathematics professionally. Much of the material in the books, particularly in *Basic Algebra* but also in some of the chapters of *Advanced Algebra*, corresponds to normal course work. The books include further topics that may be skipped in required courses but that the professional mathematician will ultimately want to learn by self-study. The test of each topic for inclusion is whether it is something that a plenary lecturer at a broad international or national meeting is likely to take as known by the audience.

The key topics and features of *Basic Algebra* are as follows:

- Linear algebra and group theory build on each other throughout the book. A small amount of linear algebra is introduced first, as the topic likely to be better known by the reader ahead of time, and then a little group theory is introduced, with linear algebra providing important examples.
- Chapters on linear algebra develop notions related to vector spaces, the theory of linear transformations, bilinear forms, classical linear groups, and multilinear algebra.
- Chapters on modern algebra treat groups, rings, fields, modules, and Galois groups, including many uses of Galois groups and methods of computation.
- Three prominent themes recur throughout and blend together at times: the analogy between integers and polynomials in one variable over a field, the interplay between linear algebra and group theory, and the relationship between number theory and geometry.
- The development proceeds from the particular to the general, often introducing examples well before a theory that incorporates them.
- More than 400 problems at the ends of chapters illuminate aspects of the text, develop related topics, and point to additional applications. A separate

90-page section "Hints for Solutions of Problems" at the end of the book gives detailed hints for most of the problems, complete solutions for many.
- Applications such as the fast Fourier transform, the theory of linear error-correcting codes, the use of Jordan canonical form in solving linear systems of ordinary differential equations, and constructions of interest in mathematical physics arise naturally in sequences of problems at the ends of chapters and illustrate the power of the theory for use in science and engineering.

Basic Algebra endeavors to show some of the interconnections between different areas of mathematics, beyond those listed above. Here are examples: Systems of orthogonal functions make an appearance with inner-product spaces. Covering spaces naturally play a role in the examination of subgroups of free groups. Cohomology of groups arises from considering group extensions. Use of the power-series expansion of the exponential function combines with algebraic numbers to prove that π is transcendental. Harmonic analysis on a cyclic group explains the mysterious method of Lagrange resolvents in the theory of Galois groups.

Algebra plays a singular role in mathematics by having been developed so extensively at such an early date. Indeed, the major discoveries of algebra even from the days of Hilbert are well beyond the knowledge of most nonalgebraists today. Correspondingly most of the subject matter of the present book is at least 100 years old. What has changed over the intervening years concerning algebra books at this level is not so much the mathematics as the point of view toward the subject matter and the relative emphasis on and generality of various topics. For example, in the 1920s Emmy Noether introduced vector spaces and linear mappings to reinterpret coordinate spaces and matrices, and she defined the ingredients of what was then called "modern algebra"—the axiomatically defined rings, fields, and modules, and their homomorphisms. The introduction of categories and functors in the 1940s shifted the emphasis even more toward the homomorphisms and away from the objects themselves. The creation of homological algebra in the 1950s gave a unity to algebraic topics cutting across many fields of mathematics. Category theory underwent a period of great expansion in the 1950s and 1960s, followed by a contraction and a return more to a supporting role. The emphasis in topics shifted. Linear algebra had earlier been viewed as a separate subject, with many applications, while group theory and the other topics had been viewed as having few applications. Coding theory, cryptography, and advances in physics and chemistry have changed all that, and now linear algebra and group theory together permeate mathematics and its applications. The other subjects build on them, and they too have extensive applications in science and engineering, as well as in the rest of mathematics.

Basic Algebra presents its subject matter in a forward-looking way that takes this evolution into account. It is suitable as a text in a two-semester advanced undergraduate or first-year graduate sequence in algebra. Depending on the graduate school, it may be appropriate to include also some material from *Advanced Algebra*. Briefly the topics in *Basic Algebra* are linear algebra and group theory, rings, fields, and modules. A full list of the topics in *Advanced Algebra* appears on page x; of these, the Wedderburn theory of semisimple algebras, homological algebra, and foundational material for algebraic geometry are the ones that most commonly appear in syllabi of first-year graduate courses.

A chart on page xvii tells the dependence among chapters and can help with preparing a syllabus. Chapters I–VII treat linear algebra and group theory at various levels, except that three sections of Chapter IV and one of Chapter V introduce rings and fields, polynomials, categories and functors, and determinants over commutative rings with identity. Chapter VIII concerns rings, with emphasis on unique factorization; Chapter IX concerns field extensions and Galois theory, with emphasis on applications of Galois theory; and Chapter X concerns modules and constructions with modules.

For a graduate-level sequence the syllabus is likely to include all of Chapters I–V and parts of Chapters VIII and IX, at a minimum. Depending on the knowledge of the students ahead of time, it may be possible to skim much of the first three chapters and some of the beginning of the fourth; then time may allow for some of Chapters VI and VII, or additional material from Chapters VIII and IX, or some of the topics in *Advanced Algebra*. For many of the topics in *Advanced Algebra*, parts of Chapter X of *Basic Algebra* are prerequisite.

For an advanced undergraduate sequence the first semester can include Chapters I through III except Section II.9, plus the first six sections of Chapter IV and as much as reasonable from Chapter V; the notion of category does not appear in this material. The second semester will involve categories very gently; the course will perhaps treat the remainder of Chapter IV, the first five or six sections of Chapter VIII, and at least Sections 1–3 and 5 of Chapter IX.

More detailed information about how the book can be used with courses can be deduced by using the chart on page xvii in conjunction with the section "Guide for the Reader" on pages xix–xxii. In my own graduate teaching, I have built one course around Chapters I–III, Sections 1–6 of Chapter IV, all of Chapter V, and about half of Chapter VI. A second course dealt with the remainder of Chapter IV, a little of Chapter VII, Sections 1–6 of Chapter VIII, and Sections 1–11 of Chapter IX.

The problems at the ends of chapters are intended to play a more important role than is normal for problems in a mathematics book. Almost all problems are solved in the section of hints at the end of the book. This being so, some blocks of problems form additional topics that could have been included in the text but were not; these blocks may either be regarded as optional topics, or they may be treated as challenges for the reader. The optional topics of this kind usually either carry out further development of the theory or introduce significant applications. For example one block of problems at the end of Chapter VII carries the theory of representations of finite groups a little further by developing the Poisson summation formula and the fast Fourier transform. For a second example blocks of problems at the ends of Chapters IV, VII, and IX introduce linear error-correcting codes as an application of the theory in those chapters.

Not all problems are of this kind, of course. Some of the problems are really pure or applied theorems, some are examples showing the degree to which hypotheses can be stretched, and a few are just exercises. The reader gets no indication which problems are of which type, nor of which ones are relatively easy. Each problem can be solved with tools developed up to that point in the book, plus any additional prerequisites that are noted.

Beyond a standard one-variable calculus course, the most important prerequisite for using *Basic Algebra* is that the reader already know what a proof is, how to read a proof, and how to write a proof. This knowledge typically is obtained from honors calculus courses, or from a course in linear algebra, or from a first junior–senior course in real variables. In addition, it is assumed that the reader is comfortable with a small amount of linear algebra, including matrix computations, row reduction of matrices, solutions of systems of linear equations, and the associated geometry. Some prior exposure to groups is helpful but not really necessary.

The theorems, propositions, lemmas, and corollaries within each chapter are indexed by a single number stream. Figures have their own number stream, and one can find the page reference for each figure from the table on pages xi–xii. Labels on displayed lines occur only within proofs and examples, and they are local to the particular proof or example in progress. Some readers like to skim or skip proofs on first reading; to facilitate this procedure, each occurrence of the word "PROOF" or "PROOF" is matched by an occurrence at the right margin of the symbol $\square$ to mark the end of that proof.

I am grateful to Ann Kostant and Steven Krantz for encouraging this project and for making many suggestions about pursuing it. I am especially indebted to an anonymous referee, who made detailed comments about many aspects of a preliminary version of the book, and to David Kramer, who did the copyediting. The typesetting was by $A_\mathcal{M}S$-TeX, and the figures were drawn with Mathematica.

I invite corrections and other comments from readers. I plan to maintain a list of known corrections on my own Web page.

A. W. KNAPP
August 2006

DEPENDENCE AMONG CHAPTERS

Below is a chart of the main lines of dependence of chapters on prior chapters. The dashed lines indicate helpful motivation but no logical dependence. Apart from that, particular examples may make use of information from earlier chapters that is not indicated by the chart.

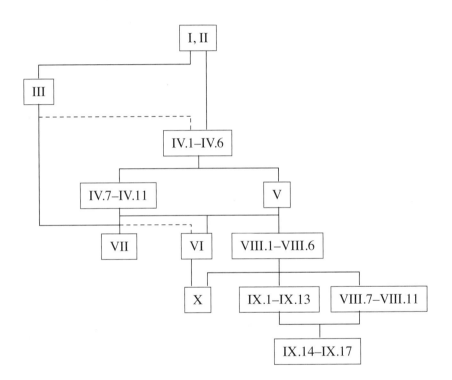

STANDARD NOTATION

See the Index of Notation, pp. 699–701, for symbols defined starting on page 1.

Item	Meaning
#S or $|S|$	number of elements in S
$\varnothing$	empty set
$\{x \in E \mid P\}$	the set of x in E such that P holds
E^c	complement of the set E
$E \cup F$, $E \cap F$, $E - F$	union, intersection, difference of sets
$\bigcup_\alpha E_\alpha$, $\bigcap_\alpha E_\alpha$	union, intersection of the sets E_α
$E \subseteq F$, $E \supseteq F$	E is contained in F, E contains F
$E \subsetneq F$, $E \supsetneq F$	E properly contained in F, properly contains F
$E \times F$, $\bigtimes_{s \in S} X_s$	products of sets
$(a_1, \ldots, a_n)$, $\{a_1, \ldots, a_n\}$	ordered n-tuple, unordered n-tuple
$f : E \to F$, $x \mapsto f(x)$	function, effect of function
$f \circ g$ or fg, $f\big|_E$	composition of g followed by f, restriction to E
$f(\,\cdot\,, y)$	the function $x \mapsto f(x, y)$
$f(E)$, $f^{-1}(E)$	direct and inverse image of a set
δ_{ij}	Kronecker delta: 1 if $i = j$, 0 if $i \neq j$
$\binom{n}{k}$	binomial coefficient
n positive, n negative	$n > 0$, $n < 0$
$\mathbb{Z}, \mathbb{Q}, \mathbb{R}, \mathbb{C}$	integers, rationals, reals, complex numbers
max (and similarly min)	maximum of a finite subset of a totally ordered set
$\sum$ or $\prod$	sum or product, possibly with a limit operation
countable	finite or in one-one correspondence with $\mathbb{Z}$
$[x]$	greatest integer $\leq x$ if x is real
Re z, Im z	real and imaginary parts of complex z
$\bar{z}$	complex conjugate of z
$|z|$	absolute value of z
1	multiplicative identity
1 or I	identity matrix or operator
1_X	identity function on X
$\mathbb{Q}^n, \mathbb{R}^n, \mathbb{C}^n$	spaces of column vectors
$\mathrm{diag}(a_1, \ldots, a_n)$	diagonal matrix
$\cong$	is isomorphic to, is equivalent to

GUIDE FOR THE READER

This section is intended to help the reader find out what parts of each chapter are most important and how the chapters are interrelated. Further information of this kind is contained in the abstracts that begin each of the chapters.

The book pays attention to at least three recurring themes in algebra, allowing a person to see how these themes arise in increasingly sophisticated ways. These are the analogy between integers and polynomials in one indeterminate over a field, the interplay between linear algebra and group theory, and the relationship between number theory and geometry. Keeping track of how these themes evolve will help the reader understand the mathematics better and anticipate where it is headed.

In Chapter I the analogy between integers and polynomials in one indeterminate over the rationals, reals, or complex numbers appears already in the first three sections. The main results of these sections are theorems about unique factorization in each of the two settings. The relevant parts of the underlying structures for the two settings are the same, and unique factorization can therefore be proved in both settings by the same argument. Many readers will already know this unique factorization, but it is worth examining the parallel structure and proof at least quickly before turning to the chapters that follow.

Before proceeding very far into the book, it is worth looking also at the appendix to see whether all its topics are familiar. Readers will find Section A1 useful at least for its summary of set-theoretic notation and for its emphasis on the distinction between range and image for a function. This distinction is usually unimportant in analysis but becomes increasingly important as one studies more advanced topics in algebra. Readers who have not specifically learned about equivalence relations and partial orderings can learn about them from Sections A2 and A5. Sections A3 and A4 concern the real and complex numbers; the emphasis is on notation and the Intermediate Value Theorem, which plays a role in proving the Fundamental Theorem of Algebra. Zorn's Lemma and cardinality in Sections A5 and A6 are usually unnecessary in an undergraduate course. They arise most importantly in Sections II.9 and IX.4, which are normally omitted in an undergraduate course, and in Proposition 8.8, which is invoked only in the last few sections of Chapter VIII.

The remainder of this section is an overview of individual chapters and pairs of chapters.

Chapter I is in three parts. The first part, as mentioned above, establishes unique factorization for the integers and for polynomials in one indeterminate over the rationals, reals, or complex numbers. The second part defines permutations and shows that they have signs such that the sign of any composition is the product of the signs; this result is essential for defining general determinants in Section II.7. The third part will likely be a review for all readers. It establishes notation for row reduction of matrices and for operations on matrices, and it uses row reduction to show that a one-sided inverse for a square matrix is a two-sided inverse.

Chapters II–III treat the fundamentals of linear algebra. Whereas the matrix computations in Chapter I were concrete, Chapters II–III are relatively abstract. Much of this material is likely to be a review for graduate students. The geometric interpretation of vectors spaces, subspaces, and linear mappings is not included in the chapter, being taken as known previously. The fundamental idea that a newly constructed object might be characterized by a "universal mapping property" appears for the first time in Chapter II, and it appears more and more frequently throughout the book. One aspect of this idea is that it is sometimes not so important what certain constructed objects are, but what they do. A related idea being emphasized is that the mappings associated with a newly constructed object are likely to be as important as the object, if not more so; at the least, one needs to stop and find what those mappings are. Section II.9 uses Zorn's Lemma and can be deferred until Chapter IX if one wants. Chapter III discusses special features of real and complex vector spaces endowed with inner products. The main result is the Spectral Theorem in Section 3. Many of the problems at the end of the chapter make contact with real analysis. The subject of linear algebra continues in Chapter V.

Chapter IV is the primary chapter on group theory and may be viewed as in three parts. Sections 1–6 form the first part, which is essential for all later chapters in the book. Sections 1–3 introduce groups and some associated constructions, along with a number of examples. Many of the examples will be seen to be related to specific or general vector spaces, and thus the theme of the interplay between group theory and linear algebra is appearing concretely for the first time. In practice, many examples of groups arise in the context of group actions, and abstract group actions are defined in Section 6. Of particular interest are group representations, which are group actions on a vector space by linear mappings. Sections 4–5 are a digression to define rings, fields, and ring homomorphisms, and to extend the theories concerning polynomials and vector spaces as presented in Chapters I–II. The immediate purpose of the digression is to make prime fields, their associated multiplicative groups, and the notion of characteristic available for the remainder of the chapter. The definition of vector space is extended to allow scalars from any field. The definition of polynomial is extended to allow coefficients from any commutative ring with identity, rather than just the rationals or reals or complex numbers, and to allow more than one indeterminate. Universal mapping properties for polynomial rings are proved. Sections 7–10 form the second part of the chapter and are a continuation of group theory. The main result is the Fundamental Theorem of Finitely Generated Abelian Groups, which is in Section 9. Section 11 forms the third part of the chapter. This section is a gentle introduction to categories and functors, which are useful for working with parallel structures in different settings within algebra. As S. Mac Lane says in his book, "Category theory asks of every type of Mathematical object: 'What are the morphisms?'; it suggests that these morphisms should be described at the same time as the objects.... This emphasis on (homo)morphisms is largely due to Emmy Noether, who emphasized the use of homomorphisms of groups and rings." The simplest parallel structure reflected in categories is that of an isomorphism. The section also discusses general notions of product and coproduct functors. Examples of products are direct products in linear algebra and in group theory. Examples of coproducts are direct sums in linear algebra and in *abelian* group theory, as well as disjoint unions in set theory. The theory in this section helps in unifying the mathematics that is to come in Chapters VI–VIII and X. The subject of group theory in continued in Chapter VII, which assumes knowledge of the material on category theory.

Chapters V and VI continue the development of linear algebra. Chapter VI uses categories, but Chapter V does not. Most of Chapter V concerns the analysis of a linear transformation carrying a finite-dimensional vector space over a field into itself. The questions are to find invariants of such transformations and to classify the transformations up to similarity. Section 2 at the start extends the theory of determinants so that the matrices are allowed to have entries in a commutative ring with identity; this extension is necessary in order to be able to work easily with characteristic polynomials. The extension of this theory is carried out by an important principle known as the "permanence of identities." Chapter VI largely concerns bilinear forms and tensor products, again in the context that the coefficients are from a field. This material is necessary in many applications to geometry and physics, but it is not needed in Chapters VII–IX. Many objects in the chapter are constructed in such a way that they are uniquely determined by a universal mapping property. Problems 18–22 at the end of the chapter discuss universal mapping properties in the general context of category theory, and they show that a uniqueness theorem is automatic in all cases.

Chapter VII continues the development of group theory, making use of category theory. It is in two parts. Sections 1–3 concern free groups and the topic of generators and relations; they are essential for abstract descriptions of groups and for work in topology involving fundamental groups. Section 3 constructs a notion of free product and shows that it is the coproduct functor for the category of groups. Sections 4–6 continue the theme of the interplay of group theory and linear algebra. Section 4 analyzes group representations of a finite group when the underlying field is the complex numbers, and Section 5 applies this theory to obtain a conclusion about the structure of finite groups. Section 6 studies extensions of groups and uses them to motivate the subject of cohomology of groups.

Chapter VIII introduces modules, giving many examples in Section 1, and then goes on to discuss questions of unique factorization in integral domains. Section 6 obtains a generalization for principal ideal domains of the Fundamental Theorem of Finitely Generated Abelian Groups, once again illustrating the first theme—similarities between the integers and certain polynomial rings. Section 7 introduces the third theme, the relationship between number theory and geometry, as a more sophisticated version of the first theme. The section compares a certain polynomial ring in two variables with a certain ring of algebraic integers that extends the ordinary integers. Unique factorization of elements fails for both, but the geometric setting has a more geometrically meaningful factorization in terms of ideals that is evidently unique. This kind of unique factorization turns out to work for the ring of algebraic integers as well. Sections 8–11 expand the examples in Section 7 into a theory of unique factorization of ideals in any integrally closed Noetherian domain whose nonzero prime ideals are all maximal.

Chapter IX analyzes algebraic extensions of fields. The first 13 sections make use only of Sections 1–6 in Chapter VIII. Sections 1–5 of Chapter IX give the foundational theory, which is sufficient to exhibit all the finite fields and to prove that certain classically proposed constructions in Euclidean geometry are impossible. Sections 6–8 introduce Galois theory, but Theorem 9.28 and its three corollaries may be skipped if Sections 14–17 are to be omitted. Sections 9–11 give a first round of applications of Galois theory: Gauss's theorem about which regular n-gons are in principle constructible with straightedge and compass, the Fundamental Theorem of Algebra, and the Abel–Galois theorem that solvability of a polynomial equation with rational coefficients in terms of radicals implies solvability of the Galois group. Sections 12–13 give a second round of applications: Gauss's method in principle for actually constructing the constructible regular n-gons and a converse to the Abel–Galois theorem. Sections 14–17 make use of Sections 7–11 of Chapter VIII, proving that π is transcendental and obtaining two methods for computing Galois groups.

Chapter X is a relatively short chapter developing further tools for dealing with modules over a ring with identity. The main construction is that of the tensor product over a ring of a unital right module and a unital left module, the result being an abelian group. The chapter makes use of material from Chapters VI and VIII, but not from Chapter IX.

Basic Algebra

CHAPTER I

Preliminaries about the Integers, Polynomials, and Matrices

Abstract. This chapter is mostly a review, discussing unique factorization of positive integers, unique factorization of polynomials whose coefficients are rational or real or complex, signs of permutations, and matrix algebra.

Sections 1–2 concern unique factorization of positive integers. Section 1 proves the division and Euclidean algorithms, used to compute greatest common divisors. Section 2 establishes unique factorization as a consequence and gives several number-theoretic consequences, including the Chinese Remainder Theorem and the evaluation of the Euler φ function.

Section 3 develops unique factorization of rational and real and complex polynomials in one indeterminate completely analogously, and it derives the complete factorization of complex polynomials from the Fundamental Theorem of Algebra. The proof of the fundamental theorem is postponed to Chapter IX.

Section 4 discusses permutations of a finite set, establishing the decomposition of each permutation as a disjoint product of cycles. The sign of a permutation is introduced, and it is proved that the sign of a product is the product of the signs.

Sections 5–6 concern matrix algebra. Section 5 reviews row reduction and its role in the solution of simultaneous linear equations. Section 6 defines the arithmetic operations of addition, scalar multiplication, and multiplication of matrices. The process of matrix inversion is related to the method of row reduction, and it is shown that a square matrix with a one-sided inverse automatically has a two-sided inverse that is computable via row reduction.

1. Division and Euclidean Algorithms

The first three sections give a careful proof of unique factorization for integers and for polynomials with rational or real or complex coefficients, and they give an indication of some first consequences of this factorization. For the moment let us restrict attention to the set $\mathbb{Z}$ of integers. We take addition, subtraction, and multiplication within $\mathbb{Z}$ as established, as well as the properties of the usual ordering in $\mathbb{Z}$.

A **factor** of an integer n is a nonzero integer k such that $n = kl$ for some integer l. In this case we say also that k **divides** n, that k is a **divisor** of n, and that n is a **multiple** of k. We write $k \mid n$ for this relationship. If n is nonzero, any product formula $n = kl_1 \cdots l_r$ is a **factorization** of n. A **unit** in $\mathbb{Z}$ is a divisor of 1, hence is either $+1$ or -1. The factorization $n = kl$ of $n \neq 0$ is called **nontrivial** if neither k nor l is a unit. An integer $p > 1$ is said to be **prime** if it has no nontrivial factorization $p = kl$.

The statement of unique factorization for positive integers, which will be given precisely in Section 2, says roughly that each positive integer is the product of primes and that this decomposition is unique apart from the order of the factors.[1] Existence will follow by an easy induction. The difficulty is in the uniqueness. We shall prove uniqueness by a sequence of steps based on the "Euclidean algorithm," which we discuss in a moment. In turn, the Euclidean algorithm relies on the following.

Proposition 1.1 (division algorithm). If a and b are integers with $b \neq 0$, then there exist unique integers q and r such that $a = bq + r$ and $0 \leq r < |b|$.

PROOF. Possibly replacing q by $-q$, we may assume that $b > 0$. The integers n with $bn \leq a$ are bounded above by $|a|$, and there exists such an n, namely $n = -|a|$. Therefore there is a largest such integer, say $n = q$. Set $r = a - bq$. Then $0 \leq r$ and $a = bq + r$. If $r \geq b$, then $r - b \geq 0$ says that $a = b(q + 1) + (r - b) \geq b(q + 1)$. The inequality $q + 1 > q$ contradicts the maximality of q, and we conclude that $r < b$. This proves existence.

For uniqueness when $b > 0$, suppose $a = bq_1 + r_1 = bq_2 + r_2$. Subtracting, we obtain $b(q_1 - q_2) = r_2 - r_1$ with $|r_2 - r_1| < b$, and this is a contradiction unless $r_2 - r_1 = 0$. $\square$

Let a and b be integers not both 0. The **greatest common divisor** of a and b is the largest integer $d > 0$ such that $d \mid a$ and $d \mid b$. Let us see existence. The integer 1 divides a and b. If b, for example, is nonzero, then any such d has $|d| \leq |b|$, and hence the greatest common divisor indeed exists. We write $d = \mathrm{GCD}(a, b)$.

Let us suppose that $b \neq 0$. The **Euclidean algorithm** consists of iterated application of the division algorithm (Proposition 1.1) to a and b until the remainder term r disappears:

$$a = bq_1 + r_1, \quad 0 \leq r_1 < b,$$
$$b = r_1 q_2 + r_2, \quad 0 \leq r_2 < r_1,$$
$$r_1 = r_2 q_3 + r_3, \quad 0 \leq r_3 < r_2,$$
$$\vdots$$
$$r_{n-2} = r_{n-1} q_n + r_n, \quad 0 \leq r_n < r_{n-1} \text{ (with } r_n \neq 0, \text{ say)},$$
$$r_{n-1} = r_n q_{n+1}.$$

---

[1] It is to be understood that the prime factorization of 1 is as the empty product.

The process must stop with some remainder term r_{n+1} equal to 0 in this way since $b > r_1 > r_2 > \cdots \geq 0$. The last nonzero remainder term, namely r_n above, will be of interest to us.

EXAMPLE. For $a = 13$ and $b = 5$, the steps read

$$13 = 5 \cdot 2 + 3,$$
$$5 = 3 \cdot 1 + 2,$$
$$3 = 2 \cdot 1 + \boxed{1},$$
$$2 = 1 \cdot 2.$$

The last nonzero remainder term is written with a box around it.

Proposition 1.2. Let a and b be integers with $b \neq 0$, and let $d = \text{GCD}(a, b)$. Then

(a) the number r_n in the Euclidean algorithm is exactly d,
(b) any divisor d' of both a and b necessarily divides d,
(c) there exist integers x and y such that $ax + by = d$.

EXAMPLE, CONTINUED. We rewrite the steps of the Euclidean algorithm, as applied in the above example with $a = 13$ and $b = 5$, so as to yield successive substitutions:

$13 = 5 \cdot 2 + 3,$ $3 = 13 - 5 \cdot 2,$
$5 = 3 \cdot 1 + 2,$ $2 = 5 - 3 \cdot 1 = 5 - (13 - 5 \cdot 2) \cdot 1 = 5 \cdot 3 - 13 \cdot 1,$
$3 = 2 \cdot 1 + \boxed{1},$ $1 = 3 - 2 \cdot 1 = (13 - 5 \cdot 2) - (5 \cdot 3 - 13 \cdot 1) \cdot 1$
$\phantom{3 = 2 \cdot 1 + \boxed{1},\ \ \ 1} = 13 \cdot 2 - 5 \cdot 5.$

Thus we see that $1 = 13x + 5y$ with $x = 2$ and $y = -5$. This shows for the example that the number r_n works in place of d in Proposition 1.2c, and the rest of the proof of the proposition for this example is quite easy. Let us now adjust this computation to obtain a complete proof of the proposition in general.

PROOF OF PROPOSITION 1.2. Put $r_0 = b$ and $r_{-1} = a$, so that

$$r_{k-2} = r_{k-1} q_k + r_k \qquad \text{for } 1 \leq k \leq n. \qquad (*)$$

The argument proceeds in three steps.

Step 1. We show that r_n is a divisor of both a and b. In fact, from $r_{n-1} = r_n q_{n+1}$, we have $r_n \mid r_{n-1}$. Let $k \leq n$, and assume inductively that r_n divides

$r_{k-1}, \ldots, r_{n-1}, r_n$. Then (∗) shows that r_n divides r_{k-2}. Induction allows us to conclude that r_n divides $r_{-1}, r_0, \ldots, r_{n-1}$. In particular, r_n divides a and b.

Step 2. We prove that $ax + by = r_n$ for suitable integers x and y. In fact, we show by induction on k for $k \leq n$ that there exist integers x and y with $ax + by = r_k$. For $k = -1$ and $k = 0$, this conclusion is trivial. If $k \geq 1$ is given and if the result is known for $k - 2$ and $k - 1$, then we have

$$ax_2 + by_2 = r_{k-2},$$
$$ax_1 + by_1 = r_{k-1} \qquad (**)$$

for suitable integers x_2, y_2, x_1, y_1. We multiply the second of the equalities of (∗∗) by q_k, subtract, and substitute into (∗). The result is

$$r_k = r_{k-2} - r_{k-1}q_k = a(x_2 - q_k x_1) + b(y_2 - q_k y_1),$$

and the induction is complete. Thus $ax + by = r_n$ for suitable x and y.

Step 3. Finally we deduce (a), (b), and (c). Step 1 shows that r_n divides a and b. If $d' > 0$ divides both a and b, the result of Step 2 shows that $d' \mid r_n$. Thus $d' \leq r_n$, and r_n is the greatest common divisor. This is the conclusion of (a); (b) follows from (a) since $d' \mid r_n$, and (c) follows from (a) and Step 2. □

Corollary 1.3. Within $\mathbb{Z}$, if c is a nonzero integer that divides a product mn and if $\text{GCD}(c, m) = 1$, then c divides n.

PROOF. Proposition 1.2c produces integers x and y with $cx + my = 1$. Multiplying by n, we obtain $cnx + mny = n$. Since c divides mn and divides itself, c divides both terms on the left side. Therefore it divides the right side, which is n. □

Corollary 1.4. Within $\mathbb{Z}$, if a and b are nonzero integers with $\text{GCD}(a, b) = 1$ and if both of them divide the integer m, then ab divides m.

PROOF. Proposition 1.2c produces integers x and y with $ax + by = 1$. Multiplying by m, we obtain $amx + bmy = m$, which we rewrite in integers as $ab(m/b)x + ab(m/a)y = m$. Since ab divides each term on the left side, it divides the right side, which is m. □

2. Unique Factorization of Integers

We come now to the theorem asserting unique factorization for the integers. The precise statement is as follows.

2. Unique Factorization of Integers

Theorem 1.5 (Fundamental Theorem of Arithmetic). Each positive integer n can be written as a product of primes, $n = p_1 p_2 \cdots p_r$, with the integer 1 being written as an empty product. This factorization is unique in the following sense: if $n = q_1 q_2 \cdots q_s$ is another such factorization, then $r = s$ and, after some reordering of the factors, $q_j = p_j$ for $1 \leq j \leq r$.

The main step is the following lemma, which relies on Corollary 1.3.

Lemma 1.6. Within $\mathbb{Z}$, if p is a prime and p divides a product ab, then p divides a or p divides b.

PROOF. Suppose that p does not divide a. Since p is prime, $\mathrm{GCD}(a, p) = 1$. Taking $m = a, n = b$, and $c = p$ in Corollary 1.3, we see that p divides b. □

PROOF OF EXISTENCE IN THEOREM 1.5. We induct on n, the case $n = 1$ being handled by an empty product expansion. If the result holds for $k = 1$ through $k = n - 1$, there are two cases: n is prime and n is not prime. If n is prime, then $n = n$ is the desired factorization. Otherwise we can write $n = ab$ nontrivially with $a > 1$ and $b > 1$. Then $a \leq n - 1$ and $b \leq n - 1$, so that a and b have factorizations into primes by the inductive hypothesis. Putting them together yields a factorization into primes for $n = ab$. □

PROOF OF UNIQUENESS IN THEOREM 1.5. Suppose that $n = p_1 p_2 \cdots p_r = q_1 q_2 \cdots q_s$ with all factors prime and with $r \leq s$. We prove the uniqueness by induction on r, the case $r = 0$ being trivial and the case $r = 1$ following from the definition of "prime." Inductively from Lemma 1.6 we have $p_r \mid q_k$ for some k. Since q_k is prime, $p_r = q_k$. Thus we can cancel and obtain $p_1 p_2 \cdots p_{r-1} = q_1 q_2 \cdots \widehat{q_k} \cdots q_s$, the hat indicating an omitted factor. By induction the factors on the two sides here are the same except for order. Thus the same conclusion is valid when comparing the two sides of the equality $p_1 p_2 \cdots p_r = q_1 q_2 \cdots q_s$. The induction is complete, and the desired uniqueness follows. □

In the product expansion of Theorem 1.5, it is customary to group factors that are equal, thus writing the positive integer n as $n = p_1^{k_1} \cdots p_r^{k_r}$ with the primes p_j distinct and with the integers k_j all ≥ 0. This kind of decomposition is unique up to order if all factors $p_j^{k_j}$ with $k_j = 0$ are dropped, and we call it a **prime factorization** of n.

Corollary 1.7. If $n = p_1^{k_1} \cdots p_r^{k_r}$ is a prime factorization of a positive integer n, then the positive divisors d of n are exactly all products $d = p_1^{l_1} \cdots p_r^{l_r}$ with $0 \leq l_j \leq k_j$ for all j.

REMARK. A general divisor of n within $\mathbb{Z}$ is the product of a unit ± 1 and a positive divisor.

PROOF. Certainly any such product divides n. Conversely if d divides n, write $n = dx$ for some positive integer x. Apply Theorem 1.5 to d and to x, form the resulting prime factorizations, and multiply them together. Then we see from the uniqueness for the prime factorization of n that the only primes that can occur in the expansions of d and x are $p_1, \ldots, p_r$ and that the sum of the exponents of p_j in the expansions of d and x is k_j. The result follows. □

If we want to compare prime factorizations for two positive integers, we can insert 0th powers of primes as necessary and thereby assume that the same primes appear in both expansions. Using this device, we obtain a formula for greatest common divisors.

Corollary 1.8. If two positive integers a and b have expansions as products of powers of r distinct primes given by $a = p_1^{k_1} \cdots p_r^{k_r}$ and $b = p_1^{l_1} \cdots p_r^{l_r}$, then

$$\text{GCD}(a, b) = p_1^{\min(k_1, l_1)} \cdots p_r^{\min(k_r, l_r)}.$$

PROOF. Let d' be the right side of the displayed equation. It is plain that d' is positive and that d' divides a and b. On the other hand, two applications of Corollary 1.7 show that the greatest common divisor of a and b is a number d of the form $p_1^{m_1} \cdots p_r^{m_r}$ with the property that $m_j \leq k_j$ and $m_j \leq l_j$ for all j. Therefore $m_j \leq \min(k_j, l_j)$ for all j, and $d \leq d'$. Since any positive divisor of both a and b is $\leq d$, we have $d' \leq d$. Thus $d' = d$. □

In special cases Corollary 1.8 provides a useful way to compute $\text{GCD}(a, b)$, but the Euclidean algorithm is usually a more efficient procedure. Nevertheless, Corollary 1.8 remains a handy tool for theoretical purposes. Here is an example: Two nonzero integers a and b are said to be **relatively prime** if $\text{GCD}(a, b) = 1$. It is immediate from Corollary 1.8 that two nonzero integers a and b are relatively prime if and only if there is no prime p that divides both a and b.

Corollary 1.9 (Chinese Remainder Theorem). Let a and b be positive relatively prime integers. To each pair (r, s) of integers with $0 \leq r < a$ and $0 \leq s < b$ corresponds a unique integer n such that $0 \leq n < ab$, a divides $n - r$, and b divides $n - s$. Moreover, every integer n with $0 \leq n < ab$ arises from some such pair (r, s).

REMARK. In notation for congruences that we introduce formally in Chapter IV, the result says that if $\text{GCD}(a, b) = 1$, then the congruences $n \equiv r \mod a$ and $n \equiv s \mod b$ have one and only one simultaneous solution n with $0 \leq n < ab$.

PROOF. Let us see that n exists as asserted. Since a and b are relatively prime, Proposition 1.2c produces integers x' and y' such that $ax' - by' = 1$. Multiplying by $s - r$, we obtain $ax - by = s - r$ for suitable integers x and y. Put $n' = ax + r = by + s$, and write by the division algorithm (Proposition 1.1) $n' = abq + n$ for some integer q and for some integer n with $0 \leq n < ab$. Then $n - r = n' - abq - r = ax - abq$ is divisible by a, and similarly $n - s$ is divisible by b.

Suppose that n and n' both have the asserted properties. Then a divides $n - n' = (n - r) - (n' - r)$, and b divides $n - n' = (n - s) - (n' - s)$. Since a and b are relatively prime, Corollary 1.4 shows that ab divides $n - n'$. But $|n - n'| < ab$, and the only integer N with $|N| < ab$ that is divisible by ab is $N = 0$. Thus $n - n' = 0$ and $n = n'$. This proves uniqueness.

Finally the argument just given defines a one-one function from a set of ab pairs (r, s) to a set of ab elements n. Its image must therefore be all such integers n. This proves the corollary. □

If n is a positive integer, we define $\varphi(n)$ to be the number of integers k with $0 \leq k < n$ such that k and n are relatively prime. The function φ is called the **Euler φ function**.

Corollary 1.10. Let $N > 1$ be an integer, and let $N = p_1^{k_1} \cdots p_r^{k_r}$ be a prime factorization of N. Then

$$\varphi(N) = \prod_{j=1}^{r} p_j^{k_j - 1}(p_j - 1).$$

REMARK. The conclusion is valid also for $N = 1$ if we interpret the right side of the formula to be the empty product.

PROOF. For positive integers a and b, let us check that

$$\varphi(ab) = \varphi(a)\varphi(b) \quad \text{if} \quad \text{GCD}(a, b) = 1. \tag{$*$}$$

In view of Corollary 1.9, it is enough to prove that the mapping $(r, s) \mapsto n$ given in that corollary has the property that $\text{GCD}(r, a) = \text{GCD}(s, b) = 1$ if and only if $\text{GCD}(n, ab) = 1$.

To see this property, suppose that n satisfies $0 \leq n < ab$ and $\text{GCD}(n, ab) > 1$. Choose a prime p dividing both n and ab. By Lemma 1.6, p divides a or p divides b. By symmetry we may assume that p divides a. If (r, s) is the pair corresponding to n under Corollary 1.9, then the corollary says that a divides $n - r$. Since p divides a, p divides $n - r$. Since p divides n, p divides r. Thus $\text{GCD}(r, a) > 1$.

Conversely suppose that (r, s) is a pair with $0 \leq r < a$ and $0 \leq s < b$ such that $\text{GCD}(r, a) = \text{GCD}(s, b) = 1$ is false. Without loss of generality, we may assume that $\mathrm{GCD}(r, a) > 1$. Choose a prime p dividing both r and a. If n is the integer with $0 \leq n < ab$ that corresponds to (r, s) under Corollary 1.9, then the corollary says that a divides $n - r$. Since p divides a, p divides $n - r$. Since p divides r, p divides n. Thus $\mathrm{GCD}(n, ab) > 1$. This completes the proof of $(*)$.

For a power p^k of a prime p with $k > 0$, the integers n with $0 \leq n < p^k$ such that $\mathrm{GCD}(n, p^k) > 1$ are the multiples of p, namely $0, p, 2p, \ldots, p^k - p$. There are p^{k-1} of them. Thus the number of integers n with $0 \leq n < p^k$ such that $\mathrm{GCD}(n, p^k) = 1$ is $p^k - p^{k-1} = p^{k-1}(p-1)$. In other words,

$$\varphi(p^k) = p^{k-1}(p-1) \qquad \text{if } p \text{ is prime and } k \geq 1. \qquad (**)$$

To prove the corollary, we induct on r, the case $r = 1$ being handled by $(**)$. If the formula of the corollary is valid for $r - 1$, then $(*)$ allows us to combine that result with the formula for $\varphi(p^{k_r})$ given in $(**)$ to obtain the formula for $\varphi(N)$. □

We conclude this section by extending the notion of greatest common divisor to apply to more than two integers. If $a_1, \ldots, a_t$ are integers not all 0, their **greatest common divisor** is the largest integer $d > 0$ that divides all of $a_1, \ldots, a_t$. This exists, and we write $d = \mathrm{GCD}(a_1, \ldots, a_t)$ for it. It is immediate that d equals the greatest common divisor of the nonzero members of the set $\{a_1, \ldots, a_t\}$. Thus, in deriving properties of greatest common divisors, we may assume that all the integers are nonzero.

Corollary 1.11. Let $a_1, \ldots, a_t$ be positive integers, and let d be their greatest common divisor. Then

(a) if for each j with $1 \leq j \leq t$, $a_j = p_1^{k_{1,j}} \cdots p_r^{k_{r,j}}$ is an expansion of a_j as a product of powers of r distinct primes $p_1, \ldots, p_r$, it follows that

$$d = p_1^{\min_{1 \leq j \leq r}\{k_{1,j}\}} \cdots p_r^{\min_{1 \leq j \leq r}\{k_{r,j}\}},$$

(b) any divisor d' of all of $a_1, \ldots, a_t$ necessarily divides d,
(c) $d = \mathrm{GCD}\big(\mathrm{GCD}(a_1, \ldots, a_{t-1}), a_t\big)$ if $t > 1$,
(d) there exist integers $x_1, \ldots, x_t$ such that $a_1 x_1 + \cdots + a_t x_t = d$.

PROOF. Part (a) is proved in the same way as Corollary 1.8 except that Corollary 1.7 is to be applied r times rather than just twice. Further application of Corollary 1.7 shows that any positive divisor d' of $a_1, \ldots, a_t$ is of the form $d' = p_1^{m_1} \cdots p_r^{m_r}$ with $m_1 \leq k_{1,j}$ for all $j, \ldots$, and with $m_r \leq k_{r,j}$ for all j. Therefore $m_1 \leq \min_{1 \leq j \leq r}\{k_{1,j}\}, \ldots$, and $m_r \leq \min_{1 \leq j \leq r}\{k_{r,j}\}$, and it follows that d' divides d. This proves (b). Conclusion (c) follows by using the formula in (a), and (d) follows by combining (c), Proposition 1.2c, and induction. □

3. Unique Factorization of Polynomials

This section establishes unique factorization for ordinary rational, real, and complex polynomials. We write $\mathbb{Q}$ for the set of rational numbers, $\mathbb{R}$ for the set of real numbers, and $\mathbb{C}$ for the set of complex numbers, each with its arithmetic operations. The rational numbers are constructed from the integers by a process reviewed in Section A3 of the appendix, the real numbers are defined from the rational numbers by a process reviewed in that same section, and the complex numbers are defined from the real numbers by a process reviewed in Section A4 of the appendix. Sections A3 and A4 of the appendix mention special properties of $\mathbb{R}$ and $\mathbb{C}$ beyond those of the arithmetic operations, but we shall not make serious use of these special properties here until nearly the end of the section—after unique factorization of polynomials has been established. Let $\mathbb{F}$ denote any of $\mathbb{Q}$, $\mathbb{R}$, or $\mathbb{C}$. The members of $\mathbb{F}$ are called **scalars**.

We work with ordinary polynomials with coefficients in $\mathbb{F}$. Informally these are expressions $P(X) = a_n X^n + \cdots + a_1 X + a_0$ with $a_n, \ldots, a_1, a_0$ in $\mathbb{F}$. Although it is tempting to think of $P(X)$ as a function with independent variable X, it is better to identify P with the sequence $(a_0, a_1, \ldots, a_n, 0, 0, \ldots)$ of coefficients, using expressions $P(X) = a_n X^n + \cdots + a_1 X + a_0$ only for conciseness and for motivation of the definitions of various operations.

The precise definition therefore is that a **polynomial** in **one indeterminate** with **coefficients** in $\mathbb{F}$ is an infinite sequence of members of $\mathbb{F}$ such that all terms of the sequence are 0 from some point on. The indexing of the sequence is to begin with 0. We may refer to a polynomial P as $P(X)$ if we want to emphasize that the indeterminate is called X. Addition, subtraction, and scalar multiplication are defined in coordinate-by-coordinate fashion:

$$(a_0, a_1, \ldots, a_n, 0, 0, \ldots) + (b_0, b_1, \ldots, b_n, 0, 0, \ldots)$$
$$= (a_0 + b_0, a_1 + b_1, \ldots, a_n + b_n, 0, 0, \ldots),$$
$$(a_0, a_1, \ldots, a_n, 0, 0, \ldots) - (b_0, b_1, \ldots, b_n, 0, 0, \ldots)$$
$$= (a_0 - b_0, a_1 - b_1, \ldots, a_n - b_n, 0, 0, \ldots),$$
$$c(a_0, a_1, \ldots, a_n, 0, 0, \ldots) = (ca_0, ca_1, \ldots, ca_n, 0, 0, \ldots).$$

Polynomial multiplication is defined so as to match multiplication of expressions $a_n X^n + \cdots + a_1 X + a_0$ if the product is expanded out, powers of X are added, and then terms containing like powers of X are collected:

$$(a_0, a_1, \ldots, 0, 0, \ldots)(b_0, b_1, \ldots, 0, 0, \ldots) = (c_0, c_1, \ldots, 0, 0, \ldots),$$

where $c_N = \sum_{k=0}^{N} a_k b_{N-k}$. We take it as known that the usual associative, commutative, and distributive laws are then valid. The set of all polynomials in the indeterminate X is denoted by $\mathbb{F}[X]$.

The polynomial with all entries 0 is denoted by 0 and is called the **zero polynomial**. For all polynomials $P = (a_0, \ldots, a_n, 0, \ldots)$ other than 0, the **degree** of P, denoted by $\deg P$, is defined to be the largest index n such that $a_n \neq 0$. The **constant polynomials** are by definition the zero polynomial and the polynomials of degree 0. If P and Q are nonzero polynomials, then

$$P + Q = 0 \quad \text{or} \quad \deg(P+Q) \leq \max(\deg P, \deg Q),$$
$$\deg(cP) = \deg P,$$
$$\deg(PQ) = \deg P + \deg Q.$$

In the formula for $\deg(P+Q)$, equality holds if $\deg P \neq \deg Q$. Implicit in the formula for $\deg(PQ)$ is the fact that PQ cannot be 0 unless $P=0$ or $Q=0$. A cancellation law for multiplication is an immediate consequence:

$$PR = QR \quad \text{with} \quad R \neq 0 \quad \text{implies} \quad P = Q.$$

In fact, $PR = QR$ implies $(P-Q)R = 0$; since $R \neq 0$, $P-Q$ must be 0.

If $P = (a_0, \ldots, a_n, 0, \ldots)$ is a polynomial and r is in $\mathbb{F}$, we can **evaluate** P at r, obtaining as a result the number $P(r) = a_n r^n + \cdots + a_1 r + a_0$. Taking into account all values of r, we obtain a mapping $P \mapsto P(\cdot)$ of $\mathbb{F}[X]$ into the set of functions from $\mathbb{F}$ into $\mathbb{F}$. Because of the way that the arithmetic operations on polynomials have been defined, we have

$$(P+Q)(r) = P(r) + Q(r),$$
$$(P-Q)(r) = P(r) - Q(r),$$
$$(cP)(r) = cP(r),$$
$$(PQ)(r) = P(r)Q(r).$$

In other words, the mapping $P \mapsto P(\cdot)$ respects the arithmetic operations. We say that r is a **root** of P if $P(r) = 0$.

Now we turn to the question of unique factorization. The definitions and the proof are completely analogous to those for the integers. A **factor** of a polynomial A is a nonzero polynomial B such that $A = BQ$ for some polynomial Q. In this case we say also that B **divides** A, that B is a **divisor** of A, and that A is a **multiple** of B. We write $B \mid A$ for this relationship. If A is nonzero, any product formula $A = BQ_1 \cdots Q_r$ is a **factorization** of A. A **unit** in $\mathbb{F}[X]$ is a divisor of 1, hence is any polynomial of degree 0; such a polynomial is a constant polynomial $A(X) = c$ with c equal to a nonzero scalar. The factorization $A = BQ$ of $A \neq 0$ is called **nontrivial** if neither B nor Q is a unit. A **prime** P in $\mathbb{F}[X]$ is a nonzero polynomial that is not a unit and has no nontrivial factorization $P = BQ$. Observe that the product of a prime and a unit is always a prime.

3. Unique Factorization of Polynomials

Proposition 1.12 (division algorithm). If A and B are polynomials in $\mathbb{F}[X]$ and if B not the 0 polynomial, then there exist unique polynomials Q and R in $\mathbb{F}[X]$ such that

(a) $A = BQ + R$ and
(b) either R is the 0 polynomial or $\deg R < \deg B$.

REMARK. This result codifies the usual method of dividing polynomials in high-school algebra. That method writes $A/B = Q + R/B$, and then one obtains the above result by multiplying by B. The polynomial Q is the quotient in the division, and R is the remainder.

PROOF OF UNIQUENESS. If $A = BQ + R = BQ_1 + R_1$, then $B(Q - Q_1) = R_1 - R$. Without loss of generality, $R_1 - R$ is not the 0 polynomial since otherwise $Q - Q_1 = 0$ also. Then

$$\deg B + \deg(Q - Q_1) = \deg(R_1 - R) \leq \max(\deg R, \deg R_1) < \deg B,$$

and we have a contradiction. $\square$

PROOF OF EXISTENCE. If $A = 0$ or $\deg A < \deg B$, we take $Q = 0$ and $R = A$, and we are done. Otherwise we induct on $\deg A$. Assume the result for degree $\leq n - 1$, and let $\deg A = n$. Write $A = a_n X^n + A_1$ with $A_1 = 0$ or $\deg A_1 < \deg A$. Let $B = b_k X^k + B_1$ with $B_1 = 0$ or $\deg B_1 < \deg B$. Put $Q_1 = a_n b_k^{-1} X^{n-k}$. Then

$$A - BQ_1 = a_n X^n + A_1 - a_n X^n - a_n b_k^{-1} X^{n-k} B_1 = A_1 - a_n b_k^{-1} X^{n-k} B_1$$

with the right side equal to 0 or of degree $< \deg A$. Then the right side, by induction, is of the form $BQ_2 + R$, and $A = B(Q_1 + Q_2) + R$ is the required decomposition. $\square$

Corollary 1.13 (Factor Theorem). If r is in $\mathbb{F}$ and if P is a polynomial in $\mathbb{F}[X]$, then $X - r$ divides P if and only if $P(r) = 0$.

PROOF. If $P = (X - r)Q$, then $P(r) = (r - r)Q(r) = 0$. Conversely let $P(r) = 0$. Taking $B(X) = X - r$ in the division algorithm (Proposition 1.12), we obtain $P = (X - r)Q + R$ with $R = 0$ or $\deg R < \deg(X - r) = 1$. Thus R is a constant polynomial, possibly 0. In any case we have $0 = P(r) = (r - r)Q(r) + R(r)$, and thus $R(r) = 0$. Since R is constant, we must have $R = 0$, and then $P = (X - r)Q$. $\square$

Corollary 1.14. If P is a nonzero polynomial with coefficients in $\mathbb{F}$ and if $\deg P = n$, then P has at most n distinct roots.

REMARKS. Since there are infinitely many scalars in any of $\mathbb{Q}$ and $\mathbb{R}$ and $\mathbb{C}$, the corollary implies that the function from $\mathbb{F}$ to $\mathbb{F}$ associated to P, namely $r \mapsto P(r)$, cannot be identically 0 if $P \neq 0$. Starting in Chapter IV, we shall allow other $\mathbb{F}$'s besides $\mathbb{Q}$ and $\mathbb{R}$ and $\mathbb{C}$, and then this implication can fail. For example, when $\mathbb{F}$ is the two-element "field" $\mathbb{F} = \{0, 1\}$ with $1 + 1 = 0$ and with otherwise the expected addition and multiplication, then $P(X) = X^2 + X$ is not the zero polynomial but $P(r) = 0$ for $r = 0$ and $r = 1$. It is thus important to distinguish polynomials in one indeterminate from their associated functions of one variable.

PROOF. Let $r_1, \ldots, r_{n+1}$ be distinct roots of $P(X)$. By the Factor Theorem (Corollary 1.13), $X - r_1$ is a factor of $P(X)$. We prove inductively on k that the product $(X - r_1)(X - r_2) \cdots (X - r_k)$ is a factor of $P(X)$. Assume that this assertion holds for k, so that $P(X) = (X - r_1) \cdots (X - r_k) Q(X)$ and

$$0 = P(r_{k+1}) = (r_{k+1} - r_1) \cdots (r_{k+1} - r_k) Q(r_{k+1}).$$

Since the r_j's are distinct, we must have $Q(r_{k+1}) = 0$. By the Factor Theorem, we can write $Q(X) = (X - r_{k+1}) R(X)$ for some polynomial $R(X)$. Substitution gives $P(X) = (X - r_1) \cdots (X - r_k)(X - r_{k+1}) R(X)$, and $(X - r_1) \cdots (X - r_{k+1})$ is exhibited as a factor of $P(X)$. This completes the induction. Consequently

$$P(X) = (X - r_1) \cdots (X - r_{n+1}) S(X)$$

for some polynomial $S(X)$. Comparing the degrees of the two sides, we find that $\deg S = -1$, and we have a contradiction. $\square$

We can use the division algorithm in the same way as with the integers in Sections 1–2 to obtain unique factorization. Within the set of integers, we defined greatest common divisors so as to be positive, but their negatives would have worked equally well. That flexibility persists with polynomials; the essential feature of any greatest common divisor of polynomials is shared by any product of that polynomial by a unit. A **greatest common divisor** of polynomials A and B with $B \neq 0$ is any polynomial D of maximum degree such that D divides A and D divides B. We shall see that D is indeed unique up to multiplication by a nonzero scalar.[2]

---

[2]For some purposes it is helpful to isolate one particular greatest common divisor by taking the coefficient of the highest power of X to be 1.

3. Unique Factorization of Polynomials

The **Euclidean algorithm** is the iterative process that makes use of the division algorithm in the form

$$A = BQ_1 + R_1, \qquad R_1 = 0 \text{ or } \deg R_1 < \deg B,$$
$$B = R_1 Q_2 + R_2, \qquad R_2 = 0 \text{ or } \deg R_2 < \deg R_1,$$
$$R_1 = R_2 Q_3 + R_3, \qquad R_3 = 0 \text{ or } \deg R_3 < \deg R_2,$$
$$\vdots$$
$$R_{n-2} = R_{n-1} Q_n + R_n, \qquad R_n = 0 \text{ or } \deg R_n < \deg R_{n-1},$$
$$R_{n-1} = R_n Q_{n+1}.$$

In the above computation the integer n is defined by the conditions that $R_n \neq 0$ and that $R_{n+1} = 0$. Such an n must exist since $\deg B > \deg R_1 > \cdots \geq 0$. We can now obtain an analog for $\mathbb{F}[X]$ of the result for $\mathbb{Z}$ given as Proposition 1.2.

Proposition 1.15. Let A and B be polynomials in $\mathbb{F}[X]$ with $B \neq 0$, and let $R_1, \ldots, R_n$ be the remainders generated by the Euclidean algorithm when applied to A and B. Then

(a) R_n is a greatest common divisor of A and B,
(b) any D_1 that divides both A and B necessarily divides R_n,
(c) the greatest common divisor of A and B is unique up to multiplication by a nonzero scalar,
(d) any greatest common divisor D has the property that there exist polynomials P and Q with $AP + BQ = D$.

PROOF. Conclusions (a) and (b) are proved in the same way that parts (a) and (b) of Proposition 1.2 are proved, and conclusion (d) is proved with $D = R_n$ in the same way that Proposition 1.2c is proved.

If D is a greatest common divisor of A and B, it follows from (a) and (b) that D divides R_n and that $\deg D = \deg R_n$. This proves (c). $\square$

Using Proposition 1.15, we can prove analogs for $\mathbb{F}[X]$ of the two corollaries of Proposition 1.2. But let us instead skip directly to what is needed to obtain an analog for $\mathbb{F}[X]$ of unique factorization as in Theorem 1.5.

Lemma 1.16. If A and B are nonzero polynomials with coefficients in $\mathbb{F}$ and if P is a prime polynomial such that P divides AB, then P divides A or P divides B.

PROOF. If P does not divide A, then 1 is a greatest common divisor of A and P, and Proposition 1.15d produces polynomials S and T such that $AS + PT = 1$. Multiplication by B gives $ABS + PTB = B$. Then P divides ABS because it divides AB, and P divides PTB because it divides P. Hence P divides B. $\square$

Theorem 1.17 (unique factorization). Every member of $\mathbb{F}[X]$ of degree ≥ 1 is a product of primes. This factorization is unique up to order and up to multiplication of each prime factor by a unit, i.e., by a nonzero scalar.

PROOF. The existence follows in the same way as the existence in Theorem 1.5; induction on the integers is to be replaced by induction on the degree. The uniqueness follows from Lemma 1.16 in the same way that the uniqueness in Theorem 1.5 follows from Lemma 1.6. $\square$

We turn to a consideration of properties of polynomials that take into account special features of $\mathbb{R}$ and $\mathbb{C}$. If $\mathbb{F}$ is $\mathbb{R}$, then $X^2 + 1$ is prime. The reason is that a nontrivial factorization of $X^2 + 1$ would have to involve two first-degree real polynomials and then $r^2 + 1$ would have to be 0 for some real r, namely for r equal to the root of either of the first-degree polynomials. On the other hand, $X^2 + 1$ is not prime when $\mathbb{F} = \mathbb{C}$ since $X^2 + 1 = (X + i)(X - i)$. The Fundamental Theorem of Algebra, stated below, implies that every prime polynomial over $\mathbb{C}$ is of degree 1. It is possible to prove the Fundamental Theorem of Algebra within complex analysis as a consequence of Liouville's Theorem or within real analysis as a consequence of the Heine–Borel Theorem and other facts about compactness. This text gives a proof of the Fundamental Theorem of Algebra in Chapter IX using modern algebra, specifically Sylow theory as in Chapter IV and Galois theory as in Chapter IX. One further fact is needed; this fact uses elementary calculus and is proved below as Proposition 1.20.

Theorem 1.18 (Fundamental Theorem of Algebra). Any polynomial in $\mathbb{C}[X]$ with degree ≥ 1 has at least one root.

Corollary 1.19. Let P be a nonzero polynomial of degree n in $\mathbb{C}[X]$, and let $r_1, \ldots, r_k$ be the distinct roots. Then there exist unique integers $m_j > 0$ for $1 \leq j \leq k$ such that $P(X)$ is a scalar multiple of $\prod_{j=1}^{k} (X - r_j)^{m_j}$. The numbers m_j have $\sum_{j=1}^{k} m_j = n$.

PROOF. We may assume that $\deg P > 0$. We apply unique factorization (Theorem 1.17) to $P(X)$. It follows from the Fundamental Theorem of Algebra (Theorem 1.18) and the Factor Theorem (Corollary 1.13) that each prime polynomial with coefficients in $\mathbb{C}$ has degree 1. Thus the unique factorization of $P(X)$ has to be of the form $c \prod_{l=1}^{n}(X - z_l)$ for some $c \neq 0$ and for some complex numbers z_l that are unique up to order. The z_l's are roots, and every root is a z_l by the Factor Theorem. Grouping like factors proves the desired factorization and its uniqueness. The numbers m_j have $\sum_{j=1}^{k} m_j = n$ by a count of degrees. $\square$

The integers m_j in the corollary are called the **multiplicities** of the roots of the polynomial $P(X)$.

We conclude this section by proving the result from calculus that will enter the proof of the Fundamental Theorem of Algebra in Chapter IX.

Proposition 1.20. Any polynomial in $\mathbb{R}[X]$ with odd degree has at least one root.

PROOF. Without loss of generality, we may take the leading coefficient to be 1. Thus let the polynomial be $P(X) = X^{2n+1} + a_{2n}X^{2n} + \cdots + a_1 X + a_0 = X^{2n+1} + R(X)$. For $|r| \geq 1$, the polynomial R satisfies $|R(r)| \leq C|r|^{2n}$, where $C = |a_{2n}| + \cdots + |a_1| + |a_0|$. Thus $|r| > \max(C, 1)$ implies $|P(r) - r^{2n+1}| \leq C|r|^{2n} < |r|^{2n+1}$, and it follows that $P(r)$ has the same sign as r^{2n+1} for $|r| > \max(C, 1)$. For $r_0 = \max(C, 1) + 1$, we therefore have $P(-r_0) < 0$ and $P(r_0) > 0$. By the Intermediate Value Theorem, given in Section A3 of the appendix, $P(r) = 0$ for some r with $-r_0 \leq r \leq r_0$. $\square$

4. Permutations and Their Signs

Let S be a finite nonempty set of n elements. A **permutation** of S is a one-one function from S onto S. The elements might be listed as $a_1, a_2, \ldots, a_n$, but it will simplify the notation to view them simply as $1, 2, \ldots, n$. We use ordinary function notation for describing the effect of permutations. Thus the value of a permutation σ at j is $\sigma(j)$, and the composition of τ followed by σ is $\sigma \circ \tau$ or simply $\sigma\tau$, with $(\sigma\tau)(j) = \sigma(\tau(j))$. Composition is automatically associative, i.e., $(\rho\sigma)\tau = \rho(\sigma\tau)$, because the effect of both sides on j, when we expand things out, is $\rho(\sigma(\tau(j)))$. The composition of two permutations is also called their **product**.

The identity permutation will be denoted by 1. Any permutation σ, being a one-one onto function, has a well-defined inverse permutation σ^{-1} with the property that $\sigma\sigma^{-1} = \sigma^{-1}\sigma = 1$. One way of describing concisely the effect of a permutation is to list its domain values and to put the corresponding range values beneath them. Thus $\sigma = \begin{pmatrix} 1 & 2 & 3 & 4 & 5 \\ 4 & 3 & 5 & 1 & 2 \end{pmatrix}$ is the permutation of $\{1, 2, 3, 4, 5\}$ with $\sigma(1) = 4$, $\sigma(2) = 3$, $\sigma(3) = 5$, $\sigma(4) = 1$, and $\sigma(5) = 2$. The inverse permutation is obtained by interchanging the two rows to obtain $\begin{pmatrix} 4 & 3 & 5 & 1 & 2 \\ 1 & 2 & 3 & 4 & 5 \end{pmatrix}$ and then adjusting the entries in the rows so that the first row is in the usual order: $\sigma^{-1} = \begin{pmatrix} 1 & 2 & 3 & 4 & 5 \\ 4 & 5 & 2 & 1 & 3 \end{pmatrix}$.

If $2 \leq k \leq n$, a k-**cycle** is a permutation σ that fixes each element in some subset of $n - k$ elements and moves the remaining elements $c_1, \ldots, c_k$ according to $\sigma(c_1) = c_2$, $\sigma(c_2) = c_3, \ldots, \sigma(c_{k-1}) = c_k$, $\sigma(c_k) = c_1$. Such a cycle may be denoted by $(c_1 \; c_2 \; \cdots \; c_{k-1} \; c_k)$ to stress its structure. For example take $n = 5$; then $\sigma = (2 \; 3 \; 5)$ is the 3-cycle given in our earlier notation by $\begin{pmatrix} 1\,2\,3\,4\,5 \\ 1\,3\,5\,4\,2 \end{pmatrix}$. The cycle (2 3 5) is the same as the cycle (3 5 2) and the cycle (5 2 3). It is sometimes helpful to speak of the identity permutation 1 as the unique 1-cycle.

A system of cycles is said to be **disjoint** if the sets that each of them moves are disjoint in pairs. Thus (2 3 5) and (1 4) are disjoint, but (2 3 5) and (1 3) are not. Any two disjoint cycles σ and τ commute in the sense that $\sigma\tau = \tau\sigma$.

Proposition 1.21. Any permutation σ of $\{1, 2, \ldots, n\}$ is a product of disjoint cycles. The individual cycles in the decomposition are unique in the sense of being determined by σ.

EXAMPLE. $\begin{pmatrix} 1\,2\,3\,4\,5 \\ 4\,3\,5\,1\,2 \end{pmatrix} = (2 \; 3 \; 5)(1 \; 4)$.

PROOF. Let us prove existence. Working with $\{1, 2, \ldots, n\}$, we show that any σ is the disjoint product of cycles in such a way that no cycle moves an element j unless σ moves j. We do so for all σ simultaneously by induction downward on the number of elements fixed by σ. The starting case of the induction is that σ fixes all n elements. Then σ is the identity, and we are regarding the identity as a 1-cycle.

For the inductive step suppose σ fixes the elements in a subset T of r elements of $\{1, 2, \ldots, n\}$ with $r < n$. Let j be an element not in T, so that $\sigma(j) \neq j$. Choose k as small as possible so that some element is repeated among $j, \sigma(j), \sigma^2(j), \ldots, \sigma^k(j)$. This condition means that $\sigma^l(j) = \sigma^k(j)$ for some l with $0 \leq l < k$. Then $\sigma^{k-l}(j) = j$, and we obtain a contradiction to the minimality of k unless $k - l = k$, i.e., $l = 0$. In other words, we have $\sigma^k(j) = j$. We may thus form the k-cycle $\gamma = (j \; \sigma(j) \; \sigma^2(j) \; \sigma^{k-1}(j))$. The permutation $\gamma^{-1}\sigma$ then fixes the $r + k$ elements of $T \cup U$, where U is the set of elements $j, \sigma(j), \sigma^2(j), \ldots, \sigma^{k-1}(j)$. By the inductive hypothesis, $\gamma^{-1}\sigma$ is the product $\tau_1 \cdots \tau_p$ of disjoint cycles that move only elements not in $T \cup U$. Since γ moves only the elements in U, γ is disjoint from each of $\tau_1, \ldots, \tau_p$. Therefore $\sigma = \gamma\tau_1 \cdots \tau_p$ provides the required decomposition of σ.

For uniqueness we observe from the proof of existence that each element j generates a k-cycle C_j for some $k \geq 1$ depending on j. If we have two decompositions as in the proposition, then the cycle within each decomposition that contains j must be C_j. Hence the cycles in the two decompositions must match. $\square$

A 2-cycle is often called a **transposition**. The proposition allows us to see quickly that any permutation is a product of transpositions.

4. Permutations and Their Signs

Corollary 1.22. Any k-cycle σ permuting $\{1, 2, \ldots, n\}$ is a product of $k - 1$ transpositions if $k > 1$. Therefore any permutation σ of $\{1, 2, \ldots, n\}$ is a product of transpositions.

PROOF. For the first statement, we observe that $(c_1 \; c_2 \; \cdots \; c_{k-1} \; c_k) = (c_1 \; c_k)(c_1 \; c_{k-1}) \cdots (c_1 \; c_3)(c_1 \; c_2)$. The second statement follows by combining this fact with Proposition 1.21. $\square$

Our final tasks for this section are to attach a sign to each permutation and to examine the properties of these signs. We begin with the special case that our underlying set S is $\{1, \ldots, n\}$. If σ is a permutation of $\{1, \ldots, n\}$, consider the numerical products

$$\prod_{1 \leq j < k \leq n} |\sigma(k) - \sigma(j)| \quad \text{and} \quad \prod_{1 \leq j < k \leq n} (\sigma(k) - \sigma(j)).$$

If (r, s) is any pair of integers with $1 \leq r < s \leq n$, then the expression $s - r$ appears once and only once as a factor in the first product. Therefore the first product is independent of σ and equals $\prod_{1 \leq j < k \leq n} (k - j)$. Meanwhile, each factor of the second product is ± 1 times the corresponding factor of the first product. Therefore we have

$$\prod_{1 \leq j < k \leq n} (\sigma(k) - \sigma(j)) = (\text{sgn } \sigma) \prod_{1 \leq j < k \leq n} (k - j),$$

where $\text{sgn } \sigma$ is $+1$ or -1, depending on σ. This sign is called the **sign** of the permutation σ.

Lemma 1.23. Let σ be a permutation of $\{1, \ldots, n\}$, let $(a \; b)$ be a transposition, and form the product $\sigma(a \; b)$. Then $\text{sgn}\,(\sigma(a \; b)) = -\text{sgn } \sigma$.

PROOF. For the pairs (j, k) with $j < k$, we are to compare $\sigma(k) - \sigma(j)$ with $\sigma(a \; b)(k) - \sigma(a \; b)(j)$. There are five cases. Without loss of generality, we may assume that $a < b$.

Case 1. If neither j nor k equals a or b, then $\sigma(a \; b)(k) - \sigma(a \; b)(j) = \sigma(k) - \sigma(j)$. Thus such pairs (j, k) make the same contribution to the product for $\sigma(a \; b)$ as to the product for σ, and they can be ignored.

Case 2. If one of j and k equals one of a and b while the other does not, there are three situations of interest. For each we compare the contributions of two such pairs together. The first situation is that of pairs (a, t) and (t, b) with $a < t < b$. These together contribute the factors $(\sigma(t) - \sigma(a))$ and $(\sigma(b) - \sigma(t))$ to the product for σ, and they contribute the factors $(\sigma(t) - \sigma(b))$ and $(\sigma(a) - \sigma(t))$ to the product for $\sigma(a \; b)$. Since

$$(\sigma(t) - \sigma(a))(\sigma(b) - \sigma(t)) = (\sigma(t) - \sigma(b))(\sigma(a) - \sigma(t)),$$

the pairs together make the same contribution to the product for $\sigma(a\ b)$ as to the product for σ, and they can be ignored.

Case 3. Continuing with matters as in Case 2, we next consider pairs (a, t) and (b, t) with $a < b < t$. These together contribute the factors $(\sigma(t) - \sigma(a))$ and $(\sigma(t) - \sigma(b))$ to the product for σ, and they contribute the factors $(\sigma(t) - \sigma(b))$ and $(\sigma(t) - \sigma(a))$ to the product for $\sigma(a\ b)$. Since

$$(\sigma(t) - \sigma(a))(\sigma(t) - \sigma(b)) = (\sigma(t) - \sigma(b))(\sigma(t) - \sigma(a)),$$

the pairs together make the same contribution to the product for $\sigma(a\ b)$ as to the product for σ, and they can be ignored.

Case 4. Still with matters as in Case 2, we consider pairs (t, a) and (t, b) with $t < a < b$. Arguing as in Case 3, we are led to an equality

$$(\sigma(a) - \sigma(t))(\sigma(b) - \sigma(t)) = (\sigma(b) - \sigma(t))(\sigma(a) - \sigma(t)),$$

and these pairs can be ignored.

Case 5. Finally we consider the pair (a, b) itself. It contributes $\sigma(b) - \sigma(a)$ to the product for σ, and it contributes $\sigma(a) - \sigma(b)$ to the product for $\sigma(a\ b)$. These are negatives of one another, and we get a net contribution of one minus sign in comparing our two product formulas. The lemma follows. □

Proposition 1.24. The signs of permutations of $\{1, 2, \ldots, n\}$ have the following properties:
(a) $\operatorname{sgn} 1 = +1$,
(b) $\operatorname{sgn} \sigma = (-1)^k$ if σ can be written as the product of k transpositions,
(c) $\operatorname{sgn}(\sigma\tau) = (\operatorname{sgn}\sigma)(\operatorname{sgn}\tau)$,
(d) $\operatorname{sgn}(\sigma^{-1}) = \operatorname{sgn}\sigma$.

PROOF. Conclusion (a) is immediate from the definition. For (b), let $\sigma = \tau_1 \cdots \tau_k$ with each τ_j equal to a transposition. We apply Lemma 1.23 recursively, using (a) at the end:

$$\operatorname{sgn}(\tau_1 \cdots \tau_k) = (-1)\operatorname{sgn}(\tau_1 \cdots \tau_{k-1}) = (-1)^2 \operatorname{sgn}(\tau_1 \cdots \tau_{k-2})$$
$$= \cdots = (-1)^{k-1} \operatorname{sgn}\tau_1 = (-1)^k \operatorname{sgn} 1 = (-1)^k.$$

For (c), Corollary 1.22 shows that any permutation is the product of transpositions. If σ is the product of k transpositions and τ is the product of l transpositions, then $\sigma\tau$ is manifestly the product of $k + l$ transpositions. Thus (c) follows from (b). Finally (d) follows from (c) and (a) by taking $\tau = \sigma^{-1}$. □

Our discussion of signs has so far attached signs only to permutations of $S = \{1, 2, \ldots, n\}$. If we are given some other set S' of n elements and we want to adapt our discussion of signs so that it applies to permutations of S', we need to identify S with S', say by a one-one onto function $\varphi : S \to S'$. If σ is a permutation of S', then $\varphi^{-1}\sigma\varphi$ is a permutation of S, and we can define $\operatorname{sgn}_\varphi(\sigma) = \operatorname{sgn}(\varphi^{-1}\sigma\varphi)$. The question is whether this definition is independent of φ.

Fortunately the answer is yes, and the proof is easy. Suppose that $\psi : S \to S'$ is a second one-one onto function, so that $\operatorname{sgn}_\psi(\sigma) = \operatorname{sgn}(\psi^{-1}\sigma\psi)$. Then $\varphi^{-1}\psi = \tau$ is a permutation of $\{1, 2, \ldots, n\}$, and (c) and (d) in Proposition 1.24 give

$$\operatorname{sgn}_\psi(\sigma) = \operatorname{sgn}(\psi^{-1}\sigma\psi) = \operatorname{sgn}(\psi^{-1}\varphi\varphi^{-1}\sigma\varphi\varphi^{-1}\psi)$$
$$= \operatorname{sgn}(\tau^{-1})\operatorname{sgn}(\varphi^{-1}\sigma\varphi)\operatorname{sgn}(\tau) = \operatorname{sgn}(\tau)\operatorname{sgn}_\varphi(\sigma)\operatorname{sgn}(\tau) = \operatorname{sgn}_\varphi(\sigma).$$

Consequently the definition of signs of permutations of $\{1, 2, \ldots, n\}$ can be carried over to give a definition of signs of permutations of any finite nonempty set of n elements, and the resulting signs are independent of the way we enumerate the set. The conclusions of Proposition 1.24 are valid for this extended definition of signs of permutations.

5. Row Reduction

This section and the next review row reduction and matrix algebra for rational, real, and complex matrices. As in Section 3 let $\mathbb{F}$ denote $\mathbb{Q}$ or $\mathbb{R}$ or $\mathbb{C}$. The members of $\mathbb{F}$ are called **scalars**.

The term "row reduction" refers to the main part of the algorithm used for solving simultaneous systems of algebraic linear equations with coefficients in $\mathbb{F}$. Such a system is of the form

$$a_{11}x_1 + a_{12}x_2 + \cdots + a_{1n}x_n = b_1,$$
$$\vdots$$
$$a_{k1}x_1 + a_{k2}x_2 + \cdots + a_{kn}x_n = b_k,$$

where the a_{ij} and b_i are known scalars and the x_j are the **unknowns**, or **variables**. The algorithm makes repeated use of three operations on the equations, each of which preserves the set of solutions $(x_1, \ldots, x_n)$ because its inverse is an operation of the same kind:

(i) interchange two equations,
(ii) multiply an equation by a nonzero scalar,
(iii) replace an equation by the sum of it and a multiple of some other equation.

The repeated writing of the variables in carrying out these steps is tedious and unnecessary, since the steps affect only the known coefficients. Instead, we can simply work with an array of the form

$$\begin{pmatrix} a_{11} & a_{12} & \cdots & a_{1n} & b_1 \\ & \ddots & & & \vdots \\ a_{k1} & a_{k2} & \cdots & a_{kn} & b_k \end{pmatrix}.$$

The individual scalars appearing in the array are called **entries**. The above operations on equations correspond exactly to operations on the rows[3] of the array, and they become

 (i) interchange two rows,
 (ii) multiply a row by a nonzero scalar,
 (iii) replace a row by the sum of it and a multiple of some other row.

Any operation of these types is called an **elementary row operation**. The vertical line in the array is handy from one point of view in that it separates the left sides of the equations from the right sides; if we have more than one set of right sides, we can include all of them to the right of the vertical line and thereby solve all the systems at the same time. But from another point of view, the vertical line is unnecessary since it does not affect which operation we perform at a particular time. Let us therefore drop it, abbreviating the system as

$$\begin{pmatrix} a_{11} & a_{12} & \cdots & a_{1n} & b_1 \\ & \ddots & & & \vdots \\ a_{k1} & a_{k2} & \cdots & a_{kn} & b_k \end{pmatrix}.$$

The main step in solving the system is to apply the three operations in succession to the array to reduce it to a particularly simple form. An array with k rows and m columns[4] is in **reduced row-echelon form** if it meets several conditions:

- Each member of the first l of the rows, for some l with $0 \leq l \leq k$, has at least one nonzero entry, and the other rows have all entries 0.
- Each of the nonzero rows has 1 as its first nonzero entry; let us say that the ith nonzero row has this 1 in its $j(i)^{\text{th}}$ entry.
- The integers $j(i)$ are to be strictly increasing as a function of i, and the only entry in the $j(i)^{\text{th}}$ column that is nonzero is to be the one in the ith row.

Proposition 1.25. Any array with k rows and m columns can be transformed into reduced row-echelon form by a succession of steps of types (i), (ii), (iii).

---

[3] "Rows" are understood to be horizontal, while "columns" are vertical.
[4] In the above displayed matrix, the array has $m = n + 1$ columns.

5. Row Reduction

In fact, the transformation in the proposition is carried out by an algorithm known as the method of **row reduction** of the array. Let us begin with an example, indicating the particular operation at each stage by a label over an arrow $\mapsto$. To keep the example from being unwieldy, we consolidate steps of type (iii) into a single step when the "other row" is the same.

EXAMPLE. In this example, $k = m = 4$. Row reduction gives

$$\begin{pmatrix} 0 & 0 & 2 & 7 \\ 1 & -1 & 1 & 1 \\ -1 & 1 & -4 & 5 \\ -2 & 2 & -5 & 4 \end{pmatrix} \stackrel{(i)}{\mapsto} \begin{pmatrix} 1 & -1 & 1 & 1 \\ 0 & 0 & 2 & 7 \\ -1 & 1 & -4 & 5 \\ -2 & 2 & -5 & 4 \end{pmatrix} \stackrel{(iii)}{\mapsto} \begin{pmatrix} 1 & -1 & 1 & 1 \\ 0 & 0 & 2 & 7 \\ 0 & 0 & -3 & 6 \\ 0 & 0 & -3 & 6 \end{pmatrix}$$

$$\stackrel{(ii)}{\mapsto} \begin{pmatrix} 1 & -1 & 1 & 1 \\ 0 & 0 & 1 & \frac{7}{2} \\ 0 & 0 & -3 & 6 \\ 0 & 0 & -3 & 6 \end{pmatrix} \stackrel{(iii)}{\mapsto} \begin{pmatrix} 1 & -1 & 0 & -\frac{5}{2} \\ 0 & 0 & 1 & \frac{7}{2} \\ 0 & 0 & 0 & \frac{33}{2} \\ 0 & 0 & 0 & \frac{33}{2} \end{pmatrix} \stackrel{(ii)}{\mapsto} \begin{pmatrix} 1 & -1 & 0 & -\frac{5}{2} \\ 0 & 0 & 1 & \frac{7}{2} \\ 0 & 0 & 0 & 1 \\ 0 & 0 & 0 & \frac{33}{2} \end{pmatrix}$$

$$\stackrel{(iii)}{\mapsto} \begin{pmatrix} 1 & -1 & 0 & 0 \\ 0 & 0 & 1 & 0 \\ 0 & 0 & 0 & 1 \\ 0 & 0 & 0 & 0 \end{pmatrix}.$$

The final matrix here is in reduced row-echelon form. In the notation of the definition, the number of nonzero rows in the reduced row-echelon form is $l = 3$, and the integers $j(i)$ are $j(1) = 1$, $j(2) = 3$, and $j(3) = 4$.

The example makes clear what the algorithm is that proves Proposition 1.25. We find the first nonzero column, apply an interchange (an operation of type (i)) if necessary to make the first entry in the column nonzero, multiply by a nonzero scalar to make the first entry 1 (an operation of type (ii)), and apply operations of type (iii) to eliminate the other nonzero entries in the column. Then we look for the next column with a nonzero entry in entries 2 and later, interchange to get the nonzero entry into entry 2 of the column, multiply to make the entry 1, and apply operations of type (iii) to eliminate the other entries in the column. Continuing in this way, we arrive at reduced row-echelon form.

Once our array, which contains both sides of our system of equations, has been transformed into reduced row-echelon form, we can read off exactly what the solutions are. It will be handy to distinguish two kinds of variables among $x_1, \ldots, x_n$ without including any of $x_{n+1}, \ldots, x_m$ in either of the classes. The **corner variables** are those x_j's for which j is $\leq n$ and is some $j(i)$ in the definition of "reduced row-echelon form," and the other x_j's with $j \leq n$ will be called **independent variables**. Let us describe the last steps of the solution technique in the setting of an example. We restore the vertical line that separated the data on the two sides of the equations.

EXAMPLE. We consider what might happen to a certain system of 4 equations in 4 unknowns. Putting the data in place for the right side makes the array have 4 rows and 5 columns. We transform the array into reduced row-echelon form and suppose that it comes out to be

$$\left(\begin{array}{cccc|c} 1 & -1 & 0 & 0 & 1 \\ 0 & 0 & 1 & 0 & 2 \\ 0 & 0 & 0 & 1 & 3 \\ 0 & 0 & 0 & 0 & 1 \text{ or } 0 \end{array} \right).$$

If the lower right entry is 1, there are no solutions. In fact, the last row corresponds to an equation $0 = 1$, which announces a contradiction. More generally, if any row of 0's to the left of the vertical line is equal to something nonzero, there are no solutions. In other words, there are no solutions to a system if the reduced row-echelon form of the entire array has more nonzero rows than the reduced row-echelon form of the part of the array to the left of the vertical line.

On the other hand, if the lower right entry is 0, then there are solutions. To see this, we restore the reduced array to a system of equations:

$$\begin{aligned} x_1 - x_2 &= 1, \\ x_3 &= 2, \\ x_4 &= 3; \end{aligned}$$

we move the independent variables (namely x_2 here) to the right side to obtain

$$\begin{aligned} x_1 &= 1 + x_2, \\ x_3 &= 2, \\ x_4 &= 3; \end{aligned}$$

and we collect everything in a tidy fashion as

$$\begin{pmatrix} x_1 \\ x_2 \\ x_3 \\ x_4 \end{pmatrix} = \begin{pmatrix} 1 \\ 0 \\ 2 \\ 3 \end{pmatrix} + x_2 \begin{pmatrix} 1 \\ 1 \\ 0 \\ 0 \end{pmatrix}.$$

The independent variables are allowed to take on arbitrary values, and we have succeeded in giving a formula for the solution that corresponds to an arbitrary set of values for the independent variables.

The method in the above example works completely generally. We obtain solutions whenever each row of 0's to the left of the vertical line is matched by a 0 on the right side, and we obtain no solutions otherwise. In the case that we are solving several systems with the same left sides, solutions exist for each of the systems if the reduced row-echelon form of the entire array has the same number of nonzero rows as the reduced row-echelon form of the part of the array to the left of the vertical line.

Let us record some observations about the method for solving systems of linear equations and then some observations about the method of row reduction itself.

Proposition 1.26. In the solution process for a system of k linear equations in n variables with the vertical line in place,
  (a) the sum of the number of corner variables and the number of independent variables is n,
  (b) the number of corner variables equals the number of nonzero rows on the left side of the vertical line and hence is $\leq k$,
  (c) when solutions exist, they are of the form

$$\text{column} + \genfrac{}{}{0pt}{}{\text{independent}}{\text{variable}} \times \text{column} + \cdots + \genfrac{}{}{0pt}{}{\text{independent}}{\text{variable}} \times \text{column}$$

  in such a way that each independent variable x_j is a free parameter in $\mathbb{F}$, the column multiplying x_j has a 1 in its jth entry, and the other columns have a 0 in that entry,
  (d) a **homogeneous system**, i.e., one with all right sides equal to 0, has a nonzero solution if the number k of equations is $<$ the number n of variables,
  (e) the solutions of an **inhomogeneous system**, i.e., one in which the right sides are not necessarily all 0, are all given by the sum of any one particular solution and an arbitrary solution of the corresponding homogeneous system.

PROOF. Conclusions (a), (b), and (c) follow immediately by inspection of the solution method. For (d), we observe that no contradictory equation can arise when the right sides are 0 and, in addition, that there must be at least one independent variable by (a) since (b) shows that the number of corner variables is $\leq k < n$. Conclusion (e) is apparent from (c), since the first column in the solution written in (c) is a column of 0's in the homogeneous case. □

Proposition 1.27. For an array with k rows and n columns in reduced row-echelon form,
  (a) the sum of the number of corner variables and the number of independent variables is n,
  (b) the number of corner variables equals the number of nonzero rows and hence is $\leq k$,

(c) when $k = n$, either the array is of the form

$$\begin{pmatrix} 1 & 0 & 0 & \cdots & 0 \\ 0 & 1 & 0 & \cdots & 0 \\ 0 & 0 & 1 & \cdots & 0 \\ & & & \ddots & \\ 0 & 0 & 0 & \cdots & 1 \end{pmatrix}$$

or else it has a row of 0's.

PROOF. Conclusions (a) and (b) are immediate by inspection. In (c), failure of the reduced row-echelon form to be as indicated forces there to be some noncorner variable, so that the number of corner variables is $< n$. By (b), the number of nonzero rows is $< n$, and hence there is a row of 0's. □

One final comment: For the special case of n equations in n variables, some readers may be familiar with a formula known as "Cramer's rule" for using determinants to solve the system when the determinant of the array of coefficients on the left side of the vertical line is nonzero. Determinants, including their evaluation, and Cramer's rule will be discussed in Chapter II. The point to make for current purposes is that the use of Cramer's rule for computation is, for n large, normally a more lengthy process than the method of row reduction. In fact, Problem 13 at the end of this chapter shows that the number of steps for solving the system via row reduction is at most a certain multiple of n^3. On the other hand, the typical number of steps for solving the system by rote application of Cramer's rule is approximately a multiple of n^4.

6. Matrix Operations

A rectangular array of scalars (i.e., members of $\mathbb{F}$) with k rows and n columns is called a k-by-n matrix. More precisely a k-**by**-n **matrix** over $\mathbb{F}$ is a function from $\{1, \ldots, k\} \times \{1, \ldots, n\}$ to $\mathbb{F}$. The expression "k-by-n" is called the **size** of the matrix. The value of the function at the ordered pair (i, j) is often indicated with subscript notation, such as a_{ij}, rather than with the usual function notation $a(i, j)$. It is called the $(i, j)^{\text{th}}$ **entry**. Two matrices are **equal** if they are the same function on ordered pairs; this means that they have the same size and their corresponding entries are equal. A matrix is called **square** if its number of rows equals its number of columns. A square matrix with all entries 0 for $i \neq j$ is called **diagonal**, and the entries with $i = j$ are the **diagonal entries**.

As the reader likely already knows, it is customary to write matrices in rectangular patterns. By convention the first index always tells the number of the row and the second index tells the number of the column. Thus a typical 2-by-3 matrix is $\begin{pmatrix} a_{11} & a_{12} & a_{13} \\ a_{21} & a_{22} & a_{23} \end{pmatrix}$. In the indication of the size of the matrix, here 2-by-3, the 2 refers to the number of rows and the 3 refers to the number of columns.

An n-dimensional **row vector** is a 1-by-n matrix, while a k-dimensional **column vector** is a k-by-1 matrix. The set of all k-dimensional column vectors is denoted by $\mathbb{F}^k$. The set $\mathbb{F}^k$ is to be regarded as the space of all ordinary garden-variety vectors. For economy of space, books often write such vectors horizontally with entries separated by commas, for example as (c_1, c_2, c_3), and it is extremely important to treat such vectors as *column* vectors, not as row vectors, in order to get matrix operations and the effect of linear transformations to correspond nicely.[5] Thus in this book, (c_1, c_2, c_3) *is to be regarded as a space-saving way of writing the column vector* $\begin{pmatrix} c_1 \\ c_2 \\ c_3 \end{pmatrix}$.

If a matrix is denoted by some letter like A, its $(i, j)^{\text{th}}$ entry will typically be denoted by A_{ij}. In the reverse direction, sometimes a matrix is assembled from its individual entries, which may be expressions depending on i and j. If some such expression a_{ij} is given for each pair (i, j), then we denote the corresponding matrix by $[a_{ij}]_{\substack{i=1,\ldots,k \\ j=1,\ldots,n}}$, or simply by $[a_{ij}]$ if there is no possibility of confusion.

Various operations are defined on matrices. Specifically let $M_{kn}(\mathbb{F})$ be the set of k-by-n matrices with entries in $\mathbb{F}$, so that $M_{k1}(\mathbb{F})$ is the same thing as $\mathbb{F}^k$. **Addition** of matrices is defined whenever two matrices have the same size, and it is defined entry by entry; thus if A and B are in $M_{kn}(\mathbb{F})$, then $A+B$ is the member of $M_{kn}(\mathbb{F})$ with $(A+B)_{ij} = A_{ij} + B_{ij}$. **Scalar multiplication** on matrices is defined entry by entry as well; thus if A is in $M_{kn}(\mathbb{F})$ and c is in $\mathbb{F}$, then cA is the member of $M_{kn}(\mathbb{F})$ with $(cA)_{ij} = cA_{ij}$. The matrix $(-1)A$ is denoted by $-A$. The k-by-n matrix with 0 in each entry is called a **zero matrix**. Ordinarily it is denoted simply by 0; if some confusion is possible in a particular situation, more precise notation will be introduced at the time. With these operations the set $M_{kn}(\mathbb{F})$ has the following properties:

(i) the operation of addition satisfies

   (a) $A + (B + C) = (A + B) + C$ for all A, B, C in $M_{kn}(\mathbb{F})$ (associative law),
   (b) $A + 0 = 0 + A = A$ for all A in $M_{kn}(\mathbb{F})$,
   (c) $A + (-A) = (-A) + 0$ for all A in $M_{kn}(\mathbb{F})$,
   (d) $A + B = B + A$ for all A and B in $M_{kn}(\mathbb{F})$ (commutative law);

---

[5]The alternatives are unpleasant. Either one is forced to write certain functions in the unnatural notation $x \mapsto (x)f$, or the correspondence is forced to involve transpose operations on frequent occasions. Unhappily, books following either of these alternative conventions may be found.

(ii) the operation of scalar multiplication satisfies
  (a) $(cd)A = c(dA)$ for all A in $M_{kn}(\mathbb{F})$ and all scalars c and d,
  (b) $1A = A$ for all A in $M_{kn}(\mathbb{F})$ and for the scalar 1;
(iii) the two operations are related by the distributive laws
  (a) $c(A + B) = cA + cB$ for all A and B in $M_{kn}(\mathbb{F})$ and for all scalars c,
  (b) $(c + d)A = cA + dA$ for all A in $M_{kn}(\mathbb{F})$ and all scalars c and d.

Since addition and scalar multiplication are defined entry by entry, all of these identities follow from the corresponding identities for members of $\mathbb{F}$.

Multiplication of matrices is defined in such a way that the kind of system of linear equations discussed in the previous section can be written as a matrix equation in the form $AX = B$, where

$$A = \begin{pmatrix} a_{11} & \cdots & a_{1n} \\ & \ddots & \\ a_{k1} & \cdots & a_{kn} \end{pmatrix}, \quad X = \begin{pmatrix} x_1 \\ \vdots \\ x_n \end{pmatrix}, \quad \text{and} \quad B = \begin{pmatrix} b_1 \\ \vdots \\ b_k \end{pmatrix}.$$

More precisely if A is a k-by-m matrix and B is an m-by-n matrix, then the product $C = AB$ is the k-by-n matrix defined by

$$C_{ij} = \sum_{l=1}^{m} A_{il} B_{lj}.$$

The $(i, j)^{\text{th}}$ entry of C is therefore the product of the ith row of A and the jth column of B.

Let us emphasize that the condition for a product AB to be defined is that the number of columns of A should equal the number of rows of B. With this definition the system of equations mentioned above is indeed of the form $AX = B$.

Proposition 1.28. Matrix multiplication has the properties that
  (a) it is associative in the sense that $(AB)C = A(BC)$, provided that the sizes match correctly, i.e., A is in $M_{km}(\mathbb{F})$, B is in $M_{mn}(\mathbb{F})$, and C is in $M_{np}(\mathbb{F})$,
  (b) it is distributive over addition in the sense that $A(B + C) = AB + AC$ and $(B + C)D = BD + CD$ if the sizes match correctly.

REMARK. Matrix multiplication is not necessarily commutative, even for square matrices. For example, $\begin{pmatrix} 1 & 0 \\ 0 & 0 \end{pmatrix}\begin{pmatrix} 0 & 1 \\ 0 & 0 \end{pmatrix} = \begin{pmatrix} 0 & 1 \\ 0 & 0 \end{pmatrix}$, while $\begin{pmatrix} 0 & 1 \\ 0 & 0 \end{pmatrix}\begin{pmatrix} 1 & 0 \\ 0 & 0 \end{pmatrix} = \begin{pmatrix} 0 & 0 \\ 0 & 0 \end{pmatrix}$.

PROOF. For (a), we have

$$((AB)C)_{ij} = \sum_{t=1}^{n} (AB)_{it} C_{tj} = \sum_{t=1}^{n} \sum_{s=1}^{m} A_{is} B_{st} C_{tj}$$

and $(A(BC))_{ij} = \sum_{s=1}^{m} A_{is} (BC)_{sj} = \sum_{s=1}^{m} \sum_{t=1}^{n} A_{is} B_{st} C_{tj},$

and these are equal. For the first identity in (b), we have

$$(A(B+C))_{ij} = \sum_{l} A_{il}(B+C)_{lj} = \sum_{l} A_{il}(B_{lj} + C_{lj})$$
$$= \sum_{l} A_{il} B_{lj} + \sum_{l} A_{il} C_{lj} = (AB)_{ij} + (AC)_{ij},$$

and the second identity is proved similarly. □

We have already defined the zero matrix 0 of a given size to be the matrix having 0 in each entry. This matrix has the property that $0A = 0$ and $B0 = 0$ if the sizes match properly. The n-by-n **identity matrix**, denoted by I or sometimes 1, is defined to be the matrix with $I_{ij} = \delta_{ij}$, where δ_{ij} is the **Kronecker delta** defined by

$$\delta_{ij} = \begin{cases} 1 & \text{if } i = j, \\ 0 & \text{if } i \neq j. \end{cases}$$

In other words, the identity matrix is the square matrix of the form

$$I = \begin{pmatrix} 1 & 0 & 0 & \cdots & 0 \\ 0 & 1 & 0 & \cdots & 0 \\ 0 & 0 & 1 & \cdots & 0 \\ & & & \ddots & \\ 0 & 0 & 0 & \cdots & 1 \end{pmatrix}.$$

It has the property that $IA = A$ and $BI = I$ whenever the sizes match properly.

Let A be an n-by-n matrix. We say that A is **invertible** and has the n-by-n matrix B as **inverse** if $AB = BA = I$. If B and C are n-by-n matrices with $AB = I$ and $CA = I$, then associativity of multiplication (Proposition 1.28a) implies that $B = IB = (CA)B = C(AB) = CI = C$. Hence an inverse for A is unique if it exists. We write A^{-1} for this inverse if it exists. Inverses of n-by-n matrices have the property that if A and D are invertible, then AD is invertible and $(AD)^{-1} = D^{-1}A^{-1}$; moreover, if A is invertible, then A^{-1} is invertible and its inverse is A.

The method of row reduction in the previous section suggests a way of computing the inverse of a matrix. Suppose that A is a square matrix to be inverted and we are seeking its inverse B. Then $AB = I$. Examining the definition of matrix multiplication, we see that this matrix equation means that the product of A and the first column of B equals the first column of I, the product of A and the second column of B equals the second column of I, and so on. We can thus think of a column of B as the unknowns in a system of linear equations, the known right sides being the entries of the column of the identity matrix. As the column index varies, the left sides of these equations do not change, since they are always given by A. So we can attempt to solve all of the systems (one for each column) simultaneously. For example, to attempt to invert $A = \begin{pmatrix} 1 & 2 & 3 \\ 4 & 5 & 6 \\ 7 & 8 & 10 \end{pmatrix}$, we set up

$$\left(\begin{array}{ccc|ccc} 1 & 2 & 3 & 1 & 0 & 0 \\ 4 & 5 & 6 & 0 & 1 & 0 \\ 7 & 8 & 10 & 0 & 0 & 1 \end{array} \right).$$

Imagine doing the row reduction. We can hope that the result will be of the form

$$\left(\begin{array}{ccc|ccc} 1 & 0 & 0 & - & - & - \\ 0 & 1 & 0 & - & - & - \\ 0 & 0 & 1 & - & - & - \end{array} \right),$$

with the identity matrix on the left side of the vertical line. If this is indeed the result, then the computation shows that the matrix on the right side of the vertical line is the only possibility for A^{-1}. But does A^{-1} in fact exist?

Actually, another question arises as well. According to Proposition 1.27c, the other possibility in applying row reduction is that the left side has a row of 0's. In this case, can we deduce that A^{-1} does not exist? Or, to put it another way, can we be sure that some row of the reduced row-echelon form has all 0's on the left side of the vertical line and something nonzero on the right side?

All of the answers to these questions are yes, and we prove them in a moment. First we need to see that elementary row operations are given by matrix multiplications.

Proposition 1.29. Each elementary row operation is given by left multiplication by an invertible matrix. The inverse matrix is the matrix of another elementary row operation.

REMARK. The square matrices giving these left multiplications are called **elementary matrices**.

PROOF. For the interchange of rows i and j, the part of the elementary matrix in the rows and columns with i or j as index is

$$\begin{array}{c} \quad i \quad j \\ \begin{array}{c} i \\ j \end{array} \begin{pmatrix} 0 & 1 \\ 1 & 0 \end{pmatrix}, \end{array}$$

and otherwise the matrix is the identity. This matrix is its own inverse.

For the multiplication of the ith row by a nonzero scalar c, the matrix is diagonal with c in the ith diagonal entry and with 1 in all other diagonal entries. The inverse matrix is of this form with c^{-1} in place of c.

For the replacement of the ith row by the sum of the ith row and the product of a times the jth row, the part of the elementary matrix in the rows and columns with i or j as index is

$$\begin{array}{c} & i \quad j \\ i \\ j \end{array} \begin{pmatrix} 1 & a \\ 0 & 1 \end{pmatrix},$$

and otherwise the matrix is the identity. The inverse of this matrix is the same except that a is replaced by $-a$. □

Theorem 1.30. The following conditions on an n-by-n square matrix A are equivalent:

(a) the reduced row-echelon form of A is the identity,
(b) A is the product of elementary matrices,
(c) A has an inverse,
(d) the system of equations $AX = 0$ with $X = \begin{pmatrix} x_1 \\ \vdots \\ x_n \end{pmatrix}$ has only the solution $X = 0$.

PROOF. If (a) holds, choose a sequence of elementary row operations that reduce A to the identity, and let $E_1, \ldots, E_r$ be the corresponding elementary matrices given by Proposition 1.29. Then we have $E_r \cdots E_1 A = I$, and hence $A = E_1^{-1} \cdots E_r^{-1}$. The proposition says that each E_j^{-1} is an elementary matrix, and thus (b) holds.

If (b) holds, then (c) holds because the elementary matrices are invertible and the product of invertible matrices is invertible.

If (c) holds and if $AX = 0$, then $X = IX = (A^{-1}A)X = A^{-1}(AX) = A^{-1}0 = 0$. Hence (d) holds.

If (d) holds, then the number of independent variables in the row reduction of A is 0. Proposition 1.26a shows that the number of corner variables is n, and parts (b) and (c) of Proposition 1.27 show that the reduced row-echelon form of A is I. Thus (a) holds. □

Corollary 1.31. If the solution procedure for finding the inverse of a square matrix A leads from $(A \mid I)$ to $(I \mid X)$, then A is invertible and its inverse is X. Conversely if the solution procedure leads to $(R \mid Y)$ and R has a row of 0's, then A is not invertible.

REMARK. Proposition 1.27c shows that this corollary addresses the only possible outcomes of the solution procedure.

PROOF. We apply the equivalence of (a) and (c) in Theorem 1.30 to settle the existence or nonexistence of A^{-1}. In the case that A^{-1} exists, we know that the solution procedure has to yield the inverse. □

Corollary 1.32. Let A be a square matrix. If B is a square matrix such that $BA = I$, then A is invertible and B is its inverse. If C is a square matrix such that $AC = I$, then A is invertible with inverse C.

PROOF. Suppose $BA = I$. Let X be a column vector with $AX = 0$. Then $X = IX = (BA)X = B(AX) = B0 = 0$. Since (d) implies (c) in Theorem 1.30, A is invertible.

Suppose $AC = I$. Applying the result of the previous paragraph to C, we conclude that C is invertible with inverse A. Therefore A is invertible with inverse C. □

7. Problems

1. What is the greatest common divisor of 9894 and 11058?
2. (a) Find integers x and y such that $11x + 7y = 1$.
   (b) How are all pairs (x, y) of integers satisfying $11x + 7y = 1$ related to the pair you found in (a)?
3. Let $\{a_n\}_{n\geq 1}$ be a sequence of positive integers, and let d be the largest integer dividing all a_n. Prove that d is the greatest common divisor of finitely many of the a_n.
4. Determine the integers n for which there exist integers x and y such that n divides $x + y - 2$ and $2x - 3y - 3$.
5. Let $P(X)$ and $Q(X)$ be the polynomials $P(X) = X^4 + X^3 + 2X^2 + X + 1$ and $Q(X) = X^5 + 2X^3 + X$ in $\mathbb{R}[X]$.
   (a) Find a greatest common divisor $D(X)$ of $P(X)$ and $Q(X)$.
   (b) Find polynomials A and B such that $AP + BQ = D$.
6. Let $P(X)$ and $Q(X)$ be polynomials in $\mathbb{R}[X]$. Prove that if $D(X)$ is a greatest common divisor of $P(X)$ and $Q(X)$ in $\mathbb{C}[X]$, then there exists a nonzero complex number c such that $cD(X)$ is in $\mathbb{R}[X]$.
7. (a) Let $P(X)$ be in $\mathbb{R}[X]$, and regard it as in $\mathbb{C}[X]$. Applying the Fundamental Theorem of Algebra and its corollary to P, prove that if z_j is a root of P, then so is $\bar{z}_j$, and z_j and $\bar{z}_j$ have the same multiplicity.
   (b) Deduce that any prime polynomial in $\mathbb{R}[X]$ has degree at most 2.

8. (a) Suppose that a polynomial $A(X)$ of degree > 0 in $\mathbb{Q}[X]$ has integer coefficients and leading coefficient 1. Show that if p/q is a root of $A(X)$ with p and q integers such that $\mathrm{GCD}(p, q) = 1$, then p/q is an integer n and n divides the constant term of $A(X)$.
   (b) Deduce that $X^2 - 2$ and $X^3 + X^2 + 1$ are prime in $\mathbb{Q}[X]$.

9. Reduce the fraction $8645/10465$ to lowest terms.

10. How many different patterns are there of disjoint cycle structures for permutations of $\{1, 2, 3, 4\}$? Give examples of each, telling how many permutations there are of each kind and what the signs are of each.

11. Prove for $n \geq 2$ that the number of permutations of $\{1, \ldots, n\}$ with sign -1 equals the number with sign $+1$.

12. Find all solutions X of the system $AX = B$ when $A = \begin{pmatrix} 1 & 2 & 3 \\ 4 & 5 & 6 \\ 7 & 8 & 9 \end{pmatrix}$ and B is given by

   (a) $B = \begin{pmatrix} 0 \\ 0 \\ 0 \end{pmatrix}$,   (b) $B = \begin{pmatrix} 5 \\ 3 \\ 2 \end{pmatrix}$,   (c) $B = \begin{pmatrix} 3 \\ 2 \\ 1 \end{pmatrix}$.

13. Suppose that a single step in the row reduction process means a single arithmetic operation or a single interchange of two entries. Prove that there exists a constant C such that any square matrix can be transformed into reduced row-echelon form in $\leq Cn^3$ steps, the matrix being of size n-by-n.

14. Compute $A + B$ and AB if $A = \begin{pmatrix} 2 & 3 \\ 4 & 5 \end{pmatrix}$ and $B = \begin{pmatrix} -4 & 8 \\ -1 & 3 \end{pmatrix}$.

15. Prove that if A and B are square matrices with $AB = BA$, then $(A + B)^n$ is given by the Binomial Theorem: $(A + B)^n = \sum_{k=0}^{n} \binom{n}{k} A^{n-k} B^k$, where $\binom{n}{k}$ is the binomial coefficient $n!/((n - k)!k!)$.

16. Find a formula for the nth power of $\begin{pmatrix} 1 & 1 & 0 \\ 0 & 1 & 1 \\ 0 & 0 & 1 \end{pmatrix}$, n being a positive integer.

17. Let D be an n-by-n diagonal matrix with diagonal entries $d_1, \ldots, d_n$, and let A be an n-by-n matrix. Compute AD and DA, and give a condition for the equality $AD = DA$ to hold.

18. Fix n, and let E_{ij} denote the n-by-n matrix that is 1 in the $(i, j)^{\mathrm{th}}$ entry and is 0 elsewhere. Compute the product $E_{kl} E_{pq}$, expressing the result in terms of matrices E_{ij} and instances of the Kronecker delta.

19. Verify that if $ad - bc \neq 0$, then $\begin{pmatrix} a & b \\ c & d \end{pmatrix}^{-1} = (ad - bc)^{-1} \begin{pmatrix} d & -b \\ -c & a \end{pmatrix}$ and that the system $\begin{pmatrix} a & b \\ c & d \end{pmatrix} \begin{pmatrix} x \\ y \end{pmatrix} = \begin{pmatrix} p \\ q \end{pmatrix}$ has the unique solution $\begin{pmatrix} x \\ y \end{pmatrix} = (ad - bc)^{-1} \begin{pmatrix} dp - bq \\ aq - cp \end{pmatrix}$.

20. Which of the following matrices A is invertible? For the invertible ones, find A^{-1}.

(a) $A = \begin{pmatrix} 1 & 2 & 3 \\ 4 & 5 & 6 \\ 7 & 8 & 9 \end{pmatrix}$,   (b) $A = \begin{pmatrix} 1 & 2 & 3 \\ 4 & 5 & 6 \\ 7 & 8 & 10 \end{pmatrix}$,   (c) $A = \begin{pmatrix} 7 & 4 & 1 \\ 6 & 4 & 1 \\ 4 & 3 & 1 \end{pmatrix}$.

21. Can a square matrix with a row of 0's be invertible? Why or why not?

22. Prove that if the product AB of two n-by-n matrices is invertible, then A and B are invertible.

23. Let A be a square matrix such that $A^k = 0$ for some positive integer n. Prove that $I + A$ is invertible.

24. Give an example of a set S and functions $f : S \to S$ and $g : S \to S$ such that the composition $g \circ f$ is the identity function but neither f nor g has an inverse function.

25. Give an example of two matrices, A of size 1-by-2 and B of size 2-by-1, such that $AB = I$, I being the 1-by-1 identity matrix. Verify that BA is not the 2-by-2 identity matrix. Give a proof for these sizes that BA can never be the identity matrix.

Problems 26–29 concern least common multiples. Let a and b be positive integers. A **common multiple** of a and b is an integer N such that a and b both divide N. The **least common multiple** of a and b is the smallest positive common multiple of a and b. It is denoted by $\text{LCM}(a, b)$.

26. Prove that a and b have a least common multiple.

27. If a has a prime factorization given by $a = p_1^{k_1} \cdots p_r^{k_r}$, prove that any positive multiple M of a has a prime factorization given by $a = p_1^{m_1} \cdots p_r^{m_r} q_1^{n_1} \cdots q_s^{n_s}$, where $q_1, \ldots, q_s$ are primes not in the list $p_1, \ldots, p_r$, where $m_j \geq k_j$ for all j, and where $n_j \geq 0$ for all j.

28. (a) Prove that if $a = p_1^{k_1} \cdots p_r^{k_r}$ and $b = p_1^{l_1} \cdots p_r^{l_r}$ are expansions of a and b as products of powers of r distinct primes $p_1, \ldots, p_r$, then $\text{LCM}(a, b) = p_1^{\max(k_1, l_1)} \cdots p_r^{\max(k_r, l_r)}$.

   (b) Prove that if N is any common multiple of a and b, then $\text{LCM}(a, b)$ divides N.

   (c) Deduce that $ab = \text{GCD}(a, b)\,\text{LCM}(a, b)$.

29. If $a_1, \ldots, a_t$ are positive integers, define their **least common multiple** to be the smallest positive integer M such that each a_j divides M. Give a formula for this M in terms of expansions of $a_1, \ldots, a_t$ as products of powers of distinct primes.

CHAPTER II

Vector Spaces over $\mathbb{Q}$, $\mathbb{R}$, and $\mathbb{C}$

Abstract. This chapter introduces vector spaces and linear maps between them, and it goes on to develop certain constructions of new vector spaces out of old, as well as various properties of determinants.

Sections 1–2 define vector spaces, spanning, linear independence, bases, and dimension. The sections make use of row reduction to establish dimension formulas for certain vector spaces associated with matrices. They conclude by stressing methods of calculation that have quietly been developed in proofs.

Section 3 relates matrices and linear maps to each other, first in the case that the linear map carries column vectors to column vectors and then in the general finite-dimensional case. Techniques are developed for working with the matrix of a linear map relative to specified bases and for changing bases. The section concludes with a discussion of isomorphisms of vector spaces.

Sections 4–6 take up constructions of new vector spaces out of old ones, together with corresponding constructions for linear maps. The four constructions of vector spaces in these sections are those of the dual of a vector space, the quotient of two vector spaces, and the direct sum and direct product of two or more vector spaces.

Section 7 introduces determinants of square matrices, together with their calculation and properties. Some of the results that are established are expansion in cofactors, Cramer's rule, and the value of the determinant of a Vandermonde matrix. It is shown that the determinant function is well defined on any linear map from a finite-dimensional vector space to itself.

Section 8 introduces eigenvectors and eigenvalues for matrices, along with their computation. Also, in this section the characteristic polynomial and the trace of a square matrix are defined, and all these notions are reinterpreted in terms of linear maps.

Section 9 proves the existence of bases for infinite-dimensional vector spaces and discusses the extent to which the material of the first eight sections extends from the finite-dimensional case to be valid in the infinite-dimensional case.

1. Spanning, Linear Independence, and Bases

This chapter develops a theory of rational, real, and complex vector spaces. Many readers will already be familiar with some aspects of this theory, particularly in the case of the vector spaces $\mathbb{Q}^n$, $\mathbb{R}^n$, and $\mathbb{C}^n$ of column vectors, where the tools developed from row reduction allow one to introduce geometric notions and to view geometrically the set of solutions to a set of linear equations. Thus we shall be brief about many of these matters, concentrating on the algebraic aspects of the theory. Let $\mathbb{F}$ denote any of $\mathbb{Q}, \mathbb{R}$, or $\mathbb{C}$. Members of $\mathbb{F}$ are called **scalars**.[1]

A **vector space** over $\mathbb{F}$ is a set V with two operations, **addition** carrying $V \times V$ into V and **scalar multiplication** carrying $\mathbb{F} \times V$ into V, with the following properties:

(i) the operation of addition, written $+$, satisfies
  (a) $v_1 + (v_2 + v_3) = (v_1 + v_2) + v_3$ for all v_1, v_2, v_3 in V (associative law),
  (b) there exists an element 0 in V with $v + 0 = 0 + v = v$ for all v in V,
  (c) to each v in V corresponds an element $-v$ in V such that $v + (-v) = (-v) + v = 0$,
  (d) $v_1 + v_2 = v_2 + v_1$ for all v_1 and v_2 in V (commutative law);

(ii) the operation of scalar multiplication, written without a sign, satisfies
  (a) $a(bv) = (ab)v$ for all v in V and all scalars a and b,
  (a) $1v = v$ for all v in V and for the scalar 1;

(iii) the two operations are related by the distributive laws
  (a) $a(v_1 + v_2) = av_1 + av_2$ for all v_1 and v_2 in V and for all scalars a,
  (b) $(a + b)v = av + bv$ for all v in V and all scalars a and b.

It is immediate from these properties that

- 0 is unique (since $0' = 0' + 0 = 0$),
- $-v$ is unique (since $(-v)' = (-v)' + 0 = (-v)' + (v + (-v)) = ((-v)' + v) + (-v) = 0 + (-v) = (-v)$),
- $0v = 0$ (since $0v = (0 + 0)v = 0v + 0v$),
- $(-1)v = -v$ (since $0 = 0v = (1 + (-1))v = 1v + (-1)v = v + (-1)v$),
- $a0 = 0$ (since $a0 = a(0 + 0) = a0 + a0$).

Members of V are called **vectors**.

EXAMPLES.

(1) $V = M_{kn}(\mathbb{F})$, the space of all k-by-n matrices. The above properties of a vector space over $\mathbb{F}$ were already observed in Section I.6. The vector space $\mathbb{F}^k$ of all k-dimensional column vectors is the special case $n = 1$, and the vector space $\mathbb{F}$ of scalars is the special case $k = n = 1$.

(2) Let S be any nonempty set, and let V be the set of all functions from S into $\mathbb{F}$. Define operations by $(f + g)(s) = f(s) + g(s)$ and $(cf)(s) = c(f(s))$. The operations on the right sides of these equations are those in $\mathbb{F}$, and the properties of a vector space follow from the fact that they hold in $\mathbb{F}$ at each s.

---

[1] All the material of this chapter will ultimately be seen to work when $\mathbb{F}$ is replaced by any "field." This point will not be important for us at this stage, and we postpone considering it further until Chapter IV.

1. Spanning, Linear Independence, and Bases   35

(3) More generally than in Example 2, let S be any nonempty set, let U be a vector space over $\mathbb{F}$, and let V be the set of all functions from S into U. Define the operations as in Example 2, but interpret the operations on the right sides of the defining equations as those in U. Then the properties of a vector space follow from the fact that they hold in U at each s.

(4) Let V be any vector space over $\mathbb{C}$, and restrict scalar multiplication to an operation $\mathbb{R} \times V \to V$. Then V becomes a vector space over $\mathbb{R}$. In particular, $\mathbb{C}$ is a vector space over $\mathbb{R}$.

(5) Let $V = \mathbb{F}[X]$ be the set of all polynomials in one indeterminate with coefficients in $\mathbb{F}$, and define addition and scalar multiplication as in Section I.3. Then V is a vector space.

(6) Let V be any vector space over $\mathbb{F}$, and let U be any nonempty subset closed under addition and scalar multiplication. Then U is a vector space over $\mathbb{F}$. Such a subset U is called a **vector subspace** of V; sometimes one says simply **subspace** if the context is unambiguous.[2]

(7) Let V be any vector space over $\mathbb{F}$, and let $U = \{v_\alpha\}$ be any subset of V. A **finite linear combination** of the members of U is any vector of the form $c_{\alpha_1} v_{\alpha_1} + \cdots + c_{\alpha_n} v_{\alpha_n}$ with each c_{α_j} in $\mathbb{F}$, each v_{α_j} in U, and $n \geq 0$. The **linear span** of U is the set of all finite linear combinations of members of U. It is a vector subspace of V and is denoted by $\mathrm{span}\{v_\alpha\}$. By convention, span $\varnothing = 0$.

(8) Many vector subspaces arise in the context of some branch of mathematics after some additional structure is imposed. For example let V be the vector space of all functions from $\mathbb{R}^3$ into $\mathbb{R}$, an instance of Example 2. The subset U of continuous members of V is a vector subspace; the closure under addition and scalar multiplication comes down to knowing that addition is a continuous function from $\mathbb{R}^3 \times \mathbb{R}^3$ into $\mathbb{R}^3$ and that scalar multiplication from $\mathbb{R} \times \mathbb{R}^3$ into $\mathbb{R}^3$ is continuous as well. Another example is the subset of twice continuously differentiable members f of V satisfying the partial differential equation $\frac{\partial^2 f}{\partial x_1^2} + \frac{\partial^2 f}{\partial x_2^2} + \frac{\partial^2 f}{\partial x_3^2} + f = 0$ on $\mathbb{R}^3$.

The associative and commutative laws in the definition of "vector space" imply certain more complicated formulas of which the stated laws are special cases. With associativity of addition, if n vectors $v_1, \ldots, v_n$ are given, then any way of inserting parentheses into the expression $v_1 + v_2 + \cdots + v_n$ leads to the same result, and a similar conclusion applies to the associativity-like formula $a(bv) = (ab)v$ for scalar multiplication. In the presence of associativity, the commutative law for addition implies that $v_1 + v_2 + \cdots + v_n = v_{\sigma(1)} + v_{\sigma(2)} + \cdots + v_{\sigma(n)}$ for any

---

[2]The word "subspace" arises also in the context of metric spaces and more general topological spaces, and the metric-topological notion of subspace is distinct from the vector notion of subspace.

permutation of $\{1, \ldots, n\}$. All these facts are proved by inductive arguments, and the details are addressed in Problems 2–3 at the end of the chapter.

Let V be a vector space over $\mathbb{F}$. A subset $\{v_\alpha\}$ of V **spans** V or is a **spanning set** for V if the linear span of $\{v_\alpha\}$, in the sense of Example 7 above, is all of V. A subset $\{v_\alpha\}$ is **linearly independent** if whenever a finite linear combination $c_{\alpha_1} v_{\alpha_1} + \cdots + c_{\alpha_n} v_{\alpha_n}$ equals the 0 vector, then all the coefficients must be 0: $c_{\alpha_1} = \cdots = c_{\alpha_n} = 0$. By subtraction we see that in this case any equality of two finite linear combinations

$$c_{\alpha_1} v_{\alpha_1} + \cdots + c_{\alpha_n} v_{\alpha_n} = d_{\alpha_1} v_{\alpha_1} + \cdots + d_{\alpha_n} v_{\alpha_n}$$

implies that the respective coefficients are equal: $c_{\alpha_j} = d_{\alpha_j}$ for $1 \leq j \leq n$.

A subset $\{v_\alpha\}$ is a **basis** if it spans V and is linearly independent. In this case each member of V has one and only one expansion as a finite linear combination of the members of $\{v_\alpha\}$.

EXAMPLE. In $\mathbb{F}^n$, the vectors

$$e_1 = \begin{pmatrix} 1 \\ 0 \\ 0 \\ \vdots \\ 0 \end{pmatrix}, \quad e_2 = \begin{pmatrix} 0 \\ 1 \\ 0 \\ \vdots \\ 0 \end{pmatrix}, \quad e_3 = \begin{pmatrix} 0 \\ 0 \\ 1 \\ \vdots \\ 0 \end{pmatrix}, \quad \ldots, \quad e_n = \begin{pmatrix} 0 \\ 0 \\ 0 \\ \vdots \\ 1 \end{pmatrix}$$

form a basis of $\mathbb{F}^n$ called the **standard basis** of $\mathbb{F}^n$.

Proposition 2.1. Let V be a vector space over $\mathbb{F}$.

(a) If $\{v_\alpha\}$ is a linearly independent subset of V that is maximal with respect to the property of being linearly independent (i.e., has the property of being strictly contained in no linearly independent set), then $\{v_\alpha\}$ is a basis of V.

(b) If $\{v_\alpha\}$ is a spanning set for V that is minimal with respect to the property of spanning (i.e., has the property of strictly containing no spanning set), then $\{v_\alpha\}$ is a basis of V.

PROOF. For (a), let v be given. We are to show that v is in the span of $\{v_\alpha\}$. Without loss of generality, we may assume that v is not in the set $\{v_\alpha\}$ itself. By the assumed maximality, $\{v_\alpha\} \cup \{v\}$ is not linearly independent, and hence $cv + c_{\alpha_1} v_{\alpha_1} + \cdots + c_{\alpha_n} v_{\alpha_n} = 0$ for some scalars $c, c_{\alpha_1}, \ldots, c_{\alpha_n}$ not all 0. Here $c \neq 0$ since $\{v_\alpha\}$ is linearly independent. Then $v = -c^{-1} c_{\alpha_1} v_{\alpha_1} - \cdots - c^{-1} c_{\alpha_n} v_{\alpha_n}$, and v is exhibited as in the linear span of $\{v_\alpha\}$.

For (b), suppose that $c_{\alpha_1} v_{\alpha_1} + \cdots + c_{\alpha_n} v_{\alpha_n} = 0$ with $c_{\alpha_1}, \ldots, c_{\alpha_n}$ not all 0. Say $c_{\alpha_1} \neq 0$. Then we can solve for v_{α_1} and see that v_{α_1} is a finite linear combination of $v_{\alpha_2}, \ldots, v_{\alpha_n}$. Substitution shows that any finite linear combination of the v_α's is a finite linear combination of the v_α's other than v_{α_1}, and we obtain a contradiction to the assumed minimality of the spanning set. $\square$

1. Spanning, Linear Independence, and Bases

Proposition 2.2. Let V be a vector space over $\mathbb{F}$. If V has a finite spanning set $\{v_1, \ldots, v_m\}$, then any linearly independent set in V has $\leq m$ elements.

PROOF. It is enough to show that no subset of $m + 1$ vectors can be linearly independent. Arguing by contradiction, suppose that $\{u_1, \ldots, u_n\}$ is a linearly independent set with $n = m + 1$. Write

$$u_1 = c_{11}v_1 + c_{21}v_2 + \cdots + c_{m1}v_m,$$
$$\vdots$$
$$u_n = c_{1n}v_1 + c_{2n}v_2 + \cdots + c_{mn}v_m.$$

The system of linear equations

$$c_{11}x_1 + \cdots + c_{1n}x_n = 0,$$
$$\vdots$$
$$c_{m1}x_1 + \cdots + c_{mn}x_n = 0,$$

is a homogeneous system of linear equations with more unknowns than equations, and Proposition 1.26d shows that it has a nonzero solution $(x_1, \ldots, x_n)$. Then we have

$$\begin{array}{ccccc}
x_1u_1 + \cdots + x_nu_n = & c_{11}x_1v_1 & + c_{21}x_1v_2 + \cdots + & c_{m1}x_1v_m \\
& + & + & + \\
& \cdots & \cdots & \cdots \\
& + & + & + \\
& c_{1n}x_nv_1 & + c_{2n}x_nv_2 + \cdots + & c_{mn}x_nv_m \\
= 0, & & &
\end{array}$$

in contradiction to the assumed linear independence of $\{u_1, \ldots, u_n\}$. $\square$

Corollary 2.3. If the vector space V has a finite spanning set $\{v_1, \ldots, v_m\}$, then

(a) $\{v_1, \ldots, v_m\}$ has a subset that is a basis,
(b) any linearly independent set in V can be extended to a basis,
(c) V has a basis,
(d) any two bases have the same finite number of elements, necessarily $\leq m$.

REMARKS. In this case we say that V is **finite-dimensional**, and the number of elements in a basis is called the **dimension** of V, written $\dim V$. If V has no finite spanning set, we say that V is infinite-dimensional. A suitable analog of the conclusion in Corollary 2.3 is valid in the infinite-dimensional case, but the proof is more complicated. We take up the infinite-dimensional case in Section 9.

PROOF. By discarding elements of the set $\{v_1, \ldots, v_m\}$ one at a time if necessary and by applying Proposition 2.1b, we obtain (a). For (b), we see from Proposition 2.2 that the given linearly independent set has $\leq m$ elements. If we adjoin elements to it one at a time so as to obtain larger linearly independent sets, Proposition 2.2 shows that there must be a stage at which we can proceed no further without violating linear independence. Proposition 2.1a then says that we have a basis. For (c), we observe that (a) has already produced a basis. Any two bases have the same number of elements, by two applications of Proposition 2.2, and this proves (d). □

EXAMPLES. The vector space $M_{kn}(\mathbb{F})$ of k-by-n matrices has dimension kn. The vector space of all polynomials in one indeterminate is infinite-dimensional because the subspace consisting of 0 and of all polynomials of degree $\leq n$ has dimension $n + 1$.

Corollary 2.4. If V is a finite-dimensional vector space with dim $V = n$, then any spanning set of n elements is a basis of V, and any linearly independent set of n elements is a basis of V. Consequently any n-dimensional vector subspace U of V coincides with V.

PROOF. These conclusions are immediate from parts (a) and (b) of Corollary 2.3 if we take part (d) into account. □

Corollary 2.5. If V is a finite-dimensional vector space and U is a vector subspace of V, then U is finite-dimensional, and dim $U \leq$ dim V.

PROOF. Let $\{v_1, \ldots, v_m\}$ be a basis of V. According to Proposition 2.2, any linearly independent set in U has $\leq m$ elements, being linearly independent in V. We can thus choose a maximal linearly independent subset of U with $\leq m$ elements, and Proposition 2.1a shows that the result is a basis of U. □

2. Vector Spaces Defined by Matrices

Let A be a member of $M_{kn}(\mathbb{F})$, thus a k-by-n matrix. The **row space** of A is the linear span of the rows of A, regarded as a vector subspace of the vector space of all n-dimensional row vectors. The **column space** of A is the linear span of the columns, regarded as a vector subspace of k-dimensional column vectors. The **null space** of A is the vector subspace of n-dimensional column vectors v for which $Av = 0$, where Av is the matrix product. The fact that this last space is a vector subspace follows from the properties $A(v_1 + v_2) = Av_1 + Av_2$ and $A(cv) = c(Av)$ of matrix multiplication.

2. Vector Spaces Defined by Matrices

We can use matrix multiplication to view the matrix A as defining a function $v \mapsto Av$ of $\mathbb{F}^n$ to $\mathbb{F}^k$. This function satisfies the properties just listed,

$$A(v_1 + v_2) = Av_1 + Av_2 \quad \text{and} \quad A(cv) = c(Av),$$

and we shall consider further functions with these two properties starting in the next section. In terms of this function, the null space of A is the set in the domain $\mathbb{F}^n$ mapped to 0. Because of these same properties and because the product Ae_j of A and the jth standard basis vector e_j in $\mathbb{F}^n$ is the jth column of A, the column space of A is the image of the function $v \mapsto Av$ as a subset of the range $\mathbb{R}^k$.

Theorem 2.6. If A is in $M_{kn}(\mathbb{F})$, then

$$\dim(\text{column space}(A)) + \dim(\text{null space}(A)) = \#(\text{columns of } A) = n.$$

PROOF. Corollary 2.5 says that the null space is finite-dimensional, being a vector subspace of $\mathbb{F}^n$, and Corollary 2.3c shows that the null space has a basis, say $\{v_1, \ldots, v_r\}$. By Corollary 2.3b we can adjoin vectors $v_{r+1}, \ldots, v_n$ so that $\{v_1, \ldots, v_n\}$ is a basis of $\mathbb{F}^n$. If v is in $\mathbb{F}^n$, we can expand v in terms of this basis as $v = c_1 v_1 + \cdots + c_n v_n$. Application of A gives

$$Av = A(c_1 v_1 + \cdots + c_n v_n) = c_1 A v_1 + \cdots + c_r A v_r + c_{r+1} A v_{r+1} + \cdots + A v_n$$
$$= c_{r+1} A v_{r+1} + \cdots + c_n A v_n.$$

Therefore the vectors $A v_{r+1}, \ldots, A v_n$ span the column space.

Let us see that they form a basis for the column space. Thus suppose that $c_{r+1} A v_{r+1} + \cdots + c_n A v_n = 0$. Then $A(c_{r+1} v_{r+1} + \cdots + c_n v_n) = 0$, and $c_{r+1} v_{r+1} + \cdots + c_n v_n$ is in the null space. Since $\{v_1, \ldots, v_r\}$ is a basis of the null space, we have

$$c_{r+1} v_{r+1} + \cdots + c_n v_n = a_1 v_1 + \cdots + a_r v_r$$

for suitable scalars $a_1, \ldots, a_r$. Therefore

$$(-a_1) v_1 + \cdots + (-a_r) v_r + c_{r+1} v_{r+1} + \cdots + c_n v_n = 0.$$

Since $v_1, \ldots, v_n$ are linearly independent, all the c_j are 0. We conclude that $A v_{r+1}, \ldots, A v_n$ are linearly independent and therefore form a basis of the column space.

As a result, we have established in the identity $r + (n - r) = n$ that $n - r$ can be interpreted as $\dim(\text{column space}(A))$ and that r can be interpreted as $\dim(\text{null space}(A))$. The theorem follows. $\square$

Proposition 2.7. If A is in $M_{kn}(\mathbb{F})$, then each elementary row operation on A preserves the row space of A.

PROOF. Let the rows of A be $r_1, \ldots, r_k$. Their span is unchanged if we interchange two of them or multiply one of them by a nonzero scalar. If we replace the row r_i by $r_i + cr_j$ with $j \neq i$, then the span is unchanged since

$$a_i r_i + a_j r_j = a_i(r_i + cr_j) + (a_j - a_i c) r_j$$

shows that any finite linear combination of the old rows is a finite linear combination of the new rows and since

$$b_i(r_i + cr_j) + b_j r_j = b_i r_i + (b_i c + b_j) r_j$$

shows the reverse. $\square$

Theorem 2.8. If A in $M_{kn}(\mathbb{F})$ has reduced row-echelon form R, then

$$\dim(\text{row space}(A)) = \dim(\text{row space}(R))$$
$$= \#(\text{nonzero rows of } R) = \#(\text{corner variables of } R)$$

and

$$\dim(\text{null space}(A)) = \dim(\text{null space}(R)) = \#(\text{independent variables of } R).$$

PROOF. The first equality in the first conclusion is immediate from Proposition 2.7, and the last equality of that conclusion is known from the method of row reduction. To see the middle inequality, we need to see that the nonzero rows of R are linearly independent. Let these rows be $r_1, \ldots, r_t$. For each i with $1 \leq i \leq t$, the index of the first nonzero entry of r_i was denoted by $j(i)$ in Section I.5. That entry has to be 1, and the other rows have to be 0 in that entry, by definition of reduced row-echelon form. If a finite linear combination $c_1 r_1 + \cdots + c_t r_t$ is 0, then inspection of the $j(i)^{\text{th}}$ entry yields the equality $c_i = 0$, and thus we conclude that all the coefficients are 0. This proves the desired linear independence.

The first equality in the second conclusion is by the solution procedure for homogeneous systems of equations in Section I.5; the set of solutions is unchanged by each row operation. To see the second equality, we recall that the form of the solution is as a finite linear combination of specific vectors, the coefficients being the independent variables. What the second equality is asserting is that these vectors form a basis of the space of solutions. We are thus to prove that they are linearly independent. Let the independent variables be certain x_j's, and let the corresponding vectors be v_j's. Then we know that the vector v_j has j^{th} entry 1 and that all the other vectors have j^{th} entry 0. If a finite linear combination of the vectors is 0, then examination of the jth entry shows that the jth coefficient is 0. The result follows. $\square$

Corollary 2.9. If A is in $M_{kn}(\mathbb{F})$, then

$$\dim(\text{row space}(A)) + \dim(\text{null space}(A)) = \#(\text{columns of } A) = n.$$

PROOF. We add the two formulas in Theorem 2.8 and see that

$$\dim(\text{row space}(A)) + \dim(\text{null space}(A))$$

equals the sum #(corner variables of R) + #(independent variables of R). Since all variables are corner variables or independent variables, this sum is n, and the result follows. □

Corollary 2.10. If A is in $M_{kn}(\mathbb{F})$, then

$$\dim(\text{row space}(A)) = \dim(\text{column space}(A)).$$

REMARK. The common value of the dimension of the row space of A and the dimension of the column space of A is called the **rank** of A. Some authors use the separate terms "row rank" and "column rank" for the two sides, and then the result is that these integers are equal.

PROOF. This follows by comparing Theorem 2.6 and Corollary 2.9. □

Although the above results may seem to have an abstract sound at first, methods of calculation for all the objects in question have quietly been carried along in the proofs, with everything rooted in the method of row reduction. All the proofs have in effect already been given that these methods of calculation do what they are supposed to do. If A is in $M_{kn}(\mathbb{F})$, the **transpose** of A, denoted by A^t, is the member of $M_{nk}(\mathbb{F})$ with entries $(A^t)_{ij} = A_{ji}$. In particular, the transpose of a row vector is a column vector, and vice versa.

METHODS OF CALCULATION.

(1) Basis of the row space of A. Row reduce A, and use the nonzero rows of the reduced row-echelon form.

(2) Basis of the column space of A. Transpose A, compute a basis of the row space of A^t by Method 1, and transpose the resulting row vectors into column vectors.

(3) Basis of the null space of A. Use the solution procedure for $Av = 0$ given in Section I.5. The set of solutions is given as all finite linear combinations of certain column vectors, the coefficients being the independent variables. The column vectors that are obtained form a basis of the null space.

(4) Basis of the linear span of the column vectors $v_1, \ldots, v_n$. Arrange the columns into a matrix A. Then the linear span is the column space of A, and a basis can be determined by Method 2.

(5) Extension of a linearly independent set $\{v_1, \ldots, v_r\}$ of column vectors in $\mathbb{F}^n$ to a basis of $\mathbb{F}^n$. Arrange the columns into a matrix, transpose, and row reduce. Adjoin additional row vectors, one for each independent variable, as follows: if x_j is an independent variable, then the row vector corresponding to x_j is to be 1 in the jth entry and 0 elsewhere. Transpose these additional row vectors so that they become column vectors, and these are vectors that may be adjoined to obtain a basis.

(6) Shrinking of a set $\{v_1, \ldots, v_r\}$ of column vectors to a subset that is a basis for the linear span of $\{v_1, \ldots, v_r\}$. For each i with $0 \leq i \leq r$, compute $d_i = \dim(\text{span}\{v_1, \ldots, v_r\})$. Retain v_i for $i \geq 0$ if $d_{i-1} < d_i$, and discard v_i otherwise.

3. Linear Maps

In this section we discuss linear maps, first in the setting of functions from $\mathbb{F}^n$ to $\mathbb{F}^k$ and then in the setting of functions between two vector spaces over $\mathbb{F}$. Much of the discussion will center on making computations for such functions by means of matrices.

We have seen that any k-by-n matrix A defines a function $L : \mathbb{F}^n$ to $\mathbb{F}^k$ by $L(v) = Av$ and that this function satisfies

$$L(u + v) = L(u) + L(v),$$
$$L(cv) = cL(v),$$

for all u and v in $\mathbb{F}^n$ and all scalars c. A function $L : \mathbb{F}^n \to \mathbb{F}^k$ satisfying these two conditions is said to be **linear**, or $\mathbb{F}$ **linear** if the scalars need emphasizing. Traditional names for such functions are **linear maps**, **linear mappings**, and **linear transformations**.[3] Thus matrices yield linear maps. Here is a converse.

Proposition 2.11. If $L : \mathbb{F}^n \to \mathbb{F}^k$ is a linear map, then there exists a unique k-by-n matrix A such that $L(v) = Av$ for all v in $\mathbb{F}^n$.

REMARK. The proof will show how to obtain the matrix A.

PROOF. For $1 \leq j \leq n$, let e_j be the jth standard basis vector of $\mathbb{F}^n$, having 1 in its jth entry and 0's elsewhere, and let the jth column of A be the k-dimensional column vector $L(e_j)$. If v is the column vector $(c_1, c_2, \ldots, c_n)$, then

$$L(v) = L\left(\sum_{j=1}^n c_j e_j\right) = \sum_{j=1}^n L(c_j e_j)$$
$$= \sum_{j=1}^n c_j L(e_j) = \sum_{j=1}^n c_j (j^{\text{th}} \text{ column of } A).$$

---

[3]The term **linear function** is particularly appropriate when the emphasis is on the fact that a certain function is linear. The term **linear operator** is used also, particularly when the context has something to do with analysis.

If $L(v)_i$ denotes the ith entry of the column vector $L(v)$, this equality says that
$$L(v)_i = \sum_{j=1}^{n} c_j A_{ij}.$$
The right side is the ith entry of Av, and hence $L(v) = Av$. This proves existence. For uniqueness we observe from the formula $L(e_j) = Ae_j$ that the jth column of A has to be $L(e_j)$ for each j, and therefore A is unique. $\square$

In the special case of linear maps from $\mathbb{F}^n$ to $\mathbb{F}^k$, the proof shows that two linear maps that agree on the members of the standard basis are equal on all vectors. We shall give a generalization of this fact as Proposition 2.13 below.

EXAMPLE 1. Let $L : \mathbb{R}^2 \to \mathbb{R}^2$ be rotation about the origin counterclockwise through the angle θ. Taking L to be defined geometrically, one finds from the parallelogram rule for addition of vectors that L is linear. Computation shows that $L\begin{pmatrix} 1 \\ 0 \end{pmatrix} = \begin{pmatrix} \cos\theta \\ \sin\theta \end{pmatrix}$ and that $L\begin{pmatrix} 0 \\ 1 \end{pmatrix} = \begin{pmatrix} -\sin\theta \\ \cos\theta \end{pmatrix}$. Applying Proposition 2.11 and the prescription for forming the matrix A given in the proof of the proposition, we see that $L(v) = \begin{pmatrix} \cos\theta & -\sin\theta \\ \sin\theta & \cos\theta \end{pmatrix} v$ for all v in $\mathbb{R}^2$.

We can add two linear maps $L : \mathbb{F}^n \to \mathbb{F}^k$ and $M : \mathbb{F}^n \to \mathbb{F}^k$ by adding their values at corresponding points: $(L + M)(v) = L(v) + M(v)$. In addition, we can multiply a linear map by a scalar by multiplying its values. Then $L + M$ and cL are linear, and it follows that the set of linear maps from $\mathbb{F}^n$ to $\mathbb{F}^k$ is a vector subspace of the vector space of all functions from $\mathbb{F}^n$ to $\mathbb{F}^k$, hence is itself a vector space. The customary notation for this vector space is $\text{Hom}_{\mathbb{F}}(\mathbb{F}^n, \mathbb{F}^k)$; the symbol Hom refers to the validity of the rule $L(u + v) = L(u) + L(v)$, and the subscript $\mathbb{F}$ refers to the validity of the additional rule $L(cv) = cL(v)$ for all c in $\mathbb{F}$.

If L corresponds to the matrix A and M corresponds to the matrix B, then $L + M$ corresponds to $A + B$ and cL corresponds to cA. The next proposition shows that composition of linear maps corresponds to multiplication of matrices.

Proposition 2.12. Let $L : \mathbb{F}^n \to \mathbb{F}^m$ be the linear map corresponding to an m-by-n matrix A, and let $M : \mathbb{F}^m \to \mathbb{F}^k$ be the linear map corresponding to a k-by-m matrix B. Then the composite function $M \circ L : \mathbb{F}^n \to \mathbb{F}^k$ is linear, and it corresponds to the k-by-n matrix BA.

PROOF. The function $M \circ L$ satisfies $(M \circ L)(u + v) = M(L(u + v)) = M(Lu + Lv) = M(Lu) + M(Lv) = (M \circ L)(u) + (M \circ L)(v)$, and similarly it satisfies $(M \circ L)(cv) = c(M \circ L)(v)$. Therefore it is linear. The correspondence of linear maps to matrices and the associativity of matrix multiplication together give $(M \circ L)(v) = M(L(v)) = (B)(Lv) = B(Av) = (BA)v$, and therefore $M \circ L$ corresponds to BA. $\square$

Now let us enlarge the setting for our discussion, treating arbitrary linear maps $L : U \to V$ between vector spaces over $\mathbb{F}$. We say that $L : U \to V$ is **linear**, or $\mathbb{F}$ **linear**, if
$$L(u + v) = L(u) + L(v),$$
$$L(cv) = cL(v),$$
for all u and v in U and all scalars c. As with the special case that $U = \mathbb{F}^n$ and $V = \mathbb{F}^k$, linear functions are called **linear maps**, **linear mappings**, and **linear transformations**. The set of all linear maps $L : U \to V$ is a vector space over $\mathbb{F}$ and is denoted by $\mathrm{Hom}_{\mathbb{F}}(U, V)$. The following result is fundamental in working with linear maps.

Proposition 2.13. Let U and V be vector spaces over $\mathbb{F}$, and let Γ be a basis of U. Then to each function $\ell : \Gamma \to V$ corresponds one and only one linear map $L : U \to V$ whose restriction to Γ has $L\big|_{\Gamma} = \ell$.

REMARK. We refer to L as the **linear extension** of ℓ.

PROOF. Suppose that $\ell : \Gamma \to V$ is given. Since Γ is a basis of U, each element of U has a unique expansion as a finite linear combination of members of Γ. Say that $u = c_{\alpha_1} u_{\alpha_1} + \cdots + c_{\alpha_r} u_{\alpha_r}$. Then the requirement of linearity on L forces $L(u) = L(c_{\alpha_1} u_{\alpha_1} + \cdots + c_{\alpha_r} u_{\alpha_r}) = c_{\alpha_1} L(u_{\alpha_1}) + \cdots + c_{\alpha_r} L(u_{\alpha_r})$, and therefore L is uniquely determined. For existence, define L by this formula. Expanding u and v in this way, we readily see that $L(u + v) = L(u) + L(v)$ and $L(cu) = cL(u)$. Therefore ℓ has a linear extension. $\square$

The definition of linearity and the proposition just proved make sense even if U and V are infinite-dimensional, but our objective for now will be to understand linear maps in terms of matrices. Thus, until further notice at a point later in this section, we shall assume that U and V are finite-dimensional. Remarks about the infinite-dimensional case appear in Section 9.

Since U and V are arbitrary finite-dimensional vector spaces, we no longer have standard bases at hand, and thus we have no immediate way to associate a matrix to a linear map $L : U \to V$. What we therefore do is fix *arbitrary bases* of U and V and work with them. It will be important to have an enumeration of each of these bases, and we therefore let

$$\Gamma = (u_1, \ldots, u_n)$$

and
$$\Delta = (v_1, \ldots, v_k)$$

be *ordered* bases of U and V, respectively.[4] If a member u of U may be expanded

---

[4]The notation $(u_1, \ldots, u_n)$ for an ordered basis, with each u_j equal to a vector, is not to be confused with the condensed notation $(c_1, \ldots, c_n)$ for a single column vector, with each c_j equal to a scalar.

in terms of Γ as $u = c_1u_1 + \cdots + c_nu_n$, we write

$$\begin{pmatrix} u \\ \Gamma \end{pmatrix} = \begin{pmatrix} c_1 \\ \vdots \\ c_n \end{pmatrix},$$

calling this the column vector expressing u in the ordered basis Γ. Using our linear map $L : U \to V$, let us define a k-by-n matrix $\begin{pmatrix} L \\ \Delta\Gamma \end{pmatrix}$ by requiring that the

$$j^{\text{th}} \text{ column of } \begin{pmatrix} L \\ \Delta\Gamma \end{pmatrix} \text{ be } \begin{pmatrix} L(u_j) \\ \Delta \end{pmatrix}.$$

The positions in which the ordered bases Δ and Γ are listed in the notation is important here; the range basis is to the left of the domain basis.[5]

EXAMPLE 2. Let V be the space of all complex-valued solutions on $\mathbb{R}$ of the differential equation $y''(t) = y(t)$. Then V is a vector subspace of functions, hence is a vector space in its own right. It is known that V is 2-dimensional with solutions $c_1 e^t + c_2 e^{-t}$. If $y(t)$ is a solution, then differentiation of the equation shows that $y'(t)$ is another solution. In other words, the derivative operator d/dt is a linear map from V to itself. One ordered basis of V is $\Gamma = (e^t, e^{-t})$, and another is $\Delta = (\cosh t, \sinh t)$, where $\cosh t = \frac{1}{2}(e^t + e^{-t})$ and $\sinh t = \frac{1}{2}(e^t - e^{-t})$. To find $\begin{pmatrix} d/dt \\ \Delta\Gamma \end{pmatrix}$, we need to express $(d/dt)(e^t)$ and $(d/dt)(e^{-t})$ in terms of $\cosh t$ and $\sinh t$. We have

$$\begin{pmatrix} (d/dt)(e^t) \\ \Delta \end{pmatrix} = \begin{pmatrix} e^t \\ \Delta \end{pmatrix} = \begin{pmatrix} \cosh t + \sinh t \\ \Delta \end{pmatrix} = \begin{pmatrix} 1 \\ 1 \end{pmatrix}$$

and

$$\begin{pmatrix} (d/dt)(e^{-t}) \\ \Delta \end{pmatrix} = \begin{pmatrix} -e^{-t} \\ \Delta \end{pmatrix} = \begin{pmatrix} -\cosh t + \sinh t \\ \Delta \end{pmatrix} = \begin{pmatrix} -1 \\ 1 \end{pmatrix}.$$

Therefore $\begin{pmatrix} d/dt \\ \Delta\Gamma \end{pmatrix} = \begin{pmatrix} 1 & -1 \\ 1 & 1 \end{pmatrix}$.

Theorem 2.14. If $L : U \to V$ is a linear map between finite-dimensional vector spaces over $\mathbb{F}$ and if Γ and Δ are ordered bases of U and V, respectively, then

$$\begin{pmatrix} L(u) \\ \Delta \end{pmatrix} = \begin{pmatrix} L \\ \Delta\Gamma \end{pmatrix} \begin{pmatrix} u \\ \Gamma \end{pmatrix}$$

for all u in U.

---

[5]This order occurs in a number of analogous situations in mathematics and has the effect of keeping the notation reasonably consistent with the notation for composition of functions.

PROOF. The two sides of the identity in question are linear in u, and Proposition 2.13 shows that it is enough to prove the identity for the members u of some ordered basis of U. We choose Γ as this ordered basis. For the basis vector u equal to the jth member u_j of Γ, use of the definition shows that $\begin{pmatrix} u_j \\ \Gamma \end{pmatrix}$ is the column vector e_j that is 1 in the jth entry and is 0 elsewhere. The product $\begin{pmatrix} L \\ \Delta\Gamma \end{pmatrix} e_j$ is the jth column of $\begin{pmatrix} L \\ \Delta\Gamma \end{pmatrix}$, which was defined to be $\begin{pmatrix} L(u_j) \\ \Delta \end{pmatrix}$. Thus the identity in question is valid for u_j, and the theorem follows. □

If we take into account Proposition 2.13, saying that linear maps on U arise uniquely from arbitrary functions on a basis of U, then Theorem 2.14 supplies a one-one correspondence of linear maps L from U to V with matrices A of the appropriate size, once we fix ordered bases in the domain and range. The correspondence is $L \leftrightarrow \begin{pmatrix} L \\ \Delta\Gamma \end{pmatrix}$.

As in the special case with linear maps between spaces of column vectors, this correspondence respects addition and scalar multiplication. Theorem 2.14 implies that under this correspondence, the image of L corresponds to the column space of A. It implies also that the vector subspace of the domain U with $L(u) = 0$, which is called the **kernel** of L and is sometimes denoted by ker L, corresponds to the null space of A. The kernel of L has the important property that

the linear map L is one-one if and only if ker $L = 0$.

Another important property comes from this association of kernel with null space and of image with column space. Namely, we apply Theorem 2.6, and we obtain the following corollary.

Corollary 2.15. If $L : U \to V$ is a linear map between finite-dimensional vector spaces over $\mathbb{F}$, then

$$\dim(\text{domain}(L)) = \dim(\text{kernel}(L)) + \dim(\text{image}(L)).$$

The next result says that composition corresponds to matrix multiplication under the correspondence of Theorem 2.14.

Theorem 2.16. Let $L : U \to V$ and $M : V \to W$ be linear maps between finite-dimensional vector spaces, and let Γ, Δ, and Ω be ordered bases of U, V, and W. Then the composition ML is linear, and the corresponding matrix is given by

$$\begin{pmatrix} ML \\ \Omega\Gamma \end{pmatrix} = \begin{pmatrix} M \\ \Omega\Delta \end{pmatrix} \begin{pmatrix} L \\ \Delta\Gamma \end{pmatrix}.$$

3. Linear Maps   47

PROOF. If u is in U, three applications of Theorem 2.14 and one application of associativity of matrix multiplication give

$$\begin{pmatrix} ML \\ \Omega\Gamma \end{pmatrix}\begin{pmatrix} u \\ \Gamma \end{pmatrix} = \begin{pmatrix} ML(u) \\ \Omega \end{pmatrix} = \begin{pmatrix} M \\ \Omega\Delta \end{pmatrix}\begin{pmatrix} L(u) \\ \Delta \end{pmatrix}$$
$$= \begin{pmatrix} M \\ \Omega\Delta \end{pmatrix}\left[\begin{pmatrix} L \\ \Delta\Gamma \end{pmatrix}\begin{pmatrix} u \\ \Gamma \end{pmatrix}\right] = \left[\begin{pmatrix} M \\ \Omega\Delta \end{pmatrix}\begin{pmatrix} L \\ \Delta\Gamma \end{pmatrix}\right]\begin{pmatrix} u \\ \Gamma \end{pmatrix}.$$

Taking u to be the jth member of Γ, we see from this equation that the jth column of $\begin{pmatrix} ML \\ \Omega\Gamma \end{pmatrix}$ equals the jth column of $\begin{pmatrix} M \\ \Omega\Delta \end{pmatrix}\begin{pmatrix} L \\ \Delta\Gamma \end{pmatrix}$. Since j is arbitrary, the theorem follows. □

A computational device that appears at first to be only of theoretical interest and then, when combined with other things, becomes of practical interest, is to change one of the ordered bases in computing the matrix of a linear map. A handy device for this purpose is a change-of-basis matrix $\begin{pmatrix} I \\ \Delta\Gamma \end{pmatrix}$ since Theorem 2.16 gives $\begin{pmatrix} L \\ \Delta\Gamma \end{pmatrix} = \begin{pmatrix} I \\ \Delta\Gamma \end{pmatrix}\begin{pmatrix} L \\ \Gamma\Gamma \end{pmatrix}$.

EXAMPLE 2, CONTINUED. Let L be d/dt as a linear map carrying the space of solutions of $y''(t) = y(t)$ to itself, with $\Gamma = (e^t, e^{-t})$ and $\Delta = (\cosh t, \sinh t)$ as before. Then $\begin{pmatrix} d/dt \\ \Gamma\Gamma \end{pmatrix} = \begin{pmatrix} 1 & 0 \\ 0 & -1 \end{pmatrix}$. Since $e^t = \cosh t + \sinh t$ and $e^{-t} = \cosh t - \sinh t$, $\begin{pmatrix} I \\ \Delta\Gamma \end{pmatrix} = \begin{pmatrix} 1 & 1 \\ 1 & -1 \end{pmatrix}$ by inspection. The product is $\begin{pmatrix} L \\ \Delta\Gamma \end{pmatrix} = \begin{pmatrix} I \\ \Delta\Gamma \end{pmatrix}\begin{pmatrix} d/dt \\ \Gamma\Gamma \end{pmatrix} = \begin{pmatrix} 1 & -1 \\ 1 & 1 \end{pmatrix}$, a result we found before with a little more effort by computing matters directly.

Often in practical applications the domain and the range are the same vector space, the domain's ordered basis equals the range's ordered basis, and the matrix of a linear map is known in this ordered basis. The problem is to determine the matrix when the ordered basis is changed in both domain and range—changed in such a way that the ordered bases in the domain and range are the same. This time we use two change-of-basis matrices $\begin{pmatrix} I \\ \Delta\Gamma \end{pmatrix}$ and $\begin{pmatrix} I \\ \Gamma\Delta \end{pmatrix}$, but these are related. Since $\begin{pmatrix} I \\ \Gamma\Delta \end{pmatrix}\begin{pmatrix} I \\ \Delta\Gamma \end{pmatrix} = \begin{pmatrix} I \\ \Gamma\Gamma \end{pmatrix} = I$, the two matrices are the inverses of one another. Thus, except for matrix algebra, the problem is to compute just one of $\begin{pmatrix} I \\ \Gamma\Delta \end{pmatrix}$ and $\begin{pmatrix} I \\ \Delta\Gamma \end{pmatrix}$.

Normally one of these two matrices can be written down by inspection. For example, if we are working with a linear map from a space of column vectors to itself, one ordered basis of interest is the standard ordered basis Σ. Another ordered basis Δ might be determined by special features of the linear map. In this case the members of Δ are given as column vectors, hence are expressed in terms of Σ. Thus $\begin{pmatrix} I \\ \Sigma\Delta \end{pmatrix}$ can be written by inspection. We shall encounter this situation later in this chapter when we use "eigenvectors" in order to understand linear maps better. Here is an example, but without eigenvectors.

EXAMPLE 1, CONTINUED. We saw that rotation L counterclockwise about the origin in $\mathbb{R}^2$ is given in the standard ordered basis $\Sigma = \left(\begin{pmatrix} 1 \\ 0 \end{pmatrix}, \begin{pmatrix} 0 \\ 1 \end{pmatrix} \right)$ by $\begin{pmatrix} L \\ \Sigma\Sigma \end{pmatrix} = \begin{pmatrix} \cos\theta & -\sin\theta \\ \sin\theta & \cos\theta \end{pmatrix}$. Let us compute the matrix of L in the ordered basis $\Delta = \left(\begin{pmatrix} 1 \\ 0 \end{pmatrix}, \begin{pmatrix} 1 \\ 1 \end{pmatrix} \right)$. The easy change-of-basis matrix to form is $\begin{pmatrix} I \\ \Sigma\Delta \end{pmatrix} = \begin{pmatrix} 1 & 1 \\ 0 & 1 \end{pmatrix}$. Hence

$$\begin{pmatrix} L \\ \Delta\Delta \end{pmatrix} = \begin{pmatrix} I \\ \Delta\Sigma \end{pmatrix} \begin{pmatrix} L \\ \Sigma\Sigma \end{pmatrix} \begin{pmatrix} I \\ \Sigma\Delta \end{pmatrix} = \begin{pmatrix} 1 & 1 \\ 0 & 1 \end{pmatrix}^{-1} \begin{pmatrix} \cos\theta & -\sin\theta \\ \sin\theta & \cos\theta \end{pmatrix} \begin{pmatrix} 1 & 1 \\ 0 & 1 \end{pmatrix},$$

and the problem is reduced to one of matrix algebra.

Our computations have proved the following proposition, which, as we shall see later, motivates much of Chapter V. The matrix C in the statement of the proposition is $\begin{pmatrix} I \\ \Gamma\Delta \end{pmatrix}$.

Proposition 2.17. Let $L : V \to V$ be a linear map on a finite-dimensional vector space, and let A be the matrix of L relative to an ordered basis Γ (in domain and range). Then the matrix of L in any other ordered basis Δ is of the form $C^{-1}AC$ for some invertible matrix C depending on Δ.

REMARK. If A is a square matrix, any square matrix of the form $C^{-1}AC$ is said to be **similar** to A. It is immediate that "is similar to" is an equivalence relation.

Now let us return to the setting in which our vector spaces are allowed to be infinite-dimensional. Two vector spaces U and V are said to be **isomorphic** if there is a one-one linear map of U onto V. In this case, the linear map in question is called an **isomorphism**, and one often writes $U \cong V$.

Here is a finite-dimensional example: If U is n-dimensional with an ordered basis Γ and V is k-dimensional with an ordered basis Δ, then $\text{Hom}_{\mathbb{F}}(U, V)$ is isomorphic to $M_{nk}(\mathbb{F})$ by the linear map that carries a member L of $\text{Hom}_{\mathbb{F}}(U, V)$ to the k-by-n matrix $\begin{pmatrix} L \\ \Delta\Gamma \end{pmatrix}$.

The relation "is isomorphic to" is an equivalence relation. In fact, it is reflexive since the identity map exhibits U as isomorphic to itself. It is transitive since Theorem 2.16 shows that the composition ML of two linear maps $L : U \to V$ and $M : V \to W$ is linear and since the composition of one-one onto functions is one-one onto. To see that it is symmetric, we need to observe that the inverse function L^{-1} of a one-one onto linear map $L : U \to V$ is linear. To see this linearity, we observe that $L(L^{-1}(v_1) + L^{-1}(v_2)) = L(L^{-1}(v_1)) + L(L^{-1}(v_2)) = v_1 + v_2 = I(v_1 + v_2) = L(L^{-1}(v_1 + v_2))$. Since L is one-one,

$$L^{-1}(v_1) + L^{-1}(v_2) = L^{-1}(v_1 + v_2).$$

Similarly the facts that $L(L^{-1}(cv)) = cv = cL(L^{-1}v) = L(c(L^{-1}(v)))$ and that L is one-one imply that

$$L^{-1}(cv) = c(L^{-1}(v)),$$

and hence L^{-1} is linear. Thus "is isomorphic to" is indeed an equivalence relation.

The vector spaces over $\mathbb{F}$ are partitioned, according to the basic result about equivalence relations in Section A2 of the appendix, into equivalence classes. Each member of an equivalence class is isomorphic to all other members of that class and to no member of any other class.

An isomorphism preserves all the vector-space structure of a vector space. Spanning sets are mapped to spanning sets, linearly independent sets are mapped to linearly independent sets, vector subspaces are mapped to vector subspaces, dimensions of subspaces are preserved, and so on. In other words, for all purposes of abstract vector-space theory, isomorphic vector spaces may be regarded as the same. Let us give a condition for isomorphism that might at first seem to trivialize all vector-space theory, reducing it to a count of dimensions, but then let us return to say why this result is not to be considered as so important.

Proposition 2.18. Two finite-dimensional vector spaces over $\mathbb{F}$ are isomorphic if and only if they have the same dimension.

PROOF. If a vector space U is isomorphic to a vector space V, then the isomorphism carries any basis of U to a basis of V, and hence U and V have the same dimension. Conversely if they have the same dimension, let $(u_1, \ldots, u_n)$ be an ordered basis of U, and let $(v_1, \ldots, v_n)$ be an ordered basis of V. Define $\ell(u_j) = v_j$ for $1 \leq j \leq n$, and let $L : U \to V$ be the linear extension of ℓ given by Proposition 2.13. Then L is linear, one-one, and onto, and hence U is isomorphic to V. $\square$

The proposition does not mean that one should necessarily be eager to make the identification of two vector spaces that are isomorphic. An important distinction is the one between "isomorphic" and "isomorphic via a canonically constructed linear map." The isomorphism of linear maps with matrices given by $L \mapsto \begin{pmatrix} L \\ \Delta \Gamma \end{pmatrix}$ is canonical since no choices are involved once Γ and Δ have been specified. This is a useful isomorphism because we can track matters down and use the isomorphism to make computations. On the other hand, it is not very useful to say merely that $\operatorname{Hom}_{\mathbb{F}}(U, V)$ and $M_{kn}(\mathbb{F})$ are isomorphic because they have the same dimension.

What tends to happen in practice is that vector spaces in applications come equipped with additional structure—some rigid geometry, or a multiplication operation, or something else. A general vector-space isomorphism has little chance of having any connection to the additional structure and thereby of being very helpful. On the other hand, a concrete isomorphism that is built by taking this additional structure into account may indeed be useful.

In the next section we shall encounter an example of an additional structure that involves neither a rigid geometry nor a multiplication operation. We shall introduce the "dual" V' of a vector space V, and we shall see that V and V' have the same dimension if V is finite-dimensional. But no particular isomorphism of V with V' is singled out as better than other ones, and it is wise not to try to identify these spaces. By contrast, the double dual V'' of V, which too will be constructed in the next section, will be seen to be isomorphic to V in the finite-dimensional case via a linear map $\iota : V \to V''$ that we define explicitly. The function ι is an example of a canonical isomorphism that we might want to exploit.

4. Dual Spaces

Let V be a vector space over $\mathbb{F}$. A **linear functional** on V is a linear map from V into $\mathbb{F}$. The space of all such linear maps, as we saw in Section 3, is a vector space. We denote it by V' and call it the **dual space** of V.

The development of Section 3 tells us right away how to compute the dual space of the space of column vectors $\mathbb{F}^n$. If Σ is the standard ordered basis of $\mathbb{F}^n$ and if 1 denotes the basis of $\mathbb{F}$ consisting of the scalar 1, then we can associate to a linear functional v' on $\mathbb{F}^n$ its matrix

$$\begin{pmatrix} v' \\ 1\Sigma \end{pmatrix} = \begin{pmatrix} v'(e_1) & v'(e_2) & \cdots & v'(e_n) \end{pmatrix},$$

which is an n-dimensional row vector. The operation of v' on a column vector

$v = \begin{pmatrix} x_1 \\ \vdots \\ x_n \end{pmatrix}$ is given by Theorem 2.14. Namely, $v'(v)$ is a multiple of the scalar 1, and the theorem tells us how to compute this multiple:

$$\begin{pmatrix} v'(v) \\ 1 \end{pmatrix} = \begin{pmatrix} v' \\ 1\Sigma \end{pmatrix} \begin{pmatrix} x_1 \\ \vdots \\ x_n \end{pmatrix} = (v'(e_1) \quad v'(e_2) \quad \cdots \quad v'(e_n)) \begin{pmatrix} x_1 \\ \vdots \\ x_n \end{pmatrix}.$$

Thus the space of all linear functionals on $\mathbb{F}^n$ may be identified with the space of all n-dimensional row vectors, and the effect of the row vector on a column vector is given by matrix multiplication. Since the standard ordered basis of $\mathbb{F}^n$ and the basis 1 of $\mathbb{F}$ are singled out as special, this identification is actually canonical, and it is thus customary to make this identification without further comment.

For a more general vector space V, no natural way of writing down elements of V' comes to mind. Indeed, if a concrete V is given, it can help considerably in understanding V to have an identification of V' that does not involve choices. For example, in real analysis one proves in a suitable infinite-dimensional setting that a (continuous) linear functional on the space of integrable functions is given by integration with a bounded function, and that fact simplifies the handling of the space of integrable functions.

In any event, the canonical identification of linear functionals that we found for $\mathbb{F}^n$ does not work once we pass to a more general finite-dimensional vector space V. To make such an identification in the absence of additional structure, we first fix an ordered basis $(v_1, \ldots, v_n)$ of V. If we do so, then V' is indeed identified with the space of n-dimensional row vectors. The members of V' that correspond to the standard basis of row vectors, i.e., the row vectors that are 1 in one entry and are 0 elsewhere, are of special interest. These are the linear functionals v'_i such that

$$v'_i(v_j) = \delta_{ij},$$

where δ_{ij} is the Kronecker delta. Since these standard row vectors form a basis of the space of row vectors, $(v'_1, \ldots, v'_n)$ is an ordered basis of V'. If the members of the ordered basis $(v_1, \ldots, v_n)$ are permuted in some way, the members of $(v'_1, \ldots, v'_n)$ are permuted in the same way. Thus the basis $\{v'_1, \ldots, v'_n\}$ depends only on the basis $\{v_1, \ldots, v_n\}$, not on the enumeration.[6] The basis $\{v'_1, \ldots, v'_n\}$ is called the **dual basis** of V relative to $\{v_1, \ldots, v_n\}$. A consequence of this discussion is the following result.

Proposition 2.19. If V is a finite-dimensional vector space with dual V', then V' is finite-dimensional with $\dim V' = \dim V$.

---

[6]Although the enumeration is not important, more structure is present here than simply an association of an unordered basis of V' to an unordered basis of V. Each member of $\{v'_1, \ldots, v'_n\}$ is matched to a particular member of $\{v_1, \ldots, v_n\}$, namely the one on which it takes the value 1.

Linear functionals play an important role in working with a vector space. To understand this role, it is helpful to think somewhat geometrically. Imagine the problem of describing a vector subspace of a given vector space. One way of describing it is from the inside, so to speak, by giving a spanning set. In this case we end up by describing the subspace in terms of parameters, the parameters being the scalar coefficients when we say that the subspace is the set of all finite linear combinations of members of the spanning set. Another way of describing the subspace is from the outside, cutting it down by conditions imposed on its elements. These conditions tend to be linear equations, saying that certain linear maps on the elements of the subspace give 0. Typically the subspace is then described as the intersection of the kernels of some set of linear maps. Frequently these linear maps will be scalar-valued, and then we are in a situation of describing the subspace by a set of linear functionals.

We know that every vector subspace of a finite-dimensional vector space V can be described from the inside in this way; we merely give all its members. A statement with more content is that we can describe it with finitely many members; we can do so because we know that every vector subspace of V has a basis.

For linear functionals really to be useful, we would like to know a corresponding fact about describing subspaces from the outside—that every vector subspace U of a finite-dimensional V can be described as the intersection of the kernels of a finite set of linear functionals. To do so is easy. We take a basis of the vector subspace U, say $\{v_1, \ldots, v_r\}$, extend it to a basis of V by adjoining vectors $v_{r+1}, \ldots, v_n$, and form the dual basis $\{v'_1, \ldots, v'_n\}$ of V'. The subspace U is then described as the set of all vectors v in V such that $v'_j(v) = 0$ for $r+1 \leq j \leq n$. The following proposition expresses this fact in ways that are independent of the choice of a basis. It uses the terminology **annihilator** of U, denoted by $\mathrm{Ann}(U)$, for the vector subspace of all members v' of V' with $v'(u) = 0$ for all u in U.

Proposition 2.20. Let V be a finite-dimensional vector space, and let U be a vector subspace of V. Then

(a) $\dim U + \dim \mathrm{Ann}(U) = \dim V$,
(b) every linear functional on U extends to a linear functional on V,
(c) whenever v_0 is a member of V that is not in U, there exists a linear functional on V that is 0 on U and is 1 on v_0.

PROOF. We retain the notation above, writing $\{v_1, \ldots, v_r\}$ for a basis of U, $v_{r+1}, \ldots, v_n$ for vectors that are adjoined to form a basis of V, and $\{v'_1, \ldots, v'_n\}$ for the dual basis of V'. For (a), we check that $\{v'_{r+1}, \ldots, v'_n\}$ is a basis of $\mathrm{Ann}(U)$. It is enough to see that they span $\mathrm{Ann}(U)$. These linear functionals are 0 on every member of the basis $\{v_1, \ldots, v_r\}$ of U and hence are in $\mathrm{Ann}(U)$. On the other hand, if v' is a member of $\mathrm{Ann}(U)$, we can certainly write $v' = c_1 v'_1 + \cdots + c_n v'_n$

for some scalars $c_1, \ldots, c_n$. Since v' is 0 on U, we must have $v'(v_i) = 0$ for $i \leq r$. Since $v'(v_i) = c_i$, we obtain $c_i = 0$ for $i \leq r$. Therefore v' is a linear combination of $v'_{r+1}, \ldots, v'_n$, and (a) is proved.

For (b), let us observe that the restrictions $v'_1\big|_U, \ldots, v'_r\big|_U$ form the dual basis of U' relative to the basis $\{v_1, \ldots, v_r\}$ of U. If u' is in U', we can therefore write $u' = c_1 v'_1\big|_U + \cdots + c_r v'_r\big|_U$ for some scalars $c_1, \ldots, c_r$. Then $v' = c_1 v'_1 + \cdots + c_r v'_r$ is the required extension of u' to all of V.

For (c), we use a special choice of basis of V in the argument above. Namely, we still take $\{v_1, \ldots, v_r\}$ to be a basis of U, and then we let $v_{r+1} = v_0$. Finally we adjoin $v_{r+2}, \ldots, v_n$ to obtain a basis $\{v_1, \ldots, v_n\}$ of V. Then v'_{r+1} has the required property. □

If $L : U \to V$ is a linear map between finite-dimensional vector spaces, then the formula

$$(L^t(v'))(u) = v'(L(u)) \qquad \text{for } u \in U \text{ and } v' \in V'$$

defines a linear map $L^t : V' \to U'$. The linear map L^t is called the **contragredient** of L. The matrix of the contragredient of L is the transpose of the matrix of L in the following sense.[7]

Proposition 2.21. Let $L : U \to V$ be a linear map between finite-dimensional vector spaces, let $L^t : V' \to U'$ be its contragredient, let Γ and Δ be respective ordered bases of U and V, and let Γ' and Δ' be their dual ordered bases. Then

$$\begin{pmatrix} L^t \\ \Gamma' \Delta' \end{pmatrix} = \begin{pmatrix} L \\ \Delta \Gamma \end{pmatrix}.$$

PROOF. Let $\Gamma = (u_1, \ldots, u_n)$, $\Delta = (v_1, \ldots, v_k)$, $\Gamma' = (u'_1, \ldots, u'_n)$, and $\Delta' = (v'_1, \ldots, v'_k)$. Write B and A for the respective matrices in the formula in question. The equations $L(u_j) = \sum_{i'=1}^k A_{i'j} v_{i'}$ and $L^t(v'_i) = \sum_{j'=1}^n B_{j'i} u'_{j'}$ imply that

$$v'_i(L(u_j)) = v'_i\left(\sum_{i'=1}^k A_{i'j} v_{i'}\right) = A_{ij}$$

and

$$L^t(v'_i)(u_j) = \sum_{j'=1}^n B_{j'i} u'_{j'}(u_j) = B_{ji}.$$

Therefore $B_{ji} = L^t(v'_i)(u_j) = v'_i(L(u_j)) = A_{ij}$, as required. □

---

[7]A general principle is involved in the definition of contragredient once we have a definition of dual vector space, and we shall see further examples of this principle in the next two sections and in later chapters: whenever a new systematic construction appears for the objects under study, it is well to look for a corresponding construction with the functions relating these new objects. In language to be introduced near the end of Chapter IV, the context for the construction will be a "category," and the principle says that it is well to see whether the construction is that of a "functor" on the category.

With V finite-dimensional, now consider $V'' = (V')'$, the double dual. In the case that $V = \mathbb{F}^n$, we saw that V' could be viewed as the space of row vectors, and it is reasonable to expect V'' to involve a second transpose and again be the space of column vectors. If so, then V gets identified with V''. In fact, this is true in all cases, and we argue as follows. If v is in V, we can define a member $\iota(v)$ of V'' by

$$\iota(v)(v') = v'(v) \qquad \text{for } v \in V \text{ and } v' \in V'.$$

This definition makes sense whether or not V is finite-dimensional. The function ι is a linear map from V into V'' called the **canonical map** of V into V''. It is independent of any choice of basis.

Proposition 2.22. If V is any finite-dimensional vector space over $\mathbb{F}$, then the canonical map $\iota : V \to V''$ is one-one onto.

REMARKS. In the infinite-dimensional case the canonical map is one-one but it is not onto. The proof that it is one-one uses the fact that V has a basis, but we have deferred the proof of this fact about infinite-dimensional vector spaces to Section 9. Problem 14 at the end of the chapter will give an example of an infinite-dimensional V for which ι does not carry V onto V''. When combined with the first corollary in Section A6 of the appendix, this example shows that ι *never* carries V onto V'' in the infinite-dimensional case.

PROOF. We saw in Section 3 that a linear map ι is one-one if and only if $\ker \iota = 0$. Thus suppose $\iota(v) = 0$. Then $0 = \iota(v)(v') = v'(v)$ for all v'. Arguing by contradiction, suppose $v \neq 0$. Then we can extend $\{v\}$ to a basis of V, and the linear functional v' that is 1 on v and is 0 on the other members of the basis will have $v'(v) \neq 0$, contradiction. We conclude that ι is one-one. By Proposition 2.19 we have

$$\dim V = \dim V' = \dim V''. \qquad (*)$$

Since ι is one-one, it carries any basis of V to a linearly independent set in V''. This linearly independent set has to be a basis, by Corollary 2.4 and the dimension formula $(*)$. □

5. Quotients of Vector Spaces

This section constructs a vector space V/U out of a vector space V and a vector subspace U. We begin with the example illustrated in Figure 2.1. In the vector space $V = \mathbb{R}^2$, let U be a line through the origin. The lines parallel to U are of the form $v + U = \{v + u \mid u \in U\}$, and we make the set of these lines into a vector space by defining $(v_1 + U) + (v_2 + U) = (v_1 + v_2) + U$ and

$c(v + U) = cv + U$. The figure suggests that if we were to take any other line W through the origin, then W would meet all the lines $v + U$, and the notion of addition of lines $v + U$ would correspond exactly to addition in W. Indeed we can successfully make such a correspondence, but the advantage of introducing the vector space of all lines $v + U$ is that it is canonical, independent of the kind of choice we have to make in selecting W. One example of the utility of having a canonical construction is the ease with which we obtain correspondence of linear maps stated in Proposition 2.25 below. Other examples will appear later.

FIGURE 2.1. The vector space of lines $v + U$ in $\mathbb{R}^2$ parallel to a given line U through the origin.

Proposition 2.23. Let V be a vector space over $\mathbb{F}$, and let U be a vector subspace. The relation defined by saying that $v_1 \sim v_2$ if $v_1 - v_2$ is in U is an equivalence relation, and the equivalence classes are all sets of the form $v + U$ with $v \in V$. The set of equivalence classes V/U is a vector space under the definitions

$$(v_1 + U) + (v_2 + U) = (v_1 + v_2) + U,$$
$$c(v + U) = cv + U,$$

and the function $q(v) = v + U$ is linear from V onto V/U with kernel U.

REMARKS. We say that V/U is the **quotient space** of V by U. The linear map $q(v) = v + U$ is called the **quotient map** of V onto V/U.

PROOF. The properties of an equivalence relation are established as follows:

$v_1 \sim v_1$	because 0 is in U,
$v_1 \sim v_2$ implies $v_2 \sim v_1$	because U is closed under negatives,
$v_1 \sim v_2$ and $v_2 \sim v_3$ together imply $v_1 \sim v_3$	because U is closed under addition.

Thus we have equivalence classes. The class of v_1 consists of all vectors v_2 such that $v_2 - v_1$ is in U, hence consists of all vectors in $v_1 + U$. Thus the equivalence classes are indeed the sets $v + U$.

Let us check that addition and scalar multiplication, as given in the statement of the proposition, are well defined. For addition let $v_1 \sim w_1$ and $v_2 \sim w_2$. Then $v_1 - w_1$ and $v_2 - w_2$ are in U. Since U is a vector subspace, the sum $(v_1 - w_1) + (v_2 - w_2) = (v_1 + v_2) - (w_1 + w_2)$ is in U. Thus $v_1 + v_2 \sim w_1 + w_2$, and addition is well defined. For scalar multiplication let $v \sim w$, and let a scalar c be given. Then $v - w$ is in U, and $c(v - w) = cv - cw$ is in U since U is a vector subspace. Hence $cv \sim cw$, and scalar multiplication is well defined.

The vector-space properties of V/U are consequences of the properties for V. To illustrate, consider associativity of addition. The argument in this case is that
$$((v_1 + U) + (v_2 + U)) + (v_3 + U) = ((v_1 + v_2) + U) + (v_3 + U)$$
$$= ((v_1 + v_2) + v_3) + U = (v_1 + (v_2 + v_3)) + U$$
$$= (v_1 + U) + ((v_2 + v_3) + U) = (v_1 + U) + ((v_2 + U) + (v_3 + U)).$$

Finally the quotient map $q : V \to V/U$ given by $q(v) = v + U$ is certainly linear. Its kernel is $\{v \mid v + U = 0 + U\}$, and this equals $\{v \mid v \in U\}$, as asserted. The map q is onto V/U since $v + U = q(v)$. $\square$

Corollary 2.24. If V is a vector space over $\mathbb{F}$ and U is a vector subspace, then
(a) $\dim V = \dim U + \dim(V/U)$,
(b) the subspace U is the kernel of some linear map defined on V.

REMARK. The first conclusion is valid even when all the spaces are not finite-dimensional. For current purposes it is sufficient to regard $\dim V$ as $+\infty$ if V is infinite-dimensional; the sum of $+\infty$ and any dimension as $+\infty$.

PROOF. Let q be the quotient map. The linear map q meets the conditions of (b). For (a), take a basis of U and extend to a basis of V. Then the images under q of the additional vectors form a basis of V/U. $\square$

Quotients of vector spaces allow for the factorization of certain linear maps, as indicated in Proposition 2.25 and Figure 2.2.

Proposition 2.25. Let $L : V \to W$ be a linear map between vector spaces over $\mathbb{F}$, let $U_0 = \ker L$, let U be a vector subspace of V contained in U_0, and let $q : V \to V/U$ be the quotient map. Then there exists a linear map $\overline{L} : V/U \to W$ such that $L = \overline{L}q$. It has the same image as L, and $\ker \overline{L} = \{u_0 + U \mid u_0 \in U_0\}$.

FIGURE 2.2. Factorization of linear maps via a quotient of vector spaces.

REMARK. One says that L **factors through** V/U or **descends** to V/U.

PROOF. The definition of $\overline{L}$ has to be $\overline{L}(v+U) = L(v)$. This forces $\overline{L}q = L$, and $\overline{L}$ will have to be linear. What needs proof is that $\overline{L}$ is well defined. Thus suppose $v_1 \sim v_2$. We are to prove that $\overline{L}(v_1 + U) = \overline{L}(v_2 + U)$, i.e., that $L(v_1) = L(v_2)$. Now $v_1 - v_2$ is in $U \subseteq U_0$, and hence $L(v_1 - v_2) = 0$. Then $L(v_1) = L(v_1 - v_2) + L(v_2) = L(v_2)$, as required. This proves that $\overline{L}$ is well defined, and the conclusions about the image and the kernel of $\overline{L}$ are immediate from the definition. □

Corollary 2.26. Let $L : V \to W$ be a linear map between vector spaces over $\mathbb{F}$, and suppose that L is onto W and has kernel U. Then V/U is canonically isomorphic to W.

PROOF. Take $U = U_0$ in Proposition 2.25, and form $\overline{L} : V/U \to W$ with $L = \overline{L}q$. The proposition shows that $\overline{L}$ is onto W and has trivial kernel, i.e., the 0 element of V/U. Having trivial kernel, $\overline{L}$ is one-one. □

Theorem 2.27 (First Isomorphism Theorem). Let $L : V \to W$ be a linear map between vector spaces over $\mathbb{F}$, and suppose that L is onto W and has kernel U. Then the map $S \mapsto L(S)$ gives a one-one correspondence between

(a) the vector subspaces S of V containing U and
(b) the vector subspaces of W.

REMARK. As in Section A1 of the appendix, we write $L(S)$ and $L^{-1}(T)$ to indicate the direct and inverse images of S and T, respectively.

PROOF. The passage from (a) to (b) is by direct image under L, and the passage from (b) to (a) will be by inverse image under L^{-1}. Certainly the direct image of a vector subspace as in (a) is a vector subspace as in (b). We are to show that the inverse image of a vector subspace as in (b) is a vector subspace as in (a) and that these two procedures invert one another.

For any vector subspace T of W, $L^{-1}(T)$ is a vector subspace of V. In fact, if v_1 and v_2 are in $L^{-1}(T)$, we can write $L(v_1) = t_1$ and $L(v_2) = t_2$ with t_1 and t_2 in T. Then the equations $L(v_1 + v_2) = t_1 + t_2$ and $L(cv_1) = cL(v_1) = ct_1$ show that $v_1 + v_2$ and cv_1 are in $L^{-1}(T)$.

Moreover, the vector subspace $L^{-1}(T)$ contains $L^{-1}(0) = U$. Therefore the inverse image under L of a vector subspace as in (b) is a vector subspace as in (a). Since L is a function, we have $L(L^{-1}(T)) = T$. Thus passing from (b) to (a) and back recovers the vector subspace of W.

If S is a vector subspace of V containing U, we still need to see that $S = L^{-1}(L(S))$. Certainly $S \subseteq L^{-1}(L(S))$. In the reverse direction let v be in $L^{-1}(L(S))$. Then $L(v)$ is in $L(S)$, i.e., $L(v) = L(s)$ for some s in S. Since L

is linear, $L(v-s) = 0$. Thus $v-s$ is in $\ker L = U$, which is contained in S by assumption. Then s and $v-s$ are in S, and hence v is in S. We conclude that $L^{-1}(L(S)) \subseteq S$, and thus passing from (a) to (b) and then back recovers the vector subspace of V containing U.

If V is a vector space and V_1 and V_2 are vector subspaces, then we write $V_1 + V_2$ for the set $V_1 + V_2$ of all sums $v_1 + v_2$ with $v_1 \in V_1$ and $v_2 \in V_2$. This is again a vector subspace of V and is called the **sum** of V_1 and V_2. If we have vector subspaces $V_1, \ldots, V_n$, we abbreviate $((\cdots(V_1 + V_2) + V_3) + \cdots + V_n)$ as $V_1 + \cdots + V_n$. $\square$

Theorem 2.28 (Second Isomorphism Theorem). Let M and N be vector subspaces of a vector space V over $\mathbb{F}$. Then the map $n + (M \cap N) \mapsto n + M$ is a well-defined canonical vector-space isomorphism

$$N/(M \cap N) \cong (M+N)/M.$$

PROOF. The function $L(n+(M\cap N)) = n+M$ is well defined since $M \cap N \subseteq M$, and L is linear. The domain of L is $\{n+(M\cap N) \mid n \in N\}$, and the kernel is the subset of this where n lies in M as well as N. For this to happen, n must be in $M \cap N$, and thus the kernel is the 0 element of $N/(M\cap N)$. Hence L is one-one.

To see that L is onto $(M+N)/M$, let $(m+n)+M$ be given. Then $n+(M\cap N)$ maps to $n+M$, which equals $(m+n)+M$. Hence L is onto. $\square$

Corollary 2.29. Let M and N be finite-dimensional vector subspaces of a vector space V over $\mathbb{F}$. Then

$$\dim(M+N) + \dim(M \cap N) = \dim M + \dim N.$$

PROOF. Theorem 2.28 and two applications of Corollary 2.24a yield

$$\dim(M+N) - \dim M = \dim((M+N)/M)$$
$$= \dim(N/(M\cap N)) = \dim N - \dim(M\cap N),$$

and the result follows. $\square$

6. Direct Sums and Direct Products of Vector Spaces

In this section we introduce the direct sum and direct product of two or more vector spaces over $\mathbb{F}$. When there are only finitely many such subspaces, these constructions come to the same thing, and we call it "direct sum." We begin with the case that two vector spaces are given.

6. Direct Sums and Direct Products of Vector Spaces

We define two kinds of direct sums. The **external direct sum** of two vector spaces V_1 and V_2 over $\mathbb{F}$, written $V_1 \oplus V_2$, is a vector space obtained as follows. The underlying set is the set-theoretic product, i.e., the set $V_1 \times V_2$ of ordered pairs (v_1, v_2) with $v_1 \in V_1$ and $v_2 \in V_2$. The operations of addition and scalar multiplication are defined coordinate by coordinate:

$$(u_1, u_2) + (v_1, v_2) = (u_1 + v_1, u_2 + v_2),$$
$$c(v_1, v_2) = (cv_1, cv_2),$$

and it is immediate that $V_1 \oplus V_2$ satisfies the defining properties of a vector space.

If $\{a_i\}$ is a basis of V_1 and $\{b_j\}$ is a basis of V_2, then it follows from the formula $(v_1, v_2) = (v_1, 0) + (0, v_2)$ that $\{(a_i, 0)\} \cup \{(0, b_j)\}$ is a basis of $V_1 \oplus V_2$. Consequently if V_1 and V_2 are finite-dimensional, then $V_1 \oplus V_2$ is finite-dimensional with

$$\dim(V_1 \oplus V_2) = \dim V_1 + \dim V_2.$$

Associated to the construction of the external direct sum of two vector spaces are four linear maps of interest:

two "projections," $p_1 : V_1 \oplus V_2 \to V_1$ with $p_1(v_1, v_2) = v_1$,
$p_2 : V_1 \oplus V_2 \to V_2$ with $p_2(v_1, v_2) = v_2$,
two "injections," $i_1 : V_1 \to V_1 \oplus V_2$ with $i_1(v_1) = (v_1, 0)$,
$i_2 : V_2 \to V_1 \oplus V_2$ with $i_2(v_2) = (0, v_2)$.

These have the properties that

$$p_r i_s = \begin{cases} I & \text{on } V_s \text{ if } r = s, \\ 0 & \text{on } V_s \text{ if } r \neq s, \end{cases}$$

$$i_1 p_1 + i_2 p_2 = I \quad \text{on } V_1 \oplus V_2.$$

The second notion of direct sum captures the idea of recognizing a situation as canonically isomorphic to an external direct sum. This is based on the following proposition.

Proposition 2.30. Let V be a vector space over $\mathbb{F}$, and let V_1 and V_2 be vector subspaces of V. Then the following conditions are equivalent:
 (a) every member v of V decomposes uniquely as $v = v_1 + v_2$ with $v_1 \in V_1$ and $v_2 \in V_2$,
 (b) $V_1 + V_2 = V$ and $V_1 \cap V_2 = 0$,
 (c) the function from the external direct sum $V_1 \oplus V_2$ to V given by $(v_1, v_2) \mapsto v_1 + v_2$ is an isomorphism of vector spaces.

REMARKS.

(1) If V is a vector space with vector subspaces V_1 and V_2 satisfying the equivalent conditions of Proposition 2.30, then we say that V is the **internal direct sum** of V_1 and V_2. It is customary to write $V = V_1 \oplus V_2$ in this case even though what we have is a canonical isomorphism of the two sides, not an equality.

(2) The dimension formula

$$\dim(V_1 \oplus V_2) = \dim V_1 + \dim V_2$$

for an internal direct sum follows, on the one hand, from the corresponding formula for external direct sums; it follows, on the other hand, by using (b) and Corollary 2.29.

(3) In the proposition it is possible to establish a fourth equivalent condition as follows: there exist linear maps $p_1 : V \to V$, $p_2 : V \to V$, $i_1 :$ image $p_1 \to V$, and $i_2 :$ image $p_2 \to V$ such that

- $p_r i_s p_s$ equals p_r if $r = s$ and equals 0 if $r \neq s$,
- $i_1 p_1 + i_2 p_2 = I$, and
- $V_1 =$ image $i_1 p_1$ and $V_2 =$ image $i_2 p_2$.

PROOF. If (a) holds, then the existence of the decomposition $v = v_1 + v_2$ shows that $V_1 + V_2 = V$. If v is in $V_1 \cap V_2$, then $0 = v + (-v)$ is a decomposition of the kind in (a), and the uniqueness forces $v = 0$. Therefore $V_1 \cap V_2 = 0$. This proves (b).

The function in (c) is certainly linear. If (b) holds and v is given in V, then the identity $V_1 + V_2 = V$ allows us to decompose v as $v = v_1 + v_2$. This proves that the linear map in (c) is onto. To see that it is one-one, suppose that $v_1 + v_2 = 0$. Then $v_1 = -v_2$ shows that v_1 is in $V_1 \cap V_2$. By (b), this intersection is 0. Therefore $v_1 = v_2 = 0$, and the linear map in (c) is one-one.

If (c) holds, then the fact that the linear map in (c) is onto V proves the existence of the decomposition in (a). For uniqueness, suppose that $v_1 + v_2 = u_1 + u_2$ with u_1 and v_1 in V_1 and with u_2 and v_2 in V_2. Then (u_1, u_2) and (v_1, v_2) have the same image under the linear map in (c). Since the function in (c) is assumed one-one, we conclude that $(u_1, u_2) = (v_1, v_2)$. This proves the uniqueness of the decomposition in (a). $\square$

If $V = V_1 \oplus V_2$ is a direct sum, then we can use the above projections and injections to pass back and forth between linear maps with V_1 and V_2 as domain or range and linear maps with V as domain or range. This passage back and forth is called the **universal mapping property** of $V_1 \oplus V_2$ and will be seen later in this section to characterize $V_1 \oplus V_2$ up to canonical isomorphism. Let us be specific about how this property works.

To arrange for V to be the range, suppose that U is a vector space over $\mathbb{F}$ and that $L_1 : U \to V_1$ and $L_2 : U \to V_2$ are linear maps. Then we can define a linear map $L : U \to V$ by $L = i_1 L_1 + i_2 L_2$, i.e., by

$$L(u) = (i_1 L_1 + i_2 L_2)(u) = (L_1(u), L_2(u)),$$

and we can recover L_1 and L_2 from L by $L_1 = p_1 L$ and $L_2 = p_2 L$.

To arrange for V to be the domain, suppose that W is a vector space over $\mathbb{F}$ and that $M_1 : V_1 \to W$ and $M_2 : V_2 \to W$ are linear maps. Then we can define a linear map $M : V \to W$ by $M = M_1 p_1 + M_2 p_2$, i.e., by

$$M(v_1, v_2) = M_1(v_1) + M_2(v_2),$$

and we can recover M_1 and M_2 from M by $M_1 = M i_1$ and $M_2 = M i_2$.

The notion of direct sum readily extends to the direct sum of n vector spaces over $\mathbb{F}$. The **external direct sum** $V_1 \oplus \cdots \oplus V_n$ is the set of ordered pairs $(v_1, \ldots, v_n)$ with each v_j in V_j and with addition and scalar multiplication defined coordinate by coordinate. In the finite-dimensional case we have

$$\dim(V_1 \oplus \cdots \oplus V_n) = \dim V_1 + \cdots + \dim V_n.$$

If $V_1, \ldots, V_n$ are given as vector subspaces of a vector space V, then we say that V is the **internal direct sum** of $V_1, \ldots, V_n$ if the equivalent conditions of Proposition 2.31 below are satisfied. In this case we write $V = V_1 \oplus \cdots \oplus V_n$ even though once again we really have a canonical isomorphism rather than an equality.

Proposition 2.31. Let V be a vector space over $\mathbb{F}$, and let $V_1, \ldots, V_n$ be vector subspaces of V. Then the following conditions are equivalent:
(a) every member v of V decomposes uniquely as $v = v_1 + \cdots + v_n$ with $v_j \in V_j$ for $1 \leq j \leq n$,
(b) $V_1 + \cdots + V_n = V$ and also $V_j \cap (V_1 + \cdots + V_{j-1} + V_{j+1} + \cdots + V_n) = 0$ for each j with $1 \leq j \leq n$,
(c) the function from the external direct sum $V_1 \oplus \cdots \oplus V_n$ to V given by $(v_1, \ldots, v_n) \mapsto v_1 + \cdots + v_n$ is an isomorphism of vector spaces.

Proposition 2.31 is proved in the same way as Proposition 2.30, and the expected analog of Remark 3 with that proposition is valid as well. Notice that the second condition in (b) is stronger than the condition that $V_i \cap V_j = 0$ for all $i \neq j$. Figure 2.3 illustrates how the condition $V_i \cap V_j = 0$ for all $i \neq j$ can be satisfied even though (b) is not satisfied and even though the vector subspaces do not therefore form a direct sum.

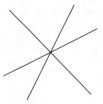

FIGURE 2.3. Three 1-dimensional vector subspaces of $\mathbb{R}^2$ such that each pair has intersection 0.

If $V = V_1 \oplus \cdots \oplus V_n$ is a direct sum, then we can define projections $p_1, \ldots, p_n$ and injections $i_1, \ldots, i_n$ in the expected way, and we again get a universal mapping property. That is, we can pass back and forth between linear maps with $V_1, \ldots, V_n$ as domain or range and linear maps with V as domain or range. The argument given above for $n = 2$ is easily adjusted to handle general n, and we omit the details.

To generalize the above notions to infinitely many vector spaces, there are two quite different ways of proceeding. Let us treat first the external constructions. Let a nonempty collection of vector spaces V_α over $\mathbb{F}$ be given, one for each $\alpha \in A$. The **direct sum** $\bigoplus_{\alpha \in A} V_\alpha$ is the set of all tuples $\{v_\alpha\}$ in the Cartesian product $\bigtimes_{\alpha \in A} V_\alpha$ with all but finitely many v_α equal to 0 and with addition and scalar multiplication defined coordinate by coordinate. For this construction we obtain a basis as the union of embedded bases of the constituent spaces. The **direct product** $\prod_{\alpha \in A} V_\alpha$ is the set of *all* tuples $\{v_\alpha\}$ in $\bigtimes_{\alpha \in A} V_\alpha$, again with addition and scalar multiplication defined coordinate by coordinate. When there are only finitely many factors $V_1, \ldots, V_n$, the direct product, which manifestly coincides with the direct sum, is sometimes denoted by $V_1 \times \cdots \times V_n$. For the direct product when there are infinitely many factors, there is no evident way to obtain a basis of the product from bases of the constituents.

The projections and injections that we defined in the case of finitely many vector spaces are still meaningful here. The universal mapping property is still valid as well, but it splinters into one form for direct sums and another form for direct products. The formulas given above for using linear maps with the V_α's as domain or range to define linear maps with the direct sum or direct product as domain or range may involve sums with infinitely many nonzero terms, and they are not directly usable. Instead, the formulas that continue to make sense are the ones for recovering linear maps with the V_α's as domain or range from linear maps with the direct sum or direct product as domain or range. These turn out to determine the formulas uniquely for the linear maps with the direct sum or direct product as domain or range. In other words, the appropriate universal mapping property uniquely determines the direct sum or direct product up to an isomorphism that respects the relevant projections and injections.

6. Direct Sums and Direct Products of Vector Spaces

Let us see to the details. We denote typical members of $\prod_{\alpha \in A} V_\alpha$ and $\bigoplus_{\alpha \in A} V_\alpha$ by $\{v_\alpha\}_{\alpha \in A}$, with the understanding that only finitely many v_α can be nonzero in the case of the direct sum. The formulas are

$$p_\beta : \prod_{\alpha \in A} V_\alpha \to V_\beta \qquad \text{with } p_\beta(\{v_\alpha\}_{\alpha \in A}) = v_\beta,$$

$$i_\beta : V_\beta \to \bigoplus_{\alpha \in A} V_\alpha \qquad \text{with } i_\beta(v_\beta) = \{w_\alpha\}_{\alpha \in A} \text{ and } w_\alpha = \begin{cases} v_\beta & \text{if } \alpha = \beta, \\ 0 & \text{if } \alpha \neq \beta. \end{cases}$$

If U is a vector space over $\mathbb{F}$ and if a linear map $L_\beta : U \to V_\beta$ is given for each $\beta \in A$, we can obtain a linear map $L : U \to \prod_{\alpha \in A} V_\alpha$ that satisfies $p_\beta L = L_\beta$ for all β. The definition that makes perfectly good sense is

$$L(u) = \{L(u)_\alpha\}_{\alpha \in A} = \{L_\alpha(u)\}_{\alpha \in A}.$$

What does not make sense is to try to express the right side in terms of the injections i_α; we cannot write the right side as $\sum_{\alpha \in A} i_\alpha(L_\alpha(u))$ because infinitely many terms might be nonzero.

If W is a vector space and a linear map $M_\beta : V_\beta \to W$ is given for each β, we can obtain a linear map $M : \bigoplus_{\alpha \in A} V_\alpha \to W$ that satisfies $M i_\beta = M_\beta$ for all β; the definition that makes perfectly good sense is

$$M(\{v_\alpha\}_{\alpha \in A}) = \sum_{\alpha \in A} M_\alpha(v_\alpha).$$

The right side is meaningful since only finitely many v_α can be nonzero. It can be misleading to write the formula as $M = \sum_{\alpha \in A} M_\alpha p_\alpha$ because infinitely many of the linear maps $M_\alpha p_\alpha$ can be nonzero functions.

In any event, we have a universal mapping property in both cases—for the direct product with the projections in place and for the direct sum with the injections in place. Let us see that these universal mapping properties characterize direct products and direct sums up to an isomorphism respecting the projections and injections, and that they allow us to define and recognize "internal" direct products and direct sums.

A **direct product** of a set of vector spaces V_α over $\mathbb{F}$ for $\alpha \in A$ consists of a vector space V and a system of linear maps $p_\alpha : V \to V_\alpha$ with the following **universal mapping property**: whenever U is a vector space and $\{L_\alpha\}$ is a system of linear maps $L_\alpha : U \to V_\alpha$, then there exists a unique linear map $L : U \to V$ such that $p_\alpha L = L_\alpha$ for all α. See Figure 2.4. The external direct product establishes existence of a direct product, and Proposition 2.32 below establishes its uniqueness up to an isomorphism of the V's that respects the p_α's. A direct product is said to be **internal** if each V_α is a vector subspace of V and if for each α, the restriction $p_\alpha|_{V_\alpha}$ is the identity map on V_α. Because of the uniqueness, this definition of internal direct product is consistent with the earlier one when there are only finitely V_α's.

FIGURE 2.4. Universal mapping property of a direct product of vector spaces.

Proposition 2.32. Let A be a nonempty set of vector spaces over $\mathbb{F}$, and let V_α be the vector space corresponding to the member α of A. If $(V, \{p_\alpha\})$ and $(V^*, \{p_\alpha^*\})$ are two direct products of the V_α's, then the linear maps $p_\alpha : V \to V_\alpha$ and $p_\alpha^* : V^* \to V_\alpha$ are onto V_α, there exists a unique linear map $L : V^* \to V$ such that $p_\alpha^* = p_\alpha L$ for all $\alpha \in A$, and L is invertible.

PROOF. In Figure 2.4 let $U = V^*$ and $L_\alpha = p_\alpha^*$. If $L : V^* \to V$ is the linear map produced by the fact that V is a direct product, then we have $p_\alpha L = p_\alpha^*$ for all α. Reversing the roles of V and V^*, we obtain a linear map $L^* : V \to V^*$ with $p_\alpha^* L^* = p_\alpha$ for all α. Therefore $p_\alpha(LL^*) = (p_\alpha L)L^* = p_\alpha^* L^* = p_\alpha$.

In Figure 2.4 we next let $U = V$ and $L_\alpha = p_\alpha$ for all α. Then the identity 1_V on V has the same property $p_\alpha 1_V = p_\alpha$ relative to all p_α that LL^* has, and the uniqueness says that $LL^* = 1_V$. Reversing the roles of V and V^*, we obtain $L^*L = 1_{V^*}$. Therefore L is invertible.

For uniqueness suppose that $\Phi : V^* \to V$ is another linear map with $p_\alpha^* = p_\alpha \Phi$ for all $\alpha \in A$. Then the argument of the previous paragraph shows that $L^*\Phi = 1_{V^*}$. Applying L on the left gives $\Phi = (LL^*)\Phi = L(L^*\Phi) = L1_{V^*} = L$. Thus $\Phi = L$.

Finally we have to show that the αth map of a direct product is onto V_α. It is enough to show that p_α^* is onto V_α. Taking V as the external direct product $\prod_{\alpha \in A} V_\alpha$ with p_α equal to the coordinate mapping, form the invertible linear map $L^* : V \to V^*$ that has just been proved to exist. This satisfies $p_\alpha = p_\alpha^* L^*$ for all $\alpha \in A$. Since p_α is onto V_α, p_α^* must be onto V_α. □

A **direct sum** of a set of vector spaces V_α over $\mathbb{F}$ for $\alpha \in A$ consists of a vector space V and a system of linear maps $i_\alpha : V_\alpha \to V$ with the following **universal mapping property**: whenever W is a vector space and $\{M_\alpha\}$ is a system of linear maps $M_\alpha : V_\alpha \to W$, then there exists a unique linear map $M : V \to W$ such that $Mi_\alpha = M_\alpha$ for all α. See Figure 2.5. The external direct sum establishes existence of a direct sum, and Proposition 2.33 below establishes its uniqueness up to isomorphism of the V's that respects the i_α's. A direct sum is said to be **internal** if each V_α is a vector subspace of V and if for each α, the map i_α is the inclusion map of V_α into V. Because of the uniqueness, this definition of internal direct sum is consistent with the earlier one when there are only finitely V_α's.

FIGURE 2.5. Universal mapping property of a direct sum of vector spaces.

Proposition 2.33. Let A be a nonempty set of vector spaces over $\mathbb{F}$, and let V_α be the vector space corresponding to the member α of A. If $(V, \{i_\alpha\})$ and $(V^*, \{i_\alpha^*\})$ are two direct sums of the V_α's, then the linear maps $i_\alpha : V_\alpha \to V$ and $i_\alpha^* : V_\alpha \to V^*$ are one-one, there exists a unique linear map $M : V \to V^*$ such that $i_\alpha^* = M i_\alpha$ for all $\alpha \in A$, and M is invertible.

PROOF. In Figure 2.5 let $W = V^*$ and $M_\alpha = i_\alpha^*$. If $M : V \to V^*$ is the linear map produced by the fact that V is a direct sum, then we have $M i_\alpha = i_\alpha^*$ for all α. Reversing the roles of V and V^*, we obtain a linear map $M^* : V^* \to V$ with $M^* i_\alpha^* = i_\alpha$ for all α. Therefore $(M^* M) i_\alpha = M^* i_\alpha^* = i_\alpha$.

In Figure 2.5 we next let $W = V$ and $M_\alpha = i_\alpha$ for all α. Then the identity 1_V on V has the same property $1_V i_\alpha = i_\alpha$ relative to all i_α that $M^* M$ has, and the uniqueness says that $M^* M = 1_V$. Reversing the roles of V and V^*, we obtain $M M^* = 1_{V^*}$. Therefore M is invertible.

For uniqueness suppose that $\Phi : V \to V^*$ is another linear map with $i_\alpha^* = \Phi i_\alpha$ for all $\alpha \in A$. Then the argument of the previous paragraph shows that $M^* \Phi = 1_V$. Applying M on the left gives $\Phi = (M M^*) \Phi = M(M^* \Phi) = M 1_V = M$. Thus $\Phi = M$.

Finally we have to show that the αth map of a direct sum is one-one on V_α. It is enough to show that i_α^* is one-one on V_α. Taking V as the external direct sum $\bigoplus_{s \in S} V_\alpha$ with i_α equal to the embedding mapping, form the invertible linear map $M^* : V^* \to V$ that has just been proved to exist. This satisfies $i_\alpha = M^* i_\alpha^*$ for all $\alpha \in A$. Since i_α is one-one, i_α^* must be one-one. □

7. Determinants

A "determinant" is a certain scalar attached initially to any square matrix and ultimately to any linear map from a finite-dimensional vector space into itself.

The definition is presumably known from high-school algebra in the case of 2-by-2 and 3-by-3 matrices:

$$\det \begin{pmatrix} a & b \\ c & d \end{pmatrix} = ad - bc,$$

$$\det \begin{pmatrix} a & b & c \\ d & e & f \\ g & h & i \end{pmatrix} = aei + bfg + cdh - afh - bdi - ceg.$$

For n-by-n square matrices the determinant function will have the following important properties:

(i) $\det(AB) = \det A \det B$,
(ii) $\det I = 1$,
(iii) $\det A = 0$ if and only if A has no inverse.

Once we have constructed the determinant function with these properties, we can then extend the function to be defined on all linear maps $L : V \to V$ with V finite-dimensional. To do so, we let Γ be any ordered basis of V, and we define $\det L = \begin{pmatrix} L \\ \Gamma\Gamma \end{pmatrix}$. If Δ is another ordered basis, then

$$\det \begin{pmatrix} L \\ \Delta\Delta \end{pmatrix} = \det \begin{pmatrix} I \\ \Delta\Gamma \end{pmatrix} \det \begin{pmatrix} L \\ \Gamma\Gamma \end{pmatrix} \det \begin{pmatrix} I \\ \Gamma\Delta \end{pmatrix},$$

and this equals $\det \begin{pmatrix} L \\ \Gamma\Gamma \end{pmatrix}$ by (i) since $\begin{pmatrix} I \\ \Delta\Gamma \end{pmatrix}$ and $\begin{pmatrix} I \\ \Gamma\Delta \end{pmatrix}$ are inverses of each other and since their determinants, by (i) and (ii), are reciprocals. Hence the definition of $\det L$ is independent of the choice of ordered basis, and determinant is well defined on the linear map $L : V \to V$. It is then immediate that the determinant function on linear maps from V into V satisfies (i), (ii), and (iii) above.

Thus it is enough to establish the determinant function on n-by-n matrices. Setting matters up in a useful way involves at least one subtle step, but much of this step has fortunately already been carried out in the discussion of signs of permutations in Section I.4. To proceed, we view det on n-by-n matrices over $\mathbb{F}$ as a function of the n rows of the matrix, rather than the matrix itself. We write V for the vector space $M_{1n}(\mathbb{F})$ of all n-dimensional row vectors. A function $f : V \times \cdots \times V \to \mathbb{F}$ defined on ordered k-tuples of members of V is called a k-**multilinear** functional or k-**linear** functional if it depends linearly on each of the k vector variables when the other $k - 1$ vector variables are held fixed. For example,

$$f((a \quad b), (c \quad d)) = ac + b(c + d) + \tfrac{1}{2}ad$$

is a 2-linear functional on $M_{12}(\mathbb{F}) \times M_{12}(\mathbb{F})$. A little more generally and more suggestively,

$$g((a \quad b), (c \quad d)) = \ell_1(a \quad b)\ell_2(c \quad d) + \ell_3(a \quad b)\ell_4(c \quad d)$$

is a 2-linear functional on $M_{12}(\mathbb{F}) \times M_{12}(\mathbb{F})$ whenever $\ell_1, \ldots, \ell_4$ are linear functionals on $M_{12}(\mathbb{F})$.

Let $\{v_1, \ldots, v_n\}$ be a basis of V. Then a k-multilinear functional as above is determined by its value on all k-tuples of basis vectors $(v_{i_1}, \ldots, v_{i_k})$. (Here $i_1, \ldots, i_k$ are integers between 1 and n.) The reason is that we can fix all but the first variable and expand out the expression by linearity so that only a basis vector remains in each term for the first variable; for each resulting term we can fix all but the second variable and expand out the expression by linearity; and so on. Conversely if we specify arbitrary scalars for the values on each such k-tuple, then we can define a k-multilinear functional assuming those values on the tuples of basis vectors.

A k-multilinear functional f on k-tuples from $M_{1n}(\mathbb{F})$ is said to be **alternating** if f is 0 whenever two of the variables are equal.

EXAMPLE. For $k = 2$ and $n = 2$, we use $\{v_1 = (1 \quad 0), v_2 = (0 \quad 1)\}$ as basis. Then a 2-linear multilinear functional f is determined by $f(v_1, v_1)$, $f(v_1, v_2)$, $f(v_2, v_1)$, and $f(v_2, v_2)$. If f is alternating, then $f(v_1, v_1) = f(v_2, v_2) = 0$. But also $f(v_1 + v_2, v_1 + v_2) = 0$, and expansion via 2-multilinearity gives

$$f(v_1, v_1) + f(v_1, v_2) + f(v_2, v_1) + f(v_2, v_2) = 0.$$

We have already seen that the first and last terms on the left side are 0, and thus $f(v_2, v_1) = -f(v_1, v_2)$. Therefore f is completely determined by $f(v_1, v_2)$.

The principle involved in the computation within the example is valid more generally: whenever a multilinear functional f is alternating and two of its arguments are interchanged, then the value of f is multiplied by -1. In fact, let us suppress all variables except for the ith and the jth. Then we have

$$0 = f(v + w, v + w) = f(v + w, v) + f(v + w, w)$$
$$= f(v, v) + f(w, v) + f(v, w) + f(w, w) = f(w, v) + f(v, w).$$

Theorem 2.34. For $M_{1n}(\mathbb{F})$, the vector space of alternating n-multilinear functionals has dimension 1, and a nonzero such functional has nonzero value on $(e_1^t, \ldots, e_n^t)$, where $\{e_1, \ldots, e_n\}$ is the standard basis of $\mathbb{F}^n$. Let f_0 be the unique such alternating n-multilinear functional taking the value 1 on $(e_1^t, \ldots, e_n^t)$. If a function $\det : M_{nn}(\mathbb{F}) \to \mathbb{F}$ is defined by

$$\det A = f_0(A_{1\cdot}, \ldots, A_{n\cdot})$$

when A has rows $A_{1\cdot}, \ldots, A_{n\cdot}$, then det has the properties that
  (a) $\det(AB) = \det A \det B$,
  (b) $\det I = 1$,
  (c) $\det A = 0$ if and only if A has no inverse,
  (d) $\det A = \sum_\sigma (\operatorname{sgn} \sigma) A_{1\sigma(1)} A_{2\sigma(2)} \cdots A_{n\sigma(n)}$, the sum being taken over all permutations σ of $\{1, \ldots, n\}$.

PROOF OF UNIQUENESS. Let f be an alternating n-multilinear functional, and let $\{u_1, \ldots, u_n\}$ be the basis of the space of row vectors defined by $u_i = e_i^t$. Since f is multilinear, f is determined by its values on all n-tuples $(u_{k_1}, \ldots, u_{k_n})$. Since f is alternating, $f(u_{k_1}, \ldots, u_{k_n}) = 0$ unless the u_{k_i} are distinct, i.e., unless $(u_{k_1}, \ldots, u_{k_n})$ is of the form $(u_{\sigma(1)}, \ldots, u_{\sigma(n)})$ for some permutation σ. We have seen that the value of f on an n-tuple of rows is multiplied by -1 if two of the rows are interchanged. Corollary 1.22 and Proposition 1.24b consequently together imply that the value of f on an n-tuple is multiplied by $\operatorname{sgn} \sigma$ if the members of the n-tuple are permuted by σ. Therefore $f(u_{\sigma(1)}, \ldots, u_{\sigma(n)}) = (\operatorname{sgn} \sigma) f(u_1, \ldots, u_n)$, and f is completely determined by its value on $(u_1, \ldots, u_n)$. We conclude that the vector space of alternating n-multilinear functionals has dimension at most 1. □

PROOF OF EXISTENCE. Define $\det A$, and therefore also f_0, by (d). Each term in this definition is the product of n linear functionals, the kth linear functional being applied to the kth argument of f_0, and f_0 is consequently n-multilinear. To see that f_0 is alternating, suppose that the ith and j^{th} rows are equal with $i \neq j$. If τ is the transposition of i and j, then $A_{1\sigma\tau(1)} A_{2\sigma\tau(2)} \cdots A_{n\sigma\tau(n)} = A_{1\sigma(1)} A_{2\sigma(2)} \cdots A_{n\sigma(n)}$, and Lemma 1.23 hence shows that

$$(\operatorname{sgn} \sigma\tau) A_{1\sigma\tau(1)} A_{2\sigma\tau(2)} \cdots A_{n\sigma\tau(n)} + (\operatorname{sgn} \sigma) A_{1\sigma(1)} A_{2\sigma(2)} \cdots A_{n\sigma(n)} = 0.$$

Thus if we compute the sum in (d) by grouping pairs of terms, the one for $\sigma\tau$ and the one for σ if $\operatorname{sgn} \sigma = +1$, we see that the whole sum is 0. Thus f_0 is alternating. Finally when A is the identity matrix I, we see that $A_{1\sigma(1)} A_{2\sigma(2)} \cdots A_{n\sigma(n)} = 0$ unless σ is the identity permutation, and then the product is 1. Since $\operatorname{sgn} 1 = +1$, $\det I = +1$. We conclude that the vector space of alternating n-multilinear functionals has dimension exactly 1. □

PROOF OF PROPERTIES OF det. Fix an n-by-n matrix B. Since f_0 is alternating n-multilinear, so is $(v_1, \ldots, v_n) \mapsto f_0(v_1 B, \ldots, v_n B)$. The vector space of alternating n-multilinear functionals has been proved to be of dimension 1, and therefore $f_0(v_1 B, \ldots, v_n B) = c(B) f_0(v_1, \ldots, v_n)$ for some scalar $c(B)$. In the notation with det, this equation reads $\det(AB) = c(B) \det A$. Putting $A = I$, we obtain $\det B = c(B) \det I$. Thus $c(B) = \det B$, and (a) follows. We have already proved (b), and (d) was the definition of $\det A$. We are left with (c). If A^{-1}

exists, then (a) and (b) give $\det(A^{-1})\det A = \det I = 1$, and hence $\det A \neq 0$. If A^{-1} does not exist, then Theorem 1.30 and Proposition 1.27c show that the reduced row-echelon form R of A has a row of 0's. We combine Proposition 1.29, conclusion (a), the invertibility of elementary matrices, and the fact that invertible matrices have nonzero determinant, and we see that $\det A$ is the product of $\det R$ and a nonzero scalar. Since det is linear as a function of each row and since R has a row of 0's, $\det R = 0$. Therefore $\det A = 0$. This completes the proof of the theorem. $\square$

The fast procedure for evaluating determinants is to use row reduction, keeping track of what happens. The effect of each kind of row operation on a determinant and the reasons the function det behaves in this way are as follows:

(i) Interchange two rows. This operation multiplies the determinant by -1 because of the alternating property.

(ii) Multiply a row by a nonzero scalar c. This operation multiplies the determinant by c because of the linearity of determinant as a function of that row.

(iii) Replace the ith row by the sum of it and a multiple of the jth row with $j \neq i$. This operation leaves the determinant unchanged. In fact, the matrix whose i^{th} row is replaced by the jth row has determinant 0 by the alternating property, and the rest follows by linearity in the ith row.

As with row reduction the number of steps required to compute a determinant this way is $\leq Cn^3$ in the n-by-n case.

A certain savings of computation is possible as compared with full-fledged row reduction. Namely, we have only to arrange for the reduced matrix to be 0 below the main diagonal, and then the determinant of the reduced matrix will be the product of the diagonal entries, by inspection of the formula in Theorem 2.34d.

EXAMPLE. For the matrix $\begin{pmatrix} 1 & 2 & 3 \\ 4 & 5 & 6 \\ 7 & 8 & 10 \end{pmatrix}$, we have

$$\det \begin{pmatrix} 1 & 2 & 3 \\ 4 & 5 & 6 \\ 7 & 8 & 10 \end{pmatrix} \stackrel{(\text{iii})}{=} \det \begin{pmatrix} 1 & 2 & 3 \\ 0 & -3 & -6 \\ 0 & -6 & -11 \end{pmatrix}$$

$$\stackrel{(\text{ii})}{=} -3 \det \begin{pmatrix} 1 & 2 & 3 \\ 0 & 1 & 2 \\ 0 & -6 & -11 \end{pmatrix} \stackrel{(\text{iii})}{=} -3 \det \begin{pmatrix} 1 & 2 & 3 \\ 0 & 1 & 2 \\ 0 & 0 & 1 \end{pmatrix} = -3.$$

We conclude this section with a number of formulas for determinants.

Proposition 2.35. If A is an n-by-n square matrix, then $\det A^t = \det A$.

PROOF. Corollary 2.9 says that the row space and the column space of A have the same dimension, and A is invertible if and only if the row space has dimension n. Thus A is invertible if and only if A^t is invertible, and Theorem 2.34c thus shows that $\det A = 0$ if and only if $\det A^t = 0$. Now suppose that $\det A$ and $\det A^t$ are nonzero. Then we can write $A = E_1 \cdots E_r$ with each E_j an elementary matrix of one of the three types. Theorem 2.34a shows that $\det A = \prod_{j=1}^r \det E_j$ and $\det A^t = \prod_{j=1}^r \det E_j^t$, and hence it is enough to prove that $\det E_j = \det E_j^t$ for each j. For E_j of either of the first two types, $E_j = E_j^t$ and there is nothing to prove. For E_j of the third type, we have $\det E_j = \det E_j^t = 1$. The result follows. □

Proposition 2.36 (expansion in cofactors). Let A be an n-by-n matrix, and let $\widehat{A_{ij}}$ be the square matrix of size $n - 1$ obtained by deleting the ith row and the jth column. Then

(a) for any j, $\det A = \sum_{i=1}^n (-1)^{i+j} A_{ij} \det \widehat{A_{ij}}$, i.e., $\det A$ may be calculated by "expansion in cofactors" about the jth column,
(b) for any i, $\det A = \sum_{j=1}^n (-1)^{i+j} A_{ij} \det \widehat{A_{ij}}$, i.e., $\det A$ may be calculated by "expansion in cofactors" about the ith row.

REMARKS. If this formula is iterated, we obtain a procedure for evaluating a determinant in about $Cn!$ steps. This procedure amounts to using the formula for $\det A$ in Theorem 2.34d and is ordinarily not of practical use. However, it *is* of theoretical use, and Corollary 2.37 will provide a simple example of a theoretical application.

PROOF. It is enough to prove (a) since (b) then follows by combining (a) and Proposition 2.35. In (a), the right side is 1 when $A = I$, and it is enough by Theorem 2.34 to prove that the right side is alternating and n-multilinear. Each term on the right side is n-multilinear, and hence so is the whole expression. To see that the right side is alternating, suppose that the kth and l^{th} rows are equal with $k < l$. The kth and l^{th} rows are both present in $\widehat{A_{ij}}$ if i is not equal to k or l, and thus each $\det \widehat{A_{ij}}$ is 0 for i not equal to k or l. We are left with showing that

$$(-1)^{k+j} A_{kj} \det \widehat{A_{kj}} + (-1)^{l+j} A_{lj} \det \widehat{A_{lj}} = 0.$$

The two matrices $\widehat{A_{kj}}$ and $\widehat{A_{lj}}$ have the same rows but in a different order. The order is

$1, \ldots, k-1, k+1, \ldots, l-1, l, l+1, \ldots, n$ in the case of $\widehat{A_{kj}}$,
$1, \ldots, k-1, k, k+1, \ldots, l-1, l+1, \ldots, n$ in the case of $\widehat{A_{lj}}$.

We can transform the first matrix into the second by transposing the index for row l to the left one step at a time until it gets to the kth position. The number of steps is $l - k - 1$, and therefore $\det \widehat{A_{lj}} = (-1)^{l-k-1} \det \widehat{A_{kj}}$. Consequently
$(-1)^{k+j} A_{kj} \det \widehat{A_{kj}} + (-1)^{l+j} \det \widehat{A_{lj}} = \left((-1)^{k+j} A_{kj} + (-1)^{2l-k-1+j} A_{lj}\right) \det \widehat{A_{kj}}$.
The right side is 0 since $A_{kj} = A_{lj}$, and the proof is complete. □

Corollary 2.37 (Vandermonde matrix and determinant). If $r_1, \ldots, r_n$ are scalars, then

$$\det \begin{pmatrix} 1 & 1 & \cdots & 1 \\ r_1 & r_2 & \cdots & r_n \\ r_1^2 & r_2^2 & \cdots & r_n^2 \\ \vdots & \vdots & \ddots & \vdots \\ r_1^{n-1} & r_2^{n-1} & \cdots & r_n^{n-1} \end{pmatrix} = \prod_{j>i} (r_j - r_i).$$

PROOF. We show that the determinant is

$$= \prod_{j>1} (r_j - r_1) \det \begin{pmatrix} 1 & \cdots & 1 \\ r_2 & \cdots & r_n \\ \vdots & \ddots & \vdots \\ r_2^{n-2} & \cdots & r_n^{n-2} \end{pmatrix},$$

and then the result follows by induction. In the given matrix, replace the nth row by the sum of it and $-r_1$ times the $(n-1)^{\text{st}}$ row, then the $(n-1)^{\text{st}}$ row by the sum of it and $-r_1$ times the $(n-2)^{\text{nd}}$ row, and so on. The resulting determinant is

$$\det \begin{pmatrix} 1 & 1 & \cdots & 1 \\ 0 & r_2 - r_1 & \cdots & r_n - r_1 \\ \vdots & \vdots & \ddots & \vdots \\ 0 & r_2^{n-2} - r_1 r_2^{n-3} & \cdots & r_n^{n-2} - r_1 r_n^{n-3} \\ 0 & r_2^{n-1} - r_1 r_2^{n-2} & \cdots & r_n^{n-1} - r_1 r_n^{n-2} \end{pmatrix}$$

$$= \det \begin{pmatrix} r_2 - r_1 & \cdots & r_n - r_1 \\ \vdots & \ddots & \vdots \\ r_2^{n-2} - r_1 r_2^{n-3} & \cdots & r_n^{n-2} - r_1 r_n^{n-3} \\ r_2^{n-1} - r_1 r_2^{n-2} & \cdots & r_n^{n-1} - r_1 r_n^{n-2} \end{pmatrix} \quad \text{by Proposition 2.36a applied with } j = 1$$

$$= (r_2 - r_1) \cdots (r_n - r_1) \det \begin{pmatrix} 1 & \cdots & 1 \\ r_2 & \cdots & r_n \\ \vdots & \ddots & \vdots \\ r_2^{n-2} & \cdots & r_n^{n-2} \end{pmatrix},$$

the last step following by multilinearity of the determinant in the columns (as a consequence of Proposition 2.35 and multilinearity in the rows). □

The **classical adjoint** of the square matrix A, denoted by A^{adj}, is the matrix with entries $A_{ij}^{\text{adj}} = (-1)^{i+j} \det \widehat{A_{ji}}$ with $\widehat{A_{kl}}$ defined as in the statement of Proposition 2.36: $\widehat{A_{kl}}$ is the matrix A with the kth row and l^{th} column deleted.

In the 2-by-2 case, we have $\begin{pmatrix} a & b \\ c & d \end{pmatrix}^{\text{adj}} = \begin{pmatrix} d & -b \\ -c & a \end{pmatrix}$. Thus we have $AA^{\text{adj}} = A^{\text{adj}}A = (\det A)I$ in the 2-by-2 case. Cramer's rule for solving simultaneous linear equations results from the n-by-n generalization of this formula.

Proposition 2.38 (Cramer's rule). If A is an n-by-n matrix, then $AA^{\text{adj}} = A^{\text{adj}}A = (\det A)I$, and thus $\det A \neq 0$ implies $A^{-1} = (\det A)^{-1}A^{\text{adj}}$. Consequently if $\det A \neq 0$, then the unique solution of the simultaneous system $Ax = b$ of n equations in n unknowns, in which $x = \begin{pmatrix} x_1 \\ \vdots \\ x_n \end{pmatrix}$ and $b = \begin{pmatrix} b_1 \\ \vdots \\ b_n \end{pmatrix}$, has

$$x_j = \frac{\det B_j}{\det A}$$

with B_j equal to the n-by-n matrix obtained from A by replacing the jth column of A by b.

REMARKS. If we think of the calculation of the determinant of an n-by-n matrix as requiring about n^3 steps, then application of Cramer's rule, at least if done in an unthinking fashion, suggests that solving an invertible system requires about $n^3(n+1)$ steps, i.e., $n+1$ determinants are involved in the explicit solution. Use of row reduction directly to solve the system is more efficient than proceeding this way. Thus Cramer's rule is more important for its theoretical applications than it is for making computations. One simple theoretical application is the observation that each entry of the inverse of a matrix is the quotient of a polynomial function of the entries divided by the determinant.

PROOF. The $(i, j)^{\text{th}}$ entry of $A^{\text{adj}}A$ is

$$(A^{\text{adj}}A)_{ij} = \sum_{k=1}^{n} A_{ik}^{\text{adj}} A_{kj} = \sum_{k=1}^{n} (-1)^{i+k} (\det \widehat{A_{ki}}) A_{kj}.$$

If $i = j$, then expansion in cofactors about the jth column (Proposition 2.36a) identifies the right side as $\det A$. If $i \neq j$, consider the matrix B obtained from A by replacing the ith column of A by the jth column. Then the ith and j^{th} columns of B are equal, and hence $\det B = 0$. Expanding $\det B$ in cofactors about the ith column (Proposition 2.36a), we obtain

$$0 = \det B = \sum_{k=1}^{n} (-1)^{i+k} (\det \widehat{B_{ki}}) B_{ki} = \sum_{k=1}^{n} (-1)^{i+k} (\det \widehat{A_{ki}}) A_{kj}.$$

Thus $AA^{\text{adj}} = (\det A)I$. A similar argument proves that $A^{\text{adj}}A = (\det A)I$.

For the application to $Ax = b$, we multiply both sides on the left by A^{adj} and obtain $(\det A)x = A^{\text{adj}}b$. Hence

$$(\det A)x_j = \sum_{i=1}^{n} (A^{\text{adj}})_{ji} b_i = \sum_{i=1}^{n} (-1)^{i+j} b_i \det \widehat{A_{ij}},$$

and the right side equals $\det B_j$ by expansion in cofactors of $\det B_j$ about the jth column (Proposition 2.36a). □

8. Eigenvectors and Characteristic Polynomials

A vector $v \neq 0$ in $\mathbb{F}^n$ is an **eigenvector** of the n-by-n matrix A if $Av = \lambda v$ for some scalar λ. We call λ the **eigenvalue** associated with v. When λ is an eigenvalue, the vector space of all v with $Av = \lambda v$, i.e., the set consisting of the eigenvectors and the 0 vector, is called the **eigenspace** for λ.

If we think of A as giving a linear map L from $\mathbb{F}^n$ to itself, an eigenvector takes on geometric significance as a vector mapped to a multiple of itself by L. Another geometric way of viewing matters is that the eigenvector yields a 1-dimensional subspace $U = \mathbb{F}v$ that is **invariant**, or **stable**, under L in the sense of satisfying $L(U) \subseteq U$.

Proposition 2.39. An n-by-n matrix A has an eigenvector with eigenvalue λ if and only if $\det(\lambda I - A) = 0$. In this case the eigenspace for λ is the kernel of $\lambda I - A$.

PROOF. We have $Av = \lambda v$ if and only if $(\lambda I - A)v = 0$, if and only if v is in $\ker(\lambda I - A)$. This kernel is nonzero if and only if $\det(\lambda I - A) = 0$. □

With A fixed, the expression $\det(\lambda I - A)$ is a polynomial in λ of degree n and is called the **characteristic polynomial**[8] of A. To see that it is at least a polynomial function of λ, let us expand $\det(\lambda I - A)$ as

$$\det \begin{pmatrix} \lambda - A_{11} & -A_{12} & \cdots & -A_{1n} \\ -A_{21} & \lambda - A_{22} & \cdots & -A_{2n} \\ \vdots & \vdots & \ddots & \vdots \\ -A_{n1} & -A_{n2} & \cdots & \lambda - A_{nn} \end{pmatrix}$$

$$= \sum_{\sigma} (\text{sgn } \sigma) \text{term}_{1,\sigma(1)} \cdots \text{term}_{n,\sigma(n)}.$$

---

[8] Some authors call $\det(A - \lambda I)$ the characteristic polynomial. This is the same polynomial as $\det(\lambda I - A)$ if n is even and is the negative of it if n is odd. The choice made here has the slight advantage of always having leading coefficient 1, which is a handy property in some situations.

The term for the permutation $\sigma = 1$ has $\sigma(k) = k$ for every k and gives $\prod_{j=1}^{n}(\lambda - A_{jj})$. All other σ's have $\sigma(k) = k$ for at most $n - 2$ values of k, and λ therefore occurs at most $n - 2$ times. Thus the above expression is

$$= \prod_{j=1}^{n}(\lambda - A_{jj}) + \left\{\begin{array}{l}\text{other terms with powers}\\ \text{of }\lambda\text{ at most }n-2\end{array}\right\}$$

$$= \lambda^n - \left(\sum_{j=1}^{n} A_{jj}\right)\lambda^{n-1} + \left\{\begin{array}{l}\text{terms with powers of}\\ \lambda\text{ from }n-2\text{ to }1\end{array}\right\} + (-1)^n \det A.$$

The constant term is $(-1)^n \det A$ as indicated because it is the value of the polynomial at $\lambda = 0$, which is $\det(-A)$. In any event, we now see that characteristic polynomials are polynomial functions and can even be treated as polynomials in an indeterminate λ in the sense of Section I.3.[9] The negative of the coefficient of λ^{n-1} is the **trace** of A, denoted by $\operatorname{Tr} A$. Thus $\operatorname{Tr} A = \sum_{j=1}^{n} A_{jj}$. Trace is a linear functional on the vector space $M_{nn}(\mathbb{F})$ of n-by-n matrices.

EXAMPLE 1. For $A = \begin{pmatrix} 4 & 1 \\ -2 & 1 \end{pmatrix}$, the characteristic polynomial is

$$\det(\lambda I - A) = \det\begin{pmatrix} \lambda - 4 & -1 \\ 2 & \lambda - 1 \end{pmatrix}$$

$$= (\lambda - 4)(\lambda - 1) + 2 = \lambda^2 - 5\lambda + 6 = (\lambda - 2)(\lambda - 3).$$

The roots, and hence the eigenvalues, are $\lambda = 2$ and $\lambda = 3$. The eigenvectors for $\lambda = 2$ are computed by solving $(2I - A)v = 0$. The method of row reduction gives

$$\begin{pmatrix} 2-4 & -1 & | & 0 \\ 2 & 2-1 & | & 0 \end{pmatrix} = \begin{pmatrix} -2 & -1 & | & 0 \\ 2 & 1 & | & 0 \end{pmatrix} \mapsto \begin{pmatrix} 1 & \frac{1}{2} & | & 0 \\ 0 & 0 & | & 0 \end{pmatrix}.$$

Thus we have $x_1 + \frac{1}{2}x_2 = 0$ and $x_1 = -\frac{1}{2}x_2$. So the eigenvectors for $\lambda = 2$ are the nonzero vectors of the form $\begin{pmatrix} x_1 \\ x_2 \end{pmatrix} = x_2 \begin{pmatrix} -\frac{1}{2} \\ 1 \end{pmatrix}$. Similarly we find the eigenvectors for $\lambda = 3$ by starting from $(3I - A)v = 0$ and solving. The result is that the eigenvectors for $\lambda = 3$ are the nonzero vectors of the form $\begin{pmatrix} x_1 \\ x_2 \end{pmatrix} = x_2 \begin{pmatrix} -1 \\ 1 \end{pmatrix}$. For this example, there is a basis of eigenvectors.

---

[9]In Chapter V we will allow determinants of matrices whose entries are from any "commutative ring with identity," $\mathbb{C}[\lambda]$ being an example. Then we can think of $\det(\lambda I - A)$ directly as involving an indeterminate λ and not initially as a function of a scalar λ.

8. Eigenvectors and Characteristic Polynomials

Corollary 2.40. An n-by-n matrix A has at most n eigenvalues.

PROOF. Since $\det(\lambda I - A)$ is a polynomial of degree n, this follows from Proposition 2.39 and Corollary 1.14. $\square$

It will later be of interest that certain matrices A have a basis of eigenvectors. Such a basis exists for A as in Example 1 but not in general. One thing that can prevent a matrix from having a basis of eigenvectors is the failure of the characteristic polynomial to factor into first-degree factors. Thus, for example, $A = \begin{pmatrix} 0 & 1 \\ -1 & 0 \end{pmatrix}$ has characteristic polynomial $\lambda^2 + 1$, which does not factor into first-degree factors when $\mathbb{F} = \mathbb{R}$. Even when we do have a factorization into first-degree factors, we can still fail to have a basis of eigenvectors, as the following example shows.

EXAMPLE 2. For $A = \begin{pmatrix} 1 & -1 \\ 0 & 1 \end{pmatrix}$, the characteristic polynomial is given by $\det(\lambda I - A) = \det \begin{pmatrix} \lambda - 1 & 1 \\ 0 & \lambda - 1 \end{pmatrix} = (\lambda - 1)^2$. When we solve for eigenvectors, we get $\begin{pmatrix} 0 & 1 & | & 0 \\ 0 & 0 & | & 0 \end{pmatrix}$, and $x_2 = 0$. Thus $\begin{pmatrix} x_1 \\ x_2 \end{pmatrix} = x_1 \begin{pmatrix} 1 \\ 0 \end{pmatrix}$, and we do not have a basis of eigenvectors.

What happens is that the presence of a factor $(\lambda - c)^k$ in the characteristic polynomial ensures the existence of an r-parameter family of eigenvectors for eigenvalue c, with $1 \leq r \leq k$, but not necessarily with $r = k$. Example 2 shows that r can be strictly less than k. For purposes of deciding whether there is a basis of eigenvectors, the positive result is that the different roots of the characteristic polynomial do not interfere with each other; this is a consequence of the following proposition.

Proposition 2.41. If A is an n-by-n matrix, then eigenvectors for distinct eigenvalues are linearly independent.

REMARK. It follows that if the characteristic polynomial of A has n distinct eigenvalues, then it has a basis of eigenvectors.

PROOF. Let $Av_1 = \lambda_1 v_1, \ldots, Av_k = \lambda_k v_k$ with $\lambda_1, \ldots, \lambda_k$ distinct, and suppose that
$$c_1 v_1 + \cdots + c_k v_k = 0.$$

Applying A repeatedly gives

$$c_1 \lambda_1 v_1 + \cdots + c_k \lambda_k v_k = 0,$$
$$c_1 \lambda_1^2 v_1 + \cdots + c_k \lambda_k^2 v_k = 0,$$
$$\vdots$$
$$c_1 \lambda_1^{k-1} v_1 + \cdots + c_k \lambda_k^{k-1} v_k = 0.$$

If the jth entry of v_i is denoted by $v_i^{(j)}$, this system of vector equations says that

$$\begin{pmatrix} 1 & \cdots & 1 \\ \lambda_1 & \cdots & \lambda_k \\ \vdots & \ddots & \vdots \\ \lambda_1^{k-1} & \cdots & \lambda_k^{k-1} \end{pmatrix} \begin{pmatrix} c_1 v_1^{(j)} \\ \vdots \\ c_k v_k^{(j)} \end{pmatrix} = \begin{pmatrix} 0 \\ \vdots \\ 0 \end{pmatrix} \quad \text{for } 1 \leq j \leq n.$$

The square matrix on the left side is a Vandermonde matrix, which is invertible by Corollary 2.37 since $\lambda_1, \ldots, \lambda_k$ are distinct. Therefore $c_i v_i^{(j)} = 0$ for all i and j. Each v_i is nonzero in some entry $v_i^{(j)}$ with j perhaps depending on i, and hence $c_i = 0$. Since all the coefficients c_i have to be 0, $v_1, \ldots, v_k$ are linearly independent. $\square$

The theory of eigenvectors and eigenvalues for square matrices allows us to develop a corresponding theory for linear maps $L : V \to V$, where V is an n-dimensional vector space over $\mathbb{F}$. If L is such a function, a vector $v \neq 0$ in V is an **eigenvector** of L if $L(v) = \lambda v$ for some scalar λ. We call λ the **eigenvalue**. When λ is an eigenvalue, the vector space of all v with $L(v) = \lambda v$ is called the **eigenspace** for λ under L. We can compute the eigenvalues and eigenvectors of L by working in any ordered basis Γ of V. The equation $L(v) = \lambda v$ becomes $\begin{pmatrix} L \\ \Gamma\Gamma \end{pmatrix} \begin{pmatrix} v \\ \Gamma \end{pmatrix} = \lambda \begin{pmatrix} v \\ \Gamma \end{pmatrix}$ and is satisfied if and only if the column vector $\begin{pmatrix} v \\ \Gamma \end{pmatrix}$ is an eigenvalue of the matrix $A = \begin{pmatrix} L \\ \Gamma\Gamma \end{pmatrix}$ with eigenvalue λ. Applying Proposition 2.39 and remembering that determinants are well defined on linear maps $L : V \to V$, we see that L has an eigenvector with eigenvalue λ if and only if $\det(\lambda I - L) = 0$ and that in this case the eigenspace is the kernel of $\lambda I - L$.

What happens if we make these computations in a different ordered basis Δ? We know from Proposition 2.17 that the matrices $A = \begin{pmatrix} L \\ \Gamma\Gamma \end{pmatrix}$ and $B = \begin{pmatrix} L \\ \Delta\Delta \end{pmatrix}$ are similar, related by $B = C^{-1}AC$, where $C = \begin{pmatrix} I \\ \Gamma\Delta \end{pmatrix}$. Computing with

A leads to $u = \begin{pmatrix} v \\ \Gamma \end{pmatrix}$ as eigenvector for the eigenvalue λ. The corresponding result for B is that $B(C^{-1}u) = C^{-1}ACC^{-1}u = C^{-1}Au = \lambda C^{-1}u$. Thus $C^{-1}u = \begin{pmatrix} I \\ \Delta\Gamma \end{pmatrix}\begin{pmatrix} v \\ \Gamma \end{pmatrix} = \begin{pmatrix} v \\ \Delta \end{pmatrix}$ is an eigenvector of B with eigenvalue λ, just as it should be.

These considerations about eigenvalues suggest some facts about similar matrices that we can observe more directly without first passing from matrices to linear maps: One is that similar matrices have the same characteristic polynomial. To see this, suppose that $B = C^{-1}AC$; then

$$\begin{aligned}\det(\lambda I - B) &= \det(\lambda I - C^{-1}AC) = \det(C^{-1}(\lambda I - A)C) \\ &= (\det C^{-1})\det(\lambda I - A)(\det C^{-1}) \\ &= (\det C^{-1})(\det C^{-1})\det(\lambda I - A) = \det(\lambda I - A).\end{aligned}$$

A second fact is that similar matrices have the same trace. In fact, the trace is the negative of the coefficient of λ^{n-1} in the characteristic polynomial, and the characteristic polynomials are the same.

Because of these considerations we are free in the future to speak of the characteristic polynomial, the eigenvalues, and the trace of a linear map from a finite-dimensional vector space to itself, as well as the determinant, and these notions do not depend on any choice of ordered basis. We can speak unambiguously also of the eigenvectors of such a linear map. For this notion the realization of the eigenvectors in an ordered basis as column vectors depends on the ordered basis, the dependence being given by the formulas two paragraphs before the present one.

One final remark is in order. When the scalars are taken to be the complex numbers $\mathbb{C}$, the Fundamental Theorem of Algebra (Theorem 1.18) is applicable: every polynomial of degree ≥ 1 has at least one root. When applied to the characteristic polynomial of a square matrix or a linear map from a finite-dimensional vector space to itself, this theorem tells us that the matrix or linear map always has at least one eigenvalue, hence an eigenvector. We shall make serious use of this fact in Chapter III.

9. Bases in the Infinite-Dimensional Case

So far in this chapter, the use of bases has been limited largely to vector spaces having a finite spanning set. In this case we know from Corollary 2.3 that the finite spanning set has a subset that is a basis, any linearly independent set can be extended to a basis, and any two bases have the same finite number of elements.

We called such spaces finite-dimensional and defined the dimension of the vector space to be the number of elements in a basis.

The first objective in this section is to prove analogs of these results in the infinite-dimensional case. We shall make use of Zorn's Lemma as in Section A5 of the appendix, as well as the notion of cardinality discussed in Section A6 of the appendix. Once these analogs are in place, we shall examine the various results that we proved about finite-dimensional spaces to see the extent to which they remain valid for infinite-dimensional spaces.

Theorem 2.42. If V is any vector space over $\mathbb{F}$, then
   (a) any spanning set in V has a subset that is a basis,
   (b) any linearly independent set in V can be extended to a basis,
   (c) V has a basis,
   (d) any two bases have the same cardinality.

REMARKS. The common cardinality mentioned in (d) is called the **dimension** of the vector space V. In many applications it is enough to use $+\infty$ in place of each infinite cardinal in dimension formulas. This was the attitude conveyed in the remark with Corollary 2.24.

PROOF. For (b), let E be the given linearly independent set, and let $\mathcal{S}$ be the collection of all linearly independent subsets of V that contain E. Partially order $\mathcal{S}$ by inclusion upward. The set $\mathcal{S}$ is nonempty because E is in $\mathcal{S}$. Let $\mathcal{T}$ be a chain in $\mathcal{S}$, and let A be the union of the members of $\mathcal{T}$. We show that A is in $\mathcal{S}$, and then A is certainly an upper bound of $\mathcal{T}$. Because of its definition, A contains E, and we are to prove that A is linearly independent. For A to fail to be linearly independent would mean that there are vectors $v_1, \ldots, v_n$ in A with $c_1 v_1 + \cdots + c_n v_n = 0$ for some system of scalars not all 0. Let v_j be in the member A_j of the chain $\mathcal{T}$. Since $A_1 \subseteq A_2$ or $A_2 \subseteq A_1$, v_1 and v_2 are both in A_1 or both in A_2. To keep the notation neutral, say they are both in A'_2. Since $A'_2 \subseteq A_3$ or $A_3 \subseteq A'_2$, all of v_1, v_2, v_3 are in A'_2 or they are all in A_3. Say they are all in A'_3. Continuing in this way, we arrive at one of the sets $A_1, \ldots, A_n$, say A'_n, such that all of $v_1, \ldots, v_n$ are all in A'_n. The members of A'_n are linearly independent by assumption, and we obtain the contradiction $c_1 = \cdots = c_n = 0$. We conclude that A is linearly independent. Thus the chain $\mathcal{T}$ has an upper bound in $\mathcal{S}$. By Zorn's Lemma, S has a maximal element, say M. By Proposition 2.1a, M is a basis of V containing E.

For (a), let E be the given spanning set, and let $\mathcal{S}$ be the collection of all linearly independent subsets of V that are contained in E. Partially order $\mathcal{S}$ by inclusion upward. The set $\mathcal{S}$ is nonempty because $\varnothing$ is in $\mathcal{S}$. Let $\mathcal{T}$ be a chain in $\mathcal{S}$, and let A be the union of the members of $\mathcal{T}$. We show that A is in $\mathcal{S}$, and then A is certainly an upper bound of $\mathcal{T}$. Because of its definition, A is contained in

E, and the same argument as in the previous paragraph shows that A is linearly independent. Thus the chain $\mathcal{T}$ has an upper bound in S. By Zorn's Lemma, S has a maximal element, say M. Proposition 2.1a is not applicable, but its proof is easily adjusted to apply here to show that M spans V and hence is a basis: Given v in V, we are to prove that v lies is the linear span of M. First suppose that v is in E. If v is in M, there is nothing to prove. Since $M \cup \{v\}$ is contained in E, the assumed maximality implies that $M \cup \{v\}$ is not linearly independent, and hence $cv + c_1 v_1 + \cdots + c_n v_n = 0$ for some scalars $c, c_1, \ldots, c_n$ not all 0 and for some vectors $v_1, \ldots, v_n$ in M. The scalar c cannot be 0 since M is linearly independent. Thus $v = -c^{-1} c_1 v_1 - \cdots - c^{-1} c_n v_n$, and v is exhibited as in the linear span of M. Consequently every member of E lies in the linear span of M. Now suppose that v is not in E. Since every member of V lies in the linear span of E, every member of V lies in the linear span of M.

Conclusion (c) follows from (a) by taking the spanning set to be V; alternatively it follows from (b) by taking the linearly independent set to be $\varnothing$.

For (d), let $A = \{v_\alpha\}$ and $B = \{w_\beta\}$ be two bases of V. Each member a of A can be written as $a = c_1 w_{\beta_1} + \cdots + c_n w_{\beta_n}$ uniquely with the scalars $c_1, \ldots, c_n$ nonzero and with each w_{β_j} in B. Let B_a be the finite subset $\{w_{\beta_1}, \ldots, w_{\beta_n}\}$. Then we have associated to each member of A a finite subset B_a of B. Let us see that $\bigcup_{a \in A} B_a = B$. If b is in B, then the linear span of $B - \{b\}$ is not all of V. Thus some v in V is not in this span. Expand v in terms of A as $v = d_1 v_{\alpha_1} + \cdots + d_m v_{\alpha_m}$ with all $d_j \neq 0$. Since v is not in the linear span of $B - \{b\}$, some $a_0 = v_{\alpha_{j_0}}$ with $1 \leq j_0 \leq m$ is not in this linear span. Then b is in B_{a_0}, and we conclude that $B = \bigcup_{a \in A} B_a$. By the corollary near the end of Section A6 of the appendix, card $B \leq$ card A. Reversing the roles of A and B, we obtain card $A \leq$ card B. By the Schroeder–Bernstein Theorem, A and B have the same cardinality. This proves (d). □

Now let us go through the results of the chapter and see how many of them extend to the infinite-dimensional case and why. It is possible but not very useful in the infinite-dimensional case to associate an infinite "matrix" to a linear map when bases or ordered bases are specified for the domain and range. Because this association is not very useful, we shall not attempt to extend any of the results concerning matrices. The facts concerning extensions of results just dealing with dimensions and linear maps are as follows:

COROLLARY 2.5. If V is any vector space and U is a vector subspace, then $\dim U \leq \dim V$.

In fact, take a basis of U and extend it to a basis of V; a basis of U is then exhibited as a subset of a basis of V, and the conclusion about cardinal-number dimensions follows.

PROPOSITION 2.13. Let U and V be vector spaces over $\mathbb{F}$, and let Γ be a basis of U. Then to each function $\ell : \Gamma \to V$ corresponds one and only one linear map $L : U \to V$ such that $L|_\Gamma = \ell$.

In fact, the proof given in Section 3 is valid with no assumption about finite dimensionality.

COROLLARY 2.15. If $L : U \to V$ is a linear map between vector spaces over $\mathbb{F}$, then
$$\dim(\mathrm{domain}(L)) = \dim(\mathrm{kernel}(L)) + \dim(\mathrm{image}(L)).$$

In fact, this formula remains valid, but the earlier proof via matrices has to be replaced. Instead, take a basis $\{v_\alpha \mid \alpha \in A\}$ of the kernel and extend it to a basis $\{v_\alpha \mid \alpha \in S\}$ of the domain. It is routine to check that $\{L(v_\alpha) \mid \alpha \in S - A\}$ is a basis of the image of L.

THEOREM 2.16 (part). The composition of two linear maps is linear.

In fact, the proof in Section 3 remains valid with no assumption about finite dimensionality.

PROPOSITION 2.18. Two vector spaces over $\mathbb{F}$ are isomorphic if and only if they have the same cardinal-number dimension.

In fact, this result follows from Proposition 2.13 just as it did in the finite-dimensional case; the only changes that are needed in the argument in Section 3 are small adjustments of the notation. Of course, one must not overinterpret this result on the basis of the remark with Theorem 2.42: two vector spaces with dimension $+\infty$ need not be isomorphic. Despite the apparent definitive sound of Proposition 2.18, one must not attach too much significance to it; vector spaces that arise in practice tend to have some additional structure, and an isomorphism based merely on equality of dimensions need not preserve the additional structure.

PROPOSITION 2.19. If V is a vector space and V' is its dual, then $\dim V \leq \dim V'$. (In the infinite-dimensional case we do not have equality.)

In fact, take a basis $\{v_\alpha\}$ of V. If for each α we define $v'_\alpha(v_\beta) = \delta_{\alpha\beta}$ and use Proposition 2.13 to form the linear extension v'_α, then the set $\{v'_\alpha\}$ is a linearly independent subset of V' that is in one-one correspondence with the basis of V. Extending $\{v'_\alpha\}$ to a basis of V', we obtain the result.

PROPOSITION 2.20. Let V be a vector space, and let U be a vector subspace of V. Then

(b) every linear functional on U extends to a linear functional on V,
(c) whenever v_0 is a member of V that is not in U, there exists a linear functional on V that is 0 on U and is 1 on v_0.

Conclusion (a) of the original Proposition 2.20, which concerns annihilators, does not extend to the infinite-dimensional case.

To prove (b) without the finite dimensionality, let u' be a given linear functional on U, let $\{u_\alpha\}$ be a basis of U, and let $\{v_\beta\}$ be a subset of V such that $\{u_\alpha\} \cup \{v_\beta\}$ is a basis of V. Define $v'(u_\alpha) = u'(u_\alpha)$ for each α and $v'(v_\beta) = 0$ for each β. Using Proposition 2.13, let v' be the linear extension to a linear functional on V. Then v' has the required properties.

To prove (c) without the finite dimensionality, we take a basis $\{u_\alpha\}$ of U and extend $\{u_\alpha\} \cup \{v_0\}$ to a basis of V. Define v' to equal 0 on each u_α, to equal 1 on v_0, and to equal 0 on the remaining members of the basis of V. Then the linear extension of v' to V is the required linear functional.

PROPOSITION 2.22. If V is any vector space over $\mathbb{F}$, then the canonical map $\iota : V \to V''$ is one-one. The canonical map is not onto V'' if V is infinite-dimensional.

The proof that it is one-one given in Section 4 is applicable in the infinite-dimensional case since we know from Theorem 2.42 that any linearly independent subset of V can be extended to a basis. For the second conclusion when V has a countably infinite basis, see Problem 31 at the end of the chapter.

PROPOSITION 2.23 THROUGH COROLLARY 2.29. For these results about quotients, the only place that finite dimensionality played a role was in the dimension formulas, Corollaries 2.24 and 2.29. We restate these two results separately.

COROLLARY 2.24. If V is a vector space over $\mathbb{F}$ and U is a vector subspace, then

(a) $\dim V = \dim U + \dim(V/U)$,
(b) the subspace U is the kernel of some linear map defined on V.

The proof in Section 5 requires no changes: Let q be the quotient map. The linear map q meets the conditions of (b). For (a), take a basis of U and extend to a basis of V. Then the images under q of the additional vectors form a basis of V/U.

COROLLARY 2.29. Let M and N be vector subspaces of a vector space V over $\mathbb{F}$. Then
$$\dim(M+N) + \dim(M \cap N) = \dim M + \dim N.$$

In fact, Corollary 2.24a gives us $\dim(M+N) = \dim((M+N)/M) + \dim M$. Substituting $\dim((M+N)/M) = \dim(N/(M \cap N))$ from Theorem 2.28 and adding $\dim(M \cap N)$ to both sides, we obtain $\dim(M+N) + \dim(M \cap N) = \dim(M \cap N) + \dim(N/(M \cap N)) + \dim M$. The first two terms on the right side add to $\dim N$ by Corollary 2.24a, and the result follows.

PROPOSITIONS 2.30 THROUGH 2.33. These results about direct products and direct sums did not assume any finite dimensionality.

The determinants of Sections 7–8 have no infinite-dimensional generalization, and Proposition 2.41 is the only result in those two sections with a valid infinite-dimensional analog. The valid analog in the infinite-dimensional case is that eigenvectors for distinct eigenvalues under a linear map are linearly independent. The proof given for Proposition 2.41 in Section 8 adapts to handle this analog, provided we interpret components $v_i^{(j)}$ of a vector v_i as the coefficients needed to expand v_i in a basis of the underlying vector space.

10. Problems

1. Determine bases of the following subsets of $\mathbb{R}^3$:
   (a) the plane $3x - 2y + 5z = 0$,
   (b) the line $\left\{\begin{array}{l} x = 2t \\ y = -t \\ z = 4t \end{array}\right\}$, where $-\infty < t < \infty$.

2. This problem shows that the associativity law in the definition of "vector space" implies certain more complicated formulas of which the stated law is a special case. Let $v_1, \ldots, v_n$ be vectors in a vector space V. The only vector-space properties that are to be used in this problem are associativity of addition and the existence of the 0 element.
   (a) Define $v_{(k)}$ inductively upward by $v_{(0)} = 0$ and $v_{(k)} = v_{(k-1)} + v_k$, and define $v^{(l)}$ inductively downward by $v^{(n+1)} = 0$ and $v^{(l)} = v_l + v^{(l+1)}$. Prove that $v_{(k)} + v^{(k+1)}$ is always the same element for $0 \le k \le n$.
   (b) Prove that the same element of V results from any way of inserting parentheses in the sum $v_1 + \cdots + v_n$ so that each step requires the addition of only two members of V.

3. This problem shows that the commutative and associative laws in the definition of "vector space" together imply certain more complicated formulas of which the stated commutative law is a special case. Let $v_1, \ldots, v_n$ be vectors in a vector space V. The only vector-space properties that are to be used in this problem are commutativity of addition and the properties in the previous problem. Because of the previous problem, $v_1 + \cdots + v_n$ is a well-defined element of V, and it is not necessary to insert any parentheses in it. Prove that $v_1 + v_2 + \cdots + v_n = v_{\sigma(1)} + v_{\sigma(2)} + \cdots + v_{\sigma(n)}$ for each permutation σ of $\{1, \ldots, n\}$.

4. For the matrix $A = \begin{pmatrix} 1 & 2 & -1 \\ 2 & 4 & 6 \\ 0 & 0 & -8 \end{pmatrix}$, find
   (a) a basis for the row space,
   (b) a basis for the column space, and

(c) the rank of the matrix.

5. Let A be an n-by-n matrix of rank one. Prove that there exists an n-dimensional column vector c and an n-dimensional row vector r such that $A = cr$.

6. Let A be a k-by-n matrix, and let R be a reduced row-echelon form of A.
   (a) Prove for each r that the rows of R whose first r entries are 0 form a basis for the vector subspace of all members of the row space of A whose first r entries are 0.
   (b) Prove that the reduced row-echelon form of A is unique in the sense that any two sequences of steps of row reduction lead to the same reduced form.

7. Let E be an finite set of N points, let V be the N-dimensional vector space of all real-valued functions on E, and let n be an integer with $0 < n \leq N$. Suppose that U is an n-dimensional subspace of V. Prove that there exists a subset D of n points in E such that the vector space of restrictions to D of the members of U has dimension n.

8. A linear map $L : \mathbb{R}^2 \to \mathbb{R}^2$ is given in the standard ordered basis by the matrix $\begin{pmatrix} -6 & -12 \\ 6 & 11 \end{pmatrix}$. Find the matrix of L in the ordered basis $\left\{ \begin{pmatrix} 3 \\ -2 \end{pmatrix}, \begin{pmatrix} -4 \\ 3 \end{pmatrix} \right\}$.

9. Let V be the real vector space of all polynomials in x of degree ≤ 2, and let $L : V \to V$ be the linear map $I - D^2$, where I is the identity and D is the differentiation operator d/dx. Prove that L is invertible.

10. Let A be in $M_{km}(\mathbb{C})$ and B be in $M_{mn}(\mathbb{C})$. Prove that

$$\text{rank}(AB) \leq \max(\text{rank } A, \text{rank } B).$$

11. Let A be in $M_{kn}(\mathbb{C})$ with $k > n$. Prove that there exists no B in $M_{nk}(\mathbb{C})$ with $AB = I$.

12. Let A be in $M_{kn}(\mathbb{C})$ and B be in $M_{nk}(\mathbb{C})$. Give an example with $k = n$ to show that $\text{rank}(AB)$ need not equal $\text{rank}(BA)$.

13. With the differential equation $y''(t) = y(t)$ in Example 2 of Section 3, two examples of linear functionals on the vector space of solutions are given by $\ell_1(y) = y(0)$ and $\ell_2(y) = y'(0)$. Find a basis of the space of solutions such that $\{\ell_1, \ell_2\}$ is the dual basis.

14. Suppose that a vector space V has a countably infinite basis. Prove that the dual V' has an uncountable linearly independent set.

15. (a) Give an example of a vector space and three vector subspaces L, M, and N such that $L \cap (M + N) \neq (L \cap M) + (L \cap N)$.
    (b) Show that inclusion always holds in one direction in (a).
    (c) Show that equality always holds in (a) if $L \supseteq M$.

16. Construct three vector subspaces M, N_1, and N_2 of a vector space V such that $M \oplus N_1 = M \oplus N_2 = V$ but $N_1 \neq N_2$. What is the geometric picture corresponding to this situation?

17. Suppose that x, y, u, and v are vectors in $\mathbb{R}^4$; let M and N be the vector subspaces of $\mathbb{R}^4$ spanned by $\{x, y\}$ and $\{u, v\}$, respectively. In which of the following cases is it true that $\mathbb{R}^4 = M \oplus N$?
    (a) $x = (1, 1, 0, 0)$, $y = (1, 0, 1, 0)$, $u = (0, 1, 0, 1)$, $v = (0, 0, 1, 1)$;
    (b) $x = (-1, 1, 1, 0)$, $y = (0, 1, -1, 1)$, $u = (1, 0, 0, 0)$, $v = (0, 0, 0, 1)$;
    (c) $x = (1, 0, 0, 1)$, $y = (0, 1, 1, 0)$, $u = (1, 0, 1, 0)$, $v = (0, 1, 0, 1)$.

18. Section 6 gave definitions and properties of projections and injections associated with the direct sum of two vector spaces. Write down corresponding definitions and properties for projections and injections in the case of the direct sum of n vector spaces, n being an integer > 2.

19. Let $T : \mathbb{R}^n \to \mathbb{R}^n$ be a linear map with $\ker T \cap \operatorname{image} T = 0$.
    (a) Prove that $\mathbb{R}^n = \ker T \oplus \operatorname{image} T$.
    (b) Prove that the condition $\ker T \cap \operatorname{image} T = 0$ is satisfied if $T^2 = T$.

20. If V_1 and V_2 are two vector spaces over $\mathbb{F}$, prove that $(V_1 \oplus V_2)'$ is canonically isomorphic to $V_1' \oplus V_2'$.

21. Suppose that M is a vector subspace of a vector space V and that $q : V \to V/M$ is the quotient map. Corresponding to each linear functional y on V/M is a linear functional z on V given by $z = yq$. Why is the correspondence $y \mapsto z$ an isomorphism between $(V/M)'$ and $\operatorname{Ann} M$?

22. Let M be a vector subspace of the vector space V, and let $q : V \to V/M$ be the quotient map. Suppose that N is a vector subspace of V. Prove that $V = M \oplus N$ if and only if the restriction of q to N is an isomorphism of N onto V/M.

23. For a square matrix A of integers, prove that the inverse has integer entries if and only if $\det A = \pm 1$.

24. Let A be in $M_{kn}(\mathbb{C})$, and let $r = \operatorname{rank} A$. Prove that r is the largest integer such that there exist r row indices $i_1, \ldots, i_r$ and r column indices $j_1, \ldots, j_r$ for which the r-by-r matrix formed from these rows and columns of A has nonzero determinant. (Educational note: This problem characterizes the subset of matrices of rank $\leq r - 1$ as the set in which all determinants of r-by-r submatrices are zero.)

25. Suppose that a linear combination of functions $t \mapsto e^{ct}$ with c real vanishes for every integer $t \geq 0$. Prove that it vanishes for every real t.

26. Find all eigenvalues and eigenvectors of $A = \begin{pmatrix} 0 & 1 \\ -6 & 5 \end{pmatrix}$.

27. Let A and C be n-by-n matrices with C invertible. By making a direct calculation with the entries, prove that $\operatorname{Tr}(C^{-1}AC) = \operatorname{Tr} A$.

28. Find the characteristic polynomial of the n-by-n matrix $\begin{pmatrix} 0 & 1 & 0 & 0 & & 0 & 0 \\ 0 & 0 & 1 & 0 & & 0 & 0 \\ 0 & 0 & 0 & 1 & & 0 & 0 \\ 0 & 0 & 0 & 0 & & 0 & 0 \\ & & & & \ddots & & \\ 0 & 0 & 0 & 0 & \cdots & 0 & 1 \\ a_0 & a_1 & a_2 & a_3 & \cdots & a_{n-2} & a_{n-1} \end{pmatrix}$.

29. Let A and B be in $M_{nn}(\mathbb{C})$.
    (a) Prove under the assumption that A is invertible that $\det(\lambda I - AB) = \det(\lambda I - BA)$.
    (b) By working with $A + \epsilon I$ and letting ϵ tend to 0, show that the assumption in (a) that A is invertible can be dropped.

30. In proving Theorem 2.42a, it is tempting to argue by considering all spanning subsets of the given set, ordering them by inclusion downward, and seeking a minimal element by Zorn's Lemma. Give an example of a chain in this ordering that has no lower bound, thereby showing that this line of argument cannot work.

Problems 31–34 concern annihilators. Let V be a vector space, let M and N be vector subspaces, and let $\iota : V \to V''$ be the canonical map.

31. If V has a countably infinite basis, how can we conclude that ι does not carry V onto V''?

32. Prove that $\text{Ann}(M + N) = \text{Ann } M \cap \text{Ann } N$.

33. Prove that $\text{Ann}(M \cap N) = \text{Ann } M + \text{Ann } N$.

34. (a) Prove that $\iota(M) \subseteq \text{Ann}(\text{Ann } M)$.
    (b) Prove that equality holds in (a) if V is finite-dimensional.
    (c) Give an infinite-dimensional example in which equality fails in (a).

Problems 35–39 concern operations by blocks within matrices.

35. Let A be a k-by-m matrix of the form $A = (A_1 \;\; A_2)$, where A_1 has size k-by-m_1, A_2 has size k-by-m_2, and $m_1 + m_2 = m$. Let B by an m'-by-n matrix of the form $B = \begin{pmatrix} B_1 \\ B_2 \end{pmatrix}$, where B_1 has size m'_1-by-n, B_2 has size m'_2-by-n, and $m'_1 + m'_2 = m'$.
    (a) If $m_1 = m'_1$ and $m_2 = m'_2$, prove that $AB = A_1 B_1 + A_2 B_2$.
    (b) If $k = n$, prove that $BA = \begin{pmatrix} B_1 A_1 & B_1 A_2 \\ B_2 A_2 & B_2 A_2 \end{pmatrix}$.
    (c) Deduce a general rule for block multiplication of matrices that are in 2-by-2 block form.

36. Let A be in $M_{kk}(\mathbb{C})$, B be in $M_{kn}(\mathbb{C})$, and D be in $M_{nn}(\mathbb{C})$. Prove that $\det \begin{pmatrix} A & B \\ 0 & D \end{pmatrix} = \det A \det D$.

37. Let $A, B, C,$ and D be in $M_{nn}(\mathbb{C})$. Suppose that A is invertible and that $AC = CA$. Prove that $\det \begin{pmatrix} A & B \\ C & D \end{pmatrix} = \det(AD - CB)$.

38. Let A be in $M_{kn}(\mathbb{C})$ and B be in $M_{nk}(\mathbb{C})$ with $k \leq n$. Let I_k be the k-by-k identity, and let I_n be the n-by-n identity. Using Problem 29, prove that $\det(\lambda I_n - BA) = \lambda^{n-k} \det(\lambda I_k - AB)$.

39. Prove the following block-form generalization of the expansion-in-cofactors formula. For each subset S of $\{1, \ldots, n\}$, let S^c be the complementary subset within $\{1, \ldots, n\}$, and let $\text{sgn}(S, S^c)$ be the sign of the permutation that carries $(1, \ldots, n)$ to the members of S in order, followed by the members of S^c in order. Fix k with $1 \leq k \leq n-1$, and let the subset S have $|S| = k$. For an n-by-n matrix A, define $A(S)$ to be the square matrix of size k obtained by using the rows of A indexed by $1, \ldots, k$ and the columns indexed by the members of S. Let $\widehat{A}(S)$ be the square matrix of size $k-1$ obtained by using the rows of A indexed by $k+1, \ldots, n$ and the columns indexed by the members of S^c. Prove that
$$\det A = \sum_{\substack{S \subseteq \{1,\ldots,n\}, \\ |S|=k}} \text{sgn}(S, S^c) \det A(S) \det \widehat{A}(S).$$

Problems 40–44 compute the determinants of certain matrices known as **Cartan matrices**. These have geometric significance in the theory of Lie groups.

40. Let A_n be the n-by-n matrix
$$\begin{pmatrix} 2 & -1 & 0 & 0 & \cdots & 0 & 0 \\ -1 & 2 & -1 & 0 & \cdots & 0 & 0 \\ 0 & -1 & 2 & -1 & \cdots & 0 & 0 \\ 0 & 0 & -1 & 2 & \cdots & 0 & 0 \\ & & & & \ddots & & \\ 0 & 0 & 0 & 0 & \cdots & 2 & -1 \\ 0 & 0 & 0 & 0 & \cdots & -1 & 2 \end{pmatrix}.$$
Using expansion in cofactors about the last row, prove that $\det A_n = 2 \det A_{n-1} - \det A_{n-2}$ for $n \geq 3$.

41. Computing $\det A_1$ and $\det A_2$ directly and using the recursion in Problem 40, prove that $\det A_n = n + 1$ for $n \geq 1$.

42. Let C_n for $n \geq 2$ be the matrix A_n except that the $(1, 2)^{\text{th}}$ entry is changed from -1 to -2.
    (a) Expanding in cofactors about the last row, prove that the argument of Problem 40 is still applicable when $n \geq 4$ and a recursion formula for $\det C_n$ results with the same coefficients.
    (b) Computing $\det C_2$ and $\det C_3$ directly and using the recursion equation in (a), prove that $\det C_n = 2$ for $n \geq 2$.

43. Let D_n for $n \geq 3$ be the matrix A_n except that the upper left 3-by-3 piece is changed from $\begin{pmatrix} 2 & -1 & 0 \\ -1 & 2 & -1 \\ 0 & -1 & 2 \end{pmatrix}$ to $\begin{pmatrix} 2 & 0 & -1 \\ 0 & 2 & -1 \\ -1 & -1 & 2 \end{pmatrix}$.

   (a) Expanding in cofactors about the last row, prove that the argument of Problem 40 is still applicable when $n \geq 5$ and a recursion formula for $\det D_n$ results with the same coefficients.

   (b) Show that D_3 can be transformed into A_3 by suitable interchanges of rows and interchanges of columns, and conclude that $\det D_3 = \det A_3 = 4$.

   (c) Computing $\det D_4$ directly and using (b) and the recursion equation in (a), prove that $\det D_n = 4$ for $n \geq 3$.

44. Let E_n for $n \geq 4$ be the matrix A_n except that the upper left 4-by-4 piece is changed from $\begin{pmatrix} 2 & -1 & 0 & 0 \\ -1 & 2 & -1 & 0 \\ 0 & -1 & 2 & -1 \\ 0 & 0 & -1 & 2 \end{pmatrix}$ to $\begin{pmatrix} 2 & -1 & 0 & 0 \\ -1 & 2 & 0 & -1 \\ 0 & 0 & 2 & -1 \\ 0 & -1 & -1 & 2 \end{pmatrix}$.

   (a) Expanding in cofactors about the last row, prove that the argument of Problem 40 is still applicable when $n \geq 6$ and a recursion formula for $\det E_n$ results with the same coefficients.

   (b) Show that E_4 can be transformed into A_4 by suitable interchanges of rows and interchanges of columns, and conclude that $\det E_4 = \det A_4 = 5$.

   (c) Show that E_5 can be transformed into D_5 by suitable interchanges of rows and interchanges of columns, and conclude that $\det E_5 = \det D_5 = 4$.

   (d) Using (b) and (c) and the recursion equation in (a), prove that $\det E_n = 9 - n$ for $n \geq 4$.

CHAPTER III

Inner-Product Spaces

Abstract. This chapter investigates the effects of adding the additional structure of an inner product to a finite-dimensional real or complex vector space.

Section 1 concerns the effect on the vector space itself, defining inner products and their corresponding norms and giving a number of examples and formulas for the computation of norms. Vector-space bases that are orthonormal play a special role.

Section 2 concerns the effect on linear maps. The inner product makes itself felt partly through the notion of the adjoint of a linear map. The section pays special attention to linear maps that are self-adjoint, i.e., are equal to their own adjoints, and to those that are unitary, i.e., preserve norms of vectors.

Section 3 proves the Spectral Theorem for self-adjoint linear maps on finite-dimensional inner-product spaces. The theorem says in part that any self-adjoint linear map has an orthonormal basis of eigenvectors. The Spectral Theorem has several important consequences, one of which is the existence of a unique positive semidefinite square root for any positive semidefinite linear map. The section concludes with the polar decomposition, showing that any linear map factors as the product of a unitary linear map and a positive semidefinite one.

1. Inner Products and Orthonormal Sets

In this chapter we examine the effect of adding further geometric structure to the structure of a real or complex vector space as defined in Chapter II. To be a little more specific in the cases of $\mathbb{R}^2$ and $\mathbb{R}^3$, the development of Chapter II amounted to working with points, lines, planes, coordinates, and parallelism, but nothing further. In the present chapter, by comparison, we shall take advantage of additional structure that captures the notions of distances and angles.

We take $\mathbb{F}$ to be $\mathbb{R}$ or $\mathbb{C}$, continuing to call its members the scalars. We do not allow $\mathbb{F}$ to be $\mathbb{Q}$ in this chapter; the main results will make essential use of additional facts about $\mathbb{R}$ and $\mathbb{C}$ beyond those of addition, subtraction, multiplication, and division. The relevant additional facts are summarized in Sections A3 and A4 of the appendix.[1]

---
[1] The theory of Chapter II will be observed in Chapter IV to extend to any "field" $\mathbb{F}$ in place of $\mathbb{Q}$ or $\mathbb{R}$ or $\mathbb{C}$, but the theory of the present chapter is limited to $\mathbb{R}$ and $\mathbb{C}$, as well as some other special fields that we shall not try to isolate.

1. Inner Products and Orthonormal Sets

Many of the results that we obtain will be limited to the finite-dimensional case. The theory of inner-product spaces that we develop has an infinite-dimensional generalization, but useful results for the generalization make use of a hypothesis of "completeness" for an inner-product space that we are not in a position to verify in examples.[2]

Let V be a vector space over $\mathbb{F}$. An **inner product** on V is a function from $V \times V$ into $\mathbb{F}$, which we here denote by $(\cdot, \cdot)$, with the following properties:

(i) the function $u \mapsto (u, v)$ of V into $\mathbb{F}$ is linear,
(ii) the function $v \mapsto (u, v)$ of V into $\mathbb{F}$ is **conjugate linear** in the sense that it satisfies $(u, v_1 + v_2) = (u, v_1) + (u, v_2)$ for v_1 and v_2 in V and $(u, cv) = \bar{c}(u, v)$ for v in V and c in $\mathbb{F}$,
(iii) $(u, v) = \overline{(v, u)}$ for u and v in V,
(iv) $(v, v) \geq 0$ for all v in V,
(v) $(v, v) = 0$ only if $v = 0$ in V.

The overbars in (ii) and (iii) indicate complex conjugation. Property (ii) reduces when $\mathbb{F} = \mathbb{R}$ to the fact that $v \mapsto (u, v)$ is linear. Properties (i) and (ii) together are summarized by saying that $(\cdot, \cdot)$ is **bilinear** if $\mathbb{F} = \mathbb{R}$ or **sesquilinear** if $\mathbb{F} = \mathbb{C}$. Property (iii) is summarized when $\mathbb{F} = \mathbb{R}$ by saying that $(\cdot, \cdot)$ is **symmetric**, or when $\mathbb{F} = \mathbb{C}$ by saying that $(\cdot, \cdot)$ is **Hermitian symmetric**.

An **inner-product space**, for purposes of this book, is a vector space over $\mathbb{R}$ or $\mathbb{C}$ with an inner product in the above sense.[3,4]

EXAMPLES.

(1) $V = \mathbb{R}^n$ with $(\cdot, \cdot)$ as the **dot product**, i.e., with $(x, y) = y^t x = x_1 y_1 + \cdots + x_n y_n$ if $x = \begin{pmatrix} x_1 \\ \vdots \\ x_n \end{pmatrix}$ and $y = \begin{pmatrix} y_1 \\ \vdots \\ y_n \end{pmatrix}$. The traditional notation for the dot product is $x \cdot y$.

(2) $V = \mathbb{C}^n$ with $(\cdot, \cdot)$ defined by $(x, y) = \bar{y}^t x = x_1 \bar{y}_1 + \cdots + x_n \bar{y}_n$ if $x = \begin{pmatrix} x_1 \\ \vdots \\ x_n \end{pmatrix}$ and $y = \begin{pmatrix} y_1 \\ \vdots \\ y_n \end{pmatrix}$. Here $\bar{y}$ denotes the entry-by-entry complex conjugate of y. The sesquilinear expression $(\cdot, \cdot)$ is different from the complex *bilinear* **dot product** $x \cdot y = x_1 y_1 + \cdots + x_n y_n$.

---

[2] A careful study in the infinite-dimensional case is normally made only after the development of a considerable number of topics in real analysis.

[3] When the scalars are complex, many books emphasize the presence of complex scalars by referring to the inner product as a "Hermitian inner product." This book does not need to distinguish the complex case very often and therefore will not use the modifier "Hermitian" with the term "inner product."

[4] Some authors, particularly in connection with mathematical physics, reverse the roles of the two variables, defining inner products to be conjugate linear in the first variable and linear in the second variable.

(3) V equal to the vector space of all complex-valued polynomials with $(f, g) = \int_0^1 f(x)\overline{g(x)}\,dx$.

Let V be an inner-product space. If v is in V, define $\|v\| = \sqrt{(v, v)}$, calling $\|\cdot\|$ the **norm** associated with the inner product. The norm of v is understood to be the nonnegative square root of the nonnegative real number (v, v) and is well defined as a consequence of (iv). In the case of $\mathbb{R}^n$, $\|x\|$ is the Euclidean distance $\sqrt{x_1^2 + \cdots + x_n^2}$ from the origin to the column vector $x = (x_1, \ldots, x_n)$. In this interpretation the dot product of two nonzero vectors in $\mathbb{R}^n$ is shown in analytic geometry to be given by $x \cdot y = \|x\|\|y\|\cos\theta$, where θ is the angle between the vectors x and y.

Direct expansion of norms squared of sums of vectors using bilinearity or sesquilinearity leads to certain formulas of particular interest. The formula that we shall use most frequently is

$$\|u + v\|^2 = \|u\|^2 + 2\operatorname{Re}(u, v) + \|v\|^2,$$

which generalizes from $\mathbb{R}^2$ a version of the **law of cosines** in trigonometry relating the lengths of the three sides of a triangle when one of the angles is known. With the additional hypothesis that $(u, v) = 0$, this formula generalizes from $\mathbb{R}^2$ the **Pythagorean Theorem**

$$\|u + v\|^2 = \|u\|^2 + \|v\|^2.$$

Another such formula is the **parallelogram law**

$$\|u + v\|^2 + \|u - v\|^2 = 2\|u\|^2 + 2\|v\|^2 \qquad \text{for all } u \text{ and } v \text{ in } V,$$

which is proved by computing $\|u + v\|^2$ and $\|u - v\|^2$ by the law of cosines and adding the results. The name "parallelogram law" is explained by the geometric interpretation in the case of the dot product for $\mathbb{R}^2$ and is illustrated in Figure 3.1. That figure uses the familiar interpretation of vectors in $\mathbb{R}^2$ as arrows, two arrows being identified if they are translates of one another; thus the arrow from v to u represents the vector $u - v$.

The parallelogram law is closely related to a formula for recovering the inner product from the norm, namely

$$(u, v) = \frac{1}{4}\sum_k i^k \|u + i^k v\|^2,$$

where the sum extends for $k \in \{0, 2\}$ if the scalars are real and extends for $k \in \{0, 1, 2, 3\}$ if the scalars are complex. This formula goes under the name

polarization. To prove it, we expand $\|u+i^k v\|^2 = \|u\|^2 + 2\operatorname{Re}(u, i^k v) + \|v\|^2 = \|u\|^2 + 2\operatorname{Re}\left((-i)^k(u,v)\right) + \|v\|^2$. Multiplying by i^k and summing on k shows that $\sum_k i^k \|u+i^k v\|^2 = 2\sum_k i^k \operatorname{Re}\left((-i)^k(u,v)\right)$. If k is even, then $i^k \operatorname{Re}((-i)^k z) = \operatorname{Re} z$ for any complex z, while if k is odd, then $i^k \operatorname{Re}((-i)^k z) = i \operatorname{Im} z$. So $2\sum_k i^k \operatorname{Re}((-i)^k z) = 4z$, and $\sum_k i^k \|u+i^k v\|^2 = 4(u,v)$, as asserted.

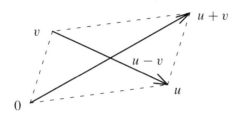

FIGURE 3.1. Geometric interpretation of the parallelogram law: the sum of the squared lengths of the four sides of a parallelogram equals the sum of the squared lengths of the diagonals.

Proposition 3.1 (Schwarz inequality). In any inner-product space V, $|(u,v)| \leq \|u\| \|v\|$ for all u and v in V.

REMARK. The proof is written so as to use properties (i) through (iv) in the definition of inner product but not (v), a situation often encountered with integrals.

PROOF. Possibly replacing u by $e^{i\theta}u$ for some real θ, we may assume that (u,v) is real. In the case that $\|v\| \neq 0$, the law of cosines gives

$$\left|u - \|v\|^{-2}(u,v)v\right|^2 = \|u\|^2 - 2\|v\|^{-2}|(u,v)|^2 + \|v\|^{-4}|(u,v)|^2\|v\|^2.$$

The left side is ≥ 0, and the right side simplifies to $\|u\|^2 - \|v\|^{-2}|(u,v)|^2$. Thus the inequality follows in this case.

In the case that $\|v\| = 0$, it is enough to prove that $(u,v) = 0$ for all u. If c is a scalar, then we have

$$\|u+cv\|^2 = \|u\|^2 + 2\operatorname{Re}\left(c(u,v)\right) + |c|^2\|v\|^2 = \|u\|^2 + 2\operatorname{Re}\left(c(u,v)\right).$$

The left side is ≥ 0 as c varies, but the right side is < 0 for a suitable choice of c unless $(u,v) = 0$. This completes the proof. □

Proposition 3.2. In any inner-product space V, the norm satisfies
 (a) $\|v\| \geq 0$ for all v in V, with equality if and only if $v = 0$,
 (b) $\|cv\| = |c| \|v\|$ for all v in V and all scalars c,
 (c) $\|u+v\| \leq \|u\| + \|v\|$ for all u and v in V.

PROOF. Conclusion (a) is immediate from properties (iv) and (v) of an inner product, and (b) follows since $\|cv\|^2 = (cv, cv) = c\bar{c}(v, v) = |c|^2 \|v\|^2$. Finally we use the law of cosines and the Schwarz inequality (Proposition 3.1) to write $\|u+v\|^2 = \|u\|^2 + 2\operatorname{Re}(u, v) + \|v\|^2 \leq \|u\|^2 + 2\|u\|\|v\| + \|v\|^2 = (\|u\| + \|v\|)^2$. Taking the square root of both sides yields (c). $\square$

Two vectors u and v in V are said to be **orthogonal** if $(u, v) = 0$, and one sometimes writes $u \perp v$ in this case. The notation is a reminder of the interpretation in the case of dot product—that dot product 0 means that the cosine of the angle between the two vectors is 0 and the vectors are therefore perpendicular. An **orthogonal set** in V is a set of vectors such that each pair is orthogonal.

The nonzero members of an orthogonal set are linearly independent. In fact, if $\{v_1, \ldots, v_k\}$ is an orthogonal set of nonzero vectors and some linear combination has $c_1 v_1 + \cdots + c_k v_k = 0$, then the inner product of this relation with v_j gives $0 = (c_1 v_1 + \cdots + c_k v_k, v_j) = c_j \|v_j\|^2$, and we see that $c_j = 0$ for each j.

A **unit vector** in V is a vector u with $\|u\| = 1$. If v is any nonzero vector, then $v/\|v\|$ is a unit vector. An **orthonormal set** in V is an orthogonal set of unit vectors. Under the assumption that V is finite-dimensional, an **orthonormal basis** of V is an orthonormal set that is a vector-space basis.[5]

EXAMPLES.

(1) In $\mathbb{R}^n$ or $\mathbb{C}^n$, the standard basis $\{e_1, \ldots, e_n\}$ is an orthonormal set.

(2) Let V be the complex inner-product space of all complex finite linear combinations, for n from $-N$ to $+N$, of the functions $x \mapsto e^{inx}$ on the closed interval $[-\pi, \pi]$, the inner product being $(f, g) = \frac{1}{2\pi} \int_{-\pi}^{\pi} f(x)\overline{g(x)}\,dx$. With respect to this inner product, the functions e^{inx} form an orthonormal set.

A simple but important exercise in an inner-product space is to resolve a vector into the sum of a multiple of a given unit vector and a vector orthogonal to the given unit vector. This exercise is solved as follows: If v is given and u is a unit vector, then v decomposes as

$$v = (v, u)u + \big(v - (v, u)u\big).$$

Here $(v, u)u$ is a multiple of u, and the two components are orthogonal since $\big(u, v - (v, u)u\big) = (u, v) - \overline{(v, u)}(u, u) = (u, v) - (u, v) = 0$. This decomposition is unique since if $v = v_1 + v_2$ with $v_1 = cu$ and $(v_2, u) = 0$, then the inner product of $v = v_1 + v_2$ with u yields $(v, u) = (cu, u) + (v_2, u) = c$. Hence

---

[5]In the infinite-dimensional theory the term "orthonormal basis" is used for an orthonormal set that spans V when *limits* of finite sums are allowed, in addition to finite sums themselves; when V is infinite-dimensional, an orthonormal basis is never large enough to be a vector-space basis.

c must be (v, u), v_1 must be $(v, u)u$, and v_2 must be $v - (v, u)u$. Figure 3.2 illustrates the decomposition, and Proposition 3.3 generalizes it by replacing the multiples of a single unit vector by the span of a finite orthonormal set.

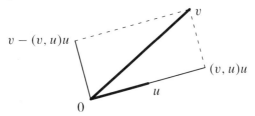

FIGURE 3.2. Resolution of v into a component $(v, u)u$ parallel to a unit vector u and a component orthogonal to u.

Proposition 3.3. Let V be an inner-product space. If $\{u_1, \ldots, u_k\}$ is an orthonormal set in V and if v is given in V, then there exists a unique decomposition

$$v = c_1 u_1 + \cdots + c_k u_k + v^\perp$$

with $v^\perp$ orthogonal to u_j for $1 \leq j \leq k$. In this decomposition $c_j = (v, u_j)$.

REMARK. The proof illustrates a technique that arises often in mathematics. We seek to prove an existence–uniqueness theorem, and we begin by making calculations toward uniqueness that narrow down the possibilities. We are led to some formulas or conditions, and we use these to define the object in question and thereby prove existence. Although it may not be so clear except in retrospect, this was the technique that lay behind proving the equivalence of various conditions for the invertibility of a square matrix in Section I.6. The technique occurred again in defining and working with determinants in Section II.7.

PROOF OF UNIQUENESS. Taking the inner product of both sides with u_j, we obtain $(v, u_j) = (c_1 u_1 + \cdots + c_k u_k + v^\perp, u_j) = c_j$ for each j. Then $c_j = (v, u_j)$ is forced, and $v^\perp$ must be given by $v - (v, u_1)u_1 - \cdots - (v, u_k)u_k$. □

PROOF OF EXISTENCE. Putting $c_j = (v, u_j)$, we need check only that the difference $v - (v, u_1)u_1 - \cdots - (v, u_k)u_k$ is orthogonal to each u_j with $1 \leq j \leq k$. Direct calculation gives

$$\left(v - \sum_i (v, u_i)u_i, \; u_j\right) = (v, u_j) - \sum_i ((v, u_i)u_i, u_j) = (v, u_j) - (v, u_j) = 0,$$

and the proof is complete. □

Corollary 3.4 (Bessel's inequality). Let V be an inner-product space. If $\{u_1, \ldots, u_k\}$ is an orthonormal set in V and if v is given in V, then $\sum_{j=1}^{k} |(v, u_j)|^2 \leq \|v\|^2$ with equality if and only if v is in span$\{u_1, \ldots, u_k\}$.

PROOF. Using Proposition 3.3, write $v = \sum_{j=1}^{k} (v, u_j) u_j + v^\perp$ with $v^\perp$ orthogonal to $u_1, \ldots, u_k$. Then

$$\begin{aligned}
\|v\|^2 &= \left(\sum_{i=1}^{k} (v, u_i) u_i + v^\perp, \; \sum_{j=1}^{k} (v, u_j) u_j + v^\perp \right) \\
&= \sum_{i,j} (v, u_i) \overline{(v, u_j)} (u_i, u_j) + \left(\sum_i (v, u_i) u_i, v^\perp \right) \\
&\quad + \left(v^\perp, \sum_j (v, u_j) u_j \right) + \|v^\perp\|^2 \\
&= \sum_{i,j} (v, u_i) \overline{(v, u_j)} \delta_{ij} + 0 + 0 + \|v^\perp\|^2 \\
&= \sum_{j=1}^{k} |(v, u_j)|^2 + \|v^\perp\|^2.
\end{aligned}$$

From Proposition 3.3 we know that v is in span$\{u_1, \ldots, u_k\}$ if and only if $v^\perp = 0$, and the corollary follows. $\square$

We shall now impose the condition of finite dimensionality in order to obtain suitable kinds of orthonormal sets. The argument will enable us to give a basis-free interpretation of Proposition 3.3 and Corollary 3.4, and we shall obtain equivalent conditions for the vector $v^\perp$ in Proposition 3.3 and Corollary 3.4 to be 0 for every v.

If an ordered set of k linearly independent vectors in the inner-product space V is given, the above proposition suggests a way of adjusting the set so that it becomes orthonormal. Let us write the formulas here and carry out the verification via Proposition 3.3 in the proof of Proposition 3.5 below. The method of adjusting the set so as to make it orthonormal is called the **Gram–Schmidt orthogonalization process**. The given linearly independent set is denoted by $\{v_1, \ldots, v_k\}$, and we define

$$u_1 = \frac{v_1}{\|v_1\|},$$
$$u_2' = v_2 - (v_2, u_1) u_1,$$
$$u_2 = \frac{u_2'}{\|u_2'\|},$$
$$u_3' = v_3 - (v_3, u_1) u_1 - (v_3, u_2) u_2,$$
$$u_3 = \frac{u_3'}{\|u_3'\|},$$
$$\vdots$$
$$u_k' = v_k - (v_k, u_1) - \cdots - (v_k, u_{k-1}) u_{k-1},$$
$$u_k = \frac{u_k'}{\|u_k'\|}.$$

Proposition 3.5. If $\{v_1, \ldots, v_k\}$ is a linearly independent set in an inner-product space V, then the Gram–Schmidt orthogonalization process replaces $\{v_1, \ldots, v_k\}$ by an orthonormal set $\{u_1, \ldots, u_k\}$ such that $\text{span}\{v_1, \ldots, v_j\} = \text{span}\{u_1, \ldots, u_j\}$ for all j.

PROOF. We argue by induction on j. The base case is $j = 1$, and the result is evident in this case. Assume inductively that $u_1, \ldots, u_{j-1}$ are well defined and orthonormal and that $\text{span}\{v_1, \ldots, v_{j-1}\} = \text{span}\{u_1, \ldots, u_{j-1}\}$. Proposition 3.3 shows that u'_j is orthogonal to $u_1, \ldots, u_{j-1}$. If $u'_j = 0$, then v_j has to be in $\text{span}\{u_1, \ldots, u_{j-1}\} = \text{span}\{v_1, \ldots, v_{j-1}\}$, and we have a contradiction to the assumed linear independence of $\{v_1, \ldots, v_k\}$. Thus $u'_j \neq 0$, and $\{u_1, \ldots, u_j\}$ is a well-defined orthonormal set. This set must be linearly independent, and hence its linear span is a j-dimensional vector subspace of the linear span of $\{v_1, \ldots, v_j\}$. By Corollary 2.4, the two linear spans coincide. This completes the induction and the proof. □

Corollary 3.6. If V is a finite-dimensional inner-product space, then any orthonormal set in a vector subspace S of V can be extended to an orthonormal basis of S.

PROOF. Extend the given orthonormal set to a basis of S by Corollary 2.3b. Then apply the Gram–Schmidt orthogonalization process. The given vectors do not get changed by the process, as we see from the formulas for the vectors u'_j and u_j, and hence the result is an extension of the given orthonormal set to an orthonormal basis. □

Corollary 3.7. If S is a vector subspace of a finite-dimensional inner-product space V, then S has an orthonormal basis.

PROOF. This is the special case of Corollary 3.6 in which the given orthonormal set is empty. □

The set of all vectors orthogonal to a subset M of the inner-product space V is denoted by $M^\perp$. In symbols,

$$M^\perp = \{u \in V \mid (u, v) = 0 \text{ for all } v \in M\}.$$

We see by inspection that $M^\perp$ is a vector subspace. Moreover, $M \cap M^\perp = 0$ since any u in $M \cap M^\perp$ must have $(u, u) = 0$. The interest in the vector subspace $M^\perp$ comes from the following proposition.

Theorem 3.8 (Projection Theorem). If S is a vector subspace of the finite-dimensional inner-product space V, then every v in V decomposes uniquely as $v = v_1 + v_2$ with v_1 in S and v_2 in $S^\perp$. In other words, $V = S \oplus S^\perp$.

REMARKS. Because of this proposition, $S^\perp$ is often called the **orthogonal complement** of the vector subspace S.

PROOF. Uniqueness follows from the fact that $S \cap S^\perp = 0$. For existence, use of Corollaries 3.7 and 3.6 produces an orthonormal basis $\{u_1, \ldots, u_r\}$ of S and extends it to an orthonormal basis $\{u_1, \ldots, u_n\}$ of V. The vectors u_j for $j > r$ are orthogonal to each u_i with $i \leq r$ and hence are in $S^\perp$. If v is given in S, we can write $v = \sum_{j=1}^{n} u_j$ as $v = v_1 + v_2$ with $v_1 = \sum_{i=1}^{r} u_i$ and $v_2 = \sum_{j=r+1}^{n} u_j$, and this decomposition for all v shows that $V = S + S^\perp$. $\square$

Corollary 3.9. If S is a vector subspace of the finite-dimensional inner-product space V, then
(a) $\dim V = \dim S + \dim S^\perp$,
(b) $S^{\perp\perp} = S$.

PROOF. Conclusion (a) is immediate from the direct-sum decomposition $V = S \oplus S^\perp$ of Theorem 3.8. For (b), the definition of orthogonal complement gives $S \subseteq S^{\perp\perp}$. On the other hand, application of (a) twice shows that S and $S^{\perp\perp}$ have the same finite dimension. By Corollary 2.4, $S^{\perp\perp} = S$. $\square$

Section II.6 introduced "projection" mappings in the setting of any direct sum of two vector spaces, and we shall use those mappings in connection with the decomposition $V = S \oplus S^\perp$ of Theorem 3.8. We make one adjustment in working with the projections, changing their ranges from the image, namely S or $S^\perp$, to the larger space V. In effect, a linear map p_1 or p_2 as in Section II.6 will be replaced by $i_1 p_1$ or $i_2 p_2$.

Specifically let $E : V \to V$ be the linear map that is the identity on S and is 0 on $S^\perp$. Then E is called the **orthogonal projection** of V on S. The linear map $I - E$ is the identity on $S^\perp$ and is 0 on S. Since $S = S^{\perp\perp}$, $I - E$ is the orthogonal projection of V on $S^\perp$. It is the linear map that picks out the $S^\perp$ component relative to the direct-sum decomposition $V = S^\perp \oplus S^{\perp\perp}$. Proposition 3.3 and Corollary 3.4 can be restated in terms of orthogonal projections.

Corollary 3.10. Let V be a finite-dimensional inner-product space, let S be a vector subspace of V, let $\{u_1, \ldots, u_k\}$ be an orthonormal basis of S, and let E be the orthogonal projection of V on S. If v is in V, then

$$E(v) = \sum_{j=1}^{k} (v, u_j) u_j$$

and

$$\|E(v)\|^2 = \sum_{j=1}^{k} |(v, u_j)|^2.$$

The vector $v^\perp$ in the expansion $v = \sum_{j=1}^{k} (v, u_j) u_j + v^\perp$ of Proposition 3.3 is equal to $(I - E)v$, and the equality of norms

$$\|v\|^2 = \sum_{j=1}^{k} |(v, u_j)|^2 + \|v^\perp\|^2$$

has the interpretations that

$$\|v\|^2 = \|E(v)\|^2 + \|(I - E)v\|^2$$

and that equality holds in Bessel's inequality if and only if $E(v) = v$.

PROOF. Write $v = \sum_{j=1}^{k} (v, u_j) u_j + v^\perp$ as in Proposition 3.3. Then each u_j is in S, and the vector $v^\perp$, being orthogonal to each member of a basis of S, is in $S^\perp$. This proves the formula for $E(v)$, and the formula for $\|E(v)\|^2$ follows by applying Corollary 3.4 to $v - v^\perp$.

Reassembling v, we now have $v = E(v) + v^\perp$, and hence $v^\perp = v - E(v) = (I - E)v$. Finally the decomposition $v = E(v) + (I - E)(v)$ is into orthogonal terms, and the Pythagorean Theorem shows that $\|v\|^2 = \|E(v)\|^2 + \|(I - E)v\|^2$. □

Theorem 3.11 (Parseval's equality). *If V is a finite-dimensional inner-product space, then the following conditions on an orthonormal set $\{u_1, \ldots, u_m\}$ are equivalent:*
  (a) *$\{u_1, \ldots, u_m\}$ is a vector-space basis of V, hence an orthonormal basis,*
  (b) *the only vector orthogonal to all of $u_1, \ldots, u_m$ is 0,*
  (c) *$v = \sum_{j=1}^{m} (v, u_j) u_j$ for all v in V,*
  (d) *$\|v\|^2 = \sum_{j=1}^{m} |(v, u_j)|^2$ for all v in V,*
  (e) *$(v, w) = \sum_{j=1}^{m} (v, u_j)\overline{(w, u_j)}$ for all v and w in V.*

PROOF. Let $S = \mathrm{span}\{u_1, \ldots, u_m\}$, and let E be the orthogonal projection of V on S. If (a) holds, then $S = V$ and $S^\perp = 0$. Thus (b) holds.

If (b) holds, then $S^\perp = 0$ and E is the identity. Thus (c) holds by Corollary 3.10.

If (c) holds, then Corollary 3.4 shows that (d) holds.

If (d) holds, we use polarization to prove (e). Let k be in $\{0, 2\}$ if $\mathbb{F} = \mathbb{R}$, or in $\{0, 1, 2, 3\}$ if $\mathbb{F} = \mathbb{C}$. Conclusion (d) gives us

$$\|v + i^k w\|^2 = \sum_{j=1}^{m} |(v + i^k w, u_j)|^2 = \|v\|^2 + \sum_{j=1}^{m} 2\,\mathrm{Re}\left((v, u_j)\overline{i^k(w, u_j)}\right) + \|w\|^2.$$

Multiplying by i^k and summing over k, we obtain

$$4(v, w) = 2 \sum_{j=1}^{m} \sum_{k} i^k \,\mathrm{Re}\left((-i)^k (v, u_j)\overline{(w, u_j)}\right).$$

In the proof of polarization, we saw that $2\sum_k i^k \operatorname{Re}((-i)^k z) = 4z$. Hence $4(v, w) = 4\sum_{j=1}^{m} (v, u_j)\overline{(w, u_j)}$. This proves (e).

If (e) holds, we take $w = v$ in (e) and apply Corollary 3.10 to see that $\|E(v)\|^2 = \|v\|^2$ for all v. Then $\|(I - E)v\|^2 = 0$ for all v, and $E(v) = v$ for all v. Hence $S = V$, and $\{u_1, \ldots, u_m\}$ is a basis. This proves (a). $\square$

Theorem 3.12 (Riesz Representation Theorem). If ℓ is a linear functional on the finite-dimensional inner-product space V, then there exists a unique v in V with $\ell(u) = (u, v)$ for all u in V.

PROOF. Uniqueness is immediate by subtracting two such expressions, since if $(u, v) = 0$ for all u, then the special case $u = v$ gives $(v, v) = 0$ and $v = 0$. Let us prove existence. If $\ell = 0$, take $v = 0$. Otherwise let $S = \ker \ell$. Corollary 2.15 shows that $\dim S = \dim V - 1$, and Corollary 3.9a then shows that $\dim S^\perp = 1$. Let w be a nonzero vector in $S^\perp$. This vector w must have $\ell(w) \neq 0$ since $S \cap S^\perp = 0$, and we let v be the member of $S^\perp$ given by

$$v = \frac{\overline{\ell(w)}}{\|w\|^2} w.$$

For any u in V, we have $\ell\left(u - \frac{\ell(u)}{\ell(w)} w\right) = 0$, and hence $u - \frac{\ell(u)}{\ell(w)} w$ is in S. Since v is in $S^\perp$, $u - \frac{\ell(u)}{\ell(w)} w$ is orthogonal to v. Thus

$$(u, v) = \left(\frac{\ell(u)}{\ell(w)} w, v\right) = \left(\frac{\ell(u)}{\ell(w)} w, \frac{\overline{\ell(w)}}{\|w\|^2} w\right) = \ell(u) \frac{\ell(w)}{\ell(w)} \frac{\|w\|^2}{\|w\|^2} = \ell(u).$$

This proves existence. $\square$

2. Adjoints

Throughout this section, V will denote a finite-dimensional inner-product space with inner product $(\cdot, \cdot)$ and with scalars from $\mathbb{F}$, with $\mathbb{F}$ equal to $\mathbb{R}$ or $\mathbb{C}$. We shall study aspects of linear maps $L : V \to V$ related to the inner product on V. The starting point is to associate to any such L another linear map $L^* : V \to V$ known as the "adjoint" of V, and then to investigate some of its properties. A tool in this investigation will be the scalar-valued function on $V \times V$ given by $(u, v) \mapsto (L(u), v)$, which captures the information in any matrix of L without requiring the choice of an ordered basis. This function determines L uniquely because an equality $(L(u), v) = (L'(u), v)$ for all u and v implies $(L(u) - L'(u), v) = 0$ for all u and v, in particular for $v = L(u) - L'(u)$; thus $\|L(u) - L'(u)\|^2 = 0$ and $L(u) = L'(u)$ for all u.

Proposition 3.13. Let $L : V \to V$ be a linear map on the finite-dimensional inner-product space V. For each u in V, there exists a unique vector $L^*(u)$ in V such that
$$(L(v), u) = (v, L^*(u)) \qquad \text{for all } v \text{ in } V.$$
As u varies, this formula defines L^* as a linear map from V to V.

REMARK. The linear map $L^* : V \to V$ is called the **adjoint** of L.

PROOF. The function $v \mapsto (L(v), u)$ is a linear functional on V, and Theorem 3.12 shows that it is given by the inner product with a unique vector of V. Thus we define $L^*(u)$ to be the unique vector of V with $(L(v), u) = (v, L^*(u))$ for all v in V.

If c is a scalar, then the uniqueness and the computation $(v, L^*(cu)) = (L(v), cu) = \bar{c}(L(v), u) = \bar{c}(v, L^*(u)) = (v, cL^*(u))$ yield $L^*(cu) = cL^*(u)$. Similarly the uniqueness and the computation
$$(v, L^*(u_1 + u_2)) = (L(v), u_1 + u_2) = (L(v), u_1) + (L(v), u_2)$$
$$= (v, L^*(u_1)) + (v, L^*(u_2)) = (v, L^*(u_1) + L^*(u_2))$$
yield $L^*(u_1 + u_2) = L^*(u_1) + L^*(u_2)$. Therefore L^* is linear. $\square$

The passage $L \mapsto L^*$ to the adjoint is a function from $\operatorname{Hom}_{\mathbb{F}}(V, V)$ to itself that is conjugate linear, and it reverses the order of multiplication: $(L_1 L_2)^* = L_2^* L_1^*$. Since the formula $(L(v), u) = (v, L^*(u))$ in the proposition is equivalent to the formula $(u, L(v)) = (L^*(u), v)$, we see that $L^{**} = L$.

All of the results in Section II.3 concerning the association of matrices to linear maps are applicable here, but our interest now will be in what happens when the bases we use are orthonormal. Recall from Section II.3 that if $\Gamma = (u_1, \ldots, u_n)$ and $\Delta = (v_1, \ldots, v_n)$ are any ordered bases of V, then the matrix $A = \begin{pmatrix} L \\ \Delta\Gamma \end{pmatrix}$ associated to the linear map $L : V \to V$ has $A_{ij} = \begin{pmatrix} L(u_j) \\ \Delta \end{pmatrix}_i$.

Lemma 3.14. If $L : V \to V$ is a linear map on the finite-dimensional inner-product space V and if $\Gamma = (u_1, \ldots, u_n)$ and $\Delta = (v_1, \ldots, v_n)$ are ordered orthonormal bases of V, then the the matrix $A = \begin{pmatrix} L \\ \Delta\Gamma \end{pmatrix}$ has $A_{ij} = (L(u_j), v_i)$.

PROOF. Applying Theorem 3.11c, we have
$$A_{ij} = \begin{pmatrix} L(u_j) \\ \Delta \end{pmatrix}_i = \begin{pmatrix} \sum_{i'} (L(u_j), v_{i'}) v_{i'} \\ \Delta \end{pmatrix}_i$$
$$= \sum_{i'} (L(u_j), v_{i'}) \begin{pmatrix} v_{i'} \\ \Delta \end{pmatrix}_i = \sum_{i'} (L(u_j), v_{i'}) \delta_{ii'} = (L(u_j), v_i). \quad \square$$

Proposition 3.15. If $L : V \to V$ is a linear map on the finite-dimensional inner-product space V and if $\Gamma = (u_1, \ldots, u_n)$ and $\Delta = (v_1, \ldots, v_n)$ are ordered orthonormal bases of V, then the matrices $A = \begin{pmatrix} L \\ \Delta\Gamma \end{pmatrix}$ and $A^* = \begin{pmatrix} L^* \\ \Gamma\Delta \end{pmatrix}$ of L and its adjoint are related by $A^*_{ij} = \overline{A_{ji}}$.

PROOF. Lemma 3.14 and the definition of L^* give $A^*_{ij} = (L^*(v_j), u_i) = (v_j, L(u_i)) = \overline{(L(u_i), v_j)} = \overline{A_{ji}}$. □

Accordingly, we define $A^* = \overline{A}^t$ for any square matrix A, sometimes calling A^* the **adjoint**[6] of A.

A linear map $L : V \to V$ is called **self-adjoint** if $L^* = L$. Correspondingly a square matrix A is **self-adjoint** if $A^* = A$. It is more common, however, to say that a matrix with $A^* = A$ is **symmetric** if $\mathbb{F} = \mathbb{R}$ or **Hermitian**[7] if $\mathbb{F} = \mathbb{C}$. A real Hermitian matrix is symmetric, and the term "Hermitian" is thus applicable also when $\mathbb{F} = \mathbb{R}$.

Any Hermitian matrix A arises from a self-adjoint linear map L. Namely, we take V to be $\mathbb{F}^n$ with the usual inner product, and we let Γ and Δ each be the standard ordered basis $\Sigma = (e_1, \ldots, e_n)$. This basis is orthonormal, and we define L by the matrix product $L(v) = Av$ for any column vector v. We know that $\begin{pmatrix} L \\ \Sigma\Sigma \end{pmatrix} = A$. Since $A^* = A$, we conclude from Proposition 3.15 that $L^* = L$. Thus we are free to deduce properties of Hermitian matrices from properties of self-adjoint linear maps.

Self-adjoint linear maps will be of special interest to us. Nontrivial examples of self-adjoint linear maps, constructed without simply writing down Hermitian matrices, may be produced by the following proposition.

Proposition 3.16. If V is a finite-dimensional inner-product space and S is a vector subspace of V, then the orthogonal projection $E : V \to V$ of V on S is self-adjoint.

PROOF. Let $v = v_1 + v_2$ and $u = u_1 + u_2$ be the decompositions of two members of V according to $V = S \oplus S^\perp$. Then we have $(v, E^*(u)) = (E(v), u) = (v_1, u_1 + u_2) = (v_1, u_1) = (v, u_1) = (v, E(u))$, and the proposition follows by the uniqueness in Proposition 3.13. □

---

[6]The name "adjoint" happens to coincide with the name for a different notion that arose in connection with Cramer's rule in Section II.7. The two notions never seem to arise at the same time, and thus no confusion need occur.

[7]The term "Hermitian" is used also for a class of linear maps in the infinite-dimensional case, but care is needed because the terms "Hermitian" and "self-adjoint" mean different things in the infinite-dimensional case.

To understand Proposition 3.16 in terms of matrices, take an ordered orthonormal basis $(u_1, \ldots, u_r)$ of S, and extend it to an ordered orthonormal basis $\Gamma = (u_1, \ldots, u_n)$ of V. Then

$$E(u_j) = \begin{cases} u_j & \text{for } j \leq r, \\ 0 & \text{for } j > r, \end{cases}$$

and hence $\begin{pmatrix} E(u_j) \\ \Gamma \end{pmatrix}$ equals the jth standard basis vector e_j if $j \leq r$ and equals 0 if $j > r$. Consequently the matrix $\begin{pmatrix} E \\ \Gamma\Gamma \end{pmatrix}$ is diagonal with 1's in the first r diagonal entries and 0's elsewhere. This matrix is equal to its conjugate transpose, as it must be according to Propositions 3.15 and 3.16.

Proposition 3.17. If V is a finite-dimensional inner-product space and $L : V \to V$ is a self-adjoint linear map, then $(L(v), v)$ is in $\mathbb{R}$ for every v in V, and consequently every eigenvalue of L is in $\mathbb{R}$. Conversely if $\mathbb{F} = \mathbb{C}$ and if $L : V \to V$ is a linear map such that $(L(v), v)$ is in $\mathbb{R}$ for every v in V, then L is self-adjoint.

REMARK. The hypothesis $\mathbb{F} = \mathbb{C}$ is essential in the converse. In fact, the 90° rotation L of $\mathbb{R}^2$ whose matrix in the standard basis is $\begin{pmatrix} 0 & 1 \\ -1 & 0 \end{pmatrix}$ is not self-adjoint but does have $L(v) \cdot v = 0$ for every v in $\mathbb{R}^2$.

PROOF. If $L = L^*$, then $(L(v), v) = (v, L^*(v)) = (v, L(v)) = \overline{(L(v), v)}$, and hence $(L(v), v)$ is real-valued. If v is an eigenvector with eigenvalue λ, then substitution of $L(v) = \lambda v$ into $(L(v), v) = \overline{(L(v), v)}$ gives $\lambda \|v\|^2 = \bar{\lambda} \|v\|^2$. Since $v \neq 0$, λ must be real.

For the converse we begin with the special case that $(L(w), w) = 0$ for all w. For $0 \leq k \leq 3$, we then have

$$(-i)^k(L(u), v) + i^k(L(v), u) = (L(u + i^k v), u + i^k v) - (L(u), u) - (L(v), v) = 0.$$

Taking $k = 0$ gives $(L(u), v) + (L(v), u) = 0$, while taking $k = 1$ gives $(L(u), v) - (L(v), u) = 0$. Hence $(L(u), v) = 0$ for all u and v. Since the function $(u, v) \mapsto L(u, v)$ determines L, we obtain $L = 0$.

In the general case, $(L(v), v)$ real-valued implies that $(L(v), v) = (L^*(v), v)$ for all v. Therefore $((L - L^*)(v), v) = 0$ for all v, and the special case shows that $L - L^* = 0$. This completes the proof. □

We conclude this section by examining one further class of linear maps having a special relationship with their adjoints.

Proposition 3.18. If V is a finite-dimensional inner-product space, then the following conditions on a linear map $L : V \to V$ are equivalent:
(a) $L^*L = I$,
(b) L carries some orthonormal basis of V to an orthonormal basis,
(c) L carries each orthonormal basis of V to an orthonormal basis,
(d) $(L(u), L(v)) = (u, v)$ for all u and v in V,
(e) $\|L(v)\| = \|v\|$ for all v in V.

REMARK. A linear map satisfying these equivalent conditions is said to be **orthogonal** if $\mathbb{F} = \mathbb{R}$ and **unitary** if $\mathbb{F} = \mathbb{C}$.

PROOF. We prove that (a), (d), and (e) are equivalent and that (b), (c), and (d) are equivalent.

If (a) holds and u and v are given in V, then $(L(u), L(v)) = (L^*L(u), v) = (I(u), v) = (u, v)$, and (d) holds. If (d) holds, then setting $u = v$ shows that (e) holds. If (e) holds, we use polarization twice to write

$$(L(u), L(v)) = \sum_k \tfrac{1}{4} i^k \|L(u) + i^k L(v)\|^2 = \sum_k \tfrac{1}{4} i^k \|L(u + i^k v)\|^2$$
$$= \sum_k \tfrac{1}{4} i^k \|u + i^k v\|^2 = (u, v).$$

Then $((L^*L - I)(u), v) = 0$ for all u and v, and we conclude that (a) holds.

Since (b) is a special case of (c) and (c) is a special case of (d), proving that (b) implies (d) will prove that (b), (c), and (d) are equivalent. Thus let $\{u_1, \ldots, u_n\}$ be an orthonormal basis of V such that $\{L(u_1), \ldots, L(u_n)\}$ is an orthonormal basis, and let u and v be given. Then

$$(L(u), L(v)) = \left(L\left(\sum_i (u, u_i) u_i\right), L\left(\sum_j (v, u_j) u_j\right)\right)$$
$$= \sum_{i,j} (u, u_i) \overline{(v, u_j)} (L(u_i), L(u_j))$$
$$= \sum_{i,j} (u, u_i) \overline{(v, u_j)} \delta_{ij} = \sum_i (u, u_i) \overline{(v, u_i)} = (u, v),$$

the last equality following from Parseval's equality (Theorem 3.11). □

As with self-adjointness, we use the geometrically meaningful definition for linear maps to obtain a definition for matrices: a square matrix A with $A^*A = I$ is said to be **orthogonal** if $\mathbb{F} = \mathbb{R}$ and **unitary** if $\mathbb{F} = \mathbb{C}$. The condition is that A is invertible and its inverse equals its adjoint. In terms of individual entries, the condition is that $\sum_k A^*_{ik} A_{kj} = \delta_{ij}$, hence that $\sum_k \overline{A_{ki}} A_{kj} = \delta_{ij}$. This is the condition that the columns of A form an orthonormal basis relative to the usual inner product on $\mathbb{R}^n$ or $\mathbb{C}^n$. A real unitary matrix is orthogonal.

If A is an orthogonal or unitary matrix, we can construct a corresponding orthogonal or unitary linear map on $\mathbb{R}^n$ or $\mathbb{C}^n$ relative to the standard ordered basis Σ. Namely, we define $L(v) = Av$, and Proposition 3.15 shows that L is orthogonal or unitary: $L^*L(v) = A^*Av = Iv = v$. Proposition 3.19 below gives a converse.

Let us notice that an orthogonal or unitary matrix A necessarily has $|\det A| = 1$. In fact, the formula $A^* = (\overline{A})^t$ implies that $\det A^* = \overline{\det A}$. Then

$$1 = \det I = \det A^*A = \det A^* \det A = \overline{\det A} \det A = |\det A|^2.$$

An orthogonal matrix thus has determinant ± 1, while we conclude for a unitary matrix only that the determinant is a complex number of absolute value 1.

EXAMPLES.

(1) The 2-by-2 orthogonal matrices of determinant $+1$ are all matrices of the form $\begin{pmatrix} \cos\theta & \sin\theta \\ \sin\theta & \cos\theta \end{pmatrix}$. The 2-by-2 orthogonal matrices of determinant -1 are the product of $\begin{pmatrix} 1 & 0 \\ 0 & -1 \end{pmatrix}$ and the 2-by-2 orthogonal matrices of determinant $+1$.

(2) The 2-by-2 unitary matrices of determinant $+1$ are all matrices of the form $\begin{pmatrix} \alpha & \beta \\ -\bar\beta & \bar\alpha \end{pmatrix}$ with $|\alpha|^2 + |\beta|^2 = 1$; these may be regarded as parametrizing the points of the unit sphere S^3 of $\mathbb{R}^4$. The 2-by-2 unitary matrices of arbitrary determinant are the products of all matrices $\begin{pmatrix} 1 & 0 \\ 0 & e^{i\theta} \end{pmatrix}$ and the 2-by-2 unitary matrices of determinant $+1$.

Proposition 3.19. If V is a finite-dimensional inner-product space, if $\Gamma = (u_1, \ldots, u_n)$ and $\Delta = (v_1, \ldots, v_n)$ are ordered orthonormal bases of V, and if $L : V \to V$ is a linear map that is orthogonal if $\mathbb{F} = \mathbb{R}$ and unitary if $\mathbb{F} = \mathbb{C}$, then the matrix $A = \begin{pmatrix} L \\ \Delta\Gamma \end{pmatrix}$ is orthogonal or unitary.

PROOF. Proposition 3.15 and Theorem 2.16 give $A^*A = \begin{pmatrix} L^* \\ \Gamma\Delta \end{pmatrix}\begin{pmatrix} L \\ \Delta\Gamma \end{pmatrix} = \begin{pmatrix} I \\ \Delta\Delta \end{pmatrix}$, and the right side is the identity matrix, as required. $\square$

One consequence of Proposition 3.19 is that any matrix $\begin{pmatrix} I \\ \Delta\Gamma \end{pmatrix}$ relative to two ordered orthonormal bases is orthogonal or unitary, since the identity function $I : V \to V$ is certainly orthogonal or unitary. Thus a change from writing the matrix of a linear map L in one ordered orthonormal basis Γ to writing the matrix of L in another ordered orthonormal basis Δ is implemented by the formula $\begin{pmatrix} L \\ \Gamma\Gamma \end{pmatrix} = C^{-1} \begin{pmatrix} L \\ \Delta\Delta \end{pmatrix} C$, where C is the orthogonal or unitary matrix $\begin{pmatrix} I \\ \Delta\Gamma \end{pmatrix}$.

Another consequence of Proposition 3.19 is that the matrix $\begin{pmatrix} L \\ \Gamma\Gamma \end{pmatrix}$ of an orthogonal or unitary linear map L in an ordered orthonormal basis Γ is an orthogonal or unitary matrix. We have defined $\det L$ to be the determinant of $\begin{pmatrix} L \\ \Gamma\Gamma \end{pmatrix}$ relative to *any* Γ, and we conclude that $|\det L| = 1$.

3. Spectral Theorem

In this section we deal with the geometric structure of certain kinds of linear maps from finite-dimensional inner-product spaces into themselves. We shall see that linear maps that are self-adjoint or unitary, among other possible conditions, have bases of eigenvectors in the sense of Section II.8. Moreover, such a basis may be taken to be orthonormal. When an ordered basis of eigenvectors is used for expressing the linear map as a matrix, the result is that the matrix is diagonal. Thus these linear maps have an especially uncomplicated structure. In terms of matrices, the result is that a Hermitian or unitary matrix A is similar to a diagonal matrix D, and the matrix C with $D = C^{-1}AC$ may be taken to be unitary. We begin with a lemma.

Lemma 3.20. If $L : V \to V$ is a self-adjoint linear map on an inner-product space V, then $v \mapsto (L(v), v)$ is real-valued, every eigenvalue of L is real, eigenvalues under L for distinct eigenvalues are orthogonal, and every vector subspace S of V with $L(S) \subseteq S$ has $L(S^\perp) \subseteq S^\perp$.

PROOF. The first two conclusions are contained in Proposition 3.17. If v_1 and v_2 are eigenvectors of L with distinct real eigenvalues λ_1 and λ_2, then

$$(\lambda_1 - \lambda_2)(v_1, v_2) = (\lambda_1 v_1, v_2) - (v_1, \lambda_2 v_2) = (L(v_1), v_2) - (v_1, L(v_2)) = 0.$$

Since $\lambda_1 \neq \lambda_2$, we must have $(v_1, v_2) = 0$. If S is a vector subspace with $L(S) \subseteq S$, then also $L(S^\perp) \subseteq S^\perp$ because $s \in S$ and $s^\perp \in S^\perp$ together imply

$$0 = (L(s), s^\perp) = (s, L(s^\perp)). \qquad \square$$

Theorem 3.21 (Spectral Theorem). Let $L : V \to V$ be a self-adjoint linear map on an inner-product space V. Then V has an orthonormal basis of eigenvectors of L. In addition, for each scalar λ, let

$$V_\lambda = \{v \in V \mid L(v) = \lambda v\},$$

so that V_λ when nonzero is the eigenspace of L for the eigenvalue λ. Then the eigenvalues of L are all real, the vector subspaces V_λ are mutually orthogonal, and any orthonormal basis of V of eigenvectors of L is the union of orthonormal bases of the V_λ's. Correspondingly if A is any Hermitian n-by-n matrix, then there exists a unitary matrix C such that $C^{-1}AC$ is diagonal with real entries. If the matrix A has real entries, then C may be taken to be an orthogonal matrix.

PROOF. Lemma 3.20 shows that the eigenvalues of L are all real and that the vector subspaces V_λ are mutually orthogonal.

To proceed further, we first assume that $\mathbb{F} = \mathbb{C}$. Applying the Fundamental Theorem of Algebra (Theorem 1.18) to the characteristic polynomial of L, we see that L has at least one eigenvalue, say λ_1. Then $L(V_{\lambda_1}) \subseteq V_{\lambda_1}$, and Lemma 3.20 shows that $L((V_{\lambda_1})^\perp) \subseteq (V_{\lambda_1})^\perp$. The vector subspace $(V_{\lambda_1})^\perp$ is an inner-product space, and the claim is that $L\big|_{(V_{\lambda_1})^\perp}$ is self-adjoint. In fact, if v_1 and v_2 are in $(V_{\lambda_1})^\perp$, then

$$\big((L\big|_{(V_{\lambda_1})^\perp})^*(v_1), v_2\big) = \big(v_1, L\big|_{(V_{\lambda_1})^\perp}(v_2)\big) = (v_1, L(v_2))$$
$$= (L(v_1), v_2) = \big(L\big|_{(V_{\lambda_1})^\perp}(v_1), v_2\big),$$

and the claim is proved. Since λ_1 is an eigenvalue of L, $\dim(V_{\lambda_1})^\perp < \dim V$. Therefore we can now set up an induction that ultimately exhibits V as an orthogonal direct sum $V = V_{\lambda_1} \oplus \cdots \oplus V_{\lambda_k}$. If v is an eigenvector of L with eigenvalue λ', then either $\lambda' = \lambda_j$ for some j in this decomposition, in which case v is in V_{λ_j}, or λ' is not equal to any λ_j, in which case v, by the lemma, is orthogonal to all vectors in $V_{\lambda_1} \oplus \cdots \oplus V_{\lambda_k}$, hence to all vectors in V; being orthogonal to all vectors in V, v must be 0. Choosing an orthonormal basis for each V_{λ_j} and taking their union provides an orthonormal basis of eigenvectors and completes the proof for L when $\mathbb{F} = \mathbb{C}$.

Next assume that A is a Hermitian n-by-n matrix. We define a linear map $L : \mathbb{C}^n \to \mathbb{C}^n$ by $L(v) = Av$, and we know from Proposition 3.15 that L is self-adjoint. The case just proved shows that L has an ordered orthonormal basis Γ of eigenvectors, all the eigenvalues being real. If Σ denotes the standard ordered basis of $\mathbb{C}^n$, then $D = \begin{pmatrix} L \\ \Gamma\Gamma \end{pmatrix}$ is diagonal with real entries and is equal to

$$\begin{pmatrix} I \\ \Gamma\Sigma \end{pmatrix} \begin{pmatrix} L \\ \Sigma\Sigma \end{pmatrix} \begin{pmatrix} I \\ \Sigma\Gamma \end{pmatrix} = C^{-1}AC,$$

where $C = \begin{pmatrix} L \\ \Sigma\Gamma \end{pmatrix}$. The matrix C is unitary by Proposition 3.19, and the formula $D = C^{-1}AC$ shows that A is as asserted.

Now let us return to L and suppose that $\mathbb{F} = \mathbb{R}$. The idea is to use the same argument as above in the case that $\mathbb{F} = \mathbb{C}$, but we need a substitute for the use of the Fundamental Theorem of Algebra. Fixing any orthonormal basis of V, let A be the matrix of L. Then A is Hermitian with real entries. The previous paragraph shows that any Hermitian matrix, whether or not real, has a characteristic polynomial that splits as a product $\prod_{j=1}^{m} (\lambda - r_j)^{m_j}$ with all r_j real. Consequently L has this property as well. Thus any self-adjoint L when $\mathbb{F} = \mathbb{R}$ has an eigenvalue. Returning to the argument for L above when $\mathbb{F} = \mathbb{C}$, we readily see that it now applies when $\mathbb{F} = \mathbb{R}$.

Finally if A is a Hermitian matrix with real entries, then we can define a self-adjoint linear map $L : \mathbb{R}^n \to \mathbb{R}^n$ by $L(v) = Av$, obtain an orthonormal basis of eigenvectors for L, and argue as above to obtain $D = C^{-1}AC$, where D is diagonal and C is unitary. The matrix C has columns that are eigenvectors in $\mathbb{R}^n$ of the associated L, and these have real entries. Thus C is orthogonal. $\square$

An important application of the Spectral Theorem is to the formation of a square root for any "positive semidefinite" linear map. We say that a linear map $L : V \to V$ on a finite-dimensional inner-product space is **positive semidefinite** if $L^* = L$ and $(L(v), v) \geq 0$ for all v in V. If $\mathbb{F} = \mathbb{C}$, then the condition $L^* = L$ is redundant, according to Proposition 3.17, but that fact will not be important for us. Similarly an n-by-n matrix A is **positive semidefinite** if $A^* = A$ and $\bar{x}^t A x \geq 0$ for all column vectors x. An example of a positive semidefinite n-by-n matrix is any matrix $A = B^*B$, where B is an arbitrary k-by-n matrix. In fact, if x is in $\mathbb{F}^n$, then $\bar{x}^t B^* B x = (\overline{Bx})^t (Bx)$, and the right side is ≥ 0, being a sum of absolute values squared.

Corollary 3.22. Let $L : V \to V$ be a positive semidefinite linear map on a finite-dimensional inner-product space, and let A be an n-by-n Hermitian matrix. Then

(a) L or A is positive semidefinite if and only if all of its eigenvalues are ≥ 0.
(b) whenever L or A is positive semidefinite, L or A is invertible if and only if $(L(v), v) > 0$ for all $v \neq 0$ or $\bar{x}^t A x > 0$ for all $x \neq 0$.
(c) whenever L or A is positive semidefinite, L or A has a unique positive semidefinite square root.

REMARKS. A positive semidefinite linear map or matrix satisfying the condition in (b) is said to be **positive definite**, and the content of (b) is that a positive semidefinite linear map or matrix is positive definite if and only if it is invertible.

PROOF. We apply the Spectral Theorem (Theorem 3.21). For each conclusion the result for a matrix A is a special case of the result for the linear map L, and it is enough to treat only L. In (a), let $(u_1, \ldots, u_n)$ be an ordered basis of eigenvectors with respective eigenvalues $\lambda_1, \ldots, \lambda_n$, not necessarily distinct. Then $(L(u_j), u_j) = \lambda_j$ shows the necessity of having $\lambda_j \geq 0$, while the computation

$$(L(v), v) = \left(L\left(\sum_i (v, u_i)u_i\right), \sum_j (v, u_j)u_j\right)$$
$$= \left(\sum_i \lambda_i (v, u_i)u_i, \sum_j (v, u_j)u_j\right)$$
$$= \sum_i \lambda_i |(v, u_i)|^2$$

shows the sufficiency.

In (b), if L fails to be invertible, then 0 is an eigenvalue for some eigenvector $v \neq 0$, and v has $(L(v), v) = 0$. Conversely if L is invertible, then all the eigenvalues λ_i are > 0 by (a), and the computation in (a) yields

$$(L(v), v) = \sum_i \lambda_i |(v, u_i)|^2 \geq \left(\min_j \lambda_j\right) \sum_i |(v, u_i)|^2 = \left(\min_j \lambda_j\right) \|v\|^2,$$

the last step following from Parseval's equality (Theorem 3.11).

For existence in (c), the Spectral Theorem says that there exists an ordered orthonormal basis $\Gamma = (u_1, \ldots, u_n)$ of eigenvectors of L, say with respective eigenvalues $\lambda_1, \ldots, \lambda_n$. The eigenvalues are all ≥ 0 by (a). The linear extension of the function P with $P(u_j) = \lambda_j^{1/2} u_j$ is given by

$$P(v) = \sum_{j=1}^n \lambda_j^{1/2} (v, u_j) u_j,$$

and it has

$$P^2(v) = \sum_j \lambda_j (v, u_j) u_j = \sum_j (v, u_j) L(u_j) = L\left(\sum_j (v, u_j) u_j\right) = L(v).$$

Thus $P^2 = L$. Relative to Γ, we have

$$\left(\frac{P}{\Gamma\Gamma}\right)_{ij} = \left((P(u_j), u_1)u_1 + \cdots + (P(u_j), u_n)u_n\right)_i = (P(u_j), u_i) = \lambda_j^{1/2} \delta_{ij},$$

and this is a Hermitian matrix; Proposition 3.15 therefore shows that $P^* = P$. Finally

$$(P(v), v) = \left(\sum_i \lambda_i^{1/2} (v, u_i) u_i, \sum_j (v, u_j) u_j\right) = \lambda_i^{1/2} |(v, u_i)|^2 \geq 0,$$

and thus P is positive semidefinite. This proves existence.

For uniqueness in (c), let P satisfy $P^* = P$ and $P^2 = L$, and suppose P is positive semidefinite. Choose an orthonormal basis of eigenvectors $u_1, \ldots, u_n$ of P, say with eigenvalues $c_1, \ldots, c_n$, all ≥ 0. Then $L(u_j) = P^2(u_j) = c_j^2 u_j$, and we see that $u_1, \ldots, u_n$ form an orthonormal basis of eigenvectors of L with eigenvalues c_j^2. On the space where L acts as the scalar λ_i, P must therefore act as the scalar $\lambda_i^{1/2}$. We conclude that P is unique. $\square$

The technique of proof of (c) allows one, more generally, to define $f(L)$ for any function $f : \mathbb{R} \to \mathbb{C}$ whenever L is self-adjoint. Actually, the function f needs to be defined only on the set of eigenvalues of L for the definition to make sense.

At the end of this section, we shall use the existence of the square root in (c) to obtain the so-called "polar decomposition" of square matrices. But before doing that, let us mine three additional easy consequences of the Spectral Theorem. The first deals with several self-adjoint linear maps rather than one, and the other two apply that conclusion to deal with single linear maps that are not necessarily self-adjoint.

Corollary 3.23. Let V be a finite-dimensional inner-product space, and let $L_1, \ldots, L_m$ be self-adjoint linear maps from V to V that commute in the sense that $L_i L_j = L_j L_i$ for all i and j. Then V has an orthonormal basis of simultaneous eigenvectors of $L_1, \ldots, L_m$. In addition, for each m-tuple of scalars $\lambda_1, \ldots, \lambda_m$, let

$$V_{\lambda_1, \ldots, \lambda_m} = \{v \in V \mid L_j(v) = \lambda_j v \text{ for } 1 \leq j \leq m\}$$

consist of 0 and the simultaneous eigenvectors of $L_1, \ldots, L_m$ corresponding to $\lambda_1, \ldots, \lambda_m$. Then all the eigenvalues λ_j are real, the vector subspaces $V_{\lambda_1, \ldots, \lambda_m}$ are mutually orthogonal, and any orthonormal basis of V of simultaneous eigenvectors of $L_1, \ldots, L_m$ is the union of orthonormal bases of the $V_{\lambda_1, \ldots, \lambda_m}$'s. Correspondingly if $A_1, \ldots, A_m$ are commuting Hermitian n-by-n matrices, then there exists a unitary matrix C such that $C^{-1} A_j C$ is diagonal with real entries for all j. If all the matrices A_j have real entries, then C may be taken to be an orthogonal matrix.

PROOF. This follows by iterating the Spectral Theorem (Theorem 3.21). In fact, let $\{V_{\lambda_1}\}$ be the system of vector subspaces produced by the theorem for L_1. For each j, the commutativity of the linear maps L_i forces

$$L_1(L_i(v)) = L_i(L_1(v)) = L_i(\lambda_1 v) = \lambda_1 L_i(v) \qquad \text{for } v \in V_{\lambda_1},$$

and thus $L_i(V_{\lambda_1}) \subseteq V_{\lambda_1}$. The restrictions of $L_1, \ldots, L_m$ to V_{λ_1} are self-adjoint and commute. Let $\{V_{\lambda_1,\lambda_2}\}$ be the system of vector subspaces produced by the Spectral Theorem for $L_2\big|_{V_{\lambda_1}}$. Each of these, by the commutativity, is carried into itself by $L_3, \ldots, L_m$, and the restrictions of $L_3, \ldots, L_m$ to V_{λ_1,λ_2} form a commuting family of self-adjoint linear maps. Continuing in this way, we arrive at the decomposition asserted by the corollary for $L_1, \ldots, L_m$. The assertion of the corollary about commuting Hermitian matrices is a special case, in the same way that the assertions in Theorem 3.21 about matrices were special cases of the assertions about linear maps. □

3. Spectral Theorem

A linear map $L : V \to V$, not necessarily self-adjoint, is said to be **normal** if L commutes with its adjoint: $LL^* = L^*L$.

Corollary 3.24. Suppose that $\mathbb{F} = \mathbb{C}$, and let $L : V \to V$ be a normal linear map on the finite-dimensional inner-product space V. Then V has an orthonormal basis of eigenvectors of L. In addition, for each complex scalar λ, let

$$V_\lambda = \{v \in V \mid L(v) = \lambda v\},$$

so that V_λ when nonzero is the eigenspace of L for the eigenvalue λ. Then the vector subspaces V_λ are mutually orthogonal, and any orthonormal basis of V of eigenvectors of L is the union of orthonormal bases of the V_λ's. Correspondingly if A is any n-by-n complex matrix such that $AA^* = A^*A$, then there exists a unitary matrix C such that $C^{-1}AC$ is diagonal.

REMARK. The corollary fails if $\mathbb{F} = \mathbb{R}$: for the linear map $L : \mathbb{R}^2 \to \mathbb{R}^2$ with $L(v) = Av$ and $A = \begin{pmatrix} 0 & 1 \\ -1 & 0 \end{pmatrix}$, $L^* = L^{-1}$ commutes with L, but L has no eigenvectors in $\mathbb{R}^2$ since the characteristic polynomial $\lambda^2 + 1$ has no first-degree factors with real coefficients.

PROOF. The point is that $L = \left(\frac{1}{2}(L+L^*)\right) + i\left(\frac{1}{2i}(L-L^*)\right)$ and that $\frac{1}{2}(L+L^*)$ and $\frac{1}{2i}(L-L^*)$ are self-adjoint. If L commutes with L^*, then $T_1 = \frac{1}{2}(L+L^*)$ and $T_2 = \frac{1}{2i}(L-L^*)$ commute with each other. We apply Corollary 3.23 to the commuting self-adjoint linear maps T_1 and T_2. The vector subspace $V_{\alpha,\beta}$ produced by Corollary 3.23 coincides with the vector subspace $V_{\alpha+i\beta}$ defined in the present corollary, and the result for L follows. The result for matrices is a special case. □

Corollary 3.25. Suppose that $\mathbb{F} = \mathbb{C}$, and let $L : V \to V$ be a unitary linear map on the finite-dimensional inner-product space V. Then V has an orthonormal basis of eigenvectors of L. In addition, for each complex scalar λ, let

$$V_\lambda = \{v \in V \mid L(v) = \lambda v\},$$

so that V_λ when nonzero is the eigenspace of L for the eigenvalue λ. Then the eigenvalues of L all have absolute value 1, the vector subspaces V_λ are mutually orthogonal, and any orthonormal basis of V of eigenvectors of L is the union of orthonormal bases of the V_λ's. Correspondingly if A is any n-by-n unitary matrix, then there exists a unitary matrix C such that $C^{-1}AC$ is diagonal; the diagonal entries of $C^{-1}AC$ all have absolute value 1.

PROOF. This is a special case of Corollary 3.24 since a unitary linear map L has $LL^* = I = L^*L$. The eigenvalues all have absolute value 1 as a consequence of Proposition 3.18e. □

Now we come to the **polar decomposition** of linear maps and of matrices. When $\mathbb{F} = \mathbb{C}$, this is a generalization of the polar decomposition $z = e^{i\theta}r$ of complex numbers. When $\mathbb{F} = \mathbb{R}$, it generalizes the decomposition $x = (\operatorname{sgn} x)|x|$ of real numbers.

Theorem 3.26 (polar decomposition). If $L : V \to V$ is a linear map on a finite-dimensional inner-product space, then L decomposes as $L = UP$, where P is positive semidefinite and U is orthogonal if $\mathbb{F} = \mathbb{R}$ and unitary if $\mathbb{F} = \mathbb{C}$. The linear map P is unique, and U is unique if L is invertible. Correspondingly any n-by-n matrix A decomposes as $A = UP$, where P is a positive semidefinite matrix and U is an orthogonal matrix if $\mathbb{F} = \mathbb{R}$ and a unitary matrix if $\mathbb{F} = \mathbb{C}$. The matrix P is unique, and U is unique if A is invertible.

REMARKS. As we have already seen in other situations, the motivation for the proof comes from the uniqueness.

PROOF OF UNIQUENESS. Let $L = UP = U'P'$. Then $L^*L = P^2 = P'^2$. The linear map L^*L is positive semidefinite since its adjoint is $(L^*L)^* = L^*L^{**} = L^*L$ and since $(L^*L(v), v) = (L(v), L(v)) \geq 0$. Therefore Corollary 3.22c shows that L^*L has a unique positive semidefinite square root. Hence $P = P'$. If L is invertible, then P is invertible and $L = UP$ implies that $U = LP^{-1}$. The same argument applies in the case of matrices. □

PROOF OF EXISTENCE. If L is given, then we have just seen that L^*L is positive semidefinite. Let P be its unique positive semidefinite square root. The proof is clearer when L is invertible, and we consider that case first. Then we can set $U = LP^{-1}$. Since $U^* = (P^{-1})^*L^* = P^{-1}L^*$, we find that $U^*U = P^{-1}L^*LP^{-1} = P^{-1}P^2P^{-1} = I$, and we conclude that U is unitary.

When L is not necessarily invertible, we argue a little differently with the positive semidefinite square root P of L^*L. The kernel K of P is the 0 eigenspace of P, and the Spectral Theorem (Theorem 3.21) shows that the image of P is the sum of all the other eigenspaces and is just $K^\perp$. Since $K \cap K^\perp = 0$, P is one-one from $K^\perp$ onto itself. Thus $P(v) \mapsto L(v)$ is a one-one linear map from $K^\perp$ into V. Call this function U, so that $U(P(v)) = L(v)$. For any v_1 and v_2 in V, we have

$$(L(v_1), L(v_2)) = (L^*L(v_1), v_2) = (P^2(v_1), v_2) = (P(v_1), P(v_2)), \quad (*)$$

and hence $U : K^\perp \to V$ preserves inner products. Let $\{u_1, \ldots, u_k\}$ be an orthonormal basis of $K^\perp$, and let $\{u_{k+1}, \ldots, u_n\}$ be an orthonormal basis of K. Since U preserves inner products and is linear, $\{U(u_1), \ldots, U(u_k)\}$ is an orthonormal basis of $U(K^\perp)$. Extend $\{U(u_1), \ldots, U(u_k)\}$ to an orthonormal basis of V by adjoining vectors $v_{k+1}, \ldots, v_n$, define $U(u_j) = v_j$ for $k + 1 \leq$

$j \leq n$, and write U also for the linear extension to all of V. Since U carries one orthonormal basis $\{u_1, \ldots, u_n\}$ of V to another, U is unitary. We have $UP = L$ on $K^\perp$, and equation (∗) with $v_1 = v_2$ shows that $\ker L = \ker P = K$. Therefore $UP = L$ everywhere. □

4. Problems

1. Let $V = M_{nn}(\mathbb{C})$, and define an inner product on V by $\langle A, B \rangle = \text{Tr}(B^*A)$. The norm $\|\cdot\|_{\text{HS}}$ obtained from this inner product is called the **Hilbert–Schmidt norm** of the matrix in question.
    (a) Prove that $\|A\|_{\text{HS}}^2 = \sum_{i,j} |A_{ij}|^2$ for A in V.
    (b) Let E_{ij} be the matrix that is 1 in the $(i, j)^{\text{th}}$ entry and is 0 elsewhere. Prove that the set of all E_{ij} is an orthonormal basis of V.
    (c) Interpret (a) in the light of (b).
    (d) Prove that the Hilbert–Schmidt norm is given on any matrix A in V by
    $$\|A\|_{\text{HS}}^2 = \sum_j \|Au_j\|^2 = \sum_{i,j} |v_i^* Au_j|^2,$$
    where $\{u_1, \ldots, u_n\}$ and $\{v_1, \ldots, v_n\}$ are any orthonormal bases of $\mathbb{C}^n$ and v^* refers to the conjugate transpose of any member v of $\mathbb{C}^n$.
    (e) Let W be the vector subspace of all diagonal matrices in V. Describe explicitly the orthogonal complement $W^\perp$, and find its dimension.

2. Let V_n be the inner-product space over $\mathbb{R}$ of all polynomials on $[0, 1]$ of degree $\leq n$ with real coefficients. (The 0 polynomial is to be included.) The Riesz Representation Theorem says that there is a unique polynomial p_n such that $f(\frac{1}{2}) = \int_0^1 f(x)p_n(x)\,dx$ for all f in V_n. Set up a system of linear equations whose solution tells what p_n is.

3. Let V be a finite-dimensional inner-product space, and suppose that L and M are self-adjoint linear maps from V to V. Show that LM is self-adjoint if and only if $LM = ML$.

4. Let V be a finite-dimensional inner-product space. If $L : V \to V$ is a linear map with adjoint L^*, prove that $\ker L = (\text{image } L^*)^\perp$.

5. Find all 2-by-2 Hermitian matrices A with characteristic polynomial $\lambda^2 + 4\lambda + 6$.

6. Let V_1 and V_2 be finite-dimensional inner-product spaces over the same $\mathbb{F}$, the inner products being $(\cdot, \cdot)_1$ and $(\cdot, \cdot)_2$.
    (a) Using the case when $V_1 = V_2$ as a model, define the adjoint of a linear map $L : V_1 \to V_2$, proving its existence. The adjoint is to be a linear map $L^* : V_2 \to V_1$.

(b) If Γ is an orthonormal basis of V_1 and Δ is an orthonormal basis of V_2, prove that the matrices of L and L^* in these bases are conjugate transposes of one another.

7. Suppose that a finite-dimensional inner-product space V is a direct sum $V = S \oplus T$ of vector subspaces. Let $E : V \to V$ be the linear map that is the identity on S and is 0 on T.
    (a) Prove that $V = S^\perp \oplus T^\perp$.
    (b) Prove that $E^* : V \to V$ is the linear map that is the identity on $T^\perp$ and is 0 on $S^\perp$.

8. **(Iwasawa decomposition)** Let g be an invertible n-by-n complex matrix. Apply the Gram–Schmidt orthogonalization process to the basis $\{ge_1, \ldots, ge_n\}$, where $\{e_1, \ldots, e_n\}$ is the standard basis, and let the resulting orthonormal basis be $\{v_1, \ldots, v_n\}$. Define an invertible n-by-n matrix k such that $k^{-1}v_j = e_j$ for $1 \leq j \leq n$. Prove that $k^{-1}g$ is upper triangular with positive diagonal entries, and conclude that $g = k(k^{-1}g)$ exhibits g as the product of a unitary matrix and an upper triangular matrix whose diagonal entries are positive.

9. Let A be an n-by-n positive definite matrix.
    (a) Prove that $\det A > 0$.
    (b) Prove for any subset of integers $1 \leq i_1 < i_2 < \cdots < i_k \leq n$ that the submatrix of A built from rows and columns indexed by $(i_1, \ldots, i_k)$ is positive definite.

10. Prove that if A is a positive definite n-by-n matrix, then there exists an n-by-n upper-triangular matrix B with positive diagonal entries such that $A = B^*B$.

11. The most general 2-by-2 Hermitian matrix is of the form $A = \begin{pmatrix} a & b \\ \bar{b} & d \end{pmatrix}$ with a and d real and with b complex. Find a diagonal matrix D and a unitary matrix U such that $D = U^{-1}AU$.

12. In the previous problem,
    (a) what conditions on A make A positive definite?
    (b) when A is positive definite, how can its positive definite square root be computed explicitly?

13. Prove that if an n-by-n real symmetric matrix A has $v^t A v = 0$ for all v in $\mathbb{R}^n$, then $A = 0$.

14. Let $L : \mathbb{C}^n \to \mathbb{C}^n$ be a self-adjoint linear map. Show for each $x \in \mathbb{C}^n$ that there is some $y \in \mathbb{C}^n$ such that $(I - L)^2(y) = (I - L)(x)$.

15. In the polar decomposition $L = UP$, prove that if P and U commute, then L is normal.

16. Let V be an n-dimensional inner-product space over $\mathbb{R}$. What is the largest possible dimension of a commuting family of self-adjoint linear maps $L : V \to V$?

4. Problems

17. Let $v_1, \ldots, v_n$ be an ordered list of vectors in an inner-product space. The associated **Gram matrix** is the Hermitian matrix of inner products given by $G(v_1, \ldots, v_n) = [(v_i, v_j)]$, and $\det G(v_1, \ldots, v_n)$ is called its **Gram determinant**.

    (a) If $c_1, \ldots, c_n$ are in $\mathbb{C}$, let $c = \begin{pmatrix} c_1 \\ \vdots \\ c_n \end{pmatrix}$. Prove that $c^t G(v_1, \ldots, v_n) \bar{c} = \|c_1 v_1 + \cdots + c_n v_n\|^2$, and conclude that $G(v_1, \ldots, v_n)$ is positive semidefinite.

    (b) Prove that $\det G(v_1, \ldots, v_n) \geq 0$ with equality if and only if $v_1, \ldots, v_n$ are linearly dependent. (This generalizes the Schwarz inequality.)

    (c) Under what circumstances does equality hold in the Schwarz inequality?

Problems 18–23 introduce the **Legendre polynomials** and establish some of their elementary properties, including their orthogonality under the inner product $\langle P, Q \rangle = \int_{-1}^{1} P(x) Q(x) \, dx$. They form the simplest family of classical orthogonal polynomials. They are uniquely determined by the conditions that the nth one P_n, for $n \geq 0$, is of degree n, they are orthogonal under $\langle \cdot, \cdot \rangle$, and they are normalized so that $P_n(1) = 1$. But these conditions are a little hard to work with initially, and instead we adopt the recursive definition $P_0(x) = 1$, $P_1(x) = x$, and

$$(n+1) P_{n+1}(x) = (2n+1) x P_n(x) - n P_{n-1}(x) \quad \text{for } n \geq 1.$$

18. (a) Prove that $P_n(x)$ has degree n, that $P_n(-x) = (-1)^n P_n(x)$, and that $P_n(1) = 1$. In particular, P_n is an even function if n is even and is an odd function if n is odd.

    (b) Let $c^{(n)}$ be the constant term of P_n if n is even and the coefficient of x if n is odd, so that $c^{(0)} = c^{(1)} = 1$. Prove that $c^{(n)} = -\frac{n-1}{n} c^{(n-2)}$ for $n \geq 2$.

19. This part establishes a useful concrete formula for $P_n(x)$. Let $D = d/dx$ and $X = x^2 - 1$, writing $X' = 2x$, $X'' = 2$, and $X''' = 0$ for the derivatives. Two parts of this problem make use of the Leibniz rule $D^n(fg) = \sum_{k=0}^{n} \binom{n}{k} (D^{n-k} f)(D^k g)$ for higher-order derivatives of a product.

    (a) Verify that $D^2(X^{n+1}) = (2n+1) D(X^n X') - n(2n+1) X'' X^n - 4n^2 X^{n-1}$.

    (b) By applying D^{n-1} to the result of (a) and rearranging terms, show that $D^{n+1}(X^{n+1}) = (2n+1) X' D^n(X^n) - 4n^2 D^{n-1}(X^{n-1})$.

    (c) Put $R_n(x) = (2^n n!)^{-1} D^n(X^n)$ for $n \geq 0$. Show that $R_0(x) = 1$, $R_1(x) = x$, and $(n+1) R_{n+1}(x) = (2n+1) x R_n(x) - n R_{n-1}(x)$ for $n \geq 1$.

    (d) **(Rodrigues's formula)** Conclude that $2^n n! P_n(x) = \left(\frac{d}{dx}\right)^n [(x^2 - 1)^n]$.

20. Using Rodrigues's formula and iterated integration by parts, prove that

$$\int_{-1}^{1} P_m(x) P_n(x) \, dx = 0 \quad \text{for } m < n.$$

Conclude that $\{P_0, P_1, \ldots, P_n\}$ is an orthogonal basis of the inner-product space of polynomials on $[-1, 1]$ of degree $\leq n$ with inner product $\langle \cdot, \cdot \rangle$.

21. Arguing as in the previous problem and taking for granted that $\int_{-1}^{1} (1-x^2)^n \, dx = \frac{2(2^n n!)^2}{(2n+1)!}$, prove that $\langle P_n, P_n \rangle = \left(n + \frac{1}{2}\right)^{-1}$.

22. This problem shows that $P_n(x)$ satisfies a certain second-order differential equation. Let $D = d/dx$. The first two parts of this problem use the Leibniz rule quoted in Problem 19. Let $X = x^2 - 1$ and $K_n = 2^n n!$, so that Rodrigues's formula says that $K_n P_n = D^n(X^n)$.
    (a) Expand $D^{n+1}[(D(X^n))X]$ by the Leibniz rule.
    (b) Observe that $(D(X^n))X = nX^n X'$, and expand $D^{n+1}[(nX^n)X']$ by the Leibniz rule.
    (c) Equating the results of the previous two parts, conclude that $y = P_n(x)$ satisfies the differential equation $(1 - x^2)y'' - 2xy' + n(n+1)y = 0$.

23. Let $P_n(x) = \sum_{k=0}^{n} c_k x^k$. Using the differential equation, show that the coefficients c_k satisfy $k(k-1)c_k = [(k-2)(k-1) - n(n+1)]c_{k-2}$ for $k \geq 2$ and that $c_k = 0$ unless $n - k$ is even.

Problems 24–28 concern the complex conjugate of an inner-product space over $\mathbb{C}$. For any finite-dimensional inner-product space V, the Riesz Representation Theorem identifies the dual V' with V, saying that each member of V' is given by taking the inner product with some member of V. When the scalars are real, this identification is linear; thus the Riesz theorem uses the inner product to construct a canonical isomorphism of V onto V'. When the scalars are complex, the identification is conjugate linear, and we do not get an isomorphism of V with V'. The **complex conjugate** of V provides a substitute result.

24. Let V be a finite-dimensional vector space over $\mathbb{C}$. Define a new complex vector space $\overline{V}$ as follows: The elements of $\overline{V}$ are the elements of V, and the definition of addition is unchanged. However, there is a change in the definition of scalar multiplication, in that if v is in V, then the product cv in $\overline{V}$ is to equal the product $\bar{c}v$ in V. Verify that $\overline{V}$ is indeed a complex vector space.

25. If V is a complex vector space and $L : V \to V$ is a linear map, define $\overline{L} : \overline{V} \to \overline{V}$ to be the same function as L. Prove that $\overline{L}$ is linear.

26. Suppose that the complex vector space V is actually a finite-dimensional inner-product space, with inner product $(\cdot, \cdot)_V$. Define $(u, v)_{\overline{V}} = (v, u)_V$. Verify that $\overline{V}$ is an inner-product space.

27. With V as in the previous problem, show that the Riesz Representation Theorem uses the inner product to set up a canonical isomorphism of V' with $\overline{V}$.

28. With V and $\overline{V}$ as in the two previous problems, let $L : V \to V$ be linear, so that $(\overline{L})^* : \overline{V} \to \overline{V}$ is linear. Under the identification of the previous problem of $\overline{V}$ with V', show that $(\overline{L})^*$ corresponds to the contragredient L^t as defined in Section II.4.

Problems 29–32 use inner-product spaces to obtain a decomposition of polynomials in several variables. A real-valued polynomial function p in $x_1, \ldots, x_n$ is said to be **homogeneous** of degree N if every monomial in p has total degree N. Let V_N be the space of real-valued polynomials in $x_1, \ldots, x_n$ homogeneous of degree N. For any homogeneous polynomial p, we define a differential operator $\partial(p)$ with constant coefficients by requiring that $\partial(\,\cdot\,)$ be linear in $(\,\cdot\,)$ and that

$$\partial(x_1^{k_1} \cdots x_n^{k_n}) = \frac{\partial^{k_1 + \cdots + k_n}}{\partial x_1^{k_1} \cdots \partial x_n^{k_n}}.$$

For example, if $|x|^2$ stands for $x_1^2 + \cdots + x_n^2$, then $\partial(|x|^2) = \Delta = \frac{\partial^2}{\partial x_1^2} + \cdots + \frac{\partial^2}{\partial x_n^2}$. If p and q are in the same V_N, then $\partial(q)p$ is a constant polynomial, and we define $\langle p, q \rangle$ to be that constant. Then $\langle \,\cdot\, , \,\cdot\, \rangle$ is bilinear.

29. (a) Prove that $\langle \,\cdot\, , \,\cdot\, \rangle$ satisfies $\langle p, q \rangle - \langle q, p \rangle$.
    (b) Prove that $\langle x_1^{k_1} \cdots x_n^{k_n}, x_1^{l_1} \cdots x_n^{l_n} \rangle$ is positive if $(k_1, \ldots, k_n) = (l_1, \ldots, l_n)$ and is 0 otherwise.
    (c) Deduce that $\langle \,\cdot\, , \,\cdot\, \rangle$ is an inner product on V_N.

30. Call $p \in V_N$ **harmonic** if $\partial(|x|^2)p = 0$, and let H_N be the vector subspace of harmonic polynomials. Prove that the orthogonal complement of $|x|^2 V_{N-2}$ in V_N relative to $\langle \,\cdot\, , \,\cdot\, \rangle$ is H_N.

31. Deduce from Problem 30 that each $p \in V_N$ decomposes uniquely as

$$p = h_N + |x|^2 h_{N-2} + |x|^4 h_{N-4} + \cdots$$

with $h_N, h_{N-2}, h_{N-4}, \ldots$ homogeneous harmonic of the indicated degrees.

32. For $n = 2$, describe a computational procedure for decomposing the element $x_1^4 + x_2^4$ of V_4 as in Problem 31.

Problems 33–34 concern products of n-by-n positive semidefinite matrices. They make use of Problem 26 in Chapter II, which says that $\det(\lambda I - CD) = \det(\lambda I - DC)$.

33. Let A and B be positive semidefinite. Using the positive definite square root of B, prove that every eigenvalue of AB is ≥ 0.

34. Let A, B, and C be positive semidefinite, and suppose that ABC is Hermitian. Under the assumption that C is invertible, introduce the positive definite square root P of C. By considering $P^{-1}ABCP^{-1}$, prove that ABC is positive semidefinite.

CHAPTER IV

Groups and Group Actions

Abstract. This chapter develops the basics of group theory, with particular attention to the role of group actions of various kinds. The emphasis is on groups in Sections 1–3 and on group actions starting in Section 6. In between is a two-section digression that introduces rings, fields, vector spaces over general fields, and polynomial rings over commutative rings with identity.

Section 1 introduces groups and a number of examples, and it establishes some easy results. Most of the examples arise either from number-theoretic settings or from geometric situations in which some auxiliary space plays a role. The direct product of two groups is discussed briefly so that it can be used in a table of some groups of low order.

Section 2 defines coset spaces, normal subgroups, homomorphisms, quotient groups, and quotient mappings. Lagrange's Theorem is a simple but key result. Another simple but key result is the construction of a homomorphism with domain a quotient group G/H when a given homomorphism is trivial on H. The section concludes with two standard isomorphism theorems.

Section 3 introduces general direct products of groups and direct sums of abelian groups, together with their concrete "external" versions and their universal mapping properties.

Sections 4–5 are a digression to define rings, fields, and ring homomorphisms, and to extend the theories concerning polynomials and vector spaces as presented in Chapters I–II. The immediate purpose of the digression is to make prime fields and the notion of characteristic available for the remainder of the chapter. The definitions of polynomials are extended to allow coefficients from any commutative ring with identity and to allow more than one indeterminate, and universal mapping properties for polynomial rings are proved.

Sections 6–7 introduce group actions. Section 6 gives some geometric examples beyond those in Section 1, it establishes a counting formula concerning orbits and isotropy subgroups, and it develops some structure theory of groups by examining specific group actions on the group and its coset spaces. Section 7 uses a group action by automorphisms to define the semidirect product of two groups. This construction, in combination with results from Sections 5–6, allows one to form several new finite groups of interest.

Section 8 defines simple groups, proves that alternating groups on five or more letters are simple, and then establishes the Jordan–Hölder Theorem concerning the consecutive quotients that arise from composition series.

Section 9 deals with finitely generated abelian groups. It is proved that "rank" is well defined for any finitely generated free abelian group, that a subgroup of a free abelian group of finite rank is always free abelian, and that any finitely generated abelian group is the direct sum of cyclic groups.

Section 10 returns to structure theory for finite groups. It begins with the Sylow Theorems, which produce subgroups of prime-power order, and it gives two sample applications. One of these classifies the groups of order pq, where p and q are distinct primes, and the other provides the information necessary to classify the groups of order 12.

Section 11 introduces the language of "categories" and "functors." The notion of category is a precise version of what is sometimes called a "context" at points in the book before this section,

1. Groups and Subgroups

Linear algebra and group theory are two foundational subjects for all of algebra, indeed for much of mathematics. Chapters II and III have introduced the basics of linear algebra, and the present chapter introduces the basics of group theory. In this section we give the definition and notation for groups and provide examples that fit with the historical development of the notion of group. Many readers will already be familiar with some group theory, and therefore we can be brief at the start.

A **group** is a nonempty set G with an operation $G \times G \to G$ satisfying the three properties (i), (ii), and (iii) below. In the absence of any other information the operation is usually called **multiplication** and is written $(a, b) \mapsto ab$ with no symbol to indicate the multiplication. The defining properties of a group are

(i) $(ab)c = a(bc)$ for all a, b, c in G (**associative** law),
(ii) there exists an element 1 in G such that $a1 = 1a = a$ for all a in G (existence of **identity**),
(iii) for each a in G, there exists an element a^{-1} in G with $aa^{-1} = a^{-1}a = 1$ (existence of **inverses**).

It is immediate from these properties that

- 1 is unique (since $1' = 1'1 = 1$),
- a^{-1} is unique (since $(a^{-1})' = (a^{-1})'1 = (a^{-1})'(a(a^{-1})) = ((a^{-1})'a)(a^{-1}) = 1(a^{-1}) = (a^{-1})$),
- the existence of a left inverse for each element implies the existence of a right inverse for each element (since $ba = 1$ and $cb = 1$ together imply $c = c(ba) = (cb)a = a$ and hence also $ab = cb = 1$),
- 1 is its own inverse (since $11 = 1$),
- $ax = ay$ implies $x = y$, and $xa = ya$ implies $x = y$ (**cancellation** laws) (since $x = 1x = (a^{-1}a)x = a^{-1}(ax) = a^{-1}(ay) = (a^{-1}a)y = 1y = y$ and since a similar argument proves the second implication).

Problem 2 at the end of Chapter II shows that the associative law extends to products of any finite number of elements of G as follows: parentheses can be inserted in any fashion in such a product, and the value of the product is unchanged; hence any expression $a_1 a_2 \cdots a_n$ in G is well defined without the use of parentheses.

The group whose only element is the identity 1 will be denoted by $\{1\}$. It is called the **trivial group**.

We come to other examples in a moment. First we make three more definitions and offer some comments. A **subgroup** H of a group G is a subset containing the identity that is closed under multiplication and inverses. Then H itself is a group because the associativity in G implies associativity in H. The intersection of any nonempty collection of subgroups of G is again a subgroup.

An **isomorphism** of a group G_1 with a group G_2 is a function $\varphi : G_1 \to G_2$ that is one-one onto and satisfies $\varphi(ab) = \varphi(a)\varphi(b)$ for all a and b in G_1. It is immediate that

- $\varphi(1) = 1$ (by taking $a = b = 1$),
- $\varphi(a^{-1}) = \varphi(a)^{-1}$ (by taking $b = a^{-1}$),
- $\varphi^{-1} : G_2 \to G_1$ satisfies $\varphi^{-1}(cd) = \varphi^{-1}(c)\varphi^{-1}(d)$ (by taking $c = \varphi(a)$ and $d = \varphi(b)$ on the right side and then observing that $\varphi(\varphi^{-1}(c)\varphi^{-1}(d)) = \varphi(ab) = \varphi(a)\varphi(b) = cd = \varphi(\varphi^{-1}(cd)))$.

The first and second of these properties show that an isomorphism respects all the structure of a group, not just products. The third property shows that the inverse of an isomorphism is an isomorphism, hence that the relation "is isomorphic to" is symmetric. Since the identity isomorphism exhibits this relation as reflexive and since the use of compositions shows that it is transitive, we see that "is isomorphic to" is an equivalence relation. Common notation for an isomorphism between G_1 and G_2 is $G_1 \cong G_2$; because of the symmetry, one can say that G_1 and G_2 are **isomorphic**.

An **abelian group** is a group G with the additional property

(iv) $ab = ba$ for all a and b in G (**commutative** law).

In an abelian group the operation is sometimes, but by no means always, called **addition** instead of "multiplication." Addition is typically written $(a, b) \mapsto a+b$, and then the identity is usually denoted by 0 and the inverse of a is denoted by $-a$, the **negative** of a. Depending on circumstances, the trivial abelian group may be denoted by $\{0\}$ or 0. Problem 3 at the end of Chapter II shows for an abelian group G with its operation written additively that n-fold sums of elements of G can be written in any order: $a_1 + a_2 + \cdots + a_n = a_{\sigma(1)} + a_{\sigma(2)} + \cdots + a_{\sigma(n)}$ for each permutation σ of $\{1, \ldots, n\}$.

Historically the original examples of groups arose from two distinct sources, and it took a while for the above definition of group to be distilled out as the essence of the matter.

One of the two sources involved number systems and vectors. Here are examples.

EXAMPLES.

(1) Additive groups of familiar number systems. The systems in question are the integers $\mathbb{Z}$, the rational numbers $\mathbb{Q}$, the real numbers $\mathbb{R}$, and the complex

1. Groups and Subgroups

numbers $\mathbb{C}$. In each case the set with its usual operation of addition forms an abelian group. The group properties of $\mathbb{Z}$ under addition are taken as known in advance in this book, as mentioned in Section A3 of the appendix, and the group properties of $\mathbb{Q}$, $\mathbb{R}$, and $\mathbb{C}$ under addition are sketched in Sections A3 and A4 of the appendix as part of the development of these number systems.

(2) Multiplicative groups connected with familiar number systems. In the cases of $\mathbb{Q}$, $\mathbb{R}$, and $\mathbb{C}$, the nonzero elements form a group under multiplication. These groups are denoted by $\mathbb{Q}^\times$, $\mathbb{R}^\times$, and $\mathbb{C}^\times$. Again the properties of a group for each of them are properties that are sketched during the development of each of these number systems in Sections A3 and A4 of the appendix. With $\mathbb{Z}$, the nonzero integers do not form a group under multiplication, because only the two units, i.e., the divisors $+1$ and -1 of 1, have inverses. The units do form a group, however, under multiplication, and the group of units is denoted by $\mathbb{Z}^\times$.

(3) Vector spaces under addition. Spaces such as $\mathbb{Q}^n$ and $\mathbb{R}^n$ and $\mathbb{C}^n$ provide us with further examples of abelian groups. In fact, the defining properties of addition in a vector space are exactly the defining properties of an abelian group. Thus every vector space provides us with an example of an abelian group if we simply ignore the scalar multiplication.

(4) Integers modulo m, under addition. Another example related to number systems is the additive group of integers modulo a positive integer m. Let us say that an integer n_1 is **congruent modulo** m to an integer n_2 if m divides $n_1 - n_2$. One writes $n_1 \equiv n_2$ or $n_1 \equiv n_2 \bmod m$ or $n_1 = n_2 \bmod m$ for this relation.[1] It is an equivalence relation, and we can write $[n]$ for the equivalence class of n when it is helpful to do so. The division algorithm (Proposition 1.1) tells us that each equivalence class has one and only one member between 0 and $m-1$. Thus there are exactly m equivalence classes, and we know a representative of each. The set of classes will be denoted by[2] $\mathbb{Z}/m\mathbb{Z}$. The point is that $\mathbb{Z}/m\mathbb{Z}$ inherits an abelian-group structure from the abelian-group structure of $\mathbb{Z}$. Namely, we attempt to define

$$[a] + [b] = [a+b].$$

To see that this formula actually defines an operation on $\mathbb{Z}/m\mathbb{Z}$, we need to check that the result is meaningful if the representatives of the classes $[a]$ and $[b]$ are changed. Thus let $[a] = [a']$ and $[b] = [b']$. Then m divides $a - a'$ and $b - b'$, and m must divide the sum $(a-a') + (b-b') = (a+b) - (a'+b')$; consequently $[a+b] = [a'+b']$, and addition is well defined. The same kind of

---

[1] This notation was anticipated in a remark explaining the classical form of the Chinese Remainder Theorem (Corollary 1.9).

[2] The notation $\mathbb{Z}/(m)$ is an allowable alternative. Some authors, particularly in topology, write $\mathbb{Z}_m$ for this set, but the notation $\mathbb{Z}_m$ can cause confusion since $\mathbb{Z}_p$ is the standard notation for the "p-adic integers" when p is prime. These are defined in *Advanced Algebra*.

argument shows that the associativity and commutativity of addition in $\mathbb{Z}$ imply associativity and commutativity in $\mathbb{Z}/m\mathbb{Z}$. The identity element is $[0]$, and group inverses (negatives) are given by $-[a] = [-a]$. Therefore $\mathbb{Z}/m\mathbb{Z}$ is an abelian group under addition, and it has m elements. If x and y are members of $\mathbb{Z}/m\mathbb{Z}$, their sum is often denoted by $x + y \bmod m$.

The other source of early examples of groups historically has the members of the group operating as transformations of some auxiliary space. Before abstracting matters, let us consider some concrete examples, ignoring some of the details of verifying the defining properties of a group.

EXAMPLES, CONTINUED.

(5) Permutations. A **permutation** of a nonempty finite set E of n elements is a one-one function from E onto itself. Permutations were introduced in Section I.4. The product of two permutations is just the composition, defined by $(\sigma\tau)(x) = \sigma(\tau(x))$ for x in E, with the symbol $\circ$ for composition dropped. The resulting operation makes the set of permutations of E into a group: we already observed in Section I.4 that composition is associative, and it is plain that the identity permutation may be taken as the group identity and that the inverse function to a permutation is the group inverse. The group is called the **symmetric group** on the n **letters** of E. It has $n!$ members for $n \geq 1$. The notation $\mathfrak{S}_n$ is often used for this group, especially when $E = \{1, \ldots, n\}$. Signs ± 1 were defined for permutations in Section I.4, and we say that a permutation is **even** or **odd** according as its sign is $+1$ or -1. The sign of a product is the product of the signs, according to Proposition 1.24, and it follows that the even permutations form a subgroup of $\mathfrak{S}_n$. This subgroup is called the **alternating group** on n letters and is denoted by $\mathfrak{A}_n$. It has $\frac{1}{2}(n!)$ members if $n \geq 2$.

(6) Symmetries of a regular polygon. Imagine a regular polygon in $\mathbb{R}^2$ centered at the origin. The plane-geometry rotations and reflections about the origin that carry the polygon to itself form a group. If the number of sides of the polygon is n, then the group always contains the rotations through all multiples of the angle $2\pi/n$. The rotations themselves form an n-element subgroup of the group of all symmetries. To consider what reflections give symmetries, we distinguish the cases n odd and n even. When n is odd, the reflection in the line that passes through any vertex and bisects the opposite side carries the polygon to itself, and no other reflections have this property. Thus the group of symmetries contains n reflections. When n is even, the reflection in the line passing through any vertex and the opposite vertex carries the polygon to itself, and so does the reflection in the line that bisects a side and also the opposite side. There are $n/2$ reflections of each kind, and hence the group of symmetries again contains n reflections. The group of symmetries thus has $2n$ elements in all cases. It is called the **dihedral**

group D_n. The group D_n is isomorphic to a certain subgroup of the permutation group $\mathfrak{S}_n$. Namely, we number the vertices of the polygon, and we associate to each member of D_n the permutation that moves the vertices the way the member of D_n does.

(7) General linear group. With $\mathbb{F}$ equal to $\mathbb{Q}$ or $\mathbb{R}$ or $\mathbb{C}$, consider any n-dimensional vector space V over $\mathbb{F}$. One possibility is $V = \mathbb{F}^n$, but we do not insist on this choice. Among all one-one functions carrying V onto itself, let G consist of the linear ones. The composition of two linear maps is linear, and the inverse of an invertible function is linear if the given function is linear. The result is a group known as the **general linear group** $GL(V)$. When $V = \mathbb{F}^n$, we know from Chapter II that we can identify linear maps from $\mathbb{F}^n$ to itself with matrices in $M_{nn}(\mathbb{F})$ and that composition corresponds to matrix multiplication. It follows that the set of all invertible matrices in $M_{nn}(\mathbb{F})$ is a group, which is denoted by $GL(n, \mathbb{F})$, and that this group is isomorphic to $GL(\mathbb{F}^n)$. The set $SL(V)$ or $SL(n, \mathbb{F})$ of all members of $GL(V)$ or $GL(n, \mathbb{F})$ of determinant 1 is a group since the determinant of a product is the product of the determinants; it is called the **special linear group**. The dihedral group D_n is isomorphic to a subgroup of $GL(2, \mathbb{R})$ since each rotation and reflection of $\mathbb{R}^2$ that fixes the origin is given by the operation of a 2-by-2 matrix.

(8) Orthogonal and unitary groups. If V is a finite-dimensional inner-product space over $\mathbb{R}$ or $\mathbb{C}$, Chapter III referred to the linear maps carrying the space to itself and preserving lengths of vectors as **orthogonal** in the real case and **unitary** in the complex case. Such linear maps are invertible. The condition of preserving lengths of vectors is maintained under composition and inverses, and it follows that the orthogonal or unitary linear maps form a subgroup $O(V)$ or $U(V)$ of the general linear group $GL(V)$. One writes $O(n)$ for $O(\mathbb{R}^n)$ and $U(n)$ for $U(\mathbb{C}^n)$. The subgroup of members of $O(V)$ or $O(n)$ of determinant 1 is called the **rotation group** $SO(V)$ or $SO(n)$. The subgroup of members of $U(V)$ or $U(n)$ of determinant 1 is called the **special unitary group** $SU(V)$ or $SU(n)$.

Before coming to Example 9, let us establish a closure property under the arithmetic operations for certain subsets of $\mathbb{C}$. We are going to use the theories of polynomials as in Chapter I and of vector spaces as in Chapter II with the rationals $\mathbb{Q}$ as the scalars. Fix a complex number θ, and form the result of evaluating at θ every polynomial in one indeterminate with coefficients in $\mathbb{Q}$. The resulting set of complex numbers comes by substituting θ for X in the members of $\mathbb{Q}[X]$, and we denote this subset of $\mathbb{C}$ by $\mathbb{Q}[\theta]$.

Suppose that θ has the property that the set $\{1, \theta, \theta^2, \ldots, \theta^n\}$ is linearly dependent over $\mathbb{Q}$ for some integer $n \geq 1$, i.e., has the property that $F_0(\theta) = 0$ for some nonzero member F_0 of $\mathbb{Q}[X]$ of degree $\leq n$. For example, if $\theta = \sqrt{2}$, then the set $\{1, \sqrt{2}, (\sqrt{2})^2\}$ is linearly dependent since $2 - (\sqrt{2})^2 = 0$; if $\theta = e^{2\pi i/5}$, then $\{1, \theta, \theta^2, \theta^3, \theta^4, \theta^5\}$ is linearly dependent since $1 - \theta^5 = 0$, or alternatively since $1 + \theta + \theta^2 + \theta^3 + \theta^4 = 0$.

Returning to the general θ, we lose no generality if we assume that the polynomial F_0 has degree exactly n. If we divide the equation $F_0(\theta) = 0$ by the leading coefficient, we obtain an equality $\theta^n = G_0(\theta)$, where G_0 is the zero polynomial or is a nonzero polynomial of degree at most $n - 1$. Then $\theta^{n+m} = \theta^m G_0(\theta)$, and we see inductively that every power θ^r with $r \geq n$ is a linear combination of the members of the set $\{1, \theta, \theta^2, \ldots, \theta^{n-1}\}$. This set is therefore a spanning set for the vector space $\mathbb{Q}[\theta]$, and we find that $\mathbb{Q}[\theta]$ is finite-dimensional, with dimension at most n. Since every positive integer power of θ lies in $\mathbb{Q}[\theta]$ and since these powers are closed under multiplication, the vector space $\mathbb{Q}[\theta]$ is closed under multiplication. More striking is that $\mathbb{Q}[\theta]$ is closed under division, as is asserted in the following proposition.

Proposition 4.1. Let θ be in $\mathbb{C}$, and suppose for some integer $n \geq 1$ that the set $\{1, \theta, \theta^2, \ldots, \theta^n\}$ is linearly dependent over $\mathbb{Q}$. Then the finite-dimensional rational vector space $\mathbb{Q}[\theta]$ is closed under taking reciprocals (of nonzero elements), as well as multiplication, and hence is closed under division.

REMARKS. Under the hypotheses of Proposition 4.1, $\mathbb{Q}[\theta]$ is called an **algebraic number field**,[3] or simply a **number field**, and θ is called an **algebraic number**. The relevant properties of $\mathbb{C}$ that are used in proving the proposition are that $\mathbb{C}$ is closed under the usual arithmetic operations, that these satisfy the usual properties, and that $\mathbb{Q}$ is a subset of $\mathbb{C}$. The deeper closure properties of $\mathbb{C}$ that are developed in Sections A3 and A4 of the appendix play no role.

PROOF. We have seen that $\mathbb{Q}[\theta]$ is closed under multiplication. If x is a nonzero member of $\mathbb{Q}[\theta]$, then all positive powers of x must be in $\mathbb{Q}[\theta]$, and the fact that $\dim \mathbb{Q}[\theta] \leq n$ forces $\{1, x, x^2, \ldots, x^n\}$ to be linearly dependent. Therefore there are integers j and k with $0 \leq j < k \leq n$ such that $c_j x^j + c_{j+1} x^{j+1} + \cdots + c_k x^k = 0$ for some rational numbers $c_j, \ldots, c_k$ with $c_k \neq 0$. Since x is assumed nonzero, we can discard unnecessary terms and arrange that $c_j \neq 0$. Then

$$1 = x(-c_j^{-1} c_{j+1} - c_j^{-1} c_{j+2} x - c_j^{-1} c_k x^{k-j-1}),$$

and the reciprocal of x has been exhibited as in $\mathbb{Q}[\theta]$. □

EXAMPLES, CONTINUED.

(9) Galois's notion of automorphisms of number fields. Let θ be a complex number as in Proposition 4.1. The subject of Galois theory, whose details will

---

[3] The definition of "algebraic number field" that is given later in the book is ostensibly more general, but the Theorem of the Primitive Element in Chapter IX will show that it amounts to the same thing as this.

be discussed in Chapter IX and whose full utility will be glimpsed only later, works in an important special case with the "automorphisms" of $\mathbb{Q}[\theta]$ that fix $\mathbb{Q}$. The automorphisms are the one-one functions from $\mathbb{Q}[\theta]$ onto itself that respect addition and multiplication and carry every element of $\mathbb{Q}$ to itself. The identity is such a function, the composition of two such functions is again one, and the inverse of such a function is again one. Therefore the automorphisms of $\mathbb{Q}[\theta]$ form a group under composition. We call this group $\mathrm{Gal}(\mathbb{Q}[\theta]/\mathbb{Q})$. Let us see that it is finite. In fact, if σ is in $\mathrm{Gal}(\mathbb{Q}[\theta]/\mathbb{Q})$, then σ is determined by its effect on θ, since we must have $\sigma(F(\theta)) = F(\sigma(\theta))$ for every F in $\mathbb{Q}[X]$. We know that there is some nonzero polynomial $F_0(X)$ such that $F_0(\theta) = 0$. Applying σ to this equality, we see that $F_0(\sigma(\theta)) = 0$. Therefore $\sigma(\theta)$ has to be a root of F_0. Viewing F_0 as in $\mathbb{C}[X]$, we can apply Corollary 1.14 and see that F_0 has only finitely many complex roots. Therefore there are only finitely many possibilities for σ, and the group $\mathrm{Gal}(\mathbb{Q}[\theta]/\mathbb{Q})$ has to be finite. Galois theory shows that this group gives considerable insight into the structure of $\mathbb{Q}[\theta]$. For example it allows one to derive the Fundamental Theorem of Algebra (Theorem 1.18) just from algebra and the Intermediate Value Theorem (Section A3 of the appendix); it allows one to show the impossibility of certain constructions in plane geometry by straightedge and compass; and it allows one to show that a quintic polynomial with rational coefficients need not have a root that is expressible in terms of rational numbers, arithmetic operations, and the extraction of square roots, cube roots, and so on. We return to these matters in Chapter IX.

Examples 5–9, which all involve auxiliary spaces, fit the pattern that the members of the group are invertible transformations of the auxiliary space and the group operation is composition. This notion will be abstracted in Section 6 and will lead to the notion of a "group action." For now, let us see why we obtained groups in each case. If X is any nonempty set, then the set of invertible functions $f : X \to X$ forms a group under composition, composition being defined by $(fg)(x) = f(g(x))$ with the usual symbol $\circ$ dropped. The associative law is just a matter of unwinding this definition:

$$((fg)h)(x) = (fg)(h(x)) = f(g(h(x))) = f((gh)(x)) = (f(gh))(x).$$

The identity function is the identity of the group, and inverse functions provide the inverse elements in the group.

For our examples, the set X was E in Example 5, $\mathbb{R}^2$ in Example 6, V or $\mathbb{F}^n$ in Example 7, V or $\mathbb{Q}^n$ or $\mathbb{R}^n$ or $\mathbb{C}^n$ in Example 8, and $\mathbb{Q}[\theta]$ in Example 9. All that was needed in each case was to know that our set G of invertible functions from X to itself formed a subgroup of the set of all invertible functions from X to itself. In other words, we had only to check that G contained the identity and was closed under composition and inversion. Associativity was automatic for G because it was valid for the group of all invertible functions from X to itself.

Actually, any group can be realized in the fashion of Examples 5–9. This is the content of the next proposition.

Proposition 4.2 (Cayley's Theorem). Any group G is isomorphic to a subgroup of invertible functions on a set X. The set X can be taken to be G itself. In particular any finite group with n elements is isomorphic to a subgroup of the symmetric group $\mathfrak{S}_n$.

PROOF. Define $X = G$, put $f_a(x) = ax$ for a in G, and let $G' = \{f_a \mid a \in G\}$. To see that G' is a group, we need G' to contain the identity and to be closed under composition and inverses. Since f_1 is the identity, the identity is indeed in G'. Since $f_{ab}(x) = (ab)x = a(bx) = f_a(bx) = f_a(f_b(x)) = (f_a f_b)(x)$, G' is closed under composition. The formula $f_a f_{a^{-1}} = f_1 = f_{a^{-1}} f_a$ then shows that $f_{a^{-1}} = (f_a)^{-1}$ and that G' is closed under inverses. Thus G' is a group.

Define $\varphi : G \to G'$ by $\varphi(a) = f_a$. Certainly φ is onto G', and it is one-one because $\varphi(a) = \varphi(b)$ implies $f_a = f_b$, $f_a(1) = f_b(1)$, and $a = b$. Also, $\varphi(ab) = f_{ab} = f_a f_b = \varphi(a)\varphi(b)$, and hence φ is an isomorphism.

In the case that G is finite with n elements, G is exhibited as isomorphic to a subgroup of the group of permutations of the members of G. Hence it is isomorphic to a subgroup of $\mathfrak{S}_n$. □

It took the better part of a century for mathematicians to sort out that two distinct notions are involved here—that of a group, as defined above, and that of a group action, as will be defined in Section 6. In sorting out these matters, mathematicians realized that it is wise to study the abstract group first and then to study the group in the context of its possible group actions. This does not at all mean ignoring group actions until after the study of groups is complete; indeed, we shall see in Sections 6, 7, and 10 that group actions provide useful tools for the study of abstract groups.

We turn to a discussion of two general group-theoretic notions—cyclic group and the direct product of two or more groups. The second of these notions will be discussed only briefly now; more detail will come in Section 3.

If a is an element of a group, we define a^n for integers $n > 0$ inductively by $a^1 = a$ and $a^n = a^{n-1}a$. Then we can put $a^0 = 1$ and $a^{-n} = (a^{-1})^n$ for $n > 0$. A little checking, which we omit, shows that the ordinary rules of exponents apply: $a^{m+n} = a^m a^n$ and $a^{mn} = (a^m)^n$ for all integers m and n. If the underlying group is abelian and additive notation is being used, these formulas read $(m+n)a = ma + na$ and $(mn)a = n(ma)$.

A **cyclic group** is a group with an element a such that every element is a power of a. The element a is called a **generator** of the group, and the group is said to be **generated** by a.

Proposition 4.3. Each cyclic group G is isomorphic either to the additive group $\mathbb{Z}$ of integers or to the additive group $\mathbb{Z}/m\mathbb{Z}$ of integers modulo m for some positive integer m.

PROOF. If all a^n are distinct, then the rule $a^{m+n} = a^m a^n$ implies that the function $n \mapsto a^n$ is an isomorphism of $\mathbb{Z}$ with G. On the other hand, if $a^k = a^l$ with $k > l$, then $a^{k-l} = 1$ and there exists a positive integer n such that $a^n = 1$. Let m be the least positive integer with $a^m = 1$. For any integers q and r, we have $a^{qm+r} = (a^m)^q a^r = a^r$. Thus the function $\varphi : \mathbb{Z}/m\mathbb{Z} \to G$ given by $\varphi([n]) = a^n$ is well defined, is onto G, and carries sums in $\mathbb{Z}/m\mathbb{Z}$ to products in G. If $0 \le l < k < m$, then $a^k \ne a^l$ since otherwise a^{k-l} would be 1. Hence φ is one-one, and we conclude that $\varphi : \mathbb{Z}/m\mathbb{Z} \to G$ is an isomorphism. □

Let us denote abstract cyclic groups by C_∞ and C_m, the subscript indicating the number of elements. Finite cyclic groups arise in guises other than as $\mathbb{Z}/m\mathbb{Z}$. For example the set of all elements $e^{2\pi i k/m}$ in $\mathbb{C}$, with multiplication as operation, forms a group isomorphic to C_m. So does the set of all rotation matrices $\begin{pmatrix} \cos 2\pi k/m & -\sin 2\pi k/m \\ \sin 2\pi k/m & \cos 2\pi k/m \end{pmatrix}$ with matrix multiplication as operation.

Proposition 4.4. Any subgroup of a cyclic group is cyclic.

PROOF. Let G be a cyclic group with generator a, and let H be a subgroup. We may assume that $H \ne \{1\}$. Then there exists a positive integer n such that a^n is in H, and we let k be the smallest such positive integer. If n is any integer such that a^n is in H, then Proposition 1.2 produces integers x and y such that $xk + yn = d$, where $d = \text{GCD}(k, n)$. The equation $a^d = (a^k)^x (a^n)^y$ exhibits a^d as in H, and the minimality of k forces $d \ge k$. Since $\text{GCD}(k, n) \le k$, we conclude that $d = k$. Hence k divides n. Consequently H consists of the powers of a^k and is cyclic. □

A notion of the direct product of two groups is definable in the same way as was done with vector spaces in Section II.6, except that a little care is needed in saying how this construction interacts with mappings. As with the corresponding construction for vector spaces, one can define an explicit "external" direct product, and one can recognize a given group as an "internal" direct product, i.e., as isomorphic to an external direct product. We postpone a fuller discussion of direct product, as well as all comments about direct sums and mappings associated with direct sums and direct products, to Section 3.

The **external direct product** $G_1 \times G_2$ of two groups G_1 and G_2 is a group whose underlying set is the set-theoretic product of G_1 and G_2 and whose group law is $(g_1, g_2)(g_1', g_2') = (g_1 g_1', g_2 g_2')$. The identity is $(1, 1)$, and the formula for inverses is $(g_1, g_2)^{-1} = (g_1^{-1}, g_2^{-1})$. The two subgroups $G_1 \times \{1\}$ and $\{1\} \times G_2$ of $G_1 \times G_2$ commute with each other.

A group G is the **internal direct product** of two subgroups G_1 and G_2 if the function from the external direct product $G_1 \times G_2$ to G given by $(g_1, g_2) \mapsto g_1 g_2$ is an isomorphism of groups. The literal analog of Proposition 2.30, which gave three equivalent definitions of internal direct product[4] of vector spaces, fails here. It is not sufficient that G_1 and G_2 be two subgroups such that $G_1 \cap G_2 = \{1\}$ and every element in G decomposes as a product $g_1 g_2$ with $g_1 \in G_1$ and $g_2 \in G_2$. For example, with $G = \mathfrak{S}_3$, the two subgroups

$$G_1 = \{1, (1\ 2)\} \quad \text{and} \quad G_2 = \{1, (1\ 2\ 3), (1\ 3\ 2)\}$$

have these properties, but G is not isomorphic to $G_1 \times G_2$ because the elements of G_1 do not commute with the elements of G_2.

Proposition 4.5. If G is a group and G_1 and G_2 are subgroups, then the following conditions are equivalent:

(a) G is the internal direct product of G_1 and G_2,
(b) every element in G decomposes uniquely as a product $g_1 g_2$ with $g_1 \in G_1$ and $g_2 \in G_2$, and every member of G_1 commutes with every member of G_2,
(c) $G_1 \cap G_2 = \{1\}$, every element in G decomposes as a product $g_1 g_2$ with $g_1 \in G_1$ and $g_2 \in G_2$, and every member of G_1 commutes with every member of G_2.

PROOF. We have seen that (a) implies (b). If (b) holds and g is in $G_1 \cap G_2$, then the formula $1 = gg^{-1}$ and the uniqueness of the decomposition of 1 as a product together imply that $g = 1$. Hence (c) holds.

If (c) holds, define $\varphi : G_1 \times G_2 \to G$ by $\varphi(g_1, g_2) = g_1 g_2$. This map is certainly onto G. To see that it is one-one, suppose that $\varphi(g_1, g_2) = \varphi(g'_1, g'_2)$. Then $g_1 g_2 = g'_1 g'_2$ and hence $g'_1{}^{-1} g_1 = g'_2 g_2^{-1}$. Since $G_1 \cap G_2 = \{1\}$, $g'_1{}^{-1} g_1 = g'_2 g_2^{-1} = 1$. Thus $(g_1, g_2) = (g'_1, g'_2)$, and φ is one-one. Finally the fact that elements of G_1 commute with elements of G_2 implies that $\varphi((g_1, g_2)(g'_1, g'_2)) = \varphi(g_1 g'_1, g_2 g'_2) = g_1 g'_1 g_2 g'_2 = g_1 g_2 g'_1 g'_2 = \varphi(g_1, g_2) \varphi(g'_1, g'_2)$. Therefore φ is an isomorphism, and (a) holds. $\square$

Here are two examples of internal direct products of groups. In each let $\mathbb{R}^+$ be the multiplicative group of positive real numbers. The first example is $\mathbb{R}^\times \cong C_2 \times \mathbb{R}^+$ with C_2 providing the sign. The second example is $\mathbb{C}^\times \cong S^1 \times \mathbb{R}^+$, where S^1 is the multiplicative group of complex numbers of absolute value 1; the isomorphism here is given by the polar-coordinate mapping $(e^{i\theta}, r) \mapsto e^{i\theta} r$.

---

[4]The direct sum and direct product of two vector spaces were defined to be the same thing in Chapter II.

1. Groups and Subgroups

We conclude this section by giving an example of a group that falls outside the pattern of the examples above and by summarizing what groups we have identified with ≤ 15 elements.

EXAMPLES, CONTINUED.

(10) Groups associated with the quaternions. The set $\mathbb{H}$ of **quaternions** is an object like $\mathbb{R}$ or $\mathbb{C}$ in that it has both an addition/subtraction and a multiplication/division, but $\mathbb{H}$ is unlike $\mathbb{R}$ and $\mathbb{C}$ in that multiplication is not commutative. We give two constructions. In one we start from $\mathbb{R}^4$ with the standard basis vectors written as $1, \mathbf{i}, \mathbf{j}, \mathbf{k}$. The multiplication table for these basis vectors is

$$\begin{array}{llll} 11 = 1, & 1\mathbf{i} = \mathbf{i}, & 1\mathbf{j} = \mathbf{j}, & 1\mathbf{k} = \mathbf{k}, \\ \mathbf{i}1 = \mathbf{i}, & \mathbf{i}\mathbf{i} = -1, & \mathbf{i}\mathbf{j} = \mathbf{k}, & \mathbf{i}\mathbf{k} = -\mathbf{j}, \\ \mathbf{j}1 = \mathbf{j}, & \mathbf{j}\mathbf{i} = -\mathbf{k}, & \mathbf{j}\mathbf{j} = -1, & \mathbf{j}\mathbf{k} = \mathbf{i}, \\ \mathbf{k}1 = \mathbf{k}, & \mathbf{k}\mathbf{i} = \mathbf{j}, & \mathbf{k}\mathbf{j} = -\mathbf{i}, & \mathbf{k}\mathbf{k} = -1, \end{array}$$

and the multiplication is extended to general elements by the usual distributive laws. The multiplicative identity is 1, and multiplicative inverses of nonzero elements are given by

$$(a1 + b\mathbf{i} + c\mathbf{j} + \mathbf{k})^{-1} = s^{-1}a1 - s^{-1}b\mathbf{i} - s^{-1}c\mathbf{j} - s^{-1}d\mathbf{k}$$

with $s = \sqrt{a^2 + b^2 + c^2 + d^2}$. Since $\mathbf{ij} = \mathbf{k}$ while $\mathbf{ji} = -\mathbf{k}$, multiplication is not commutative. What takes work to see is that multiplication is associative. To see this, we give another construction, using $M_{22}(\mathbb{C})$. Within $M_{22}(\mathbb{C})$, take

$$1 = \begin{pmatrix} 1 & 0 \\ 0 & 1 \end{pmatrix}, \quad \mathbf{i} = \begin{pmatrix} i & 0 \\ 0 & -i \end{pmatrix}, \quad \mathbf{j} = \begin{pmatrix} 0 & -1 \\ 1 & 0 \end{pmatrix}, \quad \mathbf{k} = \begin{pmatrix} 0 & -i \\ -i & 0 \end{pmatrix},$$

and define $\mathbb{H}$ to be the linear span, with real coefficients, of these matrices. The operations are the usual matrix addition and multiplication. Then multiplication is associative, and we readily verify the multiplication table for $1, \mathbf{i}, \mathbf{j}, \mathbf{k}$. A little computation verifies also the formula for multiplicative inverses. The set $\mathbb{H}^\times$ of nonzero elements forms a group under multiplication, and it is isomorphic to $\mathbb{R}^+ \times \mathrm{SU}(2)$, where

$$\mathrm{SU}(2) = \left\{ \begin{pmatrix} \alpha & \beta \\ -\bar{\beta} & \bar{\alpha} \end{pmatrix} \,\Big|\, |\alpha|^2 + |\beta|^2 = 1 \right\}$$

is the 2-by-2 special unitary group defined in Example 8. Of interest for our current purposes is the 8-element subgroup $\pm 1, \pm \mathbf{i}, \pm \mathbf{j}, \pm \mathbf{k}$, which is called the **quaternion group** and will be denoted by H_8.

The **order** of a finite group is the number of elements in the group. Let us list some of the groups we have discussed that have order at most 15:

1	C_1	9	$C_9, C_3 \times C_3$
2	C_2	10	C_{10}, D_5
3	C_3	11	C_{11}
4	$C_4, C_2 \times C_2$	12	$C_{12}, C_6 \times C_2, D_6, \mathfrak{A}_4$
5	C_5	13	C_{13}
6	C_6, D_3	14	C_{14}, D_7
7	C_7	15	C_{15}
8	$C_8, C_4 \times C_2, C_2 \times C_2 \times C_2, D_4, H_8$		

No two groups in the above table are isomorphic, as one readily checks by counting elements of each "order" in the sense of the next section. We shall see in Section 10 and in the problems at the end of the chapter that the above table is complete through order 15 except for one group of order 12. Some groups that we have discussed have been omitted from the above table because of isomorphisms with the groups above. For example, $\mathfrak{S}_2 \cong C_2, \mathfrak{A}_3 \cong C_3, C_3 \times C_2 \cong C_6, \mathfrak{S}_3 \cong D_3$, $C_5 \times C_2 \cong C_{10}, C_4 \times C_3 \cong C_{12}, D_3 \times C_2 \cong D_6, C_7 \times C_2 \cong C_{14}$, and $C_5 \times C_3 \cong C_{15}$.

2. Quotient Spaces and Homomorphisms

Let G be a group, and let H be a subgroup. For purposes of this paragraph, say that g_1 in G is equivalent to g_2 in G if $g_1 = g_2 h$ for some h in H. The relation "equivalent" is an equivalence relation: it is reflexive because 1 is in H, it is symmetric since H is closed under inverses, and it is transitive since H is closed under products. The equivalence classes are called **left cosets** of H in G. The left coset containing an element g of G is the set $gH = \{gh \mid h \in H\}$.

EXAMPLES.

(1) When $G = \mathbb{Z}$ and $H = m\mathbb{Z}$, the left cosets are the sets $r + m\mathbb{Z}$, i.e., the sets $\{x \in \mathbb{Z} \mid x \equiv r \bmod m\}$ for the various values of r.

(2) When $G = \mathfrak{S}_3$ and $H = \{(1), (1\ 3)\}$, there are three left cosets: H, $(1\ 2)H = \{(1\ 2), (1\ 3\ 2)\}$, and $(2\ 3)H = \{(2\ 3), (1\ 2\ 3)\}$.

Similarly one can define the **right cosets** Hg of H in G. When G is nonabelian, these need not coincide with the left cosets; in Example 2 above with $G = \mathfrak{S}_3$ and $H = \{(1), (1\ 3)\}$, the right coset $H(1\ 2) = \{(1\ 2), (1\ 2\ 3)\}$ is not a left coset.

2. Quotient Spaces and Homomorphisms

Lemma 4.6. If H is a subgroup of the group G, then any two left cosets of H in G have the same cardinality, namely card H.

REMARKS. We shall be especially interested in the case that card H is finite, and then we write $|H| = $ card H for the number of elements in H.

PROOF. If $g_1 H$ and $g_2 H$ are given, then the map $g \mapsto g_2 g_1^{-1} g$ is one-one on G and carries $g_1 H$ onto $g_2 H$. Hence $g_1 H$ and $g_2 H$ have the same cardinality. Taking $g_1 = 1$, we see that this common cardinality is card H. □

We write G/H for the set $\{gH\}$ of all left cosets of H in G, calling it the **quotient space** or **left-coset space** of G by H. The set $\{Hg\}$ of right cosets is denoted by $H\backslash G$.

Theorem 4.7 (Lagrange's Theorem). If G is a finite group, then $|G| = |G/H||H|$. Consequently the order of any subgroup of G divides the order of G.

PROOF. Lemma 4.6 shows that each left coset has $|H|$ elements. The left cosets are disjoint and exhaust G, and there are $|G/H|$ left cosets. Thus G has $|G/H||H|$ elements. □

If a is an element of a group G, then we have seen that the powers a^n of a form a cyclic subgroup of G that is isomorphic either to $\mathbb{Z}$ or to some group $\mathbb{Z}/m\mathbb{Z}$ for a positive integer m. We say that a has **finite order** m when the cyclic group is isomorphic to $\mathbb{Z}/m\mathbb{Z}$. Otherwise a has **infinite order**. In the finite-order case the order of a is thus the least positive integer n such that $a^n = 1$.

Corollary 4.8. If G is a finite group, then each element a of G has finite order, and the order of a divides the order of G.

PROOF. The order of a equals $|H|$ if $H = \{a^n \mid n \in \mathbb{Z}\}$, and Corollary 4.8 is thus a special case of Theorem 4.7. □

Corollary 4.9. If p is a prime, then the only group of order p, up to isomorphism, is the cyclic group C_p, and it has no subgroups other than $\{1\}$ and C_p itself.

PROOF. Suppose that G is a finite group of order p and that $H \neq \{1\}$ is a subgroup of G. Let $a \neq 1$ be in H, and let $P = \{a^n \mid n \in \mathbb{Z}\}$. Since $a \neq 1$, Corollary 4.8 shows that the order of a is an integer > 1 that divides p. Since p is prime, the order of a must equal p. Then $|P| = p$. Since $P \subseteq H \subseteq G$ and $|G| = p$, we must have $P = G$. □

Let G_1 and G_2 be groups. We say that $\varphi : G_1 \to G_2$ is a **homomorphism** if $\varphi(ab) = \varphi(a)\varphi(b)$ for all a and b in G. In other words, φ is to respect products, but it is not assumed that φ is one-one or onto. Any homomorphism φ automatically respects the identity and inverses, in the sense that

- $\varphi(1) = 1$ (since $\varphi(1) = \varphi(11) = \varphi(1)\varphi(1)$),
- $\varphi(a^{-1}) = \varphi(a)^{-1}$ (since $1 = \varphi(1) = \varphi(aa^{-1}) = \varphi(a)\varphi(a^{-1})$ and similarly $1 = \varphi(a^{-1})\varphi(a)$).

EXAMPLES. The following functions are homomorphisms: any isomorphism, the function $\varphi : \mathbb{Z} \to \mathbb{Z}/m\mathbb{Z}$ given by $\varphi(k) = k \bmod m$, the function $\varphi : \mathfrak{S}_n \to \{\pm 1\}$ given by $\varphi(\sigma) = \operatorname{sgn} \sigma$, the function $\varphi : \mathbb{Z} \to G$ given for fixed a in G by $\varphi(n) = a^n$, and the function $\varphi : GL(n, \mathbb{F}) \to \mathbb{F}^\times$ given by $\varphi(A) = \det A$.

The **image** of a homomorphism $\varphi : G_1 \to G_2$ is just the image of φ considered as a function. It is denoted by $\operatorname{image} \varphi = \varphi(G_1)$ and is necessarily a subgroup of G_2 since if $\varphi(g_1) = g_2$ and $\varphi(g_1') = g_2'$, then $\varphi(g_1 g_1') = g_2 g_2'$ and $\varphi(g_1^{-1}) = g_2^{-1}$.

The **kernel** of a homomorphism $\varphi : G_1 \to G_2$ is the set $\ker \varphi = \varphi^{-1}(\{1\}) = \{x \in G_1 \mid \varphi(x) = 1\}$. This is a subgroup since if $\varphi(x) = 1$ and $\varphi(y) = 1$, then $\varphi(xy) = \varphi(x)\varphi(y) = 1$ and $\varphi(x^{-1}) = \varphi(x)^{-1} = 1$.

The homomorphism $\varphi : G_1 \to G_2$ is one-one if and only if $\ker \varphi$ is the trivial group $\{1\}$. The necessity follows since 1 is already in $\ker \varphi$, and the sufficiency follows since $\varphi(x) = \varphi(y)$ implies that $\varphi(xy^{-1}) = 1$ and therefore that xy^{-1} is in $\ker \varphi$.

The kernel H of a homomorphism $\varphi : G_1 \to G_2$ has the additional property of being a **normal subgroup** of G_1 in the sense that ghg^{-1} is in H whenever g is in G_1 and h is in H, i.e., $gHg^{-1} = H$. In fact, if h is in $\ker \varphi$ and g is in G_1, then $\varphi(ghg^{-1}) = \varphi(g)\varphi(h)\varphi(g)^{-1} = \varphi(g)\varphi(g)^{-1} = 1$ shows that ghg^{-1} is in $\ker \varphi$.

EXAMPLES.

(1) Any subgroup H of an abelian group G is normal since $ghg^{-1} = gg^{-1}h = h$. The alternating subgroup $\mathfrak{A}_n$ of the symmetric group $\mathfrak{S}_n$ is normal since $\mathfrak{A}_n$ is the kernel of the homomorphism $\sigma \mapsto \operatorname{sgn} \sigma$.

(2) The subgroup $H = \{1, (1\ 3)\}$ of $\mathfrak{S}_3$ is not normal since $(1\ 2)H(1\ 2)^{-1} = \{1, (2\ 3)\}$.

(3) If a subgroup H of a group G has just two left cosets, then H is normal even if G is an infinite group. In fact, suppose $G = H \cup g_0 H$ whenever g_0 is not in H. Taking inverses of all elements of G, we see that $G = H \cup Hg_1$ whenever g_1 is not in H. If g in G is given, then either g is in H and $gHg^{-1} = H$, or g is not in H and $gH = Hg$, so that $gHg^{-1} = H$ in this case as well.

Let H be a subgroup of G. Let us look for the circumstances under which G/H inherits a multiplication from G. The natural definition is

$$(g_1H)(g_2H) \stackrel{?}{=} g_1g_2H,$$

but we have to check that this definition makes sense. The question is whether we get the same left coset as product if we change the representatives of g_1H and g_2H from g_1 and g_2 to g_1h_1 and g_2h_2. Since our prospective definition makes $(g_1h_1H)(g_2h_2H) = g_1h_1g_2h_2H$, the question is whether $g_1h_1g_2h_2H$ equals g_1g_2H. That is, we ask whether $g_1h_1g_2h_2 = g_1g_2h$ for some h in H. If this equality holds, then $h_1g_2h_2 = g_2h$, and hence $g_2^{-1}h_2g_2$ equals hh_2^{-1}, which is an element of H. Conversely if every expression $g_2^{-1}h_2g_2$ is in H, then we can go backwards and see that $g_1h_1g_2h_2 = g_1g_2h$ for some h in H, hence see that G/H indeed inherits a multiplication from G. Thus *a necessary and sufficient condition for G/H to inherit a multiplication from G is that the subgroup H is normal*. According to the next proposition, the multiplication inherited by G/H when this condition is satisfied makes G/H into a group.

Proposition 4.10. If H is a normal subgroup of a group G, then G/H becomes a group under the inherited multiplication $(g_1H)(g_2H) = (g_1g_2)H$, and the function $q : G \to G/H$ given by $q(g) = gH$ is a homomorphism of G onto G/H with kernel H. Consequently every normal subgroup of G is the kernel of some homomorphism.

REMARKS. When H is normal, the group G/H is called a **quotient group** of G, and the homomorphism $q : G \to G/H$ is called the **quotient homomorphism**.[5] In the special case that $G = \mathbb{Z}$ and $H = m\mathbb{Z}$, the construction reduces to the construction of the additive group of integers modulo m and accounts for using the notation $\mathbb{Z}/m\mathbb{Z}$ for that group.

PROOF. The coset $1H$ is the identity, and $(gH)^{-1} = g^{-1}H$. Also, the computation $(g_1Hg_2H)g_3H = g_1g_2g_3H = g_1H(g_2Hg_3H)$ proves associativity. Certainly q is onto G/H. It is a homomorphism since $q(g_1g_2) = g_1g_2H = g_1Hg_2H = q(g_1)q(g_2)$. □

In analogy with what was shown for vector spaces in Proposition 2.25, quotients in the context of groups allow for the factorization of certain homomorphisms of groups. The appropriate result is stated as Proposition 4.11 and is pictured in Figure 4.1. We can continue from there along the lines of Section II.5.

---

[5]Some authors call G/H a "factor group." A "factor set," however, is something different.

Proposition 4.11. Let $\varphi : G_1 \to G_2$ be a homomorphism between groups, let $H_0 = \ker \varphi$, let H be a normal subgroup of G_1 contained in H_0, and define $q : G_1 \to G_1/H$ to be the quotient homomorphism. Then there exists a homomorphism $\overline{\varphi} : G_1/H \to G_2$ such that $\varphi = \overline{\varphi} \circ q$, i.e, $\overline{\varphi}(g_1 H) = \varphi(g_1)$. It has the same image as φ, and $\ker \overline{\varphi} = \{h_0 H \mid h_0 \in H_0\}$.

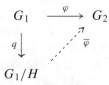

FIGURE 4.1. Factorization of homomorphisms of groups via the quotient of a group by a normal subgroup.

REMARK. One says that φ **factors through** G_1/H or **descends to** G_1/H. See Figure 4.1.

PROOF. We will have $\overline{\varphi} \circ q = \varphi$ if and only if $\overline{\varphi}$ satisfies $\overline{\varphi}(g_1 H) = \varphi(g_1)$. What needs proof is that $\overline{\varphi}$ is well defined. Thus suppose that g_1 and g_1' are in the same left coset, so that $g_1' = g_1 h$ with h in H. Then $\varphi(g_1') = \varphi(g_1)\varphi(h) = \varphi(g_1)$ since $H \subseteq \ker \varphi$, and $\overline{\varphi}$ is therefore well defined.

The computation $\overline{\varphi}(g_1 H g_2 H) = \overline{\varphi}(g_1 g_2 H) = \varphi(g_1 g_2) = \varphi(g_1)\varphi(g_2) = \overline{\varphi}(g_1 H)\overline{\varphi}(g_2 H)$ shows that $\overline{\varphi}$ is a homomorphism. Since image $\overline{\varphi}$ = image φ, $\overline{\varphi}$ is onto image φ. Finally $\ker \overline{\varphi}$ consists of all $g_1 H$ such that $\overline{\varphi}(g_1 H) = 1$. Since $\overline{\varphi}(g_1 H) = \varphi(g_1)$, the condition that g_1 is to satisfy is that g_1 be in $\ker \varphi = H_0$. Hence $\ker \overline{\varphi} = \{h_0 H \mid h_0 \in H_0\}$, as asserted. □

Corollary 4.12. Let $\varphi : G_1 \to G_2$ be a homomorphism between groups, and suppose that φ is onto G_2 and has kernel H. Then φ exhibits the group G_1/H as canonically isomorphic to G_2.

PROOF. Take $H = H_0$ in Proposition 4.11, and form $\overline{\varphi} : G_1/H \to G_2$ with $\varphi = \overline{\varphi} \circ q$. The proposition shows that $\overline{\varphi}$ is onto G_2 and has trivial kernel, i.e., the identity element of G_1/H. Having trivial kernel, $\overline{\varphi}$ is one-one. □

Theorem 4.13 (First Isomorphism Theorem). Let $\varphi : G_1 \to G_2$ be a homomorphism between groups, and suppose that φ is onto G_2 and has kernel K. Then the map $H_1 \mapsto \varphi(H_1)$ gives a one-one correspondence between

(a) the subgroups H_1 of G_1 containing K and
(b) the subgroups of G_2.

Under this correspondence normal subgroups correspond to normal subgroups. If H_1 is normal in G_1, then $g H_1 \mapsto \varphi(g)\varphi(H_1)$ is an isomorphism of G_1/H_1 onto $G_2/\varphi(H_1)$.

REMARK. In the special case of the last statement that $\varphi : G_1 \to G_2$ is a quotient map $q : G \to G/K$ and H is a normal subgroup of G containing K, the last statement of the theorem asserts the isomorphism

$$G/H \cong (G/K)/(H/K).$$

PROOF. The passage from (a) to (b) is by direct image under φ, and the passage from (b) to (a) will be by inverse image under φ^{-1}. Certainly the direct image of a subgroup as in (a) is a subgroup as in (b). To prove the one-one correspondence, we are to show that the inverse image of a subgroup as in (b) is a subgroup as in (a) and that these two constructions invert one another.

For any subgroup H_2 of G_2, $\varphi^{-1}(H_2)$ is a subgroup of G_1. In fact, if g_1 and g_1' are in $\varphi^{-1}(H_2)$, we can write $\varphi(g_1) = h_2$ and $\varphi(g_1') = h_2'$ with h_2 and h_2' in H_2. Then the equations $\varphi(g_1 g_1') = h_2 h_2'$ and $\varphi(g_1^{-1}) = \varphi(g_1)^{-1} = h_2^{-1}$ show that $h_2 h_2'$ and h_2^{-1} are in $\varphi^{-1}(H_2)$.

Moreover, the subgroup $\varphi^{-1}(H_2)$ contains $\varphi^{-1}(\{1\}) = K$. Therefore the inverse image under φ of a subgroup as in (b) is a subgroup as in (a). Since φ is a function, we have $\varphi(\varphi^{-1}(H_2)) = H_2$. Thus passing from (b) to (a) and back recovers the subgroup of G_2.

If H_1 is a subgroup of G_1 containing K, we still need to see that $H_1 = \varphi^{-1}(\varphi(H_1))$. Certainly $H_1 \subseteq \varphi^{-1}(\varphi(H_1))$. For the reverse inclusion let g_1 be in $\varphi^{-1}(\varphi(H_1))$. Then $\varphi(g_1)$ is in $\varphi(H_1)$, i.e., $\varphi(g_1) = \varphi(h_1)$ for some h_1 in H_1. Since φ is a homomorphism, $\varphi(g_1 h_1^{-1}) = 1$. Thus $g_1 h_1^{-1}$ is in $\ker \varphi = K$, which is contained in H_1 by assumption. Then h_1 and $g_1 h_1^{-1}$ are in H_1, and hence their product $(g_1 h_1^{-1}) h_1 = g_1$ is in H_1. We conclude that $\varphi^{-1}(\varphi(H_1)) \subseteq H_1$, and thus passing from (a) to (b) and then back recovers the subgroup of G_1 containing K.

Next let us show that normal subgroups correspond to normal subgroups. If H_2 is normal in G_2, let H_1 be the subgroup $\varphi^{-1}(H_2)$ of G_1. For h_1 in H_1 and g_1 in G_1, we can write $\varphi(h_1) = h_2$ with h_2 in H_2, and then $\varphi(g_1 h_1 g_1^{-1}) = \varphi(g_1) h_2 \varphi(g_1)^{-1}$ is in $\varphi(g_1) H_2 \varphi(g_1)^{-1} = H_2$. Hence $g_1 h_1 g_1^{-1}$ is in $\varphi^{-1}(H_2) = H_1$. In the reverse direction let H_1 be normal in G_1, and let g_2 be in G_2. Since φ is onto G_2, we can write $g_2 = \varphi(g_1)$ for some g_1 in G_1. Then $g_2 \varphi(H_1) g_2^{-1} = \varphi(g_1) \varphi(H_1) \varphi(g_1)^{-1} = \varphi(g_1 H_1 g_1^{-1}) = \varphi(H_1)$. Thus $\varphi(H_1)$ is normal.

For the final statement let $H_2 = \varphi(H_1)$. We have just proved that this image is normal, and hence G_2/H_2 is a group. The mapping $\Phi : G_1 \to G_2/H_2$ given by $\Phi(g_1) = \varphi(g_1) H_2$ is the composition of two homomorphisms and hence is a homomorphism. Its kernel is

$$\{g_1 \in G_1 \mid \varphi(g_1) \in H_2\} = \{g_1 \in G_1 \mid \varphi(g_1) \in \varphi(H_1)\} = \varphi^{-1}(\varphi(H_1)),$$

and this equals H_1 by the first conclusion of the theorem. Applying Corollary 4.12 to Φ, we obtain the required isomorphism $\overline{\Phi} : G_1/H_1 \to G_2/\varphi(H_1)$. □

Theorem 4.14 (Second Isomorphism Theorem). Let H_1 and H_2 be subgroups of a group G with H_2 normal in G. Then $H_1 \cap H_2$ is a normal subgroup of H_1, the set $H_1 H_2$ of products is a subgroup of G with H_2 as a normal subgroup, and the map $h_1(H_1 \cap H_2) \mapsto h_1 H_2$ is a well-defined canonical isomorphism of groups

$$H_1/(H_1 \cap H_2) \cong (H_1 H_2)/H_2.$$

PROOF. The set $H_1 \cap H_2$ is a subgroup, being the intersection of two subgroups. For h_1 in H_1, we have $h_1(H_1 \cap H_2)h_1^{-1} \subseteq h_1 H_1 h_1^{-1} \subseteq H_1$ since H_1 is a subgroup and $h_1(H_1 \cap H_2)h_1^{-1} \subseteq h_1 H_2 h_1^{-1} \subseteq H_2$ since H_2 is normal in G. Therefore $h_1(H_1 \cap H_2)h_1^{-1} \subseteq H_1 \cap H_2$, and $H_1 \cap H_2$ is normal in H_1.

The set $H_1 H_2$ of products is a subgroup since $h_1 h_2 h_1' h_2' = h_1 h_1' (h_1'^{-1} h_2 h_1') h_2'$ and since $(h_1 h_2)^{-1} = (h_2^{-1} h_1^{-1} h_2) h_2^{-1}$, and H_2 is normal in $H_1 H_2$ since H_2 is normal in G.

The function $\varphi(h_1(H_1 \cap H_2)) = h_1 H_2$ is well defined since $H_1 \cap H_2 \subseteq H_2$, and φ respects products. The domain of φ is $\{h_1(H_1 \cap H_2) \mid h_1 \in H_1\}$, and the kernel is the subset of this such that h_1 lies in H_2 as well as H_1. For this to happen, h_1 must be in $H_1 \cap H_2$, and thus the kernel is the identity coset of $H_1/(H_1 \cap H_2)$. Hence φ is one-one.

To see that φ is onto $(H_1 H_2)/H_2$, let $h_1 h_2 H_2$ be given. Then $h_1(H_1 \cap H_2)$ maps to $h_1 H_2$, which equals $h_1 h_2 H_2$. Hence φ is onto. $\square$

3. Direct Products and Direct Sums

We return to the matter of direct products and direct sums of groups, direct products having been discussed briefly in Section 1. In a footnote in Section II.4 we mentioned a general principle in algebra that "whenever a new systematic construction appears for the objects under study, it is well to look for a corresponding construction with the functions relating these new objects." This principle will be made more precise in Section 11 of the present chapter with the aid of the language of "categories" and "functors."

Another principle that will be relevant for us is that constructions in one context in algebra often recur, sometimes in slightly different guise, in other contexts. One example of the operation of this principle occurs with quotients. The construction and properties of the quotient of a vector space by a vector subspace, as in Section II.5, is analogous in this sense to the construction and properties of the quotient of a group by a *normal* subgroup, as in Section 2 in the present chapter. The need for the subgroup to be normal is an example of what is meant by "slightly different guise." Anyway, this principle too will be made more precise in Section 11 of the present chapter using the language of categories and functors.

3. Direct Products and Direct Sums

Let us proceed with an awareness of both these principles in connection with direct products and direct sums of groups, looking for analogies with what happened for vector spaces and expecting our work to involve constructions with homomorphisms as well as with groups.

The external direct product $G_1 \times G_2$ was defined as a group in Section 1 to be the set-theoretic product with coordinate-by-coordinate multiplication. There are four homomorphisms of interest connected with $G_1 \times G_2$, namely

$$i_1 : G_1 \to G_1 \times G_2 \quad \text{given by} \quad i_1(g_1) = (g_1, 1),$$
$$i_2 : G_2 \to G_1 \times G_2 \quad \text{given by} \quad i_2(g_2) = (1, g_2),$$
$$p_1 : G_1 \times G_2 \to G_1 \quad \text{given by} \quad p_1(g_1, g_2) = g_1,$$
$$p_2 : G_1 \times G_2 \to G_2 \quad \text{given by} \quad p_2(g_1, g_2) = g_2.$$

Recall from the discussion before Proposition 4.5 that Proposition 2.30 for the direct product of two vector spaces does not translate directly into an analog for the direct product of groups; instead that proposition is replaced by Proposition 4.5, which involves some condition of commutativity.

Warned by this anomaly, let us work with mappings rather than with groups and subgroups, and let us use mappings in formulating a definition of the direct product of groups. As with the direct product of two vector spaces, the mappings to use are p_1 and p_2 but not i_1 and i_2. The way in which p_1 and p_2 enter is through the effect of the direct product on homomorphisms. If $\varphi_1 : H \to G_1$ and $\varphi_2 : H \to G_2$ are two homomorphisms, then $h \mapsto (\varphi_1(h), \varphi_2(h))$ is the corresponding homomorphism of H into $G_1 \times G_2$. In order to state matters fully, let us give the definition with an arbitrary number of factors.

Let S be an arbitrary nonempty set of groups, and let G_s be the group corresponding to the member s of S. The **external direct product** of the G_s's consists of a group $\prod_{s \in S} G_s$ and a system of group homomorphisms. The group as a set is $\times_{s \in S} G_s$, whose elements are arbitrary functions from S to $\bigcup_{s \in S} G_s$ such that the value of the function at s is in G_s, and the group law is $(\{g_s\}_{s \in S})(\{g'_s\}_{s \in S}) = \{g_s g'_s\}_{s \in S}$. The group homomorphisms are the coordinate mappings $p_{s_0} : \prod_{s \in S} G_s \to G_{s_0}$ with $p_{s_0}(\{g_s\}_{s \in S}) = g_{s_0}$. The individual groups G_s are called the **factors**, and a direct product of n groups may be written as $G_1 \times \cdots \times G_n$ instead of with the symbol $\prod$. The group $\prod_{s \in S} G_s$ has the **universal mapping property** described in Proposition 4.15 and pictured in Figure 4.2.

Proposition 4.15 (universal mapping property of external direct product). Let $\{G_s \mid s \in S\}$ be a nonempty set of groups, and let $\prod_{s \in S} G_s$ be the external direct product, the associated group homomorphisms being the coordinate mappings $p_{s_0} : \prod_{s \in S} G_s \to G_{s_0}$. If H is any group and $\{\varphi_s \mid s \in S\}$ is a system of group homomorphisms $\varphi_s : H \to G_s$, then there exists a unique group homomorphism $\varphi : H \to \prod_{s \in S} G_s$ such that $p_{s_0} \circ \varphi = \varphi_{s_0}$ for all $s_0 \in S$.

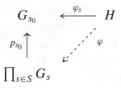

FIGURE 4.2. Universal mapping property of an external direct product of groups.

PROOF. Existence of φ is proved by taking $\varphi(h) = \{\varphi_s(h)\}_{s \in S}$. Then $p_{s_0}(\varphi(h)) = p_{s_0}(\{\varphi_s(h)\}_{s \in S}) = \varphi_{s_0}(h)$ as required. For uniqueness let $\varphi' : H \to \prod_{s \in S} G_s$ be a homomorphism with $p_{s_0} \circ \varphi' = \varphi_{s_0}$ for all $s_0 \in S$. For each h in H, we can write $\varphi'(h) = \{\varphi'(h)_s\}_{s \in S}$. For s_0 in S, we then have $\varphi_{s_0}(h) = (p_{s_0} \circ \varphi')(h) = p_{s_0}(\varphi'(h)) = \varphi'(h)_{s_0}$, and we conclude that $\varphi' = \varphi$. □

Now we give an abstract definition of direct product that allows for the possibility that the direct product is "internal" in the sense that the various factors are identified as subgroups of a given group. The definition is by means of the above universal mapping property and will be seen to characterize the direct product up to canonical isomorphism. Let S be an arbitrary nonempty set of groups, and let G_s be the group corresponding to the member s of S. A **direct product** of the G_s's consists of a group G and a system of group homomorphisms $p_s : G \to G_s$ for $s \in S$ with the following **universal mapping property**: whenever H is a group and $\{\varphi_s \mid s \in S\}$ is a system of group homomorphisms $\varphi_s : H \to G_s$, then there exists a unique group homomorphism $\varphi : H \to G$ such that $p_s \circ \varphi = \varphi_s$ for all $s \in S$. Proposition 4.15 proves existence of a direct product, and the next proposition addresses uniqueness. A direct product is **internal** if each G_s is a subgroup of G and each restriction $p_s|_{G_s}$ is the identity map.

FIGURE 4.3. Universal mapping property of a direct product of groups.

Proposition 4.16. Let S be a nonempty set of groups, and let G_s be the group corresponding to the member s of S. If $(G, \{p_s\})$ and $(G', \{p'_s\})$ are two direct products, then the homomorphisms $p_s : G \to G_s$ and $p'_s : G' \to G_s$ are onto G_s, there exists a unique homomorphism $\Phi : G' \to G$ such that $p'_s = p_s \circ \Phi$ for all $s \in S$, and Φ is an isomorphism.

PROOF. In Figure 4.3 let $H = G'$ and $\varphi_s = p'_s$. If $\Phi : G' \to G$ is the homomorphism produced by the fact that G is a direct product, then we have

$p_s \circ \Phi = p'_s$ for all s. Reversing the roles of G and G', we obtain a homomorphism $\Phi' : G \to G'$ with $p'_s \circ \Phi' = p_s$ for all s. Therefore $p_s \circ (\Phi \circ \Phi') = p'_s \circ \Phi' = p_s$.

In Figure 4.3 we next let $H = G$ and $\varphi_s = p_s$ for all s. Then the identity 1_G on G has the same property $p_s \circ 1_G = p_s$ relative to all p_s that $\Phi \circ \Phi'$ has, and the uniqueness says that $\Phi \circ \Phi' = 1_G$. Reversing the roles of G and G', we obtain $\Phi' \circ \Phi = 1_{G'}$. Therefore Φ is an isomorphism.

For uniqueness suppose that $\Psi : G' \to G$ is another homomorphism with $p'_s = p_s \circ \Psi$ for all $s \in S$. Then the argument of the previous paragraph shows that $\Phi' \circ \Psi = 1_{G'}$. Applying Φ on the left gives $\Psi = (\Phi \circ \Phi') \circ \Psi = \Phi \circ (\Phi' \circ \Psi) = \Phi \circ 1_{G'} = \Phi$. Thus $\Psi = \Phi$.

Finally we have to show that the sth mapping of a direct product is onto G_s. It is enough to show that p'_s is onto G_s. Taking G as the external direct product $\prod_{s \in S} G_s$ with p_s equal to the coordinate mapping, form the isomorphism $\Phi' : G \to G'$ that has just been proved to exist. This satisfies $p_s = p'_s \circ \Phi'$ for all $s \in S$. Since p_s is onto G_s, p'_s must be onto G_s. □

Let us turn to direct sums. Part of what we seek is a definition that allows for an abstract characterization of direct sums in the spirit of Proposition 4.16. In particular, the interaction with homomorphisms is to be central to the discussion. In the case of two factors, we use i_1 and i_2 rather than p_1 and p_2. If $\varphi_1 : G_1 \to H$ and $\varphi_2 : G_2 \to H$ are two homomorphisms, then the corresponding homomorphism φ of $G_1 \oplus G_2$ to H is to satisfy $\varphi_1 = \varphi \circ i_1$ and $\varphi_2 = \varphi \circ i_2$. With $G_1 \oplus G_2$ defined, as expected, to be the same group as $G_1 \times G_2$, we are led to the formula

$$\varphi(g_1, g_2) = \varphi(g_1, 1)\varphi(1, g_2) = \varphi_1(g_1)\varphi_2(g_2).$$

The images of commuting elements under a homomorphism have to commute, and hence H had better be abelian. Then in order to have an analog of Proposition 4.16, we will want to specialize H at some point to $G_1 \oplus G_2$, and therefore G_1 and G_2 had better be abelian. With these observations in place, we are ready for the general definition.

Let S be an arbitrary nonempty set of *abelian* groups, and let G_s be the group corresponding to the member s of S. We shall use additive notation for the group operation in each G_s. The **external direct sum** of the G_s's consists of an abelian group $\bigoplus_{s \in S} G_s$ and a system of group homomorphisms i_s for $s \in S$. The group is the subgroup of $\prod_{s \in S} G_s$ of all elements that are equal to 0 in all but finitely many coordinates. The group homomorphisms are the mappings $i_{s_0} : G_{s_0} \to \bigoplus_{s \in S} G_s$ carrying a member g_{s_0} of G_{s_0} to the element that is g_{s_0} in coordinate s_0 and is 0 at all other coordinates. The individual groups are called the **summands**, and a direct sum of n abelian groups may be written as $G_1 \oplus \cdots \oplus G_n$. The group $\bigoplus_{s \in S} G_s$ has the **universal mapping property** described in Proposition 4.17 and pictured in Figure 4.4.

Proposition 4.17 (universal mapping property of external direct sum). Let $\{G_s \mid s \in S\}$ be a nonempty set of abelian groups, and let $\bigoplus_{s \in S} G_s$ be the external direct sum, the associated group homomorphisms being the embedding mappings $i_{s_0} : G_{s_0} \to \bigoplus_{s \in S} G_s$. If H is any abelian group and $\{\varphi_s \mid s \in S\}$ is a system of group homomorphisms $\varphi_s : G_s \to H$, then there exists a unique group homomorphism $\varphi : \bigoplus_{s \in S} G_s \to H$ such that $\varphi \circ i_{s_0} = \varphi_{s_0}$ for all $s_0 \in S$.

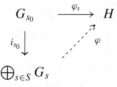

FIGURE 4.4. Universal mapping property of an external direct sum of abelian groups.

PROOF. Existence of φ is proved by taking $\varphi(\{g_s\}_{s \in S}) = \sum_s \varphi_s(g_s)$. The sum on the right side is meaningful since the element $\{g_s\}_{s \in S}$ of the direct sum has only finitely many nonzero coordinates. Since H is abelian, the computation

$$\varphi(\{g_s\}_{s \in S}) + \varphi(\{g'_s\}_{s \in S}) = \sum_s \varphi_s(g_s) + \sum_s \varphi_s(g'_s)$$
$$= \sum_s (\varphi_s(g_s) + \varphi_s(g'_s)) = \sum_s \varphi_s(g_s + g'_s)$$
$$= \varphi(\{g_s + g'_s\}_{s \in S}) = \varphi(\{g_s\}_{s \in S} + \{g'_s\}_{s \in S})$$

shows that φ is a homomorphism. If g_{s_0} is given and $\{g_s\}_{s \in S}$ denotes the element that is g_{s_0} in the $s_0{}^{\text{th}}$ coordinate and is 0 elsewhere, then $\varphi(i_{s_0}(g_{s_0})) = \varphi(\{g_s\}_{s \in S}) = \sum_s \varphi_s(g_s)$, and the right side equals $\varphi_{s_0}(g_{s_0})$ since $g_s = 0$ for all other s's. Thus $\varphi \circ i_{s_0} = \varphi_{s_0}$.

For uniqueness let $\varphi' : \bigoplus_{s \in S} G_s \to H$ be a homomorphism with $\varphi' \circ i_{s_0} = \varphi_{s_0}$ for all $s_0 \in S$. Then the value of φ' is determined at all elements of $\bigoplus_{s \in S} G_s$ that are 0 in all but one coordinate. Since the most general member of $\bigoplus_{s \in S} G_s$ is a finite sum of such elements, φ' is determined on all of $\bigoplus_{s \in S} G_s$. □

Now we give an abstract definition of direct sum that allows for the possibility that the direct sum is "internal" in the sense that the various constituents are identified as subgroups of a given group. Again the definition is by means of a universal mapping property and will be seen to characterize the direct sum up to canonical isomorphism. Let S be an arbitrary nonempty set of *abelian* groups, and let G_s be the group corresponding to the member s of S. A **direct sum** of the G_s's consists of an abelian group G and a system of group homomorphisms $i_s : G_s \to G$ for $s \in S$ with the following **universal mapping property**: whenever H is an abelian group and $\{\varphi_s \mid s \in S\}$ is a system of group homomorphisms

$\varphi_s : G_s \to H$, then there exists a unique group homomorphism $\varphi : G \to H$ such that $\varphi \circ i_s = \varphi_s$ for all $s \in S$. Proposition 4.17 proves existence of a direct sum, and the next proposition addresses uniqueness. A direct sum is **internal** if each G_s is a subgroup of G and each mapping i_s is the inclusion mapping.

FIGURE 4.5. Universal mapping property of a direct sum of abelian groups.

Proposition 4.18. Let S be a nonempty set of abelian groups, and let G_s be the group corresponding to the member s of S. If $(G, \{i_s\})$ and $(G', \{i'_s\})$ are two direct sums, then the homomorphisms $i_s : G_s \to G$ and $i'_s : G_s \to G'$ are one-one, there exists a unique homomorphism $\Phi : G \to G'$ such that $i'_s = \Phi \circ i_s$ for all $s \in S$, and Φ is an isomorphism.

PROOF. In Figure 4.5 let $H = G'$ and $\varphi_s = i'_s$. If $\Phi : G \to G'$ is the homomorphism produced by the fact that G is a direct sum, then we have $\Phi \circ i_s = i'_s$ for all s. Reversing the roles of G and G', we obtain a homomorphism $\Phi' : G' \to G$ with $\Phi' \circ i'_s = i_s$ for all s. Therefore $(\Phi' \circ \Phi) \circ i_s = \Phi' \circ i'_s = i_s$.

In Figure 4.5 we next let $H = G$ and $\varphi_s = i_s$ for all s. Then the identity 1_G on G has the same property $1_G \circ i_s = i_s$ relative to all i_s that $\Phi' \circ \Phi$ has, and the uniqueness says that $\Phi' \circ \Phi = 1_G$. Reversing the roles of G and G', we obtain $\Phi \circ \Phi' = 1_{G'}$. Therefore Φ is an isomorphism.

For uniqueness suppose that $\Psi : G \to G'$ is another homomorphism with $i'_s = \Psi \circ i_s$ for all $s \in S$. Then the argument of the previous paragraph shows that $\Phi' \circ \Psi = 1_G$. Applying Φ on the left gives $\Psi = (\Phi \circ \Phi') \circ \Psi = \Phi \circ (\Phi' \circ \Psi) = \Phi \circ 1_G = \Phi$. Thus $\Psi = \Phi$.

Finally we have to show that the sth mapping of a direct sum is one-one on G_s. It is enough to show that i'_s is one-one on G_s. Taking G as the external direct sum $\bigoplus_{s \in S} G_s$ with i_s equal to the embedding mapping, form the isomorphism $\Phi' : G' \to G$ that has just been proved to exist. This satisfies $i_s = \Phi' \circ i'_s$ for all $s \in S$. Since i_s is one-one, i'_s must be one-one. □

EXAMPLE. The group $\mathbb{Q}^\times$ is the direct sum of copies of $\mathbb{Z}$, one for each prime, plus one copy of $\mathbb{Z}/2\mathbb{Z}$. If p is a prime, the mapping $i_p : \mathbb{Z} \to \mathbb{Q}^\times$ is given by $i_p(n) = p^n$. The remaining coordinate gives the sign. The isomorphism results from unique factorization, only finitely many primes being involved for any particular nonzero rational number.

4. Rings and Fields

In this section we begin a two-section digression in order to develop some more number theory beyond what is in Chapter I and to make some definitions as new notions arise. In later sections of the present chapter, some of this material will yield further examples of concrete groups and tools for working with them.

We begin with the additive group $\mathbb{Z}/m\mathbb{Z}$ of integers modulo a positive integer m. We continue to write $[a]$ for the equivalence class of the integer a when it is helpful to do so. Our interest will be in the multiplication structure that $\mathbb{Z}/m\mathbb{Z}$ inherits from multiplication in $\mathbb{Z}$. Namely, we attempt to define

$$[a][b] = [ab].$$

To see that this formula is meaningful in $\mathbb{Z}/m\mathbb{Z}$, we need to check that the same equivalence class results on the right side if the representatives of $[a]$ and $[b]$ are changed. Thus let $[a] = [a']$ and $[b] = [b']$. Then m divides $a - a'$ and $b - b'$ and must divide the sum of products $(a - a')b + a'(b - b') = ab - a'b'$. Consequently $[ab] = [a'b']$, and multiplication is well defined. If x and y are in $\mathbb{Z}/m\mathbb{Z}$, their product is often denoted by $xy \bmod m$.

The same kind of argument as just given shows that the associativity of multiplication in $\mathbb{Z}$ and the distributive laws imply corresponding facts about $\mathbb{Z}/m\mathbb{Z}$. The result is that $\mathbb{Z}/m\mathbb{Z}$ is a "commutative ring with identity" in the sense of the following definitions.

A **ring** is a set R with two operations $R \times R \to R$, usually called **addition** and **multiplication** and often denoted by $(a, b) \mapsto a + b$ and $(a, b) \mapsto ab$, such that

  (i) R is an abelian group under addition,
  (ii) multiplication is associative in the sense that $a(bc) = (ab)c$ for all a, b, c in R,
  (iii) the two distributive laws

  $$a(b + c) = (ab) + (ac) \quad \text{and} \quad (b + c)a = (ba) + (ca)$$

  hold for all a, b, c in R.

The additive identity is denoted by 0, and the additive inverse of a is denoted by $-a$. A sum $a + (-b)$ is often abbreviated $a - b$. By convention when parentheses are absent, multiplications are to be carried out before additions and subtractions. Thus the distributive laws may be rewritten as

$$a(b + c) = ab + ac \quad \text{and} \quad (b + c)a = ba + ca.$$

A ring R is called a **commutative ring** if multiplication satisfies the commutative law

  (iv) $ab = ba$ for all a and b in R.

A ring R is called a **ring with identity**[6] if there exists an element 1 such that $1a = a1 = a$ for all a in R. It is immediate from the definitions that

- $0a = 0$ and $a0 = 0$ in any ring (since, in the case of the first formula, $0 = 0a - 0a = (0+0)a - 0a = 0a + 0a - 0a = 0a$),
- the multiplicative identity is unique in a ring with identity (since $1' = 1'1 = 1$),
- $(-1)a = -a = a(-1)$ in any ring with identity (partly since $0 = 0a = (1 + (-1))a = 1a + (-1)a = a + (-1)a$).

In a ring with identity, it will be convenient not to insist that the identity be different from the zero element 0. If 1 and 0 do happen to coincide in R, then it readily follows that 0 is the only element of R, and R is said to be the **zero ring**.

The set $\mathbb{Z}$ of integers is a basic example of a commutative ring with identity. Returning to $\mathbb{Z}/m\mathbb{Z}$, suppose now that m is a prime p. If $[a]$ is in $\mathbb{Z}/p\mathbb{Z}$ with a in $\{1, 2, \ldots, p-1\}$, then $\text{GCD}(a, p) = 1$ and Proposition 1.2 produces integers r and s with $ar + ps = 1$. Modulo p, this equation reads $[a][r] = [1]$. In other words, $[r]$ is a multiplicative inverse of $[a]$. The result is that $\mathbb{Z}/p\mathbb{Z}$, when p is a prime, is a "field" in the sense of the following definition.

A **field** $\mathbb{F}$ is a commutative ring with identity such that $\mathbb{F} \neq 0$ and such that

(v) to each $a \neq 0$ in $\mathbb{F}$ corresponds an element a^{-1} in $\mathbb{F}$ such that $aa^{-1} = 1$.

In other words, $\mathbb{F}^\times = \mathbb{F} - \{0\}$ is an abelian group under multiplication. Inverses are necessarily unique as a consequence of one of the properties of groups.

When p is prime, we shall write $\mathbb{F}_p$ for the field $\mathbb{Z}/p\mathbb{Z}$. Its multiplicative group $\mathbb{F}_p^\times$ has order $p - 1$, and Lagrange's Theorem (Corollary 4.8) immediately implies that $a^{p-1} \equiv 1 \bmod p$ whenever a and p are relatively prime. This result is known as **Fermat's Little Theorem**.[7]

For general m, certain members of $\mathbb{Z}/m\mathbb{Z}$ have multiplicative inverses. The product of two such elements is again one, and the inverse of one is again one. Thus, even though $\mathbb{Z}/m\mathbb{Z}$ need not be a field, the subset $(\mathbb{Z}/m\mathbb{Z})^\times$ of members of $\mathbb{Z}/m\mathbb{Z}$ with multiplicative inverses is a group. The same argument as when m is prime shows that the class of a has an inverse if and only if $\text{GCD}(a, m) = 1$. The number of such classes was defined in Chapter I in terms of the Euler φ function as $\varphi(m)$, and a formula for $\varphi(m)$ was obtained in Corollary 1.10. The

---

[6]Some authors, particularly when discussing only algebra, find it convenient to incorporate the existence of an identity into the definition of a ring. However, in real analysis some important natural rings do not have an identity, and the theory is made more complicated by forcing an identity into the picture. For example the space of integrable functions on $\mathbb{R}$ forms a very natural ring, with convolution as multiplication, and there is no identity; forcing an identity into the picture in such a way that the space remains stable under translations makes the space large and unwieldy. The distinction between working with rings and working with rings with identity will be discussed further in Section 11.

[7]As opposed to Fermat's Last Theorem, which lies deeper.

conclusion is that $(\mathbb{Z}/m\mathbb{Z})^\times$ is an abelian group of order $\varphi(m)$. Application of Lagrange's Theorem yields Euler's generalization of Fermat's Little Theorem, namely that $a^{\varphi(m)} \equiv 1 \bmod m$ for every positive integer m and every integer a relatively prime to m.

More generally, in any ring R with identity, a **unit** is defined to be any element a such that there exists an element a^{-1} with $aa^{-1} = a^{-1}a = 1$. The element a^{-1} is unique if it exists[8] and is called the **multiplicative inverse** of a. The units of R form a group denoted by $R^\times$. For example the group $\mathbb{Z}^\times$ consists of $+1$ and -1, and the zero ring R has $R^\times = \{0\}$. If R is a nonzero ring, then 0 is not in $R^\times$.

Here are some further examples of fields.

EXAMPLES OF FIELDS.

(1) $\mathbb{Q}, \mathbb{R}$, and $\mathbb{C}$. These are all fields.

(2) $\mathbb{Q}[\theta]$. This was introduced between Examples 8 and 9 of Section 1. It is assumed that θ is a complex number and that there exists an integer $n > 0$ such that the complex numbers $1, \theta, \theta^2, \ldots, \theta^n$ are linearly dependent over $\mathbb{Q}$. The set $\mathbb{Q}[\theta]$ is defined to be the linear span over $\mathbb{Q}$ of all powers $1, \theta, \theta^2, \ldots$ of θ, which is the same as the linear span of the finite set $1, \theta, \theta^2, \ldots, \theta^{n-1}$. The set $\mathbb{Q}[\theta]$ was shown in Proposition 4.1 to be a subset of $\mathbb{C}$ that is closed under the arithmetic operations, including the passage to reciprocals in the case of the nonzero elements. It is therefore a field.

(3) A field of 4 elements. Let $\mathbb{F}_4 = \{0, 1, \theta, \theta+1\}$, where θ is some symbol not standing for 0 or 1. Define addition in $\mathbb{F}_4$ and multiplication in $\mathbb{F}_4^\times$ by requiring that $a + 0 = 0 + a = a$ for all a, that

$$1 + 1 = 0, \qquad 1 + \theta = (\theta + 1), \qquad 1 + (\theta + 1) = \theta,$$
$$\theta + 1 = (\theta + 1), \qquad \theta + \theta = 0, \qquad \theta + (\theta + 1) = 1,$$
$$(\theta + 1) + 1 = \theta, \qquad (\theta + 1) + \theta = 1, \qquad (\theta + 1) + (\theta + 1) = 0,$$

and that

$$1 \cdot 1 = 1, \qquad 1\theta = \theta, \qquad 1(\theta + 1) = (\theta + 1),$$
$$\theta 1 = \theta, \qquad \theta\theta = (\theta + 1), \qquad \theta(\theta + 1) = 1,$$
$$(\theta + 1)1 = (\theta + 1), \qquad (\theta + 1)\theta = 1, \qquad (\theta + 1)(\theta + 1) = \theta.$$

The result is a field. With this direct approach a certain amount of checking is necessary to verify all the properties of a field. We shall return to this matter in Chapter IX when we consider finite fields more generally, and we shall then have a way of constructing $\mathbb{F}_4$ that avoids tedious checking.

---

[8] In fact, if b and c exist with $ab = ca = 1$, then a is a unit with $a^{-1} = b = c$ because $b = 1b = (ca)b = c(ab) = c1 = c$.

In analogy with the theory of groups, we define a **subring** of a ring to be a nonempty subset that is closed under addition, negation, and multiplication. The set $2\mathbb{Z}$ of even integers is a subring of the ring $\mathbb{Z}$ of integers. A **subfield** of a field is a subset containing 0 and 1 that is closed under addition, negation, multiplication, and multiplicative inverses for its nonzero elements. The set $\mathbb{Q}$ of rationals is a subfield of the field $\mathbb{R}$ of reals.

Intermediate between rings and fields are two kinds of objects—integral domains and division rings—that arise frequently enough to merit their own names.

The setting for the first is a commutative ring R. A nonzero element a of R is called a **zero divisor** if there is some nonzero b in R with $ab = 0$. For example the element 2 in the ring $\mathbb{Z}/6\mathbb{Z}$ is a zero divisor because $2 \cdot 3 = 0$. An **integral domain** is a *nonzero* commutative ring with identity having no zero divisors. Fields have no zero divisors since if a and b are nonzero, then $ab = 0$ would force $b = 1b = (a^{-1}a)b = a^{-1}(ab) = a^{-1}0 = 0$ and would give a contradiction; therefore every field is an integral domain. The ring of integers $\mathbb{Z}$ is another example of an integral domain, and the polynomial rings $\mathbb{Q}[X]$ and $\mathbb{R}[X]$ and $\mathbb{C}[X]$ introduced in Section I.3 are further examples. A cancellation law for multiplication holds in any integral domain:

$$ab = ac \text{ with } a \neq 0 \quad \text{implies} \quad b = c.$$

In fact, $ab = ac$ implies $a(b - c) = 0$; since $a \neq 0$, $b - c$ must be 0.

The other object with its own name is a **division ring**, which is a nonzero ring with identity such that every nonzero element is a unit. The commutative division rings are the fields, and we have encountered only one noncommutative division ring so far. That is the set $\mathbb{H}$ of quaternions, which was introduced in Section 1. Division rings that are not fields will play only a minor role in this book but are of importance in *Advanced Algebra*.

Let us turn to mappings. A function $\varphi : R \to R'$ between two rings is an **isomorphism** of rings if φ is one-one onto and satisfies $\varphi(a + b) = \varphi(a) + \varphi(b)$ and $\varphi(ab) = \varphi(a)\varphi(b)$ for all a and b in R. In other words, φ is to be an isomorphism of the additive groups and to satisfy $\varphi(ab) = \varphi(a)\varphi(b)$. Such a mapping carries the identity, if any, in R to the identity of R'. The relation "is isomorphic to" is an equivalence relation. Common notation for an isomorphism of rings is $R \cong R'$; because of the symmetry, one can say that R and R' are **isomorphic**.

A function $\varphi : R \to R'$ between two rings is a **homomorphism** of rings if φ satisfies $\varphi(a + b) = \varphi(a) + \varphi(b)$ and $\varphi(ab) = \varphi(a)\varphi(b)$ for all a and b in R. In other words, φ is to be a homomorphism of the additive groups and to satisfy $\varphi(ab) = \varphi(a)\varphi(b)$.

EXAMPLES OF HOMOMORPHISMS OF RINGS.

(1) The mapping $\varphi : \mathbb{Z} \to \mathbb{Z}/m\mathbb{Z}$ given by $\varphi(k) = k \bmod m$.

(2) The evaluation mapping $\varphi : \mathbb{R}[X] \to \mathbb{R}$ given by $P(X) \mapsto P(r)$ for some fixed r in $\mathbb{R}$.

(3) Mappings with the direct product $\mathbb{Z} \times \mathbb{Z}$. The additive group $\mathbb{Z} \times \mathbb{Z}$ becomes a commutative ring with identity under coordinate-by-coordinate multiplication, namely $(a, a') + (b, b') = (a + b, a' + b')$. The identity is $(1, 1)$. Projection $(a, a') \mapsto a$ to the first coordinate is a homomorphism of rings $\mathbb{Z} \times \mathbb{Z} \to \mathbb{Z}$ that carries identity to identity. Inclusion $a \mapsto (a, 0)$ of $\mathbb{Z}$ into the first coordinate is a homomorphism of rings $\mathbb{Z} \to \mathbb{Z} \times \mathbb{Z}$ that does not carry identity to identity.[9]

Proposition 4.19. If R is a ring with identity 1_R, then there exists a unique homomorphism of rings $\varphi_1 : \mathbb{Z} \to R$ such that $\varphi(1) = 1_R$.

PROOF. The formulas for manipulating exponents of an element in a group, when translated into the additive notation for addition in R, say that $n \mapsto nr$ satisfies $(m + n)r = mr + nr$ and $(mn)r = m(nr)$ for all r in R and all integers m and n. The first of these formulas implies, for any r in R, that $\varphi_r(n) = nr$ is a homomorphism between the additive groups of $\mathbb{Z}$ and R, and it is certainly uniquely determined by its value for $n = 1$. The distributive laws imply that $\psi_r(r') = r'r$ is another homomorphism of additive groups. Hence $\psi_r \circ \varphi_{r'}$ and $\varphi_{r'r}$ are homomorphisms between the additive groups of $\mathbb{Z}$ and R. Since $(\psi_r \circ \varphi_{r'})(1) = \psi_r(r') = r'r = \varphi_{r'r}(1)$, we must have $(\psi_r \circ \varphi_{r'})(m) = \varphi_{r'r}(m)$ for all integers m. Thus $(mr')r = m(r'r)$ for all m. Putting $r = n1_R$ and $r' = 1_R$ proves the fourth equality of the computation

$$\varphi_1(mn) = (mn)1_R = m(n1_R)$$
$$= m(1_R(n1_R)) = (m1_R)(n1_R) = \varphi_1(m)\varphi_1(n),$$

and shows that φ_1 is in fact a homomorphism of rings. □

The image of a homomorphism $\varphi : R \to R'$ of rings is a subring of R', as is easily checked. The kernel turns out to be more than just of subring of R. If a is in the kernel and b is any element of R, then $\varphi(ab) = \varphi(a)\varphi(b) = 0\varphi(b) = 0$ and similarly $\varphi(ba) = 0$. Thus the kernel of a ring homomorphism is closed under products of members of the kernel with arbitrary members of R. Adapting a definition to this circumstance, one says that an **ideal** I of R (or **two-sided ideal** in case of ambiguity) is an additive subgroup such that ab and ba are in I whenever a is in I and b is in R. Briefly then, the kernel of a homomorphism of rings is an ideal.

Conversely suppose that I is an ideal in a ring R. Since I is certainly an additive subgroup of an abelian group, we can form the additive quotient group

---

[9]Sometimes authors who build the existence of an identity into the definition of "ring" insist as a matter of definition that homomorphisms of rings carry identity to identity. Such authors would then exclude this particular mapping from consideration as a homomorphism.

R/I. It is customary to write the individual cosets in additive notation, thus as $r + I$. In analogy with Proposition 4.10, we have the following result for the present context.

Proposition 4.20. If I is an ideal in a ring R, then a well-defined operation of multiplication is obtained within the additive group R/I by the definition $(r_1 + I)(r_2 + I) = r_1 r_2 + I$, and R/I becomes a ring. If R has an identity 1, then $1 + I$ is an identity in R/I. With these definitions the function $q : R \to R/I$ given by $q(r) = r + I$ is a ring homomorphism of R onto R/I with kernel I. Consequently every ideal of R is the kernel of some homomorphism of rings.

REMARKS. When I is an ideal, the ring R/I is called a **quotient ring**[10] of R, and the homomorphism $q : R \to R/I$ is called the **quotient homomorphism**. In the special case that $R = \mathbb{Z}$ and $I = m\mathbb{Z}$, the construction of R/I reduces to the construction of $\mathbb{Z}/m\mathbb{Z}$ as a ring at the beginning of this section.

PROOF. If we change the representatives of the cosets from r_1 and r_2 to $r_1 + i_1$ and $r_2 + i_2$ with i_1 and i_2 in I, then $(r_1 + i_1)(r_2 + i_2) = r_1 r_2 + (i_1 r_1 + r_1 i_2 + i_1 i_2)$ is in $r_1 r_2 + I$ by the closure properties of I. Hence multiplication is well defined.

The associativity of this multiplication follows from the associativity of multiplication in R because

$$\big((r_1 + I)(r_2 + I)\big)(r_3 + I) = (r_1 r_2 + I)(r_3 + I) = (r_1 r_2)r_3 + I = r_1(r_2 r_3) + I$$
$$= (r_1 + I)(r_2 r_3 + I) = (r_1 + I)\big((r_2 + I)(r_3 + I)\big).$$

Similarly the computation

$$(r_1 + I)\big((r_2 + I) + (r_3 + I)\big) = r_1(r_2 + r_3) + I = (r_1 r_2 + r_1 r_3) + I$$
$$= (r_1 + I)(r_2 + I) + (r_1 + I)(r_3 + I)$$

yields one distributive law, and the other distributive law is proved in the same way. If R has an identity 1, then $(1 + I)(r + I) = 1r + I = r + I$ and $(r + I)(1 + I) = r1 + I = r + I$ show that $1 + I$ is an identity in R/I.

Finally we know that the quotient map $q : R \to R/I$ is a homomorphism of additive groups, and the computation $q(r_1 r_2) = r_1 r_2 + I = (r_1 + I)(r_2 + I) = q(r_1)q(r_2)$ shows that q is a homomorphism of rings. □

EXAMPLES OF IDEALS.

(1) The ideals in the ring $\mathbb{Z}$ coincide with the additive subgroups and are the sets $m\mathbb{Z}$; the reason each $m\mathbb{Z}$ is an ideal is that if a and b are integers and m divides a, then m divides ab.

---

[10]Quotient rings are known also as "factor rings." A "ring of quotients," however, is something different.

(2) The ideals in a field $\mathbb{F}$ are 0 and $\mathbb{F}$ itself, no others; in fact, if $a \neq 0$ is in an ideal and b is in $\mathbb{F}$, then the equality $b = (ba^{-1})a$ shows that b is in the ideal and that the ideal therefore contains all elements of $\mathbb{F}$.

(3) If R is $\mathbb{Q}[X]$ or $\mathbb{R}[X]$ or $\mathbb{C}[X]$, then every ideal I is of the form $I = Rf(X)$ for some polynomial $f(X)$. In fact, we can take $f(X) = 0$ if $I = 0$. If $I \neq 0$, let $f(X)$ be a nonzero member of I of lowest possible degree. If $A(X)$ is in I, then Proposition 1.12 shows that $A(X) = f(X)B(X) + C(X)$ with $C(X) = 0$ or $\deg C < \deg f$. The equality $C(X) = A(X) - f(X)B(X)$ shows that $C(X)$ is in I, and the minimality of $\deg f$ implies that $C(X) = 0$. Thus $A(X) = f(X)B(X)$.

(4) In a ring R with identity 1, an ideal I is a proper subset of R if and only if 1 is not in I. In fact, I is certainly a proper subset if 1 is not in I. In the converse direction if 1 is in I, then every element $r = r1$, for r in R, lies in I. Hence $I = R$, and I is not a proper subset.

In analogy with what was shown for vector spaces in Proposition 2.25 and for groups in Proposition 4.11, quotients in the context of rings allow for the factorization of certain homomorphisms of rings. The appropriate result is stated as Proposition 4.21 and is pictured in Figure 4.6.

Proposition 4.21. Let $\varphi : R_1 \to R_2$ be a homomorphism of rings, let $I_0 = \ker \varphi$, let I be an ideal of R_1 contained in I_0, and let $q : R_1 \to R_1/I$ be the quotient homomorphism. Then there exists a homomorphism of rings $\overline{\varphi} : R_1/I \to R_2$ such that $\varphi = \overline{\varphi} \circ q$, i.e., $\overline{\varphi}(r_1 + I) = \varphi(r_1)$. It has the same image as φ, and $\ker \overline{\varphi} = \{r + I \mid r \in I_0\}$.

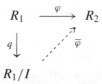

FIGURE 4.6. Factorization of homomorphisms of rings via the quotient of a ring by an ideal.

REMARK. One says that φ **factors through** R_1/I or **descends to** R_1/I.

PROOF. Proposition 4.11 shows that φ descends to a homomorphism $\overline{\varphi}$ of the additive group of R_1/I into the additive group of R_2 and that all the other conclusions hold except possibly for the fact that $\overline{\varphi}$ respects multiplication. To see that $\overline{\varphi}$ respects multiplication, we just compute that $\overline{\varphi}((r + I)(r' + I)) = \overline{\varphi}(rr' + I) = \varphi(rr') = \varphi(r)\varphi(r') = \overline{\varphi}(r + I)\overline{\varphi}(r' + I)$. □

An example of special interest occurs when φ is a homomorphism of rings $\varphi : \mathbb{Z} \to R$ and the ideal $m\mathbb{Z}$ of $\mathbb{Z}$ is contained in the kernel of φ. Then the proposition says that φ descends to a homomorphism of rings $\overline{\varphi} : \mathbb{Z}/m\mathbb{Z} \to R$. We shall make use of this result shortly. But first let us state a different special case as a corollary.

Corollary 4.22. Let $\varphi : R_1 \to R_2$ be a homomorphism of rings, and suppose that φ is onto R_2 and has kernel I. Then φ exhibits the ring R_1/I as canonically isomorphic to R_2.

PROOF. Take $I = I_0$ in Proposition 4.21, and form $\overline{\varphi} : R_1/I \to R_2$ with $\varphi = \overline{\varphi} \circ q$. The proposition shows that $\overline{\varphi}$ is onto R_2 and has trivial kernel, i.e., the identity element of R_1/I. Having trivial kernel, $\overline{\varphi}$ is one-one. $\square$

Proposition 4.23. Any field $\mathbb{F}$ contains a subfield isomorphic to the rationals $\mathbb{Q}$ or to some field $\mathbb{F}_p$ with p prime.

REMARKS. The subfield in the proposition is called the **prime field** of $\mathbb{F}$. The **characteristic** of $\mathbb{F}$ is defined to be 0 if the prime field is isomorphic to $\mathbb{Q}$ and to be p if the prime field is isomorphic to $\mathbb{F}_p$.

PROOF. Proposition 4.19 produces a homomorphism of rings $\varphi_1 : \mathbb{Z} \to \mathbb{F}$ with $\varphi_1(1) = 1$. The kernel of φ_1 is an ideal, necessarily of the form $m\mathbb{Z}$ with m an integer ≥ 0, and the image of φ_1 is a commutative subring with identity in $\mathbb{F}$. Let $\overline{\varphi}_1 : \mathbb{Z}/m\mathbb{Z} \to \mathbb{F}$ be the descended homomorphism given by Proposition 4.21. The integer m cannot factor nontrivially, say as $m = rs$, because otherwise $\overline{\varphi}_1(r)$ and $\overline{\varphi}_1(s)$ would be nonzero members of $\mathbb{F}$ with $\overline{\varphi}_1(r)\overline{\varphi}_1(s) = \overline{\varphi}_1(rs) = \overline{\varphi}_1(0) = 0$, in contradiction to the fact that a field has no zero divisors.

Thus m is prime or m is 0. If m is a prime p, then $\mathbb{Z}/p\mathbb{Z}$ is a field, and the image of $\overline{\varphi}_1$ is the required subfield of $\mathbb{F}$. Thus suppose that $m = 0$. Then φ_1 is one-one, and $\mathbb{F}$ contains a subring with identity isomorphic to $\mathbb{Z}$. Define a function $\Phi_1 : \mathbb{Q} \to \mathbb{F}$ by saying that if k and l are integers with $l \neq 0$, then $\Phi_1(kl^{-1}) = \varphi_1(k)\varphi_1(l)^{-1}$. This is well defined because $\varphi_1(l) \neq 0$ and because $k_1 l_1^{-1} = k_2 l_2^{-1}$ implies $k_1 l_2 = k_2 l_1$ and hence $\varphi_1(k_1)\varphi_1(l_2) = \varphi_1(k_2)\varphi_1(l_1)$ and $\varphi_1(k_1)\varphi_1(l_1)^{-1} = \varphi_1(k_2)\varphi_1(l_2)^{-1}$. We readily check that Φ_1 is a homomorphism with kernel 0. Then $\mathbb{F}$ contains the subfield $\Phi_1(\mathbb{Q})$ isomorphic to $\mathbb{Q}$. $\square$

5. Polynomials and Vector Spaces

In this section we complete the digression begun in Section 4. We shall be using the elementary notions of rings and fields established in Section 4 in order to work with (i) polynomials over any commutative ring with identity and (ii) vector spaces over arbitrary fields.

It is an important observation that a good deal of what has been proved so far in this book concerning polynomials when $\mathbb{F}$ is $\mathbb{Q}$ or $\mathbb{R}$ or $\mathbb{C}$ remains valid when $\mathbb{F}$ is any field. Specifically all the results in Section I.3 through Theorem 1.17 on the topic of polynomials in one indeterminate remain valid as long as the coefficients are from a field. The theory breaks down somewhat when one tries to extend it by allowing coefficients that are not in a field or by allowing more than one indeterminate. Because of this circumstance and because we have not yet announced a universal mapping property for polynomial rings and because we have not yet addressed the several-variable case, we shall briefly review matters now while extending the reach of the theory that we have.

Let R be a nonzero commutative ring with identity, so that $1 \neq 0$. A polynomial in one indeterminate is to be an expression $P(X) = a_n X^n + \cdots + a_2 X^2 + a_1 X + a_0$ in which X is a symbol, not a variable. Nevertheless, the usual kinds of manipulations with polynomials are to be valid. This description lacks precision because X has not really been defined adequately. To make a precise definition, we remove X from the formalism and simply define the polynomial to be the tuple $(a_0, a_1, \ldots, a_n, 0, 0, \ldots)$ of its coefficients. Thus a **polynomial** in **one indeterminate** with **coefficients** in R is an infinite sequence of members of R such that all terms of the sequence are 0 from some point on. The indexing of the sequence is to begin with 0, and X is to refer to the polynomial $(0, 1, 0, 0, \ldots)$. We may refer to a polynomial P as $P(X)$ if we want to emphasize that the indeterminate is called X. Addition and negation of polynomials are defined in coordinate-by-coordinate fashion by

$$(a_0, a_1, \ldots, a_n, 0, 0, \ldots) + (b_0, b_1, \ldots, b_n, 0, 0, \ldots)$$
$$= (a_0 + b_0, a_1 + b_1, \ldots, a_n + b_n, 0, 0, \ldots),$$
$$-(a_0, a_1, \ldots, a_n, 0, 0, \ldots) = (-a_0, -a_1, \ldots, -a_n, 0, 0, \ldots),$$

and the set $R[X]$ of polynomials is then an abelian group isomorphic to the direct sum of infinitely many copies of the additive group of R. As in Section I.3, X^n is to be the polynomial whose coefficients are 1 in the nth position, with $n \geq 0$, and 0 in all other positions. Polynomial multiplication is then defined so as to match multiplication of expressions $a_n X^n + \cdots + a_1 X + a_0$ if the product is expanded out, powers of X are added, and the terms containing like powers of X are collected. Thus the precise definition is that

$$(a_0, a_1, \ldots, 0, 0, \ldots)(b_0, b_1, \ldots, 0, 0, \ldots) = (c_0, c_1, \ldots, 0, 0, \ldots),$$

where $c_N = \sum_{k=0}^{N} a_k b_{N-k}$. It is a simple matter to check that this multiplication makes $R[X]$ into a commutative ring.

The polynomial with all entries 0 is denoted by 0 and is called the **zero polynomial**. For all polynomials $P = (a_0, \ldots, a_n, 0, \ldots)$ other than 0, the **degree** of P, denoted by $\deg P$, is defined to be the largest index n such that $a_n \neq 0$. In this case, a_n is called the **leading coefficient**, and $a_n X^n$ is called the **leading term**; if $a_n = 1$, the polynomial is called **monic**. The usual convention with the 0 polynomial is either to leave its degree undefined or to say that the degree is $-\infty$; let us follow the latter approach in this section in order not to have to separate certain formulas into cases.

There is a natural one-one homomorphism of rings $\iota : R \to R[X]$ given by $\iota(c) = (c, 0, 0, \ldots)$ for c in R. This sends the identity of R to the identity of $R[X]$. Thus we can identify R with the **constant polynomials**, i.e., those of degree ≤ 0.

If P and Q are nonzero polynomials, then

$$\deg(P + Q) \leq \max(\deg P, \deg Q).$$

In this formula equality holds if $\deg P \neq \deg Q$. In the case of multiplication, let P and Q have respective leading terms $a_m X^m$ and $b_n X^n$. All the coefficients of PQ are 0 beyond the $(m+n)^{\text{th}}$, and the $(m+n)^{\text{th}}$ is $a_m b_n$. This in principle could be 0 but is nonzero if R is an integral domain. Thus P and Q nonzero implies

$$\deg(PQ) \begin{cases} \leq \deg P + \deg Q & \text{for general } R, \\ = \deg P + \deg Q & \text{if } R \text{ is an integral domain.} \end{cases}$$

It follows in particular that $R[X]$ is an integral domain if R is.

Normally we shall write out specific polynomials using the informal notation with powers of X, using the more precise notation with tuples only when some ambiguity might otherwise result.

In the special case that R is a field, Section I.3 introduced the notion of evaluation of a polynomial $P(X)$ at a point r in the field, thus providing a mapping $P(X) \mapsto P(r)$ from $R[X]$ to R for each r in R. We listed a number of properties of this mapping, and they can be summarized in our present language by the statement that the mapping is a homomorphism of rings. Evaluation is a special case of a more sweeping property of polynomials given in the next proposition as a **universal mapping property** of $R[X]$.

Proposition 4.24. Let R be a nonzero commutative ring with identity, and let $\iota : R \to R[X]$ be the identification of R with constant polynomials. If T is any commutative ring with identity, if $\varphi : R \to T$ is a homomorphism of rings sending 1 into 1, and if t is in T, then there exists a unique homomorphism of rings $\Phi : R[X] \to T$ carrying identity to identity such that $\Phi(\iota(r)) = \varphi(r)$ for all $r \in R$ and $\Phi(X) = t$.

REMARKS. The mapping Φ is called the **substitution homomorphism** extending φ and substituting t for X, and the mapping is written $P(X) \mapsto P^\varphi(t)$. The notation means that φ is to be applied to the coefficients of P and then X is to be replaced by t. A diagram of this homomorphism as a universal mapping property appears in Figure 4.7. In the special case that $T = R$ and φ is the identity, Φ reduces to **evaluation** at t, and the mapping is written $P(X) \mapsto P(t)$, just as in Section I.3.

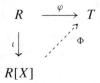

FIGURE 4.7. Substitution homomorphism for polynomials in one indeterminate.

PROOF. Define $\Phi(a_0, a_1, \ldots, a_n, 0, \ldots) = \varphi(a_0) + \varphi(a_1)t + \cdots + \varphi(a_n)t^n$. It is immediate that Φ is a homomorphism of rings sending the identity $\iota(1) = (1, 0, 0, \ldots)$ of $R[X]$ to the identity $\varphi(1)$ of T. If r is in R, then $\Phi(\iota(r)) = \Phi(r, 0, 0, \ldots) = \varphi(r)$. Also, $\Phi(X) = \Phi(0, 1, 0, 0, \ldots) = t$. This proves existence. Uniqueness follows since $\iota(R)$ and X generate $R[X]$ and since a homomorphism defined on $R[X]$ is therefore determined by its values on $\iota(R)$ and X. □

The formulation of the proposition with the general $\varphi : R \to T$, rather than just the identity mapping on R, allows several kinds of applications besides the routine evaluation mapping. An example of one kind occurs when $R = \mathbb{C}[X]$ and $\varphi : \mathbb{C} \to \mathbb{C}[X]$ is the composition of complex conjugation on $\mathbb{C}$ followed by the identification of complex numbers with constant polynomials in $\mathbb{C}[X]$; the proposition then says that complex conjugation of the coefficients of a member of $\mathbb{C}[X]$ is a ring homomorphism. This observation simplifies the solution of Problem 7 in Chapter I. Similarly one can set up matters so that the proposition shows the passage from $\mathbb{Z}[X]$ to $(\mathbb{Z}/m\mathbb{Z})[X]$ by reduction of coefficients modulo m to be a ring homomorphism.

Still a third kind of application is to take T in the proposition to be a ring with the same kind of universal mapping property that $R[X]$ has, and the consequence is an abstract characterization of $R[X]$. We carry out the details below as Proposition 4.25. This result will be applied later in this section to the several-indeterminate case to show that introducing several indeterminates at once yields the same ring, up to canonical isomorphism, as introducing them one at a time.

Proposition 4.25. Let R and S be nonzero commutative rings with identity, let X' be an element of S, and suppose that $\iota' : R \to S$ is a one-one ring homomorphism of R into S carrying 1 to 1. Suppose further that (S, ι', X') has the following property: whenever T is a commutative ring with identity, $\varphi : R \to T$ is a homomorphism of rings sending 1 into 1, and t is in T, then there exists a unique homomorphism $\Phi' : S \to T$ carrying identity to identity such that $\Phi'(\iota'(r)) = \varphi(r)$ for all $r \in R$ and $\Phi'(X') = t$. Then there exists a unique homomorphism of rings $\Psi : R[X] \to S$ such that $\Psi \circ \iota = \iota'$ and $\Psi(X) = X'$, and Ψ is an isomorphism.

REMARK. A somewhat weaker conclusion than in the proposition is that any triple (S, ι', X') having the same universal mapping property as $(R[X], \iota, X)$ is isomorphic to (S, ι', X'), the isomorphism being unique.

PROOF. In the universal mapping property for S, take $T = R[X]$, $\varphi = \iota$, and $t = X$. The hypothesis gives us a ring homomorphism $\Phi' : S \to R[X]$ with $\Phi'(1) = 1$, $\Phi' \circ \iota' = \iota$, and $\Phi'(X') = X$. Next apply Proposition 4.24 with $T = S$, $\varphi = \iota'$, and $t = X'$. We obtain a ring homomorphism $\Phi : R[X] \to S$ with $\Phi(1) = 1$, $\Phi \circ \iota = \iota'$, and $\Phi(X) = X'$. Then $\Phi' \circ \Phi$ is a ring homomorphism from $R[X]$ to itself carrying 1 to 1, fixing X, and having $\Phi' \circ \Phi\big|_{\iota(R)} = \iota$. From the uniqueness in Proposition 4.24 when $T = R[X]$, $\varphi = \iota$, and $t = X$, we see that $\Phi' \circ \Phi$ is the identity on $R[X]$. Reversing the roles of Φ and Φ' and applying the uniqueness in the universal mapping property for S, we see that $\Phi \circ \Phi'$ is the identity on S. Therefore Φ may be taken as the isomorphism Ψ in the statement of the proposition. This proves existence for Ψ, and uniqueness follows since $\iota(R)$ and X together generate $R[X]$ and since Ψ is a homomorphism. $\square$

If P is a polynomial over R in one indeterminate and r is in R, then r is a **root** of P if $P(r) = 0$. We know as a consequence of Corollary 1.14 that for any prime p, any polynomial in $\mathbb{F}_p[X]$ of degree $n \geq 1$ has at most n roots. This result does not extend to $\mathbb{Z}/m\mathbb{Z}$ for all positive integers m: when $m = 8$, the polynomial $X^2 - 1$ has 4 roots, namely 1, 3, 5, 7. This result about $\mathbb{F}_p[X]$ has the following consequence.

Proposition 4.26. If $\mathbb{F}$ is a field, then any finite subgroup of the multiplicative group $\mathbb{F}^\times$ is cyclic.

PROOF. Let C be a subgroup of $\mathbb{F}^\times$ of finite order n. Lagrange's Theorem (Corollary 4.8) shows that the order of each element of C divides n. With h defined as the maximum order of an element of C, it is enough to show that $h = n$. Let a be an element of order h. The polynomial $X^h - 1$ has at most h roots by Corollary 1.14, and a is one of them, by definition of "order." If $h < n$, then it follows that some member b of C is not a root of $X^h - 1$. The order h' of b is then a divisor of n but cannot be a divisor of h since otherwise we would have $b^h = (b^{h'})^{h/h'} = 1^{h/h'} = 1$. Consequently there exists a prime p such that some power p^r of p divides h' but not h. Let $s < r$ be the exact power of p dividing h, and write $h = mp^s$, so that $\text{GCD}(m, p^r) = 1$ and $a' = a^{p^s}$ has order m. Put $q = h'/p^r$, so that $b' = b^q$ has order p^r. The proof will be completed by showing that $c = a'b'$ has order $mp^r = hp^{r-s} > h$, in contradiction to the maximality of h.

Let t be the order of c. On the one hand, from $c^{mp^r} = (a')^{mp^r}(b')^{mp^r} = a^{hp^{r-s}}b^{mp^r q} = a^{hp^{r-s}}b^{mh'} = (a^h)^{p^{r-s}}(b^{h'})^m = 1$, we see that t divides mp^r. On the other hand, $1 = c^t$ says that $(a')^t = (b')^{-t}$. Raising both sides to the p^r power gives $1 = ((b')^{p^r})^{-t} = (a')^{tp^r}$, and hence m divides tp^r; by Corollary 1.3, m divides t. Raising both sides of $(a')^t = (b')^{-t}$ to the mth power gives $1 = ((a')^m)^t = (b')^{-tm}$, and hence p^r divides tm; by Corollary 1.3, p^r divides t. Applying Corollary 1.4, we conclude that mp^r divides t. Therefore $t = mp^r$, and the proof is complete. □

Corollary 4.27. The multiplicative group of a finite field is cyclic.

PROOF. This is a special case of Proposition 4.26. □

A finite field $\mathbb{F}$ can have a nonzero polynomial that is 0 at every element of $\mathbb{F}$. Indeed, every element of $\mathbb{F}_p$ is a root of $X^p - X$, as a consequence of Fermat's Little Theorem. It is for this reason that it is unwise to confuse a polynomial in an indeterminate with a "polynomial function."

Let us make the notion of a polynomial function of one variable rigorous. If $P(X)$ is a polynomial with coefficients in the commutative ring R with identity, then Proposition 4.24 gives us an evaluation homomorphism $P \mapsto P(r)$ for each r in R. The function $r \mapsto P(r)$ from R into R is the **polynomial function** associated to the polynomial P. This function is a member of the commutative ring of all R-valued functions on R, and the mapping $P \mapsto (r \mapsto P(r))$ is a homomorphism of rings. What we know from Corollary 1.14 is that this homomorphism is one-one if R is an infinite field. A negative result is that if R is a finite commutative ring with identity, then $\prod_{r \in R}(X - r)$ is a polynomial that maps to the 0 function, and hence the homomorphism is not one-one. A more general positive result than the one above for infinite fields is the following.

Proposition 4.28.

(a) If R is a nonzero commutative ring with identity and $P(X)$ is a member of $R[X]$ with a root r, then $P(X) = (X - r)Q(X)$ for some $Q(X)$ in $R[X]$.

(b) If R is an integral domain, then a nonzero member of $R[X]$ of degree n has at most n roots.

(c) If R is an infinite integral domain, then the ring homomorphism of $R[X]$ to the ring of polynomial functions from R to R, given by evaluation, is one-one.

PROOF. For (a), we proceed by induction on the degree of P, the base case of the induction being degree ≤ 0. If the conclusion has been proved for degree $< n$ with $n \geq 1$, let the leading term of P be $a_n X^n$. Then $P(X) = a_n(X-r)^n + A(X)$ with $\deg A < n$. Evaluation at r gives, by virtue of Proposition 4.24, $0 = 0 + A(r)$. By the inductive hypothesis, $A(X) = (X-r)B(X)$. Then $P(X) = (X-r)Q(X)$ with $Q(X) = a_n(X-r)^{n-1} + B(X)$, and the induction is complete.

For (b), let $P(X)$ have degree n with at least $n+1$ distinct roots $r_1, \ldots, r_{n+1}$. Part (a) shows that $P(X) = (X - r_1)P_1(X)$ with $\deg P_1 = n - 1$. Also, $0 = P(r_2) = (r_2 - r_1)P_1(r_2)$. Since $r_2 - r_1 \neq 0$ and since R has no zero divisors, $P_1(r_2) = 0$. Part (a) then shows that $P_1(X) = (X - r_2)P_2(X)$, and substitution gives $P(X) = (X - r_1)(X - r_2)P_2(X)$. Continuing in this way, we obtain $P(X) = (X - r_1)\cdots(X - r_n)P_n(X)$ with $\deg P_n = 0$. Since $P \neq 0$, $P_n \neq 0$. So P_n is a nonzero constant polynomial $P_n(X) = c \neq 0$. Evaluating at r_{n+1}, we obtain $0 = (r_{n+1} - r_1)\cdots(r_{n+1} - r_n)c$ with each factor nonzero, in contradiction to the fact that R is an integral domain.

For (c), a polynomial in the kernel of the ring homomorphism has every member of R as a root. If R is infinite, (b) shows that such a polynomial is necessarily the zero polynomial. Thus the kernel is 0, and the ring homomorphism has to be one-one. □

Let us turn our attention to polynomials in several indeterminates. Fix the nonzero commutative ring R with identity, and let n be a positive integer. Informally a polynomial over R in n indeterminates is to be a finite sum

$$\sum_{j_1 \geq 0, \ldots, j_n \geq 0} r_{j_1, \ldots, j_n} X_1^{j_1} \cdots X_n^{j_n}$$

with each $r_{j_1, \ldots, j_n}$ in R. To make matters precise, we work just with the system of coefficients, just as in the case of one indeterminate.

Let J be the set of integers ≥ 0, and let J^n be the set of n-tuples of elements of J. A member of J^n may be written as $j = (j_1, \ldots, j_n)$. Addition of members of J^n is defined coordinate by coordinate. Thus $j + j' = (j_1 + j'_1, \ldots, j_n + j'_n)$ if $j = (j_1, \ldots, j_n)$ and $j' = (j'_1, \ldots, j'_n)$. A **polynomial** in n **indeterminates** with **coefficients** in R is a function $f : J^n \to R$ such that $f(j) \neq 0$ for only finitely many $j \in J^n$. Temporarily let us write S for the set of all such polynomials for a particular n. If f and g are two such polynomials, their sum h and product k are the polynomials defined by

$$h(j) = f(j) + g(j),$$
$$k(i) = \sum_{j+j'=i} f(j)g(j').$$

Under these definitions, S is a commutative ring.

Define a mapping $\iota : R \to S$ by

$$\iota(r)(j) = \begin{cases} r & \text{if } j = (0, \ldots, 0), \\ 0 & \text{otherwise.} \end{cases}$$

Then ι is a one-one homomorphism of rings, $\iota(0)$ is the zero element of S and is called simply 0, and $\iota(1)$ is a multiplicative identity for S. The polynomials in the image of ι are called the **constant polynomials**.

For $1 \le k \le n$, let e_k be the member of J^n that is 1 in the kth place and is 0 elsewhere. Define X_k to be the polynomial that assigns 1 to e_k and assigns 0 to all other members of J^n. We say that X_k is an **indeterminate**. If $j = (j_1, \ldots, j_n)$ is in J^n, define X^j to be the product

$$X^j = X_1^{j_1} \cdots X_n^{j_n}.$$

If r is in R, we allow ourselves to abbreviate $\iota(r)X^j$ as rX^j, and any such polynomial is called a **monomial**. The monomial rX^j is the polynomial that assigns r to j and assigns 0 to all other members of J^n. Then it follows immediately from the definitions that each polynomial has a unique expansion as a finite sum of nonzero monomials. Thus the most general member of S is of the form $\sum_{j \in J^n} r_j X^j$ with only finitely many nonzero terms. This is called the **monomial expansion** of the given polynomial.

We may now write $R[X_1, \ldots, X_n]$ for S. A polynomial $\sum_{j \in J^n} r_j X^j$ may be conveniently abbreviated as P or as $P(X)$ or as $P(X_1, \ldots, X_n)$ when its monomial expansion is either understood or irrelevant.

The **degree** of the 0 polynomial is defined for this section to be $-\infty$, and the degree of any monomial rX^j with $r \ne 0$ is defined to be the integer

$$|j| = j_1 + \cdots + j_n \qquad \text{if } j = (j_1, \ldots, j_n).$$

Finally the **degree** of any nonzero polynomial P, denoted by $\deg P$, is defined to be the maximum of the degrees of the terms in its monomial expansion. If all the nonzero monomials in the monomial expansion of a polynomial P have the same degree d, then P is said to be **homogeneous** of degree d. Under these definitions the 0 polynomial has degree $-\infty$ but is homogeneous of every degree. If P and Q are homogeneous polynomials of degrees d and d', then PQ is homogeneous of degree dd' (and possibly equal to the 0 polynomial).

In any event, by grouping terms in the monomial expansion of a polynomial according to their degree, we see that every polynomial is uniquely the sum of nonzero homogeneous polynomials of distinct degrees. Let us call this the **homogeneous-polynomial expansion** of the given polynomial. Let us expand two such nonzero polynomials P and Q in this fashion, writing $P = P_{d_1} + \cdots + P_{d_k}$

and $Q = Q_{d'_1} + \cdots + Q_{d'_l}$ with $d_1 < \cdots < d_k$ and $d'_1 < \cdots < d'_l$. Then we see directly that
$$\deg(P + Q) \leq \max(\deg P, \deg Q),$$
$$\deg(PQ) \leq \deg P + \deg Q.$$

In the formula for $\deg(P + Q)$, the term that is potentially of largest degree is $P_{d_k} + Q_{d'_l}$, and it is of degree $\max(\deg P, \deg Q)$ if $\deg P \neq \deg Q$. In the formula for $\deg(PQ)$, the term that is potentially of largest degree is $P_{d_k} Q_{d'_l}$. It is homogeneous of degree $d_k + d'_l$, but it could be 0. Some proof is required that it is not 0 if R is an integral domain, as follows.

Proposition 4.29. If R is an integral domain, then $R[X_1, \ldots, X_n]$ is an integral domain.

PROOF. Let P and Q be nonzero homogeneous polynomials with $\deg P = d$ and $\deg Q = d'$. We are to prove that $PQ \neq 0$. We introduce an ordering on the set of all members j of J^n, saying $j = (j_1, \ldots, j_n) > j' = (j'_1, \ldots, j'_n)$ if there is some k such that $j_i = j'_i$ for $i < k$ and $j_k > j'_k$. In the monomial expansion of P as $P(X) = \sum_{|j|=d} a_j X^j$, let i be the largest n-tuple j in the ordering such that $a_j \neq 0$. Similarly with $Q(X) = \sum_{|j'|=d'} b_{j'} X^{j'}$, let i' be the largest n-tuple j' in the ordering such that $b_{j'} \neq 0$. Then

$$P(X)Q(X) = a_i b_{i'} X^{i+i'} + \sum_{\substack{j, j' \text{ with} \\ (j, j') \neq (i, i')}} a_j b_{j'} X^{j+j'},$$

and all terms in the sum $\sum_{j, j'}$ on the right side have $j + j' < i + i'$. Thus $a_i b_{i'} X^{i+i'}$ is the only term in the monomial expansion of $P(X)Q(X)$ involving the monomial $X^{i+i'}$. Since R is an integral domain and a_i and $b_{i'}$ are nonzero, $a_i b_{i'}$ is nonzero. Thus $P(X)Q(X)$ is nonzero. □

Proposition 4.30. Let R be a nonzero commutative ring with identity, let $R[X_1, \ldots, X_n]$ be the ring of polynomials in n indeterminates, and define $\iota : R \to R[X_1, \ldots, X_n]$ to be the identification of R with constant polynomials. If T is any commutative ring with identity, if $\varphi : R \to T$ is a homomorphism of rings sending 1 into 1, and if $t_1, \ldots, t_n$ are in T, then there exists a unique homomorphism $\Phi : R[X_1, \ldots, X_n] \to T$ carrying identity to identity such that $\Phi(\iota(r)) = \varphi(r)$ for all $r \in R$ and $\Phi(X_j) = t_j$ for $1 \leq j \leq n$.

REMARKS. The mapping Φ is called the **substitution homomorphism** extending φ and substituting t_j for X_j for $1 \leq j \leq n$, and the mapping is written $P(X_1, \ldots, X_n) \mapsto P^\varphi(t_1, \ldots, t_n)$. The notation means that φ is to be applied to each coefficient of P and then $X_1, \ldots, X_n$ are to be replaced by $t_1, \ldots, t_n$.

A diagram of this homomorphism as a **universal mapping property** appears in Figure 4.8. In the special case that $T = R \times \cdots \times R$ (cf. Example 3 of homomorphisms in Section 4) and φ is the identity, Φ reduces to **evaluation** at $(t_1, \ldots, t_n)$, and the mapping is written $P(X_1, \ldots, X_n) \mapsto P(t_1, \ldots, t_n)$.

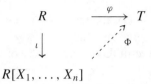

$$R[X_1, \ldots, X_n]$$

FIGURE 4.8. Substitution homomorphism for polynomials in n indeterminates.

PROOF. If $P(X_1, \ldots, X_n) = \sum_{j_1 \geq 0, \ldots, j_n \geq 0} a_{j_1, \ldots, j_n} X_1^{j_1} \cdots X_n^{j_n}$ is the monomial expansion of a member P of $R[X_1, \ldots, X_n]$, then $\Phi(P)$ is defined to be the corresponding finite sum $\sum_{j_1 \geq 0, \ldots, j_n \geq 0} a_{j_1, \ldots, j_n} t_1^{j_1} \cdots t_n^{j_n}$. Existence readily follows, and uniqueness follows since $\iota(R)$ and $X_1, \ldots, X_n$ generate $R[X_1, \ldots, X_n]$ and since Φ is a homomorphism. □

Corollary 4.31. If R is a nonzero commutative ring with identity, then $R[X_1, \ldots, X_{n-1}][X_n]$ is isomorphic as a ring to $R[X_1, \ldots, X_n]$.

REMARK. The proof will show that the isomorphism is the expected one.

PROOF. In the notation with n-tuples and J^n, any $(n-1)$-tuple may be identified with an n-tuple by adjoining 0 as its nth coordinate, and in this way, every monomial in $R[X_1, \ldots, X_{n-1}]$ can be regarded as a monomial in $R[X_1, \ldots, X_n]$. The extension of this mapping to sums gives us a one-one homomorphism of rings $\iota' : R[X_1, \ldots, X_{n-1}] \to R[X_1, \ldots, X_n]$. We are going to use Proposition 4.25 to prove the isomorphism of rings $R[X_1, \ldots, X_{n-1}][X_n] \cong R[X_1, \ldots, X_n]$. In the notation of that proposition, the role of R is played by $R[X_1, \ldots, X_{n-1}]$, we take $S = R[X_1, \ldots, X_n]$, and we have constructed ι'. We are to show that (S, ι', X_n) satisfies a certain universal mapping property. Thus suppose that T is a commutative ring with identity, that t is in T, and that $\varphi' : R[X_1, \ldots, X_{n-1}] \to T$ is a homomorphism of rings carrying identity to identity.

We shall apply Proposition 4.30 in order to obtain the desired homomorphism $\Phi' : S \to T$. Let $\iota_{n-1} : R \to R[X_1, \ldots, X_{n-1}]$ be the identification of R with constant polynomials in $R[X_1, \ldots, X_{n-1}]$, and let $\iota_n = \iota' \circ \iota_{n-1}$ be the identification of R with constant polynomials in S. Define $\varphi : R \to T$ by $\varphi = \varphi' \circ \iota_{n-1}$, and take $t_n = t$ and $t_j = \varphi'(X_j)$ for $1 \leq j \leq n-1$. Then Proposition 4.30 produces a homomorphism of rings $\Phi' : S \to T$ with $\Phi'(\iota_n(r)) = \varphi(r)$ for $r \in R$, $\Phi'(\iota'(X_j)) = \varphi'(X_j)$ for $1 \leq j \leq n-1$, and $\Phi'(X_n) = t_n$. The equations

$$\Phi'(\iota'(\iota_{n-1}(r))) = \Phi'(\iota_n(r)) = \varphi(r) = \varphi'(\iota_{n-1}(r))$$

and $\qquad \Phi'(\iota'(X_j)) = \varphi'(X_j)$

show that $\Phi' \circ \iota' = \varphi'$ on $R[X_1, \ldots, X_n]$. Also, $\Phi'(X_n) = t_n = t$. Thus the mapping Φ' sought by Proposition 4.25 exists. It is unique since $R[X_1, \ldots, X_{n-1}]$ and X_n together generate S. The conclusion from Proposition 4.25 is that S is isomorphic to $R[X_1, \ldots, X_{n-1}][X_n]$ via the expected isomorphism of rings. □

We conclude the discussion of polynomials in several variables by making the notion of a polynomial function of several variables rigorous. If $P(X_1, \ldots, X_n)$ is a polynomial in n indeterminates with coefficients in the commutative ring R with identity, then Proposition 4.30 gives us an evaluation homomorphism $P \mapsto P(r_1, \ldots, r_n)$ for each n-tuple $(r_1, \ldots, r_n)$ of members of R. The function $(r_1, \ldots, r_n) \mapsto P(r_1, \ldots, r_n)$ from $R \times \cdots \times R$ into R is the **polynomial function** associated to the polynomial P. This function is a member of the commutative ring of all R-valued functions on $R \times \cdots \times R$, and the mapping $P \mapsto \big((r_1, \ldots, r_n) \mapsto P(r_1, \ldots, r_n)\big)$ is a homomorphism of rings.

Corollary 4.32. If R is an infinite integral domain, then the ring homomorphism of $R[X_1, \ldots, X_n]$ to polynomial functions from $R \times \cdots \times R$ to R, given by evaluation, is one-one.

REMARK. This result extends Proposition 4.28 to several indeterminates.

PROOF. We proceed by induction on n, the case $n = 1$ being handled by Proposition 4.28. Assume the result for $n - 1$ indeterminates. If $P \neq 0$ is in $R[X_1, \ldots, X_n]$, Corollary 4.31 allows us to write

$$P(X_1, \ldots, X_n) = \sum_{i=1}^{k} P_i(X_1, \ldots, X_{n-1}) X_n^i$$

for some k, with each P_i in $R[X_1, \ldots, X_{n-1}]$ and with $P_k(X_1, \ldots, X_{n-1}) \neq 0$. By the inductive hypothesis, $P_k(r_1, \ldots, r_{n-1})$ is nonzero for some elements $r_1, \ldots, r_{n-1}$ of R. So the polynomial $\sum_{i=0}^{k} P_i(r_1, \ldots, r_{n-1}) X_n^i$ in $R[X_n]$ is not the 0 polynomial, and Proposition 4.28 shows that it is not 0 when evaluated at some r_n. Then $P(r_1, \ldots, r_n) \neq 0$. □

It is possible also to introduce polynomial rings in infinitely many variables. These will play roles only as counterexamples in this book, and thus we shall not stop to treat them in detail.

We complete this section with some remarks about vector spaces. The definition of a **vector space** over a general field $\mathbb{F}$ remains the same as in Section II.1, where $\mathbb{F}$ is assumed to be $\mathbb{Q}$ or $\mathbb{R}$ or $\mathbb{C}$. We shall make great use of the fact that all the results in Chapter II concerning vector spaces remain valid when $\mathbb{Q}$ or $\mathbb{R}$ or

$\mathbb{C}$ is replaced by a general field $\mathbb{F}$. The proofs need no adjustments, and it is not necessary to write out the details. For the moment we make only the following application of vector spaces over general fields, but the extended theory of vector spaces will play an important role in most of the remaining chapters of this book.

Proposition 4.33. If $\mathbb{F}$ is a finite field, then the number of elements in $\mathbb{F}$ is a power of a prime.

REMARK. We return to this matter in Chapter IX, showing at that time that for each prime power $p^n > 1$, there is one and only one field with p^n elements, up to isomorphism.

PROOF. The characteristic of $\mathbb{F}$ cannot be 0 since $\mathbb{F}$ is finite, and hence it is some prime p. Denote the prime field of $\mathbb{F}$ by $\mathbb{F}_p$. By restricting the multiplication so that it is defined only on $\mathbb{F}_p \times \mathbb{F}$, we make $\mathbb{F}$ into a vector space over $\mathbb{F}_p$, necessarily finite-dimensional. Proposition 2.18 shows that $\mathbb{F}$ is isomorphic as a vector space to the space $(\mathbb{F}_p)^n$ of n-dimensional column vectors for some n, and hence $\mathbb{F}$ must have p^n elements. $\square$

6. Group Actions and Examples

Let X be a nonempty set, let $\mathcal{F}(X)$ be the group of invertible functions from X onto itself, the group operation being composition, and let G be a group. A **group action** of G on X is a homomorphism of G into $\mathcal{F}(X)$. Examples 5–9 of groups in Section 1 were in fact subgroups of various groups $\mathcal{F}(X)$ and are therefore examples of group actions. Thus every group of permutations of $\{1, \ldots, n\}$, every dihedral group acting on $\mathbb{R}^2$, and every general linear group or subgroup acting on a finite-dimensional vector space over $\mathbb{Q}$ or $\mathbb{R}$ or $\mathbb{C}$ or an arbitrary field $\mathbb{F}$ provides an example. So do the orthogonal and unitary groups acting on $\mathbb{R}^n$ and $\mathbb{C}^n$, as well as the automorphism group of any number field.

We saw an indication in Section 1 that many early examples of groups arose in this way. One source of examples that is of some importance and was not listed in Section 1 occurs in the geometry of $\mathbb{R}^2$. The translations in $\mathbb{R}^2$, together with the rotations about arbitrary points of $\mathbb{R}^2$ and the reflections about arbitrary lines in $\mathbb{R}^2$, form a group G of rigid motions of the plane.[11] This group G is a subgroup of $\mathcal{F}(\mathbb{R}^2)$, and thus G acts on $\mathbb{R}^2$. More generally, whenever a nonempty set X has a notion of distance, the set of **isometries** of X, i.e., the distance-preserving members of $\mathcal{F}(X)$, forms a subgroup of $\mathcal{F}(X)$, and thus the group of isometries of X acts on X.

---

[11]One can show that G is the full group of rigid motions of $\mathbb{R}^2$, but this fact will not concern us.

At any rate a group action τ of G on X, being a homomorphism of G into $\mathcal{F}(X)$, is of the form $g \mapsto \tau_g$, where τ_g is in $\mathcal{F}(X)$ and $\tau_{g_1 g_2} = \tau_{g_1} \tau_{g_2}$. There is an equivalent way of formulating matters that does not so obviously involve the notion of a homomorphism. Namely, we write $\tau_g(x) = gx$. In this notation the group action becomes a function $G \times X \to X$ with $(g, x) \mapsto gx$ such that

(i) $(g_1 g_2)x = g_1(g_2 x)$ for all g_1 and g_2 in G and for all x in X (from the fact that $\tau_{g_1 g_2} = \tau_{g_1} \tau_{g_2}$),

(ii) $1x = x$ for all x in X (from the fact that $\tau_1 = 1$).

Conversely if $G \times X \to X$ satisfies (i) and (ii), then the formulas $x = 1x = (gg^{-1})x = g(g^{-1}x)$ and $x = 1x = (g^{-1}g)x = g^{-1}(gx)$ show that the function $x \mapsto gx$ from X to itself is invertible with inverse $x \mapsto g^{-1}x$. Consequently the definition $\tau_g(x) = gx$ makes $g \mapsto \tau_g$ a function from G into $\mathcal{F}(X)$, and (i) shows that τ is a homomorphism. Thus (i) and (ii) indeed give us an equivalent formulation of the notion of a group action. Both formulations are useful.

Quite often the homomorphism $G \to \mathcal{F}(X)$ of a group action is one-one, and then G can be regarded as a subgroup of $\mathcal{F}(X)$. Here is an important geometric example in which the homomorphism is not one-one.

EXAMPLE. Linear fractional transformations. Let $X = \mathbb{C} \cup \{\infty\}$, a set that becomes the **Riemann sphere** in complex analysis. The group $G = \mathrm{GL}(2, \mathbb{C})$ acts on X by the **linear fractional transformations**

$$\begin{pmatrix} a & b \\ c & d \end{pmatrix}(z) = \frac{az+b}{cz+d},$$

the understanding being that the image of ∞ is ac^{-1} and the image of $-dc^{-1}$ is ∞, just as if we were to pass to a limit in each case. Property (ii) of a group action is clear. To verify (i), we simply calculate that

$$\begin{pmatrix} a' & b' \\ c' & d' \end{pmatrix}\left(\begin{pmatrix} a & b \\ c & d \end{pmatrix}(z)\right) = \frac{a'\left(\frac{az+b}{cz+d}\right)+b'}{c'\left(\frac{az+b}{cz+d}\right)+d'}$$

$$= \frac{(a'a+b'c)z+(a'b+b'd)}{(c'a+d'c)z+(c'b+d'd)}$$

$$= \left(\begin{pmatrix} a' & b' \\ c' & d' \end{pmatrix}\begin{pmatrix} a & b \\ c & d \end{pmatrix}\right)(z),$$

and indeed we have a group action. Let $\mathrm{SL}(2, \mathbb{R})$ be the subgroup of real matrices in $\mathrm{GL}(2, \mathbb{C})$ of determinant 1, and let Y be the subset of X where $\mathrm{Im}\, z > 0$, not including ∞. The members of $SL(2, \mathbb{R})$ carry the subset Y into itself, as we see from the computation

$$\operatorname{Im} \frac{az+b}{cz+d} = \operatorname{Im} \frac{(az+b)(c\bar{z}+d)}{|cz+d|^2} = \operatorname{Im} \frac{adz+bc\bar{z}}{|cz+d|^2}$$

$$= \frac{(ad-bc)\operatorname{Im} z}{|cz+d|^2} = \frac{\operatorname{Im} z}{|cz+d|^2}.$$

Since the effect of a matrix g^{-1} is to invert the effect of g, and since both g and g^{-1} carry Y to itself, we conclude that $SL(2, \mathbb{R})$ acts on $Y = \{z \in \mathbb{C} \mid \operatorname{Im} z > 0\}$ by linear fractional transformations. In similar fashion one can verify that the subgroup

$$SU(1,1) = \left\{ \begin{pmatrix} \alpha & \beta \\ \bar{\beta} & \bar{\alpha} \end{pmatrix} \,\middle|\, \alpha \in \mathbb{C},\ \beta \in \mathbb{C},\ |\alpha|^2 - |\beta|^2 = 1 \right\}$$

of $GL(2, \mathbb{C})$ acts on $\{z \in \mathbb{C} \mid |z| < 1\}$ by linear fractional transformations.

One group action can yield many others. For example, from an action of G on X, we can construct an action on the space of all complex-valued functions on X. The definition is $(gf)(x) = f(g^{-1}x)$, the use of the inverse being necessary in order to verify property (i) of a group action:

$$((g_1g_2)f)(x) = f((g_1g_2)^{-1}x) = f((g_2^{-1}g_1^{-1})x)$$
$$= f(g_2^{-1}(g_1^{-1}x)) = (g_2f)(g_1^{-1}x) = (g_1(g_2f))(x).$$

There is nothing special about the complex numbers as range for the functions here. We can allow any set as range, and we can even allow G to act on the range, as well as on the domain.[12] If G acts on X and Y, then the set of functions from X to Y inherits a group action under the definition

$$(gf)(x) = g(f(g^{-1}x)),$$

as is easily checked. In other words, we are to use g^{-1} where the domain enters the formula and we are to use g where the range enters the formula.

If V is a vector space over a field $\mathbb{F}$, a **representation** of G on V is a group action of G on V by *linear* functions. Specifically for each $g \in G$, τ_g is to be a member of the group of linear maps from V into itself. Usually one writes $\tau(g)$ instead of τ_g in representation theory, and thus the condition is that $\tau(g)$ is to be linear for each $g \in G$ and we are to have $\tau(1) = 1$ and $\tau(g_1g_2) = \tau(g_1)\tau(g_2)$ for all g_1 and g_2. There are interesting examples both when V is finite-dimensional and when V is infinite-dimensional.[13]

---

[12]When $\mathbb{C}$ was used as range in the previous display, the group action of G on $\mathbb{C}$ was understood to be **trivial** in the sense that $gz = z$ for every g in G and z in $\mathbb{C}$.

[13]In some settings a continuity assumption may be added to the definition of a representation, or the field $\mathbb{F}$ may be restricted in some way. We impose no such assumption here at this time.

6. Group Actions and Examples

EXAMPLES OF REPRESENTATIONS.

(1) If $m \geq 1$, then the additive group $\mathbb{Z}/m\mathbb{Z}$ acts linearly on $\mathbb{R}^2$ by

$$\tau(k) = \begin{pmatrix} \cos \frac{2\pi k}{m} & -\sin \frac{2\pi k}{m} \\ \sin \frac{2\pi k}{m} & \cos \frac{2\pi k}{m} \end{pmatrix}, \qquad k \in \{0, 1, 2, \ldots, m-1\}.$$

Each $\tau(k)$ is a rotation matrix about the origin through an angle that is a multiple of $2\pi/m$. These transformations of $\mathbb{R}^2$ form a subgroup of the group of symmetries of a regular k-gon centered at the origin in $\mathbb{R}^2$.

(2) The dihedral group D_3 acts linearly on $\mathbb{R}^2$ with

$$\tau(1) = \begin{pmatrix} 1 & 0 \\ 0 & 1 \end{pmatrix}, \quad \tau(2\ 3) = \begin{pmatrix} 1 & 0 \\ 0 & -1 \end{pmatrix}, \quad \tau(1\ 2) = \begin{pmatrix} -\frac{1}{2} & \frac{\sqrt{3}}{2} \\ \frac{\sqrt{3}}{2} & \frac{1}{2} \end{pmatrix}, \quad \tau(1\ 3) = \begin{pmatrix} -\frac{1}{2} & -\frac{\sqrt{3}}{2} \\ -\frac{\sqrt{3}}{2} & \frac{1}{2} \end{pmatrix},$$

$$\tau(1\ 2\ 3) = \begin{pmatrix} -\frac{1}{2} & -\frac{\sqrt{3}}{2} \\ \frac{\sqrt{3}}{2} & -\frac{1}{2} \end{pmatrix}, \quad \tau(1\ 3\ 2) = \begin{pmatrix} -\frac{1}{2} & \frac{\sqrt{3}}{2} \\ -\frac{\sqrt{3}}{2} & -\frac{1}{2} \end{pmatrix}.$$

Each of these matrices carries into itself the equilateral triangle with center at the origin and one vertex at $(1, 0)$. To obtain these matrices, we number the vertices #1, #2, #3 counterclockwise with the vertex at $(1, 0)$ as #1.

(3) The symmetric group $\mathfrak{S}_n$ acts linearly on $\mathbb{R}^n$ by permuting the indices of standard basis vectors. For example, with $n = 3$, we have $(1\ 3)e_1 = e_3$, $(1\ 3)e_2 = e_2$, etc. The matrices may be computed by the techniques of Section II.3. With $n = 3$, we obtain, for example,

$$(1\ 3) \mapsto \begin{pmatrix} 0 & 0 & 1 \\ 0 & 1 & 0 \\ 1 & 0 & 0 \end{pmatrix} \quad \text{and} \quad (1\ 2\ 3) \mapsto \begin{pmatrix} 0 & 0 & 1 \\ 1 & 0 & 0 \\ 0 & 1 & 0 \end{pmatrix}.$$

(4) If G acts on a set X, then the corresponding action $(gf)(x) = f(g^{-1}x)$ on complex-valued functions is a representation on the vector space of all complex-valued functions on X. This vector space is infinite-dimensional if X is an infinite set. The linearity of the action on functions follows from the definitions of addition and scalar multiplication of functions. In fact, let functions f_1 and f_2 be given, and let c be a scalar. Then

$$(g(f_1 + f_2))(x) = (f_1 + f_2)(g^{-1}x) = f_1(g^{-1}x) + f_2(g^{-1}x)$$
$$= (gf_1)(x) + (gf_2)(x) = (gf_1 + gf_2)(x)$$

and

$$(g(cf_1))(x) = (cf_1)(g^{-1}x) = c(f_1(g^{-1}x)) = c((gf_1)(x)) = (c(gf_1))(x).$$

One more important class of group actions consists of those that are closely related to the structure of the group itself. Two simple ones are the action of G on itself by left translations $(g_1, g_2) \mapsto g_1 g_2$ and the action of G on itself by right translations $(g_1, g_2) \mapsto g_2 g_1^{-1}$. More useful is the action of G on a quotient space G/H, where H is a subgroup. This action is given by $(g_1, g_2 H) \mapsto g_1 g_2 H$. There are still others, and some of them are particularly handy in analyzing finite groups. We give some applications in the present section and the next, and we postpone others to Section 10. Before describing some of these actions in detail, let us make some general definitions and establish two easy results.

Let $G \times X \to X$ be a group action. If p is in X, then $G_p = \{g \in G \mid gp = p\}$ is a subgroup of G called the **isotropy subgroup** at p. This is not always a normal subgroup; however, the subgroup $\bigcap_{p \in G} G_p$ that fixes all points of X is the kernel of the homomorphism $G \to \mathcal{F}(X)$ defining the group action, and such a kernel has to be normal.

Let p and q be in X. We say that p is equivalent to q for the purposes of this paragraph if $p = gq$ for some $g \in G$. The result is an equivalence relation: it is reflexive since $p = 1p$, it is symmetric since $p = gq$ implies $g^{-1}p = q$, and it is transitive since $p = gq$ and $q = g'r$ together imply $p = (gg')r$. The equivalence classes are called **orbits** of the group action. The orbit of a point p in X is $Gp = \{gp \mid g \in G\}$. If $Y = Gp$ is an orbit, or more generally if Y is any subset of X carried to itself by every element of G, then $G \times Y \to Y$ is a group action. In fact, each function $y \mapsto gy$ is invertible on Y with $y \mapsto g^{-1}y$ as the inverse function, and properties (i) and (ii) of a group action follow from the same properties for X.

A group action $G \times X \to X$ is said to be **transitive** if there is just one orbit, hence if $X = Gp$ for each p in X. It is **simply transitive** if it is transitive and if for each p and q in X, there is just one element g of G with $gp = q$.

Proposition 4.34. Let $G \times X \to X$ be a group action, let p be in X, and let H be the isotropy subgroup at p. Then the map $G \to Gp$ given by $g \mapsto gp$ descends to a well-defined map $G/H \to Gp$ that is one-one from G/H onto the orbit Gp and respects the group actions.

REMARK. In other words, a group action of G on a single orbit is always **isomorphic as a group action** to the action of G on some quotient space G/H.

PROOF. Let $\varphi : G \to Gp$ be defined by $\varphi(g) = gp$. For h in $H = G_p$, $\varphi(gh) = (gh)p = g(hp) = gp = \varphi(g)$ shows that φ descends to a well-defined function $\overline{\varphi} : G/H \to Gp$, and $\overline{\varphi}$ is certainly onto Gp. If $\overline{\varphi}(g_1 H) = \overline{\varphi}(g_2 H)$, then $g_1 p = \varphi(g_1 p) = \varphi(g_2 p) = g_2 p$, and hence $g_2^{-1} g_1 p = p$, $g_2^{-1} g_1$ is in H, g_1 is in $g_2 H$, and $g_1 H = g_2 H$. Thus $\overline{\varphi}$ is one-one.

Respecting the group action means that $\overline{\varphi}(gg'H) = g\overline{\varphi}(g'H)$, and this identity holds since $g\overline{\varphi}(g'H) = g\varphi(g') = g(g'p) = (gg')p = \varphi(gg') = \overline{\varphi}(gg'H)$. □

A simple consequence is the following important **counting formula** in the case of a group action by a finite group.

Corollary 4.35. Let G be a finite group, let $G \times X \to X$ be a group action, let p be in X, and G_p be the isotropy group at p, and let Gp be the orbit of p. Then $|G| = |Gp| |G_p|$.

PROOF. Proposition 4.34 shows that the action of G on some G/G_p is the most general group action on a single orbit, G_p being the isotropy subgroup. Thus the corollary follows from Lagrange's Theorem (Theorem 4.7) with $H = G_p$ and $G/H = Gp$. □

We turn to applications of group actions to the structure of groups. If H is a subgroup of a group G, the **index** of H in G is the number of elements in G/H, finite or infinite. The first application notes a situation in which a subgroup of a finite group is automatically normal.

Proposition 4.36. Let G be a finite group, and let p be the smallest prime dividing the order of G. If H is a subgroup of G of index p, then H is normal.

REMARKS. The most important case is $p = 2$: any subgroup of index 2 is automatically normal, and this conclusion is valid even if G is infinite, as was already pointed out in Example 3 of Section 2. If G is finite and if 2 divides the order of G, there need not, however, be any subgroup of index 2; for example, the alternating group $\mathfrak{A}_4$ has order 12, and Problem 11 at the end of the chapter shows that $\mathfrak{A}_4$ has no subgroup of order 6.

PROOF. Let $X = G/H$, and restrict the group action $G \times X \to X$ to an action $H \times X \to X$. The subset $\{1H\}$ is a single orbit under H, and the remaining $p-1$ members of G/H form a union of orbits. Corollary 4.35 shows that the number of elements in an orbit has to be a divisor of $|H|$, and the smallest divisor of $|H|$ other than 1 is $\geq p$ since the smallest divisor of $|G|$ other than 1 equals p and since $|H|$ divides $|G|$. Hence any orbit of H containing more than one element has at least p elements. Since only $p - 1$ elements are left under consideration, each orbit under H contains only one element. Therefore $hgH = gH$ for all h in H and g in G. Then $g^{-1}hg$ is in H, and we conclude that H is normal. □

If G is a group, the **center** Z_G of G is the set of all elements x such that $gx = xg$ for all g in G. The center of G is a subgroup (since $gx = xg$ and $gy = yg$ together imply $g(xy) = xgy = (xy)g$ and $xg^{-1} = g^{-1}(gx)g^{-1} = g^{-1}(xg)g^{-1} = g^{-1}x$), and every subgroup of the center is normal since $x \in Z_G$ and $g \in G$ together imply $gxg^{-1} = x$. Here are examples: the center of a group G is G itself if and only if G is abelian, the center of the quaternion group H_8 is $\{\pm 1\}$, and the center of any symmetric group $\mathfrak{S}_n$ with $n \geq 3$ is $\{1\}$.

If x is in G, the **centralizer** of x in G, denoted by $Z_G(x)$, is the set of all g such that $gx = xg$. This is a subgroup of G, and it equals G itself if and only if x is in the center of G. For example the centralizer of **i** in H_8 is the 4-element subgroup $\{\pm 1, \pm \mathbf{i}\}$.

Having made these definitions, we introduce a new group action of G on G, namely $(g, x) \mapsto gxg^{-1}$. The orbits are called the **conjugacy classes** of G. If x and y are two elements of G, we say that x is **conjugate** to y if x and y are in the same conjugacy class. In other words, x is conjugate to y if there is some g in G with $gxg^{-1} = y$. The result is an equivalence relation. Let us write $C\ell(x)$ for the conjugacy class of x. We can easily compute the isotropy subgroup G_x at x under this action; it consists of all $g \in G$ such that $gxg^{-1} = x$ and hence is exactly the centralizer $Z_G(x)$ of x in G. In particular, $C\ell(x) = \{x\}$ if and only if x is in the center Z_G. Applying Corollary 4.35, we immediately obtain the following result.

Proposition 4.37. If G is a finite group, then $|G| = |C\ell(x)| |Z_G(x)|$ for all x in G.

Thus $|C\ell(x)|$ is always a divisor of $|G|$, and it equals 1 if and only if x is in the center Z_G. Let us apply these considerations to groups whose order is a power of a prime.

Corollary 4.38. If G is a finite group whose order is a positive power of a prime, then the center Z_G is not $\{1\}$.

PROOF. Let $|G| = p^n$ with p prime and with $n > 0$. The conjugacy classes of G exhaust G, and thus the sum of all $|C\ell(x)|$'s equals $|G|$. Since $|C\ell(x)| = 1$ if and only if x is in Z_G, the sum of $|Z_G|$ and all the $|C\ell(x)|$'s that are not 1 is equal to $|G|$. All the terms $|C\ell(x)|$ that are not 1 are positive powers of p, by Proposition 4.37, and so is $|G|$. Therefore p divides $|Z_G|$. □

Corollary 4.39. If G is a finite group of order p^2 with p prime, then G is abelian.

PROOF. From Corollary 4.38 we see that either $|Z_G| = p^2$, in which case G is abelian, or $|Z_G| = p$. We show that the latter is impossible. If fact, if x is not in Z_G, then $Z_G(x)$ is a subgroup of G that contains Z_G and the element x. It must then have order p^2 and be all of G. Hence every element of G commutes with x, and x is in Z_G, contradiction. □

Corollary 4.40. If G is a finite group whose order is a positive power p^n of a prime p, then there exist normal subgroups G_k of G for $0 \le k \le n$ such that $|G| = p^k$ for all $k \le n$ and such that $G_k \subseteq G_{k+1}$ for all $k < n$.

PROOF. We proceed by induction on n. The base case of the induction is $n = 1$ and is handled by Corollary 4.9. Assume inductively that the result holds for n, and let G have order p^{n+1}. Corollary 4.39 shows that $Z_G \neq \{1\}$. Any element $\neq 1$ in Z_G must have order a power of p, and some power of it must therefore have order p. Thus let a be an element of Z_G of order p, and let H be the subgroup consisting of the powers of a. Then H is normal and has order p. Let $G' = G/H$ be the quotient group, and let $\varphi : G \to G'$ be the quotient homomorphism. The group G' has order p^n, and the inductive hypothesis shows that G' has normal subgroups G'_k for $0 \leq k \leq n$ such that $|G'_k| = p^k$ for $k \leq n$ and $G'_k \subseteq G'_{k+1}$ for $k \leq n-1$. For $1 \leq k \leq n+1$, define $G_k = \varphi^{-1}(G'_{k-1})$, and let $G_0 = \{1\}$. The First Isomorphism Theorem (Theorem 4.13) shows that each G_k for $k \geq 1$ is a normal subgroup of G containing H and that $\varphi(G_k) = G'_{k-1}$. Then $\varphi|_{G_k}$ is a homomorphism of G_k onto G'_{k-1} with kernel H, and hence $|G_k| = |G'_{k-1}||H| = p^{k-1}p = p^k$. Therefore the G_k's will serve as the required subgroups of G. □

It is not always so easy to determine the conjugacy classes in a particular group. For example, in GL$(n, \mathbb{C})$ the question of conjugacy is the question whether two matrices are similar in the sense of Section II.3; this will be one of the main problems addressed in Chapter V. By contrast, the problem of conjugacy in symmetric groups has a simple answer. Recall that every permutation is uniquely the product of disjoint cycles. The **cycle structure** of a permutation consists of the number of cycles of each length in this decomposition.

Lemma 4.41. Let σ and τ be members of the symmetric group $\mathfrak{S}_n$. If σ is expressed as the product of disjoint cycles, then $\tau\sigma\tau^{-1}$ has the same cycle structure as σ, and the expression for $\tau\sigma\tau^{-1}$ as the product of disjoint cycles is obtained from that for σ by substituting $\tau(k)$ for k throughout.

REMARK. For example, if $\sigma = (a\ b)(c\ d\ e)$, then $\tau\sigma\tau^{-1}$ decomposes as $\big(\tau(a)\ \tau(b)\big)\big(\tau(c)\ \tau(d)\ \tau(e)\big)$.

PROOF. Because the conjugate of a product equals the product of the conjugates, it is enough to handle a cycle $\gamma = (a_1\ a_2\ \cdots\ a_n)$ appearing in σ. The corresponding cycle $\gamma' = \tau\gamma\tau^{-1}$ is asserted to be $\gamma' = (\tau(a_1)\ \tau(a_2)\ \cdots\ \tau(a_n))$. Application of τ^{-1} to $\tau(a_j)$ yields a_j, application of σ to this yields a_{j+1} if $j < n$ and a_1 if $j = n$, and application of τ to the result yields $\tau(a_{j+1})$ or $\tau(a_1)$. For each of the symbols b not in the list $\{a_1, \ldots, a_n\}$, $\tau\gamma\tau^{-1}(\tau(b)) = \tau(b)$ since $\gamma(b) = b$. Thus $\tau\gamma\tau^{-1} = \gamma'$, as asserted. □

Proposition 4.42. Let H be a subgroup of a symmetric group $\mathfrak{S}_n$. If $C\ell(x)$ denotes a conjugacy class in H, then all members of $C\ell(x)$ have the same cycle structure. Conversely if $H = \mathfrak{S}_n$, then the conjugacy class of a permutation σ consists of all members of $\mathfrak{S}_n$ having the same cycle structure as σ.

PROOF. The first conclusion is immediate from Lemma 4.41. For the second conclusion, let σ and σ' have the same cycle structure, and let τ be the permutation that moves, for each k, the kth symbol appearing in the disjoint-cycle expansion of σ into the kth symbol in the corresponding expansion of σ'. Define τ on the remaining symbols in any fashion at all. Application of the lemma shows that $\tau \sigma \tau^{-1} = \sigma'$. Thus any two permutations with the same cycle structure are conjugate. $\square$

7. Semidirect Products

One more application of group actions to the structure theory of groups will be to the construction of "semidirect products" of groups. If H is a group, then an isomorphism of H with itself is called an **automorphism**. The set of automorphisms of H is a group under composition, and we denote it by $\text{Aut}\, H$. We are going to be interested in "group actions by automorphisms," i.e., group actions of a group G on a space X when X is itself a group and the action by each member of G is an automorphism of the group structure of X; the group action is therefore a homomorphism of the form $\tau : G \to \text{Aut}\, X$.

EXAMPLE 1. In $\mathbb{R}^2$, we can identify the additive group of the underlying vector space with the group of translations $\ell_v(w) = v + w$; the identification associates a translation ℓ with the member $\ell(0)$ of $\mathbb{R}^2$. Let H be the group of translations. The rotations about the origin in $\mathbb{R}^2$, namely the linear maps with matrices $\begin{pmatrix} \cos\theta & \sin\theta \\ -\sin\theta & \cos\theta \end{pmatrix}$, form a group $G = \text{SO}(2)$ that acts on $\mathbb{R}^2$, hence acts on the set H of translations. The linearity of the rotations says that the action of $G = \text{SO}(2)$ on the translations is by automorphisms of H, i.e., that each rotation, in its effect on G, is in $\text{Aut}\, H$. Out of these data—the two groups G and H and a homomorphism of G into $\text{Aut}\, H$—we will construct below what amounts to the group of all rotations (about any point) and translations of $\mathbb{R}^2$. The construction is that of a "semidirect product."

EXAMPLE 2. Take any group G, and let G act on $X = G$ by conjugation. Each conjugation $x \mapsto gxg^{-1}$ is an automorphism of G, and thus the action of G on itself by conjugation is an action by automorphisms.

Let G and H be groups. Suppose that a group action $\tau : G \to \mathcal{F}(H)$ is given with G acting on H by automorphisms. That is, suppose that each map $h \to \tau_g(h)$ is an automorphism of H. We define a group $G \times_\tau H$ whose underlying set will be the Cartesian product $G \times H$. The motivation for the definition of multiplication comes from Example 2, in which $\tau_g(h) = ghg^{-1}$. We want to write a product $g_1 h_1 g_2 h_2$ in the form $g'h'$, and we can do so using the formula

$$g_1 h_1 g_2 h_2 = g_1 g_2 (g_2^{-1} h_1 g_2) h_2 = (g_1 g_2)\big((\tau_{g_2^{-1}}(h_1)) h_2\big).$$

Similarly the formula for inverses is motivated by the formula

$$(gh)^{-1} = h^{-1} g^{-1} = g^{-1}(g h^{-1} g^{-1}) = g^{-1} \tau_g(h^{-1}).$$

Proposition 4.43. Let G and H be groups, and let τ be a group action of G on H by automorphisms. Then the set-theoretic product $G \times H$ becomes a group $G \times_\tau H$ under the definitions

$$(g_1, h_1)(g_2, h_2) = (g_1 g_2, (\tau_{g_2^{-1}}(h_1)) h_2)$$

and

$$(g, h)^{-1} = (g^{-1}, \tau_g(h^{-1})).$$

The mappings $i_1 : G \to G \times_\tau H$ and $i_2 : H \to G \times_\tau H$ given by $i_1(g) = (g, 1)$ and $i_2(h) = (1, h)$ are one-one homomorphisms, and $p_2 : G \times_\tau H \to G$ given by $p_2(g, h) = g$ is a homomorphism onto G. The images $G' = i_1(G)$ and $H' = i_2(H)$ are subgroups of $G \times_\tau H$ with H' normal such that $G' \cap H' = \{1\}$, such that every element of $G \times_\tau H$ is the product of an element of G' and an element of H', and such that conjugation of G' on H' is given by $i_1(g) i_2(h) i_1(g)^{-1} = i_2(\tau_g(h))$.

REMARK. The group $G \times_\tau H$ is called the **external semidirect product**[14] of G and H with respect to τ.

PROOF. For associativity we compute directly that

$$\big((g_1, h_1)(g_2, h_2)\big)(g_3, h_3) = (g_1 g_2 g_3, \tau_{g_3^{-1}}(\tau_{g_2^{-1}}(h_1)) h_2) h_3)$$

and $\quad (g_1, h_1)\big((g_2, h_2)(g_3, h_3)\big) = (g_1 g_2 g_3, \tau_{g_3^{-1} g_2^{-1}}(h_1) \tau_{g_3^{-1}}(h_2) h_3).$

Since

$$\tau_{g_3^{-1}}(\tau_{g_2^{-1}}(h_1) h_2) = (\tau_{g_3^{-1}} \tau_{g_2^{-1}}(h_1)) \tau_{g_3^{-1}}(h_2) = \tau_{g_3^{-1} g_2^{-1}}(h_1) \tau_{g_3^{-1}}(h_2),$$

we have a match. It is immediate that $(1, 1)$ is a two-sided identity. Since $(g, h)(g^{-1}, \tau_g(h^{-1})) = (1, \tau_g(h) \tau_g(h^{-1})) = (1, \tau_g(h h^{-1})) = (1, \tau_g(1)) = (1, 1)$ and $(g^{-1}, \tau_g(h^{-1}))(g, h) = (1, \tau_{g^{-1}}(\tau_g(h^{-1})) h) = (1, \tau_1(h^{-1}) h) = (1, 1)$, $(g^{-1}, \tau_g(h^{-1}))$ is indeed a two-sided inverse of (g, h). It is immediate from the definition of multiplication that i_1, i_2, and p_2 are homomorphisms, that i_1 and i_2 are one-one, that p_2 is onto, that $G' \cap H' = \{1\}$, and that $G \times_\tau H = G'H'$. Since i_1 and i_2 are homomorphisms, G' and H' are subgroups. Since H' is the kernel of p_2, H' is normal. Finally the definition of multiplication gives $i_1(g) i_2(h) i_1(g)^{-1} = (g, h)(g^{-1}, 1) = (1, (\tau_g(h)) 1) = i_2(\tau_g(h))$, and the proof is complete. □

---

[14]The notation $\ltimes$ is used by some authors in place of $\times_\tau$.

Proposition 4.44. Let S be a group, let G and H be subgroups with H normal, and suppose that $G \cap H = \{1\}$ and that every element of S is the product of an element of G and an element of H. For each $g \in G$, define an automorphism τ_g of H by $\tau_g(h) = ghg^{-1}$. Then τ is a group action of G on H by automorphisms, and the mapping $G \times_\tau H \to S$ given by $(g, h) \mapsto gh$ is an isomorphism of groups.

REMARKS. In this case we call S an **internal semidirect product** of G and H with respect to τ. We shall not attempt to write down a universal mapping property that characterizes internal semidirect products.

PROOF. Since $\tau_{g_1 g_2}(h) = g_1 g_2 h g_2^{-1} g_1^{-1} = g_1 \tau_{g_2}(h) g_1^{-1} = \tau_{g_1} \tau_{g_2}(h)$ and since each τ_g is an automorphism of H, τ is an action by automorphisms. Proposition 4.43 therefore shows that $G \times_\tau H$ is a well-defined group. The function φ from $G \times_\tau H$ to S given by $\varphi(g, h) = gh$ is a homomorphism by the same computation that motivated the definition of multiplication in a semidirect product, and φ is onto S since every element of S lies in the set GH of products. If $gh = 1$, then $g = h^{-1}$ exhibits g as in $G \cap H = \{1\}$. Hence $g = 1$ and $h = 1$. Therefore φ is one-one and must be an isomorphism. □

EXAMPLE 1. Dihedral groups D_n. We show that D_n is the internal semidirect product of a 2-element group and the rotation subgroup. Let H be the group of rotations about the origin through multiples of the angle $2\pi/n$. This group is cyclic of order n, and it is normal in D_n because it is of index 2. If s is any of the reflections in D_n, then $G = \{1, s\}$ is a subgroup of D_n of order 2 with $G \cap H = \{1\}$. Counting the elements, we see that every element of D_n is of the form r^k or sr^k, in other words that the set of products GH is all of D_n. Thus Proposition 4.44 shows that D_n is an (internal) semidirect product of G and H with respect to some $\tau : G \to \text{Aut } H$. To understand the homomorphism τ, let us write the members of H as the powers of r, where r is rotation counterclockwise about the origin through the angle $2\pi/n$. For the reflection s (or indeed for any reflection in D_n), a look at the geometry shows that $sr^k s^{-1} = r^{-k}$ for all k. In other words, the automorphism $\tau(1)$ leaves each element of H fixed while $\tau(s)$ sends each $k \mod n$ to $-k \mod n$. The map that sends each element of a cyclic group to its group inverse is indeed an automorphism of the cyclic group, and thus τ is indeed a homomorphism of G into $\text{Aut } H$.

EXAMPLE 2. Construction of a nonabelian group of order 21. Let $H = C_7$, written multiplicatively with generator a, and let $G = C_3$, written multiplicatively with generator b. To arrange for G to act on H by automorphisms, we make use of a nontrivial automorphism of H of order 3. Such a mapping is $a^k \mapsto a^{2k}$. In fact, there is no doubt that this mapping is an automorphism, and we have to see that it has order 3. The effect of applying it twice is $a^k \mapsto a^{4k}$, and the effect of applying it three times is $a^k \mapsto a^{8k}$. But $a^{8k} = a^k$ since $a^7 = 1$, and thus the mapping $a^k \mapsto a^{2k}$ indeed has order 3. We send b^n into the nth power of this automorphism, and the result is a homomorphism $\tau : G \to \operatorname{Aut} H$. The semidirect product $G \times_\tau H$ is certainly a group of order $3 \times 7 = 21$. To see that it is nonabelian, we observe from the group law in Proposition 4.43 that $ab = b\tau_{b^{-1}}(a) = ba^4$. Thus $ab \neq ba$, and $G \times_\tau H$ is nonabelian.

It is instructive to generalize the construction in Example 2 a little bit. To do so, we need a lemma.

Lemma 4.45. If p is a prime, then the automorphisms of the additive group of the field $\mathbb{F}_p$ are the multiplications by the members of the multiplicative group $\mathbb{F}_p^\times$, and consequently $\operatorname{Aut} C_p$ is isomorphic to a cyclic group C_{p-1}.

PROOF. Let us write $\operatorname{Aut} \mathbb{F}_p$ for the automorphism group of the additive group of $\mathbb{F}_p$. Each function $\varphi_a : \mathbb{F}_p \to \mathbb{F}_p$ given by $\varphi_a(n) = na$, taken modulo p, is in $\operatorname{Aut} \mathbb{F}_p$ as a consequence of the distributive law. We define a function $\Phi : \operatorname{Aut} \mathbb{F}_p \to \mathbb{F}_p^\times$ by $\Phi(\varphi) = \varphi(1)$ for $\varphi \in \operatorname{Aut} \mathbb{F}_p$. Again by the distributive law $\varphi(n) = n\varphi(1)$ for every integer n. Thus if φ_1 and φ_2 are in $\operatorname{Aut} \mathbb{F}_p$, then $\Phi(\varphi_1 \circ \varphi_2) = (\varphi_1 \circ \varphi_2)(1) = \varphi_1(\varphi_2(1)) = \varphi_2(1)\varphi_1(1)$, and consequently Φ is a homomorphism. If a member φ of $\operatorname{Aut} \mathbb{F}_p$ has $\Phi(\varphi) = 1$ in $\mathbb{F}_p^\times$, then $\varphi(1) = 1$ and therefore $\varphi(n) = n\varphi(1) = n$ for all n. Therefore φ is the identity in $\operatorname{Aut} \mathbb{F}_p$. We conclude that Φ is one-one. If a is given in $\mathbb{F}_p^\times$, then $\Phi(\varphi_a) = \varphi_a(1) = a$, and hence Φ is onto $\mathbb{F}_p^\times$. Therefore Φ is an isomorphism of $\operatorname{Aut} \mathbb{F}_p$ and $\mathbb{F}_p^\times$. By Corollary 4.27, Φ exhibits $\operatorname{Aut} \mathbb{F}_p$ as isomorphic to the cyclic group C_{p-1}. $\square$

Proposition 4.46. If p and q are primes with $p < q$ such that p divides $q - 1$, then there exists a nonabelian group of order pq.

REMARKS. For $p = 2$, the divisibility condition is automatic, and the proof will yield the dihedral group D_q. For $p = 3$ and $q = 7$, the condition is that 3 divides $7 - 1$, and the constructed group will be the group in Example 2 above.

PROOF. Let $G = C_p$ with generator a, and let $H = C_q$. Lemma 4.45 shows that $\operatorname{Aut} C_q \cong C_{q-1}$. Let b be a generator of $\operatorname{Aut} C_q$. Since p divides $q - 1$, $b^{(q-1)/p}$ has order p. Then the map $a^k \to b^{k(q-1)/p}$ is a well-defined homomorphism τ of G into $\operatorname{Aut} H$, and it determines a semidirect product $S = G \times_\tau H$, by Proposition 4.43. The order of S is pq, and the multiplication is nonabelian since for $h \in H$, we have $(a, 1)(1, h) = (a, h)$ and $(1, h)(a, 1) = (a, \tau_{a^{-1}}(h)) = (a, b^{-(q-1)/p}(h))$, but $b^{-(q-1)/p}$ is not the identity automorphism of H because it has order p. $\square$

8. Simple Groups and Composition Series

A group $G \neq \{1\}$ is said to be **simple** if its only normal subgroups are $\{1\}$ and G.

Among abelian groups the simple ones are the cyclic groups of prime order. Indeed, a cyclic group C_p of prime order has no nontrivial subgroups at all, by Corollary 4.9. Conversely if G is abelian and simple, let $a \neq 1$ be in G. Then $\{a^n\}$ is a cyclic subgroup and is normal since G is abelian. Thus $\{a^n\}$ is all of G, and G is cyclic. The group $\mathbb{Z}$ is not simple, having the nontrivial subgroup $2\mathbb{Z}$, and the group $\mathbb{Z}/(rs)\mathbb{Z}$ with $r > 1$ and $s > 1$ is not simple, having the multiples of r as a nontrivial subgroup. Thus G has to be cyclic of prime order.

The interest is in nonabelian simple groups. We shall establish that the alternating groups $\mathfrak{A}_n$ are simple for $n \geq 5$, and some other simple groups will be considered in Problems 55–62 at the end of the chapter.

Theorem 4.47. The alternating group $\mathfrak{A}_n$ is simple if $n \geq 5$.

PROOF. Let $K \neq \{1\}$ be a normal subgroup of $\mathfrak{A}_n$. Choose σ in K with $\sigma \neq 1$ such that $\sigma(i) = i$ for the maximum possible number of integers i with $1 \leq i \leq n$. The main step is to show that σ is a 3-cycle. Arguing by contradiction, suppose that σ is not a 3-cycle. Then there are two cases.

The first case is that the decomposition of σ as the product of disjoint cycles contains a k-cycle for some $k \geq 3$. Without loss of generality, we may take the cycle in question to be $\gamma = (1\ 2\ 3\ \cdots)$, and then $\sigma = \gamma \rho = (1\ 2\ 3\ \cdots)\rho$ with ρ equal to a product of disjoint cycles not containing the symbols appearing in γ. Being even and not being a 3-cycle, σ moves at least two other symbols besides the three listed ones, say 4 and 5. Put $\tau = (3\ 4\ 5)$. Lemma 4.41 shows that $\sigma' = \tau \sigma \tau^{-1} = \gamma' \rho' = (1\ 2\ 4\ \cdots)\rho'$ with ρ' not containing any of the symbols appearing in γ'. Thus $\sigma'\sigma^{-1}$ moves 3 into 4 and cannot be the identity. But $\sigma'\sigma^{-1}$ is in K and fixes all symbols other than 1, 2, 3, 4, 5 that are fixed by σ. In addition, $\sigma'\sigma^{-1}$ fixes 2, and none of 1, 2, 3, 4, 5 is fixed by σ. Thus $\sigma'\sigma^{-1}$ is a member of K other than the identity that fixes fewer symbols than σ, and we have arrived at a contradiction.

The second case is that σ is a product $\sigma = (1\ 2)(3\ 4)\cdots$ of disjoint transpositions. There must be at least two factors since σ is even. Put $\tau = (1\ 2)(4\ 5)$, the symbol 5 existing since the group $\mathfrak{A}_n$ in question has $n \geq 5$. Then $\sigma' = (1\ 2)(3\ 5)\cdots$. Since $\sigma'\sigma^{-1}$ carries 4 into 5, $\sigma'\sigma^{-1}$ is a member of K other than the identity. It fixes all symbols other than 1, 2, 3, 4, 5 that are fixed by σ, and in addition it fixes 1 and 2. Thus $\sigma'\sigma^{-1}$ fixes more symbols than σ does, and again we have arrived at a contradiction.

We conclude that K contains a 3-cycle, say $(1\ 2\ 3)$. If i, j, k, l, m are five arbitrary symbols, then we can construct a permutation τ with $\tau(1) = i, \tau(2) = j$, $\tau(3) = k, \tau(4) = l$, and $\tau(5) = m$. If τ is odd, we replace τ by $\tau(l\ m)$, and the result is even. Thus we may assume that τ is in $\mathfrak{A}_n$ and has $\tau(1) = i, \tau(2) = j$, and $\tau(3) = k$. Lemma 4.41 shows that $\tau\sigma\tau^{-1} = (i\ j\ k)$. Since K is normal, we conclude that K contains all 3-cycles.

To complete the proof, we show for $n \geq 3$ that every element of $\mathfrak{A}_n$ is a product of 3-cycles. If σ is in $\mathfrak{A}_n$, we use Corollary 1.22 to decompose σ as a product of transpositions. Since σ is even, we can group these in pairs. If the members of a pair of transpositions are not disjoint, then their product is a 3-cycle. If they are disjoint, then the identity $(1\ 2)(3\ 4) = (1\ 2\ 3)(2\ 3\ 4)$ shows that their product is a product of 3-cycles. This completes the proof. $\square$

Let G be a group. A descending sequence
$$G_n \supseteq G_{n-1} \supseteq \cdots \supseteq G_1 \supseteq G_0$$
of subgroups of G with $G_n = G$, $G_0 = \{1\}$, and each G_{k-1} normal in G_k is called a **normal series** for G. The normal series is called a **composition series** if each inclusion $G_k \supseteq G_{k-1}$ is proper and if each consecutive quotient G_k/G_{k-1} is simple.

EXAMPLES.

(1) Let G be a cyclic group of order N. A normal series for G consists of certain subgroups of G, all necessarily cyclic by Proposition 4.4. Their respective orders $N_n, N_{n-1}, \ldots, N_1, N_0$ have $N_n = N$, $N_0 = 1$, and $N_{k-1} \mid N_k$ for all k. The series is a composition series if and only if each quotient N_k/N_{k-1} is prime. In this case the primes that occur are exactly the prime divisors of N, and a prime p occurs r times if p^r is the exact power of p that divides N. Thus the consecutive quotients from a composition series of this G, up to isomorphisms, are independent of the particular composition series — though they may arise in a different order.

(2) For $G = \mathbb{Z}$, a normal series is of the form
$$\mathbb{Z} \supseteq m_1\mathbb{Z} \supseteq m_1 m_2\mathbb{Z} \supseteq m_1 m_2 m_3 \mathbb{Z} \supseteq \cdots \supseteq 0.$$
The group $G = \mathbb{Z}$ has no composition series.

(3) For the symmetric group $G = \mathfrak{S}_4$, let $C_2 \times C_2$ refer to the 4-element subgroup $\{1, (1\ 2)(3\ 4), (1\ 3)(2\ 4), (1\ 4)(2\ 3)\}$. The series
$$\mathfrak{S}_4 \supseteq \mathfrak{A}_4 \supseteq C_2 \times C_2 \supseteq \{1, (1\ 2)\} \supseteq \{1\}$$
is a composition series, the consecutive quotients being C_2, C_3, C_2, C_2. Each term in the composition series except for $\{1, (1\ 2)\}$ is actually normal in the whole group G, but there is no way to choose the 2-element subgroup to make it normal in G. The other two possible choices of 2-element subgroup, which lead to different composition series but with isomorphic consecutive quotients, are obtained by replacing $\{1, (1\ 2)\}$ by $\{1, (1\ 3)\}$ and again by $\{1, (1\ 4)\}$.

(4) For the symmetric group $G = \mathfrak{S}_5$, the series

$$\mathfrak{S}_5 \supseteq \mathfrak{A}_5 \supseteq \{1\}$$

is a composition series, the consecutive quotients being C_2 and $\mathfrak{A}_5$.

(5) Let G be a finite group of order p^n with p prime. Corollary 4.40 produces a composition series, and this time all the subgroups are normal in G. The successive normal subgroups have orders p^k for $k = n, n-1, \ldots, 0$, and each consecutive quotient is isomorphic to C_p.

Historically the Jordan–Hölder Theorem addressed composition series for groups, showing that the consecutive quotients, up to isomorphisms, are independent of the particular composition series. They can then consistently be called the **composition factors** of the group. Finding the composition factors of a particular group may be regarded as a step toward understanding the structure of the group. A generalization of the Jordan–Hölder Theorem due to Zassenhaus and Schreier applies to normal series in situations in which composition series might not exist, such as Example 2 above. We prove the Zassenhaus–Schreier Theorem, and the Jordan–Hölder Theorem is then a special case.

Two normal series

$$G_m \supseteq G_{m-1} \supseteq \cdots \supseteq G_1 \supseteq G_0$$
and
$$H_n \supseteq H_{n-1} \supseteq \cdots \supseteq H_1 \supseteq H_0$$

for the same group G are said to be **equivalent normal series** if $m = n$ and the order of the consecutive quotients $G_m/G_{m-1}, G_{m-1}/G_{m-2}, \ldots, G_1/G_0$ may be rearranged so that they are respectively isomorphic to $H_m/H_{m-1}, H_{m-1}/H_{m-2}, \ldots, H_1/H_0$. One normal series is said to be a **refinement** of another if the subgroups appearing in the second normal series all appear as subgroups in the first normal series.

Lemma 4.48 (Zassenhaus). Let G_1, G_2, G_1', and G_2' be subgroups of a group G with $G_1' \subseteq G_1$ and $G_2' \subseteq G_2$, G_1' normal in G_1, and G_2' normal in G_2. Then $(G_1 \cap G_2')G_1'$ is normal in $(G_1 \cap G_2)G_1'$, $(G_1' \cap G_2)G_2'$ is normal in $(G_1 \cap G_2)G_2'$, and

$$((G_1 \cap G_2)G_1')/((G_1 \cap G_2')G_1') \cong ((G_1 \cap G_2)G_2')/((G_1' \cap G_2)G_2').$$

PROOF. Let us check that $(G_1 \cap G_2')G_1'$ is normal in $(G_1 \cap G_2)G_1'$. Handling conjugation by members of $G_1 \cap G_2$ is straightforward: If g is in $G_1 \cap G_2$, then $g(G_1 \cap G_2')g^{-1} = G_1 \cap G_2'$ since g is in G_1 and $gG_2'g^{-1} = G_2'$. Also, $gG_1'g^{-1} = G_1'$ since g is in G_1. Hence $g(G_1 \cap G_2')G_1'g^{-1} = (G_1 \cap G_2')G_1'$.

Handling conjugation by members of G_1' requires a little trick: Let g be in G_1' and let hg' be in $(G_1 \cap G_2')G_1'$. Then $g(hg')g^{-1} = h(h^{-1}gh)g'g^{-1}$. The left factor h is in $G_1 \cap G_2'$. The remaining factors are in G_1'; for g' and g^{-1}, this is a matter of definition, and for $h^{-1}gh$, it follows because h is in G_1 and g is in G_1'. Thus $g(G_1 \cap G_2')G_1'g^{-1} = (G_1 \cap G_2')G_1'$, and $(G_1 \cap G_2')G_1'$ is normal in $(G_1 \cap G_2)G_1'$. The other assertion about normal subgroups holds by symmetry in the indexes 1 and 2.

By the Second Isomorphism Theorem (Theorem 4.14),

$$(G_1 \cap G_2)/(((G_1 \cap G_2')G_1') \cap (G_1 \cap G_2))$$
$$\cong ((G_1 \cap G_2)(G_1 \cap G_2')G_1')/((G_1 \cap G_2')G_1') \quad (*)$$
$$= ((G_1 \cap G_2)G_1')/((G_1 \cap G_2')G_1').$$

Since we have

$$((G_1 \cap G_2')G_1') \cap (G_1 \cap G_2) = ((G_1 \cap G_2')G_1') \cap G_2 = (G_1 \cap G_2')(G_1' \cap G_2),$$

we can rewrite the conclusion of $(*)$ as

$$(G_1 \cap G_2)/((G_1 \cap G_2')(G_1' \cap G_2)) \cong ((G_1 \cap G_2)G_1')/((G_1 \cap G_2')G_1'). \quad (**)$$

The left side of $(**)$ is symmetric under interchange of the indices 1 and 2. Hence so is the right side, and the lemma follows. □

Theorem 4.49 (Schreier). *Any two normal series of a group G have equivalent refinements.*

PROOF. Let the two normal series be

$$G_m \supseteq G_{m-1} \supseteq \cdots \supseteq G_1 \supseteq G_0,$$
$$H_n \supseteq H_{n-1} \supseteq \cdots \supseteq H_1 \supseteq H_0, \quad (*)$$

and define

$$G_{ij} = (G_i \cap H_j)G_{i+1} \quad \text{for } 0 \leq j \leq n,$$
$$H_{ji} = (G_i \cap H_j)H_{j+1} \quad \text{for } 0 \leq i \leq m. \quad (**)$$

Then we obtain respective refinements of the two normal series $(*)$ given by

$$G = G_{00} \supseteq G_{01} \supseteq \cdots \supseteq G_{0n}$$
$$\supseteq G_{10} \supseteq G_{11} \supseteq \cdots \supseteq G_{1n} \cdots \supseteq G_{m-1,n} = \{1\},$$
$$G = H_{00} \supseteq H_{01} \supseteq \cdots \supseteq H_{0m} \quad (\dagger)$$
$$\supseteq H_{10} \supseteq H_{11} \supseteq \cdots \supseteq H_{1m} \cdots \supseteq H_{n-1,m} = \{1\}.$$

The containments $G_{in} \supseteq G_{i+1,0}$ and $H_{jm} \supseteq H_{j+1,0}$ are equalities in (†), and the only nonzero consecutive quotients are therefore of the form $G_{ij}/G_{i,j+1}$ and $H_{ji}/H_{j,i+1}$. For these we have

$$\begin{aligned} G_{ij}/G_{i,j+1} &= ((G_i \cap H_j)G_{i+1})/((G_i \cap H_{j+1})G_{i+1}) & \text{by (**)} \\ &\cong ((G_i \cap H_j)H_{j+1})/((G_{i+1} \cap H_j)H_{j+1}) & \text{by Lemma 4.48} \\ &= H_{ji}/H_{j,i+1} & \text{by (**),} \end{aligned}$$

and thus the refinements (†) are equivalent. □

Corollary 4.50 (Jordan–Hölder Theorem). Any two composition series of a group G are equivalent as normal series.

PROOF. Let two composition series be given. Theorem 4.49 says that we can insert terms in each so that the refined series have the same length and are equivalent. Since the given series are composition series, the only way to insert a new term is by repeating some term, and the repetition results in a consecutive quotient of $\{1\}$. Because of Theorem 4.49 we know that the quotients $\{1\}$ from the two refined series must match. Thus the number of terms added to each series is the same. Also, the quotients that are not $\{1\}$ must match in pairs. Thus the given composition series are equivalent. □

9. Structure of Finitely Generated Abelian Groups

A set of **generators** for a group G is a set such that each element of G is a finite product of generators and their inverses. (A generator and its inverse are allowed to occur multiple times in a product.)

In this section we shall study *abelian* groups having a finite set of generators. Such groups are said to be **finitely generated abelian groups**, and our goal is to classify them up to isomorphism. We use additive notation for all our abelian groups in this section. We begin by introducing an analog $\mathbb{Z}^n$ for the integers $\mathbb{Z}$ of the vector space $\mathbb{R}^n$ for the reals $\mathbb{R}$, and along with it a generalization.

A **free abelian group** is any abelian group isomorphic to a direct sum, finite or infinite, of copies of the additive group $\mathbb{Z}$ of integers. The external direct sum of n copies of $\mathbb{Z}$ will be denoted by $\mathbb{Z}^n$. Let us use Proposition 4.17 to see that we can recognize groups isomorphic to free abelian groups by means of the following condition: an abelian group G is isomorphic to a free abelian group if and only if it has a $\mathbb{Z}$ **basis**, i.e., a subset that generates G and is such that no nontrivial linear combination, with integer coefficients, of the members of the subset is equal to the 0 element of the group. It will be helpful to use terminology adapted from the theory of vector spaces for this latter condition — that the subset is to be **linearly independent** over $\mathbb{Z}$.

Let us give the proof that the condition is necessary and sufficient for G to be free abelian. In one direction if G is an external direct sum of copies of $\mathbb{Z}$, then the members of G that are 1 in one coordinate and are 0 elsewhere form a $\mathbb{Z}$ basis. Conversely if $\{g_s\}_{s \in S}$ is a $\mathbb{Z}$ basis, let G_{s_0} be the subgroup of multiples of g_{s_0}, and let φ_{s_0} be the inclusion homomorphism of G_{s_0} into G. Proposition 4.17 produces a unique group homomorphism $\varphi : \bigoplus_{s \in S} G_s \to G$ such that $\varphi \circ i_{s_0} = \varphi_{s_0}$ for all $s_0 \in S$. The spanning condition for the $\mathbb{Z}$ basis says that φ is onto G, and the linear independence condition for the $\mathbb{Z}$ basis says that φ has 0 kernel.

The similarity between vector-space bases and $\mathbb{Z}$ bases suggests further comparison of vector spaces and abelian groups. With vector spaces over a field, every vector space has a basis over the field. However, it is exceptional for an abelian group to have a $\mathbb{Z}$ basis. Two examples that hint at the difficulty are the additive group $\mathbb{Z}/m\mathbb{Z}$ with $m > 1$ and the additive group $\mathbb{Q}$. The group $\mathbb{Z}/m\mathbb{Z}$ has no nonempty linearly independent set, while the group $\mathbb{Q}$ has a linearly independent set of one element, no spanning set of one element, and no linearly independent set of more than one element. Here are two positive examples.

EXAMPLES.

(1) The additive group of all points in $\mathbb{R}^n$ whose coordinates are integers. The standard basis of $\mathbb{R}^n$ is a $\mathbb{Z}$ basis.

(2) The additive group of all points (x, y) in $\mathbb{R}^2$ with x and y both in $\mathbb{Z}$ or both in $\mathbb{Z} + \frac{1}{2}$. The set $\{(1, 0), (\frac{1}{2}, \frac{1}{2})\}$ is a $\mathbb{Z}$ basis.

Next we take a small step that eliminates technical complications from the discussion, proving that any subgroup of a finitely generated abelian group is finitely generated.

Lemma 4.51. Let $\varphi : G \to H$ be a homomorphism of abelian groups. If $\ker \varphi$ and $\operatorname{image} \varphi$ are finitely generated, then G is finitely generated.

PROOF. Let $\{x_1, \ldots, x_m\}$ and $\{y_1, \ldots, y_n\}$ be respective finite sets of generators for $\ker \varphi$ and $\operatorname{image} \varphi$. For $1 \leq j \leq n$, choose x'_j in G with $\varphi(x'_j) = y_j$. We shall prove that $\{x_1, \ldots, x_m, x'_1, \ldots, x'_n\}$ is a set of generators for G. Thus let x be in G. Since $\varphi(x)$ is in $\operatorname{image} \varphi$, there exist integers $a_1, \ldots, a_n$ with $\varphi(x) = a_1 y_1 + \cdots + a_n y_n$. The element $x' = a_1 x'_1 + \cdots + a_n x'_n$ of G has $\varphi(x') = a_1 y_1 + \cdots + a_n y_n = \varphi(x)$. Therefore $\varphi(x - x') = 0$, and there exist integers $b_1, \ldots, b_m$ with $x - x' = b_1 x_1 + \cdots + b_m x_m$. Hence

$$x = b_1 x_1 + \cdots + b_m x_m + x' = b_1 x_1 + \cdots + b_m x_m + a_1 x'_1 + \cdots + a_n x'_n. \quad \square$$

Proposition 4.52. Any subgroup of a finitely generated abelian group is finitely generated.

PROOF. Let G be finitely generated with a set $\{g_1, \ldots, g_n\}$ of n generators, and define $G_k = \mathbb{Z}g_1 + \cdots + \mathbb{Z}g_k$ for $1 \leq k \leq n$. If H is any subgroup of G, define $H_k = H \cap G_k$ for $1 \leq k \leq n$. We shall prove by induction on k that every H_k is finitely generated, and then the case $k = n$ gives the proposition. For $k = 1$, $G_1 = \mathbb{Z}g_1$ is a cyclic group, and any subgroup of it is cyclic by Proposition 4.4 and hence is finitely generated.

Assume inductively that every subgroup of G_k is known to be finitely generated. Let $q : G_{k+1} \to G_{k+1}/G_k$ be the quotient homomorphism, and let $\varphi = q\big|_{H_{k+1}}$, mapping H_{k+1} into G_{k+1}/G_k. Then $\ker \varphi = H_{k+1} \cap G_k$ is a subgroup of G_k and is finitely generated by the inductive hypothesis. Also, image φ is a subgroup of G_{k+1}/G_k, which is a cyclic group with generator equal to the coset of g_{k+1}. Since a subgroup of a cyclic group is cyclic, image φ is finitely generated. Applying Lemma 4.51 to φ, we see that H_{k+1} is finitely generated. This completes the induction and the proof. $\square$

A free abelian group has **finite rank** if it has a finite $\mathbb{Z}$ basis, hence if it is isomorphic to $\mathbb{Z}^n$ for some n. The first theorem is that the integer n is determined by the group.

Theorem 4.53. The number of $\mathbb{Z}$ summands in a free abelian group of finite rank is independent of the direct-sum decomposition of the group.

We define this number to be the **rank** of the free abelian group. Actually, "rank" is a well-defined cardinal in the infinite-rank case as well, because the rank coincides in that case with the cardinality of the group. In any event, Theorem 4.53 follows immediately by two applications of the following lemma.

Lemma 4.54. If G is a free abelian group with a finite $\mathbb{Z}$ basis $x_1, \ldots, x_n$, then any linearly independent subset of G has $\leq n$ elements.

PROOF. Let $\{y_1, \ldots, y_m\}$ be a linearly independent set in G. Since $\{x_1, \ldots, x_n\}$ is a $\mathbb{Z}$ basis, we can define an m-by-n matrix C of integers by $y_i = \sum_{j=1}^{n} C_{ij}x_j$. As a matrix in $M_{mn}(\mathbb{Q})$, C has rank $\leq n$. Consequently if $m > n$, then the rows are linearly dependent over $\mathbb{Q}$, and we can find rational numbers $q_1, \ldots, q_m$ not all 0 such that $\sum_{i=1}^{m} q_i C_{ij} = 0$ for all j. Multiplying by a suitable integer to clear fractions, we obtain integers $k_1, \ldots, k_m$ not all 0 such that $\sum_{i=1}^{m} k_i C_{ij} = 0$ for all j. Then we have

$$\sum_{i=1}^{m} k_i y_i = \sum_{i=1}^{m} k_i \sum_{j=1}^{n} C_{ij}x_j = \sum_{j=1}^{n} \Big(\sum_{i=1}^{m} k_i C_{ij}\Big)x_j = \sum_{j=1}^{n} 0 x_j = 0,$$

in contradiction to the linear independence of $\{y_1, \ldots, y_m\}$ over $\mathbb{Z}$. Therefore $m \leq n$. $\square$

9. Structure of Finitely Generated Abelian Groups

Now we come to the two main results of this section. The first is a special case of the second by Proposition 4.52 and Lemma 4.54. The two will be proved together, and it may help to regard the proof of the first as a part of the proof of the second.

Theorem 4.55. A subgroup H of a free abelian group G of finite rank n is free abelian of rank $\leq n$.

REMARK. This result persists in the case of infinite rank, but we do not need the more general result and will not give a proof.

Theorem 4.56 (Fundamental Theorem of Finitely Generated Abelian Groups). Every finitely generated abelian group is a finite direct sum of cyclic groups. The cyclic groups may be taken to be copies of $\mathbb{Z}$ and various C_{p^k} with p prime, and in this case the cyclic groups are unique up to order and to isomorphism.

REMARKS. The main conclusion of the theorem is the decomposition of each finitely generated abelian group into the direct sum of cyclic groups. An alternative decomposition of the given group that forces uniqueness is as the direct sum of copies of $\mathbb{Z}$ and finite cyclic groups $C_{d_1}, \ldots, C_{d_r}$ such that $d_1 \mid d_2, d_2 \mid d_3, \ldots, d_{r-1} \mid d_r$. A proof of the additional statement appears in the problems at the end of Chapter VIII. The integers $d_1, \ldots, d_r$ are sometimes called the **elementary divisors** of the group.

Let us establish the setting for the proof of Theorem 4.56. Let G be the given group, and say that it has a set of n generators. Proposition 4.17 produces a homomorphism $\varphi : \mathbb{Z}^n \to G$ that carries the standard generators $x_1, \ldots, x_n$ of $\mathbb{Z}^n$ to the generators of G, and φ is onto G. Let H be the kernel of φ. As a subgroup of $\mathbb{Z}^n$, H is finitely generated, by Proposition 4.52. Let $y_1, \ldots, y_m$ be generators. Theorem 4.55 predicts that H is in fact free abelian, hence that $\{y_1, \ldots, y_m\}$ could be taken to be linearly independent over $\mathbb{Z}$ with $m \leq n$, but we do not assume that knowledge in the proof of Theorem 4.56.

The motivation for the main part of the proof of Theorem 4.56 comes from the elementary theory of vector spaces, particularly from the method of using a basis for a finite-dimensional vector space to find a basis of a vector subspace when we know a finite spanning set for the vector subspace. Thus let V be a finite-dimensional vector space over $\mathbb{R}$, with basis $\{x_j\}_{j=1}^n$, and let U be a vector subspace with spanning set $\{y_i\}_{i=1}^m$. To produce a vector-space basis for U, we imagine expanding the y_i's as linear combinations of $x_1, \ldots, x_n$. We can think symbolically of this expansion as expressing each y_i as the product of a row vector of real numbers times the formal "column vector" $\begin{pmatrix} x_1 \\ \vdots \\ x_n \end{pmatrix}$. The entries of this column vector are vectors, but there is no problem in working with it since this is all just a matter of notation anyway. Then the formal column vector $\begin{pmatrix} y_1 \\ \vdots \\ y_m \end{pmatrix}$ of m members of U equals the product of an m-by-n matrix of real numbers times the formal column vector $\begin{pmatrix} x_1 \\ \vdots \\ x_n \end{pmatrix}$. We know from Chapter II that the procedure for finding a basis of U is to row reduce this matrix of real numbers. The nonzero rows of the result determine a basis of the span of the m vectors we have used, and this basis is related tidily to the given basis for V. We can compare the two bases to understand the relationship between U and V. To prove Theorem 4.56, we would like to use the same procedure, but we have to work with an integer matrix and avoid division. This means that only two of the three usual row operations are fully available for the row reduction; division of a row by an integer is allowable only when the integer is ± 1. A partial substitute for division comes by using the steps of the Euclidean algorithm via the division algorithm (Proposition 1.1), but even that is not enough. For example, if the m-by-n matrix is $\begin{pmatrix} 2 & 1 & 1 \\ 0 & 0 & 3 \end{pmatrix}$, no further row reduction is possible with integer operations. However, the equations tell us that H is the subgroup of $\mathbb{Z}^3$ generated by $(2, 1, 1)$ and $(0, 0, 3)$, and it is not at all clear how to write $\mathbb{Z}^3/H$ as a direct sum of cyclic groups.

The row operations have the effect of changing the set of generators of H while maintaining the fact that they generate H. What is needed is to allow also column reduction with integer operations. Steps of this kind have the effect of changing the $\mathbb{Z}$ basis of $\mathbb{Z}^n$. When steps of this kind are allowed, we can produce new generators of H and a new basis of $\mathbb{Z}^n$ so that the two can be compared. With the example above, suitable column operations are

$$\begin{pmatrix} 2 & 1 & 1 \\ 0 & 0 & 3 \end{pmatrix} \mapsto \begin{pmatrix} 1 & 2 & 1 \\ 0 & 0 & 3 \end{pmatrix} \mapsto \begin{pmatrix} 1 & 0 & 0 \\ 0 & 0 & 3 \end{pmatrix} \mapsto \begin{pmatrix} 1 & 0 & 0 \\ 0 & 3 & 0 \end{pmatrix}.$$

The equations with the new generators say that $y_1' = x_1'$ and $y_2' = 3x_2'$. Thus H is the subgroup $\mathbb{Z} \oplus 3\mathbb{Z} \oplus 0\mathbb{Z}$, nicely aligned with $\mathbb{Z}^3 = \mathbb{Z} \oplus \mathbb{Z} \oplus \mathbb{Z}$. The quotient is $(\mathbb{Z}/\mathbb{Z}) \oplus (\mathbb{Z}/3\mathbb{Z}) \oplus (\mathbb{Z}/0\mathbb{Z}) \cong C_3 \oplus \mathbb{Z}$.

The proof of Theorem 4.56 will make use of an algorithm that uses row and column operations involving only allowable divisions and that converts the matrix C of coefficients so that its nonzero entries are the **diagonal entries** C_{ii} for $1 \leq i \leq r$ and no other entries. The algorithm in principle can be very slow, and it may be helpful to see what it does in an ordinary example.

EXAMPLE. Suppose that the relationship between generators y_1, y_2, y_3 of H

and the standard $\mathbb{Z}$ basis $\{x_1, x_2\}$ of $\mathbb{Z}^2$ is

$$\begin{pmatrix} y_1 \\ y_2 \\ y_3 \end{pmatrix} = C \begin{pmatrix} x_1 \\ x_2 \end{pmatrix}, \qquad \text{where } C = \begin{pmatrix} 3 & 5 \\ 7 & 13 \\ 5 & 9 \end{pmatrix}.$$

In row reduction in vector-space theory, we would start by dividing the first row of C by 3, but division by 3 is not available in the present context. Our target for the upper-left entry is $\text{GCD}(3, 7, 5) = 1$, and we use the division algorithm one step at a time to get there. To begin with, it says that $7 = 2 \cdot 3 + 1$ and hence $7 - 2 \cdot 3 = 1$. The first step of row reduction is then to replace the second row by the difference of it and 2 times the first row. The result can be achieved by left multiplication by

$$\begin{pmatrix} 1 & 0 & 0 \\ -2 & 1 & 0 \\ 0 & 0 & 1 \end{pmatrix} \qquad \text{and is} \qquad \begin{pmatrix} 3 & 5 \\ 1 & 3 \\ 5 & 9 \end{pmatrix}.$$

We write this step as

$$\begin{pmatrix} 3 & 5 \\ 7 & 13 \\ 5 & 9 \end{pmatrix} \xrightarrow{\text{left by } \begin{pmatrix} 1 & 0 & 0 \\ -2 & 1 & 0 \\ 0 & 0 & 1 \end{pmatrix}} \begin{pmatrix} 3 & 5 \\ 1 & 3 \\ 5 & 9 \end{pmatrix}.$$

The entry 1 in the first column is our target for this stage since $\text{GCD}(3, 7, 5) = 1$. The next step interchanges two rows to move the 1 to the upper left entry, and the subsequent step uses the 1 to eliminate the other entries of the first column:

$$\begin{pmatrix} 3 & 5 \\ 1 & 3 \\ 5 & 9 \end{pmatrix} \xrightarrow{\text{left by } \begin{pmatrix} 0 & 1 & 0 \\ 1 & 0 & 0 \\ 0 & 0 & 1 \end{pmatrix}} \begin{pmatrix} 1 & 3 \\ 3 & 5 \\ 5 & 9 \end{pmatrix} \xrightarrow{\text{left by } \begin{pmatrix} 1 & 0 & 0 \\ -3 & 1 & 0 \\ -5 & 0 & 1 \end{pmatrix}} \begin{pmatrix} 1 & 3 \\ 0 & -4 \\ 0 & -6 \end{pmatrix}.$$

The algorithm next seeks to eliminate the off-diagonal entry 3 in the first row. This is done by a column operation:

$$\begin{pmatrix} 1 & 3 \\ 0 & -4 \\ 0 & -6 \end{pmatrix} \xrightarrow{\text{right by } \begin{pmatrix} 1 & -3 \\ 0 & 1 \end{pmatrix}} \begin{pmatrix} 1 & 0 \\ 0 & -4 \\ 0 & -6 \end{pmatrix}.$$

With two further row operations we are done:

$$\begin{pmatrix} 1 & 0 \\ 0 & -4 \\ 0 & -6 \end{pmatrix} \xrightarrow{\text{left by } \begin{pmatrix} 1 & 0 & 0 \\ 0 & 1 & -1 \\ 0 & 0 & 1 \end{pmatrix}} \begin{pmatrix} 1 & 0 \\ 0 & 2 \\ 0 & -6 \end{pmatrix} \xrightarrow{\text{left by } \begin{pmatrix} 1 & 0 & 0 \\ 0 & 1 & 0 \\ 0 & 3 & 1 \end{pmatrix}} \begin{pmatrix} 1 & 0 \\ 0 & 2 \\ 0 & 0 \end{pmatrix}.$$

Our steps are summarized by the fact that the matrix A with

$$A = \begin{pmatrix} 1 & 0 & 0 \\ 0 & 1 & 0 \\ 0 & 3 & 1 \end{pmatrix} \begin{pmatrix} 1 & 0 & 0 \\ 0 & 1 & -1 \\ 0 & 0 & 1 \end{pmatrix} \begin{pmatrix} 1 & 0 & 0 \\ -3 & 1 & 0 \\ -5 & 0 & 1 \end{pmatrix} \begin{pmatrix} 0 & 1 & 0 \\ 1 & 0 & 0 \\ 0 & 0 & 1 \end{pmatrix} \begin{pmatrix} 1 & 0 & 0 \\ -2 & 1 & 0 \\ 0 & 0 & 1 \end{pmatrix}$$

has

$$AC \begin{pmatrix} 1 & -3 \\ 0 & 1 \end{pmatrix} = \begin{pmatrix} 1 & 0 \\ 0 & 2 \\ 0 & 0 \end{pmatrix}$$

and by the fact that the integer matrices to the left and right of C have determinant ± 1. The determinant condition ensures that A^{-1} and $\begin{pmatrix} 1 & -3 \\ 0 & 1 \end{pmatrix}^{-1}$ have integer entries, according to Cramer's rule (Proposition 2.38).

Lemma 4.57. If C is an m-by-n matrix of integers, then there exist an m-by-m matrix A of integers with determinant ± 1 and an n-by-n matrix B of integers with determinant ± 1 such that for some $r \geq 0$, the nonzero entries of $D = ACB$ are exactly the diagonal entries $D_{11}, D_{22}, \ldots, D_{rr}$.

PROOF. Given C, choose (i, j) with $|C_{ij}| \neq 0$ but $|C_{ij}|$ as small as possible. (If $C = 0$, the algorithm terminates.) Possibly by interchanging two rows and/or then two columns (a left multiplication with determinant -1 and then a right multiplication with determinant -1), we may assume that $(i, j) = (1, 1)$. By the division algorithm write, for each i,

$$C_{i1} = q_i C_{11} + r_i \qquad \text{with } 0 \leq r_i < |C_{11}|,$$

and replace the ith row by the difference of the ith row and q_i times the first row (a left multiplication). If some r_i is not 0, the result will leave a nonzero entry in the first column that is $< |C_{11}|$ in absolute value. Permute the least such $r_i \neq 0$ to the upper left and repeat the process. Since the least absolute value is going down, this process at some point terminates with all r_i equal to 0. The first column then has a nonzero diagonal entry and is otherwise 0.

Now consider C_{1j} and apply the division algorithm and column operations in similar fashion in order to process the first row. If we get a smaller nonzero remainder, permute the smallest one to the first column. Repeat this process until the first row is 0 except for entry C_{11}. Continue alternately with row and column operations in this fashion until both $C_{1j} = 0$ for $j > 1$ and $C_{i1} = 0$ for $i > 1$.

Repeat the algorithm for the $(m-1)$-by-$(n-1)$ matrix consisting of rows 2 through m and columns 2 through n, and continue inductively. The algorithm terminates when either the reduced-in-size matrix is empty or is all 0. At this point the original matrix has been converted into the desired "diagonal form." □

Lemma 4.58. Let $G_1, \ldots, G_n$ be abelian groups, and for $1 \leq j \leq n$, let H_j be a subgroup of G_j. Then

$$(G_1 \oplus \cdots \oplus G_n)/(H_1 \oplus \cdots \oplus H_n) \cong (G_1/H_1) \oplus \cdots \oplus (G_n/H_n).$$

PROOF. Let $\varphi : G_1 \oplus \cdots \oplus G_n \to (G_1/H_1) \oplus \cdots \oplus (G_n/H_n)$ be the homomorphism defined by $\varphi(g_1, \ldots, g_n) = (g_1 H_1, \ldots, g_n H_n)$. The mapping φ is onto $(G_1/H_1) \oplus \cdots \oplus (G_n/H_n)$, and the kernel is $H_1 \oplus \cdots \oplus H_n$. Then Corollary 4.12 shows that φ descends to the required isomorphism. $\square$

PROOF OF THEOREM 4.55 AND MAIN CONCLUSION OF THEOREM 4.56. Given G with n generators, we set up matters as indicated immediately after the statement of Theorem 4.56, writing

$$\begin{pmatrix} y_1 \\ \vdots \\ y_m \end{pmatrix} = C \begin{pmatrix} x_1 \\ \vdots \\ x_n \end{pmatrix},$$

where $x_1, \ldots, x_n$ are the standard generators of $\mathbb{Z}^n$, $y_1, \ldots, y_m$ are the generators of the kernel of the homomorphism from $\mathbb{Z}^n$ onto G, and C is a matrix of integers. Applying Lemma 4.57, let A and B be square integer matrices of determinant ± 1 such that $D = ACB$ is diagonal as in the statement of the lemma. Define

$$\begin{pmatrix} z_1 \\ \vdots \\ z_m \end{pmatrix} = A \begin{pmatrix} y_1 \\ \vdots \\ y_m \end{pmatrix} \quad \text{and} \quad \begin{pmatrix} u_1 \\ \vdots \\ u_n \end{pmatrix} = B^{-1} \begin{pmatrix} x_1 \\ \vdots \\ x_n \end{pmatrix}.$$

Substitution gives

$$\begin{pmatrix} z_1 \\ \vdots \\ z_m \end{pmatrix} = A \begin{pmatrix} y_1 \\ \vdots \\ y_m \end{pmatrix} = (ACB)B^{-1} \begin{pmatrix} x_1 \\ \vdots \\ x_n \end{pmatrix} = ACB \begin{pmatrix} u_1 \\ \vdots \\ u_n \end{pmatrix}.$$

If $(c_1 \; \cdots \; c_n)$ and $(d_1 \; \cdots \; d_n) = (c_1 \; \cdots \; c_n)B^{-1}$ are row vectors, then the formula

$$c_1 u_1 + \cdots + c_n u_n = (c_1 \; \cdots \; c_n)\begin{pmatrix} u_1 \\ \vdots \\ u_n \end{pmatrix} = (d_1 \; \cdots \; d_n)\begin{pmatrix} x_1 \\ \vdots \\ x_n \end{pmatrix}$$

$$= d_1 x_1 + \cdots + d_n x_n \tag{$*$}$$

shows that $\{u_1, \ldots, u_n\}$ generates the same subset of $\mathbb{Z}^n$ as $\{x_1, \ldots, x_n\}$. Since $(c_1 \; \cdots \; c_n)$ is nonzero if and only if $(d_1 \; \cdots \; d_n)$ is nonzero, the formula $(*)$ shows also that the linear independence of $\{x_1, \ldots, x_n\}$ implies that of $\{u_1, \ldots, u_n\}$. Hence $\{u_1, \ldots, u_n\}$ is a $\mathbb{Z}$ basis of $\mathbb{Z}^n$. Similarly $\{y_1, \ldots, y_m\}$ and $\{z_1, \ldots, z_m\}$

generate the same subgroup H of $\mathbb{Z}^n$. Therefore we can compare H and $\mathbb{Z}^n$ using $\{z_1, \ldots, z_m\}$ and $\{u_1, \ldots, u_n\}$. Since D is diagonal, the equations relating $\{z_1, \ldots, z_m\}$ and $\{u_1, \ldots, u_n\}$ are $z_j = D_{jj}u_j$ for $j \leq \min(m, n)$ and $z_j = 0$ for $\min(m, n) < j \leq m$. If $q = \min(m, n)$, then we see that

$$H = \sum_{i=1}^m \mathbb{Z}z_i = \sum_{i=1}^q D_{ii}\mathbb{Z}u_i + \sum_{i=q+1}^m \mathbb{Z}z_i = \sum_{i=1}^q D_{ii}\mathbb{Z}u_i.$$

Since the set $\{u_1, \ldots, u_q\}$ is linearly independent over $\mathbb{Z}$, this sum exhibits H as given by

$$H = D_{11}\mathbb{Z} \oplus \cdots \oplus D_{qq}\mathbb{Z}$$

with $D_{11}u_1, \ldots, D_{qq}u_q$ as a $\mathbb{Z}$ basis. Consequently H has been exhibited as free abelian of rank $\leq q \leq n$. This proves Theorem 4.55. Applying Lemma 4.58 to the quotient $\mathbb{Z}^n/H$ and letting $D_{11}, \ldots, D_{rr}$ be the nonzero diagonal entries of D, we see that H has rank r, and we obtain an expansion of G in terms of cyclic groups as

$$G = C_{D_{11}} \oplus \cdots \oplus C_{D_{rr}} \oplus \mathbb{Z}^{n-r}.$$

This proves the main conclusion of Theorem 4.56. □

PROOF OF THE DECOMPOSITION WITH CYCLIC GROUPS OF PRIME-POWER ORDER. It is enough to prove that if $m = \prod_{j=1}^N p_j^{k_j}$ with the p_j equal to distinct primes, then $\mathbb{Z}/m\mathbb{Z} \cong (\mathbb{Z}/p_1^{k_1}\mathbb{Z}) \oplus \cdots \oplus (\mathbb{Z}/p_N^{k_N}\mathbb{Z})$. This is a variant of the Chinese Remainder Theorem (Corollary 1.9). For the proof let

$$\varphi : \mathbb{Z} \to (\mathbb{Z}/p_1^{k_1}\mathbb{Z}) \oplus \cdots \oplus (\mathbb{Z}/p_N^{k_N}\mathbb{Z})$$

be the homomorphism given by $\varphi(s) = \big(s \bmod p_1^{k_1}, \ldots, s \bmod p_N^{k_N}\big)$ for $s \in \mathbb{Z}$. Since $\varphi(m) = (0, \ldots, 0)$, φ descends to a homomorphism

$$\overline{\varphi} : \mathbb{Z}/m\mathbb{Z} \to (\mathbb{Z}/p_1^{k_1}\mathbb{Z}) \oplus \cdots \oplus (\mathbb{Z}/p_N^{k_N}\mathbb{Z}).$$

The map $\overline{\varphi}$ is one-one because if $\varphi(s) = 0$, then $p_j^{k_j}$ divides s for all j. Since the $p_j^{k_j}$ are relatively prime in pairs, their product m divides s. Since m divides s, $s \equiv 0 \bmod m$. The map $\overline{\varphi}$ is onto since it is one-one and since the finite sets $\mathbb{Z}/m\mathbb{Z}$ and $(\mathbb{Z}/p_1^{k_1}\mathbb{Z}) \oplus \cdots \oplus (\mathbb{Z}/p_N^{k_N}\mathbb{Z})$ both have m elements. □

PROOF OF UNIQUENESS OF THE DECOMPOSITION. Write $G = \mathbb{Z}^s \oplus T$, where

$$T = (\mathbb{Z}/p_1^{l_1}\mathbb{Z}) \oplus \cdots \oplus (\mathbb{Z}/p_M^{l_M}\mathbb{Z})$$

and the p_j's are not necessarily distinct. The subgroup T is the subgroup of elements of finite order in G, and it is well defined independently of the decomposition of G as the direct sum of cyclic groups. The quotient $G/T \cong \mathbb{Z}^s$ is free abelian of finite rank, and its rank s is well defined by Theorem 4.53. Thus the number s of factors of $\mathbb{Z}$ in the decomposition of G is uniquely determined, and we need only consider uniqueness of the decomposition of the finite abelian group T.

For p prime the elements of T of order p^a for some a are those in the sum of the groups $\mathbb{Z}/p_j^{l_j}\mathbb{Z}$ for which $p_j = p$, and we are reduced to considering a group

$$H = \mathbb{Z}/p^{l_1}\mathbb{Z} \oplus \cdots \oplus \mathbb{Z}/p^{l_{M'}}\mathbb{Z}$$

with p fixed and $l_1 \leq \cdots \leq l_{M'}$. The set of p^j powers of elements of H is a subgroup of H and is given by $\mathbb{Z}/p^{l_t-j}\mathbb{Z} \oplus \cdots \oplus \mathbb{Z}/p^{l_{M'}-j}\mathbb{Z}$ if l_t is the first index $\geq j$, while the set of p^{j+1} powers of elements of H is given by $\mathbb{Z}/p^{l_{t'}-j-1}\mathbb{Z} \oplus \cdots \oplus \mathbb{Z}/p^{l_{M'}-j}\mathbb{Z}$ if $l_{t'}$ is the first index $\geq j+1$. Therefore Lemma 4.58 gives

$$p^j H/p^{j+1} H \cong (\mathbb{Z}/p^{l_{t'}-j-1}\mathbb{Z})/(\mathbb{Z}/p^{l_{t'}-j}\mathbb{Z}) \oplus \cdots \oplus (\mathbb{Z}/p^{l_{M'}-j-1}\mathbb{Z})/(\mathbb{Z}/p^{l_{M'}-j}\mathbb{Z}).$$

Each term of $p^j H/p^{j+1} H$ has order p, and thus

$$|p^j H/p^{j+1} H| = p^{|\{i \mid l_i > j\}|}.$$

Hence H determines the integers $l_1, \ldots, l_{M'}$, and uniqueness is proved. $\square$

10. Sylow Theorems

This section continues the use of group actions to obtain results concerning structure theory for abstract groups. We shall prove the three Sylow Theorems, which are a starting point for investigations of the structure of finite groups that are deeper than those in Sections 6 and 7. We state the three theorems as the parts of Theorem 4.59.

Theorem 4.59 (Sylow Theorems). Let G be a finite group of order $p^m r$, where p is prime and p does not divide r. Then
  (a) G contains a subgroup of order p^m, and any subgroup of G of order p^l with $0 \leq l < m$ is contained in a subgroup of order p^m,
  (b) any two subgroups of order p^m in G are conjugate in G, i.e., any two such subgroups P_1 and P_2 have $P_2 = aP_1a^{-1}$ for some $a \in G$,
  (c) the number of subgroups of order p^m is of the form $pk + 1$ and divides r.

REMARK. A subgroup of order p^m as in the theorem is called a **Sylow p-subgroup** of G.

Before coming to the proof, let us carefully give two simple applications to structure theory. The applications combine Theorem 4.59, some results of Sections 6 and 7, and Problems 35–38 and 45–48 at the end of the chapter.

Proposition 4.60. If p and q are primes with $p < q$, then there exists a nonabelian group of order pq if and only if p divides $q - 1$, and in this case the nonabelian group is unique up to isomorphism. It may be taken to be a semidirect product of the cyclic groups C_p and C_q with C_q normal.

REMARK. It follows from Theorem 4.56 that the only abelian group of order pq, up to isomorphism, is $C_p \times C_q \cong C_{pq}$. If $p = 2$ in the proposition, then q is odd and p divides $q - 1$; the proposition yields the dihedral group D_q. For $p > 2$, the divisibility condition may or may not hold: For $pq = 15$, the condition does not hold, and hence every group of order 15 is cyclic. For $pq = 21$, the condition does hold, and there exists a nonabelian group of order 21; this group was constructed explicitly in Example 2 in Section 7.

PROOF. Existence of a nonabelian group of order pq, together with the semidirect-product structure, is established by Proposition 4.46 if p divides $q-1$. Let us see uniqueness and the necessity of the condition that p divide $q - 1$.

If G has order pq, Theorem 4.59a shows that G has a Sylow p-subgroup H_p and a Sylow q-subgroup H_q. Corollary 4.9 shows that these two groups are cyclic. The conjugates of H_q are Sylow q-subgroups, and Theorem 4.59c shows that the number of such conjugates is of the form $kq + 1$ and divides p. Since $p < q$, $k = 0$. Therefore H_q is normal. (Alternatively, one can apply Proposition 4.36 to see that H_q is normal.)

Each element of G is uniquely a product ab with a in H_p and b in H_q. For the uniqueness, if $a_1 b_1 = a_2 b_2$, then $a_2^{-1} a_1 = b_2 b_1^{-1}$ is an element of $H_p \cap H_q$. Its order must divide both p and q and hence must be 1. Thus the pq products ab with a in H_p and b in H_q are all different. Since the number of them equals the order of G, every member of G is such a product. By Proposition 4.44, G is a semidirect product of H_p and H_q.

If the action of H_p on H_q is nontrivial, then Problem 37 at the end of the chapter shows that p divides $q - 1$, and Problem 38 shows that the group is unique up to isomorphism. On the other hand, if the action is trivial, then G is certainly abelian. □

Proposition 4.61. If G is a group of order 12, then G contains a subgroup H of order 3 and a subgroup K of order 4, and at least one of them is normal. Consequently there are exactly five groups of order 12, up to isomorphism—two abelian and three nonabelian.

REMARK. The second statement follows from the first, as a consequence of Problems 45–48 at the end of the chapter. Those problems show how to construct the groups.

PROOF. Theorem 4.59a shows that H may be taken to be a Sylow 3-subgroup and K may be taken to be a Sylow 2-subgroup. We have to prove that either H or K is normal.

Suppose that H is not normal. Theorem 4.59c shows that the number of Sylow 3-subgroups is of the form $3k + 1$ and divides 4. The subgroup H, not being normal, fails to equal one of its conjugates, which will be another Sylow 3-subgroup; hence $k > 0$. Therefore $k = 1$, and there are four Sylow 3-subgroups. The intersection of any two such subgroups is a subgroup of both and must be trivial since 3 is prime. Thus the set-theoretic union of the Sylow 3-subgroups accounts for $4 \cdot 2 + 1$ elements. None of these elements apart from the identity lies in K, and thus K contributes 3 further elements, for a total of 12. Thus every element of G lies in K or a conjugate of H. Consequently K equals every conjugate of K, and K is normal. □

Let us see where we are with classifying finite groups of certain orders, up to isomorphism. A group of order p is cyclic by Corollary 4.9, and a group of order p^2 is abelian by Corollary 4.39. Groups of order pq are settled by Proposition 4.60. Thus for p and q prime, we know the structure of all groups of order p, p^2, and pq. Problems 39–44 at the end of the chapter tell us the structure of the groups of order 8, and Proposition 4.61 and Problems 45–48 tell us the structure of the groups of order 12. In particular, the table at the end of Section 1, which gives examples of nonisomorphic groups of order at most 15, is complete except for the one group of order 12 that is discussed in Problem 48.

Problems 30–34 and 49–54 at the end of the chapter go in the direction of classifying finite groups of certain other orders.

Now we return to Theorem 4.59. The proof of the theorem makes use of the theory of group actions as in Section 6. In fact, the proof of existence of Sylow p-subgroups is just an elaboration of the argument used to prove Corollary 4.38, saying that a group of prime-power order has a nontrivial center. The relevant action for the existence part of the proof is the one $(g, x) \mapsto gxg^{-1}$ given by conjugation of the elements of the group, the orbit of x being the conjugacy class $C\ell(x)$. Proposition 4.37 shows that $|G| = |C\ell(x)||Z_G(x)|$, where $Z_G(x)$ is the centralizer of x. Since the disjoint union of the conjugacy classes is all of $|G|$, we have

$$|G| = |Z_G| + \sum_{\substack{\text{representatives } x_j \\ \text{of each conjugacy class} \\ \text{with } |C\ell(x)| \neq 1}} |G|/|Z_G(x_j)|,$$

a formula sometimes called the **class equation** of G.

PROOF OF EXISTENCE OF SYLOW p-SUBGROUPS IN THEOREM 4.59a. We induct on $|G|$, the base case being $|G| = 1$. Suppose that existence holds for groups of order $< |G|$. Without loss of generality suppose that $m > 0$, so that p divides $|G|$.

First suppose that p does not divide $|Z_G|$. Referring to the class equation of G, we see that p must fail to divide some integer $|G|/|Z_G(x_j)|$ for which $|Z_G(x_j)| < |G|$. Since p^m is the exact power of p dividing $|G|$, we conclude that p^m divides this $|Z_G(x_j)|$ and p^{m+1} does not. Since $|Z_G(x_j)| < |G|$, the inductive hypothesis shows that $Z_G(x_j)$ has a subgroup of order p^m, and this is a Sylow p-subgroup of G.

Now suppose that p divides $|Z_G|$. The group Z_G is finitely generated abelian, hence is a direct sum of cyclic groups by Theorem 4.56. Thus Z_G contains an element c of order p. The cyclic group C generated by c then has order p. Being a subgroup of Z_G, C is normal in G. The group G/C has order $p^{m-1}r$, and the inductive hypothesis implies that G/C has a subgroup H of order p^{m-1}. If $\varphi : G \to G/C$ denotes the quotient map, then $\varphi^{-1}(H)$ is a subgroup of G of order $|H||\ker \varphi| = p^{m-1}p = p^m$. □

For the remaining parts of Theorem 4.59, we make use of a different group action. If Γ denotes the set of all subgroups of G, then G acts on Γ by conjugation: $(g, H) \mapsto gHg^{-1}$. The orbit of a subgroup of H consists of all subgroups conjugate to H in G, and the isotropy subgroup at the point H in Γ is

$$\{g \in G \mid gHg^{-1} = H\}.$$

This is a subgroup $N(H)$ of G known as the **normalizer** of H in G. It has the properties that $N(H) \supseteq H$ and that H is a normal subgroup of $N(H)$. The counting formula of Corollary 4.35 gives

$$\left|\{gHg^{-1} \mid g \in G\}\right| = |G/N(H)|.$$

Meanwhile, application of Lagrange's Theorem (Theorem 4.7) to the three quotients $G/H, G/N(H)$, and $N(H)/H$ shows that

$$|G/H| = |G/N(H)||N(H)/H|,$$

with all three factors being integers.

Now assume as in the statement of Theorem 4.59 that $|G| = p^m r$ with p prime and p not dividing r. In this setting we have the following lemma.

10. Sylow Theorems

Lemma 4.62. If P is a Sylow p-subgroup of G and if H is a subgroup of the normalizer $N(P)$ whose order is a power of p, then $H \subseteq P$.

PROOF. Since $H \subseteq N(P)$ and P is normal in $N(P)$, the set HP of products is a group, by the same argument as used for $H_p H_q$ in the proof of Proposition 4.60. Then $HP/P \cong H/(H \cap P)$ by the Second Isomorphism Theorem (Theorem 4.14), and hence $|HP/P|$ is some power p^k of p. By Lagrange's Theorem (Theorem 4.7), $|HP| = p^{m+k}$ with $k \geq 0$. Since no subgroup of G can have order p^l with $l > m$, we must have $k = 0$. Thus $HP = P$ and $H \subseteq P$. □

PROOF OF THE REMAINDER OF THEOREM 4.59. Within the set Γ of all subgroups of G, let Π be the set of all subgroups of G of order p^m. We have seen that Π is not empty. Since the conjugate of a subgroup has the same order as the subgroup, Π is the union of orbits of Γ under conjugation by G. Thus we can restrict the group action by conjugation from $G \times \Gamma \to \Gamma$ to $G \times \Pi \to \Pi$.

Let P and P' be members of Π, and let Σ and Σ' be the G orbits of P and P' under conjugation. Suppose that Σ and Σ' are distinct orbits of G. Let us restrict the group action by conjugation from $G \times \Pi \to \Pi$ to $P \times \Pi \to \Pi$. The G orbits Σ and Σ' then break into P orbits, and the counting formula Corollary 4.35 says for each orbit that

$$p^m = |P| = \#\{\text{subgroups in a } P \text{ orbit}\} \times \big|\text{isotropy subgroup within } P\big|.$$

Hence the number of subgroups in a P orbit is of the form p^l for some $l \geq 0$.

Suppose that $l = 0$. Then the P orbit is some singleton set $\{P''\}$, and the corresponding isotropy subgroup within P is all of P:

$$P = \{p \in P \mid pP''p^{-1} = P''\} \subseteq N(P'').$$

Lemma 4.62 shows that $P \subseteq P''$, and therefore $P = P''$. Thus $l = 0$ only for the P orbit $\{P\}$. In other words, the number of elements in any P orbit other than $\{P\}$ is divisible by p. Consequently $|\Sigma| \equiv 1 \bmod p$ while $|\Sigma'| \equiv 0 \bmod p$, the latter because Σ and Σ' are assumed distinct. But this conclusion is asymmetric in the G orbits Σ and Σ', and we conclude that Σ and Σ' must coincide. Hence there is only one G orbit in Π, and it has $kp + 1$ members for some k. This proves parts (b) and (c) except for the fact that $kp + 1$ divides r.

For this divisibility let us apply the counting formula Corollary 4.35 to the orbit Σ of G. The formula gives $|G| = |\Sigma| |\text{isotropy subgroup}|$, and hence $|\Sigma|$ divides $|G| = p^m r$. Since $|\Sigma| = kp + 1$, we have $\gcd(|\Sigma|, p) = 1$ and also $\gcd(|\Sigma|, p^m) = 1$. By Corollary 1.3, $kp + 1$ divides r.

Finally we prove that any subgroup H of G of order p^l lies in some Sylow p-subgroup. Let $\Sigma = \Pi$ again be the G orbit in Γ of subgroups of order p^m, and restrict the action by conjugation from $G \times \Sigma \to \Sigma$ to $H \times \Sigma \to \Sigma$. Each H orbit in Σ must have p^a elements for some a, by one more application of the counting formula Corollary 4.35. Since $|\Sigma| \equiv 1 \bmod p$, some H orbit has one element, say the H orbit of P. Then the isotropy subgroup of H at the point P is all of H, and $H \subseteq N(P)$. By Lemma 4.62, $H \subseteq P$. This completes the proof of Theorem 4.59. □

11. Categories and Functors

The mathematics thus far in the book has taken place in several different contexts, and we have seen that the same notions sometimes recur in more than one context, possibly with variations. For example we have worked with vector spaces, inner-product spaces, groups, rings, and fields, and we have seen that each of these areas has its own definition of isomorphism. In addition, the notion of direct product or direct sum has arisen in more than one of these contexts, and there are other similarities. In this section we introduce some terminology to make the notion of "context" precise and to provide a setting for discussing similarities between different contexts.

A **category** $\mathcal{C}$ consists of three things:

- a class of **objects**, denoted by $\mathrm{Obj}(\mathcal{C})$,
- for any two objects A and B in the category, a set $\mathrm{Morph}(A, B)$ of **morphisms**,
- for any three objects A, B, and C in the category, a **law of composition** for morphisms, i.e., a function carrying $\mathrm{Morph}(A, B) \times \mathrm{Morph}(B, C)$ into $\mathrm{Morph}(A, C)$, with the image of (f, g) under composition written as gf,

and these are to satisfy certain properties that we list in a moment. When more than one category is under discussion, we may use notation like $\mathrm{Morph}_{\mathcal{C}}(A, B)$ to distinguish between the categories.

We are to think initially of the objects as the sets we are studying with a particular kind of structure on them; the morphisms are then the functions from one object to another that respect this additional structure, and the law of composition is just composition of functions. Indeed, the defining conditions that are imposed on general categories are arranged to be obvious for this special kind of category, and this setting accounts for the order in which we write the composition of two morphisms. But the definition of a general category is not so restrictive, and it is important not to restrict the definition in this way.

The properties that are to be satisfied to have a category are as follows:

(i) the sets $\mathrm{Morph}(A_1, B_1)$ and $\mathrm{Morph}(A_2, B_2)$ are disjoint unless $A_1 = A_2$ and $B_1 = B_2$ (because two functions are declared to be different

unless their domains match and their ranges match, as is underscored in Section A1 of the appendix),
(ii) the law of composition satisfies the associativity property $h(gf) = (hg)f$ for $f \in \text{Morph}(A, B)$, $g \in \text{Morph}(B, C)$, and $h \in \text{Morph}(C, D)$,
(iii) for each object A, there is an **identity morphism** 1_A in $\text{Morph}(A, A)$ such that $f1_A = f$ and $1_A g = g$ for $f \in \text{Morph}(A, B)$ and $g \in \text{Morph}(C, A)$.

A **subcategory** $\mathcal{S}$ of a category $\mathcal{C}$ by definition is a category with $\text{Obj}(\mathcal{S}) \subseteq \text{Obj}(\mathcal{C})$ and $\text{Morph}_{\mathcal{S}}(A, B) \subseteq \text{Morph}_{\mathcal{C}}(A, B)$ whenever A and B are in $\text{Obj}(\mathcal{S})$, and it is assumed that the laws of composition in $\mathcal{S}$ and $\mathcal{C}$ are consistent when both are defined.

Here are several examples in which the morphisms are functions and the law of composition is ordinary composition of functions. They are usually identified in practice just by naming their objects, since the morphisms are understood to be all functions from one object to another respecting the additional structure on the objects.

EXAMPLES OF CATEGORIES.

(1) The category of all sets. An object A is a set, and a morphism in the set $\text{Morph}(A, B)$ is a function from A into B.

(2) The category of all vector spaces over a field $\mathbb{F}$. The morphisms are linear maps.

(3) The category of all groups. The morphisms are group homomorphisms.

(4) The category of all abelian groups. The morphisms again are group homomorphisms. This is a subcategory of the previous example.

(5) The category of all rings. The morphisms are all ring homomorphisms. The kernel and the image of a morphism are necessarily objects of the category.

(6) The category of all rings with identity. The morphisms are all ring homomorphisms carrying identity to identity. This is a subcategory of the previous example. The image of a morphism is necessarily an object of the category, but the kernel of a morphism is usually not in the category.

(7) The category of all fields. The morphisms are as in Example 6, and the result is a subcategory of Example 6. In this case any morphism is necessarily one-one and carries inverses to inverses.

(8) The category of all group actions by a particular group G. If G acts on X and on Y, then a morphism from the one space to the other is a G **equivariant mapping** from X to Y, i.e., a function $\varphi : X \to Y$ such that $\varphi(gx) = g\varphi(x)$ for all x in X.

(9) The category of all representations by a particular group G on a vector space over a particular field $\mathbb{F}$. The morphisms are the linear G equivariant functions. This is a subcategory of the previous example.

Readers who are familiar with point-set topology will recognize that one can impose topologies on everything in the above examples, insisting that the functions be continuous, and again we obtain examples of categories. For example the category of all topological spaces consists of objects that are topological spaces and morphisms that are continuous functions. The category of all continuous group actions by a particular topological group has objects that are group actions $G \times X \to X$ that are continuous functions, and the morphisms are the equivariant functions that are continuous.

Readers who are familiar with manifolds will recognize that another example is the category of all smooth manifolds, which consists of objects that are smooth manifolds and morphisms that are smooth functions.

The morphisms in a category need not be functions in the usual sense. An important example is the "opposite category" $\mathcal{C}^{\text{opp}}$ to a category $\mathcal{C}$, which is a handy technical device and is discussed in Problems 78–80 at the end of the chapter.

In all of the above examples of categories, the class of objects fails to be a set. This behavior is typical. However, it does not cause problems in practice because in any particular argument involving categories, we can restrict to a subcategory for which the objects do form a set.[15]

If $\mathcal{C}$ is a category, a morphism $\varphi \in \text{Morph}(A, B)$ is said to be an **isomorphism** if there exists a morphism $\psi \in \text{Morph}(B, A)$ such that $\psi\varphi = 1_A$ and $\varphi\psi = 1_B$. In this case we say that A is **isomorphic** to B in the category $\mathcal{C}$. Let us check that the morphism ψ is unique if it exists. In fact, if ψ' is a member of $\text{Morph}(B, A)$ with $\psi'\varphi = 1_A$ and $\varphi\psi' = 1_B$, then $\psi = 1_A\psi = (\psi'\varphi)\psi = \psi'(\varphi\psi) = \psi'1_B = \psi'$. We can therefore call ψ the **inverse** to φ.

The relation "is isomorphic to" is an equivalence relation.[16] In fact, the relation is symmetric by definition, and it is reflexive because $1_A \in \text{Morph}(A, A)$ has 1_A as inverse. For transitivity let $\varphi_1 \in \text{Morph}(A, B)$ and $\varphi_2 \in \text{Morph}(B, C)$ be isomorphisms, with respective inverses $\psi_1 \in \text{Morph}(B, A)$ and $\psi_2 \in \text{Morph}(C, B)$. Then $\varphi_2\varphi_1$ is in $\text{Morph}(A, C)$, and $\psi_1\psi_2$ is in $\text{Morph}(C, A)$. Calculation gives $(\psi_1\psi_2)(\varphi_2\varphi_1) = \psi_1(\psi_2(\varphi_2\varphi_1)) = \psi_1((\psi_2\varphi_2)\varphi_1) = \psi_1(1_B\varphi_1) = \psi_1\varphi_1 = 1_A$, and similarly $(\varphi_2\varphi_1)(\psi_1\psi_2) = 1_C$. Therefore $\varphi_2\varphi_1 \in \text{Morph}(A, C)$ is an isomorphism, and "is isomorphic to" is an equivalence relation. When A is isomorphic to B, it is permissible to say that A and B are **isomorphic**.

The next step is to abstract a frequent kind of construction that we have

---

[15]For the interested reader, a book that pays closer attention to the inherent set-theoretic difficulties in the theory is Mac Lane's *Categories for the Working Mathematician*.

[16]Technically one considers relations only when they are defined on sets, and the class of objects in a category is typically not a set. However, just as with vector spaces, groups, and so on, we can restrict attention in any particular situation to a subcategory for which the objects do form a set, and then there is no difficulty.

used with our categories. If $\mathcal{C}$ and $\mathcal{D}$ are two categories, a **covariant functor** $F : \mathcal{C} \to \mathcal{D}$ associates to each object A in $\mathrm{Obj}(\mathcal{C})$ an object $F(A)$ in $\mathrm{Obj}(\mathcal{D})$ and to each pair of objects A and B and morphism f in $\mathrm{Morph}_\mathcal{C}(A, B)$ a morphism $F(f)$ in $\mathrm{Morph}_\mathcal{D}(F(A), F(B))$ such that

(i) $F(gf) = F(g)F(f)$ for $f \in \mathrm{Morph}_\mathcal{C}(A, B)$ and $g \in \mathrm{Morph}_\mathcal{C}(B, C)$,
(ii) $F(1_A) = 1_{F(A)}$ for A in $\mathrm{Obj}(\mathcal{C})$.

EXAMPLES OF COVARIANT FUNCTORS.

(1) Inclusion of a subcategory into a category is a covariant functor.

(2) Let $\mathcal{C}$ be the category of all sets. If F carries each set X to the set 2^X of all subsets of X, then F is a covariant functor as soon as its effect on functions between sets, i.e., its effect on morphisms, is defined in an appropriate way. Namely, if $f : X \to Y$ is a function, then $F(f)$ is to be a function from $F(X) = 2^X$ to $F(Y) = 2^Y$. That is, we need a definition of $F(f)(A)$ as a subset of Y whenever A is a subset of X. A natural way of making such a definition is to put $F(f)(A) = f(A)$, and then F is indeed a covariant functor.

(3) Let $\mathcal{C}$ be any of Examples 2 through 6 of categories above, and let $\mathcal{D}$ be the category of all sets, as in Example 1 of categories. If F carries an object A in $\mathcal{C}$ (i.e., a vector space, group, ring, etc.) into its underlying set and carries each morphism into its underlying function between two sets, then F is a covariant functor and furnishes an example of what is called a **forgetful functor**.

(4) Let $\mathcal{C}$ be the category of all vector spaces over a field $\mathbb{F}$, let U be a vector space over $\mathbb{F}$, and let $F : \mathcal{C} \to \mathcal{C}$ be defined on a vector space to be the vector space of linear maps $F(V) = \mathrm{Hom}_\mathbb{F}(U, V)$. The set of morphisms $\mathrm{Morph}_\mathcal{C}(V_1, V_2)$ is $\mathrm{Hom}_\mathbb{F}(V_1, V_2)$. If f is in $\mathrm{Morph}_\mathcal{C}(V_1, V_2)$, then $F(f)$ is to be in $\mathrm{Morph}_\mathcal{C}\big(\mathrm{Hom}_\mathbb{F}(U, V_1), \mathrm{Hom}_\mathbb{F}(U, V_2)\big)$, and the definition is that $F(f)(L) = f \circ L$ for $L \in \mathrm{Hom}_\mathbb{F}(U, V_1)$. Then F is a covariant functor: to check that $F(gf) = F(g)F(f)$ when g is in $\mathrm{Morph}_\mathcal{C}(V_2, V_3)$, we write $F(gf)(L) = gf \circ L = g \circ fL = g \circ F(f) = F(g)F(f)$.

(5) Let $\mathcal{C}$ be the category of all groups, let $\mathcal{D}$ be the category of all sets, let G be a group, and let $F : \mathcal{C} \to \mathcal{D}$ be the functor defined as follows. For a group H, $F(H)$ is the set of all group homomorphisms from G into H. The set of morphisms $\mathrm{Morph}_\mathcal{C}(H_1, H_2)$ is the set of group homomorphisms from H_1 into H_2. If f is in $\mathrm{Morph}_\mathcal{C}(H_1, H_2)$, then $F(f)$ is to be a function with domain the set of homomorphisms from G into H_1 and with range the set of homomorphisms from G into H_2. Let $F(f)(\varphi) = \varphi \circ f$. Then F is a covariant functor.

(6) Let $\mathcal{C}$ be the category of all sets, and let $\mathcal{D}$ be the category of all abelian groups. To a set S, associate the free abelian group $F(S)$ with S as $\mathbb{Z}$ basis. If $f : S \to S'$ is a function, then the universal mapping property of external

direct sums of abelian groups (Proposition 4.17) yields a corresponding group homomorphism from $F(S)$ to $F(S')$, and we define this group homomorphism to be $F(f)$. Then F is a covariant functor.

(7) Let $\mathcal{C}$ be the category of all finite sets, fix a commutative ring R with identity, and let $\mathcal{D}$ be the category of all commutative rings with identity. To a finite set S, associate the commutative ring $F(S) = R[\{X_s \mid s \in S\}]$. If $f : S \to S'$ is a function, then the properties of substitution homomorphisms give us a corresponding homomorphism of rings with identity carrying $F(S)$ to $F(S')$, and the result is a covariant functor.

There is a second kind of functor of interest to us. If $\mathcal{C}$ and $\mathcal{D}$ are two categories, a **contravariant functor** $F : \mathcal{C} \to \mathcal{D}$ associates to each object A in $\text{Obj}(\mathcal{C})$ an object $F(A)$ in $\text{Obj}(\mathcal{D})$ and to each pair of objects A and B and morphism f in $\text{Morph}_\mathcal{C}(A, B)$ a morphism $F(f)$ in $\text{Morph}_\mathcal{D}(F(B), F(A))$ such that

(i) $F(gf) = F(f)F(g)$ for $f \in \text{Morph}_\mathcal{C}(A, B)$ and $g \in \text{Morph}_\mathcal{D}(B, C)$,
(ii) $F(1_A) = 1_{F(A)}$ for A in $\text{Obj}(\mathcal{C})$.

EXAMPLES OF CONTRAVARIANT FUNCTORS.

(1) Let $\mathcal{C}$ be the category of all vector spaces over a field $\mathbb{F}$, let W be a vector space over $\mathbb{F}$, and let $F : \mathcal{C} \to \mathcal{C}$ be defined on a vector space to be the vector space of linear maps $F(V) = \text{Hom}_\mathbb{F}(V, W)$. The set of morphisms $\text{Morph}_\mathcal{C}(V_1, V_2)$ is $\text{Hom}_\mathbb{F}(V_1, V_2)$. If f is in $\text{Morph}_\mathcal{C}(V_1, V_2)$, then $F(f)$ is to be in $\text{Morph}_\mathcal{C}\bigl(\text{Hom}_\mathbb{F}(V_2, W), \text{Hom}_\mathbb{F}(V_1, W)\bigr)$, and the definition is that $F(f)(L) = L \circ f$ for $L \in \text{Hom}_\mathbb{F}(V_1, W)$. Then F is a contravariant functor: to check that $F(gf) = F(f)F(g)$ when g is in $\text{Morph}_\mathcal{C}(V_2, V_3)$, we write $F(gf)(L) = L \circ gf = Lg \circ f = F(f)(Lg) = F(f)F(g)$.

(2) Let $\mathcal{C}$ be the category of all vector spaces over a field $\mathbb{F}$, define F of a vector space V to be the dual vector space V', and define F of a linear mapping f between two vector spaces V and W to be the contragredient f^t carrying W' into V', defined by $f^t(w')(v) = w'(f(v))$. This is the special case of Example 1 of contravariant functors in which $W = \mathbb{F}$. Hence F is a contravariant functor.

(3) Let $\mathcal{C}$ be the category of all groups, let $\mathcal{D}$ be the category of all sets, let G be a group, and let $F : \mathcal{C} \to \mathcal{D}$ be the functor defined as follows. For a group H, $F(H)$ is the set of all group homomorphisms from H into G. The set of morphisms $\text{Morph}_\mathcal{C}(H_1, H_2)$ is the set of group homomorphisms from H_1 into H_2. If f is in $\text{Morph}_\mathcal{C}(H_1, H_2)$, then $F(f)$ is to be a function with domain the set of homomorphisms from H_2 into G and with range the set of homomorphisms from H_1 into G. The definition is $F(f)(\varphi) = f \circ \varphi$. Then F is a contravariant functor.

11. Categories and Functors

It is an important observation about functors that the composition of two functors is a functor. This is immediate from the definition. If the two functors are both covariant or both contravariant, then the composition is covariant. If one of them is covariant and the other is contravariant, then the composition is contravariant.

$$\begin{array}{ccc} A & \xrightarrow{\alpha} & B \\ \beta \downarrow & & \downarrow \gamma \\ C & \xrightarrow{\delta} & D \end{array}$$

FIGURE 4.9. A square diagram. The square commutes if $\gamma\alpha = \delta\beta$.

In the subject of category theory, a great deal of information is conveyed by "commutative diagrams" of objects and morphisms. By a **diagram** is meant a directed graph, usually but not necessarily planar, in which the vertices represent some relevant objects in a category and the arrows from one vertex to another represent morphisms of interest between pairs of these objects. Often the vertices and arrows are labeled, but in fact labels on the vertices can be deduced from the labels on the arrows since any morphism determines its "domain" and "range" as a consequence of defining property (i) of categories. A diagram is said to be **commutative** if for each pair of vertices A and B and each directed path from A to B, the compositions of the morphisms along each path are the same. For example a square as in Figure 4.9 is commutative if $\gamma\alpha = \delta\beta$. The triangular diagrams in Figures 4.1 through 4.8 are all commutative.

$$\begin{array}{ccc} F(A) & \xrightarrow{F(\alpha)} & F(B) \\ F(\beta) \downarrow & & \downarrow F(\gamma) \\ F(C) & \xrightarrow{F(\delta)} & F(D) \end{array} \quad \text{and} \quad \begin{array}{ccc} G(A) & \xleftarrow{G(\alpha)} & G(B) \\ G(\beta) \uparrow & & \uparrow G(\gamma) \\ G(C) & \xleftarrow{G(\delta)} & G(D) \end{array}$$

FIGURE 4.10. Diagrams obtained by applying a covariant functor F and a contravariant functor G to the diagram in Figure 4.9.

Functors can be applied to diagrams, yielding new diagrams. For example, suppose that Figure 4.9 is a diagram in the category $\mathcal{C}$, that $F : \mathcal{C} \to \mathcal{D}$ is a covariant functor, and that $G : \mathcal{C} \to \mathcal{D}$ is a contravariant functor. Then we can apply F and G to the diagram in Figure 4.9, obtaining the two diagrams in the category $\mathcal{D}$ that are pictured in Figure 4.10. *If the diagram in Figure 4.9 is commutative, then so are the diagrams in Figure 4.10,* as a consequence of the effect of functors on compositions of morphisms.

The subject of category theory seeks to analyze functors that make sense for all categories, or at least all categories satisfying some additional properties. The most important investigation of this kind is concerned with homology and cohomology, as well as their ramifications, for "abelian categories," which include several important examples affecting algebra, topology, and several complex variables. The topic in question is called "homological algebra" and is discussed further in *Advanced Algebra*.

There are a number of other functors that are investigated in category theory, and we mention four:

- products, including direct products,
- coproducts, including direct sums,
- direct limits, also called inductive limits,
- inverse limits, also called projective limits.

We discuss general products and coproducts in the present section, omitting a general discussion of direct limits and inverse limits. Inverse limits will arise in *Advanced Algebra* for one category in connection with Galois groups, but we shall handle that one situation on its own without attempting a generalization. An attempt in the 1960s to recast as much mathematics as possible in terms of category theory is now regarded by many mathematicians as having been overdone, and it seems wiser to cast bodies of mathematics in the framework of category theory only when doing so can be justified by the amount of time saved by eliminating redundant arguments.

When a category C and a nonempty set S are given, we can define a category C^S. The objects of C^S are functions on S with the property that the value of the function at each s in S is in $\text{Obj}(C)$, two such functions being regarded as the same if they consist of the same ordered pairs.[17] Let us refer to such a function as an S-**tuple** of members of $\text{Obj}(C)$, denoting it by an expression like $\{X_s\}_{s \in S}$. A morphism in $\text{Morph}_{C^S}\big(\{X_s\}_{s \in S}, \{Y_s\}_{s \in S}\big)$ is an S-tuple $\{f_s\}_{s \in S}$ of morphisms of C such that f_s lies in $\text{Morph}_C(X_s, Y_s)$ for all s, and the law of composition of such morphisms takes place coordinate by coordinate.

Let $\{X_s\}_{s \in S}$ be an object in C^S. A **product** of $\{X_s\}_{s \in S}$ is a pair $(X, \{p_s\}_{s \in S})$ such that X is in $\text{Obj}(C)$ and each p_s is in $\text{Morph}_C(X, X_s)$ with the following **universal mapping property**: whenever A in $\text{Obj}(C)$ is given and a morphism $\varphi_s \in \text{Morph}_C(A, X_s)$ is given for each s, then there exists a unique morphism $\varphi \in \text{Morph}_C(A, X)$ such that $p_s \varphi = \varphi_s$ for all s. The relevant diagram is pictured in Figure 4.11.

---

[17] In other words, the range of such a function is considered as irrelevant. We might think of the range as $\text{Obj}(C)$ except for the fact that a function is supposed to have a *set* as range and $\text{Obj}(C)$ need not be a set.

11. Categories and Functors

FIGURE 4.11. Universal mapping property of a product in a category.

EXAMPLES OF PRODUCTS.

(1) Products exist in the category of vector spaces over a field $\mathbb{F}$. If vector spaces V_s indexed by a nonempty set S are given, then their product exists in the category, and an example is their external direct product $\prod_{s \in S} V_s$, according to Figure 2.4 and the discussion around it.

(2) Products exist in the category of all groups. If groups G_s indexed by a nonempty set S are given, then their product exists in the category, and an example is their external direct product $\prod_{s \in S} G_s$, according to Figure 4.2 and Proposition 4.15. If the groups G_s are abelian, then $\prod_{s \in S} G_s$ is abelian, and it follows that products exist in the category of all abelian groups.

(3) Products exist in the category of all sets. If sets X_s indexed by a nonempty set S are given, then their product exists in the category, and an example is their Cartesian product $\bigtimes_{s \in S} X_s$, as one easily checks.

(4) Products exist in the category of all rings and in the category of all rings with identity. If objects R_s in the category indexed by a nonempty set S are given, then their product may be taken as an abelian group to be the external direct product $\prod_{s \in S} R_s$, with multiplication defined coordinate by coordinate, and the group homomorphisms p_s are easily checked to be morphisms in the category.

A product of objects in a category need not exist in the category. An artificial example may be formed as follows: Let $\mathcal{C}$ be a category with one object G, namely a group of order 2, and let $\text{Morph}(G, G) = \{0, 1_G\}$, the law of composition being the usual composition. Let S be a 2-element set, and let the corresponding objects be $X_1 = G$ and $X_2 = G$. The claim is that the product $X_1 \times X_2$ does not exist in $\mathcal{C}$. In fact, take $A = G$. There are four S-tuples of morphisms (φ_1, φ_2) meeting the conditions of the definition. Yet the only possibility for the product is $X = G$, and then there are only two possible φ's in $\text{Morph}(A, X)$. Hence we cannot account for all possible S-tuples of morphisms, and the product cannot exist.

The thing that category theory addresses is the uniqueness. A product is always unique up to canonical isomorphism, according to Proposition 4.63. We proved uniqueness for products in the special cases of Examples 1 and 2 above in Propositions 2.32 and 4.16.

Proposition 4.63. Let $\mathcal{C}$ be a category, and let S be a nonempty set. If $\{X_s\}_{s \in S}$ is an object in $\mathcal{C}^S$ and if $(X, \{p_s\})$ and $(X', \{p'_s\})$ are two products, then there exists a unique morphism $\Phi : X' \to X$ such that $p'_s = p_s \circ \Phi$ for all $s \in S$, and Φ is an isomorphism.

REMARK. There is no assertion that p_s is onto X_s. In fact, "onto" has no meaning for a general category.

PROOF. In Figure 4.11 let $A = X'$ and $\varphi_s = p'_s$. If $\Phi \in \text{Morph}(X', X)$ is the morphism produced by the fact that X is a direct product, then we have $p_s \Phi = p'_s$ for all s. Reversing the roles of X and X', we obtain a morphism $\Phi' \in \text{Morph}(X, X')$ with $p'_s \Phi' = p_s$ for all s. Therefore $p_s(\Phi \Phi') = (p_s \Phi) \Phi' = p'_s \Phi' = p_s$.

In Figure 4.11 we next let $A = X$ and $\varphi_s = p_s$ for all s. Then the identity 1_X in $\text{Morph}(X, X)$ has the same property $p_s 1_X = p_s$ relative to all p_s that $\Phi \Phi'$ has, and the uniqueness in the statement of the universal mapping property implies that $\Phi \Phi' = 1_X$. Reversing the roles of X and X', we obtain $\Phi' \Phi = 1_{X'}$. Therefore Φ is an isomorphism.

For uniqueness suppose that $\Psi \in \text{Morph}(X', X)$ is another morphism with $p'_s = p_s \Psi$ for all $s \in S$. Then the argument of the previous paragraph shows that $\Phi' \Psi = 1_{X'}$. Consequently $\Psi = 1_X \Psi = (\Phi \Phi') \Psi = \Phi(\Phi' \Psi) = \Phi 1_{X'} = \Phi$, and $\Psi = \Phi$. $\square$

If products always exist in a particular category, they are not unique, only unique up to canonical isomorphism. Such a product is commonly denoted by $\prod_{s \in S} X_s$, even though it is not uniquely defined. *It is customary to treat the product over S as a covariant functor $F : \mathcal{C}^S \to \mathcal{C}$*, the effect of the functor on objects being given by $F(\{X_s\}_{s \in S}) = \prod_{s \in S} X_s$. For a well-defined functor we have to fix a choice of product for each object under consideration[18] in $\text{Obj}(\mathcal{C}^S)$. For the effect of F on morphisms, we argue with the universal mapping property. Thus let $\{X_s\}_{s \in S}$ and $\{Y_s\}_{s \in S}$ be objects in $\mathcal{C}^S$, let f_s be in $\text{Morph}_\mathcal{C}(X_s, Y_s)$ for all s, and let the products in question be $\left(\prod_{s \in S} X_s, \{p_s\}_{s \in S}\right)$ and $\left(\prod_{s \in S} Y_s, \{q_s\}_{s \in S}\right)$. Then $f_{s_0} p_{s_0}$ is in $\text{Morph}_\mathcal{C}\left(\prod_{s \in S} X_s, Y_{s_0}\right)$ for each s_0, and the universal mapping property gives us f in $\text{Morph}_\mathcal{C}\left(\prod_{s \in S} X_s, \prod_{s \in S} Y_s\right)$ such that $q_s f = f_s p_s$ for all s. We define this f to be $F(\{f_s\}_{s \in S})$, and we readily check that F is a functor.

We turn to coproducts, which include direct sums. Let $\{X_s\}_{s \in S}$ be an object in $\mathcal{C}^S$. A **coproduct** of $\{X_s\}_{s \in S}$ is a pair $(X, \{i_s\}_{s \in S})$ such that X is in $\text{Obj}(\mathcal{C})$ and each i_s is in $\text{Morph}_\mathcal{C}(X_s, X)$ with the following **universal mapping property**: whenever A in $\text{Obj}(\mathcal{C})$ is given and a morphism $\varphi_s \in \text{Morph}_\mathcal{C}(X_s, A)$ is given

---

[18]Since $\text{Obj}(\mathcal{C}^S)$ need not be a set, it is best to be wary of applying the Axiom of Choice when the indexing of sets is given by $\text{Obj}(\mathcal{C}^S)$. Instead, one makes the choice only for all objects in some set of objects large enough for a particular application.

for each s, then there exists a unique morphism $\varphi \in \operatorname{Morph}_\mathcal{C}(X, A)$ such that $\varphi i_s = \varphi_s$ for all s. The relevant diagram is pictured in Figure 4.12.

FIGURE 4.12. Universal mapping property of a coproduct in a category.

EXAMPLES OF COPRODUCTS.

(1) Coproducts exist in the category of vector spaces over a field $\mathbb{F}$. If vector spaces V_s indexed by a nonempty set S are given, then their coproduct exists in the category, and an example is their external direct sum $\bigoplus_{s \in S} V_s$, according to Figure 2.5 and the discussion around it.

(2) Coproducts exist in the category of all abelian groups. If abelian groups G_s indexed by a nonempty set S are given, then their coproduct exists in the category, and an example is their external direct sum $\bigoplus_{s \in S} G_s$, according to Figure 4.4 and Proposition 4.17.

(3) Coproducts exist in the category of all sets. If sets X_s indexed by a nonempty set S are given, then their coproduct exists in the category, and an example is their disjoint union $\bigcup_{s \in S} \{(x_s, s) \mid x_s \in X_s\}$. The verification appears as Problem 74 at the end of the chapter.

(4) Coproducts exist in the category of all groups. Suppose that groups G_s indexed by a nonempty set S are given. It will be shown in Chapter VII that the coproduct is the "free product" $*_{s \in S} G_s$ that is defined in that chapter. In the special case that each G_s is the group $\mathbb{Z}$ of integers, the free product coincides with the free group on S. Therefore, even if all the groups G_s are abelian, their coproduct need not be a subgroup of the direct product and need not even be abelian. In particular it need not coincide with the direct sum.

A coproduct of objects in a category need not exist in the category. Problem 76 at the end of the chapter offers an example that the reader is invited to check.

Proposition 4.64. Let $\mathcal{C}$ be a category, and let S be a nonempty set. If $\{X_s\}_{s \in S}$ is an object in $\mathcal{C}^S$ and if $(X, \{i_s\})$ and $(X', \{i'_s\})$ are two coproducts, then there exists a unique morphism $\Phi : X \to X'$ such that $i'_s = \Phi \circ i_s$ for all $s \in S$, and Φ is an isomorphism.

REMARKS. There is no assertion that i_s is one-one. In fact, "one-one" has no meaning for a general category. This proposition may be derived quickly from Proposition 4.63 by a certain duality argument that is discussed in Problems

78–80 at the end of the chapter. Here we give a direct argument without taking advantage of duality.

PROOF. In Figure 4.12 let $A = X'$ and $\varphi_s = i'_s$. If $\Phi \in \text{Morph}(X, X')$ is the morphism produced by the fact that X is a coproduct, then we have $\Phi i_s = i'_s$ for all s. Reversing the roles of X and X', we obtain a morphism $\Phi' \in \text{Morph}(X', X)$ with $\Phi' i'_s = i_s$ for all s. Therefore $(\Phi'\Phi)i_s = \Phi' i'_s = i_s$.

In Figure 4.12 we next let $A = X$ and $\varphi_s = i_s$ for all s. Then the identity 1_X in $\text{Morph}(X, X)$ has the same property $1_X i_s = i_s$ relative to all i_s that $\Phi'\Phi$ has, and the uniqueness says that $\Phi'\Phi = 1_X$. Reversing the roles of X and X', we obtain $\Phi\Phi' = 1_{X'}$. Therefore Φ is an isomorphism.

For uniqueness suppose that $\Psi \in \text{Morph}(X, X')$ is another morphism with $i'_s = \Psi i_s$ for all $s \in S$. Then the argument of the previous paragraph shows that $\Phi'\Psi = 1_X$. Consequently $\Psi = 1_{X'}\Psi = (\Phi\Phi')\Psi = \Phi(\Phi'\Psi) = \Phi 1_X = \Phi$, and $\Psi = \Phi$. □

If coproducts always exist in a particular category, they are not unique, only unique up to canonical isomorphism. Such a coproduct is commonly denoted by $\coprod_{s \in S} X_s$, even though it is not uniquely defined. As with product, *it is customary to treat the coproduct over S as a covariant functor $F : \mathcal{C}^S \to \mathcal{C}$*, the effect of the functor on objects being given by $F(\{X_s\}_{s \in S}) = \coprod_{s \in S} X_s$. For a well-defined functor we have to fix a choice of coproduct for each object under consideration in $\text{Obj}(\mathcal{C}^S)$. For the effect of F on morphisms, we argue with the universal mapping property. Thus let $\{X_s\}_{s \in S}$ and $\{Y_s\}_{s \in S}$ be objects in $\mathcal{C}^S$, let f_s be in $\text{Morph}_{\mathcal{C}}(X_s, Y_s)$ for all s, and let the coproducts in question be $\left(\coprod_{s \in S} X_s, \{i_s\}_{s \in S}\right)$ and $\left(\coprod_{s \in S} Y_s, \{j_s\}_{s \in S}\right)$. Then $j_{s_0} f_{s_0}$ is in $\text{Morph}_{\mathcal{C}}\left(X_{s_0}, \coprod_{s \in S} Y_s\right)$ for each s_0, and the universal mapping property gives us f in $\text{Morph}_{\mathcal{C}}\left(\coprod_{s \in S} X_s, \coprod_{s \in S} Y_s\right)$ such that $f i_s = j_s f_s$ for all s. We define this f to be $F(\{f_s\}_{s \in S})$, and we readily check that F is a functor.

Universal mapping properties occur in other contexts than for products and coproducts. We have already seen them in connection with homomorphisms on free abelian groups and with substitution homomorphisms on polynomial rings, and more such properties will occur in the development of tensor products in Chapter VI. A general framework for discussing universal mapping properties appears in the problems at the end of Chapter VI.

12. Problems

1. Let G be a group in which all elements other than the identity have order 2. Prove that G is abelian.
2. The dihedral group D_4 of order 8 can be viewed as a subgroup of the symmetric group $\mathfrak{S}_4$ of order 8. Find 8 explicit permutations in $\mathfrak{S}_4$ forming a subgroup isomorphic to D_4.

3. Suppose G is a finite group, H is a subgroup, and $a \in G$ is an element with a^l in H for some integer l with $\gcd(l, |G|) = 1$. Prove that a is in H.

4. Let G be a group, and define a new group G' to have the same underlying set as G but to have multiplication given by $a \circ b = ba$. Prove that G' is a group and that it is isomorphic to G.

5. Prove that if G is an abelian group and n is an integer, then $a \mapsto a^n$ is a homomorphism of G. Give an example of a nonabelian group for which $a \mapsto a^2$ is not a homomorphism.

6. Suppose that G is a group and that H and K are *normal* subgroups of G with $H \cap K = \{1\}$. Verify that the set HK of products is a subgroup and that this subgroup is isomorphic as a group to the external direct product $H \times K$.

7. Take as known that 8191 is prime, so that $\mathbb{F}_{8191}$ is a field. Without carrying through the computations and without advocating trial and error, describe what steps you would carry out to solve for x mod 8191 such that $1234x \equiv 1 \bmod 8191$.

8. **(Wilson's Theorem)** Let p be an odd prime. Starting from the fact that $1, \ldots, p-1$ are roots of the polynomial $X^{p-1} - 1 \equiv 0 \bmod p$ in $\mathbb{F}_p$, prove that $(p-1)! \equiv -1 \bmod p$.

9. Classify, up to isomorphism, all groups of order p^2 if p is a prime.

10. This problem concerns conjugacy classes in a group G.
    (a) Prove that all elements of a conjugacy class have the same order.
    (b) Prove that if ab is in a conjugacy class, so is ba.

11. (a) Find explicitly all the conjugacy classes in the alternating group $\mathfrak{A}_4$.
    (b) For each conjugacy class in $\mathfrak{A}_4$, find the centralizer of one element in the class.
    (c) Prove that $\mathfrak{A}_4$ has no subgroup isomorphic to C_6 or $\mathfrak{S}_3$.

12. Prove that the alternating group $\mathfrak{A}_5$ has no subgroup of order 30.

13. Let G be a nonabelian group of order p^n, where p is prime. Prove that any subgroup of order p^{n-1} is normal.

14. Let G be a finite group, and let H be a normal subgroup. If $|H| = p$ and p is the smallest prime dividing $|G|$, prove that H is contained in the center of G.

15. Let G be a group. An automorphism of G of the form $x \mapsto gxg^{-1}$ is called an **inner automorphism**. Prove that the set of inner automorphisms is a normal subgroup of the group Aut G of all automorphisms and is isomorphic to G/Z_G.

16. (a) Prove that Aut C_m is isomorphic to $(\mathbb{Z}/m\mathbb{Z})^\times$.
    (b) Find a value of m for which Aut C_m is not cyclic.

17. Fix $n \geq 2$. In the symmetric group $\mathfrak{S}_n$, for each integer k with $1 \leq k \leq n/2$, let C_k be the set of elements in $\mathfrak{S}_n$ that are products of k disjoint transpositions.
    (a) Prove that if τ is an automorphism of $\mathfrak{S}_n$, then $\tau(C_1) = C_k$ for some k.
    (b) Prove that $|C_k| = \binom{n}{2k} \frac{(2k)!}{2^k k!}$.
    (c) Prove that $|C_k| \neq |C_1|$ unless $k = 1$ or $n = 6$. (Educational note: From this, it follows that $\tau(C_1) = C_1$ except possibly when $n = 6$. One can deduce as a consequence that every automorphism of $\mathfrak{S}_n$ is inner except possibly when $n = 6$.)

18. Give an example: G is a group with a normal subgroup N, N has a subgroup M that is normal in M, yet M is not normal in G.

19. Show that the cyclic group C_{rs} is isomorphic to $C_r \times C_s$ if and only if $\text{GCD}(r, s) = 1$.

20. How many abelian groups, up to isomorphism, are there of order 27?

21. Let G be the free abelian group with $\mathbb{Z}$ basis $\{x_1, x_2, x_3\}$. Let H be the subgroup of G generated by $\{u_1, u_2, u_3\}$, where
$$u_1 = 3x_1 + 2x_2 + 5x_3,$$
$$u_2 = x_2 + 3x_3,$$
$$u_3 = x_2 + 5x_3.$$
Express G/H as a direct sum of cyclic groups.

22. Let $\{e_1, e_2, e_3, e_4\}$ be the standard basis of $\mathbb{R}^4$. Let G be the additive subgroup of $\mathbb{R}^4$ generated by the four elements
$$e_1, \quad e_1 + e_2, \quad \tfrac{1}{2}(e_1 + e_2 + e_3 + e_4), \quad \tfrac{1}{2}(e_1 + e_2 + e_3 - e_4),$$
and let H be the subgroup of G generated by the four elements
$$e_1 - e_2, \quad e_2 - e_3, \quad e_3 - e_4, \quad e_3 + e_4.$$
Identify the abelian group G/H as a direct sum of cyclic groups.

23. Let G be the free abelian group with $\mathbb{Z}$ basis $\{x_1, \ldots, x_n\}$, and let H be the subgroup generated by $\{u_1, \ldots, u_m\}$, where $\begin{pmatrix} u_1 \\ \vdots \\ u_m \end{pmatrix} = C \begin{pmatrix} x_1 \\ \vdots \\ x_n \end{pmatrix}$ for an m-by-n matrix C of integers. Prove that the number of summands $\mathbb{Z}$ in the decomposition of G/H into cyclic groups is equal to the rank of the matrix C when C is considered as in $M_{mn}(\mathbb{Q})$.

24. Prove that every abelian group is the homomorphic image of a free abelian group.

25. Let G be a group, and let H and K be subgroups.
    (a) For x and y in G, prove that $xH \cap yK$ is empty or is a coset of $H \cap K$.
    (b) Deduce from (a) that if H and K have finite index in G, then so does $H \cap K$.

26. Let G be a free abelian group of finite rank n, and let H be a free abelian subgroup of rank n. Prove that H has finite index in G.

27. Let $G = \mathfrak{S}_4$ be the symmetric group on four letters.
    (a) Find a Sylow 2-subgroup of G. How many Sylow 2-subgroups are there, and why?
    (b) Find a Sylow 3-subgroup of G. How many Sylow 3-subgroups are there, and why?

28. Let H be a subgroup of a group G. Prove or disprove that the normalizer $N(H)$ of H in G is a normal subgroup of G.

29. How many elements of order 7 are there in a simple group of order 168?

30. Let G be a group of order pq^2, where p and q are primes with $p < q$. Let S_p and S_q be Sylow subgroups for the primes p and q. Prove that G is a semidirect product of S_p and S_q with S_q normal.

31. Suppose that G is a finite group and that H is a subgroup whose index in G is a prime p. By considering the action of G on the set of subgroups conjugate to H and considering the possibilities for the normalizer $N(H)$, determine the possibilities for the number of subgroups conjugate to H.

32. Let G be a group of order 24, let H be a subgroup of order 8, and assume that H is not normal.
    (a) Using the Sylow Theorems, explain why H has exactly 3 conjugates in G, counting H itself as one.
    (b) Show how to use the conjugates in (a) to define a homomorphism of G into the symmetric group $\mathfrak{S}_3$ on three letters.
    (c) Use the homomorphism of (b) to conclude that G is not simple.

33. Let G be a group of order 36. Arguing in the style of the previous problem, show that there is a nontrivial homomorphism of G into the symmetric group $\mathfrak{S}_4$.

34. Let G be a group of order $2pq$, where p and q are primes with $2 < p < q$.
    (a) Prove that if $q + 1 \neq 2p$, then a Sylow q-subgroup is normal.
    (b) Suppose that $q + 1 = 2p$, let H be a Sylow p-subgroup, and let K be a Sylow q-subgroup. Prove that at least one of H and K is normal, that the set HK of products is a subgroup, and that the subgroup HK is cyclic of index 2 in G.

Problems 35–38 concern the detection of isomorphisms among semidirect products. For the first two of the problems, let H and K be groups, and let $\varphi_1 : H \to \operatorname{Aut} K$ and $\varphi_2 : H \to \operatorname{Aut} K$ be homomorphisms.

35. Suppose that $\varphi_2 = \varphi_1 \circ \varphi$ for some automorphism φ of H. Define $\psi : H \times_{\varphi_2} K \to H \times_{\varphi_1} K$ by $\psi(h, k) = (\varphi(h), k)$. Prove that ψ is an isomorphism.

36. Suppose that $\varphi_2 = \varphi \circ \varphi_1$ for some inner automorphism φ of Aut K in the sense of Problem 15, i.e., $\varphi :$ Aut $K \to$ Aut K is to be given by $\varphi(x) = axa^{-1}$ with a in Aut K. Define $\psi : H \times_{\varphi_1} K \to H \times_{\varphi_2} K$ by $\psi(h, k) = (h, a(k))$. Prove that ψ is an isomorphism.

37. Suppose that p and q are primes and that the cyclic group C_p acts on C_q by automorphisms with a nontrivial action. Prove that p divides $q - 1$.

38. Suppose that p and q are primes such that p divides $q - 1$. Let τ_1 and τ_2 be nontrivial homomorphisms from C_p to Aut C_q. Prove that $C_p \times_{\tau_1} C_q \cong C_p \times_{\tau_2} C_q$, and conclude that there is only one nonabelian semidirect product $C_p \times_\tau C_q$ up to isomorphism.

Problems 39–44 discuss properties of groups of order 8, obtaining a classification of these groups up to isomorphism.

39. Prove that the five groups $C_8, C_4 \times C_2, C_2 \times C_2 \times C_2, D_4$, and H_8 are mutually nonisomorphic and that the first three exhaust the abelian groups of order 8, apart from isomorphisms.

40. (a) Find a composition series for the 8-element dihedral group D_4.
    (b) Find a composition series for the 8-element quaternion group H_8.

41. (a) Prove that every subgroup of the quaternion group H_8 is normal.
    (b) Identify the conjugacy classes in H_8.
    (c) Compute the order of Aut H_8.

42. Suppose that G is a nonabelian group of order 8. Prove that G has an element of order 4 but no element of order 8.

43. Let G be a nonabelian group of order 8, and let K be the copy of C_4 generated by some element of order 4. If G has some element of order 2 that is not in K, prove that $G \cong D_4$.

44. Let G be a nonabelian group of order 8, and let K be the copy of C_4 generated by some element of order 4. If G has no element of order 2 that is not in K, prove that $G \cong H_8$.

Problems 45–48 classify groups of order 12, making use of Proposition 4.61, Problem 15, and Problems 35–38. Let G be a group of order 12, let H be a Sylow 3-subgroup, and let K be a Sylow 2-subgroup. Proposition 4.61 says that at least one of H and K is normal. Consequently there are three cases, and these are addressed by the first three of the problems.

45. Verify that there are only two possibilities for G up to isomorphism if G is abelian.

46. Suppose that K is normal, so that $G \cong H \times_\tau K$. Prove that either
    (i) τ is trivial or
    (ii) τ is nontrivial and $K \cong C_2 \times C_2$,
    and deduce that G is abelian if (i) holds and that $G \cong \mathfrak{A}_4$ if (ii) holds.

47. Suppose that H is normal, so that $G = K \times_\tau H$. Prove that one of the conditions
    - (i) τ is trivial,
    - (ii) $K \cong C_2 \times C_2$ and τ is nontrivial,
    - (iii) $K \cong C_4$ and τ is nontrivial

    holds, and deduce that G is abelian if (i) holds, that $G \cong D_6$ if (ii) holds, and that G is nonabelian and is not isomorphic to $\mathfrak{A}_4$ or D_6 if (iii) holds.

48. In the setting of the previous problem, prove that there is one and only one group, up to isomorphism, satisfying condition (iii), and find the order of each of its elements.

Problems 49–52 assume that p and q are primes with $p < q$. The problems go in the direction of classifying finite groups of order $p^2 q$.

49. If G is a group of order $p^2 q$, prove that either $p^2 q = 12$ or a Sylow q-subgroup is normal.

50. If p^2 divides $q - 1$, exhibit three nonabelian groups of order $p^2 q$ that are mutually nonisomorphic.

51. If p divides $q - 1$ but p^2 does not divide $q - 1$, exhibit two nonabelian groups of order $p^2 q$ that are not isomorphic.

52. If p does not divide $q - 1$, prove that any group of order $p^2 q$ is abelian.

Problems 53–54 concern nonabelian groups of order 27.

53. (a) Show that multiplication by the elements 1, 4, 7 mod 9 defines a nontrivial action of $\mathbb{Z}/3\mathbb{Z}$ on $\mathbb{Z}/9\mathbb{Z}$ by automorphisms.
    (b) Show from (a) that there exists a nonabelian group of order 27.
    (c) Show that the group in (b) is generated by elements a and b that satisfy

$$a^9 = b^3 = b^{-1}aba^{-4} = 1.$$

54. Show that any nonabelian group of order 27 having a subgroup H isomorphic to C_9 and an element of order 3 not lying in H is isomorphic to the group constructed in the previous problem.

Problems 55–62 give a construction of infinitely many simple groups, some of them finite and some infinite. Let $\mathbb{F}$ be a field. For $n \geq 2$, let $\mathrm{SL}(n, \mathbb{F})$ be the special linear group for the space $\mathbb{F}^n$ of n-dimensional column vectors. The center Z of $\mathrm{SL}(n, \mathbb{F})$ consists of the scalar multiples of the identity, the scalar being an nth root of 1. Let $\mathrm{PSL}(n, \mathbb{F}) = \mathrm{SL}(n, \mathbb{F})/Z$. It is known that $\mathrm{PSL}(n, \mathbb{F})$ is simple except for $\mathrm{PSL}(2, \mathbb{F}_2)$ and $\mathrm{PSL}(2, \mathbb{F}_3)$. These problems will show that $\mathrm{PSL}(2, \mathbb{F})$ is simple if $|\mathbb{F}| > 5$ and $\mathbb{F}$ is not of characteristic 2. Most of the argument will consider $\mathrm{SL}(2, \mathbb{F})$, and the passage to PSL will occur only at the very end. In Problems 56–61, G denotes a normal subgroup of $\mathrm{SL}(2, \mathbb{F})$ that is not contained in the center Z, and it is to be proved that $G = \mathrm{SL}(2, \mathbb{F})$.

55. Suppose that $\mathbb{F}$ is a finite field with q elements.
    (a) By considering the possibilities for the first column of a matrix and then considering the possibilities for the second column when the first column is fixed, compute $|GL(2, \mathbb{F})|$ as a function of q.
    (b) By using the determinant homomorphism, compute $|SL(2, \mathbb{F})|$ in terms of $|GL(2, \mathbb{F})|$.
    (c) Taking into account that $\mathbb{F}$ does not have characteristic 2, prove that $|PSL(2, \mathbb{F})| = \frac{1}{2}|SL(2, \mathbb{F})|$.
    (d) Show for a suitable finite field $\mathbb{F}$ with more than 5 elements that $PSL(2, \mathbb{F})$ has order 168.

56. Let M be a member of G that is not in Z. Since M is not scalar, there exists a column vector u with Mu not a multiple of u. Define $v = Mu$, so that (u, v) is an ordered basis of $\mathbb{F}^2$. By rewriting all matrices with the ordered basis (u, v), show that there is no loss in generality in assuming that G contains a matrix $A = \begin{pmatrix} 0 & -1 \\ 1 & c \end{pmatrix}$ if it is ultimately shown that $G = SL(2, \mathbb{F})$.

57. Let a be a member of the multiplicative group $\mathbb{F}^\times$ to be chosen shortly, and let B be the member $\begin{pmatrix} ca & a^{-1} \\ -a & 0 \end{pmatrix}$ of $SL(2, \mathbb{F})$. Prove that
    (a) $B^{-1}A^{-1}BA$ is upper triangular and is in G,
    (b) $B^{-1}A^{-1}BA$ has unequal diagonal entries if $a^4 \neq 1$,
    (c) the condition in (b) can be satisfied for a suitable choice of a under the assumption that $|\mathbb{F}| > 5$.

58. Suppose that $C = \begin{pmatrix} x & y \\ 0 & x^{-1} \end{pmatrix}$ is a member of G for some $x \neq \pm 1$ and some y. Taking $D = \begin{pmatrix} 1 & 1 \\ 0 & 1 \end{pmatrix}$ and forming $CDC^{-1}D^{-1}$, show that G contains a matrix $E = \begin{pmatrix} 1 & \lambda \\ 0 & 1 \end{pmatrix}$ with $\lambda \neq 0$.

59. By conjugating E by $\begin{pmatrix} \alpha & 0 \\ 0 & \alpha^{-1} \end{pmatrix}$, show that the set of λ in $\mathbb{F}$ such that $\begin{pmatrix} 1 & \lambda \\ 0 & 1 \end{pmatrix}$ is in G is closed under multiplication by squares and under addition and subtraction.

60. Using the identity $x = \frac{1}{4}(x+1)^2 - \frac{1}{4}(x-1)^2$, deduce from Problems 56–59 that G contains all matrices $\begin{pmatrix} 1 & \lambda \\ 0 & 1 \end{pmatrix}$ with $\lambda \in \mathbb{F}$.

61. Show that $\begin{pmatrix} 1 & \lambda \\ 0 & 1 \end{pmatrix}$ is conjugate to $\begin{pmatrix} 1 & 0 \\ -\lambda & 1 \end{pmatrix}$, and show that the set of all matrices $\begin{pmatrix} 1 & \lambda \\ 0 & 1 \end{pmatrix}$ and $\begin{pmatrix} 1 & 0 \\ \lambda' & 1 \end{pmatrix}$ generates $SL(2, \mathbb{F})$. Conclude that $G = SL(2, \mathbb{F})$.

62. Using the First Isomorphism Theorem, conclude that the only normal subgroup of $PSL(2, \mathbb{F})$ other than $\{1\}$ is $PSL(2, \mathbb{F})$ itself.

Problems 63–73 briefly introduce the theory of error-correcting codes. Let $\mathbb{F}$ be the finite field $\mathbb{Z}/2\mathbb{Z}$. The vector space $\mathbb{F}^n$ over $\mathbb{F}$ will be called **Hamming space**, and its members are regarded as "words" (potential messages consisting of 0's and 1's). The **weight** $\text{wt}(c)$ of a word c is the number of nonzero entries in c. The **Hamming**

distance $d(a, b)$ between words $a = (a_1, \ldots, a_n)$ and $b = (b_1, \ldots, b_n)$ is the weight of $a - b$, i.e., the number of indices i with $1 \le i \le n$ and $a_i \ne b_i$. A **code** is a nonempty subset C of $\mathbb{F}^n$, and the **minimal distance** $\delta(C)$ of a code is the smallest value of $d(a, b)$ for a and b in C with $a \ne b$. By convention if $|C| = 1$, take $\delta(C) = n + 1$. One imagines that members of C, which are called code words, are allowable messages, i.e., words that can be stored and retrieved, or transmitted and received. A code with minimal distance δ can then detect up to $\delta - 1$ errors in a word ostensibly from C that has been retrieved from storage or has been received in a transmission. The code can correct up to $(\delta - 1)/2$ errors because no word of $\mathbb{F}^n$ can be at distance $\le (\delta - 1)/2$ from more than one word in C, by Problem 63 below. The interest is in **linear codes**, those for which C is a vector subspace. It is desirable that each message have a high percentage of content and a relatively low percentage of further information used for error correction; thus a fundamental theoretical problem for linear codes is to find the maximum dimension of a linear code if n and a lower bound on the minimal distance for the code are given. As a practical matter, information is likely to be processed in packets of a standard length, such as some power of 2. In many situations packets can be reprocessed if they have been found to have errors. The initial interest is therefore in codes that can recognize and possibly correct a small number of errors. The problems in this set are continued at the ends of Chapters VII and IX.

63. Prove that the Hamming distance satisfies $d(a, b) \le d(a, c) + d(c, b)$, and conclude that if a word w in $\mathbb{F}^n$ is at distance $\le (D - 1)/2$ from two distinct members of the linear code C, then $\delta(C) < D$.

64. Explain why the minimal distance $\delta(C)$ of a linear code $C \ne \{0\}$ is given by the minimal weight of the nonzero words in C.

65. Fix $n \ge 2$. List $\delta(C)$ and $\dim C$ for the following elementary linear codes:
    (a) $C = 0$.
    (b) $C = \mathbb{F}^n$.
    (c) **(Repetition code)** $C = \{0, (1, 1, \ldots, 1)\}$.
    (d) **(Parity-check code)** $C = \{c \in \mathbb{F}^n \mid \text{wt}(c) \text{ is even}\}$. (Educational note: To use this code, one sends the message in the first $n - 1$ bits and adjusts the last bit so that the word is in C. If there is at most one error in the word, this parity bit will tell when there is an error, but it will not tell where the error occurs.)

66. One way to get a sense of what members of a linear code C in $\mathbb{F}^n$ have small weight starts by making a basis for the code into the row vectors of a matrix and row reducing the matrix.
    (a) Taking into account the distinction between corner variables and independent variables in the process of row reduction, show that every basis vector of C has weight at most the sum of 1 and the number of independent variables. Conclude that $\dim C + \delta(C) \le n + 1$.
    (b) Give an example of a linear code with $\delta(C) = 2$ for which equality holds.

(c) Examining the argument for (a) more closely, show that $2 \leq \dim C \leq n-2$ implies $\dim C + \delta(C) \leq n$.

67. Let C be a linear code with a basis consisting of the rows of $\begin{pmatrix} 1 & 0 & 0 & 1 & 1 & 0 \\ 0 & 1 & 0 & 1 & 0 & 1 \\ 0 & 0 & 1 & 0 & 1 & 1 \end{pmatrix}$. Show that $\delta(C) = 3$. Educational note: Thus for $n = 6$ and $\delta(C) = 3$, we always have $\dim C \leq 3$, and equality is possible.

68. **(Hamming codes)** The Hamming code C_7 of order 7 is a certain linear code having $\dim C_7 = 4$ that will be seen to have $\delta(C_7) = 3$. The code words of a basis, with their commas removed, may be taken as

$$1110000,\ 1001100,\ 0101010,\ 1101001.$$

The basis may be described as follows. Bits 1, 2, 4 are used as checks. The remaining bits are used to form the standard basis of $\mathbb{F}^4$. What is put in bits 1, 2, 4 is the binary representation of the position of the nonzero entry in positions 3, 5, 6, 7. When all 16 members of C_7 are listed in the order dictated by the bits in positions 3, 5, 6, 7, the resulting list is

Decimal value in 3, 5, 6, 7	Code word	Decimal value in 3, 5, 6, 7	Code word
0	0000000	8	1110000
1	1101001	9	0011001
2	0101010	10	1011010
3	1000011	11	0110011
4	1001100	12	0111100
5	0100101	13	1010101
6	1100110	14	0010110
7	0001111	15	1111111

For the general members of C_7, not just the basis vectors, the check bits in positions 1, 2, 4 may be described as follows: the bit in position 1 is a parity bit for the positions among 3, 5, 6, 7 having a 1 in their binary expansions, the bit in position 2 is a parity bit for the positions among 3, 5, 6, 7 having a 2 in their binary expansions, and the bit in position 4 is a parity bit for the positions among 3, 5, 6, 7 having a 4 in their binary expansions. The Hamming code C_8 of order 8 is obtained from C_7 by adjoining a parity bit in position 8.

(a) Prove that $\delta(C_7) = 3$. (Educational note: Thus for $n = 7$ and $\delta(C) = 3$, we always have $\dim C \leq 4$, and equality is possible.)

(b) Prove that $\delta(C_8) = 4$.

(c) Describe how to form a generalization that replaces $n = 8$ by $n = 2^r$ with $r \geq 3$. The Hamming codes that are obtained will be called C_{2^r-1} and C_{2^r}.

(d) Prove that $\dim C_{2^r-1} = \dim C_{2^r} = 2^r - r - 1$, $\delta(C_{2^r-1}) = 3$, and $\delta(C_{2^r}) = 4$.

69. The matrix $H = \begin{pmatrix} 1 & 0 & 1 & 0 & 1 & 0 & 1 \\ 0 & 1 & 1 & 0 & 0 & 1 & 1 \\ 0 & 0 & 0 & 1 & 1 & 1 & 1 \end{pmatrix}$, when multiplied by any column vector c in the Hamming code C_7, performs the three parity checks done by bits 1, 2, 4 and described in the previous problem. Therefore such a c must have $Hc = 0$.
   (a) Prove that the condition works in the reverse direction as well—that $Hc = 0$ only if c is in C_7.
   (b) Deduce that if a received word r is not in C_7 and if r is assumed to match some word of C_7 except in the ith position, then Hr matches the ith column of H and this fact determines the integer i. (Educational note: Thus there is a simple procedure for testing whether a received word is a code word and for deciding, in the case that it is not a code word, what unique bit to change to convert it into a code word.)

70. Let $r \geq 4$. Prove for $2^{r-1} \leq n \leq 2^r - 1$ that any linear code C in $\mathbb{F}^n$ with $\delta(C) \geq 3$ has $\dim C \leq n - r$. Observe that equality holds for $C = C_{2^r-1}$.

71. The **weight enumerator polynomial** of a linear code C is the polynomial $W_C(X, Y)$ in $\mathbb{Z}[X, Y]$ given by $W_C(X, Y) = \sum_{k=0}^{n} N_k(C) X^{n-k} Y^k$, where $N_k(C)$ is the number of words of weight k in C.
   (a) Compute $W_C(X, Y)$ for the following linear codes C: the 0 code, the code $\mathbb{F}^n$, the repetition code, the parity code, the code in Problem 67, the Hamming code C_7, and the Hamming code C_8.
   (b) Why is the coefficient of X^n in $W_C(X, Y)$ necessarily equal to 1?
   (c) Show that $W_C(X, Y) = \sum_{c \in C} X^{n - \text{wt}(c)} Y^{\text{wt}(c)}$.

72. **(Cyclic redundancy codes)** Cyclic redundancy codes treat blocks of data as coefficients of polynomials in $\mathbb{F}[X]$. With the size n of data blocks fixed, one fixes a monic **generating polynomial** $G(X) = 1 + a_1 X + \cdots + a_{g-1} X^{g-1} + X^g$ with a nonzero constant term and with degree g suitably less than n. Data to be transmitted are provided as members $(b_0, b_1, \ldots, b_{n-g-1})$ of $\mathbb{F}^{n-g}$ and are converted into polynomials $B(X) = b_0 + b_1 X + \cdots + b_{n-g-1} X^{n-g-1}$. Then the n-tuple of coefficients of $G(X)B(X)$ is transmitted. To decode a polynomial $P(X)$ that is received, one writes $P(X) = G(X)Q(X) + R(X)$ via the division algorithm. If $R(X) = 0$, it is assumed that $P(X)$ is a code word. Otherwise $R(X)$ is definitely not a code word. Thus the code C amounts to the system of coefficients of all polynomials $G(X)B(X)$ with $B(X) = 0$ or $\deg B(X) \leq n - g - 1$. A basis of C is obtained by letting $B(X)$ run through the monomials $1, X, \ldots, X^{n-g-1}$, and therefore $\dim C = n - g$. Take $G(X) = 1 + X + X^2 + X^4$ and $n \geq 8$. Prove that $\delta(C) = 2$.

73. **(CRC-8)** The cyclic redundancy code C bearing the name CRC-8 has $G(X) = 1 + X + X^2 + X^8$. Prove that if $8 \leq n \leq 19$, then $\delta(C) = 4$. (Educational note: It will follow from the theory of finite fields in Chapter IX, together with the problems on coding theory at the end of that chapter, that $n = 255$ plays a special role for this code, and $\delta(C) = 4$ in that case.)

Problems 74–77 concern categories and functors. Problem 75 assumes knowledge of point-set topology.

74. Let $\mathcal{C}$ be the category of all sets, the morphisms being the functions between sets. Verify that the disjoint union of sets is a coproduct.

75. Let $\mathcal{C}$ be the category of all topological spaces, the morphisms being the continuous functions. Let S be a nonempty set, and let X_s be a topological space for each s in S.
    (a) Show that the Cartesian product of the spaces X_s, with the product topology, is a product of the X_s's.
    (b) Show that the disjoint union of the spaces X_s, topologized so that a set E is open if and only if its intersection with each X_s is open, is a coproduct of the X_s's.

76. Taking a cue from the example of a category in which products need not exist, exhibit a category in which coproducts need not exist.

77. Let $\mathcal{C}$ be a category having just one object, say X, and suppose that every member of Morph(X, X) is an isomorphism. Prove that Morph(X, X) is a group under the law of composition for the category. Can every group be realized in this way, up to isomorphism?

Problems 78–80 introduce a notion of **duality** in category theory and use it to derive Proposition 4.64 from Proposition 4.63. If $\mathcal{C}$ is a category, then the **opposite category** $\mathcal{C}^{\text{opp}}$ is defined to have Obj($\mathcal{C}^{\text{opp}}$) = Obj($\mathcal{C}$) and Morph$_{\mathcal{C}^{\text{opp}}}(A, B)$ = Morph$_\mathcal{C}(B, A)$. If $\circ$ denotes the law of composition in $\mathcal{C}$, then the law of composition $\circ^{\text{opp}}$ in $\mathcal{C}^{\text{opp}}$ is defined by $g \circ^{\text{opp}} f = f \circ g$ for $f \in$ Morph$_{\mathcal{C}^{\text{opp}}}(A, B)$ and $g \in$ Morph$_{\mathcal{C}^{\text{opp}}}(B, C)$.

78. Verify that $\mathcal{C}^{\text{opp}}$ is indeed a category, that $(\mathcal{C}^{\text{opp}})^{\text{opp}} = \mathcal{C}$, and that to pass from a diagram involving objects and morphisms in $\mathcal{C}$ to a corresponding diagram involving the same objects and morphisms considered as in $\mathcal{C}^{\text{opp}}$, one leaves all the vertices and labels alone and reverses the directions of all the arrows. Verify also that the diagram of $\mathcal{C}$ commutes if and only if the diagram in $\mathcal{C}^{\text{opp}}$ commutes.

79. Let $\mathcal{C}$ be the category of all sets, the morphisms in Morph$_\mathcal{C}(A, B)$ being all functions from A to B. Show that the morphisms in Morph$_{\mathcal{C}^{\text{opp}}}(A, B)$ cannot necessarily all be regarded as functions from A to B.

80. Suppose that S is a nonempty set and that $\{X_s\}_{s \in S}$ is an object in $\mathcal{C}$.
    (a) Prove that if $(X, \{p_s\}_{s \in S})$ is a product of $\{X_s\}_{s \in S}$ in $\mathcal{C}$, then $(X, \{p_s\}_{s \in S})$ is a coproduct of $\{X_s\}_{s \in S}$ in $\mathcal{C}^{\text{opp}}$, and that if $(X, \{p_s\}_{s \in S})$ is a coproduct of $\{X_s\}_{s \in S}$ in $\mathcal{C}$, then $(X, \{p_s\}_{s \in S})$ is a product of $\{X_s\}_{s \in S}$ in $\mathcal{C}^{\text{opp}}$.
    (b) Show that Proposition 4.64 for $\mathcal{C}$ follows from the validity of Proposition 4.63 for $\mathcal{C}^{\text{opp}}$.

CHAPTER V

Theory of a Single Linear Transformation

Abstract. This goal of this chapter is to find finitely many canonical representatives of each similarity class of square matrices with entries in a field and correspondingly of each isomorphism class of linear maps from a finite-dimensional vector space to itself.

Section 1 frames the problem in more detail. Section 2 develops the theory of determinants over a commutative ring with identity in order to be able to work easily with characteristic polynomials $\det(\lambda I - A)$. The discussion is built around the principle of "permanence of identities," which allows for passage from certain identities with integer coefficients to identities with coefficients in the ring in question.

Section 3 introduces the minimal polynomial of a square matrix or linear map. The Cayley–Hamilton Theorem establishes that such a matrix satisfies its characteristic equation, and it follows that the minimal polynomial divides the characteristic polynomial. It is proved that a matrix is similar to a diagonal matrix if and only if its minimal polynomial is the product of distinct factors of degree 1. In combination with the fact that two diagonal matrices are similar if and only if their diagonal entries are permutations of one another, this result solves the canonical-form problem for matrices whose minimal polynomial is the product of distinct factors of degree 1.

Section 4 introduces general projection operators from a vector space to itself and relates them to vector-space direct-sum decompositions with finitely many summands. The summands of a direct-sum decomposition are invariant under a linear map if and only if the linear map commutes with each of the projections associated to the direct-sum decomposition.

Section 5 concerns the Primary Decomposition Theorem, whose subject is the operation of a linear map $L : V \to V$ with V finite-dimensional. The statement is that if L has minimal polynomial $P_1(\lambda)^{l_1} \cdots P_k(\lambda)^{l_k}$ with the $P_j(\lambda)$ distinct monic prime, then V has a unique direct-sum decomposition in which the respective summands are the kernels of the linear maps $P_j(L)^{l_j}$, and moreover the minimal polynomial of the restriction of L to the jth summand is $P_j(\lambda)^{l_j}$.

Sections 6–7 concern Jordan canonical form. For the case that the prime factors of the minimal polynomial of a square matrix all have degree 1, the main theorem gives a canonical form under similarity, saying that a given matrix is similar to one in "Jordan form" and that the Jordan form is completely determined up to permutation of the constituent blocks. The theorem applies to all square matrices if the field is algebraically closed, as is the case for $\mathbb{C}$. The theorem is stated and proved in Section 6, and Section 7 shows how to make computations in two different ways.

1. Introduction

This chapter will work with vector spaces over a common field of "scalars," which will be called $\mathbb{K}$. As was observed near the end of Section IV.5, all the results concerning vector spaces in Chapter II remain valid when the scalars are taken from $\mathbb{K}$ rather than just $\mathbb{Q}$ or $\mathbb{R}$ or $\mathbb{C}$. The ring of polynomials in one indeterminate X over $\mathbb{K}$ will be denoted by $\mathbb{K}[X]$.

For the field $\mathbb{C}$ of complex numbers, every nonconstant polynomial in $\mathbb{C}[X]$ has a root, according to the Fundamental Theorem of Algebra (Theorem 1.18). Because of this fact some results in this chapter will take an especially simple form when $\mathbb{K} = \mathbb{C}$, and this simple form will persist for any field with this same property. Accordingly, we make a definition. Let us say that a field $\mathbb{K}$ is **algebraically closed** if every nonconstant polynomial in $\mathbb{K}[X]$ has a root. We shall work hard in Chapter IX to obtain examples of algebraically closed fields beyond $\mathbb{K} = \mathbb{C}$, but let us mention now what a few of them are.

EXAMPLES.

(1) The subset of $\mathbb{C}$ of all roots of polynomials with rational coefficients is an algebraically closed field.

(2) For each prime p, we have seen that any finite field of characteristic p has p^n elements for some n. It turns out that there is one and only one field of p^n elements, up to isomorphism, for each n. If we align them suitably for fixed p and take their union on n, then the result is an algebraically closed field.

(3) If $\mathbb{K}$ is any field, then there exists an algebraically closed field having $\mathbb{K}$ as a subfield. We shall prove this existence in Chapter IX by means of Zorn's Lemma (which appears in Section A5 of the appendix).

The general problem to be addressed in this chapter is to find "canonical forms" for linear maps from finite-dimensional vector spaces to themselves, special ways of realizing the linear maps that bring out some of their properties. Let us phrase a specific problem of this kind completely in terms of linear algebra at first. Then we can rephrase it in terms of a combination of linear algebra and group theory, and we shall see how it fits into a more general context.

In terms of matrices, the specific problem is to find a way of deciding whether two square matrices represent the same linear map in different bases. We know from Proposition 2.17 that if $L : V \to V$ is linear on the finite-dimensional vector space V and if A is the matrix of L relative to a particular ordered basis in domain and range, then the matrix B of L in another ordered basis is of the form $B = C^{-1}AC$ for some invertible matrix C, i.e., A and B are similar. Thus one kind of solution to the problem would be to specify one representative of each similarity class of square matrices. But this is not a convenient kind of answer to look for; in fact, the matrices $A = \begin{pmatrix} 1 & 0 \\ 0 & 2 \end{pmatrix}$ and $B = \begin{pmatrix} 2 & 0 \\ 0 & 1 \end{pmatrix}$ are similar via $C = \begin{pmatrix} 0 & 1 \\ 1 & 0 \end{pmatrix}$, but there is no particular reason to prefer one of A or B to the other. Thus a "canonical form" for detecting similarity will allow more than one repre-

1. Introduction

sentative of each similarity class (but typically only finitely many such representatives), and a supplementary statement will tell us when two such are similar.

So far, the best information that we have about solving this problem concerning square matrices comes from Section II.8. In that section the discussion of eigenvalues gave us some necessary conditions for similarity, but we did not obtain a useful necessary and sufficient condition.

In terms of linear maps, what we seek for a linear $L : V \to V$ is to use the geometry of L to construct an ordered basis of V such that L acts in a particularly simple way on that ordered basis. Ideally the description of how L acts on the ordered basis is to be detailed enough so that the matrix of L in that ordered basis is completely determined by the description, even though the ordered basis may not be determined by it. For example, if L were to have a basis of eigenvectors, then the description could be that "L has an ordered basis of eigenvectors with eigenvalues $x_1, \ldots, x_n$." In any ordered basis with this property, the matrix of L would then be diagonal with diagonal entries $x_1, \ldots, x_n$.

Suppose then that we have this kind of detailed description of how a linear map L acts on some ordered basis. To what extent is L completely determined? The answer is that L is determined up to an isomorphism of the underlying vector space. In fact, suppose that L and M are linear maps from V to itself such that $\begin{pmatrix} L \\ \Gamma\Gamma \end{pmatrix} = A = \begin{pmatrix} M \\ \Delta\Delta \end{pmatrix}$ for some ordered bases Γ and Δ. Then

$$\begin{pmatrix} L \\ \Gamma\Gamma \end{pmatrix} = A = \begin{pmatrix} M \\ \Delta\Delta \end{pmatrix} = \begin{pmatrix} I \\ \Delta\Gamma \end{pmatrix} \begin{pmatrix} M \\ \Gamma\Gamma \end{pmatrix} \begin{pmatrix} I \\ \Gamma\Delta \end{pmatrix}$$

$$= \begin{pmatrix} S \\ \Gamma\Gamma \end{pmatrix}^{-1} \begin{pmatrix} M \\ \Gamma\Gamma \end{pmatrix} \begin{pmatrix} S \\ \Gamma\Gamma \end{pmatrix} = \begin{pmatrix} S^{-1}MS \\ \Gamma\Gamma \end{pmatrix},$$

where $S : V \to V$ is the invertible linear map defined by $\begin{pmatrix} S \\ \Gamma\Gamma \end{pmatrix} = \begin{pmatrix} I \\ \Gamma\Delta \end{pmatrix}$. Hence $L = S^{-1}MS$ and $SL = MS$. In other words, if we think of having two copies of V, one called V_1 and the other called V_2, that are isomorphic via $S : V_1 \to V_2$, then the effect of M in V_2 corresponds under S to the effect of L in V_1. In this sense, L is determined up to an isomorphism of V.

Thus we are looking for a geometric description that determines linear maps up to isomorphism. Two linear maps L and M that are related in this way have $L = S^{-1}MS$ for some invertible linear map S. Passing to matrices with respect to some basis, we see that the matrices of L and M are to be similar. Consequently our two problems, one to characterize similarity for matrices and the other to characterize isomorphism for linear maps, come to the same thing.

These two problems have an interpretation in terms of group theory. In the case of n-by-n matrices, the group $GL(n, \mathbb{K})$ of invertible matrices acts on the set of all square matrices of size n by conjugation via $(g, x) \mapsto gxg^{-1}$; the similarity classes are exactly the orbits of this group action, and the canonical form is to single out finitely many representatives from each orbit. In the case of linear maps, the group $GL(V)$ of invertible linear maps on the finite-dimensional vector space V acts by conjugation on the set of all linear maps from V into itself; the isomorphism classes of linear maps on V are the orbits, and the canonical form is to single out finitely many representatives from each orbit.

The above problem, whether for matrices or for linear maps, does not have a unique acceptable solution. Nevertheless, the text of this chapter will ultimately concentrate on one such solution, known as the "Jordan canonical form."

Now that we have brought group theory into the statement of the problem, we can put matters in a more general context: The situation is that some "important" group G acts in an important way on an "interesting" vector space of matrices. The **canonical-form problem** for this situation is to single out finitely many representatives of each orbit and give a way of deciding, in terms of these representatives, whether two of the given matrices lie in the same orbit. We shall not pursue the more general problem in the text at this time. However, Problem 1 at the end of the chapter addresses one version beyond the one concerning similarity: to find a canonical form for the action of $GL(m, \mathbb{K}) \times GL(n, \mathbb{K})$ on m-by-n matrices by $((g, h), x) = gxh^{-1}$. Some other groups that are important in this sense, besides products of general linear groups, are introduced in Chapter VI, and a problem at the end of Chapter VI reinterprets two theorems of that chapter as further canonical-form theorems under the action of a general linear group.

Let us return to the canonical-form problems for similarity of matrices and isomorphism of linear maps. The basic tool in studying these problems is the characteristic polynomial of a matrix or a linear map, as in Chapter II. However, we subtly used a special feature of $\mathbb{Q}$ and $\mathbb{R}$ and $\mathbb{C}$ in working with characteristic polynomials in Chapter II: we passed back and forth between the characteristic polynomial $\det(\lambda I - A)$ as a polynomial in one indeterminate (defined by its expression after expanding it out) and as a polynomial function of λ, defined for each value of λ in $\mathbb{Q}$ or $\mathbb{R}$ or $\mathbb{C}$, one value at a time. This passage was legitimate because the homomorphism of the ring of polynomials in one indeterminate over a field to the ring of polynomial functions is one-one when the field is infinite, by Proposition 4.28c or Corollary 1.14. Some care is required, however, in working with general fields, and we begin by supplying the necessary details for justifying manipulations with determinants in a more general setting than earlier.

2. Determinants over Commutative Rings with Identity

Throughout this section let R be a commutative ring with identity. The main case of interest for us at this time will be that $R = \mathbb{K}[\lambda]$ is the polynomial ring in one indeterminate λ over a field $\mathbb{K}$.

2. Determinants over Commutative Rings with Identity

The set of n-by-n matrices with entries in R is an abelian group under entry-by-entry addition, and matrix multiplication makes it into a ring with identity. Following tradition, we shall usually write $M_n(R)$ rather than $M_{nn}(R)$ for this ring. In this section we shall define a determinant function $\det : M_n(R) \to R$ and establish some of its properties. For the case that R is a field, some of our earlier proofs concerning determinants used vector-space concepts—bases, dimensions, and so forth—and these are not available for general R. Yet most of the properties of determinants remain valid for general R because of a phenomenon known as **permanence of identities**. We shall not try to state a general theorem about this principle but instead will be content to observe a pattern in how the relevant identities are proved.

If A is in $M_n(R)$, we define its **determinant** to be

$$\det A = \sum_{\sigma \in \mathfrak{S}_n} (\operatorname{sgn} \sigma) A_{1\sigma(1)} A_{2\sigma(2)} \cdots A_{n\sigma(n)},$$

in effect converting into a definition the formula obtained in Theorem 2.34d when R is a field.

A sample of the kind of identity we have in mind is the formula

$$\det(AB) = \det A \det B \quad \text{for } A \text{ and } B \text{ in } M_n(R).$$

The key is that this formula says that two polynomials in $2n^2$ variables, with integer coefficients, are equal whenever arbitrary members of R are substituted for the variables. Thus let us introduce $2n^2$ indeterminates $X_{11}, X_{12}, \ldots, X_{nn}$ and $Y_{11}, Y_{12}, \ldots, Y_{nn}$ to correspond to these variables. Forming the commutative ring $S = \mathbb{Z}[X_{11}, X_{12}, \ldots, X_{nn}, Y_{11}, Y_{12}, \ldots, Y_{nn}]$, we assemble the matrices $X = [X_{ij}]$, $Y = [Y_{ij}]$, and $XY = \left[\sum_k X_{ik} Y_{kj}\right]$ in $M_n(S)$. Consider the two members of S given by

$\det X \det Y$
$$= \Big(\sum_{\sigma \in \mathfrak{S}_n} (\operatorname{sgn} \sigma) X_{1\sigma(1)} X_{2\sigma(2)} \cdots X_{n\sigma(n)}\Big) \Big(\sum_{\sigma \in \mathfrak{S}_n} (\operatorname{sgn} \sigma) Y_{1\sigma(1)} Y_{2\sigma(2)} \cdots Y_{n\sigma(n)}\Big)$$

and
$$\det(XY) = \sum_{\sigma \in \mathfrak{S}_n} (\operatorname{sgn} \sigma)(XY)_{1\sigma(1)}(XY)_{2\sigma(2)} \cdots (XY)_{n\sigma(n)},$$

where $(XY)_{ij} = \sum_k X_{ik} Y_{kj}$. If we fix arbitrary elements $x_{11}, x_{12}, \ldots, x_{nn}$ and $y_{11}, y_{12}, \ldots, y_{nn}$ of $\mathbb{Z}$, then Proposition 4.30 gives us a unique substitution homomorphism $\Psi : S \to \mathbb{Z}$ such that $\Psi(1) = 1$, $\Psi(X_{ij}) = x_{ij}$, and $\Psi(Y_{ij}) = y_{ij}$ for all i and j. Writing $x = [x_{ij}]$ and $y = [y_{ij}]$ and using that matrices with integer entries have $\det(xy) = \det x \det y$ because $\mathbb{Z}$ is a subset of the field $\mathbb{Q}$, we

see that $\Psi(\det(XY)) = \Psi(\det X \det Y)$ for each choice of x and y. Since $\mathbb{Z}$ is an infinite integral domain and since x and y are arbitrary, Corollary 4.32 allows us to deduce that
$$\det(XY) = \det X \det Y$$
as an equality in S.

Now we pass from an identity in S to an identity in R. Let 1_R be the identity in R. Proposition 4.19 gives us a unique homomorphism of rings $\varphi_1 : \mathbb{Z} \to R$ such that $\varphi_1(1) = 1_R$. If we fix arbitrary elements $A_{11}, A_{12}, \ldots, A_{nn}$ and $B_{11}, B_{12}, \ldots, B_{nn}$ of R, then Proposition 4.30 gives us a unique substitution homomorphism $\Phi : S \to R$ such that $\Phi(1) = \varphi_1(1) = 1_R$, $\Phi(X_{ij}) = A_{ij}$ for all i and j, and $\Phi(Y_{ij}) = B_{ij}$ for all i and j. Applying Φ to the equality $\det(XY) = \det X \det Y$, we obtain the identity we sought, namely
$$\det(AB) = \det A \det B \qquad \text{for } A \text{ and } B \text{ in } M_n(R).$$

Proposition 5.1. If R is a commutative ring with identity, then the determinant function $\det : M_n(R) \to R$ has the following properties:
(a) $\det(AB) = \det A \det B$,
(b) $\det I = 1$,
(c) $\det A^t = \det A$,
(d) $\det C = \det A + \det B$ if A, B, and C match in all rows but the jth and if the jth row of C is the sum of the jth rows of A and B,
(e) $\det B = r \det A$ if A and B match in all rows but the jth and if the jth row of B is equal entry by entry to r times the jth row of A for some r in R,
(f) $\det A = 0$ if A has two equal rows,
(g) $\det \begin{pmatrix} A & B \\ 0 & D \end{pmatrix} = \det A \det D$ if A is in $M_k(R)$, D is in $M_l(R)$, and $k + l = n$.

REMARK. Properties (d), (e), and (f) imply that usual steps in manipulating determinants by row reduction continue to be valid.

PROOF. Part (a) was proved above, and parts (c) through (f) may be proved in the same way from the corresponding facts about integer matrices in Section II.7. Part (b) is immediate from the definition.

For (g), we first prove the result when the entries are in $\mathbb{Q}$, and then we argue in the same way as with (a) above. When the entries are in $\mathbb{Q}$, row reduction of D allows us to reduce to the case either that D has a row of 0's or that D is the identity. If D has a row of 0's, then $\det \begin{pmatrix} A & B \\ 0 & D \end{pmatrix}$ and $\det A \det D$ are both 0 and hence are equal. If D is the identity, then further row reduction shows that $\det \begin{pmatrix} A & B \\ 0 & I \end{pmatrix} = \det \begin{pmatrix} A & 0 \\ 0 & I \end{pmatrix}$, and the right side equals $\det A = \det A \det I$, as required. $\square$

2. Determinants over Commutative Rings with Identity

Proposition 5.2 (expansion in cofactors). Let R be a commutative ring with identity, let A be in $M_n(R)$, and let $\widehat{A}_{ij}$ be the member of $M_{n-1}(R)$ obtained by deleting the ith row and the jth column from A. Then

(a) for any j, $\det A = \sum_{i=1}^{n}(-1)^{i+j}A_{ij}\det\widehat{A}_{ij}$, i.e., $\det A$ may be calculated by "expansion in cofactors" about the jth column,

(b) for any i, $\det A = \sum_{j=1}^{n}(-1)^{i+j}A_{ij}\det\widehat{A}_{ij}$, i.e., $\det A$ may be calculated by "expansion in cofactors" about the ith row.

PROOF. This may be derived in the same way from Proposition 2.36 by using the principle of permanence of identities. □

Corollary 5.3 (Vandermonde matrix and determinant). If $r_1,\ldots,r_n$ lie in a commutative ring R with identity, then

$$\det\begin{pmatrix} 1 & 1 & \cdots & 1 \\ r_1 & r_2 & \cdots & r_n \\ r_1^2 & r_2^2 & \cdots & r_n^2 \\ \vdots & \vdots & \ddots & \vdots \\ r_1^{n-1} & r_2^{n-1} & \cdots & r_n^{n-1} \end{pmatrix} = \prod_{j>i}(r_j - r_i).$$

PROOF. The derivation of this from Proposition 5.2 is the same as the derivation of Corollary 2.37 from Proposition 2.35. □

Proposition 5.4 (Cramer's rule). Let R be a commutative ring with identity, let A be in $M_n(R)$, and define A^{adj} in $M_n(R)$ to be the classical adjoint of A, namely the matrix with entries $A_{ij}^{\text{adj}} = (-1)^{i+j}\det\widehat{A}_{ji}$, where $\widehat{A}_{kl}$ defined as in the statement of Proposition 5.2. Then $AA^{\text{adj}} = A^{\text{adj}}A = (\det A)I$.

PROOF. This may be derived from Proposition 2.38 in the same way as for Propositions 5.1 and 5.2 using the principle of permanence of identities. □

Corollary 5.5. Let R be a commutative ring with identity, and let A be in $M_n(R)$. If $\det A$ is a unit in R, then A has a two-sided inverse in $M_n(R)$. Conversely if A has a one-sided inverse in $M_n(R)$, then $\det A$ is a unit in R.

REMARK. If R is a field, then A and any associated linear map are often called **nonsingular** if invertible, **singular** otherwise. When R is not a field, terminology varies for what to call a noninvertible matrix whose determinant is not 0.

PROOF. If $\det A$ is a unit in R, let r be its multiplicative inverse. Then Proposition 5.4 shows that $r^{-1}A^{\text{adj}}$ is a two-sided inverse of A. Conversely if A has, say, a left inverse B, then $BA = I$ implies $(\det B)(\det A) = \det I = 1$, and $\det B$ is an inverse for $\det A$. A similar argument applies if A has a right inverse. □

3. Characteristic and Minimal Polynomials

Again let $\mathbb{K}$ be a field. If A is in $M_n(\mathbb{K})$, the **characteristic polynomial** of A is defined to be the member of the ring $\mathbb{K}[\lambda]$ of polynomials in one indeterminate λ given by $F(\lambda) = \det(\lambda I - A)$. The material of Section 2 shows that $F(\lambda)$ is well defined, being the determinant of a member of $M_n(\mathbb{K}[\lambda])$. It is apparent from the definition of determinant in Section 2 that $F(\lambda)$ is a monic polynomial of degree n with coefficient $-\operatorname{Tr} A = -\sum_{j=1}^{n} A_{jj}$ for λ^{n-1}. Evaluating $F(\lambda)$ at 0, we see that the constant term is $(-1)^n \det A$.

Since the determinant of a product in $M_n(\mathbb{K}[\lambda])$ is the product of the determinants (Proposition 5.1a) and since $C^{-1}(\lambda I - A)C = \lambda I - C^{-1}AC$, we have

$$\det(\lambda I - C^{-1}AC) = (\det C)^{-1} \det(\lambda I - A)(\det C) = \det(\lambda I - A).$$

Thus similar matrices have equal characteristic polynomials. If V is an n-dimensional vector space over $\mathbb{K}$ and $L : V \to V$ is linear, then the matrices of L in any two ordered bases of V (the domain basis being assumed equal to the range basis) are similar, and their characteristic polynomials are the same. Consequently we can define the **characteristic polynomial** of L to be the characteristic polynomial of any matrix of L.

The development of characteristic polynomials has thus be redone in a way that is valid over any field $\mathbb{K}$ without making use of the ring homomorphism from polynomials in one indeterminate over $\mathbb{K}$ to polynomial functions from $\mathbb{K}$ into itself. The discussion in Section II.8 of eigenvectors and eigenvalues for members A of $M_n(\mathbb{K})$ and for linear maps $L : V \to V$ with V finite-dimensional over $\mathbb{K}$ is now meaningful, and there is no need to repeat it.

In particular, the eigenvalues of A and L are exactly the roots of their characteristic polynomial, no matter what $\mathbb{K}$ is. If $\mathbb{K}$ is algebraically closed, then the characteristic polynomial has a root, and consequently A and L each have at least one eigenvalue.

If $L : V \to V$ is linear and V is finite-dimensional, then a vector subspace U of V is said to be **invariant** under L if $L(U) \subseteq U$. In this case $L\big|_U$ is a well-defined linear map from U to itself. Since $L(U) \subseteq U$, Proposition 2.25 shows that $L : V \to V$ factors through V/U as a linear map $\overline{L} : V/U \to V/U$. We shall use this construction, the existence of eigenvalues in the algebraically closed case, and an induction to prove the following.

Proposition 5.6. If $\mathbb{K}$ is an algebraically closed field, if V is a finite-dimensional vector space over $\mathbb{K}$, and if $L : V \to V$ is linear, then V has an ordered basis in which the matrix of L is upper triangular. Consequently any member of $M_n(\mathbb{K})$ is similar to an upper triangular matrix.

3. Characteristic and Minimal Polynomials

REMARKS. For an upper triangular matrix $A = \begin{pmatrix} c_1 & & * \\ & \ddots & \\ 0 & & c_n \end{pmatrix}$ in $M_n(\mathbb{K})$, the characteristic polynomial is $\prod_{j=1}^{n}(\lambda - c_j)$ because the only nonzero term in the definition of $\det(\lambda I - A)$ is the one corresponding to the identity permutation. Triangular form is not yet the canonical form we seek for a square matrix because a particular square matrix may be similar to infinitely many matrices in triangular form.

PROOF. We proceed by induction on $n = \dim V$, with the base case $n = 1$ being clear. Suppose that the result holds for all linear maps from spaces of dimension $< n$ to themselves. Given $L : V \to V$ with $\dim V = n$, let v_1 be an eigenvector of L. This exists by the remarks before the proposition since $\mathbb{K}$ is algebraically closed. Let U be the vector subspace $\mathbb{K}v_1$. Then $L(U) \subseteq U$, and Proposition 2.25 shows that $L : V \to V$ factors through V/U as a linear map $\overline{L} : V/U \to V/U$. Since $\dim V/U = n - 1$, the inductive hypothesis produces an ordered basis $(\bar{v}_2, \ldots, \bar{v}_n)$ of V/U such that the matrix of $\overline{L}$ is upper triangular in this basis. This condition means that $\overline{L}(\bar{v}_j) = \sum_{i=2}^{j} c_{ij}\bar{v}_i$ for $j \geq 2$. Select coset representatives $v_2, \ldots, v_n$ of $\bar{v}_2, \ldots, \bar{v}_n$ so that $\bar{v}_j = v_j + U$ for $j \geq 2$. Then $L(v_j + U) = \sum_{i=2}^{j} c_{ij}(v_i + U)$ for $j \geq 2$, and hence $L(v_j)$ lies in the coset $\sum_{i=2}^{j} c_{ij}v_i + U$ for $j \geq 2$. For each $j \geq 1$, we then have $L(v_j) = \sum_{i=2}^{j} c_{ij}v_i + c_{1j}v_1$ for some scalar c_{1j}, and we see that $(v_1, \ldots, v_n)$ is the required ordered basis. $\square$

Let us return to the situation in which $\mathbb{K}$ is any field. For a matrix A in $M_n(\mathbb{K})$ and a polynomial P in $\mathbb{K}[\lambda]$, it is meaningful to form $P(A)$. We can do so by two equivalent methods, both useful. The concrete way of forming $P(A)$ is as $P(A) = c_n A^n + \cdots + c_1 A + c_0 I$ if $P(\lambda) = c_n \lambda^n + \cdots + c_1 \lambda + c_0$. The abstract way is to form the subring T of $M_n(A)$ generated by $\mathbb{K}I$ and A. This subring is commutative. We let $\varphi : \mathbb{K} \to T$ be given by $\varphi(c) = cI$. Then the universal mapping property of $\mathbb{K}[\lambda]$ given in Proposition 4.24 produces a unique ring homomorphism $\Phi : \mathbb{K}[\lambda] \to T$ such that $\Phi(c) = cI$ for all $c \in \mathbb{K}$ and $\Phi(\lambda) = A$. The value of $P(A)$ is the element $\Phi(P)$ of T.

For A in $M_n(\mathbb{K})$, let us study all polynomials P such that $P(A) = 0$. For any polynomial P and any invertible matrix C, we have

$$P(C^{-1}AC) = C^{-1}P(A)C$$

because if $P(\lambda) = c_n\lambda^n + \cdots + c_1\lambda + c_0$, then

$$P(C^{-1}AC) = c_n(C^{-1}AC)^n + \cdots + c_1 C^{-1}AC + c_0 I$$
$$= C^{-1}(c_n A^n + \cdots + c_1 A + c_0 I)C.$$

Consequently if $P(A) = 0$, then $P(C^{-1}AC) = 0$, and the set of matrices with $P(A) = 0$ is closed under similarity. We shall make use of this observation a little later in this section.

Proposition 5.7. If A is in $M_n(\mathbb{K})$, then there exists a nonzero polynomial P in $\mathbb{K}[\lambda]$ such that $P(A) = 0$.

PROOF. The $\mathbb{K}$ vector space $M_n(\mathbb{K})$ has dimension n^2. Therefore the $n^2 + 1$ matrices $I, A, A^2, \ldots, A^{n^2}$ are linearly dependent, and we have

$$c_0 + c_1 A + c_2 A^2 + \cdots + c_{n^2} A^{n^2} = 0$$

for some set of scalars not all 0. Then $P(A) = 0$ for the polynomial $P(\lambda) = c_0 + c_1 \lambda + c_2 \lambda^2 + \cdots + c_{n^2} \lambda^{n^2}$; this P is not the 0 polynomial since at least one of the coefficients is not 0. □

ALTERNATIVE PROOF IF $\mathbb{K}$ IS ALGEBRAICALLY CLOSED. Since the set of polynomials P with $P(A) = 0$ depends only on the similarity class of A, Proposition 5.6 shows that there is no loss of generality in assuming that A is upper triangular, say of the form $\begin{pmatrix} \lambda_1 & * \\ & \ddots & \\ 0 & & \lambda_n \end{pmatrix}$. Then $A - \lambda_j I$ is upper triangular with 0 in the jth diagonal entry, and $\prod_{j=1}^{n} (A - \lambda_j I)$ is upper triangular with 0 in all diagonal entries. Therefore $\left(\prod_{j=1}^{n} (A - \lambda_j I) \right)^n = 0$. □

With A fixed, we continue to consider the set of all polynomials $P(\lambda)$ such that $P(A) = 0$. Let us think of $P(A)$ as being computed by the abstract procedure described above, namely as the image of A under the ring homomorphism $\Phi : \mathbb{K}[\lambda] \to T$ such that $\Phi(c) = cI$ for all $c \in \mathbb{K}$ and $\Phi(\lambda) = A$, where T is the commutative subring of $M_n(\mathbb{K})$ generated by $\mathbb{K}I$ and A. Then the set of all polynomials $P(\lambda)$ with $P(A) = 0$ is the kernel of the ring homomorphism Φ. This set is therefore an ideal, and Proposition 5.7 shows that the ideal is nonzero. We shall apply the following proposition to this ideal.

Proposition 5.8. If I is a nonzero ideal in $\mathbb{K}[\lambda]$, then there exists a unique monic polynomial of lowest degree in I, and every member of I is the product of this particular polynomial by some other polynomial.

PROOF. Let $B(\lambda)$ be a nonzero member of I of lowest possible degree; adjusting B by a scalar factor, we may assume that B is monic. If A is in I, then Proposition 1.12 produces polynomials Q and R such that $A = BQ + R$ and either $R = 0$ or $\deg R < \deg B$. Since I is an ideal, BQ is in I and hence $R = A - BQ$ is in I. From minimality of the degree of B, we conclude that $R = 0$. Hence $A = BQ$,

3. Characteristic and Minimal Polynomials

and A is exhibited as the product of B and some other polynomial Q. If B_1 is a second monic polynomial of lowest degree in I, then we can take $A = B_1$ to see that $B_1 = QB$. Since $\deg B_1 = \deg B$, we conclude that $\deg Q = 0$. Thus Q is a constant polynomial. Comparing the leading coefficients of B and B_1, we see that $Q(\lambda) = 1$. □

With A fixed in $M_n(\mathbb{K})$, let us apply Proposition 5.8 to the ideal of all polynomials P in $\mathbb{K}[\lambda]$ with $P(A) = 0$. The unique monic polynomial of lowest degree in this ideal is called the **minimal polynomial** of A. Let us try to identify this minimal polynomial.

Theorem 5.9 (Cayley–Hamilton Theorem). If A is in $M_n(\mathbb{K})$ and if $F(\lambda) = \det(\lambda I - A)$ is its characteristic polynomial, then $F(A) = 0$.

PROOF. Let T be the commutative subring of $M_n(\mathbb{K})$ generated by $\mathbb{K}I$ and A, and define a member $B(\lambda)$ of the ring $T[\lambda]$ by $B(\lambda) = \lambda I - A$. The $(i, j)^{\text{th}}$ entry of $B(\lambda)$ is $B_{ij}(\lambda) = \delta_{ij}\lambda - A_{ij}$, and $F(\lambda) = \det B(\lambda)$.

Let $C(\lambda) = B(\lambda)^{\text{adj}}$ denote the classical adjoint of $B(\lambda)$ as a member of $T[\lambda]$; the form of $C(\lambda)$ is given in the statement of Cramer's rule (Proposition 5.4), and that proposition says that

$$B(\lambda)C(\lambda) = (\det B(\lambda))I = F(\lambda)I.$$

The equality in the $(i, j)^{\text{th}}$ entry is the equality $\delta_{ij}F(\lambda) = \sum_j B_{ik}(\lambda)C_{kj}(\lambda)$ of members of $\mathbb{K}[\lambda]$. Application of the substitution homomorphism $\lambda \mapsto A$ gives

$$\delta_{ij}F(A) = \sum_k B_{ik}(A)C_{kj}(A) = \sum_k (\delta_{ik}A - A_{ik}I)C_{kj}(A).$$

Multiplying on the right by the ith standard basis vector e_i and summing on i, we obtain the equality of vectors

$$F(A)e_j = \sum_i \sum_k (\delta_{ik}Ae_i - A_{ik}e_i)C_{kj}(A) = \sum_k C_{kj}(A)\Big(\sum_i (\delta_{ik}Ae_i - A_{ik}e_i)\Big)$$

since $C_{kj}(A)$ is a scalar. But $\sum_i (\delta_{ik}Ae_i - A_{ik}e_i) = Ae_k - \sum_i A_{ik}e_i = 0$ for all k, and therefore $F(A)e_j = 0$. Since j is arbitrary, $F(A) = 0$. □

Corollary 5.10. If A is in $M_n(\mathbb{K})$, then the minimal polynomial of A divides the characteristic polynomial of A.

PROOF. Theorem 5.9 shows that the characteristic polynomial of A lies in the ideal of all polynomials vanishing on A. Then the corollary follows from Proposition 5.8. □

For our matrix A in $M_n(\mathbb{K})$, let $F(\lambda)$ be the characteristic polynomial, and let $M(\lambda)$ be the minimal polynomial. By unique factorization (Theorem 1.17), the monic polynomial $F(\lambda)$ has a factorization into powers of distinct prime monic polynomials of the form

$$F(\lambda) = P_1(\lambda)^{k_1} \cdots P_r(\lambda)^{k_r},$$

and this factorization is unique up to the order of the factors. Since $M(\lambda)$ is a monic polynomial dividing $F(\lambda)$, we must have

$$M(\lambda) = P_1(\lambda)^{l_1} \cdots P_r(\lambda)^{l_r}$$

with $l_1 \leq k_1, \ldots, l_r \leq k_r$, by the same argument that deduced Corollary 1.7 from unique factorization in the ring of integers. We shall see shortly that $k_j > 0$ implies $l_j > 0$ if $P_j(\lambda)$ is of degree 1, i.e., if $P_j(\lambda)$ is of the form $\lambda - \lambda_0$; in other words, if λ_0 is an eigenvalue of A, then $\lambda - \lambda_0$ divides its minimal polynomial. We return to this point in a moment. Problem 31 at the end of the chapter will address the same question when $P_j(\lambda)$ has degree > 1.

EXAMPLES.

(1) In the 2-by-2 case, $\begin{pmatrix} c & 0 \\ 0 & c \end{pmatrix}$ has minimal polynomial $M(\lambda) = \lambda - c$, and $\begin{pmatrix} c & 1 \\ 0 & c \end{pmatrix}$ has $M(\lambda) = (\lambda - c)^2$. Both matrices have characteristic polynomial $F(\lambda) = (\lambda - c)^2$.

(2) The k-by-k matrix

$$\begin{pmatrix} c & 1 & 0 & \cdots & 0 & 0 \\ 0 & c & 1 & \cdots & 0 & 0 \\ & & \ddots & & & \\ 0 & 0 & 0 & \cdots & c & 1 \\ 0 & 0 & 0 & \cdots & 0 & c \end{pmatrix}$$

with c in every diagonal entry, with 1 in every entry just above the diagonal, and with 0 elsewhere has minimal polynomial $M(\lambda) = (\lambda - c)^k$ and characteristic polynomial $F(\lambda) = (\lambda - c)^k$.

(3) If a matrix A is made up exclusively of several blocks of the type in Example 2 with the same c in each case, the ith block being of size k_i, then the minimal polynomial is $M(\lambda) = (\lambda - c)^{\max_i k_i}$, and the characteristic polynomial is $F(\lambda) = (\lambda - c)^{\sum_i k_i}$.

(4) If A is made up exclusively of several blocks as in Example 3 but with c different for each block, then the minimal and characteristic polynomials for A are obtained by multiplying the minimal and characteristic polynomials obtained from Example 3 for the various c's.

3. Characteristic and Minimal Polynomials

To proceed further, let us change our point of view, working with linear maps $L : V \to V$, where V is a finite-dimensional vector space over $\mathbb{K}$. We have already defined the characteristic polynomial of L to be the characteristic polynomial of the matrix of L in any ordered basis; this is well defined because similar matrices have the same characteristic polynomial. In analogous fashion we can define the **minimal polynomial** of L to be the minimal polynomial of the matrix of L in any ordered basis; this is well defined since, as we have seen, the set of polynomials P in one indeterminate with $P(A) = 0$ is the same as the set with $P(C^{-1}AC) = 0$ if C is invertible.

Another way of approaching the matter of the minimal polynomial of L is to define $P(L)$ for any polynomial P in one indeterminate. As with matrices, we can define $P(L)$ either concretely by substituting L for λ in the expression for $P(\lambda)$, or we can define $P(L)$ abstractly by appealing to the universal mapping property in Proposition 4.24. For the latter we work with the subring T' of linear maps from V to itself generated by $\mathbb{K}I$ and L. This subring is commutative. We let $\varphi : \mathbb{K} \to T'$ be given by $\varphi(c) = cI$, and we use Proposition 4.24 to obtain the unique ring homomorphism $\Phi : \mathbb{K}[\lambda] \to T'$ such that $\Phi(c) = cI$ for all $c \in \mathbb{K}$ and $\Phi(\lambda) = L$. Then $P(L)$ is the element $\Phi(P)$ of T'. Once $P(L)$ is defined, we observe that the set of polynomials $P(\lambda)$ such that $P(L) = 0$ is a nonzero ideal in $\mathbb{K}[\lambda]$; Proposition 5.8 yields a unique monic polynomial of lowest degree in this ideal, and that is the minimal polynomial of L.

Linear maps enable us to make convenient use of invariant subspaces. Recall from earlier in the section that a vector subspace U of V is said to be invariant under the linear map $L : V \to V$ if $L(U) \subseteq U$; in this case we obtain associated linear maps $L|_U : U \to U$ and $\overline{L} : V/U \to V/U$. Relationships among the characteristic polynomials and minimal polynomials of these linear maps are given in the next two propositions.

Proposition 5.11. Let V be a finite-dimensional vector space over $\mathbb{K}$, let $L : V \to V$ be linear, let U be a proper nonzero invariant subspace under L, and let $\overline{L} : V/U \to V/U$ be the induced linear map on V/U. Then the characteristic polynomials of $L, L|_U$, and $\overline{L}$ are related by

$$\det(\lambda I - L) = \det(\lambda I - L|_U) \det(\lambda I - \overline{L}).$$

PROOF. Let $\Gamma_U = (v_1, \ldots, v_k)$ be an ordered basis of U, and extend Γ_U to an ordered basis $\Gamma = (v_1, \ldots, v_n)$ of V. Then $\overline{\Gamma} = (v_{k+1} + U, \ldots, v_n + U)$ is an ordered basis of V/U. Since U is invariant under L, the matrix of L in the ordered basis Γ is of the form $\begin{pmatrix} A & B \\ 0 & D \end{pmatrix}$, where A is the matrix of $L|_U$ in the ordered basis Γ_U and D is the matrix of $\overline{L}$ in the ordered basis $\overline{\Gamma}$. Passing to the characteristic polynomials and applying Proposition 5.1g, we obtain the desired conclusion. $\square$

Proposition 5.12. Let V be a finite-dimensional vector space over $\mathbb{K}$, let $L : V \to V$ be linear, let U be a proper nonzero invariant subspace under L, and let $\overline{L} : V/U \to V/U$ be the induced linear map on V/U. Then the minimal polynomials of $L\big|_U$ and $\overline{L}$ divide the minimal polynomial of L.

PROOF. Let $N(\lambda)$ be the minimal polynomial of $L\big|_U$. Then $N(\lambda)$ is the unique monic polynomial of lowest degree in the ideal of all polynomials $P(\lambda)$ such that $P(L)u = 0$ for all u in U. The minimal polynomial $M(\lambda)$ of L has this property because $M(\lambda)v = 0$ for all v in V. Therefore $M(\lambda)$ is in the ideal and is the product of $N(\lambda)$ and some other polynomial.

Among linear maps S from V into V carrying U into itself, the function $S \mapsto \overline{S}$ sending S to the linear map $\overline{S}$ induced on V/U is a homomorphism of rings. It follows that if $P(\lambda)$ is a polynomial with $P(L) = 0$, then $P(\overline{L}) = 0$. Taking $P(\lambda)$ to be the minimal polynomial of L, we see that the minimal polynomial of L is in the ideal of polynomials vanishing on $\overline{L}$. Therefore it is the product of the minimal polynomial of $\overline{L}$ and some other polynomial. $\square$

Let us come back to the unproved assertion before the examples—that $k_j > 0$ implies $l_j > 0$ if $P_r(\lambda)$ has degree 1. We prove the linear-function version of this statement as a corollary of Proposition 5.12.

Corollary 5.13. If $L : V \to V$ is linear on a finite-dimensional vector space over $\mathbb{K}$ and if a first-degree polynomial $\lambda - \lambda_0$ divides the characteristic polynomial of L, then $\lambda - \lambda_0$ divides the minimal polynomial of L.

PROOF. If $\lambda - \lambda_0$ divides the characteristic polynomial, then λ_0 is an eigenvalue of L, say with v as an eigenvector. Then $U = \mathbb{K}v$ is an invariant subspace under L, and the characteristic and minimal polynomials of $L\big|_U$ are both $\lambda - \lambda_0$. By Proposition 5.12, $\lambda - \lambda_0$ divides the minimal polynomial of L. $\square$

Theorem 5.14. If $L : V \to V$ is linear on a finite-dimensional vector space over $\mathbb{K}$, then L has a basis of eigenvectors if and only if the minimal polynomial $M(\lambda)$ of L is the product of distinct factors of degree 1; in this case, $M(\lambda)$ equals $(\lambda - \lambda_1) \cdots (\lambda - \lambda_k)$, where $\lambda_1, \ldots, \lambda_k$ are the distinct eigenvalues of L. Consequently a matrix A in $M_n(\mathbb{K})$ is similar to a diagonal matrix if and only if its minimal polynomial is the product of distinct factors of degree 1.

PROOF. The easy direction is that $v_1, \ldots, v_n$ are the members of a basis of eigenvectors for L with respective eigenvalues $\mu_1, \ldots, \mu_n$. In this case, let $\lambda_1, \ldots, \lambda_k$ be the distinct members of the set of eigenvalues, with $\mu_i = \lambda_{j(i)}$ for some function $j : \{1, \ldots, n\} \to \{1, \ldots, k\}$. Then $(L - \lambda_j I)(v) = 0$ for v equal to any v_i with $j(i) = j$. Since the linear maps $L - \lambda_j I$ commute as j varies, $\prod_{j=1}^k (L - \lambda_j I)(v) = 0$ for v equal to each of $v_1, \ldots, v_n$, hence for all v. Therefore the minimal polynomial $M(\lambda)$ of L divides $\prod_{j=1}^{k}(\lambda - \lambda_j)$. On the other hand, Corollary 5.13 shows that the deg $M(\lambda) \geq k$. Hence $M(\lambda) = \prod_{j=1}^{k}(\lambda - \lambda_j)$.

Conversely suppose that $M(\lambda) = \prod_{j=1}^{k}(\lambda - \lambda_j)$ with the λ_j distinct. If S_1 is the linear map $S_1 = \prod_{j=2}^{k}(L - \lambda_j I)$, then the formula for $M(\lambda)$ shows that $(L - \lambda_1 I)S_1(v) = 0$ for all v in V, and hence image S_1 is a vector subspace of the eigenspace of L for the eigenvalue λ_1. If v is in ker $S_1 \cap$ image S_1, we then have $0 = S_1(v) = \prod_{j=2}^{k}(L - \lambda_j I)(v) = \prod_{j=2}^{k}(\lambda_1 - \lambda_j)v$. Since λ_1 is distinct from $\lambda_2, \ldots, \lambda_k$, we conclude that $v = 0$, hence that ker $S_1 \cap$ image $S_1 = 0$. Since dim ker S_1 + dim image S_1 = dim V, Corollary 2.29 therefore gives

$$\dim V = \dim \ker S_1 + \dim \operatorname{image} S_1$$
$$= \dim(\ker S_1 + \operatorname{image} S_1) + \dim(\ker S_1 \cap \operatorname{image} S_1)$$
$$= \dim(\ker S_1 + \operatorname{image} S_1).$$

Hence $V = \ker S_1 + \operatorname{image} S_1$. Since $\ker S_1 \cap \operatorname{image} S_1 = 0$, we conclude that $V = \ker S_1 \oplus \operatorname{image} S_1$.

Actually, the same calculation of $S_1(v)$ as above shows that image S_1 is the full eigenspace of L for the eigenvalue λ_1. In fact, if $L(v) = \lambda_1 v$, then $S_1(v) = \prod_{j=2}^{k}(\lambda_1 - \lambda_j)v$, and hence v equals the image under S_1 of $\left(\prod_{j=2}^{k}(\lambda_1 - \lambda_j)\right)^{-1}v$.

Next, since L commutes with S_1, $\ker S_1$ is an invariant subspace under L, and λ_1 is not an eigenvalue of $L\big|_{\ker S_1}$. Thus $\lambda - \lambda_1$ does not divide the minimal polynomial of $L\big|_{\ker S_1}$. On the other hand, S_1 vanishes on the eigenspaces of L for eigenvalues $\lambda_2, \ldots, \lambda_k$, and Corollary 5.13 shows for $j \geq 2$ that $\lambda - \lambda_j$ divides the minimal polynomial of $L\big|_{\ker S_1}$. Taking Proposition 5.12 into account, we conclude that $L\big|_{\ker S_1}$ has minimal polynomial $\prod_{j=2}^{k}(\lambda - \lambda_j)$. We have succeeded in splitting off the eigenspace of L under λ_1 as a direct summand and reducing the proposition to the case of $k - 1$ eigenvalues. Thus induction shows that V is the direct sum of its eigenspaces for the eigenvalues $\lambda_2, \ldots, \lambda_k$, and L thus has a basis of eigenvectors. $\square$

Theorem 5.14 comes close to solving the canonical-form problem for similarity in the case of one kind of square matrices: if the minimal polynomial of A is the product of distinct factors of degree 1, then A is similar to a diagonal matrix. To complete the solution for this case, all we have to do is to say when two diagonal matrices are similar to each other; this step is handled by the following easy proposition.

Proposition 5.15. Two diagonal matrices A and A' in $M_n(\mathbb{K})$ with respective diagonal entries $d_1, \ldots, d_n$ and $d'_1, \ldots, d'_n$ are similar if and only if there is a permutation σ in $\mathfrak{S}_n$ such that $d'_j = d_{\sigma(j)}$ for all j.

PROOF. The respective characteristic polynomials are $\prod_{j=1}^{n} (\lambda - d_j)$ and $\prod_{j=1}^{n} (\lambda - d'_j)$. If A and A' are similar, then the characteristic polynomials are equal, and unique factorization (Theorem 1.17) shows that the factors $\lambda - d'_j$ match the factors $\lambda - d_j$ up to order. Conversely if there is a permutation σ in $\mathfrak{S}_n$ such that $d'_j = d_{\sigma(j)}$ for all j, then the matrix C whose j^{th} column is $e_{\sigma(j)}$ has the property that $A' = C^{-1}AC$. $\square$

To proceed further with obtaining canonical forms for matrices under similarity and for linear maps under isomorphism, we shall use linear maps in ways that we have not used them before. In particular, it will be convenient to be able to recognize direct-sum decompositions from properties of linear maps. We take up this matter in the next section.

4. Projection Operators

In this section we shall see how to recognize direct-sum decompositions of a vector space V from the associated projection operators, and we shall relate these operators to invariant subspaces under a linear map $L : V \to V$.

If $V = U_1 \oplus U_2$, then the function E_1 defined by $E_1(u_1 + u_2) = u_1$ when u_1 is in U_1 and u_2 is in U_2 is linear, satisfies $E_1^2 = E_1$, and has image $E_1 = U_1$ and $\ker E_1 = U_2$. We call E_1 the **projection** of V on U_1 along U_2. A decomposition of V as the direct sum of two vector spaces, when the first of the two spaces is singled out, therefore determines a projection operator uniquely. A converse is as follows.

Proposition 5.16. If V is a vector space and $E_1 : V \to V$ is a linear map such that $E_1^2 = E_1$, then there exists a direct-sum decomposition $V = U_1 \oplus U_2$ such that E_1 is the projection of V on U_1 along U_2. In this case, $(I - E_1)^2 = I - E_1$, and $I - E_1$ is the projection of V on U_2 along U_1.

PROOF. Define $U_1 = \text{image } E_1$ and $U_2 = \ker E_1$. If v is in image $E_1 \cap \ker E_1$, then $E_1(v) = 0$ since v is in $\ker E_1$ and $v = E_1(w)$ for some w in V since v is in image E_1. Then $0 = E_1(v) = E_1^2(w) = E_1(w) = v$, and therefore image $E_1 \cap \ker E_1 = 0$.

If $v \in V$ is given, write $v = E_1(v) + (I - E_1)(v)$. Then $E_1(v)$ is in image E_1, and the computation $E_1(I - E_1)(v) = (E_1 - E_1^2)(v) = (E_1 - E_1)(v) = 0$ shows that $(I - E_1)(v) = 0$. Consequently $V = \text{image } E_1 + \ker E_1$, and we conclude that $V = \text{image } E_1 \oplus \ker E_1$.

Hence $V = U_1 \oplus U_2$, where $U_1 = \text{image } E_1$ and $U_2 = \ker E_1$. In this notation, E_1 is 0 on U_2. If v is in U_1, then $v = E_1(w)$ for some w, and we have

$v = E_1(w) = E_1^2(w) = E_1(E_1(w)) = E_1(v)$. Thus E_1 is the identity on U_1 and is the projection as asserted.

For $(I - E_1)^2$, we have $(I - E_1)^2 = I - 2E_1 + E_1^2 = I - 2E_1 + E_1 = I - E_1$, and $I - E_1$ is a projection. It is 1 on U_2 and is 0 on U_1, hence is the projection of V on U_2 along U_1. □

Let us generalize these considerations to the situation that V is the direct sum of r vector subspaces. The following facts about the situation in Proposition 5.16, with the definition $E_2 = I - E_1$, are relevant to formulating the generalization:
   (i) E_1 and E_2 have $E_1^2 = E_1$ and $E_2^2 = E_2$,
   (ii) $E_1 E_2 = E_2 E_1 = 0$,
   (iii) $E_1 + E_2 = I$.

Suppose that $V = U_1 \oplus \cdots \oplus U_r$. Define $E_j(u_1 + \cdots + u_r) = u_j$. Then E_j is linear from V to itself with $E_j^2 = E_j$, and Proposition 5.16 shows that E_j is the projection of V on U_j along the direct sum of the remaining U_i's. The linear maps $E_1, \ldots, E_r$ then satisfy
   (i') $E_j^2 = E_j$ for $1 \leq j \leq r$,
   (ii') $E_j E_i = 0$ if $i \neq j$,
   (iii') $E_1 + \cdots + E_r = I$.

A converse is as follows.

Proposition 5.17. If V is a vector space and $E_j : V \to V$ for $1 \leq j \leq r$ are linear maps such that
   (a) $E_j E_i = 0$ if $i \neq j$, and
   (b) $E_1 + \cdots + E_r = I$,

then $E_j^2 = E_j$ for $1 \leq j \leq r$ and the vector subspaces $U_j = \text{image } E_j$ have the properties that $V = U_1 \oplus \cdots \oplus U_r$ and that E_j is the projection of V on U_j along the direct sum of all U_i but U_j.

PROOF. Multiplying (b) through by E_j on the left and applying (a) to each term on the left side except the jth, we obtain $E_j^2 = E_j$. Therefore, for each j, E_j is a projection on U_j along some vector subspace depending on j.

If v is in V, then (b) gives $v = E_1(v) + \cdots + E_r(v)$ and shows that $V = U_1 + \cdots + U_r$. Suppose that v is in the intersection of U_j with the sum of the other U_i's. Write $v = \sum_{i \neq j} u_i$ with $u_i = E_i(w_i)$ in U_i. Applying E_j and using the fact that v is in U_j, we obtain $v = E_j(v) = \sum_{i \neq j} E_j E_i(w_i)$. Every term of the right side is 0 by (a), and hence $v = 0$. Thus $V = U_1 \oplus \cdots \oplus U_r$.

Since $E_j E_i = 0$ for $i \neq j$, E_j is 0 on each U_i for $i \neq j$. Therefore the sum of all U_i except U_j is contained in the kernel of E_j. Since the image and kernel of E_j intersect in 0, the sum of all U_i except U_j is exactly equal to the kernel of E_j. This completes the proof. □

226   V. Theory of a Single Linear Transformation

Proposition 5.18. Suppose that a vector space V is a direct sum $V = U_1 \oplus \cdots \oplus U_r$ of vector subspaces, that $E_1, \ldots, E_r$ are the corresponding projections, and that $L : V \to V$ is linear. Then all the subspaces U_j are invariant under L if and only if $LE_j = E_j L$ for all j.

PROOF. If $L(U_j) \subseteq U_j$ for all j, then $i \neq j$ implies $E_i L(U_j) \subseteq E_i(U_j) = 0$ and $LE_i(U_j) = L(0) = 0$. Also, $v \in U_j$ implies $E_j L(v) = L(v) = LE_j(v)$. Hence $E_i L = E_i L$ for all i.

Conversely if $E_j L = LE_j$ and if v is in U_j, then $E_j L(v) = LE_j(v) = L(v)$ shows that $L(v)$ is in U_j. Therefore $L(U_j) \subseteq U_j$ for all j. $\square$

5. Primary Decomposition

For the case that the minimal polynomial of a linear map $L : V \to V$ is the product of distinct factors of degree 1, Theorem 5.14 showed that V is a direct sum of its eigenspaces. The proof used elementary vector-space techniques from Chapter II but did not take full advantage of the machinery developed in the present chapter for passing back and forth between polynomials in one indeterminate and the values of polynomials on L. Let us therefore rework the proof of that proposition, taking into account the discussion of projections in Section 4.

We seek an eigenspace decomposition $V = V_{\lambda_1} \oplus \cdots \oplus V_{\lambda_k}$ relative to L. Proposition 5.17 suggests looking for the corresponding decomposition of the identity operator as a sum of projections: $I = E_1 + \cdots + E_k$. According to that proposition, we obtain a direct-sum decomposition as soon as we obtain this kind of sum of linear maps such that $E_i E_j = 0$ for $i \neq j$. The E_j's will automatically be projections.

The proof of Theorem 5.14 showed that $S_1 = \prod_{j=2}^{k} (L - \lambda_j I)$ has image equal to the kernel of $L - \lambda_1 I$, i.e., equal to the eigenspace for eigenvalue λ_1. If v is in this eigenspace, then $S_1(v) = \prod_{j=2}^{k} (\lambda_1 - \lambda_j) v$. Hence $E_1 = c_1 S_1$, where $c_1^{-1} = \prod_{j=2}^{k} (\lambda_1 - \lambda_j)$. The linear map S_1 equals $Q_1(L)$, where $Q_1(\lambda) = \prod_{j=2}^{k} (\lambda - \lambda_j)$. Thus $E_1 = c_1 Q_1(L)$. Similar remarks apply to the other eigenspaces, and therefore the required decomposition of the identity operator has to be of the form $I = c_1 Q_1(L) + \cdots + c_k Q_k(L)$ with $c_1, \ldots, c_k$ equal to certain scalars.

The polynomials $Q_1(\lambda), \ldots, Q_l(\lambda)$ are at hand from the start, each containing all but one factor of the minimal polynomial. Moreover, $i \neq j$ implies that

$$Q_i(L) Q_j(L) = \Big(\prod_{l=1}^{k} (L - \lambda_l I) \Big) \Big(\prod_{l \neq i, j} (L - \lambda_l I) \Big).$$

The first factor on the right side is the value of the minimal polynomial of L with L substituted for λ. Hence the right side is 0, and we see that our linear maps $E_1, \ldots, E_k$ have $E_i E_j = 0$ for $i \neq j$.

As soon as we allow nonconstant coefficients in place of the c_j's in the above argument, we obtain a generalization of Theorem 5.14 to the situation that the minimal polynomial of L is arbitrary. The prime factors of the minimal polynomial need not even be of degree 1. Hence the theorem applies to all L's even if $\mathbb{K}$ is not algebraically closed.

Theorem 5.19 (Primary Decomposition Theorem). Let $L : V \to V$ be linear on a finite-dimensional vector space over $\mathbb{K}$, and let $M(\lambda) = P_1(\lambda)^{l_1} \cdots P_k(\lambda)^{l_k}$ be the unique factorization of the minimal polynomial $M(\lambda)$ of L into the product of powers of distinct monic prime polynomials $P_j(\lambda)$. Define $U_j = \ker(P_j(L)^{l_j})$ for $1 \leq j \leq k$. Then

(a) $V = U_1 \oplus \cdots \oplus U_k$,
(b) the projection E_j of V on U_j along the sum of the other U_i's is of the form $T_j(L)$ for some polynomial T_j,
(c) each vector subspace U_j is invariant under L,
(d) any linear map from V to itself that commutes with L carries each U_j into itself,
(e) any vector subspace W invariant under L has the property that
$$W = (W \cap U_1) \oplus \cdots \oplus (W \cap U_k),$$
(f) the minimal polynomial of $L_j = L\big|_{U_j}$ is $P_j(\lambda)^{l_j}$.

REMARKS. The decomposition in (a) is called the **primary decomposition** of V under L, and the vector subspaces U_j are called the **primary subspaces** of V under L.

PROOF. For $1 \leq j \leq k$, define $Q_j(\lambda) = M(\lambda)/P_j(\lambda)^{l_j}$. The ideal in $\mathbb{K}[\lambda]$ generated by $Q_1(\lambda), \ldots, Q_k(\lambda)$ consists of all products of a single monic polynomial $D(\lambda)$ by arbitrary polynomials, according to Proposition 5.8, and $D(\lambda)$ has to divide each $Q_j(\lambda)$. Since $Q_j(\lambda) = \prod_{i \neq j} P_i(\lambda)^{l_i}$, $D(\lambda)$ cannot be divisible by any $P_j(\lambda)$, and consequently $D(\lambda) = 1$. Thus there exist polynomials $R_1(\lambda), \ldots, R_k(\lambda)$ such that

$$1 = Q_1(\lambda) R_1(\lambda) + \cdots + Q_k(\lambda) R_k(\lambda).$$

Define $E_j = Q_j(L) R_j(L)$, so that $E_1 + \cdots + E_k = I$. If $i \neq j$, then $Q_i(\lambda) Q_j(\lambda) = M(\lambda) \prod_{r \neq i, j} P_r(\lambda)^{l_r}$. Since $M(L) = 0$, we see that $E_i E_j = 0$.

Proposition 5.17 says that each E_j is a projection. Also, it says that if U_j denotes image E_j, then $V = U_1 \oplus \cdots \oplus U_k$, and E_j is the projection on U_j along the sum of the other U_i's. With this definition of the U_j's (rather than the one in the statement of the theorem), we have therefore shown that (a) and (b) hold.

Let us see that conclusions (c), (d), and (e) follow from (b). Conclusion (c) holds by Proposition 5.18 since L commutes with $T_j(L)$ whenever T_j is a polynomial. For (d), if $J : V \to V$ is a linear map commuting with L, then J commutes with each E_j since (b) shows that each E_j is of the form $T_j(L)$. From Proposition 5.18 we conclude that each U_j is invariant under J. For (e), the subspace W certainly contains $(W \cap U_1) \oplus \cdots \oplus (W \cap U_k)$. For the reverse containment suppose w is in W. Since E_j is of the form $T_j(L)$ and since W is invariant under L, $E_j(w)$ is in W. But also $E_j(w)$ is in U_j. Therefore the expansion $w = \sum_j E_j(w)$ exhibits w as the sum of members of the spaces $W \cap U_j$.

Next let us prove that U_j, as we have defined it, is given also by the definition in the statement of the theorem. In other words, let us prove that

$$\text{image } E_j = \ker(P_j(L)^{l_j}). \tag{$*$}$$

We need a preliminary fact. The polynomial $P_j(\lambda)^{l_j}$ has the property that $M(\lambda) = P_j(\lambda)^{l_j} Q_j(\lambda)$. Hence $P_j(L)^{l_j} Q_j(L) = M(L) = 0$. Multiplying by $R_j(L)$, we obtain

$$P_j(L)^{l_j} E_j = 0. \tag{$**$}$$

Now suppose that v is in image E_j. Then $P_j(L)^{l_j}(v) = P_j(L)^{l_j} E_j(v) = 0$ by $(**)$, and hence image $E_j \subseteq \ker(P_j(L)^{l_j})$. For the reverse inclusion, let v be in $\ker(P_j(L)^{l_j})$. For $i \neq j$, $Q_i(\lambda) R_i(\lambda) = \left(\prod_{r \neq i, j} P_r(\lambda)^{l_r}\right) R_i(\lambda) P_j(\lambda)^{l_j}$ and hence

$$E_i(v) = \left(\prod_{r \neq i, j} P_r(L)^{l_r}\right) R_i(L) P_j(L)^{l_j}(v) = 0.$$

Writing $v = E_1(v) + \cdots + E_k(v)$, we see that $v = E_j(v)$. Thus $\ker(P_j(L)^{l_j}) \subseteq$ image E_j. Therefore $(*)$ holds, and U_j is as in the statement of the theorem.

Finally let us prove (f). Let $M_j(\lambda)$ be the minimal polynomial of $L_j = L\big|_{U_j}$. From $(**)$ we see that $P_j(L_j)^{l_j} = 0$. Hence $M_j(\lambda)$ divides $P_j(\lambda)^{l_j}$. For the reverse divisibility we have $M_j(L_j) = 0$. Then certainly $M_j(L_j) Q_j(L_j) R_j(L_j)$, which equals $M_j(L) E_j$ on U_j, is 0 on U_j. Consider $M_j(L) E_j$ on $U_i = $ image E_i when $i \neq j$. Since $E_j E_i = 0$, $M_j(L) E_j$ equals 0 on all U_i other than U_j. We conclude that $M_j(L) E_j$ equals 0 on V, i.e., $M_j(L) Q_j(L) R_j(L) = 0$. Since $M(\lambda)$ is the minimal polynomial of L, $M(\lambda)$ divides

$$M_j(\lambda) Q_j(\lambda) R_j(\lambda) = M_j(\lambda)\Big(1 - \sum_{i \neq j} Q_i(\lambda) R_i(\lambda)\Big), \tag{$\dagger$}$$

and the factor $P_j(\lambda)^{l_j}$ of $M(\lambda)$ must divide the right side of $(\dagger)$. On that right side, $P_j(\lambda)^{l_j}$ divides each $Q_i(\lambda)$ with $i \neq j$. Since $P_j(\lambda)$ does not divide 1, $P_j(\lambda)$

does not divide the factor $1 - \sum_{i \neq j} Q_i(\lambda) R_i(\lambda)$. Since $P_j(\lambda)$ is prime, $P_j(\lambda)^{l_j}$ and $1 - \sum_{i \neq j} Q_i(\lambda) R_i(\lambda)$ are relatively prime. We know that $P_j(\lambda)^{l_j}$ divides the product of $M_j(\lambda)$ and $1 - \sum_{i \neq j} Q_i(\lambda) R_i(\lambda)$, and consequently $P_j(\lambda)^{l_j}$ divides $M_j(\lambda)$. This proves the reverse divisibility and completes the proof of (f). □

6. Jordan Canonical Form

Now we can return to the canonical-form problem for similarity of square matrices and isomorphism of linear maps from a finite-dimensional vector space to itself. The answer obtained in this section will solve the problem completely if $\mathbb{K}$ is algebraically closed but only partially if $\mathbb{K}$ fails to be algebraically closed. Problems 32–40 at the end of the chapter extend the content of this section to give a complete answer for general $\mathbb{K}$.

The present theorem is most easily stated in terms of matrices. A square matrix is called a **Jordan block** if it is of the form

$$\begin{pmatrix} c & 1 & 0 & 0 & \cdots & 0 & 0 \\ & c & 1 & 0 & \cdots & 0 & 0 \\ & & c & 1 & \cdots & 0 & 0 \\ & & & \ddots & \ddots & \vdots & \vdots \\ & & & & c & 1 & 0 \\ & & & & & c & 1 \\ & & & & & & c \end{pmatrix},$$

of some size and for some c in $\mathbb{K}$, as in Example 2 of Section 3, with 0 everywhere below the diagonal. A square matrix is in **Jordan form**, or **Jordan normal form**, if it is block diagonal and each block is a Jordan block. One can insist on grouping the blocks for which the constant c is the same and arranging the blocks for given c in some order, but these refinements are inessential.

Theorem 5.20 (Jordan canonical form).

(a) If the field $\mathbb{K}$ is algebraically closed, then every square matrix over $\mathbb{K}$ is similar to a matrix in Jordan form, and two matrices in Jordan form are similar to each other if and only if their Jordan blocks can be permuted so as to match exactly.

(b) For a general field $\mathbb{K}$, a square matrix A is similar to a matrix in Jordan form if and only if each prime factor of its minimal polynomial has degree 1. Two matrices in Jordan form are similar to each other if and only if their Jordan blocks can be permuted so as to match exactly.

The first step in proving existence of a matrix in Jordan form similar to a given matrix is to use the Primary Decomposition Theorem (Theorem 5.19). We think of the matrix A as operating on the space $\mathbb{K}^n$ of column vectors in the usual way. The primary subspaces are uniquely defined vector subspaces of $\mathbb{K}^n$, and we introduce an ordered basis, yet to be specified in full detail, within each primary subspace. The union of these ordered bases gives an ordered basis of $\mathbb{K}^n$, and we change from the standard basis to this one. The result is that the given matrix has been conjugated so that its appearance is block diagonal, each block having minimal polynomial equal to a power of a prime polynomial and the prime polynomials all being different. Let us call these blocks **primary blocks**. The effect of Theorem 5.19 has been to reduce matters to a consideration of each primary block separately. The hypothesis either that $\mathbb{K}$ is algebraically closed or, more generally, that the prime divisors of the minimal polynomial all have degree 1 means that the minimal polynomial of the primary block under study may be taken to be $(\lambda - c)^l$ for some c in $\mathbb{K}$ and some integer $l \geq 1$. In terms of Jordan form, we have isolated, for each c in $\mathbb{K}$, what will turn out to be the subspace of $\mathbb{K}^n$ corresponding to Jordan blocks with c in every diagonal entry.

Let us write B for a primary block with minimal polynomial $(\lambda - c)^l$. We certainly have $(B - cI)^l = 0$, and it follows that the matrix $N = B - cI$ has $N^l = 0$. A matrix N with $N^l = 0$ for some integer $l \geq 0$ is said to be **nilpotent**. To prove the existence part of Theorem 5.20, it is enough to prove the following theorem.

Theorem 5.21. For any field $\mathbb{K}$, each nilpotent matrix N in $M_n(\mathbb{K})$ is similar to a matrix in Jordan form.

The proof of Theorem 5.21 and of the uniqueness statements in Theorem 5.20 will occupy the remainder of this section. It is implicit in Theorem 5.21 that a nilpotent matrix in $M_n(\mathbb{K})$ has 0 as a root of its characteristic polynomial with multiplicity n, in particular that the only prime polynomials dividing the characteristic polynomial are the ones dividing the minimal polynomial. We proved such a fact about divisibility earlier for general square matrices when the prime factor has degree 1, but we did not give a proof for general degree. We pause for a moment to give a direct proof in the nilpotent case.

Lemma 5.22. If N is a nilpotent matrix in $M_n(K)$, then N has characteristic polynomial λ^n and satisfies $N^n = 0$.

PROOF. If $N^l = 0$, then
$$(\lambda I - N)(\lambda^{l-1}I + \lambda^{l-2}N + \cdots + \lambda^2 N^{l-3} + \lambda N^{l-2} + N^{l-1}) = \lambda^l I - N^l = \lambda^l I.$$
Taking determinants and using Proposition 5.1 in the ring $R = \mathbb{K}[\lambda]$, we obtain
$$\det(\lambda I - N)\det(\text{other factor}) = \det(\lambda^l I) = \lambda^{ln}.$$

Thus $\det(\lambda I - N)$ divides λ^{ln}. By unique factorization in $\mathbb{K}[\lambda]$, $\det(\lambda I - N)$ is a constant times a power of λ. Then we must have $\det(\lambda I - N) = \lambda^n$. Applying the Cayley–Hamilton Theorem (Theorem 5.9), we obtain $N^n = 0$. □

Let us now prove the uniqueness statements in Theorem 5.20; this step will in fact help orient us for the proof of Theorem 5.21. In (b), one thing we are to prove is that if A is similar to a matrix in Jordan form, then every prime polynomial dividing the minimal polynomial has degree 1. Since characteristic and minimal polynomials are unchanged under similarity, we may assume that A is itself in Jordan form. The characteristic and minimal polynomials of A are computed in the four examples of Section 3. Since the minimal polynomial is the product of polynomials of degree 1, the only primes dividing it have degree 1.

In both (a) and (b) of Theorem 5.20, we are to prove that the Jordan form is unique up to permutation of the Jordan blocks. The matrix A determines its characteristic polynomial, which determines the roots of the characteristic polynomial, which are the diagonal entries of the Jordan form. Thus the sizes of the primary blocks within the Jordan form are determined by A. Within each primary block, we need to see that the sizes of the various Jordan blocks are completely determined.

Thus we may assume that N is nilpotent and that $C^{-1}NC = J$ is in Jordan form with 0's on the diagonal. Although we shall make statements that apply in all cases, the reader may be helped by referring to the particular matrix J in Figure 5.1 and its powers in Figure 5.2.

$$J = \begin{pmatrix} 0 & 1 & 0 & 0 & & & & & & \\ 0 & 0 & 1 & 0 & & & & & & \\ 0 & 0 & 0 & 1 & & & & & & \\ 0 & 0 & 0 & 0 & & & & & & \\ & & & & 0 & 1 & 0 & & & \\ & & & & 0 & 0 & 1 & & & \\ & & & & 0 & 0 & 0 & & & \\ & & & & & & & 0 & 1 & \\ & & & & & & & 0 & 0 & \\ & & & & & & & & & 0 & 1 \\ & & & & & & & & & 0 & 0 \\ & & & & & & & & & & & 0 \end{pmatrix}.$$

FIGURE 5.1. Example of a nilpotent matrix in Jordan form.

Each block of the Jordan form J contributes 1 to the dimension of the kernel (or null space really) of J via the first column of the block, and hence

$$\dim(\ker J) = \#\{\text{Jordan blocks in } J\}.$$

In Figure 5.1 this number is 5.

$$J^2 = \begin{pmatrix} 0&0&1&0 \\ 0&0&0&1 \\ 0&0&0&0 \\ 0&0&0&0 \\ & & & & 0&0&1 \\ & & & & 0&0&0 \\ & & & & 0&0&0 \\ & & & & & & & 0&0 \\ & & & & & & & 0&0 \\ & & & & & & & & & 0&0 \\ & & & & & & & & & 0&0 \\ & & & & & & & & & & & 0 \end{pmatrix} \quad \text{and} \quad J^3 = \begin{pmatrix} 0&0&0&1 \\ 0&0&0&0 \\ 0&0&0&0 \\ 0&0&0&0 \\ & & & & 0&0&0 \\ & & & & 0&0&0 \\ & & & & 0&0&0 \\ & & & & & & & 0&0 \\ & & & & & & & 0&0 \\ & & & & & & & & & 0&0 \\ & & & & & & & & & 0&0 \\ & & & & & & & & & & & 0 \end{pmatrix}$$

FIGURE 5.2. Powers of the nilpotent matrix in Figure 5.1.

When J is squared, the 1's in J move up and to the right one more step beyond the diagonal except that blocks of size 2 become 0. When J is cubed, the 1's in J move up and to the right one further step except that blocks of size 3 become 0. Each time J is raised to a new power one higher than before, each block that is nonzero in the old power contributes an additional 1 to the dimension of the kernel. Thus we have

$$\dim(\ker J^2) - \dim(\ker J) = \#\{\text{Jordan blocks of size} \geq 2\}$$

and $\dim(\ker J^3) - \dim(\ker J^2) = \#\{\text{Jordan blocks of size} \geq 3\};$

in the general case,

$$\dim(\ker J^k) - \dim(\ker J^{k-1}) = \#\{\text{Jordan blocks of size} \geq k\} \qquad \text{for } k \geq 1.$$

Lemma 5.22 says that $J^k = 0$ when k is $\geq$ the size of J, and the differences need not be computed beyond that point.

For Figure 5.2 the values by inspection are $\dim(\ker J^2) = 9$ and $\dim(\ker J^3) = 11$; also $J^4 = 0$ and hence $\dim(\ker J^4) = 12$. The numbers of Jordan blocks of size $\geq k$ for $k = 1, 2, 3, 4$ are 5, 4, 2, 1, and these numbers indeed match the differences $5 - 0$, $9 - 5$, $11 - 9$, $12 - 11$, as predicted by the above formula.

Since $C^{-1}NC = J$, we have $C^{-1}N^kC = J^k$ and $N^kC = CJ^k$. The matrix C is invertible, and therefore $\dim(\ker J^k) = \dim(\ker CJ^k) = \dim(\ker N^kC) = \dim(\ker N^k)$. Hence

$$\dim(\ker N^k) - \dim(\ker N^{k-1}) = \#\{\text{Jordan blocks of size} \geq k\} \qquad \text{for } k \geq 1,$$

and the number of Jordan blocks of each size is uniquely determined by properties of N. This completes the proof of all the uniqueness statements in Theorem 5.20.

6. Jordan Canonical Form

Now let us turn to the proof of Theorem 5.21, first giving the idea. The argument involves a great many choices, and it may be helpful to understand it in the context of Figures 5.1 and 5.2. Let $\Sigma = (e_1, \ldots, e_{12})$ be the standard ordered basis of $\mathbb{K}^{12}$. The matrix J, when operating by multiplication on the left, moves basis vectors to other basis vectors or to 0. Namely,

$$Je_1 = 0, \quad Je_2 = e_1, \quad Je_3 = e_2, \quad Je_4 = e_3,$$
$$Je_5 = 0, \quad Je_6 = e_5, \quad Je_7 = e_6,$$
$$Je_8 = 0, \quad Je_9 = e_8,$$
$$Je_{10} = 0, \quad Je_{11} = e_{10},$$
$$Je_{12} = 0,$$

with each line describing what happens for a single Jordan block. Let us think of the given nilpotent matrix N as equal to $\begin{pmatrix} L \\ \Sigma\Sigma \end{pmatrix}$ for some linear map L. We want to find a new ordered basis $\Gamma = (v_1, \ldots, v_{12})$ in which the matrix of L is J. In the expression $C^{-1}NC = J$, the matrix C equals $\begin{pmatrix} I \\ \Sigma\Gamma \end{pmatrix}$, and its columns are expressions for $v_1, \ldots, v_{12}$ in the basis Σ, i.e., $Ce_i = v_i$. For each index i, we have $Je_i = Je_{i-1}$ or $Je_i = 0$. The formula $NC = CJ$, when applied to e_i, therefore says that

$$Nv_i = NCe_i = CJe_i = \begin{cases} Ce_{i-1} = v_{i-1} & \text{if } Je_i = e_{i-1}, \\ 0 & \text{if } Je_i = 0. \end{cases}$$

Thus we are looking for an ordered basis such that N sends each member of the basis either into the previous member or into 0. The procedure in this example will be to pick out v_4 as a vector not annihilated by N^3, obtain v_3, v_2, v_1, from it by successively applying N, pick out v_7 as a vector not annihilated by N^2 and independent of what has been found, obtain v_6, v_5 from it by successively applying N, and so on. It is necessary to check that the appropriate linear independence can be maintained, and that step will be what the proof is really about.

The proof of Theorem 5.21 will now be given in the general case. The core of the argument concerns linear maps and appears as three lemmas. Afterward the results of the lemmas will be interpreted in terms of matrices. For all the lemmas let V be an n-dimensional vector space over $\mathbb{K}$, and let $N : V \to V$ be linear with $N^n = 0$. Define $K_j = \ker N^j$, so that

$$0 = K_0 \subseteq K_1 \subseteq K_2 \subseteq \cdots \subseteq K_n = V.$$

Lemma 5.23. Suppose $j \geq 1$ and suppose S_j is any vector subspace of V such that $K_{j+1} = K_j \oplus S_j$. Then N is one-one from S_j into K_j and $N(S_j) \cap K_{j-1} = 0$.

PROOF. Since $N(\ker N^{j+1}) \subseteq \ker N^j$, we obtain $N(S_j) \subseteq K_j$; thus N indeed sends S_j into K_j. To see that N is one-one from S_j into K_j, suppose that s is a member of S_j with $N(s) = 0$. Then s is in K_1. Since $j \geq 1$, $K_1 \subseteq K_j$. Thus s is in K_j. Since $K_j \cap S_j = 0$, s is 0. Hence N is one-one from S_j into K_j. To see that $N(S_j) \cap K_{j-1} = 0$, suppose s is a member of S_j with $N(s)$ in K_{j-1}. Then $0 = N^{j-1}(N(s)) = N^j(s)$ shows that s is in K_j. Since $K_j \cap S_j = 0$, s equals 0. □

Lemma 5.24. Define $U_n = W_n = 0$. For $0 \leq j \leq n - 1$, there exist vector subspaces U_j and W_j of K_{j+1} such that

$$K_{j+1} = K_j \oplus U_j \oplus W_j,$$
$$U_j = N(U_{j+1} \oplus W_{j+1}),$$

and $\qquad N : U_{j+1} \oplus W_{j+1} \to U_j \quad \text{is one-one.}$

PROOF. Define $U_{n-1} = N(U_n \oplus W_n) = 0$, and let W_{n-1} be a vector subspace such that $V = K_n = K_{n-1} \oplus W_{n-1}$. Put $S_{n-1} = U_{n-1} \oplus W_{n-1}$. Proceeding inductively downward, suppose that $U_n, U_{n-1}, \ldots, U_{j+1}, W_n, W_{n-1}, \ldots, W_{j+1}$ have been defined so that $U_k = N(U_{k+1} \oplus W_{k+1})$, $N : U_{k+1} \oplus W_{k+1} \to U_k$ is one-one, and $K_{k+1} = K_k \oplus U_k \oplus W_k$ whenever k satisfies $j < k \leq n - 1$. We put $S_k = U_k \oplus W_k$ for these values of k, and then S_k satisfies the hypothesis of Lemma 5.23 whenever k satisfies $j < k \leq n - 1$. We now construct U_j and W_j. We put $U_j = N(S_{j+1})$. Since S_{j+1} satisfies the hypothesis of Lemma 5.23, we see that $U_j \subseteq K_{j+1}$, N is one-one from S_{j+1} into U_j, and $U_j \cap K_j = 0$. Thus we can find a vector subspace W_j with $K_{j+1} = K_j \oplus U_j \oplus W_j$, and the inductive construction is complete. □

Lemma 5.25. The vector subspaces of Lemma 5.24 satisfy

$$V = U_0 \oplus W_0 \oplus U_1 \oplus W_1 \oplus \cdots \oplus U_{n-1} \oplus W_{n-1}.$$

PROOF. Iterated use of Lemma 5.24 gives

$$V = K_n = K_{n-1} \oplus (U_{n-1} \oplus W_{n-1})$$
$$= K_{n-2} \oplus (U_{n-2} \oplus W_{n-2}) \oplus (U_{n-1} \oplus W_{n-1})$$
$$= \cdots = K_0 \oplus (U_0 \oplus W_0) \oplus \cdots \oplus (U_{n-1} \oplus W_{n-1})$$
$$= (U_0 \oplus W_0) \oplus \cdots \oplus (U_{n-1} \oplus W_{n-1}),$$

the last step holding since $K_0 = 0$, K_0 being the kernel of the identity function. □

PROOF OF THEOREM 5.21. We regard N as acting on $V = \mathbb{K}^n$ by multiplication on the left, and we describe an ordered basis in which the matrix of N is in Jordan form. For $0 \leq j \leq n-1$, form a basis of the vector subspace W_j of Lemma 5.24, and let $v^{(j)}$ be a typical member of this basis. Each $v^{(j)}$ will be used as the last basis vector corresponding to a Jordan block of size $j+1$. The full ordered basis for that Jordan block will therefore be $N^j v^{(j)}, N^{j-1} v^{(j)}, \ldots, N v^{(j)}, v^{(j)}$. The theorem will be proved if we show that the union of these sets as j and $v^{(j)}$ vary is a basis of $\mathbb{K}^n$ and that $N^{j+1} v^{(j)} = 0$ for all j and $v^{(j)}$.

From the first conclusion of Lemma 5.24 we see for $j \geq 0$ that $W_j \subseteq K_{j+1}$, and hence $N^{j+1}(W_j) = 0$. Therefore $N^{j+1} v^{(j)} = 0$ for all j and $v^{(j)}$.

Let us prove by induction downward on j that a basis of $U_j \oplus W_j$ consists of all $v^{(j)}$ and all $N^k v^{(j+k)}$ for $k > 0$. The base case of the induction is $j = n-1$, and the statement holds in that case since $U_{n-1} = 0$ and since the vectors $v^{(n-1)}$ form a basis of W_{n-1}. The inductive hypothesis is that all $v^{(j+1)}$ and all $N^k v^{(j+1+k)}$ for $k > 0$ together form a basis of $U_{j+1} \oplus W_{j+1}$. The second and third conclusions of Lemma 5.24 together show that all $N v^{(j+1)}$ and all $N^{k+1} v^{(j+1+k)}$ for $k > 0$ together form a basis of U_j. In other words, all $N^k v^{(j+k)}$ with $k > 0$ together form a basis of U_j. The vectors $v^{(j)}$ by construction form a basis of W_j, and $U_j \cap W_j = 0$. Therefore the union of these separate bases is a basis for $U_j \oplus W_j$, and the induction is complete.

Taking the union of the bases of $U_j \oplus W_j$ for all j and applying Lemma 5.25, we see that we have a basis of $V = \mathbb{K}^n$. This shows that the desired set is a basis of $\mathbb{K}^n$ and completes the proof of Theorem 5.21. $\square$

7. Computations with Jordan Form

Let us illustrate the computation of Jordan form and the change-of-basis matrix with a few examples. We are given a matrix A and we seek J and C with $J = C^{-1} A C$. We regard A as the matrix of some linear L in the standard ordered basis Σ, and we regard J as the matrix of L in some other ordered basis Γ. Then $C = \begin{pmatrix} I \\ \Sigma \Gamma \end{pmatrix}$, and so the columns of C give the members of Γ written as ordinary column vectors (in the standard ordered basis).

EXAMPLE 1. This example will be a nilpotent matrix, and we shall compute J and C merely by interpreting the proof of Theorem 5.21 in concrete terms. Let

$$A = \begin{pmatrix} -1 & 1 & 0 \\ -1 & 1 & 0 \\ -1 & 1 & 0 \end{pmatrix}.$$

The first step is to compute the characteristic polynomial, which is

$$\det(\lambda I - A) = \det \begin{pmatrix} \lambda+1 & -1 & 0 \\ 1 & \lambda-1 & 0 \\ 1 & -1 & \lambda \end{pmatrix} = \lambda \det \begin{pmatrix} \lambda+1 & -1 \\ 1 & \lambda-1 \end{pmatrix} = \lambda^3.$$

Then $A^3 = 0$ by the Cayley–Hamilton Theorem (Theorem 5.9), and A is indeed nilpotent. The diagonal entries of J are thus all 0, and we have to compute the sizes of the various Jordan blocks. To do so, we compute the dimension of the kernel of each power of A. The dimension of the kernel of a matrix equals the number of independent variables when we solve $AX = 0$ by row reduction. With the first power of A, the variable x_1 is dependent, and x_2 and x_3 are independent. Also, $A^2 = 0$. Thus

$$\dim(\ker A^0) = 0, \quad \dim(\ker A) = 2, \quad \text{and} \quad \dim(\ker A^2) = 3.$$

Hence

$$\#\{\text{Jordan blocks of size } \geq 1\} = \dim(\ker A) - \dim(\ker A^0) = 2 - 0 = 2,$$
$$\#\{\text{Jordan blocks of size } \geq 2\} = \dim(\ker A^2) - \dim(\ker A) = 3 - 2 = 1.$$

From these equalities we see that one Jordan block has size 2 and the other has size 1. Thus

$$J = \begin{pmatrix} 0 & 1 & \\ 0 & 0 & \\ & & 0 \end{pmatrix}.$$

We want to set up vector subspaces as in Lemma 5.24 so that $K_{j+1} = K_j \oplus U_j \oplus W_j$ and $U_j = A(U_{j+1} \oplus W_{j+1})$ for $0 \leq j \leq 2$. Since $K_3 = K_2$, the equations begin with $K_2 = \cdots$ and are

$$K_2 = K_1 \oplus 0 \oplus W_1, \quad U_0 = A(0 \oplus W_1), \quad K_1 = K_0 \oplus U_0 \oplus W_0.$$

Here $K_2 = \mathbb{K}^3$ and K_1 is the subspace of all $X = \begin{pmatrix} x_1 \\ x_2 \\ x_3 \end{pmatrix}$ such that $AX = 0$. The space W_1 is to satisfy $K_2 = K_1 \oplus W_1$, and we see that W_1 is 1-dimensional. Let $\{v^{(1)}\}$ be a basis of the 1-dimensional vector subspace W_1. Then U_0 is 1-dimensional with basis $\{Av^{(1)}\}$. The subspace K_1 is 2-dimensional and contains U_0. The space W_0 is to satisfy $K_1 = U_0 \oplus W_0$, and we see that W_0 is 1-dimensional. Let $\{v^{(0)}\}$ be a basis of W_0. Then the respective columns of C may be taken to be

$$Av^{(1)}, \quad v^{(1)}, \quad v^{(0)}.$$

Let us compute these vectors.

7. Computations with Jordan Form   237

If we extend a basis of K_1 to a basis of K_2, then W_1 may be taken to be the linear span of the added vector. To obtain a basis of K_1, we compute that the reduced row-echelon form of A is $\begin{pmatrix} 1 & -1 & 0 \\ 0 & 0 & 0 \\ 0 & 0 & 0 \end{pmatrix}$, and the resulting system consists of the single equation $x_1 - x_2 = 0$. Thus $x_1 = x_2$, and

$$\begin{pmatrix} x_1 \\ x_2 \\ x_3 \end{pmatrix} = x_2 \begin{pmatrix} 1 \\ 1 \\ 0 \end{pmatrix} + x_3 \begin{pmatrix} 0 \\ 0 \\ 1 \end{pmatrix}.$$

The coefficients of x_2 and x_3 on the right side form a basis of K_1, and we are to choose a vector that is not a linear combination of these. Thus we can take $v^{(1)} = \begin{pmatrix} 1 \\ 0 \\ 0 \end{pmatrix}$ as the basis vector of W_1. Then $U_0 = A(W_1)$ has $Av^{(1)} = A\begin{pmatrix} 1 \\ 0 \\ 0 \end{pmatrix} = \begin{pmatrix} -1 \\ -1 \\ -1 \end{pmatrix}$ as a basis, and the basis of W_0 may be taken as any vector in K_1 but not U_0. We can take this basis to consist of $v^{(0)} = \begin{pmatrix} 0 \\ 0 \\ 1 \end{pmatrix}$.

Lining up our three basis vectors as the columns of C gives us $C = \begin{pmatrix} -1 & 1 & 0 \\ -1 & 0 & 0 \\ -1 & 0 & 1 \end{pmatrix}$. Computation gives $C^{-1} = \begin{pmatrix} 0 & -1 & 0 \\ 1 & -1 & 0 \\ 0 & -1 & 1 \end{pmatrix}$, and we readily check that $C^{-1}AC = J$.

EXAMPLE 2. We continue with A and J as in Example 1, but we compute the columns of C without directly following the proof of Theorem 5.21. The method starts from the fact that each Jordan block corresponds to a 1-dimensional space of eigenvectors, and then we backtrack to find vectors corresponding to the other columns. For this particular A, we know that the three columns of C are to be of the form $v_1 = Av^{(1)}$, $v_2 = v^{(1)}$, and $v_3 = v^{(0)}$. The vectors v_1 and v_3 together span the 0 eigenspace of A. We find all the 0 eigenvectors, writing them as a two-parameter family. This eigenspace is just $K_1 = \ker A$, and we found in Example 1 that $K_1 = \left\{ \begin{pmatrix} x_2 \\ x_2 \\ x_3 \end{pmatrix} \right\}$. One of these vectors is to be v_1, and it has to equal Av_2. Thus we solve $Av_2 = \begin{pmatrix} x_2 \\ x_2 \\ x_3 \end{pmatrix}$. Applying the solution procedure yields

$$\left(\begin{array}{ccc|c} 1 & -1 & 0 & -x_2 \\ 0 & 0 & 0 & 0 \\ 0 & 0 & 0 & x_3 - x_2 \end{array} \right).$$

This system has no solutions unless $x_3 - x_2 = 0$. If we take $x_2 = x_3 = -1$, then we obtain the same first two columns of C as in Example 1, and any vector in K_1 independent of $\begin{pmatrix} -1 \\ -1 \\ -1 \end{pmatrix}$ may be taken as the third column.

EXAMPLE 3. Let

$$A = \begin{pmatrix} 2 & 1 & 0 \\ -1 & 4 & 0 \\ -1 & 2 & 2 \end{pmatrix}.$$

Direct calculation shows that the characteristic polynomial is $\det(\lambda I - A) = \lambda^3 - 8\lambda^2 + 21\lambda - 18 = (\lambda - 2)(\lambda - 3)^2$. The possibilities for J are therefore

$$\begin{pmatrix} 3 & 0 & 0 \\ 0 & 3 & 0 \\ 0 & 0 & 2 \end{pmatrix} \quad \text{and} \quad \begin{pmatrix} 3 & 1 & 0 \\ 0 & 3 & 0 \\ 0 & 0 & 2 \end{pmatrix};$$

the first one will be correct if the dimension of the eigenspace for the eigenvalue 3 is 2, and the second one will be correct if that dimension is 1.

The third column of C corresponds to an eigenvector for the eigenvalue 2, hence to a nonzero solution of $(A - 2I)v = 0$. The solutions are $v = k\begin{pmatrix} 0 \\ 0 \\ 1 \end{pmatrix}$, and we can therefore use $\begin{pmatrix} 0 \\ 0 \\ 1 \end{pmatrix}$.

For the first two columns of C, we have to find $\ker(A - 3I)$ no matter which of the methods we use, the one in Example 1 or the one in Example 2. Solving the system of equations, we obtain all vectors in the space $\left\{ z \begin{pmatrix} 1 \\ 1 \\ 1 \end{pmatrix} \right\}$. The dimension of the space is 1, and the second possibility for the Jordan form is the correct one.

Following the method of Example 1 to find the columns of C means that we pick a basis of this kernel and extend it to a basis of $\ker(A - 3I)^2$. A basis of $\ker(A - 3I)$ consists of the vector $\begin{pmatrix} 1 \\ 1 \\ 1 \end{pmatrix}$. The matrix $(A - 3I)^2$ is $\begin{pmatrix} 0 & 0 & 0 \\ 0 & 0 & 0 \\ 0 & -1 & 1 \end{pmatrix}$, and the solution procedure leads to the formula

$$\begin{pmatrix} a \\ b \\ c \end{pmatrix} = a \begin{pmatrix} 1 \\ 0 \\ 0 \end{pmatrix} + c \begin{pmatrix} 0 \\ 1 \\ 1 \end{pmatrix}$$

for its kernel. The vector $\begin{pmatrix} 1 \\ 1 \\ 1 \end{pmatrix}$ arises from $a = 1$ and $c = 1$. We are to make an independent choice, say $a = 1$ and $c = 0$. Then the second basis vector to use is $\begin{pmatrix} 1 \\ 0 \\ 0 \end{pmatrix}$. This becomes the second column of C, and the first column then has to be

$(A - 3I)\begin{pmatrix} 1 \\ 0 \\ 0 \end{pmatrix} = \begin{pmatrix} -1 \\ -1 \\ -1 \end{pmatrix}$. The result is that $C = \begin{pmatrix} -1 & 1 & 0 \\ -1 & 0 & 0 \\ -1 & 0 & 1 \end{pmatrix}$.

Following the method of Example 2 for this example means that we retain the entire kernel of $A - 3I$, namely all vectors $v_1 = z\begin{pmatrix} 1 \\ 1 \\ 1 \end{pmatrix}$, as candidates for the first column of C. The second column is to satisfy $(A - 3I)v_2 = v_1$. Solving leads to $v_2 = z \begin{pmatrix} -1 \\ 0 \\ 0 \end{pmatrix} + c \begin{pmatrix} 1 \\ 1 \\ 1 \end{pmatrix}$. In contrast to Example 2, there is no potential contradictory equation. So we choose z and then c. If we take $z = 1$ and $c = 0$, we find that the first two columns of C are to be $\begin{pmatrix} 1 \\ 1 \\ 1 \end{pmatrix}$ and $\begin{pmatrix} -1 \\ 0 \\ 0 \end{pmatrix}$. Then

$$C = \begin{pmatrix} 1 & -1 & 0 \\ 1 & 0 & 0 \\ 1 & 0 & 1 \end{pmatrix}.$$

For any example in which we can factor the characteristic polynomial exactly, either of the two methods used above will work. The first method appears complicated but uses numbers throughout; it tends to be more efficient with large examples involving high-degree minimal polynomials. The second method appears direct but requires solving equations with symbolic variables; it tends to be more efficient for relatively simple examples.

8. Problems

In Problems 1–25 all vector spaces are assumed finite-dimensional, and all linear transformations are assumed defined from such spaces into themselves. Unless information is given to the contrary, the underlying field $\mathbb{K}$ is assumed arbitrary.

1. Let $M_{mn}(\mathbb{C})$ be the vector space of m-by-n complex matrices. The group $\text{GL}(m, \mathbb{C}) \times \text{GL}(n, \mathbb{C})$ acts on $M_{mn}(\mathbb{C})$ by $((g, h), x) \mapsto gxh^{-1}$, where gxh^{-1} denotes a matrix product.
   (a) Verify that this is indeed a group action.
   (b) Prove that two members of $M_{mn}(\mathbb{C})$ lie in the same orbit if and only if they have the same rank.
   (c) For each possible rank, give an example of a member of $M_{mn}(\mathbb{C})$ with that rank.

2. Prove that a member of $M_n(\mathbb{K})$ is invertible if and only if the constant term of its minimal polynomial is different from 0.

3. Suppose that $L : V \to V$ is a linear map with minimal polynomial $M(\lambda) = P_1(\lambda)^{l_1} \cdots P_k(\lambda)^{l_k}$ and that $V = U \oplus W$ with U and W both invariant under L. Let $P_1(\lambda)^{r_1} \cdots P_k(\lambda)^{r_k}$ and $P_1(\lambda)^{s_1} \cdots P_k(\lambda)^{s_k}$ be the respective minimal polynomials of $L\big|_U$ and $L\big|_W$. Prove that $l_j = \max(r_j, s_j)$ for $1 \leq j \leq k$.

4. (a) If A and B are in $M_n(\mathbb{K})$, if $P(\lambda)$ is a polynomial such that $P(AB) = 0$, and if $Q(\lambda) = \lambda P(\lambda)$, prove that $Q(BA) = 0$.
   (b) What can be inferred from (a) about the relationship between the minimal polynomials of AB and of BA?

5. (a) Suppose that D and D' are in $M_n(\mathbb{K})$, are similar to diagonal matrices, and have $DD' = D'D$. Prove that there is a matrix C such that $C^{-1}DC$ and $C^{-1}D'C$ are both diagonal.
   (b) Give an example of two nilpotent matrices N and N' in $M_n(\mathbb{K})$ with $NN' = N'N$ such that there is no C with $C^{-1}NC$ and $C^{-1}N'C$ both in Jordan form.

6. (a) Prove that the matrix of a projection is similar to a diagonal matrix. What are the eigenvalues?
   (b) Give a necessary and sufficient condition for two projections involving the same V to be given by similar matrices.

7. Let $E : V \to V$ and $F : V \to V$ be projections. Prove that E and F have
   (a) the same image if and only if $EF = F$ and $FE = E$,
   (b) the same kernel if and only if $EF = E$ and $FE = F$.

8. Let $E : V \to V$ and $F : V \to V$ be projections. Prove that EF is a projection if $EF = FE$. Prove or disprove a converse.

9. An **involution** on V is a linear map $U : V \to V$ such that $U^2 = I$. Show that the equation $U = 2E - 1$ establishes a one-one correspondence between all projections E and all involutions U.

10. Let $L : V \to V$ be linear. Prove that there exist vector subspaces U and W of V such that
    (i) $V = U \oplus W$,
    (ii) $L(U) \subseteq U$ and $L(W) \subseteq W$,
    (iii) L is nilpotent on U,
    (iv) L is nonsingular on W.

11. Prove that the vector subspaces U and W in the previous problem are uniquely characterized by (i) through (iv).

12. Let $L : V \to V$ be a linear map, and suppose that its minimal polynomial is of the form $M(\lambda) = \prod_{j=1}^{k} (\lambda - \lambda_j)^{l_j}$ with the λ_j distinct. Let $V = U_1 \oplus \cdots \oplus U_k$ be the corresponding primary decomposition of V, and define $D : V \to V$ by $D = \lambda_1 E_1 + \cdots + \lambda_k E_k$, where $E_1, \ldots, E_k$ are the projections associated with the primary decomposition. Finally put $N = L - D$. Prove that
    (a) $L = D + N$,
    (b) D has a basis of eigenvectors,
    (c) N is nilpotent,
    (d) $DN = ND$.
    (e) D and N are given by polynomials in L,
    (f) the minimal polynomial of D is $\prod_{j=1}^{k} (\lambda - \lambda_j)$,
    (g) the minimal polynomial of N is $\lambda^{\max l_j}$.

13. In the previous problem with L given, prove that the decomposition $L = D + N$ is uniquely determined by properties (a) through (d).

14. Let A be a nilpotent square matrix. Prove that $\det(I + A) = 1$.

15. For the complex matrix $A = \begin{pmatrix} -5 & 9 \\ -4 & 7 \end{pmatrix}$, find a Jordan-form matrix J and an invertible matrix C such that $J = C^{-1}AC$.

16. For the complex matrix $A = \begin{pmatrix} 4 & 1 & -1 \\ -8 & -2 & 2 \\ 8 & 2 & -2 \end{pmatrix}$, find a Jordan-form matrix J and an invertible matrix C such that $J = C^{-1}AC$.

17. For the upper triangular matrix

$$A = \begin{pmatrix} 2 & 0 & 0 & 1 & 1 & 0 & 0 \\ & 2 & 0 & 0 & 0 & 1 & 1 \\ & & 2 & 0 & 1 & 0 & 0 \\ & & & 2 & 0 & 1 & 2 \\ & & & & 2 & 1 & 1 \\ & & & & & 2 & 1 \\ & & & & & & 3 \end{pmatrix},$$

find a Jordan-form matrix J and an invertible matrix C such that $J = C^{-1}AC$.

18. (a) For $M_3(\mathbb{C})$, prove that any two matrices with the same minimal polynomial and the same characteristic polynomial must be similar.
    (b) Is the same thing true for $M_4(\mathbb{C})$?

19. Suppose that $\mathbb{K}$ has characteristic 0 and that J is a Jordan block with nonzero eigenvalue and with size > 1. Prove that there is no $n \geq 1$ such that J^n is diagonal.

20. Classify up to similarity all members A of $M_n(\mathbb{C})$ with $A^n = I$.

21. How many similarity classes are there of 3-by-3 matrices A with entries in $\mathbb{C}$ such that $A^3 = A$? Explain.

22. Let $n \geq 2$, and let N be a member of $M_n(\mathbb{K})$ with $N^n = 0$ but $N^{n-1} \neq 0$. Prove that there is no n-by-n matrix A with $A^2 = N$.

23. For a Jordan block J, prove that J^t is similar to J.

24. Prove that if A is in $M_n(\mathbb{C})$, then A^t is similar to A.

25. Let N be the 2-by-2 matrix $\begin{pmatrix} 0 & 1 \\ 0 & 0 \end{pmatrix}$, and let A and B be the 4-by-4 matrices $A = \begin{pmatrix} N & 0 \\ 0 & N \end{pmatrix}$ and $B = \begin{pmatrix} N & N \\ 0 & N \end{pmatrix}$. Prove that A and B are similar.

Problems 26–31 concern cyclic vectors. Fix a linear map $L : V \to V$ from a finite-dimensional vector space V to itself. For v in V, let $\mathcal{P}(v)$ denote the set of all vectors $Q(L)(v)$ in V for $Q(\lambda)$ in $\mathbb{K}[\lambda]$; $\mathcal{P}(v)$ is a vector subspace and is invariant under L. If U is an invariant subspace of V, we say that U is a **cyclic subspace** if there is some

v in U such that $\mathcal{P}(v) = U$; in this case, v is said to be a **cyclic vector** for U, and U is called the **cyclic subspace generated** by v. For v in V, let $\mathcal{I}_v$ be the ideal of all polynomials $Q(\lambda)$ in $\mathbb{K}[\lambda]$ with $Q(L)v = 0$. The **monic generator** of v is the unique monic polynomial $M_v(\lambda)$ such that $M_v(\lambda)$ divides every member of $\mathcal{I}_v$.

26. For $v \in V$, explain why $\mathcal{I}_v$ is nonzero and why $M_v(\lambda)$ therefore exists.

27. For $v \in V$, prove that
    (a) the degree of the monic generator $M_v(\lambda)$ equals the dimension of the cyclic subspace $\mathcal{P}(v)$,
    (b) the vectors $v, L(v), L^2(v), \ldots, L^{\deg M_v - 1}(v)$ form a vector-space basis of $\mathcal{P}(v)$,
    (c) the minimal and characteristic polynomials of $L|_{\mathcal{P}(v)}$ are both equal to $M_v(\lambda)$.

28. Suppose that $M_v(\lambda) = c_0 + c_1\lambda + \cdots + c_{d-1}\lambda^{d-1} + \lambda^d$. Prove that the matrix of $L|_{\mathcal{P}(v)}$ in a suitable ordered basis is

$$\begin{pmatrix} -c_{d-1} & 1 & 0 & \cdots & & & \\ -c_{d-2} & 0 & 1 & & & & \\ -c_{d-3} & 0 & 0 & & & & \\ \vdots & & & \ddots & \ddots & & \vdots \\ -c_2 & 0 & 0 & \cdots & 0 & 1 & 0 \\ -c_1 & 0 & 0 & \cdots & 0 & 0 & 1 \\ -c_0 & 0 & 0 & \cdots & & 0 & 0 \end{pmatrix}.$$

29. Suppose that v is in V, that $M_v(\lambda)$ is a power of a prime polynomial $P(\lambda)$, and that $Q(\lambda)$ is a nonzero polynomial with $\deg Q(\lambda) < \deg P(\lambda)$. Prove that $P(Q(L)(v)) = \mathcal{P}(v)$.

30. Let $P(\lambda)$ be a prime polynomial.
    (a) Prove by induction on $\dim V$ that if the minimal polynomial of L is $P(\lambda)$, then the characteristic polynomial of L is a power of $P(\lambda)$.
    (b) Prove by induction on l that if the minimal polynomial of L is $P(\lambda)^l$, then the characteristic polynomial of L is a power of $P(\lambda)$.
    (c) Conclude that if the minimal polynomial of L is a power of $P(\lambda)$, then $\deg P(\lambda)$ divides $\dim V$.

31. (a) Prove that every prime factor of the characteristic polynomial of L divides the minimal polynomial of L.
    (b) In Problem 12 prove that D and L have the same characteristic polynomial.

Problems 32–40 continue the study of cyclic vectors begun in Problems 26–31, using the same notation. The goal is to obtain a canonical-form theorem like Theorem 5.20 for L but with no assumption on $\mathbb{K}$ or $P(\lambda)$, namely that each primary subspace for L is the direct sum of cyclic subspaces and the resulting decomposition is unique up to isomorphism. This result and the Fundamental Theorem of Finitely Generated

8. Problems 243

Abelian Groups (Theorem 4.56) will be seen in Chapter VIII to be special cases of a single more general theorem. Still another canonical form for matrices and linear maps is an analog of the result with elementary divisors mentioned in the remarks with Theorem 4.56 and is valid here; it is called **rational canonical form**, but we shall not pursue it until the problems at the end of Chapter VIII. The proof in Problems 32–40 uses ideas similar to those used for Theorem 5.21 except that the hypothesis will now be that the minimal polynomial of L is $P(\lambda)^l$ with $P(\lambda)$ prime, rather than just λ^l. Define $K_j = \ker(P(L)^j)$ for $j \geq 0$, so that $K_0 = 0$, $K_j \subseteq K_{j+1}$ for all j, $K_l = V$, and each K_j is an invariant subspace under L. Define $d = \deg P(\lambda)$.

32. Suppose $j \geq 1$, and suppose S_j is any vector subspace of V such that $K_{j+1} = K_j \oplus S_j$. Prove that $P(L)$ is one-one from S_j into K_j and $P(L)(S_j) \cap K_{j-1} = 0$.

33. Define $U_l = W_l = 0$. For $0 \leq j \leq l-1$, prove that there exist vector subspaces U_j and W_j of K_{j+1} such that
$$K_{j+1} = K_j \oplus U_j \oplus W_j,$$
$$U_j = P(L)(U_{j+1} \oplus W_{j+1}),$$
$$P(L) : U_{j+1} \oplus W_{j+1} \to U_j \quad \text{is one-one}.$$

34. Prove that the vector subspaces of the previous problem satisfy
$$V = U_0 \oplus W_0 \oplus U_1 \oplus W_1 \oplus \cdots \oplus U_{l-1} \oplus W_{l-1}.$$

35. For $v \neq 0$ in W_j, prove that the set of all $L^r P(L)^s(v)$ with $0 \leq r \leq d-1$ and $0 \leq s \leq j$ is a vector-space basis of $\mathcal{P}(v)$.

36. Going back over the construction in Problem 33, prove that each W_j can be chosen to have a basis consisting of vectors $L^r(v_i^{(j)})$ for $1 \leq i \leq (\dim W_j)/d$ and $0 \leq r \leq d-1$.

37. Let the index i used in the previous problem with j be denoted by i_j for $1 \leq i_j \leq (\dim W_j)/d$. Prove that a vector-space basis of $U_j \oplus W_j$ consists of all $L^r P(L)^k(v_{i_{j+k}}^{(j+k)})$ for $0 \leq r \leq d-1$, $k \geq 0$, $1 \leq i_{j+k} \leq (\dim W_{j+k})/d$.

38. Prove that V is the direct sum of cyclic subspaces under L. Prove specifically that each $v_{i_j}^{(j)}$ generates a cyclic subspace and that the sum of all these vector subspaces, with $0 \leq j \leq l$ and $1 \leq i_j \leq (\dim W_j)/d$, is a direct sum and equals V.

39. In the decomposition of the previous problem, each cyclic subspace generated by some $v_{i_j}^{(j)}$ has minimal polynomial $P(\lambda)^{j+1}$. Prove that
$$\#\begin{cases} \text{direct summands with minimal polynomial} \\ P(\lambda)^k \text{ for some } k \geq j+1 \end{cases} = (\dim K_{j+1} - \dim K_j)/d.$$

40. Prove that the formula of the previous problem persists for any decomposition of V as the direct sum of cyclic subspaces, and conclude from Problem 28 that the decomposition into cyclic subspaces is unique up to isomorphism.

Problems 41–46 concern systems of ordinary differential equations with constant coefficients. The underlying field is taken to be $\mathbb{C}$, and differential calculus is used. For A in $M_n(\mathbb{C})$ and t in $\mathbb{R}$, define $e^{tA} = \sum_{k=0}^{\infty} \frac{t^k A^k}{k!}$. Take for granted that the series defining e^{tA} converges entry by entry, that the series may be differentiated term by term to yield $\frac{d}{dt}(e^{tA}) = Ae^{tA} = e^{tA}A$, and that $e^{sA+tB} = e^{sA}e^{tB}$ if A and B commute.

41. Calculate e^{tA} for A equal to

    (a) $\begin{pmatrix} 0 & 1 \\ -1 & 0 \end{pmatrix}$,

    (b) $\begin{pmatrix} 0 & 1 \\ 1 & 0 \end{pmatrix}$,

    (c) the diagonal matrix with diagonal entries $d_1, \ldots, d_n$.

42. (a) Calculate e^{tJ} when J is a nilpotent n-by-n Jordan block.

    (b) Use (a) to calculate e^{tJ} when J is a general n-by-n Jordan block.

43. Let $y_1, \ldots, y_n$ be unknown functions from $\mathbb{R}$ to $\mathbb{C}$, and let y be the vector-valued function formed by arranging $y_1, \ldots, y_n$ in a column. Suppose that A is in $M_n(\mathbb{C})$. Prove for each vector $v \in \mathbb{C}^n$ that $y(t) = e^{tA}v$ is a solution of the system of differential equations $\frac{dy}{dt} = Ay(t)$.

44. With notation as in the previous problem and with v fixed in $\mathbb{C}^n$, use $e^{-tA}y(t)$ to show, for each open interval of t's containing 0, that the only solution of $\frac{dy}{dt} = Ay(t)$ on that interval such that $y(0) = v$ is $y(t) = e^{tA}v$.

45. For C invertible, prove that $e^{tC^{-1}AC} = C^{-1}e^{tA}C$, and deduce a relationship between solutions of $\frac{dy}{dt} = Ay(t)$ and solutions of $\frac{dy}{dt} = (C^{-1}AC)y(t)$.

46. Let $A = \begin{pmatrix} 2 & 1 & 0 \\ -1 & 4 & 0 \\ -1 & 2 & 2 \end{pmatrix}$. Taking into account Example 3 in Section 7 and Problems 42 through 45 above, find all solutions for t in $(-1, 1)$ to the system $\frac{dy}{dt} = Ay(t)$ such that $y(0) = \begin{pmatrix} 1 \\ 2 \\ 3 \end{pmatrix}$.

CHAPTER VI

Multilinear Algebra

Abstract. This chapter studies, in the setting of vector spaces over a field, the basics concerning multilinear functions, tensor products, spaces of linear functions, and algebras related to tensor products.

Sections 1–5 concern special properties of bilinear forms, all vector spaces being assumed to be finite-dimensional. Section 1 associates a matrix to each bilinear form in the presence of an ordered basis, and the section shows the effect on the matrix of changing the ordered basis. It then addresses the extent to which the notion of "orthogonal complement" in the theory of inner-product spaces applies to nondegenerate bilinear forms. Sections 2–3 treat symmetric and alternating bilinear forms, producing bases for which the matrix of such a form is particularly simple. Section 4 treats a related subject, Hermitian forms when the field is the complex numbers. Section 5 discusses the groups that leave some particular bilinear and Hermitian forms invariant.

Section 6 introduces the tensor product of two vector spaces, working with it in a way that does not depend on a choice of basis. The tensor product has a universal mapping property—that bilinear functions on the product of the two vector spaces extend uniquely to linear functions on the tensor product. The tensor product turns out to be a vector space whose dual is the vector space of all bilinear forms. One particular application is that tensor products provide a basis-independent way of extending scalars for a vector space from a field to a larger field. The section includes a number of results about the vector space of linear mappings from one vector space to another that go hand in hand with results about tensor products. These have convenient formulations in the language of category theory as "natural isomorphisms."

Section 7 begins with the tensor product of three and then n vector spaces, carefully considering the universal mapping property and the question of associativity. The section defines an algebra over a field as a vector space with a bilinear multiplication, not necessarily associative. If E is a vector space, the tensor algebra $T(E)$ of E is the direct sum over $n \geq 0$ of the n-fold tensor product of E with itself. This is an associative algebra with a universal mapping property relative to any linear mapping of E into an associative algebra A with identity: the linear map extends to an algebra homomorphism of $T(E)$ into A carrying 1 into 1.

Sections 8–9 define the symmetric and exterior algebras of a vector space E. The symmetric algebra $S(E)$ is a quotient of $T(E)$ with the following universal mapping property: any linear mapping of E into a commutative associative algebra A with identity extends to an algebra homomorphism of $S(E)$ into A carrying 1 into 1. The symmetric algebra is commutative. Similarly the exterior algebra $\bigwedge(E)$ is a quotient of $T(E)$ with this universal mapping property: any linear mapping l of E into an associative algebra A with identity such that $l(v)^2 = 0$ for all $v \in E$ extends to an algebra homomorphism of $\bigwedge(E)$ into A carrying 1 into 1.

The problems at the end of the chapter introduce some other algebras that are of importance in applications, and the problems relate some of these algebras to tensor, symmetric, and exterior algebras. Among the objects studied are Lie algebras, universal enveloping algebras, Clifford algebras, Weyl algebras, Jordan algebras, and the division algebra of octonions.

1. Bilinear Forms and Matrices

This chapter will work with vector spaces over a common field of "scalars," which will be called $\mathbb{K}$. In Section 6 a field containing $\mathbb{K}$ as a subfield will briefly play a role, and that will be called $\mathbb{L}$.

If V is a vector space over $\mathbb{K}$, a **bilinear form** on V is a function from $V \times V$ into $\mathbb{K}$ that is linear in each variable when the other variable is held fixed.

EXAMPLES.

(1) For general $\mathbb{K}$, take $V = \mathbb{K}^n$. Any matrix A in $M_n(\mathbb{K})$ determines a bilinear form by the rule $\langle v, w \rangle = v^t A w$.

(2) For $\mathbb{K} = \mathbb{R}$, let V be an inner-product space, in the sense of Chapter III, with inner product $(\cdot, \cdot)$. Then $(\cdot, \cdot)$ is a bilinear form on V.

Multilinear functionals on a vector space of row vectors, also called k-linear functionals or k-multilinear functionals, were defined in the course of working with determinants in Section II.7, and that definition transparently extends to general vector spaces. A bilinear form on a general vector space is then just a 2-linear functional. From the point of view of definitions, the words "functional" and "form" are interchangeable here, but the word "form" is more common in the bilinear case because of a certain homogeneity that it suggests and that comes closer to the surface in Corollary 6.12 and in Section 7.

For the remainder of this section, all vector spaces will be finite-dimensional.

Bilinear forms, i.e., 2-linear functionals, are of special interest relative to k-linear functionals for general k because of their relationships with matrices and linear mappings. To begin with, each bilinear form, in the presence of an ordered basis, is given by a matrix. In more detail let V be a finite-dimensional vector space, and let $\langle \cdot, \cdot \rangle$ be a bilinear form on V. If an ordered basis $\Gamma = (v_1, \ldots, v_n)$ of V is specified, then the bilinear form determines the matrix B with entries $B_{ij} = \langle v_i, v_j \rangle$. Conversely we can recover the bilinear form from B as follows: Write $v = \sum_i a_i v_i$ and $w = \sum_j b_j v_j$. Then

$$\langle v, w \rangle = \left\langle \sum_i a_i v_i, \sum_j b_j v_j \right\rangle = \sum_{i,j} a_i \langle v_i, v_j \rangle b_j.$$

In other words, $\langle v, w \rangle = a^t B b$, where $a = \begin{pmatrix} v \\ \Gamma \end{pmatrix}$ and $b = \begin{pmatrix} w \\ \Gamma \end{pmatrix}$ in the notation of Section II.3. Therefore

$$\langle v, w \rangle = \begin{pmatrix} v \\ \Gamma \end{pmatrix}^t B \begin{pmatrix} w \\ \Gamma \end{pmatrix}.$$

1. Bilinear Forms and Matrices

Consequently we see that all bilinear forms on a finite-dimensional vector space reduce to Example 1 above—once we choose an ordered basis.

Let us examine the effect of a change of ordered basis. Suppose that $\Gamma = (v_1, \ldots, v_m)$ and $\Delta = (w_1, \ldots, w_n)$, and let B and C be the matrices of the bilinear form in these two ordered bases: $B_{ij} = \langle v_i, v_j \rangle$ and $C_{ij} = \langle w_i, w_j \rangle$. Let the two bases be related by $w_j = \sum_i a_{ij} v_i$, i.e., let $[a_{ij}] = \begin{pmatrix} I \\ \Gamma\Delta \end{pmatrix}$. Then we have

$$C_{ij} = \langle w_i, w_j \rangle = \Big\langle \sum_k a_{ki} v_k, \sum_l a_{lj} v_l \Big\rangle = \sum_{k,l} a_{ki} a_{lj} \langle v_k, v_l \rangle = \sum_{k,l} a_{ki} B_{kl} a_{lj}.$$

Translating this formula into matrix form, we obtain the following proposition.

Proposition 6.1. Let $\langle \cdot, \cdot \rangle$ be a bilinear form on a finite-dimensional vector space V, let Γ and Δ be ordered bases of V, and let B and C be the respective matrices of $\langle \cdot, \cdot \rangle$ relative to Γ and Δ. Then

$$C = \begin{pmatrix} I \\ \Gamma\Delta \end{pmatrix}^t B \begin{pmatrix} I \\ \Gamma\Delta \end{pmatrix}.$$

The qualitative conclusion about the matrices may be a little unexpected. It is not that they are similar but that they are related by $C = S^t B S$ for some nonsingular square matrix S. In particular, B and C need not have the same determinant.

Guided by the circle of ideas around the Riesz Representation Theorem for inner products (Theorem 3.12), let us examine what happens when we fix one of the variables of a bilinear form and work with the resulting linear map. Thus again let $\langle \cdot, \cdot \rangle$ be a bilinear form on V. For fixed u in V, $v \mapsto \langle u, v \rangle$ is a linear functional on V, thus a member of the dual space V' of V. If we write $L(u)$ for this linear functional, then L is a function from V to V' satisfying $L(u)(v) = \langle u, v \rangle$. The formula for L shows that L is in fact a linear function. We define the **left radical**, lrad, of $\langle \cdot, \cdot \rangle$ to be the kernel of L; thus

$$\text{lrad}\,(\langle \cdot, \cdot \rangle) = \{u \in V \mid \langle u, v \rangle = 0 \text{ for all } v \in V\}.$$

Similarly we let $R : V \to V'$ be the linear map $R(v)(u) = \langle u, v \rangle$, and we define the **right radical**, rrad, of $\langle \cdot, \cdot \rangle$ to be the kernel of R; thus

$$\text{rrad}\,(\langle \cdot, \cdot \rangle) = \{v \in V \mid \langle u, v \rangle = 0 \text{ for all } u \in V\}.$$

EXAMPLE 1, CONTINUED. The vector space V is the space $\mathbb{K}^n$ of n-dimensional column vectors, the dual V' is the space of n-dimensional row vectors, A is an n-by-n matrix with entries in $\mathbb{K}$, and $\langle \cdot, \cdot \rangle$ is given by $\langle u, v \rangle = u^t A v = L(u)(v) = R(v)(u)$ for u and v in $\mathbb{K}^n$. Explicit formulas for L and R are given by

$$L(u) = u^t A = (A^t u)^t$$

and

$$R(v) = (Av)^t.$$

Thus

$$\mathrm{lrad}\,(\langle \cdot, \cdot \rangle) = \ker L = \text{null space}(A^t),$$
$$\mathrm{rrad}\,(\langle \cdot, \cdot \rangle) = \ker R = \text{null space}(A).$$

Since A is square and since the row rank and column rank of A are equal, the dimensions of the null spaces of A and A^t are equal. Hence

$$\dim \mathrm{lrad}\,(\langle \cdot, \cdot \rangle) = \dim \mathrm{rrad}\,(\langle \cdot, \cdot \rangle).$$

This equality of dimensions for the case of $\mathbb{K}^n$ extends to general V, as is noted in the next proposition.

Proposition 6.2. If $\langle \cdot, \cdot \rangle$ is any bilinear form on a finite-dimensional vector space V, then
$$\dim \mathrm{lrad}\,(\langle \cdot, \cdot \rangle) = \dim \mathrm{rrad}\,(\langle \cdot, \cdot \rangle).$$

PROOF. We saw above that computations with bilinear forms of V reduce, once we choose an ordered basis for V, to computations with matrices, row vectors, and column vectors. Thus the argument just given in the continuation of Example 1 is completely general, and the proposition is proved. $\square$

A bilinear form $\langle \cdot, \cdot \rangle$ is said to be **nondegenerate** if its left radical is 0. In view of the Proposition 6.2, it is equivalent to require that the right radical be 0. When the radicals are 0, the associated linear maps L and R from V to V' are one-one. Since $\dim V = \dim V'$, it follows that L and R are onto V'. Thus a nondegenerate bilinear form on V sets up two canonical isomorphisms of V with its dual V'.

For definiteness let us work with the linear mapping $L : V \to V'$ given by $L(u)(v) = \langle u, v \rangle$. If $U \subseteq V$ is a vector subspace, define

$$U^\perp = \{u \in V \mid \langle u, v \rangle = 0 \text{ for all } v \in U\}.$$

It is apparent from the definitions that

$$\boxed{U \cap U^\perp = \mathrm{lrad}\,(\langle \cdot, \cdot \rangle)\big|_{U \times U}.}$$

In contrast to the special case that $\mathbb{K} = \mathbb{R}$ and the bilinear form is an inner product, $U \cap U^\perp$ may be nonzero even if $\langle \cdot, \cdot \rangle$ is nondegenerate. For example let $V = \mathbb{R}^2$, define
$$\left\langle \begin{pmatrix} x_1 \\ x_2 \end{pmatrix}, \begin{pmatrix} y_1 \\ y_2 \end{pmatrix} \right\rangle = x_1 y_1 - x_2 y_2,$$
and suppose that U is the 1-dimensional vector subspace $U = \left\{ \begin{pmatrix} x_1 \\ x_1 \end{pmatrix} \right\}$. The matrix of the bilinear form in the standard ordered basis is $\begin{pmatrix} 1 & 0 \\ 0 & -1 \end{pmatrix}$; since the matrix is nonsingular, the bilinear form is nondegenerate. Direct calculation shows that $U^\perp = \left\{ \begin{pmatrix} y_1 \\ y_1 \end{pmatrix} \right\} = U$, so that $U \cap U^\perp \neq 0$. Nevertheless, in the nondegenerate case the dimensions of U and $U^\perp$ behave as if $U^\perp$ were an orthogonal complement. The precise result is as follows.

Proposition 6.3. If $\langle \cdot, \cdot \rangle$ is a nondegenerate bilinear form on the finite-dimensional vector space V and if U is a vector subspace of V, then
$$\dim V = \dim U + \dim U^\perp.$$

PROOF. Define $\ell : V \to U'$ by $\ell(v)(u) = \langle v, u \rangle$ for $v \in V$ and $u \in U$. The definition of $U^\perp$ shows that $\ker \ell = U^\perp$. To see that image $\ell = U'$, choose a vector subspace U_1 of V with $V = U \oplus U_1$, let u' be in U', and define v' in V' by
$$v' = \begin{cases} u' & \text{on } U, \\ 0 & \text{on } U_1. \end{cases}$$
Since $\langle \cdot, \cdot \rangle$ is nondegenerate, the linear mapping $L : V \to V'$ is onto V'. Thus we can choose $v \in V$ with $L(v) = v'$. Then
$$\ell(v)(u) = \langle v, u \rangle = L(v)(u) = v'(u) = u'(u)$$
for all u in U, and hence $\ell(v) = u'$. Therefore image $\ell = U'$, and we conclude that
$$\dim V = \dim(\ker \ell) + \dim(\text{image } \ell) = \dim U^\perp + \dim U' = \dim U^\perp + \dim U.$$
$\square$

Corollary 6.4. If $\langle \cdot, \cdot \rangle$ is a nondegenerate bilinear form on the finite-dimensional vector space V and if U is a vector subspace of V, then $V = U \oplus U^\perp$ if and only if $\langle \cdot, \cdot \rangle \big|_{U \times U}$ is nondegenerate.

PROOF. Corollary 2.29 and Proposition 6.3 together give
$$\dim(U + U^\perp) + \dim(U \cap U^\perp) = \dim U + \dim U^\perp = \dim V.$$
Thus $U + U^\perp = V$ if and only if $U \cap U^\perp = 0$, if and only if $\langle \cdot, \cdot \rangle \big|_{U \times U}$ is nondegenerate. The result therefore follows from Proposition 2.30. $\square$

2. Symmetric Bilinear Forms

We continue with the setting in which $\mathbb{K}$ is a field and all vector spaces of interest are defined over $\mathbb{K}$ and are finite-dimensional.

A bilinear form $\langle\,\cdot\,,\,\cdot\,\rangle$ on V is said to be **symmetric** if $\langle u, v\rangle = \langle v, u\rangle$ for all u and v in V, **skew-symmetric** if $\langle u, v\rangle = -\langle v, u\rangle$ for all u and v in V, and **alternating** if $\langle u, u\rangle = 0$ for all u in V.

"Alternating" always implies "skew-symmetric." In fact, if $\langle\,\cdot\,,\,\cdot\,\rangle$ is alternating, then $0 = \langle u+v, u+v\rangle = \langle u, u\rangle + \langle u, v\rangle + \langle v, u\rangle + \langle v, v\rangle = \langle u, v\rangle + \langle v, u\rangle$; thus $\langle\,\cdot\,,\,\cdot\,\rangle$ is skew-symmetric. If $\mathbb{K}$ has characteristic different from 2, then the converse is valid: "skew-symmetric" implies "alternating." In fact, if $\langle\,\cdot\,,\,\cdot\,\rangle$ is skew-symmetric, then $\langle u, u\rangle = -\langle u, u\rangle$ and hence $2\langle u, u\rangle = 0$; thus $\langle u, u\rangle = 0$, and $\langle\,\cdot\,,\,\cdot\,\rangle$ is alternating.

Let us examine further the effect of the characteristic of $\mathbb{K}$. If, on the one hand, $\mathbb{K}$ has characteristic different from 2, the most general bilinear form $\langle\,\cdot\,,\,\cdot\,\rangle$ is the sum of the symmetric form $\langle\,\cdot\,,\,\cdot\,\rangle_s$ and the alternating form $\langle\,\cdot\,,\,\cdot\,\rangle_a$ given by

$$\langle u, v\rangle_s = \tfrac{1}{2}(\langle u, v\rangle + \langle v, u\rangle),$$
$$\langle u, v\rangle_a = \tfrac{1}{2}(\langle u, v\rangle - \langle v, u\rangle).$$

In this sense the symmetric and alternating bilinear forms are the extreme cases among all bilinear forms, and we shall study the two cases separately.

If, on the other hand, $\mathbb{K}$ has characteristic 2, then "alternating" implies "skew-symmetric" but not conversely. "Alternating" is a serious restriction, and we shall be able to deal with it. However, "symmetric" and "skew-symmetric" are equivalent since $1 = -1$, and thus neither condition is much of a restriction; we shall not attempt to say anything insightful in these cases.

In this section we study symmetric bilinear forms, obtaining results when $\mathbb{K}$ has characteristic different from 2. From the symmetry it is apparent that the left and right radicals of a symmetric bilinear form are the same, and we call this vector subspace the **radical** of the form. By way of an example, here is a continuation of Example 1 from the previous section.

EXAMPLE. Let $V = \mathbb{K}^n$, let A be a **symmetric** n-by-n matrix (i.e., one with $A^t = A$), and let $\langle u, v\rangle = u^t A v$. The computation $\langle v, u\rangle = v^t A u = (v^t A u)^t = u^t A^t v = u^t A v = \langle u, v\rangle$ shows that the bilinear form $\langle\,\cdot\,,\,\cdot\,\rangle$ is symmetric; the second equality $v^t A u = (v^{\text{th}} A u)^t$ holds since $v^t A u$ is a 1-by-1 matrix.

Again the example is completely general. In fact, if $\Gamma = (v_1, \ldots, v_n)$ is an ordered basis of a vector space V and if $\langle\,\cdot\,,\,\cdot\,\rangle$ is a given symmetric bilinear form on V, then the matrix of the form has entries $A_{ij} = \langle v_i, v_j\rangle$, and these evidently satisfy $A_{ij} = A_{ji}$. So A is a symmetric matrix, and computations with the bilinear form are reduced to those used in the example.

2. Symmetric Bilinear Forms

Theorem 6.5 (Principal Axis Theorem). Suppose that $\mathbb{K}$ has characteristic different from 2.

(a) If $\langle \cdot, \cdot \rangle$ is a symmetric bilinear form on a finite-dimensional vector space V, then there exists an ordered basis of V in which the matrix of $\langle \cdot, \cdot \rangle$ is diagonal.

(b) If A is an n-by-n symmetric matrix, then there exists a nonsingular n-by-n matrix M such that $M^t A M$ is diagonal.

REMARKS. Because computations with general symmetric bilinear forms reduce to computations in the special case of a symmetric matrix and because Proposition 6.1 tells the effect of a change of ordered basis, (a) and (b) amount to the same result; nevertheless, we give two proofs of Theorem 6.5—a proof via matrices and a proof via linear maps. A hint of the validity of the theorem comes from the case that $\mathbb{K} = \mathbb{R}$. For the field $\mathbb{R}$ when the bilinear form is an inner product, the Spectral Theorem (Theorem 3.21) says that there is an orthonormal basis of eigenvectors and hence that (a) holds. When $\mathbb{K} = \mathbb{R}$, the same theorem says that there exists an orthogonal matrix M with $M^{-1} A M$ diagonal; since any orthogonal matrix M satisfies $M^{-1} = M^t$, the Spectral Theorem is saying that (b) holds.

PROOF VIA MATRICES. If A is an n-by-n symmetric matrix, we seek a nonsingular M with $M^t A M$ diagonal. We induct on the size of A, the base case of the induction being $n = 1$, where there is nothing to prove. Assume the result to be known for size $n - 1$, and write the given n-by-n matrix A in block form as $A = \begin{pmatrix} a & b \\ b^t & d \end{pmatrix}$ with d of size 1-by-1. If $d \neq 0$, let x be the column vector $-d^{-1}b$. Then

$$\begin{pmatrix} I & x \\ 0 & 1 \end{pmatrix} \begin{pmatrix} a & b \\ b^t & d \end{pmatrix} \begin{pmatrix} I & 0 \\ x^t & 1 \end{pmatrix} = \begin{pmatrix} * & 0 \\ 0 & d \end{pmatrix},$$

and the induction goes through. If $d = 0$, we argue in a different way. We may assume that $b \neq 0$ since otherwise the result is immediate by induction. Say $b_i \neq 0$ with $1 \leq i \leq n - 1$. Let y be an $(n - 1)$-dimensional row vector with i^{th} entry a member δ of $\mathbb{K}$ to be specified and with other entries 0. Then

$$\begin{pmatrix} I & 0 \\ y & 1 \end{pmatrix} \begin{pmatrix} a & b \\ b^t & 0 \end{pmatrix} \begin{pmatrix} I & y^t \\ 0 & 1 \end{pmatrix} = \begin{pmatrix} * & * \\ * & yay^t + b^t y^t + yb \end{pmatrix} = \begin{pmatrix} * & * \\ * & \delta^2 a_{ii} + 2\delta b_i \end{pmatrix}.$$

Since $\mathbb{K}$ has characteristic different from 2, $2b_i$ is not 0; thus there is some value of δ for which $\delta^2 a_{ii} + 2\delta b_i \neq 0$. Then we are reduced to the case $d \neq 0$, which we have already handled, and the induction goes through. □

PROOF VIA LINEAR MAPS. We may assume that the given symmetric bilinear form is not identically 0, since otherwise any basis will do. Let the radical of the form be denoted by $\text{rad} = \text{rad}(\langle \cdot, \cdot \rangle)$. Choose a vector subspace S of V such that $V = \text{rad} \oplus S$, and put $[\cdot, \cdot] = \langle \cdot, \cdot \rangle|_{S \times S}$. Then $[\cdot, \cdot]$ is a symmetric bilinear form on S, and it is nondegenerate. In fact, $[u, \cdot] = 0$ means $\langle u, v \rangle = 0$ for all $v \in S$; since $\langle u, v \rangle = 0$ for v in rad anyway, $\langle u, v \rangle = 0$ for all $v \in V$, u is in rad as well as S, and $u = 0$.

Since $\langle \cdot, \cdot \rangle$ is not identically 0, the subspace S is not 0. Thus the nondegenerate symmetric bilinear form $[\cdot, \cdot]$ on S is not 0. Since

$$[u, v] = \tfrac{1}{2}\bigl([u+v, u+v] - [u, u] - [v, v]\bigr),$$

it follows that $[v, v] \neq 0$ for some v in S. Put $U_1 = \mathbb{K}v$. Then $[\cdot, \cdot]\big|_{U_1 \times U_1}$ is nondegenerate, and Corollary 6.4 implies that $S = U_1 \oplus U_1^\perp$. Applying the converse direction of the same corollary to $U_1^\perp$, we see that $[\cdot, \cdot]\big|_{U_1^\perp \times U_1^\perp}$ is nondegenerate. Repeating this construction with $U^\perp$ and iterating, we obtain

$$V = \mathrm{rad} \oplus U_1 \oplus \cdots \oplus U_k$$

with $\langle U_i, U_j \rangle = 0$ for $i \neq j$ and with $\dim U_i = 1$ for all i. This completes the proof. $\square$

Theorem 6.5 fails in characteristic 2. Problem 2 at the end of the chapter illustrates the failure.

Let us examine the matrix version of Theorem 6.5 more closely when $\mathbb{K}$ is $\mathbb{C}$ or $\mathbb{R}$. The theorem says that if A is n-by-n symmetric, then we can find a nonsingular M with $B = M^t A M$ diagonal. Taking D diagonal and forming $C = D^t B D$, we see that we can adjust the diagonal entries of B by arbitrary nonzero squares. Over $\mathbb{C}$, we can therefore arrange that C is of the form $\mathrm{diag}(1, \ldots, 1, 0, \ldots, 0)$. The number of 1's equals the rank, and this has to be the same as the rank of the given matrix A. The form is nondegenerate if and only if there are no 0's. Thus we understand everything about the diagonal form.

Over $\mathbb{R}$, matters are more subtle. We can arrange that C is of the form $\mathrm{diag}(\pm 1, \ldots, \pm 1, 0, \ldots, 0)$, the various signs ostensibly not being correlated. Replacing C by $P^t C P$ with P a permutation matrix, we may assume that our diagonal matrix is of the form $\mathrm{diag}(+1, \ldots, +1, -1, \ldots, -1, 0, \ldots, 0)$. The number of $+1$'s and -1's together is again the rank of A, and the form is nondegenerate if and only if there are no 0's. But what about the separate numbers of $+1$'s and -1's? The triple given by

$$(p, m, z) = \bigl(\#(+1)\text{'s}, \#(-1)\text{'s}, \#(0)\text{'s}\bigr)$$

is called the **signature** of A when $\mathbb{K} = \mathbb{R}$. A similar notion can be defined in the case of a symmetric bilinear form over $\mathbb{R}$.

Theorem 6.6 (Sylvester's Law). The signature of an n-by-n symmetric matrix over $\mathbb{R}$ is well defined.

PROOF. The integer $p + m$ is the rank, which does not change under a transformation $A \mapsto M^t A M$ if M is nonsingular. Thus we may take z as known. Let (p', m', z) and (p, m, z) be two signatures for a symmetric matrix A, with $p' \leq p$. Define the corresponding symmetric bilinear form on $\mathbb{R}^n$ by $\langle u, v \rangle = u^t A v$. Let $(v'_1, \ldots, v'_n)$ and $(v_1, \ldots, v_n)$ be ordered bases of $\mathbb{R}^n$ diagonalizing the bilinear form and exhibiting the resulting signature, i.e., having $\langle v'_i, v'_j \rangle = \langle v_i, v_j \rangle = 0$ for $i \neq j$ and having

$$\langle v'_j, v'_j \rangle = \begin{cases} +1 & \text{for } 1 \leq j \leq p', \\ -1 & \text{for } p'+1 \leq j \leq n-z, \\ 0 & \text{for } n-z+1 \leq j \leq n, \end{cases}$$

$$\langle v_j, v_j \rangle = \begin{cases} +1 & \text{for } 1 \leq j \leq p, \\ -1 & \text{for } p+1 \leq j \leq n-z, \\ 0 & \text{for } n-z+1 \leq j \leq n. \end{cases}$$

We shall prove that $\{v_1, \ldots, v_p, v'_{p'+1}, \ldots, v'_n\}$ is linearly independent, and then we must have $p' \geq p$. Reversing the roles of p and p', we see that $p' = p$ and $m' = m$, and the theorem is proved. Thus suppose we have a linear dependence:

$$a_1 v_1 + \cdots + a_p v_p = b_{p'+1} v'_{p'+1} + \cdots + b_n v'_n.$$

Let v be the common value of the two sides of this equation. Then

$$\langle v, v \rangle = \langle a_1 v_1 + \cdots + a_p v_p, a_1 v_1 + \cdots + a_p v_p \rangle = \sum_{j=1}^{p} a_j^2 \geq 0$$

and

$$\langle v, v \rangle = \langle b_{p'+1} v'_{p'+1} + \cdots + b_n v'_n, b_{p'+1} v'_{p'+1} + \cdots + b_n v'_n \rangle = - \sum_{j=p'+1}^{n-z} b_j^2 \leq 0.$$

We conclude that $\langle v, v \rangle = 0$, $\sum_{j=1}^{p} a_j^2 = 0$, and $a_1 = \cdots = a_p = 0$. Thus $v = 0$ and $b_{p'+1} v'_{p'+1} + \cdots + b_n v'_n = 0$. Since $\{v'_{p'+1}, \ldots, v'_n\}$ is linearly independent, we obtain also $b_{p'+1} = \cdots = b_n = 0$. Therefore $\{v_1, \ldots, v_p, v'_{p'+1}, \ldots, v'_n\}$ is a linearly independent set, and the proof is complete. $\square$

3. Alternating Bilinear Forms

We continue with the setting in which $\mathbb{K}$ is a field and all vector spaces of interest are defined over $\mathbb{K}$ and are finite-dimensional.

In this section we study alternating bilinear forms, imposing no restriction on the characteristic of $\mathbb{K}$. From the skew symmetry of any alternating bilinear form it is apparent that the left and right radicals of such a form are the same, and we call this vector subspace the **radical** of the form. First let us consider examples given in terms of matrices. Temporarily let us separate matters according to the characteristic.

EXAMPLE 1 OF SECTION 1 WITH $\mathbb{K}$ OF CHARACTERISTIC $\neq 2$. Let $V = \mathbb{K}^n$, let A be a **skew-symmetric** n-by-n matrix (i.e., one with $A^t = -A$), and let $\langle u, v \rangle = u^t A v$. The computation $\langle v, u \rangle = v^t A u = (v^t A u)^t = u^t A^t v = -u^t A v = -\langle u, v \rangle$ shows that the bilinear form $\langle \cdot, \cdot \rangle$ is skew-symmetric, hence alternating.

EXAMPLE 1 OF SECTION 1 WITH $\mathbb{K}$ OF CHARACTERISTIC $= 2$. Let $V = \mathbb{K}^n$, let A be an n-by-n matrix, and define $\langle u, v \rangle = u^t A v$. We suppose that A is skew-symmetric; it is the same to assume that A is symmetric since the characteristic is 2. In order to have $\langle e_i, e_i \rangle = 0$ for each standard basis vector, we shall assume that $A_{ii} = 0$ for all i. If u is a column vector with entries $u_1, \ldots, u_n$, then $\langle u, u \rangle = u^t A u = \sum_{i,j} u_i A_{ij} u_j = \sum_{i \neq j} u_i A_{ij} u_j = \sum_{i<j} (A_{ij} u_i u_j + A_{ji} u_i u_j) = \sum_{i<j} 2 A_{ij} u_i u_j = 0$. Hence the bilinear form $\langle \cdot, \cdot \rangle$ is alternating.

Again the examples are completely general. In fact, if $\Gamma = (v_1, \ldots, v_n)$ is an ordered basis of a vector space V and if $\langle \cdot, \cdot \rangle$ is a given alternating bilinear form, then the matrix of the form has entries $A_{ij} = \langle v_i, v_j \rangle$ that evidently satisfy $A_{ij} = -A_{ji}$ and $A_{ii} = 0$. So A is a skew-symmetric matrix with 0's on the diagonal, and computations with the bilinear form are reduced to those used in the examples. To keep the terminology parallel, let us say that a square matrix is **alternating** if it is skew-symmetric and has 0's on the diagonal.

Theorem 6.7.

(a) If $\langle \cdot, \cdot \rangle$ is an alternating bilinear form on a finite-dimensional vector space V, then there exists an ordered basis of V in which the matrix of $\langle \cdot, \cdot \rangle$ has the form

$$\begin{pmatrix} \boxed{\begin{matrix} 0 & 1 \\ -1 & 0 \end{matrix}} & & & & & \\ & \boxed{\begin{matrix} 0 & 1 \\ -1 & 0 \end{matrix}} & & & & \\ & & \ddots & & & \\ & & & \boxed{\begin{matrix} 0 & 1 \\ -1 & 0 \end{matrix}} & & \\ & & & & 0 & \\ & & & & & \ddots \\ & & & & & & 0 \end{pmatrix}$$

If $\langle\,\cdot\,,\,\cdot\,\rangle$ is nondegenerate, then dim V is even.

(b) If A is an n-by-n alternating matrix, then there exists a nonsingular n-by-n matrix M such that $M^t A M$ is as in (a).

PROOF. It is enough to prove (a). Let rad be the radical of the given form $\langle\,\cdot\,,\,\cdot\,\rangle$, and choose a vector subspace S of V with $V = \text{rad} \oplus S$. The restriction of $\langle\,\cdot\,,\,\cdot\,\rangle$ to S is then alternating and nondegenerate. We may now proceed by induction on dim V under the assumption that $\langle\,\cdot\,,\,\cdot\,\rangle$ is nondegenerate. For dim $V = 1$, the form is degenerate. For dim $V = 2$, we can find u and v with $\langle u, v \rangle \neq 0$, and we can normalize one of the vectors to make $\langle u, v \rangle = 1$. Then (u, v) is the required ordered basis.

Assuming the result in the nondegenerate case for dimension $< n$, suppose that dim $V = n$. Again choose u and v with $\langle u, v \rangle = 1$, and define $U = \mathbb{K}u \oplus \mathbb{K}v$. Then $\langle\,\cdot\,,\,\cdot\,\rangle|_{U \times U}$ has matrix $\begin{pmatrix} 0 & 1 \\ -1 & 0 \end{pmatrix}$ and is nondegenerate. By Corollary 6.4, $V = U \oplus U^\perp$, and an application of the converse of the corollary shows that $\langle\,\cdot\,,\,\cdot\,\rangle|_{U^\perp \times U^\perp}$ is nondegenerate. The induction hypothesis applies to $U^\perp$, and we obtain the desired matrix for the given form. $\square$

4. Hermitian Forms

In this section the field will be $\mathbb{C}$, and V will be a finite-dimensional vector space over $\mathbb{C}$.

A **sesquilinear form** $\langle\,\cdot\,,\,\cdot\,\rangle$ on V is a function from $V \times V$ into $\mathbb{C}$ that is linear in the first variable and conjugate linear in the second.[1] Sesquilinear forms do not make sense for general fields because of the absence of a universal analog of complex conjugation, and we shall consequently work only with the field $\mathbb{C}$ in this section.[2]

A sesquilinear form $\langle\,\cdot\,,\,\cdot\,\rangle$ is **Hermitian** if $\langle u, v \rangle = \overline{\langle v, u \rangle}$ for all u and v in V. The form is **skew-Hermitian** if instead $\langle u, v \rangle = -\overline{\langle v, u \rangle}$ for all u and v in V. Hermitian and skew-Hermitian forms are the extreme types of sesquilinear forms since any sesquilinear form $\langle\,\cdot\,,\,\cdot\,\rangle$ is the sum of a Hermitian form $\langle\,\cdot\,,\,\cdot\,\rangle_{\text{h}}$ and a skew-Hermitian form $\langle\,\cdot\,,\,\cdot\,\rangle_{\text{sh}}$ given by

$$\langle u, v \rangle_{\text{h}} = \tfrac{1}{2}(\langle u, v \rangle + \overline{\langle v, u \rangle}),$$
$$\langle u, v \rangle_{\text{sh}} = \tfrac{1}{2}(\langle u, v \rangle - \overline{\langle v, u \rangle}).$$

---

[1] Some authors, particularly in mathematical physics, reverse the roles of the two variables and assume the conjugate linearity in the first variable instead of the second.

[2] Sesquilinear forms make sense in number fields like $\mathbb{Q}[\sqrt{2}]$ that have an automorphism of order 2 (see Section IV.1), but sesquilinear forms in this kind of setting will not concern us here.

In addition, any skew-Hermitian form becomes a Hermitian form simply by multiplying by i. Specifically if $\langle \cdot, \cdot \rangle_{sh}$ is skew-Hermitian, then $i\langle \cdot, \cdot \rangle_{sh}$ is sesquilinear and Hermitian, as is readily checked. Consequently the study of skew-Hermitian forms immediately reduces to the study of Hermitian forms.

EXAMPLE. Let $V = \mathbb{C}^n$, and let A be a **Hermitian** matrix, i.e., one with $A^* = A$, where A^* is the conjugate transpose of A. Then it is a simple matter to check that $\langle u, v \rangle = v^* A u$ defines a Hermitian form on $\mathbb{C}^n$.

Again the example with a matrix is completely general. In fact, let $\langle \cdot, \cdot \rangle$ be a Hermitian form on V, let $\Gamma = (v_1, \ldots, v_n)$ be an ordered basis of V, and define $A_{ij} = \langle v_i, v_j \rangle$. Then A is a Hermitian matrix, and $\langle u, v \rangle = u^t A \bar{v}$, where $\bar{v}$ is the entry-by-entry complex conjugate of v.

If $\Delta = (w_1, \ldots, w_n)$ is a second ordered basis, then the formula for changing basis may be derived as follows: Write $w_j = \sum_i c_{ij} v_i$, so that $[c_{ij}]$ is the matrix $\begin{pmatrix} I \\ \Gamma \Delta \end{pmatrix}$. If $B_{ij} = \langle w_i, w_j \rangle$, then $B_{ij} = \langle w_i, w_j \rangle = \sum_{kl} c_{ki} \langle v_k, v_l \rangle \bar{c}_{lj}$, and hence

$$B = \begin{pmatrix} I \\ \Gamma \Delta \end{pmatrix}^t A \overline{\begin{pmatrix} I \\ \Gamma \Delta \end{pmatrix}}.$$

Thus two Hermitian matrices A and B represent the same Hermitian form in different bases if and only if $B = M^* A M$ for some nonsingular matrix M.

Proposition 6.8.
(a) If $\langle \cdot, \cdot \rangle$ is a Hermitian form on a finite-dimensional vector space V over $\mathbb{C}$, then there exists an ordered basis of V in which the matrix of $\langle \cdot, \cdot \rangle$ is diagonal with real entries.
(b) If A is an n-by-n Hermitian matrix, then there exists a nonsingular n-by-n matrix M such that $M^* A M$ is diagonal.

PROOF. The above considerations show that (a) and (b) are reformulations of the same result. Hence it is enough to prove (b). By the Spectral Theorem (Theorem 3.21), there exists a unitary matrix U such that $U^{-1} A U$ is diagonal with real entries. Since U is unitary, $U^{-1} = U^*$. Thus we can take $M = U$ to prove (b). □

Just as with symmetric bilinear forms over $\mathbb{R}$, we can do a little better than Proposition 6.8 indicates. If B is Hermitian and diagonal with diagonal entries b_i, and if D is diagonal with positive entries d_i, then $C = D^* B D$ is diagonal with diagonal entries $d_i^2 b_i$. Choosing D suitably and then replacing C by $P^t C P$ for a suitable permutation matrix P, we may assume that $P^t C P$ is of the form diag$(+1, \ldots, +1, -1, \ldots, -1, 0, \ldots, 0)$. The number of $+1$'s and -1's together is the rank of A, and the form is nondegenerate if and only if there are no 0's. The triple given by

$$(p, m, z) = \big(\#(+1)\text{'s}, \#(-1)\text{'s}, \#(0)\text{'s}\big)$$

is again called the **signature** of A. A similar notion can be defined in the case of a Hermitian form, as opposed to a Hermitian matrix.

Theorem 6.9 (Sylvester's Law). The signature of an n-by-n Hermitian matrix is well defined.

The proof is the same as for Theorem 6.6 except for adjustments in notation.

5. Groups Leaving a Bilinear Form Invariant

Although it is not logically necessary to do so, we digress in this section to introduce some important groups that are defined by means of bilinear or Hermitian forms. These groups arise in many areas of mathematics, both pure and applied, and their detailed structure constitutes a topic in the fields of Lie groups, algebraic groups, and finite groups that is beyond the scope of this book. Thus the best place to define them seems to be now.

We limit our comments on applications to just these: When the underlying field in the definition of these groups is $\mathbb{R}$ or $\mathbb{C}$, the group is quite often a "simple Lie group," one of the basic building blocks of the theory of the continuous groups that so often arise in topology, geometry, differential equations, and mathematical physics. When the underlying field is a number field in the sense of Example 9 of Section IV.1, the group quite often plays a role in algebraic number theory. When the underlying field is a finite field, the group is often closely related to a finite simple group; an example of this relationship occurred in Problems 55–62 at the end of Chapter IV, where it was shown that the group PSL$(2, \mathbb{K})$, built in an easy way from the general linear group GL$(2, \mathbb{K})$, is simple if the field $\mathbb{K}$ has more than 5 elements. More general examples of finite simple groups produced by analogous constructions are said to be of "Lie type." A celebrated theorem of the late twentieth century classified the finite simple groups—establishing that the only such groups are the cyclic groups of prime order, the alternating groups on 5 or more letters, the simple groups of Lie type, and 26 so-called sporadic simple groups.

If $\langle \,\cdot\,, \,\cdot\, \rangle$ is a bilinear form on an n-dimensional vector space V over a field $\mathbb{K}$, a nonsingular linear map $g : V \to V$ is said to **leave the bilinear form invariant** if

$$\langle g(u), g(v) \rangle = \langle u, v \rangle$$

for all u and v in V. Fix an ordered basis Γ of V, let A be the matrix of the bilinear form in this basis, let $g' = \begin{pmatrix} g \\ \Gamma\Gamma \end{pmatrix}$ be the member of $GL(n, \mathbb{K})$ corresponding to g, and abbreviate $\begin{pmatrix} w \\ \Gamma \end{pmatrix}$ as w' for any w in V. To translate the invariance condition into one concerning matrices, we use the formula $\langle u, v \rangle = u'^t A v'$, the corresponding formula for $\langle g(u), g(v) \rangle$, and the formula $g(w)' = g'(w')$ from Theorem 2.14. Then we obtain $u'^t g'^t A g' v' = u'^t A v'$. Taking u to be the ith member of the ordered basis Γ and v to be the jth member, we obtain equality of the $(i, j)^{\text{th}}$ entry of the two matrices $g'^t A g'$ and A. Thus the matrix form of the invariance condition is that a nonsingular matrix g' satisfy

$$g'^t A g' = A.$$

We know that changing the ordered basis Γ amounts to replacing A by $M^t A M$ for some nonsingular matrix M. If g' satisfies the invariance condition $g'^t A g' = A$ relative to A, then $M^{-1} g' M$ satisfies

$$(M^{-1} g' M)^t (M^t A M)(M^{-1} g' M) = M^t A M.$$

Thus we are led to a conjugate subgroup within $GL(n, \mathbb{K})$. A conjugate subgroup is not something substantially new, and thus we might as well make a convenient choice of basis so that A looks particularly special.

The interesting cases are that the given bilinear form is symmetric or alternating, hence that the matrix A is symmetric or alternating. Let us restrict our attention to them. The left and right radicals coincide in these cases, and the first thing to do is to take the two-sided radical into account. Returning to the original bilinear form, we write $V = \text{rad} \oplus S$, where rad is the radical and S is some vector subspace of S, and we choose an ordered basis $(v_1, \ldots, v_p, v_{p+1}, \ldots, v_n)$ such that $v_1, \ldots, v_p$ are in S and $v_{p+1}, \ldots, v_n$ are in rad. Then $\langle v_i, v_j \rangle = 0$ if $i > p$ or $j > p$, and consequently A has its only nonzero entries in the upper left p-by-p block. The same argument as in the proofs of Theorems 6.5 and 6.7 shows that the restriction of the bilinear form to S is nondegenerate, and consequently the upper left p-by-p block of A is nonsingular. Changing notation slightly, suppose that g is an n-by-n matrix written in block form as $g = \begin{pmatrix} g_{11} & g_{12} \\ g_{21} & g_{22} \end{pmatrix}$ with g_{11} of size p-by-p, suppose that $\begin{pmatrix} A & 0 \\ 0 & 0 \end{pmatrix}$ is another matrix written in the same block form, suppose that the p-by-p matrix A is nonsingular, and suppose that $g^t \begin{pmatrix} A & 0 \\ 0 & 0 \end{pmatrix} g = \begin{pmatrix} A & 0 \\ 0 & 0 \end{pmatrix}$. Making a brief computation, we find that necessary and sufficient conditions on g are that g_{11} be nonsingular and have $g_{11}^t A g_{11} = A$, that $g_{12} = 0$, that g_{22} be arbitrary nonsingular, and that g_{21} be arbitrary. In other words, the only interesting condition $g_{11}^t A g_{11} = A$ is a reflection of what happens in the nonsingular case.

Consequently the interesting cases are that the given bilinear form is nondegenerate, as well as either symmetric or alternating. If A is symmetric and nonsingular, then the group of all nonsingular matrices g such that $g^t A g = A$ is called the **orthogonal group** relative to A. If A is alternating and nonsingular, then the group of all nonsingular matrices g such that $g^t A g = A$ is called the **symplectic group** relative to A.

For the symplectic case it is customary to invoke Theorem 6.7 and take A to be

$$ J = \begin{pmatrix} \begin{array}{|cc|} \hline 0 & 1 \\ -1 & 0 \\ \hline \end{array} & & & \\ & \begin{array}{|cc|} \hline 0 & 1 \\ -1 & 0 \\ \hline \end{array} & & \\ & & \ddots & \\ & & & \begin{array}{|cc|} \hline 0 & 1 \\ -1 & 0 \\ \hline \end{array} \end{pmatrix}, $$

except possibly for a permutation of the rows and columns and possibly for a multiplication by -1. Two conflicting notations are in common use for the symplectic group, namely $Sp(n, \mathbb{K})$ and $Sp(\frac{1}{2}n, \mathbb{K})$, and one always has to check a particular author's definitions.

For the orthogonal case the notation is less standardized. Theorem 6.5 says that we may take A to be diagonal except when $\mathbb{K}$ has characteristic 2. But the theorem does not tell us exactly which A's are representative of the same bilinear form. When $\mathbb{K} = \mathbb{C}$, we know that we can take A to be the identity matrix I. The group is known as the complex orthogonal group and is denoted by $O(n, \mathbb{C})$. When $\mathbb{K} = \mathbb{R}$, we can take A to be diagonal with diagonal entries ± 1. Sylvester's Law (Theorem 6.6) says that the form determines the number of $+1$'s and the number of -1's. The groups are called indefinite orthogonal groups and are denoted by $O(p, q)$, where p is the number of $+1$'s and q is the number of -1's. When $q = 0$, we obtain the ordinary orthogonal group of matrices relative to an inner product.

A similar analysis applies to Hermitian forms. The field is now $\mathbb{C}$, the invariance condition with the form is still $\langle g(u), g(v) \rangle = \langle u, v \rangle$, and the corresponding condition with matrices is $g^t A \bar{g} = A$. The interesting case is that the Hermitian form is nondegenerate. Proposition 6.8 and Sylvester's Law (Theorem 6.9) together show that we may take A to be diagonal with diagonal entries ± 1 and that the Hermitian form determines the number of $+1$'s and the number of -1's. The groups are the indefinite unitary groups and are denoted by $U(p, q)$, where p is the number of $+1$'s and q is the number of -1's. When $q = 0$, we obtain the ordinary unitary group of matrices relative to an inner product.

6. Tensor Product of Two Vector Spaces

If E is a vector space over $\mathbb{K}$, then the set of all bilinear forms on E is a vector space under addition and scalar multiplication of the values, i.e., it is a vector subspace of the set of all functions from $E \times E$ into $\mathbb{K}$. In this section we introduce a vector space called the "tensor product" of E with itself, whose dual, even if E is infinite-dimensional, is canonically isomorphic to this vector space of bilinear forms.

Matters will be clearer if we work initially with something slightly more general than bilinear forms on a single vector space E. Thus fix a field $\mathbb{K}$, and let E and F be vector spaces over $\mathbb{K}$. A function from $E \times F$ into a vector space U over $\mathbb{K}$ is said to be **bilinear** if it is linear in each of the two variables when the other one is held fixed. Such a space of bilinear functions is a vector space over $\mathbb{K}$ under addition and scalar multiplication of the values. The bilinear functions are called **bilinear forms** when the range space U is $\mathbb{K}$ itself. More generally, if $E_1, \ldots, E_k$ are vector spaces over $\mathbb{K}$, a function from $E_1 \times \cdots \times E_k$ into a vector space over $\mathbb{K}$ is said to be k-**linear** or k-**multilinear** if it is linear in each of its k variables when the other $k - 1$ variables are held fixed. Again the word "form" is used in the scalar-valued case, and all of these spaces of multilinear functions are vector spaces over $\mathbb{K}$.

In this section we shall introduce the tensor product of two vector spaces E and F over $\mathbb{K}$, ultimately denoting it by $E \otimes_{\mathbb{K}} F$. The dual of this tensor product will be canonically isomorphic to the vector space of bilinear forms on $E \times F$. More generally the space of linear functions from the tensor product into a vector space U will be canonically isomorphic to the vector space of bilinear functions on $E \times F$ with values in U.

Following the habit encouraged by Chapter IV, we want to arrange that tensor product is a functor. If $\mathcal{V}$ denotes the category of vector spaces over $\mathbb{K}$ and if $\mathcal{V} \times \mathcal{V}$ denotes the category described in Section IV.11 as $\mathcal{V}^S$ for a two-element set S, then tensor product is to be a functor from $\mathcal{V} \times \mathcal{V}$ into $\mathcal{V}$. Hence we will want to examine the effect of tensor products on morphisms, i.e., on linear maps.

As in similar constructions in Chapter IV, the effect of tensor product on linear maps is captured by defining the tensor product by means of a universal mapping property. The appropriate universal mapping property rephrases the statement above that the space of linear functions from the tensor product into any vector space U is canonically isomorphic to the vector space of bilinear functions on $E \times F$ with values in U.

If E and F are vector spaces over $\mathbb{K}$, a **tensor product** of E and F is a pair (V, ι) consisting of a vector space V over $\mathbb{K}$ together with a bilinear function $\iota : E \times F \to V$, with the following **universal mapping property**: whenever b is a bilinear mapping of $E \times F$ into a vector space U over $\mathbb{K}$, then there exists a unique linear mapping B of V into U such that the diagram in Figure 6.1 commutes, i.e., such that $B\iota = b$ holds in the diagram. When ι is understood, one frequently refers to V itself as the tensor product. The linear mapping $B : V \to U$ is called the **linear extension** of b to the tensor product.

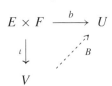

FIGURE 6.1. Universal mapping property of a tensor product.

Theorem 6.10. If E and F are vector spaces over $\mathbb{K}$, then a tensor product of E and F exists and is unique up to canonical isomorphism in this sense: if (V_1, ι_1) and (V_2, ι_2) are tensor products, then there exists a unique linear mapping $B : V_2 \to V_1$ with $B\iota_2 = \iota_1$, and B is an isomorphism. Any tensor product is spanned linearly by the image of $E \times F$ in it.

REMARKS. As usual, uniqueness will follow readily from the universal mapping property. What is really needed is a proof of existence. This will be carried out by an explicit construction. Later, in Chapter X, we shall reintroduce tensor products, taking the basic construction to be that of the tensor product of two abelian groups, and then the tensor product of two vector spaces will in effect be obtained in a slightly different way. However, the exact construction does not matter, only the existence; the uniqueness allows us to match the results of any two constructions.

FIGURE 6.2. Diagrams for uniqueness of a tensor product.

PROOF OF UNIQUENESS. Let (V_1, ι_1) and (V_2, ι_2) be tensor products. Set up the diagrams in Figure 6.2, and use the universal mapping property to obtain linear maps $B_2 : V_1 \to V_2$ and $B_1 : V_2 \to V_1$ extending ι_2 and ι_1. Then $B_1 B_2 : V_1 \to V_1$ has $B_1 B_2 \iota_1 = B_1 \iota_2 = \iota_1$, and $1_{V_1} : V_1 \to V_1$ has $(1_{V_1})\iota_1 = \iota_1$. By the assumed uniqueness within the universal mapping property, $B_1 B_2 = 1_{V_1}$ on V_1. Similarly $B_2 B_1 = 1_{V_2}$ on V_2. Then $B_1 : V_2 \to V_1$ gives the canonical isomorphism. Because of the isomorphism the image of $E \times F$ will span an arbitrary tensor product if it spans some particular tensor product. □

PROOF OF EXISTENCE. Let $V_1 = \bigoplus_{(e,f)} \mathbb{K}(e, f)$, the direct sum being taken over all ordered pairs (e, f) with $E \in E$ and $f \in F$. Then V_1 is a vector space over $\mathbb{K}$ with a basis consisting of all ordered pairs (e, f). We think of all identities that the elements of V_1 must satisfy to be a tensor product, writing each as some expression set equal to 0, and then we assemble those expressions into a vector subspace to factor out from V_1. Namely, let V_0 be the vector subspace of V_1 generated by all elements of any of the kinds

$$(e_1 + e_2, f) - (e_1, f) - (e_2, f),$$
$$(ce, f) - c(e, f),$$
$$(e, f_1 + f_2) - (e, f_1) - (e, f_2),$$
$$(e, cf) - c(e, f),$$

the understanding being that c is in $\mathbb{K}$, the elements e, e_1, e_2 are in E, and the elements f, f_1, f_2 are in F. Define $V = V_1/V_0$, and define $\iota : E \times F \to V_1/V_0$ by $\iota(e, f) = (e, f) + V_0$. We shall prove that (V, ι) is a tensor product of E and F. The definitions show that the image of ι spans V linearly.

Let $b : E \times F \to U$ be given as in Figure 6.1. To see that a linear extension B exists and is unique, define B_1 on V_1 by

$$B_1\Big(\sum_{\text{(finite)}} c_i(e_i, f_i)\Big) = \sum_{\text{(finite)}} c_i b(e_i, f_i).$$

The bilinearity of b shows that B_1 maps V_0 to 0. By Proposition 2.25, B_1 descends to a linear map $B : V_1/V_0 \to U$, and we have $B\iota = b$. Hence B exists as required.

To check uniqueness of B, we observe again that the cosets $(e, f) + V_0$ within V_1/V_0 span V; since commutativity of the diagram in Figure 6.1 forces

$$B((e, f) + V_0) = B(\iota(e, f)) = b(e, f),$$

B is unique. This completes the proof. □

A tensor product of E and F is denoted by $(E \otimes_{\mathbb{K}} F, \iota)$, with the bilinear map ι given by $\iota(e, f) = e \otimes f$; the map ι is frequently dropped from the notation when there is no chance of ambiguity. The tensor product that was constructed in the proof of existence in Theorem 6.10 is not given any special notation to distinguish it from any other tensor product. The elements $e \otimes f$ span $E \otimes_{\mathbb{K}} F$, as was noted in the statement of the theorem. Elements of the form $e \otimes f$ are sometimes called **pure tensors**.

Not every element need be a pure tensor, but every element in $E \otimes_{\mathbb{K}} F$ is a finite sum of pure tensors. We shall see in Proposition 6.14 that if $\{u_i\}$ is a basis of E and $\{v_j\}$ is a basis of F, then the pure tensors $u_i \otimes v_j$ form a basis of $E \otimes_{\mathbb{K}} F$. In particular the dimension of the tensor product is the product of the dimensions of the factors. We could have defined the tensor product in this way—by taking bases and declaring that $u_i \otimes v_j$ is to be a basis of the desired space. The difficulty is that we would be forever wedded to our choice of those particular bases, or we would constantly have to prove that our definitions are independent of bases. The definition by means of Theorem 6.10 avoids this difficulty.

To make tensor product $(E, F) \mapsto E \otimes_{\mathbb{K}} F$ into a functor, we have to describe the effect on linear mappings. To aid in that discussion, let us reintroduce some notation first used in Chapter II: if U and V are vector spaces over $\mathbb{K}$, then $\text{Hom}_{\mathbb{K}}(U, V)$ is defined to be the vector space of $\mathbb{K}$ linear maps from U to V.

Corollary 6.11. If E, F, and V are vector spaces over $\mathbb{K}$, then the vector space $\text{Hom}_{\mathbb{K}}(E \otimes_{\mathbb{K}} F, V)$ is canonically isomorphic (via restriction to pure tensors) to the vector space of all V-valued bilinear functions on $E \times F$.

PROOF. Restriction is a linear mapping from $\text{Hom}_{\mathbb{K}}(E \otimes_{\mathbb{K}} F, V)$ to the vector space of all V-valued bilinear functions on $E \times F$, and it is one-one since the image of $E \times F$ in $E \otimes_{\mathbb{K}} F$ spans $E \otimes_{\mathbb{K}} F$. It is onto since any bilinear function from $E \times F$ to V has a linear extension to $E \otimes_{\mathbb{K}} F$, by Theorem 6.10. □

Corollary 6.12. If E and F are vector spaces over $\mathbb{K}$, then the vector space of all bilinear forms on $E \times F$ is canonically isomorphic to $(E \otimes_{\mathbb{K}} F)'$, the dual of the vector space $E \otimes_{\mathbb{K}} F$.

PROOF. This is the special case of Corollary 6.11 in which $V = \mathbb{K}$. □

Corollary 6.13. If E, F, and V are vector spaces over $\mathbb{K}$, then there is a canonical $\mathbb{K}$ linear isomorphism Φ of left side to right side in

$$\text{Hom}_{\mathbb{K}}(E \otimes_{\mathbb{K}} F, V) \cong \text{Hom}_{\mathbb{K}}(E, \text{Hom}_{\mathbb{K}}(F, V))$$

such that

$$\Phi(\varphi)(e)(f) = \varphi(e \otimes f)$$

for all $\varphi \in \text{Hom}_{\mathbb{K}}(E \otimes_{\mathbb{K}} F, V), e \in E$, and $f \in F$.

REMARK. This result is just a restatement of Corollary 6.11, but let us prove it anyway, writing the proof in the language of the statement.

PROOF. The map Φ is well defined and $\mathbb{K}$ linear, and it carries the left side to the right side. For ψ in the right side, define $\Psi(\psi)(e, f) = \psi(e)(f)$. Then $\Psi(\psi)$ is a bilinear map from $E \times F$ into V, and we let $\widetilde{\Psi}(\psi)$ be the linear extension from $E \otimes_{\mathbb{K}} F$ into V given in Theorem 6.10. Then $\widetilde{\Psi}$ is a two-sided inverse to Φ, and the corollary follows. □

Let us now make $(E, F) \mapsto E \otimes_\mathbb{K} F$ into a covariant functor. If (E_1, F_1) and (E_2, F_2) are objects in $\mathcal{V} \times \mathcal{V}$, i.e., if they are two ordered pairs of vector spaces, then a morphism from the first to the second is a pair (L, M) of linear maps of the form $L : E_1 \to E_2$ and $M : F_1 \to F_2$. To (L, M), we are to associate a linear map from $E_1 \otimes_\mathbb{K} F_1$ into $E_2 \otimes_\mathbb{K} F_2$; this linear map will be denoted by $L \otimes M$. We use Corollary 6.11 to define $L \otimes M$ as the member of $\mathrm{Hom}_\mathbb{K}(E_1 \otimes_\mathbb{K} F_1, E_2 \otimes_\mathbb{K} F_2)$ that corresponds under restriction to the bilinear map $(e_1, f_1) \mapsto L(e_1) \otimes M(f_1)$ of $E_1 \times F_1$ into $E_2 \otimes_\mathbb{K} F_2$. In terms of pure tensors, the map $L \otimes M$ satisfies

$$(L \otimes M)(e_1 \otimes f_1) = L(e_1) \otimes M(f_1),$$

and this formula completely determines $L \otimes M$ because of the uniqueness of linear extensions of bilinear maps.

To check that this definition of the effect of tensor product on pairs of linear maps makes $(E, F) \mapsto E \otimes_\mathbb{K} F$ into a covariant functor, we have to check the effect on the identity map and the effect on composition. For the effect on the identity map $(1_{E_1}, 1_{F_1})$ when $E_1 = E_2$ and $F_1 = F_2$, we see from the above displayed formula that $(1_{E_1} \otimes 1_{F_1})(e_1 \otimes f_1) = 1_{E_1}(e_1) \otimes 1_{F_1}(f_1) = e_1 \otimes f_1 = 1_{E_1 \otimes_\mathbb{K} F_1}(e_1 \otimes f_1)$. Since elements of the form $e_1 \otimes f_1$ span $E_1 \otimes_\mathbb{K} F_1$, we conclude that $1_{E_1} \otimes 1_{F_1} = 1_{E_1 \otimes_\mathbb{K} F_1}$.

For the effect on composition, let $(L_1, M_1) : (E_1, F_1) \to (E_2, F_2)$ and $(L_2, M_2) : (E_2, F_2) \to (E_3, F_3)$ be given. Then we have

$$(L_2 \otimes M_2)(L_1 \otimes M_1)(e_1 \otimes f_1) = (L_2 \otimes M_2)(L_1(e_1) \otimes M_1(f_1))$$
$$= (L_2 L_1)(e_1) \otimes (M_2 M_1)(f_1) = (L_2 L_1 \otimes M_2 M_1)(e_1 \otimes f_1).$$

Since elements of the form $e_1 \otimes f_1$ span $E_1 \otimes_\mathbb{K} F_1$, we conclude that

$$(L_2 \otimes M_2)(L_1 \otimes M_1) = L_2 L_1 \otimes M_2 M_1.$$

Therefore $(E, F) \mapsto E \otimes_\mathbb{K} F$ is a covariant functor.

In particular, $E \mapsto E \otimes_\mathbb{K} F$ and $F \mapsto E \otimes_\mathbb{K} F$ are covariant functors from $\mathcal{V}$ into itself. For these two functors from $\mathcal{V}$ into itself, the effect on linear mappings is especially nice, namely that

$$L_1 \mapsto L_1 \otimes M_1 \quad \begin{cases} \text{is } \mathbb{K} \text{ linear from } \mathrm{Hom}_\mathbb{K}(E_1, E_2) \\ \text{into } \mathrm{Hom}_\mathbb{K}(E_1 \otimes_\mathbb{K} F_1, E_2 \otimes_\mathbb{K} F_2), \end{cases}$$

$$M_1 \mapsto L_1 \otimes M_1 \quad \begin{cases} \text{is } \mathbb{K} \text{ linear from } \mathrm{Hom}_\mathbb{K}(F_1, F_2) \\ \text{into } \mathrm{Hom}_\mathbb{K}(E_1 \otimes_\mathbb{K} F_1, E_2 \otimes_\mathbb{K} F_2). \end{cases}$$

To prove the first of these assertions, for example, we observe that the sum of the linear extensions of

$$(e_1, f_1) \mapsto L_1(e_1) \otimes M_1(f_1) \quad \text{and} \quad (e_1, f_1) \mapsto L_1'(e_1) \otimes M_1(f_1)$$

is a linear extension of $(e_1, f_1) \mapsto (L_1+L_1')(e_1) \otimes M_1(f_1)$, and the uniqueness in the universal mapping property implies that $(L_1+L_1') \otimes M_1 = L_1 \otimes M_1 + L_1' \otimes M_1$. Similar remarks apply to multiplication by scalars.

Let us mention some identities satisfied by $\otimes_\mathbb{K}$. There is a canonical isomorphism

$$E \otimes_\mathbb{K} F \cong F \otimes_\mathbb{K} E$$

given by taking the linear extension of $(e, f) \mapsto f \otimes e$ as the map from left to right. The linear extension of $(f, e) \mapsto e \otimes f$ gives a two-sided inverse. Category theory has a way of capturing the idea that this isomorphism is systematic, rather than randomly dependent on E and F. The two sides of the above isomorphism may be regarded as the values of the covariant functors $(E, F) \mapsto E \otimes_\mathbb{K} F$ and $(E, F) \mapsto F \otimes_\mathbb{K} E$. The notion in category theory capturing "systematic" is called "naturality." It makes precise the fact that the system of isomorphisms respects linear maps, as well as the vector spaces. Here is the general definition. Its usefulness will be examined later in this section.

Let $\mathcal{C}$ and $\mathcal{D}$ be two categories, and let $\Phi : \mathcal{C} \to \mathcal{D}$ and $\Psi : \mathcal{C} \to \mathcal{D}$ be covariant functors. Suppose that for each X in $\text{Obj}(\mathcal{C})$, a morphism T_X in $\text{Morph}_\mathcal{D}(\Phi(X), \Psi(X))$ is given. Then the system $\{T_X\}$ is called a **natural transformation** of Φ into Ψ if for each pair of objects X_1 and X_2 in $\mathcal{C}$ and each h in $\text{Morph}_\mathcal{C}(X_1, X_2)$, the diagram in Figure 6.3 commutes. If furthermore each T_X is an isomorphism, then it is immediate that the system $\{T_X^{-1}\}$ is a natural transformation of Ψ into Φ, and we say that $\{T_X\}$ is a **natural isomorphism**.

$$\begin{array}{ccc} \Phi(X_1) & \xrightarrow{\Phi(h)} & \Phi(X_2) \\ T_{X_1} \downarrow & & \downarrow T_{X_2} \\ \Psi(X_1) & \xrightarrow{\Psi(h)} & \Psi(X_2) \end{array}$$

FIGURE 6.3. Commutative diagram of a natural transformation $\{T_X\}$.

If Φ and Ψ are contravariant functors, then the system $\{T_X\}$ is called a **natural transformation** of Φ into Ψ if the diagram obtained from Figure 6.3 by reversing the horizontal arrows commutes. The system is a **natural isomorphism** if furthermore each T_x is an isomorphism.

In the case we are studying, we have $\mathcal{C} = \mathcal{V} \times \mathcal{V}$ and $\mathcal{D} = \mathcal{V}$. Objects X in $\mathcal{C}$ are pairs (E, F) of vector spaces, and Φ and Ψ are the covariant functors with $\Phi(E, F) = E \otimes_\mathbb{K} F$ and $\Psi(E, F) = F \otimes_\mathbb{K} E$. The mapping $T_{(E,F)} : E \otimes_\mathbb{K} F \to F \otimes_\mathbb{K} E$ is uniquely determined by the condition that $T_{(E,F)}(e \otimes f) = f \otimes e$ for all $e \in E$ and $f \in F$. A morphism of pairs from (E_1, F_1) to (E_2, F_2) is of the form $h = (L, M)$ with $L \in \operatorname{Hom}_{\mathbb{K}}(E_1, E_2)$ and $M \in \operatorname{Hom}_{\mathbb{K}}(F_1, F_2)$. Our constructions above show that

$$\Phi(L, M) = L \otimes M \in \operatorname{Hom}_{\mathbb{K}}(E_1 \otimes_{\mathbb{K}} F_1, E_2 \otimes_{\mathbb{K}} F_2)$$

and
$$\Psi(L, M) = M \otimes L \in \operatorname{Hom}_{\mathbb{K}}(F_1 \otimes_{\mathbb{K}} E_1, F_2 \otimes_{\mathbb{K}} E_2).$$

In Figure 6.3 the two routes from top left to bottom right in the diagram have

$$T_{(E_2, F_2)} \Phi(L, M)(e_1 \otimes f_1) = T_{(E_2, F_2)}(L \otimes M)(e_1 \otimes f_1)$$
$$= T_{(E_2, F_2)}(L(e_1) \otimes M(f_1)) = M(f_1) \otimes L(e_1)$$

and

$$\Psi(L, M) T_{(E_1, F_1)}(e_1 \otimes f_1) = \Psi(L, M)(f_1 \otimes e_1)$$
$$= (M \otimes L)(f_1 \otimes e_1) = M(f_1) \otimes L(e_1).$$

The results are equal, and therefore the diagram commutes. Consequently the isomorphism

$$E \otimes_{\mathbb{K}} F \cong F \otimes_{\mathbb{K}} E$$

is natural in the pair (E, F).

Another canonical isomorphism of interest is

$$E \otimes_{\mathbb{K}} \mathbb{K} \cong E.$$

Here the map from left to right is the linear extension of $(e, c) \mapsto ce$, while the map from right to left is $e \mapsto e \otimes 1$. In view of the previous canonical isomorphism, we have $\mathbb{K} \otimes_{\mathbb{K}} E \cong E$ also. Each of these isomorphisms is natural in E.

Next let us consider how $\otimes_{\mathbb{K}}$ interacts with direct sums. The result is that tensor product distributes over direct sums, even infinite direct sums:

$$E \otimes_{\mathbb{K}} \left(\bigoplus_{s \in S} F_s \right) \cong \bigoplus_{s \in S} (E \otimes_{\mathbb{K}} F_s).$$

The map from left to right is the linear extension of the bilinear map $(e, \{f_s\}_{s \in S}) \mapsto \{e \otimes f_s\}_{s \in S}$. For the definition of the inverse, the constructions of Section II.6 show that we have only to define the map on each $E \otimes_{\mathbb{K}} F_s$, where it is the linear extension of $(e, f_s) \mapsto e \otimes \{i_s(f_s))\}_{s \in S}$; here $i_{s_0} : F_{s_0} \to \bigoplus_s F_s$ is the one-one linear map carrying the s_0^{th} vector space into the direct sum. Once again it is possible to prove that the isomorphism is natural; we omit the details.

It follows from the displayed isomorphism and the isomorphism $E \otimes_{\mathbb{K}} \mathbb{K} \cong E$ that if $\{x_i\}$ is a basis of E and $\{y_j\}$ is a basis of F, then $\{x_i \otimes y_j\}$ is a basis of $E \otimes_{\mathbb{K}} F$. This proves the following result.

6. Tensor Product of Two Vector Spaces

Proposition 6.14. If E and F are vector spaces over $\mathbb{K}$, then

$$\dim(E \otimes_\mathbb{K} F) = (\dim E)(\dim F).$$

If $\{y_j\}$ is a basis of F, then the most general member of $E \otimes_\mathbb{K} F$ is of the form $\sum_j e_j \otimes y_j$ with all e_j in E.

We turn to a consideration of $\operatorname{Hom}_\mathbb{K}$ from the point of view of functors. In the examples in Section IV.11, we saw that $V \mapsto \operatorname{Hom}_\mathbb{K}(U, V)$ is a covariant functor from $\mathcal{V}$ to itself and that $U \mapsto \operatorname{Hom}_\mathbb{K}(U, V)$ is a contravariant functor from $\mathcal{V}$ to itself. If we are not squeamish about mixing the two types—covariant and contravariant—then we can consider $(U, V) \mapsto \operatorname{Hom}_\mathbb{K}(U, V)$ as a functor[3] from $\mathcal{V} \times \mathcal{V}$ to $\mathcal{V}$. At any rate if L is in $\operatorname{Hom}_\mathbb{K}(U_1, U_2)$ and M is in $\operatorname{Hom}_\mathbb{K}(V_1, V_2)$, then $\operatorname{Hom}(L, M)$ carries $\operatorname{Hom}_\mathbb{K}(U_2, V_1)$ into $\operatorname{Hom}_\mathbb{K}(U_1, V_2)$ and is given by

$$\operatorname{Hom}(L, M)(h) = MhL \quad \text{for } h \in \operatorname{Hom}_\mathbb{K}(U_2, V_1).$$

It is evident that the result is $\mathbb{K}$ linear as a function of h, and hence

$$\operatorname{Hom}(L, M) \quad \text{is in } \operatorname{Hom}_\mathbb{K}\big(\operatorname{Hom}_\mathbb{K}(U_2, V_1), \operatorname{Hom}_\mathbb{K}(U_1, V_2)\big).$$

When we look for analogs for the functor $\operatorname{Hom}_\mathbb{K}$ of the identity $E \otimes_\mathbb{K} \mathbb{K} \cong E$ for the functor $\otimes_\mathbb{K}$, we are led to two identities. One is just the definition of the dual of a vector space:

$$\operatorname{Hom}_\mathbb{K}(U, \mathbb{K}) = U'.$$

The other is the natural isomorphism

$$\operatorname{Hom}_\mathbb{K}(\mathbb{K}, V) \cong V.$$

In the proof of the latter identity, the mapping from left to right is given by sending a linear $h : \mathbb{K} \to V$ to $h(1)$, and the mapping from right to left is given by sending v in V to h with $h(c) = cv$.

Next let us consider how $\operatorname{Hom}_\mathbb{K}$ interacts with direct sums and direct products. The construction $\operatorname{Hom}_\mathbb{K}(U, V)$ distributes over finite direct sums in each variable, but the situation with infinite direct sums or direct products is more subtle. Valid identities are

$$\operatorname{Hom}_\mathbb{K}\Big(\bigoplus_{s \in S} U_s, V\Big) \cong \prod_{s \in S} \operatorname{Hom}_\mathbb{K}(U_s, V)$$

and

$$\operatorname{Hom}_\mathbb{K}\Big(U, \prod_{s \in S} V_s\Big) \cong \prod_{s \in S} \operatorname{Hom}_\mathbb{K}(U, V_s),$$

---

[3]Readers who prefer to be careful about this point can regard U as in the category $\mathcal{V}^{\text{opp}}$ defined in Problems 78–80 at the end of Chapter IV. Then $(U, V) \mapsto \operatorname{Hom}_\mathbb{K}(U, V)$ is a covariant functor from $\mathcal{V}^{\text{opp}} \times \mathcal{V}$ into $\mathcal{V}$.

and these are natural isomorphisms. Proofs of these identities for all S and counterexamples related to them when S is infinite appear in Problems 7–8 at the end of the chapter.

We have already checked that the isomorphism $E \otimes_{\mathbb{K}} F \cong F \otimes_{\mathbb{K}} E$ is natural in (E, F), and we have asserted naturality in some other situations in which it is easy to check. The next proposition asserts naturality for the identity of Corollary 6.13, which combines $\otimes_{\mathbb{K}}$ and $\mathrm{Hom}_{\mathbb{K}}$ in a nontrivial way. After the proof of the result, we shall digress for a moment to indicate the usefulness of natural isomorphisms.

Proposition 6.15. Let E, F, V, E_1, F_1, and V_1 be vector spaces over $\mathbb{K}$, and let $L_{E_1} : E_1 \to E$, $L_{F_1} : F_1 \to F$, and $L_V : V \to V_1$ be $\mathbb{K}$ linear maps. Then the isomorphism Φ of Corollary 6.13 is natural in the sense that the diagram

$$\begin{array}{ccc} \mathrm{Hom}_{\mathbb{K}}(E \otimes_{\mathbb{K}} F, V) & \xrightarrow{\Phi} & \mathrm{Hom}_{\mathbb{K}}(E, \mathrm{Hom}_{\mathbb{K}}(F, V)) \\ {\scriptstyle \mathrm{Hom}(L_{E_1} \otimes L_{F_1}, L_V)} \Big\downarrow & & \Big\downarrow {\scriptstyle \mathrm{Hom}(L_{E_1}, \mathrm{Hom}(L_{F_1}, L_V))} \\ \mathrm{Hom}_{\mathbb{K}}(E_1 \otimes_{\mathbb{K}} F_1, V_1) & \xrightarrow{\Phi} & \mathrm{Hom}_{\mathbb{K}}(E_1, \mathrm{Hom}_{\mathbb{K}}(F_1, V_1)) \end{array}$$

commutes.

REMARKS. Observe that the first two linear maps L_{E_1} and L_{F_1} go in the opposite direction to the two vertical maps, while L_V goes in the same direction as the vertical maps. This is a reflection of the fact that both sides of the identity in Corollary 6.13 are contravariant in the first two variables and covariant in the third variable.

PROOF. For φ in $\mathrm{Hom}_{\mathbb{K}}(E \otimes_{\mathbb{K}} F, V)$, e_1 in E_1, and f_1 in F_1, we have

$$(\mathrm{Hom}(L_{E_1}, \mathrm{Hom}(L_{F_1}, L_V)) \circ \Phi)(\varphi)(e_1)(f_1)$$
$$= (\mathrm{Hom}(L_{F_1}, L_V) \circ \Phi(\varphi) \circ L_{E_1})(e_1)(f_1)$$
$$= (\mathrm{Hom}(L_{F_1}, L_V) \circ (\Phi(\varphi) \circ L_{E_1}))(e_1)(f_1)$$
$$= L_V(\Phi(\varphi)(L_{E_1}(e_1))(L_{F_1}(f_1)))$$
$$= L_V(\varphi(L_{E_1}(e_1) \otimes L_{F_1}(f_1)))$$
$$= (L_V \circ \varphi \circ (L_{E_1} \otimes L_{F_1}))(e_1 \otimes f_1)$$
$$= (\mathrm{Hom}(L_{E_1} \otimes L_{F_1}, L_V)(\varphi))(e_1 \otimes f_1)$$
$$= \Phi(\mathrm{Hom}(L_{E_1} \otimes L_{F_1}, L_V) \circ \varphi)(e_1)(f_1).$$

This proves the proposition. $\square$

Let us now discuss naturality in a wider context. In a general category $\mathcal{D}$, if we have two objects U and U' such that $\text{Morph}(U, V)$ and $\text{Morph}(U', V)$ have the same cardinality for each object V, then we cannot really say anything about the relationship between U and U'. But under a hypothesis that the isomorphism of sets has a certain naturality to it, then, according to Proposition 6.16 below, U and U' are isomorphic objects. Thus naturality of a system of weak-looking set-theoretic isomorphisms can lead to a much stronger-looking isomorphism. Corollary 6.17 goes on to make a corresponding assertion about functors. The assertion about functors in the corollary is a helpful tool for establishing natural isomorphisms of functors, and an example appears below in Proposition 6.20'.

Proposition 6.16. Let $\mathcal{D}$ be a category, and suppose that U and U' are objects in $\mathcal{D}$ with the following property: to each object V in $\mathcal{D}$ corresponds a one-one onto function
$$T_V : \text{Morph}(U, V) \to \text{Morph}(U', V)$$
with the system $\{T_V\}$ natural in V in the sense that whenever σ is in $\text{Morph}(V, V')$, then the diagram

$$\begin{array}{ccc} \text{Morph}(U, V) & \xrightarrow{T_V} & \text{Morph}(U', V) \\ \text{left-by-}\sigma \downarrow & & \downarrow \text{left-by-}\sigma \\ \text{Morph}(U, V') & \xrightarrow{T_{V'}} & \text{Morph}(U', V') \end{array}$$

commutes. Then U is isomorphic to U' as an object in $\mathcal{D}$, an isomorphism from U to U' being the member $T_{U'}^{-1}(1_{U'})$ of $\text{Morph}(U, U')$.

REMARKS.

(1) Another way of formulating this result is as follows: Let $\mathcal{D}$ be any category, let $\mathcal{S}$ be the category of sets, and let U and U' be objects in $\mathcal{D}$. Define a covariant functor $H_U : \mathcal{D} \to \mathcal{S}$ by $H_U(V) = \text{Morph}_{\mathcal{D}}(U, V)$ and $H_U(\sigma) = \text{left-by-}\sigma$ for $\sigma \in \text{Morph}_{\mathcal{D}}(V, V')$, and define $H_{U'}$ similarly. If H_U and $H_{U'}$ are naturally isomorphic functors, then U and U' are isomorphic objects in $\mathcal{D}$.

(2) A similar result is valid when H_U and $H_{U'}$ are contravariant functors, H_U being defined by $H_U(V) = \text{Hom}_{\mathcal{D}}(V, U)$ and $H_U(\sigma) = \text{right-by-}\sigma$ for $\sigma \in \text{Morph}_{\mathcal{D}}(V, V')$. The result in this case follows immediately by applying Proposition 6.16 to the opposite category $\mathcal{D}^{\text{opp}}$ of $\mathcal{D}$ as defined in Problems 78–80 at the end of Chapter IV.

PROOF. Let φ be the element $T_{U'}^{-1}(1_{U'})$ of $\text{Morph}(U, U')$, and let ψ be the element $T_U(1_U)$ of $\text{Morph}(U', U)$. To prove the proposition, it is enough to show that $\varphi\psi = 1_{U'}$ and $\psi\varphi = 1_U$.

For σ in $\text{Morph}(V, V')$, form the commutative diagram in the statement of the proposition. The commutativity says that

$$\sigma T_V(h) = T_{V'}(\sigma h) \qquad \text{for } h \in \text{Morph}(U, V). \tag{$*$}$$

Taking $V = U$, $V' = U'$, $\sigma = \varphi$, and $h = 1_U$ in $(*)$ proves the second equality of the chain

$$\varphi\psi = \varphi T_U(1_U) = T_{U'}(\varphi 1_U) = T_{U'}(\varphi) = 1_{U'}.$$

Taking $V = U'$, $V' = U$, $\sigma = \psi$, and $h = \varphi$ in $(*)$ proves the first equality of the chain

$$T_U(\psi\varphi) = \psi T_{U'}(\varphi) = \psi 1_{U'} = \psi = T_U(1_U);$$

Applying T_U^{-1}, we obtain $\psi\varphi = 1_U$, as required. $\square$

Corollary 6.17. Let $\mathcal{C}$ and $\mathcal{D}$ be categories, and let $F : \mathcal{C} \to \mathcal{D}$ and $G : \mathcal{C} \to \mathcal{D}$ be covariant functors. Suppose that to each pair of objects (A, V) in $\mathcal{C} \times \mathcal{D}$ corresponds a one-one onto function

$$T_{A,V} : \text{Morph}(F(A), V) \to \text{Morph}(G(A), V)$$

with the system $\{T_{A,V}\}$ natural in (A, V). Then the functors F and G are naturally isomorphic.

REMARKS. A similar result is valid if $T_{A,V}$ carries $\text{Morph}(V, F(A))$ to $\text{Morph}(V, G(A))$ and/or if F and G are contravariant. To handle these situations, we apply the corollary to the opposite categories $\mathcal{D}^{\text{opp}}$ and/or $\mathcal{C}^{\text{opp}}$, as defined in Problems 78–80 at the end of Chapter IV, instead of to the categories $\mathcal{D}$ and/or $\mathcal{C}$.

PROOF. By Proposition 6.16 and the hypotheses, the member $T_{A,G(A)}^{-1}(1_{G(A)})$ of $\text{Morph}_\mathcal{D}(F(A), G(A))$ is an isomorphism. We are to prove that the system $\{T_{A,G(A)}\}$ is natural in A. If σ in $\text{Morph}_\mathcal{C}(A, A')$ is given, then the naturality of $T_{A,V}$ in the V variable implies that the diagram

$$\begin{array}{ccc}
\text{Morph}_\mathcal{D}(F(A), G(A)) & \xrightarrow{T_{A,G(A)}} & \text{Morph}_\mathcal{D}(G(A), G(A)) \\
\text{left-by-}G(\sigma) \downarrow & & \downarrow \text{left-by-}G(\sigma) \\
\text{Morph}_\mathcal{D}(F(A), G(A')) & \xrightarrow{T_{A,G(A')}} & \text{Morph}_\mathcal{D}(G(A), G(A'))
\end{array}$$

commutes. Evaluating at $T_{A,G(A)}^{-1}(1_{G(A)}) \in \text{Morph}_\mathcal{D}(F(A), G(A))$ the two equal compositions in the diagram, we obtain

$$G(\sigma) = G(\sigma)1_{G(A)} = T_{A,G(A')}\bigl(G(\sigma)T_{A,G(A)}^{-1}(1_{G(A)})\bigr). \tag{$*$}$$

6. Tensor Product of Two Vector Spaces

With σ as above, the naturality of $T_{A,V}$ in the A variable implies that the diagram

$$
\begin{array}{ccc}
\mathrm{Morph}_{\mathcal{D}}(F(A'), G(A')) & \xrightarrow{T_{A',G(A')}} & \mathrm{Morph}_{\mathcal{D}}(G(A'), G(A')) \\
\text{right-by-}F(\sigma) \downarrow & & \downarrow \text{right-by-}G(\sigma) \\
\mathrm{Morph}_{\mathcal{D}}(F(A), G(A')) & \xrightarrow{T_{A,G(A')}} & \mathrm{Morph}_{\mathcal{D}}(G(A), G(A'))
\end{array}
$$

commutes. Evaluating at $T^{-1}_{A',G(A')}(1_{G(A')}) \in \mathrm{Morph}_{\mathcal{D}}(F(A'), G(A'))$ the two equal compositions in the diagram, we obtain

$$G(\sigma) = 1_{G(A')}G(\sigma) = T_{A,G(A')}\bigl(T^{-1}_{A',G(A')}(1_{G(A')})F(\sigma)\bigr). \qquad (**)$$

Equations $(*)$ and $(**)$, together with the fact that $T_{A,G(A')}$ is invertible, say that

$$G(\sigma)T^{-1}_{A,G(A)}(1_{G(A)}) = T^{-1}_{A',G(A')}(1_{G(A')})F(\sigma).$$

In other words, the isomorphism $\widetilde{T}_A \in \mathrm{Morph}_{\mathcal{D}}(F(A), G(A))$ given by $\widetilde{T}_A = T^{-1}_{A,G(A)}(1_{G(A)})$ makes the diagram

$$
\begin{array}{ccc}
F(A) & \xrightarrow{\widetilde{T}_A} & G(A) \\
F(\sigma) \downarrow & & \downarrow G(\sigma) \\
F(A') & \xrightarrow{\widetilde{T}_{A'}} & G(A')
\end{array}
$$

commute. Thus F is naturally isomorphic to G. $\square$

Tensor product provides a device for converting a real vector space canonically into a complex vector space, so that a basis over $\mathbb{R}$ in the original space becomes a basis over $\mathbb{C}$ in the new space. If E is the given real vector space, then the complex vector space, called the **complexification** of E, is the space $E^{\mathbb{C}} = E \otimes_{\mathbb{R}} \mathbb{C}$ with multiplication by a complex number c in $E^{\mathbb{C}}$ defined to be $1 \otimes (z \mapsto cz)$.

This construction works more generally when we have any inclusion of fields $\mathbb{K} \subseteq \mathbb{L}$. In this situation, $\mathbb{L}$ becomes a vector space over $\mathbb{K}$ if scalar multiplication $\mathbb{K} \times \mathbb{L} \to \mathbb{L}$ is defined as the restriction of the multiplication $\mathbb{L} \times \mathbb{L} \to \mathbb{L}$ within $\mathbb{L}$. For any vector space E over $\mathbb{K}$, we define $E^{\mathbb{L}} = E \otimes_{\mathbb{K}} \mathbb{L}$, initially as a vector space over $\mathbb{K}$. For $c \in \mathbb{L}$, we then define

(multiplication by c in $E \otimes_{\mathbb{K}} \mathbb{L}$) = $1 \otimes$ (multiplication by c in $\mathbb{L}$).

The above identities concerning tensor products of linear maps allow one easily to prove the following identities:

$$c_1(c_2 v) = (c_1 c_2)v,$$
$$c(u + v) = cu + cv,$$
$$(c_1 + c_2)v = c_1 v + c_2 v,$$
$$1v = v.$$

Together these identities say that $E^{\mathbb{L}} = E \otimes_{\mathbb{K}} \mathbb{L}$, with its vector-space addition and the above definition of multiplication by scalars in $\mathbb{L}$, is a vector space over $\mathbb{L}$. The further identity

$$c(e \otimes 1) = ce \otimes 1 \qquad \text{if } c \text{ is in } \mathbb{K} \text{ and } e \text{ is in } E$$

shows that its scalar multiplication is consistent with scalar multiplication in E when the scalars are in $\mathbb{K}$ and E is identified with the subset $E \otimes 1$ of $E^{\mathbb{L}}$.

Let us say that the pair $(E^{\mathbb{L}}, \iota)$, where $\iota : E \to E^{\mathbb{L}}$ is the mapping $e \mapsto e \otimes 1$, is obtained by **extension of scalars**. This construction is characterized by a universal mapping property as follows.

Proposition 6.18. Let $\mathbb{K} \subseteq \mathbb{L}$ be an inclusion of fields, and let E be a vector space over $\mathbb{K}$.

(a) If $(E^{\mathbb{L}}, \iota)$ is formed by extension of scalars, then $(E^{\mathbb{L}}, \iota)$ has the following universal mapping property: whenever U is a vector space over $\mathbb{L}$ and $\varphi : E \to U$ is a $\mathbb{K}$ linear map, there exists a unique $\mathbb{L}$ linear map $\Phi : E^{\mathbb{L}} \to U$ such that $\Phi \iota = \varphi$.

(b) Suppose that (V, j) is any pair in which V is a vector space over $\mathbb{L}$ and $j : E \to V$ is a $\mathbb{K}$ linear function such that the following universal mapping property holds: whenever U is a vector space over $\mathbb{L}$ and $\varphi : E \to U$ is a $\mathbb{K}$ linear map, there exists a unique $\mathbb{L}$ linear map $\Phi : V \to U$ such that $\Phi j = \varphi$. Then there exists a unique isomorphism $\Psi : E^{\mathbb{L}} \to V$ of $\mathbb{L}$ vector spaces such that $\Psi \iota = j$.

PROOF. In (a), for the uniqueness of Φ, we must have $\Phi(e \otimes c) = c\Phi(e \otimes 1) = c(\Phi \iota)(e) = c\varphi(e)$. Hence Φ is determined by φ on pure tensors in $E \otimes_{\mathbb{K}} \mathbb{L}$ and therefore everywhere.

For existence let $\Phi : E \otimes_{\mathbb{K}} \mathbb{L} \to U$ be the $\mathbb{K}$ linear extension of the $\mathbb{K}$ bilinear function of $E \times \mathbb{L}$ into U given by

$$(e, c) \mapsto c\varphi(e) \qquad \text{for } e \in E \text{ and } c \in \mathbb{L}.$$

In the $\mathbb{L}$ vector space $E \otimes_{\mathbb{K}} \mathbb{L}$, multiplication by a member c_0 of $\mathbb{L}$ is defined to be $1 \otimes$ (multiplication by c_0). On a pure tensor $e \otimes c$, we therefore have

$$\Phi(c_0(e \otimes c)) = \Phi(e \otimes c_0 c) = (c_0 c)\varphi(e) = c_0(c\varphi(e)) = c_0(\Phi(e \otimes c)).$$

Since $E \otimes_{\mathbb{K}} \mathbb{L}$ is generated by pure tensors, Φ is $\mathbb{L}$ linear. By the construction of Φ, $\varphi(e) = \Phi(e \otimes 1) = (\Phi \iota)(e)$. Thus Φ has the required properties.

In (b), let (V, j) have the same universal mapping property as $(E^{\mathbb{L}}, \iota)$. We apply the universal mapping property of $(E^{\mathbb{L}}, \iota)$ to the $\mathbb{K}$ linear map $j : E \to V$ to obtain an $\mathbb{L}$ linear $\Phi : E^{\mathbb{L}} \to V$ with $\Phi \iota = j$, and we apply the universal mapping property of (V, j) to the $\mathbb{K}$ linear map $\iota : E \to E^{\mathbb{L}}$ to obtain an $\mathbb{L}$ linear $\Phi' : V \to E^{\mathbb{L}}$ with $\Phi' j = \iota$. From $(\Phi'\Phi)\iota = \Phi' j = \iota$ and $1_{E^{\mathbb{L}}} \iota = \iota$, the uniqueness in the universal mapping property for $(E^{\mathbb{L}}, \iota)$ implies $\Phi'\Phi = 1_{E^{\mathbb{L}}}$. Arguing similarly, we obtain $\Phi\Phi' = 1_V$. Thus Φ is an isomorphism with the required properties.

If $\Psi : E^{\mathbb{L}} \to V$ is another isomorphism with $\Psi \iota = j$, then the argument just given shows that $\Phi'\Psi = 1_{E^{\mathbb{L}}}$ and $\Psi\Phi' = 1_V$. Hence $\Psi = (\Phi')^{-1} = \Phi$, and Ψ is unique. $\square$

To make $E \mapsto E^{\mathbb{L}}$ into a covariant functor from vector spaces over $\mathbb{K}$ to vector spaces over $\mathbb{L}$, we must examine the effect on linear maps. The tool is Proposition 6.18a. Thus let E and F be two vector spaces over $\mathbb{K}$, and let $M : E \to F$ be a $\mathbb{K}$ linear map between them. We extend scalars for E and F. The proposition applies to the composition $E \to F \to F^{\mathbb{L}}$ and shows that the composition extends uniquely to an $\mathbb{L}$ linear map from $E^{\mathbb{L}}$ to $F^{\mathbb{L}}$. A quick look at the proof shows that this $\mathbb{L}$ linear map is $M \otimes 1$. Actually, we can see directly that $M \otimes 1$ is indeed linear over $\mathbb{L}$ and not just over $\mathbb{K}$: we just use our identity for compositions of tensor products to write

$$(M \otimes 1)(I \otimes (\text{multiplication by } c)) = M \otimes (\text{multiplication by } c)$$
$$= (I \otimes (\text{multiplication by } c))(M \otimes 1).$$

In any event, the explicit form of the extended linear map as $M \otimes 1$ shows immediately that the identity linear map goes to the identity and that compositions go to compositions. Thus $E \mapsto E^{\mathbb{L}}$ is a covariant functor.

In the special case that the vector spaces are $\mathbb{K}^n$ and $\mathbb{K}^m$, extension of scalars has a particularly simple interpretation. The new spaces may be viewed as $\mathbb{L}^n$ and $\mathbb{L}^m$. Thus column vectors with entries in $\mathbb{K}$ get replaced by column vectors with entries in $\mathbb{L}$. What happens with linear mappings is even more transparent. A linear map $M : E \to F$ is given by an m-by-n matrix A with entries in $\mathbb{K}$, and the linear map $M \otimes 1 : E^{\mathbb{L}} \to F^{\mathbb{L}}$ is the one given by the same matrix A. Now the entries of A are to be regarded as members of the larger field $\mathbb{L}$. Viewed this way, extension of scalars might look as if it is dependent on choices of bases, but the tensor-product formalism shows that it is not.

A related notion to extension of scalars is that of **restriction of scalars**. Again with an inclusion $\mathbb{K} \subseteq \mathbb{L}$ of fields, a vector space E over the larger field $\mathbb{L}$ becomes a vector space $E_{\mathbb{K}}$ over the smaller field $\mathbb{K}$ by ignoring unnecessary scalar multiplications. Although this notion is related to extension of scalars, it is not inverse to it. For example, if the two fields are $\mathbb{R}$ and $\mathbb{C}$ and if we start with an n-dimensional vector space E over $\mathbb{R}$, then $E^{\mathbb{C}}$ is a complex vector space of dimension n and $(E^{\mathbb{C}})_{\mathbb{R}}$ is a real vector space of dimension $2n$. We thus do not get back to the original space E.

7. Tensor Algebra

Just as polynomial rings are often used in the construction of more general commutative rings, so "tensor algebras" are often used in the construction of more general rings that may not be commutative. In this section we construct the "tensor algebra" of a vector space as a direct sum of iterated tensor products of the vector space with itself, and we establish its properties. We shall proceed with care, in order to provide a complete proof of the associativity of the multiplication.

Let A, B, and C be vector spaces over a field $\mathbb{K}$. A **triple tensor product** $V = A \otimes_{\mathbb{K}} B \otimes_{\mathbb{K}} C$ is a vector space over $\mathbb{K}$ with a 3-linear map $\iota : A \times B \times C \to V$ having the following universal mapping property: whenever t is a 3-linear mapping of $A \times B \times C$ into a vector space U over $\mathbb{K}$, then there exists a linear mapping T of V into U such that the diagram in Figure 6.4 commutes.

$$
\begin{array}{ccc}
A \times B \times C & \xrightarrow{t} & U \\
{\scriptstyle \iota}\downarrow & \nearrow {\scriptstyle T} & \\
V = A \otimes_{\mathbb{K}} B \otimes_{\mathbb{K}} C & &
\end{array}
$$

FIGURE 6.4. Commutative diagram of a triple tensor product.

The usual argument with universal mapping properties shows that there is at most one triple tensor product up to a well-determined isomorphism, and one can give an explicit construction of it that is similar to the one for ordinary tensor products $E \otimes_{\mathbb{K}} F$. We shall not need that particular proof of existence since Proposition 6.19a below will give us an alternative argument. Once we have that statement, we shall use the uniqueness of triple tensor products to establish in Proposition 6.19b an associativity formula for ordinary iterated tensor products.

A shorter proof of Proposition 6.19b, which avoids Proposition 6.19a and uses naturality, will be given after the proof of Proposition 6.20.

Proposition 6.19. If $\mathbb{K}$ is a field and A, B, C are vector spaces over $\mathbb{K}$, then
(a) $(A \otimes_{\mathbb{K}} B) \otimes_{\mathbb{K}} C$ and $A \otimes_{\mathbb{K}} (B \otimes_{\mathbb{K}} C)$ are triple tensor products.
(b) there exists a unique $\mathbb{K}$ isomorphism Φ from left to right in

$$(A \otimes_{\mathbb{K}} B) \otimes_{\mathbb{K}} C \cong A \otimes_{\mathbb{K}} (B \otimes_{\mathbb{K}} C)$$

such that $\Phi((a \otimes b) \otimes c) = a \otimes (b \otimes c)$ for all $a \in A, b \in B$, and $c \in C$.

PROOF. In (a), consider $(A \otimes_{\mathbb{K}} B) \otimes_{\mathbb{K}} C$. Let $t : A \times B \times C \to U$ be 3-linear. For $c \in C$, define $t_c : A \times B \to U$ by $t_c(a, b) = t(a, b, c)$. Then t_c is bilinear and hence extends to a linear $T_c : A \otimes_{\mathbb{K}} B \to U$. Since t is 3-linear, $t_{c_1+c_2} = t_{c_1} + t_{c_2}$ and $t_{xc} = xt_c$ for scalar x; thus uniqueness of the linear extension forces $T_{c_1+c_2} = T_{c_1} + T_{c_2}$ and $T_{xc} = xT_c$. Consequently

$$t' : (A \otimes_{\mathbb{K}} B) \times C \to U$$

given by $t'(d, c) = T_c(d)$ is bilinear and therefore extends to a linear $T : (A \otimes_{\mathbb{K}} B) \otimes_{\mathbb{K}} C \to U$. This T proves existence of the linear extension of the given t. Uniqueness is trivial, since the elements $(a \otimes b) \otimes c$ span $(A \otimes_{\mathbb{K}} B) \otimes_{\mathbb{K}} C$. So $(A \otimes_{\mathbb{K}} B) \otimes_{\mathbb{K}} C$ is a triple tensor product. In a similar fashion, $A \otimes_{\mathbb{K}} (B \otimes_{\mathbb{K}} C)$ is a triple tensor product.

For (b), set up the diagram of the universal mapping property for a triple tensor product, using $V = (A \otimes_{\mathbb{K}} B) \otimes_{\mathbb{K}} C$, $U = A \otimes_{\mathbb{K}} (B \otimes_{\mathbb{K}} C)$, and $t(a, b, c) = a \otimes (b \otimes c)$. We have just seen in (a) that V is a triple tensor product with $\iota(a, b, c) = (a \otimes b) \otimes c$. Thus there exists a linear $T : V \to U$ with $T\iota(a, b, c) = t(a, b, c)$. This equation means that $T((a \otimes b) \otimes c) = a \otimes (b \otimes c)$. Interchanging the roles of $(A \otimes_{\mathbb{K}} B) \otimes_{\mathbb{K}} C$ and $A \otimes_{\mathbb{K}} (B \otimes_{\mathbb{K}} C)$, we obtain a two-sided inverse for T. Thus T will serve as Φ in (b), and existence is proved. Uniqueness is trivial, since the elements $(a \otimes b) \otimes c$ span $(A \otimes_{\mathbb{K}} B) \otimes_{\mathbb{K}} C$. □

When there is no danger of confusion, Proposition 6.19 allows us to write a triple tensor product without parentheses as $A \otimes_{\mathbb{K}} B \otimes_{\mathbb{K}} C$. The same argument as in Corollaries 6.11 and 6.12 shows that the vector space of 3-linear forms on $A \times B \times C$ is canonically isomorphic to the dual of the vector space $A \otimes_{\mathbb{K}} B \otimes_{\mathbb{K}} C$.

Just as with Corollary 6.13 and Proposition 6.15, the result of Proposition 6.19 can be improved by saying that the isomorphism is natural in the variables A, B, and C, as follows.

Proposition 6.20. Let A, B, C, A_1, B_1, and C_1 be vector spaces over a field $\mathbb{K}$, and let $L_A : A \to A_1$, $L_B : B \to B_1$, and $L_C : C \to C_1$ be linear maps. Then the isomorphism Φ of Proposition 6.19b is natural in the triple (A, B, C) in the sense that the diagram

$$\begin{array}{ccc} (A \otimes_{\mathbb{K}} B) \otimes_{\mathbb{K}} C & \xrightarrow{\Phi} & A \otimes_{\mathbb{K}} (B \otimes_{\mathbb{K}} C) \\ {\scriptstyle (L_A \otimes L_B) \otimes L_C} \downarrow & & \downarrow {\scriptstyle L_A \otimes (L_B \otimes L_C)} \\ (A_1 \otimes_{\mathbb{K}} B_1) \otimes_{\mathbb{K}} C_1 & \xrightarrow{\Phi} & A_1 \otimes_{\mathbb{K}} (B_1 \otimes_{\mathbb{K}} C_1) \end{array}$$

commutes.

PROOF. We have
$$((L_A \otimes (L_B \otimes L_C)) \circ \Phi)((a \otimes b) \otimes c)$$
$$= (L_A \otimes (L_B \otimes L_C))(a \otimes (b \otimes c))$$
$$= L_A a \otimes (L_B \otimes L_C)(b \otimes c)$$
$$= L_A a \otimes (L_B b \otimes L_C c)$$
$$= \Phi((L_A a \otimes L_B b) \otimes L_C c)$$
$$= \Phi((L_A \otimes L_B)(a \otimes b) \otimes L_C c)$$
$$= (\Phi \circ ((L_A \otimes L_B) \otimes L_C))((a \otimes b) \otimes c),$$

and the proposition follows. $\square$

The treatment of Propositions 6.19 and 6.20 can be shortened if we are willing to bypass the notion of a triple tensor product and use what was proved about naturality in the previous section. The result and the proof are as follows.

Proposition 6.20'. Let A, B, and C be vector spaces over a field $\mathbb{K}$. Then there is an isomorphism $\Phi : (A \otimes_{\mathbb{K}} B) \otimes_{\mathbb{K}} C \to A \otimes_{\mathbb{K}} (B \otimes_{\mathbb{K}} C)$ that is natural in the triple (A, B, C) and satisfies $\Phi(a \otimes (b \otimes c)) = a \otimes (b \otimes c)$.

PROOF. Writing $\cong$ for "naturally isomorphic in all variables" and applying Proposition 6.15 and other natural isomorphisms of the previous section repeatedly, we have

$$\operatorname{Hom}_{\mathbb{K}} \big((A \otimes_{\mathbb{K}} B) \otimes_{\mathbb{K}} C, V \big) \cong \operatorname{Hom}_{\mathbb{K}} \big(A \otimes_{\mathbb{K}} B, \operatorname{Hom}_{\mathbb{K}}(C, V) \big)$$
$$\cong \operatorname{Hom}_{\mathbb{K}} \big(B, \operatorname{Hom}_{\mathbb{K}}(A, \operatorname{Hom}_{\mathbb{K}}(C, V)) \big)$$
$$\cong \operatorname{Hom}_{\mathbb{K}} \big(B, \operatorname{Hom}_{\mathbb{K}}(A \otimes_{\mathbb{K}} C, V) \big)$$
$$\cong \operatorname{Hom}_{\mathbb{K}} \big(B, \operatorname{Hom}_{\mathbb{K}}(C \otimes_{\mathbb{K}} A, V) \big)$$
$$\cong \operatorname{Hom}_{\mathbb{K}} \big((C \otimes_{\mathbb{K}} B) \otimes_{\mathbb{K}} A, V \big) \quad \text{by symmetry}$$
$$\cong \operatorname{Hom}_{\mathbb{K}} \big(A \otimes_{\mathbb{K}} (C \otimes_{\mathbb{K}} B), V \big)$$
$$\cong \operatorname{Hom}_{\mathbb{K}} \big(A \otimes_{\mathbb{K}} (B \otimes_{\mathbb{K}} C), V \big).$$

Then the existence of the natural isomorphism follows from Corollary 6.17. Using the explicit formula for the isomorphism in Proposition 6.16 and tracking matters down, we see that $\Phi(a \otimes (b \otimes c)) = a \otimes (b \otimes c)$. □

There is no difficulty in generalizing matters to n-fold tensor products by induction. An n-**fold tensor product** is to be universal for n-multilinear maps. Again it is unique up to canonical isomorphism, as one proves by an argument that runs along familiar lines. A direct construction of an n-fold tensor product is possible in the style of the proof for ordinary tensor products, but such a construction will not be needed. Instead, we can form an n-fold tensor product as the $(n-1)$-fold tensor product of the first $n-1$ spaces, tensored with the nth space. Proposition 6.19b allows us to regroup parentheses (inductively) in any fashion we choose, and the same argument as in Corollaries 6.11 and 6.12 yields the following proposition.

Proposition 6.21. If $E_1, \ldots, E_n$, and V are vector spaces over $\mathbb{K}$, then the vector space $\text{Hom}_{\mathbb{K}}(E_1 \otimes_{\mathbb{K}} \cdots \otimes_{\mathbb{K}} E_n, V)$ is canonically isomorphic (via restriction to pure tensors) to the vector space of all V-valued n-multilinear functions on $E_1 \times \cdots \times E_n$. In particular the vector space of all n-multilinear forms on $E_1 \times \cdots \times E_n$ is canonically isomorphic to $(E_1 \otimes_{\mathbb{K}} \cdots \otimes_{\mathbb{K}} E_n)'$.

Iterated application of Proposition 6.20 shows that we get also a well-defined notion of a linear map $L_1 \otimes \cdots \otimes L_n$, the tensor product of n linear maps. Thus $(E_1, \ldots, E_n) \mapsto E_1 \otimes_{\mathbb{K}} \cdots \otimes_{\mathbb{K}} E_n$ is a functor. There is no need to write out the details.

We turn to the question of defining a multiplication operation on tensors. If $\mathbb{K}$ is a field, an **algebra**[4] over $\mathbb{K}$ is a vector space V over $\mathbb{K}$ with a **multiplication** or **product** operation $V \times V \to V$ that is $\mathbb{K}$ bilinear. The additive part of the $\mathbb{K}$ bilinearity means that the product operation satisfies the distributive laws

$$a(b+c) = ab + ac \quad \text{and} \quad (b+c)a = ba + ca \quad \text{for all } a, b, c \text{ in } V,$$

and the scalar-multiplication part of the $\mathbb{K}$ bilinearity means that

$$(ka)b = k(ab) = a(kb) \quad \text{for all } k \text{ in } \mathbb{K} \text{ and } a, b \text{ in } V.$$

Within the text of the book, we shall work mostly just with **associative algebras**, i.e., those algebras satisfying the usual associative law

$$a(bc) = (ab)c \quad \text{for all } a, b, c \text{ in } V.$$

---
[4]Some authors use the term "algebra" to mean what we shall call an "associative algebra."

An associative algebra is therefore a ring and a vector space, the scalar multiplication and the ring multiplication being linked by the requirement that $(ka)b = k(ab) = a(kb)$ for all scalars k. Some commutative examples of associative algebras over $\mathbb{K}$ are any field $\mathbb{L}$ containing $\mathbb{K}$, the polynomial algebra $\mathbb{K}[X_1, \ldots, X_n]$, and the algebra of all $\mathbb{K}$-valued functions on a nonempty set S. Two noncommutative examples of associative algebras over $\mathbb{K}$ are the matrix algebra $M_n(\mathbb{K})$, with matrix multiplication as its product, and $\text{Hom}_\mathbb{K}(V, V)$ for any vector space V, with composition as its product. The division ring $\mathbb{H}$ of quaternions (Example 10 in Section IV.1) is another example of a noncommutative associative algebra over $\mathbb{R}$.

Despite our emphasis on algebras that are associative, certain kinds of nonassociative algebras are of great importance in applications, and consequently several problems at the end of the chapter make use of nonassociative algebras. A nonassociative algebra is determined by its vector-space structure and the multiplication table for the members of a $\mathbb{K}$ basis. There is no restriction on the multiplication table; all multiplication tables define algebras. Perhaps the best-known nonassociative algebra is the 3-dimensional algebra over $\mathbb{R}$ determined by **vector product** in $\mathbb{R}^3$. A basis is $\{\mathbf{i}, \mathbf{j}, \mathbf{k}\}$, the multiplication operation is denoted by $\times$, and the multiplication table is

$$\begin{array}{lll} \mathbf{i} \times \mathbf{i} = 0, & \mathbf{i} \times \mathbf{j} = \mathbf{k}, & \mathbf{i} \times \mathbf{k} = -\mathbf{j}, \\ \mathbf{j} \times \mathbf{i} = -\mathbf{k}, & \mathbf{j} \times \mathbf{j} = 0, & \mathbf{j} \times \mathbf{k} = \mathbf{i}, \\ \mathbf{k} \times \mathbf{i} = \mathbf{j}, & \mathbf{k} \times \mathbf{j} = -\mathbf{i}, & \mathbf{k} \times \mathbf{k} = 0. \end{array}$$

Since $\mathbf{i} \times (\mathbf{i} \times \mathbf{k}) = \mathbf{i} \times (-\mathbf{j}) = -\mathbf{k}$ and $(\mathbf{i} \times \mathbf{i}) \times \mathbf{k} = 0$, vector product is not associative. The vector-product algebra is a special case of a Lie algebra; Lie algebras are defined in Problems 31–35 at the end of the chapter.

Tensor algebras, which we shall now construct, will be associative algebras. Fix a vector space E over $\mathbb{K}$, and for integers $n \geq 1$, let $T^n(E)$ be the n-fold tensor product of E with itself. In the case $n = 0$, we let $T^0(E)$ be the field $\mathbb{K}$. Define, initially as a vector space, $T(E)$ to be the direct sum

$$T(E) = \bigoplus_{n=0}^{\infty} T^n(E).$$

The elements that lie in one or another $T^n(E)$ are called **homogeneous**. We define a bilinear multiplication on homogeneous elements

$$T^m(E) \times T^n(E) \to T^{m+n}(E)$$

to be the restriction of the canonical isomorphism

$$T^m(E) \otimes_\mathbb{K} T^n(E) \to T^{m+n}(E)$$

7. Tensor Algebra

resulting from iterating Proposition 6.19b. This multiplication, denoted by $\otimes$, is associative, as far as it goes, because the restriction of the $\mathbb{K}$ isomorphism

$$T^l(E) \otimes_{\mathbb{K}} (T^m(E) \otimes_{\mathbb{K}} T^n(E)) \to (T^l(E) \otimes_{\mathbb{K}} T^m(E)) \otimes_{\mathbb{K}} T^n(E)$$

to $T^l(E) \times (T^m(E) \times T^n(E))$ factors through the map

$$T^l(E) \times (T^m(E) \times T^n(E)) \to (T^l(E) \times T^m(E)) \times T^n(E)$$

given by $(r, (s, t)) \mapsto ((r, s), t)$.

This much tells how to multiply homogeneous elements in $T(E)$. Since each element t in $T(E)$ has a unique expansion as a finite sum $t = \sum_{k=0}^{n} t_k$ with $t_k \in T^k(E)$, we can define the product of this t and the element $t' = \sum_{k=0}^{n'} t'_k$ to be the element $t \otimes t' = \sum_{l=0}^{n+n'} \sum_{k+k'=l} (t_k \otimes t'_{k'})$; the expression $\sum_{k+k'=l} (t_k \otimes t'_{k'})$ is the component of the product in $T^l(E)$.

Multiplication is thereby well defined in $T(E)$, and it satisfies the distributive laws and is associative. Thus $T(E)$ becomes an associative algebra with a (two-sided) identity, namely the element 1 in $T^0(E)$. In the presence of the identification $\iota : E \to T^1(E)$, $T(E)$ is known as the **tensor algebra** of E. The pair $(T(E), \iota)$ has the **universal mapping property** given in Proposition 6.22 and pictured in Figure 6.5.

FIGURE 6.5. University mapping property of a tensor algebra.

Proposition 6.22. The pair $(T(E), \iota)$ has the following universal mapping property: whenever $l : E \to A$ is a linear map from E into an associative algebra with identity, then there exists a unique associative algebra homomorphism $L : T(E) \to A$ with $L(1) = 1$ such that the diagram in Figure 6.5 commutes.

PROOF. Uniqueness is clear, since E and 1 generate $T(E)$ as an algebra. For existence we define $L^{(n)}$ on $T^n(E)$ to be the linear extension of the n-multilinear map

$$(v_1, v_2, \ldots, v_n) \mapsto l(v_1) l(v_2) \cdots l(v_n),$$

and we let $L = \bigoplus L^{(n)}$ in obvious notation. Let $u_1 \otimes \cdots \otimes u_m$ be in $T^m(E)$ and $v_1 \otimes \cdots \otimes v_n$ be in $T^n(E)$. Then we have

$$L^{(m)}(u_1 \otimes \cdots \otimes u_m) = l(u_1) \cdots l(u_m),$$
$$L^{(n)}(v_1 \otimes \cdots \otimes v_n) = l(v_1) \cdots l(v_n),$$
$$L^{(m+n)}(u_1 \otimes \cdots \otimes u_m \otimes v_1 \otimes \cdots \otimes v_n) = l(u_1) \cdots l(u_m) l(v_1) \cdots l(v_n).$$

Hence

$$L^{(m)}(u_1 \otimes \cdots \otimes u_m) L^{(n)}(v_1 \otimes \cdots \otimes v_n) = L^{(m+n)}(u_1 \otimes \cdots \otimes u_m \otimes v_1 \otimes \cdots \otimes v_n).$$

Taking linear combinations, we see that L is a homomorphism. $\square$

Proposition 6.22 allows us to make $E \mapsto T(E)$ into a functor from the category of vector spaces over $\mathbb{K}$ to the category of associative algebras with identity over $\mathbb{K}$. To carry out the construction, we suppose that $\varphi : E \to F$ is a linear map between two vector spaces over $\mathbb{K}$. If $i : E \to T(E)$ and $j : F \to T(F)$ are the inclusion maps, then $j\varphi$ is a linear map from E into $T(F)$, and Proposition 6.22 produces a unique algebra homomorphism $\Phi : T(E) \to T(F)$ carrying 1 to 1 and satisfying $\Phi i = \varphi$. Then the tensor-product functor is defined to carry the linear map φ to the homomorphism Φ of associative algebras with identity.

For the situation in which R is a commutative ring with identity, Section IV.5 introduced the ring $R[X_1, \ldots, X_n]$ of polynomials in n commuting indeterminates with coefficients in R. This ring was characterized by a universal mapping property saying that if a ring homomorphism of R into a commutative ring with identity were given and if n elements $t_1, \ldots, t_n$ were given, then the ring homomorphism of R could be extended uniquely to a ring homomorphism of $R[X_1, \ldots, X_n]$ carrying X_j into t_j for each j.

Proposition 6.22 yields a noncommutative version of this result, except that the ring of coefficients is assumed this time to be a field $\mathbb{K}$. To arrange for $X_1, \ldots, X_n$ to be *noncommuting* indeterminates, we form a vector space with $\{X_1, \ldots, X_n\}$ as a basis. Thus we let $E = \bigoplus_{j=1}^{n} \mathbb{K} X_j$. If $t_1, \ldots, t_n$ are arbitrary elements of an associative algebra A with identity, then the formulas $l(X_j) = t_j$ for $1 \le j \le n$ define a linear map $l : E \to A$. The associative-algebra homomorphism $L : T(E) \to A$ produced by the proposition extends the inclusion of $\mathbb{K}$ into the subfield $\mathbb{K}1$ of A and carries each X_j to t_j.

8. Symmetric Algebra

We continue to allow $\mathbb{K}$ to be an arbitrary field. Let E be a vector space over $\mathbb{K}$, and let $T(E)$ be the tensor algebra. We begin by defining the symmetric algebra $S(E)$. This is to be a version of $T(E)$ in which the elements, which are called symmetric tensors, commute with one another. It will not be canonically an algebra of polynomials, as we shall see presently, and thus we make no use of polynomial rings in the construction.

Just as the vector space of n-multilinear forms $E \times \cdots \times E \to \mathbb{K}$ is canonically the dual of $T^n(E)$, so the vector space of **symmetric** n-multilinear forms will be canonically the dual of $S^n(E)$. Here "symmetric" means that $f(x_1, \ldots, x_n) = f(x_{\tau(1)}, \ldots, x_{\tau(n)})$ for every permutation τ in the symmetric group $\mathfrak{S}_n$.

Since tensor algebras are supposed to be universal devices for constructing associative algebras over $\mathbb{K}$, whether commutative or not, we seek to form $S(E)$ as a quotient of $T(E)$. If q is the quotient homomorphism, we want to have $q(u \otimes v) = q(v \otimes u)$ in $S(E)$ whenever u and v are in $\iota(E) = T^1(E)$. Hence every element $u \otimes v - v \otimes u$ is to be in the kernel of the homomorphism. On the other hand, we do not want to impose any unnecessary conditions on our quotient, and so we factor out only what the elements $u \otimes v - v \otimes u$ force us to factor out. Thus we define the **symmetric algebra** by

$$S(E) = T(E)/I,$$

where $I = \begin{pmatrix} \text{two-sided ideal generated by all} \\ u \otimes v - v \otimes u \text{ with } u \text{ and } v \\ \text{in } T^1(E) \end{pmatrix}.$

Then $S(E)$ is an associative algebra with identity.

Let us see that the fact that the generators of the ideal I are homogeneous elements (all being in $T^2(E)$) implies that

$$I = \bigoplus_{n=0}^{\infty} (I \cap T^n(E)).$$

In fact, each $I \cap T^n(E)$ is contained in I, and hence I contains the right side. On the other hand, if x is any element of I, then x is a sum of terms of the form $a \otimes (u \otimes v - v \otimes u) \otimes b$, and we may assume that each a and b is homogeneous. Any individual term $a \otimes (u \otimes v - v \otimes u) \otimes b$ is in some $I \cap T^n(E)$, and x is exhibited as a sum of members of the various intersections $I \cap T^n(E)$.

An ideal with the property $I = \bigoplus_{n=0}^{\infty} (I \cap T^n(E))$ is said to be **homogeneous**. Since I is homogeneous,

$$S(E) = \bigoplus_{n=0}^{\infty} T^n(E)/(I \cap T^n(E)).$$

We write $S^n(E)$ for the nth summand on the right side, so that

$$S(E) = \bigoplus_{n=0}^{\infty} S^n(E).$$

Since $I \cap T^1(E) = 0$, the map of $E \to T^1(E) \to S^1(E)$ into first-order elements is one-one onto. The product operation in $S(E)$ is written without a product sign, the image in $S^n(E)$ of $v_1 \otimes \cdots \otimes v_n$ in $T^n(E)$ being written as $v_1 \cdots v_n$. If a is in $S^m(E)$ and b is in $S^n(E)$, then ab is in $S^{m+n}(E)$. Moreover, $S^n(E)$ is generated by elements $v_1 \cdots v_n$ with all v_j in $S^1(E) \cong E$, since $T^n(E)$ is generated by corresponding elements $v_1 \otimes \cdots \otimes v_n$. The defining relations for $S(E)$ make $v_i v_j = v_j v_i$ for v_i and v_j in $S^1(E)$, and it follows that the associative algebra $S(E)$ is commutative. □

Proposition 6.23. Let E be a vector space over the field $\mathbb{K}$.

(a) Let ι be the n-multilinear function $\iota(v_1, \ldots, v_n) = v_1 \cdots v_n$ of $E \times \cdots \times E$ into $S^n(E)$. Then $(S^n(E), \iota)$ has the following **universal mapping property**: whenever l is any symmetric n-multilinear map of $E \times \cdots \times E$ into a vector space U, then there exists a unique linear map $L : S^n(E) \to U$ such that the diagram

$$\begin{array}{ccc} E \times \cdots \times E & \xrightarrow{l} & U \\ {\scriptstyle \iota}\downarrow & \nearrow{\scriptstyle L} & \\ S^n(E) & & \end{array}$$

commutes.

(b) Let ι be the one-one linear function that embeds E as $S^1(E) \subseteq S(E)$. Then $(S(E), \iota)$ has the following **universal mapping property**: whenever l is any linear map of E into a commutative associative algebra A with identity, then there exists a unique algebra homomorphism $L : S(E) \to A$ with $L(1) = 1$ such that the diagram

commutes.

PROOF. In both cases uniqueness is trivial. For existence we use the universal mapping properties of $T^n(E)$ and $T(E)$ to produce $\widetilde{L}$ on $T^n(E)$ or $T(E)$. If we can show that $\widetilde{L}$ annihilates the appropriate subspace so as to descend to $S^n(E)$ or $S(E)$, then the resulting map can be taken as L, and we are done. For (a), we have $\widetilde{L} : T^n(E) \to U$, and we are to show that $\widetilde{L}(T^n(E) \cap I) = 0$, where I is generated by all $u \otimes v - v \otimes u$ with u and v in $T^1(E)$. A member of $T^n(E) \cap I$ is thus of the form $\sum a_i \otimes (u_i \otimes v_i - v_i \otimes u_i) \otimes b_i$ with each term in $T^n(E)$. Each term here is a sum of pure tensors

$$x_1 \otimes \cdots \otimes x_r \otimes u_i \otimes v_i \otimes y_1 \otimes \cdots \otimes y_s - x_1 \otimes \cdots \otimes x_r \otimes v_i \otimes u_i \otimes y_1 \otimes \cdots \otimes y_s \quad (*)$$

with $r + 2 + s = n$. Since l by assumption takes equal values on

$$x_1 \times \cdots \times x_r \times u_i \times v_i \times y_1 \times \cdots \times y_s$$

and
$$x_1 \times \cdots \times x_r \times v_i \times u_i \times y_1 \times \cdots \times y_s,$$

$\widetilde{L}$ vanishes on (∗), and it follows that $\widetilde{L}(T^n(E) \cap I) = 0$.

For (b) we are to show that $\widetilde{L} : T(E) \to A$ vanishes on I. Since $\ker \widetilde{L}$ is an ideal, it is enough to check that $\widetilde{L}$ vanishes on the generators of I. But $\widetilde{L}(u \otimes v - v \otimes u) = l(u)l(v) - l(v)l(u) = 0$ by the commutativity of A, and thus $L(I) = 0$. □

Corollary 6.24. If E and F are vector spaces over the field $\mathbb{K}$, then the vector space $\text{Hom}_{\mathbb{K}}(S^n(E), F)$ is canonically isomorphic (via restriction to pure tensors) to the vector space of all F-valued symmetric n-multilinear functions on $E \times \cdots \times E$.

PROOF. Restriction is linear and one-one. It is onto by Proposition 6.23a. □

Corollary 6.25. If E is a vector space over the field $\mathbb{K}$, then the dual $(S^n(E))'$ of $S^n(E)$ is canonically isomorphic (via restriction to pure tensors) to the vector space of symmetric n-multilinear forms on $E \times \cdots \times E$.

PROOF. This is a special case of Corollary 6.24. □

If $\varphi : E \to F$ is a linear map between vector spaces, then we can use Proposition 6.23b to define a corresponding homomorphism $\Phi : S(E) \to S(F)$ of associative algebras with identity. In this way, we can make $E \mapsto S(E)$ into a functor from the category of vector spaces over $\mathbb{K}$ to the category of commutative associative algebras with identity over $\mathbb{K}$. The details appear in Problem 14 at the end of the chapter.

Next we shall identify a basis for $S^n(E)$ as a vector space. The union of such bases as n varies will then be a basis of $S(E)$. Let $\{u_i\}_{i \in A}$ be a basis of E, possibly infinite. As noted in Section A5 of the appendix, a **simple ordering** on the index set A is a partial ordering in which every pair of elements is comparable and in which $a \le b$ and $b \le a$ together imply $a = b$.

Proposition 6.26. Let E be a vector space over the field $\mathbb{K}$, let $\{u_i\}_{i \in A}$ be a basis of E, and suppose that a simple ordering has been imposed on the index set A. Then the set of all monomials $u_{i_1}^{j_1} \cdots u_{i_k}^{j_k}$ with $i_1 < \cdots < i_k$ and $\sum_m j_m = n$ is a basis of $S^n(E)$.

REMARK. In particular if E is finite-dimensional with $(u_1, \ldots, u_N)$ as an ordered basis, then the monomials $u_1^{j_1} \cdots u_N^{j_N}$ of total degree n form a basis of $S^n(E)$.

PROOF. Since $S(E)$ is commutative and since n-fold products of elements $\iota(u_i)$ in $T^1(E)$ span $T^n(E)$, the indicated set of monomials spans $S^n(E)$. Let us see that the set is linearly independent. Take any finite subset $F \subseteq A$ of indices. The map $\sum_{i \in A} c_i u_i \mapsto \sum_{i \in F} c_i X_i$ of E into the polynomial algebra $\mathbb{K}[\{X_i\}_{i \in F}]$ is linear into a commutative algebra with identity. Its extension via Proposition 6.23b maps all monomials in the u_i for $i \in F$ into distinct monomials in $\mathbb{K}[\{X_i\}_{i \in F}]$, which are necessarily linearly independent. Hence any finite subset of the monomials in the statement of the proposition is linearly independent, and the whole set must be linearly independent. Therefore our spanning set is a basis. □

The proof of Proposition 6.26 shows that $S(E)$ may be identified with polynomials in indeterminates identified with members of E once a basis has been chosen, but this identification depends on the choice of basis. Indeed, if we think of E as specified in advance, then the isomorphism was set up by mapping the set $\{X_i\}_{i \in A}$ to the specified basis of E, and the result certainly depended on what basis was used. Nevertheless, if E is finite-dimensional, there is still an isomorphism that is independent of basis; it is between $S(E')$, where E' is the *dual* of E, and a natural basis-free notion of "polynomials" *on* E. We return to this point after one application of Proposition 6.26.

Corollary 6.27. Let E be a finite-dimensional vector space over $\mathbb{K}$ of dimension N. Then

(a) $\dim S^n(E) = \binom{n+N-1}{N-1}$ for $0 \leq n < \infty$,

(b) $S^n(E')$ is canonically isomorphic to $S^n(E)'$ in such a way that

$$(f_1 \cdots f_n)(w_1 \cdots w_n) = \sum_{\tau \in \mathfrak{S}_n} \prod_{j=1}^n f_j(w_{\tau(j)}),$$

for any $f_1, \ldots, f_n$ in E' and any $w_1, \ldots, w_n$ in E, provided $\mathbb{K}$ has characteristic 0; here $\mathfrak{S}_n$ is the symmetric group on n letters.

PROOF. For (a), a basis has been described in Proposition 6.26. To see its cardinality, we recognize that picking out $N-1$ objects from $n+N-1$ to label as dividers is a way of assigning exponents to the u_j's in an ordered basis; thus the cardinality of the indicated basis is $\binom{n+N-1}{N-1}$.

8. Symmetric Algebra

For (b), let $f_1, \ldots, f_n$ be in E' and $w_1, \ldots, w_n$ be in E, and define

$$l_{f_1,\ldots,f_n}(w_1, \ldots, w_n) = \sum_{\tau \in \mathfrak{S}_n} \prod_{j=1}^{n} f_j(w_{\tau(j)}).$$

Then $l_{f_1,\ldots,f_n}$ is symmetric n-multilinear from $E \times \cdots \times E$ into $\mathbb{K}$ and extends by Proposition 6.23a to a linear $L_{f_1,\ldots,f_n} : S^n(E) \to \mathbb{K}$. Thus $l(f_1, \ldots, f_n) = L_{f_1,\ldots,f_n}$ defines a symmetric n-multilinear map of $E' \times \cdots \times E'$ into $S^n(E)'$. Its linear extension L maps $S^n(E')$ into $S^n(E)'$.

To complete the proof, we shall show that L carries basis to basis. Let $u_1, \ldots, u_N$ be an ordered basis of E, and let $u'_1, \ldots, u'_N$ be the dual basis. Part (a) shows that the elements $(u'_1)^{j_1} \cdots (u'_N)^{j_N}$ with $\sum_m j_m = n$ form a basis of $S^n(E')$ and that the elements $(u_1)^{k_1} \cdots (u_N)^{k_N}$ with $\sum_m k_m = n$ form a basis of $S^n(E)$. We show that L of the basis of $S^n(E')$ is the dual basis of the basis of $S^n(E)$, except for positive-integer factors. Thus let all of $f_1, \ldots, f_{j_1}$ be u'_1, let all of $f_{j_1+1}, \ldots, f_{j_1+j_2}$ be u'_2, and so on. Similarly let all of $w_1, \ldots, w_{k_1}$ be u_1, let all of $w_{k_1+1}, \ldots, w_{k_1+k_2}$ be u_2, and so on. Then

$$L((u'_1)^{j_1} \cdots (u'_N)^{j_N})((u_1)^{k_1} \cdots (u_N)^{k_N}) = L(f_1 \cdots f_n)(w_1 \cdots w_n)$$
$$= l(f_1, \ldots, f_n)(w_1 \cdots w_n)$$
$$= \sum_{\tau \in \mathfrak{S}_n} \prod_{i=1}^{n} f_i(w_{\tau(i)}).$$

For given τ, the product on the right side is 0 unless, for each index i, an inequality $j_{m-1} + 1 \leq i \leq j_m$ implies that $k_{m-1} + 1 \leq \tau(i) \leq k_m$. In this case the product is 1; so the right side counts the number of such τ's. For given τ, obtaining a nonzero product forces $k_m = j_m$ for all m. And when $k_m = j_m$ for all m, the choice $\tau = 1$ does lead to product 1. Hence the members of L of the basis are positive-integer multiples of the members of the dual basis, as asserted. □

Let us return to the question of introducing a basis-free notion of polynomials on the vector space E under the assumption that E is finite-dimensional. We take a cue from Corollary 4.32, which tells us that the evaluation homomorphism carrying $\mathbb{K}[X_1, \ldots, X_n]$ to the algebra of $\mathbb{K}$-valued polynomial functions of $(t_1, \ldots, t_n)$ is one-one if $\mathbb{K}$ is an infinite field. We regard the latter as the algebra of polynomial functions on $\mathbb{K}^n$, and we check what happens when we carry the vector space E over to $\mathbb{K}^n$ by fixing a basis. Let $\Gamma = \{x_1, \ldots, x_n\}$ be a basis of E, and let $\Gamma' = \{x'_1, \ldots, x'_n\}$ be the dual basis of E'. If $e = t_1 x_1 + \cdots + t_n x_n$ is the expansion of a member of E in terms of Γ, then we have $x'_j(e) = t_j$. Thus the polynomial functions t_j are given by the members of the dual basis. The vector space of all homogeneous first-degree polynomial functions is the set of linear combinations of the t_j's, and these are given by arbitrary linear functionals on E. Thus the vector space of homogeneous first-degree polynomial functions on E is just the dual space E', and this conclusion does not depend on the choice of basis. The algebra of all polynomial functions on E is then the algebra of all $\mathbb{K}$-valued functions on E generated by E' and the constant functions.

This discussion tells us unambiguously what polynomial *functions* on E are to be, and we want to backtrack to handle abstract polynomials on E. Although the evaluation homomorphism from $\mathbb{K}[X_1, \ldots, X_n]$ to the algebra of polynomial functions on $\mathbb{K}^n$ may fail to be one-one if $\mathbb{K}$ is a finite field, its restriction to homogeneous first-degree polynomials *is* one-one. Thus, whatever we might mean by the vector space of homogeneous first-degree polynomials on E, the evaluation mapping should exhibit this space as isomorphic to E'.

Armed with these clues, we define the **polynomial algebra** $P(E)$ on E to be the symmetric algebra $S(E')$ if E is finite-dimensional. We need an evaluation mapping for each point e of E, and we obtain this from the universal mapping property of symmetric algebras (Proposition 6.23b): With e fixed, we have a linear map l from the vector space E' to the commutative associative algebra $\mathbb{K}$ given with $l(e') = e'(e)$. The universal mapping property gives us a unique algebra homomorphism $L : S(E') \to \mathbb{K}$ that extends l and carries 1 to 1. The algebra homomorphism L is then a multiplicative linear functional on $P(E) = S(E')$ that carries 1 to 1 and agrees with evaluation at e on homogeneous first-degree polynomials. We write this homomorphism as $p \mapsto p(e)$, and we define $P^n(E) = S^n(E')$; this is the vector space of homogeneous n^{th}-degree polynomials on E. A confirmation that $P(E)$ is indeed to be regarded as the algebra of abstract polynomials on E comes from the following.

Proposition 6.28. If E is a finite-dimensional vector space over the field $\mathbb{K}$, then the system of evaluation homomorphisms $P(E) \to \mathbb{K}$ on polynomials given by $p \mapsto \{p(e)\}_{e \in E}$ is an algebra homomorphism of $P(E)$ onto the algebra of $\mathbb{K}$-valued polynomial functions on E that carries the identity to the constant function 1, and it is one-one if $\mathbb{K}$ is an infinite field.

PROOF. Certainly $p \mapsto \{p(e)\}_{e \in E}$ is an algebra homomorphism of $P(E)$ into the algebra of $\mathbb{K}$-valued polynomial functions on E, and it carries the identity to the constant function 1. We have seen that the image of $P^1(E)$ is exactly E', and hence the image of $P(E)$ is the algebra of $\mathbb{K}$-valued functions on E generated by E' and the constants. This is exactly the algebra of all $\mathbb{K}$-valued polynomial functions, and hence the mapping is onto.

Suppose that $\mathbb{K}$ is infinite. The restriction of $p \mapsto \{p(e)\}_{e \in E}$ to the finite-dimensional subspace $P^n(E)$ of $P(E)$ maps into the finite-dimensional subspace of all polynomial functions on E homogeneous of degree n, and this restriction must therefore be onto. We can read off the dimension of the space of all polynomial functions on E homogeneous of degree n from Corollary 4.32 and Corollary 6.27a. This dimension matches the dimension of $P^n(E)$, according to Corollary 6.27a. Since the mapping is onto and the finite dimensions match, the restricted mapping is one-one. Hence $p \mapsto \{p(e)\}_{e \in E}$ is one-one. □

We have defined the symmetric algebra $S(E)$ as a quotient of the tensor algebra $T(E)$. Now let us suppose that $\mathbb{K}$ has characteristic 0. With this hypothesis we shall be able to identify an explicit vector subspace of $T(E)$ that maps one-one onto $S(E)$ during the passage to the quotient. This subspace of $T(E)$ can therefore be viewed as a version of $S(E)$ for some purposes.

We define an n-multilinear function from $E \times \cdots \times E$ into $T^n(E)$ by

$$(v_1, \ldots, v_n) \mapsto \frac{1}{n!} \sum_{\tau \in \mathfrak{S}_n} v_{\tau(1)} \otimes \cdots \otimes v_{\tau(n)},$$

and let $\sigma : T^n(E) \to T^n(E)$ be its linear extension. We call σ the **symmetrizer** operator. The image of σ in $T(E)$ is denoted by $\widetilde{S}^n(E)$, and the members of this subspace are called **symmetrized** tensors.

Proposition 6.29. Let the field $\mathbb{K}$ have characteristic 0, and let E be a vector space over $\mathbb{K}$. Then the symmetrizer operator σ satisfies $\sigma^2 = \sigma$. The kernel of σ on $T^n(E)$ is exactly $T^n(E) \cap I$, and therefore

$$T^n(E) = \widetilde{S}^n(E) \oplus (T^n(E) \cap I).$$

REMARK. In view of this corollary, the quotient map $T^n(E) \to S^n(E)$ carries $\widetilde{S}^n(E)$ one-one onto $S^n(E)$. Thus $\widetilde{S}^n(E)$ can be viewed as a copy of $S^n(E)$ embedded as a direct summand of $T^n(E)$.

PROOF. We have

$$\sigma^2(v_1 \otimes \cdots \otimes v_n) = \frac{1}{(n!)^2} \sum_{\rho, \tau \in \mathfrak{S}_n} v_{\rho\tau(1)} \otimes \cdots \otimes v_{\rho\tau(n)}$$

$$= \frac{1}{(n!)^2} \sum_{\rho \in \mathfrak{S}_n} \sum_{\substack{\omega \in \mathfrak{S}_n, \\ (\omega = \rho\tau)}} v_{\omega(1)} \otimes \cdots \otimes v_{\omega(n)}$$

$$= \frac{1}{n!} \sum_{\rho \in \mathfrak{S}_n} \sigma(v_1 \otimes \cdots \otimes v_n)$$

$$= \sigma(v_1 \otimes \cdots \otimes v_n).$$

Hence $\sigma^2 = \sigma$. Thus σ fixes any member of image σ, and it follows that image $\sigma \cap \ker \sigma = 0$. Consequently $T^n(E)$ is the direct sum of image σ and $\ker \sigma$. We are left with identifying $\ker \sigma$ as $T^n(E) \cap I$.

The subspace $T^n(E) \cap I$ is spanned by elements

$$x_1 \otimes \cdots \otimes x_r \otimes u \otimes v \otimes y_1 \otimes \cdots \otimes y_s - x_1 \otimes \cdots \otimes x_r \otimes v \otimes u \otimes y_1 \otimes \cdots \otimes y_s$$

with $r + 2 + s = n$, and the symmetrizer σ certainly vanishes on such elements. Hence $T^n(E) \cap I \subseteq \ker \sigma$. Suppose that the inclusion is strict, say with t in $\ker \sigma$ but t not in $T^n(E) \cap I$. Let q be the quotient map $T^n(E) \to S^n(E)$. The kernel of q is $T^n(E) \cap I$, and thus $q(t) \neq 0$. From Proposition 6.26 the $T(E)$ monomials in basis elements from E with increasing indices map onto a basis of $S(E)$. Since $\mathbb{K}$ has characteristic 0, the symmetrized versions of these monomials map to nonzero multiples of the images of the initial monomials. Consequently q carries $\widetilde{S}^n(E) = \text{image} \, \sigma$ onto $S^n(E)$. Thus choose $t' \in \widetilde{S}^n(E)$ with $q(t') = q(t)$. Then $t' - t$ is in $\ker q = T^n(E) \cap I \subseteq \ker \sigma$. Since $\sigma(t) = 0$, we see that $\sigma(t') = 0$. Consequently t' is in $\ker \sigma \cap \text{image} \, \sigma = 0$, and we obtain $t' = 0$ and $q(t) = q(t') = 0$, contradiction. □

9. Exterior Algebra

We turn to a discussion of the exterior algebra. Let $\mathbb{K}$ be an arbitrary field, and let E be a vector space over $\mathbb{K}$. The construction, results, and proofs for the exterior algebra $\bigwedge(E)$ are similar to those for the symmetric algebra $S(E)$. The elements of $\bigwedge(E)$ are to be all the alternating tensors (= skew-symmetric if $\mathbb{K}$ has characteristic $\neq 2$), and so we want to force $v \otimes v = 0$. Thus we define the **exterior algebra** by

$$\bigwedge(E) = T(E)/I',$$

where $I' = \begin{pmatrix} \text{two-sided ideal generated by all} \\ v \otimes v \text{ with } v \text{ in } T^1(E) \end{pmatrix}$.

Then $\bigwedge(E)$ is an associative algebra with identity.

It is clear that I' is homogeneous: $I' = \bigoplus_{n=0}^{\infty} (I' \cap T^n(E))$. Thus we can write

$$\bigwedge(E) = \bigoplus_{n=0}^{\infty} T^n(E)/(I' \cap T^n(E)).$$

We write $\bigwedge^n(E)$ for the nth summand on the right side, so that

$$\bigwedge(E) = \bigoplus_{n=0}^{\infty} \bigwedge^n(E).$$

9. Exterior Algebra

Since $I' \cap T^1(E) = 0$, the map of E into first-order elements $\bigwedge^1(E)$ is one-one onto. The product operation in $\bigwedge(E)$ is denoted by $\wedge$ rather than $\otimes$, the image in $\bigwedge^n(E)$ of $v_1 \otimes \cdots v_n$ in $T^n(E)$ being denoted by $v_1 \wedge \cdots \wedge v_n$. If a is in $\bigwedge^m(E)$ and b is in $\bigwedge^n(E)$, then $a \wedge b$ is in $\bigwedge^{m+n}(E)$. Moreover, $\bigwedge^n(E)$ is generated by elements $v_1 \wedge \cdots \wedge v_n$ with all v_j in $\bigwedge^1(E) \cong E$, since $T^n(E)$ is generated by corresponding elements $v_1 \otimes \cdots \otimes v_n$. The defining relations for $\bigwedge(E)$ make $v_i \wedge v_j = -v_j \wedge v_i$ for v_i and v_j in $\bigwedge^1(E)$, and it follows that

$$a \wedge b = (-1)^{mn} b \wedge a \qquad \text{for } a \in \bigwedge^m(E) \text{ and } b \in \bigwedge^n(E).$$

Proposition 6.30. Let E be a vector space over the field $\mathbb{K}$.

(a) Let ι be the n-multilinear function $\iota(v_1, \ldots, v_n) = v_1 \wedge \cdots \wedge v_n$ of $E \times \cdots \times E$ into $\bigwedge^n(E)$. Then $(\bigwedge^n(E), \iota)$ has the following **universal mapping property**: whenever l is any alternating n-multilinear map of $E \times \cdots \times E$ into a vector space U, then there exists a unique linear map $L : \bigwedge^n(E) \to U$ such that the diagram

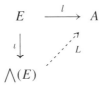

commutes.

(b) Let ι be the function that embeds E as $\bigwedge^1(E) \subseteq \bigwedge(E)$. Then $(\bigwedge(E), \iota)$ has the following **universal mapping property**: whenever l is any linear map of E into an associative algebra A with identity such that $l(v)^2 = 0$ for all $v \in E$, then there exists a unique algebra homomorphism $L : \bigwedge(E) \to A$ with $L(1) = 1$ such that the diagram

$$\begin{array}{ccc} E & \xrightarrow{l} & A \\ {\scriptstyle \iota}\downarrow & \nearrow{\scriptstyle L} & \\ \bigwedge(E) & & \end{array}$$

commutes.

PROOF. The proof is completely analogous to the proof of Proposition 6.23. □

Corollary 6.31. If E and F are vector spaces over the field $\mathbb{K}$, then the vector space $\operatorname{Hom}_\mathbb{K}(\bigwedge^n(E), F)$ is canonically isomorphic (via restriction to pure tensors) to the vector space of all F-valued alternating n-multilinear functions on $E \times \cdots \times E$.

PROOF. Restriction is linear and one-one. It is onto by Proposition 6.30a. □

Corollary 6.32. If E is a vector space over the field $\mathbb{K}$, then the dual $(\bigwedge^n(E))'$ of $\bigwedge^n(E)$ is canonically isomorphic (via restriction to pure tensors) to the vector space of alternating n-multilinear forms on $E \times \cdots \times E$.

PROOF. This is a special case of Corollary 6.31. □

If $\varphi : E \to F$ is a linear map between vector spaces, then we can use Proposition 6.30b to define a corresponding homomorphism $\Phi : \bigwedge(E) \to \bigwedge(F)$ of associative algebras with identity. In this way, we can make $E \mapsto \bigwedge(E)$ into a functor from the category of vector spaces over $\mathbb{K}$ to the category of commutative associative algebras with identity over $\mathbb{K}$. We omit the details, which are similar to those for symmetric tensors.

Next we shall identify a basis for $\bigwedge^n(E)$ as a vector space. The union of such bases as n varies will then be a basis of $\bigwedge(E)$.

Proposition 6.33. Let E be a vector space over the field $\mathbb{K}$, let $\{u_i\}_{i \in A}$ be a basis of E, and suppose that a simple ordering has been imposed on the index set A. Then the set of all monomials $u_{i_1} \wedge \cdots \wedge u_{i_n}$ with $i_1 < \cdots < i_n$ is a basis of $\bigwedge^n(E)$.

PROOF. Since multiplication in $\bigwedge(E)$ satisfies $a \wedge b = (-1)^{mn} b \wedge a$ for $a \in \bigwedge^m(E)$ and $b \in \bigwedge^n(E)$ and since monomials span $T^n(E)$, the indicated set spans $\bigwedge^n(E)$. Let us see that the set is linearly independent. For $i \in A$, let u'_i be the member of E' with $u'_i(u_j)$ equal to 1 for $j = i$ and equal to 0 for $j \neq i$. Fix $r_1 < \cdots < r_n$, and define

$$l(w_1, \ldots, w_n) = \det\{u'_{r_i}(w_j)\} \qquad \text{for } w_1, \ldots, w_n \text{ in } E.$$

Then l is alternating n-multilinear from $E \times \cdots \times E$ into $\mathbb{K}$ and extends by Proposition 6.30a to $L : \bigwedge^n(E) \to \mathbb{K}$. If $k_1 < \cdots < k_n$, then

$$L(u_{k_1} \wedge \cdots \wedge u_{k_n}) = l(u_{k_1}, \ldots, u_{k_n}) = \det\{u'_{r_i}(u_{k_j})\},$$

and the right side is 0 unless $r_1 = k_1, \ldots, r_n = k_n$, in which case it is 1. This proves that the $u_{r_1} \wedge \cdots \wedge u_{r_n}$ are linearly independent in $\bigwedge^n(E)$. □

Corollary 6.34. Let E be a finite-dimensional vector space over $\mathbb{K}$ of dimension N. Then

(a) $\dim \bigwedge^n(E) = \binom{N}{n}$ for $0 \leq n \leq N$ and $= 0$ for $n > N$,

(b) $\bigwedge^n(E')$ is canonically isomorphic to $\bigwedge^n(E)'$ by

$$(f_1 \wedge \cdots \wedge f_n)(w_1, \ldots, w_n) = \det\{f_i(w_j)\}.$$

9. Exterior Algebra

PROOF. Part (a) is an immediate consequence of Proposition 6.33, and (b) is proved in the same way as Corollary 6.27b, using Proposition 6.30a as a tool. The "positive-integer multiples" that arise in the proof of Corollary 6.27b are all 1 in the current proof, and hence no restriction on the characteristic of $\mathbb{K}$ is needed. $\square$

Now let us suppose that $\mathbb{K}$ has characteristic 0. We define an n-multilinear function from $E \times \cdots \times E$ into $T^n(E)$ by

$$(v_1, \ldots, v_n) \mapsto \frac{1}{n!} \sum_{\tau \in \mathfrak{S}_n} (\text{sgn } \tau) v_{\tau(1)} \otimes \cdots \otimes v_{\tau(n)},$$

and let $\sigma' : T^n(E) \to T^n(E)$ be its linear extension. We call σ' the **antisymmetrizer** operator. The image of σ' in $T(E)$ is denoted by $\widetilde{\bigwedge}^n(E)$, and the members of this subspace are called **antisymmetrized** tensors.

Proposition 6.35. Let the field $\mathbb{K}$ have characteristic 0, and let E be a vector space over $\mathbb{K}$. Then the antisymmetrizer operator σ' satisfies $\sigma'^2 = \sigma'$. The kernel of σ' on $T^n(E)$ is exactly $T^n(E) \cap I'$, and therefore

$$T^n(E) = \widetilde{\bigwedge}^n(E) \oplus (T^n(E) \cap I').$$

REMARK. In view of this corollary, the quotient map $T^n(E) \to \bigwedge^n(E)$ carries $\widetilde{\bigwedge}^n(E)$ one-one onto $\bigwedge^n(E)$. Thus $\widetilde{\bigwedge}^n(E)$ can be viewed as a copy of $\bigwedge^n(E)$ embedded as a direct summand of $T^n(E)$.

PROOF. We have

$$\sigma'^2(v_1 \otimes \cdots \otimes v_n) = \frac{1}{(n!)^2} \sum_{\rho, \tau \in \mathfrak{S}_n} (\text{sgn } \rho\tau) v_{\rho\tau(1)} \otimes \cdots \otimes v_{\rho\tau(n)}$$

$$= \frac{1}{(n!)^2} \sum_{\rho \in \mathfrak{S}_n} \sum_{\substack{\omega \in \mathfrak{S}_n, \\ (\omega = \rho\tau)}} (\text{sgn } \omega) v_{\omega(1)} \otimes \cdots \otimes v_{\omega(n)}$$

$$= \frac{1}{n!} \sum_{\rho \in \mathfrak{S}_n} \sigma'(v_1 \otimes \cdots \otimes v_n)$$

$$= \sigma'(v_1 \otimes \cdots \otimes v_n).$$

Hence $\sigma'^2 = \sigma'$. Consequently $T^n(E)$ is the direct sum of image σ' and $\ker \sigma'$, and we are left with identifying $\ker \sigma'$ as $T^n(E) \cap I'$.

The subspace $T^n(E) \cap I'$ is spanned by elements

$$x_1 \otimes \cdots \otimes x_r \otimes v \otimes v \otimes y_1 \otimes \cdots \otimes y_s$$

with $r+2+s = n$, and the antisymmetrizer σ' certainly vanishes on such elements. Hence $T^n(E) \cap I' \subseteq \ker \sigma'$. Suppose that the inclusion is strict, say with t in $\ker \sigma'$ but t not in $T^n(E) \cap I'$. Let q be the quotient map $T^n(E) \to \bigwedge^n(E)$. The kernel of q is $T^n(E) \cap I'$, and thus $q(t) \neq 0$. From Proposition 6.33 the $T(E)$ monomials with strictly increasing indices map onto a basis of $\bigwedge(E)$. Since $\mathbb{K}$ has characteristic 0, the antisymmetrized versions of these monomials map to nonzero multiples of the images of the initial monomials. Consequently q carries $\widetilde{\bigwedge}^n(E) = \text{image } \sigma'$ onto $\bigwedge^n(E)$. Thus choose $t' \in \widetilde{\bigwedge}^n(E)$ with $q(t') = q(t)$. Then $t' - t$ is in $\ker q = T^n(E) \cap I' \subseteq \ker \sigma'$. Since $\sigma'(t) = 0$, we see that $\sigma'(t') = 0$. Consequently t' is in $\ker \sigma' \cap \text{image } \sigma' = 0$, and we obtain $t' = 0$ and $q(t) = q(t') = 0$, contradiction. $\square$

10. Problems

1. Let V be a vector space over a field $\mathbb{K}$, and let $\langle \cdot, \cdot \rangle$ be a nondegenerate bilinear form on V.
   (a) Prove that every member v' of V is of the form $v'(w) = \langle v, w \rangle$ for one and only one member v of V.
   (b) Suppose that $(\cdot, \cdot)$ is another bilinear form on V. Prove that there is some linear function $L : V \to V$ such that $(v, w) = \langle L(v), w \rangle$ for all v and w in V.

2. The matrix $A = \begin{pmatrix} 0 & 1 \\ 1 & 0 \end{pmatrix}$ with entries in $\mathbb{F}_2$ is symmetric. Prove that there is no nonsingular M with $M^t A M$ diagonal.

3. This problem shows that one possible generalization of Sylvester's Law to other fields is not valid. Over the field $\mathbb{F}_3$, show that there is a nonsingular matrix M such that $\begin{pmatrix} -1 & 0 \\ 0 & -1 \end{pmatrix} = M^t \begin{pmatrix} 1 & 0 \\ 0 & 1 \end{pmatrix} M$. Conclude that the number of squares in $\mathbb{K}^\times$ among the diagonal entries of the diagonal form in Theorem 6.5 is not an invariant of the symmetric matrix.

4. Let V be a complex n-dimensional vector space, let $(\cdot, \cdot)$ be a Hermitian form on V, let $V_\mathbb{R}$ be the $2n$-dimensional real vector space obtained from V by restricting scalar multiplication to real scalars, and define $\langle \cdot, \cdot \rangle = \text{Im}(\cdot, \cdot)$. Prove that
   (a) $\langle \cdot, \cdot \rangle$ is an alternating bilinear form on $V_\mathbb{R}$,
   (b) $\langle J(v_1), J(v_2) \rangle = \langle v_1, v_2 \rangle$ for all v_1 and v_2 if $J : V_\mathbb{R} \to V_\mathbb{R}$ is what multiplication by i becomes when viewed as a linear map from $V_\mathbb{R}$ to itself,
   (c) $\langle \cdot, \cdot \rangle$ is nondegenerate on $V_\mathbb{R}$ if and only if $(\cdot, \cdot)$ is nondegenerate on V.

5. Let W be a $2n$-dimensional real vector space, and let $\langle \cdot, \cdot \rangle$ be a nondegenerate alternating bilinear form on W. Suppose that $J : W \to W$ is a linear map such

that $J^2 = -I$ and $\langle J(w_1), J(w_2)\rangle = \langle w_1, w_2\rangle$ for all w_1 and w_2 in W. Prove that W equals $V_{\mathbb{R}}$ for some n-dimensional complex vector space V possessing a Hermitian form whose imaginary part is $\langle \cdot, \cdot \rangle$.

6. This problem sharpens the result of Theorem 6.7 in the nondegenerate case. Let $\langle \cdot, \cdot \rangle$ be a nondegenerate alternating bilinear form on a $2n$-dimensional vector space V over $\mathbb{K}$. A vector subspace S of V is called an **isotropic** subspace if $\langle u, v \rangle = 0$ for all u and v in S. Prove that
   (a) any isotropic subspace of V that is maximal under inclusion has dimension n,
   (b) for any maximal isotropic subspace S_1, there exists a second maximal isotropic subspace S_2 such that $S_1 \cap S_2 = 0$.
   (c) if S_1 and S_2 are maximal isotropic subspaces of V such that $S_1 \cap S_2 = 0$, then the linear map $S_2 \to S_1'$ given by $s_2 \mapsto \langle \cdot, s_2\rangle\big|_{S_1}$ is an isomorphism of S_2 onto the dual space S_1'.
   (d) if S_1 and S_2 are maximal isotropic subspaces of V such that $S_1 \cap S_2 = 0$, then there exist bases $\{p_1, \ldots, p_n\}$ of S_1 and $\{q_1, \ldots, q_n\}$ of S_2 such that $\langle p_i, p_j \rangle = \langle q_i, q_j \rangle = 0$ and $\langle p_i, q_j \rangle = \delta_{ij}$ for all i and j. (The resulting basis $\{p_1, \ldots, p_n, q_1, \ldots, q_n\}$ of V is called a **Weyl basis** of V.)

7. Let S be a nonempty set, and let $\mathbb{K}$ be a field. For s in S, let U_s and V_s be vector spaces over $\mathbb{K}$, and let U and V be two further vector spaces over $\mathbb{K}$.
   (a) Prove that $\mathrm{Hom}_{\mathbb{K}}\left(\bigoplus_{s\in S} U_s, V\right) \cong \prod_{s\in S} \mathrm{Hom}_{\mathbb{K}}(U_s, V)$.
   (b) Prove that $\mathrm{Hom}_{\mathbb{K}}\left(U, \prod_{s\in S} V_s\right) \cong \prod_{s\in S} \mathrm{Hom}_{\mathbb{K}}(U, V_s)$.
   (c) Give examples to show that neither isomorphism in (a) and (b) need remain valid if all three direct products are changed to direct sums.

8. This problem continues Problem 1 at the end of Chapter V, which established a canonical-form theorem for an action of $GL(m, \mathbb{K}) \times GL(n, \mathbb{K})$ on m-by-n matrices. For the present problem, the group $GL(n, \mathbb{K})$ acts on $M_n(\mathbb{K})$ by $(g, x) \mapsto gxg^t$.
   (a) Verify that this is indeed a group action and that the vector subspaces $A_{nn}(\mathbb{K})$ of alternating matrices and $S_{nn}(\mathbb{K})$ of symmetric matrices are mapped into themselves under the group action.
   (b) Prove that two members of $A_{nn}(\mathbb{K})$ lie in the same orbit if and only if they have the same rank, and that the rank must be even. For each even rank $\leq n$, find an example of a member of $A_{nn}(\mathbb{K})$ with that rank.
   (c) Prove that two members of $S_{nn}(\mathbb{C})$ lie in the same orbit if and only if they have the same rank, and for each rank $\leq n$, find an example of a member of $S_{nn}(\mathbb{C})$ with that rank.

9. Let U and V be vector spaces over $\mathbb{K}$, and let U' be the dual of U. The bilinear map $(u', v) \mapsto u'(\cdot)v$ of $U' \times V$ into $\mathrm{Hom}_{\mathbb{K}}(U, V)$ extends to a linear map $T_{UV} : U' \otimes_{\mathbb{K}} V \to \mathrm{Hom}_{\mathbb{K}}(U, V)$.

(a) Prove that T_{UV} is one-one.
(b) Prove that T_{UV} is onto $\text{Hom}_{\mathbb{K}}(U, V)$ if U is finite-dimensional.
(c) Give an example for which T_{UV} is not onto $\text{Hom}_{\mathbb{K}}(U, V)$.
(d) Let $\mathcal{C}$ be the category of all vector spaces over $\mathbb{K}$, and let Φ and Ψ be the functors from $\mathcal{C} \times \mathcal{C}$ into $\mathcal{C}$ whose effects on objects are $\Phi(U, V) = U' \otimes_{\mathbb{K}} V$ and $\Psi(U, V) = \text{Hom}_{\mathbb{K}}(U, V)$. Prove that the system $\{T_{UV}\}$ is a natural transformation of Φ into Ψ.
(e) In view of (c), can the system $\{T_{UV}\}$ be a natural isomorphism?

10. Let $\mathbb{K} \subseteq \mathbb{L}$ be an inclusion of fields, and let $\mathcal{V}_{\mathbb{K}}$ and $\mathcal{V}_{\mathbb{L}}$ be the categories of vector spaces over $\mathbb{K}$ and $\mathbb{L}$. Section 6 of the text defined extension of scalars as a covariant functor $\Phi(E) = E \otimes_{\mathbb{K}} \mathbb{L}$. Another definition of extension of scalars is $\Psi(E) = \text{Hom}_{\mathbb{K}}(\mathbb{L}, E)$ with $(l\varphi)(l') = \varphi(ll')$. Verify that $\Psi(E)$ is a vector space over $\mathbb{L}$ and that Ψ is a functor.

11. A linear map $L : E \to F$ between finite-dimensional complex vector spaces becomes a linear map $L_{\mathbb{R}} : E_{\mathbb{R}} \to F_{\mathbb{R}}$ when we restrict attention to real scalars. Explain how to express a matrix for $L_{\mathbb{R}}$ in terms of a matrix for L.

12. **(Kronecker product of matrices)** Let $L : E_1 \to E_2$ and $M : F_1 \to F_2$ be linear maps between finite-dimensional vector spaces over $\mathbb{K}$, let Γ_1 and Γ_2 be ordered bases of E_1 and E_2, and let Δ_1 and Δ_2 be ordered bases of F_1 and F_2. Define matrices A and B by $A = \begin{pmatrix} L \\ \Gamma_2 \Gamma_1 \end{pmatrix}$ and $B = \begin{pmatrix} M \\ \Delta_2 \Delta_1 \end{pmatrix}$. Use $\Gamma_1, \Gamma_2, \Delta_1$, and Δ_2 to define ordered bases Ω_1 and Ω_2 of $E_1 \otimes_{\mathbb{K}} F_1$ and $E_2 \otimes_{\mathbb{K}} F_2$, and describe how the matrix $C = \begin{pmatrix} L \otimes M \\ \Omega_2 \Omega_1 \end{pmatrix}$ is related to A and B.

13. Let $\mathbb{K}$ be a field, and let E be the vector space $\mathbb{K}X \oplus \mathbb{K}Y$. Prove that the subalgebra of $T(E)$ generated by $1, Y$, and $X^2 + XY + Y^2$ is isomorphic as an algebra with identity to $T(F)$ for some vector space F.

Problems 14–17 concern the functors $E \mapsto T(E)$, $E \mapsto S(E)$, and $E \mapsto \bigwedge E$ defined for vector spaces over a field $\mathbb{K}$.

14. If $\varphi : E \to F$ is a linear map between vector spaces over $\mathbb{K}$, Section 8 of the text indicated how to define a corresponding homomorphism $\Phi : S(E) \to S(F)$ of associative algebras with identity over $\mathbb{K}$, using Proposition 6.23b.
(a) Fill in the details of this application of Proposition 6.23b.
(b) Establish the appropriate conditions on mappings that complete the proof that $E \mapsto S(E)$ is a functor.
(c) Verify that Φ carries $S^n(E)$ linearly into $S^n(F)$ for all integers $n \geq 0$.

15. Suppose that a linear map $\varphi : E \to E$ is given. Let $\Phi : S(E) \to S(E)$ and $\widetilde{\Phi} : T(E) \to T(E)$ be the associated algebra homomorphisms of $S(E)$ into itself and of $T(E)$ into itself, and let $q : T(E) \to S(E)$ be the quotient homomorphism appearing in the definition of $S(E)$. These mappings are related by the equation $\Phi q(x) = q\widetilde{\Phi}(x)$ for x in $T(E)$. Proposition 6.29 shows for each $n \geq 0$ that

$T^n(E) = \widetilde{S}^n(E) \oplus (T^n(E) \cap I)$, where $\widetilde{S}^n(E)$ is the image of $T^n(E)$ under the symmetrizer mapping. The remark with the proposition observes that q carries $\widetilde{S}^n(E)$ one-one onto $S^n(E)$. Prove that $\widetilde{\Phi}$ carries $\widetilde{S}^n(E)$ into itself and that $\widetilde{\Phi}|_{\widetilde{S}^n(E)}$ matches $\Phi|_{S^n(E)}$ in the sense that $q\widetilde{\Phi}(x) = \Phi q(x)$ for all x in $\widetilde{S}^n(E)$.

16. With E finite-dimensional let $\varphi : E \to E$ be a linear mapping, and define $\Phi : \bigwedge E \to \bigwedge E$ to be the corresponding algebra homomorphism of $\bigwedge E$ sending 1 into 1. This carries each $\bigwedge^n E$ into itself. Prove that Φ acts as multiplication by the scalar $\det \varphi$ on the 1-dimensional space $\bigwedge^{\dim E}(E)$.

17. Suppose that G is a group, that the vector space E over $\mathbb{K}$ is finite-dimensional, and that $\varphi : G \to \mathrm{GL}(E)$ is a representation of G on E. The functors $E \mapsto T(E)$, $E \mapsto S(E)$, and $E \mapsto \bigwedge E$ yield, for each $\varphi(g)$, algebra homomorphisms of $T(E)$ into itself, $S(E)$ into itself, and $\bigwedge E$ into itself.
    (a) Show that as g varies, the result in each case is a representation of G.
    (b) Suppose that $E = \mathbb{K}^n$. Give a formula for the representation of G on a member of $P(\mathbb{K}^n) = S((\mathbb{K}^n)')$.

Problems 18–22 concern universal mapping properties. Let $\mathcal{A}$ and $\mathcal{V}$ be two categories, and let $\mathcal{F} : \mathcal{A} \to \mathcal{V}$ be a covariant functor. (In practice, $\mathcal{F}$ tends to be a relatively simple functor, such as one that simply ignores some of the structure of $\mathcal{A}$.) Let E be in Obj($\mathcal{V}$). A pair (S, ι) with S in Obj($\mathcal{A}$) and ι in Morph$_\mathcal{V}(E, \mathcal{F}(S))$ is said to have the **universal mapping property** relative to E and $\mathcal{F}$ if the following condition is satisfied: whenever A is in Obj($\mathcal{A}$) and a member l of Morph$_\mathcal{V}(E, \mathcal{F}(A))$ is given, there exists a unique member L of Morph$_\mathcal{A}(S, A)$ such that $\mathcal{F}(L)\iota = l$.

18. (a) By suitably specializing $\mathcal{A}$, $\mathcal{V}$, $\mathcal{F}$, etc., show that the universal mapping property of the symmetric algebra of a vector space over $\mathbb{K}$ is an instance of what has been described.
    (b) How should the answer to (a) be adjusted so as to account for the universal mapping property of the exterior algebra of a vector space over $\mathbb{K}$?
    (c) How should the answer to (a) be adjusted so as to account for the universal mapping property of the coproduct of $\{X_j\}_{j\in J}$ in a category $\mathcal{C}$, the universal mapping property being as in Figure 4.12? (Educational note: For the *product* of $\{X_j\}_{j\in J}$ in $\mathcal{C}$, the above description does not apply directly because the morphisms go the wrong way. Instead, one applies the above description to the opposite categories $\mathcal{A}^\mathrm{opp}$ and $\mathcal{V}^\mathrm{opp}$, defined as in Problems 78–80 at the end of Chapter IV.)

19. If (S, ι) and (S', ι') are two pairs that each have the universal mapping property relative to E and $\mathcal{F}$, prove that S and S' are canonically isomorphic as objects in $\mathcal{A}$. More specifically prove that there exists a unique L in Morph$_\mathcal{A}(S, S')$ such that $\mathcal{F}(L)\iota = \iota'$ and that L is an isomorphism whose inverse L' in Morph$_\mathcal{A}(S', S)$ has $\mathcal{F}(L')\iota' = \iota$.

20. Suppose that the pair (S, ι) has the universal mapping property relative to E and $\mathcal{F}$. Let $\mathcal{S}$ be the category of sets, and define functors $F : \mathcal{A} \to \mathcal{S}$ and $G : \mathcal{A} \to \mathcal{S}$ by $F(A) = \text{Morph}_{\mathcal{A}}(S, A)$, $F(\varphi)$ equals composition on the left by φ for $\varphi \in \text{Morph}_{\mathcal{A}}(A, A')$, $G(A) = \text{Morph}_{\mathcal{V}}(E, \mathcal{F}(A))$, and $G(\varphi)$ equals composition on the left by $\mathcal{F}(\varphi)$. Let $T_A : \text{Morph}_{\mathcal{A}}(S, A) \to \text{Morph}_{\mathcal{V}}(E, \mathcal{F}(A))$ be the one-one onto map given by the universal mapping property. Show that the system $\{T_A\}$ is a natural isomorphism of F into G.

21. Suppose that (S', ι) is a second pair having the universal mapping property relative to E and $\mathcal{F}$. Define $F' : \mathcal{A} \to \mathcal{S}$ by $F'(A) = \text{Morph}_{\mathcal{A}}(S', A)$. Combining the previous problem and Proposition 6.16, obtain a second proof (besides the one in Problem 19) that S and S' are canonically isomorphic.

22. Suppose that for each E in $\text{Obj}(\mathcal{V})$, there is some pair (S, ι) with the universal mapping property relative to E and $\mathcal{F}$. Fix such a pair (S, ι) for each E, calling it $(S(E), \iota_E)$. Making an appropriate construction for morphisms and carrying out the appropriate verifications, prove that $E \mapsto S(E)$ is a functor.

Problems 23–28 introduce the **Pfaffian** of a $(2n)$-by-$(2n)$ alternating matrix $X = [x_{ij}]$ with entries in a field $\mathbb{K}$. This is the polynomial in the entries of X with integer coefficients given by

$$\text{Pfaff}(X) = \sum_{\substack{\text{some } \tau\text{'s} \\ \text{in } \mathfrak{S}_{2n}}} (\text{sgn } \tau) \prod_{k=1}^{n} x_{\tau(2k-1), \tau(2k)},$$

where the sum is taken over those permutations τ such that $\tau(2k - 1) < \tau(2k)$ for $1 \leq k \leq n$ and such that $\tau(1) < \tau(3) < \cdots < \tau(2n - 1)$. It will be seen that $\det X$ is the square of this polynomial. Examples of Pfaffians are

$$\text{Pfaff}\begin{pmatrix} 0 & x \\ -x & 0 \end{pmatrix} = x \quad \text{and} \quad \text{Pfaff}\begin{pmatrix} 0 & a & b & c \\ -a & 0 & d & e \\ -b & -d & 0 & f \\ -c & -e & -f & 0 \end{pmatrix} = af - be + cd.$$

The problems in this set will be continued at the end of Chapter VIII.

23. For the matrix J in Section 5, show that $\text{Pfaff}(J) = 1$.

24. In the expansion $\det X = \sum_{\sigma \in \mathfrak{S}_{2n}} (\text{sgn } \sigma) \prod_{l=1}^{2n} x_{l, \sigma(l)}$, prove that the value of the right side with X as above is not changed if the sum is extended only over those σ's whose expansion in terms of disjoint cycles involves only cycles of even length (and in particular no cycles of length 1).

25. Define $\sigma \in \mathfrak{S}_{2n}$ to be "good" if its expansion in terms of disjoint cycles involves only cycles of even length. If σ is good, show that there uniquely exist two disjoint subsets A and B of n elements each in $\{1, \ldots, 2n\}$ such that A contains the smallest-numbered index in each cycle and such that σ maps each set onto the other.

26. In the notation of the previous problem with σ good, let $y(\sigma)$ be the product of the monomials x_{ab} such that a is in A and $b = \sigma(a)$. For each factor x_{ij} of $y(\sigma)$ with $i > j$, replace the factor by $-x_{ji}$. In the resulting product, arrange the factors in order so that their first subscripts are increasing, and denote this expression by $s x_{i_1 i_2} x_{i_3 i_4} \cdots x_{i_{2n-1} i_{2n}}$, where s is a sign. Let τ be the permutation that carries each r to i_r, and define $s(\tau)$ to be the sign s. Similarly let $z(\sigma)$ be the product of the monomials x_{ba} such that b is in B and $a = \sigma(b)$. For each factor x_{ij} of $z(\sigma)$ with $i > j$, replace the factor by $-x_{ji}$. In the resulting product, arrange the factors in order so that their first subscripts are increasing, and denote this expression by $s' x_{j_1 j_2} x_{j_3 j_4} \cdots x_{j_{2n-1} j_{2n}}$, where s' is a sign. Let τ' be the permutation that carries each r to j_r, and define $s'(\tau')$ to be the sign s'. Prove, apart from signs, that the σth term in the expansion of $\det X$ matches the product of the τth term of $\text{Pfaff}(X)$ and the τ'^{th} term of $\text{Pfaff}(X)$.

27. In the previous problem, take the signs $s(\tau)$ and $s'(\tau')$ into account and show that the signs of σ, τ, and τ' work out so that the σth term in the expansion of $\det X$ is the product of the τth and τ'^{th} terms of $\text{Pfaff}(X)$.

28. Show that every term of the product of $\text{Pfaff}(X)$ with itself is accounted for once and only once by the construction in the previous three problems, and conclude that the alternating matrix X has $\det X = (\text{Pfaff}(X))^2$.

Problems 29–30 concern filtrations and gradings. A vector space V over $\mathbb{K}$ is said to be **filtered** when an increasing sequence of subspaces $V_0 \subseteq V_1 \subseteq V_2 \subseteq \cdots$ is specified with union V. In this case we put $V_{-1} = 0$ by convention. The space V is **graded** if a sequence of subspaces $V^0, V^1, V^2, \ldots$ is specified such that

$$V = \bigoplus_{n=0}^{\infty} V^n.$$

When V is graded, there is a natural filtration of V given by $V_n = \bigoplus_{k=0}^n V^k$. Examples of graded vector spaces are any tensor algebra $V = T(E)$, symmetric algebra $S(E)$, exterior algebra $\bigwedge(E)$, and polynomial algebra $P(E)$, the nth subspace of the grading consisting of those elements that are homogeneous of degree n. Any polynomial algebra $\mathbb{K}[X_1, \ldots, X_n]$ is another example of a graded vector space, the grading being by total degree.

29. When V is a filtered vector space as in (A.34), the **associated graded vector space** is $\text{gr } V = \bigoplus_{n=0}^{\infty} V_n / V_{n-1}$. Let V and $V^\#$ be two filtered vector spaces, and let φ be a linear map between them such that $\varphi(V_n) \subseteq V_n^\#$ for all n. Since the restriction of φ to V_n carries V_{n-1} into $V_{n-1}^\#$, this restriction induces a linear map $\text{gr}^n \varphi : (V_n / V_{n-1}) \to (V_n^\# / V_{n-1}^\#)$. The direct sum of these linear maps is then a linear map $\text{gr } \varphi : \text{gr } V \to \text{gr } V^\#$ called the **associated graded map** for φ. Prove that if $\text{gr } \varphi$ is a vector-space isomorphism, then φ is a vector-space isomorphism.

30. Let A be an associative algebra over $\mathbb{K}$ with identity. If A has a filtration $A_0, A_1, \ldots$ of vector subspaces with $1 \in A_0$ such that $A_m A_n \subseteq A_{m+n}$ for all m and n, then one says that A is a **filtered associative algebra**; similarly if A is graded as $A = \bigoplus_{n=0}^{\infty} A^n$ in such a way that $A^m A^n \subseteq A^{m+n}$ for all m and n, then one says that A is a **graded associative algebra**. If A is a filtered associative algebra with identity, prove that the graded vector space gr A acquires a multiplication in a natural way, making it into a graded associative algebra with identity.

Problems 31–35 concern Lie algebras and their universal enveloping algebras. If $\mathbb{K}$ is a field, a **Lie algebra** $\mathfrak{g}$ over $\mathbb{K}$ is a nonassociative algebra whose product, called the **Lie bracket** and written $[x, y]$, is alternating as a function of the pair (x, y) and satisfies the **Jacobi identity** $[x, [y, z]] + [y, [z, x]] + [z, [x, y]] = 0$ for all x, y, z in $\mathfrak{g}$. The **universal enveloping algebra** $U(\mathfrak{g})$ of $\mathfrak{g}$ is the quotient $T(\mathfrak{g})/I''$, where I'' is the two-sided ideal generated by all elements $x \otimes y - y \otimes x - [x, y]$ with x and y in $T^1(\mathfrak{g})$. The grading for $T(\mathfrak{g})$ makes $U(\mathfrak{g})$ into a filtered associate algebra with identity. The product of x and y in $U(\mathfrak{g})$ is written xy.

31. If A is an associative algebra over $\mathbb{K}$, prove that A becomes a Lie algebra if the Lie bracket is defined by $[x, y] = xy - yx$. In particular, observe that $M_n(\mathbb{K})$ becomes a Lie algebra in this way.

32. Fix a matrix $A \in M_n(\mathbb{K})$, and let $\mathfrak{g}$ be the vector subspace of all members x of $M_n(\mathbb{K})$ with $x^t A + Ax = 0$.
    (a) Prove that $\mathfrak{g}$ is closed under the bracket operation of the previous problem and is therefore a Lie subalgebra of $M_n(\mathbb{K})$.
    (b) Deduce as a special case of (a) that the vector space of all skew-symmetric matrices in $M_n(\mathbb{K})$ is a Lie subalgebra of $M_n(\mathbb{K})$.

33. Let $\mathfrak{g}$ be a Lie algebra over $\mathbb{K}$, and let ι be the linear map obtained as the composition of $\mathfrak{g} \to T^1(\mathfrak{g})$ and the passage to the quotient $U(\mathfrak{g})$. Prove that $(U(\mathfrak{g}), \iota)$ has the following universal mapping property: whenever l is any linear map of $\mathfrak{g}$ into an associative algebra A with identity satisfying the condition of being a Lie algebra homomorphism, namely $l[x, y] = l(x)l(y) - l(y)l(x)$ for all x and y in $\mathfrak{g}$, then there exists a unique associative algebra homomorphism $L : U(\mathfrak{g}) \to A$ with $L(1) = 1$ such that $L \circ \iota = l$.

34. Let $\mathfrak{g}$ be a Lie algebra over $\mathbb{K}$, let $\{u_i\}_{i \in A}$ be a vector-space basis of $\mathfrak{g}$, and suppose that a simple ordering has been imposed on the index set A. Prove that the set of all monomials $u_{i_1}^{j_1} \cdots u_{i_k}^{j_k}$ with $i_1 < \cdots < i_k$ and $\sum_m j_m$ arbitrary is a spanning set for $U(\mathfrak{g})$.

35. For a Lie algebra $\mathfrak{g}$ over $\mathbb{K}$, the **Poincaré–Birkhoff–Witt Theorem** says that the spanning set for $U(\mathfrak{g})$ in the previous problem is actually a basis. Assuming this theorem, prove that gr $U(\mathfrak{g})$ is isomorphic as a graded algebra to $S(\mathfrak{g})$.

Problems 36–40 introduce Clifford algebras. Let $\mathbb{K}$ be a field of characteristic $\neq 2$, let E be a finite-dimensional vector space over $\mathbb{K}$, and let $\langle \cdot , \cdot \rangle$ be a symmetric bilinear form on E. The **Clifford algebra** $\mathrm{Cliff}(E, \langle \cdot , \cdot \rangle)$ is the quotient $T(E)/I''$, where I'' is the two-sided ideal generated by all elements[5] $v \otimes v + \langle v, v \rangle$ with v in E. The grading for $T(E)$ makes $\mathrm{Cliff}(E, \langle \cdot , \cdot \rangle)$ into a filtered associative algebra with identity. Products in $\mathrm{Cliff}(E, \langle \cdot , \cdot \rangle)$ are written as ab with no special symbol.

36. Let ι be the composition of the inclusion $E \subseteq T^1(E)$ and the passage to the quotient modulo I''. Prove that $(\mathrm{Cliff}(E, \langle \cdot , \cdot \rangle), \iota)$ has the following universal mapping property: whenever l is any linear map of E into an associative algebra A with identity such that $l(v)^2 = -\langle v, v \rangle 1$ for all $v \in E$, then there exists a unique algebra homomorphism $L : \mathrm{Cliff}(E, \langle \cdot , \cdot \rangle) \to A$ with $L(1) = 1$ and such that $L \circ \iota = l$.

37. Let $\{u_1, \ldots, u_n\}$ be a basis of E. Prove that the 2^n elements of $\mathrm{Cliff}(E, \langle \cdot , \cdot \rangle)$ given by $u_{i_1} u_{i_2} \cdots u_{i_k}$ with $i_1 < \cdots < i_k$ form a spanning set of $\mathrm{Cliff}(E, \langle \cdot , \cdot \rangle)$.

38. Using the Principal Axis Theorem, fix a basis $\{e_1, \ldots, e_n\}$ of E such that $\langle e_i, e_j \rangle = d_i \delta_{ij}$ for all j. Introduce an algebra C over $\mathbb{K}$ of dimension 2^n with generators $e_1, \ldots, e_n$ and with a basis parametrized by subsets of $\{1, \ldots, n\}$ and given by all elements

$$e_{i_1} e_{i_2} \cdots e_{i_k} \quad \text{with} \quad i_1 < i_2 < \cdots < i_k,$$

with the multiplication that is implicit in the rules

$$e_i^2 = -d_i \quad \text{and} \quad e_i e_j = -e_j e_i \quad \text{if } i \neq j,$$

namely, to multiply two monomials $e_{i_1} e_{i_2} \cdots e_{i_k}$ and $e_{j_1} e_{j_2} \cdots e_{j_l}$, put them end to end, replace any occurrence of two e_k's by the scalar $-d_k$, and then permute the remaining e_k's until their indices are in increasing order, introducing a minus sign each time two distinct e_k's are interchanged. Prove that the algebra C is associative.

39. Prove that the associative algebra C of the previous problem is isomorphic as an algebra to $\mathrm{Cliff}(E, \langle \cdot , \cdot \rangle)$.

40. Prove that $\mathrm{gr}\,\mathrm{Cliff}(E, \langle \cdot , \cdot \rangle)$ is isomorphic as a graded algebra to $\bigwedge(E)$.

Problems 41–48 introduce finite-dimensional Heisenberg Lie algebras and the corresponding Weyl algebras. They make use of Problems 31–35 concerning Lie algebras and universal enveloping algebras. Let V be a finite-dimensional vector space over the field $\mathbb{K}$, and let $\langle \cdot , \cdot \rangle$ be a nondegenerate alternating bilinear form on $V \times V$. Write $2n$ for the dimension of V. Introduce an indeterminate X_0. The **Heisenberg Lie algebra** $H(V)$ on V is a Lie algebra whose underlying vector space is $\mathbb{K}X_0 \oplus V$ and whose Lie bracket is given by $[(cX_0, u), (dX_0, v)] = \langle u, v \rangle X_0$. Let $U(H(V))$ be its universal enveloping algebra. The **Weyl algebra** $W(V)$ on V is the quotient of the tensor algebra $T(V)$ by the two-sided ideal generated by all $u \otimes v - v \otimes u - \langle u, v \rangle 1$ with u and v in V; as such, it is a filtered associative algebra.

---

[5]Some authors factor out the elements $v \otimes v - \langle v, v \rangle$ instead. There is no generally accepted convention.

41. Verify when the field is $\mathbb{K} = \mathbb{R}$ that an example of a $2n$-dimensional V with its nondegenerate alternating bilinear form $\langle \cdot, \cdot \rangle$ is $V = \mathbb{C}^n$ with $\langle u, v \rangle = \text{Im}(u, v)$, where $(\cdot, \cdot)$ is the usual inner product on $\mathbb{C}^n$. For this V, exhibit a Lie-algebra isomorphism of $H(V)$ with the Lie algebra of all complex $(n+1)$-by-$(n+1)$ matrices of the form $\begin{pmatrix} 0 & \bar{z}^t & ir \\ 0 & 0 & z \\ 0 & 0 & 0 \end{pmatrix}$ with $z \in \mathbb{C}^n$ and $r \in \mathbb{R}$.

42. In the general situation show that the linear map $\iota(cX_0, v) = c1 + v$ is a Lie algebra homomorphism of $H(V)$ into $W(V)$ and that its extension to an associative algebra homomorphism $\widetilde{\iota}: U(H(V)) \to W(V)$ is onto and has kernel equal to the two-sided ideal in $U(H(V))$ generated by $X_0 - 1$.

43. Prove that $W(V)$ has the following universal mapping property: whenever $\varphi: H(V) \to A$ is a Lie algebra homomorphism of $H(V)$ into an associative algebra A with identity such that $\varphi(X_0) = 1$, then there exists a unique associative algebra homomorphism $\widetilde{\varphi}$ of $W(V)$ into A such that $\varphi = \widetilde{\varphi} \circ \iota$.

44. Let $v_1, \ldots, v_{2n}$ be any vector space basis of V. Prove that the elements $v_1^{k_1} \cdots v_{2n}^{k_{2n}}$ with integer exponents ≥ 0 span $W(V)$.

45. For $\mathbb{K} = \mathbb{R}$, let S be the vector space of all real-valued functions $P(x)e^{-\pi|x|^2}$, where $P(x)$ is a polynomial in n real variables. Show that S is mapped into itself by the linear operators $\partial/\partial x_i$ and $m_j =$ (multiplication by x_j).

46. With $\mathbb{K} = \mathbb{R}$, let $\{p_1, \ldots, p_n, q_1, \ldots, q_n\}$ be a Weyl basis of V in the terminology of Problem 6. In the notation of Problem 45, let $\varphi: V \to \text{Hom}_\mathbb{R}(S, S)$ be the linear map given by $\varphi(p_i) = \partial/\partial x_i$ and $\varphi(q_j) = m_j$. Use Problem 43 to extend φ to an algebra homomorphism $\widetilde{\varphi}: W(V) \to \text{Hom}_\mathbb{R}(S, S)$ with $\widetilde{\varphi}(1) = 1$, and use Problem 42 to obtain a representation of $H(V)$ on S. Prove that this representation of $H(V)$ is irreducible in the sense that there is no proper nonzero vector subspace carried to itself by all members of $\widetilde{\varphi}(H(V))$.

47. In Problem 46 with $\mathbb{K} = \mathbb{R}$, prove that the associative algebra homomorphism $\widetilde{\varphi}: W(V) \to \text{Hom}_\mathbb{R}(S, S)$ is one-one. Conclude for $\mathbb{K} = \mathbb{R}$ that the elements $v_1^{k_1} \cdots v_{2n}^{k_{2n}}$ of Problem 44 form a vector-space basis of $W(V)$.

48. For $\mathbb{K} = \mathbb{R}$, prove that $\text{gr } W(V)$ is isomorphic as a graded algebra to $S(V)$.

Problems 49–51 deal with Jordan algebras. Let $\mathbb{K}$ be a field of characteristic $\neq 2$. An algebra J over $\mathbb{K}$ with multiplication $a \cdot b$ is called a **Jordan algebra** if the identities $a \cdot b = b \cdot a$ and $a^2 \cdot (b \cdot a) = (a^2 \cdot b) \cdot a$ are always satisfied; here a^2 is an abbreviation for $a \cdot a$.

49. Let A be an associative algebra, and define $a \cdot b = \frac{1}{2}(ab + ba)$. Prove that A becomes a Jordan algebra under this new multiplication.

50. In the situation of the previous problem, suppose that $a \mapsto a^t$ is a one-one linear mapping of A onto itself such that $(ab)^t = b^t a^t$ for all a and b. (For example, $a \mapsto a^t$ could be the transpose mapping if $A = M_n(\mathbb{K})$.) Prove that the vector subspace of all a with $a^t = a$ is carried to itself by the Jordan product $a \cdot b$ and hence is a Jordan algebra.

51. Let V be a finite-dimensional vector space over $\mathbb{K}$, and let $\langle \cdot, \cdot \rangle$ be a symmetric bilinear form on V. Define $A = \mathbb{K}1 \oplus V$ as a vector space, and define a multiplication in A by $(c1, x) \cdot (d1, y) = \big((cd + \langle x, y \rangle)1, cy + dx\big)$. Prove that A is a Jordan algebra under this definition of multiplication.

Problems 52–56 deal with the algebra $\mathbb{O}$ of real **octonions**, sometimes known as the **Cayley numbers**. This is a certain 8-dimensional nonassociative algebra with identity over $\mathbb{R}$ with an inner product such that $\|ab\| = \|a\|\|b\|$ for all a and b and such that the left and right multiplications by any element $a \neq 0$ are always invertible.

52. Let A be an algebra over $\mathbb{R}$. Let $[a, b] = ab - ba$ and $[a, b, c] = (ab)c - a(bc)$.
    (a) The 3-multilinear function $(a, b, c) \mapsto [a, b, c]$ from $A \times A \times A$ to A is called the **associator** in A. Observe that it is 0 if and only if A is associative. Show that it is alternating if and only if A always satisfies the limited associativity laws

    $$(aa)b = a(ab), \qquad (ab)a = a(ba), \qquad (ba)a = b(aa).$$

    In this case, A is said to be **alternative**.
    (b) Show that A is alternative if the first and third of the limited associativity laws in (a) are always satisfied.

53. **(Cayley–Dickson construction)** Suppose that A is an algebra over $\mathbb{R}$ with a two-sided identity 1, and suppose that there is an $\mathbb{R}$ linear function $*$ from A to itself (called "conjugation") such that $1^* = 1$, $a^{**} = a$, and $(ab)^* = b^* a^*$ for all a and b in A. Define an algebra B over $\mathbb{R}$ to have the underlying real vector-space structure of $A \oplus A$ and to have multiplication and conjugation given by

    $$(a, b)(c, d) = (ac - db^*, a^*d + cb) \qquad \text{and} \qquad (a, b)^* = (a^*, -b).$$

    (a) Prove that $(1, 0)$ is a two-sided identity in B and that the operation $*$ in B satisfies the required properties of a conjugation.
    (b) Prove that if $a^* = a$ for all $a \in A$, then A is commutative.
    (c) Prove that if $a^* = a$ for all $a \in A$, then B is commutative.
    (d) Prove that if A is commutative and associative, then B is associative.
    (e) Verify the following outcomes of the above construction $A \to B$:
        (i) $A = \mathbb{R}$ yields $B = \mathbb{C}$,
        (ii) $A = \mathbb{C}$ yields $B = \mathbb{H}$, the algebra of quaternions.

54. Suppose that A is an algebra over $\mathbb{R}$ with an identity and a conjugation as in the previous problem. Say that A is **nicely normed** if
   (i) $a + a^*$ is always of the form $r1$ with r real and
   (ii) aa^* always equals a^*a and for $a \neq 0$, is of the form $r1$ with r real and positive.

   (a) Prove that if A is nicely normed, then so is the algebra B of the previous problem.
   (b) Prove that if A is nicely normed, then $(a, b) = \frac{1}{2}(ab^* + ba^*)$ is an inner product on A with norm $\|a\| = (aa^*)^{1/2} = (a^*a)^{1/2}$.
   (c) Prove that if A is associative and nicely normed, then the algebra B of the previous problem is alternative.

55. Starting from the real algebra $A = \mathbb{H}$, apply the construction of Problem 53, and let the resulting 8-dimensional real algebra be denoted by $\mathbb{O}$, the algebra of octonions.

   (a) Prove that $\mathbb{O}$ is an alternative algebra and is nicely normed.
   (b) Prove that $(xx^*)y = x(x^*y)$ and $x(yy^*) = (xy)y^*$ within $\mathbb{O}$.
   (c) Prove that $\|ab\|^2 a = \|a\|^2 \|b\|^2 a$ within $\mathbb{O}$.
   (d) Conclude from (c) that the operations of left and right multiplication by any $a \neq 0$ within $\mathbb{O}$ are invertible.
   (e) Show that the inverse operators are left and right multiplication by $\|a\|^{-2} a^*$.
   (f) Denote the usual basis vectors of $\mathbb{H}$ by **1, i, j, k**. Write down a multiplication table for the eight basis vectors of $\mathbb{O}$ given by $(x, 0)$ and $(0, y)$ as x and y run through the basis vectors of $\mathbb{H}$.

56. What prevents the construction of Problem 53, when applied with $A = \mathbb{O}$, from yielding a 16-dimensional algebra B in which $\|ab\|^2 = \|a\|^2 \|b\|^2$ and therefore in which the operations of left and right multiplication by any $a \neq 0$ within B are invertible?

CHAPTER VII

Advanced Group Theory

Abstract. This chapter continues the development of group theory begun in Chapter IV, the main topics being the use of generators and relations, representation theory for finite groups, and group extensions. Representation theory uses linear algebra and inner-product spaces in an essential way, and a structure-theory theorem for finite groups is obtained as a consequence. Group extensions introduce the subject of cohomology of groups.

Sections 1–3 concern generators and relations. The context for generators and relations is that of a free group on the set of generators, and the relations indicate passage to a quotient of this free group by a normal subgroup. Section 1 constructs free groups in terms of words built from an alphabet and shows that free groups are characterized by a certain universal mapping property. This universal mapping property implies that any group may be defined by generators and relations. Computations with free groups are aided by the fact that two reduced words yield the same element of a free group if and only if the reduced words are identical. Section 2 obtains the Nielsen–Schreier Theorem that subgroups of free groups are free. Section 3 enlarges the construction of free groups to the notion of the free product of an arbitrary set of groups. Free product is what coproduct is for the category of groups; free groups themselves may be regarded as free products of copies of the integers.

Sections 4–5 introduce representation theory for finite groups and give an example of an important application whose statement lies outside representation theory. Section 4 contains various results giving an analysis of the space $C(G, \mathbb{C})$ of all complex-valued functions on a finite group G. In this analysis those functions that are constant on conjugacy classes are shown to be linear combinations of the characters of the irreducible representations. Section 5 proves Burnside's Theorem as an application of this theory—that any finite group of order $p^a q^b$ with p and q prime and with $a+b > 1$ has a nontrivial normal subgroup.

Section 6 introduces cohomology of groups in connection with group extensions. If N is to be a normal subgroup of G and Q is to be isomorphic to G/N, the first question is to parametrize the possibilities for G up to isomorphism. A second question is to parametrize the possibilities for G if G is to be a semidirect product of N and Q.

1. Free Groups

This section and the next two introduce some group-theoretic notions that in principle apply to all groups but in practice are used with countable groups, often countably infinite groups that are nonabelian. The material is especially useful in applications in topology, particularly in connection with fundamental groups and covering spaces. But the formal development here will be completely algebraic, not making use of any definitions or theorems from topology.

In the case of abelian groups, every abelian group G is a quotient of a suitable free abelian group, i.e., a suitable direct sum of copies of the additive group $\mathbb{Z}$ of integers.[1] Recall the discussion of Section IV.9: We introduce a copy $\mathbb{Z}_g$ of $\mathbb{Z}$ for each g in G, define $\widetilde{G} = \bigoplus_{g \in G} \mathbb{Z}_g$, let $i_g : \mathbb{Z}_g \to \widetilde{G}$ be the standard embedding, and let $\varphi_g : \mathbb{Z}_g \to G$ be the group homomorphism written additively as $\varphi_g(n) = ng$. The universal mapping property of direct sums that was stated as Proposition 4.17 produces a unique group homomorphism $\varphi : \widetilde{G} \to G$ such that $\varphi \circ i_g = \varphi_g$ for all g, and φ is the required homomorphism of a free abelian group onto G.

The goal in this section is to carry out an analogous construction for groups that are not necessarily abelian. The constructed groups, to be called "free groups," are to be rather concrete, and the family of all of them is to have the property that every group is the quotient of some member of the family.

If S is any set, we construct a "free group $F(S)$ on the set S." Let us speak of S as a set of "symbols" or as the members of an "alphabet," possibly infinite, with which we are working. If S is empty, the group $F(S)$ is taken to be the one-element trivial group, and we shall therefore now assume that S is not empty. If a is a symbol in S, we introduce a new symbol a^{-1} corresponding to it, and we let S^{-1} denote the set of all such symbols a^{-1} for $a \in S$. Define $S' = S \cup S^{-1}$. A **word** is a finite string of symbols from S', i.e., an ordered n-tuple for some n of members of S' with repetitions allowed. Words that are n-tuples are said to have **length** n. The empty word, with length 0, will be denoted by 1. Other words are usually written with the symbols juxtaposed and all commas omitted, as in $abca^{-1}cb^{-1}$. The set of words will be denoted by $W(S')$. We introduce a multiplication $W(S') \times W(S') \to W(S')$ by writing end-to-end the words that are to be multiplied: $(abca^{-1}, cb^{-1}) \mapsto abca^{-1}cb^{-1}$. The length of a product is the sum of the lengths of the factors. It is plain that this multiplication is associative and that 1 is a two-sided identity. It is not a group operation, however, since most elements of $W(S')$ do not have inverses: multiplication never decreases length, and thus the only way that 1 can be a product of two elements is as the product 11. To obtain a group from $W(S')$, we shall introduce an equivalence relation in $W(S')$.

Two words are said to be **equivalent** if one of the words can be obtained from the other by a finite succession of insertions and deletions of expressions aa^{-1} or $a^{-1}a$ within the word; here a is assumed to be an element of S. It will be convenient to refer to the pairs aa^{-1} and $a^{-1}a$ together; therefore when $b = a^{-1}$ is in S^{-1}, let us define $b^{-1} = (a^{-1})^{-1}$ to be a. Then two words are equivalent if one of the words can be obtained from the other by a finite succession of insertions and deletions of expressions of the form bb^{-1} with b in S'. This definition is

---
[1] Direct sum here is what coproduct, in the sense of Section IV.11, amounts to in the category of all abelian groups.

1. Free Groups

arranged so that "equivalent" is an equivalence relation. We write $x \sim y$ if x and y are words that are equivalent. The underlying set for the free group $F(S)$ will be taken to be the set of equivalence classes of members of $W(S')$.

Theorem 7.1. If S is a set and $W(S')$ is the corresponding set of words built from $S' = S \cup S^{-1}$, then the product operation defined on $W(S')$ descends in a well-defined fashion to the set $F(S)$ of equivalence classes of members of $W(S')$, and $F(S)$ thereby becomes a group. Define $\iota : S \to F(S)$ to be the composition of the inclusion into words of length one followed by passage to equivalence classes. Then the pair $(F(S), \iota)$ has the following universal mapping property: whenever G is a group and $\varphi : S \to G$ is a function, then there exists a unique group homomorphism $\widetilde{\varphi} : F(S) \to G$ such that $\varphi = \widetilde{\varphi} \circ \iota$.

REMARK. The group $F(S)$ is called the **free group** on S. Figure 7.1 illustrates its universal mapping property. The brief form in words of the property is that any function from S into a group G extends uniquely to a group homomorphism of $F(S)$ into G. This universal mapping property actually characterizes $F(S)$, as will be seen in Proposition 7.2.

FIGURE 7.1. Universal mapping property of a free group.

PROOF. Let us denote equivalence classes by brackets. We want to define multiplication in $F(S)$ by $[w_1][w_2] = [w_1 w_2]$. To see that this formula makes sense in $F(S)$, let x_1, x_2, and y be words, and let b be in S'. Define $x = x_1 x_2$ and $x' = x_1 b b^{-1} x_2$, so that $x' \sim x$. Then it is evident that $x'y \sim xy$ and $yx' \sim yx$. Iteration of this kind of relationship shows that $w_1' \sim w_1$ and $w_2' \sim w_2$ implies $w_1' w_2' \sim w_1 w_2$, and hence multiplication of equivalence classes is well defined.

Since multiplication in $W(S')$ is associative, we have $[w_1]([w_2][w_3]) = [w_1][w_2 w_3] = [w_1(w_2 w_3)] = [(w_1 w_2)w_3] = [w_1 w_2][w_3] = ([w_1][w_2])[w_3]$. Thus multiplication is associative in $F(S)$. The class $[1]$ of the empty word 1 is a two-sided identity. If $b_1, \ldots, b_n$ are in S', then $b_n^{-1} \cdots b_2^{-1} b_1^{-1} b_1 b_2 \cdots b_n$ is equivalent to 1, and so is $b_1 b_2 \cdots b_n b_n^{-1} \cdots b_2^{-1} b_1^{-1}$. Consequently $[b_n^{-1} \cdots b_2^{-1} b_1^{-1}]$ is a two-sided inverse of $[b_1 b_2 \cdots b_n]$, and $F(S)$ is a group.

Now we address the universal mapping property, first proving the stated uniqueness of the homomorphism. Every member of $F(S)$ is the product of classes $[b]$ with b in S'. In turn, if b is of the form a^{-1} with a in S, then $[b] = [a]^{-1}$. Hence $F(S)$ is generated by all classes $[a]$ with a in S, i.e., by $\iota(S)$. Any homomorphism of a group is determined by its values on the members of a generating set, and uniqueness therefore follows from the formula $\widetilde{\varphi}([a]) = \widetilde{\varphi}(\iota(a)) = \varphi(a)$.

For existence we begin by defining a function $\Phi : W(S') \to G$ such that

$$\Phi(a) = \varphi(a) \qquad \text{for } a \text{ in } S,$$
$$\Phi(a^{-1}) = \varphi(a)^{-1} \qquad \text{for } a^{-1} \text{ in } S^{-1},$$
$$\Phi(w_1 w_2) = \Phi(w_1)\Phi(w_2) \qquad \text{for } w_1 \text{ and } w_2 \text{ in } W(S').$$

We use the formulas $\Phi(a) = \varphi(a)$ for a in S and $\Phi(a^{-1}) = \varphi(a)^{-1}$ for a^{-1} in S^{-1} as a definition of $\Phi(b)$ for b in S'. Any member of $W(S')$ can be written uniquely as $b_1 \cdots b_n$ with each b_j in S', and we set $\Phi(b_1 \cdots b_n) = \Phi(b_1) \cdots \Phi(b_n)$. (If $n = 0$, the understanding is that $\Phi(1) = 1$.) Then Φ has the required properties.

Let us show that $w' \sim w$ implies $\Phi(w') = \Phi(w)$. If $b_1, \ldots, b_n$ are in S' and b is in S', then the question is whether

$$\Phi(b_1 \cdots b_k b b^{-1} b_{k+1} \cdots b_n) \stackrel{?}{=} \Phi(b_1 \cdots b_k b_{k+1} \cdots b_n).$$

If g and g' denote the elements $\Phi(b_1) \cdots \Phi(b_k)$ and $\Phi(b_{k+1}) \cdots \Phi(b_n)$ of G, then the two sides of the queried formula are

$$g\Phi(b)\Phi(b^{-1})g' \qquad \text{and} \qquad gg'.$$

Thus the question is whether $\Phi(b)\Phi(b^{-1})$ always equals 1 in G. If $b = a$ is in S, this equals $\varphi(a)\varphi(a)^{-1} = 1$, while if $b = a^{-1}$ is in S^{-1}, it equals $\varphi(a)^{-1}\varphi(a) = 1$. We conclude that $w' \sim w$ implies $\Phi(w') = \Phi(w)$.

We may therefore define $\widetilde{\varphi}([w]) = \Phi(w)$ for $[w]$ in $F(S)$. Since $\widetilde{\varphi}([w][w']) = \widetilde{\varphi}([ww']) = \Phi(ww') = \Phi(w)\Phi(w') = \widetilde{\varphi}([w])\widetilde{\varphi}([w'])$, $\widetilde{\varphi}$ is a homomorphism of $F(S)$ into G. For a in S, we have $\widetilde{\varphi}([a]) = \Phi(a) = \varphi(a)$. In other words, $\widetilde{\varphi}(\iota(a)) = \varphi(a)$. This completes the proof of existence. $\square$

Proposition 7.2. Let S be a set, F be a group, and $\iota' : S \to F$ be a function. Suppose that the pair (F, ι') has the following universal mapping property: whenever G is a group and $\varphi : S \to G$ is a function, then there exists a unique group homomorphism $\widetilde{\varphi} : F \to G$ such that $\varphi = \widetilde{\varphi} \circ \iota'$. Then there exists a unique group homomorphism $\Phi : F(S) \to F$ such that $\iota' = \Phi \circ \iota$, and it is a group isomorphism.

REMARKS. Chapter VI is not a prerequisite for the present chapter. However, readers who have been through Chapter VI will recognize that Proposition 7.2 is a special case of Problem 19 at the end of that chapter.

PROOF. We apply the universal mapping property of $(F(S), \iota)$, as stated in Theorem 7.1, to the group $G = F$ and the function $\varphi = \iota'$, obtaining a group homomorphism $\Phi : F(S) \to F$ such that $\iota' = \Phi \circ \iota$. Then we apply the given universal mapping property of (F, ι') to the group $G = F(S)$ and the function $\varphi = \iota$, obtaining a group homomorphism $\Psi : F \to F(S)$ such that $\iota = \Psi \circ \iota'$.

The group homomorphism $\Psi \circ \Phi : F(S) \to F(S)$ has the property that $(\Psi \circ \Phi) \circ \iota = \Psi \circ (\Phi \circ \iota) = \Psi \circ \iota' = \iota$, and the identity $1_{F(S)}$ has this same property. By the uniqueness of the group homomorphism in Theorem 7.1, $\Psi \circ \Phi = 1_{F(S)}$.

Similarly the group homomorphism $\Phi \circ \Psi : F \to F$ has the property that $(\Phi \circ \Psi) \circ \iota' = \iota'$, and the identity 1_F has this same property. By the uniqueness of the group homomorphism in the assumed universal mapping property of F, $\Phi \circ \Psi = 1_F$.

Therefore Φ is a group isomorphism. We know that $\iota(S)$ generates $F(S)$. If $\Phi' : F(S) \to F$ is another group isomorphism with $\iota' = \Phi' \circ \iota$, then Φ' and Φ agree on $\iota(S)$ and therefore have to agree everywhere. Hence Φ is unique. □

Proposition 7.2 raises the question of recognizing candidates for the set $T = \iota'(S)$ in a given group F so as to be in a position to exhibit F as isomorphic to the free group $F(S)$. Certainly T has to generate F. But there is also an independence condition. The idea is that if we form words from the members of T, then two words are to lead to equal members of F only if they can be transformed into one another by the same rules that are allowed with free groups.

What this problem amounts to in the case that $F = F(S)$ is that we want a decision procedure for telling whether two given words are equivalent. This is the so-called **word problem** for the free group. If we think about the matter for a moment, not much is instantly obvious. If a_1 and a_2 are two members of S and if they are considered as words of length 1, are they equivalent? Equivalence allows for inserting pairs bb^{-1} with b in S', as well as deleting them. Might it be possible to do some complicated iterated insertion and deletion of pairs to transform a_1 into a_2? Although the negative answer can be readily justified in this situation by a parity argument, it can be justified even more easily by the universal mapping property: there exist groups G with more than one element; we can map a_1 to one element of G and a_2 to another element of G, extend to a homomorphism $\widetilde{\varphi} : F(S) \to G$, see that $\widetilde{\varphi}(\iota(a_1)) \neq \widetilde{\varphi}(\iota(a_2))$, and conclude that $\iota(a_1) \neq \iota(a_2)$. But what about the corresponding problem for two more-complicated words in a free group? Fortunately there is a decision procedure for the word problem in a free group. It involves the notion of "reduced" words. A word in $W(S')$ is said to be **reduced** if it contains no consecutive pair bb^{-1} with b in S'.

Proposition 7.3 (solution of the word problem for free groups). Let S be a set, let $S' = S \cup S^{-1}$, and let $W(S')$ be the corresponding set of words. Then each word in $W(S')$ is equivalent to one and only one reduced word.

REMARK. To test whether two words are equivalent, the proposition says to delete pairs bb^{-1} with $b \in S'$ as much as possible from each given word, and to check whether the resulting reduced words are identical.

PROOF. Removal of a pair bb^{-1} with $b \in S'$ decreases the length of a word by 2, and the length has to remain ≥ 0. Thus the process of successively removing such pairs has to stop after finitely many steps, and the result is a reduced word. This proves that each equivalence class contains a reduced word.

For uniqueness we shall associate to each word a finite sequence of reduced words such that the last member of the sequence is unchanged when we insert or delete within the given word any expression bb^{-1} with $b \in S'$. Specifically if $w = b_1 \cdots b_n$, with each b_i in S', is a given word, we associate to w the sequence of words $x_0, x_1, \ldots, x_n$ defined inductively by

$$x_0 = 1,$$
$$x_1 = b_1,$$
$$x_i = \begin{cases} x_{i-1} b_i & \text{if } i \geq 2 \text{ and } x_{i-1} \text{ does not end in } b_i^{-1}, \\ y_{i-2} & \text{if } i \geq 2 \text{ and } x_{i-1} = y_{i-2} b_i^{-1}, \end{cases} \quad (*)$$

and we define $r(w) = x_n$. Let us see, by induction on $i \geq 0$, that x_i is reduced. The base cases $i = 0$ and $i = 1$ are clear from the definition. Suppose that $i \geq 2$ and that $x_0, \ldots, x_{i-1}$ are reduced. If $x_{i-1} = y_{i-2} b_i^{-1}$ for some y_{i-2}, then x_{i-1} reduced forces y_{i-2} to be reduced, and hence $x_i = y_{i-2}$ is reduced. If x_{i-1} does not end in b_i^{-1}, then the last two symbols of $x_i = x_{i-1} b_i$ do not cancel, and no earlier pair can cancel since x_{i-1} is assumed reduced; hence x_i is reduced. This completes the induction and shows that x_i is reduced for $0 \leq i \leq n$.

If the word $w = b_1 \cdots b_n$ is reduced, then each x_i for $i \geq 2$ is determined by the first of the two choices in $(*)$, and hence $x_i = b_1 \cdots b_i$ for all i. Consequently $r(w) = w$ if w is reduced. If we can prove for a general word $b_1 \cdots b_n$ that

$$r(b_1 \cdots b_n) = r(b_1 \cdots b_k b b^{-1} b_{k+1} \cdots b_n), \quad (**)$$

then it follows that every word w' equivalent to a word w has $r(w') = r(w)$. Since $r(w) = w$ for w reduced, there can be only one reduced word in an equivalence class.

To prove $(**)$, let $x_0, \ldots, x_n$ be the finite sequence associated with $b_1 \cdots b_n$, and let $x'_0, \ldots, x'_{n+2}$ be the sequence associated with $b_1 \cdots b_k b b^{-1} b_{k+1} \cdots b_n$. Certainly $x_i = x'_i$ for $i \leq k$. Let us compute x'_{k+1} and x'_{k+2}. From $(*)$ we see that

$$x'_{k+1} = \begin{cases} x_k b & \text{if } x_k \text{ does not end in } b^{-1}, \\ y & \text{if } x_k = y b^{-1}. \end{cases}$$

In the first of these cases, x'_{k+1} ends in b, and (∗) says therefore that $x'_{k+2} = x_k$. In the second of the cases, the fact that x_k is reduced implies that y does not end in b; hence (∗) says that $x'_{k+2} = yb^{-1} = x_k$. In other words, $x'_{k+2} = x_k$ in both cases. Since the inductive definition of any x_i depends only on x_{i-1}, and similarly for x'_i, we see that $x'_{k+2+i} = x_{k+i}$ for $0 \le i \le n-k$. Therefore $x'_{n+2} = x_n$, and (∗∗) follows. This proves the proposition. □

Let us return to the problem of recognizing candidates for the set $T = \iota'(S)$ in a given group F so that the subgroup generated by T is a free group. Using the universal mapping property for the free group $F(T)$, we form the group homomorphism of $F(T)$ into F that extends the identity mapping on T. We want this homomorphism to be one-one, i.e., to have the property that the only way a word in F built from the members of T can equal the identity is if it comes from the identity. Because of Proposition 7.3 the only reduced word in $F(T)$ that yields the identity is the empty word. Thus the condition that the homomorphism be one-one is that the only image in F of a reduced word in $F(T)$ that can equal the identity is the image of the empty word. Making this condition into a definition, we say that a subset $S = \{g_t \mid t \in T\}$ of F not containing 1 is **free** if no nonempty product $h_1 h_2 \cdots h_m$ in which each h_i or h_i^{-1} is in S and each h_{i+1} is different from h_i^{-1} can be the identity. A free set in F that generates F is called a **free basis** for F.

EXAMPLE. Within the free group $F(\{x, y\})$ on two generators x and y, consider the subgroup generated by $u = x^2$, $v = y^2$, and $w = xy$. The claim is that the subset $\{u, v, w\}$ is free, so that the subgroup generated by u, v, and w is isomorphic to a free group $F(\{u, v, w\})$ on three generators. We are to check that no nonempty reduced word in $u, v, w, u^{-1}, v^{-1}, w^{-1}$ can reduce to the empty word after substitution in terms of x and y. We induct on the length of the u, v, w word, the base case being length 0. Suppose that $v = y^2$ occurs somewhere in our reduced u, v, w word that collapses to the empty word after substitution. Consider what is needed for the left-hand factor of y in the y^2 to cancel. The cancellation must result from the presence of some y^{-1}. Suppose that this y^{-1} occurs to the left of y^2. Since passing to a reduced word need involve only deletions and not insertions of pairs, everything between y^{-1} and y^2 must cancel. If the y^{-1} has resulted from $w^{-1} = y^{-1} x^{-1}$, then the number of x, y symbols between y^{-1} and y^2 is odd, and an odd number of factors can never cancel. So the y^{-1} must arise from the right-hand y^{-1} in a factor $v^{-1} = y^{-2}$. The symbols between y^{-2} and y^2 come from some reduced u, v, w word, and induction shows that this word must be trivial. Then y^{-2} and y^2 are adjacent, contradiction. Thus the left factor of y^2 must cancel because of some y^{-1} on the right of y^2. If the y^{-1} is part of $w^{-1} = y^{-1} x^{-1}$ or is the left y^{-1} in $v^{-1} = y^{-2}$, then the number of x, y

symbols between the left y and the y^{-1} is odd, and we cannot get cancellation. So the y^{-1} must be the right-hand y^{-1} in a factor y^{-2}. Then we have an expression $y(y \cdots y^{-1})y^{-1}$ in which the symbols in parentheses cancel. The symbols $\cdots$ must cancel also; since these represent some reduced u, v, w word, induction shows that $\cdots$ is empty. We conclude that y^2 and y^{-2} are adjacent, contradiction. Thus our reduced u, v, w word contains no factor v. Similarly examination of the right-hand factor x in an occurrence of x^2 shows that our reduced u, v, w word contains no factor u. It must therefore be a product of factors w or a product of factors w^{-1}. Substitution of $w = xy$ leads directly without any cancellation to an x, y reduced word, and we conclude that the u, v, w word is empty. Thus the subset $\{u, v, w\}$ is free.

If G is any group, the **commutator subgroup** G' of G is the subgroup generated by all elements $xyx^{-1}y^{-1}$ with $x \in G$ and $y \in G$.

Proposition 7.4. If G is a group, then the commutator subgroup is normal, and G/G' is abelian. If $\varphi : G \to H$ is any homomorphism of G into an abelian group H, then $\ker \varphi \supseteq G'$.

PROOF. The computation

$$axyx^{-1}y^{-1}a^{-1} = (axa^{-1})(aya^{-1})(axa^{-1})^{-1}(aya^{-1})^{-1}$$

shows that G' is normal. If $\psi : G \to G/G'$ is the quotient homomorphism, then $\psi(x)\psi(y) = xyG' = xy(y^{-1}x^{-1}yx)G' = yxG' = \psi(y)\psi(x)$, and therefore G/G' is abelian. Finally if $\varphi : G \to H$ is a homomorphism of G into an abelian group H, then the computation $\varphi(xyx^{-1}y^{-1}) = \varphi(x)\varphi(y)\varphi(x)^{-1}\varphi(y)^{-1} = \varphi(x)\varphi(x)^{-1}\varphi(y)\varphi(y)^{-1} = 1$ shows that $G' \subseteq \ker \varphi$. □

Corollary 7.5. If F is the free group on a set S and if F' is the commutator subgroup of F, then F/F' is isomorphic to the free abelian group $\bigoplus_{s \in S} \mathbb{Z}_s$.

PROOF. Let $H = \bigoplus_{s \in S} \mathbb{Z}_s$, and let $\varphi : S \to H$ be the function with $\varphi(s) = 1_s$, i.e., $\varphi(s)$ is to be the member of H that is 1 in the sth coordinate and is 0 elsewhere. Application of the universal mapping property of F as given in Theorem 7.1 yields a group homomorphism $\widetilde{\varphi} : F \to H$ such that $\widetilde{\varphi} \circ \iota = \varphi$. Since the elements $\varphi(s)$, with s in S, generate H, $\widetilde{\varphi}$ carries F onto H. Since H is abelian, Proposition 7.4 shows that $\ker \widetilde{\varphi} \supseteq F'$. Proposition 4.11 shows that $\widetilde{\varphi}$ descends to a homomorphism $\widetilde{\varphi}_0 : F/F' \to H$, and $\widetilde{\varphi}_0$ has to be onto H.

To complete the proof, we show that $\widetilde{\varphi}_0$ is one-one. Let x be a member of F. Since the products of the elements $\iota(s)$ and their inverses generate F and since F/F' is abelian, we can write $xF' = s_{i_1}^{j_1} \cdots s_{i_n}^{j_n} F'$, where s_{i_1} occurs a total of j_1 times in $x, \ldots$, and s_{i_n} occurs a total of j_n times in x; it is understood that an occurrence of $s_{i_1}^{-1}$ is to contribute -1 toward j_1. Then we have $\widetilde{\varphi}_0(xF') = j_1\varphi(s_{i_1}) + \cdots + j_n\varphi(s_{i_n})$. If $\widetilde{\varphi}_0(xF') = 0$, we obtain $j_1\varphi(s_{i_1}) + \cdots + j_n\varphi(s_{i_n}) = 0$, and then $j_1 = \cdots = j_n = 0$ since the elements $\varphi(s_{i_1}), \ldots, \varphi(s_{i_n})$ are members of a $\mathbb{Z}$ basis of H. Hence $xF' = F'$, x is in F', and $\widetilde{\varphi}_0$ is one-one. □

Corollary 7.6. If F_1 and F_2 are isomorphic free groups on sets S_1 and S_2, respectively, then S_1 and S_2 have the same cardinality.

PROOF. Corollary 7.5 shows that an isomorphism of F_1 with F_2 induces an isomorphism of the free abelian groups $\bigoplus_{s \in S_1} \mathbb{Z}_{s_1}$ and $\bigoplus_{s \in S_2} \mathbb{Z}_{s_2}$. The rank of a free abelian group is a well-defined cardinal, and the result follows—almost.

We did not completely prove this fact about the rank of a free abelian group in Section IV.9. Theorem 4.53 did prove, however, that rank is well defined for finitely generated free abelian groups. Thus the corollary follows if S_1 and S_2 are finite. If S_1 or S_2 is uncountable, then the cardinality of the corresponding free abelian group matches the cardinality of its $\mathbb{Z}$ basis; hence the corollary follows if S_1 or S_2 is uncountable. The only remaining case to eliminate is that one of S_1 and S_2, say the first of them, has a countably infinite $\mathbb{Z}$ basis and the other has finite rank n. The first of the groups then has a linearly independent set of $n + 1$ elements, and Lemma 4.54 shows that the span of these elements cannot be isomorphic to a subgroup of a free abelian group of rank n. This completes the proof in all cases. □

Because of Corollary 7.6, it is meaningful to speak of the **rank** of a free group; it is the cardinality of any free basis. We shall see in the next section that any subgroup of a free group is free. In contrast to the abelian case, however, the rank may actually increase in passing from a free group to one of its subgroups: the example earlier in this section exhibited a free group of rank 3 as a subgroup of a free group of rank 2.

We turn to a way of describing general groups, particularly groups that are at most countable. The method uses "generators," which we already understand, and "relations," which are defined in terms of free groups. Let S be a set, let R be a subset of $F(S)$, and let $N(R)$ be the smallest *normal* subgroup of $F(S)$ containing R. The group $G = F(S)/N(R)$ is sometimes written as $G = \langle S; R \rangle$ or as

$$G = \langle \text{elements of } S; \text{ elements of } R \rangle,$$

with the elements of S and R listed rather than grouped as a set. Either of these expressions is called a **presentation** of G. The set S is a set of **generators**, and the set R is the corresponding set of **relations**. The following result implicit in the universal mapping property of Theorem 7.1 shows the scope of this definition.

Proposition 7.7. Each group G is the homomorphic image of a free group.

PROOF. Let S be a set of generators for G; for example, S can be taken to be G itself. Let $\varphi : S \to G$ be the inclusion of the set of generators into G, and let $\widetilde{\varphi} : F(S) \to G$ be the group homomorphism of Theorem 7.1 such that $\widetilde{\varphi}(\iota(s)) = \varphi(s)$ for all s in S. The image of $\widetilde{\varphi}$ is a subgroup of G that contains the generating set S and is therefore equal to all of G. Thus $\widetilde{\varphi}$ is the required homomorphism. □

If G is any group and $\widetilde{\varphi} : F(S) \to G$ is the homomorphism given in Proposition 7.7, then the subgroup $R = \ker \widetilde{\varphi}$ has the property that $G \cong \langle S; R \rangle$. Consequently *every* group can be given by generators and relations.

For example the proof of the proposition shows that one possibility is to take $S = G$ and R equal to the set of all members of the multiplication table, but with the multiplication table entry $ss' = s''$ rewritten as the left side $ss'(s'')^{-1}$ of an equation $ss'(s'')^{-1} = 1$ specifying a combination of generators that maps to 1. This is of course not a very practical example. Generators and relations are most useful when S and R are fairly small. One says that G is **finitely generated** if S can be chosen to be finite, **finitely presented** if both S and R can be chosen to be finite.

A frequently used device in working with generators and relations is the following simple proposition.

Proposition 7.8. Let $G = \langle S; R \rangle$ be a group given by generators and relations, let G' be a second group, let φ be a one-one function φ from S onto a set of generators for G', and let $\Phi : F(S) \to G'$ be the extension of φ to a group homomorphism. If $\Phi(r) = 1$ for every member r of R, then Φ descends to a homomorphism of G onto G'. In particular, if $G = \langle S; R \rangle$ and $G' = \langle S; R' \rangle$ are groups given by generators and relations with $R \subseteq R'$, then the natural homomorphism of $F(S)$ onto G' descends to a homomorphism of G onto G'.

PROOF. The proposition follows immediately from the universal mapping property in Theorem 7.1 in combination with Proposition 4.11. □

Now let us consider some examples of groups given by generators and relations. The case of one generator is something we already understand: the group has to be cyclic. A presentation of $\mathbb{Z}$ is as $\langle a; \ \rangle$, and a presentation of C_n is as $\langle a; a^n \rangle$. But other presentations are possible with one generator, such as $\langle a; a^6, a^9 \rangle$ for C_3. Here is an example with two generators.

EXAMPLE. Let us prove that $D_n \cong \langle x, y; \ x^n, y^2, (xy)^2 \rangle$, where D_n is the dihedral group of order $2n$. Concretely let us work with D_n as the group of 2-by-2 real matrices generated by $\begin{pmatrix} \cos 2\pi/n & -\sin 2\pi/n \\ \sin 2\pi/n & \cos 2\pi/n \end{pmatrix}$ and $\begin{pmatrix} 1 & 0 \\ 0 & -1 \end{pmatrix}$. The generated group indeed has order $2n$. If we identify

$$x \text{ with } \begin{pmatrix} \cos 2\pi/n & -\sin 2\pi/n \\ \sin 2\pi/n & \cos 2\pi/n \end{pmatrix} \quad \text{and} \quad y \text{ with } \begin{pmatrix} 1 & 0 \\ 0 & -1 \end{pmatrix},$$

then $y^2 = 1$, and the formula

$$\begin{pmatrix} \cos 2\pi/n & -\sin 2\pi/n \\ \sin 2\pi/n & \cos 2\pi/n \end{pmatrix}^k = \begin{pmatrix} \cos 2\pi k/n & -\sin 2\pi k/n \\ \sin 2\pi k/n & \cos 2\pi k/n \end{pmatrix}$$

shows that $x^n = 1$. In addition, $xy = \begin{pmatrix} \cos 2\pi/n & \sin 2\pi/n \\ \sin 2\pi/n & -\cos 2\pi/n \end{pmatrix}$, and the square of this is the identity. By Proposition 7.8, D_n is a homomorphic image of $\widetilde{D}_n = \langle x, y; \ x^n, y^2, (xy)^2 \rangle$. To complete the identification, it is enough to show that the order of $\widetilde{D}_n$ is $\leq 2n$ because the homomorphism of $\widetilde{D}_n$ onto D_n must then be one-one. In $\langle x, y; \ x^n, y^2, (xy)^2 \rangle$, we compute that $y^{-1} = y$ and that $x(yx)y = 1$ implies $yx = x^{-1}y^{-1} = x^{-1}y$. Induction then yields $yx^k = x^{-k}y$ for $k > 0$. Multiplying left and right by y gives $yx^{-k} = x^k y$ for $k > 0$. So $yx^l = x^{-l}y$ for every integer l. This means that every element is of the form x^m or $x^m y$, and we may take $0 \leq m \leq n-1$. Hence there are at most $2n$ elements.

Without trying to be too precise, let us mention that the **word problem** for finitely presented groups is to give an algorithm for deciding whether two words represent the same element of the group. It is known that there is no such algorithm applicable to all finitely presented groups. Of course, there can be such an algorithm for certain special classes of presentations. For example, if there are no relations in the presentation, then the group is a free group, and Proposition 7.3 gives a solution in this case. There tends to be a solution for a class of groups if the groups all correspond rather concretely to some geometric situation, such as a tiling of Euclidean space or some other space. The example above with D_n is of this kind.

By way of a concrete class of examples, one can identify any doubly generated group of the form $\langle x, y; \ x^a, y^b, (xy)^c \rangle$ if a, b, c are integers > 1, and one can describe what words represent what elements in these groups. These groups all correspond to tilings in 2 dimensions. In fact, let $\gamma = a^{-1} + b^{-1} + c^{-1}$. If $\gamma > 1$, the tiling is of the Riemann sphere, and the group is finite. If $\gamma = 1$, the tiling is of the Euclidean plane $\mathbb{R}^2$, and the group is infinite. If $\gamma < 1$, the tiling is of the hyperbolic plane, and the group is infinite. In all cases one starts from a triangle in the appropriate geometry with angles $\pi/a, \pi/b$, and π/c, and a basic tile consists of the double of this triangle obtained by reflecting the triangle about any of its sides. The group elements x, y, and xy are rotations, suitably oriented, about the vertices of the triangle through respective angles $2\pi/a$, $2\pi/b$, and $2\pi/c$. Further information about the cases $\gamma > 1$ and $\gamma = 1$ is obtained in Problems 37–46 at the end of the chapter.

We conclude with one further example of a presentation whose group we can readily identify concretely.

Proposition 7.9. Let S be a set, and let $R = \{sts^{-1}t^{-1} \mid s \in S, t \in S\}$. Then the smallest normal subgroup of the free group $F(S)$ containing R is the commutator subgroup $F(S)'$, and therefore $\langle S; R \rangle$ is isomorphic to the free abelian group $\bigoplus_{s \in S} \mathbb{Z}_s$.

PROOF. The members of R are in $F(S)'$, the product of two members of $F(S)'$ is in $F(S)'$, and any conjugate of a member of $F(S)'$ is in $F(S)'$. Therefore the smallest normal subgroup $N(R)$ containing R has $N(R) \subseteq F(S)'$. Let $\varphi : F(S) \to F(S)/N(R)$ be the quotient homomorphism. Elements of the quotient $F(S)/N(R)$ may be expressed as words in the elements $\varphi(s)$ and $\varphi(s)^{-1}$ for s in S, and the factors commute because of the definition of R. Therefore $F(S)/N(R)$ is abelian. By Proposition 7.4, $N(R) \supseteq F(S)'$. Therefore $N(R) = F(S)'$. This proves the first conclusion, and the second conclusion follows from Corollary 7.5. □

2. Subgroups of Free Groups

The main result of this section is that any subgroup of a free group is a free group. An example in the previous section shows that the rank can actually increase in the process of passing to the subgroup.

The proof of the main result is ostensibly subtle but is relatively easy to understand in topological terms. Although we shall give the topological interpretation, we shall not pursue it further, and the proof that we give may be regarded as a translation of the topological proof into the language of algebra, combined with some steps of beautification.

For purposes of the topological argument, let us think of the given free group for the moment as finitely generated, and let us suppose that the subgroup has finite index. A free group on n symbols is the fundamental group of a bouquet of n circles, all joined at a single point, which we take as the base point. By the theory of covering spaces, any subgroup of index k is the fundamental group of some k-sheeted covering space of the bouquet of circles. This covering space is a 1-dimensional simplicial complex, and one can prove with standard tools that the fundamental group of any 1-dimensional simplicial complex is a free group. The theorem follows.

2. Subgroups of Free Groups

If the special hypotheses are dropped that the given free group is finitely generated and the subgroup has finite index, then the same proof is applicable as long as one allows a suitable generalization of the notion of simplicial complex. Thus the topological argument is completely general.

The theorem then is as follows.

Theorem 7.10 (Nielsen–Schreier Theorem). Every subgroup of a free group is a free group.

REMARKS. The algebraic proof will occupy the remainder of the section but will occasionally be interrupted by comments about the example in the previous section.

Let the given free group be F, let the subgroup be H, and form the right cosets Hg in F. Let C be a set of representatives for these cosets, with 1 chosen as the representative of the identity coset; we shall impose further conditions on C shortly.

EXAMPLE, continued. For the example in the previous section, we were given a free group F with two generators x, y, and the subgroup H is taken to have generators x^2, xy, y^2. In fact, one readily checks that H is the subgroup formed from all words of even length, and we shall think of it that way. The set C of coset representatives may be taken to be $\{1, x\}$ in this case. The argument we gave that H is free has points of contact with the proof we give of Theorem 7.10 but is not an exact special case of it. One point of contact is that within each generator of H that we identify, there is some particular factor that does not cancel when that generator appears in a word representing a member of the subgroup.

We define a function $\rho : F \to C$ by taking $\rho(x)$ to be the coset representative of the member x of F. This function has the property that $\rho(hx) = \rho(x)$ for all h in H and x in F. Also, $x \mapsto x\rho(x)^{-1}$ is a function from F to H, and it is the identity function on H. The first lemma shows that a relatively small subset of the elements $x\rho(x)^{-1}$ is a set of generators of H.

Lemma 7.11. Let S be the set of generators of F, and let $S' = S \cup S^{-1}$. Every element of H is a product of elements of the form $gb\rho(gb)^{-1}$ with g in C and b in S'. Furthermore the element $g' = \rho(gb)$ of C has the properties that $g = \rho(g'b^{-1})$ and that $gb^{-1}\rho(gb^{-1})^{-1}$ is of the form $\left(g'b\rho(g'b)^{-1}\right)^{-1}$. Consequently the elements $ga\rho(ga)^{-1}$ with g in C and a in S form a set of generators of H.

EXAMPLE, continued. In the example, we are taking $C = \{1, x\}$ and $S = \{x, y\}$. The elements $gb\rho(gb)^{-1}$ obtained with $g=1$ and b equal to x, y, x^{-1}, y^{-1} are $1, yx^{-1}, x^{-1}x^{-1}$, and $y^{-1}x^{-1}$. The elements $gb\rho(gb)^{-1}$ obtained with $g = x$ and b equal to x, y, x^{-1}, y^{-1} are $xx, xy, 1$, and xy^{-1}. The lemma says that $1, yx^{-1}, xx$, and xy form a set of generators of H and that the elements $x^{-1}x^{-1}, y^{-1}x^{-1}, 1$, and xy^{-1} are inverses of these generators in some order.

REMARK. The lemma needs no hypothesis that F is free. A nontrivial application of the lemma with F not free appears in Problem 43 at the end of the chapter.

PROOF. Any h in F can be written as a product $h = b_1 \cdots b_n$ with each b_j in S'. Define $r_0 = 1$ and $r_k = \rho(b_1 \cdots b_k)$ for $1 \leq k \leq n$. Then

$$hr_n^{-1} = (r_0 b_1 r_1^{-1})(r_1 b_2 r_2)^{-1} \cdots (r_{n-1} b_n r_n^{-1}). \qquad (*)$$

Since

$$r_k = \rho(b_1 \cdots b_k) = \rho(b_1 \cdots b_{k-1} b_k) = \rho(\rho(b_1 \cdots b_{k-1}) b_k) = \rho(r_{k-1} b_k),$$

we have $r_{k-1} b_k r_k^{-1} = gb\rho(gb)^{-1}$ with $g = r_{k-1}$ and $b = b_k$. Thus $(*)$ exhibits hr_n^{-1} as a product of elements as in the first conclusion of the lemma. Since $r_n = \rho(b_1 \cdots b_n) = \rho(h)$, $r_n = 1$ if h is in H. Therefore in this case, h itself is a product of elements as in the statement of that conclusion, and that conclusion is now proved.

For the other conclusion, let $gb^{-1}\rho(gb^{-1})^{-1}$ be given, and put $g' = \rho(gb^{-1})$, so that $gb^{-1}g'^{-1} = h$ is in H. This equation implies that $g'b = h^{-1}g$. Hence $\rho(g'b) = \rho(h^{-1}g) = \rho(g) = g$, and it follows that $gb^{-1}\rho(gb^{-1})^{-1} = gb^{-1}g'^{-1} = (g'bg^{-1})^{-1} = \left(g'b\rho(g'b)^{-1}\right)^{-1}$. This proves the lemma. □

Lemma 7.12. With F free it is possible to choose the set C of coset representatives in such a way that all of its members have expansions in terms of S' as $g = b_1 \cdots b_n$ in which

(a) $g = b_1 b_2 \cdots b_n$ is a reduced word as written,
(b) $b_1 b_2 \cdots b_{n-1}$ is also a member of C.

REMARKS. It is understood from the case of $n = 1$ in (b) that 1 is the representative of the identity coset. When C is chosen as in this lemma, C is said to be a **Schreier set**. In the example, $C = \{1, x\}$ is a Schreier set. So is $C = \{1, y\}$, and hence the selection of a Schreier set may involve a choice.

PROOF. If S' is finite or countably infinite, we enumerate it. In the uncountable case (which is of less practical interest), we introduce a well ordering in S' by means of Zermelo's Well-Ordering Theorem as in Section A5 of the appendix.

2. Subgroups of Free Groups

The ordering of S' will be used to define a lexicographic ordering of the set of all reduced words in the members of S'. If

$$x = b_1 \cdots b_m \quad \text{and} \quad y = b'_1 \cdots b'_n \qquad (*)$$

are reduced words with $m \leq n$, we say that $x < y$ if any of the following hold:

(i) $m < n$,
(ii) $m = n$ and $b_1 < b'_1$,
(iii) $m = n$, and for some $k < m$, $b_1 = b'_1, \ldots, b_k = b'_k$, and $b_{k+1} < b'_{k+1}$.

With this definition the set of reduced words is well ordered, and hence any nonempty subset of reduced words has a least element.

Let us observe that if x, y, z are reduced words with $x < y$ and if yz is reduced as written, then $xz < yz$ after xz has been reduced. In fact, let us assume that x and y are as in $(*)$ and that the length of z is r. The assumption is that yz has length $n + r$, and the length of xz is at most $m + r$. If $m < n$, then certainly $xz < yz$. If $m = n$ and xz fails to be reduced, then the length of xz is less than the length of yz, and $xz < yz$. If $m = n$ and xz is reduced, then the first inequality $b_k < b'_k$ with x and y shows that $xz < yz$.

To define the set C of coset representatives, let the representative of Hg be the least member of the set Hg, each element being written as a reduced word. Since the length of the empty word is 0, the representative of the identity coset H is 1 under this definition. Thus all we have to check is that an initial segment of a member of C is again in C.

Suppose that $b_1 \cdots b_n$ is in C, so that $b_1 \cdots b_n$ is the least element of $Hb_1 \cdots b_n$. Denote the least element of $Hb_1 \cdots b_{n-1}$ by g. If $g = b_1 \cdots b_{n-1}$, we are done. Otherwise $g < b_1 \cdots b_{n-1}$, and then the fact that $b_1 \cdots b_n$ is reduced implies that $gb_n < b_1 \cdots b_n$. But gb_n is in $Hb_1 \cdots b_n$, and this inequality contradicts the minimality of $b_1 \cdots b_n$ in that coset. Thus we conclude that $g = b_1 \cdots b_{n-1}$. This proves the lemma. □

For the remainder of the proof of Theorem 7.10, we assume, as we may by Lemma 7.12, that the set C of coset representatives is a Schreier set. Typical elements of S will be denoted by a, and typical elements of $S' = S \cup S^{-1}$ will be denoted by b. Let us write u for a typical element $ga\rho(ga)^{-1}$ with g in C, and let us write v for a typical element $gb\rho(gb)^{-1}$ with g in C. The elements u generate H by Lemma 7.11, and each element v is either an element u or the inverse of an element u, according to the lemma. We shall prove that the elements u not equal to 1 are distinct and form a free basis of H.

First we prove that each of the elements $v = gb\rho(gb)^{-1}$ either is reduced as written or is equal to 1. Put $g' = \rho(gb)$, so that $v = gbg'^{-1}$. Since g and g' are in the Schreier set C, they are reduced as written, and hence so are g and g'^{-1}. Thus the only possible cancellation in v occurs because the last factor of g is b^{-1} or the last factor of g' is b. If the last factor of g is b^{-1}, then gb is an initial segment of g and hence is in the Schreier set C; thus $\rho(gb) = gb$ and $v = gb\rho(gb)^{-1} = 1$. Similarly if the last factor of g' is b, then $g'b^{-1}$ is an initial segment of g' and hence is in the Schreier set C; thus $\rho(g'b^{-1}) = g'b^{-1}$, and Lemma 7.11 gives $v^{-1} = (gb\rho(gb)^{-1})^{-1} = g'b^{-1}\rho(g'b^{-1})^{-1} = 1$. Thus $v = gb\rho(gb)^{-1}$ either is reduced as written or is equal to 1.

Next let us see that the elements v other than 1 are distinct. Suppose that $v = gb\rho(gb)^{-1} = g'b'\rho(g'b')^{-1}$ is different from 1. Remembering that each of these expressions is reduced as written, we see that if g is shorter than g', then gb is an initial segment of g'. Since C is a Schreier set, gb is in C and $\rho(gb) = gb$; thus $v = gb\rho(gb)^{-1}$ equals 1, contradiction. Similarly g' cannot be shorter than g. So g and g' must have the same length l. In this case the first $l + 1$ factors must match in the two equal reduced words, and we conclude that $g = g'$ and $b = b'$. This proves the uniqueness.

We know that each v is either some u or some u'^{-1}, and this uniqueness shows that it cannot be both unless $v = 1$. Therefore the nontrivial u's are distinct, and the nontrivial v's consist of the u's and their inverses, each appearing once.

Since an element v not equal to 1 therefore determines its g and b, let us refer to the factor b of $v = gb\rho(gb)^{-1}$ as the **significant factor** of v. This is the part that will not cancel out when we pass from a product of v's to its reduced form.

Specifically suppose that we have $v = gb\rho(gb)^{-1}$ and $\bar{v} = \bar{g}\bar{b}\rho(\bar{g}\bar{b})^{-1}$, that neither of these is 1, and that $\bar{v} \neq v^{-1}$. Put $g' = \rho(gb)$ and $\bar{g}' = \rho(\bar{g}\bar{b})$. The claim is that the cancellation in forming $v\bar{v} = gbg'^{-1}\bar{g}\bar{b}\bar{g}'^{-1}$ does not extend to either of the significant factors b and $\bar{b}$. If it does, then one of three things happens:

(i) the b in bg'^{-1} gets canceled because the last factor of g' is b, in which case $g'b^{-1}$ is an initial segment of g', $g'b^{-1} = \rho(g'b^{-1}) = g$, and $v = gbg'^{-1} = 1$, or

(ii) the $\bar{b}$ in $\bar{g}\bar{b}$ gets canceled because the last factor of $\bar{g}$ is $\bar{b}^{-1}$, in which case $\bar{g}\bar{b}$ is an initial segment of $\bar{g}$, $\bar{g}\bar{b} = \rho(\bar{g}\bar{b}) = \bar{g}'$, and $\bar{v} = \bar{g}\bar{b}\bar{g}'^{-1} = 1$, or

(iii) $g'^{-1}\bar{g} = 1$ and $b\bar{b} = 1$, in which case $\bar{g} = g'$, $\bar{b} = b^{-1}$, and the middle conclusion of Lemma 7.11 allows us to conclude that $\bar{v} = v^{-1}$.

All three of these possibilities have been ruled out by our assumptions, and therefore neither of the significant factors in $v\bar{v}$ cancels.

As a consequence of this noncancellation, we can see that in any product $v_1 \cdots v_m$ of v's in which no v_k is 1 and no v_{k+1} equals v_k^{-1}, none of the significant factors cancel. In fact, the previous paragraph shows that the significant factors of v_1 and v_2 survive in forming v_1v_2, the significant factors of v_2 and v_3 survive in right multiplying by v_3, and so on. Since the nontrivial u's are distinct and the nontrivial v's consist of the u's and their inverses, each appearing once, we conclude that the set of nontrivial u's is a free subset of F. Lemma 7.11 says that the u's generate H, and therefore the set of nontrivial u's is a free basis of H.

3. Free Products

The free abelian group on an index set S, as constructed in Section IV.9, has a universal mapping property that allows arbitrary functions from S into any target abelian group to be extended to homomorphisms of the free abelian group into the target group. The construction of free groups in Section 1 was arranged to adapt the construction so that the target group in the universal mapping property could be any group, abelian or nonabelian.

In this section we make a similar adaptation of the construction of a direct sum of abelian groups so that the result is applicable in a context of arbitrary groups. Proposition 4.17 gave the universal mapping property of the external direct sum $\bigoplus_{s\in S} G_s$ of a set of abelian groups with associated embedding homomorphisms $i_{s_0} : G_{s_0} \to \bigoplus_{s\in S} G_s$. The statement is that if H is any abelian group and $\{\varphi_s \mid s \in S\}$ is a system of group homomorphisms $\varphi_s : G_s \to H$, then there exists a unique group homomorphism $\varphi : \bigoplus_{s\in S} G_s \to H$ such that $\varphi \circ i_{s_0} = \varphi_{s_0}$ for all $s_0 \in S$. Example 2 of coproducts in Section IV.11 shows that direct sum is therefore the coproduct functor in the category of all abelian groups.

This universal mapping property of $\bigoplus_{s\in S} G_s$ fails when H is a nonabelian group such as the symmetric group $\mathfrak{S}_3$. In fact, $\mathfrak{S}_3$ has an element of order 2 and an element of order 3 and hence admits nontrivial homomorphisms $\varphi_2 : C_2 \to \mathfrak{S}_3$ and $\varphi_3 : C_3 \to \mathfrak{S}_3$. But there is no homomorphism $\varphi : C_2 \oplus C_3 \to \mathfrak{S}_3$ such that $\varphi \circ i_2 = \varphi_2$ and $\varphi \circ i_3 = \varphi_3$ because the image of φ has to be abelian but the images of φ_2 and φ_3 do not commute. Consequently direct sum cannot extend to a coproduct functor in the category of all groups.

Instead, the appropriate group constructed from C_2 and C_3 for this kind of universal mapping property is the "free product" of C_2 and C_3, denoted by $C_2 * C_3$. In this section we construct the free product of any set of groups, finite or infinite. Also, we establish its universal mapping property and identify it in terms of generators and relations. The prototype of a free product is the free group $F(S)$, which equals a free product of copies of $\mathbb{Z}$ indexed by S. A free product is always an infinite group if at least two of the factors are not 1-element groups.

An important application of free products occurs in the theory of the fundamental group in topology: if X is a topological space for which the theory of covering spaces is applicable, and if A and B are open subsets of X with $X = A \cup B$ such that $A \cap B$ is nonempty, connected, and simply connected, then the fundamental group of X is the free product of the fundamental group of A and the fundamental group of B. This result, together with a generalization that no longer requires $A \cap B$ to be simply connected, is known as the **Van Kampen Theorem**.

Let S be a nonempty set of groups G_s for s in S. The set S is allowed to be infinite, but in practice it often has just two elements. We shall describe the group defined to be the free product $G = *_{s \in S} G_s$. We start from the set $W(\{G_s\})$ of all words built from the groups G_s. This consists of all finite sequences $g_1 \cdots g_n$ with each g_i in some G_s depending on i. The length of a word is the number of factors in it. The empty word is denoted by 1. We multiply two words by writing them end to end, and the resulting operation of multiplication is associative. A word is said to be equivalent to a second word if the first can be obtained from the second by a finite sequence of steps of the following kinds and their inverses:

(i) drop a factor for which g_i is the identity element of the group in which it lies,
(ii) collapse two factors $g_i g_{i+1}$ to a single one g_i^* if g_i and g_{i+1} lie in the same G_s and their product in that group is g_i^*.

The result is an equivalence relation, and the set of equivalence classes is the underlying set of $*_{s \in S} G_s$.

Theorem 7.13. If S is a nonempty set of groups G_s and $W(\{G_s\})$ is the set of all words from the groups G_s, then the product operation defined on $W(\{G_s\})$ descends in a well-defined fashion to the set $*_{s \in S} G_s$ of equivalence classes of members of $W(\{G_s\})$, and $*_{s \in S} G_s$ thereby becomes a group. For each s_0 in S, define $i_{s_0} : G_{s_0} \to *_{s \in S} G_s$ to be the group homomorphism obtained as the composition of the inclusion of G_{s_0} into words of length 1 followed by passage to equivalence classes. Then the pair $\left(*_{s \in S} G_s, \{i_s\} \right)$ has the following universal mapping property: whenever H is a group and $\{\varphi_s \mid s \in S\}$ is a system of group homomorphisms $\varphi_s : G_s \to H$, then there exists a unique group homomorphism $\varphi : *_{s \in S} G_s \to H$ such that $\varphi \circ i_{s_0} = \varphi_{s_0}$ for all $s_0 \in S$.

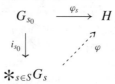

FIGURE 7.2. Universal mapping property of a free product.

REMARKS. The group $*_{s \in S} G_s$ is called the **free product** of the groups G_s. Figure 7.2 illustrates its universal mapping property. This universal mapping property actually characterizes $*_{s \in S} G_s$, as will be seen in Proposition 7.14. One often writes $G_1 * \cdots * G_n$ when the set S is finite; the order of listing the groups is immaterial. The proof of Theorem 7.13 is rather similar to the proof of Theorem 7.1, and we shall skip some details.

PROOF. Let us write $\sim$ for the equivalence relation on words, and let us denote equivalence classes by brackets. We want to define multiplication in $*_{s\in S}G_s$ by $[w_1][w_2] = [w_1w_2]$. To see that this formula makes sense in $*_{s\in S}G_s$, let x, x', and y be words in $W(\{G_s\})$, and suppose that x and x' differ by only one operation of type (i) or type (ii) as above. Then $x \sim x'$, and it is evident that $x'y \sim xy$ and $yx' \sim yx$. Iteration of this kind of relationship shows that $w'_1 \sim w_1$ and $w'_2 \sim w_2$ implies $w'_1w'_2 \sim w_1w_2$, and hence multiplication is well defined.

The associativity of multiplication in $W(\{G_s\})$ implies that multiplication in $*_{s\in S}G_s$ is associative, and [1] is a two-sided identity. We readily check that if $g = g_1 \cdots g_n$ is a word, then the word $g^{-1} = g_n^{-1} \cdots g_1^{-1}$ has the property that $[g^{-1}]$ is a two-sided inverse to $[g]$. Therefore $*_{s\in S}G_s$ is a group.

The uniqueness of the homomorphism φ in the universal mapping property is no problem since all words are products of words of length 1 and since the subgroups $i_{s_0}(G_{s_0})$ together generate $*_{s\in S}G_s$.

For existence of φ, we begin by defining a function $\Phi : W(\{G_s\}) \to H$ such that

$$\Phi(g_s) = \varphi_s(g_s) \quad \text{for } g_s \text{ in } G_s \text{ when viewed as a word of length 1,}$$
$$\Phi(w_1w_2) = \Phi(w_1)\Phi(w_2) \quad \text{for } w_1 \text{ and } w_2 \text{ in } W(\{G_s\}).$$

We take the formulas $\Phi(g_s) = \varphi(g_s)$ for g_s in G_s as a definition of Φ on words of length 1. Any member of $W(\{G_s\})$ can be written uniquely as $g_1 \cdots g_n$ with each g_i in G_{s_i}, and we set $\Phi(g_1 \cdots g_n) = \Phi(g_1) \cdots \Phi(g_n)$. (If $n = 0$, the understanding is that $\Phi(1) = 1$.) Then Φ has the required properties.

Let us show that $w' \sim w$ implies $\Phi(w') = \Phi(w)$. The questions are whether

(i) if $g_1, \ldots, g_n$ are in various G_s's with g_i equal to the identity 1_{s_i} of G_{s_i}, then

$$\Phi(g_1 \cdots g_{i-1} 1_{s_i} g_{i+1} \cdots g_n) \stackrel{?}{=} \Phi(g_1 \cdots g_{i-1} g_{i+1} \cdots g_n),$$

(ii) if $g_1, \ldots, g_n$ are in various G_s's with $G_{s_i} = G_{s_{i+1}}$ and if $g_i g_{i+1} = g_i^*$ in G_{s_i}, then

$$\Phi(g_1 \cdots g_{i-1} g_i g_{i+1} g_{i+2} \cdots g_n) \stackrel{?}{=} \Phi(g_1 \cdots g_{i-1} g_i^* g_{i+2} \cdots g_n).$$

In the case of (i), the question comes down to whether a certain $h\Phi(1_{s_i})h'$ in H equals hh', and this is true because $\Phi(1_{s_i}) = \varphi_{s_i}(1_{s_i})$ is the identity of H. In the case of (ii), the question comes down to whether $h\Phi(g_i)\Phi(g_{i+1})h'$ equals $h\Phi(g_i^*)h'$ if $G_{s_i} = G_{s_{i+1}}$ and $g_i g_{i+1} = g_i^*$ in G_{s_i}, and this is true because $\Phi(g_i)\Phi(g_{i+1}) = \varphi_{s_i}(g_i)\varphi_{s_i}(g_{i+1}) = \varphi_{s_i}(g_i g_{i+1}) = \varphi_{s_i}(g_i^*) = \Phi(g_i^*)$. We conclude that $w' \sim w$ implies $\Phi(w') = \Phi(w)$.

We may therefore define $\varphi([w]) = \Phi(w)$ for $[w]$ in $F(\{G_s\})$, and φ is a homomorphism of $F(\{G_s\})$ into H as a consequence of the property $\Phi(w_1 w_2) = \Phi(w_1)\Phi(w_2)$ of Φ on $W(\{G_s\})$. For g_s in G_s, we have $\varphi([g_s]) = \Phi(g_s) = \varphi_s(g_s)$, i.e., $\varphi(i(g_s)) = \varphi_s(g_s)$. This completes the proof of existence. □

Proposition 7.14. Let S be a nonempty set of groups G_s. Suppose that G' is a group and that $i'_s : G_s \to G'$ for $s \in S$ is a system of group homomorphisms with the following universal mapping property: whenever H is a group and $\{\varphi_s \mid s \in S\}$ is a system of group homomorphisms $\varphi_s : G_s \to H$, then there exists a unique group homomorphism $\varphi : G' \to H$ such that $\varphi \circ i'_s = \varphi_s$ for all $s \in S$. Then there exists a unique group homomorphism $\Phi : *_{s \in S} G_s \to G'$ such that $i'_s = \Phi \circ i_s$ for all $s \in S$. Moreover, Φ is a group isomorphism, and the homomorphisms $i'_s : G_s \to G'$ are one-one.

REMARKS. As was true with Proposition 7.2, readers who have been through Chapter VI will recognize that Proposition 7.14 is a special case of Problem 19 at the end of that chapter.

PROOF. Put $G = *_{s \in S} G_s$. In the universal mapping property of Theorem 7.13, let $H = G'$ and $\varphi_s = i'_s$, and let $\Phi : G \to G'$ be the homomorphism φ produced by that theorem. Then Φ satisfies $\Phi \circ i_s = i'_s$ for all s. Reversing the roles of G and G', we obtain a homomorphism $\Phi' : G' \to G$ with $\Phi' \circ i'_s = i_s$ for all s. Therefore $(\Phi' \circ \Phi) \circ i_s = \Phi' \circ i'_s = i_s$.

Comparing $\Phi' \circ \Phi$ with the identity 1_G and applying the uniqueness in the universal mapping property for G, we see that $\Phi' \circ \Phi = 1_G$. Similarly the uniqueness in the universal mapping property of G' gives $\Phi \circ \Phi' = 1_{G'}$. Thus Φ is a group isomorphism. It is uniquely determined by the given properties since the various subgroups $i_s(G_s)$ generate G. Since $i'_s = \Phi \circ i_s$ and since Φ and i_s are one-one, i'_s is one-one. □

As was the case for free groups, we want a decision procedure for telling whether two given words in $W(\{G_s\})$ are equivalent. This is the so-called **word problem** for the free product. Solving it allows us to use free products concretely, just as Proposition 7.3 allowed us to use free groups concretely. A word in $W(\{G_s\})$ is said to be **reduced** if it

(i) contains no factor for which g_i is the identity element of the group G_s in which it lies,

(ii) contains no two consecutive factors g_i and g_{i+1} taken from the same group G_s.

Proposition 7.15. (solution of the word problem for free products). If S is a nonempty set of groups G_s and $W(\{G_s\})$ is the set of all words from the groups G_s, then each word in $W(\{G_s\})$ is equivalent to one and only one reduced word.

EXAMPLE. Consider the free product $C_2 * C_2$ of two cyclic groups, one with x as generator and the other with y as generator. Words consist of a finite sequence of factors of x, y, the identity of the first factor, and the identity of the second factor. A word is reduced if no factor is an identity and if no two x's are adjacent and no two y's are adjacent. Thus the reduced words consist of finite sequences whose terms are alternately x and y. Those of length ≤ 3 are $1, x, y, xy, yx, xyx, yxy$, and in general there are two of each length > 0. The proposition tells us that all these reduced words give distinct members of $C_2 * C_2$. In particular, the group is infinite.

REMARK. More generally, to test whether two words are equivalent, the proposition says to eliminate factors of the identity and multiply consecutive factors in each word when they come from the same group, and repeat these steps until it is no longer possible to do either of these operations on either word. Then each of the given words has been replaced by a reduced word, and the two given words are equivalent if and only if the two reduced words are identical. Problems 37–46 at the end of the chapter concern $C_2 * C_3$, and some of these problems make use of the result of this proposition—that distinct reduced words are inequivalent.

PROOF OF PROPOSITION 7.15. Both operations—eliminating factors of the identity and multiplying consecutive factors in each word when they come from the same group—reduce the length of a word. Since the length has to remain ≥ 0, the process of successively carrying out these two operations as much as possible has to stop after finitely many steps, and the result is a reduced word. This proves that each equivalence class of words contains a reduced word.

For uniqueness of the reduced word in an equivalence class, we proceed somewhat as with Proposition 7.3, associating to each word a finite sequence of reduced words such that the last member of the sequence is unchanged when we apply an operation to the word that preserves equivalence. However, there are considerably more details to check this time.

If $w = g_1 \cdots g_n$ is a given word with each g_i in G_{s_i}, then we associate to w the sequence of reduced words $x_0, x_1, \ldots, x_n$ defined inductively by

$$x_0 = 1,$$

$$x_1 = \begin{cases} g_1 & \text{if } g_1 \text{ is not the identity of } G_{s_1}, \\ 1 & \text{if } g_1 \text{ is the identity of } G_{s_1}, \end{cases}$$

and the following formula for $i \geq 2$ if x_{i-1} is of the reduced form $h_1 \cdots h_k$ with h_j in G_{t_j}:

$$x_i = \begin{cases} h_1 \cdots h_k g_i & \text{if } G_{s_i} \neq G_{t_k} \text{ and } g_i \text{ is not the identity } 1_{G_{s_i}} \text{ of } G_{s_i}, \\ h_1 \cdots h_k & \text{if } g_i \text{ is the identity } 1_{G_{s_i}} \text{ of } G_{s_i}, \\ h_1 \cdots h_{k-1} & \text{if } G_{t_k} = G_{s_i} \text{ with } h_k g_i = 1_{G_{s_i}}, \\ h_1 \cdots h_{k-1} g_i^* & \text{if } G_{t_k} = G_{s_i} \text{ with } h_k g_i = g_i^* \neq 1_{G_{s_i}}. \end{cases}$$

Put $r(w) = x_n$. We check inductively for $i \geq 0$ that each x_i is reduced. In fact, x_i for $i \geq 2$ begins in every case with $h_1 \cdots h_{k-1}$, which is assumed reduced. The only possible reduction for x_i thus comes from factors that are adjoined or from interference with h_{k-1}, and all possibilities are addressed in the above choices. Thus $r(w) = x_n$ is necessarily reduced for each word w.

If $g_1 \cdots g_n$ is reduced as given, then x_i is determined by the first possible choice $h_1 \cdots h_k g_i$ every time, and hence $x_i = g_1 \cdots g_i$ for all i. Therefore we obtain $r(w) = w$ if w is reduced.

Now consider the equivalent words

$$w = g_1 \cdots g_j g_{j+1} \cdots g_n \quad \text{and} \quad w' = g_1 \cdots g_j 1_{G_s} g_{j+1} \cdots g_n.$$

Form $x_0, \ldots, x_n$ for w and $x'_0, \ldots, x'_{n+1}$ for w'. Then we have $x'_j = x_j$; let $h_1 \cdots h_k$ be a reduced form of x'_j. The formula for x'_{j+1} is governed by the second choice in the display, and $x'_{j+1} = h_1 \cdots h_k = x_j$. Then $x'_{j+i+1} = x_{j+i}$ for $1 \leq i \leq n - j$ as well. Hence $x'_{n+1} = x_n$, and $r(w') = r(w)$.

Next suppose that $g_j^* = g_j g_{j+1}$ in G_{s_j}, and consider the equivalent words

$$w = g_1 \cdots g_{j-1} g_j^* g_{j+2} \cdots g_n \quad \text{and} \quad w' = g_1 \cdots g_{j-1} g_j g_{j+1} g_{j+2} \cdots g_n.$$

As above, form $x_0, \ldots, x_n$ for w and $x'_0, \ldots, x'_{n+1}$ for w'. Then we have $x_{j-1} = x'_{j-1}$, and we let $h_1 \cdots h_k$ be a reduced form of x_{j-1}. There are cases, subcases, and subsubcases.

First assume $G_{t_k} \neq G_{s_j}$. Then x_j equals $h_1 \cdots h_k g_j^*$ or $h_1 \cdots h_k$ in the two subcases $g_j^* \neq 1_{G_{s_j}}$ and $g_j^* = 1_{G_{s_j}}$. In the first subcase, we have $g_j^* \neq 1_{G_{s_j}}$ and $x_j = h_1 \cdots h_k g_j^*$. Then x'_j equals $h_1 \cdots h_k g_j$ or $h_1 \cdots h_k$ in the two subsubcases $g_j \neq 1_{G_{s_j}}$ and $g_j = 1_{G_{s_j}}$. In the first subsubcase, $x'_{j+1} = h_1 \cdots h_k g_j^* = x_j$ whether or not $g_{j+1} = 1_{G_{s_j}}$. In the second subsubcase, $g_j^* = g_j g_{j+1}$ cannot be $1_{G_{s_j}}$, and therefore $x'_{j+1} = h_1 \cdots h_k g_j^* = x_j$.

In the second subcase of the case $G_{t_k} \neq G_{s_j}$, we have $g_j^* = 1_{G_{s_j}}$ and $x_j = x_{j-1} = h_1 \cdots h_k$. Then x'_j equals $h_1 \cdots h_k g_j$ or $h_1 \cdots h_k$ in the two subsubcases $g_j \neq 1_{G_{s_j}}$ and $g_j = 1_{G_{s_j}}$. In both subsubcases, $x'_{j+1} = h_1 \cdots h_k$, so that $x'_{j+1} = x_j$.

Now assume $G_{t_k} = G_{s_j}$. Then x_j equals $h_1 \cdots h_{k-1} h_k^*$ or $h_1 \cdots h_{k-1}$ in the two subcases $h_k g_j^* = h_k^* \neq 1_{G_{s_j}}$ and $h_k g_j^* = 1_{G_{s_j}}$. In the first subcase, we have $h_k g_j^* = h_k^* \neq 1_{G_{s_j}}$ and $x_j = h_1 \cdots h_{k-1} h_k^*$. Then x_j' equals $h_1 \cdots h_{k-1} h_k'$ or $h_1 \cdots h_{k-1}$ in the two subsubcases $h_k g_j = h_k' \neq 1_{G_{s_j}}$ and $h_k g_j = 1_{G_{s_j}}$. In the first subsubcase, $h_k' g_{j+1} = h_k g_j g_{j+1} = h_k g_j^* = h_k^*$ implies $x_{j+1}' = h_1 \cdots h_{k-1} h_k^* = x_j$. In the second subsubcase, we know that h_k^* cannot be $1_{G_{s_i}}$ and hence that $g_{j+1} = h_k g_j g_{j+1} = h_k g_j^* = h_k^*$ cannot be $1_{G_{s_j}}$; thus $x_{j+1}' = h_1 \cdots h_{k-1} h_k^* = x_j$.

In the second subcase of the case $G_{t_k} = G_{s_j}$, we have $h_k g_j^* = 1_{G_{s_j}}$ and $x_j = h_1 \cdots h_{k-1}$. Then x_j' equals $h_1 \cdots h_{k-1} h_k^{*'}$ or $h_1 \cdots h_{k-1}$ in the two subsubcases $h_k g_j = h_k^{*'} \neq 1_{G_{s_j}}$ and $h_k g_j = 1_{G_{s_j}}$. In the first subsubcase, g_{j+1} cannot be $1_{G_{s_j}}$ but $h_k^{*'} g_{j+1} = h_k g_j g_{j+1} = h_k g_j^* = 1_{G_{s_j}}$; hence $x_{j+1}' = h_1 \cdots h_{k-1} = x_j$. In the second subsubcase, $x_j' = h_1 \cdots h_{k-1}$ and $g_{j+1} = 1_{G_{s_j}}$, so that $x_{j+1}' = h_1 \cdots h_{k-1} = x_j$.

We conclude that $x_{j+1}' = x_j$ in all cases. Hence $x_{j+i+1}' = x_{j+i}$ for $0 \leq i \leq n-j$, $x_{n+1}' = x_n$, and $r(w') = r(w)$. Consequently the only reduced word that is equivalent to w is $r(w)$. $\square$

Proposition 7.16. Let S be a nonempty set of groups G_s, and suppose that $\langle S_s; R_s \rangle$ is a presentation of G_s, the sets S_s being understood to be disjoint for $s \in S$. Then $\langle \bigcup_{s \in S} S_s; \bigcup_{s \in S} R_s \rangle$ is a presentation of the free product $*_{s \in S} G_s$.

REMARK. One effect of this proposition is to make Proposition 7.8 available as a tool for use with free products. Using Proposition 7.8 may be easier than appealing to the universal mapping property in Theorem 7.13.

PROOF. Put $S = \bigcup_{s \in S} S_s$ and $R = \bigcup_{s \in S} R_s$, and define G to be a group given by generators and relations as $G = \langle S; R \rangle$. Consider the function from S_s into the quotient group $G = F(S)/N(R)$ given by carrying x in S_s into the word x in S and then passing to $F(S)$ and its quotient G. Because of the universal mapping property of free groups, this function extends to a group homomorphism $\widetilde{i_s} : F(S_s) \to G$. If r is a reduced word relative to S_s representing a member of R_s, then r is carried by $\widetilde{i_s}$ into a member of the larger set R and then into the identity of G. Since $\ker \widetilde{i_s}$ is normal in $F(S_s)$, $\ker \widetilde{i_s}$ contains the smallest normal subgroup $N(R_s)$ in $F(S_s)$ that contains R_s. Proposition 4.11 shows that $\widetilde{i_s}$ descends to a group homomorphism $i_s : G_s \to G$.

We shall prove that G and the system $\{i_s\}$ have the universal mapping property of Proposition 7.14 that characterizes a free product. Then it will follow from that proposition that $G \cong *_{s \in S} G_s$, and the proof will be complete.

Thus let H be a group, and let $\{\varphi_s \mid s \in S\}$ be a system of group homomorphisms $\varphi_s : G_s \to H$. We are to produce a homomorphism $\Phi : G \to H$ such that $\Phi \circ i_s = \varphi_s$ for all s, and we are to prove that such a homomorphism is unique. Let $q_s : F(S_s) \to G_s$ be the quotient homomorphism, and define $\widetilde{\varphi}_s : F(S_s) \to H$ by $\widetilde{\varphi}_s = \varphi_s \circ q_s$. Now define $\widetilde{\Phi} : S \to H$ as follows: if x is in S, then x is in a set S_s for a unique s and thereby defines a member of $F(S_s)$ for that unique s; $\widetilde{\Phi}(x)$ is taken to be $\widetilde{\varphi}_s(x)$. The universal mapping property of the free group $F(S)$ allows us to extend $\widetilde{\Phi}$ to a group homomorphism, which we continue to call $\widetilde{\Phi}$, of $F(S)$ into H. Let r be a nontrivial relation in $R \subseteq F(S)$. Then r, by hypothesis of disjointness for the sets S_s, lies in a unique R_s. Hence $\widetilde{\Phi}(r) = \widetilde{\varphi}_s(r) = \varphi_s(q_s(r)) = \varphi_s(1_s) = 1_H$. Consequently the kernel of $\widetilde{\Phi}$ contains the smallest normal subgroup $N(R)$ of $F(S)$ containing R, and $\widetilde{\Phi}$ descends to a homomorphism $\Phi : G \to H$. This Φ satisfies

$$\Phi \circ i_s \circ q_s = \Phi \circ \widetilde{i}_s = \widetilde{\Phi}\big|_{F(S_s)} = \widetilde{\varphi}_s = \varphi_s \circ q_s.$$

Since the quotient homomorphism q_s is onto G_s, we obtain $\Phi \circ i_s = \varphi_s$, and existence of the homomorphism Φ is established.

For uniqueness, we observe that the identities $\Phi \circ i_s = \varphi_s$ imply that Φ is uniquely determined on the subgroup of G generated by the images of all i_s. Since q_s is onto G_s, this subgroup is the same as the subgroup generated by the images of all $\widetilde{i}_s$. This subgroup contains the image in G of every generator of $F(S)$ and hence is all of G. Thus Φ is uniquely determined. $\square$

4. Group Representations

Group representations were defined in Section IV.6 as group actions on vector spaces by invertible linear functions. The underlying field of the vector space will be taken to be $\mathbb{C}$ in this section and the next, and the theory will then be especially tidy. The subject of group representations is one that uses a mix of linear algebra and group theory to reveal hidden structure within group actions. It has broad applications to algebra and analysis, but we shall be most interested in an application to finite groups known as Burnside's Theorem that will be proved in the next section.

Let us begin with the abelian case, taking G for the moment to be a finite abelian group. A **multiplicative character** of G is a homomorphism $\chi : G \to S^1 \subseteq \mathbb{C}^\times$ of G into the multiplicative group of complex numbers of absolute value 1. The multiplicative characters form an abelian group $\widehat{G}$ under pointwise multiplication of their complex values: $(\chi \chi')(g) = \chi(g)\chi'(g)$. The identity of $\widehat{G}$ is the multiplicative character that is identically 1 on G, and the inverse of χ is the complex conjugate of χ.

The notion of multiplicative character adapts to the case of a finite group the familiar exponential functions $x \mapsto e^{inx}$ on the line, which can be regarded as multiplicative characters of the additive group $\mathbb{R}/2\pi\mathbb{Z}$ of real numbers modulo 2π. These functions have long been used to resolve a periodic function of

4. Group Representations 327

time into its component frequencies: The device is the Fourier series of the function f. If f is periodic of period 2π, then the **Fourier coefficients** of f are $c_n = \frac{1}{2\pi} \int_{-\pi}^{\pi} f(x) e^{-inx} \, dx$, and the **Fourier series** of f is the infinite series $\sum_{n=-\infty}^{\infty} c_n e^{inx}$. A portion of the subject of Fourier series looks for senses in which $f(x)$ is actually equal to the sum of its Fourier series. This is the problem of **Fourier inversion**.

A similar problem can be formulated when $\mathbb{R}/2\pi\mathbb{Z}$ is replaced by the finite abelian group G. The exponential functions are replaced by the multiplicative characters. One can form an analog of Fourier coefficients for the vector space $C(G, \mathbb{C})$ of complex-valued functions[2] defined on G, and then one can form the analog of the Fourier series of the function. The problem of Fourier inversion becomes one of linear algebra, once we take into account the known structure of all finite abelian groups (Theorem 4.56). The result is as follows.

Theorem 7.17 (Fourier inversion formula for finite abelian groups). Let G be a finite abelian group, and introduce an inner product on the complex vector space $C(G, \mathbb{C})$ of all functions from G to $\mathbb{C}$ by the formula

$$\langle F, F' \rangle = \sum_{g \in G} F(g) \overline{F'(g)},$$

the corresponding norm being $\|F\| = \langle F, F \rangle^{1/2}$. Then the members of $\widehat{G}$ form an orthogonal basis of $C(G, \mathbb{C})$, each χ in $\widehat{G}$ satisfying $\|\chi\|^2 = |G|$. Consequently $|\widehat{G}| = |G|$, and any function $F : G \to \mathbb{C}$ is given by the "sum of its Fourier series":

$$F(g) = \frac{1}{|G|} \sum_{\chi \in \widehat{G}} \left(\sum_{h \in G} F(h) \overline{\chi(h)} \right) \chi(g).$$

REMARKS. This theorem is one of the ingredients in the proof in *Advanced Algebra* of Dirichlet's theorem that if a and b are positive relatively prime integers, then there are infinitely many primes of the form $an + b$. In applications to engineering, the ordinary Fourier transform on the line is often approximated, for computational purposes, by a Fourier series on a large cyclic group, and then Theorem 7.17 is applicable. Such a Fourier series can be computed with unexpected efficiency using a special grouping of terms; this device is called the **fast Fourier transform** and is described in Problems 29–31 at the end of the chapter.

---

[2]The notation $C(G, \mathbb{C})$ is to be suggestive of what happens for $G = S^1$ and for $G = \mathbb{R}^1$, where one works in part with the space of *continuous* complex-valued functions vanishing off a bounded set. In any event, pointwise multiplication makes $C(G, \mathbb{C})$ into a commutative ring. Later in the section we introduce a second multiplication, called "convolution," that makes $C(G, \mathbb{C})$ into a ring in a different way. In Chapter VIII we shall introduce the "complex group algebra" $\mathbb{C}G$ of G. The vector space $C(G, \mathbb{C})$ is the dual vector space of $\mathbb{C}G$. However, $C(G, \mathbb{C})$ and $\mathbb{C}G$ are canonically isomorphic because they have distinguished bases, and the isomorphism respects the multiplication structures—convolution in $C(G, \mathbb{C})$ and the group-algebra multiplication in $\mathbb{C}G$.

PROOF. For orthogonality let χ and χ' be distinct members of $\widehat{G}$, and put $\chi'' = \chi\overline{\chi'} = \chi{\chi'}^{-1}$. Choose g_0 in G with $\chi''(g_0) \neq 1$. Then

$$\chi''(g_0)\left(\sum_{g \in G} \chi''(g)\right) = \sum_{g \in G} \chi''(g_0 g) = \sum_{g \in G} \chi''(g),$$

so that
$$[1 - \chi''(g_0)] \sum_{g \in G} \chi''(g) = 0$$

and therefore
$$\sum_{g \in G} \chi''(g) = 0.$$

Consequently

$$\langle \chi, \chi' \rangle = \sum_{g \in G} \chi(g)\overline{\chi'(g)} = \sum_{g \in G} \chi''(g) = 0.$$

The orthogonality implies that the members of $\widehat{G}$ are linearly independent, and we obtain $|\widehat{G}| \leq \dim C(G, \mathbb{C}) = |G|$. Certainly $\|\chi\|^2 = \sum_{g \in G} |\chi(g)|^2 = \sum_{g \in G} 1 = |G|$.

To see that the members of $\widehat{G}$ are a basis of $C(G, \mathbb{C})$, we write G as a direct sum of cyclic groups, by Theorem 4.56. A summand $\mathbb{Z}/m\mathbb{Z}$ has at least m distinct multiplicative characters, given by $j \bmod m \mapsto e^{2\pi i j r/m}$ for $0 \leq r \leq m-1$, and these characters extend to G as 1 on the other direct summands of G. Taking products of such multiplicative characters from the different summands of G, we see that $|\widehat{G}| \geq |G|$. Therefore $|\widehat{G}| = |G|$, and $\widehat{G}$ is an orthogonal basis by Corollary 2.4. The formula for $F(g)$ in the statement of the theorem follows by applying Theorem 3.11c. $\square$

Now suppose that the finite group G is not necessarily abelian. Since S^1 is abelian, Proposition 7.4 shows that χ takes the value 1 on every member of the commutator subgroup G' of G. Consequently there is no way that the multiplicative characters can form a basis for the vector space $C(G, \mathbb{C})$ of complex-valued functions on G. The above analysis thus breaks down, and some adjustment is needed in order to extend the theory.

The remedy is to use representations, as defined in Section IV.6, on complex vector spaces of dimension > 1. We shall assume in the text that the vector space is finite-dimensional. The sense in which representations extend the theory of multiplicative characters is that any multiplicative character χ gives a representation R on the 1-dimensional vector space $\mathbb{C}$ by $R(g)(z) = \chi(g)z$ for g in G and z in $\mathbb{C}$. Conversely any 1-dimensional representation gives a multiplicative character: if R is the representation on the 1-dimensional vector space V and if $v_0 \neq 0$ is in V, then $\chi(g)$ is the scalar such that $R(g)v_0 = \chi(g)v_0$. It is enough to observe that the only elements of finite order in the multiplicative group $\mathbb{C}^\times$ are certain members of the circle S^1, and then it follows that χ takes values in S^1.

In the higher-dimensional case, the analog of the multiplicative character χ in passing to a 1-dimensional representation R is a "matrix representation." A **matrix representation** of G is a function $g \mapsto [\rho(g)_{ij}]$ from G into invertible square matrices of some given size such that $\rho(g_1 g_2)_{ij} = \sum_{k=1}^{n} \rho(g_1)_{ik} \rho(g_2)_{kj}$. If a representation R acts on the finite-dimensional complex vector space V, then the choice of an ordered basis Γ for V leads to a matrix representation by the formula

$$[\rho(g)_{ij}] = \begin{pmatrix} R(g) \\ \Gamma\Gamma \end{pmatrix}.$$

Conversely if a matrix representation $g \mapsto [\rho(g)_{ij}]$ and an ordered basis Γ of V are given, then the same formula may be used to obtain a representation R of G on V.

In contrast to the 1-dimensional case, the matrices that occur with a matrix representation of dimension > 1 need not be unitary. The correspondence between unitary linear maps and unitary matrices was discussed in Chapter III. When the finite-dimensional vector space V has an inner product, a linear map was defined to be unitary if it satisfies the equivalent conditions of Proposition 3.18. A complex square matrix A was defined to be unitary if $A^* A = I$. The matrix of a unitary linear map relative to an ordered orthonormal basis is unitary, and conversely when a unitary matrix and an ordered orthonormal basis are given, the associated linear map is unitary. We can thus speak of **unitary representations** and **unitary matrix representations**.

Some examples of representations appear in Section IV.6. One further pair of examples will be of interest to us. With the finite group G fixed but not necessarily abelian, we continue to let $C(G, \mathbb{C})$ be the complex vector space of all functions $f : G \to \mathbb{C}$. We define two representations of G on $C(G, \mathbb{C})$: the **left regular representation** ℓ given by $(\ell(g)f)(x) = f(g^{-1}x)$ and the **right regular representation** r given by $(r(g)f)(x) = f(xg)$. The reason for the presence of an inverse in one case and not the other was discussed in Section IV.6. Relative to the inner product

$$(f_1, f_2) = \sum_{x \in G} f_1(x) \overline{f_2(x)},$$

both ℓ and r are unitary. The argument for ℓ is that

$$(\ell(g)f_1, \ell(g)f_2) = \sum_{x \in G} (\ell(g)f_1)(x) \overline{(\ell(g)f_2)(x)} = \sum_{x \in G} f_1(g^{-1}x) \overline{f_2(g^{-1}x)}$$

$$\stackrel{\text{under } y = g^{-1}x}{=} \sum_{y \in G} f_1(y) \overline{f_2(y)} = (f_1, f_2),$$

and the argument for r is completely analogous.

It will be convenient to abbreviate "representation R on V" as "representation (R, V)." Let (R, V) be a representation of the finite group G on a finite-dimensional complex vector space. An **invariant subspace** U of V is a vector subspace such that $R(g)U \subseteq U$ for all g in G. The representation is **irreducible** if $V \neq 0$ and if V has no invariant subspaces other than 0 and V.

Two representations (R_1, V_1) and (R_2, V_2) on finite-dimensional complex vector spaces are **equivalent** if there exists a linear invertible function $A : V_1 \to V_2$ such that $AR_1(g) = R_2(g)A$ for all g in G. In the terminology of Section IV.11, "equivalent" is the notion of "is isomorphic to" in the category of all finite-dimensional representations of G.

In more detail a morphism from (R_1, V_1) to (R_2, V_2) in this category is an **intertwining operator**, namely a linear map $A : V_1 \to V_2$ such that $AR_1(g) = R_2(g)A$ for all g in G. The condition for this equality to hold is that the diagram in Figure 7.3 commute.

$$\begin{array}{ccc} V_1 & \xrightarrow{A} & V_2 \\ {\scriptstyle R_1(g)}\downarrow & & \downarrow{\scriptstyle R_2(g)} \\ V_1 & \xrightarrow{A} & V_2 \end{array}$$

FIGURE 7.3. An intertwining operator for two representations, i.e., a morphism in the category of finite-dimensional representations of G.

An example of a pair of representations that are equivalent is the left and right regular representations of G on $C(G, \mathbb{C})$: in fact, if we define $(Af)(x) = f(x^{-1})$, then

$$(\ell(g)Af)(x) = (Af)(g^{-1}x) = f(x^{-1}g) = (r(g)f)(x^{-1}) = (Ar(g)f)(x).$$

Proposition 7.18 (Schur's Lemma). If (R_1, V_1) and (R_2, V_2) are irreducible representations of the finite group G on finite-dimensional complex vector spaces and if $A : V_1 \to V_2$ is an intertwining operator, then A is invertible (and hence exhibits R_1 and R_2 as equivalent) or else $A = 0$. If $(R_1, V_1) = (R_2, V_2)$ and $A : V_1 \to V_2$ is an intertwining operator, then A is scalar.

REMARK. The conclusion that A is scalar makes essential use of the fact that the underlying field is $\mathbb{C}$.

PROOF. The equality $R_2(g)Av_1 = AR_1(g)v$ shows that ker A and image A are invariant subspaces. By the assumed irreducibility, ker A equals 0 or V_1, and image A equals 0 or V_2. The first statement follows. When $(R_1, V_1) = (R_2, V_2)$, the identity $I : V_1 \to V_2$ is an intertwining operator. If λ is an eigenvalue of A, then $A - \lambda I$ is another intertwining operator. Since $A - \lambda I$ is not invertible when λ is an eigenvalue of A, A must be 0. $\square$

Corollary 7.19. Every irreducible finite-dimensional representation of a finite abelian group G is 1-dimensional.

PROOF. If (R, V) is given, then the linear map $A = R(g)$ satisfies $AR(\tilde{g}) = R(x\tilde{g}) = R(\tilde{g}x) = R(\tilde{g})A$ for all x in G. By Schur's Lemma (Proposition 7.18), $A = R(g)$ is scalar. Since g is arbitrary, every vector subspace of V is invariant. Irreducibility therefore implies that V is 1-dimensional. $\square$

Let R be a representation of the finite group G on the finite-dimensional complex vector space V, let $(\,\cdot\,,\,\cdot\,)_0$ be any inner product on V, and define

$$(v_1, v_2) = \sum_{x \in G} (R(x)v_1, R(x)v_2)_0.$$

Then we have

$$\begin{aligned}
(R(g)v_1, R(g)v_2) &= \sum_{x \in G} (R(x)R(g)v_1, R(x)R(g)v_2)_0 \\
&= \sum_{x \in G} (R(xg)v_1, R(xg)v_2)_0 \\
&= \sum_{y \in G} (R(y)v_1, R(y)v_2)_0 \quad \text{by the change } y = xg \\
&= (v_1, v_2).
\end{aligned}$$

With respect to the inner product $(\,\cdot\,,\,\cdot\,)$, the representation (R, V) is therefore unitary. In other words, we are always free to introduce an inner product to make a given finite-dimensional representation unitary. The significance of this construction is noted in the following proposition.

Proposition 7.20. If (R, V) is a finite-dimensional representation of the finite group G and if an inner product is introduced in V that makes the representation unitary, then the orthogonal complement of an invariant subspace is invariant.

PROOF. Let U be an invariant subspace. If u is in U and $u^\perp$ is in $U^\perp$, then $(R(g)u^\perp, u) = (R(g)^{-1}R(g)u^\perp, R(g)^{-1}u) = (u^\perp, R(g)^{-1}u) = 0$. Thus $u^\perp$ in $U^\perp$ implies $R(g)u^\perp$ is in $U^\perp$. $\square$

Corollary 7.21. Any finite-dimensional representation of the finite group G is a direct sum of irreducible representations.

REMARK. That is, we can find a system of invariant subspaces such that the action of G is irreducible on each of these subspaces and such that the whole vector space is the direct sum of these subspaces.

PROOF. This is immediate by induction on the dimension. For dimension 0, the representation is the empty direct sum of irreducible representations. If the decomposition is known for dimension $< n$ and if U is an invariant subspace under R of smallest possible dimension ≥ 1, then U is irreducible under R, and Proposition 7.20 says that the subspace $U^\perp$, which satisfies $V = U \oplus U^\perp$, is invariant. It is therefore enough to decompose $U^\perp$, and induction achieves such a decomposition. □

Proposition 7.22 (Schur orthogonality). For finite-dimensional representations of a finite group G in which inner products have been introduced to make the representations unitary,

(a) if (R_1, V_1) and (R_2, V_2) are inequivalent and irreducible, then

$$\sum_{x \in G}(R_1(x)v_1, v_1')\overline{(R_2(x)v_2, v_2')} = 0 \quad \text{for all } v_1, v_2 \in V_1 \text{ and } v_2, v_2' \in V_2.$$

(b) if (R, V) is irreducible, then

$$\sum_{x \in G}(R(x)v_1, v_1')\overline{(R(x)v_2, v_2')} = \frac{|G|(v_1, v_2)\overline{(v_1', v_2')}}{\dim V} \quad \text{for } v_1, v_2, v_1', v_2' \in V.$$

REMARKS. If G is abelian, then V_1 and V_2 in (a) are 1-dimensional, and the conclusion of (a) reduces to the statement that the multiplicative characters are orthogonal. Conclusion (b) in this case reduces to a trivial statement.

PROOF. For (a), let $l : V_2 \to V_1$ be any linear map, and form the linear map

$$L = \sum_{x \in G} R_1(x) l R_2(x^{-1}).$$

Multiplying on the left by $R_1(g)$ and on the right by $R_2(g^{-1})$ and changing variables in the sum, we obtain $R_1(g)LR_2(g^{-1}) = L$, so that $R_1(g)L = LR_2(g)$ for all $g \in G$. By Schur's Lemma (Proposition 7.18) and the assumed irreducibility and inequivalence, $L = 0$. Thus $(Lv_2', v_1') = 0$. For the particular choice of l as $l(w_2) = (w_2, v_2)v_1$, we have

$$0 = (Lv_2', v_1') = \sum_{x \in G}(R_1(x)lR_2(x^{-1})v_2', v_1')$$

$$= \sum_{x \in G}\big(R_1(x)(R_2(x^{-1})v_2', v_2)v_1, v_1'\big) = \sum_{x \in G}(R_1(x)v_1, v_1')(R_2(x^{-1})v_2', v_2),$$

and (a) results since $(R_2(x^{-1})v_2', v_2) = \overline{(R_2(x)v_2, v_2')}$.

For (b), we proceed in the same way, starting from $l : V \to V$, and we obtain $L = \lambda I$ from Schur's Lemma. Taking the trace of both sides, we find that

4. Group Representations 333

$$\lambda \dim V = \operatorname{Tr} L = |G| \operatorname{Tr} l.$$

Therefore $\lambda = |G|(\operatorname{Tr} l)/\dim V$. Since $L = \lambda I$,

$$(Lv_2', v_1') = \frac{|G| \operatorname{Tr} l}{\dim V} \overline{(v_1', v_2')}.$$

Again we make the particular choice of l as $l(w_2) = (w_2, v_2)v_1$. Since $\operatorname{Tr} l = (v_1, v_2)$, we obtain

$$\begin{aligned}
\frac{(v_1, v_2)\overline{(v_1', v_2')}}{\dim V} &= \frac{\operatorname{Tr} l}{\dim V} \overline{(v_1', v_2')} = |G|^{-1}(Lv_2', v_1') \\
&= |G|^{-1} \sum_{x \in G} (R(x)l R(x^{-1}) v_2', v_1') \\
&= |G|^{-1} \sum_{x \in G} \left(R(x)(R(x^{-1})v_2', v_2)v_1, v_1' \right) \\
&= |G|^{-1} \sum_{x \in G} (R(x)v_1, v_1')(R(x^{-1})v_2', v_2),
\end{aligned}$$

and (b) results since $(R(x^{-1})v_2', v_2) = \overline{(R(x)v_2, v_2')}$. $\square$

Let us interpret Proposition 7.22 as a statement about the left and right regular representations ℓ and r of G on the inner-product space $C(G, \mathbb{C})$, the inner product being $\langle f, f' \rangle = \sum_{g \in G} f(g)\overline{f'(g)}$. Let R be an irreducible representation of G on the finite-dimensional vector space V, and introduce an inner product to make it unitary. A member of $C(G, \mathbb{C})$ of the form $g \mapsto (R(g)v, v')$ is called a **matrix coefficient** of R. Let $v_1', \ldots, v_n'$ be an orthonormal basis of V. The matrix representation of G that corresponds to R and this choice of orthonormal basis has $\rho(g)_{ij} = (R(g)v_j, v_i)$, and hence the entries of $[\rho(g)_{ij}]$, as functions on G, provide examples of matrix coefficients. These particular matrix coefficients are orthogonal, according to Proposition 7.22b, with

$$\sum_{x \in G} |\rho(x)_{ij}|^2 = \sum_{x \in G} (R(g)v_j, v_i)\overline{(R(g)v_j, v_i)} = \frac{|G|(v_j, v_j)\overline{(v_i, v_i)}}{\dim V} = \frac{|G|}{\dim V}.$$

Thus the functions $\sqrt{|G|^{-1} \dim V} \, \rho(x)_{ij}$ form an orthonormal basis of an n^2-dimensional subspace V_R of $C(G, \mathbb{C})$, where $n = \dim V$. The vector subspace V_R has the following properties:

(i) All matrix coefficients of R are in V_R, as is seen by expanding $v = \sum_j c_j v_j$ and $v' = \sum_i d_i v_i$ and obtaining $(R(g)v, v') = \sum_{i,j} c_j \bar{d}_i (R(g)v_j, v_i) = \sum_{i,j} c_j \bar{d}_i \rho(g)_{ij}$.

(ii) V_R is invariant under ℓ and r because
$$\ell(g)(R(\cdot)v, v')(x) = (R(g^{-1}x)v, v') = (R(x)v, R(g)v'),$$
$$r(g)(R(\cdot)v, v')(x) = (R(xg)v, v') = (R(x)R(g)v, v').$$

(iii) Any representation R' equivalent to R has $V_{R'} = V_R$.

Let us see how V_R decomposes into irreducible subspaces under r. The computation with r in (ii) above shows, for each i, that the vector space of all functions $x \mapsto (R(x)v, v_i)$ for $v \in V$ is invariant under r. This is the linear span of the matrix coefficients obtained from the ith row of $[\rho(x)_{ij}]$. Define a linear map A from V into this vector space by $Av = (R(\cdot)v, v_i)$. It is evident that A is one-one onto, and moreover $AR(g)v = (R(\cdot)R(g)v, v_i) = r(g)(R(\cdot)v, v_i) = r(g)Av$. Thus A exhibits this space, with r as representation, as equivalent to (R, V). The space V_R is the direct sum of these spaces on i, and the summands are orthogonal, according to Proposition 7.22b. Thus V_R decomposes under r as the direct sum of dim V irreducible subspaces, each one equivalent to (R, V).

One can make a similar analysis with ℓ, using columns in place of rows. However, this analysis is a little more subtle since V_R, acted upon by ℓ, is the direct sum of dim V copies of the "contragredient" of (R, V), rather than (R, V) itself. The details are left to Problems 32–36 at the end of the chapter.

As R varies over inequivalent representations, these vector spaces V_R are orthogonal, according to Proposition 7.22a. The claim is that their direct sum is the space $C(G, \mathbb{C})$ of all functions on G. In fact, the sum is invariant under r, and if it is nonzero, then we can find a nonzero vector subspace $U = \{f(\cdot)\}$ of $C(G, \mathbb{C})$ orthogonal to all the spaces V_R such that U is invariant and irreducible under r. Let $u_1, \ldots, u_m$ be an orthonormal basis of U. Then each function $x \mapsto (r(x)u_j, u_i)$ is orthogonal to U by construction, i.e.,
$$0 = \sum_{x \in G} (r(x)u_j, u_i)\overline{f(x)} \qquad \text{for all } f \text{ in } U.$$

Applying the Riesz Representation Theorem (Theorem 3.12), choose a member e of U such that $f(1) = (f, e)$ for all f in U. By definition of $r(x)$ and e, we find that
$$u(x) = (r(x)u)(1) = (r(x)u, e)$$
for all u in U. Substitution and use once more of Proposition 7.22b gives
$$0 = \sum_{x \in G} (r(x)u_j, u_i)\overline{(r(x)u, e)} = \frac{|G|(u_j, u)\overline{(u_i, e)}}{\dim U}$$
for all i and j. Since we can take $u = u_j = u_1$ and since i is arbitrary, this equation forces $e = 0$ and gives a contradiction. We conclude that the sum of all the spaces V_R is all of $C(G, \mathbb{C})$. Let us state the result as a theorem.

Theorem 7.23. For the finite group G, let $\{(R_\alpha, U_\alpha)\}$ be a complete set of inequivalent irreducible finite-dimensional representations of G, and let V_{R_α} be the linear span of the matrix coefficients of R_α. Then

(a) the spaces V_{R_α} are mutually orthogonal and are invariant under the left and right regular representations ℓ and r,
(b) the representation (r, V_{R_α}) is equivalent to the direct sum of dim U_α copies of (R_α, U_α),
(c) the direct sum of the spaces V_{R_α} is the space $C(G, \mathbb{C})$ of all complex-valued functions on G.

Moreover,

(d) the number of R_α's is finite,
(e) $\dim V_{R_\alpha} = (\dim U_\alpha)^2$,
(f) any irreducible subspace of $(r, C(G, \mathbb{C}))$ that is equivalent to (R_α, U_α) is contained in V_{R_α}.

Corollary 7.24. Let $\{(R_\alpha, U_\alpha)\}$ be a complete set of inequivalent irreducible finite-dimensional representations of the finite group G, and let $d_\alpha = \dim U_\alpha$. In each U_α, introduce an inner product making (R_α, U_α) unitary. For each α, let $\{u_1^{(\alpha)}, \ldots, u_{d_\alpha}^{(\alpha)}\}$ be an orthonormal basis of U_α. Then the functions in $C(G, \mathbb{C})$ given by $\sqrt{|G|^{-1}d_\alpha}\,(R_\alpha(x)v_j^{(\alpha)}, v_i^{(\alpha)})$ form an orthonormal basis of $C(G, \mathbb{C})$. Consequently every f in $C(G, \mathbb{C})$ satisfies

$$f(x) = \frac{1}{|G|} \sum_\alpha d_\alpha \sum_{i,j} \left(\sum_{y \in G} f(y) \overline{(R_\alpha(y)v_j^{(\alpha)}, v_i^{(\alpha)})} \right) (R_\alpha(x)v_j^{(\alpha)}, v_i^{(\alpha)})$$

and

$$\sum_{x \in G} |f(x)|^2 = \frac{1}{|G|} \sum_\alpha d_\alpha \sum_{i,j} \left| \sum_{y \in G} f(y) \overline{(R_\alpha(y)v_j^{(\alpha)}, v_i^{(\alpha)})} \right|^2.$$

REMARKS. The first displayed formula is the **Fourier inversion formula** for an arbitrary finite group G and generalizes Theorem 7.17, which gives the result in the abelian case; in the abelian case all the dimensions d_α equal 1, and the functions $(R_\alpha(x)v_j^{(\alpha)}, v_i^{(\alpha)})$ are just the multiplicative characters of G. The second displayed formula is known as the **Plancherel formula**, a result incorporating the conclusion about norms in Parseval's equality (Theorem 3.11d).

PROOF. This follows form (a), (c), and (e) in Theorem 7.23, together with Theorem 3.11 and the remarks made before the statement of Theorem 7.23. □

Corollary 7.25. Let $\{(R_\alpha, U_\alpha)\}$ be a complete set of inequivalent irreducible finite-dimensional representations of the finite group G, and let $d_\alpha = \dim U_\alpha$. Then $\sum_\alpha d_\alpha^2 = |G|$.

PROOF. This follows by counting the number of members listed in the orthonormal basis of $C(G, \mathbb{C})$ given in Corollary 7.24. □

We shall make use of a second multiplication on the vector space $C(G, \mathbb{C})$ besides the pointwise multiplication that itself makes $C(G, \mathbb{C})$ into a ring. The new multiplication is called **convolution** and is defined by

$$(f_1 * f_2)(x) = \sum_{y \in G} f_1(y) f_2(y^{-1}x) = \sum_{y \in G} f_1(xy^{-1}) f_2(y),$$

the two expressions on the right being equal by a change of variables. The first of the expressions on the right equals the value of the function $\sum_{y \in G} f_1(y) \ell(y) f_2$ at x and shows that the convolution is an average of the left translates of f_2 weighted by f_1. Convolution is associative because

$$(f_1 * (f_2 * f_3))(x) = \sum_y f_1(y)(f_2 * f_3)(y^{-1}x) = \sum_{y,z} f_1(y) f_2(y^{-1}xz^{-1}) f_3(z)$$
$$= \sum_z (f_1 * f_2)(xz^{-1}) f_3(z) = ((f_1 * f_2) * f_3)(x),$$

and one readily checks that $C(G, \mathbb{C})$ becomes a ring when convolution is used as the multiplication.

For any finite-dimensional representation (R, V) and any v in V, let us define $R(f)v = \sum_{x \in G} f(x) R(x) v$. Convolution has the property that

$$R(f_1 * f_2) = R(f_1) R(f_2)$$

because

$$R(f_1 * f_2)v = \sum_x (f_1 * f_2)(x) R(x) v = \sum_{x,y} f_1(xy^{-1}) f_2(y) R(x) v$$
$$= \sum_{x,y} f_1(x) f_2(y) R(xy) v = \sum_x f_1(x) R(x) \Big(\sum_y f_2(y) R(y) v \Big)$$
$$= \sum_x f_1(x) R(x) R(f_2) v = R(f_1) R(f_2) v.$$

We shall combine the notion of convolution with the notion of a "character." If (R, V) is a finite-dimensional representation of G, then the **character** of (R, V) is the function χ_R given by

$$\chi_R(x) = \operatorname{Tr} R(x),$$

with Tr denoting the trace. Equivalent representations have the same character since $\operatorname{Tr}(A R(x) A^{-1}) = \operatorname{Tr} R(x)$ if A is invertible. Characters have the additional properties that

(i) $\chi_R(gxg^{-1}) = \chi_R(x)$ because $\operatorname{Tr} R(gxg^{-1}) = \operatorname{Tr}(R(g) R(x) R(g)^{-1}) = \operatorname{Tr} R(x)$,

(ii) $\chi_{R_1 \oplus \cdots \oplus R_n} = \chi_{R_1} + \cdots + \chi_{R_n}$ since the trace of a block-diagonal matrix is the sum of the traces of the blocks.

The character of a 1-dimensional representation is the associated multiplicative character. Here is an example of a character for a representation on a space of dimension more than 1; its values are not all in S^1.

EXAMPLE. The dihedral group D_n with $2n$ elements, defined in Section IV.1, is isomorphic to the matrix group generated by

$$x = \begin{pmatrix} \cos 2\pi/n & -\sin 2\pi/n \\ \sin 2\pi/n & \cos 2\pi/n \end{pmatrix} \quad \text{and} \quad y = \begin{pmatrix} 1 & 0 \\ 0 & -1 \end{pmatrix}.$$

The map carrying each matrix of the group to itself is a representation of D_n on $\mathbb{C}^2$. The value of the character of this representation is $2\cos 2\pi k/n$ on x^k for $0 \leq k \leq n-1$, and the value of the character is 0 on y and on the remaining $n-1$ elements of the group.

Computations with characters are sometimes aided by the use of inner products. If an inner product is imposed on a finite-dimensional complex vector space V and if $\{v_i\}$ is an orthonormal basis, then the trace of a linear $A : V \to V$ is given by $\text{Tr } A = \sum_i (Av_i, v_i)$. If R is a representation on V, we consequently have $\chi_R(x) = \sum_i (R(x)v_i, v_i)$.

Proposition 7.26. Let R, R_1, and R_2 be irreducible finite-dimensional representations of a finite group G. Then their characters satisfy

(a) $\sum_{x \in G} |\chi_R(x)|^2 = |G|$,
(b) $\sum_{x \in G} \chi_{R_1}(x)\overline{\chi_{R_2}(x)} = 0$ if R_1 and R_2 are inequivalent.

PROOF. These follow from Schur orthogonality (Proposition 7.22): For (a), let R act on the vector space V, let $d = \dim V$, introduce an inner product with respect to which R is unitary, and let $\{v_i\}$ be an orthonormal basis of V. Then Proposition 7.22b gives

$$\sum_x |\chi_R(x)|^2 = \sum_x \left(\sum_i (R(x)v_i, v_i)\right)\overline{\left(\sum_j (R(x)v_j, v_j)\right)}$$
$$= \sum_{i,j} \sum_x (R(x)v_i, v_i)\overline{(R(x)v_j, v_j)}$$
$$= \sum_{i,j} |G|d^{-1}\delta_{ij}\delta_{ij} = \sum_i |G|d^{-1} = |G|.$$

Part (b) is proved in the same fashion, using Proposition 7.22a. □

Let us now bring together the notions of convolution and character. A **class function** on G is a function f in $C(G, \mathbb{C})$ with $f(gxg^{-1}) = f(x)$ for all g and x in G. That is, class functions are the ones that are constant on each conjugacy class of the group. Every character is an example of a class function. The class functions form a vector subspace of $C(G, \mathbb{C})$, and the dimension of this vector subspace equals the number of conjugacy classes in G. Class functions are closed under convolution because if f_1 and f_2 are class functions, then

$$(f_1 * f_2)(gxg^{-1}) = \sum_y f_1(gxg^{-1}y^{-1})f_2(y) = \sum_y f_1(xg^{-1}y^{-1}g)f_2(g^{-1}yg)$$
$$= \sum_z f_1(xz^{-1})f_2(z) = (f_1 * f_2)(x).$$

On an abelian group every member of $C(G, \mathbb{C})$ is a class function.

Theorem 7.27 (Fourier inversion formula for class functions). For the finite group G, let $\{(R_\alpha, U_\alpha)\}$ be a complete set of inequivalent irreducible finite-dimensional representations of G. If f is a class function on G, then

$$f(x) = \frac{1}{|G|} \sum_\alpha \left(\sum_{y \in G} f(y) \overline{\chi_{R_\alpha}(y)} \right) \chi_{R_\alpha}(x).$$

REMARK. This result may be regarded as a second way (besides the one in Corollary 7.24) of generalizing Theorem 7.17 to the nonabelian case.

PROOF. Using the result and notation of Corollary 7.24, we have

$$f(x) = |G|^{-1} \sum_\alpha d_\alpha \sum_{i,j} \left(\sum_{y \in G} f(y) \overline{(R_\alpha(y)v_i^{(\alpha)}, v_j^{(\alpha)})} \right) (R_\alpha(x)v_i^{(\alpha)}, v_j^{(\alpha)}).$$

Replace $f(y)$ by $f(gyg^{-1})$ since f is a class function, and then change variables and sum over g in G to see that $|G|f(x)$ is equal to

$$|G|^{-1} \sum_\alpha d_\alpha \sum_{i,j} \left(\sum_{g,y} f(y) \overline{(R_\alpha(y)R_\alpha(g)v_i^{(\alpha)}, R_\alpha(g)v_j^{(\alpha)})} \right) (R_\alpha(x)v_i^{(\alpha)}, v_j^{(\alpha)}).$$

Within this expression we have

$$\sum_g \overline{(R_\alpha(y)R_\alpha(g)v_i^{(\alpha)}, R_\alpha(g)v_j^{(\alpha)})}$$

$$= \sum_{g,k} \overline{\left(R_\alpha(y)(R_\alpha(g)v_i^{(\alpha)}, v_k^{(\alpha)})v_k^{(\alpha)}, R_\alpha(g)v_j^{(\alpha)}\right)}$$

$$= \sum_{g,k} \overline{(R_\alpha(g)v_i^{(\alpha)}, v_k^{(\alpha)})(R_\alpha(g)v_j^{(\alpha)}, R_\alpha(y)v_k^{(\alpha)})}$$

$$= \frac{|G|}{d_\alpha} \sum_k (v_j^{(\alpha)}, v_i^{(\alpha)}) \overline{(R_\alpha(y)v_k^{(\alpha)}, v_k^{(\alpha)})} \qquad \text{by Schur orthogonality}$$

$$= \frac{|G|}{d_\alpha} (v_j^{(\alpha)}, v_i^{(\alpha)}) \overline{\chi_{R_\alpha}(y)}$$

$$= \frac{|G|}{d_\alpha} \delta_{ij} \overline{\chi_{R_\alpha}(y)}.$$

Substituting, we obtain the formula of the theorem. $\square$

4. Group Representations

Corollary 7.28. If G is a finite group, then the number of irreducible finite-dimensional representations of G, up to equivalence, equals the number of conjugacy classes of G.

PROOF. Theorem 7.27 shows that the irreducible characters span the vector space of class functions. Proposition 7.26b shows that the irreducible characters are orthogonal and hence are linearly independent. Thus the number of irreducible characters equals the dimension of the space of class functions, which equals the number of conjugacy classes. □

EXAMPLE. The above information already gives us considerable control over finding a complete set of inequivalent irreducible finite-dimensional representations of elementary groups. We know that the number of such representations equals the number of conjugacy classes and that the sum of the squares of their dimensions equals $|G|$. For the symmetric group $\mathfrak{S}_3$ of order 6, for example, the conjugacy classes are given by the cycle structures of the possible permutations, namely the cycle structures of (1), (1 2), and (1 2 3). Hence there are three inequivalent irreducible representations. The sum of the squares of the three dimensions is to be 6; thus we have two of dimension 1 and one of dimension 2. The multiplicative characters 1 and sgn are the two of dimension 1, and the one of dimension 2 can be taken to be the 2-dimensional representation of D_3 whose character was computed in the example preceding Proposition 7.26.

One final constraint on the dimensions of the irreducible representations of a finite group G is as follows.

Proposition 7.29. If G is a finite group and (R, V) is an irreducible finite-dimensional representation of G, then $\dim V$ divides $|G|$.

For example, if $|G| = p^2$ with p prime, then it follows from Propositions 7.29 and 7.25 that every irreducible finite-dimensional representation of G has dimension 1, and one can easily conclude from this fact that G is abelian. (See Problem 14 at the end of the chapter.) Thus we recover as an immediate consequence the conclusion of Corollary 4.39 that groups of order p^2 are abelian.

The proof of Proposition 7.29 is surprisingly subtle. We shall obtain the theorem as a consequence of Theorem 7.31 below, a theorem that will be used also in the proof of Burnside's Theorem in the next section. Theorem 7.31 gives a little taste of the usefulness of algebraic number theory, and we shall see more of this usefulness in Chapter IX. The application to Burnside's Theorem will use the Fundamental Theorem of Galois Theory, whose proof is deferred to Chapter IX.

An **algebraic integer** is any complex number that is a root of a monic polynomial with coefficients in $\mathbb{Z}$. For example, $\sqrt{2}$ and $\frac{1}{2}(1 + i\sqrt{3})$ are algebraic integers because they are roots of $X^2 - 2$ and $X^2 - X + 1$, respectively. Any root of unity is an algebraic integer, being a root of some polynomial $X^n - 1$. The set of algebraic integers will be denoted in this chapter by $\mathcal{O}$. Before stating Theorem 7.31, let us establish two elementary facts about $\mathcal{O}$.

Lemma 7.30. The set $\mathcal{O}$ of algebraic integers is a ring, and $\mathcal{O} \cap \mathbb{Q} = \mathbb{Z}$.

PROOF. Suppose that x and y are complex numbers satisfying the polynomial equations $x^m + a_{m-1}x^{m-1} + \cdots + a_1 x + a_0 = 0$ and $y^n + b_{n-1}y^{n-1} + \cdots + b_1 y + b_0 = 0$, each with integer coefficients. Form the subset of $\mathbb{C}$ given by

$$M = \sum_{k=0}^{m-1} \sum_{l=0}^{n-1} \mathbb{Z} x^k y^l.$$

This is a finitely generated subgroup of the abelian group $\mathbb{C}$ under addition. It satisfies

$$xM = \sum_{k=1}^{m} \sum_{l=0}^{n-1} \mathbb{Z} x^k y^l \subseteq M + \sum_{l=0}^{n-1} \mathbb{Z} y^l x^m$$

$$= M + \sum_{l=0}^{n-1} \mathbb{Z} y^l (-a_{m-1} x^{m-1} - \cdots - a_1 x - a_0) \subseteq M,$$

and similarly $yM \subseteq M$. Hence $(x \pm y)M \subseteq M$ and $xy \subseteq M$.

To prove that $\mathcal{O}$ is a ring, it is enough to show that if N is a nonzero finitely generated subgroup of the abelian group $\mathbb{C}$ under addition and if z is a complex number with $zN \subseteq N$, then z is an algebraic integer. By Theorem 4.56, N is a direct sum of cyclic groups. Since every nonzero member of $\mathbb{C}$ has infinite order additively, these cyclic groups must be copies of $\mathbb{Z}$. So N is free abelian. Let $z_1, \ldots, z_n$ be a $\mathbb{Z}$ basis of N. Here $n > 0$. Since $zN \subseteq N$, we can find unique integers c_{ij} such that

$$zz_i = \sum_{j=1}^{n} c_{ij} z_j \quad \text{for } 1 \leq i \leq n.$$

This equation says that the matrix $C = [c_{ij}]$ has $\begin{pmatrix} z_1 \\ \vdots \\ z_n \end{pmatrix}$ as an eigenvector with eigenvalue z. Therefore the matrix $zI - C$ is singular, and $\det(zI - C) = 0$. Since $\det(zI - C)$ is a monic polynomial expression in z with integer coefficients, z is an algebraic integer.

To see that $\mathcal{O} \cap \mathbb{Q} = \mathbb{Z}$, let p and q be relatively prime integers with $q > 0$, and suppose that p/q is a root of $X^n + a_{n-1}X^{n-1} + \cdots + a_1 X + a_0$ with $a_{n-1}, \ldots, a_0$ in $\mathbb{Z}$. Substituting p/q for X, setting the expression equal to 0, and clearing fractions, we obtain $p^n + a_{n-1} p^{n-1} q + \cdots + a_1 p q^{n-1} + a_0 q^n = 0$. Since q divides every term here after the first, we conclude that q divides p^n. Since $\gcd(p, q) = 1$, we conclude that $q = 1$. Thus p/q is in $\mathbb{Z}$. $\square$

Lemma 7.30 allows us to see that if G is a finite group and χ is the irreducible character corresponding to an irreducible finite-dimensional representation R, then $\chi(x)$ is an algebraic integer for each x in G. In fact, the subgroup H of G generated by x is cyclic and is in particular abelian. Corollary 7.21 says that $R\big|_H$ is the direct sum of irreducible representations of H, and Corollary 7.19 says that each such irreducible representation is 1-dimensional. Thus in a suitable basis, $R\big|_H$ is diagonal. The diagonal entries must be roots of unity (in fact, N^{th} roots of unity if x has order N), and $\chi(x)$ is thus a sum of roots of unity. By Lemma 7.30, $\chi(x)$ is an algebraic integer.

Theorem 7.31. Let G be a finite group, (R, V) be an irreducible finite-dimensional representation of G, χ be the character of R, and C be a conjugacy class in G. Denote by $\chi(C)$ the constant value of χ on the conjugacy class C. Then $|C|\chi(C)/\dim V$ is an algebraic integer.

PROOF. If f is any class function on G, then $R(f)$ commutes with each $R(x)$ for x in G because $R(f) = \sum_y f(y) R(y)$ yields

$$R(x) R(f) R(x)^{-1} = \sum_y f(y) R(x) R(y) R(x)^{-1} = \sum_y f(y) R(xyx^{-1})$$
$$= \sum_z f(x^{-1}zx) R(z) = \sum_z f(z) R(z) = R(f).$$

By Schur's Lemma (Proposition 7.18), $R(f)$ is scalar. If C is a conjugacy class, then the function I_C that is 1 on C and is 0 elsewhere is a class function, and hence $R(I_C)$ is a scalar λ_C. As C varies, the functions I_C form a vector-space basis of the space of class functions. The formula $(I_C * I_{C'})(x) = \sum_y I_C(y) I_{C'}(y^{-1}x)$ shows that $I_C * I_{C'}$ is integer-valued, and we have seen that the convolution of two class functions is a class function. Therefore $I_C * I_{C'} = \sum_{C''} n_{CC'C''} I_{C''}$ for suitable integers $n_{CC'C''}$. Application of R gives $\lambda_C \lambda_{C'} = \sum_{C''} n_{CC'C''} \lambda_{C''}$. If we fix C and let A be the square matrix with entries $A_{C'C''} = n_{CC'C''}$, we obtain

$$\lambda_C \lambda_{C'} = \sum_{C''} A_{C'C''} \lambda_{C''}.$$

This equation says that the matrix A has the column vector with entries $\lambda_{C''}$ as an eigenvector with eigenvalue λ_C. Therefore the matrix $\lambda_C I - A$ is singular, and $\det(\lambda_C I - A) = 0$. Since $\det(\lambda_C I - A)$ is a monic polynomial expression in λ_C with integer coefficients, λ_C is an algebraic integer. Taking the trace of the equation $R(I_C) = \lambda_C I$, we obtain $\sum_{x \in C} \chi(x) = \lambda_C \dim V$. Since $\chi(x) = \chi(C)$ for x in C, the result is that $|C|\chi(C)/\dim V = \lambda_C$. Since λ_C is an algebraic integer, $|C|\chi(C)/\dim V$ is an algebraic integer. □

PROOF THAT THEOREM 7.31 IMPLIES PROPOSITION 7.29. Proposition 7.26a gives

$$\frac{|G|}{\dim V} = \frac{\sum_{x \in G} |\chi(x)|^2}{\dim V} = \frac{\sum_C \sum_{x \in C} |\chi(x)|^2}{\dim V} = \sum_C \left(\frac{|C|\chi(C)}{\dim V} \right) \overline{\chi(C)}.$$

Each term in parentheses on the right side is an algebraic integer, according to Theorem 7.31, and therefore Lemma 7.30 shows that $|G|/\dim V$ is an algebraic integer. Since $|G|/\dim V$ is in $\mathbb{Q}$, Lemma 7.30 shows that $|G|/\dim V$ is in $\mathbb{Z}$. □

5. Burnside's Theorem

The theorem of this section is as follows.

Theorem 7.32 (Burnside's Theorem). If G is a finite group of order $p^a q^b$ with p and q prime and with $a + b > 1$, then G has a nontrivial normal subgroup.

The argument will use the result Theorem 7.31 from algebraic number theory, and also it will make use of a special case of the Fundamental Theorem of Galois Theory, whose proof is deferred to Chapter IX. That special case is the following statement, whose context was anticipated in Section IV.1, where groups of automorphisms of certain fields were discussed briefly. Since the set $\{1, e^{2\pi i/n}, e^{2 \cdot 2\pi i/n}, e^{3 \cdot 2\pi i/n}, \dots\}$ is linearly dependent over $\mathbb{Q}$, Proposition 4.1 in that section implies that the subring $\mathbb{Q}[e^{2\pi i/n}]$ of $\mathbb{C}$ generated by $\mathbb{Q}$ and $e^{2\pi i/n}$ is a subfield and is a finite-dimensional vector space over $\mathbb{Q}$. According to Example 9 of that section, the group $\Gamma = \text{Gal}(\mathbb{Q}[e^{2\pi i/n}]/\mathbb{Q})$ of automorphisms of $\mathbb{Q}[e^{2\pi i/n}]$ fixing every element of $\mathbb{Q}$ is a finite group.

Proposition 7.33 (special case of the Fundamental Theorem of Galois Theory). Let $n > 0$ be an integer, and put $K = \mathbb{Q}[e^{2\pi i/n}]$. Let Γ be the finite group of field automorphisms of K fixing every element of $\mathbb{Q}$. Then the only members β of K such that $\sigma(\beta) = \beta$ for every σ in Γ are the members of $\mathbb{Q}$.

Lemma 7.34. Let G be a finite group, (R, V) be an irreducible finite-dimensional representation of G, χ be the character of R, and C be a conjugacy class in G. If $\text{GCD}(|C|, \dim V) = 1$ and if x is in C, then either $R(x)$ is scalar or $\chi(x) = 0$.

PROOF. Define $\chi(C)$ to be the constant value of χ on C, and put $\alpha = \chi(x)/\dim V = \chi(C)/\dim V$. Since $\text{GCD}(|C|, \dim V) = 1$, we can choose integers m and n with $m|C| + n \dim V = 1$. Multiplication by α yields

$$\frac{m|C|\chi(C)}{\dim V} + n\chi(C) = \alpha.$$

Theorem 7.31 shows that the coefficients $\frac{|C|\chi(C)}{\dim V}$ and $\chi(C)$ of m and n on the left side are algebraic integers, and therefore α is an algebraic integer. As we observed toward the end of the previous section, $\chi(x) = \chi(C)$ is the sum of $\dim V$ roots of unity. Since $\alpha = \chi(C)/\dim V$, we see that $|\alpha| \leq 1$ with equality only if all the roots of unity are equal, in which case $R(x)$ is scalar. In view of the hypothesis, we may assume that $|\alpha| < 1$. We shall show that $\alpha = 0$.

Let $K = \mathbb{Q}[e^{2\pi i/|G|}]$ be the smallest subfield of $\mathbb{C}$ containing $\mathbb{Q}$ and the complex number $e^{2\pi i/|G|}$, and let Γ be the group of field automorphisms of K that fix every element of $\mathbb{Q}$. We know that K is finite-dimensional over $\mathbb{Q}$ and that Γ is a finite group, and Proposition 7.33 shows that the only members of K fixed by every element of Γ are the members of $\mathbb{Q}$.

Our element x of G has $x^{|G|} = 1$. Thus every root of unity contributing to $\chi(x)$ is a $|G|^{\text{th}}$ root of unity and is in K. Therefore the algebraic integer α is in K. If σ is in Γ, each of the $|G|^{\text{th}}$ roots of unity is mapped by σ to some complex number x satisfying $x^{|G|} = 1$, and hence the member $\sigma(\alpha)$ of K satisfies $|\sigma(\alpha)| \leq 1$. Also, $\sigma(\alpha)$ is an algebraic integer, as we see by applying σ to the monic equation with integer coefficients satisfied by α, and we are assuming that $|\alpha| < 1$. Consequently $\beta = \prod_{\sigma \in \Gamma} \sigma(\alpha)$ is an algebraic integer and has absolute value < 1. A change of variables in the product shows that β is fixed by every member of Γ, and we see from the previous paragraph that β is in $\mathbb{Q}$. By Lemma 7.30, β is in $\mathbb{Z}$. Being of absolute value less than 1, it is 0. Thus $\alpha = 0$, and $\chi(x) = 0$. □

Lemma 7.35. Let G be a finite group, and let C be a conjugacy class in G such that $|C| = p^k$ for some prime p and some integer $k > 0$. Then there exists an irreducible finite-dimensional representation $R \neq 1$ of G with $R(x)$ scalar for every x in C. Consequently G is not simple.

PROOF. The conjugacy class C cannot be $\{1\}$ because $|\{1\}| \neq p^k$ with $k > 0$. Let χ_{reg} be the character of the right regular representation r of G on $C(G, \mathbb{C})$. If I_g denotes the function that is 1 at g and is 0 elsewhere, then the functions I_g form an orthonormal basis of $C(G, \mathbb{C})$, and therefore $\chi_{\text{reg}}(x) = \sum_{g \in G} (r(x)I_g, I_g) = \sum_{g \in G} (I_{gx^{-1}}, I_g)$. Every term on the right side is 0 if $x \neq 1$, and thus Theorem 7.23 gives

$$0 = \chi_{\text{reg}}(x) = 1 + \sum_{\chi \neq 1} d_\chi \chi(x) \qquad \text{for } x \in C, \qquad (*)$$

the sum being taken over all irreducible characters other than 1, with d_χ being the dimension of an irreducible representation corresponding to χ. Let R_χ be an irreducible representation with character χ. Any χ such that p does not divide d_χ has $\gcd(|C|, d_\chi) = 1$ since $|C|$ is assumed to be a power of p. Arguing by contradiction, we may assume that no such χ has $R_\chi(x)$ scalar, and then Lemma 7.34 says that $\chi(x) = 0$ for all such χ. Hence (*) simplifies to

$$0 = 1 + \sum_{\chi \neq 1,\ p \text{ divides } d_\chi} d_\chi \chi(x) \qquad \text{for } x \in C. \qquad (**)$$

Since $\chi(x)$ is an algebraic integer, Lemma 7.30 shows that this equation is of the form $1 + p\beta = 0$, where β is an algebraic integer. Then $\beta = -1/p$ shows that $-1/p$ is an algebraic integer. Since $-1/p$ is in $\mathbb{Q}$, Lemma 7.30 shows that it must be in $\mathbb{Z}$, and we have arrived at a contradiction. Thus there must have been some χ with $R_\chi(x)$ scalar for x in C.

The set of g in G for which this R_χ has $R_\chi(g)$ scalar is a normal subgroup of G that contains x and cannot therefore be $\{1\}$. Assume by way of contradiction that G is simple. Then $R_\chi(g)$ is scalar for all g in G. Since R_χ is irreducible, R_χ is 1-dimensional. Then the commutator subgroup G' of G is contained in the kernel of R_χ. Since $R_\chi \neq 1$, G' is not all of G. Since G' is normal, $G' = \{1\}$, and we conclude that G is abelian. But the given G has a conjugacy class with more than one element, and we have arrived at a contradiction. □

PROOF OF THEOREM 7.32. Corollary 4.38 shows that a group of prime-power order has a center different from $\{1\}$, and we may therefore assume that $p \neq q$, $a > 0$, and $b > 0$. Let H be a Sylow q-subgroup. Applying Corollary 4.38, let x be a member of the center Z_H of H other than 1. The centralizer $Z_G(\{x\})$ is a subgroup containing H, and it therefore has order $p^{a'}q^b$. If $a' = a$, then x is in the center of G, and the powers of x form the desired proper normal subgroup of G. Thus $a' < a$. By Proposition 4.37 the conjugacy class C of x has $|G|/p^{a'}q^b = p^{a-a'}$ elements with $a - a' > 0$. By Lemma 7.35, G is not simple. □

6. Extensions of Groups

In Section IV.8 we examined composition series for finite groups. For a given finite group, a composition series consists of a decreasing sequence of subgroups starting with the whole group and ending with $\{1\}$, each normal in the next larger one, such that the successive quotient groups are simple. The Jordan–Hölder Theorem (Corollary 4.50) assured us that the set of successive quotients, up to isomorphism, is independent of the choice of composition series. This theorem raises the question of reconstructing the whole group from data of this kind. Consider a single step of the process. If we know the normal subgroup and the simple quotient that it yields at a certain stage, what are the possibilities for the next-larger subgroup? We study this question and some of its ramifications in this section, dropping any hypotheses that are not helpful in the analysis. Here is an example that we shall carry along.

EXAMPLE. Suppose that the normal subgroup is the cyclic group C_4 and that the quotient is the cyclic group C_2. The whole group has to be of order 8, and the classification of groups of order 8 done in Problems 39–44 at the end of Chapter IV tells us that there are four different possibilities for the whole group: the abelian groups $C_4 \times C_2$ and C_8, the dihedral group D_4, and the quaternion group H_8.

Let us establish a framework for the general problem. We start with a group E, a normal subgroup N, and the quotient $G = E/N$. We seek data that determine the group law in E in terms of N and G. For each member u of G, fix a coset representative $\bar{u}$ in E such that $\bar{u}N = u$. Since N is normal, the element $\bar{u}$ of E yields an automorphism $(\,\cdot\,)^u$ of N defined by $x^u = \bar{u}x\bar{u}^{-1}$. In addition, the fact that G is a group says that any two of our representatives $\bar{u}$ and $\bar{v}$ have

$$\bar{u}\bar{v} = a(u,v)\overline{uv} \quad \text{for some unique } a(u,v) \text{ in } N.$$

The set of all elements $a(u,v)$ for this choice of coset representatives is called a **factor set**, and E is called a **group extension** of N by the group[3] G.

The automorphisms and the factor set constructed above have to satisfy two compatibility conditions, as follows:

(i) $(x^v)^u = a(u,v)x^{uv}a(u,v)^{-1}$ because $(x^u)^v = \bar{u}(x^v)\bar{u}^{-1} = \bar{u}\bar{v}x\bar{v}^{-1}\bar{u}^{-1}$
$= (a(u,v)\overline{uv})x(a(u,v)\overline{uv})^{-1} = a(u,v)x^{uv}a(u,v)^{-1}$,

(ii) $a(v,w)^u a(u,vw) = a(u,v)a(uv,w)$ because $(\bar{u}\bar{v})\bar{w} = a(u,v)\overline{uv}\bar{w}$
$= a(u,v)a(uv,w)\overline{uvw}$ and $\bar{u}(\bar{v}\bar{w}) = \bar{u}a(v,w)\overline{vw} = a(v,w)^u \bar{u}\overline{vw} = a(v,w)^u a(u,vw)\overline{uvw}$.

Then the multiplication law in E is given in terms of the automorphisms and the factor set by the formula

(iii) $(x\bar{u})(y\bar{v}) = xy^u a(u,v)\overline{uv}$ by the computation $(x\bar{u})(y\bar{v}) = xy^u \bar{u}\bar{v} = xy^u a(u,v)\overline{uv}$.

Conversely, according to the proposition below, such data determine a group E with a normal subgroup isomorphic to N and a quotient E/N isomorphic to G.

Proposition 7.36 (Schreier). Let two groups N and G be given, along with a family of automorphisms $x \mapsto x^u$ of N parametrized by u in G, as well as a function $a : G \times G \to N$ such that

(a) $(x^v)^u = a(u,v)x^{uv}a(u,v)^{-1}$ for all u and v in G,
(b) $a(v,w)^u a(u,vw) = a(u,v)a(uv,w)$ for all u, v, w in G.

Then the set $N \times G$ becomes a group E under the multiplication

(c) $(x,u)(y,v) = (xy^u a(u,v), uv)$,

---

[3] *Warning:* Some authors say "group extension of G by N."

and this group has a normal subgroup isomorphic to N with quotient group isomorphic to G. More particularly, the identity of E is $(a(1, 1)^{-1}, 1)$, the map $x \mapsto (xa(1, 1)^{-1}, 1)$ of N into E is a one-one homomorphism that exhibits N as a normal subgroup of E, and the map $(x, u) \mapsto u$ of E onto G is a homomorphism that exhibits G as isomorphic to E/N.

PROOF. Reverting to the earlier notation, let us write $x\bar{u}$ in place of (x, u) for elements of E. Associativity of multiplication follows from the computation

$$\begin{aligned}
(x\bar{u}y\bar{v})(z\bar{w}) &= \left(xy^u a(u, v)\overline{uv}\right)z\bar{w} & \text{by (c)} \\
&= xy^u a(u, v) z^{uv} a(uv, w) \overline{uvw} & \text{by (c)} \\
&= xy^u a(u, v) z^{uv} a(u, v)^{-1} a(u, v) a(uv, w) \overline{uvw} \\
&= xy^u a(u, v) z^{uv} a(u, v)^{-1} a(v, w)^u a(u, vw) \overline{uvw} & \text{by (b)} \\
&= x\left(yz^v a(v, w)\right)^u a(u, vw) \overline{uvw} & \text{by (a)} \\
&= (x\bar{u})\left(yz^v a(v, w)\overline{vw}\right) & \text{by (c)} \\
&= (x\bar{u})(y\bar{v}z\bar{w}) & \text{by (c).}
\end{aligned}$$

The identity is to be $\bar{1}a(1, 1)^{-1}$. Before checking this assertion, we prove three preliminary identities. Setting $u = v = 1$ in (a) and replacing x^1 by x gives[4]

$$x^1 = a(1, 1) x a(1, 1)^{-1} \qquad \text{for all } x \in N. \tag{$*$}$$

Setting $v = w = 1$ in (b) gives $a(1, 1)^u a(u, 1) = a(u, 1)a(u, 1)$ and hence

$$a(1, 1)^u = a(u, 1) \qquad \text{for all } u \in G. \tag{$\dagger$}$$

Meanwhile, setting $u = v = 1$ in (b) gives $a(1, w)^1 a(1, w) = a(1, 1) a(1, w)$ and hence $a(1, w)^1 = a(1, 1)$ for all $w \in G$. The left side $a(1, w)^1$ of this last equality is equal to $a(1, 1) a(1, w) a(1, 1)^{-1}$ by $(*)$; canceling $a(1, 1)$ yields

$$a(1, w) = a(1, 1) \qquad \text{for all } w \in G. \tag{$\dagger\dagger$}$$

Using these identities, we check that $a(1, 1)^{-1}\bar{1}$ is a two-sided identity by making the computations

$$\begin{aligned}
(x\bar{u})(a(1, 1)^{-1}\bar{1}) &= x(a(1, 1)^{-1})^u a(u, 1)\bar{u} & \text{by (c)} \\
&= x(a(1, 1)^{-1})^u a(1, 1)^u \bar{u} & \text{by ($\dagger$)} \\
&= x\bar{u}
\end{aligned}$$

---

[4]The effect of the automorphism $x \mapsto x^1$ is not necessarily trivial since the coset representative $\bar{1}$ of 1 is not assumed to be the identity. Thus we must distinguish between x^1 and x.

and

$$(a(1,1)^{-1}\bar{1})(y\bar{v}) = a(1,1)^{-1}y^1 a(1,v)\bar{v} \quad \text{by (c)}$$
$$= ya(1,1)^{-1}a(1,v)\bar{v} \quad \text{by (*)}$$
$$= y\bar{v} \quad \text{by (††).}$$

Let us check that a left inverse for $x\bar{u}$ is $a(1,1)^{-1}a(u^{-1},u)^{-1}(x^{u^{-1}})^{-1}\overline{u^{-1}}$. In fact,

$$\left(a(1,1)^{-1}a(u^{-1},u)^{-1}(x^{u^{-1}})^{-1}\overline{u^{-1}}\right)(x\bar{u})$$
$$= a(1,1)^{-1}a(u^{-1},u)^{-1}(x^{u^{-1}})^{-1}x^{u^{-1}}a(u^{-1},u)\bar{1} \quad \text{by (c)}$$
$$= a(1,1)^{-1}\bar{1},$$

as required. Thus multiplication is associative, there is a two-sided identity, and every element has a left inverse. It follows that E is a group.

The map $x\bar{u} \mapsto u$ of E into G is a homomorphism by (c), and it is certainly onto G. Its kernel is evidently the subgroup of all elements $xa(1,1)^{-1}\bar{1}$ in E. Since

$$(xa(1,1)^{-1}\bar{1})(ya(1,1)^{-1}\bar{1}) = xa(1,1)^{-1}(ya(1,1)^{-1})^1 a(1,1)\bar{1} \quad \text{by (c)}$$
$$= xa(1,1)^{-1}a(1,1)(ya(1,1)^{-1})\bar{1} \quad \text{by (*)}$$
$$= xya(1,1)^{-1}\bar{1},$$

the one-one map $x \mapsto xa(1,1)^{-1}\bar{1}$ of N onto the kernel respects the group structures and is therefore an isomorphism. In other words, the embedded version of N is the kernel. Being a kernel, it is a normal subgroup. □

EXAMPLE, CONTINUED. Let $N = C_4 = \{1, r, r^2, r^3\}$ and $G = C_2 = \{1, u_0\}$ with $u_0^2 = 1$. The group N has two automorphisms, the nontrivial one fixing 1 and r^2 while interchanging r and r^3. The automorphism of N from $1 \in G$ has to be trivial, while the automorphism of N from $u_0 \in G$ can be trivial or nontrivial. In fact,

$$\text{the automorphism is} \begin{cases} \text{trivial} & \text{for } E = C_4 \times C_2 \text{ and } E = C_8, \\ \text{nontrivial} & \text{for } E = D_4 \text{ and } E = H_8. \end{cases}$$

In each case the automorphism does not depend on the choice of coset representatives. The factor sets do depend on the choice of representatives, however. Let us fix $\bar{1}$ as the identity of E and make a particular choice of $\bar{u_0}$ for each E. Then the definition of factor set shows that $a(1, 1) = a(u_0, 1) = a(1, u_0) = 1$, and the only part of the factor set yet to be determined is $a(u_0, u_0)$. Let us consider matters group by group. For $C_4 \times C_2$, we can take $\overline{u_0}$ to be the generator of the C_2 factor; this has square 1, and hence $a(u_0, u_0) = 1$. For $C_8 = \{1\theta, \theta^2, \ldots, \theta^7\}$, let us think of N as embedded in E with $r = \theta^2$. The element $\overline{u_0}$ can be any odd power of θ; if we take $\overline{u_0} = \theta$, then $(\overline{u_0})^2 = \theta^2 = r$, and hence $a(u_0, u_0) = r$. For $E = D_4$, the example following Proposition 7.8 shows that we may view the elements as the rotations $1, r, r^2, r^3$ and the reflections s, rs, r^2s, r^3s for particular choices of r and s. We can take $\overline{u_0}$ to be any of the reflections, and then $(\overline{u_0})^2 = 1$ and $a(u_0, u_0) = 1$. Finally for $E = H_8 = \{\pm 1, \pm\mathbf{i}, \pm\mathbf{j}, \pm\mathbf{k}\}$, let us say that N is embedded as $\{\pm 1, \pm\mathbf{i}\}$. Then $\overline{u_0}$ can be any of the four elements $\pm\mathbf{j}$ and $\pm\mathbf{k}$. Each of these has square -1, and hence $a(u_0, u_0) = -1$. For the choices we have made, we therefore have

$$a(u_0, u_0) = \begin{cases} 1 & \text{for } E = C_4 \times C_2 \text{ and } E = D_4, \\ r & \text{for } E = C_8, \\ -1 & \text{for } E = H_8. \end{cases}$$

The formula of Proposition 7.36a reduces to $(x^v)^u = x^{uv}$ since N is abelian, and it is certainly satisfied. The formula for Proposition 7.36b is $a(v, w)^u a(u, vw) = a(u, v)a(uv, w)$. This is satisfied for $E = C_4 \times C_2$ and $E = D_4$ since $a(\cdot, \cdot)$ is identically 1. For the other two cases the values of $a(\cdot, \cdot)$ lie in the 2-element subgroup of N that is fixed by the nontrivial automorphism, and hence $a(v, w)^u = a(v, w)$ in every case. The formula to be checked reduces to $a(v, w)a(1, 1) = a(1, 1)a(v, w)$ by (††) if $u = 1$, to $a(1, 1)a(u, w) = a(1, 1)a(u, w)$ by (†) and (††) if $v = 1$, and to $a(1, 1)a(u, v) = a(u, v)a(1, 1)$ by (†) if $w = 1$. Thus all that needs checking is the case that $u = v = w = u_0$, and then the formula in question reduces to $a(u_0, u_0)a(1, 1) = a(u_0, u_0)a(1, 1)$ by (†) and (††).

Let us examine for a particular extension the dependence of the automorphisms and factor set on the choice of coset representatives. Returning to our original construction, suppose that we change the coset representatives of the members of G, associating a member $\widetilde{u}$ to $u \in G$ in place of $\bar{u}$. We then obtain a new automorphism of N corresponding to u, and we write it as $x \mapsto x^{u^*} = \widetilde{u}x\widetilde{u}^{-1}$ instead of $x \mapsto x^u = \bar{u}x\bar{u}^{-1}$. To quantify matters, we observe that $\widetilde{u}$ lies in the same coset of N as does $\bar{u}$. Thus $\widetilde{u} = \alpha(u)\bar{u}$ for some function $\alpha : G \to N$, and the function α can be absolutely arbitrary. In terms of this function α, the two automorphisms are related by

$$x^{u^*} = \widetilde{u}x\widetilde{u}^{-1} = \alpha(u)\bar{u}x\bar{u}^{-1}\alpha(u)^{-1} = \alpha(u)x^u\alpha(u)^{-1}.$$

If the factor set for the system $\{\widetilde{u}\}$ of coset representatives is denoted by $\{b(u, v)\}$, then we have $b(u, v)\alpha(uv)\overline{uv} = b(u, v)\widetilde{uv} = \widetilde{u}\widetilde{v} = \alpha(u)\bar{u}\alpha(v)\bar{v} =$

$\alpha(u)\alpha(v)^u a(u, v)\overline{uv}$. Equating coefficients of $\overline{uv}$, we obtain

$$b(u, v) = \alpha(u)\alpha(v)^u a(u, v)\alpha(uv)^{-1}.$$

Accordingly we say that a group extension of N by G determined by automorphisms $x \mapsto x^u$ and a factor set $a(u, v)$ is **equivalent**, or **isomorphic**, to a group extension of N by G determined by automorphisms $x \mapsto x^{u^*}$ and a factor set $b(u, v)$ if there is a function $\alpha : G \to N$ such that

$$x^{u^*} = \alpha(u) x^u \alpha(u)^{-1} \quad \text{and} \quad b(u, v) = \alpha(u)\alpha(v)^u a(u, v)\alpha(uv)^{-1}$$

for all u and v in G. It is immediate that equivalence of group extensions is an equivalence relation.

Proposition 7.37. Suppose that E_1 and E_2 are group extensions of N by G with respective inclusions $i_1 : N \to E_1$ and $i_2 : N \to E_2$ and with respective quotient homomorphisms $\varphi_1 : E_1 \to G$ and $\varphi_2 : E_2 \to G$. If there exists a group isomorphism $\Phi : E_1 \to E_2$ such that the two squares in Figure 7.4 commute, then the two group extensions are equivalent. Conversely if the two group extensions are equivalent, then there exists a group isomorphism $\Phi : E_1 \to E_2$ such that the two squares in Figure 7.4 commute.

$$\begin{array}{ccccc} N & \xrightarrow{i_1} & E_1 & \xrightarrow{\varphi_1} & G \\ \parallel & & \downarrow{\Phi} & & \parallel \\ N & \xrightarrow{i_2} & E_2 & \xrightarrow{\varphi_2} & G \end{array}$$

FIGURE 7.4. Equivalent group extensions.

REMARKS. The commutativity of the squares is important. Just because two group extensions of N by G are isomorphic as groups does not imply that they are equivalent group extensions. An example is given in Problem 19 at the end of the chapter.

PROOF. For the direct part, suppose that Φ exists. For each u in G, select $\bar{u}$ in E_1 with $\varphi_1(\bar{u}) = u$. Then we can form the extension data $\{x \mapsto x^u\}$ and $\{a(u, v)\}$ for E_1 relative to the normal subgroup $i_1(N)$ and the system $\{\bar{u} \mid u \in G\}$ of coset representatives. When reinterpreted in terms of N, E_1, and G, these data become $\{i_1^{-1}(x) \mapsto i_1^{-1}(x^u)\}$ and $\{i_1^{-1}(a(u, v))\}$.

Application of Φ to the coset $i_1(N)\bar{u}$ yields $i_2(N)\Phi(\bar{u})$ since $\Phi i_1 = i_2$, and $\Phi(\bar{u})$ is a member of E_2 with $\varphi_2(\Phi(\bar{u})) = \varphi_1(\bar{u}) = u$. Setting $\tilde{u} = \Phi(\bar{u})$, we see that $\Phi(i_1(N)\bar{u})$ is the coset $i_2(N)\tilde{u}$ of $i_2(N)$ in E_2. Thus we can determine extension data for E_2 relative to $i_2(N)$ and the system $\{\widetilde{u} \mid u \in G\}$, and we can transform them by i_2^{-1} to obtain data relative to N, E_2, and G.

The claim is that the data relative to N, E_2, and G match those for N, E_1, and G. The automorphisms of N from E_2 are the maps $i_2^{-1}(x') \mapsto i_2^{-1}(x'^{u^*})$, where $x'^{u^*} = \widetilde{u}x'\widetilde{u}^{-1}$. From $i_2 = \Phi i_1$ and the fact that each of these maps is one-one, we obtain $i_2^{-1} = i_1^{-1}\Phi^{-1}$ on $i_2(N)$. Substitution shows that the automorphisms of N from E_2 are

$$i_1^{-1}(\Phi^{-1}(x')) \mapsto i_1^{-1}(\Phi^{-1}(x'^{u^*})) = i_1^{-1}(\Phi^{-1}(\widetilde{u}x'\widetilde{u}^{-1}))$$
$$= i_1^{-1}(\bar{u}\Phi^{-1}(x')\bar{u}^{-1}) = i_1^{-1}((\Phi^{-1}(x'))^u).$$

If we set $x' = \Phi(x)$ with x in $i_1(N)$, then the automorphisms of N from E_2 take the form $i_1^{-1}(x) \mapsto i_1^{-1}(x^u)$. Thus they match the automorphisms of N from E_1.

In the case of the factor sets, we have $\bar{u}\bar{v} = a(u, v)\overline{uv}$. Application of Φ gives $\widetilde{u}\widetilde{v} = \Phi(a(u, v))\widetilde{uv}$. Thus the factor set for E_2 relative to N is $\{i_2^{-1}\Phi(a(u, v))\}$. Since $i_2^{-1}\Phi = i_1^{-1}$, this matches the factor set for E_1 relative to N.

We turn to the converse part. Suppose that the multiplication law in E_1 is $(i_1(x)\bar{u})(i_1(y)\bar{v}) = i_1(x)i_1(y)^u i_1(a(u, v))\overline{uv}$ for x and y in N, and that the multiplication law in E_2 is $(i_2(x)\widetilde{u})(i_2(y)\widetilde{v}) = i_2(x)i_2(y)^{u^*} i_2(b(u, v))\widetilde{uv}$. Here $\bar{u}$ and $\bar{v}$ are preimages of u and v under φ_1, and $\widetilde{u}$ and $\widetilde{v}$ are preimages of u and v under φ_2. Define automorphisms of N by $x^u = i_1^{-1}(i_1(x)^u)$ and $x^{u^*} = i_2^{-1}(i_2(x)^{u^*})$. We can then rewrite the multiplication laws as

$$(i_1(x)\bar{u})(i_1(y)\bar{v}) = i_1(xy^u a(u, v))\overline{uv}$$
and
$$(i_2(x)\widetilde{u})(i_2(y)\widetilde{v}) = i_2(xy^{u^*} b(u, v))\widetilde{uv}.$$

The assumption that E_1 is equivalent to E_2 as an extension of N by G means that there exists a function $\alpha : G \to N$ such that

$$x^{u^*} = \alpha(u)x^u\alpha(u)^{-1} \quad \text{and} \quad b(u, v) = \alpha(u)\alpha(v)^u a(u, v)\alpha(uv)^{-1}$$

for all u and v in G. Define $\Phi : E_1 \to E_2$ by

$$\Phi(i_1(x)\bar{u}) = i_2(x\alpha(u)^{-1})\widetilde{u}.$$

Certainly Φ is one-one onto. It remains to check that Φ is a group homomorphism and that the squares commute in Figure 7.4.

To check that $\Phi : E_1 \to E_2$ is a group homomorphism, we compare

$$\Phi(i_1(x)\bar{u}i_1(y)\bar{v}) = \Phi(i_1(xy^u a(u, v))\overline{uv}) = i_2(xy^u a(u, v)\alpha(uv)^{-1})\widetilde{uv}$$

with the product

$$\Phi(i_1(x)\bar{u})\Phi(i_1(y)\bar{v}) = i_2(x\alpha(u)^{-1})\widetilde{u}i_2(y\alpha(v)^{-1})\widetilde{v}$$
$$= i_2(x\alpha(u)^{-1}(y\alpha(v)^{-1})^{u^*}b(u,v))\widetilde{uv}.$$

Since

$$\alpha(u)^{-1}(y\alpha(v)^{-1})^{u^*}b(u,v) = \alpha(u)^{-1}(y\alpha(v)^{-1})^{u^*}\alpha(u)\alpha(v)^u a(u,v)\alpha(uv)^{-1}$$
$$= (y\alpha(v)^{-1})^u \alpha(v)^u a(u,v)\alpha(uv)^{-1}$$
$$= y^u a(u,v)\alpha(uv)^{-1},$$

these expressions are equal, and Φ is a group homomorphism. Thus Φ is a group isomorphism.

Now we check the commutativity of the squares. The computation

$$\varphi_2 \Phi(i_1(x)\bar{u}) = \varphi_2(i_2(x\alpha(u)^{-1})\widetilde{u}) = u = \varphi_1(i_1(x)\bar{u})$$

shows that the right-hand square commutes.

For the left-hand square we use the fact recorded in the statement of Proposition 7.36 that $i_1(a(1,1)^{-1})\widetilde{1}$ is the identity of E_1 and $i_2(b(1,1)^{-1})\widetilde{1}$ is the identity of E_2. Therefore $\Phi i_1(x) = \Phi(i_1(xa(1,1)^{-1})\widetilde{1}) = i_2(xa(1,1)^{-1}\alpha(1)^{-1})\widetilde{1}$. Since $i_2(x) = xb(1,1)^{-1}\widetilde{1}$, the left-hand square commutes if $b(1,1) = \alpha(1)a(1,1)$. This formula follows from (∗) in the proof of Proposition 7.36 by the computation

$$b(1,1) = \alpha(1)\alpha(1)^1 a(1,1)\alpha(1)^{-1} = \alpha(1)a(1,1)\alpha(1)\alpha(1)^{-1} = \alpha(1)a(1,1),$$

and thus the left-hand square indeed commutes. □

For the remainder of this section, *let us assume that N is abelian*. In this case Proposition 7.36a reduces to the identity $(x^v)^u = x^{uv}$ for all u and v in G independently of the choice of representatives, just as it does in the example we studied with $N = C_4$ and $G = C_2$. In the terminology of Section IV.7, G acts on N by automorphisms.[5] Suppose we fix such an action $\tau : G \to \text{Aut } N$ by automorphisms and consider all extensions of N by G built from τ. In our example we are thus to consider E equal to $C_4 \times C_2$ or C_8, which are built with the trivial τ, or else E equal to D_4 or H_8, which are built with the nontrivial τ (in which the nontrivial element of G acts by the nontrivial automorphism of N).

Since N is abelian, let us switch to additive notation for N and to ordinary function notation for $\tau(w)$, rewriting the formula of Proposition 7.36b as

$$\tau(u)a(v,w) + a(u,vw) = a(u,v) + a(uv,w).$$

---

[5] The formula $(x^v)^u = x^{uv}$ correctly corresponds to a group action with the group on the left as in Section IV.7.

This condition is preserved under addition of factor sets as long as τ does not change, it is satisfied by the 0 factor set, and the negative of a factor set is again a factor set. Therefore the factor sets for this τ form an abelian group.

Two factor sets for this τ are equivalent (in the sense of yielding equivalent group extensions) if and only if their difference is equivalent to 0, and $a(u, v)$ is equivalent to 0 if and only if

$$a(u, v) = \alpha(uv) - \alpha(u) - \tau(u)\alpha(v)$$

for some function $\alpha : G \to N$. The set of factor sets for this τ that are equivalent to 0 is thus a subgroup,[6] and we arrive at the following result.

Proposition 7.38. Let G and N be groups with N abelian, and suppose that $\tau : G \to \operatorname{Aut} N$ is a homomorphism. Then the set of equivalence classes of group extensions of N by G corresponding to the action $\tau : G \to \operatorname{Aut} N$ is parametrized by the quotient of the abelian group of factor sets by the subgroup of factor sets equivalent to 0.

The extension E corresponding to the 0 factor set is of special interest. In this case the multiplication law for the coset representatives is $\bar{u}\bar{v} = \overline{uv}$ since the member $a(u, v) = 0$ of N is to be interpreted multiplicatively in this product formula. Consequently the map $u \mapsto \bar{u}$ of G into E is a group homomorphism, necessarily one-one, and we can regard G as a subgroup of E. Proposition 4.44 allows us to conclude that E is the semidirect product $G \times_\tau N$. The multiplication law for general elements of E, with multiplicative notation used for N, is

$$(x\bar{u})(y\bar{v}) = x(\tau(u)y)\overline{uv}.$$

It is possible also to describe explicitly the extension one obtains from the sum of two factor sets corresponding to the same τ, but we leave this matter to Problems 20–23 at the end of the chapter. The operation on extensions that corresponds to addition of factor sets in this way is called **Baer multiplication**. What we saw in the previous paragraph says that the group identity under Baer multiplication is the semidirect product.

The two conditions, the compatibility condition on a factor set given in Proposition 7.36b and the condition with α in it for equivalence to 0, are of a combinatorial type that occurs in many contexts in mathematics and is captured by the ideas of "homology" and "cohomology." For the current situation the notion is that of **cohomology of groups**, and we shall define it now. The subject of homological

---

[6]One can legitimately ask whether an arbitrary $\alpha : G \to N$ leads to a factor set under the definition $a(u, v) = \alpha(uv) - \tau(v)\alpha(u) - \alpha(v)$, and one easily checks that the answer is yes. Alternatively, one can refer to the case $n = 2$ in the upcoming Proposition 7.39.

algebra, which is introduced in *Advanced Algebra*, puts cohomology of groups in a wider context and explains some of its mystery.

We fix an abelian group N, a group G, and a group action τ of G on N by automorphisms. It is customary to suppress τ in the notation for the group action, and we shall follow that convention. For integers $n \geq 0$, one begins with the abelian group $C^n(G, N)$ of n-**cochains** of G with coefficients in N. This is defined by

$$C^n(G, N) = \begin{cases} N & \text{if } n = 0, \\ \{f : \prod_{k=1}^{n} G \to M\} & \text{if } n > 0. \end{cases}$$

In words, $C^n(G, N)$ is the set of all functions into M from the n-fold direct product of G with itself. The **coboundary map** $\delta_n : C^n(G, N) \to C^{n+1}(G, N)$ is the homomorphism of abelian groups defined by

$$(\delta_0 f)(g_1) = g_1 f - f$$

and by

$$(\delta_n f)(g_1, \ldots, g_{n+1}) = g_1(f(g_2, \ldots, g_{n+1}))$$
$$+ \sum_{i=1}^{n} (-1)^i f(g_1, \ldots, g_{i-1}, g_i g_{i+1}, g_{i+2}, \ldots, g_{n+1})$$
$$+ (-1)^{n+1} f(g_1, \ldots, g_n)$$

for $n > 0$. We postpone to the end of this section the proof of the following result.

Proposition 7.39. $\delta_n \delta_{n-1} = 0$ for all $n \geq 1$.

It follows from Proposition 7.39 that image $\delta_{n-1} \subseteq \ker \delta_n$ for all $n \geq 1$. Thus if we define abelian groups by

$$Z^n(G, N) = \ker \delta_n,$$

$$B^n(G, N) = \begin{cases} 0 & \text{for } n = 0, \\ \text{image } \delta_{n-1} & \text{for } n > 0, \end{cases}$$

then $B^n(G, N) \subseteq Z^n(G, N)$ for all n, and it makes sense to define the abelian groups

$$H^n(G, N) = Z^n(G, N) / B^n(G, N) \quad \text{for } n \geq 0.$$

The elements of $Z^n(G, N)$ are called n-**cocycles**, the elements of $B^n(G, N)$ are called n-**coboundaries**, and $H^n(G, N)$ is called the nth **cohomology group** of G with coefficients in N.

EXAMPLES IN LOW DEGREE.

DEGREE 0. Here $(\delta_0 f)(u) = uf - f$ with f in N and u in G. The cocycle condition is that this is 0 for all u. Thus f is to be fixed by G. We say that an f

fixed by G is an **invariant** of the group action. The space of invariants is denoted by N^G. By convention above, we are taking $B^0(G, N) = 0$. Thus
$$H^0(G, N) = N^G.$$

DEGREE 1. Here $(\delta_1 f)(u, v) = u(f(v)) - f(uv) + f(u)$ with f a function from G to N. The cocycle condition is that
$$f(uv) = f(u) + u(f(v)) \qquad \text{for all } u, v \in G.$$
A function f satisfying this condition is called a **crossed homomorphism** of G into N. A coboundary is a function $f : G \to N$ of the form $f(u) = (\delta_0 x)(u) = ux - x$ for some $x \in N$. Then $H^1(G, N)$ is the quotient of the group of crossed homomorphisms by this subgroup. In the special case that the action of G on N is trivial, the crossed homomorphisms reduce to ordinary homomorphisms of G into N, and every coboundary is 0. Thus $H^1(G, N)$ is the group of homomorphisms of G into N if G acts trivially on N.

DEGREE 2. Here f is a function from $G \times G$ into N, and
$$(\delta_2 f)(u, v, w) = u(f(v, w)) - f(uv, w) + f(u, vw) - f(u, v).$$
The cocycle condition is that
$$u(f(v, w)) + f(u, vw) = f(uv, w) + f(u, v) \qquad \text{for all } u, v, w \in G.$$
This is the same as the condition that $\{f(u, v)\}$ be a factor set for extensions of N by G relative to the given action of G on N by automorphisms. A coboundary is a function $f : G \times G \to N$ of the form
$$f(u, v) = (\delta_0 \alpha)(u, v) = u(\alpha(v)) - \alpha(uv) + \alpha(u) \qquad \text{for some } \alpha : G \to N.$$
This is the same as the condition that $\{-f(u, v)\}$ be a factor set equivalent to 0. Thus we can restate Proposition 7.38 as follows.

Proposition 7.40. Let G and N be groups with N abelian, and suppose that $\tau : G \to \operatorname{Aut} N$ is a homomorphism. Then the set of equivalence classes of group extensions of N by G corresponding to the action $\tau : G \to \operatorname{Aut} N$ is parametrized by $H^2(G, N)$.

Since group extensions have such a nice interpretation in terms of cohomology groups H^2, it is reasonable to look for a nice interpretation for H^1 as well. Indeed, H^1 has an interpretation in terms of uniqueness up to inner isomorphisms for semidirect-product decompositions. We continue with the abelian group N, a group G, and a group action τ of G on N by automorphisms. A semidirect product $E = G \times_\tau N$ is an allowable extension. Since G embeds as a subgroup of E, we are given a one-one group homomorphism $u \mapsto \bar{u}$ of G into E. The construction at the beginning of this section works with the set $\bar{u}$ of coset representatives, and they have $\bar{u}\bar{v} = \overline{uv}$.

6. Extensions of Groups

Suppose that the semidirect product can be formed by a second one-one group homomorphism $u \mapsto \widetilde{u}$ of G into E. If we write $\widetilde{u} = \alpha(u)\bar{u}$ for a function $\alpha : G \to N$, then we know from earlier in the section that the extensions formed from $\{\bar{u}\}$ and from $\{\widetilde{u}\}$ are equivalent. Because G maps homomorphically into E for both systems, the factor sets are 0 in both cases. Consequently the function α must satisfy

$$\alpha(uv) - \alpha(u) - \tau(u)\alpha(v) = 0.$$

This is exactly the condition that $\alpha : G \to N$ be a 1-cocyle. Thus the group $Z^1(G, N)$ parametrizes all ways that we can embed G as a complementary subgroup to N in the semidirect product $E = G \times_\tau N$.

A relatively trivial way to construct a one-one group homomorphism $u \mapsto \widetilde{u}$ from $u \mapsto \bar{u}$ is to form, in the usual multiplicative notation, $\widetilde{u} = x_0^{-1}\bar{u}x_0$ for some $x_0 \in N$. Then $\widetilde{u} = x_0^{-1}\bar{u}x_0\bar{1} = x_0^{-1}(\tau(u)(x_0))\bar{u}$, and the additive notation for $\alpha(u)$ has $\alpha(u) = \tau(u)(x_0) - x_0$. Referring to our earlier computations in degree 1, we see that α is in the group $B^1(G, N)$ of coboundaries.

The conclusion is that $H^1(G, N)$ parametrizes all ways, modulo relatively trivial ways, that we can embed G as a complementary subgroup to N in the semidirect product $E = G \times_\tau N$.

As promised, we now return to the proof of Proposition 7.39.

PROOF OF PROPOSITION 7.39. For $n = 1$, we have

$$(\delta_1\delta_0 f)(u, v) = u((\delta_0 f)(v)) - (\delta_0 f)(uv) + (\delta_0 f)(u)$$
$$= u(vf - f) - (uvf - f) + (uf - f) = 0.$$

For $n > 1$, we begin with

$$(\delta_n\delta_{n-1} f)(g_1, \ldots, g_{n+1}) = g_1((\delta_{n-1} f)(g_2, \ldots, g_{n+1}))$$
$$+ \sum_{i=1}^{n}(-1)^i(\delta_{n-1} f)(g_1, \ldots, g_ig_{i+1}, \ldots, g_{n+1})$$
$$+ (-1)^{n+1}(\delta_{n-1} f)(g_1, \ldots, g_n)$$
$$= \mathrm{I} + \mathrm{II} + \mathrm{III}.$$

Here

$$\mathrm{I} = g_1 g_2(f(g_3, \ldots, g_{n+1})) + \sum_{i=2}^{n}(-1)^{i-1}g_1(f(g_2, \ldots, g_ig_{i+1}, \ldots, g_{n+1}))$$
$$+ (-1)^n g_1(f(g_2, \ldots, g_n)) = \mathrm{IA} + \mathrm{IB} + \mathrm{IC},$$

$$\mathrm{II} = -(\delta_{n-1} f)(g_1g_2, g_3, \ldots, g_n) + \sum_{i=2}^{n}(-1)^i(\delta_{n-1} f)(g_1, \ldots, g_ig_{i+1}, \ldots, g_{n+1})$$
$$= \mathrm{IIA} + \mathrm{IIB},$$

$$\text{III} = (-1)^{n+1} g_1(f(g_2, \ldots, g_n)) + (-1)^{n+1}(-1) f(g_1 g_2, g_3, \ldots, g_n)$$
$$+ (-1)^{n+1} \sum_{i=2}^{n-1} (-1)^i f(g_1, \ldots, g_i g_{i+1}, \ldots, g_n)$$
$$+ (-1)^{n+1}(-1)^n f(g_1, \ldots, g_{n-1})$$
$$= \text{IIIA} + \text{IIIB} + \text{IIIC} + \text{IIID}.$$

Terms IIA and IIB decompose further as

$$\text{IIA} = -g_1 g_2(f(g_3, \ldots, g_{n+1})) + f(g_1 g_2 g_3, g_4, \ldots, g_{n+1})$$
$$- \sum_{i=3}^{n} (-1)^{i+1} f(g_1 g_2, \ldots, g_i g_{i+1}, \ldots, g_{n+1}) - (-1)^n f(g_1 g_2, g_3, \ldots, g_n)$$
$$= \text{IIAa} + \text{IIAb} + \text{IIAc} + \text{IIAd},$$

$$\text{IIB} = \sum_{i=2}^{n} (-1)^i g_1(f(g_2, \ldots, g_i g_{i+1}, \ldots, g_{n+1}))$$
$$+ (-1)^2(-1) f(g_1 g_2 g_3, g_4, \ldots, g_{n+1})$$
$$+ \sum_{i=3}^{n} (-1)^i (-1) f(g_1 g_2, \ldots, g_i g_{i+1}, \ldots, g_{n+1})$$
$$+ \sum_{i=2}^{n} (-1)^i \sum_{j=2}^{i-2} (-1)^j f(g_1, \ldots, g_j g_{j+1}, \ldots, g_i g_{i+1}, \ldots, g_{n+1})$$
$$+ \sum_{i=3}^{n} (-1)^i (-1)^{i-1} f(g_1, \ldots, g_{i-1} g_i g_{i+1}, \ldots, g_{n+1})$$
$$+ \sum_{i=2}^{n-1} (-1)^i (-1)^i f(g_1, \ldots, g_i g_{i+1} g_{i+2}, \ldots, g_{n+1})$$
$$+ \sum_{i=2}^{n-2} (-1)^i \sum_{j=i+2}^{n} (-1)^{j-1} f(g_1, \ldots, g_i g_{i+1}, \ldots, g_j g_{j+1}, \ldots, g_{n+1})$$
$$+ \sum_{i=2}^{n-1} (-1)^i (-1)^n f(g_1, \ldots, g_i g_{i+1}, \ldots, g_n)$$
$$+ (-1)^n (-1)^n f(g_1, \ldots, g_{n-1})$$
$$= \text{IIBa} + \text{IIBb} + \text{IIBc} + \text{IIBd} + \text{IIBe} + \text{IIBf} + \text{IIBg} + \text{IIBh} + \text{IIBi}.$$

Inspection shows that we have cancellation between term IA and term IIAa, term IB and term IIBa, term IC and term IIIA, term IIAb and term IIBb, term IIAc and term IIBc, term IIAd and term IIIB, term IIBd and term IIBg, term IIBe and term IIBf, term IIBh and term IIIC, and term IIBi and term IIID. All the terms cancel, and we conclude that $\delta_n \delta_{n-1} f = 0$. $\square$

7. Problems

1. Using Burnside's Theorem and Problem 34 at the end of Chapter IV, show that 60 is the smallest possible order of a nonabelian simple group.

2. A **commutator** in a group is any element of the form $xyx^{-1}y^{-1}$.
   (a) Prove that the inverse of a commutator is a commutator.
   (b) Prove that any conjugate of a commutator is a commutator.

3. Let a and b be elements of a group G. Prove that the subgroup generated by a and b is the same as the subgroup generated by bab^2 and bab^3.

4. A subgroup H of a group G is said to be **characteristic** if it is carried into itself by every automorphism of G.
   (a) Prove that characteristic implies normal.
   (b) Prove that the center Z_G of G is a characteristic subgroup.
   (c) Prove that the commutator subgroup G' of G is a characteristic subgroup.

5. In the terminology of the previous problem, which subgroups of the quaternion subgroup H_8 are characteristic?

6. Is every finite group finitely presented? Why or why not?

7. Let $G = \mathrm{SL}(2, \mathbb{R})$, and let G' be the commutator subgroup.
   (a) Prove that every element $\begin{pmatrix} 1 & t \\ 0 & 1 \end{pmatrix}$ is in G'.
   (b) Prove that $G' = G$.
   (c) Prove that $\begin{pmatrix} -1 & 0 \\ 0 & -1 \end{pmatrix}$ is not a commutator even though it is in G'.

8. Problem 53 at the end of Chapter IV produced a group G of order 27 generated by two elements a and b satisfying $a^9 = b^3 = b^{-1}aba^{-4} = 1$. Prove that G is given by generators and relations as
$$G = \langle a, b;\ a^9, b^3, b^{-1}aba^{-4} \rangle.$$

9. Let G_n be given by generators and a single relation as
$$G_n = \langle x_1, y_1, \ldots, x_n, y_n;\ x_1 y_1 x_1^{-1} y_1^{-1} \cdots x_n y_n x_n^{-1} y_n^{-1} \rangle.$$

    Prove that G_n/G'_n is free abelian of rank $2n$, and conclude that the groups G_n are mutually nonisomorphic as n varies. (Educational note related to topology: The group G_n may be shown to be the fundamental group of a compact orientable 2-dimensional manifold without boundary and with n handles.)

10. Prove that a free group of finite rank n cannot be generated by fewer than n elements.

11. Let F be the free group on generators a, b, c, and let H be the subgroup generated by all words of length 2.
    (a) Find coset representatives g such that G is the disjoint union of the cosets Hg.
    (b) Find a free basis of H.

12. For the free group on generators x and y, prove that the elements y, xyx^{-1}, $x^2yx^{-2}, x^3yx^{-3}, \ldots$, constitute a free basis of the subgroup that they generate. Conclude that a free group of rank 2 has a free subgroup of infinite rank.

13. Let $G = C_2 * C_2$. Prove that the only quotient groups of G, up to isomorphism, are G itself, $\{1\}, C_2, C_2 \times C_2$, and the dihedral groups D_n for $n \geq 3$.

14. Prove that if every irreducible finite-dimensional representation of a finite group G is 1-dimensional, then G is abelian.

15. Let G be a finitely generated group, and let H be a subgroup of finite index. Prove that H is finitely generated.

16. Let N be an abelian group, let G be a group, let τ be an action of G on N by automorphisms, and let $n > 0$ be an integer.
    (a) Prove that if every element of N has finite order dividing an integer m, then every member of $H^n(G, N)$ has finite order dividing m.
    (b) Suppose that G is finite and that f is an n-cocycle. Define an $(n-1)$-cochain F by
    $$F(g_1, \ldots, g_{n-1}) = \sum_{g \in G} f(g_1, \ldots, g_{n-1}, g).$$
    By summing the cocycle condition for f over the last variable, express $|G|f(g_1, \ldots, g_n)$ in terms of F, and deduce that $|G|f$ is a coboundary. Conclude that every member of $H^n(G, N)$ has order dividing $|G|$.

17. Let G be a finite group. Suppose that G has a normal abelian subgroup N, and suppose that $\text{GCD}(|N|, |G/N|) = 1$. Prove that there exists a subgroup H of G such that G is the semidirect product of H and N.

18. Let N be the cyclic group C_2, and let G be an arbitrary group of order 4. Identify up to equivalence all group extensions of N by G.

19. Let $N = C_2$, and let $E = \bigoplus_{n=1}^{\infty} (C_2 \oplus C_4)$. Regard E as an extension of N in two ways—first by embedding N as one of the summands C_2 of E and then by embedding N as a subgroup of one of the summands C_4 of E. Show that the quotient groups E/N in the two cases are isomorphic, that E/N acts trivially on N in both cases, and that the two group extensions are not equivalent.

Problems 20–23 concern **Baer multiplication** of extensions. Let N be an abelian group, let G be a group, let τ be an action of G on N by automorphisms, and let E_1 and E_2 be two extensions of N by G relative to τ. Write $\varphi_1 : E_1 \to G$ and $\varphi_2 : E_2 \to G$ for the quotient mappings. Let (E, E') denote the subgroup of all members (e_1, e_2) of $E_1 \times E_2$ for which $\varphi_1(e_1) = \varphi_2(e_2)$. Writing the operation in N multiplicatively, let $Q = \{(x, x^{-1}) \in E_1 \times E_2 \mid x \in N\}$. The Baer product of E_1 and E_2 is defined to be the quotient $(E_1, E_2)/Q$. A typical coset of the Baer product will be denoted by $(e_1, e_2)Q$.

20. Prove that the homomorphism $x \mapsto (x, 1)Q$ is one-one from N into $(E_1, E_2)/Q$, that the homomorphism $\varphi : (E_1, E_2) \to G$ defined by $\varphi(e_1, e_2) = \varphi_1(e_1)$ has image G and descends to the quotient $(E_1, E_2)/Q$, and that the kernel of the descended φ is the embedded copy of N. (Therefore $(E_1, E_2)/Q$ is an extension of N by G, evidently relative to τ.)

21. For each $u \in G$, select $\bar{u} \in E_1$ and $\tilde{u} \in E_2$ with $\varphi_1(\bar{u}) = u = \varphi_2(\tilde{u})$, and define $a(u, v)$ and $b(u, v)$ for u and v in G by $(x\bar{u})(y\bar{v}) = a(u, v)\overline{uv}$ and $(x\tilde{u})(y\tilde{v}) = b(u, v)\widetilde{b(u, v)}$. Show that $(\bar{u}, \tilde{u})Q$ has $\varphi((\bar{u}, \tilde{u})Q) = u$ and that the associated 2-cocycle for $(E_1, E_2)/Q$ is $a(u, v)b(u, v)$ if the group operation in N is written multiplicatively.

22. Prove that Baer multiplication descends to a well-defined multiplication of equivalence classes of extensions of N by G relative to τ, in the following sense: Suppose that E_1 and E_1' are equivalent extensions and that E_2 and E_2' are equivalent extensions. Let $(E_1, E_2)/Q$ and $(E_1', E_2')/Q'$ be the Baer products. Then $(E_1, E_2)/Q$ is equivalent to $(E_1', E_2')/Q'$. Conclude that if Baer multiplication is imposed on equivalence classes of extensions of N by G relative to τ, then the correspondence stated in Proposition 7.40 of equivalence classes to members of $H^2(G, N)$ is a group isomorphism.

Problems 23–24 derive the Poisson summation formula for finite abelian groups. If G is a finite abelian group and $\widehat{G}$ is its group of multiplicative characters, then the **Fourier coefficient** at $\chi \in \widehat{G}$ of a function f in $C(G, \mathbb{C})$ is $\widehat{f}(\chi) = \sum_{g \in G} f(g)\overline{\chi(g)}$. The Fourier inversion formula in Theorem 7.17 says that $f(g) = |G|^{-1} \sum_{\chi \in \widehat{G}} \widehat{f}(\chi)\chi(g)$.

23. Let G be a finite abelian group, let H be a subgroup, and let G/H be the quotient group. If t is in G, write $\dot{t}$ for the coset of t in G/H. Let f be in $C(G, \mathbb{C})$ and define $F(\dot{t}) = \sum_{h \in H} f(t + h)$ as a function on G/H. Suppose that χ is a member of $\widehat{G}$ that is identically 1 on H, so that χ descends to a member $\dot{\chi}$ of $\widehat{G/H}$. Prove that $\widehat{f}(\chi) = \widehat{F}(\dot{\chi})$.

24. **(Poisson summation formula)** With f and F as in the previous problem, apply the Fourier inversion formula for G/H to the function F, and derive the formula

$$\sum_{h \in H} f(t+h) = \frac{1}{|G/H|} \sum_{\omega \in \widehat{G},\ \omega|_H = 1} \widehat{f}(\omega)\omega(t).$$

(Educational note: This formula is often applied with $t = 0$, in which case it reduces to $\sum_{h \in H} f(h) = \frac{1}{|G/H|} \sum_{\omega \in \widehat{G},\ \omega|_H = 1} \widehat{f}(\omega)$.)

Problems 25–28 continue the introduction to error-correcting codes begun in Problems 63–73 at the end of Chapter IV, combining those results with the Poisson summation formula in the problems above and with notions from Section VI.1. Let $\mathbb{F}$ be the field $\mathbb{Z}/2\mathbb{Z}$, and form the Hamming space $\mathbb{F}^n$. Define a nondegenerate bilinear form on $\mathbb{F}^n$ by $(a, c) = \sum_{i=1}^{n} a_i c_i$ for a and c in $\mathbb{F}^n$. Recall from Chapter IV that a linear code C is a vector subspace of $\mathbb{F}^n$. For such a C, let $C^\perp$ as in Section VI.1 be the set of all $a \in \mathbb{F}^n$ such that $(a, c) = 0$ for all $c \in C$; the linear code $C^\perp$ is called the **dual code**. A linear code is **self dual** if $C^\perp = C$.

25. (a) Show that the codes 0 and $\mathbb{F}^n$ are dual to each other.
    (b) Show that the repetition code and the parity-check code are dual to each other.
    (c) Show that the Hamming code of order 8 is self dual.
    (d) Show that any self-dual linear code C has $\dim C = n/2$, and conclude that the Hamming code of order 2^r with $r > 3$ is not self dual.
    (e) Show that any member c of a self-dual linear code C has even weight.
    (f) Show that if a linear code C has $C \subseteq C^\perp$ and if every member c of C has even weight, then $c \mapsto \frac{1}{2}\text{wt}(c) \bmod 2$ is a group homomorphism of C into $\mathbb{Z}/2\mathbb{Z}$. Here wt(c) denotes the weight of c.

26. Regard $\mathbb{F}^n$ as an additive group G to which the Fourier inversion formula of Section 4 can be applied.
    (a) Show that one can map $\widehat{G}$ to $\mathbb{F}^n$ by $\chi \mapsto a_\chi$ with $\chi(c) = (-1)^{(a_\chi, c)}$ and that the result is a group isomorphism. (Therefore if f is in $C(\mathbb{F}^n, \mathbb{C})$, we can henceforth regard $\widehat{f}$ as a function on $\mathbb{F}^n$.)
    (b) Show under the identification in (a) that if f is in $C(\mathbb{F}^n, \mathbb{C})$, then $\widehat{f}(a) = \sum_{c \in \mathbb{F}^n} f(c)(-1)^{(a,c)}$ for a in $\mathbb{F}^n$.
    (c) Suppose that the function $f \in C(\mathbb{F}^n, \mathbb{C})$ is of the special form $f(c) = \prod_{i=1}^{n} f_i(c_i)$ whenever $c = (c_1, \ldots, c_n)$. Here each f_i is a function on the 2-element group $\mathbb{F}$. Prove that $\widehat{f}(a) = \prod_{i=1}^{n} \widehat{f_i}(a_i)$ whenever $a = (a_1, \ldots, a_n)$. Here $\widehat{f_i}$ is given by the formula of (b) for the case $n = 1$: $\widehat{f_i}(a_i) = \sum_{c_i \in \mathbb{F}} f_i(c_i)(-1)^{a_i c_i}$.

27. Fix two complex numbers x and y. Define $f_0 : \mathbb{F} \to \mathbb{C}$ to be the function with $f_0(0) = x$ and $f_0(1) = y$. Define $f : \mathbb{F} \to \mathbb{C}$ to be the function with $f(c) = \prod_{i=1}^{n} f_0(c_i) = x^{n-\text{wt}(c)} y^{\text{wt}(c)}$ where wt(c) is the weight of c.
    (a) Show that $\widehat{f_0}(0) = x + y$ and $\widehat{f_0}(1) = x - y$.
    (b) Show that $\widehat{f}(a) = (x+y)^{n-\text{wt}(a)}(x-y)^{\text{wt}(a)}$.

28. Let C be a linear code in $\mathbb{F}^n$. Take G to be the additive group of $\mathbb{F}^n$ and H to be the additive group of C. Regard $C^\perp$ as an additive group also.
    (a) Map $\widehat{G/H}$ to $C^\perp$ by $\chi \mapsto a_\chi$ with $\chi(c) = (-1)^{(a_\chi, c)}$. Show that this mapping is a group isomorphism.

(b) Applying the Poisson summation formula of Problem 24, prove that
$$\sum_{h \in C} f(h) = \frac{1}{|C^\perp|} \sum_{a \in C^\perp} \widehat{f}(a)$$
for all f in $C(\mathbb{F}^n, \mathbb{C})$.

(c) **(MacWilliams identity)** Let $W_C(X, Y) = \sum_{k=0}^n N_k(C) X^{n-k} Y^k$, where $N_k(C)$ is the number of members of C with weight k, be the weight-enumerator polynomial of C, and let $W_{C^\perp}(X, Y)$ be defined similarly. By applying (b) to the function f in the previous problem, prove that $W_C(x, y) = |C^\perp|^{-1} W_{C^\perp}(x + y, x - y)$ for each x and y. Conclude from Corollary 4.32 that weight-enumerator polynomials satisfy $W_C(X, Y) = |C^\perp|^{-1} W_{C^\perp}(X + Y, X - Y)$.

(d) The polynomials $W_C(X, Y)$ were seen in Chapter IV to be X^n for the 0 code, $(X + Y)^n$ for the code $\mathbb{F}^n$, $X^n + Y^n$ for the repetition code, $\frac{1}{2}((X+Y)^n + (X-Y)^n)$ for the parity-check code, and $X^8 + 14X^4Y^4 + Y^8$ for the Hamming code of order 8. Using relationships established in Problem 25, verify the result of (c) for each of these codes.

(e) Suppose that C is a self-dual linear code. Applying (c) in this case, exhibit $W_C(X, Y)$ as being invariant under a copy of the dihedral group D_8 of order 16. (Educational note: If the polynomial $W_C(X, Y)$ is invariant also under $X \mapsto iX$, as is true for the Hamming code of order 8, then $W_C(X, Y)$ is invariant under the group generated by D_8 and this transformation, which can be shown to have order 192.)

Problems 29–31 concern an unexpectedly fast method of computation of Fourier coefficients in the context of finite abelian groups, particularly in the context of cyclic groups. They show for a cyclic group of order $m = pq$ that the use of the idea behind the Poisson summation formula of Problem 24 makes it possible to compute the Fourier coefficients of a function in about $pq(p+q)$ steps rather than the expected $m^2 = p^2 q^2$ steps. This savings may be iterated in the case of a cyclic group of order 2^n so that the Fourier coefficients are computed in about $n2^n$ steps rather than the expected 2^{2n} steps. An organized algorithm to implement this method of computation is known as the **fast Fourier transform**. Write the cyclic group C_m as the set $\{0, 1, 2, \ldots, m-1\}$ of integers modulo m under addition, and let $\zeta_m = e^{2\pi i/m}$. For k in C_m define a multiplicative character χ_n of C_m by $\chi_n(k) = (\zeta_m^n)^k$. The resulting m multiplicative characters satisfy $\chi_n \chi_{n'} = \chi_{n+n'}$, and they exhaust $\widehat{C_m}$ since distinct multiplicative characters are orthogonal. It will be convenient to identify χ_n with $\chi_n(1) = \zeta_m^n$.

29. In the setting of Problem 23, suppose that $G = C_m$ with $m = pq$; here p and q need not be relatively prime. Let $H = \{0, q, 2q, \ldots, (p-1)q\}$ be the subgroup of G isomorphic to C_p, so that $G/H = \{0, 1, 2, \ldots, q-1\}$ is isomorphic to C_q. Prove that the characters χ of G identified with $\zeta_m^0, \zeta_m^p, \zeta_m^{2p}, \ldots, \zeta_m^{(q-1)p}$

are the ones that are identically 1 on H and therefore descend to characters of G/H. Verify that the descended characters $\dot{\chi}$ are the ones identified with $\zeta_q^0, \zeta_q^1, \zeta_q^2, \ldots, \zeta_q^{q-1}$. Consequently the formula $\widehat{f}(\chi) = \widehat{F}(\dot{\chi})$ of Problem 23 provides a way of computing $\widehat{f}$ at $\zeta_m^0, \zeta_m^p, \zeta_m^{2p}, \ldots, \zeta_m^{(q-1)p}$ from the values of $\widehat{F}$. Show that if $\widehat{F}$ is computed from the definition of Fourier coefficients, then the number of steps involved in its computation is about q^2, apart from a constant factor. Show therefore that the total number of steps in computing $\widehat{f}$ at these special values of χ is therefore on the order of $q^2 + pq$.

30. In the previous problem show for each k with $0 \leq k \leq p-1$ that the value of $\widehat{f}$ at $\zeta_m^k, \zeta_m^{p+k}, \zeta_m^{2p+k}, \ldots, \zeta_m^{(q-1)p+k}$ can be handled in the same way with a different F by replacing f by a suitable variant of f. Doing so for each k requires p times the number of steps detected in the previous problem, and therefore all of $\widehat{f}$ can be computed in about $p(q^2 + pq) = pq(p+q)$ steps.

31. Show how iteration of this process to compute the Fourier coefficients of each F, together with further iteration of this process, allows one to compute the Fourier coefficients for a function on $C_{m_1 m_2 \cdots m_r}$ in about $m_1 m_2 \cdots m_r (m_1 + m_2 + \cdots + m_r)$ steps.

Problems 32–36 concern contragredient representations and the decomposition of the left regular representation of a finite group G. They make use of Problems 24–28 in Chapter III, which introduce the complex conjugate $\overline{V}$ of a complex vector space V. In the case that V is an inner-product space, those problems define $(u, v)_{\overline{V}} = (v, u)_V$, and they show that if $\ell_v \in V'$ is given by $\ell_v(u) = (u, v)_V = (v, u)_{\overline{V}}$, then the mapping $\ell_v \leftrightarrow v$ is an isomorphism of V' with $\overline{V}$.

32. Show that the definition $(\ell_{v_1}, \ell_{v_2})_{V'} = (v_1, v_2)_{\overline{V}}$ makes the isomorphism of V' with $\overline{V}$ preserve inner products.

33. If R is a unitary representation of G on the finite-dimensional complex vector space V, define the **contragredient** representation R^c of G on V' by $R^c(x) = R(x^{-1})^t$. Prove that $R^c(x)\ell_v = \ell_{R(x)v}$ and that R^c is unitary on V'.

34. Show that the matrix coefficients of R^c are the complex conjugates of those of R and that the characters satisfy $\chi_{R^c} = \overline{\chi_R}$.

35. Give an example of an irreducible representation of a finite group G that is not equivalent to its contragredient.

36. Let ℓ be the left regular representation of G on $C(G, \mathbb{C})$, and let V_R be the linear span in $C(G, \mathbb{C})$ of the matrix coefficients of an irreducible representation R of dimension d. Prove that the representation (ℓ, V_R) of G is equivalent to the direct sum of d copies of the contragredient R^c.

Problems 37–46 concern the free product $C_2 * C_3$ and its quotients. The problems make use of the group of matrices $\mathrm{SL}(2, \mathbb{Z}/m\mathbb{Z})$ of determinant 1 over the commutative ring $\mathbb{Z}/m\mathbb{Z}$, as discussed in Section V.2. One of the quotients of $C_2 * C_3$

will be PSL(2, $\mathbb{Z}$) = SL(2, $\mathbb{Z}$)/{scalar matrices}, and these problems show that the quotient mapping can be arranged to be an isomorphism. Other quotients will be the groups $G_m = \langle X, Y;\ X^2, Y^3, (XY)^m \rangle$ with $m \geq 2$. These arise in connection with tilings in 2-dimensional geometry. The isomorphism $C_2 * C_3 \cong \text{PSL}(2, \mathbb{Z})$ leads to a homomorphism that will be called σ_m carrying G_m onto PSL(2, $\mathbb{Z}/m\mathbb{Z}$) = SL(2, $\mathbb{Z}/m\mathbb{Z}$)/{scalar matrices}, the image group being finite. The problems show that the homomorphism $\sigma_m : G_m \to \text{PSL}(2, \mathbb{Z}/m\mathbb{Z})$ is an isomorphism for the cases in which G_m arises from spherical geometry, namely for $2 \leq m \leq 5$, and that the homomorphism is not an isomorphism for $m = 6$, the case in which G_m arises from Euclidean geometry.

37. Show that the elements $\begin{pmatrix} 0 & -1 \\ 1 & 0 \end{pmatrix}$ and $\begin{pmatrix} 0 & 1 \\ -1 & -1 \end{pmatrix}$ generate SL(2, $\mathbb{Z}$) by arguing as follows: if the subgroup Γ of SL(2, $\mathbb{Z}$) generated by these two elements is not SL(2, $\mathbb{Z}$), choose an element $\begin{pmatrix} a & b \\ c & d \end{pmatrix}$ outside Γ having max($|a|, |b|$) as small as possible, and derive a contradiction by showing that a suitable right multiple of it by elements of Γ is in Γ.

38. By mapping $X \mapsto x = \begin{pmatrix} 0 & -1 \\ 1 & 0 \end{pmatrix}$ mod $\pm I$ and $Y \mapsto y = \begin{pmatrix} 0 & 1 \\ -1 & -1 \end{pmatrix}$ mod $\pm I$, produce a group homomorphism Φ of $C_2 * C_3 = \langle X, Y;\ X^2, Y^3 \rangle$ onto PSL(2, $\mathbb{Z}$).

39. Let x, y, and $\Phi : C_2 * C_3 \to \text{PSL}(2, \mathbb{Z})$ be as in the previous problem.
    (a) For any member $\begin{pmatrix} a & b \\ c & d \end{pmatrix}$ mod $\pm I$ of PSL(2, $\mathbb{Z}$), define $\mu\left(\begin{pmatrix} a & b \\ c & d \end{pmatrix} \text{mod} \pm I\right)$ = max($|a|, |b|$) and $\nu\left(\begin{pmatrix} a & b \\ c & d \end{pmatrix} \text{mod} \pm I\right)$ = max($|c|, |d|$). Prove that if $z = \begin{pmatrix} a & b \\ c & d \end{pmatrix}$ mod $\pm I$ in PSL(2, $\mathbb{Z}$) has $ab \leq 0$, then $\mu(zyx) \geq \mu(z)$ and $\mu(zy^{-1}x) \geq \mu(z)$, while if $cd \leq 0$, then $\nu(zyx) \geq \nu(z)$ and $\nu(zy^{-1}x) \geq \nu(z)$.
    (b) Prove that $\mu(zx) = \mu(z)$ and $\nu(zx) = \nu(z)$ for all z in PSL(2, $\mathbb{Z}$).
    (c) Show that there are only 10 members z of PSL(2, $\mathbb{Z}$) for which the two conditions $\mu(z) = 1$ and $\nu(z) = 1$ both hold.
    (d) A reduced word in $C_2 * C_3$ is a finite sequence of factors X, Y, and Y^{-1}, with no two consecutive factors equal and with no two consecutive factors YY^{-1} or $Y^{-1}Y$. Prove for any reduced word $a_1 \cdots a_n$ in $C_2 * C_3$, where each a_j is one of X, Y, and Y^{-1}, that $\mu(\Phi(a_1 \cdots a_n)) \geq \mu(\Phi(a_1 \cdots a_{n-1}))$ and that $\nu(\Phi(a_1 \cdots a_n)) \geq \nu(\Phi(a_1 \cdots a_{n-1}))$.
    (e) Deduce that the homomorphism Φ is an isomorphism.

40. Let $\Gamma(m)$ be the group of all matrices M in SL(2, $\mathbb{Z}$) such that every entry of $M - I$ is divisible by m.
    (a) Prove that passage from a matrix in SL(2, $\mathbb{Z}$) to the same matrix with its entries considered modulo m gives a homomorphism $\widetilde{\sigma}_m : \text{SL}(2, \mathbb{Z}) \to \text{SL}(2, \mathbb{Z}/m\mathbb{Z})$ with ker $\widetilde{\sigma}_m = \Gamma(m)$.

(b) Prove that if α, β, and m are positive integers with $\mathrm{GCD}(\alpha, \beta, m) = 1$, then there exists an integer r such that $\mathrm{GCD}(\alpha + mr, \beta) = 1$. (One way of proceeding is to use Dirichlet's theorem on primes in arithmetic progressions.)

(c) Prove that image $\tilde{\sigma}_m = \mathrm{SL}(2, \mathbb{Z}/m\mathbb{Z})$, i.e., $\tilde{\sigma}_m$ is onto.

41. Let $\Phi_m : C_2 * C_3 \to G_m$ be the homomorphism defined by the conditions $X \mapsto X$ and $Y \mapsto Y$. Let H_m be the smallest normal subgroup of $\mathrm{PSL}(2, \mathbb{Z})$ containing $(xy)^m$ mod $\pm I$. Let $\tilde{\sigma}_m : \mathrm{SL}(2, \mathbb{Z}) \to \mathrm{SL}(2, \mathbb{Z}/m\mathbb{Z})$ be the homomorphism of the previous problem.

(a) Why is Φ_m well defined?

(b) Why is $H_m = \Phi(\ker \Phi_m)$?

(c) Define $\mathrm{PSL}(\mathbb{Z}/m\mathbb{Z}) = \mathrm{SL}(2, \mathbb{Z}/m\mathbb{Z})/\{\text{scalar matrices}\}$. Why does the composition of $\tilde{\sigma}_m$ followed by passage to the quotient descend to a homomorphism σ_m of $\mathrm{PSL}(2, \mathbb{Z})$ onto $\mathrm{PSL}(2, \mathbb{Z}/m\mathbb{Z})$?

(d) If $K \subseteq \mathrm{PSL}(2, \mathbb{Z})$ is the kernel of σ_m, why is $H_m \subseteq K_m$?

(e) Show that if t is any integer, then the following members of K_m lie in the subgroup H_m: $\begin{pmatrix} 1 & tm \\ 0 & 1 \end{pmatrix}$ mod $\pm I$, $\begin{pmatrix} 1 & 0 \\ tm & 1 \end{pmatrix}$ mod $\pm I$, $\begin{pmatrix} 1+tm & tm \\ -tm & 1-tm \end{pmatrix}$ mod $\pm I$, and $\begin{pmatrix} 1+tm & -tm \\ tm & 1-tm \end{pmatrix}$ mod $\pm I$.

42. With G_m defined as above, exhibit homomorphisms of various groups G_m onto the following finite groups:

(a) $\mathfrak{S}_3$ when $m = 2$ by sending $X \mapsto (1\ 2)$ and $Y \mapsto (1\ 2\ 3)$.

(b) $\mathfrak{A}_4$ when $m = 3$ by sending $X \mapsto (1\ 2)(3\ 4)$ and $Y \mapsto (1\ 2\ 3)$.

(c) $\mathfrak{S}_4$ when $m = 4$ by sending $X \mapsto (1\ 2)$ and $Y \mapsto (2\ 3\ 4)$.

(d) $\mathfrak{A}_5$ when $m = 5$ by sending $X \mapsto (1\ 2)(3\ 4)$ and $Y \mapsto (1\ 3\ 5)$.

43. This problem shows how to prove that $H_m = K_m$ for $2 \leq m \leq 5$, and it asks that the steps be carried out for $m = 2$ and $m = 3$. Recall from the remark with Lemma 7.11 that Lemma 7.11 is valid for *all* groups in determining a set of generators of a subgroup from generators of the whole group and a system of coset representatives. The lemma is to be applied to the group $\mathrm{PSL}(2, \mathbb{Z})$ and the subgroup K_m. Generators of $\mathrm{PSL}(2, \mathbb{Z})$ are taken as $b_1 = x$ mod $\pm I$ and $b_2 = y$ mod $\pm I$.

(a) For the case $m = 2$, find members $g_1, \ldots, g_6$ of $\mathrm{PSL}(2, \mathbb{Z})$ such that the six cosets of $\mathrm{PSL}(2, \mathbb{Z})/K_2$ are exactly $K_2 g_1, \ldots, K_2 g_6$.

(b) Still for the case $m = 2$, find $g_j b_i \rho(g_j b_i)^{-1}$ for $1 \leq i \leq 2$ and $1 \leq j \leq 6$. Lemma 7.11 says that these 12 elements generate K_2.

(c) Using Problem 41e and any necessary variations of it, show that each of the 12 generators of K_2 in (b) lies in the subgroup H_2, and conclude that $H_2 = K_2$.

(d) Repeat steps (a), (b), and (c) for $m = 3$. There are 12 cosets $K_3 g_j$ of PSL$(2, \mathbb{Z})/K_3$. (Educational note: There are 24 cosets for PSL$(2, \mathbb{Z})/K_4$ and 60 cosets for PSL$(2, \mathbb{Z})/K_5$.)

44. Take for granted that $H_m = K_m$ for $2 \leq m \leq 5$. Deduce the isomorphisms
    (a) $G_2 \cong \text{PSL}(2, \mathbb{Z}/2\mathbb{Z}) \cong \mathfrak{S}_3$.
    (b) $G_3 \cong \text{PSL}(2, \mathbb{Z}/3\mathbb{Z}) \cong \mathfrak{A}_4$. (This group is called the **tetrahedral group**.)
    (c) $G_4 \cong \text{PSL}(2, \mathbb{Z}/4\mathbb{Z}) \cong \mathfrak{S}_4$. (This group is called the **octahedral group**.)
    (d) $G_5 \cong \text{PSL}(2, \mathbb{Z}/5\mathbb{Z}) \cong \mathfrak{A}_5$. (This group is called the **icosahedral group**.)

45. A translation in the Euclidean plane $\mathbb{R}^2$ is any function $T_{(a,b)}(x, y) = (a + x, b + y)$, the rotation about the origin clockwise through the angle θ is the linear map R_θ given by the matrix $\begin{pmatrix} \cos\theta & -\sin\theta \\ \sin\theta & \cos\theta \end{pmatrix}$, and the rotation about (x_0, y_0) clockwise through the angle θ is the linear map given by $(x, y) \mapsto R_\theta(x - x_0, y - y_0) + (x_0, y_0)$.
    (a) Prove that $R_\theta T_{(a,b)} R_\theta^{-1} = T_{R_\theta(a,b)}$.
    (b) Prove that the union of the set of translations and all the sets of rotations about points of $\mathbb{R}^2$ is a group by showing that it is the semidirect product of the subgroup of rotations about the origin and the normal subgroup of translations.

46. Fix a triangle T in the Euclidean plane with vertices arranged counterclockwise at a, b, c and with angles $\pi/2$ at a, $\pi/3$ at b, and $\pi/6$ at c. Let r_a be rotation clockwise through π at a, r_b be rotation clockwise through $2\pi/3$ at b, and r_c be rotation *counter*clockwise through $\pi/3$ at c.
    (a) Show that $r_a^2 = 1, r_b^3 = 1, r_c^6 = 1$, and $r_c = r_a r_b$.
    (b) Show that the member $r_b r_a r_b r_a r_b$ of the group generated by r_a and r_b is a nontrivial translation and therefore that the generated group is infinite.
    (c) Conclude that $G_6 \not\cong \text{PSL}(2, \mathbb{Z}/6\mathbb{Z})$. (Educational note: If $\widetilde{T}$ denotes the union of T and the reflection of T in one of the sides of T, it can be shown that the group generated by r_a and r_b is isomorphic to G_6 and tiles the plane with copies of $\widetilde{T}$.)

Problems 47–52 establish a harmonic analysis for arbitrary representations of finite groups on complex vector spaces, whether finite-dimensional or infinite-dimensional. Let G be a finite group, and let V be a complex vector space. For any representation R of G on V, one defines $R(f)v = \sum_{x \in G} f(x) R(x) v$ for f in $C(G, \mathbb{C})$ and v in V, just as in the case that V is finite-dimensional. The same computation as in Section VII.4 shows that the formula $R(f_1 * f_2) = R(f_1) R(f_2)$ remains valid when V is infinite-dimensional.

47. Let (R_1, V_1) and (R_2, V_2) be irreducible finite-dimensional representations of G on complex vector spaces, and let χ_{R_1} and χ_{R_2} be their characters. Using Schur orthogonality, prove that
    (a) $\chi_{R_1} * \chi_{R_2} = 0$ if R_1 and R_2 are inequivalent,

(b) $\chi_{R_1} * \chi_{R_1} = |G| d_{R_1}^{-1} \chi_{R_1}$, where $d_{R_1} = \dim V_R$.

48. With (R, V) given, let (R_α, V_α) be any irreducible finite-dimensional representation of G, and define $E_\alpha : V \to V$ by $E_\alpha = |G|^{-1} d_\alpha R(\overline{\chi_\alpha})$, where χ_α is the character of R_α and where $d_\alpha = \dim V_\alpha$.
    (a) Prove that $E_\alpha^2 = E_\alpha$.
    (b) Prove that $E_\alpha E_\beta = E_\beta E_\alpha = 0$ if (R_β, V_β) is an irreducible finite-dimensional representation of G such that R_α and R_β are inequivalent.

49. Observe for each v in V that $\{R(x)v \mid x \in G\}$ spans a finite-dimensional invariant subspace of V. By Corollary 7.21, each v in V lies in a finite direct sum of finite-dimensional invariant subspaces of V on each of which R acts irreducibly. Using Zorn's Lemma, prove that V is the direct sum of finite-dimensional subspaces on each of which R acts irreducibly. (If V is infinite-dimensional, there will of course be infinitely many such subspaces.)

50. Suppose that V_0 is a finite-dimensional invariant subspace of V such that $R|_{V_0}$ is equivalent to some R_α, where R_α is as in Problem 48. Prove that E_α is the identity on V_0.

51. Deduce that if $\{(R_\beta, V_\beta)\}$ is a maximal collection of inequivalent finite-dimensional irreducible representations of G, then $\sum_\beta E_\beta = I$ on V and the image of E_α is the set of all sums of vectors in V lying in some finite-dimensional invariant subspace V_0 of V such that $R|_{V_0}$ is equivalent to R_α. (Educational note: Consequently V is exhibited as the finite direct sum of the spaces image E_α, each space image E_α is the direct sum of finite-dimensional irreducible invariant subspaces, and the restriction of R to any finite-dimensional irreducible invariant subspace of image E_α is equivalent with R_α.

52. Suppose that (R_α, V_α) is a 1-dimensional representation of G given by a multiplicative character ω. Prove that the image of E_α consists of all vectors v in V such that $R(x)v = \omega(x)v$ for all x in G.

CHAPTER VIII

Commutative Rings and Their Modules

Abstract. This chapter amplifies the theory of commutative rings that was begun in Chapter IV, and it introduces modules for any ring. Emphasis is on the topic of unique factorization.

Section 1 gives many examples of rings, some commutative and some noncommutative, and introduces the notion of a module for a ring.

Sections 2–4 discuss some of the tools related to questions of factorization in integral domains. Section 2 defines the field of fractions for an integral domain and gives its universal mapping property. Section 3 defines prime and maximal ideals and relates quotients of them to integral domains and fields. Section 4 introduces principal ideal domains, which are shown to have unique factorization, and it defines Euclidean domains as a special kind of principal ideal domain for which greatest common divisors can be obtained constructively.

Section 5 proves that if R is an integral domain with unique factorization, then so is the polynomial ring $R[X]$. This result is a consequence of Gauss's Lemma, which addresses what happens to the greatest common divisor of the coefficients when one multiplies two members of $R[X]$. Gauss's Lemma has several other consequences that relate factorization in $R[X]$ to factorization in $F[X]$, where F is the field of fractions of R. Still another consequence is Eisenstein's irreducibility criterion, which gives a sufficient condition for a member of $R[X]$ to be irreducible.

Section 6 contains the theorem that every finitely generated unital module over a principal ideal domain is a direct sum of cyclic modules. The cyclic modules may be assumed to be primary in a suitable sense, and then the isomorphism types of the modules appearing in the direct-sum decomposition, together with their multiplicities, are uniquely determined. The main results transparently generalize the Fundamental Theorem for Finitely Generated Abelian Groups, and less transparently they generalize the existence and uniqueness of Jordan canonical form for square matrices with entries in an algebraically closed field.

Sections 7–11 contain foundational material related to factorization for the two subjects of algebraic number theory and algebraic geometry. Both these subjects rely heavily on the theory of commutative rings. Section 7 is a section of motivation, showing the analogy between a situation in algebraic number theory and a situation in algebraic geometry. Sections 8–10 introduce Noetherian rings, integral closures, and localizations. Section 11 uses this material to establish unique factorization of ideals for Dedekind domains, as well as some other properties.

1. Examples of Rings and Modules

Sections 4–5 of Chapter IV introduced rings and fields, giving a small number of examples of each. In the present section we begin by recalling those examples and giving further ones. Although Chapters VI and VII are not prerequisite for the present chapter, our list of examples will include some rings and fields that arose in those two chapters.

The theory to be developed in this chapter is intended to apply to commutative rings, especially to questions related to unique factorization in such rings. Despite this limitation it seems wise to include examples of noncommutative rings in the list below.

In the conventions of this book, a ring need not have an identity. Many rings that arise only in the subject of algebra have an identity, but there are important rings in the subject of real analysis that do not. From the point of view of category theory, one therefore distinguishes between the category of all rings, with ring homomorphisms as morphisms, and the category of all rings with identity, with ring homomorphisms carrying 1 to 1 as morphisms. In the latter case one may want to exclude the zero ring from being an object in the category under certain circumstances.

EXAMPLES OF RINGS.

(1) Basic commutative rings from Chapter IV. All of the structures $\mathbb{Z}$, $\mathbb{Q}$, $\mathbb{R}$, $\mathbb{C}$, $\mathbb{Z}/m\mathbb{Z}$, and $2\mathbb{Z}$ are commutative rings. All but the last have an identity. Of these, $\mathbb{Q}$, $\mathbb{R}$, and $\mathbb{C}$ are fields, and so is $\mathbb{F}_p = \mathbb{Z}/p\mathbb{Z}$ if p is a prime number. The others are not fields.

(2) Polynomial rings. Let R be a nonzero commutative ring with identity. In Section IV.5 we defined the commutative ring $R[X_1, \ldots, X_n]$ of polynomials over R in n indeterminates. It has a universal mapping property with respect to substitution for the indeterminates and use of a homomorphism on the coefficients. Making substitutions from R itself and mapping the coefficients by the identity homomorphism, we are led to the ring of all functions $(r_1, \ldots, r_n) \mapsto f(r_1, \ldots, r_n)$ for $r_1, \ldots, r_n$ in R and $f(X_1, \ldots, X_n)$ in $R[X_1, \ldots, X_n]$; this is called the ring of all polynomial functions in n variables on R. Polynomials may be considered also in infinitely many variables, but we did not treat this case in any detail.

(3) Matrix rings over commutative rings. Let R be a nonzero commutative ring with identity. The set $M_n(R)$ of all n-by-n matrices with entries in R is a ring under entry-by-entry addition and the usual definition of matrix multiplication: $(AB)_{ij} = \sum_{k=1}^{n} A_{ik} B_{kj}$. It has an identity, namely the identity matrix I with $I_{ij} = \delta_{ij}$. In this setting, Section V.2 introduced a theory of determinants, and it was proved that a matrix has a one-sided inverse if and only if it has a two-sided inverse, if and only if its determinant is a member of the group $R^\times$ of units in R, i.e., elements of R invertible under multiplication. The matrix ring $M_n(R)$ is always noncommutative if $n > 1$.

(4) Matrix rings over noncommutative rings. If R is any ring, we can still make the set $M_n(R)$ of all n-by-n matrices with entries in R into a ring. However, if

R has no identity, $M_n(R)$ will have no identity. The theory of determinants does not directly apply if R is noncommutative or if R fails to have an identity,[1] and as a consequence, questions about the invertibility of matrices are more subtle than with the previous example.

(5) Spaces of linear maps from a vector space into itself. Let V be a vector space over a field $\mathbb{K}$. The vector space $\mathrm{End}_{\mathbb{K}}(V) = \mathrm{Hom}_{\mathbb{K}}(V, V)$ of all $\mathbb{K}$ linear maps from V to itself is initially a vector space over $\mathbb{K}$. Composition provides a multiplication that makes $\mathrm{End}_{\mathbb{K}}(V)$ into a ring with identity. In fact, associativity of multiplication is automatic for any kind of function, and so is the distributive law $(L_1+L_2)L_3 = L_1L_3+L_2L_3$. The distributive law $L_1(L_2+L_3) = L_1L_2+L_1L_3$ follows from the fact that L_1 is linear. This ring is isomorphic as a ring to $M_n(\mathbb{K})$ if V is n-dimensional, an isomorphism being determined by specifying an ordered basis of V.

(6) Associative algebras over fields. These were defined in Section VI.7, knowledge of which is not being assumed now. Thus we repeat the definition. If $\mathbb{K}$ is a field, then an associative algebra over $\mathbb{K}$, or associative $\mathbb{K}$ algebra, is a ring A that is also a vector space over $\mathbb{K}$ such that the multiplication $A \times A \to A$ is $\mathbb{K}$-linear in each variable. The conditions of linearity concerning multiplication have two parts to them: an additive part saying that the usual distributive laws are valid and a scalar-multiplication part saying that

$$(ka)b = k(ab) = a(kb) \qquad \text{for all } k \text{ in } \mathbb{K} \text{ and } a, b \text{ in } A.$$

If A has an identity, the displayed condition says that all scalar multiples of the identity lie in the **center** of A, i.e., commute with every element of A. In Examples 2 and 3, when R is a field $\mathbb{K}$, the polynomial rings and matrix rings over $\mathbb{K}$ provide examples of associative algebras over $\mathbb{K}$; scalar multiplication is to be done in entry-by-entry fashion. Example 5 is an associative algebra as well. If $\mathbb{L}$ is any field such that $\mathbb{K}$ is a subfield, then $\mathbb{L}$ may be regarded as an associative algebra over $\mathbb{K}$. An interesting commutative associative algebra over $\mathbb{C}$ without identity is the algebra $C_{\mathrm{com}}(\mathbb{R})$ of all continuous complex-valued functions on $\mathbb{R}$ that vanish outside a bounded interval; the vector-space operations are the usual pointwise operations, and the operation of multiplication is given by **convolution**

$$(f * g)(x) = \int_{\mathbb{R}} f(x - y)g(y) \, dy.$$

Section VII.4 worked with an analog $C(G, \mathbb{C})$ of this algebra in the context that $\mathbb{R}$ is replaced by a finite group G.

---

[1] A limited theory of determinants applies in the noncommutative case, but it will not be helpful for our purposes.

(7) Division rings. A division ring is a nonzero ring with identity such that every element has a two-sided inverse under multiplication. A commutative division ring is just a field. The ring $\mathbb{H}$ of quaternions is the only explicit noncommutative division ring that we have encountered so far. It is an associative algebra over $\mathbb{R}$. More generally, if A is a division ring, then we can easily check that the center $\mathbb{K}$ of A is a field and that A is an associative algebra over $\mathbb{K}$.[2]

(8) Tensor, symmetric, and exterior algebras. If E is a vector space over a field $\mathbb{K}$, Chapter VI defined the tensor, symmetric, and exterior algebras of E over $\mathbb{K}$, as well as the polynomial algebra on E in the case that E is finite-dimensional. These are all associative algebras with identity. Symmetric algebras and polynomial algebras are commutative. None of these algebras will be discussed further in this chapter.

(9) A field of 4 elements. This was constructed in Section IV.4. Further finite fields beyond the field of 4 elements and the fields $\mathbb{F}_p = \mathbb{Z}/p\mathbb{Z}$ with p prime will be constructed in Chapter IX.

(10) Algebraic number fields $\mathbb{Q}[\theta]$. These were discussed in Sections IV.1 and IV.4. In defining $\mathbb{Q}[\theta]$, we assume that θ is a complex number and that there exists an integer $n > 0$ such that the complex numbers $1, \theta, \theta^2, \ldots, \theta^n$ are linearly dependent over $\mathbb{Q}$. The set $\mathbb{Q}[\theta]$ is defined to be the subset of $\mathbb{C}$ obtained by substitution of θ into all members of $\mathbb{Q}[X]$. It coincides with the linear span over $\mathbb{Q}$ of $1, \theta, \theta^2, \ldots, \theta^{n-1}$. Proposition 4.1 shows that it is closed under the arithmetic operations, including passage to multiplicative inverses of nonzero elements, and it is therefore a subfield of $\mathbb{C}$. This example ties in with the notion of minimal polynomial in Chapter V because the members of $\mathbb{Q}[X]$ with θ as a root are all multiples of one nonzero such polynomial that exhibits the linear dependence. We return to this example occasionally later in this chapter, particularly in Sections 7–11, and then we treat it in more detail in Chapter IX.

(11) Algebraic integers in a number field $\mathbb{Q}[\theta]$. Algebraic integers were defined in Section VII.5 as the roots in $\mathbb{C}$ of monic polynomials in $\mathbb{Z}[X]$, and they were shown to form a commutative ring with identity. The set of algebraic integers in $\mathbb{Q}[\theta]$ is therefore a commutative ring with identity, and it plays somewhat the same role for $\mathbb{Q}[\theta]$ that $\mathbb{Z}$ plays for $\mathbb{Q}$. We discuss this example further in Sections 7–11.

(12) Integral group rings. If G is a group, then we can make the free abelian group $\mathbb{Z}G$ on the elements of G into a ring by defining multiplication to be $\left(\sum_i m_i g_i\right)\left(\sum_j n_j h_j\right) = \sum_{i,j} (m_i n_j)(g_i h_j)$ when the m_i and n_j are in $\mathbb{Z}$ and the g_i and h_j are in G. It is immediate that the result is a ring with identity, and $\mathbb{Z}G$

---

[2]Use of the term "division algebra" requires some care. Some mathematicians understand division algebras to be associative, and others do not. The real algebra $\mathbb{O}$ of octonions, as defined in Problems 52–56 at the end of Chapter VI, is not associative, but it does have division.

1. Examples of Rings and Modules

is called the **integral group ring** of G. The group G is embedded as a subgroup of the group $(\mathbb{Z}G)^\times$ of units of $\mathbb{Z}G$, each element of g being identified with a sum $\iota(g) = \sum m_i g_i$ in which the only nonzero term is $1g$. The ring $\mathbb{Z}G$ has the universal mapping property illustrated in Figure 8.1 and described as follows: whenever $\varphi : G \to R$ is a group homomorphism of G into the group $R^\times$ of units of a ring R, then there exists a unique ring homomorphism $\Phi : \mathbb{Z}G \to R$ such that $\Phi \iota = \varphi$. The existence of Φ as a homomorphism of additive groups follows from the universal mapping property of free abelian groups, and then one readily checks that Φ respects multiplication.[3]

FIGURE 8.1. Universal mapping property of the integral group ring of G.

(13) Quotient rings. If R is a ring and I is a two-sided ideal, then we saw in Section IV.4 that the additive quotient R/I has a natural multiplication that makes it into a ring called a **quotient ring** of R. This in effect was the construction that obtained the ring $\mathbb{Z}/m\mathbb{Z}$ from the ring $\mathbb{Z}$.

(14) Direct product of rings. If $\{R_s \mid s \in S\}$ is a nonempty set of rings, then a **direct product** $\prod_{s \in S} R_s$ is a ring whose additive group is any direct product of the underlying additive groups and whose multiplication is given in entry-by-entry fashion. The resulting ring and the associated ring homomorphisms $p_{s_0} : \prod_{s \in S} R_s \to R_{s_0}$ amount to the product functor for the category of rings; if each R_s has an identity, the result amounts also to the product functor for the category of rings with identity.

We give further examples of rings near the end of this section after we have defined modules and given some examples.

Informally *a module is a vector space over a ring*. But let us be more precise. If R is a ring, then a **left R module**[4] M is an abelian group with the additional structure of a "scalar multiplication" $R \times M \to M$ such that

(i) $r(r'm) = (rr')m$ for r and r' in R and m in M,

---

[3]Universal mapping properties are discussed systematically in Problems 18–22 at the end of Chapter VI. The subject of such a property, here the pair $(\mathbb{Z}G, \iota)$, is always unique up to canonical isomorphism in a given category, but its existence has to be proved.

[4]Many algebra books write "R-module," using a hyphen. However, when R is replaced by an expression, particularly in applications of the theory, the hyphen is often dropped. For an example, see "module" in Hall's *The Theory of Groups*. The present book omits the hyphen in all cases in order to be consistent.

(ii) $(r+r')m = rm + r'm$ and $r(m+m') = rm + rm'$ if r and r' are in R and m and m' are in M.

In addition, if R has an identity, we say that M is **unital** if

(iii) $1m = m$ for all m in M.

One may also speak of **right R modules**. For these the scalar multiplication is usually written as mr with m in M and r in R, and the expected analogs of (i) and (ii) are to hold.

When R is commutative, it is immaterial which side is used for the scalar multiplication, and one speaks simply of an R **module**.

Let R be a ring, and let M and N be two left R modules. A **homomorphism of left R modules**, or more briefly an R **homomorphism**, is an additive group homomorphism $\varphi : M \to N$ such that $\varphi(rm) = r\varphi(m)$ for all r in R. Then we can form a category for fixed R in which the objects are the left R modules and the morphisms are the R homomorphisms from one left R module to another. Similarly the right R modules, along with the corresponding kind of R homomorphisms, form a category. If R has an identity, then the unital R modules form a subcategory in each case. These categories are fundamental to the subject of homological algebra, which we take up in *Advanced Algebra*.

EXAMPLES OF MODULES.

(1) Vector spaces. If R is a field, *the unital R modules are exactly the vector spaces over R.*

(2) Abelian groups. *The unital $\mathbb{Z}$ modules are exactly the abelian groups.* Scalar multiplication is given in the expected way: If n is a positive integer, the product nx is the n-fold sum of x with itself. If $n = 0$, the product nx is 0. If $n < 0$, the product nx is $-((-n)x)$.

(3) Vector spaces as unital modules for the polynomial ring $\mathbb{K}[X]$. Let V be a finite-dimensional vector space over the field $\mathbb{K}$, and fix L be in $\text{End}_{\mathbb{K}}(V)$. Then V becomes a unital $\mathbb{K}[X]$ module under the definition $A(X)v = A(L)(v)$ whenever $A(X)$ is a polynomial in $\mathbb{K}[X]$; here $A(L)$ is the member of $\text{End}_{\mathbb{K}}(V)$ defined as in Section V.3. In Section 6 in this chapter we shall see that some of the deeper results in the theory of a single linear transformation, as developed in Chapter V, follow from the theory of unital $\mathbb{K}[X]$ modules that will emerge from the present chapter.

(4) Modules in the context of algebraic number fields. Let $\mathbb{Q}[\theta]$ be a subfield of $\mathbb{C}$ as in Example 10 of rings earlier in this section. It is assumed that the $\mathbb{Q}$ vector space $\mathbb{Q}[\theta]$ is finite-dimensional. Let L be the member of $\text{End}_{\mathbb{Q}}(\mathbb{Q}[\theta])$ given as left multiplication by θ on $\mathbb{Q}[\theta]$. As in the previous example, $\mathbb{Q}[\theta]$ becomes a unital $\mathbb{Q}[X]$ module. Chapter V defines a minimal polynomial for L, as well as a characteristic polynomial. These objects play a role in the study

to be carried out in Chapter IX of fields like $\mathbb{Q}[\theta]$. If θ is an algebraic integer as in Example 11 of rings earlier in this section, then we can get more refined information by replacing $\mathbb{Q}$ by $\mathbb{Z}$ in the above analysis; this technique plays a role in the theory to be developed in Sections 7–11.

(5) Rings and their quotients. If R is a ring, then R is a left R module and also a right R module. If I is a two-sided ideal in I, then the quotient ring R/I, as defined in Proposition 4.20, is a left R module and also a right R module. These modules are automatically unital if R has an identity. Later in this section we shall consider quotients of R by "one-sided ideals."

(6) Spaces of rectangular matrices. If R is a ring, then the space $M_{mn}(R)$ of m-by-n matrices with entries in R is an abelian group under addition and becomes a left R module when multiplication by the scalar r is defined as left multiplication by r in each entry. Also, if we put $S = M_m(R)$, then $M_{mn}(R)$ is a left S module under the usual definition of matrix multiplication: $(sv)_{ij} = \sum_{k=1}^{n} s_{ik} v_{kj}$, where s is in S and v is in $M_{mn}(R)$.

(7) Direct product of R modules. If S is a nonempty set and $\{M_s\}_{s \in S}$ is a corresponding system of left R modules, then a **direct product** $\prod_{s \in S} M_s$ is obtained as an additive group by forming any direct product of the underlying additive groups of the M_s's and defining scalar multiplication by members of R to be scalar multiplication in each coordinate. The associated abelian-group homomorphisms $p_{s_0} : \prod_{s \in S} M_s \to M_{s_0}$ become R homomorphisms under this definition of scalar multiplication on the direct product. Direct product amounts to the product functor for the category of left R modules; we omit the easy verification, which makes use of the corresponding fact about abelian groups. As in the case of abelian groups, we can speak of an **external** direct product as the result of a construction that starts with the product of the sets M_s, and we can speak of recognizing a direct product as **internal** when the M_s's are contained in the direct product and the restriction of each p_s to M_s is the identity function.

(8) Direct sum of R modules. If S is a nonempty set and $\{M_s\}_{s \in S}$ is a corresponding system of left R modules, then a **direct sum** $\bigoplus_{s \in S} M_s$ is obtained as an additive group by forming any direct sum of the underlying additive groups of the M_s's and defining scalar multiplication by members of R to be scalar multiplication in each coordinate. The associated abelian-group homomorphisms $i_{s_0} : M_{s_0} \to \bigoplus_{s \in S} M_s$ become R homomorphisms under this definition of scalar multiplication on the direct sum. Direct sum amounts to the coproduct functor for the category of left R modules; we omit the easy verification, which makes use of the corresponding fact about abelian groups. As in the case of abelian groups, we can speak of an **external** direct sum as the result of a construction that starts with a subset of the product of the sets M_s, and we can speak of recognizing a

direct sum as **internal** when the M_s's are contained in the direct sum and each i_s is the inclusion mapping.

(9) Free R modules. Let R be a nonzero ring with identity, and let S be a nonempty set. As in Example 5, let us regard R as a unital left R module. Then the left R module given as the direct sum $F(S) = \bigoplus_{s \in S} R$ is called a **free R module**, or free left R module. We define $\iota : S \to F(S)$ by $\iota(s) = i_s(1)$, where i_s is the usual embedding map for the direct sum of R modules. The left R module $F(S)$ has a universal mapping property similar to the corresponding property of free abelian groups. This is illustrated in Figure 8.2 and is described as follows: whenever M is a unital left R module and $\varphi : S \to M$ is a function, then there exists a unique R homomorphism $\Phi : F(S) \to M$ such that $\Phi \iota = \varphi$. The existence of Φ as an R homomorphism follows from the universal mapping property of direct sums (Example 8) as soon as the property is demonstrated for S equal to a singleton set. Thus let A be any left R module, and let $a \in A$ be given; then it is evident that $r \mapsto ra$ is the unique R homomorphism of the left R module R into A carrying 1 to a.

FIGURE 8.2. Universal mapping property of a free left R module.

If R is a ring and M is a left R module, then an R **submodule** N of M is an additive subgroup of M that is closed under scalar multiplication, i.e., has rm in N when r is in R and m is in N. In situations in which there is no ambiguity, the use of "left" in connection with R submodules is not necessary.

EXAMPLES OF SUBMODULES. If V is a vector space over a field $\mathbb{K}$, then a $\mathbb{K}$ submodule of V is a vector subspace of V. If M is an abelian group, then a $\mathbb{Z}$ submodule of M is a subgroup. In Example 6 of modules, in which $S = M_m(R)$, then an example of a left S submodule of $M_{mn}(R)$ is all matrices with 0 in every entry of a specified subset of the n columns.

If the ring R has an identity and M is a unital left R module, then the R submodule of M **generated** by $m \in M$, i.e., the smallest R submodule containing m, is Rm, the set of products rm with r in R. In fact, the set of all rm is an abelian group since $(r \pm s)m = rm \pm sm$, it is closed under scalar multiplication since $s(rm) = (sr)m$, and it contains m since $1m = m$. However, if the left R module M is not unital, then the R submodule generated by m may not equal Rm, and it was for that reason that R modules were assumed to be unital in the construction of free R modules in Example 9 of modules above. More generally the R submodule of M **generated** by a finite set $\{m_1, \ldots, m_n\}$ in M is $Rm_1 + \cdots + Rm_n$ if the left R module M is unital.

Example 5 of modules treated R as a left R module. In this setting the left R submodules are called **left ideals** in R. That is, a left ideal I is an additive subgroup of R such that ri is in I whenever r is in R and i is in I. As a special case of what was said in the previous paragraph, if the ring R has an identity, then the left R module R is automatically unital, and the left ideal of R generated by an element a is Ra, the set of all products ra with r in R.

Similarly a **right ideal** in R is an additive subgroup I such that ir is in I whenever r is in R and i is in I. The right ideals are the right R submodules of the right R module R. If R is commutative, then left ideals, right ideals, and two-sided ideals are all the same.

Suppose that $\varphi : M \to N$ is an R homomorphism of left R modules. In this situation we readily verify that the kernel of φ, denoted by $\ker \varphi$ as usual, is an R submodule of M, and the image of φ, denoted by image φ as usual, is an R submodule of N. The R homomorphism φ is one-one if and only if $\ker \varphi = 0$, as a consequence of properties of homomorphisms of abelian groups. A one-one R homomorphism of one left R module onto another is called an R **isomorphism**; its inverse is automatically an R isomorphism, and "is R isomorphic to" is an equivalence relation.

Still with R as a ring, suppose that M is a left R submodule and N is an R submodule. Then we can form the quotient M/N of abelian groups. This becomes a left R module under the definition $r(m + N) = rm + N$, as we readily check. We call M/N a **quotient module**. The quotient mapping $m \mapsto m + N$ of M to M/N is an R homomorphism onto. A particular example of a quotient module is R/I, where I is a left ideal in R.

We can now go over the results on quotients of abelian groups in Section IV.2, specifically Proposition 4.11 through Theorem 4.14, and check that they extend immediately to results about left R modules. The statements appear below. The arguments are all routine, and there is no point in repeating them. In the special case that R is a field and the R modules are vector spaces, these results specialize to results proved in Sections II.5 and II.6.

Proposition 8.1. Let R be a ring, let $\varphi : M_1 \to M_2$ be an R homomorphism between left R modules, let $N_0 = \ker \varphi$, let N be an R submodule of M_1 contained in N_0, and define $q : M_1 \to M_1/N$ to be the R module quotient map. Then there exists an R homomorphism $\overline{\varphi} : M_1/N \to M_2$ such that $\varphi = \overline{\varphi} q$, i.e, $\overline{\varphi}(m_1 + N) = \varphi(m_1)$. It has the same image as φ, and $\ker \overline{\varphi} = \{h_0 N \mid h_0 \in N_0\}$.

REMARK. As with groups, one says that φ **factors through** M_1/N or **descends to** M_1/N. Figure 8.3 illustrates matters.

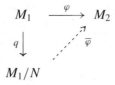

FIGURE 8.3. Factorization of R homomorphisms via a quotient of R modules.

Corollary 8.2. Let R be a ring, let $\varphi : M_1 \to M_2$ be an R homomorphism between left R modules, and suppose that φ is onto M_2 and has kernel N. Then φ exhibits the left R module M_1/N as canonically R isomorphic to M_2.

Theorem 8.3 (First Isomorphism Theorem). Let R be a ring, let $\varphi : M_1 \to M_2$ be an R homomorphism between left R modules, and suppose that φ is onto M_2 and has kernel K. Then the map $N_1 \mapsto \varphi(N_1)$ gives a one-one correspondence between

(a) the R submodules N_1 of M_1 containing K and
(b) the R submodules of M_2.

Under this correspondence the mapping $m + N_1 \mapsto \varphi(m) + \varphi(N_1)$ is an R isomorphism of M_1/N_1 onto $M_2/\varphi(N_1)$.

REMARK. In the special case of the last statement that $\varphi : M_1 \to M_2$ is an R module quotient map $q : M \to M/K$ and N is an R submodule of M containing K, the last statement of the theorem asserts the R isomorphism $M/N \cong (M/K)/(N/K)$.

Theorem 8.4 (Second Isomorphism Theorem). Let R be a ring, let M be a left R module, and let N_1 and N_2 be R submodules of M. Then $N_1 \cap N_2$ is an R submodule of N_1, the set $N_1 + N_2$ of sums is an R submodule of M, and the map $n_1 + (N_1 \cap N_2) \mapsto n_1 + N_2$ is a well-defined canonical R isomorphism

$$N_1/(N_1 \cap N_2) \cong (N_1 + N_2)/N_2.$$

A quotient of a direct sum of R modules by the direct sum of R submodules is the direct sum of the quotients, according to the following proposition. The result generalizes Lemma 4.58, which treats the special case of abelian groups (unital $\mathbb{Z}$ modules).

Proposition 8.5. Let R be a ring, let $M = \bigoplus_{s \in S} M_s$ be a direct sum of left R modules, and for each s in S, let N_s be a left R submodule of M_s. Then the natural map of $\bigoplus_{s \in S} M_s$ to the direct sum of quotients descends to an R isomorphism

$$\bigoplus_{s \in S} M_s \Big/ \bigoplus_{s \in S} N_s \cong \bigoplus_{s \in S} (M_s/N_s).$$

PROOF. Let $\varphi : \bigoplus_{s \in S} M_s \to \bigoplus_{s \in S} (M_s/N_s)$ be the R homomorphism defined by $\varphi(\{m_s\}_{s \in S}) = \{m_s + N_s\}_{s \in S}$. The mapping φ is onto $\bigoplus_{s \in S} (M_s/N_s)$, and the kernel is $\bigoplus_{s \in S} N_s$. Then Corollary 8.2 shows that φ descends to the required R isomorphism. $\square$

EXAMPLES OF RINGS, CONTINUED.

(15) Associative algebras over commutative rings with identity. These directly generalize Example 6 of rings. Let R be a nonzero commutative ring with identity. An **associative algebra over** R, or **associative R algebra**, is a ring A that is also a left R module such that multiplication $A \times A \to A$ is R linear in each variable. The conditions of R linearity in each variable mean that addition satisfies the usual distributive laws for a ring and that the following condition is to be satisfied relating multiplication and scalar multiplication:

$$(ra)b = r(ab) = a(rb) \qquad \text{for all } r \text{ in } R \text{ and } a, b \in A.$$

If A has an identity, the displayed condition says that all scalar multiples of the identity lie in the **center** of A, i.e., commute with every element of A. Examples 2 and 3, treating polynomial rings and matrix rings whose scalars lie in a commutative ring with identity, furnish examples. Every ring R is an associative $\mathbb{Z}$ algebra when the $\mathbb{Z}$ action is defined so as to make the abelian group underlying the additive structure of R into a $\mathbb{Z}$ module. All that needs to be checked is the displayed formula. For $n = 1$, we have $(1a)b = 1(ab) = a(1b)$ since the $\mathbb{Z}$ module R is unital. If we also have $(na)b = n(ab) = a(nb)$ for a positive integer n, then we can add and use the appropriate distributive laws to obtain $((n+1)a)b = (n+1)(ab) = a((n+1)b)$. Induction therefore gives $(na)b = n(ab) = a(nb)$ for all positive integers n, and this equality extends to all integers n by using additive inverses. The associative R algebras form a category in which the morphisms from one such algebra to another are the ring homomorphisms that are also R homomorphisms. The product functor for this category is the direct product as in Example 14 with an overlay of scalar multiplication as in Example 7 of modules. The coproduct functor in the category of commutative associative R algebras with identity is more subtle and involves a tensor product over R, a notion we postpone introducing until Chapter X.

(16) Group algebra RG over R. If G is a group and R is a commutative ring with identity, then we can introduce a multiplication in the free R module RG on the elements of G by the definition $\left(\sum_i r_i g_i\right)\left(\sum_j s_j h_j\right) = \sum_{i,j} (r_i s_j)(g_i h_j)$ when the r_i and s_j are in R and the g_i and h_j are in G. It is immediate that this multiplication makes the free R module into an associative R algebra with identity, and RG is called the **group algebra** of G over R. The special case $R = \mathbb{Z}$ leads to the integral group ring as in Example 12. The group G is embedded as a

subgroup of the group $(RG)^\times$ of units of RG, each element of g being identified with a sum $\iota(g) = \sum r_i g_i$ in which the only nonzero term is $1g$. The associative R algebra RG has a universal mapping property similar to that in Figure 8.1 and given in Figure 8.4 as follows: whenever $\varphi : G \to A$ is a group homomorphism of G into the group $A^\times$ of units of an associative R algebra A, then there exists a unique associative R algebra homomorphism $\Phi : RG \to A$ such that $\Phi\iota = \varphi$.

FIGURE 8.4. Universal mapping property of the group algebra RG.

(17) Scalar-valued functions of finite support on a group, with convolution as multiplication. If G is a group and R is a commutative ring with identity, denote by $C(G, R)$ the R module of all functions from G into R that are of **finite support** in the sense that each function is 0 except on a finite subset of G. This R module readily becomes an associative R algebra if ring multiplication is taken to be pointwise multiplication, but the interest here is in a different definition of multiplication. Instead, multiplication is defined to be **convolution** with

$$(f_1 * f_2)(x) = \sum_{y \in G} f_1(xy^{-1}) f_2(y) = \sum_{y \in G} f_1(y) f_2(y^{-1}x).$$

The sums in question are finite because of the finite support of f_1 and f_2, and the sums are equal by a change of variables. This multiplication was introduced in the special case $R = \mathbb{C}$ in Section VII.4, and the argument for associativity given there in the special case works in general. With convolution as multiplication, $C(G, R)$ becomes an associative R algebra with identity. Problem 14 at the end of the chapter asks for a verification that the mapping $g \mapsto f_g$ with

$$f_g(x) = \begin{cases} 1 & \text{for } x = g, \\ 0 & \text{for } x \neq g, \end{cases}$$

extends to an R algebra isomorphism of RG onto $C(G, R)$.

2. Integral Domains and Fields of Fractions

For the remainder of the chapter we work with commutative rings only. In several of the sections, including this one, the commutative ring will be an integral domain, i.e., a nonzero commutative ring with identity and with no zero divisors.

2. Integral Domains and Fields of Fractions

In this section we show how an integral domain can be embedded canonically in a field. This embedding is handy for recognizing certain facts about integral domains as consequences of facts about fields. For example Proposition 4.28b established that if R is a nonzero integral domain and if $A(X)$ is a polynomial in $R[X]$ of degree $n > 0$, then $A(X)$ has at most n roots. Since the coefficients of the polynomial can be considered to be members of the larger field that contains R, this result is an immediate consequence of the corresponding fact about fields (Corollary 1.14).

The prototype is the construction of the field $\mathbb{Q}$ of rationals from the integral domain $\mathbb{Z}$ of integers as in Section A3 of the appendix, in which one thinks of $\frac{a}{b}$ as a pair (a, b) with $b \neq 0$ and then identifies pairs by saying that $\frac{a}{b} = \frac{c}{d}$ if and only if $ad = bc$.

We proceed in the same way in the general case. Thus let R be a nonzero integral domain, form the set

$$\widetilde{F} = \{(a, b) \mid a \in R, \ b \in R, \ b \neq 0\},$$

and impose the equivalence relation $(a, b) \sim (c, d)$ if $ad = bc$. The relation $\sim$ is certainly reflexive and symmetric. To see that it is transitive, suppose that $(a, b) \sim (c, d)$ and $(c, d) \sim (e, f)$. Then $ad = bc$ and $cf = de$, and these together force $adf = bcf = bde$. In turn, this implies $af = be$ since R is an integral domain and d is assumed $\neq 0$. Thus $\sim$ is transitive and is an equivalence relation. Let F be the set of equivalence classes.

The definition of addition in $\widetilde{F}$ is $(a, b)+(c, d) = (ad+bc, bd)$, the expression we get by naively clearing fractions, and we want to see that addition is consistent with the equivalence relation. In checking this, we need change only one of the pairs at a time. Thus suppose that $(a', b') \sim (a, b)$ and that (c, d) is given. We know that $a'b = ab'$, and we want to see that $(ad + bc, bd) \sim (a'd + b'c, b'd)$, i.e., that $(ad + bc)b'd = (a'd + b'c)bd$. In other words, we are to check that $adb'd = a'dbd$; we see immediately that this equality is valid since $ab' = a'b$. Consequently addition is consistent with the equivalence relation and descends to be defined on the set F of equivalence classes.

Taking into account the properties satisfied by members of an integral domain, we check directly that addition is commutative and associative on $\widetilde{F}$, and it follows that addition is commutative and associative on F.

The element $(0, 1)$ is a two-sided identity for addition in $\widetilde{F}$, and hence the class of $(0, 1)$ is a two-sided identity for addition in F. We denote this class by 0. Let us identify this class. A pair (a, b) is in the class of $(0, 1)$ if and only if $0 \cdot b = 1 \cdot a$, hence if and only if $a = 0$. In other words, the class of $(0, 1)$ consists of all $(0, b)$ with $b \neq 0$.

In $\widetilde{F}$, we have $(a, b) + (-a, b) = (ab + b(-a), bb) = (0, b^2) \sim (0, 1)$, and therefore the class of $(-a, b)$ is a two-sided inverse to the class of (a, b) under addition. Consequently F is an abelian group under addition.

The definition of multiplication in $\widetilde{F}$ is $(a,b)(c,d) = (ac, bd)$, and it is routine to see that this definition is consistent with the equivalence relation. Therefore multiplication descends to be defined on F. We check by inspection that multiplication is commutative and associative on $\widetilde{F}$, and it follows that it is commutative and associative on F. The element $(1, 1)$ is a two-sided identity for multiplication in $\widetilde{F}$, and the class of $(1, 1)$ is therefore a two-sided identity for multiplication in F. We denote this class by 1.

If (a, b) is not in the class 0, then $a \neq 0$, as we saw above. Then $ab \neq 0$, and we have $(a, b)(b, a) = (ab, ab) \sim (1, 1) = 1$. Hence the class of (b, a) is a two-sided inverse of the class of (a, b) under multiplication. Consequently the nonzero elements of F form an abelian group under multiplication.

For one of the distributive laws, the computation

$$(a, b)((c, d) + (e, f)) = (a, b)(cf + de, df) = (a(cf + de), bdf)$$
$$= (acf + ade, bdf) \sim (acbf + bdae, b^2df)$$
$$= (ac, bd) + (ae, bf) = (a, b)(c, d) + (a, b)(e, f)$$

shows that the classes of $(a, b)((c, d) + (e, f))$ and of $(a, b)(c, d) + (a, b)(e, f)$ are equal. The other distributive law follows from this one since F is commutative under multiplication. Therefore F is a field.

The field F is called the **field of fractions** of the integral domain R. The function $\eta : R \to F$ defined by saying that $\eta(r)$ is the class of $(\eta, 1)$ is easily checked to be a homomorphism of rings sending 1 to 1. It is one-one. Let us call it the canonical embedding of R into F. The pair (F, η) has the universal mapping property stated in Proposition 8.6 and illustrated in Figure 8.5.

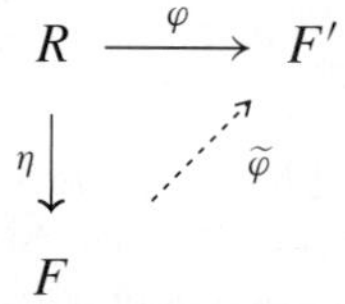

$$\begin{array}{ccc} R & \xrightarrow{\varphi} & F' \\ \eta \downarrow & \nearrow \widetilde{\varphi} & \\ F & & \end{array}$$

FIGURE 8.5. Universal mapping property of the field of fractions of R.

Proposition 8.6. Let R be a nonzero integral domain, let F be its field of fractions, and let η be the canonical embedding of R into F. Whenever φ is a one-one ring homomorphism of R into a field F' carrying 1 to 1, then there exists a unique ring homomorphism $\widetilde{\varphi} : F \to F'$ such that $\varphi = \widetilde{\varphi}\eta$, and $\widetilde{\varphi}$ is one-one as a homomorphism of fields.

REMARK. We say that $\widetilde{\varphi}$ is the extension of φ from R to F. Once this proposition has been proved, it is customary to drop η from the notation and regard R as a subring of its field of fractions.

PROOF. If (a, b) with $b \neq 0$ is a pair in $\widetilde{F}$, we define $\Phi(a, b) = \varphi(a)\varphi(b)^{-1}$. This is well defined since $b \neq 0$ and since φ, being one-one, cannot have $\varphi(b) = 0$. Let us see that Φ is consistent with the equivalence relation, i.e., that $(a, b) \sim (a', b')$ implies $\Phi(a, b) = \Phi(a', b')$. Since $(a, b) \sim (a', b')$, we have $ab' = a'b$ and therefore also $\varphi(a)\varphi(b') = \varphi(a')\varphi(b)$ and $\Phi(a, b) = \varphi(a)\varphi(b)^{-1} = \varphi(a')\varphi(b')^{-1} = \Phi(a', b')$, as required.

We can thus define $\widetilde{\varphi}$ of the class of (a, b) to be $\Phi(a, b)$, and $\widetilde{\varphi}$ is well defined as a function from F to F'. If r is in R, then $\widetilde{\varphi}(\eta(r)) = \widetilde{\varphi}(\text{class of } (r, 1)) = \Phi(r, 1) = \varphi(r)\varphi(1)^{-1}$, and this equals $\varphi(r)$ since φ is assumed to carry 1 into 1. Therefore $\widetilde{\varphi}\eta = \varphi$.

For uniqueness, let the class of (a, b) be given in F. Since b is nonzero, this class is the same as the class of $(a, 1)(b, 1)^{-1}$, which equals $\eta(a)\eta(b)^{-1}$. Since $(\widetilde{\varphi}\eta)(a) = \varphi(a)$ and $(\widetilde{\varphi}\eta)(b) = \varphi(b)$, we must have $\widetilde{\varphi}(\text{class of }(a, b)) = \widetilde{\varphi}(\eta(a))\widetilde{\varphi}(\eta(b))^{-1} = \varphi(a)\varphi(b)^{-1}$. Therefore φ uniquely determines $\widetilde{\varphi}$. □

If $\mathbb{K}$ is a field, then $R = \mathbb{K}[X]$ is an integral domain, and Proposition 8.6 applies to this R. The field of fractions consists in effect of formal rational expressions $P(X)Q(X)^{-1}$ in the indeterminate X, with the expected identifications made. We write $\mathbb{K}(X)$ for this field of fractions. More generally the field of fractions of the integral domain $\mathbb{K}[X_1, \ldots, X_n]$ consists of formal rational expressions in the indeterminates $X_1, \ldots, X_n$, with the expected identifications made, and is denoted by $\mathbb{K}(X_1, \ldots, X_n)$.

3. Prime and Maximal Ideals

In this section, R will denote a commutative ring, not necessarily having an identity. We shall introduce the notions of "prime ideal" and "maximal ideal," and we shall investigate relationships between these two notions.

A proper ideal I in R is **prime** if $ab \in I$ implies $a \in I$ or $b \in I$. The ideal $I = R$ is not prime, by convention.[5] We give three examples of prime ideals; a fourth example will be given in a proposition immediately afterward.

EXAMPLES.

(1) For $\mathbb{Z}$, it was shown in an example just before Proposition 4.21 that each ideal is of the form $m\mathbb{Z}$ for some integer m. We may assume that $m \geq 0$. The prime ideals are 0 and all $p\mathbb{Z}$ with p prime. To see this latter fact, consider $m\mathbb{Z}$ with $m \geq 2$. If $m = ab$ nontrivially, then neither a nor b is in I, but ab is in I; hence I is not prime. Conversely if m is prime, and if ab is in $I = m\mathbb{Z}$, then

---

[5]This convention is now standard. Books written before about 1960 usually regarded $I = R$ as a prime ideal. Correspondingly they usually treated the zero ring as an integral domain.

$ab = mc$ for some integer c. Since m is prime, Lemma 1.6 shows that m divides a or m divides b. Hence a is in I or b is in I. Therefore I is prime.

(2) If $\mathbb{K}$ is a field, then each ideal in $R = \mathbb{K}[X]$ is of the form $A(X)\mathbb{K}[X]$ with $A(X)$ in $\mathbb{K}[X]$, and $A(X)\mathbb{K}[X]$ is prime if and only if $A(X)$ is 0 or is a prime polynomial. In fact, each ideal is of the form $A(X)\mathbb{K}[X]$ by Proposition 5.8. If $A(X)$ is not a constant polynomial, then the argument that $A(X)\mathbb{K}[X]$ is prime if and only if the polynomial $A(X)$ is prime proceeds as in Example 1, using Lemma 1.16 in place of Lemma 1.6.

(3) In $R = \mathbb{Z}[X]$, the structure of the ideals is complicated, and we shall not attempt to list all ideals. Let us observe simply that the ideal $I = X\mathbb{Z}[X]$ is prime. In fact, if $A(X)B(X)$ is in $X\mathbb{Z}[X]$, then $A(X)B(X) = XC(X)$ for some $C(X)$ in $\mathbb{Z}[X]$. If the constant terms of $A(X)$ and $B(X)$ are a_0 and b_0, this equation says that $a_0 b_0 = 0$. Therefore $a_0 = 0$ or $b_0 = 0$. In the first case, $A(X) = XP(X)$ for some $P(X)$, and then $A(X)$ is in I; in the second case, $B(X) = XQ(X)$ for some $Q(X)$, and then $B(X)$ is in I. We conclude that I is prime.

Proposition 8.7. An ideal I in the commutative ring R is prime if and only if R/I is an integral domain.

PROOF. If a proper ideal I fails to be prime, choose ab in I with $a \notin I$ and $b \notin I$. Then $a + I$ and $b + I$ are nonzero in R/I and have product $0 + I$. So R/I is nonzero and has a zero divisor; by definition, R/I fails to be an integral domain.

Conversely if R/I (is nonzero and) has a zero divisor, choose $a + I$ and $b + I$ nonzero with product $0 + I$. Then neither a nor b is in I but ab is in I. Since I is certainly proper, I is not prime. □

A proper ideal I in the commutative ring R is said to be **maximal** if R has no proper ideal J with $I \subsetneq J$. If the commutative ring R has an identity, a simple way of testing whether an ideal I is proper is to check whether 1 is in I; in fact, if 1 is in I, then $I \supseteq RI \supseteq R1 = R$ implies $I = R$. Maximal ideals exist in abundance when R is nonzero and has an identity, as a consequence of the following result.

Proposition 8.8. In a commutative ring R with identity, any proper ideal is contained in a maximal ideal.

PROOF. This follows from Zorn's Lemma (Section A5 of the appendix). Specifically let I be the given proper ideal, and form the set S of all proper ideals that contain I. This set is nonempty, containing I as a member, and we order it by inclusion upward. If we have a chain in S, then the union of the members of the chain is an ideal that contains all the ideals in the chain, and it is proper since it does not contain 1. Therefore the union of the ideals in the chain is an upper bound for the chain. By Zorn's Lemma the set S has a maximal element, and any such maximal element is a maximal ideal containing I. □

Lemma 8.9. If R is a nonzero commutative ring with identity, then R is a field if and only if the only proper ideal in R is 0.

PROOF. If R is a field and I is a nonzero ideal in R, let $a \neq 0$ be in I. Then $1 = aa^{-1}$ is in I, and consequently $I = R$. Conversely if the only ideals in R are 0 and R, let $a \neq 0$ be given in R, and form the ideal $I = aR$. Since 1 is in R, a is in I. Thus $I \neq 0$. Then I must be R. So there exists some b in R with $1 = ba$, and a is exhibited as having the inverse b. □

Proposition 8.10. If R is a commutative ring with identity, then an ideal I is maximal if and only if R/I is a field.

REMARK. One can readily give a direct proof, but it seems instructive to give a proof reducing the result to Lemma 8.9.

PROOF. We consider R and R/I as unital R modules, the ideals for each of R and R/I being the R submodules. The quotient ring homomorphism $R \to R/I$ is an R homomorphism. By the First Isomorphism Theorem for modules (Theorem 8.3), there is a one-one correspondence between the ideals in R containing I and the ideals in R/I. Then the result follows immediately from Lemma 8.9. □

Corollary 8.11. If R is a commutative ring with identity, then every maximal ideal is prime.

PROOF. If I is maximal, then R/I is a field by Proposition 8.10. Hence R/I is an integral domain, and I must be prime by Proposition 8.7. □

In the converse direction nonzero prime ideals need not be maximal, as the following example shows. However, Proposition 8.12 will show that nonzero prime ideals *are* necessarily maximal in certain important rings.

EXAMPLE. In $R = \mathbb{Z}[X]$, we have seen that $I = X\mathbb{Z}[X]$ is a prime ideal. But I is not maximal since $X\mathbb{Z}[X] + 2\mathbb{Z}[X]$ is a proper ideal that strictly contains I.

Proposition 8.12. In $R = \mathbb{Z}$ or $R = \mathbb{K}[X]$ with $\mathbb{K}$ a field, every nonzero prime ideal is maximal.

PROOF. Examples 1 and 2 at the beginning of this section show that every nonzero prime ideal is of the form $I = pR$ with p prime. If such an I is given and if J is any ideal strictly containing I, choose a in J with a not in I. Since a

is not in $I = pR$, it is not true that p divides a. So p and a are relatively prime, and there exist elements x and y in R with $xp + ya = 1$, by Proposition 1.2c or 1.15d. Since p and a are in J, so is 1. Therefore $J = R$, and I is not strictly contained in any proper ideal. So I is maximal. □

EXAMPLE. Algebraic number fields $\mathbb{Q}[\theta]$. These were introduced briefly in Chapter IV and again in Section 1 as the $\mathbb{Q}$ linear span of all powers $1, \theta, \theta^2, \ldots$. Here θ is a nonzero complex number, and we make the assumption that $\mathbb{Q}[\theta]$ is a finite-dimensional vector space over $\mathbb{Q}$. Proposition 4.1 showed that $\mathbb{Q}[\theta]$ is then indeed a field. Let us see how this conclusion relates to the results of the present section. In fact, write a nontrivial linear dependence of $1, \theta, \theta^2, \ldots$ over $\mathbb{Q}$ in the form $c_0 + c_1\theta + c_2\theta^2 + \cdots + c_{n-1}\theta^{n-1} + \theta^n = 0$. Without loss of generality, suppose that this particular linear dependence has n as small as possible among all such relations. Then θ is a root of

$$P(X) = c_0 + c_1 X + c_2 X^2 + \cdots + c_{n-1} X^{n-1} + X^n.$$

Consider the substitution homomorphism $E : \mathbb{Q}[X] \to \mathbb{C}$ given by $E(A(X)) = A(\theta)$. This ring homomorphism carries $\mathbb{Q}[X]$ onto the ring $\mathbb{Q}[\theta]$, and the kernel is some ideal I. Specifically I consists of all polynomials $A(X)$ with $A(\theta) = 0$, and $P(X)$ is one of these of lowest possible degree. Proposition 5.8 shows that I consists of all multiples of some polynomial, and that polynomial may be taken to be $P(X)$ by minimality of the integer n. Proposition 8.1 therefore shows that $\mathbb{Q}[\theta] \cong \mathbb{Q}[X]/P(X)\mathbb{Q}[X]$ as a ring. If $P(X)$ were to have a nontrivial factorization as $P(X) = Q_1(X)Q_2(X)$, then $P(\theta) = 0$ would imply $Q_1(\theta) = 0$ or $Q_2(\theta) = 0$, and we would obtain a contradiction to the minimality of n. Therefore $P(X)$ is prime. By Example 2 earlier in the section, $I = P(X)\mathbb{Q}[X]$ is a nonzero prime ideal, and Proposition 8.12 shows that it is maximal. By Proposition 8.10 the quotient ring $\mathbb{Q}[\theta] = \mathbb{Q}[X]/P(X)\mathbb{Q}[X]$ is a field. These computations with $\mathbb{Q}[\theta]$ underlie the first part of the theory of fields that we shall develop in Chapter IX.

4. Unique Factorization

We have seen that the positive members of $\mathbb{Z}$ and the nonzero members of $\mathbb{K}[X]$, when $\mathbb{K}$ is a field, factor into the products of "primes" and that these factorizations are unique up to order and up to adjusting each of the prime factors in $\mathbb{K}[X]$ by a unit. In this section we shall investigate this idea of unique factorization more generally. Zero divisors are problematic from the point of view of factorization, and it will be convenient to exclude them. Therefore we work exclusively with integral domains.

4. Unique Factorization

The first observation is that unique factorization is not a completely general notion for integral domains. Let us consider an example in detail.

EXAMPLE. $R = \mathbb{Z}[\sqrt{-5}]$. This is the subring of $\mathbb{C}$ whose members are of the form $a + b\sqrt{-5}$ with a and b integers. Since $(a + b\sqrt{-5})(c + d\sqrt{-5}) = (ac - 5cd) + (ad + bc)\sqrt{-5}$, R is closed under multiplication and is indeed a subring. Define $N(a + b\sqrt{-5}) = a^2 + 5b^2 = (a + b\sqrt{-5})\overline{(a + b\sqrt{-5})}$. This is a nonnegative-integer-valued function on R and is 0 only on the 0 element of R. Since complex conjugation is an automorphism of $\mathbb{C}$, we check immediately that

$$N\big((a + b\sqrt{-5})(c + d\sqrt{-5})\big) = N(a + b\sqrt{-5})N(c + d\sqrt{-5}).$$

The group of units of R, i.e., of elements with inverses under multiplication, is denoted by $R^\times$ as usual. If r is in $R^\times$, then $rr^{-1} = 1$, and so $N(r)N(r^{-1}) = N(1) = 1$. Consequently the units r of R all have $N(r) = 1$. Setting $a^2 + 5b^2 = 1$, we see that the units are ± 1. The product formula for N shows that if we start factoring a member of R, then factor its factors, and so on, and if we forbid factorizations into two factors when one is a unit, then the process of factorization has to stop at some point. So complete factorization makes sense. Now consider the equality

$$6 = (1 + \sqrt{-5})(1 - \sqrt{-5}) = 2 \cdot 3.$$

The factors here have $N(1 + \sqrt{-5}) = N(1 - \sqrt{-5}) = 6$, $N(2) = 4$, and $N(3) = 9$. Considering the possible values of $a^2 + 5b^2$, we see that $N(\cdot)$ does not take on either of the values 2 and 3 on R. Consequently $1 + \sqrt{-5}$, $1 - \sqrt{-5}$, 2, and 3 do not have nontrivial factorizations. On the other hand, consideration of the values of $N(\cdot)$ shows that 2 and 3 are not products of either of $1 \pm \sqrt{-5}$ with units. We conclude that the displayed factorizations of 6 show that unique factorization has failed.

Thus unique factorization is not universal for integral domains. It is time to be careful about terminology. With $\mathbb{Z}$ and $\mathbb{K}[X]$, we have referred to the individual factors in a complete factorization as "primes." Their defining property in Chapter I was that they could not be factored further in nontrivial fashion. Primes in these rings were shown to have the additional property that if a prime divides a product then it divides one of the factors. It is customary to separate these two properties for general integral domains. Let us say that a nonzero element a **divides** b if $b = ac$ for some c. In this case we say also that a is a **factor** of b. In an integral domain R, a nonzero element r that is not a unit is said to be **irreducible** if every factorization $r = r_1 r_2$ in R has the property that either r_1 or r_2 is a unit. Nonzero nonunits that are not irreducible are said to be **reducible**. A nonzero element p that is not a unit is said to be **prime**[6] if the condition that p divides a product ab always implies that p divides a or p divides b.

Prime implies irreducible. In fact, if p is a prime that is reducible, let us write $p = r_1 r_2$ with neither r_1 nor r_2 equal to a unit. Since p is prime, p divides r_1 or r_2, say r_1. Then $r_1 = pc$ with c in R, and we obtain $p = r_1 r_2 = p c r_2$. Since R is an integral domain, $1 = c r_2$, and r_2 is exhibited as a unit with inverse c, in contradiction to the assumption that r_2 is not a unit.

On the other hand, irreducible does not imply prime. In fact, we saw in $\mathbb{Z}[\sqrt{-5}]$ that $1 + \sqrt{-5}$ is irreducible. But $1 + \sqrt{-5}$ divides $2 \cdot 3$, and $1 + \sqrt{-5}$ does not divide either of 2 or 3. Therefore $1 + \sqrt{-5}$ is not prime.

We shall see in a moment that the distinction between "irreducible" and "prime" lies at the heart of the question of unique factorization. Let us make a definition that helps identify our problem precisely. We say that an integral domain R is a **unique factorization domain** if R has the two properties

(UFD1) every nonzero nonunit of R is a finite product of irreducible elements,

(UFD2) the factorization in (UFD1) is always unique up to order and to multiplication of the factors by units.

The problem that arises for us for a given R is to decide whether R is a unique factorization domain. The following proposition shows the relevance of the distinction between "irreducible" and "prime."

Proposition 8.13. In an integral domain R in which (UFD1) holds, the condition (UFD2) is equivalent to the condition

(UFD2') every irreducible element is prime.

REMARKS. In fact, showing that irreducible implies prime was the main step in Chapter I in proving unique factorization for positive integers and for $\mathbb{K}[X]$ when $\mathbb{K}$ is a field. The mechanism for carrying out the proof that irreducible implies prime for those settings will be abstracted in Theorems 8.15 and 8.17.

PROOF. Suppose that (UFD2) holds, that p is an irreducible element, and that p divides ab. We are to show that p divides a or p divides b. We may assume that $ab \neq 0$. Write $ab = pc$, and let $a = \prod_i p_i$, $b = \prod_j p'_j$, and $c = \prod_k q_k$ be factorizations via (UFD1) into products of irreducible elements.

---

[6]This definition enlarges the definition of "prime" in $\mathbb{Z}$ to include the negatives of the usual prime numbers. Unique factorization immediately extends to nonzero integers of either sign, but the prime factors are now determined only up to factors of ± 1. In cases where confusion about the sign of an integer prime might arise, the text will henceforth refer to "primes of $\mathbb{Z}$" or "integer primes" when both signs are allowed, and to "positive primes" or "prime numbers" when the primes are understood to be as in Chapter I.

Then $\prod_{i,j} p_i p'_j = p \prod_k q_k$. By (UFD2) one of the factors on the left side is εp for some unit ε. Then p either is of the form $\varepsilon^{-1} p_i$ and then p divides a, or is of the form $\varepsilon^{-1} p'_j$ and then p divides b. Hence (UFD2′) holds.

Conversely suppose that (UFD2′) holds. Let the nonzero nonunit r have two factorizations into irreducible elements as $r = p_1 p_2 \cdots p_m = \varepsilon_0 q_1 q_2 \cdots q_n$ with $m \leq n$ and with ε_0 a unit. We prove the uniqueness by induction on m, the case $m = 0$ being trivial and the case $m = 1$ following from the definition of "irreducible." Inductively from (UFD2′) we know that p_m divides q_k for some k. Since q_k is irreducible, $q_k = \varepsilon p_m$ for some unit ε. Thus we can cancel q_k and obtain $p_1 p_2 \cdots p_{m-1} = \varepsilon_0 \varepsilon q_1 q_2 \cdots \widehat{q_k} \cdots q_n$, the hat indicating an omitted factor. By induction the factors on the two sides here are the same except for order and units. Thus the same conclusion is valid when comparing the two sides of the equality $p_1 p_2 \cdots p_m = \varepsilon_0 q_1 q_2 \cdots q_n$. The induction is complete, and (UFD2) follows. $\square$

It will be convenient to simplify our notation for ideals. In any commutative ring R with identity, if a is in R, we let (a) denote the ideal Ra generated by a. An ideal of this kind with a single generator is called a **principal ideal**. More generally, if $a_1, \ldots, a_n$ are members of R, then $(a_1, \ldots, a_n)$ denotes the ideal $Ra_1 + \cdots + Ra_n$ generated by $a_1, \ldots, a_n$. For example, in $\mathbb{Z}[X]$, $(2, X)$ denotes the ideal $2\mathbb{Z} + X\mathbb{Z}$ of all polynomials whose constant term is even. The following condition explains a bit the mystery of what it means for an element to be prime.

Proposition 8.14. A nonzero element p in an integral domain R is prime if and only if the ideal (p) in R is prime.

PROOF. Suppose that the element p is prime. Then the ideal (p) is not R; in fact, otherwise 1 would have to be of the form $1 = rp$ for some $r \in R$, r would be a multiplicative inverse of p, and p would be a unit. Now suppose that a product ab is in the ideal (p). Then $ab = pr$ for some r in R, and p divides ab. Since p is prime, p divides a or p divides b. Therefore the ideal (p) is prime.

Conversely suppose that (p) is a prime ideal with $p \neq 0$. Since $(p) \neq R$, p is not a unit. If p divides the product ab, then $ab = pc$ for some c in R. Hence ab is in (p). Since (p) is assumed prime, either a is in (p) or b is in (p). In the first case, p divides a, and in the second case, p divides b. Thus the element p is prime. $\square$

An integral domain R is called a **principal ideal domain** if every ideal in R is principal. At the beginning of Section 3, we saw a reminder that $\mathbb{Z}$ is a principal ideal domain and that so is $\mathbb{K}[X]$ whenever $\mathbb{K}$ is a field. It turns out that unique factorization for these cases is a consequence of this fact.

Theorem 8.15. Every principal ideal domain is a unique factorization domain.

REMARKS. Let R be the given principal ideal domain. Proposition 8.13 shows that it is enough to show that (UFD1) and (UFD2′) hold in R.

PROOF OF (UFD1). Let a_1 be a nonzero nonunit of R. If a_1 is not irreducible, then a_1 has a factorization $a_1 = a_2 b_2$ in which neither a_2 nor b_2 is a unit. If a_2 is not irreducible, then a_2 has a factorization $a_2 = a_3 b_3$ in which neither a_3 nor b_3 is a unit. We continue in this way as long as it is possible to do so. Let us see that this process cannot continue indefinitely. Assume the contrary. The equality $a_1 = a_2 b_2$ with b_2 not a unit says that a_1 is in the ideal (a_2) and a_2 is not in the ideal (a_1). Arguing in this way with a_2, a_3, and so on, we obtain

$$(a_1) \subsetneq (a_2) \subsetneq (a_3) \subsetneq \cdots .$$

Let $I = \bigcup_{n=1}^\infty (a_n)$. Then I is an ideal. Since R is a principal ideal domain, $I = (a)$ for some a. This element a must be in (a_k) for some k, and then we have $(a_k) = (a_{k+1}) = \cdots = (a)$. This is a contradiction, and hence the process does not continue indefinitely.

Therefore some irreducible element c_1, namely the element a_k in the above argument, divides a_1. Write $a_1 = c_1 a_2$, and repeat the above argument with a_2. Iterating this construction, we obtain $a_n = c_n a_{n+1}$ for each n with c_n irreducible. Thus $a_1 = c_1 c_2 \cdots c_n a_{n+1}$ with $c_1, \ldots, c_n$ irreducible. Let us see that this process cannot continue indefinitely. Assuming the contrary, we are led to the strict inclusions

$$(a_1) \subsetneq (a_2) \subsetneq (a_3) \subsetneq \cdots .$$

Again we cannot have such an infinite chain of strict inclusions in a principal ideal domain, and we must have $(a_n) = (a_{n+1})$ at some stage. Then c_n has to be a unit, contradiction. Thus a_n has no nontrivial factorization, and $a_1 = c_1 \cdots c_{n-1} a_n$ is the desired factorization. This proves (UFD1). □

PROOF OF (UFD2′). If p is an irreducible element, we prove that the ideal (p) is maximal. Corollary 8.11 shows that (p) is prime, and Proposition 8.14 shows that p is prime. Thus (UFD2′) will follow.

The element p, being irreducible, is not a unit. Thus (p) is proper. Suppose that I is an ideal with $I \supsetneq (p)$. Since R is a principal ideal domain, $I = (c)$ for some c. Then $p = rc$ for some r in R. Since $I \neq (p)$, r cannot be a unit. Therefore the irreducibility of p implies that c is a unit. Then $I = (c) = (1) = R$, and we conclude that (p) is maximal. □

Let us record what is essentially a corollary of the proof.

4. Unique Factorization

Corollary 8.16. In a principal ideal domain, every nonzero prime ideal is maximal.

PROOF. Let (p) be a nonzero prime ideal. Proposition 8.14 shows that p is prime, and prime elements are automatically irreducible. The proof of the uniqueness part of Theorem 8.15 then deduces in the context of a principal ideal domain that (p) is maximal. □

Principal ideal domains arise comparatively infrequently, and recognizing them is not necessarily easy. The technique that was used with $\mathbb{Z}$ and $\mathbb{K}[X]$ generalizes slightly, and we take up that generalization now. An integral domain R is called a **Euclidean domain** if there exists a function $\delta : R \to \{\text{integers} \geq 0\}$ such that whenever a and b are in R with $b \neq 0$, there exist q and r in R with $a = bq + r$ and $\delta(r) < \delta(b)$. The ring $\mathbb{Z}$ of integers is a Euclidean domain if we take $\delta(n) = |n|$, and the ring $\mathbb{K}[X]$ for $\mathbb{K}$ a field is a Euclidean domain if we take $\delta(P(X))$ to be $2^{\deg P}$ if $P(X) \neq 0$ and to be 0 if $P(X) = 0$.

Another example of a Euclidean domain is the ring $\mathbb{Z}[\sqrt{-1}] = \mathbb{Z} + \mathbb{Z}\sqrt{-1}$ of **Gaussian integers**. It has $\delta(a+b\sqrt{-1}) = (a+b\sqrt{-1})(a-b\sqrt{-1}) = a^2 + b^2$, a and b being integers. Let us abbreviate $\sqrt{-1}$ as i. To see that δ has the required property, we first extend δ to $\mathbb{Q}[i]$, writing $\delta(x+yi) = (x+yi)(x-yi) = x^2+y^2$ if x and y are rational. We use the fact that

$$\delta(zz') = \delta(z)\delta(z') \quad \text{for } z \text{ and } z' \text{ in } \mathbb{Q}[i],$$

which follows from the computation $\delta(zz') = zz' \cdot \overline{zz'} = z\bar{z}z'\bar{z'} = \delta(z)\delta(z')$. For any real number u, let $[u]$ be the greatest integer $\leq u$. Every real u satisfies $\left|[u+\tfrac{1}{2}] - u\right| \leq \tfrac{1}{2}$. Given $a + ib$ and $c + di$ with $c + di \neq 0$, we write

$$\frac{a+bi}{c+di} = \frac{(a+bi)(c-di)}{c^2+d^2} = \frac{ac+bd}{c^2+d^2} + \frac{bc-ad}{c^2+d^2}i.$$

Put $p = \left[\frac{ac+bd}{c^2+d^2} + \tfrac{1}{2}\right]$, $q = \left[\frac{bc-ad}{c^2+d^2} + \tfrac{1}{2}\right]$, and $r+si = (a+bi)-(c+di)(p+qi)$. Then

$$a+bi = (c+di)(p+qi) + (r+si),$$

and

$$\delta(r+si) = \delta\big((a+bi)-(c+di)(p+qi)\big) = \delta(c+di)\delta\left(\frac{a+bi}{c+di} - (p+qi)\right).$$

The complex number $x + yi = \frac{a+bi}{c+di} - (p+qi) = \left(\frac{ac+bd}{c^2+d^2} - p\right) + \left(\frac{bc-ad}{c^2+d^2} - q\right)i$ has $|x| \leq \tfrac{1}{2}$ and $|y| \leq \tfrac{1}{2}$, and therefore $\delta(x+yi) = x^2 + y^2 \leq \tfrac{1}{4} + \tfrac{1}{4} = \tfrac{1}{2}$. Hence $\delta(r+si) < \delta(c+di)$, as required.

Some further examples of this kind appear in Problems 13 and 25–26 at the end of the chapter. The matter is a little delicate. The ring $\mathbb{Z}[\sqrt{-5}]$ may seem superficially similar to $\mathbb{Z}[\sqrt{-1}]$. But $\mathbb{Z}[\sqrt{-5}]$ does not have unique factorization, and the following theorem, in combination with Theorem 8.15, assures us that $\mathbb{Z}[\sqrt{-5}]$ cannot be a Euclidean domain.

Theorem 8.17. Every Euclidean domain is a principal ideal domain.

PROOF. Let I be an ideal in R. We are to show that I is principal. Without loss of generality, we may assume that $I \neq 0$. Choose $b \neq 0$ in I with $\delta(b)$ as small as possible. Certainly $I \supseteq (b)$. If $a \neq 0$ is in I, write $a = bq + r$ with $\delta(r) < \delta(b)$. Then $r = a - bq$ is in I with $\delta(r) < \delta(b)$. The minimality of b forces $r = 0$ and $a = bq$. Thus $I \subseteq (b)$, and we conclude that $I = (b)$. □

5. Gauss's Lemma

In the previous section we saw that every principal ideal domain has unique factorization. In the present section we shall establish that certain additional integral domains have unique factorization, namely any integral domain $R[X]$ for which R is a unique factorization domain. A prototype is $\mathbb{Z}[X]$, which will be seen to have unique factorization even though there exist nonprincipal ideals like $(2, X)$ in the ring. An important example for applications, particularly in algebraic geometry, is $\mathbb{K}[X_1, \ldots, X_n]$, where $\mathbb{K}$ is a field; in this case our result is to be applied inductively, making use of the isomorphism $\mathbb{K}[X_1, \ldots, X_n] \cong \mathbb{K}[X_1, \ldots, X_{n-1}][X_n]$ given in Corollary 4.31.

For the conclusion that $R[X]$ has unique factorization if R does, the heart of the proof is application of a result known as Gauss's Lemma, which we shall prove in this section. Gauss's Lemma has additional consequences for $R[X]$ beyond unique factorization, and we give them as well.

Before coming to Gauss's Lemma, let us introduce some terminology and prove one preliminary result. In any integral domain R, we call two nonzero elements a and b **associates** if $a = b\varepsilon$ for some ε in the group $R^\times$ of units. The property of being associates is an equivalence relation because $R^\times$ is a group.

Still with the nonzero integral domain R, let us define a **greatest common divisor** of two nonzero elements a and b to be any element c of R such that c divides both a and b and such that any divisor of a and b divides c. Any associate of a greatest common divisor of a and b is another greatest common divisor of a and b. Conversely if a and b have a greatest common divisor, then any two greatest common divisors are associates. In fact, if c and c' are greatest common divisors, then each of them divides both a and b, and the definition forces each of them to divide the other. Thus $c' = c\varepsilon$ and $c = c'\varepsilon'$, and then $c' = c'\varepsilon'\varepsilon$ and $1 = \varepsilon'\varepsilon$. Consequently ε is a unit, and c and c' are associates.

If R is a unique factorization domain, then any two nonzero elements a and b have a greatest common divisor. In fact, we decompose a and b into the product of a unit by powers of nonassociate irreducible elements as $a = \varepsilon \prod_{i=1}^{m} p_i^{k_i}$ and $b = \varepsilon' \prod_{j=1}^{n} p_j'^{l_j}$. For each p_j' such that p_j' is associate to some p_i, we replace p_j' by p_i in the factorization of b, adjusting ε' as necessary, and then we reorder the factors of a and b so that the common p_i's are the ones for $1 \leq i \leq r$. Then $c = \prod_{i=1}^{r} p_i^{\min(k_i, l_i)}$ is a greatest common divisor of a and b. We write $\mathrm{GCD}(a, b)$ for a greatest common divisor of a and b; as we saw above, this is well defined up to a factor of a unit.[7]

One should not read too much into the notation. In a principal ideal domain if a and b are nonzero, then, as we shall see momentarily, $\mathrm{GCD}(a, b)$ is defined by the condition on ideals that

$$(\mathrm{GCD}(a, b)) = (a, b).$$

This condition implies that there exist elements x and y in R such that

$$xa + yb = \mathrm{GCD}(a, b).$$

However, in the integral domain $\mathbb{Z}[X]$, in which $\mathrm{GCD}(2, X) = 1$, there do not exist polynomials $A(X)$ and $B(X)$ with $A(X)2 + B(X)X = 1$.

To prove that $(\mathrm{GCD}(a, b)) = (a, b)$ in a principal ideal domain, write (c) for the principal ideal (a, b); c satisfies $c = xa + yb$ for some x and y in R. Since a and b lie in (c), $a = rc$ and $b = r'c$. Hence c divides both a and b. In the reverse direction if d divides a and b, then $ds = a$ and $ds' = b$. Hence $c = xa + yb = (xs + ys')d$, and d divides c. So c is indeed a greatest common divisor of a and b.

In a unique factorization domain the definition of greatest common divisor immediately extends to apply to n nonzero elements, rather than just two. We readily check up to a unit that

$$\mathrm{GCD}(a_1, \ldots, a_n) = \mathrm{GCD}\big(\mathrm{GCD}(a_1, \ldots, a_{n-1}), a_n\big).$$

Moreover, we can allow any of $a_2, \ldots, a_n$ to be 0, and there is no difficulty. In addition, we have

$$\mathrm{GCD}(da_1, \ldots, da_n) = d\,\mathrm{GCD}(a_1, \ldots, a_n) \qquad \text{up to a unit}$$

if d and a_1 are not 0.

Let R be a unique factorization domain. If $A(X)$ is a nonzero element of $R[X]$, we say that $A(X)$ is **primitive** if the GCD of its coefficients is a unit. In this case no prime of R divides all the coefficients of $A(X)$.

---

[7] Greatest common divisors can exist for certain integral domains that fail to have unique factorization, but we shall not have occasion to work with any such domains.

Theorem 8.18 (Gauss's Lemma). If R is a unique factorization domain, then the product of primitive polynomials is primitive.

PROOF #1. Arguing by contradiction, let $A(X) = a_m X^m + \cdots + a_0$ and $B(X) = b_n X^n + \cdots + b_0$ be primitive polynomials such that every coefficient of $A(X)B(X)$ is divisible by some prime p. Since $A(X)$ and $B(X)$ are primitive, we may choose k and l as small as possible such that p does not divide a_k and does not divide b_l. The coefficient of X^{k+l} in $A(X)B(X)$ is

$$a_0 b_{k+l} + a_1 b_{k+l-1} + \cdots + a_k b_l + \cdots + a_{k+l} b_0$$

and is divisible by p. Then all the individual terms, and their sum, are divisible by p except possibly for $a_k b_l$, and we conclude that p divides $a_k b_l$. Since p is prime and p divides $a_k b_l$, p must divide a_k or b_l, contradiction. □

PROOF #2. Arguing by contradiction, let $A(X)$ and $B(X)$ be primitive polynomials such that every coefficient of $A(X)B(X)$ is divisible by some prime p. Proposition 8.14 shows that the ideal (p) is prime, and Proposition 8.7 shows that $R' = R/(p)$ is an integral domain. Let $\varphi : R \to R'[X]$ be the composition of the quotient homomorphism $R \to R'$ and the inclusion of R' into constant polynomials in $R'[X]$, and let $\Phi : R[X] \to R'[X]$ be the corresponding substitution homomorphism of Proposition 4.24 that carries X to X. Since $A(X)$ and $B(X)$ are primitive, $\Phi(A(X))$ and $\Phi(B(X))$ are not zero. Their product $\Phi(A(X))\Phi(B(X)) = \Phi(A(X)B(X))$ is 0 since p divides every coefficient of $A(X)B(X)$, and this conclusion contradicts the assertion of Proposition 4.29 that $R'[X]$ is an integral domain. □

Let F be the field of fractions of the unique factorization domain R. The consequences of Theorem 8.18 exploit a simple relationship between $R[X]$ and $F[X]$, which we state below as Proposition 8.19. Once that proposition is in hand, we can state the consequences of Theorem 8.18. If $A(X)$ is a nonzero polynomial in $R[X]$, let $c(A)$ to be the greatest common divisor of the coefficients, i.e.,

$$c(A) = \text{GCD}(a_n, \ldots, a_1, a_0) \qquad \text{if } A(X) = a_n X^n + \cdots + a_1 X + a_0.$$

The element $c(A)$ is well defined up to a factor of a unit. In this notation the definition of "primitive" becomes, $A(X)$ is primitive if and only if $c(A)$ is a unit.

If $A(X)$ is not necessarily primitive, then at least $c(A)$ divides each coefficient of $A(X)$, and hence $c(A)^{-1} A(X)$ is in $R[X]$, say with coefficients $b_n, \ldots, b_1, b_0$. Then we have

$$c(A) = \text{GCD}(a_n, \ldots, a_1, a_0) = \text{GCD}(c(A)b_n, \ldots, c(A)b_1, c(A)b_0)$$
$$= c(A)\text{GCD}(b_n, \ldots, b_1, b_0) = c(A)c\bigl(c(A)^{-1} A(X)\bigr)$$

up to a unit factor, and hence $c\big(c(A)^{-1}A(X)\big)$ is a unit. We conclude that

$$A(X) \in R[X] \quad \text{implies that} \quad c(A)^{-1}A(X) \text{ is primitive.}$$

Proposition 8.19. Let R be a unique factorization domain, and let F be its field of fractions. If $A(X)$ is any nonzero polynomial in $F[X]$, then there exist α in F and $A_0(X)$ in $R[X]$ such that $A(X) = \alpha A_0(X)$ with $A_0(X)$ primitive. The scalar α and the polynomial $A_0(X)$ are unique up to multiplication by units in R.

REMARK. We call $A_0(X)$ the **associated primitive polynomial** to $A(X)$. According to the proposition, it is unique up to a unit factor in R.

PROOF. Let $A(X) = c_n X^n + \cdots + c_1 X + c_0$ with each c_k in F. We can write each c_k as $a_k b_k^{-1}$ with a_k and b_k in R and $b_k \neq 0$. We clear fractions. That is, we let $\beta = \prod_{k=0}^n b_k$. Then the kth coefficient of $\beta A(X)$ is $a_k \prod_{l \neq k} b_l$ and is in R. Hence $\beta A(X)$ is in $R[X]$. The observation just before the proposition shows that $c(\beta A)^{-1}\beta A$ is primitive. Thus $A(X) = \alpha A_0(X)$ with $\alpha = \beta^{-1}c(\beta A)$ and $A_0(X) = c(\beta A)^{-1}\beta A(X)$, $A_0(X)$ being primitive. This proves existence.

If $\alpha_1 A_1(X) = \alpha_2 A_2(X)$ with α_1 and α_2 in F and with $A_1(X)$ and $A_2(X)$ primitive, choose $r \neq 0$ in R such that $r\alpha_1$ and $r\alpha_2$ are in R. Up to unit factors in R, we then have $r\alpha_1 = r\alpha_1 c(A_1) = c(r\alpha_1 A_1) = c(r\alpha_2 A_2) = r\alpha_2 c(A_2) = r\alpha_2$. Hence, up to a unit factor in R, we have $\alpha_1 = \alpha_2$. This proves uniqueness. □

Corollary 8.20. Let R be a unique factorization domain, and let F be its field of fractions.

(a) Let $A(X)$ and $B(X)$ be nonzero polynomials in $R[X]$, and suppose that $B(X)$ is primitive. If $B(X)$ divides $A(X)$ in $F[X]$, then it divides $A(X)$ in $R[X]$.

(b) If $A(X)$ is an irreducible polynomial in $R[X]$ of degree > 0, then $A(X)$ is irreducible in $F[X]$.

(c) If $A(X)$ is a monic polynomial in $R[X]$ and if $B(X)$ is a monic factor of $A(X)$ within $F[X]$, then $B(X)$ is in $R[X]$.

(d) If $A(X)$, $B(X)$, and $C(X)$ are in $R[X]$ with $A(X)$ primitive and with $A(X) = B(X)C(X)$, then $B(X)$ and $C(X)$ are primitive.

PROOF. In (a), write $A(X) = B(X)Q(X)$ in $F(X)$, and let $Q(X) = \rho Q_0(X)$ be a decomposition of $Q(X)$ as in Proposition 8.19. Since $c(A)^{-1}A(X)$ is primitive, the corresponding decomposition of $A(X)$ is $A(X) = c(A)\big(c(A)^{-1}A(X)\big)$. The equality $A(X) = \rho B(X)Q_0(X)$ then reads $c(A)(c(A)^{-1}A(X)) = \rho B(X)Q_0(X)$. Since $B(X)Q_0(X)$ is primitive according to Theorem 8.18, the uniqueness in Proposition 8.19 shows that $c(A)^{-1}A(X) = B(X)Q_0(X)$ except possibly for a unit factor in R. Then $B(X)$ divides $A(X)$ with quotient $c(A)Q_0(X)$, apart from a unit factor in R. Since $c(A)Q_0(X)$ is in $R[X]$, (a) is proved.

In (b), the condition that $\deg A(X) > 0$ implies that $A(X)$ is not a unit in $F[X]$. Arguing by contradiction, suppose that $A(X) = B(X)Q(X)$ in $F[X]$ with neither of $B(X)$ and $Q(X)$ of degree 0. Let $B(X) = \beta B_0(X)$ be a decomposition of $B(X)$ as in Proposition 8.19. Then we have $A(X) = B_0(X)(\beta Q(X))$, and (a) shows that $\beta Q(X)$ is in $R[X]$, in contradiction to the assumed irreducibility of $A(X)$ in $R[X]$.

In (c), write $A(X) = B(X)Q(X)$, and let $B(X) = \beta B_0(X)$ be a decomposition of $B(X)$ as in Proposition 8.19. Then we have $A(X) = B_0(X)(\beta Q(X))$ with $\beta Q(X)$ in $F[X]$. Conclusion (a) shows that $\beta Q(X)$ is in $R[X]$. If $b \in R$ is the leading coefficient of $B_0(X)$ and if $q \in R$ is the leading coefficient of $\beta Q(X)$, then we have $1 = bq$, and consequently b and q are units in R. Since $B(X) = \beta B_0(X)$ and $B(X)$ is monic, $1 = \beta b$, and therefore $\beta = b^{-1}$ is a unit in R. Hence $B(X)$ is in $R[X]$.

In (d), we argue along the same lines as in (a). We may take $B(X) = c(B)(c(B)^{-1}B(X))$ and $C(X) = c(C)(c(C)^{-1}C(X))$ as decompositions of $B(X)$ and $C(X)$ according to Proposition 8.19. Then we have $A(X) = (c(B)c(C))[c(B)^{-1}B(X)c(C)^{-1}C(X)]$. Theorem 8.18 says that the factor in brackets is primitive, and the uniqueness in Proposition 8.19 shows that $1 = c(B)c(C)$, up to unit factors. Therefore $c(B)$ and $c(C)$ are units in R, and $B(X)$ and $C(X)$ are primitive. □

Corollary 8.21. If R is a unique factorization domain, then the ring $R[X]$ is a unique factorization domain.

REMARK. As was mentioned at the beginning of the section, $\mathbb{Z}[X]$ and $\mathbb{K}[X_1, \ldots, X_n]$, when $\mathbb{K}$ is a field, are unique factorization domains as a consequence of this result.

PROOF. We begin with the proof of (UFD1). Suppose that $A(X)$ is a nonzero member of $R[X]$. We may take its decomposition according to Proposition 8.19 to be $A(X) = c(A)(c(A)^{-1}A(X))$. Consider divisors of $c(A)^{-1}A(X)$ in $R[X]$. These are all primitive, according to (d). Hence those of degree 0 are units in R. Thus any nontrivial factorization of $c(A)^{-1}A(X)$ is into two factors of strictly lower degree, both primitive. In a finite number of steps, this process of factorization with primitive factors has to stop. We can then factor $c(A)$ within R. Combining the factorizations of $c(A)$ and $c(A)^{-1}A(X)$, we obtain a factorization of $A(X)$.

For (UFD2′), let $P(X)$ be irreducible in $R[X]$. Since the factorization $P(X) = c(P)(c(P)^{-1}P(X))$ has to be trivial, either $c(P)$ is a unit, in which case $P(X)$ is primitive, or $c(P)^{-1}P(X)$ is a unit, in which case $P(X)$ has degree 0. In either case, suppose that $P(X)$ divides a product $A(X)B(X)$.

In the first case, $P(X)$ is primitive. Since $F[X]$ is a principal ideal domain, hence a unique factorization domain, either $P(X)$ divides $A(X)$ in $F[X]$ or $P(X)$

divides $B(X)$ in $F[X]$. By symmetry we may assume that $P(X)$ divides $A(X)$ in $F[X]$. Then (a) shows that $P(X)$ divides $A(X)$ in $R[X]$.

In the second case, $P(X) = P$ has degree 0 and is prime in R. Write $A(X)B(X) = PQ(X)$ with $Q(X)$ in $R[X]$. Once more we argue along the same lines as in (a). We may take $A(X) = c(A)(c(A)^{-1}A(X))$, $B(X) = c(B)(c(B)^{-1}B(X))$, and $Q(X) = c(Q)(c(Q)^{-1}Q(X))$ as the decompositions of $A(X), B(X),$ and $Q(X)$ according to Proposition 8.19. Then we have

$$\bigl(c(A)c(B)\bigr)\bigl[c(A)^{-1}A(X)c(B)^{-1}B(X)\bigr] = Pc(Q)\bigl(c(Q)^{-1}Q(X)\bigr).$$

Theorem 8.18 shows that the product in brackets is primitive, and the uniqueness in Proposition 8.19 shows that we have $c(A)c(B) = Pc(Q)$ up to factors of units in R. Since P is prime in R, P divides $c(A)$ or P divides $c(B)$. By symmetry we may assume that P divides $c(A)$. Then P divides $A(X)$ since $c(A)$ divides every coefficient of $A(X)$. □

The final application, Eisenstein's irreducibility criterion, is proved somewhat in the style of Gauss's Lemma (Theorem 8.18). We shall give only the analog of Proof #1 of Gauss's Lemma, leaving the analog of Proof #2 to Problem 21 at the end of the chapter.

Corollary 8.22 (Eisenstein's irreducibility criterion). Let R be a unique factorization domain, let F be its field of fractions, and let p be a prime in R. If $A(X) = a_N X^N + \cdots + a_1 X + a_0$ is a polynomial of degree ≥ 1 in $R[X]$ such that p divides $a_{N-1}, \ldots, a_0$ but not a_N and such that p^2 does not divide a_0, then $A(X)$ is irreducible in $F[X]$.

REMARK. The polynomial $A(X)$ will be irreducible in $R[X]$ also unless all its coefficients are divisible by some nonunit of R.

PROOF. Without loss of generality, we may replace $A(X)$ by $c(A)^{-1}A(X)$ and thereby assume that $A(X)$ is primitive. Corollary 8.20b shows that it is enough to prove irreducibility in $R[X]$. Assuming the contrary, suppose that $A(X)$ factors in $R[X]$ as $A(X) = B(X)C(X)$ with $B(X) = b_m X^m + \cdots + b_1 X + b_0$, $C(X) = c_n X^n + \cdots + c_1 X + c_0$, and neither of $B(X)$ and $C(X)$ equal to a unit. Corollary 8.20d shows that $B(X)$ and $C(X)$ are primitive. In particular, $B(X)$ and $C(X)$ have to be nonconstant polynomials. Define $a_k = 0$ for $k > N$, $b_k = 0$ for $k > m$, and $c_k = 0$ for $k > n$. Since p divides $a_0 = b_0 c_0$ and p is prime, p divides either b_0 or c_0. Without loss of generality, suppose that p divides b_0. Since p^2 does not divide a_0, p does not divide c_0.

We show, by induction on k, that p divides b_k for every $k < N$. The case $k = 0$ is the base case of the induction. If p divides b_j for $j < k$, then we have

$$a_k = b_0 c_k + b_1 c_{k-1} + \cdots + b_{k-1} c_1 + b_k c_0.$$

Since $k < N$, the left side is divisible by p. The inductive hypothesis shows that p divides every term on the right side except possibly the last. Consequently p divides $b_k c_0$. Since p does not divide c_0, p divides b_k. This completes the induction.

Since $C(X)$ is nonconstant, the degree of $B(X)$ is $< N$, and therefore we have shown that every coefficient of $B(X)$ is divisible by p. Then $c(B)$ is divisible by p, in contradiction to the fact that $B(X)$ is primitive. $\square$

EXAMPLES.

(1) Cyclotomic polynomials in $\mathbb{Q}[X]$. Let us see for each prime number p that the polynomial $\Phi(X) = X^{p-1} + X^{p-2} + \cdots + X + 1$ is irreducible in $\mathbb{Q}[X]$. We have $X^p - 1 = (X - 1)\Phi(X)$. Replacing $X - 1$ by Y gives $(Y + 1)^p - 1 = Y\Phi(Y + 1)$. The left side, by the Binomial Theorem, is $\sum_{k=1}^{p} \binom{p}{k} Y^k$. Hence $\Phi(Y + 1) = \sum_{k=1}^{p} \binom{p}{k} Y^{k-1}$. The binomial coefficient $\binom{p}{k}$ is divisible by p for $1 \leq k \leq p - 1$ since p is prime, and therefore the polynomial $\Psi(Y) = \Phi(Y + 1)$ satisfies the condition of Corollary 8.22 for the ring $\mathbb{Z}$. Hence $\Psi(Y)$ is irreducible over $\mathbb{Q}[Y]$. A nontrivial factorization of $\Phi(X)$ would yield a nontrivial factorization of $\Psi(Y)$, and hence $\Phi(X)$ is irreducible over $\mathbb{Q}[X]$.

(2) Certain polynomials in $\mathbb{K}[X, Y]$ when $\mathbb{K}$ is a field. Since $\mathbb{K}[X, Y] \cong \mathbb{K}[X][Y]$, it follows that $\mathbb{K}[X, Y]$ is a unique factorization domain, and any member of $\mathbb{K}[X, Y]$ can be written as $A(X, Y) = a_n(X)Y^n + \cdots + a_1(X)Y + a_0(X)$. The polynomial X is prime in $\mathbb{K}[X, Y]$, and Corollary 8.22 therefore says that $A(X, Y)$ is irreducible in $\mathbb{K}(X)[Y]$ if X does not divide $a_n(X)$ in $\mathbb{K}[X]$, X divides $a_{n-1}(X), \ldots, a_0(X)$ in $\mathbb{K}[X]$, and X^2 does not divide $a_0(X)$ in $\mathbb{K}[X]$. The remark with the corollary points out that $A(X, Y)$ is irreducible in $\mathbb{K}[X, Y]$ if also there is no nonconstant polynomial in $\mathbb{K}[X]$ that divides every $a_k(X)$. For example, $Y^5 + XY^2 + XY + X$ is irreducible in $\mathbb{K}[X, Y]$.

6. Finitely Generated Modules

The Fundamental Theorem of Finitely Generated Abelian Groups (Theorem 4.56) says that every finitely generated abelian group is a direct sum of cyclic groups. If we think of abelian groups as $\mathbb{Z}$ modules, we can ask whether this theorem has some analog in the context of R modules. The answer is yes—the theorem readily extends to the case that $\mathbb{Z}$ is replaced by an arbitrary principal ideal domain. The surprising addendum to the answer is that we have already treated a second special case of the generalized theorem. That case arises when the principal ideal domain is $\mathbb{K}[X]$ for some field $\mathbb{K}$. If V is a finite-dimensional vector space over $\mathbb{K}$ and $L : V \to V$ is a $\mathbb{K}$ linear map, then V becomes a $\mathbb{K}[X]$ module under the definition $Xv = L(v)$. This module is finitely generated even without the X present because V is finite-dimensional, and the generalized theorem that we prove in this section recovers the analysis of L that we carried out in Chapter V. When $\mathbb{K}$ is algebraically closed, we obtain the Jordan canonical form; for general $\mathbb{K}$, we obtain a different canonical form involving cyclic subspaces that was worked out in Problems 32–40 at the end of Chapter V.

The definitions for the generalization of Theorem 4.56 are as follows. Let R be a principal ideal domain. A subset S of an R module M is called a set of **generators** of M if M is the smallest R submodule of M containing all the members of S. If $\{m_s \mid s \in S\}$ is a subset of M, then the set of all finite sums $\sum_{s \in S} r_s m_s$ is an R submodule, but it need not contain the elements m_s and therefore need not be the R submodule generated by all the m_s. However, if M is unital, then taking $r_{s_0} = 1$ and all other r_s equal to 0 exhibits m_{s_0} as in the R submodule of all finite sums $\sum_{s \in S} r_s m_s$. For this reason we shall insist that all the R submodules in this section be unital.

We say that the R module M is **finitely generated** if it has a finite set of generators. The main theorem gives the structure of unital finitely generated R modules when R is a principal ideal domain. We need to take a small preliminary step that eliminates technical complications from the discussion, the same step that was carried out in Lemma 4.51 and Proposition 4.52 in the case of $\mathbb{Z}$ modules, i.e., abelian groups.

Lemma 8.23. Let R be a commutative ring with identity, and let $\varphi : M \to N$ be a homomorphism of unital R modules. If $\ker \varphi$ and $\operatorname{image} \varphi$ are finitely generated, then M is finitely generated.

PROOF. Let $\{x_1, \ldots, x_m\}$ and $\{y_1, \ldots, y_n\}$ be respective finite sets of generators for $\ker \varphi$ and $\operatorname{image} \varphi$. For $1 \leq j \leq n$, choose x'_j in M with $\varphi(x'_j) = y_j$. We shall prove that $\{x_1, \ldots, x_m, x'_1, \ldots, x'_n\}$ is a set of generators for M. Thus let x be in M. Since $\varphi(x)$ is in image φ, there exist $r_1, \ldots, r_n$ in R with $\varphi(x) = r_1 y_1 + \cdots + r_n y_n$. The element $x' = r_1 x'_1 + \cdots + r_n x'_n$ of M has $\varphi(x') = r_1 y_1 + \cdots + r_n y_n = \varphi(x)$. Therefore $\varphi(x - x') = 0$, and there exist $s_1, \ldots, s_m$ in R such that $x - x' = s_1 x_1 + \cdots + s_m x_m$. Consequently

$$x = s_1 x_1 + \cdots + s_m x_m + x' = s_1 x_1 + \cdots + s_m x_m + r_1 x'_1 + \cdots + r_n x'_n. \quad \square$$

Proposition 8.24. If R is a principal ideal domain, then any R submodule of a finitely generated unital R module is finitely generated. Moreover, any R submodule of a singly generated unital R module is singly generated.

PROOF. Let M be unital and finitely generated with a set $\{m_1, \ldots, m_n\}$ of n generators, and define $M_k = Rm_1 + \cdots + Rm_k$ for $1 \leq k \leq n$. Then $M_n = M$ since M is unital. We shall prove by induction on k that every R submodule of M_k is finitely generated. The case $k = n$ then gives the proposition. For $k = 1$, suppose that S is an R submodule of $M_1 = Rm_1$. Since S is an R submodule and every member of S lies in Rm_1, the subset I of all r in R with rm_1 in S is an ideal with $Im_1 = S$. Since every ideal in R is singly generated, we can write $I = (r_0)$. Then $S = Im_1 = Rr_0m_1$, and the single element r_0m_1 generates S.

Assume inductively that every R submodule of M_k is known to be finitely generated, and let N_{k+1} be an R submodule of M_{k+1}. Let $q : M_{k+1} \to M_{k+1}/M_k$ be the quotient R homomorphism, and let φ be the restriction $q|_{N_{k+1}}$, mapping N_{k+1} into M_{k+1}/M_k. Then $\ker \varphi = N_{k+1} \cap M_k$ is an R submodule of M_k and is finitely generated by the inductive hypothesis. Also, image φ is an R submodule of M_{k+1}/M_k, which is singly generated with generator equal to the coset of m_{k+1}. Since an R submodule of a singly generated unital R module was shown in the previous paragraph to be singly generated, image φ is finitely generated. Applying Lemma 8.23 to φ, we see that N_{k+1} is finitely generated. This completes the induction and the proof. $\square$

According to the definition in Example 9 of modules in Section 1, a free R module is a direct sum, finite or infinite, of copies of the R module R. A free R module is said to have **finite rank** if *some* direct sum is a finite direct sum. A unital R module M is said to be **cyclic** if it is singly generated, i.e., if $M = Rm_0$ for some m_0 in M. In this case, we have an R isomorphism $M \cong R/I$, where I is the ideal $\{r \in R \mid rm_0 = 0\}$.

Before coming to the statement of the theorem and the proof, let us discuss the heart of the matter, which is related to row reduction of matrices. We regard the space $M_{1n}(R)$ of all 1-row matrices with n entries in R as a free R module. Suppose that R is a principal ideal domain, and suppose that we have a particular 2-by-n matrix with entries in R and with the property that the two rows have nonzero elements a and b, respectively, in the first column. We can regard the set of R linear combinations of the two rows of our particular matrix as an R submodule of the free R module $M_{1n}(R)$. Let $c = \mathrm{GCD}(a, b)$. This member of R is defined only up to multiplication by a unit, but we make a definite choice of it. The idea is that we can do a kind of invertible row-reduction step that simultaneously replaces the two rows of our 2-by-n matrix by a first row whose first entry is c and a second row whose first entry is 0; in the process the corresponding R submodule of $M_{1n}(R)$ will be unchanged. In fact, we saw in the previous section that the hypothesis on R implies that there exist members x and y of R with $xa + yb = c$. Since c divides a and b, we can rewrite this equality as $x(ac^{-1}) + y(bc^{-1}) = 1$. Then the 2-by-2 matrix $M = \begin{pmatrix} x & y \\ -bc^{-1} & ac^{-1} \end{pmatrix}$ with entries in R has the property that

$$\begin{pmatrix} x & y \\ -bc^{-1} & ac^{-1} \end{pmatrix} \begin{pmatrix} a & * \\ b & * \end{pmatrix} = \begin{pmatrix} c & * \\ 0 & * \end{pmatrix}.$$

This equation shows explicitly that the rows of $\begin{pmatrix} c & * \\ 0 & * \end{pmatrix}$ lie in the R linear span of the rows of $\begin{pmatrix} a & * \\ b & * \end{pmatrix}$. The key fact about M is that its determinant $x(ac^{-1}) + y(bc^{-1})$ is 1 and that M is therefore invertible with entries in R: the inverse is just $M^{-1} = \begin{pmatrix} ac^{-1} & -y \\ bc^{-1} & x \end{pmatrix}$. This invertibility shows that the rows of $\begin{pmatrix} a & * \\ b & * \end{pmatrix}$ lie in the R linear span of $\begin{pmatrix} c & * \\ 0 & * \end{pmatrix}$. Consequently the R linear span of the rows of our given 2-by-n matrix is preserved under left multiplication by M.

In effect we can do the same kind of row reduction of matrices over R as we did with matrices over $\mathbb{Z}$ in the proof of Theorem 4.56. The only difference is that this time we do not see constructively how to find the x and y that relate a, b, and c. Thus we would lack some information if we actually wanted to follow through and calculate a particular example. We were able to make calculations to imitate the proof of Theorem 4.56 because we were able to use the Euclidean algorithm to arrive at what x and y are. In the present context we would be able to make explicit calculations if R were a Euclidean domain.

Theorem 8.25 (Fundamental Theorem of Finitely Generated Modules). If R is a principal ideal domain, then

(a) the number of R summands in a free R module of finite rank is independent of the direct-sum decomposition,

(b) any R submodule of a free R module of finite rank n is a free R module of rank $\leq n$,

(c) any finitely generated unital R module is the finite direct sum of cyclic modules.

REMARK. Because of (a), it is meaningful to speak of the **rank** of a free R module of finite rank; it is the number of R summands. By convention the 0 module is a free R module of rank 0. Then the statement of (b) makes sense. Statement (c) will be amplified in Corollary 8.29 below.

PROOF. Let M be a free R module of the form $Rx_1 \oplus \cdots \oplus Rx_n$, and suppose that $y_1, \ldots, y_m$ are elements of M such that no nontrivial combination $r_1 y_1 + \cdots + r_m y_m$ is 0. Define an m-by-n matrix C with entries in R by $y_i = \sum_{j=1}^n C_{ij} x_j$ for $1 \leq i \leq m$. If F is the field of fractions of R, then we can regard C as a matrix with entries in F. As such, the matrix has rank $\leq n$. If $m > n$, then the rows are linearly dependent, and we can find members $q_1, \ldots, q_m$ of F, not all 0, such that $\sum_{i=1}^m q_i C_{ij} = 0$ for $1 \leq j \leq n$. Clearing fractions, we obtain members $r_1, \ldots, r_m$ of R, not all 0, such that $\sum_{i=1}^m r_i C_{ij} = 0$ for $1 \leq j \leq n$. Then

$$\sum_{i=1}^m r_i y_i = \sum_{i=1}^m r_i \left(\sum_{j=1}^n C_{ij} x_j \right) = \sum_{j=1}^n \left(\sum_{i=1}^m r_i C_{ij} \right) x_j = \sum_{j=1}^n 0 x_j = 0,$$

in contradiction to the assumed independence property of $y_1, \ldots, y_m$. Therefore we must have $m \leq n$.

If we apply this conclusion to a set $x_1, \ldots, x_n$ that exhibits M as free and to another set, possibly infinite, that does the same thing, we find that the second set has $\leq n$ members. Reversing the roles of the two sets, we find that they both have n members. This proves (a).

For (b) and (c), we shall reduce the result to a lemma saying that a certain kind of result can be achieved by row and column reduction of matrices with entries in R. Let M be a free R module of rank n, defined by a subset $x_1, \ldots, x_n$ of M, and let N be an R submodule of M. Proposition 8.24 shows that N is finitely generated. We let $y_1, \ldots, y_m$ be generators, not necessarily with any independence property. Define an m-by-n matrix C with entries in R by $y_i = \sum_{j=1}^{n} C_{ij} x_j$. We can recover M as the set of R linear combinations of $x_1, \ldots, x_n$, and we can recover N as the set of R linear combinations of $y_1, \ldots, y_m$.

If B is an n-by-n matrix with entries in R and with determinant in the group $R^\times$ of units, then Corollary 5.5 shows that B^{-1} exists and has entries in R. If we define $x'_i = \sum_{j=1}^{n} B_{ij} x_j$, then any R linear combination of $x'_1, \ldots, x'_n$ is an R linear combination of $x_1, \ldots, x_n$. Also, the computation $\sum_{i=1}^{n} (B^{-1})_{ki} x'_i = \sum_{i,j} (B^{-1})_{ki} B_{ij} x_j = \sum_j \delta_{kj} x_j = x_k$ shows that any R linear combination of $x_1, \ldots, x_n$ is an R linear combination of $x'_1, \ldots, x'_n$. Thus we can recover the same M and N if we replace C by CB. Arguing in the same way with $y_1, \ldots, y_m$ and $y'_1, \ldots, y'_m$, we see that we can recover the same M and N if we replace CB by ACB, where A is an m-by-m matrix with entries in R and with determinant in $R^\times$.

Lemma 8.26 below will say that we can find A and B such that the nonzero entries of $D = ACB$ are exactly the diagonal ones D_{kk} for $1 \leq k \leq l$, where l is a certain integer with $0 \leq l \leq \min(m, n)$.

That is, the resulting equations restricting $y'_1, \ldots, y'_m$ in terms of $x'_1, \ldots, x'_n$ will be of the form

$$y'_k = \begin{cases} D_{kk} x'_k & \text{for } 1 \leq k \leq l, \\ 0 & \text{for } l+1 \leq k \leq m. \end{cases} \qquad (*)$$

Now let us turn to (b) and (c). For (b), the claim is that the elements y'_k with $1 \leq k \leq l$ exhibit N as a free R module. We know that $y'_1, \ldots, y'_m$ generate N and hence that $y'_1, \ldots, y'_l$ generate N. For the independence, suppose we can find members $r_1, \ldots, r_l$ not all 0 in R such that $\sum_{k=1}^{l} r_k y'_k = 0$. Then substitution gives $\sum_{k=1}^{l} r_k D_{kk} x'_k = 0$, and the independence of $x'_1, \ldots, x'_l$ forces $r_k D_{kk} = 0$ for $1 \leq k \leq l$. Since R is an integral domain, $r_k = 0$ for such k. Thus indeed the elements y'_k with $1 \leq k \leq l$ exhibit N as a free R module. Since $l \leq \min(m, n)$, the rank of N is at most the rank of M.

For (c), let Q be a finitely generated unital R module, say with n generators. By the universal mapping property of free R modules (Example 9 in Section 1), there exists a free R module M of rank n with Q as quotient. Let $x_1, \ldots, x_n$ be generators of M that exhibit M as free, and let N be the kernel of the quotient R homomorphism $M \to Q$, so that $Q \cong M/N$. Then (b) shows that N is a free R module of rank $m \leq n$. Let $y_1, \ldots, y_m$ be generators of N that exhibit N as free, and define an m-by-n matrix C with entries in R by $y_i = \sum_{j=1}^n C_{ij} x_j$ for $1 \leq i \leq m$. The result is that we are reduced to the situation we have just considered, and we can obtain equations of the form (∗) relating their respective generators, namely $y'_1, \ldots, y'_m$ for N and $x'_1, \ldots, x'_n$ for M.

For $1 \leq k \leq n$, define $M_k = Rx'_k$ and

$$N_k = \begin{cases} Ry'_k = RD_{kk}x'_k & \text{for } 1 \leq k \leq l, \\ 0 & \text{for } l+1 \leq k \leq n, \end{cases}$$

so that $N \cong N_1 \oplus \cdots \oplus N_n$. Then M_k/N_k is R isomorphic to the cyclic R module $R/(D_{kk})$ if $1 \leq k \leq l$, while $M_k/N_k = M_k$ is isomorphic to the cyclic R module R if $l+1 \leq k \leq n$. Applying Proposition 8.5, we obtain

$$M/N \cong (M_1 \oplus \cdots \oplus M_n)/(N_1 \oplus \cdots \oplus N_n) \cong (M_1/N_1) \oplus \cdots \oplus (M_n/N_n).$$

Thus M/N is exhibited as a direct sum of cyclic R modules. □

To complete the proof of Theorem 8.25, we are left with proving the following lemma, which is where row and column reduction take place.

Lemma 8.26. Let R be a principal ideal domain. If C is an m-by-n matrix with entries in R, then there exist an m-by-m matrix A with entries in R and with determinant in $R^\times$ and an n-by-n matrix B with entries in R and with determinant in $R^\times$ such that for some l with $0 \leq l \leq \min(m, n)$, the nonzero entries of $D = ACB$ are exactly the diagonal entries $D_{11}, D_{22}, \ldots, D_{ll}$.

PROOF. The matrices A and B will be constructed as products of matrices of determinant ± 1, and then $\det A$ and $\det B$ equal ± 1 by Proposition 5.1a. The matrix A will correspond to row operations on C, and B will correspond to column operations. Each factor will be the identity except in some 2-by-2 block. Among the row and column operations of interest are the interchange of two rows or two columns, in which the 2-by-2 block is $\begin{pmatrix} 0 & 1 \\ 1 & 0 \end{pmatrix}$. Another row operation of interest replaces two rows having respective j^{th} entries a and b by R linear combinations of them in which a and b are replaced by $c = \text{GCD}(a, b)$ and 0. If $x(ac^{-1}) + y(bc^{-1}) = 1$, then the 2-by-2 block is $\begin{pmatrix} x & y \\ -bc^{-1} & ac^{-1} \end{pmatrix}$. A similar operation is possible with columns.

The reduction involves an induction that successively constructs the entries $D_{11}, D_{22}, \ldots, D_{ll}$, stopping when the part of C involving rows and columns numbered $\geq l+1$ has been replaced by 0. We start by interchanging rows and columns to move a nonzero entry into position (1, 1). By a succession of row operations as in the previous paragraph, we can reduce the entry in position (1, 1) to the greatest common divisor of the entries of C in the first column, while reducing the remaining entries of the first column to 0. Next we do the same thing with column operations, reducing the entry in position (1, 1) to the greatest common divisor of the members of the first row, while reducing the remaining entries of the first row to 0. Then we go back and repeat the process with row operations and with column operations as many times as necessary until all the entries of the first row and column other than the one in position (1, 1) are 0. We need to check that this process indeed terminates at some point. If the entries that appear in position (1, 1) as the iterations proceed are $c_1, c_2, c_3, \ldots$, then we have $(c_1) \subseteq (c_2) \subseteq (c_3) \subseteq \cdots$. The union of these ideals is an ideal, necessarily a principal ideal of the form (c), and c occurs in one of the ideals in the union; the chain of ideals must be constant after that stage. Once the corner entry becomes constant, the matrices $\begin{pmatrix} x & y \\ -bc^{-1} & ac^{-1} \end{pmatrix}$ for the row operations can be chosen to be of the form $\begin{pmatrix} 1 & 0 \\ -ba^{-1} & 1 \end{pmatrix}$, and the result is that the row operations do not change the entries of the first row. Similar remarks apply to the matrices for the column operations. The upshot is that we can reduce C in this way so that all entries of the first row and column are 0 except the one in position (1, 1). This handles the inductive step, and we can proceed until at some l^{th} stage we have only the 0 matrix to process. $\square$

This completes the proof of Theorem 8.25. In Theorem 4.56, in which we considered the special case of abelian groups, we obtained a better conclusion than in Theorem 8.25c: we showed that the direct sum of cyclic groups could be written as the direct sum of copies of $\mathbb{Z}$ and of cyclic groups of prime-power order, and that in this case the decomposition was unique up to the order of the summands. We shall now obtain a corresponding better conclusion in the setting of Theorem 8.25.

The existence of the decomposition into cyclic modules of a special kind uses a very general form of the Chinese Remainder Theorem, whose classical statement appears as Corollary 1.9. The generalization below makes use of the following operations of addition and multiplication of ideals in a commutative ring with identity: if I and J are ideals, then $I + J$ denotes the set of sums $x + y$ with $x \in I$ and $y \in J$, and IJ denotes the set of all finite sums of products xy with $x \in I$ and $y \in J$; the sets $I + J$ and IJ are ideals.

Theorem 8.27 (Chinese Remainder Theorem). Let R be a commutative ring with identity, and let $I_1, \ldots, I_n$ be ideals in R such that $I_i + I_j = R$ whenever $i \neq j$.

(a) If elements $x_1, \ldots, x_n$ of R are given, then there exists x in R such that $x \equiv x_j \bmod I_j$, i.e., $x - x_j$ is in I_j, for all j. The element x is unique if $I_1 \cap \cdots \cap I_n = 0$.

(b) The map $\varphi : R \to \prod_{j=1}^{n} R/I_j$ given by $\varphi(r) = (\ldots, r + I_j, \ldots)$ is an onto ring homomorphism, its kernel is $\bigcap_{j=1}^{n} I_j$, and the homomorphism descends to a ring isomorphism

$$R \Big/ \bigcap_{j=1}^{n} I_j \cong R/I_1 \times \cdots \times R/I_n.$$

(c) The intersection $\bigcap_{j=1}^{n} I_j$ and the product $I_1 \cdots I_n$ coincide.

PROOF. For existence in (a) when $n = 1$, we take $x = x_1$. For existence when $n = 2$, the assumption $I_1 + I_2 = R$ implies that there exist $a_1 \in I_1$ and $a_2 \in I_2$ with $a_1 + a_2 = 1$. Given x_1 and x_2, we put $x = x_1 a_2 + x_2 a_1$, and then $x \equiv x_1 a_2 \equiv x_1 \bmod I_1$ and $x \equiv x_2 a_1 \equiv x_2 \bmod I_2$.

For general n, the assumption $I_1 + I_j = R$ for $j \geq 2$ implies that there exist $a_j \in I_1$ and $b_j \in I_j$ with $a_j + b_j = 1$. If we expand out the product $1 = \prod_{j=2}^{n}(a_j + b_j)$, then all terms but one on the right side involve some a_j and are therefore in I_1. That one term is $b_2 b_2 \cdots b_n$, and it is in $\bigcap_{j=2}^{n} I_j$. Thus $I_1 + \bigcap_{j=2}^{n} I_j = R$. The case $n = 2$, which was proved above, yields an element y_1 in R such that

$$y_1 \equiv 1 \bmod I_1 \quad \text{and} \quad y_1 \equiv 0 \bmod \bigcap_{j \neq 1} I_j.$$

Repeating this process for index i and using the assumption $I_i + I_j = R$ for $j \neq i$, we obtain an element y_i in R such that

$$y_i \equiv 1 \bmod I_i \quad \text{and} \quad y_i \equiv 0 \bmod \bigcap_{j \neq i} I_j.$$

If we put $x = x_1 y_1 + \cdots + x_n y_n$, then we have $x \equiv x_i y_i \bmod I_i \equiv x_i \bmod I_i$ for each i, and the proof of existence is complete.

For uniqueness in (a), if we have two elements x and x' satisfying the congruences, then their difference $x - x'$ lies in I_j for every j, hence is 0 under the assumption that $I_1 \cap \cdots \cap I_n = 0$.

In (b), the map φ is certainly a ring homomorphism. The existence result in (a) shows that φ is onto, and the proof of the uniqueness result identifies the kernel. The isomorphism follows.

For (c), consider the special case that I and J are ideals with $I + J = R$. Certainly $IJ \subseteq I \cap J$. For the reverse inclusion, choose $x \in I$ and $y \in J$ with $x + y = 1$; this is possible since $I + J = R$. If z is in $I \cap J$, then $z = zx + zy$ with zx in JI and zy in IJ. Thus z is exhibited as in IJ.

Consequently $I_1 I_2 = I_1 \cap I_2$. Suppose inductively that $I_1 \cdots I_k = I_1 \cap \cdots \cap I_k$. We saw in the proof of (a) that $I_{k+1} + \bigcap_{j \neq k+1} I_j = R$, and thus we certainly have

$I_{k+1} + \bigcap_{j=1}^{k} I_j = R$. The special case in the previous paragraph, in combination with the inductive hypothesis, shows that $I_{k+1}I_1 \cdots I_k = I_{k+1} \cdot (\bigcap_{j=1}^{k} I_j) = \bigcap_{j=1}^{k+1} I_j$. This completes the induction and the proof. □

Corollary 8.28. Let R be a principal ideal domain, and let $a = \varepsilon p_1^{k_1} \cdots p_n^{k_n}$ be a factorization of a nonzero nonunit element a into the product of a unit and powers of nonassociate primes. Then there is a ring isomorphism

$$R/(a) \cong R/(p_1^{k_1}) \times \cdots \times R/(p_n^{k_n}).$$

PROOF. Let $I_j = (p_j^{k_j})$ in Theorem 8.27. For $i \neq j$, we have $\text{GCD}(p_i^{k_i}, p_j^{k_j}) = 1$. Since R is a principal ideal domain, there exist a and b in R with $ap_i^{k_i} + bp_j^{k_j} = 1$, and consequently $(p_i^{k_i}) + (p_j^{k_j}) = R$. The theorem applies, and the corollary follows. □

Corollary 8.29. If R is a principal ideal domain, then any finitely generated unital R module M is the direct sum of a nonunique free R submodule $\bigoplus_{i=1}^{s} R$ of a well-defined finite rank $s \geq 0$ and the R submodule T of all members m of M such that $rm = 0$ for some $r \neq 0$ in R. In turn, the R submodule T is isomorphic to a direct sum

$$T \cong \bigoplus_{j=1}^{n} R/(p_j^{k_j}),$$

where the p_j are primes in R and the ideals $(p_j^{k_j})$ are not necessarily distinct. The number of summands (p^k) for each class of associate primes p and each positive integer k is uniquely determined by M.

PROOF. Theorem 8.25c gives $M = F \oplus \bigoplus_{j=1}^{n} Ra_j$, where F is a free R submodule of some finite rank s and the a_j's are nonzero members of M that are each annihilated by some nonzero member of R. The set T of all m with $rm = 0$ for some $r \neq 0$ in R is exactly $\bigoplus_{j=1}^{n} Ra_j$. Then F is R isomorphic to M/T, hence is isomorphic to the same free R module independently of what direct-sum decomposition of M is used. By Theorem 8.25a, s is well defined.

The cyclic R module Ra_j is isomorphic to $R/(b_j)$, where (b_j) is the ideal of all elements r in R with $ra_j = 0$. The ideal (b_j) is nonzero by assumption and is not all of R since the element $r = 1$ has $1a_j = a_j \neq 0$. Applying Corollary 8.28 for each j and adding the results, we obtain $T \cong \bigoplus_{i=1}^{n} R/(p_i^{k_i})$ for suitable primes p_i and powers k_i. The isomorphism in Corollary 8.28 is given as a ring isomorphism, and we are reinterpreting it as an R isomorphism. The primes p_i that arise for fixed (b_j) are distinct, but there may be repetitions in the pairs (p_i, k_i) as j varies. This proves existence of the decomposition.

If p is a prime in R, then the elements m of T such that $p^k m = 0$ for some k are the ones corresponding to the sum of the terms in $\bigoplus_{j=1}^{n} R/(p_j^{k_j})$ in which p_j is an associate of p. Thus, to complete the proof, it is enough to show that the R isomorphism class of the R module

$$N = R/(p^{l_1}) \oplus \cdots \oplus R/(p^{l_m})$$

with p fixed and with $0 < l_1 \leq \cdots \leq l_m$ completely determines the integers $l_1, \ldots, l_m$.

For any unital R module L, we can form the sequence of R submodules $p^j L$. The element p carries $p^j L$ into $p^{j+1} L$, and thus each $p^j L / p^{j+1} L$ is an R module on which p acts as 0. Consequently each $p^j L / p^{j+1} L$ is an $R/(p)$ module. Corollary 8.16 and Proposition 8.10 together show that $R/(p)$ is a field, and therefore we can regard each $p^j L / p^{j+1} L$ as an $R/(p)$ vector space.

We shall show that the dimensions $\dim_{R/(p)}(p^j N / p^{j+1} N)$ of these vector spaces determine the integers $l_1, \ldots, l_m$. We start from

$$p^j N = p^j R/(p^{l_1}) \oplus \cdots \oplus p^j R/(p^{l_m}).$$

The term $p^j R/(p^{l_k})$ is 0 if $j \geq l_k$. Thus

$$p^j N = \bigoplus_{j < l_k} p^j R/(p^{l_k}) = \bigoplus_{j < l_k} p^j R/p^{l_k} R.$$

Similarly

$$p^{j+1} N = \bigoplus_{j < l_k} p^{j+1} R/(p^{l_k}) = \bigoplus_{j < l_k} p^{j+1} R/p^{l_k} R.$$

Proposition 8.5 and Theorem 8.3 give us the R isomorphisms

$$p^j N / p^{j+1} N \cong \bigoplus_{j < l_k} \left(p^j R/p^{l_k} R\right) / \left(p^{j+1} R/p^{l_k} R\right) \cong \bigoplus_{j < l_k} p^j R / p^{j+1} R,$$

and these must descend to $R/(p)$ isomorphisms. Consequently

$$\dim_{R/(p)}(p^j N / p^{j+1} N) = \#\{k \mid l_k > j\} \dim_{R/(p)}(p^j R / p^{j+1} R).$$

The coset $p^j + p^{j+1} R$ of $p^j R / p^{j+1} R$ has the property that multiplication by arbitrary elements of R yields all of $p^j R / p^{j+1} R$. Therefore $\dim_{R/(p)}(p^j R / p^{j+1} R) = 1$, and we obtain

$$\dim_{R/(p)}(p^j N / p^{j+1} N) = \#\{k \mid l_k > j\}.$$

Thus the R module N determines the integers on the right side, and these determine the number of l_k's equal to each positive integer j. This proves uniqueness. $\square$

Let us apply Theorem 8.25 and Corollary 8.29 to the principal ideal domain $R = \mathbb{K}[X]$, where $\mathbb{K}$ is a field. The particular unital module of interest is a finite-dimensional vector space V over $\mathbb{K}$, and the scalar multiplication by $\mathbb{K}[X]$ is given by $A(X)v = A(L)(v)$ for each polynomial $A(X)$, where L is a fixed linear map $L : V \to V$. Let us see that the results of this section recover the structure theory of L as developed in Chapter V.

Since V is finite-dimensional over $\mathbb{K}$, V is certainly finitely generated over $R = \mathbb{K}[X]$. Theorem 8.25 gives

$$V \cong R/(A_1(X)) \oplus \cdots \oplus R/(A_n(X)) \oplus R \oplus \cdots \oplus R$$

as R modules and in particular as vector spaces over $\mathbb{K}$. Each summand R is infinite-dimensional as a vector space, and consequently no summand R can be present. Corollary 8.29 refines the decomposition to the form

$$V \cong R/(P_1(X)^{k_1}) \oplus \cdots \oplus R/(P_m(X)^{k_m})$$

as R modules, the polynomials $P_j(X)$ being prime but not necessarily distinct. Since the R isomorphism is in particular an isomorphism of $\mathbb{K}$ vector spaces, each $R/(P_j(X)^{k_j})$ corresponds to a vector subspace V_j, and $V = V_1 \oplus \cdots \oplus V_m$. Since the R isomorphism respects the action by X, we have $L(V_j) \subseteq V_j$ for each j. Thus the direct sum decompositions of Theorem 8.25 and Corollary 8.29 are yielding a decomposition of V into a direct sum of vector subspaces invariant under L. Since the jth summand is of the form $R/(P_j(X)^{k_j})$, L acts on V_j in a particular way, which we have to analyze.

Let us carry out this analysis in the case that $\mathbb{K}$ is algebraically closed (as for example when $\mathbb{K} = \mathbb{C}$), seeing that each V_j yields a Jordan block of the Jordan canonical form (Theorem 5.20a) of L. For the case of general $\mathbb{K}$, the analysis can be seen to lead to the corresponding more general results that were obtained in Problems 32–40 at the end of Chapter V.

Since $\mathbb{K}$ is algebraically closed, any polynomial in $\mathbb{K}[X]$ of degree ≥ 1 has a root in $\mathbb{K}$ and therefore has a first-degree factor $X - c$. Consequently all primes in $\mathbb{K}[X]$ are of the form $X - c$, up to a scalar factor, with c in $\mathbb{K}$. To understand the action of L on V_j, we are to investigate $\mathbb{K}[X]/((X - c)^k)$.

Suppose that $A(X)$ is in $\mathbb{K}[X]$ and is of degree $n \geq 1$. Expanding the monomials of $A(X)$ by the Binomial Theorem as

$$X^j = ((X - c) + c)^j = \sum_{i=0}^{j} \binom{j}{i} c^{j-i}(X - c)^i,$$

we see that $A(X)$ has an expansion as

$$A(X) = a_0 + a_1(X - c) + \cdots + a_n(X - c)^n$$

for suitable coefficients $a_0, \ldots, a_n$ in $\mathbb{K}$. Let the invariant subspace that we are studying be $V_{j_0} \subseteq V$. Since V_{j_0} is isomorphic as an R module to $\mathbb{K}[X]/((X-c)^k)$, $(X-c)^k$ acts on V_{j_0} as 0. So does every higher power of $X-c$, and hence

$$A(X) \quad \text{acts as} \quad a_0 + a_1(X-c) + \cdots + a_{k-1}(X-c)^{k-1}.$$

The polynomials on the right, as their coefficients vary, represent distinct cosets of $\mathbb{K}[X]/((X-c)^k)$: in fact, if two were to be in the same coset, we could subtract and see that $(X-c)^k$ could not divide the difference unless it were 0. The distinct cosets match in one-one $\mathbb{K}$ linear fashion with the members of V_{j_0}, and thus $\dim V_{j_0} = k$. Let us write down this match. Let v_0 be the member of V_{j_0} that is to correspond to the coset 1 of $\mathbb{K}[X]/(X-c)^k$. On V_{j_0}, $\mathbb{K}[X]$ is acting with $Xv = L(v)$. We define recursively vectors $v_1, \ldots, v_{k-1}$ of V_{j_0} by

$$v_1 = (L - cI)v_0 = (X-c)v_0 \quad \longleftrightarrow \quad (X-c) \cdot 1 = X-c,$$
$$v_2 = (L - cI)v_1 = (X-c)v_1 \quad \longleftrightarrow \quad (X-c) \cdot (X-c) = (X-c)^2,$$
$$\vdots$$
$$v_{k-1} = (L - cI)v_{k-2} = (X-c)v_{k-2} \quad \longleftrightarrow \quad (X-c) \cdot (X-c)^{k-2} = (X-c)^{k-1},$$
$$(L - cI)v_{k-1} = (X-c)v_{k-1} \quad \longleftrightarrow \quad (X-c) \cdot (X-c)^{k-1} = (X-c)^k \equiv 0.$$

We conclude from this correspondence that the vectors $v_0, v_1, \ldots, v_{k-1}$ form a basis of V_{j_0} and that the matrix of $L - cI$ in the ordered basis $v_{k-1}, \ldots, v_1, v_0$ is

$$\begin{pmatrix} 0 & 1 & 0 & 0 & \cdots & 0 & 0 \\ & 0 & 1 & 0 & \cdots & 0 & 0 \\ & & 0 & 1 & \cdots & 0 & 0 \\ & & & \ddots & \ddots & \vdots & \vdots \\ & & & & 0 & 1 & 0 \\ & & & & & 0 & 1 \\ & & & & & & 0 \end{pmatrix}.$$

Hence the matrix of L in the same ordered basis is

$$\begin{pmatrix} c & 1 & 0 & 0 & \cdots & 0 & 0 \\ & c & 1 & 0 & \cdots & 0 & 0 \\ & & c & 1 & \cdots & 0 & 0 \\ & & & \ddots & \ddots & \vdots & \vdots \\ & & & & c & 1 & 0 \\ & & & & & c & 1 \\ & & & & & & c \end{pmatrix},$$

i.e., is a Jordan block. Thus Theorem 8.25 and Corollary 8.29 indeed establish the existence of Jordan canonical form (Theorem 5.20a) when $\mathbb{K}$ is algebraically closed. It is easy to check that Corollary 8.29 establishes also the uniqueness statement in Theorem 5.20a.

7. Orientation for Algebraic Number Theory and Algebraic Geometry

The remainder of the chapter introduces material on commutative rings with identity that is foundational for both algebraic number theory and algebraic geometry. Historically algebraic number theory grew out of Diophantine equations, particularly from two problems—from Fermat's Last Theorem and from representation of integers by binary quadratic forms. Algebraic geometry grew out of studying the geometry of solutions of equations and out of studying Riemann surfaces.

These two subjects can be studied on their own, but they also have a great deal in common. The discovery that the plane could be coordinatized and that geometry could be approached through algebra was one of the great advances of all time for mathematics. Since then, fundamental connections between algebraic number theory and algebraic geometry have been discovered at a deeper level, and the distinction between the two subjects is more and more just a question of one's point of view. The emphasis in the remainder of this chapter will be on one aspect of this relationship, the theory that emerged from trying to salvage something in the way of unique factorization.

By way of illustration, let us examine an analogy between what happens with a certain ring of "algebraic integers" and what happens with a certain "algebraic curve." The ring of algebraic integers in question was introduced already in Section 4. It is $R = \mathbb{Z}[\sqrt{-5}] = \mathbb{Z} + \mathbb{Z}\sqrt{-5}$. The units are ± 1. Our investigation of unique factorization was aided by the function

$$N(a + b\sqrt{-5}) = (a + b\sqrt{-5})(a - b\sqrt{-5}) = a^2 + 5b^2,$$

which has the property that

$$N((a + b\sqrt{-5})(c + d\sqrt{-5})) = N(a + b\sqrt{-5})N(c + d\sqrt{-5}).$$

With this function we could determine candidates for factors of particular elements. In connection with the equality $2 \cdot 3 = (1 + \sqrt{-5})(1 - \sqrt{-5})$, we saw that the two factors on the left side and the two factors on the right side are all irreducible. Moreover, neither factor on the left is the product of a unit and a factor on the right. Therefore R is not a unique factorization domain. As a consequence it cannot be a principal ideal domain. In fact, $(2, 1 + \sqrt{-5})$ is an example of an ideal that is not principal. We shall return shortly to examine this ring further.

Now we introduce the algebraic curve. Consider $y^2 = (x - 1)x(x + 1)$ as an equation in two variables x and y. To fix the ideas, we think of a solution as a pair (x, y) of complex numbers. Although the variables in this discussion are complex, it is convenient to be able to draw pictures of the solutions, and one does this by showing only the solutions (x, y) with x and y in $\mathbb{R}$. Figure 8.6 indicates the set of solutions in $\mathbb{R}^2$ for this particular curve. We can study these solutions for a while, looking for those pairs (x, y) with x and y rationals or integers, but a different level of understanding comes from studying functions on the locus of complex solutions. The functions of interest are polynomial functions in the pair (x, y), and we identify two of them if they agree on the locus. Thus we introduce the ring

$$R' = \mathbb{C}[x, y]/(y^2 - (x - 1)x(x + 1)).$$

There is a bit of a question whether this is indeed the space of restrictions, but that can be settled affirmatively by the "Nullstellensatz" in *Advanced Algebra* and a verification that the principal ideal $(y^2 - (x - 1)x(x + 1))$ is prime.[8] The ring R' is called the "affine coordinate ring" of the curve, and the curve itself is an example of an "affine algebraic curve."

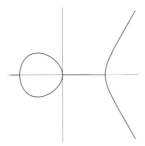

FIGURE 8.6. Real points of the curve $y^2 = (x - 1)x(x + 1)$.

We can recover the locus of the curve from the ring R' as follows. If (x_0, y_0) is a point of the curve, then it is meaningful to evaluate members of R' at (x_0, y_0), and we let $I_{(x_0, y_0)}$ be the ideal of all members of R' vanishing at (x_0, y_0). Evaluation at (x_0, y_0) exhibits the ring $R'/I_{(x_0, y_0)}$ as isomorphic to $\mathbb{C}$, which is a field. Thus $I_{(x_0, y_0)}$ is a maximal ideal and is in particular prime. It turns out for this example that all nonzero prime ideals are of this form.[9] We return to make use of this geometric interpretation of prime ideals in a moment.

Now let us consider factorization in R'. Every element of R' can be written uniquely as $A(x) + B(x)y$, where $A(x)$ and $B(x)$ are polynomials. The analog

---

[8] The polynomial $y^2 - (x - 1)x(x + 1)$ is prime since $(x - 1)x(x + 1)$ is not a square, or since Eisenstein's criterion applies. The principal ideal $(y^2 - (x - 1)x(x + 1))$ is therefore prime by Proposition 8.14. What the Nullstellensatz says when the underlying field is algebraically closed is that the only polynomials vanishing on the zero locus of a prime ideal are the members of the ideal.

[9] In Section 9, Example 3 of integral closures in combination with Proposition 8.45 shows that every nonzero prime ideal of R' is maximal. (In algebraic geometry one finds that this property of prime ideals is a reflection of the 1-dimensional nature of the curve.) The Nullstellensatz says that the maximal ideals are all of the form $I_{(x_0, y_0)}$.

in R' of the quantity $N(a+b\sqrt{-5})$ in the ring R is the quantity

$$N(A(x)+B(x)y) = (A(x)+B(x)y)(A(x)-B(x)y)$$
$$= A(x)^2 - B(x)^2 y^2$$
$$= A(x)^2 - B(x)^2(x^3-x).$$

Easy computation shows that

$$N((A(x)+B(x)y)(C(x)+D(x)y)) = N(A(x)+B(x)y)N(C(x)+D(x)y),$$

and hence $N(\cdot)$ gives us a device to use to check whether elements of R' are irreducible. We find in the equation

$$(x+y)(x-y) = x^2 - (x^3-x) = -x(x - \tfrac{1}{2}(1+\sqrt{5}))(x - \tfrac{1}{2}(1-\sqrt{5}))$$

that the two elements on the left side and the three elements on the right side are irreducible. Therefore unique factorization fails in R'.

Although unique factorization fails for the elements of R', there is a notion of factorization for ideals in R' that behaves well algebraically and has a nice geometric interpretation. Recall that the nonzero prime ideals correspond to the points of the locus $y^2 = (x-1)x(x+1)$ via passage to the zero locus, the ideal corresponding to (x_0, y_0) being called $I_{(x_0, y_0)}$. For any two ideals I and J, we can form the product ideal IJ whose elements are the sums of products of a member of I and a member of J. Then $I_{(x_0, y_0)}^k$ may be interpreted as the ideal of all members of R' vanishing at (x_0, y_0) to order k or higher, and $I_{(x_1, y_1)}^{k_1} \cdots I_{(x_n, y_n)}^{k_n}$ becomes the ideal of all members of R' vanishing at each (x_j, y_j) to order at least k_j. We shall see in Section 11 that every nonzero proper ideal I in R' factors in this way. The points (x_j, y_j) and the integers k_j have a geometric interpretation in terms of I and are therefore uniquely determined: the (x_j, y_j)'s form the locus of common zeros of the members of I, and the integer k_j is the greatest integer such that the vanishing at (x_j, y_j) is always at least to order k_j. In a sense, factorization of elements was the wrong thing to consider; the right thing to consider is factorization of ideals, which is unique because of the associated geometric interpretation.

Returning to the ring $R = \mathbb{Z}[\sqrt{-5}]$, we can ask whether factorization of ideals is a useful notion in R. Again IJ is to be the set of all sums of products of an element in I and an element in J. For $I = (2, 1+\sqrt{-5})$ and $J = (2, 1-\sqrt{-5})$, we get all sums of expressions $(2a+b(1+\sqrt{-5}))(2c+d(1-\sqrt{-5}))$ in which a, b, c, d are in $\mathbb{Z}$, hence all sums of expressions

$$2(2ac+3bd) + 2(bc+ad) + 2\sqrt{-5}(bc-ad).$$

All such elements are divisible by 2. Two examples come by taking $a = c = 1$ and $b = d = 0$ and by taking $a = c = 0$ and $b = d = 1$; these give 4 and 6. Subtracting, we see that 2 is a sum of products. Thus $IJ = (2)$. The element 2 is irreducible and not prime, and we know from Proposition 8.14 that the ideal (2) therefore cannot be prime. What we find is that the ideal (2) factors even though the element 2 does not factor. It turns out that R has unique factorization of ideals, just the way R' does.

The prime ideals of the ring R have a certain amount of structure in terms of the primes or prime ideals of $\mathbb{Z}$. To understand what to expect, let us digress for a moment to discuss what happens with the ring $R'' = \mathbb{Z}[i] = \mathbb{Z} + \mathbb{Z}\sqrt{-1}$ of Gaussian integers. This too was introduced in Section 4, and it is a Euclidean domain, hence a principal ideal domain. It has unique factorization. Its appropriate $N(\cdot)$ function is $N(a+ib) = a^2 + b^2$. Problems 27–31 at the end of the chapter ask one to verify that the primes of R'', up to multiplication by one of the units ± 1 and $\pm i$, are members of R'' of any of the three kinds

$p = 4n + 3$ that is prime in $\mathbb{Z}$ and has $n \geq 0$,
$p = a \pm ib$ with $a^2 + b^2$ prime in $\mathbb{Z}$ of the form $4n + 1$ with $n \geq 0$,
$p = 1 \pm i$ (these are associates).

These three kinds may be distinguished by what happens to the function $N(\cdot)$. In the first case $N(p) = p^2$ is the square of a prime of $\mathbb{Z}$ and is the square of a prime of R'', in the second case $N(p)$ is a prime of $\mathbb{Z}$ that is the product of two distinct primes of R'', and in the third case $N(p)$ is a prime of $\mathbb{Z}$ that is the square of a prime of R'', apart from a unit factor. The nonzero prime ideals of R'' are the principal ideals generated by the prime elements of R'', and they fall into three types as well. Each nonzero prime ideal P has a prime p of $\mathbb{Z}$ attached to it, namely the one with $(p) = \mathbb{Z} \cap P$, and the type of the ideal corresponds to the nature of the factorization of the ideal pR'' of R''. Specifically in the first case pR'' is a prime ideal in R'', in the second case pR'' is the product of two distinct prime ideals in R'', and in the third case pR'' is the square of a prime ideal in R''.

The structure of the prime ideals in R is of the same nature as with R''. Each nonzero prime ideal P has a prime p of $\mathbb{Z}$ attached to it, again given by $(p) = \mathbb{Z} \cap P$, and the three kinds correspond to the factorization of the ideal pR of R. Let us be content to give examples of the three possible behaviors:

$11R$ is prime in R,
$2R$ is the product of two distinct prime ideals in R,
$5R$ is the square of the prime ideal $(\sqrt{-5})$ in R.

We have already seen the decomposition of $2R$, and the decomposition of $5R$ is easy to check. With $11R$, the idea is to show that 11 is a prime element in R. Thus let 11 divide a product in R. Then $N(11) = 11^2$ divides the product of the $N(\cdot)$'s, 11 divides the product of the $N(\cdot)$'s, and 11 must divide one of the $N(\cdot)$'s. Say that 11 divides $N(a+b\sqrt{-5})$, i.e., that $a^2 + 5b^2 \equiv 0 \bmod 11$. If 11 divides one of a or b, then this congruence shows that 11 divides the other of them; then 11 divides $a + b\sqrt{-5}$, as we wanted to show. The other possibility is that 11 divides neither a nor b. Then $(ab^{-1})^2 \equiv -5 \bmod 11$ says that -5 is a square modulo 11, and we readily check that it is not. The conclusion is that 11 is indeed prime in R.

This structure for the prime ideals of R has an analog with the curve and its ring R'. The analogs for the curve case of $\mathbb{Z}$ and $\sqrt{-5}$ for the number-theoretic case are $\mathbb{C}[x]$ and y. The primes of $\mathbb{C}[x]$ are nonzero scalars times polynomials $x - c$ with c complex, and the relevant question for R' is how the ideal $(x - c)R'$ decomposes into prime ideals. We can think about this problem algebraically or geometrically. Algebraically, the ideal of all polynomials vanishing at (x_0, y_0) is $I_{(x_0, y_0)} = (x - x_0, y - y_0)$, the set of all $(x - x_0)A(x) - y_0 B(x) + y B(x)$ with $A(x)$ and $B(x)$ in $\mathbb{C}[x]$. The intersection with $\mathbb{C}[x]$ consists of all $(x - x_0)A(x)$ and is therefore the principal ideal $(x - x_0)$. We want to factor the ideal $(x - x_0)R'$.

If we pause for a moment and think about the problem geometrically, the answer is fairly clear. Ideals correspond to zero loci with multiplicities. The question is the factorization of the ideal of all polynomials vanishing when $x = x_0$. For most values of the complex number x_0, there are two choices of the complex y such that (x_0, y) is on the locus since y is given by a quadratic equation, namely $y^2 = (x_0 - 1)x_0(x_0 + 1)$. Thus for most values of x_0, $(x - x_0)R'$ is the product of two distinct prime ideals. The geometry thus suggests that

$$(x - x_0)R' = (x - x_0, y - y_0)(x - x_0, y + y_0),$$

where $y_0^2 = (x_0 - 1)x_0(x_0 + 1)$ and it is assumed that $y_0 \neq 0$. We can verify this algebraically: The members of the product ideal are the polynomials

$$\big((x - x_0)A(x) + (y - y_0)B(x)\big)\big((x - x_0)C(x) + (y + y_0)D(x)\big)$$
$$= (x - x_0)^2 A(x)C(x) + (x - x_0)\big(A(x)(y + y_0)D(x)) + C(x)(y - y_0)B(x)\big)$$
$$+ (y^2 - y_0^2)B(x)D(x).$$

The last term on the right side is $\big((x^3 - x) - (x_0^3 - x_0)\big)B(x)D(x)$ and is divisible by $x - x_0$. Therefore every member of the product ideal lies in the principal ideal $(x - x_0)$. On the other hand, the product ideal contains $(x - x_0)(x - x_0)$ and also $(y^2 - y_0^2) = (x^3 - x_0^3) - (x - x_0) = (x - x_0)(x^2 + xx_0 + x_0^2)$. Since $\gcd\big((x - x_0), (x^2 + xx_0 + x_0^2)\big) = 1$, the product ideal contains $x - x_0$. Therefore the product ideal equals $(x - x_0)$.

The exceptional values of x_0 are $-1, 0, +1$, where the locus has $y_0 = 0$. The geometry of the factorization is not so clear in this case, but the algebraic

7. Orientation for Algebraic Number Theory and Algebraic Geometry

computation remains valid. Thus we have $(x - x_0)R' = (x - x_0, y)^2$ if x_0 equals $-1, 0$, or $+1$. The conclusion is that the nonzero prime ideals of R' are of two types, with $(x - x_0)R'$ equal to

the product of two distinct prime ideals in R' if x_0 is not in $\{-1, 0, +1\}$,

the square of a prime ideal in R' if x_0 is in $\{-1, 0, +1\}$.

The third type, with $(x - x_0)R'$ prime in R', does not arise. Toward the end of Chapter IX we shall see how we could have anticipated the absence of the third type.

That is enough of a comparison for now. Certain structural results useful in both algebraic number theory and algebraic geometry are needed even before we get started at factoring ideals, and those are some of the topics for the remainder of this chapter. In Section 11 we conclude by establishing unique factorization of ideals for a class of examples that includes the examples above. In the examples above, the rings we considered were $\mathbb{Z}[X]/(X^2+5) = \mathbb{Z}[\sqrt{-5}]$ and $\mathbb{C}[x, y]/(y^2 - (x - 1)x(x + 1)) \cong \mathbb{C}[x][\sqrt{(x-1)x(x+1)}]$. In each case the notation $[\,\cdot\,]$ refers to forming the ring generated by the coefficients and the expression or expressions in brackets.

First we establish a result saying that ideals in the rings of interest are not too wild. For example, in algebraic geometry, one wants to consider the set of restrictions of the members of $\mathbb{K}[X_1, \ldots, X_n]$, $\mathbb{K}$ being a field, to the locus of common zeros of a set of polynomials. The general tool will tell us that any ideal in $\mathbb{K}[X_1, \ldots, X_n]$ is finitely generated; thus a description of what polynomials vanish on the locus under study is not completely out of the question. The tool is the Hilbert Basis Theorem and is the main result of Section 8.

Second we need a way of understanding, in a more general setting, the relationship that we used in the above examples between $\mathbb{Z}$ and $\mathbb{Z}[\sqrt{-5}]$, and between $\mathbb{C}[x]$ and $\mathbb{C}[x][\sqrt{(x-1)x(x+1)}]$. The tool is the notion of integral closure and is the subject of Section 9.

Third we need a way of isolating the behavior of prime ideals, of eliminating the influence of algebraic or geometric factors that have nothing to do with the prime ideal under study. The tool is the notion of localization and is the subject of Section 10.

In Section 11 we make use of these three tools to establish unique factorization of ideals for a class of integral domains known as "Dedekind domains." It is easy to see that principal ideal domains are Dedekind domains, and we shall show that many other integral domains, including the examples above, are Dedekind domains. A refined theorem producing Dedekind domains will be obtained toward the end of Chapter IX once we have introduced the notion of a "separable" extension of fields.

8. Noetherian Rings and the Hilbert Basis Theorem

In this section, R will be a commutative ring with identity, and all R modules will be assumed unital. We begin by introducing three equivalent conditions on a unital R module.

Proposition 8.30. If R is a commutative ring with identity and M is a unital R module, then the following conditions on R submodules of M are equivalent:
  (a) (**ascending chain condition**) every strictly ascending chain of R submodules $M_1 \subsetneq M_2 \subsetneq \cdots$ terminates in finitely many steps,
  (b) (**maximum condition**) every nonempty collection of R submodules has a maximal element under inclusion,
  (c) (**finite basis condition**) every R submodule is finitely generated.

PROOF. To see that (a) implies (b), let $\mathcal{C}$ be a nonempty collection of R submodules of M. Take M_1 in $\mathcal{C}$. If M_1 is not maximal, choose M_2 in $\mathcal{C}$ properly containing M_1. If M_2 is not maximal, choose M_3 in $\mathcal{C}$ properly containing M_2. Continue in this way. By (a), this process must terminate, and then we have found a maximal R submodule in $\mathcal{C}$.

To see that (b) implies (c), let N be an R submodule of M, and let $\mathcal{C}$ be the collection of all finitely generated R submodules of N. This collection is nonempty since 0 is in it. By (b), $\mathcal{C}$ has a maximal element, say N'. If x is in N but x is not in N', then $N' + Rx$ is a finitely generated R submodule of N that properly contains N' and therefore gives a contradiction. We conclude that $N' = N$, and therefore N is finitely generated.

To see that (c) implies (a), let $M_1 \subsetneq M_2 \subsetneq \cdots$ be given, and put $N = \bigcup_{n=1}^{\infty} M_n$. By (c), N is finitely generated. Since the M_n are increasing with n, we can find some M_{n_0} containing all the generators. Then the sequence stops no later than at M_{n_0}. $\square$

Let us apply Proposition 8.30 with M taken to be the unital R module R. As always, the R submodules of R are the ideals of R.

Corollary 8.31. If R is a commutative ring with identity, then the following conditions on R are equivalent:
  (a) ascending chain condition for ideals: every strictly ascending chain of ideals in R is finite,
  (b) maximum condition for ideals of R: every nonempty collection of ideals in R has a maximal element under inclusion,
  (c) finite basis condition for ideals: every ideal in R is finitely generated.

The corollary follows immediately from Proposition 8.30. A commutative ring with identity satisfying the equivalent conditions of Corollary 8.31 is said to be a **Noetherian** commutative ring.

EXAMPLES.

(1) Principal ideal domains, such as $\mathbb{Z}$ and $\mathbb{K}[X]$ when $\mathbb{K}$ is a field. The finite basis condition for ideals is satisfied since every ideal is singly generated. The fact that (c) implies (a) has already been proved manually for principal ideal domains twice in this chapter—once in the proof of (UFD1) for a principal ideal domain in Theorem 8.15 and once in the proof of Lemma 8.26.

(2) Any homomorphic image R' of a Noetherian commutative ring R, provided 1 maps to 1. In fact, if $I' \subseteq R'$ is an ideal, its inverse image I is an ideal in R; the image of a finite set of generators of I is a finite set of generators of I'.

(3) $\mathbb{K}[X_1, \ldots, X_n]$ when $\mathbb{K}$ is a field. This commutative ring is Noetherian by application of the Hilbert Basis Theorem (Theorem 8.32 below) and induction on n. This ring is also a unique factorization domain, as we saw in Section 5.

(4) $\mathbb{Z}[X]$. This commutative ring is Noetherian, also by the Hilbert Basis Theorem below. Example 2 shows therefore that the quotient $\mathbb{Z}[\sqrt{-5}] = \mathbb{Z}[X]/(X^2 + 5)$ is Noetherian. This ring is an integral domain, and we have seen that it is not a unique factorization domain.

Theorem 8.32 (Hilbert Basis Theorem). If R is a nonzero Noetherian commutative ring, then so is $R[X]$.

PROOF. If I is an ideal in $R[X]$ and if $k \geq 0$ is an integer, let $L_k(I)$ be the union of $\{0\}$ and the set of all nonzero elements of R that appear as the coefficient of X^k in some element of degree k in I. First let us see that $\{L_k(I)\}_{k \geq 0}$ is an increasing sequence of ideals in R. In fact, if $A(X)$ and $B(X)$ are polynomials of degree k in I with leading terms $a_k X^k$ and $b_k X^k$, then $A(X) + B(X)$ has degree k if $b_k \neq -a_k$, and hence $a_k + b_k$ is in $L_k(I)$ in every case. Similarly if r is in R and $r a_k \neq 0$, then $rA(X)$ has degree k, and hence $r a_k$ is in $L_k(I)$ in every case. Consequently $L_k(I)$ is an ideal in R. Since I is closed under multiplication by X, $L_k(I) \subseteq L_{k+1}(I)$ for all $k \geq 0$.

Next let us prove that if J is any ideal in $R[X]$ such that $I \subseteq J$ and $L_k(I) = L_k(J)$ for all $k \geq 0$, then $I = J$. Let $B(X)$ be in J with $\deg B(X) = k$. Arguing by contradiction, we may suppose that $B(X)$ is not in I and that k is the smallest possible degree of a polynomial in J but not in I. Since $L_k(I) = L_k(J)$, we can find $A(X)$ in I whose leading term is the same as the leading term of $B(X)$. Since $B(X)$ is not in I, $B(X) - A(X)$ is not in I. Since $I \subseteq J$, $B(X) - A(X)$ is in J. Since $\deg(B(X) - A(X)) \leq k - 1$, we have arrived at a contradiction to the defining property of k. We conclude that $I = J$.

Now let $\{I_j\}_{j\geq 0}$ be an ascending chain of ideals in $R[X]$, and form $L_i(I_j)$ for each i. When i or j is fixed, these ideals are increasing as a function of the other index, j or i. By the maximum condition in R, $L_i(I_j) \subseteq L_p(I_q)$ for some p and q and all i and j. For $i \geq p$ and $j \geq q$, we have $L_i(I_j) \supseteq L_p(I_q)$ and thus $L_i(I_j) = L_p(I_q)$. The case $j = q$ gives $L_p(I_q) = L_i(I_q)$, and therefore $L_i(I_j) = L_i(I_q)$ for $i \geq p$ and $j \geq q$. For any fixed i, the ascending chain condition on ideals gives $L_i(I_j) = L_i(I_{n(i)})$ for $j \geq n(i)$, and the above argument shows that we may take $n(i) = q$ if $i \geq p$. Hence $n(i)$ may be taken to be bounded in i, say by n_0, and $L_i(I_j) = L_i(I_{n_0})$ for all $i \geq 0$ and $j \geq n_0$. By the result of the previous paragraph, $I_j = I_{n_0}$ for $j \geq n_0$, and hence the ascending chain condition has been verified for ideals in $R[X]$. □

Proposition 8.33. In a Noetherian integral domain, every nonzero nonunit is a product of irreducible elements.

PROOF. In other words, the assertion is that the condition (UFD1) of Section 4 is satisfied. The proof of the assertion is essentially the same as the proof of (UFD1) for Theorem 8.15. □

Proposition 8.34. If R is a Noetherian commutative ring, then any R submodule of a finitely generated unital R module is finitely generated.

REMARK. The proof follows the lines of the argument for Proposition 8.24.

PROOF. Let M be a unital finitely generated R module with a set $\{m_1, \ldots, m_n\}$ of n generators, and define $M_k = Rm_1 + \cdots + Rm_k$ for $1 \leq k \leq n$. Then $M_n = M$ since M is unital. We shall prove by induction on k that every R submodule of M_k is finitely generated. The case $k = n$ then gives the proposition. For $k = 1$, suppose that S is an R submodule of $M_1 = Rm_1$. Let I be the subset of all r in R with rm_1 in S. Since S is an R submodule, I is an ideal in R, necessarily finitely generated since R is Noetherian. Let $I = (r_1, \ldots, r_l)$. Then $S = Im_1 = Rr_1m_1 + Rr_2m_1 + \cdots + Rr_lm_1$, and the elements $r_1m_1, r_2m_1, \ldots, r_lm_1$ form a finite set of generators of S.

Assume inductively that every R submodule of M_k is known to be finitely generated, and let N_{k+1} be an R submodule of M_{k+1}. Let $q : M_{k+1} \to M_{k+1}/M_k$ be the quotient R homomorphism, and let φ be the restriction $q|_{N_{k+1}}$, mapping N_{k+1} into M_{k+1}/M_k. Then $\ker \varphi = N_{k+1} \cap M_k$ is an R submodule of M_k and is finitely generated by the inductive hypothesis. Also, image φ is an R submodule of M_{k+1}/M_k, which is singly generated with generator equal to the coset of m_{k+1}. Since an R submodule of a singly generated unital R module was shown in the previous paragraph to be finitely generated, image φ is finitely generated. Applying Lemma 8.23 to φ, we see that N_{k+1} is finitely generated. This completes the induction and the proof. □

9. Integral Closure

In this section, we let R be an integral domain, F be its field of fractions, and K be a any field containing F. Sometimes we shall assume also that $\dim_F K$ is finite. The main cases of interest are as follows.

EXAMPLES OF GREATEST INTEREST.

(1) $R = \mathbb{Z}$, $F = \mathbb{Q}$, and $\dim_F K < \infty$. In Chapter IX we shall see in this case from the "Theorem of the Primitive Element" that K is necessarily of the form $\mathbb{Q}[\theta]$ as already described in Section 1 and in Chapter IV. This is the setting we used in Section 7 as orientation for certain problems in algebraic number theory.

(2) $R = \mathbb{K}[X]$ for a field $\mathbb{K}$, $F = \mathbb{K}(X)$ is the field of fractions of R, and K is a field containing F with $\dim_F K < \infty$. In the special case $\mathbb{K} = \mathbb{C}$, this is the setting we used in Section 7 as orientation for treating curves in algebraic geometry.

Proposition 8.35. Let R be an integral domain, F be its field of fractions, and K be any field containing F. Then the following conditions on an element x of K are equivalent:

(a) x is a root of a monic polynomial in $R[X]$,
(b) the subring $R[x]$ of K generated by R and x is a finitely generated R module,
(c) there exists a finitely generated nonzero unital R module $M \subseteq K$ such that $xM \subseteq M$.

REMARK. When the equivalent conditions of the proposition are satisfied, we say that x is **integral** over R or x is **integrally dependent** on R. In this terminology, in Section VII.5 and in Section 1 of the present chapter, we defined an **algebraic integer** to be any member of $\mathbb{C}$ that is integral over $\mathbb{Z}$. The equivalence of (a) and (c) in this setting allowed us to prove that the set of algebraic integers is a subring of $\mathbb{C}$.

PROOF. If (a) holds, we can write $x^n + a_{n-1}x^{n-1} + \cdots + a_1 x + a_0 = 0$ for suitable coefficients in R. Solving for x^n and substituting, we see that the subring $R[x]$, which equals $R + Rx + Rx^2 + \cdots$, is actually given by $R[x] = R + Rx + \cdots + Rx^{n-1}$. Therefore $R[x]$ is a finitely generated R module, and (b) holds.

If (b) holds, then we can take $M = R[x]$ to see that (c) holds.

If (c) holds, let $m_1, \ldots, m_k$ be generators of M as an R module. Then we can find members a_{ij} of R for which

$$xm_1 = a_{11}m_1 + \cdots + a_{1k}m_k,$$
$$\vdots$$
$$xm_k = a_{k1}m_1 + \cdots + a_{kk}m_k.$$

This set of equations, regarded as a single matrix equation over K, becomes

$$\begin{pmatrix} x-a_{11} & -a_{12} & \cdots & -a_{1k} \\ -a_{21} & x-a_{22} & \cdots & -a_{2k} \\ & & \ddots & \\ -a_{k1} & -a_{k2} & \cdots & x-a_{kk} \end{pmatrix} \begin{pmatrix} m_1 \\ m_2 \\ \vdots \\ m_k \end{pmatrix} = \begin{pmatrix} 0 \\ 0 \\ \vdots \\ 0 \end{pmatrix}.$$

The k-by-k matrix on the left is therefore not invertible, and its determinant, which is a member of the field K, must be 0. Expanding the determinant and replacing x by an indeterminate X, we obtain a monic polynomial of degree k in $R[X]$ for which x is a root. Thus (a) holds. $\square$

If R, F, and K are as above, the **integral closure** of R in K is the set of all members of K that are integral over R. In Corollary 8.38 we shall prove that the integral closure of R in K is a subring of K.

EXAMPLES OF INTEGRAL CLOSURES.

(1) The integral closure of $\mathbb{Z}$ in $\mathbb{Q}$ is $\mathbb{Z}$ itself. This fact amounts to the statement that a rational root of a monic polynomial with integer coefficients is an integer; this was proved[10] in the course of Lemma 7.30. Recall the argument: If $x = p/q$ is a rational number in lowest terms that satisfies $x^n + a_{n-1}x^{n-1} + \cdots + a_1 a + a_0 = 0$, then we clear fractions and obtain $p^n + a_{n-1}p^{n-1}q + \cdots + a_1 pq^{n-1} + a_0 q^n = 0$. Examining divisibility by q, we see that q divides p^n. Hence any prime factor of q divides p and shows that p/q cannot be in lowest terms. Therefore q has no prime factors, and p/q is an integer.

(2) Let us determine the integral closure of $\mathbb{Z}$ in $\mathbb{Q}(\sqrt{m})$, where m is a square-free integer other than 0 or 1. The result is going to be that the integral closure consists of all $a + b\sqrt{m}$ with

$$a \text{ and } b \begin{cases} \text{both in } \mathbb{Z} & \text{if } m \not\equiv 1 \bmod 4, \\ \text{both in } \mathbb{Z} \text{ or both in } \mathbb{Z} + \tfrac{1}{2} & \text{if } m \equiv 1 \bmod 4. \end{cases}$$

---

[10]It is not assumed that the reader has looked at Chapter VII. A result that implies Lemma 7.30 will be obtained below as Corollary 8.38, which makes no use of material from Chapter VII.

In other words, the integral closure is

$$\begin{cases} \mathbb{Z}[\sqrt{m}\,] & \text{if } m \not\equiv 1 \bmod 4, \\ \mathbb{Z}[\tfrac{1}{2}(1+\sqrt{m}\,)] & \text{if } m \equiv 1 \bmod 4. \end{cases} \quad (*)$$

In fact, consider the polynomial

$$P(X) = X^2 - 2aX + (a^2 - mb^2),$$

whose roots are exactly $a \pm b\sqrt{m}$. If a and b are in $\mathbb{Z}$, then $P(X)$ has coefficients in $\mathbb{Z}$, and hence both of $a \pm b\sqrt{m}$ are in the integral closure. If $m \equiv 1 \bmod 4$ and a and b are both in $\mathbb{Z} + \tfrac{1}{2}$, write $a = c/2$ and $b = d/2$ with c and d in $2\mathbb{Z} + 1$. Since $a^2 - mb^2 = \tfrac{1}{4}(c^2 - md^2)$, we have

$$c^2 - md^2 \equiv c^2 - d^2 \bmod 4 \equiv 1 - 1 \bmod 4 \equiv 0 \bmod 4,$$

and therefore $\tfrac{1}{4}(c^2 - md^2) = a^2 - mb^2$ is in $\mathbb{Z}$. Consequently the polynomial $P(X)$ exhibits $a + b\sqrt{m}$ as in the integral closure.

For the reverse inclusion, suppose that $z = a + b\sqrt{m}$ is in the integral closure and is not in $\mathbb{Z}$. Then z is a root of some monic polynomial $A(X)$ in $\mathbb{Z}[X]$. In addition, z is a root of $P(X)$ above, and $P(X)$ is a monic prime polynomial in $\mathbb{Q}[X]$ because it has no rational first-degree factor. Writing $A(X) = B(X)P(X) + R(X)$ in $\mathbb{Q}[X]$ with $R(X) = 0$ or $\deg R(X) < \deg P(X) = 2$ and substituting z for X, we see that $R(z) = 0$, and we conclude that $R(X) = 0$. Thus $P(X)$ divides $A(X)$. By Corollary 8.20c, $P(X)$ is in $\mathbb{Z}[X]$. Hence $2a$ and $a^2 - mb^2$ are in $\mathbb{Z}$. One case is that a is in $\mathbb{Z}$, and then mb^2 is in $\mathbb{Z}$; since m is square free, there are no candidates for primes dividing the denominator of b, and so b is in $\mathbb{Z}$. The other case is that a is in $\mathbb{Z} + \tfrac{1}{2}$, and then mb^2 is in $\mathbb{Z} + \tfrac{1}{4}$. So $m(2b)^2$ is in $4\mathbb{Z} + 1$. Since m is square free, there are no candidates for primes dividing the denominator of $2b$, and $2b$ is an integer. Since $m(2b)^2$ is in $4\mathbb{Z} + 1$, $m \equiv 1 \bmod 4$ and $2b \equiv 1 \bmod 2$ are forced. This completes the proof that the integral closure is given by $(*)$.

(3) Under the assumption that the characteristic of the field $\mathbb{K}$ is not 2, let us determine the integral closure T of $R = \mathbb{K}[x]$ in $K = \mathbb{K}(x)[\sqrt{P(x)}\,] = \mathbb{K}(x)[y]/(y^2 - P(x))$, where $P(X)$ is a square-free polynomial in $\mathbb{K}[x]$. Parenthetically we need to check that K is a field. Since $\mathbb{K}(x)$ is a field, $\mathbb{K}(x)[y]$ is a principal ideal domain, and the question is whether $(y^2 - P(x))$ is a prime ($=$ maximal) ideal. We have only to observe that $y^2 - P(x)$ is irreducible because $P(x)$ is not a square, and then it follows that K is a field. Thus the situation for this example fits the setting of Proposition 8.35 with $R = \mathbb{K}[x]$, $F = \mathbb{K}(x)$, and $K = F(y)/(y^2 - P)$. We are going to show that the integral closure T of R in

K consists of all $A(x) + B(x)\sqrt{P(x)}$ with $A(x)$ and $B(x)$ both in $R = \mathbb{K}[x]$. It follows that the integral closure will be

$$T = \mathbb{K}[x][\sqrt{P(x)}\,] = \mathbb{K}[x] + \mathbb{K}[x]\sqrt{P(x)}. \qquad (*)$$

To see this, first let $A(x)$ and $B(x)$ be in $\mathbb{K}[x]$, and consider the monic polynomial

$$Q(y) = y^2 - 2Ay + (A^2 - PB^2) \qquad (**)$$

in $\mathbb{K}[x][y]$. Its roots in K are exactly $A(x) \pm B(x)\sqrt{P(x)}$, and thus we see that both of $A(x) \pm B(X)P(x)$ are in T. Conversely let $z = A(x) + B(x)\sqrt{P(x)}$ be in T but not R. Here $A(x)$ and $B(x)$ are in $\mathbb{K}(x)$. Then z is a root in K of some monic polynomial $M(y)$ whose coefficients are in $\mathbb{K}[x]$. In addition, z is a root of the member $Q(y)$ of $\mathbb{K}(x)[y]$ defined in $(**)$. The division algorithm gives $M(y) = N(y)Q(y) + W(y)$ in $\mathbb{K}(x)[y]$ with $W = 0$ or $\deg W < \deg Q = 2$. Substituting $z \in T$ for y, we obtain

$$0 = M(z) = N(z)Q(z) + W(z) = N(z)0 + W(z).$$

Thus $W(z) = 0$. If $\deg W = 1$, then z is in F, and the same argument as in Example 1 shows that z is in R; since we are assuming that z is not in R, we conclude that $W = 0$. Therefore $Q(y)$ divides $M(y)$. By Corollary 8.20c, $M(y)$ is in $\mathbb{K}[x][y]$. Hence $2A$ and $A^2 - PB^2$ are in $\mathbb{K}[x]$. Since the characteristic of $\mathbb{K}$ is not 2, A is in $\mathbb{K}[x]$. Then PB^2 is in $\mathbb{K}[x]$, and B must be in $\mathbb{K}[x]$ since P is square free. Thus T is given as in $(*)$.

From these examples we can extract a rough description of the situation that will interest us. We start with a ring R such as $\mathbb{Z}$ or $\mathbb{K}[x]$, along with its field of fractions F. We assume that the integral closure of R in F is R itself, as is the case with $\mathbb{Z}$ in $\mathbb{Q}$ and as we shall see is the case with $\mathbb{K}[x]$ in $\mathbb{K}(x)$. Let K be a field containing F with $\dim_F K < \infty$. We are interested in an analog T of integral elements relative to K, and what works as T is the integral closure of R in K.

Lemma 8.36. If A, B, and C are integral domains with $A \subseteq B \subseteq C$ such that C is a finitely generated B module and B is a finitely generated A module, then C is a finitely generated A module.

PROOF. Let C be generated over B by $c_1, \ldots, c_r$, and let B be generated over A by $b_1, \ldots, b_s$. Then C is generated over A by the sr elements $b_j c_i$ for $1 \leq i \leq r$ and $1 \leq j \leq s$. □

9. Integral Closure

Proposition 8.37. Let R be an integral domain, F be its field of fractions, and K be any field containing F. If $x_1, \ldots, x_r$ are members of K integral over R, then the subring $R[x_1, \ldots, x_r]$ of K generated by R and $x_1, \ldots, x_r$ is a finitely generated R module.

REMARKS. The ring $R[x_1, \ldots, x_r]$ is certainly finitely generated over R as a ring. The proposition asserts more—that it is finitely generated as an R module. This means that all products of powers of the x_j's are in the R linear span of finitely many of them.

PROOF. We induct on r. Since x_1 is assumed integral over R, the case $r = 1$ follows from Proposition 8.35b. For the inductive step, suppose that $R[x_1, \ldots, x_s]$ is a finitely generated R module. Since x_{s+1} is integral over R, it is certainly integral over $R[x_1, \ldots, x_s]$. Thus Proposition 8.35b shows that $R[x_1, \ldots, x_{s+1}]$ is a finitely generated $R[x_1, \ldots, x_s]$ module. Taking $A = R$, $B = R[x_1, \ldots, x_s]$, and $C = R[x_1, \ldots, x_{s+1}]$ in Lemma 8.36, we see that $R[x_1, \ldots, x_{s+1}]$ is a finitely generated R module. □

Corollary 8.38. Let R be an integral domain, F be its field of fractions, and K be any field containing F. Then the integral closure of R in K is a subring of K.

REMARK. A special case of this corollary appears in somewhat different language as Lemma 7.30.

PROOF. Let x and y be integral over R. Then $R[x, y]$ is a finitely generated R module by Proposition 8.37. We have $(x \pm y)R[x, y] \subseteq R[x, y]$ and $(xy)R[x, y] \subseteq R[x, y]$. Taking $M = R[x, y]$ in Proposition 8.35c and using the implication that (c) implies (a) in that proposition, we see that $x \pm y$ and xy are integral over R. □

Corollary 8.39. Let A, B, and C be integral domains with $A \subseteq B \subseteq C$. If every member of B is integral over A and if every member of C is integral over B, then every member of C is integral over A.

PROOF. Let K be the field of fractions of C, and regard C as a subring of K. If x is in C, then x is a root of some monic polynomial with coefficients in B, say $x^n + b_{n-1}x^{n-1} + \cdots + b_0 = 0$. By Proposition 8.37 the subring $D = A[b_{n-1}, \ldots, b_0]$ of C is a finitely generated A module. Since x is integral over D, $D[x]$ is a finitely generated D module, by a second application of Proposition 8.37. Lemma 8.36 shows that $D[x]$ is a finitely generated A module. By Proposition 8.35, x is integral over A. □

We say that the integral domain R is **integrally closed** if R equals its integral closure in its field of fractions. Example 1 above in essence observed that the ring $\mathbb{Z}$ of integers is integrally closed. Example 2 above showed, for the case $m = -3$, that the integral closure of $\mathbb{Z}$ in $\mathbb{Q}[\sqrt{-3}\,]$ is something other than the ring $\mathbb{Z}[\sqrt{-3}\,]$; consequently $\mathbb{Z}[\sqrt{-3}\,]$ cannot be integrally closed. A more direct argument is to observe that the element $x = \frac{1}{2}(-1+\sqrt{-3})$ of $\mathbb{Q}[\sqrt{-3}\,]$ satisfies $x^2 + x + 1 = 0$ but is not in $\mathbb{Z}[\sqrt{-3}\,]$.

Corollary 8.40. Let R be an integral domain, F be its field of fractions, and K be any field containing F. Then the integral closure T of R in K is integrally closed.

PROOF. Corollary 8.38 shows that T is a subring of K. Let C be the integral closure of T in K. We apply Corollary 8.39 to the integral domains $R \subseteq T \subseteq C$. The corollary says that every member of C is integral over R, and hence $C \subseteq T$. That is, $C = T$. Let $\eta : T \to L$ be the one-one homomorphism of T into its field of fractions, and let $\varphi : T \to K$ be the inclusion. By Proposition 8.6, there exists a unique ring homomorphism $\widetilde{\varphi} : L \to K$ such that $\varphi = \widetilde{\varphi}\eta$. Identifying L with $\widetilde{\varphi}(L) \subseteq K$, we can treat L as a subfield of K containing T. Since the only elements of K integral over T have been shown to be the members of T, the only elements of the subfield L integral over T are the members of T. Therefore T is integrally closed. $\square$

Proposition 8.41. If R is a unique factorization domain, then R is integrally closed.

PROOF. Suppose that $y^{-1}x$ is a member of the field of fractions F of R, with x and y in R and $y \neq 0$, and suppose that $y^{-1}x$ satisfies the equation

$$(y^{-1}x)^n + a_{n-1}(y^{-1}x)^{n-1} + \cdots + a_1(y^{-1}x) + a_0 = 0$$

with coefficients in R. Clearing fractions and moving x^n over to one side by itself, we have

$$x^n = -y(a_{n-1}x^{n-1} + \cdots + a_1xy^{n-2} + a_0y^{n-1}).$$

If a prime p in R divides y, then it divides x^n and must divide x. If R is a unique factorization domain, this says that we cannot arrange for $\text{GCD}(x, y)$ to equal 1 unless no prime divides y. In this case, y is a unit in R. Consequently $y^{-1}x$ is in R. $\square$

Since $\mathbb{Z}$ is a unique factorization domain, Proposition 8.41 gives a new proof that $\mathbb{Z}$ is integrally closed. We see also that $\mathbb{K}[x]$ is integrally closed when $\mathbb{K}$ is a field.

We saw above that the ring $\mathbb{Z}[\sqrt{-3}]$ is not integrally closed; consequently it cannot be a unique factorization domain. Another way of drawing this conclusion is to verify in the equality $(1 + \sqrt{-3})(1 - \sqrt{-3}) = 2 \cdot 2$ that the two elements on the left are irreducible and are not associates of the irreducible element 2 on the right.

A more significant example, taking advantage of the contrapositive of Proposition 8.41, is that any polynomial ring $\mathbb{K}[X_1, \ldots, X_n]$ over a field $\mathbb{K}$ is integrally closed. In fact, we know from Section 5 that $\mathbb{K}[X_1, \ldots, X_n]$ has unique factorization.

Proposition 8.42. Let R be an integral domain, F be its field of fractions, and K be any field containing F. If $\dim_F K < \infty$, then any x in K has the property that there is some $c \neq 0$ in R such that cx is integral over R.

REMARKS. Consequently K may be regarded as the field of fractions of the integral closure T of R in K. In fact, let $\{x_i\}$ be a basis of K over F, and choose $c_i \neq 0$ in R for each i such that $y_i = c_i x_i$ is integral over R. Then $\{y_i\}$ is a basis for K over F consisting of members of T, and it follows that every member of K is the quotient of a member of T by a member of R. Proposition 8.6 supplies a one-one ring homomorphism of the field of fractions for T into K, and the description just given for the elements of K shows that this homomorphism is onto K. Therefore K may be regarded as the field of fractions of T.

PROOF. Since $\dim_F K < \infty$, the elements $1, x, x^2, \ldots$ of K are linearly dependent over F. Therefore $a_n x^n + \cdots + a_1 x + a_0 = 0$ for a suitable n and for suitable members of F with $a_n \neq 0$. Clearing fractions, we may assume that $a_n, \ldots, a_1, a_0$ are in R and that $a_n \neq 0$. Multiplying the equation by a_n^{n-1}, we obtain

$$(a_n x)^n + a_{n-1}(a_n x)^{n-1} + \cdots + a_1 a_n^{n-2}(a_n x) + a_0 a_n^{n-1} = 0.$$

Thus we can take $c = a_n$. □

In the base rings $\mathbb{Z}$ and $\mathbb{K}[x]$ of our examples, every nonzero prime ideal is maximal because the rings are principal ideal domains. In Section 7 we mentioned that every nonzero prime ideal in $\mathbb{Z}[\sqrt{-5}]$ is maximal even though $\mathbb{Z}[\sqrt{-5}]$ is not a principal ideal domain. The remainder of this section, particularly Proposition 8.45, shows that the feature that every nonzero prime ideal is maximal is always preserved in our passage from R to T.

Proposition 8.43. Let R be an integral domain, F be its field of fractions, K be any field containing F, and T be the integral closure of R in K. If Q is a nonzero prime ideal of T, then $P = R \cap Q$ is a nonzero prime ideal of R.

REMARKS. Corollary 8.38 shows that T is a ring. A construction for prime ideals that goes in the reverse direction, from R to T, appears below as Proposition 8.53.

PROOF. Let Q be a nonzero prime ideal of T, and put $P = R \cap Q$. The ideal P is proper since 1 is not in Q and cannot be in P. It is prime since $xy \in P$ implies that xy is in Q, x or y is in Q, and x or y is in $R \cap Q = P$. To see that P is nonzero, take $t \neq 0$ in Q. Since t is integral over R, t satisfies some monic polynomial equation $t^n + a_{n-1} t^{n-1} + \cdots + a_1 t + a_0 = 0$ with coefficients in R. Without loss of generality, $a_0 \neq 0$ since otherwise we could divide the equation by a positive power of t. Then $a_0 = t(-t^{n-1} - a_{n-1} t^{n-2} - \cdots - a_1)$ exhibits a_0 as in Q as well as in R. Thus P is nonzero. $\square$

Lemma 8.44. Let R and T be integral domains with $R \subseteq T$ and with every element of T integral over R. If T' is an integral domain and $\varphi : T \to T'$ is a homomorphism of rings onto T', then every member of T' is integral over $\varphi(R)$.

PROOF. If t is in T, then t satisfies some monic polynomial equation of the form $t^n + a_{n-1} t^{n-1} + \cdots + a_1 t + a_0 = 0$ with coefficients in R. Applying φ to this equation, we see that $\varphi(t)$ satisfies a monic polynomial equation with coefficients in $\varphi(R)$. $\square$

Proposition 8.45. Let R be an integral domain, F be its field of fractions, K be any field containing F, and T be the integral closure of R in K. If every nonzero prime ideal of R is maximal, then every nonzero prime ideal of T is maximal.

REMARK. As with Proposition 8.43, Corollary 8.38 shows that T is a ring.

PROOF. Let Q be a nonzero prime ideal in T, and let $P = R \cap Q$. Since P is a nonzero prime ideal of R by Proposition 8.43, the hypotheses say that P is maximal in R. We shall apply Lemma 8.44 to the quotient homomorphism $T \to T/Q$. The lemma says that every element of the integral domain T/Q is integral over the subring $(R + Q)/Q$. Composing the inclusion homomorphism $R \to T$ with the homomorphism $T \to T/Q$ yields a ring homomorphism $R \to T/Q$ that carries P into the 0 coset. Since $P = R \cap Q$, this ring homomorphism descends to a one-one ring homomorphism $R/P \to T/Q$. The Second Isomorphism Theorem (for abelian groups) identifies the image of R/P with $(R + Q)/Q$. Since P is maximal as an ideal in R, R/P is a field. The ring isomorphism $R/P \cong (R+Q)/Q$ thus shows that every element of T/Q is integral over a field.

Let us write k for this field isomorphic to R/P, and let k' be the field of fractions of T/Q. We can now argue as in the proof of Proposition 4.1. If $x \neq 0$ is in T/Q, then x satisfies a monic polynomial equation $x^m + c_{m-1}x^{m-1} + \cdots + c_1 x + c_0 = 0$ with coefficients in k, and we may assume that $c_0 \neq 0$. Then the equality $x^{-1} = -c_0^{-1}(c_1 + \cdots + a_{m-1}x^{m-2} + x^{m-1})$ shows that the member x^{-1} of k' is in fact in T/Q. Therefore T/Q is a field, and the ideal Q is maximal in T. □

10. Localization and Local Rings

In this section, R denotes a commutative ring with identity. The objective is to enlarge or at least adjust R so as to make further elements of R become invertible under multiplication. The prototype is the construction of the field of fractions for an integral domain. A subset S of R is called a **multiplicative system** if 1 is in S and if the product of any two members of S is in S. The multiplicative system will be used as a set of new allowable denominators, and the new ring will be denoted[11] by $S^{-1}R$.

The construction proceeds along the same lines as in Section 2, except that some care is needed to take into account the possibility of zero divisors in R and even in S. We begin with an intermediate set

$$\widetilde{R} = \{(r, s) \mid r \in R, \ s \in S\}$$

and impose the relation $(r, s) \sim (r', s')$ if $t(rs' - sr') = 0$ for some $t \in S$. To check transitivity, suppose that $(r, s) \sim (r', s')$ and $(r', s') \sim (r'', s'')$. Then we have $t(rs' - sr') = 0$ and $t'(r's'' - s'r'') = 0$ for some t and t' in S, and hence

$$s' t t' (rs'' - sr'') = s'' t' \big(t(rs' - sr') \big) + st \big(t'(r's'' - s'r'') \big) = 0.$$

Since $s'tt'$ is in S, $(r, s) \sim (r'', s'')$. Thus $\sim$ is an equivalence relation.

The set of equivalence classes is denoted by $S^{-1}R$ and is called the **localization[12] of R with respect to S**. Addition and multiplication are defined in $\widetilde{R}$ by $(r, s) + (r', s') = (rs' + sr', ss')$ and $(r, s)(r', s') = (rr', ss')$. Simple variants of the arguments in Section 2 show that these operations descend to operations on $S^{-1}R$. For example, with addition let (r, s), (r', s'), and (r'', s'') be in $\widetilde{R}$ with

---

[11] Some authors write R_S instead of $S^{-1}R$.

[12] Some authors use a term like "ring of fractions" or "ring of quotients" in connection with localization in the general case or in some special cases. We shall not use these terms. In any event, "ring of quotients" is emphatically not to be confused with "quotient ring" as in Chapter IV, which is the coset space of a ring modulo an ideal.

$(r', s') \sim (r'', s'')$, i.e., with $t'(r's'' - s'r'') = 0$ for some $t' \in S$. Then the equivalence

$$(r, s) + (r', s') = (rs' + sr', ss') \sim (rs'' + sr'', ss'') = (r, s) + (r'', s'')$$

holds because

$$t'\big((rs' + sr')ss'' - (rs'' + sr'')ss'\big) = s^2 t'(r's'' - s'r'') = 0.$$

Similarly multiplication is well defined.

The result is that $S^{-1}R$ is a commutative ring with identity and that the mapping $r \mapsto r^*$, where r^* is the class of $(r, 1)$, is a ring homomorphism of R into $S^{-1}R$ carrying 1 to 1. Let us observe the following simple properties of $S^{-1}R$:

(i) $S^{-1}R = 0$ if and only if 0 is in S, since $S^{-1}R = 0$ if and only if $(1, 1) \sim (0, 1)$, if and only if $t(1 \cdot 1 - 1 \cdot 0) = 0$ for some $t \in S$.

(ii) $r \mapsto r^*$ is one-one if and only if S contains no zero divisors, since $r^* = 0$ if and only if $(r, 1) \sim (0, 1)$, if and only if $tr = 0$ for some $t \in S$.

(iii) s^* is a unit in $S^{-1}R$ for each $s \in S$, since the class of $(1, s)$ is a multiplicative inverse for s^*.

(iv) every member of $S^{-1}R$ is of the form $(s^*)^{-1}r^*$ for some $r \in R$ and $s \in S$, since $(r, s) = (r, 1)(1, s)$ is the class of $r^*(s^*)^{-1}$.

(v) $S^{-1}R$ is an integral domain if R is an integral domain and 0 is not in S.

In working with localizations, we shall normally drop the superscript $*$ on the image r^* in $S^{-1}R$ of an element r of R.

Localizations arise in algebraic number theory and in algebraic geometry. In applications to algebraic number theory, the ring R typically is an integral domain, and therefore the map $r \mapsto r^*$ is one-one. In applications to algebraic geometry, S may have zero divisors.

EXAMPLES OF LOCALIZATIONS.

(1) R is arbitrary, and $S = \{1\}$. Then $S^{-1}R = R$.

(2) R is arbitrary, and $S = \{\text{nonzero elements that are not zero divisors in } R\}$. Then every nonzero element of $S^{-1}R$ is a zero divisor or is a unit. In this example when S consists of all members of R other than 0, then R is an integral domain and $S^{-1}R$ is the field of fractions of R.

(3) R is arbitrary, P is a prime ideal in R, and S is the set-theoretic complement of P. The identity is in S since P is proper. The prime nature of P is used in checking that S is a multiplicative system: if s and t are in S, then neither is in P, by definition, and their product st cannot be in P since P is prime; thus the product st is in S. With these definitions,

$$S^{-1}R \quad \text{is often denoted by } R_P$$

and is called the **localization of R at the prime** P. In practice this is the most important example of a localization,[13] directly generalizing the construction of the field of fractions of an integral domain as the localization at the prime ideal 0. Here are some special cases, $\mathbb{K}$ being a field in the cases in which it occurs:

(a) When $R = \mathbb{Z}$ and $P = (p)$ for a prime number p, the set S consists of nonzero integers not divisible by p, and R_P is the subset of all members of $\mathbb{Q}$ whose denominators are not divisible by p.

(b) When $R = \mathbb{K}[X]$ and $P = (X-c)$, the set S consists of all polynomials that are nonvanishing at c, and R_P is the set of formal rational expressions in X that are finite at c.

(c) When $R = \mathbb{K}[X,Y]$ and $P = (X-c, Y-d)$, the set S consists of all polynomials in X and Y that are nonvanishing at (c,d), and R_P is the set of formal rational expressions in X and Y that are finite at (c,d).

(d) When $R = \mathbb{K}[X,Y]$ and $P = (X)$, the set S consists of all polynomials in X and Y that are not divisible by X, and R_P is the set of formal rational expressions in X and Y that are meaningful as rational expressions in Y when X is set equal to 0. For example, $1/(X+Y)$ is in R_P, but $1/X$ is not.

(4) R is arbitrary, $\{P_\alpha\}$ is a nonempty collection of prime ideals, and S is the set of all elements of R that lie in none of the ideals P_α. Then $S^{-1}R$ may be regarded as the localization of R at the set of all primes P_α.

(5) R is arbitrary, u is an element of R, and $S = \{1, u, u^2, \dots\}$. For example, if $R = \mathbb{Z}/(p^2)$, where p is a prime, and if $u = p$, then 0 is in S, and observation (i) shows that $S^{-1}R = 0$.

(6) R is a Noetherian integral domain, E is an arbitrary set of nonzero elements of R, and S is the set of all finite products of members of E, including the element 1 as the empty product. Let us see that the same $S^{-1}R$ results when E is replaced by a certain set E' of units and irreducible elements of R, namely the union of $R^\times$ and the set of all irreducible elements x in R such that x^{-1} is in $S^{-1}R$. Define T to be the set of all finite products of members of E'. We show that $S^{-1}R = T^{-1}R$. If e is in E', then either e is a unit in R, in which case e^{-1} lies in R and therefore also $S^{-1}R$, or e is irreducible in R with e^{-1} in $S^{-1}R$. Passing to finite products of members of E', we see that $T^{-1} \subseteq S^{-1}R$. Hence $T^{-1}R \subseteq S^{-1}R$. Now let s be in S, and use Proposition 8.33 to write s as a product of irreducible elements $s = s_1 \cdots s_n$. Then $s_j^{-1} = s^{-1}(s_1 \cdots \widehat{s_j} \cdots s_n)$, with $\widehat{s_j}$ indicating a missing factor. By construction, each s_j is in E'. Therefore each s_j is in T, and s is in T. Consequently $S \subseteq T$, and $S^{-1}R \subseteq T^{-1}R$.

---

[13]Beware of confusing R_P with R/P. The ring R_P is obtained by suitably enlarging R, at least in the case that R is an integral domain, whereas the ring R/P is obtained by suitably factoring something out from R.

The localization of R at S is characterized up to canonical isomorphism by the same kind of universal mapping property that characterizes the field of fractions of an integral domain. To formulate a proposition, let us write η for the homomorphism $r \mapsto r^*$ of R into $S^{-1}R$. Then the pair $(S^{-1}R, \eta)$ has the universal mapping property stated in Proposition 8.46 and illustrated in Figure 8.7.

$$\begin{array}{ccc} R & \xrightarrow{\varphi} & T \\ \eta \downarrow & \nearrow \widetilde{\varphi} & \\ S^{-1}R & & \end{array}$$

FIGURE 8.7. Universal mapping property of the localization of R at S.

Proposition 8.46. Let R be a commutative ring with identity, let S be a multiplicative system in R, let $S^{-1}R$ be the localization of R at S, and let η be the canonical homomorphism of R into $S^{-1}R$. Whenever φ is a ring homomorphism of R into a commutative ring T with identity such that $\varphi(1) = 1$ and such that $\varphi(s)$ is a unit in T for each $s \in S$, then there exists a unique ring homomorphism $\widetilde{\varphi} : S^{-1}R \to T$ such that $\varphi = \widetilde{\varphi}\eta$.

PROOF. If (r, s) with $s \in S$ is a pair in $\widetilde{R}$, we define $\Phi(r, s) = \varphi(r)\varphi(s)^{-1}$. This is well defined since $\varphi(s)$ is assumed to be a unit in T. Let us see that Φ is consistent with the equivalence relation, i.e., that $(r, s) \sim (r', s')$ implies $\Phi(r, s) = \Phi(r', s')$. Since $(r, s) \sim (r', s')$, we have $u(rs' - r's) = 0$ for some $u \in S$, and therefore also $\varphi(u)(\varphi(r)\varphi(s') - \varphi(r')\varphi(s)) = 0$. Since $\varphi(u)$ is a unit, $\varphi(r)\varphi(s') = \varphi(r')\varphi(s)$. Hence $\Phi(r, s) = \varphi(r)\varphi(s)^{-1} = \varphi(r')\varphi(s')^{-1} = \Phi(r', s')$, as required.

We can thus define $\widetilde{\varphi}$ of the class of (r, s) to be $\Phi(r, s)$, and $\widetilde{\varphi}$ is well defined as a function from $S^{-1}R$ to T. It is a routine matter to check that $\widetilde{\varphi}$ is a ring homomorphism. If r is in R, then $\widetilde{\varphi}(\eta(r)) = \widetilde{\varphi}(\text{class of } (r, 1)) = \Phi(r, 1) = \varphi(r)\varphi(1)^{-1}$, and this equals $\varphi(r)$ since φ is assumed to carry 1 into 1. Therefore $\widetilde{\varphi}\eta = \varphi$.

For uniqueness, observation (iv) shows that the most general element of $S^{-1}R$ is of the form $\eta(r)\eta(s)^{-1}$ with $r \in R$ and $s \in S$. Since $(\widetilde{\varphi}\eta)(r) = \varphi(r)$ and $(\widetilde{\varphi}\eta)(s) = \varphi(s)$, we must have $\widetilde{\varphi}(\eta(r)\eta(s)^{-1}) = \widetilde{\varphi}(\eta(r))\widetilde{\varphi}(\eta(s))^{-1} = \varphi(r)\varphi(s)^{-1}$. Therefore φ uniquely determines $\widetilde{\varphi}$. □

We shall examine the relationship between ideals in R and ideals in the localization $S^{-1}R$. If I is an ideal in R, then $S^{-1}I = \{s^{-1}i \mid s \in S, i \in I\}$ is easily checked to be an ideal in $S^{-1}R$ and is called the **extension** of I to $S^{-1}R$. If J is an ideal in $S^{-1}R$, then $R \cap J$, i.e., the inverse image of J under the canonical homomorphism $\eta : R \to S^{-1}R$, is an ideal in R and is called the **contraction** of J.

Proposition 8.47. Let R be a commutative ring with identity, and let $S^{-1}R$ be a localization. If J is an ideal in $S^{-1}R$, then $S^{-1}(R \cap J) = J$. Consequently the mapping $I \mapsto S^{-1}I$ is a one-one mapping of the set of all ideals I in R of the form $I = R \cap J$ onto the set of all ideals in $S^{-1}R$, and this mapping respects intersections and inclusions.

REMARKS. As in the definition of contraction, $R \cap J$ means $\eta^{-1}(J)$, where $\eta : R \to S^{-1}R$ is the canonical homomorphism. The map $I \mapsto S^{-1}I$ that carries arbitrary ideals of R to ideals of $S^{-1}R$ need not be one-one; the localization could for example be the field of fractions of an integral domain and have only trivial ideals. The proposition says that the map $I \mapsto S^{-1}I$ is one-one, however, when restricted to ideals of the form $I = R \cap J$.

PROOF. From the facts that $R \cap J \subseteq J$ and J is an ideal in $S^{-1}R$, we obtain $S^{-1}(R \cap J) \subseteq S^{-1}J \subseteq J$. For the reverse inclusion let x be in J, and write $x = s^{-1}r$ with r in R and s in S. Then $sx = r$ is in $R \cap J$, and therefore x is in $S^{-1}(R \cap J)$.

For the conclusion about the mapping $I \mapsto S^{-1}I$, the mapping is one-one because $S^{-1}(R \cap J_1) = S^{-1}(R \cap J_2)$ implies $J_1 = J_2$ by what we have just shown; hence $R \cap J_1 = R \cap J_2$. The mapping is onto because if J is given, then $J = S^{-1}(R \cap J)$ by what has already been shown. To see that the mapping respects the intersection of ideals, let ideals $R \cap J_\alpha$ be given for α in some nonempty set. Then

$$S^{-1}\left(\bigcap_\alpha (R \cap J_\alpha)\right) = S^{-1}(R \cap \bigcap_\alpha J_\alpha) = \bigcap_\alpha J_\alpha = \bigcap_\alpha S^{-1}(R \cap J_\alpha).$$

Finally the fact that the mapping respects the intersection of two ideals implies that it respects inclusions. □

Corollary 8.48. Let R be a commutative ring with identity, and let $S^{-1}R$ be a localization.

(a) If R is Noetherian, then $S^{-1}R$ is Noetherian.

(b) If every nonzero prime ideal in R is maximal, then the same thing is true in $S^{-1}R$.

(c) If R is an integral domain that is integrally closed and if $S^{-1}R$ is not zero, then $S^{-1}R$ is integrally closed.

(d) If I is an ideal in R, then the ideal $S^{-1}I$ of $S^{-1}R$ is proper if and only if $I \cap S = \emptyset$.

PROOF. For (a), let $\{J_\alpha\}$ be a nonempty collection of ideals in $S^{-1}R$. Contraction of ideals is one-one by the first conclusion of Proposition 8.47, and it respects inclusions because it is given by the inverse image of a function. Since R is Noetherian, Corollary 8.31b produces a maximal element $R \cap J$ from among the ideals $R \cap J_\alpha$ of R. The first and second conclusions of Proposition 8.47 together show that $J = S^{-1}(R \cap J) \supseteq S^{-1}(R \cap J_\alpha) = J_\alpha$ for all α. Hence J is maximal among the J_α.

For (b), let J_1 be a nonzero prime ideal in $S^{-1}R$. Arguing by contradiction, suppose that J_2 is an ideal in $S^{-1}R$ with $J_1 \subsetneqq J_2 \subsetneqq S^{-1}R$. Then $R \cap J_1 \subseteq R \cap J_2 \subseteq R$. If either of these inclusions were an equality, then use of the second conclusion of Proposition 8.47 would give a corresponding equality for J_1, J_2, R, and there is no such equality. Hence $R \cap J_1 \subsetneqq R \cap J_2 \subsetneqq R$.

If J_1 is prime in $S^{-1}R$, then $R \cap J_1$ is prime in R: In fact, if a and b are members of R such that ab is in $R \cap J_1$, then ab is in J_1, and either a or b must be in J_1 since J_1 is prime. Since a and b are both in R, one of a and b is in $R \cap J_1$. Thus $R \cap J_1$ is prime.[14]

By assumption for (b), $R \cap J_1$ is then maximal in R, and this conclusion contradicts the fact that $R \cap J_1 \subsetneqq R \cap J_2 \subsetneqq R$. The assumption that J_2 exists has thus led us to a contradiction. Consequently there can be no such J_2, and J_1 is a maximal ideal in $S^{-1}R$.

For (c), let F be the field of fractions of R, so that $R \subseteq S^{-1}R \subseteq F$. The field of fractions of $S^{-1}R$ is the field F as a consequence of Proposition 8.6. If x is a member of F that is integral over $S^{-1}R$ and if x satisfies $x^n + b_{n-1}x^{n-1} + \cdots + b_0 = 0$ with coefficients in $S^{-1}R$, then we can find a common element s of S and rewrite this equation as

$$x^n + (s^{-1}a_{n-1})x^{n-1} + \cdots + (s^{-1}a_0) = 0$$

with $a_{n-1}, \ldots, a_0$ in R. Multiplying by s^n, we obtain

$$(sx)^n + a_{n-1}(sx)^{n-1} + \cdots + a_1 s^{n-2}(sx) + a_0 s^{n-1} = 0.$$

Therefore sx is integral over R. Since R is integrally closed, sx is in R. Write $r = sx$. Then $x = s^{-1}r$ with r in R and s in S. Hence x is exhibited as in $S^{-1}R$, and we conclude that $S^{-1}R$ is integrally closed.

For (d), suppose that $I \cap S$ is nonempty. If s is in $I \cap S$, then $1 = s^{-1}s$ is in $S^{-1}I$ and the ideal $S^{-1}I$ equals $S^{-1}R$. Conversely if $S^{-1}I = S^{-1}R$, then 1 is in $S^{-1}I = \{s^{-1}i \mid s \in S, \ i \in I\}$, and hence $1 = s^{-1}i$ for some s and i; consequently $I \cap S$ contains the element $i = s$. □

A **local ring** is a commutative ring with identity having a unique maximal ideal. An equivalent definition is given in Proposition 8.49 below, and then it follows that the localization $S^{-1}R$ of Example 2 earlier in this section is a local

---

[14]Problem 9 at the end of the chapter puts this argument in a broader context.

ring. Corollary 8.50 below will produce a more useful example: localization with respect to a prime ideal, as in Example 3 earlier, always yields a local ring.[15]

Proposition 8.49. A nonzero commutative ring R with identity is a local ring if and only if the nonunits of R form an ideal.

REMARK. The zero ring is not local, having no proper ideals, and its set of nonunits is empty, hence is not an ideal.

PROOF. If the nonunits of R form an ideal, then that ideal is a unique maximal ideal since a proper ideal cannot contain a unit; hence R is local. Conversely suppose that R is local and that M is the unique maximal ideal. If x is any nonunit, then the principal ideal (x) is a proper ideal since 1 is not of the form xr. By Proposition 8.8, (x) is contained in some maximal ideal, and we must have $(x) \subseteq M$ since M is the unique maximal ideal. Then x is in M, and we conclude that every nonunit is contained in M. □

Corollary 8.50. Let R be an integral domain, let P be a prime ideal of R, let S be the set-theoretic complement of P, and let $R_P = S^{-1}R$ be the localization of R at P. Then R_P is a local ring, its unique maximal ideal is $M = S^{-1}P$, and P can be recovered from M as $P = R \cap M$. If Q is any prime ideal of R that is not contained in P, then $S^{-1}Q = S^{-1}R$.

PROOF. The subset $S^{-1}P$ of $S^{-1}R$ is an ideal by Proposition 8.47, and Corollary 8.48d shows that it is proper. Every member of $S^{-1}R$ that is not in $S^{-1}P$ is of the form $s'^{-1}s$ with s and s' in S and hence is a unit. Since no unit lies in any proper ideal, $S^{-1}R$ has $M = S^{-1}P$ as its unique maximal ideal, and $S^{-1}R$ is local by Proposition 8.49.

The contraction $R \cap M$ consists of all elements in R of the form $s^{-1}p$ with s in S and p in P. Let us see that the contraction equals P. Certainly $R \cap M \supseteq P$. For the reverse inclusion the equation $s^{-1}p = r$ says that $p = rs$. If r is not in P, then the facts that s is not in P and P is prime imply that $p = rs$ is not in P, contradiction. Thus r is in P, and we conclude that P can be recovered from M as $P = R \cap M$.

If Q is any prime ideal of R that is not contained in P, then $S^{-1}Q = S^{-1}R$. In fact, any element q of Q that is not in P is in S; therefore 1 is in the ideal $S^{-1}Q$, and $S^{-1}Q = S^{-1}R$. □

The construction of R_P in the corollary reduces to the construction of the field of fractions of R if $P = 0$. Other interesting and typical cases occur for

---

[15]For Example 3 with $R = \mathbb{K}[X]$ and $P = (X - c)$, the sense in which the ring R_P is "local" has a geometric interpretation: the only spot in $\mathbb{K}$ where we can regard members of R_P as $\mathbb{K}$-valued functions is "near" the point c, with "near" depending on the element of R_P. See the discussion after the proof of Corollary 8.50 below.

suitable nonzero P's when $R = \mathbb{K}[X, Y]$, $\mathbb{K}$ being a field. One such prime ideal is $P = (X - c, Y - d)$; then, as was mentioned in connection with Example 3 above, the localization of R at P consists of the rational expressions $f(X, Y)$ that are well defined at (c, d). The maximal ideal in this case consists of all such rational expressions that are 0 at (c, d). Another example of a nonzero prime ideal in $R = \mathbb{K}[X, Y]$ is $P = (X)$; then the localization of R at P consists of the rational expressions $f(X, Y)$ whose denominators are not divisible by X, and the maximal ideal consists of all such rational expressions $f(X, Y)$ whose numerators are divisible by X if f is written in lowest terms.

A number-theoretic analog of the localizations of the previous paragraph is the localization of $R = \mathbb{Z}$ at (p), where p is a prime number. The discussion with Example 3 above mentioned that the localization consists of all members of $\mathbb{Q}$ with no factor of p in the denominator. In this case the maximal ideal consists of those rationals q whose numerators are divisible by p if q is written in lowest terms.

We conclude this section with introductory remarks about a product operation on ideals. Let R be a nonzero commutative ring with identity. If I and J are ideals in R, then once again IJ denotes[16] the set of all sums of products of a member of I by a member of J. Certainly IJ is closed under addition and negatives, and the fact that $r(IJ) = (rI)J \subseteq IJ$ for $r \in R$ shows that IJ is an ideal. Localization with respect to a prime ideal is a handy tool for extracting information about products of ideals. We illustrate with Propositions 8.52 and 8.53 below. The first of these will play an important role in Section 11.

Lemma 8.51 (Nakayama's Lemma). Let R be a commutative ring with identity, let I be an ideal of R contained in all maximal ideals, and let M be a finitely generated unital R module. If $IM = M$, then $M = 0$.

REMARK. Here IM means the set of sums of products of a member of I by a member of M. The lemma applies to no ideals if $R = 0$.

PROOF. We induct on the number of generators of M. If M is singly generated, say by a generator m, then the hypothesis $IM = M$ implies that $rm = m$ for some r in I. Thus $(1 - r)m = 0$. If $1 - r$ is a unit, then we can multiply by its inverse and obtain $m = 0$; we conclude that $M = 0$. If $1 - r$ is not a unit, then it lies in some maximal ideal P, by application of Proposition 8.8 to the proper principal ideal $(1 - r)$. Since r lies in P by hypothesis, 1 lies in P, and we have a contradiction to the fact that P is proper.

---

[16]Sometimes, such as in the equality $S^{-1}S^{-1} = S^{-1}$, the product notation is meant to refer only to the set of all products, not to all sums of products. With ideals we are to allow sums of products. The applicable convention will normally be clear from the context, but we shall be explicit when there might be a possibility of confusion.

Suppose that the lemma holds for $n-1$ or fewer generators, and let M be generated by $m_1, \ldots, m_n$. Since $IM = M$, we have $\sum_{j=1}^{n} r_j m_j = m_1$ for suitable $r_1, \ldots, r_n$ in I. Then $(1 - r_1)m_1 = \sum_{j=2}^{n} r_j m_j$. If $1 - r_1$ is a unit, then we can multiply by its inverse and see that the generator m_1 is unnecessary; we conclude that $M = 0$ by induction. If $1 - r_1$ is not a unit, then it lies in some maximal ideal P. Since r_1 lies in P by hypothesis, 1 lies in P, and we have a contradiction. $\square$

Proposition 8.52. Let R be a Noetherian commutative ring, and let I and P be ideals in R with P prime. If $IP = I$, then $I = 0$.

PROOF. Let us localize with respect to the prime ideal P. If we write S for the set-theoretic complement of P in R, then $R_P = S^{-1}R$ is a local ring by Corollary 8.50, and its unique maximal ideal is $S^{-1}P$. Since $(S^{-1}I)(S^{-1}R) = S^{-1}IR = S^{-1}I$, $S^{-1}I$ is an ideal in R_P. Also, $(S^{-1}I)(S^{-1}P) = S^{-1}IP = S^{-1}I$, and $S^{-1}I$ has to be proper. In Nakayama's Lemma (Lemma 8.51), let us take M to be the $S^{-1}R$ module $S^{-1}I$. Since $S^{-1}P$ is the only maximal ideal in $S^{-1}R$, M is contained in all maximal ideals of $S^{-1}R$. Since R is Noetherian, Corollary 8.48a shows $S^{-1}R$ to be Noetherian, and the ideal $S^{-1}I$ is a finitely generated $S^{-1}R$ module by Corollary 8.31c. The lemma applies since $(S^{-1}P)(S^{-1}I) = S^{-1}I$, and the conclusion is that $S^{-1}I = 0$. Then the subset I of $S^{-1}I$ must be 0. $\square$

Proposition 8.53. Let R be an integral domain, F be its field of fractions, K be any field containing F, and T be the integral closure of R in K. If P is a maximal ideal in R, then $PT \neq T$, and there exists a maximal ideal Q of T with $P = R \cap Q$.

REMARKS. This result inverts the construction of Proposition 8.43, of course not necessarily uniquely. The examples in Section 7 illustrate what can happen in simple cases. More detailed analysis of what can happen in general requires some field theory and is postponed to Chapter IX, specifically when we discuss "splitting of prime ideals in extensions."

PROOF. If $PT \neq T$, then Proposition 8.8 supplies a maximal ideal Q of T with $PT \subseteq Q$. Since 1 is not in Q, we then have $P \subseteq R \cap Q \subsetneq R$. Consequently the maximality of P implies that $P = R \cap Q$.

Arguing by contradiction, we now assume that $PT = T$. Localizing, let S be the set-theoretic complement of P in R, so that $S^{-1}P$ is the unique maximal ideal of $S^{-1}R$ by Corollary 8.50. From $PT = T$, we can write

$$1 = a_1 t_1 + \cdots + a_n t_n \qquad (*)$$

with each a_i in P and each t_i in T. If we define T_0 to be the subring $R[t_1, \cdots, t_n]$ of T, then T_0 is a finitely generated R module by Proposition 8.37, and $S^{-1}T_0$

is therefore a finitely generated $S^{-1}R$ module. Equation ($*$) shows that 1 lies in PT_0. Multiplying by an arbitrary element of T_0, we see that $PT_0 = T_0$. Since $S^{-1}S^{-1} = S^{-1}$, we obtain $(S^{-1}P)(S^{-1}T_0) = S^{-1}T_0$. Nakayama's Lemma (Lemma 8.51) allows us to conclude that $S^{-1}T_0 = 0$. Since 1 lies in T_0, we have arrived at a contradiction. $\square$

11. Dedekind Domains

A **Dedekind domain** is an integral domain with the following three properties:
  (i) it is Noetherian,
  (ii) it is integrally closed,
  (iii) every nonzero prime ideal is maximal.

Every principal ideal domain R is a Dedekind domain. In fact, (i) every ideal in R is singly generated, (ii) R is integrally closed by Proposition 8.41, and (iii) every nonzero prime ideal in R is maximal by Corollary 8.16.

We shall be interested in Dedekind domains that are obtained by enlarging a principal ideal domain suitably. The general theorem in this direction is that if R is a Dedekind domain with field of fractions F and if K is a field containing F with $\dim_F K$ finite, then the integral closure of R in K is a Dedekind domain. Let us state something less sweeping.

Theorem 8.54. If R is a Dedekind domain with field of fractions F and if K is a field containing F with $\dim_F K$ finite, then the integral closure T of R in K is a Dedekind domain if any of the following three conditions holds:
  (a) T is Noetherian,
  (b) T is finitely generated as an R module,
  (c) the field extension $F \subseteq K$ is "separable."

REMARKS. The term "separable" will be defined in Chapter IX, and the fact that (c) implies (b) will be proved at that time. It will be proved also that characteristic 0 implies separable. For now, we shall be content with showing that (b) implies (a) and that (a) implies that T is a Dedekind domain.

PROOF. We are given that R satisfies conditions (i), (ii), (iii) above, and we are to verify the conditions for T. Condition (ii) holds for T by Corollary 8.40, and Proposition 8.45 shows that (iii) holds. If (a) holds, then T satisfies the defining conditions of a Dedekind domain.

Let us see that (b) implies (a). If (b) holds, then Proposition 8.34 shows that every R submodule of T is finitely generated. Since $T \supseteq R$, every T submodule of T is finitely generated. That is, every ideal of T is finitely generated, and T is Noetherian. Thus (a) holds, and the proof is complete. $\square$

Example 2 of integral closures in Section 9 showed that the integral closure of $\mathbb{Z}$ in $\mathbb{Q}(\sqrt{m})$ is doubly generated as a $\mathbb{Z}$ module, a set of generators being either $\{1, \sqrt{m}\}$ or $\{1, \frac{1}{2}(1 + \sqrt{m})\}$, depending on the value of m. Example 3 showed, under the assumption that $\mathbb{K}$ has characteristic different from 2, that the integral closure of $\mathbb{K}[x]$ in $\mathbb{K}(x)[\sqrt{P(x)}]$ is doubly generated as a $\mathbb{K}[x]$ module, a set of generators being $\{1, \sqrt{P(x)}\}$. Since $\mathbb{Z}$ and $\mathbb{K}[x]$ are principal ideal domains and hence Dedekind domains, these examples give concrete cases in which hypothesis (b) in Theorem 8.54 is satisfied. Consequently in each case the theorem asserts that a certain explicit integral closure is a Dedekind domain.

Theorem 8.55 (unique factorization of ideals). If R is a Dedekind domain, then each nonzero proper ideal I in R decomposes as a finite product $\prod_{j=1}^{n} P_j^{k_j}$, where the P_j's are distinct nonzero prime ideals and the k_j's are positive integers. Moreover,

(a) the decomposition into positive powers of distinct nonzero prime ideals is unique up to the order of the factors,
(b) the power P^k of a nonzero prime ideal P appearing in the decomposition of I is characterized as the unique nonnegative integer such that P^k contains I and P^{k+1} does not contain I (with $k = 0$ interpreted as saying that P is not one of the P_j),
(c) whenever I, J_1, J_2 are nonzero ideals with $IJ_1 = IJ_2$, then $J_1 = J_2$,
(d) whenever I and J_1 are two nonzero proper ideals with $I \subseteq J_1$, then there exists a nonzero ideal J_2 with $I = J_1 J_2$.

Let us say that a nonzero ideal J_1 **divides** a nonzero ideal I if $I = J_1 J_2$ for some ideal J_2. We say also that J_1 is a **factor** of I. Conclusion (d), once it is established, is an important principle for working with ideals in a Dedekind domain: *to contain is to divide*.

Thinking along these lines leads us to expect that prime ideals play some special role with respect to containment. Such a role is captured by the following lemma.

Lemma 8.56. In an integral domain, if P is a prime ideal such that $P \supseteq I_1 \cdots I_n$ for the product of the ideals $I_1, \ldots, I_n$, then $P \supseteq I_j$ for some j.

PROOF. By induction it is enough to handle $n = 2$. Thus suppose $P \supseteq I_1 I_2$. We are to show that $P \supseteq I_1$ or $P \supseteq I_2$. Arguing by contradiction, suppose on the contrary that $x \in I_1$ and $y \in I_2$ are elements with $x \notin P$ and $y \notin P$. Then xy cannot be in P since P is prime, but xy is in $I_1 I_2 \subseteq P$, and we have a contradiction. $\square$

Lemma 8.57. Let R be a Dedekind domain, and let I be a nonzero ideal of R. Then there exists a finite product $P_1 \cdots P_k$ of nonzero prime ideals, possibly empty and not necessarily having distinct factors, such that $P_1 \cdots P_k \subseteq I$.

PROOF. We argue by contradiction. Among all nonzero ideals for which there is no such finite product, choose one, say J, that is maximal under inclusion. This choice is possible since R is Noetherian. The ideal J cannot be prime since otherwise $J \subseteq J$ would be the containment asserted by the lemma. Thus we can choose elements a_1 and a_2 in J with $a_1 a_2 \in J$, $a_1 \notin J$, and $a_2 \notin J$. Define ideals I_1 and I_2 by $I_1 = J + Ra_1$ and $I_2 = J + Ra_2$. These strictly contain J, and their product manifestly has $I_1 I_2 \subseteq J$. By maximality of J, we can find products $P_1 \cdots P_k$ and $Q_1 \cdots Q_l$ of nonzero prime ideals with $P_1 \cdots P_k \subseteq I_1$ and $Q_1 \cdots Q_l \subseteq I_2$. Then $P_1 \cdots P_k Q_1 \cdots Q_l \subseteq I_1 I_2 \subseteq J$, contradiction. $\square$

Lemma 8.58. Let R be a Dedekind domain, regard R as embedded in its field of fractions F, let P be a nonzero prime ideal in R, and define

$$P^{-1} = \{x \in F \mid xP \subseteq R\}.$$

Then the set PP^{-1} of sums of products equals R.

PROOF. By definition of P^{-1}, $P \subseteq PP^{-1} \subseteq R$. Since P is an ideal and PP^{-1} is closed under addition and negatives, PP^{-1} is an ideal. Property (iii) of Dedekind domains shows that P is a maximal ideal in R, and therefore $PP^{-1} = P$ or $PP^{-1} = R$. We are to rule out the first alternative.

Thus suppose that $PP^{-1} = P$. Since R is Noetherian by (i), P is a finitely generated R submodule of F. The equality $PP^{-1} = P$ implies that each member x of P^{-1} has $xP \subseteq P$, and Proposition 8.35c implies that each such x is integral over R. Since R is integrally closed by (ii), x is in R. Thus $P^{-1} \subseteq R$, and the definition of P^{-1} shows that $P^{-1} = R$.

Fix a nonzero element a of P. Applying Lemma 8.57, find a product of nonzero prime ideals such that $P_1 \cdots P_k \subseteq (a) \subseteq P$. Without loss of generality, we may assume that k is as small as possible among all such inclusions. Since P is prime and $P_1 \cdots P_k \subseteq P$, Lemma 8.56 shows that P contains some P_j, say P_1. By (iii), P_1 is maximal, and therefore $P = P_1$. Form the product $P_2 \cdots P_k$, taking this product to be R if $k = 1$. Then $P_2 \cdots P_k$ is not a subset of (a), by minimality of k, and there exists a member b of $P_2 \cdots P_k$ that is not in (a). On the other hand, $PP_2 \cdots P_k \subseteq (a)$ shows that $Pb \subseteq (a)$, hence that $a^{-1}bP \subseteq R$. Thus $a^{-1}b$ is in P^{-1}, which we are assuming is R. In other words, $a^{-1}b$ is in R, and b is in $aR = (a)$, contradiction. $\square$

PROOF OF THEOREM 8.55. Arguing by contradiction, we may assume because R is Noetherian that I is maximal among the nonzero proper ideals that do not decompose as products of prime ideals. Then certainly I is not prime. Application of Proposition 8.8 produces a maximal ideal P containing I, and P is prime by Corollary 8.11. Multiplying $I \subseteq P$ by P^{-1} as in Lemma 8.58, we obtain $I \subseteq P^{-1}I \subseteq P^{-1}P = R$, the equality holding by Lemma 8.58. Hence $P^{-1}I$ is an ideal. An equality $I = P^{-1}I$ would imply that $PI = PP^{-1}I = I$ by Lemma 8.58, and then Proposition 8.52 would yield $I = 0$, a contradiction to the hypothesis that I is nonzero. An equality $P^{-1}I = R$ would imply $I = PP^{-1}I = PR = P$ by Lemma 8.58, in contradiction to the fact that I is not prime. We conclude that $I \subsetneq P^{-1}I \subsetneq R$. The maximal choice of I shows that $P^{-1}I$ decomposes as a product $P^{-1}I = P_1 \cdots P_r$ of prime ideals, not necessarily distinct. One more application of Lemma 8.58 yields $I = PP^{-1}I = PP_1 \cdots P_r$, and we have a contradiction. We conclude that every nonzero proper ideal decomposes as a product of prime ideals. Grouping equal factors, we can write the decomposition as in the statement of the theorem.

Next let us establish uniqueness as in (a). Suppose that we have two equal decompositions $P_1 \cdots P_r = Q_1 \cdots Q_s$ as the product of prime ideals, and suppose that $r \leq s$. We show by induction on r that $r = s$ and that the factors on the two sides match, apart from their order. The base case of the induction is $r = 0$, and then it is evident that $s = 0$. Assume the uniqueness for $r - 1$. Since P_1 is prime and $P_1 \supseteq Q_1 \cdots Q_s$, $P_1 \supseteq Q_j$ for some j by Lemma 8.56. By (iii) for Dedekind domains, Q_j is a maximal ideal, and therefore $P_1 = Q_j$. Multiplying the equality $P_1 \cdots P_r = Q_1 \cdots Q_s$ by P_1^{-1} and applying Lemma 8.58 to each side, we obtain $P_2 \cdots P_r = Q_1 \cdots Q_{j-1}Q_{j+1} \cdots Q_s$. The inductive hypothesis implies that $r - 1 = s - 1$ and the factors on the two sides match, apart from their order. Then we can conclude about the equality $P_1 \cdots P_r = Q_1 \cdots Q_s$ that $r = s$ and that the factors on the two sides match, apart from their order. This proves (a).

Let us establish the formula in (b) for k_j. Suppose that P is a prime ideal. By (a), we can write $I = P^n J$ for a certain integer $n \geq 0$ in such a way that P does not appear in the unique decomposition of J. Certainly $P^k \supseteq I$ for $k \leq n$ because $P^k \supseteq P^k P^{n-k} = P^n \supseteq P^n J = I$. Suppose $P^{n+1} \supseteq I$. Multiplying $P^{n+1} \supseteq I = P^n J$ by n factors of P^{-1} and using Lemma 8.58 repeatedly, we obtain $P \supseteq P^{-n}I = J$. Since P is prime, Lemma 8.56 shows that P must contain one of the factors when J is decomposed as the product of prime ideals, and we have a contradiction to the maximality of this factor unless this factor is P itself. In this case, P appears in the decomposition of J, and again we have a contradiction.

For (c), if $IJ_1 = IJ_2$, substitute the unique decompositions as products of prime ideals for I, J_1, and J_2, and use (a) to cancel the factors from I on each side, obtaining $J_1 = J_2$.

For (d), suppose that I and J_1 are two nonzero proper ideals with $I \subseteq J_1$. If

$P_i^{k_i}$ is the largest power of a prime ideal P_i appearing in the decomposition of J_1, then $P_i^{k_i} \supseteq J_1 \supseteq I$, and (b) shows that $P_i^{k_i}$ appears in the decomposition of I. In other words, if l_i is the largest power of P_i appearing in the decomposition of I, then $l_i \geq k_i$. Let $J_2 = \prod_i P_i^{l_i - k_i}$. Then we obtain $I = J_1 J_2$, and (d) is proved. □

Corollary 8.59. Let R be a Dedekind domain, and let P be a nonzero prime ideal in R. Then there exists an element π in P such that π is not in P^2, and any such element has the property that π^k is not in P^{k+1} for any $k \geq 1$.

PROOF. Proposition 8.52 shows that P^2 is a proper subset of P, and therefore we can find an element π in P that is not in P^2. Since the principal ideal (π) has $(\pi) \subseteq P$ and $(\pi) \subsetneq P^2$, the factorization of (π) involves P but not P^2. Thus we can use Theorem 8.55 to write $(\pi) = P Q_1 \cdots Q_n$ for prime ideals $Q_1, \ldots, Q_n$ different from P. Then $(\pi^k) = (\pi)^k = P^k Q_1^k \cdots Q_n^k$, and (b) of the theorem says that P^{k+1} does not contain (π^k). □

Corollary 8.60. Let R be a Dedekind domain, and let P be a nonzero prime ideal in R. For any integer $e \geq 1$, the natural action of R on powers of P makes P^{e-1}/P^e into a vector space over the field R/P, and this vector space is 1-dimensional.

REMARKS. This technical-sounding corollary will be used late in Chapter IX and again in *Advanced Algebra*.

PROOF. Since $R(P^{e-1}) \subseteq P^{e-1}$ and $P(P^{e-1}) \subseteq P^e$, we obtain
$$(R/P)(P^{e-1}/P^e) \subseteq P^{e-1}/P^e.$$
Thus P^{e-1}/P^e is a unital R/P module, i.e., a vector space over the field R/P. We show that it has dimension 1. Corollary 8.59 shows that there exists a member π of P not in P^2, and it shows that π^k is not in P^{k+1} for any k. This element π has the property that $(\pi) = P Q_1 \cdots Q_r$ for nonzero prime ideals $Q_1, \ldots, Q_r$ distinct from P, and thus
$$R\pi^{e-1} = (\pi^{e-1}) = (\pi)^{e-1} = P^{e-1} Q_1^{e-1} \cdots Q_r^{e-1}.$$
Hence
$$R\pi^{e-1} + P^e = P^{e-1}(Q_1^{e-1} \cdots Q_r^{e-1} + P).$$
The ideal in parentheses on the right side strictly contains P since the failure of P to divide $Q_1^{e-1} \cdots Q_r^{e-1}$ means that P does not contain $Q_1^{e-1} \cdots Q_r^{e-1}$ (by Theorem 8.55d). Since P is maximal, the ideal in parentheses is R, and we see that $R(\pi^{e-1} + P^e) = P^{e-1}/P^e$. Therefore $(R/P)(\pi^{e-1} + P^e) = P^{e-1}/P^e$. This formula says that P^{e-1}/P^e consists of all scalar multiples of a certain element, and it follows that P^{e-1}/P^e is 1-dimensional. □

Lemma 8.61. If P and Q are distinct maximal ideals in an integral domain R and if k and l are positive integers, then $P^k + Q^l = R$.

PROOF. We know that $P^k + Q^l$ is an ideal. Arguing by contradiction, assume that it is proper. Then we can find a maximal ideal M with $M \supseteq P^k + Q^l$. This M satisfies $M \supseteq P^k$ and $M \supseteq Q^l$. By Lemma 8.56, $M \supseteq P$ and $M \supseteq Q$. Since P and Q are distinct and maximal, we obtain $P = M = Q$, contradiction. □

Corollary 8.62. If R is a Dedekind domain with only finitely many prime ideals, then R is a principal ideal domain.

REMARKS. Corollary 8.48 may be used to produce examples to which Corollary 8.62 is applicable. All we have to do is to take one of our standard Dedekind domains R and localize with respect to a nonzero prime ideal P. The corollary says that the result R_P is a Dedekind domain, and it has a unique maximal ideal, hence a unique nonzero prime ideal. The conclusion is that R_P is a principal ideal domain.

PROOF. Let $P_1, \ldots, P_n$ be the distinct nonzero prime ideals. Theorem 8.55 shows that any nonzero ideal I in R factors uniquely as $I = P_1^{k_1} \cdots P_n^{k_n}$ with each $k_j \geq 0$. For $1 \leq i \leq n$, Corollary 8.59 produces π_i in P_i such that π_i is not in P_i^2, and it shows that π_i^m is not in P_i^{m+1}.

Lemma 8.61 gives $P_i^{k_i} + P_j^{k_j} = R$ if $i \neq j$. Applying the Chinese Remainder Theorem (Theorem 8.27a), we can find an element a in R with $a \equiv \pi_i^{k_i} \mod P_i^{k_i+1}$ for $1 \leq i \leq n$. Using Theorem 8.55 again, let $(a) = P_1^{l_1} \cdots P_n^{l_n}$ be the unique factorization of the principal ideal (a). The defining property of a shows that a is in $P_i^{k_i}$ but not $P_i^{k_i+1}$ for each i. Thus (a) is contained in $P_i^{k_i}$ but not in $P_i^{k_i+1}$. By Theorem 8.55b, $l_i = k_i$ for each i. Hence the ideal $I = P_1^{k_1} \cdots P_n^{k_n} = (a)$ is exhibited as principal. □

Corollary 8.63. If R is a Dedekind domain and if $I = \prod_{j=1}^n P_j^{k_j}$ is the unique factorization of a nonzero proper ideal I as the product of positive powers of distinct prime ideals P_j, then the map $r \mapsto \prod_{j=1}^n P_j^{k_j}$ defined on R by $r \mapsto (\ldots, r + P_j^{k_j}, \ldots)$ descends to a ring isomorphism

$$R/I \cong R/P_1^{k_1} \times \cdots \times R/P_n^{k_n}.$$

PROOF. Lemma 8.61 shows that $P_i^{k_i} + P_j^{k_j} = R$ if $i \neq j$. Then the result follows immediately from the Chinese Remainder Theorem (Theorem 8.27). □

12. Problems

1. This problem examines ring homomorphisms of the field of real numbers into itself that carry 1 into 1. Let φ be such a homomorphism.
   (a) Prove that φ is the identity on $\mathbb{Q}$.

(b) Prove that φ maps squares into squares.
(c) Prove that φ respects the ordering of $\mathbb{R}$, i.e., that $a \leq b$ implies $\varphi(a) \leq \varphi(b)$.
(d) Prove that φ is the identity on $\mathbb{R}$.

2. An element r in a commutative ring with identity is called **nilpotent** if $r^n = 0$ for some integer n. Prove that if r is nilpotent, then $1 + r$ is a unit.

3. If R is a field, prove that the embedding of R in its field of fractions exhibits R as isomorphic to its field of fractions.

4. Prove that X is prime in $R[X]$ if R is an integral domain.

5. Suppose that R is an integral domain that is not a field.
   (a) Prove that there is a nonzero prime ideal in $R[X]$ that is not maximal.
   (b) Prove that there is an ideal in $R[X]$ that is not principal.

6. This problem makes use of real-analysis facts concerning closed bounded intervals of the real line. Let R be the ring of all continuous functions from $[0, 1]$ into $\mathbb{R}$, with pointwise multiplication as the ring multiplication.
   (a) Prove for each x_0 in $[0, 1]$ that the set I_{x_0} of members of R that vanish at x_0 is a maximal ideal of R.
   (b) Prove that any maximal ideal I of R that is not some I_{x_0} contains finitely many members $f_1, \ldots, f_n$ of R that have no common zero on $[0, 1]$.
   (c) By considering $f_1^2 + \cdots + f_n^2$ in (b), prove that every maximal ideal of R is of the form I_{x_0} for some x_0 in $[0, 1]$.

7. Let R be the ring of all bounded continuous functions from $\mathbb{R}$ into $\mathbb{R}$, with pointwise multiplication as the ring multiplication. Say that a member f of R vanishes at infinity if for each $\epsilon > 0$, there is some N such that $|f(x)| < \epsilon$ whenever $|x| \geq N$.
   (a) Show that the subset I_∞ of all members of R that vanish at infinity is an ideal but not a maximal ideal.
   (b) Why must R have at least one maximal ideal I that contains I_∞?
   (c) Why can there be no x_0 in $\mathbb{R}$ such that the maximal ideal I of (b) consists of all members of R that vanish at x_0?

8. Let I be a nonzero ideal in $\mathbb{Z}[\sqrt{-5}]$.
   (a) Prove that I contains some positive integer.
   (b) Prove that I, as an abelian group under addition, is free abelian of rank 2.
   (c) If n denotes the least positive integer in I, prove that I has a $\mathbb{Z}$ basis of the form $\{n, a + b\sqrt{-5}\}$ for a suitable member $a + b\sqrt{-5}$ of $\mathbb{Z}[\sqrt{-5}]$.

9. Let $\varphi : R \to R'$ be a homomorphism of commutative rings with identity such that $\varphi(1) = 1$. Prove that if P' is a prime ideal in R', then $P = \varphi^{-1}(P')$ is a prime ideal in R.

10. Determine the maximal ideals of each of the following rings:
    (a) $\mathbb{R} \times \mathbb{R}$,
    (b) $\mathbb{R}[X]/(X^2)$,
    (c) $\mathbb{R}[X]/(X^2 - 3X + 2)$,
    (d) $\mathbb{R}[X]/(X^2 + X + 1)$.

11. (a) Prove or disprove: If I is a nonzero prime ideal in $\mathbb{Q}[X]$, then $\mathbb{Q}[X]/I$ is a unique factorization domain.
    (b) Prove or disprove: If I is a nonzero prime ideal in $\mathbb{Z}[X]$, then $\mathbb{Z}[X]/I$ is a unique factorization domain.

12. (**Partial fractions**) Let R be a principal ideal domain, and let F be its field of fractions.
    (a) Let n be a nonzero member of R with a factorization $n = cd$ such that $\text{GCD}(c, d) = 1$. Prove for each m in R that the member mn^{-1} of F has a decomposition as $mn^{-1} = ac^{-1} + bd^{-1}$ with a and b in R.
    (b) Let n be a nonzero member of R with a factorization $n = p_1^{k_1} \cdots p_r^{k_r}$, the elements p_j being nonassociate primes in R. Prove for each m in R that the member mn^{-1} of F has a decomposition as $mn^{-1} = q_1 p_1^{-k_1} + \cdots + q_r p_r^{-k_r}$ with all q_j in R.

13. (a) By adapting the proof that the ring of Gaussian integers forms a Euclidean domain, prove that the function $\delta(a + b\sqrt{-2}) = a^2 + 2b^2$ satisfies $\delta(rr') = \delta(r)\delta(r')$ and exhibits $\mathbb{Z}[\sqrt{-2}]$ as a Euclidean domain.
    (b) It was shown in Section 9 that $\mathbb{Z}[\sqrt{-3}\,]$ is not a unique factorization domain, hence cannot be a Euclidean domain. What goes wrong with continuing the adaptation in the previous problem so that it applies to $\mathbb{Z}[\sqrt{-3}\,]$?

14. Let G be a group, and let R be a commutative ring with identity. Examples 16 and 17 in Section 1 defined the group algebra RG and the R algebra $C(G, R)$ of functions from G into R, convolution being the multiplication in $C(G, R)$. Prove that the mapping $g \mapsto f_g$ described with Example 17 extends to an R algebra isomorphism of RG onto $C(R, G)$.

15. Let I be an ideal in $\mathbb{Z}[X]$, and suppose that the lowest degree of a nonzero polynomial in I is n and that I contains some monic polynomial of degree n. Prove that I is a principal ideal.

16. For each integer $n > 0$, exhibit an ideal I_n in $\mathbb{Z}[X]$ that cannot be written with fewer than n generators.

17. Let φ be the substitution homomorphism $\varphi : \mathbb{K}[x, y] \to \mathbb{K}[t]$ defined by $x \mapsto t^2$, $y \mapsto t^3$, and $\varphi(c) = c$ for $c \in \mathbb{K}$.
    (a) Prove that $\ker \varphi$ is the principal ideal $(y^2 - x^3)$.
    (b) What is image φ?

18. Let $R = \mathbb{Z}[i]$.
    (a) Show that each unital R module M may be regarded as an abelian group with an abelian-group homomorphism $\varphi : M \to M$ for which φ^2 is the mapping $m \mapsto -m$.
    (b) Show conversely that if M is an abelian group and there exists an abelian-group homomorphism $\varphi : M \to M$ for which φ^2 is the mapping $m \mapsto -m$, then M may be regarded as a unital R module.

19. Let R be a unique factorization domain, and let F be its field of fractions. Let $A(X)$ and $B(X)$ be nonzero polynomials in $F[X]$, let $A_0(X)$ and $B_0(X)$ be their associated primitive polynomials, and suppose that $B(X)$ divides $A(X)$ in $F[X]$. Prove that $B_0(X)$ divides $A_0(X)$ in $R[X]$.

20. Prove that an integral domain with finitely many elements is a field.

21. Two proofs of Theorem 8.18 were given, one using direct multiplication of polynomials and the other using polynomials with coefficients taken modulo (p), and it was stated that proofs in both these styles could be given for Corollary 8.22. A proof in the first style was supplied in the text. Supply a proof in the second style.

22. Let $\mathbb{K}$ be a field.
    (a) Prove that $\det \begin{pmatrix} W & X \\ Y & Z \end{pmatrix}$, when considered as a polynomial in $\mathbb{K}[W, X, Y, Z]$, is irreducible.
    (b) Let X_{ij} be indeterminates for i and j from 1 to n. Doing an induction, prove that the polynomial $\det[X_{ij}]$ is irreducible in $\mathbb{K}[X_{11}, X_{12}, \ldots, X_{nn}]$.

23. Prove that two members of $\mathbb{Z}[X]$ are relatively prime in $\mathbb{Q}[X]$ if and only if the ideal they generate in $\mathbb{Z}[X]$ contains an integer.

24. Let V be the $\mathbb{Z}[i]$ module with two generators u_1, u_2 related by the conditions $(1+i)u_1 + (2-i)u_2 = 0$ and $3u_1 + 5iu_2 = 0$. Express V as the direct sum of cyclic $\mathbb{Z}[i]$ modules.

Problems 25–26 concern the ring $R = \mathbb{Z}[\frac{1}{2}(1 + \sqrt{-m})]$, where m is a square-free integer > 1 with $m \equiv 3 \bmod 4$. Let $F = \mathbb{Q}[\sqrt{-m}]$ be the field of fractions of R.

25. For $z = x + y\sqrt{-m}$ in F, define $\delta(z) = x^2 + my^2$.
    (a) Show that $\delta(zw) = \delta(z)\delta(w)$.
    (b) Show that if for each z in F there is some r in R with $\delta(z - r) < 1$, then δ exhibits R as a Euclidean domain.

26. Prove that the condition of part (b) of the previous problem is satisfied for $m = 3$, 7, and 11, and conclude that $\mathbb{Z}[\frac{1}{2}(1 + \sqrt{-m})]$ is a Euclidean domain for these values of m.

Problems 27–31 classify the primes in the ring $\mathbb{Z}[i]$ of Gaussian integers. This ring is a Euclidean domain and therefore is a unique factorization domain. Members of this ring will be written as $a + bi$, and it is understood that a and b are in $\mathbb{Z}$. Put $N(a + bi) = (a + bi)(a - bi) = a^2 + b^2$.

27. Let $a + bi$ be prime in $\mathbb{Z}[i]$. Prove that
    (a) $a - bi$ is prime.
    (b) $N(a + bi)$ is a power of some positive prime p in $\mathbb{Z}$.
    (c) $N(a + bi)$ equals p or p^2 when p is as in (b).
    (d) $N(a + bi) = p^2$ in (c) forces $a + bi = p$, apart from a unit factor.

28. Prove that no prime $a + bi$ in $\mathbb{Z}[i]$ has $N(a + bi) = p$ with p of the form $4n + 3$. Conclude that every positive prime in $\mathbb{Z}$ of the form $4n + 3$ is a prime in $\mathbb{Z}[i]$.

29. Prove that the only primes $a + bi$ of $\mathbb{Z}[i]$ for which $N(a + bi)$ equals 2 or 2^2 are $1 + i$ and its associates, for which $N(a + bi) = 2$.

30. Prove that if p is a positive prime in $\mathbb{Z}$ of the form $4n + 1$, then -1 is a square in the finite field $\mathbb{F}_p$.

31. Let p be a positive prime in $\mathbb{Z}$ of the form $4n + 1$.
    (a) Prove that there exist ring homomorphisms φ_1 of $\mathbb{Z}[X]$ onto $\mathbb{F}_p[X]/(X^2 + 1)$ and φ_2 of $\mathbb{Z}[X]$ onto $\mathbb{Z}[i]/(p)$.
    (b) Prove that $\ker \varphi_1$ and $\ker \varphi_2$ are both equal to the ideal $(p, X^2 + 1)$ in $\mathbb{Z}[X]$, and deduce a ring isomorphism $\mathbb{Z}[i]/(p) \cong \mathbb{F}_p[X]/(X^2 + 1)$.
    (c) Taking into account the results of Problems 27 and 30, show that p is not prime in $\mathbb{Z}[i]$ and is therefore of the form $p = N(a + bi) = a^2 + b^2$ for some prime $a + bi$ in $\mathbb{Z}[i]$.
    (d) Prove a uniqueness result for the decomposition $p = a^2 + b^2$, that if also $p = a'^2 + b'^2$, then $a' + b'i$ is an associate either of $a + bi$ or of $a - bi$.

Problems 32–35 establish a theory of **elementary divisors**. This theory provides a different uniqueness result, beyond the one in Corollary 8.28, to accompany the Fundamental Theorem of Finitely Generated Modules over a Principal Ideal Domain. When specialized to $\mathbb{K}[X]$ for a field $\mathbb{K}$, the theory yields the **rational canonical form** of a member of $M_n(\mathbb{K})$. Let R be a nonzero principal ideal domain. If C and D are members of $M_{mn}(R)$, let us say that C and D are **equivalent** if there exist A in $M_m(R)$ and B in $M_n(R)$ with $\det A$ in $R^\times$, $\det B$ in $R^\times$, and $D = ACB$. Fix m and n, and put $k = \min(m, n)$. If C is a member of $M_{mn}(R)$, its **diagonal entries** are the entries $C_{11}, C_{22}, \ldots, C_{kk}$. The matrix C will be called **diagonal** if its only nonzero entries are diagonal entries. Problems 26–31 of Chapter V are relevant for Problem 34.

32. (a) Suppose that C is a diagonal matrix in $M_{mn}(R)$ with $C_{11} \neq 0$. Show that C is equivalent to a matrix C' described as follows: all entries of C' are the same as those of C except possibly for the entries $C'_{21}, \ldots, C'_{k1}$ in the first column, and these satisfy $C'_{j1} = C_{jj}$.

(b) By applying the algorithm of Lemma 8.26 to the matrix C' in (a), prove that any nonzero diagonal matrix C in $M_{mn}(R)$ is equivalent to a diagonal matrix C'' such that C''_{11} divides all the diagonal entries of C''.

(c) By iterating the construction in (a) and (b), prove that any diagonal matrix C in $M_{mn}(R)$ is equivalent to a diagonal matrix D having the following properties: The nonzero diagonal entries of D are the entries D_{jj} with $1 \leq j \leq l$ for some integer l with $0 \leq l \leq k$. For each j with $1 \leq j < l$, D_{jj} divides $D_{j+1,j+1}$.

33. Establish the following uniqueness theorem: Let D and E be diagonal matrices in $M_{mn}(R)$ whose diagonal entries satisfy the divisibility property in (c) of the previous problem. Prove that if D and E are equivalent, then they have the same number of nonzero entries, and their corresponding diagonal entries are associates.

34. (**Rational canonical form**) Let $\mathbb{K}$ be a field, and let $L : V \to V$ be a $\mathbb{K}$ linear mapping from a finite-dimensional $\mathbb{K}$ vector space V to itself. By applying Theorem 8.25 and the results of the previous problems to V as a $\mathbb{K}[X]$ module with $Xv = L(v)$, prove the following: V can be written as the direct sum of cyclic subspaces $V_1, \ldots, V_r$ under L in such a way that the minimal polynomial of L on V_j divides the minimal polynomial of L on V_{j+1} for $1 \leq j < r$; moreover, the integer r and the minimal polynomials are uniquely determined by L, and any two linear mappings with the same r and matching minimal polynomials are similar over $\mathbb{K}$.

35. Let $\mathbb{K}$ and $\mathbb{L}$ be fields with $\mathbb{K} \subseteq \mathbb{L}$, and suppose that two members of $M_n(\mathbb{K})$ are conjugate via $GL(n, \mathbb{L})$. Prove that they are conjugate via $GL(n, \mathbb{K})$.

Problems 36–39 concern symmetric polynomials in n indeterminates over a field. Let F be a field, and let $R = F[X_1, \ldots, X_n]$. If $\sigma \in \mathfrak{S}_n$ is a permutation, then there is a corresponding substitution homomorphism of rings $\sigma^* : R \to R$ fixing F and carrying each X_j into $X_{\sigma(j)}$. A **symmetric polynomial** A in R is a member of R for which $\sigma^* A = A$ for every permutation σ. The symmetric polynomials form a subring of R containing the constants. The main result about symmetric polynomials is that every symmetric polynomial is a polynomial in the "elementary symmetric polynomials"; these will be defined below.

36. Prove that the ring homomorphisms σ^* satisfy $(\sigma\tau)^* = \sigma^*\tau^*$. Deduce that each $\sigma^* : R \to R$ is an isomorphism.

37. Prove that the homogeneous-polynomial expansion of any symmetric polynomial is into symmetric polynomials.

38. For each permutation σ, let σ^{**} be the substitution homomorphism of $R[X] \cong F[X_1, \ldots, X_n, X]$ acting as σ^* on R and carrying X to itself.
(a) Prove that $(\sigma\tau)^{**} = \sigma^{**}\tau^{**}$ and that each σ^{**} is a ring isomorphism of $R[X]$.

(b) Prove that each coefficient in $R[X]$ of any polynomial fixed by all σ^{**} is a symmetric polynomial in R.
(c) The polynomial $(X - X_1)(X - X_2) \cdots (X - X_n)$ is fixed by all σ^{**}, and its coefficients are called the **elementary symmetric polynomials**. Show that they are

$$E_1 = \sum_i X_i, \quad E_2 = \sum_{i<j} X_i X_j, \quad E_3 = \sum_{i<j<k} X_i X_j X_j, \ldots, \quad E_n = X_1 X_2 \cdots X_n.$$

39. Order the monomials of total degree m by saying that the monomial $aX_1^{k_1} \cdots X_n^{k_n}$ with $a \neq 0$ and $\sum k_j = m$ is greater than the monomial $a'X_1^{l_1} \cdots X_n^{l_n}$ with $a' \neq 0$ and $\sum l_j = m$ if the first j for which $k_j \neq l_j$ has $k_j > l_j$.
    (a) If $A(X_1, \ldots, X_n)$ is a nonzero symmetric polynomial homogeneous of degree m and if $aX_1^{k_1} \cdots X_n^{k_n}$ is its nonzero monomial that is highest in the above order, why must it be true that $k_1 \geq k_2 \geq \cdots \geq k_n$?
    (b) Verify that the largest monomial in $E_1^{c_1} \cdots E_n^{c_n}$ in the ordering is

$$X_1^{c_1+c_2+\cdots+c_n} X_2^{c_2+\cdots+c_n} \cdots X_n^{c_n}.$$

    (c) Show that if $A(X_1, \ldots, X_n)$ is a nonzero symmetric polynomial homogeneous of degree m, then there exist a symmetric polynomial $M = E_1^{c_1} \cdots E_n^{c_n}$ homogeneous of degree m and a scalar r such that the largest monomials in A and rM are equal.
    (d) With notation as in (c), show that $A - rM$ equals 0 or else the largest monomial of A is greater than the largest monomial of $A - rm$.
    (e) Deduce that every symmetric polynomial is a polynomial in the elementary symmetric polynomials.

Problems 40–43 concern the Pfaffian of a $(2n)$-by-$(2n)$ alternating matrix $X = [x_{ij}]$ with entries in a field $\mathbb{K}$. Here "alternating" means that $x_{ij} = -x_{ji}$ for all i and j and $x_{ii} = 0$ for all i. The Pfaffian is the polynomial in the entries of X with integer coefficients given by

$$\text{Pfaff}(X) = \sum_{\substack{\text{certain } \tau\text{'s} \\ \text{in } \mathfrak{S}_{2n}}} (\text{sgn } \tau) \prod_{k=1}^{n} x_{\tau(2k-1), \tau(2k)},$$

where the sum is taken over those permutations τ such that $\tau(2k-1) < \tau(2k)$ for $1 \leq k \leq n$ and such that $\tau(1) < \tau(3) < \cdots < \tau(2n-1)$. The Pfaffian was introduced in Problems 23–28 at the end of Chapter VI. It was shown in those problems that the Pfaffian satisfies $\det X = (\text{Pfaff}(X))^2$. The present problems will make use of that result but of no other results from Chapter VI. They will also make use of facts concerning continuous functions and connected open subsets of Euclidean space.

40. Prove by induction on m that the open subset of $\mathbb{C}^m$ on which a nonzero polynomial function $P(z_1, \ldots, z_m)$ is nonzero is pathwise connected and therefore connected.

41. For this problem let $\mathbb{K} = \mathbb{C}$.
    (a) For any two matrices A and X in $M_{2n}(\mathbb{C})$ with X alternating, prove that $\text{Pfaff}(A^t X A) = \pm(\det A)\text{Pfaff}(X)$ with the sign depending on A and X.
    (b) Fix X, and allow A to vary. Using Problem 40, prove that the sign is always positive in (a). That is, prove that $\text{Pfaff}(A^t X A) = (\det A)\text{Pfaff}(X)$.

42. For this problem let $\mathbb{K}$ be any field. By regarding the expressions $\text{Pfaff}(A^t X A)$ and $(\det A)\text{Pfaff}(X)$ as polynomials with coefficients in $\mathbb{Z}$ in the indeterminates A_{ij} for all i and j and the indeterminates X_{ij} for $i < j$, and using the principle of permanence of identities in Section V.2, prove that $\text{Pfaff}(A^t X A) = (\det A)\text{Pfaff}(X)$ whenever A and X are in $M_{2n}(\mathbb{K})$ and X is alternating.

43. Section VI.5 defines a particular alternating matrix J for which $\text{Pfaff}(J) = 1$. A **symplectic matrix** g over $\mathbb{K}$ is one for which $g^t J g = J$. Prove that every symplectic matrix has determinant 1.

Problems 44–47 concern Dedekind domains. Let R be such a domain. It is to be proved that each nonzero ideal I is doubly generated in the sense that $I = Ra + Rb$ for suitable members a and b of R.

44. Let $R_1, \ldots, R_n$ be nonzero commutative rings with identity, not necessarily integral domains. Prove that if every ideal of each R_j is principal, then every ideal in $R_1 \times \cdots \times R_n$ is principal.

45. Let P be a nonzero prime ideal, and let k be a positive integer.
    (a) Prove that the only nonzero proper ideals in R/P^k are P/P^k, $P^2/P^k, \ldots, P^{k-1}/P^k$.
    (b) Using the element π in the statement of Corollary 8.59, prove that each of the ideals in (a) is principal.

46. Combining Corollary 8.63 with Problems 44 and 45, conclude that the quotient of R by any nonzero proper ideal has only principal ideals.

47. Let I be a nonzero proper ideal in R. By letting a be any nonzero element of I and by applying (c) in the previous problem to the ideal $I/(a)$ of $R/(a)$, prove that $I = Ra + Rb$ for a suitable b in I.

Problems 48–53 introduce and classify "fractional ideals" in Dedekind domains. Let R be a Dedekind domain, regarded as a subring of its field of fractions F. A **fractional ideal** in F is a finitely generated R submodule of F.

48. Prove that the fractional ideals in F that lie in R are exactly the ordinary ideals of R.

12. Problems

49. Prove for any fractional ideal M that there exists a nonzero member a of F such that aM lies in R and hence is an ordinary ideal. Conclude that the product of two fractional ideals is a fractional ideal.

50. Prove that if I is a nonzero ideal of R and if I^{-1} is defined by

$$I^{-1} = \{x \in F \mid xR \subseteq I\},$$

then I^{-1} is a fractional ideal in F. Conclude that if P is a prime ideal in R, then P^{-1} as defined in Lemma 8.58 is a fractional ideal in F.

51. Prove, by arguing with an ideal that is maximal among those for which the statement is false, that to any nonzero ideal I in R corresponds some fractional ideal M of F such that $IM = R$.

52. Prove in the notation of the previous two problems that $M = I^{-1}$.

53. Deduce that every nonzero fractional ideal is of the form IJ^{-1}, where I and J are nonzero ideals. Conclude that
    (a) the nonzero fractional ideals are exactly all products $\prod_{i=1}^{n} P_i^{k_i}$, where the P_i are distinct nonzero prime ideals and the k_i are arbitrary nonzero integers, positive or negative,
    (b) the nonzero fractional ideals form a group.

CHAPTER IX

Fields and Galois Theory

Abstract. This chapter develops some general theory for field extensions and then goes on to study Galois groups and their uses. More than half the chapter illustrates by example the power and usefulness of the theory of Galois groups. Prerequisite material from Chapter VIII consists of Sections 1–6 for Sections 1–13 of the present chapter, and it consists of all of Chapter VIII for Sections 14–17 of the present chapter.

Sections 1–2 introduce field extensions. These are inclusions of a base field in a larger field. The fundamental construction is of a simple extension, algebraic or transcendental, and the next construction is of a splitting field. An algebraic simple extension is made by adjoining a root of an irreducible polynomial over the base field, and a splitting field is made by adjoining all the roots of such a polynomial. For both constructions, there are existence and uniqueness theorems.

Section 3 classifies finite fields. For each integer q that is a power of some prime number, there exists one and only one finite field of order q, up to isomorphism. One finite field is an extension of another, apart from isomorphisms, if and only if the order of the first field is a power of the order of the second field.

Section 4 concerns algebraic closure. Any field has an algebraic extension in which each nonconstant polynomial over the extension field has a root. Such a field exists and is unique up to isomorphism.

Section 5 applies the theory of Sections 1–2 to the problem of constructibility with straightedge and compass. First the problem is translated into the language of field theory. Then it is shown that three desired constructions from antiquity are impossible: "doubling a cube," trisecting an arbitrary constructible angle, and "squaring a circle." The full proof of the impossibility of squaring a circle uses the fact that π is transcendental over the rationals, and the proof of this property of π is deferred to Section 14. Section 5 concludes with a statement of the theorem of Gauss identifying integers n such that a regular n-gon is constructible and with some preliminary steps toward its proof.

Sections 6–8 introduce Galois groups and develop their theory. The theory applies to a field extension with three properties—that it is finite-dimensional, separable, and normal. Such an extension is called a "finite Galois extension." The Fundamental Theorem of Galois Theory says in this case that the intermediate extensions are in one-one correspondence with subgroups of the Galois group, and it gives formulas relating the corresponding intermediate fields and Galois subgroups.

Sections 9–11 give three standard initial applications of Galois groups. The first is to proving the theorem of Gauss about constructibility of regular n-gons, the second is to deriving the Fundamental Theorem of Algebra from the Intermediate Value Theorem, and the third is to proving the necessity of the condition of Abel and Galois for solvability of polynomial equations by radicals—that the Galois group of the splitting field of the polynomial have a composition series with abelian quotients.

Sections 12–13 begin to derive quantitative information, rather than qualitative information, from Galois groups. Section 12 shows how an appropriate Galois group points to the specific steps in the construction of a regular n-gon when the construction is possible. Section 13 introduces a tool known as Lagrange resolvents, a precursor of modern harmonic analysis. Lagrange resolvents are used first to show that Galois extensions in characteristic 0 with cyclic Galois group of prime order p are simple extensions obtained by adjoining a pth root, provided all the pth roots of 1 lie in the base field. Lagrange resolvents and this theorem about cyclic Galois groups combine to yield a derivation of Cardan's formula for solving general cubic equations.

Section 14 begins the part of the chapter that depends on results in the later sections of Chapter VIII. Section 14 itself contains a proof that π is transcendental; the proof is a nice illustration of the interplay of algebra and elementary real analysis.

Section 15 introduces the field polynomial of an element in a finite-dimensional extension field. The determinant and trace of this polynomial are called the norm and trace of the element. The section gives various formulas for the norm and trace, including formulas involving Galois groups. With these formulas in hand, the section concludes by completing the proof of Theorem 8.54 about extending Dedekind domains, part of the proof having been deferred from Section VIII.11.

Section 16 discusses how prime ideals split when one passes, for example, from the integers to the algebraic integers in a number field. The topic here was broached in the motivating examples for algebraic number theory and algebraic geometry as introduced in Section VIII.7, and it was the main topic of concern in that section. The present results put matters into a wider context.

Section 17 gives two tools that sometimes help in identifying Galois groups, particularly of splitting fields of monic polynomials with integer coefficients. One tool uses the discriminant of the polynomial. The other uses reduction of the coefficients modulo various primes.

1. Algebraic Elements

If $\mathbb{K}$ and $\mathbb{k}$ are fields such that $\mathbb{k}$ is a subfield of $\mathbb{K}$, we say that $\mathbb{K}$ is a **field extension** of $\mathbb{k}$. When it is necessary to refer to this situation in some piece of notation, we often write $\mathbb{K}/\mathbb{k}$ to indicate the field extension. In this section we shall study field extensions in a general way, and in the next section we shall discuss constructions and uniqueness results involving them.

If $\mathbb{K}$ and $\mathbb{K}'$ are two fields and if φ is a ring homomorphism of $\mathbb{K}$ into $\mathbb{K}'$ with $\varphi(1) = 1$, then φ is automatically one-one since $\mathbb{K}$ has no nontrivial ideals. We refer to φ as a **field map** or **field mapping**.[1] If $\mathbb{K}$ and $\mathbb{K}'$ are both field extensions of a field $\mathbb{k}$ and if the restriction of a field map φ to $\mathbb{k}$ is the identity, then φ is called a $\mathbb{k}$ **field map** or a **field map fixing** $\mathbb{k}$. The terminology "$\mathbb{k}$ field map" is consistent with the view that $\mathbb{K}$ and $\mathbb{K}'$ are two R algebras for $R = \mathbb{k}$ in the sense of Examples 6 and 15 in Section VIII.1, and that the isomorphism in question is just an R algebra isomorphism.

If a field map $\varphi : \mathbb{K} \to \mathbb{K}'$ is onto $\mathbb{K}'$, then φ is a **field isomorphism**; it is a $\mathbb{k}$ **field isomorphism** if $\mathbb{K}$ and $\mathbb{K}'$ are extensions of $\mathbb{k}$ and φ is the identity on $\mathbb{k}$. When $\mathbb{K} = \mathbb{K}'$ and φ is onto $\mathbb{K}'$, φ is called an **automorphism** of $\mathbb{K}$; if also φ is the identity on a subfield $\mathbb{k}$, then φ is called a $\mathbb{k}$ **automorphism** of $\mathbb{K}$.

---

[1]This is the notion of morphism in the category of fields.

Throughout this section we let $\mathbb{K}/\mathbb{k}$ be a field extension. If $x_1, \ldots, x_n$ are members of $\mathbb{K}$, we let

$$\mathbb{k}[x_1, \ldots, x_n] = \text{subring of } \mathbb{K} \text{ generated by } 1 \text{ and } x_1, \ldots, x_n,$$

$$\mathbb{k}(x_1, \ldots, x_n) = \text{subfield of } \mathbb{K} \text{ generated by } 1 \text{ and } x_1, \ldots, x_n.$$

The latter, in more detail, means the set of all quotients ab^{-1} with a and b in $\mathbb{k}[x_1, \ldots, x_n]$ and with $b \neq 0$. It is referred to as the **field obtained by adjoining** $x_1, \ldots, x_n$ to $\mathbb{k}$. Because of this description of the elements of $\mathbb{k}(x_1, \ldots, x_n)$, the field $\mathbb{k}(x_1, \ldots, x_n)$ can be regarded as the field of fractions $\mathbb{F}$ of $\mathbb{k}[x_1, \ldots, x_n]$. In fact, we argue as follows: let $\eta : \mathbb{k}[x_1, \ldots, x_n] \to \mathbb{F}$ be the natural ring homomorphism $a \mapsto$ class of $(a, 1)$ of $\mathbb{k}[x_1, \ldots, x_n]$ into its field of fractions; then the universal mapping property of $\mathbb{F}$ stated in Proposition 8.6 gives a factorization of the inclusion $\iota : \mathbb{k}[x_1, \ldots, x_n] \to \mathbb{k}(x_1, \ldots, x_n)$ as $\iota = \widetilde{\iota}\eta$, and the field mapping $\widetilde{\iota}$ has to be onto $\mathbb{k}(x_1, \ldots, x_n)$ since the class of (a, b) maps to the member ab^{-1} of $\mathbb{k}(x_1, \ldots, x_n)$.

As in Chapter IV and elsewhere, we let $\mathbb{k}[X]$ be the ring of polynomials in the indeterminate X with coefficients in $\mathbb{k}$. For each x in $\mathbb{K}$, we have a unique substitution homomorphism $\varphi_x : \mathbb{k}[X] \to \mathbb{k}[x]$ carrying $\mathbb{k}$ to itself and carrying X to x. We say that x is **algebraic** over $\mathbb{k}$ if φ_x is not one-one, i.e., if x is a root of some nonzero polynomial in $\mathbb{k}[X]$, and that x is **transcendental** over $\mathbb{k}$ if φ_x is one-one.

EXAMPLES.

(1) If $\mathbb{k} = \mathbb{R}$, if $\mathbb{K} = \mathbb{C}$, and if x is the usual element $i = \sqrt{-1}$, then $\varphi_i(X^2 + 1) = 0$, and i is algebraic over $\mathbb{R}$.

(2) If $\mathbb{k} = \mathbb{Q}$, if $\mathbb{K} = \mathbb{C}$, and if θ is a complex number with the property that $\theta^n + c_{n-1}\theta^{n-1} + \cdots + c_1\theta + c_0 = 0$ for some n and for some coefficients in $\mathbb{Q}$, then θ is algebraic over $\mathbb{Q}$. This situation was the subject of Proposition 4.1, of Example 2 in Section IV.4, and of Example 10 in Section VIII.1.

(3) Let $\mathbb{k} = \mathbb{Q}$ and $\mathbb{K} = \mathbb{C}$. For π equal to the usual trigonometric constant, given as the least positive real such that $e^{i\pi} = -1$ when $e^z = \sum_{n=0}^{\infty} z^n/n!$, it will be proved in Section 14 that there is no polynomial $F(X)$ in $\mathbb{Q}[X]$ with $F(\pi) = 0$, and π is consequently transcendental over $\mathbb{Q}$.

(4) If $\mathbb{k} = \mathbb{Z}/2\mathbb{Z}$ and $\mathbb{K}$ is the 4-element field constructed in Example 3 of fields in Section IV.4, then any element of $\mathbb{K}$ is algebraic over $\mathbb{k}$.

(5) If $\mathbb{k} = \mathbb{C}(X)$ and if $\mathbb{K} = \mathbb{C}(X)[\sqrt{(X-1)X(X+1)}\,]$ as with the ring R' in Section VIII.7 and as in Example 3 of integral closures in Section VIII.9, then $\sqrt{(X-1)X(X+1)}$ is algebraic over $\mathbb{C}(X)$.

1. Algebraic Elements

Suppose that x in $\mathbb{K}$ is algebraic over $\mathbb{k}$. Then

$$\ker \varphi_x = \{F(X) \in \mathbb{k}[X] \mid F(x) = 0\}$$

is an ideal in $\mathbb{k}[X]$ that is necessarily nonzero and principal. A generator is determined up to a constant factor as any nonzero polynomial in the ideal that has lowest possible degree, and we might as well take this polynomial to be monic. Thus $\ker \varphi_x$ is of the form $(F_0(X))$ for some unique monic polynomial $F_0(X)$, and this polynomial $F_0(X)$ is called the **minimal polynomial** of x over $\mathbb{k}$. Review of the example at the end of Section VIII.3 may help motivate the first five results below.

Proposition 9.1 If $x \in \mathbb{K}$ is algebraic over $\mathbb{k}$, then the minimal polynomial of x over $\mathbb{k}$ is prime as a polynomial in $\mathbb{K}[X]$.

PROOF. Suppose that $F(X)$ factors nontrivially as $F(X) = G(X)H(X)$. Since $F(x) = 0$, either $G(x) = 0$ or $H(x) = 0$, and then we have a contradiction to the fact that F has minimal degree among all polynomials vanishing at x. $\square$

Theorem 9.2. If $x \in \mathbb{K}$ is algebraic over $\mathbb{k}$, then the field $\mathbb{k}(x)$ coincides with the ring $\mathbb{k}[x]$. Moreover, if the minimal polynomial of x over $\mathbb{k}$ has degree n, then each element of $\mathbb{k}(x)$ has a unique expansion as

$$c_{n-1}x^{n-1} + c_{n-2}x^{n-2} + \cdots + c_1 x + c_0 \qquad \text{with all } c_i \in \mathbb{k}.$$

PROOF. Since the substitution ring homomorphism φ_x carries $\mathbb{k}[X]$ onto $\mathbb{k}[x]$, we have an isomorphism of rings $\mathbb{k}[x] \cong \mathbb{k}[X]/\ker \varphi_x = \mathbb{k}[X]/(F_0(X))$, where $F_0(X)$ is the minimal polynomial of x over $\mathbb{k}$. Since F_0 is prime, $(F_0(X))$ is a nonzero prime ideal and hence is maximal. Thus $\mathbb{k}[x]$ is a field. Consequently $\mathbb{k}(x) = \mathbb{k}[x]$.

Any element in $\mathbb{k}[x]$, hence in $\mathbb{k}(x)$, is a polynomial in x. Since $F_0(x) = 0$, we can solve $F_0(x) = 0$ for its leading term, say x^n, obtaining $x^n = G(x)$, where $G(X) = 0$ or $\deg G(X) \leq n - 1$. Thus the expansions in the statement of the theorem yield all the members of $\mathbb{k}[x]$. If an element has two such expansions, we subtract them and obtain a nonzero polynomial $H(X)$ of degree at most $n - 1$ with $H(x) = 0$, in contradiction to the minimality of the degree of $F_0(X)$. $\square$

Corollary 9.3. If $x \in \mathbb{K}$ is algebraic over $\mathbb{k}$, then the field $\mathbb{k}(x)$, regarded as a vector space over $\mathbb{k}$, is of dimension n, where n is the degree of the minimal polynomial of x over $\mathbb{k}$. The elements $1, x, x^2, \ldots, x^{n-1}$ form a basis of $\mathbb{k}(x)$ over $\mathbb{k}$.

PROOF. This is just a restatement of the second conclusion of Theorem 9.2. $\square$

We say that the field extension $\mathbb{K}/\mathbb{k}$ is an **algebraic extension** if every element of $\mathbb{K}$ is algebraic over $\mathbb{k}$.

Proposition 9.4. If the vector-space dimension of $\mathbb{K}$ over $\mathbb{k}$ is some finite n, then $\mathbb{K}$ is an algebraic extension of $\mathbb{k}$, and each element x of $\mathbb{K}$ has some nonzero polynomial $F(X)$ in $\mathbb{k}[X]$ of degree at most n for which $F(x) = 0$.

PROOF. This is immediate since the elements $1, x, x^2, \ldots, x^n$ of $\mathbb{K}$ have to be linearly dependent over $\mathbb{k}$. $\square$

When $\mathbb{K}/\mathbb{k}$ is a field extension, we write $[\mathbb{K} : k]$ for the vector-space dimension $\dim_{\mathbb{k}} \mathbb{K}$, and we call this the **degree** of $\mathbb{K}$ over $\mathbb{k}$. If $[\mathbb{K} : \mathbb{k}]$ is finite, we say that $\mathbb{K}$ is a **finite extension** of $\mathbb{k}$, or **finite algebraic extension** of $\mathbb{k}$, the condition "algebraic" being automatic by Proposition 9.4.

Corollary 9.5. If x is in $\mathbb{K}$, then x is algebraic over $\mathbb{k}$ if and only if $\mathbb{k}(x)$ is a finite algebraic extension of $\mathbb{k}$. In this case the minimal polynomial of x over $\mathbb{k}$ has degree $[\mathbb{k}(x) : \mathbb{k}]$.

PROOF. If x is algebraic over $\mathbb{k}$, then $[\mathbb{k}(x) : \mathbb{k}]$ is finite and is the degree of the minimal polynomial of x over $\mathbb{k}$, by Corollary 9.3. Proposition 9.4 shows in this case that $\mathbb{k}(x)$ is a finite algebraic extension. If x is transcendental over $\mathbb{k}$, then the substitution homomorphism φ_x is one-one, and $\dim_{\mathbb{k}} \mathbb{k}(x) \geq \dim_{\mathbb{k}} \mathbb{k}[X] = +\infty$.
$\square$

Theorem 9.6. Let $\mathbb{k}, \mathbb{K}$, and $\mathbb{L}$ be fields with $\mathbb{k} \subseteq \mathbb{K} \subseteq \mathbb{L}$, and suppose that $[\mathbb{K} : \mathbb{k}] = n$ and $[\mathbb{L} : \mathbb{K}] = m$, finite or infinite. Let $\{\omega_1, \omega_2, \ldots\}$ be a vector-space basis of $\mathbb{K}$ over $\mathbb{k}$, and let $\{\xi_1, \xi_2, \ldots\}$ be a vector-space basis of $\mathbb{L}/\mathbb{K}$. Then the mn products $\omega_i \xi_j$ form a basis of $\mathbb{L}$ over $\mathbb{k}$.

PROOF OF SPANNING. If ξ is in $\mathbb{L}$, write $\xi = \sum_j a_j \xi_j$ with each a_j in $\mathbb{K}$ and with only finitely many a_j's not 0. Then expand each a_j in terms of the ω_i's, and substitute. $\square$

PROOF OF LINEAR INDEPENDENCE. Let $\sum_{i,j} c_{ij} \omega_i \xi_j = 0$ with the c_{ij}'s in $\mathbb{k}$. Since the members ξ_j of $\mathbb{L}$ are linearly independent over $\mathbb{K}$, $\sum_i c_{ij} \omega_i = 0$ for each j. Since the members ω_i of $\mathbb{K}$ are linearly independent over $\mathbb{k}$, $c_{ij} = 0$ for all i and j. $\square$

Corollary 9.7. If $\mathbb{k}, \mathbb{K}$, and $\mathbb{L}$ are fields with $\mathbb{k} \subseteq \mathbb{K} \subseteq \mathbb{L}$, then

$$\boxed{[\mathbb{L} : \mathbb{k}] = [\mathbb{L} : \mathbb{K}][\mathbb{K} : \mathbb{k}].}$$

PROOF. This is immediate by counting basis elements in Theorem 9.6. $\square$

Theorem 9.8. If $\mathbb{K}/\mathbb{k}$ is a field extension and if $x_1, \ldots, x_n$ are members of $\mathbb{K}$ that are algebraic over $\mathbb{k}$, then $\mathbb{k}(x_1, \ldots, x_n)$ is a finite algebraic extension of $\mathbb{k}$.

REMARK. If a finite algebraic extension of $\mathbb{k}$ turns out to be of the form $\mathbb{k}(x)$ for some x, we say that the extension is a **simple algebraic extension**.

PROOF. Since x_i is algebraic over $\mathbb{k}$, it is algebraic over $\mathbb{k}(x_1, \ldots, x_{i-1})$. Hence $[\mathbb{k}(x_1, \ldots, x_i) : \mathbb{k}(x_1, \ldots, x_{i-1})]$ is finite. Applying Corollary 9.7 repeatedly, we see that $\mathbb{k}(x_1, \ldots, x_n)$ is a finite extension of $\mathbb{k}$. Proposition 9.4 shows that it is a finite algebraic extension. □

EXAMPLE. The sum $\sqrt{2} + \sqrt[3]{2}$ is algebraic over $\mathbb{Q}$, as a consequence of Theorem 9.8. This fact suggests Corollary 9.9 below.

Corollary 9.9 If $\mathbb{K}/\mathbb{k}$ is a field extension, then the elements of $\mathbb{K}$ that are algebraic over $\mathbb{k}$ form a field.

PROOF. If x and y in $\mathbb{K}$ are algebraic over $\mathbb{k}$, then $\mathbb{k}(x, y)$ is a finite algebraic extension of $\mathbb{k}$, according to Theorem 9.8. This extension contains $x \pm y$ and xy, and it contains x^{-1} if $x \neq 0$. The corollary therefore follows from Proposition 9.4. □

For the special case of Corollary 9.9 in which $\mathbb{K} = \mathbb{C}$ and $\mathbb{k} = \mathbb{Q}$, this subfield of $\mathbb{C}$ is called the field of **algebraic numbers**, and any finite algebraic extension of $\mathbb{Q}$ within $\mathbb{C}$ is called a **number field**, or an **algebraic number field**. The seeming discrepancy between this definition and the definition given in remarks with Proposition 4.1 (that in essence a "number field" is any simple algebraic extension of $\mathbb{Q}$) will be resolved by the Theorem of the Primitive Element (Theorem 9.34 below).

2. Construction of Field Extensions

In this section, $\mathbb{k}$ denotes any field. Our interest will be in constructing extension fields for $\mathbb{k}$ and in addressing the question of uniqueness under additional hypotheses. We begin with a kind of converse to Proposition 9.1 that generalizes the method described in Section A4 of the appendix for constructing $\mathbb{C} = \mathbb{R}(\sqrt{-1})$ from $\mathbb{R}$ and the polynomial $X^2 + 1$.

Theorem 9.10 (existence theorem for simple algebraic extensions). If $F(X)$ is a monic prime polynomial in $\mathbb{k}[X]$, then there exists a simple algebraic extension $\mathbb{K} = \mathbb{k}(x)$ of $\mathbb{k}$ such that x is a root of $F(X)$. Moreover, $F(X)$ is the minimal polynomial of x over $\mathbb{k}$.

PROOF. Define $\mathbb{K} = \Bbbk[X]/(F(X))$ as a ring. Since $F(X)$ is prime, $(F(X))$ is a nonzero prime ideal, hence maximal. Therefore $\mathbb{K}$ is a field, an extension field of $\Bbbk$. Define x to be the coset $X + (F(X))$. Then $F(x) = F(X) + (F(X)) = 0 + (F(X))$, and x is therefore algebraic over $\Bbbk$. It is immediate that $\mathbb{K} = \Bbbk[x]$, and Theorem 9.2 shows that $\mathbb{K} = \Bbbk(x)$. If $G(x) = 0$ for some $G(X)$ in $\Bbbk[X]$, then $G(X)$ is in $(F(X))$. We conclude that $F(X)$ has minimal degree among all polynomials with x as a root, and $F(X)$ is therefore the minimal polynomial. □

Theorem 9.11 (uniqueness theorem for simple algebraic extensions). If $F(X)$ is a monic prime polynomial in $\Bbbk[X]$ and if $\mathbb{K} = \Bbbk(x)$ and $\mathbb{K}' = \Bbbk(y)$ are two simple algebraic extensions such that x and y are roots of $F(X)$, then there exists a field isomorphism φ of $\mathbb{K}$ onto $\mathbb{K}'$ fixing $\Bbbk$ and carrying x to y.

EXAMPLE. The monic polynomial $F(X) = X^3 - 2$ is prime in $\mathbb{Q}[X]$, and $x = \sqrt[3]{2}$ and $y = e^{2\pi i/3}\sqrt[3]{2}$ are roots of it within $\mathbb{C}$. The fields $\mathbb{Q}(x)$ and $\mathbb{Q}(y)$ are subfields of $\mathbb{C}$ and are distinct because $\mathbb{Q}(x)$ is contained in $\mathbb{R}$ and $\mathbb{Q}(y)$ is not. Nevertheless, these fields are $\mathbb{Q}$ isomorphic, according to the theorem.

PROOF. In view of the proof of Theorem 9.10, there is no loss of generality in assuming that $\mathbb{K} = \Bbbk[X]/(F(X))$. Since y is algebraic over $\Bbbk$, we can form the substitution homomorphism $\varphi_y : \Bbbk[X] \to \Bbbk(y)$. This is a $\Bbbk$ algebra homomorphism. Its kernel is the ideal $(F(X))$ since $F(X)$ is the minimal polynomial of y, and φ_y therefore descends to a one-one $\Bbbk$ algebra homomorphism $\overline{\varphi_y} : \Bbbk(x) \to \Bbbk(y)$. Since $\dim \Bbbk(x)$ and $\dim \Bbbk(y)$ both match the degree of $F(X)$, $\overline{\varphi_y}$ is onto $\Bbbk(y)$ and is therefore the required $\Bbbk$ isomorphism. □

We say that a nonconstant polynomial $F(X)$ in $\Bbbk[X]$ **splits** in a given extension field if $F(X)$ factors completely into degree-one factors over that extension field. A **splitting field** over $\Bbbk$ for a nonconstant polynomial $F(X)$ in $\Bbbk[X]$ is an extension field $\mathbb{L}$ of $\Bbbk$ such that $F(X)$ splits in $\mathbb{L}$ and such that $\mathbb{L}$ is generated by $\Bbbk$ and the roots of $F(X)$ in $\mathbb{L}$.

EXAMPLES. Let $\Bbbk = \mathbb{Q}$. Then $\mathbb{Q}(\sqrt{-1})$ is a splitting field for $X^2 + 1$, because $\pm\sqrt{-1}$ are both in $\mathbb{Q}(\sqrt{-1})$ and they generate $\mathbb{Q}(\sqrt{-1})$ over $\mathbb{Q}$. But $\mathbb{Q}(\sqrt[3]{2})$ is not a splitting field for $X^3 - 2$ because $\mathbb{Q}(\sqrt[3]{2})$ does not contain the two nonreal roots of $X^3 - 2$.

Theorem 9.12 (existence of splitting field). If $F(X)$ is a nonconstant polynomial in $\Bbbk[X]$, then there exists a splitting field of $F(X)$ over $\Bbbk$.

PROOF. We begin by constructing a certain extension field $\mathbb{K}$ of $\Bbbk$ in which $F(X)$ factors completely into degree-one factors in $\mathbb{K}[X]$. We do so by induction on $n = \deg F(X)$. For $n = 1$, there is nothing to prove. For general n, let $G(X)$

2. Construction of Field Extensions

be a prime factor of $F(X)$, and apply Theorem 9.10 to obtain a simple algebraic extension $\mathbb{k}_1 = \mathbb{k}(x_1)$ over $\mathbb{k}$ such that $G(x_1) = 0$. Then $F(x_1) = 0$, and the Factor Theorem (Corollary 1.13) gives $F(X) = (X - x_1)H(X)$ for some $H(X)$ in $\mathbb{k}_1(X)$ of degree $n - 1$. Since $\deg H(X) = n - 1 < \deg F(X)$, the inductive hypothesis produces an extension $\mathbb{K}$ of $\mathbb{k}_1$ such that $H(X)$ is a constant multiple of $(X - x_2) \cdots (X - X_n)$ with all x_i in $\mathbb{K}$. Then $F(X)$ factors into degree-one factors in $\mathbb{K}[X]$, and the induction is complete.

Within the constructed field $\mathbb{K}$, let $\mathbb{L}$ be the subfield $\mathbb{L} = \mathbb{k}(x_1, \ldots, x_n)$. Then $F(X)$ still factors completely into degree-one factors in $\mathbb{L}(X)$, and $\mathbb{L}$ is generated by $\mathbb{k}$ and the x_i. Hence $\mathbb{L}$ is a splitting field. □

EXAMPLES OF SPLITTING FIELDS.

(1) $\mathbb{k} = \mathbb{Q}$ and $F(X) = X^3 - 2$. The proof of Theorem 9.12 takes $\mathbb{k}_1 = \mathbb{Q}(\sqrt[3]{2})$ and writes $X^3 - 2 = (X - \sqrt[3]{2})(X^2 + \sqrt[3]{2}X + (\sqrt[3]{2})^2)$. Then the proof adjoins one root θ (hence both roots) of $X^2 + \sqrt[3]{2}X + (\sqrt[3]{2})^2$, setting $\mathbb{K} = \mathbb{Q}(\sqrt[3]{2}, \theta)$. With this choice of $\mathbb{K}$, the splitting field turns out to be $\mathbb{L} = \mathbb{K}$. In fact, to see that $\mathbb{L}$ is not a proper subfield of $\mathbb{K}$, we observe that $6 = [\mathbb{K} : \mathbb{k}] = [\mathbb{K} : \mathbb{L}][\mathbb{L} : \mathbb{Q}]$ by Corollary 9.7 and that the proper containment $\mathbb{L} \supsetneq \mathbb{Q}(\sqrt[3]{2})$ implies $[\mathbb{L} : \mathbb{Q}] > 3$. Since $[\mathbb{L} : \mathbb{Q}]$ is a divisor of 6 greater than 3, $[\mathbb{L} : \mathbb{Q}] = 6$. Thus $[\mathbb{K} : \mathbb{L}] = 1$, and $\mathbb{K} = \mathbb{L}$.

(2) $\mathbb{k} = \mathbb{Q}$ and $F(X) = X^3 - X - \frac{1}{3}$. Application of Corollary 8.20c to the polynomial $G(X) = -3X^2 F(1/X) = X^3 + 3X^2 - 3$ shows that $G(X)$ has no degree-one factor and hence is irreducible over $\mathbb{Q}$. Then it follows that $F(X)$ is irreducible over $\mathbb{Q}$. The proof of Theorem 9.12 takes $\mathbb{k}_1 = \mathbb{Q}(r)$, where $r^3 - r - \frac{1}{3} = 0$. Then division gives

$$X^3 - X - \tfrac{1}{3} = (X - r)(X^2 + rX + (r^2 - 1)).$$

The discriminant $b^2 - 4ac$ of the quadratic factor is

$$r^2 - 4(r^2 - 1) = 4 - 3r^2 = \frac{r^2}{(1 + 2r)^2},$$

the right-hand equality following from direct computation. This discriminant is a square in $\mathbb{k}_1 = \mathbb{Q}(r)$, and hence $X^2 + rX + (r^2 - 1)$ factors into degree-one factors in $\mathbb{Q}(r)$ without passing to an extension field. Therefore $\mathbb{L} = \mathbb{Q}(r)$ with $[\mathbb{L} : \mathbb{Q}] = 3$.

Theorem 9.13 (uniqueness of splitting field). If $F(X)$ is a nonconstant polynomial in $\mathbb{k}[X]$, then any two splitting fields of $F(X)$ over $\mathbb{k}$ are $\mathbb{k}$ isomorphic.

The idea of the proof is simple enough, but carrying out the idea runs into a technical complication. The idea is to proceed by induction, using the uniqueness result for simple algebraic extensions (Theorem 9.11) repeatedly until all the roots have been addressed. The difficulty is that after one step the coefficients of the two quotient polynomials end up in two distinct but $\mathbb{k}$ isomorphic fields. Thus at the second step Theorem 9.11 does not apply directly. What is needed is the reformulated version given below as Theorem 9.11', which lends itself to this kind of induction. In addition, as soon as the induction involves at least three steps, the above statement of Theorem 9.13 does not lend itself to a direct inductive proof. For this reason we shall instead prove a reformulated version Theorem 9.13' of Theorem 9.13 that is ostensibly more general than Theorem 9.13.

Recall from Proposition 4.24 that a general substitution homomorphism that starts from a polynomial ring can have two ingredients. One is the substitution of some element, such as x, for the indeterminate X, and the other is a homomorphism that is made to act on the coefficients. If the homomorphism is σ, let us write $F^\sigma(X)$ to indicate the polynomial obtained by applying σ to each coefficient of $F(X)$.

Theorem 9.11'. Let $\mathbb{k}$ and $\mathbb{k}'$ be fields, and let $\sigma : \mathbb{k} \to \mathbb{k}'$ be a field isomorphism. Suppose that $F(X)$ is a monic prime polynomial in $\mathbb{k}[X]$ and that $\mathbb{K} = \mathbb{k}(x)$ and $\mathbb{K}' = \mathbb{k}'(x')$ are simple algebraic extensions such that $F(x) = 0$ and $F^\sigma(x') = 0$. Then there exists a field isomorphism $\varphi : \mathbb{k}(x) \to \mathbb{k}'(x')$ such that $\varphi\big|_\mathbb{k} = \sigma$ and $\varphi(x) = x'$.

PROOF. The argument is essentially unchanged from the proof of Theorem 9.11. We start from the substitution homomorphism $G(X) \mapsto G^\sigma(x')$ that replaces X by x' and that operates by σ on the coefficients. This descends to a field map of $\mathbb{k}[x]$ into $\mathbb{k}'[x']$, and the homomorphism must be onto $\mathbb{k}'[x']$ by a count of dimensions. □

Theorem 9.13'. Let $\mathbb{k}$ and $\mathbb{k}'$ be fields, and let $\sigma : \mathbb{k} \to \mathbb{k}'$ be a field isomorphism. If $F(X)$ is a nonconstant polynomial in $\mathbb{k}[X]$ and if $\mathbb{L}$ and $\mathbb{L}'$ are respective splitting fields for $F(X)$ over $\mathbb{k}$ and for $F^\sigma(X)$ over $\mathbb{k}'$, then there exists a field isomorphism $\varphi : \mathbb{L} \to \mathbb{L}'$ such that $\varphi\big|_\mathbb{k} = \sigma$ and such that φ sends the set of roots of $F(X)$ to the set of roots of $F^\sigma(X)$.

PROOF. We proceed by induction on $n = \deg F(X)$, the case $n = 1$ being evident. Assume the result for degree $n - 1$. Let $G(X)$ be a prime factor of $F(X)$ over $\mathbb{k}$. Then $G^\sigma(X)$ is a prime factor of $F^\sigma(X)$ over $\mathbb{k}'$. The polynomials $G(X)$ and $G^\sigma(X)$ have roots in $\mathbb{L}$ and $\mathbb{L}'$, respectively. Fix one such root for each, say x_1 and x_1'. By Theorem 9.11', there exists a field isomorphism $\sigma_1 : \mathbb{k}(x_1) \to \mathbb{k}'(x_1')$ extending σ and satisfying $\sigma_1(x_1) = x_1'$. Write $F(X) = (X - x_1)H(X)$ with coefficients in $\mathbb{k}(x_1)$, by the Factor Theorem (Corollary 1.13). Applying σ_1 to the coefficients, we obtain $F^\sigma(X) = (X - x_1')H^{\sigma_1}(X)$ with coefficients in $\Bbbk'(x_1')$. Then $\mathbb{L}$ and $\mathbb{L}'$ are splitting fields for $H(X)$ and $H^{\sigma_1}(X)$ over $\Bbbk(x_1)$ and $\Bbbk'(x_1')$, respectively. By induction we can extend σ_1 to an isomorphism $\varphi : \mathbb{L} \to \mathbb{L}'$, and the theorem readily follows. $\square$

3. Finite Fields

In this section we shall use the results on splitting fields in Section 2 to classify finite fields up to isomorphism. So far, the examples of finite fields that we have encountered are the prime fields $\mathbb{F}_p = \mathbb{Z}/p\mathbb{Z}$ with p elements, p being any prime number, and the field of 4 elements in Example 3 of fields in Section IV.4. Every finite field has to contain a subfield isomorphic to one of the prime fields $\mathbb{F}_p$, and Proposition 4.33 observed as a consequence that any finite field necessarily has p^n elements for some prime number p and some integer $n > 0$.

Theorem 9.14. For each p^n with p a prime number and with n a positive integer, there exists up to isomorphism one and only one field with p^n elements. Such a field is a splitting field for $X^{p^n} - X$ over the prime field $\mathbb{F}_p$.

If $q = p^n$, it is customary to denote by $\mathbb{F}_q$ a field of order q. The theorem says that $\mathbb{F}_q$ exists and is unique up to isomorphism. Some authors refer to finite fields as **Galois fields**.

Some preparation is needed before we can come to the proof of the theorem. We need to carry over the simplest aspects of differential calculus to polynomials with coefficients in an arbitrary field $\Bbbk$. First we give an informal definition of the **derivative** of a polynomial; then we give a more precise definition. For any polynomial $F(X) = \sum_{j=0}^{n} c_j X^j$ in $\Bbbk[X]$, we informally define the derivative to be the polynomial

$$F'(X) = \sum_{j=1}^{n} j c_j X^{j-1} = \sum_{j=0}^{n-1} (j+1) c_{j+1} X^j.$$

The more precise definition uses the definition of members of $\Bbbk[X]$ as infinite sequences of members of $\Bbbk$ whose terms are 0 from some point on. In this notation if $F = (c_0, c_1, \ldots, c_n, 0, \ldots)$ with c_j in the jth position for $j \leq n$ and with 0 in the jth position for $j > n$, then $F' = (c_1, 2c_2, \ldots, nc_n, 0, \ldots)$ with $(j+1)c_{j+1}$ in the jth position for $j \leq n-1$ and with 0 in the jth position for $j > n-1$. In any event, the mapping $F \mapsto F'$ is $\Bbbk$ linear from $\Bbbk[X]$ to itself. The operation is called **differentiation**.

Proposition 9.15. Differentiation on $\Bbbk[X]$ satisfies the product rule: $F = GH$ implies $F' = G'H + GH'$.

PROOF. Because of the $\Bbbk$ linearity, it is enough to prove the result for monomials. Thus let $G(X) = X^m$ and $H(X) = X^n$, so that $F(X) = X^{m+n}$. Then $F'(X) = (m+n)X^{m+n-1}$, $G'(X)H(X) = mX^{m+n-1}$, and $G(X)H'(X) = nX^{m+n-1}$. Hence we indeed have $F'(X) = G'(X)H(X) + G(X)H'(X)$. □

Corollary 9.16. If n is a positive integer, if r is in $\Bbbk$, and if $F(X) = (X-r)^n$ in $\Bbbk[X]$, then $F'(X) = n(X-r)^{n-1}$.

PROOF. This is immediate by induction from Proposition 9.15 since the derivative of $X - r$ is 1. □

Corollary 9.17. Let r be in $\Bbbk$, and let $F(X)$ be in $\Bbbk[X]$. If $(X-r)^2$ divides $F(X)$, then $F(r) = F'(r) = 0$.

PROOF. Write $F(X) = (X-r)^2 G(X)$. If we substitute r for X, we see that $F(r) = 0$. If instead we differentiate, using Proposition 9.15 and Corollary 9.16, then we obtain $F'(X) = 2(X-r)G(X) + (X-r)^2 G'(X)$. Substituting r for X, we obtain $F'(r) = 0 + 0 = 0$. □

Lemma 9.18. If $\Bbbk$ is a field of characteristic $p \neq 0$, then the map $\varphi : \Bbbk \to \Bbbk$ given by $\varphi(x) = x^p$ is a field mapping.

REMARK. The map $x \mapsto x^p$ is often called the **Frobenius** map. If $\Bbbk$ is a finite field, then it must carry $\Bbbk$ onto $\Bbbk$ since one-one implies onto for functions from a finite set to itself; in this case the map is an automorphism of $\Bbbk$.

PROOF. The computation $\varphi(uv) = (uv)^p = u^p v^p = \varphi(u)\varphi(v)$ shows that φ respects products. If u and v are in $\Bbbk$, then

$$\varphi(u+v) = (u+v)^p = \varphi(u) + \sum_{j=1}^{p-1} \binom{p}{j} u^{p-j} v^j + \varphi(v) = \varphi(u) + \varphi(v),$$

the last equality holding since the binomial coefficient $\binom{p}{j}$ has a p in the numerator for $1 \leq j \leq p-1$. Thus φ is a ring homomorphism. Since $\varphi(1) = 1$, φ is a field mapping. □

PROOF OF UNIQUENESS IN THEOREM 9.14. Let $\Bbbk$ be a finite field, say of characteristic p, and let $\mathbb{P}$ be the prime field of order p within $\Bbbk$. We know that $\mathbb{P}$ is isomorphic to $\mathbb{F}_p = \mathbb{Z}/p\mathbb{Z}$. Since $\Bbbk$ is a finite-dimensional vector space over $\mathbb{P}$, we know also that $\Bbbk$ has order $q = p^n$ for some integer $n > 0$. The multiplicative group $\Bbbk^\times$ of $\Bbbk$ thus has order $q - 1$, and every $x \neq 0$ in $\Bbbk$ therefore satisfies

$x^{q-1} = 1$. Taking $x = 0$ into account, we see that every member of $\Bbbk$ satisfies $x^q = x$. Forming the polynomial $X^q - X$ in $\mathbb{P}[X]$, we see that every member of $\Bbbk$ is a root of this polynomial. Iterated application q times of the Factor Theorem (Corollary 1.13) shows that $X^q - X$ factors into degree-one factors in $\Bbbk$. Since every member of $\Bbbk$ is a root of $X^q - X$, $\Bbbk$ is a splitting field of $X^q - X$ over $\mathbb{P}$. Then the uniqueness of the prime field up to isomorphism, in combination with the uniqueness of the splitting field of $X^q - X$ given in Theorem 9.13′, shows that $\Bbbk$ is uniquely determined up to isomorphism. □

PROOF OF EXISTENCE IN THEOREM 9.14. Let $q = p^n$ be given, and define $\Bbbk$ to be a splitting field of $X^q - X$ over $\mathbb{F}_p = \mathbb{Z}/p\mathbb{Z}$. The field $\Bbbk$ exists by Theorem 9.12, and it has characteristic p. Since $X^q - X$ is monic of degree q, the definition of splitting field says that we can write

$$X^q - X = (X - u_1)(X - u_2) \cdots (X - u_q) \qquad \text{with all } u_j \in \Bbbk.$$

Because of Lemma 9.18, the map $\varphi(u) = u^q$, which is the nth power of the map $u \mapsto u^p$, is a field mapping of $\Bbbk$ into itself. The members of $\Bbbk$ fixed by φ form a subfield of $\Bbbk$, and these elements of $\Bbbk$ are exactly the members of the set $S = \{u_1, \ldots, u_q\}$. Therefore S is a subfield of $\Bbbk$, necessarily containing $\mathbb{F}_p = \mathbb{Z}/p\mathbb{Z}$. Since $X^q - X$ splits in S and since the roots of $X^q - X$ generate S, S is a splitting field of $X^q - X$ over $\mathbb{F}_p$. In other words, $S = \Bbbk$. To complete the proof, it is enough to show that the elements $u_1, \ldots, u_q$ are distinct, and then $\Bbbk$ will be a field of q elements. The question is therefore whether some root of $X^q - X$ has multiplicity at least 2, i.e., whether $(X - r)^2$ divides $X^q - X$ for some r in $\Bbbk$. Corollary 9.17 gives a necessary condition for this divisibility, saying that the derivative of $X^q - X$ must have r as a root. However, the derivative of $X^q - X$ is $qX^{q-1} - 1 = -1$, and the constant polynomial -1 has no roots. We conclude that $\Bbbk$ has q elements. □

Corollary 9.19. If q and r are integers with $2 \leq q \leq r$, then the finite field $\mathbb{F}_q$ is isomorphic to a subfield of the finite field $\mathbb{F}_r$ if and only if $r = q^n$ for some integer $n \geq 1$.

PROOF. If $\mathbb{F}_q$ is isomorphic to a subfield of $\mathbb{F}_r$, then we may consider $\mathbb{F}_r$ as a vector space over $\mathbb{F}_q$, say of dimension n. In this case, $\mathbb{F}_r$ has q^n elements.

Conversely let $r = q^n$, and regard $\mathbb{F}_r$ as a splitting field of $X^{q^n} - X$ over the prime field $\mathbb{F}_p$, by Theorem 9.14. Let S be the subset of $\mathbb{F}_r$ of all roots of $X^q - X$. Putting $a = q - 1$ and $k = \frac{q^n - 1}{q - 1} = q^{n-1} + q^{n-2} + \cdots + 1$, we have

$$X^{ka} - 1 = (X^a - 1)(X^{(k-1)a} + X^{(k-2)a} + \cdots + 1).$$

Multiplying by X, we see that $X^q - X$ is a factor of $X^{q^n} - X$. Since $X^{q^n} - X$ splits in $\mathbb{F}_r$ and has distinct roots, the same is true of $X^q - X$. Therefore $|S| = q$.

Let $q = p^m$. The mth power of the homomorphism of Lemma 9.18 on $\mathbb{k} = \mathbb{F}_r$ is $x \mapsto x^q$, and the subset of $\mathbb{F}_r$ fixed by this homomorphism is a subfield. Thus S is a subfield, and it has q elements. □

4. Algebraic Closure

Algebraically closed fields—those for which every nonconstant polynomial with coefficients in the field has a root in the field—were introduced in Section V.1, and it was mentioned at that time that every field is a subfield of some algebraically closed field. We shall prove that existence theorem in this section in a form lending itself to a uniqueness result.

Throughout this section let $\mathbb{k}$ be a field. We begin by giving further descriptions of algebraically closed fields that take the theory of Sections 1–2 into account.

Proposition 9.20. The following conditions on the field $\mathbb{k}$ are equivalent:
  (a) $\mathbb{k}$ has no nontrivial algebraic extensions,
  (b) every irreducible polynomial in $\mathbb{k}[X]$ has degree 1,
  (c) every polynomial in $\mathbb{k}[X]$ of positive degree has at least one root in $\mathbb{k}$,
  (d) every polynomial in $\mathbb{k}[X]$ of positive degree factors over $\mathbb{k}$ into polynomials of degree 1.

PROOF. If (a) holds, then (b) holds since any irreducible polynomial of degree greater than 1 would give a nontrivial simple algebraic extension (Theorem 9.10). If (b) holds and a polynomial of positive degree is given, apply (b) to an irreducible factor to see that the given polynomial has a root; thus (c) holds. Condition (c) implies condition (d) by induction and the Factor Theorem. If (d) holds and if $\mathbb{K}$ is an algebraic extension of $\mathbb{k}$, let x be in $\mathbb{K}$, and let $F(X)$ be the minimal polynomial of x over $\mathbb{k}$. Then $F(X)$ is irreducible over $\mathbb{k}$, and (d) says that $F(X)$ has degree 1. Hence x is in $\mathbb{k}$, and we conclude that $\mathbb{K} = \mathbb{k}$. □

A field satisfying the equivalent conditions of Proposition 9.20 is said to be **algebraically closed**.

EXAMPLES OF ALGEBRAICALLY CLOSED FIELDS.

(1) The Fundamental Theorem of Algebra (Theorem 1.18) says that $\mathbb{C}$ is algebraically closed. This theorem was not proved in Chapter I, but a proof will be given in this chapter in Section 10.

(2) Let $\mathbb{K}$ be the subset of all members of $\mathbb{C}$ that are algebraic over $\mathbb{Q}$. By Corollary 9.9, $\mathbb{K}$ is a subfield of $\mathbb{C}$. Example 1 shows that every polynomial in $\mathbb{Q}[X]$ splits in $\mathbb{K}$, and Lemma 9.21 below then allows us to conclude that $\mathbb{K}$ is algebraically closed.

(3) Fix a prime number p, and start with $\Bbbk_0 = \mathbb{F}_p$ as the prime field $\mathbb{Z}/p\mathbb{Z}$. Enumerate the members of $\mathbb{F}_p[X]$, letting $F_n(X)$ be the nth such polynomial. We construct $\Bbbk_n$ by induction on n so that $\Bbbk_n$ is a splitting field for $F_n(X)$ over $\Bbbk_{n-1}$ when $n \geq 1$. Then $\Bbbk_0 \subseteq \Bbbk_1 \subseteq \Bbbk_2 \subseteq \cdots$ is an increasing sequence of fields containing $\mathbb{F}_p$. Let $\mathbb{K}$ be the union. Any two elements of $\mathbb{K}$ lie in a single $\Bbbk_n$, and it follows that $\mathbb{K}$ is closed under the field operations. Any three elements lie in a single $\Bbbk_n$, and it follows that any of the defining properties of a field is valid in $\mathbb{K}$ because it is valid in $\Bbbk_n$. Therefore $\mathbb{K}$ is a field. This field is an extension of $\mathbb{F}_p$, and every polynomial in $\mathbb{F}_p[X]$ splits in $\mathbb{K}$. As in Example 2, Lemma 9.21 below shows that $\mathbb{K}$ is algebraically closed.

Lemma 9.21. If $\mathbb{K}/\Bbbk$ is an algebraic extension of fields and if every nonconstant polynomial in $\Bbbk[X]$ splits into degree-one factors in $\mathbb{K}$, then $\mathbb{K}$ is algebraically closed.

PROOF. Let $\mathbb{K}'$ be an algebraic extension of $\mathbb{K}$, and let x be in $\mathbb{K}'$. Let $G(X)$ be the minimal polynomial of x over $\mathbb{K}$, and write $G(X)$ as

$$G(X) = X^n + c_{n-1}X^{n-1} + \cdots + c_0 \quad \text{with all } c_i \in \mathbb{K}.$$

Then x is algebraic over $\Bbbk(c_{n-1}, \ldots, c_0)$, which is a finite extension of $\Bbbk$ by Theorem 9.8. By Corollary 9.7, x lies in a finite extension of $\Bbbk$. Thus Proposition 9.4 shows that x is algebraic over $\Bbbk$. Let $F(X)$ be the minimal polynomial of x over $\Bbbk$. By assumption this splits over $\mathbb{K}$, say as

$$F(X) = (X - x_1) \cdots (X - x_m) \quad \text{with all } x_i \in \mathbb{K}.$$

Evaluating at x and using the fact that $F(x) = 0$, we see that $x = x_j$ for some j. Therefore x is in $\mathbb{K}$, and $\mathbb{K}$ is algebraically closed. $\square$

An extension field $\mathbb{K}/\Bbbk$ is an **algebraic closure** of $\Bbbk$ if $\mathbb{K}$ is algebraic over $\Bbbk$ and if $\mathbb{K}$ is algebraically closed. Example 2 of algebraically closed fields above gives an algebraic closure of $\mathbb{Q}$, and Example 3 gives an algebraic closure of $\mathbb{F}_p$.

Theorem 9.22 (Steinitz). Every field $\Bbbk$ has an algebraic closure, and this is unique up to $\Bbbk$ isomorphism.

REMARKS. The proof of existence is modeled on the argument for Example 3 of algebraic closures. However, we are not free in general to use a simple union of a sequence of fields and have to work harder. Because there is no evident set of possibilities within which we are forming extension fields, Zorn's Lemma is inconvenient to use and tends to result in an unintuitive construction. Instead, we use Zermelo's Well-Ordering Theorem, whose use more closely parallels the inductive construction in Example 3.

PROOF OF EXISTENCE. With $\Bbbk$ as the given field, let S be the set of nonconstant polynomials $s(X)$ in $\Bbbk[X]$, and introduce a well ordering into S by means of Zermelo's Well-Ordering Theorem (Section A5 of the appendix). Let us write $\prec$ for "strictly precedes in the ordering" and $\precsim$ for "equals or strictly precedes." For each $s \in S$, let $\bar{s}$ be the successor of s, i.e., the first element among all elements t with $s \prec t$. We write s_0 for the first element of S. Without loss of generality, we may assume that S has a last element s_∞. The idea is to construct simultaneously two kinds of things:

(i) an algebraic extension field $\Bbbk_s/\Bbbk$ for each $s \in S$ such that $\Bbbk_{s_0} = \Bbbk$ and such that $\Bbbk_{\bar{s}}$ is a splitting field for $s(X)$ over $\Bbbk_s$ whenever $s \prec s_\infty$,
(ii) a field mapping $\varphi_{ut} : \Bbbk_t \to \Bbbk_u$ for each ordered pair of elements t and u in S having $t \precsim u$, such that $\varphi_{tt} = 1$ for all t and such that $t \precsim u \precsim v$ implies $\varphi_{vt} = \varphi_{vu}\varphi_{ut}$.

These extension fields and mappings are to be such that $\Bbbk_s = \bigcup_{t \prec s} \varphi_{st}(\Bbbk_t)$ whenever s is not a successor and is not s_0. If such a system of extension fields and field homomorphisms exists, then Lemma 9.21 applies to a splitting field over $\Bbbk_{s_\infty}$ of the nonconstant polynomial $s_\infty(X)$ and shows that this splitting field is algebraically closed; since this splitting field is an algebraic extension of $\Bbbk$, it is an algebraic closure of $\Bbbk$.

A partial such system through t_0 means a system consisting of fields $\Bbbk_s$ with $s \precsim t_0$ and field homomorphisms φ_{ut} with $t \precsim u \precsim t_0$ such that the above conditions hold as far as they are applicable. A partial system exists through the first member s_0 of S because we can take $\Bbbk_{s_0} = \Bbbk$ and $\varphi_{s_0 s_0} = 1$. Arguing by contradiction, we suppose that such a system of extension fields and field homomorphisms fails to exist through some member of S. Let t_0 be the first member of S such that there is no partial system through t_0.

Suppose that t_0 is the successor of some element t_1 in S. We know that a partial system exists through t_1. If we let $\Bbbk_{t_0}$ be a splitting field for $t_1(X)$ over $\Bbbk_{t_1}$, and if we define

$$\varphi_{t_0 t} = \begin{cases} \varphi_{t_0 t_1} \varphi_{t_1 t} & \text{for } t \precsim t_1, \\ 1 & \text{for } t = t_0, \end{cases}$$

then the enlarged system is a partial system through t_0, contradiction. Thus t_0 cannot be the successor of some element of S.

When t_0 is not a successor, at least $\Bbbk_t$ is defined for $t \prec t_0$ and φ_{ut} is defined for $t \precsim u \prec t_0$. We want to form a union, but we have to keep the field operations aligned properly in the process. Define a "t-allowable tuple" to be a function $u \mapsto x_u$ defined for $t \precsim u \prec t_0$ such that x_u is in $\Bbbk_u$ and $\varphi_{vu}(x_u) = x_v$ whenever $t \precsim u \precsim v \prec t_0$. If x is in $\Bbbk_t$, then an example of a t-allowable tuple is given by $u \mapsto \varphi_{ut}(x)$ for $t \precsim u \prec t_0$.

If $t \prec t_0$ and $t' \prec t_0$, then we can apply field operations to the t-allowable tuple $u \mapsto x_u$ and to the t'-allowable tuple $u \mapsto y_u$, obtaining $\max(t, t')$-allowable

4. Algebraic Closure   463

tuples $u \mapsto x_u + y_u$, $u \mapsto -x_u$, $u \mapsto x_u y_u$, and $x_u \mapsto x_u^{-1}$ as long as $x_t \neq 0$. These operations are meaningful since each φ_{vu} is a field mapping.

If $t \prec t_0$ and $t' \prec t_0$, we say that the t-allowable tuple $u \mapsto x_u$ is equivalent to the t'-allowable tuple $u \mapsto y_u$ if $x_u = y_u$ for $\max(t, t') \precsim u \prec t_0$. The result is an equivalence relation, and the equivalence relation respects the field operations in the previous paragraph. We define $\Bbbk_{t_0}$ to be the set of equivalence classes of allowable tuples with the inherited field operations. The 0 element is the class of the s_0-allowable tuple $u \mapsto 0$, and the multiplicative identity is the class of the s_0-allowable tuple $u \mapsto 1$. It is a routine matter to check that $\Bbbk_{t_0}$ is a field.

If $t \prec t_0$ is given, we define the function $\varphi_{t_0 t} : \Bbbk_t \to \Bbbk_{t_0}$ as follows: if x is in $\Bbbk_t$, we form the t-allowable tuple $u \mapsto \varphi_{ut}(x)$ and take its equivalence class, which is a member of $\Bbbk_{t_0}$, as $\varphi_{t_0 t}(x)$. Then $\varphi_{t_0 t}$ is evidently a field mapping. It is evident also that $\varphi_{t_0 v} \varphi_{vu} = \varphi_{t_0 u}$ when $u \precsim v \prec t_0$. Defining $\varphi_{t_0 t_0}$ to be the identity, we have a complete system of field mappings φ_{vu} for $\Bbbk_{t_0}$.

The final step is to check that $\Bbbk_{t_0}$ is the union of the images of the $\varphi_{t_0 t}$ for $t \prec t_0$. Thus choose a representative of an equivalence class in $\Bbbk_{t_0}$. Let the representative be a t-allowable tuple $u \mapsto x_u$ for $t \precsim u \prec t_0$. The element x_t is in $\Bbbk_t$, and the condition $x_u = \varphi_{ut}(x_t)$ is just the condition that the class of $u \mapsto x_u$ be the image of x_t under $\varphi_{t_0 t}$. Hence every member of $\Bbbk_{t_0}$ is in the image of some $\varphi_{t_0 t}$ with $t \prec t_0$, and we have a contradiction to the hypothesis that a partial system through t_0 does not exist. This completes the proof of existence.   □

For the uniqueness in Theorem 9.22, we again need a serious application of the Axiom of Choice, but here Zorn's Lemma can be applied fairly routinely. The proof will show a little more than is needed, and in fact the uniqueness in Theorem 9.22 will be derived as a consequence of Theorem 9.23 below.

Theorem 9.23. Let $\mathbb{K}'$ be an algebraically closed field, and let $\mathbb{K}$ be an algebraic extension of a field $\Bbbk$. If φ is a field mapping of $\Bbbk$ into $\mathbb{K}'$, then φ can be extended to a field mapping of $\mathbb{K}$ into $\mathbb{K}'$.

PROOF OF UNIQUENESS IN THEOREM 9.22 USING THEOREM 9.23. Let $\mathbb{K}$ and $\mathbb{K}'$ be algebraic closures of $\Bbbk$, and let $\varphi : \Bbbk \to \mathbb{K}'$ be the inclusion mapping. Theorem 9.23 supplies a field mapping $\Phi : \mathbb{K} \to \mathbb{K}'$ such that $\Phi|_\Bbbk = \varphi$, i.e., such that Φ fixes $\Bbbk$. Since $\mathbb{K}$ is an algebraic closure of $\Bbbk$, so is $\Phi(\mathbb{K})$. Then $\mathbb{K}'$ is an algebraic extension of the algebraically closed field $\Phi(\mathbb{K})$, and we must have $\Phi(\mathbb{K}) = \mathbb{K}'$. Thus Φ is a $\Bbbk$ isomorphism of $\mathbb{K}$ onto $\mathbb{K}'$.

PROOF OF THEOREM 9.23. Let S be the set of all triples $(\mathbb{L}, \mathbb{L}', \psi)$ such that $\mathbb{L}$ is a field with $\Bbbk \subseteq \mathbb{L} \subseteq \mathbb{K}$ and ψ is a field mapping of $\mathbb{L}$ onto the subfield $\mathbb{L}'$ of $\mathbb{K}'$ with $\psi|_\Bbbk = \varphi$. The set S is nonempty since $(\Bbbk, \varphi(\Bbbk), \varphi)$ is a member of it. Defining $(\mathbb{L}_1, \mathbb{L}'_1, \psi_1) \subseteq (\mathbb{L}_2, \mathbb{L}'_2, \psi_2)$ to mean that $\mathbb{L}_1 \subseteq \mathbb{L}_2$, that $\mathbb{L}_1' \subseteq \mathbb{L}_2'$, and that ψ_1 as a set of ordered pairs is a subset of ψ_2 as a set of ordered pairs, we partially order S by inclusion upward. If $\{(\mathbb{L}_\alpha, \mathbb{L}_\alpha', \psi_\alpha)\}$ is a nonempty chain in S, form the triple $\left(\bigcup_\alpha \mathbb{L}_\alpha, \bigcup_\alpha \mathbb{L}_\alpha', \bigcup_\alpha \psi_\alpha\right)$, and put $\psi = \bigcup_\alpha \psi_\alpha$. Then $\psi\left(\bigcup_\alpha \mathbb{L}_\alpha\right) = \bigcup_\alpha \mathbb{L}_\alpha'$, and consequently $\left(\bigcup_\alpha \mathbb{L}_\alpha, \bigcup_\alpha \mathbb{L}_\alpha', \bigcup_\alpha \psi_\alpha\right)$ is an upper bound in S for the chain. By Zorn's Lemma, S has a maximal element $(\mathbb{L}_0, \mathbb{L}_0', \psi_0)$. We shall prove that $\mathbb{L}_0 = \mathbb{K}$, and the proof will be complete.

Fix x in $\mathbb{K}$, and let $F(X)$ be the minimal polynomial of x over $\mathbb{L}_0$. The minimal polynomial of $\psi_0(x)$ over $\mathbb{L}_0'$ is then $F^{\psi_0}(X)$. Since $\mathbb{K}'$ is algebraically closed, $F^{\psi_0}(X)$ has a root x' in $\mathbb{K}'$. By Theorem 9.11', $\psi_0 : \mathbb{L}_0 \to \mathbb{L}'$ can be extended to an isomorphism $\Psi_0 : \mathbb{L}_0(x) \to \mathbb{L}_0'(x')$ such that $\psi_0(x) = x'$. Then $(\mathbb{L}_0(x), \mathbb{L}_0'(x'), \Psi_0)$ is in S and contains $(\mathbb{L}_0, \mathbb{L}_0', \psi_0)$. This containment, if strict, would contradict the fact that $(\mathbb{L}_0, \mathbb{L}_0', \psi_0)$ is a maximal element of S. Thus equality must hold: $\mathbb{L}_0(x) = \mathbb{L}_0$. Therefore x is in $\mathbb{L}_0$, and we conclude that $\mathbb{L}_0 = \mathbb{K}$. $\square$

5. Geometric Constructions by Straightedge and Compass

Classical Euclidean geometry attached a certain emphasis to constructions in the Euclidean plane that could be made by straightedge and compass. These are often referred to casually as constructions by "ruler and compass," but one is not allowed to use the markings on a ruler. Thus "straightedge and compass" is a more accurate description.

In these constructions the starting configuration may be regarded as a line with two points marked on the line. Allowable constructions are the following: to form the line through a given point different from finitely many other lines through that point, to form the line through two distinct points, to form a circle with a given center and a radius different from that of finitely many other circles through the point, and to form a circle with a given center and radius. Intersections of a line or a circle with previous lines and circles establish new points for continuing the construction.

For example a line perpendicular to a given line at a given point can be constructed by drawing any circle centered at the point, using the two intersection points as centers of new circles, drawing those circles so as to have radius larger than the first circle, and forming the line between their two points of intersection. An angle at the point P of intersection between two intersecting lines A and B may be bisected by drawing any circle centered at P, selecting one of the points of intersection on each line so that P and the two new points Q and R describe the angle, drawing circles with that same radius centered at Q and R, and forming the line between the points of intersection of the two circles. And so on.

5. Geometric Constructions by Straightedge and Compass

Three notable problems remained unsolved in antiquity:

(i) how to double a cube, i.e., how to construct the side of a cube of double the volume of a given cube,
(ii) how to trisect any constructible angle, i.e., how to divide the angle into three equal parts by means of constructed lines,
(iii) how to square a circle, i.e., how to construct the side of a square whose area equals that of a given disk.

In this section we shall use the elementary field theory of Sections 1–2 to show that doubling a cube and trisecting a 60-degree angle are impossible with straightedge and compass. As to (iii), we shall reduce a proof of the impossibility of squaring the circle to a proof that π is transcendental over $\mathbb{Q}$. This latter proof we give in Section 14.

The first step is to translate the problem of geometric constructibility into a statement in algebra. Since we are given two points on a line, we can introduce Cartesian coordinates for the Euclidean plane, taking one of the points to be $(0, 0)$ and the other point to be $(1, 0)$. Points in the Euclidean plane are now determined by their Cartesian coordinates, which determine all distances. Distances in turn can be laid off on the x-axis from $(0, 0)$. Thus the question becomes, what points on the x-axis can be constructed?

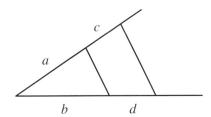

FIGURE 9.1. Closure of positive constructible x coordinates under multiplication and division.

Let $\mathcal{C}$ be the set of constructible x coordinates. We are given that 0 and 1 are in $\mathcal{C}$. Closure of $\mathcal{C}$ under addition and subtraction is evident; the straightedge is not even necessary for this step. Figure 9.1 indicates why the positive elements of $\mathcal{C}$ are closed under multiplication and division. In more detail we take two intersecting lines and mark three known positive members of $\mathcal{C}$ as the distances a, b, c in the figure. Then we form the line through the two points marking a and b, and we form a line parallel to that line through the point marked off by the distance c. The intersection of this parallel line with the other original line defines a distance d. Then $a/b = c/d$, and so $d = bc/a$. By taking $a = 1$, we see that we can multiply any two members b and c in $\mathcal{C}$, obtaining a result in $\mathcal{C}$.

By instead taking $c = 1$, we see that we can divide. The conclusion is that $\mathcal{C}$ is a field.

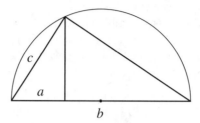

FIGURE 9.2. Closure of positive constructible x coordinates under square roots.

Figure 9.2 indicates why the positive elements of $\mathcal{C}$ are closed under taking square roots. In more detail let a and b be positive members of $\mathcal{C}$ with $a < b$. By forming a circle whose diameter is a segment of length b and by forming a line perpendicular to that line at the point marked by a, we determine the pictured right triangle with a side c satisfying $a/c = c/b$. Then $c = \sqrt{ab}$. By taking one of a and b to be 1, we see that the square root of the other of a and b is in $\mathcal{C}$. This completes the proof of the direct part of the following theorem.

Theorem 9.24. The set $\mathcal{C}$ of x coordinates that can be constructed from $x = 1$ and $x = 0$ by straightedge and compass forms a subfield of $\mathbb{R}$ such that the square root of any positive element of the field lies in the field. Conversely the members of $\mathcal{C}$ are those real numbers lying in some subfield F_n of $\mathbb{R}$ of the form

$$F_1 = \mathbb{Q}(\sqrt{a_0}), \quad F_2 = F_1(\sqrt{a_1}), \quad \ldots, \quad F_n = F_{n-1}(\sqrt{a_{n-1}})$$

with each a_j in F_j and with $a_0, \ldots, a_{n-1}$ all ≥ 0.

PROOF OF CONVERSE. Suppose we have a subfield $F = F_n$ of $\mathbb{R}$ of the kind described in the statement of the theorem. The possibilities for obtaining a new constructible point from F by an additional construction arise from three situations: the intersection of two lines, each passing through two points of F; the intersection of a line and a circle, each determined by data from F; and the intersection of two circles, each determined by data from F.

In the case of two intersecting lines, each line is of the form $ax + by = c$ for suitable coefficients a, b, c in F, and the intersection is a point (x, y) in $F \times F$. So intersections of lines do not force us to enlarge F.

For a line and a circle, we assume that the line is given by $ax + by = c$ with a, b, c in F, that the circle has radius in F and center in $F \times F$, and that the lines and the circle actually intersect. The circle is then given by $(x-h)^2 + (y-k)^2 = r^2$ with h, k, r in F. Substitution of the equation of the line into the equation of the

5. Geometric Constructions by Straightedge and Compass

circle gives us a quadratic equation either for x, and x then determines y, or for y, and y then determines x. The quadratic equation has real roots, and thus its discriminant is ≥ 0. The result is that x and y are in a field $F(\sqrt{l})$ for some $l \geq 0$ in F.

For two circles, without loss of generality, we may take their equations to be

$$x^2 + y^2 = r^2 \quad \text{and} \quad (x-h)^2 + (y-h)^2 = s^2$$

with r, h, k, s in F. Subtracting gives $2xh + 2yk = h^2 + k^2 - s^2 + r^2$. With this equation and with $x^2 + y^2 = r^2$, we again have a line and circle that are being intersected. Thus the same remarks apply as in the previous paragraph.

The conclusion is that any new single construction of points of intersection by straightedge and compass leads from F to $F(\sqrt{l})$ for some $l \geq 0$ in F. Thus every member of the set $\mathcal{C}$ is as described in the theorem. □

To apply the theorem to prove the impossibility of the three never-accomplished constructions that were described earlier in the section, we observe that $[F_i : F_{i-1}]$ in the theorem equals 1 or 2 for each i. Consequently every member of the constructible set $\mathcal{C}$ lies in a finite algebraic extension of $\mathbb{Q}$ of degree 2^k for some k.

For the problem of doubling a cube, the question amounts to constructing $\sqrt[3]{2}$. We argue by contradiction. If $\sqrt[3]{2}$ lies in F_n as in the theorem, then $\mathbb{Q}(\sqrt[3]{2}) \subseteq F_n$. With k as the integer $\leq n$ such that $[F_n : \mathbb{Q}] = 2^k$, Corollary 9.7 gives

$$2^k = [F_n : \mathbb{Q}] = [F_n : \mathbb{Q}(\sqrt[3]{2})][\mathbb{Q}(\sqrt[3]{2}) : \mathbb{Q}] = 3[F_n : \mathbb{Q}(\sqrt[3]{2})].$$

Thus 3 must divide a power of 2, and we have arrived at a contradiction. We conclude that it is not possible to double a cube with straightedge and compass.

For the problem of trisecting any constructible angle, let us show that a $60°$ angle cannot be trisected. A $60°$ angle is itself constructible, being the angle between two sides in an equilateral triangle. Trisecting a $60°$ angle amounts to constructing $\cos 20°$; $\sin 20°$ is then $(1 - \cos^2 20°)^{1/2}$. To proceed, we derive an equation satisfied by $\cos 20°$, starting from

$$(\cos 20° + i \sin 20°)^3 = \cos 60° + i \sin 60° = \tfrac{1}{2} + \tfrac{i\sqrt{3}}{2}.$$

We expand the left side and extract the real part of both sides to obtain

$$\cos^3 20° - 3\cos 20° \sin^2 20° = \tfrac{1}{2}.$$

Substituting $\sin^2 20° = 1 - \cos^2 20°$ and simplifying, we see that $r = \cos 20°$ satisfies

$$4r^3 - 3r - \tfrac{1}{2} = 0.$$

Arguing with Corollary 8.20 as in Example 2 of splitting fields in Section 2, we readily check that $4X^3 - 3X - \frac{1}{2}$ is irreducible over $\mathbb{Q}$. Hence $[\mathbb{Q}(\cos 20°) : \mathbb{Q}] = 3$, and we are led to the same contradiction as for the problem of doubling the cube. Therefore it is not possible to trisect a 60° angle with straightedge and compass.

For the problem of squaring a circle, let A be the area of the circle, and let r be the radius. If the square has side x, then $x^2 = A = \pi r^2$, with r given. Thus $x = r\sqrt{\pi}$, and the essence of the matter is to construct $\sqrt{\pi}$. However, π is known to be transcendental by a theorem of F. Lindemann (1882); we give a proof in Section 14. Since π is transcendental, $\sqrt{\pi}$ is transcendental.

A fourth notable problem, which leads to further insights, concerns the construction of a regular polygon of outer radius 1 with n sides. This construction is easy with straightedge and compass when n is a power of 2 or is 3 times a power of 2, and Euclid showed that a construction is possible for $n = 5$. But a construction cannot be managed with straightedge and compass for $n = 9$, for example, because a central angle in this case is 40° and the constructibility of $\cos 40°$ would imply the constructibility of $\cos 20°$. Thus the question is, for what values of n can a regular n-gon be constructed with straightedge and compass?

The remarkable answer was given by Gauss. By a **Fermat number** is meant any integer of the form $2^{2^N} + 1$. A **Fermat prime** is a Fermat number that is prime. The Fermat numbers for $N = 0, 1, 2, 3, 4$ are $3, 5, 17, 257, 65537$, and each is a Fermat prime. No larger Fermat primes are known.[2] The answer given by Gauss, which we shall prove in stages in Sections 6–9, is as follows.

Theorem 9.25 (Gauss).[3] A regular n-gon is constructible with straightedge and compass if and only if n is the product of distinct Fermat primes and a power of 2.

We can show the relevance of Fermat primes right now, and we can give an indication that if n is a prime number, then a regular n-gon can be constructed if and only if n is a Fermat prime. But a full proof even of this statement will make use of Galois groups, which we take up in the next three sections.

For the necessity let n be prime, and suppose that a regular n-gon is constructible. Returning from degrees to radians, we observe that each central angle is $2\pi/n$. Thus the constructibility implies the constructibility of $\cos 2\pi/n$, and it

---

[2]Many Fermat numbers for $N \geq 5$ are known not to be prime, sometimes by the discovery of an explicit factor and sometimes by a verification that 3 to the power 2^{2^N-1} is not congruent to -1 modulo $2^{2^N} + 1$. (Cf. Lemma 9.46.) For example Euler discovered that 641 divides $2^{2^5} + 1$.

[3]Gauss announced both the necessity and the sufficiency in this theorem in his *Disquisitiones Arithmeticae* in 1801, but he included a proof of only the sufficiency (partly in his articles 336 and 365). A proof of the necessity appeared in a paper of Pierre-Laurent Wantzel in 1837.

follows that $e^{2\pi i/n} = \cos 2\pi/n + i \sin 2\pi/n$ is in the field $\mathcal{C} + i\mathcal{C}$ of constructible points in the complex plane. We have the factorization

$$X^n - 1 = (X - 1)(X^{n-1} + X^{n-2} + \cdots + X + 1).$$

and $e^{2\pi i/n}$ is a root of the second factor. The first example of Eisenstein's criterion (Corollary 8.22) in Section VIII.5 shows that the second factor is irreducible. According to the results of Section 1, $\mathbb{Q}(e^{2\pi i/n})$ is a simple algebraic extension of $\mathbb{Q}$ of degree $n - 1$.

Applying Theorem 9.24, we see that $n - 1$ must be a power of two. Let us write $n - 1 = 2^m$. Suppose $m = a2^N$ with a odd. If $a > 1$, then the equality $n = 2^{a2^N} + 1 = (2^{2^N})^a + 1^a$ exhibits n as the sum of two a^{th} powers, necessarily divisible by $2^{2^N} + 1$. Since n is assumed prime, we conclude that $a = 1$. Therefore $n = 2^{2^N} + 1$, and n is a Fermat prime.

We do not quite succeed in proving the converse at this point. If n is the Fermat prime $2^{2^N} + 1$, then the above argument shows that the degree of $\mathbb{Q}(e^{2\pi i/n})$ over $\mathbb{Q}$ is 2^{2^N}. However, we cannot yet conclude that $\mathbb{Q}(e^{2\pi i/n})$ can be built from $\mathbb{Q}$ by successively adjoining 2^N square roots, and thus the converse part of Theorem 9.24 is not immediately applicable. Once we have the theory of Galois groups in hand, we shall see that the existence of these intermediate extensions involving square roots is ensured, and then the constructibility follows.

6. Separable Extensions

The **Galois group** $\text{Gal}(\mathbb{K}/\mathbb{k})$ of a field extension $\mathbb{K}/\mathbb{k}$ is defined to be the set

$$\text{Gal}(\mathbb{K}/\mathbb{k}) = \{\mathbb{k} \text{ automorphisms of } \mathbb{K}\}$$

with composition as group operation. An instance of this group was introduced in the context of Example 9 of Section IV.1; in this example the field $\mathbb{k}$ was the field $\mathbb{Q}$ of rationals and the field $\mathbb{K}$ was a number field $\mathbb{Q}[\theta]$, where θ is algebraic over $\mathbb{Q}$. In studying $\text{Gal}(\mathbb{K}/\mathbb{k})$ in this chapter, we ordinarily assume that $\dim_\mathbb{k} \mathbb{K} < \infty$, but there will be instances where we do not want to make such an assumption.

Beginning in this section, we take up a study of Galois groups in general. We shall be interested in relationships between fields $\mathbb{L}$ with $\mathbb{k} \subseteq \mathbb{L} \subseteq \mathbb{K}$ and subgroups of $\text{Gal}(\mathbb{K}/\mathbb{k})$. If H is a subgroup of $\text{Gal}(\mathbb{K}/\mathbb{k})$, then

$$K^H = \{x \in \mathbb{K} \mid \varphi(x) = x \text{ for all } \varphi \in H\}$$

is a field called the **fixed field** of H; it provides an example of an intermediate field $\mathbb{L}$ and gives a hint of the relationships we shall investigate. We begin with some examples; in each case the base field $\mathbb{k}$ is the field $\mathbb{Q}$ of rationals.

EXAMPLES OF GALOIS GROUPS.

(1a) $\mathbb{K} = \mathbb{Q}(\sqrt{-1})$. If φ is in $\mathrm{Gal}(\mathbb{K}/\mathbb{Q})$, then we must have $\varphi|_{\mathbb{Q}} = 1$, and $\varphi(\sqrt{-1})$ must be a root of $X^2 + 1$. Thus $\varphi(\sqrt{-1}) = \pm\sqrt{-1}$. Since $\mathbb{Q}$ and $\sqrt{-1}$ generate $\mathbb{Q}(\sqrt{-1})$, there are at most two such φ's. On the other hand, $\mathbb{Q}(\sqrt{-1})$ and $\mathbb{Q}(-\sqrt{-1})$ are simple extensions of $\mathbb{Q}$ such that $\sqrt{-1}$ and $-\sqrt{-1}$ have the same minimal polynomial. Theorem 9.11 therefore produces a $\mathbb{Q}$ automorphism of $\mathbb{Q}(\sqrt{-1})$ with $\varphi(\sqrt{-1}) = -\sqrt{-1}$, namely complex conjugation. We conclude that $\mathrm{Gal}(\mathbb{K}/\mathbb{Q})$ has order 2, hence that $\mathrm{Gal}(\mathbb{K}/\mathbb{Q}) \cong C_2$.

(1b) $\mathbb{K} = \mathbb{Q}(\sqrt{2})$. The same argument applies as in Example 1a, and the conclusion is that $\mathrm{Gal}(\mathbb{K}/\mathbb{Q}) \cong C_2$. The nontrivial element of the Galois group carries $\sqrt{2}$ into $-\sqrt{2}$ and is different from complex conjugation.

(2) $\mathbb{K} = \mathbb{Q}(\sqrt[3]{2})$. If φ is in $\mathrm{Gal}(\mathbb{K}/\mathbb{Q})$, then $\varphi|_{\mathbb{Q}} = 1$, and $\varphi(\sqrt[3]{2})$ has to be a root of $X^3 - 2$. But $\mathbb{K}$ is a subfield of $\mathbb{R}$, and there is only one root of $X^3 - 2$ in $\mathbb{R}$. Hence $\varphi(\sqrt[3]{2}) = \sqrt[3]{2}$. Since $\mathbb{Q}$ and $\sqrt[3]{2}$ generate $\mathbb{Q}(\sqrt[3]{2})$ as a field, we see that $\varphi = 1$. We conclude that $\mathrm{Gal}(\mathbb{K}/\mathbb{Q})$ has order 1, i.e., is the trivial group.

(3) $\mathbb{K} = \mathbb{Q}(r)$, where r is a root of $X^3 - X - \frac{1}{3}$. Any φ in $\mathrm{Gal}(\mathbb{K}/\mathbb{Q})$ fixes $\mathbb{Q}$ and sends r to a root of $X^3 - X - \frac{1}{3}$. In Example 2 of splitting fields in Section 2, we saw that all three complex roots of $X^3 - X - \frac{1}{3}$ lie in $\mathbb{K}$. Arguing as in Example 1a, we see that $\mathrm{Gal}(\mathbb{K}/\mathbb{Q})$ has order 3, hence that $\mathrm{Gal}(\mathbb{K}/\mathbb{Q}) \cong C_3$.

(4) $\mathbb{K} = \mathbb{Q}(e^{2\pi i/17})$. According to Section 5, this is the field we need to consider in addressing the constructibility of a regular 17-gon. We saw in that section that $[\mathbb{K} : \mathbb{Q}] = 16$ and that the minimal polynomial of $e^{2\pi i/17}$ over $\mathbb{Q}$ is $X^{16} + X^{15} + \cdots + X + 1$. The other roots of the minimal polynomial in $\mathbb{C}$ are $e^{2\pi i l/17}$ for $2 \le l \le 16$, and these all lie in $\mathbb{K}$. Theorem 9.11 therefore gives us a $\mathbb{Q}$ automorphism φ_l of $\mathbb{K}$ sending $e^{2\pi i/17}$ into $e^{2\pi i l/17}$ for each l with $1 \le l \le 16$. Since $\mathbb{Q}$ and $e^{2\pi i/17}$ generate $\mathbb{K}$, a $\mathbb{Q}$ automorphism of $\mathbb{K}$ is completely determined by its effect on $e^{2\pi i/17}$. Thus the order of $\mathrm{Gal}(\mathbb{K}/\mathbb{Q})$ is 16. Let us determine the group structure. Since φ_l sends $e^{2\pi i/17}$ into $e^{2\pi i l/17}$, it sends $e^{2\pi i r/17} = (e^{2\pi i/17})^r$ into $(e^{2\pi i l/17})^r = e^{2\pi i l r/17}$. If we drop the exponential from the notation, we can think of φ_l as defined on the integers modulo 17, the formula being $\varphi_l(r) = rl \bmod 17$. From this viewpoint φ_l is an automorphism of the additive group of $\mathbb{F}_{17}$. Lemma 4.45 shows that the group of additive automorphisms of $\mathbb{F}_{17}$ is isomorphic to $\mathbb{F}_{17}^\times$, and it follows from Corollary 4.27 that $\mathrm{Gal}(\mathbb{K}/\mathbb{Q}) \cong C_{16}$. For our application of constructibility of a regular 17-gon, we would like to know whether the elements of $\mathbb{K}$ are constructible. Taking Theorem 9.24 into account, we therefore seek an intermediate field $\mathbb{L}$ of which $\mathbb{K}$ is a quadratic extension. Since we know that $\mathrm{Gal}(\mathbb{K}/\mathbb{Q})$ is cyclic, we can let $H \subseteq \mathrm{Gal}(\mathbb{K}/\mathbb{Q}) \cong C_{16}$ be the 2-element subgroup, and it is natural to try the fixed field $\mathbb{L} = \mathbb{K}^H$. To understand this fixed field, we need to understand the

isomorphism $\mathbb{F}_{17}^\times \cong C_{16}$ better. Modulo 17, we have

$$3^2 = 9, \quad 3^4 = -2^2, \quad 3^8 = 2^4 = -1, \quad 3^{16} = 1.$$

Consequently 3 is a generator of the cyclic group $\mathbb{F}_{17}^\times$. Then $H = \{3^8, 1\} = \{\pm 1\}$, and $L = \{x \in \mathbb{K} \mid \varphi_{-1}(x) = \varphi_{+1}(x) = x\}$. Since $\varphi_{-1}(e^{2\pi i r/17}) = e^{-2\pi i r/17} = \overline{e^{2\pi i r/17}}$ with the overbar indicating complex conjugation, we see that

$$\mathbb{L} = \mathbb{K}^H = \{x \in \mathbb{K} \mid x = \bar{x}\}.$$

It is not hard to check that indeed $[\mathbb{K} : \mathbb{L}] = 2$. Next we need a subfield $\mathbb{L}'$ of $\mathbb{L}$ with $[\mathbb{L} : \mathbb{L}'] = 2$. We try $\mathbb{L}' = \mathbb{K}^{H'}$ with H' equal to the 4-element cyclic subgroup of $\mathrm{Gal}(\mathbb{K}/\mathbb{Q})$. Here we have a harder time checking whether $\mathbb{L}$ is indeed a quadratic extension of $\mathbb{L}'$, but we shall see in Section 8 that it is.[4] We continue in this way, and ultimately we end up with the chain of subfields that exhibits the members of $\mathbb{K}$ as constructible.

We seek to formulate the kind of argument in the above examples as a general theorem. We have to rule out the bad behavior of $\mathbb{Q}(\sqrt[3]{2})$, where one root of the minimal polynomial lies in the field but others do not, and we shall do this by assuming that the extension field is a "normal" extension, in a sense to be defined in Section 7. In addition, our style of argument shows that we might run into trouble if our irreducible polynomials over $\Bbbk$ can have repeated roots in $\mathbb{K}$. We shall rule out this bad behavior by insisting that the extension be "separable," a condition that we introduce now. The extension will automatically be separable if $\mathbb{K}$ has characteristic 0.

For the remainder of this section, fix the base field $\Bbbk$. An irreducible polynomial $F(X)$ in $\Bbbk[X]$ is called **separable** if it splits into distinct degree-one factors in its splitting field, i.e., if

$$f(X) = a_n(X - x_1) \cdots (X - x_n) \qquad \text{with } x_i \neq x_j \text{ for } i \neq j.$$

Once this splitting into distinct degree-one factors occurs in the splitting field, it occurs in any larger field as well.

Lemma 9.26. A polynomial $F(X)$ in $\Bbbk[X]$ has no repeated roots in its splitting field $\mathbb{K}$ if and only if $\mathrm{GCD}(F, F') = 1$, where $F'(X)$ is the derivative of $F(X)$.

---

[4]Actually, Section 8 will point out how Corollary 9.36 in Section 7 already handles this step. In fact, Corollary 9.37 handles this step with no supplementary argument.

PROOF. The polynomial $F(X)$ has repeated roots in $\mathbb{K}$ if and only if $F(X)$ is divisible by $(X - r)^2$ for some $r \in \mathbb{K}$, if and only if some $r \in \mathbb{K}$ has $F(r) = F'(r) = 0$ (by Corollary 9.17), if and only if some $r \in \mathbb{K}$ has $(X - r)$ dividing $F(X)$ and also $F'(X)$ (by the Factor Theorem), if and only if some $r \in \mathbb{K}$ has $(X - r)$ dividing $GCD(F, F')$ when the GCD is computed in $\mathbb{K}$, if and only if $GCD(F, F') \neq 1$ when the GCD is computed in $\mathbb{K}$ (by unique factorization in $\mathbb{K}[X]$). However, the Euclidean algorithm calculates $GCD(F, F')$ without reference to the field, and the GCD is therefore the same when computed in $\mathbb{K}$ as it is when computed in $\mathbb{k}$. The lemma follows. □

Proposition 9.27. An irreducible polynomial $F(X)$ in $\mathbb{k}[X]$ is separable if and only if $F'(X) \neq 0$. In particular, every irreducible (necessarily nonconstant) polynomial is separable if $\mathbb{k}$ has characteristic 0.

PROOF. Since the polynomial $F(X)$ is irreducible and $GCD(F, F')$ divides $F(X)$, $GCD(F, F')$ equals 1 or $F(X)$ in all cases. If $F'(X) = 0$, then $GCD(F, F') = F(X)$, and Lemma 9.26 implies that $F(X)$ is not separable. Conversely if $F'(X) \neq 0$, then the facts that $GCD(F, F')$ divides $F'(X)$ and that $\deg F' < \deg F$ together imply that $GCD(F, F')$ cannot equal $F(X)$. So $GCD(F, F') = 1$, and Lemma 9.26 implies that $F(X)$ is separable. □

Fix an algebraic extension $\mathbb{K}$ of $\mathbb{k}$. We say that an element x of $\mathbb{K}$ is **separable** over $\mathbb{k}$ if the minimal polynomial of x over $\mathbb{k}$ is separable. We say that $\mathbb{K}$ is a **separable extension** of $\mathbb{k}$ if every x in $\mathbb{K}$ is separable over $\mathbb{k}$.

EXAMPLES OF SEPARABLE EXTENSIONS AND EXTENSIONS NOT SEPARABLE.

(1) In characteristic 0, every algebraic extension $\mathbb{K}$ of $\mathbb{k}$ is separable, by Proposition 9.27.

(2) Every algebraic extension $\mathbb{K}$ of a finite field $\mathbb{k}$ is separable. In fact, if x is in $\mathbb{K}$, then $[\mathbb{k}(x) : \mathbb{k}]$ is finite. Hence $\mathbb{k}(x)$ is a finite field. Then we may assume that $\mathbb{K}$ is a finite field, say of order $q = p^n$ with p prime. Since the multiplicative group $\mathbb{K}^\times$ has order $q - 1$, every nonzero element of $\mathbb{K}$ is a root of $X^{q-1} - 1$, and every root of $\mathbb{K}$ is therefore a root of $X^q - X$. The minimal polynomial $F(X)$ of x over $\mathbb{k}$ must then divide $X^q - X$. However, we know that $X^q - X$ splits over $\mathbb{K}$ and has no repeated roots. Thus $F(X)$ splits over $\mathbb{K}$ and has no repeated roots. Then $F(X)$ is separable over $\mathbb{k}$, and x is separable over $\mathbb{k}$.

(3) Let $\mathbb{k} = \mathbb{F}_p(x)$ be a transcendental extension of the finite field $\mathbb{F}_p$. Because this extension is transcendental, $X^p - x$ is irreducible over $\mathbb{k}$. Let $\mathbb{K}$ be the simple algebraic extension $\mathbb{k}[X]/(X^p - x)$, which we can write more simply as $\mathbb{k}(x^{1/p})$. The minimal polynomial of $x^{1/p}$ over $\mathbb{k}$ is $X^p - x$, and its derivative is $pX^{p-1} = 0$ since the derivative of the constant x is 0. By Proposition 9.27, $x^{1/p}$ is not separable over $\mathbb{k}$.

6. Separable Extensions

The way that separability enters considerations with Galois groups is through the following theorem, explicitly or implicitly. One of the corollaries of the theorem is that if $\mathbb{K}/\mathbb{k}$ is an algebraic extension, then the set of elements in $\mathbb{K}$ separable over $\mathbb{k}$ is a subfield of $\mathbb{K}$.

Theorem 9.28. Let $\mathbb{k} \subseteq \mathbb{L} \subseteq \mathbb{K}$ be an inclusion of fields such that $\mathbb{K}$ is a simple algebraic extension of $\mathbb{L}$ of the form $\mathbb{K} = \mathbb{L}(\alpha)$, let $\overline{\mathbb{K}}$ be an algebraic closure of $\mathbb{K}$, and let $M(X)$ be the minimal polynomial of α over $\mathbb{L}$. Then the number of field mappings of $\mathbb{K}$ into $\overline{\mathbb{K}}$ fixing $\mathbb{k}$ is the product of the number of distinct roots of $M(X)$ in $\overline{\mathbb{K}}$ by the number of field mappings of $\mathbb{L}$ into $\overline{\mathbb{K}}$ fixing $\mathbb{k}$.

REMARKS. An algebraic closure $\overline{\mathbb{K}}$ of $\mathbb{K}$ exists by Theorem 9.22. Because $\overline{\mathbb{K}}$ is known to exist, the present theorem reduces to Theorem 9.11 when $\mathbb{L} = \mathbb{k}$.

PROOF. Any field mapping $\varphi : \mathbb{K} \to \overline{\mathbb{K}}$ is uniquely determined by $\varphi\big|_{\mathbb{L}}$ and $\varphi(\alpha)$. If $\sigma = \varphi\big|_{\mathbb{L}}$, then the equality $M(\alpha) = 0$ implies that $M^\sigma(\varphi(\alpha)) = 0$, and thus $\varphi(\alpha)$ has to be a root of $M^\sigma(X)$. The number of distinct roots of $M^\sigma(X)$ in $\overline{\mathbb{K}}$ equals the number of distinct roots of $M(X)$ in $\overline{\mathbb{K}}$; hence the number of possibilities for $\varphi(\alpha)$ is at most the number of distinct roots of $M(X)$ in $\overline{\mathbb{K}}$. Consequently the number of such φ's fixing $\mathbb{k}$ is bounded above by the product of the number of distinct roots of $M(X)$ in $\overline{\mathbb{K}}$ times the number of field mappings σ of $\mathbb{L}$ into $\overline{\mathbb{K}}$ fixing $\mathbb{k}$.

For an inequality in the reverse direction, let $\sigma : \mathbb{L} \to \overline{\mathbb{K}}$ be any field mapping of $\mathbb{L}$ into $\overline{\mathbb{K}}$ fixing $\mathbb{k}$, put $\mathbb{L}' = \sigma(\mathbb{L})$, let x be any root of $M^\sigma(X)$, and form the subfield $\mathbb{L}'(x)$ of $\overline{\mathbb{K}}$. Theorem 9.11' shows that there exists a field isomorphism $\varphi : \mathbb{L}(\alpha) \to \mathbb{L}'(x)$ with $\varphi\big|_{\mathbb{L}} = \sigma$ and $\varphi(\alpha) = x$, and we can regard φ as a field mapping of $\mathbb{K}$ into $\overline{\mathbb{K}}$ fixing $\mathbb{k}$, extending σ, and having $\varphi(\alpha) = x$. Thus the number of field mappings $\varphi : \mathbb{K} \to \overline{\mathbb{k}}$ fixing $\mathbb{k}$ is bounded below by the product of the number of distinct roots of $M(X)$ in $\overline{\mathbb{K}}$ times the number of field homomorphisms σ of $\mathbb{L}$ into $\overline{\mathbb{K}}$ fixing $\mathbb{k}$. $\square$

Corollary 9.29. Let $\mathbb{K} = \mathbb{k}(\alpha_1, \ldots, \alpha_n)$ be a finite algebraic extension of the field $\mathbb{k}$, and let $\overline{\mathbb{K}}$ be an algebraic closure of $\mathbb{K}$. Then the number of field mappings of $\mathbb{K}$ into $\overline{\mathbb{K}}$ fixing $\mathbb{k}$ is $\leq [\mathbb{K} : \mathbb{k}]$. Moreover, the following conditions are equivalent:

(a) the number of field mappings of $\mathbb{K}$ into $\overline{\mathbb{K}}$ fixing $\mathbb{k}$ equals $[\mathbb{K} : \mathbb{k}]$,
(b) each α_j is separable over $\mathbb{k}(\alpha_1, \ldots, \alpha_{j-1})$ for $1 \leq j \leq n$,
(c) each α_j is separable over $\mathbb{k}$ for $1 \leq j \leq n$.

PROOF. For $1 \leq j \leq n$, let $M_j(X)$ be the minimal polynomial of α_j over $\mathbb{k}(\alpha_1, \ldots, \alpha_{j-1})$, let d_j be the degree of $M_j(X)$, and let s_j be the number of distinct roots of $M_j(X)$ in $\overline{\mathbb{K}}$. Then $s_j \leq d_j$ with equality for a particular j if and only if

α_j is separable over $\mathbb{k}(\alpha_1, \ldots, \alpha_{j-1})$, by definition. Also, $[\mathbb{K} : \mathbb{k}] = \prod_{j=1}^n d_j$ by Corollary 9.7, and the number of field mappings of $\mathbb{K}$ into $\overline{\mathbb{K}}$ fixing $\mathbb{k}$ is $\prod_{j=1}^n s_j$ by iterated application of Theorem 9.28. From these facts, the first conclusion of the corollary is immediate, and so is the equivalence of (a) and (b).

Condition (a) is independent of the order of enumeration of $\alpha_1, \ldots, \alpha_n$. Since we can always take any particular α_j to be first, we obtain the equivalence of (a) and (c). □

Corollary 9.30. Let $\mathbb{K} = \mathbb{k}(\alpha_1, \ldots, \alpha_n)$ be a finite algebraic extension of the field $\mathbb{k}$. If each α_j for $1 \leq j \leq n$ is separable over $\mathbb{k}$, then $\mathbb{K}/\mathbb{k}$ is a separable extension.

PROOF. Let β be in $\mathbb{K}$, We apply the equivalence of (a) and (c) in Corollary 9.29 once to the set of generators $\{\alpha_1, \ldots, \alpha_n\}$ and once to the set of generators $\{\beta, \alpha_1, \ldots, \alpha_n\}$, and the result is immediate. □

Corollary 9.31. If $\mathbb{K}/\mathbb{k}$ is an algebraic field extension, then the subset $\mathbb{L}$ of elements of $\mathbb{K}$ that are separable over $\mathbb{k}$ is a subfield of $\mathbb{K}$.

PROOF. If α and β are given in $\mathbb{L}$, we apply Corollary 9.30 to the extension $\mathbb{k}(\alpha, \beta)$ of $\mathbb{k}$ to see that $\mathbb{L}$ contains the subfield generated by $\mathbb{k}$ and the elements α and β. □

Proposition 9.32. If $\mathbb{K}/\mathbb{k}$ is a separable algebraic extension and if $\mathbb{L}$ is a field with $\mathbb{k} \subseteq \mathbb{L} \subseteq \mathbb{K}$, then $\mathbb{K}$ is separable over $\mathbb{L}$, and $\mathbb{L}$ is separable over $\mathbb{k}$.

PROOF. The separability assertion about $\mathbb{L}/\mathbb{k}$ says the same thing about elements of $\mathbb{L}$ that separability of $\mathbb{K}/\mathbb{k}$ says about those same elements, and it is therefore immediate that $\mathbb{L}/\mathbb{k}$ is separable.

Next let us consider $\mathbb{K}/\mathbb{L}$. If x is in $\mathbb{K}$, let $F(X)$ be its minimal polynomial over $\mathbb{k}$, and let $G(X)$ be its minimal polynomial over $\mathbb{L}$. Since $F(X)$ is in $\mathbb{L}[X]$ and $F(x) = G(x) = 0$, $G(X)$ divides $F(X)$. Since $\mathbb{K}/\mathbb{k}$ is separable, $F(X)$ splits into distinct degree-one factors in its splitting field $\mathbb{F}$. The field $\mathbb{F}$ contains a splitting field of $G(X)$, and thus the degree-one factors of $G(X)$ in $\mathbb{F}[X]$ are a subset of the degree-one factors of $F(X)$ in $\mathbb{F}[X]$. There are no repeated factors for $F(X)$, and there can be no repeated factors for $G(X)$. Thus x is separable over $\mathbb{L}$, and $\mathbb{K}/\mathbb{L}$ is a separable extension. □

In studying Galois groups, we shall be chiefly interested in the following situation in Corollary 9.29: $\mathbb{K}$ is an algebraic field extension $\mathbb{K} = \mathbb{k}(\alpha_1, \ldots, \alpha_n)$ of $\mathbb{k}$ for which every field mapping of $\mathbb{K}$ into an algebraic closure that fixes $\mathbb{k}$ actually carries $\mathbb{K}$ into itself. We seek conditions under which this situation arises,

6. Separable Extensions

and then we mine the consequences. As we did in the study begun in Theorem 9.28, we begin with the case of a simple algebraic extension.

Let $\mathbb{K} = \mathbb{k}(\gamma)$ be a simple algebraic extension of $\mathbb{k}$, and let $F(X)$ be the minimal polynomial of γ over $\mathbb{k}$. Any member φ of the Galois group $\text{Gal}(\mathbb{K}/\mathbb{k})$ carries γ to another root γ' of $F(X)$, and φ is uniquely determined by γ' since $\mathbb{k}$ and γ generate the field $\mathbb{K}$. An element φ of $\text{Gal}(\mathbb{K}/\mathbb{k})$ carrying γ to γ' can exist only if γ' is in $\mathbb{K}$. If γ' is in $\mathbb{K}$, then $\mathbb{k}(\gamma) \supseteq \mathbb{k}(\gamma')$, and the equal finite dimensionality of $\mathbb{k}(\gamma)$ and $\mathbb{k}(\gamma')$ forces $\mathbb{k}(\gamma) = \mathbb{k}(\gamma')$. In other words, if γ' is in $\mathbb{K}$, then the unique $\mathbb{k}$ isomorphism $\mathbb{k}(\gamma) \to \mathbb{k}(\gamma')$ of Theorems 9.10 and 9.11 carrying γ to γ' is a member of $\text{Gal}(\mathbb{K}/\mathbb{k})$. Making a count of what happens to all the elements γ', we see that we have proved the following.

Proposition 9.33. Let $\mathbb{K} = \mathbb{k}(\gamma)$ be a simple algebraic extension of $\mathbb{k}$, and let $F(X)$ be the minimal polynomial of γ. Then

$$|\text{Gal}(\mathbb{K}/\mathbb{k})| \leq [\mathbb{K} : \mathbb{k}]$$

with equality if and only if $F(X)$ is a separable polynomial and $\mathbb{K}$ is a splitting field of $F(X)$ over $\mathbb{k}$.

EXAMPLE. For $\mathbb{K} = \mathbb{Q}(\sqrt[3]{2})$ with minimal polynomial $F(X)$, we know that $F(X)$ does not split in $\mathbb{K}$; the nonreal roots of $F(X)$ do not lie in $\mathbb{K}$. Proposition 9.33 gives us $|\text{Gal}(\mathbb{K}/\mathbb{Q})| < [\mathbb{K} : \mathbb{Q}] = 3$, and a glance at the argument preceding Proposition 9.33 shows that $|\text{Gal}(\mathbb{K}/\mathbb{Q})|$ has to be 1.

It is possible to investigate the case of several generators directly, but it is more illuminating to reduce it to the case of a single generator as in Proposition 9.33. The tool for doing so is the following important theorem.

Theorem 9.34 (Theorem of the Primitive Element). Let $\mathbb{K}/\mathbb{k}$ be a separable algebraic extension with $[\mathbb{K} : \mathbb{k}] < \infty$. Then there exists an element γ in $\mathbb{K}$ such that $\mathbb{K} = \mathbb{k}(\gamma)$.

PROOF. We can write $\mathbb{K} = \mathbb{k}(x_1, \ldots, x_n)$, and we proceed by induction on n, the case $n = 1$ being trivial. For general n, let $\mathbb{L} = \mathbb{k}(x_1, \ldots, x_{n-1})$, so that $\mathbb{K} = \mathbb{L}(x_n)$. By the inductive hypothesis, $\mathbb{L}$ is of the form $\mathbb{L} = \mathbb{k}(\alpha)$ for some α in $\mathbb{K}$, and thus $\mathbb{K} = \mathbb{k}(\alpha, x_n)$. Changing notation, we see that it is enough to prove that whenever $\mathbb{K}$ is a separable algebraic extension of the form $\mathbb{K} = \mathbb{k}(\alpha, \beta)$, then $\mathbb{K}$ is of the form $\mathbb{K} = \mathbb{k}(\gamma)$ for some γ. We shall show this for some γ of the form $\gamma = \beta + c\alpha$ with c in $\mathbb{k}$. Because every finite extension of a finite field is separable (by Example 2 of separable extensions), we may assume that $\mathbb{k}$ is an infinite field.

Let $F(X)$ and $G(X)$ be the minimal polynomials of α and β over $\Bbbk$, and let $\mathbb{K}'$ be an extension in which $F(X)G(X)$ splits, i.e., in which $F(X)$ and $G(X)$ both split. Let $\alpha_1 = \alpha, \alpha_2, \ldots, \alpha_m$ and $\beta_1 = \beta, \beta_2, \ldots, \beta_n$ be the roots of $F(X)$ and $G(X)$ in $\mathbb{K}'$, in each case necessarily distinct by definition of separability of α and β. Define $\mathbb{L} = \Bbbk(\gamma)$ with $\gamma = \beta + c\alpha$, where c is a member of $\Bbbk$ to be specified. For suitable c, we shall show that α is in $\mathbb{L}$. Then $\beta = \gamma - c\alpha$ must be in $\mathbb{L}$, and we obtain $\mathbb{K} \subseteq \mathbb{L}$. Since γ is in $\mathbb{K}$, the reverse inclusion is built into the construction, and thus we will have $\mathbb{K} = \mathbb{L}$.

We shall compute the minimal polynomial of α over $\mathbb{L}$. We know that α is a root of $F(X)$, and we put $H(X) = G(\gamma - cX)$. Then $H(X)$ is in $\mathbb{L}[X] \subseteq \mathbb{K}'[X]$, and $G(\beta) = 0$ implies $H(\alpha) = 0$. Therefore $X - \alpha$ divides both $F(X)$ and $H(X)$ in the ring $\mathbb{K}'[X]$. Let us determine $\mathrm{GCD}(F, H)$ in $\mathbb{K}'[X]$. The separability of α says that $X - \alpha$ divides $F(X)$ only once. Since $F(X)$ splits in $\mathbb{K}'[x]$, any other prime divisor of $\mathrm{GCD}(F, H)$ in $\mathbb{K}'[X]$ has to be of the form $X - \alpha_j$ with $j \neq 1$. The definition of $H(X)$ gives $H(\alpha_j) = G(\gamma - c\alpha_j)$. If $G(\gamma - c\alpha_j) = 0$, then $\gamma - c\alpha_j = \beta_i$ for some i, with the consequence that $\beta + c\alpha - c\alpha_j = \beta_i$ and $c = (\beta_i - \beta)(\alpha - \alpha_j)^{-1}$. Since $\Bbbk$ is an infinite field, some choice of c in $\mathbb{K}$ makes $\mathrm{GCD}(F, H) = X - \alpha$ in $\mathbb{K}'[X]$. Then $\mathrm{GCD}(F, H) = X - \alpha$, up to a scalar factor, in $\mathbb{L}[X]$ since $F(X)$ and $H(X)$ are in $\mathbb{L}[X]$ and since the GCD can be computed without reference to the field containing both elements. The ratio of the constant term to the coefficient of X has to be in $\mathbb{L}$ independently of the scalar factor multiplying $X - \alpha$, and therefore α is in $\mathbb{L}$. This completes the proof. $\square$

7. Normal Extensions

Proposition 9.33 suggests that the failure of equality to hold in the inequality $|\mathrm{Gal}(\mathbb{K}/\Bbbk)| \leq [\mathbb{K} : \Bbbk]$ has something to do with the failure of polynomials over $\Bbbk$ to split fully in $\mathbb{K}$ once they have at least one root in $\mathbb{K}$. In the case of the extension $\mathbb{Q}(\sqrt[3]{2})/\mathbb{Q}$, where equality fails, the Galois group is trivial and therefore gives us no information about the extension. Thus it makes sense to regard the failure of equality to hold as an undesirable situation. Accordingly, we make a definition, choosing among several equivalent conditions one that is easy to apply.

A finite separable[5] algebraic extension $\mathbb{K}$ of a field $\Bbbk$ is said to be **normal** over $\Bbbk$ if $\mathbb{K}$ is the splitting field of *some* $F(X)$ in $\Bbbk[X]$. The following proposition asserts some powerful consequences of this condition.

---

[5] A more advanced treatment might proceed without the assumption of separability for as long as possible. But it is unnecessary to do so in this volume, and the assumption of separability makes the Theorem of the Primitive Element available to us.

Proposition 9.35. Let $\mathbb{K}$ be a finite separable algebraic extension of a field $\mathbb{k}$, so that $|\operatorname{Gal}(\mathbb{K}/\mathbb{k})| \leq [\mathbb{K}:\mathbb{k}]$. Then the following are equivalent.
- (a) $\mathbb{K}$ is the splitting field of *some* $F(X)$ in $\mathbb{k}[X]$, i.e., $\mathbb{K}$ is normal over $\mathbb{k}$,
- (b) *every* irreducible polynomial $F(X)$ in $\mathbb{k}[X]$ with a root in $\mathbb{K}$ splits in $\mathbb{K}$, i.e., $\mathbb{K}$ contains a splitting field for each such $F(X)$,
- (c) $|\operatorname{Gal}(\mathbb{K}/\mathbb{k})| = [\mathbb{K}:\mathbb{k}]$,
- (d) $\mathbb{k} = \mathbb{K}^G$ for $G = \operatorname{Gal}(\mathbb{K}/\mathbb{k})$.

REMARKS. We prove that (a) and (c) are equivalent, that the equivalent (a) and (c) imply (d), that (d) implies (b), and that (b) implies (a).

PROOF. By separability and Theorem 9.34 we can write $\mathbb{K} = \mathbb{k}(\gamma)$ throughout the proof for some γ in $\mathbb{K}$. Let $M(X)$ be the minimal polynomial of γ over $\mathbb{k}$.

Suppose (a) holds. We prove (c). Write $\mathbb{K} = \mathbb{k}(\gamma) = \mathbb{k}(\alpha_1, \ldots, \alpha_n)$, where $\alpha_1, \ldots, \alpha_n$ are the roots of some $F(X)$ in $\mathbb{k}[X]$ that splits over $\mathbb{k}$. We may assume that $F(X)$ has no repeated prime factors and therefore, by separability of $\mathbb{K}/\mathbb{k}$, that $\alpha_1, \ldots, \alpha_n$ are distinct. Then $\gamma = H(\alpha_1, \ldots, \alpha_n)$ for some H in $\mathbb{k}[X_1, \ldots, X_n]$. Proposition 9.33 will establish (c) if we show that $M(X)$ splits over $\mathbb{K}$. Let $\mathbb{K}' \supseteq \mathbb{K}$ be a finite extension in which $M(X)$ splits, and let γ' be a root of $M(X)$ in $\mathbb{K}'$. We are to show that γ' is in $\mathbb{K}$. Theorem 9.11′ produces a $\mathbb{k}$ isomorphism $\varphi : \mathbb{k}(\gamma) \to \mathbb{k}(\gamma')$ with $\varphi(\gamma) = \gamma'$. Since $\varphi(\alpha_i)$ is a root of $F(X)$ for each i, $\varphi(\alpha_i) = \alpha_{j(i)}$ for some $j = j(i)$ that is unique since $\alpha_1, \ldots, \alpha_n$ are distinct. Thus φ permutes $\{\alpha_1, \ldots, \alpha_n\}$, and

$$\gamma' = \varphi(\gamma) = \varphi(H(\alpha_1, \ldots, \alpha_n)) = H_1(\alpha_1, \ldots, \alpha_n)$$

for some H_1 in $\mathbb{k}[X_1, \ldots, X_n]$. Therefore γ' is in $\mathbb{k}(\alpha_1, \ldots, \alpha_n) = \mathbb{K}$. This proves (c).

Suppose (c) holds. We prove (a). Proposition 9.33, in the presence of condition (c) and the given separability of $\mathbb{K}/\mathbb{k}$, implies that $\mathbb{K}$ is a splitting field of $M(X)$ over $\mathbb{k}$. Thus (a) holds.

Suppose (a) holds. We prove (d). Let $\mathbb{k}' = \mathbb{K}^G$. Since every member of $\operatorname{Gal}(\mathbb{K}/\mathbb{k})$ fixes $\mathbb{k}'$, $\operatorname{Gal}(\mathbb{K}/\mathbb{k}) \subseteq \operatorname{Gal}(\mathbb{K}/\mathbb{k}')$. Meanwhile, (a) for $\mathbb{K}/\mathbb{k}$ implies (a) for $\mathbb{K}/\mathbb{k}'$, and $\mathbb{K}$ is separable over $\mathbb{k}'$ by Proposition 9.32. Since (a) implies (c), (c) holds for both $\mathbb{k}'$ and $\mathbb{k}$, and we have

$$[\mathbb{K}:\mathbb{k}] = |\operatorname{Gal}(\mathbb{K}/\mathbb{k})| \leq |\operatorname{Gal}(\mathbb{K}/\mathbb{k}')| = [\mathbb{K}:\mathbb{k}'].$$

Since $\mathbb{k}' \supseteq \mathbb{k}$, the inequality of dimensions implies that $\mathbb{k}' = \mathbb{k}$. Thus (d) holds.

Suppose (d) holds. We prove (b). Let $F(X)$ be an irreducible polynomial in $\mathbb{k}[X]$ having a root r in $\mathbb{K}$. The polynomial $F(X)$ is necessarily the minimal polynomial of r over $\mathbb{k}$. Enumerate $\{\varphi(r) \mid \varphi \in \operatorname{Gal}(\mathbb{K}/\mathbb{k})\}$ as $r_1, \ldots, r_n$, with any possible repetitions included. If $J(X)$ is defined to be $\prod_{i=1}^{n}(X - r_i)$, then expansion of the product gives

$$J(X) = X^{|G|} - \Big(\sum_i r_i\Big) X^{|G|-1} + \Big(\sum_{i<j} r_i r_j\Big) X^{|G|-2} - \cdots \pm \Big(\prod_i r_i\Big).$$

Each member φ of $\mathrm{Gal}(\mathbb{K}/\mathbb{k})$ carries each coefficient of $J(X)$ into itself since φ permutes the elements r_i. Since $\mathbb{K}^G = \mathbb{k}$, we see therefore that $J(X)$ is in $\mathbb{k}[X]$. Since $J(r) = 0$ and $F(X)$ is the minimal polynomial of r, $F(X)$ divides $J(X)$. Over $\mathbb{K}$, $J(X)$ splits because of its definition. By unique factorization, $F(X)$ must split too. Thus (b) holds.

Finally if (b) holds, then $M(X)$, being irreducible over $\mathbb{k}$ and having γ as a root in $\mathbb{K}$, splits in $\mathbb{K}$. Thus $\mathbb{K}$ is a splitting field for $M(X)$ over $\mathbb{k}$, and (a) holds.
□

Corollary 9.36. If $\mathbb{K}$ is a finite normal separable extension of $\mathbb{k}$ and if $\mathbb{L}$ is a field with $\mathbb{k} \subseteq \mathbb{L} \subseteq \mathbb{K}$, then $\mathbb{K}$ is a finite normal separable extension of $\mathbb{L}$, and the subgroup $H = \mathrm{Gal}(\mathbb{K}/\mathbb{L})$ of $\mathrm{Gal}(\mathbb{K}/\mathbb{k})$ has

$$\boxed{|H| \cdot [\mathbb{L} : \mathbb{k}] = |\mathrm{Gal}(\mathbb{K}/\mathbb{k})|.}$$

PROOF. The field $\mathbb{K}$ is a separable extension of the intermediate field $\mathbb{L}$ by Proposition 9.32, and it is a normal extension by Proposition 9.35a. Therefore Proposition 9.35c gives $|\mathrm{Gal}(\mathbb{K}/\mathbb{L})| = [\mathbb{K} : \mathbb{L}]$, and we have

$$|H| \cdot [\mathbb{L} : \mathbb{k}] = |\mathrm{Gal}(\mathbb{K}/\mathbb{L})| \cdot [\mathbb{L} : \mathbb{k}] = [\mathbb{K} : \mathbb{L}] \cdot [\mathbb{L} : \mathbb{k}] = [\mathbb{K} : \mathbb{k}] = |\mathrm{Gal}(\mathbb{K}/\mathbb{k})|,$$

the last two equalities holding by Corollary 9.7 and Proposition 9.35c.
□

Corollary 9.37. Let $\mathbb{K}/\mathbb{k}$ be a separable algebraic extension, and suppose that H is a finite subgroup of $\mathrm{Gal}(\mathbb{K}/\mathbb{k})$. Then $\mathbb{K}/\mathbb{K}^H$ is a finite normal separable extension, H is the subgroup $\mathrm{Gal}(\mathbb{K}/\mathbb{K}^H)$ of $\mathrm{Gal}(\mathbb{K}/\mathbb{k})$, and $[\mathbb{K} : \mathbb{K}^H] = |H|$.

PROOF. Proposition 9.32 shows that $\mathbb{K}$ is separable over $\mathbb{K}^H$. For an arbitrary element x of $\mathbb{K}$, form the polynomial in $\mathbb{K}[X]$ given by

$$F(X) = \prod_{\varphi \in H} (X - \varphi(x)).$$

If φ_0 is in H, then F^{φ_0} is given by replacing each $\varphi(x)$ by $\varphi_0 \varphi(x)$, and the product is unchanged. Therefore $F(X) = F^{\varphi_0}(X)$, and $F(X)$ is in $\mathbb{K}^H[X]$. Thus $F(X)$ is a polynomial in $\mathbb{K}^H[X]$ that has x as a root and splits in $\mathbb{K}$. The minimal polynomial $M(X)$ of x over $\mathbb{K}^H$ must divide $F(X)$, and it too has x as a root.

By unique factorization in $\mathbb{K}[X]$, $M(X)$ must split in $\mathbb{K}$. Thus $\mathbb{K}/\mathbb{K}^H$ will be a normal extension if it is shown that $[\mathbb{K} : \mathbb{K}^H] < \infty$.

The element x has $[\mathbb{K}^H(x) : \mathbb{K}^H] = \deg M(X) \leq \deg F(X) = |H|$, and the claim is that $[\mathbb{K} : \mathbb{K}^H] \leq |H|$. Assuming the contrary, we would at some point have an inequality $[\mathbb{K}^H(x_1, \ldots, x_n) : \mathbb{K}^H] > |H|$ because every element of $\mathbb{K}$ is algebraic over $\Bbbk$. By the Theorem of the Primitive Element (Theorem 9.34), $\mathbb{K}^H(x_1, \ldots, x_n) = \mathbb{K}^H(z)$ for some element z, and therefore $[\mathbb{K}^H(x_1, \ldots, x_n) : \mathbb{K}^H] = [\mathbb{K}^H(z) : \mathbb{K}^H] \leq |H|$, contradiction. We conclude that $[\mathbb{K} : \mathbb{K}^H] \leq |H|$. From the previous paragraph, $\mathbb{K}/\mathbb{K}^H$ is a finite separable normal extension.

The definition of $\mathbb{K}^H$ shows that $H \subseteq \text{Gal}(\mathbb{K}/\mathbb{K}^H)$, and Proposition 9.35c gives $|\text{Gal}(\mathbb{K}/\mathbb{K}^H)| = [\mathbb{K} : \mathbb{K}^H]$. Putting these facts together with the inequality $[\mathbb{K} : \mathbb{K}^H] \leq |H|$ from the previous paragraph, we have

$$|H| \leq |\text{Gal}(\mathbb{K}/\mathbb{K}^H)| = [\mathbb{K} : \mathbb{K}^H] \leq |H|$$

with equality on the left only if $H = \text{Gal}(\mathbb{K}/\mathbb{K}^H)$. Equality must hold throughout the displayed line since the ends are equal, and therefore $H = \text{Gal}(\mathbb{K}/\mathbb{K}^H)$. □

8. Fundamental Theorem of Galois Theory

We are now in a position to obtain the main result in Galois theory.

Theorem 9.38 (Fundamental Theorem of Galois Theory). If $\mathbb{K}$ is a finite normal separable extension of $\Bbbk$, then there is a one-one inclusion-reversing correspondence between the subgroups H of $\text{Gal}(\mathbb{K}/\Bbbk)$ and the subfields $\mathbb{L}$ of $\mathbb{K}$ that contain $\Bbbk$, corresponding elements H and $\mathbb{L}$ being given by

$$\mathbb{L} = \mathbb{K}^H \qquad \text{and} \qquad H = \text{Gal}(\mathbb{K}/\mathbb{L}).$$

The effect of the theorem is to take an extremely difficult problem, namely finding intermediate fields, and reduce it to a problem that is merely difficult, namely finding the Galois group. For example the finiteness of $\text{Gal}(\mathbb{K}/\Bbbk)$ implies that there are only finitely many subgroups of $\text{Gal}(\mathbb{K}/\Bbbk)$, and the theorem therefore implies that there are only finitely many intermediate fields; this finiteness of the number of intermediate fields is not so obvious without the theorem.

As a reminder of the availability of Theorem 9.38, Proposition 9.35, and Corollary 9.36, it is customary to refer to a finite normal separable extension as a **finite Galois extension**.

Before coming to the proof of the theorem, let us examine what the theorem says for the examples in Section 6. In each case the field $\Bbbk$ is the field $\mathbb{Q}$ of rationals. The extensions are separable because the characteristic is 0.

EXAMPLES.

(1a) $\mathbb{K} = \mathbb{Q}(\sqrt{-1})$. This is a splitting field for $X^2 + 1$. Proposition 9.33 gives $|\mathrm{Gal}(\mathbb{K}/\mathbb{Q})| = [\mathbb{K} : \mathbb{Q}] = 2$. Thus $\mathrm{Gal}(\mathbb{K}/\mathbb{Q}) \cong C_2$. There are no nontrivial subgroups, and there are consequently no intermediate fields. We knew this already since there cannot be any intermediate $\mathbb{Q}$ vector spaces between $\mathbb{Q}$ and $\mathbb{K}$. Thus the theorem tells us nothing new.

(1b) $\mathbb{K} = \mathbb{Q}(\sqrt{2})$. Similar remarks apply.

(2) $\mathbb{K} = \mathbb{Q}(\sqrt[3]{2})$. This extension is not normal, and the theorem does not apply to $\mathbb{K}$. If we adjoin r to $\mathbb{K}$ with $r^2 + (\sqrt[3]{2})r + (\sqrt[3]{2})^2 = 0$, we obtain a splitting field $\mathbb{K}'$ for $X^3 - 2$ over $\mathbb{Q}$. Then $\mathbb{K}'$ is a normal extension of $\mathbb{Q}$, and the theorem applies. Since each element of $\mathrm{Gal}(\mathbb{K}'/\mathbb{Q})$ permutes the three roots of $X^3 - 2$ and is determined by its effect on these roots, $\mathrm{Gal}(\mathbb{K}'/\mathbb{Q})$ is isomorphic to a subgroup of the symmetric group $\mathfrak{S}_3$. The Galois group $\mathrm{Gal}(\mathbb{K}'/\mathbb{Q})$ has order $[\mathbb{K}' : \mathbb{Q}] = 6$ and hence is isomorphic to the whole symmetric group $\mathfrak{S}_3$. The group $\mathfrak{S}_3$ has three subgroups of order 2 and one subgroup of order 3. Therefore $\mathbb{K}$ has three intermediate fields of degree 3 and one of degree 2. The intermediate fields of degree 3 are the three fields generated by $\mathbb{Q}$ and one of the three roots of $X^3 - 2$. The intermediate field of degree 2 corresponds to the alternating subgroup of order 3 and is the subfield generated by $\mathbb{Q}$ and the cube roots of 1. It is a splitting field for $X^2 + X + 1$ over $\mathbb{Q}$.

(3) $\mathbb{K} = \mathbb{Q}(r)$, where r is a root of $X^3 - X - \frac{1}{3}$. We know from Section 2 that $X^3 - X - \frac{1}{3}$ is irreducible over $\mathbb{Q}$ and splits in $\mathbb{K}$, and $\mathbb{K}$ by definition is therefore normal. Proposition 9.33 tells us that $\mathrm{Gal}(\mathbb{K}/\mathbb{Q})$ has order 3 and hence is isomorphic to C_3. There are no nontrivial subgroups, and Theorem 9.38 tells us that there are no intermediate fields. We could have seen in more elementary fashion that there are no intermediate fields by using Corollary 9.7, since the corollary tells us that the degree of an intermediate field would have to divide 3.

(4) $\mathbb{K} = \mathbb{Q}(e^{2\pi i/17})$. We have seen that $[\mathbb{K} : \mathbb{Q}] = 16$ and that $\mathrm{Gal}(\mathbb{K}/\mathbb{Q}) \cong \mathbb{F}_{17}^\times \cong C_{16}$. Let c be a generator of the cyclic Galois group. Let $H_2 = \{1, c^8\}$, $H_4 = \{1, c^4, c^8, c^{12}\}$, and $H_8 = \{1, c^2, c^4, c^6, c^8, c^{10}, c^{12}, c^{14}\}$. Then put

$$\mathbb{L}_2 = \mathbb{K}^{H_2}, \qquad \mathbb{L}_4 = \mathbb{K}^{H_4}, \qquad \mathbb{L}_8 = \mathbb{K}^{H_8}.$$

The inclusions among our subgroups are

$$\{1\} \subseteq H_2 \subseteq H_4 \subseteq H_8 \subseteq \mathrm{Gal}(\mathbb{K}/\mathbb{Q}),$$

and the theorem says that the correspondence with intermediate fields reverses inclusions. Then we have

$$\mathbb{K} \supseteq \mathbb{L}_2 \supseteq \mathbb{L}_4 \supseteq \mathbb{L}_8 \supseteq \mathbb{Q}.$$

8. Fundamental Theorem of Galois Theory   481

Applying Corollary 9.36, we see that each of these subfields is a quadratic extension of the next-smaller one. Theorem 9.24 says that the members of $\mathbb{K}$ are therefore constructible with straightedge and compass. Consequently a regular 17-gon is constructible with straightedge and compass. The constructibility or nonconstructibility of regular n-gons for general n will be settled in similar fashion in the next section. In Section 12 we return to the question of using Galois theory to guide us through the actual steps of the construction when it is possible.

PROOF OF THEOREM 9.38. The function $\mathbb{L} \mapsto \text{Gal}(\mathbb{K}/\mathbb{L})$ has domain the set of all intermediate fields and range the set of all subgroups of $\text{Gal}(\mathbb{K}/\Bbbk)$, since an element in $\text{Gal}(\mathbb{K}/\mathbb{L})$ is necessarily in $\text{Gal}(\mathbb{K}/\Bbbk)$. Each such extension $\mathbb{K}/\mathbb{L}$ is separable by Proposition 9.32 and is normal by Proposition 9.35a. Thus Proposition 9.35d applies to each $\mathbb{K}/\mathbb{L}$ and shows that $\mathbb{L} = \mathbb{K}^{\text{Gal}(\mathbb{K}/\mathbb{L})}$. Consequently the function $\mathbb{L} \mapsto \text{Gal}(\mathbb{K}/\mathbb{L})$ is one-one. If H is a subgroup of $\text{Gal}(\mathbb{K}/\Bbbk)$, then Corollary 9.37 shows that $\mathbb{L} = \mathbb{K}^H$ is an intermediate field for which $H = \text{Gal}(\mathbb{K}/\mathbb{L})$, and therefore the function $\mathbb{L} \mapsto \text{Gal}(\mathbb{K}/\mathbb{L})$ is onto.

It is immediate from the definition of Galois group that $\mathbb{L}_1 \subseteq \mathbb{L}_2$ implies $\text{Gal}(\mathbb{K}/\mathbb{L}_1) \supseteq \text{Gal}(\mathbb{K}/\mathbb{L}_2)$, and it is immediate from the formula $\mathbb{L} = \mathbb{K}^{\text{Gal}(\mathbb{K}/\mathbb{L})}$ that $\text{Gal}(\mathbb{K}/\mathbb{L}_1) \supseteq \text{Gal}(\mathbb{K}/\mathbb{L}_2)$ implies $\mathbb{L}_1 \subseteq \mathbb{L}_2$. This completes the proof. □

Corollary 9.39. If $\mathbb{K}$ is a finite Galois extension of $\Bbbk$ and if $\mathbb{L}$ is a subfield of $\mathbb{K}$ that contains $\Bbbk$, then $\mathbb{L}$ is a normal extension of $\Bbbk$ if and only if $\text{Gal}(\mathbb{K}/\mathbb{L})$ is a normal subgroup of $\text{Gal}(\mathbb{K}/\Bbbk)$. In this case, the map $\text{Gal}(\mathbb{K}/\Bbbk) \to \text{Gal}(\mathbb{L}/\Bbbk)$ given by restriction from $\mathbb{K}$ to $\mathbb{L}$ is a group homomorphism that descends to a group isomorphism

$$\text{Gal}(\mathbb{K}/\Bbbk)\big/\text{Gal}(\mathbb{K}/\mathbb{L}) \cong \text{Gal}(\mathbb{L}/\Bbbk).$$

PROOF. Let $\mathbb{L}$ correspond to $H = \text{Gal}(\mathbb{K}/\mathbb{L})$ in Theorem 9.38, so that $\mathbb{L} = \mathbb{K}^H$. If φ is in $\text{Gal}(\mathbb{K}/\Bbbk)$, then

$$\begin{aligned}\mathbb{K}^{\varphi H \varphi^{-1}} &= \{k \in \mathbb{K} \mid \varphi h \varphi^{-1}(k) = k \text{ for all } h \in H\} \\ &= \{\varphi(k') \in \mathbb{K} \mid \varphi h(k') = \varphi(k') \text{ for all } h \in H\} \\ &= \{\varphi(k') \in \mathbb{K} \mid h(k') = k' \text{ for all } h \in H\} \\ &= \varphi(\mathbb{K}^H) = \varphi(\mathbb{L}).\end{aligned}$$

Since the correspondence of Theorem 9.38 is one-one onto, $\varphi H \varphi^{-1} = H$ if and only if $\varphi(\mathbb{L}) = \mathbb{L}$. Therefore H is a normal subgroup of $\text{Gal}(\mathbb{K}/\Bbbk)$ if and only if $\varphi(\mathbb{L}) = \mathbb{L}$ for all $\varphi \in \text{Gal}(\mathbb{K}/\Bbbk)$.

Now suppose that H is a normal subgroup of $\text{Gal}(\mathbb{K}/\Bbbk)$. We have just seen that $\varphi(\mathbb{L}) = \mathbb{L}$ for all $\varphi \in \text{Gal}(\mathbb{K}/\Bbbk)$. Then each φ defines by restriction a member

$\bar\varphi = \varphi|_\mathbb{L}$ of $\mathrm{Gal}(\mathbb{L}/\mathbb{k})$, and $\varphi \mapsto \bar\varphi$ is certainly a group homomorphism. The kernel of $\varphi \mapsto \bar\varphi$ is the subgroup of $\mathrm{Gal}(\mathbb{K}/\mathbb{k})$ given by

$$\{\varphi \in \mathrm{Gal}(\mathbb{K}/\mathbb{k}) \mid \varphi|_\mathbb{L} = 1\},$$

and this is just $\mathrm{Gal}(\mathbb{K}/\mathbb{L})$. Thus $\varphi \mapsto \bar\varphi$ descends to a one-one homomorphism of $\mathrm{Gal}(\mathbb{K}/\mathbb{k}) / \mathrm{Gal}(\mathbb{K}/\mathbb{L})$ into $\mathrm{Gal}(\mathbb{L}/\mathbb{k})$, and we have

$$|\mathrm{Gal}(\mathbb{K}/\mathbb{k})|/|\mathrm{Gal}(\mathbb{K}/\mathbb{L})| \leq |\mathrm{Gal}(\mathbb{L}/\mathbb{k})|.$$

We make use of Corollary 9.7 relating degrees of extensions. Applying Proposition 9.35c to $\mathbb{K}/\mathbb{k}$ and $\mathbb{K}/\mathbb{L}$, as well as Proposition 9.33 to $\mathbb{L}/\mathbb{k}$, we obtain

$$\begin{aligned}[\mathbb{L} : \mathbb{k}] &= [\mathbb{K} : \mathbb{k}]/[\mathbb{K} : \mathbb{L}] \\ &= |\mathrm{Gal}(\mathbb{K}/\mathbb{k})|/|\mathrm{Gal}(\mathbb{K}/\mathbb{L})| \\ &\leq |\mathrm{Gal}(\mathbb{L}/\mathbb{k})| \leq [\mathbb{L} : \mathbb{k}],\end{aligned}$$

with equality at the first $\leq$ sign only if $\varphi \mapsto \bar\varphi$ is onto $\mathrm{Gal}(\mathbb{L}/\mathbb{k})$ and with equality at the second $\leq$ sign only if $\mathbb{L}$ is the splitting field over $\mathbb{k}$ of the minimal polynomial of a certain element γ of $\mathbb{L}$. Equality must hold in both cases because the end members of the display are equal, and we conclude that $\varphi \mapsto \bar\varphi$ is onto and that $\mathbb{L}/\mathbb{k}$ is a normal extension.

We are left with proving that if $\mathbb{L}/\mathbb{k}$ is a normal extension, then H is a normal subgroup of $\mathrm{Gal}(\mathbb{K}/\mathbb{k})$. Thus let $\mathbb{L}/\mathbb{k}$ be normal. In view of the conclusion of the first paragraph of the proof, it is enough to prove that $\varphi(\mathbb{L}) = \mathbb{L}$ for all $\varphi \in \mathrm{Gal}(\mathbb{K}/\mathbb{k})$. By definition of normal extension, $\mathbb{L}$ is the splitting field of some polynomial $F(X)$ in $\mathbb{k}[X]$. We may assume that $F(X)$ is monic. Let us write

$$F(X) = (X - x_1) \cdots (X - x_n) \qquad \text{with all } x_j \text{ in } \mathbb{L}.$$

Applying a given member φ of $\mathrm{Gal}(\mathbb{K}/\mathbb{k})$ to the coefficients, we obtain

$$F(X) = (X - \varphi(x_1)) \cdots (X - \varphi(x_n)),$$

and here the $\varphi(x_j)$'s are known only to be in $\mathbb{K}$. By unique factorization in $\mathbb{K}[X]$, $\varphi(x_i) = x_{j(i)}$ for some $j = j(i)$. Therefore $\varphi(x_i)$ is in $\mathbb{L}$ for all i. Since $\mathbb{L}$ is the splitting field of $F(X)$ over $\mathbb{k}$, $\mathbb{L} = \mathbb{k}(x_1, \ldots, x_n)$. Thus φ maps $\mathbb{L}$ into $\mathbb{L}$. $\square$

The examples of Galois groups given in Section 6 all involved fields that are finite extensions of the rationals $\mathbb{Q}$. As we shall see in Section 17, it is important for the understanding of Galois groups of finite extensions of $\mathbb{Q}$ to be able to identify Galois groups of finite extensions of *finite* fields. This matter is addressed in the following proposition.

Proposition 9.40. Let $\mathbb{K}$ be a finite extension of the finite field $\mathbb{F}_q$, where $q = p^a$ and p is prime, and suppose that $[\mathbb{K} : \mathbb{F}_q] = n$. Then $\mathbb{K}$ is a Galois extension of $\mathbb{F}_q$, the Galois group $\text{Gal}(\mathbb{K}/\mathbb{F}_q)$ is cyclic of order n, and a generator is the ath-power Frobenius automorphism $x \mapsto x^q = x^{p^a}$.

PROOF. Theorem 9.14 shows that $\mathbb{K}$ is a splitting field for $X^{q^n} - X$ over $\mathbb{F}_p$. Hence it is a splitting field for $X^{q^n} - X$ over $\mathbb{F}_q$, and $\mathbb{K}/\mathbb{F}_q$ is a normal extension. The polynomial $X^{q^n} - X$ has no multiple roots, and it follows that $\mathbb{K}/\mathbb{F}_q$ is a separable extension.

Define φ by $\varphi(x) = x^q$. Lemma 9.18 shows that φ is an automorphism of $\mathbb{K}$. Since every member of $\mathbb{F}_q^\times$ has order dividing $q - 1$, every nonzero element of $\mathbb{F}_q$ is fixed by φ. The map φ certainly carries 0 to 0, and thus φ is in $\text{Gal}(\mathbb{K}/\mathbb{F}_q)$. By a similar argument, φ^n fixes every element of $\mathbb{K}$, and hence $\varphi^n = 1$. Corollary 4.27 shows that $\mathbb{K}^\times$ is cyclic, hence that there exists an element y in $\mathbb{K}^\times$ such that $y^l \neq 1$ for $1 \leq l < q^n - 1$. This y has $y^l \neq y$ for $2 \leq l \leq q^n - 1$. Then $\varphi^k(y) = y^{q^k}$ cannot be 1 for $1 \leq k \leq n - 1$, and φ must have order exactly n. This shows that φ generates a cyclic subgroup of order n in $\text{Gal}(\mathbb{K}/\mathbb{F}_q)$. Since n is an upper bound for the order of $\text{Gal}(\mathbb{K}/\mathbb{F}_q)$ by Proposition 9.33, this cyclic subgroup exhausts the Galois group. □

EXAMPLE. Suppose that we are given a polynomial with coefficients in $\mathbb{F}_p$ and we want to find the Galois group of a splitting field. Since there are efficient computer programs for factoring the polynomial into irreducible polynomials, let us take that factorization as done. The Galois group will be cyclic of some order with generator the Frobenius automorphism $x \mapsto x^p$. For an irreducible polynomial of degree n, the splitting field has degree n, and the smallest power of $x \mapsto x^p$ that gives the identity is the nth power. The conclusion is that the Galois group is cyclic of order equal to the least common multiple of the degrees of the irreducible constituents, a generator being the Frobenius automorphism.

9. Application to Constructibility of Regular Polygons

In this section we use Galois theory to give a proof of Theorem 9.25 concerning the constructiblity of regular n-gons. Let us recall the statement.

THEOREM 9.25 (Gauss). A regular n-gon is constructible with straightedge and compass if and only if n is the product of distinct Fermat primes and a power of 2.

PROOF OF SUFFICIENCY. First suppose that n is a Fermat prime $n = 2^{2^N} + 1$. Let $\mathbb{K} = \mathbb{Q}(e^{2\pi i/n})$. We saw in Section 5 that the degree $[\mathbb{K} : \mathbb{Q}]$ is 2^{2^N}, hence is a power of 2. Furthermore we know that $\mathbb{K}$ is a separable extension of $\mathbb{Q}$, being of characteristic 0, and it is normal, being the splitting field for $X^n - 1$ over $\mathbb{Q}$. In Section 6 we saw that the Galois group $\mathrm{Gal}(\mathbb{K}/\mathbb{Q})$ is cyclic of order 2^{2^N}. Let c be a generator of this group. For each integer k with $0 \le k \le 2^N$, let H_{2^k} be the unique cyclic subgroup of $\mathrm{Gal}(\mathbb{K}/\mathbb{Q})$ of order 2^k. For this subgroup, $c^{2^{2^N-k}}$ is a generator. Put $\mathbb{L}_{2^k} = \mathbb{K}^{H_{2^k}}$. Then we have inclusions

$$\{1\} \subseteq H_2 \subseteq H_{2^2} \subseteq \cdots H_{2^k} \subseteq \cdots \subseteq H_{2^{2^N-1}} \subseteq H_{2^{2^N}} = \mathrm{Gal}(\mathbb{K}/\mathbb{Q}),$$

the index being 2 at each stage. Theorem 9.38 says that the correspondence with intermediate fields reverses inclusions and that the degree of each consecutive extension of subfields matches the index of the corresponding consecutive subgroups. The intermediate fields are therefore of the form

$$\mathbb{K} \supseteq \mathbb{L}_2 \supseteq \mathbb{L}_{2^2} \supseteq \cdots \mathbb{L}_{2^k} \supseteq \cdots \supseteq \mathbb{L}_{2^{2^N-1}} \supseteq \mathbb{L}_{2^{2^N}} = \mathbb{Q},$$

and the degree in each case is 2. In view of the formula for the roots of a quadratic polynomial, each extension is obtained by adjoining some square root. By Theorem 9.24 the members of $\mathbb{K}$ are constructible with straightedge and compass. In particular, $e^{2\pi i/n}$ is constructible, and a regular n-gon is constructible.

Next suppose that $e^{2\pi i/r}$ and $e^{2\pi i/s}$ are both constructible and that $\mathrm{GCD}(r, s) = 1$. Choose integers a and b with $ar + bs = 1$, so that $\frac{a}{s} + \frac{b}{r} = \frac{1}{rs}$. Then the equality $(e^{2\pi i/s})^a (e^{2\pi i/r})^b = e^{2\pi i/(rs)}$ shows that $e^{2\pi i/(rs)}$ is constructible. This proves the sufficiency for any product of distinct Fermat primes. Bisection of an angle is always possible with straightedge and compass, as was observed in the third paragraph of Section 5, and the proof of the sufficiency in Theorem 9.25 is therefore complete. □

REMARKS. The above proof shows that the construction is possible, but it gives little clue how to carry out the construction. We shall address this matter further in Section 12.

We turn our attention to the necessity—that n has to be the product of distinct Fermat primes and a power of 2 if a regular n-gon is constructible. For the moment let $n \ge 1$ be any integer. Let us consider the distinct n^{th} roots of 1 in $\mathbb{C}$, which are $e^{k2\pi i/n}$ for $0 \le k < n$. The order of each of these elements divides n, and the order is exactly n if and only if $\mathrm{GCD}(k, n) = 1$. In this case we say that $e^{k2\pi i/n}$ is a **primitive** n^{th} root of 1. Define the **cyclotomic polynomial** $\Phi_n(X)$ by

$$\Phi_n(X) = \prod_{\substack{\mathrm{GCD}(k,n)=1, \\ 0 \le k < n}} (X - e^{k2\pi i/n}).$$

9. Application to Constructibility of Regular Polygons

Each such polynomial is monic by inspection. The splitting field $\mathbb{Q}(e^{2\pi i/n})$ in $\mathbb{C}$ is called a **cyclotomic field**. Since the complex roots of $X^n - 1$ are exactly the numbers $e^{k2\pi i/n}$, we have

$$X^n - 1 = \prod_{d|n} \Phi_d(X),$$

the product being taken over the positive divisors d of n.

Lemma 9.41. Each cyclotomic polynomial $\Phi_n(X)$ lies in $\mathbb{Z}[X]$, and the degree of $\Phi_n(X)$ is $\varphi(n)$, where φ is the Euler φ function defined just before Corollary 1.10.

PROOF. We know that $\Phi_n(X)$ is in $\mathbb{C}[X]$, and we begin by showing by induction on n that $\Phi_n(X)$ is in $\mathbb{Q}[X]$. For $n = 1$, we have $\Phi_1[X] = X - 1$, and the assertion is true. If it is true for all d with $1 \leq d < n$, then the formula $X^n - 1 = \prod_{d|n} \Phi_d(X)$ and induction show that $X^n - 1 = \Phi_n(X)F(X)$ for some $F(X)$ in $\mathbb{Q}[X]$. By the division algorithm, $X^n - 1 = F(X)Q(X) + R(X)$ for polynomials $Q(X)$ and $R(X)$ in $\mathbb{Q}[X]$ with $R(X) = 0$ or $\deg R(X) < \deg F(X)$. Subtraction gives $F(X)(\Phi_n(X) - Q(X)) = -R(X)$ in $\mathbb{C}[X]$. If $R(X)$ is not 0, then $\deg R(X) < \deg F(X)$ gives a contradiction. Therefore $R(X) = 0$ and $F(X)(\Phi_n(X) - Q(X)) = 0$. Since $\mathbb{C}[X]$ is an integral domain, $\Phi_n(X) = Q(X)$. Thus $\Phi_n(X)$ is in $\mathbb{Q}[X]$, and the induction is complete.

To see that $\Phi_n(X)$ is in $\mathbb{Z}[X]$, we again induct, the case $n = 1$ being clear. The formula $X^n - 1 = \prod_{d|n} \Phi_d(X)$ and induction show that $X^n - 1 = \Phi_n(X)F(X)$ for some $F(X)$ in $\mathbb{Z}[X]$. Since $\Phi_n(X)$ is known to be in $\mathbb{Q}[X]$, Corollary 8.20c shows that $\Phi_n(X)$ is in $\mathbb{Z}[X]$, and the induction is complete. □

Lemma 9.42. Each cyclotomic polynomial $\Phi_n(X)$ is irreducible as a member of $\mathbb{Q}[X]$.

PROOF. Let ζ be a primitive n^{th} root of 1, let p be a prime number not dividing n, let $F(X)$ be the minimal polynomial of ζ over $\mathbb{Q}$, and let $G(X)$ be the minimal polynomial of ζ^p. The main step is to show that $F(X) = G(X)$.

To carry out this step, we observe that $F(\zeta) = G(\zeta^p) = 0$ and that $F(X)$ and $G(X)$ must divide $\Phi_n(X)$. Arguing by contradiction, suppose that $F(X) \neq G(X)$. Then $\text{GCD}(F, G) = 1$ since $F(X)$ and $G(X)$ are irreducible over $\mathbb{Q}$, and therefore $F(X)G(X)$ divides $\Phi_n(X)$. Hence we can write

$$X^n - 1 = F(X)G(X)H(X),$$

and $H(X)$ is a monic member of $\mathbb{Z}[X]$ by Lemma 9.41 and Corollary 8.20c. Since ζ is a root of $G(X^p)$, we must have $G(X^p) = F(X)M(X)$ for some monic polynomial $M(X)$ in $\mathbb{Z}[X]$. We apply the substitution homomorphism to $\mathbb{Z}[X] \to \mathbb{F}_p[X]$ that carries X to X and reduces the coefficients modulo p; the mapping on the coefficients will be denoted by a bar. Then we have

$$X^n - \bar{1} = \overline{F}(X)\overline{G}(X)\overline{H}(X) \quad \text{and} \quad \overline{G}(X)^p = \overline{G}(X^p) = \overline{F}(X)\overline{M}(X),$$

the equality $\overline{G}(X)^p = \overline{G}(X^p)$ following from Lemma 9.18. If $\overline{Q}(X)$ is a prime factor of $\overline{F}(X)$, then $\overline{Q}(X)$ divides $\overline{G}(X)^p$ and therefore must divide $\overline{G}(X)$. So $\overline{Q}(X)^2$ divides $X^n - \bar{1}$. Therefore $X^n - \bar{1}$ has multiple roots in its splitting field, in contradiction to Corollary 9.17 and the fact that the derivative of $X^n - \bar{1}$ is nonzero at each nonzero member of $\mathbb{F}_p$ (since $\mathrm{GCD}(p, n) = 1$ by assumption). We conclude that $F(X) = G(X)$.

Now suppose that r is a positive integer with $\mathrm{GCD}(r, n) = 1$. Then we can write $r = p_1 \cdots p_l$ with each p_j not dividing n, and we see inductively that ζ^r has $F(X)$ as minimal polynomial. Thus $F(X)$ has at least $\varphi(n)$ roots. Since $F(X)$ divides $\Phi_n(X)$, we must have $F(X) = \Phi_n(X)$. Therefore $\Phi_n(X)$ is irreducible over $\mathbb{Q}$. □

PROOF OF NECESSITY IN THEOREM 9.25. Theorem 9.24 shows that the degree $[\mathbb{Q}(e^{2\pi i/n}) : \mathbb{Q}]$ must be a power of 2 if a regular n-gon is constructible. Since $e^{2\pi i/n}$ is a root of $\Phi_n(X)$ and since Lemma 9.42 shows $\Phi_n(X)$ to be irreducible over $\mathbb{Q}$, $\Phi_n(X)$ is the minimal polynomial of $e^{2\pi i/n}$ over $\mathbb{Q}$. By Lemma 9.41 the degree in question is given by $[\mathbb{Q}(e^{2\pi i/n}) : \mathbb{Q}] = \varphi(n)$, where φ is the Euler φ function. Corollary 1.10 shows that if $n = p_1^{k_1} \cdots p_r^{k_r}$ is a prime factorization of n into distinct prime powers with each $k_j > 0$, then

$$\varphi(n) = \prod_{j=1}^{r} p_j^{k_j - 1}(p_j - 1).$$

For constructibility this must be a power of 2. Then each p_j dividing n must be 1 more than a power of 2, i.e., must be 2 or a Fermat prime, and the only p_j allowed to have p_j^2 dividing n is $p_j = 2$. □

10. Application to Proving the Fundamental Theorem of Algebra

In this section we use Galois theory to give a proof of the Fundamental Theorem of Algebra. Let us recall the statement.

THEOREM 1.18 (Fundamental Theorem of Algebra). Any polynomial in $\mathbb{C}[X]$ with degree ≥ 1 has at least one root.

10. Application to Proving the Fundamental Theorem of Algebra

We begin with a lemma that handles three easy special cases.

Lemma 9.43. There are no finite extensions of $\mathbb{R}$ of odd degree greater than 1, the only extension of $\mathbb{R}$ of degree 2 up to $\mathbb{R}$ isomorphism is $\mathbb{C}$, and there are no finite extensions of $\mathbb{C}$ of degree 2.

PROOF. If $\mathbb{K}$ is a finite extension of $\mathbb{R}$ of odd degree and if x is in $\mathbb{K}$, then $[\mathbb{R}(x) : \mathbb{R}]$ is odd, and consequently the minimal polynomial $F(X)$ of x over $\mathbb{R}$ is irreducible of odd degree. By Proposition 1.20, which is derived from the Intermediate Value Theorem of Section A3 of the appendix, $F(X)$ has at least one root in $\mathbb{R}$. Therefore $F(X)$ has degree 1, and x is in $\mathbb{R}$.

If $F(X)$ is an irreducible polynomial in $\mathbb{R}[X]$ of degree 2, then $F(X)$ splits in $\mathbb{C}$ by the quadratic formula, and hence the only extension of $\mathbb{R}$ of degree 2 is $\mathbb{C}$, up to $\mathbb{R}$ isomorphism, by the uniqueness of splitting fields (Theorem 9.13).

Let $G(X) = X^2 + bX + c$ be a polynomial in $\mathbb{C}[X]$ of degree 2. Then $G(X)$ has a root x in $\mathbb{C}$ given by the quadratic formula since every member of $\mathbb{C}$ has a square root[6] in $\mathbb{C}$, and $G(X)$ cannot be irreducible. Since any finite extension of $\mathbb{C}$ of degree 2 would have to be of the form $\mathbb{C}(x)$, with x equal to a root of an irreducible quadratic polynomial over $\mathbb{C}$, there can be no such extension. □

PROOF OF THEOREM 1.18. First let us show that every irreducible member $F(X)$ of $\mathbb{R}[X]$ splits over $\mathbb{C}$. Let $\mathbb{K}$ be a splitting field for $F(X)$. Say that $[\mathbb{K} : \mathbb{R}] = 2^m N$ with N odd. Then $\mathbb{K}$ is a Galois extension of $\mathbb{R}$, and $|\operatorname{Gal}(\mathbb{K}/\mathbb{R})| = 2^m N$. By the Sylow Theorems (particularly Theorem 4.59a), let H be a Sylow 2-subgroup of $\operatorname{Gal}(\mathbb{K}/\mathbb{R})$. This H has $|H| = 2^m$. The field $\mathbb{L} = \mathbb{K}^H$ that corresponds to H under Theorem 9.38 has $[\mathbb{L} : \mathbb{R}] = N$ with N odd, and the first conclusion of Lemma 9.43 shows that $N = 1$. Thus $|\operatorname{Gal}(\mathbb{K}/\mathbb{R})| = 2^m$. Corollary 4.40 shows that $\operatorname{Gal}(\mathbb{K}/\mathbb{R})$ has nested subgroups of all orders 2^{m-k} with $0 \leq k \leq m$, and Theorem 9.38 says that the corresponding fixed fields are nested and have respective degrees 2^k with $0 \leq k \leq m$. The extension field of $\mathbb{R}$ for $k = 1$ is necessarily $\mathbb{C}$ by Lemma 9.43, and Lemma 9.43 shows that there are no quadratic extensions of $\mathbb{C}$. Therefore $m = 0$ or $m = 1$, and the possible splitting fields for $F(X)$ are $\mathbb{R}$ and $\mathbb{C}$ in the two cases.

To complete the proof, suppose that $\mathbb{K}$ is a finite algebraic extension of $\mathbb{C}$ of degree n. Then $\mathbb{K}$ is a finite algebraic extension of $\mathbb{R}$ of degree $2n$. The Theorem of the Primitive Element allows us to write $\mathbb{K} = \mathbb{R}(x)$ for some $x \in \mathbb{K}$, and the minimal polynomial of x over $\mathbb{R}$ necessarily has degree $2n$. The previous paragraph shows that this polynomial splits in $\mathbb{C}$. Thus x is in $\mathbb{C}$, and $\mathbb{K} = \mathbb{C}$. This completes the proof. □

---

[6]To see that every member of $\mathbb{C}$ has a square root in $\mathbb{C}$, let $c + di$ be given with c and d real and with $d \neq 0$. Let a and b be real numbers with $a^2 = \frac{1}{2}(c + \sqrt{c^2 + d^2})$, $b^2 = \frac{1}{2}(-c + \sqrt{c^2 + d^2})$, and $\operatorname{sgn}(ab) = \operatorname{sgn} d$. Then $(a + bi)^2 = c + di$.

11. Application to Unsolvability of Polynomial Equations with Nonsolvable Galois Group

The quadratic formula for finding the roots of a quadratic polynomial has in principle been known since the time of the Babylonians about 400 B.C.[7] The corresponding problem of finding roots of cubics was unsolved until the sixteenth century, and **Cardan's formula** was discovered at that time. The original formula assumes real coefficients and was in two parts, a first case corresponding to what we now view as one real root and two complex roots, the second case corresponding to what we view as three real roots.[8] There is a similar formula, but more complicated, for solving quartics. Further centuries passed with no progress on finding a corresponding formula for the roots of a polynomial of degree 5 or higher. The introduction of Galois theory in the early nineteenth century made it possible to prove a surprising negative statement about all degrees beyond 4.

Suppose that we are given a polynomial equation with coefficients in the field $\mathbb{Q}$ or a more general field $\mathbb{k}$ of characteristic 0. In this section we use Galois theory to address the question whether the roots of the equation in a splitting field can be expressed in terms of $\mathbb{k}$ and the adjunction of finitely many n^{th} roots to the field, for various values of n. For the moment let us say in this case that the roots are "expressible in terms of the members of $\mathbb{k}$ and radicals." We shall make this notion more precise shortly.

Recall from Section IV.8 that with a finite group G, we can find a strictly decreasing sequence of subgroups starting with G and ending with $\{1\}$ such that each subgroup is normal in the next larger one and each quotient group is simple. Such a series was defined to be a composition series for G. The Jordan–Hölder Theorem (Corollary 4.50) says that the respective consecutive quotients are isomorphic for any two composition series, apart from the order in which they appear. We define the finite group G to be **solvable** if each of the consecutive quotients is cyclic of prime order, rather than nonabelian. It is enough that the group have a normal series for which each of the consecutive quotients is abelian.

Examples of solvable and nonsolvable groups are obtainable from the calculations in Section IV.8: abelian groups and groups of prime-power order are always solvable, the symmetric group $\mathfrak{S}_4$ and each of its subgroups are solvable, and the

---

[7] The Babylonians did not actually have equations but had an algorithmic method that amounted to completing the square.

[8] Cardan's name was Girolamo Cardano. The solution in the first case of the cubic seems to have been discovered by Scipione dal Ferro and later by Nicolo Tartaglia. Dal Ferro died in 1526 and passed the secret method to his student Antonio Fior. In 1535 Fior engaged in a public contest with Tartaglia at solving cubics, and he lost. Cardano wheedled the solution method in the first case from Tartaglia, published it in 1539, and discovered and published the solution in the second case. Cardano's student Lodovico Ferrari discovered how to solve quartics, and Cardano published that solution as well. See "St. Andrews" in the Selected References for more information.

symmetric group $\mathfrak{S}_5$ is not solvable since a composition series is $\mathfrak{S}_5 \supseteq \mathfrak{A}_5 \supseteq \{1\}$ and the group $\mathfrak{A}_5$ is simple (Theorem 4.47).

Modulo a precise definition for a field $\Bbbk$ of the words "expressible in terms of the members of $\Bbbk$ and radicals," the answer to our main question is as follows.

Theorem 9.44 (Abel, Galois).[9] Let $\Bbbk$ be a field of characteristic 0, let $F(X)$ be in $\Bbbk[X]$, and let $\mathbb{K}$ be a splitting field of $F(X)$ over $\Bbbk$. Then the roots of $F(X)$ are expressible in terms of the members of $\Bbbk$ and radicals if and only if the group $\text{Gal}(\mathbb{K}/\Bbbk)$ is solvable.

EXAMPLE. With $\Bbbk = \mathbb{Q}$, let $F(X)$ be the polynomial $F(X) = X^5 - 5X + 1$ in $\mathbb{Q}[X]$. We shall show that

(i) $F(X)$ is irreducible over $\mathbb{Q}$,
(ii) $F(X)$ has three roots in $\mathbb{R}$ and one pair of conjugate complex roots in $\mathbb{C}$,
(iii) the splitting field $\mathbb{K}$ over $\mathbb{Q}$ of any polynomial of degree 5 for which (i) and (ii) hold has Galois group with $\text{Gal}(\mathbb{K}/\mathbb{Q}) \cong \mathfrak{S}_5$.

We know that from Theorem 4.47 that $\mathfrak{S}_5$ is not solvable, and Theorem 9.44 therefore allows us to conclude that the roots of $X^5 - 5X + 1$ are not expressible in terms of the members of $\mathbb{Q}$ and radicals.

To prove (i), we apply Eisenstein's criterion (Corollary 8.22) to the polynomial $F(X-1) = X^5 - 5X^4 + 10X^3 - 10X^2 + 5$ and to the prime $p = 5$, and the irreducibility is immediate.

To prove (ii), we observe that $F(-2) < 0$, $F(0) > 0$, $F(1) < 0$, $F(2) > 0$. Applying the Intermediate Value Theorem (Section A3 of the appendix), we see that there are at least three roots in $\mathbb{R}$. Since $F'(X) = 5(X^4 - 1)$ has exactly the two roots ± 1 in $\mathbb{R}$, $F(X)$ has at most three roots in $\mathbb{R}$ by an application of the Mean Value Theorem.

To prove (iii), label the roots 1, 2, 3, 4, 5 with 1 and 2 denoting the nonreal roots. Each member of the Galois group permutes the roots and is determined by its effect on the roots. Thus $\text{Gal}(\mathbb{K}/\mathbb{Q})$ may be regarded as a subgroup of $\mathfrak{S}_5$. Since $F(X)$ is irreducible over $\mathbb{Q}$, 5 divides $[\mathbb{K}:\mathbb{Q}]$ and 5 divides $|\text{Gal}(\mathbb{K}/\mathbb{Q})|$. By the Sylow Theorems, $\text{Gal}(\mathbb{K}/\mathbb{Q})$ contains an element of order 5, hence a 5-cycle. Some power of this 5-cycle carries root 1 to root 2. So we may assume that the 5-cycle is (1 2 3 4 5). Also, $\text{Gal}(\mathbb{K}/\mathbb{Q})$ contains complex conjugation, which acts as (1 2). Then $\text{Gal}(\mathbb{K}/\mathbb{Q})$ contains

$$(1\ 2\ 3\ 4\ 5)(1\ 2)(1\ 2\ 3\ 4\ 5)^{-1} = (2\ 3),$$
$$(1\ 2\ 3\ 4\ 5)(2\ 3)(1\ 2\ 3\ 4\ 5)^{-1} = (3\ 4),$$
$$(1\ 2\ 3\ 4\ 5)(3\ 4)(1\ 2\ 3\ 4\ 5)^{-1} = (4\ 5).$$

---

[9]Abel proved that there is no general solution via radicals that gives the roots of polynomials of degree 5. Galois found the present theorem, which shows how to decide the question for each individual polynomial of degree 5.

Since the set $\{(1\ 2), (2\ 3), (3\ 4), (4\ 5)\}$ of transpositions is easily shown from Corollary 1.22 to generate $\mathfrak{S}_5$, $\mathrm{Gal}(\mathbb{K}/\mathbb{Q}) = \mathfrak{S}_5$.

Let $\mathbb{K}'$ be a finite extension of the given field $\mathbb{k}$. A **root tower** for $\mathbb{K}'$ over $\mathbb{k}$ is a finite sequence of extensions

$$\mathbb{k} = \mathbb{K}'_0 \subseteq \mathbb{K}'_1 \subseteq \cdots \subseteq \mathbb{K}'_{l-1} \subseteq \mathbb{K}'_l = \mathbb{K}'$$

such that for each i with $0 \leq i \leq l-1$, there is a prime number $n_i > 1$ and there is an element r_i in $\mathbb{K}'_{i+1}$ with $a_i = r_i^{n_i}$ in $\mathbb{K}'_i$ and r_i not in $\mathbb{K}'_i$. Then it follows that r_i^k is not in $\mathbb{K}'_i$ for any k with $0 < k < n_i$.

(If we write $a_i = r_i^{n_i}$, then we might think of writing $\mathbb{K}'_{i+1} = \mathbb{K}'_i(\sqrt[n_i]{a_i})$, but this formulation is less precise at the moment since it does not specify precisely which choice of $\sqrt[n_i]{a_i}$ is to be used.)

With "root tower" now well defined, we can make a precise definition and thereby complete the precise formulation of Theorem 9.44. Let $\mathbb{k}$ be the given field of characteristic 0, let $F(X)$ be in $\mathbb{k}[X]$, and let $\mathbb{K}$ be a splitting field of $F(X)$ over $\mathbb{k}$. We say that the roots of $F(X)$ are **expressible in terms of members of $\mathbb{k}$ and radicals** if there exists some finite extension $\mathbb{K}'$ of $\mathbb{K}$ having a root tower over $\mathbb{k}$.

The statement of Theorem 9.44 is now completely precise, and the remainder of the section will be devoted to the proof of one direction of the theorem: if the roots are expressible in terms of members of $\mathbb{k}$ and radicals, then the Galois group is solvable. The proof of the converse direction of the theorem is postponed to Section 13. We begin with a lemma.

Lemma 9.45. Let $\mathbb{k}$ be a field of any characteristic, and let p be a prime number. If a is a member of $\mathbb{k}$ such that $X^p - a$ has no root in $\mathbb{k}$, then $X^p - a$ is irreducible in $\mathbb{k}$.

PROOF. First suppose that p is different from the characteristic. Let $\mathbb{L}$ be a splitting field for $X^p - a$. The derivative of $X^p - a$, evaluated at any root of $X^p - a$ in $\mathbb{L}$, is nonzero, and Corollary 9.17 shows that $X^p - a$ splits as the product of p distinct linear factors in $\mathbb{L}$. The quotient of any two roots of $X^p - a$ is a pth root of 1. Fixing one of these two roots of $X^p - a$ and letting the other vary, we obtain p distinct p^{th} roots of 1. Thus $\mathbb{L}$ contains all p of the pth roots of 1. Proposition 4.26 shows that the group of pth roots of 1 is cyclic. Let ζ be a generator. If $a^{1/p}$ denotes one of the roots of $X^p - a$ in $\mathbb{L}$, then the set of all the roots is given by $\{a^{1/p}\zeta^k \mid 0 \leq k \leq p-1\}$.

Now suppose that $X^p - a$ has a nontrivial factorization $X^p - a = F(X)G(X)$ in $\mathbb{k}[X]$. Possibly by adjusting the leading coefficients of $F(X)$ and $G(X)$, we may assume that $F(X)$ and $G(X)$ are both monic. Unique factorization in $\mathbb{L}[X]$

11. Application to Unsolvability of Polynomial Equations with Nonsolvable Group

then implies that there is a nonempty subset S of $\{k \mid 0 \leq k \leq p-1\}$ with a nonempty complement S^c such that

$$F(X) = \prod_{k \in S}(X - \zeta^k a^{1/p}) \quad \text{and} \quad G(X) = \prod_{k \in S^c}(X - \zeta^k a^{1/p}).$$

If S has m elements, then the constant term of $F(X)$ is $(-a^{1/p})^m \omega$, where ω is some p^{th} root of 1. Thus $x = (a^{1/p})^m \omega$ is in $\Bbbk$. Since $\mathrm{GCD}(m, p) = 1$, we can choose integers c and d with $cm + dp = 1$. Since x is in $\Bbbk$, so is $x^c a^d = (a^{1/p})^{mc+dp} \omega^c = a^{1/p} \omega^c$. But $a^{1/p} \omega^c$ is a root of $X^p - a$, in contradiction to the hypothesis that no root of $X^p - a$ lies in $\Bbbk$. Hence $X^p - a$ is irreducible.

If p equals the characteristic of $\Bbbk$, then Lemma 9.18 gives the factorization $X^p - a = (X - a^{1/p})^p$, where $a^{1/p}$ is one root of $X^p - a$ in $\mathbb{K}$. Then we can argue as above except that ζ and ω are to be replaced by 1 throughout. This completes the proof of the lemma. □

PROOF OF NECESSITY IN THEOREM 9.44 THAT $\mathrm{Gal}(\mathbb{K}/\Bbbk)$ BE SOLVABLE. We are to prove that if some finite extension $\mathbb{K}'$ of $\mathbb{K}$ has a root tower over $\Bbbk$, then $\mathrm{Gal}(\mathbb{K}/\Bbbk)$ is solvable.

Step 1. We enlarge each field in the given root tower to obtain a root tower

$$\Bbbk \subseteq \mathbb{K}''_0 \subseteq \mathbb{K}''_1 \subseteq \cdots \subseteq \mathbb{K}''_{l-1} \subseteq \mathbb{K}''_l = \mathbb{K}''$$

of a finite extension $\mathbb{K}''$ of $\mathbb{K}'$ in such a way that $\mathbb{K}''_0$ is the normal extension of $\Bbbk$ obtained by adjoining all n^{th} roots of 1 for a suitably large n and such that each $\mathbb{K}''_{i+1}$ is the normal extension of $\mathbb{K}''_i$ for $0 \leq i \leq l-1$ obtained by adjoining all n_i^{th} roots of the member a_i of $\mathbb{K}'_i$. Using Theorem 9.22, choose an algebraic closure $\overline{\mathbb{K}'}$ of $\mathbb{K}'$. Let n be the product of the integers $n_0, n_1, \ldots, n_{l-1}$. Let $\zeta_1, \ldots, \zeta_{n-1}$ be the nth roots of 1 in $\overline{\mathbb{K}'}$ other than 1 itself, define subfields of $\overline{\mathbb{K}'}$ by

$$\mathbb{K}''_i = \mathbb{K}'_i(\zeta_1, \ldots, \zeta_{n-1}) \quad \text{for } 0 \leq i \leq l,$$

and put $\mathbb{K}'' = \mathbb{K}'_l$. The field $\mathbb{K}''_0$ is a splitting field for $X^n - 1$ over $\Bbbk$ and is therefore a normal extension. The field $\mathbb{K}''_{i+1}$ is given by $\mathbb{K}''_{i+1} = \mathbb{K}''_i(r_i)$, where r_i is a root in $\mathbb{K}''_{i+1}$ of the polynomial $X^{n_i} - a_i$ in $\mathbb{K}''_i[X]$. Here n_i is prime. Lemma 9.45 shows that either r_i is in $\mathbb{K}''_i[X]$ or $X^{n_i} - a_i$ is irreducible in $\mathbb{K}''_i[X]$. In the first case, $\mathbb{K}''_{i+1} = \mathbb{K}''_i$, and we have a normal extension. In the second case, $\mathbb{K}''_{i+1}$ is a splitting field for $X^{n_i} - a_i$ over $\mathbb{K}''_i$ because it is generated by $\mathbb{K}''_i$ and one root of $X^{n_i} - a_i$ and because all n_i^{th} roots of 1 already lie in $\mathbb{K}''_0$; thus again we have a normal extension.

Step 2. The Galois group of $\mathbb{K}_0''$ over $\mathbb{k}$ is abelian. In fact, Proposition 4.26 shows that the group of nth roots of 1 in $\mathbb{K}_0''$ is cyclic. Let ζ be a generator, and let $U = \{\zeta^k\}_{k=0}^{n-1}$. The map of $\text{Gal}(\mathbb{K}_0''/\mathbb{k})$ into $\text{Aut}\,U$ given by $\varphi \mapsto \varphi|_U$ is a one-one homomorphism, and $\text{Aut}\,U$ is isomorphic to $(\mathbb{Z}/n\mathbb{Z})^\times$. Since $(\mathbb{Z}/n\mathbb{Z})^\times$ is abelian, it follows that $\text{Gal}(\mathbb{K}_0''/\mathbb{k})$ is abelian.

Step 3. The Galois group of $\mathbb{K}_{i+1}''$ over $\mathbb{K}_i''$ is trivial or is cyclic of order n_i. In fact, the Galois group is trivial if $\mathbb{K}_{i+1}'' = \mathbb{K}_i''$. The contrary case is that $[\mathbb{K}_{i+1}'' : \mathbb{K}_i''] = n_i$, and then $\text{Gal}(\mathbb{K}_{i+1}''/\mathbb{K}_i'')$ has order n_i, which is prime. Every group of order n_i is cyclic, and hence $\text{Gal}(\mathbb{K}_{i+1}''/\mathbb{K}_i'')$ is cyclic.

Step 4. We extend the root tower to a larger field $\mathbb{L} \supseteq \mathbb{K}''$ that is a normal extension of $\mathbb{k}$. The resulting root tower of $\mathbb{L}$ will be written as

$$\mathbb{k} \subseteq \mathbb{L}_0 = \mathbb{K}_0'' \subseteq \mathbb{L}_1 = \mathbb{K}_1'' \subseteq \cdots$$
$$\subseteq \mathbb{L}_{k-1} = \mathbb{K}_{l-1}'' \subseteq \mathbb{L}_l = \mathbb{K}'' \subseteq \mathbb{L}_{l+1} \subseteq \cdots \subseteq \mathbb{L}_t = \mathbb{L}.$$

As it is, we cannot say that $\mathbb{K}''$ is the splitting field over $\mathbb{k}$ for the product of the minimal polynomials used in Step 1, because the elements a_i are not assumed to lie in $\mathbb{k}$. To adjust the tower to correct this problem, write $\mathbb{K}''$ as

$$\mathbb{K}'' = \mathbb{k}(r_0, r_1, \ldots, r_{l-1}, \zeta) = \mathbb{k}(x_0, \ldots, x_l),$$

with ζ as in Step 2. Here $r_0, \ldots, r_{l-1}$ are the given elements that define the original root tower, and we define $x_l = \zeta$ and $x_j = r_j$ for $0 \leq j < l$. Since $\mathbb{K}''$ is a finite extension of $\mathbb{k}$, each x_j has a minimal polynomial $G_j(X)$ over $\mathbb{k}$. Define $G(X) = \prod_{j=0}^{l} G_j(X)$, and let $\mathbb{L}$ be the splitting field of $G(X)$ in the algebraic closure $\overline{\mathbb{K}'}$. The field $\mathbb{L}$ is a normal extension of $\mathbb{k}$. The roots of $G(X)$ are the members of $\mathbb{L}$ that are roots of some $G_j(X)$. Each x_j is a root of its own $G_j(X)$. If x_j' is another root of $G_j(X)$, then there is a $\mathbb{k}$ isomorphism of $\mathbb{k}(x_j)$ onto $\mathbb{k}(x_j')$, and we know by the uniqueness of splitting fields (Theorem 9.13′)[10] that this extends to a $\mathbb{k}$ isomorphism of $\mathbb{L}$ onto $\mathbb{L}$. Hence to each root θ of $G(X)$ in $\mathbb{L}$ corresponds some x_j and some $\varphi \in \text{Gal}(\mathbb{K}/\mathbb{k})$ with $\varphi(x_j) = \theta$. Thus

$$\mathbb{L} = \mathbb{k}(\{\varphi(x_j) \mid 0 \leq j \leq l \text{ and } \varphi \in \text{Gal}(\mathbb{L}/\mathbb{k})\}).$$

For any φ in $\text{Gal}(\mathbb{L}/\mathbb{k})$ and any $j \leq l - 1$, the element $\varphi(x_j)$ of $\mathbb{L}$ satisfies

$$(\varphi(x_j))^{n_j} = \varphi(x_j^{n_j}) = \varphi(a_j),$$

---

[10]The theorem is to be applied to $\sigma : \mathbb{k}(x_j) \to \mathbb{k}(x_j')$ with $F(X) = F^\sigma(X) = G(X)$ and with $\mathbb{L}' = \mathbb{L}$.

and the element on the right is in $\varphi(K_j'')$. Any element $\varphi(\zeta)$ is an nth root of 1 and hence is already in $\mathbb{K}_0''$; such elements are redundant for $\varphi \neq 1$. Enumerate $\text{Gal}(\mathbb{L}/\mathbb{k})$ as $\varphi_1, \ldots, \varphi_s$ with $\varphi_1 = 1$. The tower for $\mathbb{K}''$ is to be continued with the fields obtained by adjoining one at a time the elements

$$\varphi_2(r_0), \ldots, \varphi_2(r_{l-1}), \varphi_3(r_0), \ldots, \varphi_3(r_{l-1}), \ldots, \varphi_s(r_0), \ldots, \varphi_s(r_{l-1}).$$

The final field is $\mathbb{L}$, and then we have an enlarged tower as asserted.

Step 5. $\text{Gal}(\mathbb{L}/\mathbb{k})$ is a solvable group. In fact, first we prove by induction downward on i that $\text{Gal}(\mathbb{L}/\mathbb{L}_i)$ is solvable, the case $i = t$ being the case of the trivial group. Let $i < t$ be given. We have arranged that $\mathbb{L}_{i+1}$ is a normal extension of $\mathbb{L}_i$. Since $\mathbb{L}$ is normal over all the smaller fields by Step 4, Corollary 9.39 therefore gives $\text{Gal}(\mathbb{L}_{i+1}/\mathbb{L}_i) \cong \text{Gal}(\mathbb{L}/\mathbb{L}_i)/\text{Gal}(\mathbb{L}/\mathbb{L}_{i+1})$. The group on the left side is cyclic by Step 3 or the analogous proof with some r_j replaced by a suitable $\varphi(r_j)$, and thus a normal series with abelian quotients for $\text{Gal}(\mathbb{L}/\mathbb{L}_{i+1})$ may be extended by including the term $\text{Gal}(\mathbb{L}/\mathbb{L}_i)$, and the result is still a normal series with abelian quotients. Thus $\text{Gal}(\mathbb{L}/\mathbb{L}_i)$ is solvable. This completes the induction and shows that $\text{Gal}(\mathbb{L}/\mathbb{L}_0)$ is solvable. To complete the proof we use the isomorphism $\text{Gal}(\mathbb{L}_0/\mathbb{k}) \cong \text{Gal}(\mathbb{L}/\mathbb{k})/\text{Gal}(\mathbb{L}/\mathbb{L}_0)$ given by Corollary 9.39. The group on the left side is abelian by Step 2, and thus a normal series with abelian quotients for $\text{Gal}(\mathbb{L}/\mathbb{L}_0)$ may be extended by including the term $\text{Gal}(\mathbb{L}/\mathbb{k})$, and the result is still a normal series with abelian quotients. Thus $\text{Gal}(\mathbb{L}/\mathbb{k})$ is solvable.

Step 6. $\text{Gal}(\mathbb{K}/\mathbb{k})$ is a solvable group. We have $\mathbb{L} \supseteq \mathbb{K} \supseteq \mathbb{k}$ with $\mathbb{L}/\mathbb{k}$ normal by Step 4 and with $\mathbb{K}/\mathbb{k}$ normal since $\mathbb{K}$ is a splitting field of $F(X)$ over $\mathbb{k}$. Applying Corollary 9.39, we obtain an isomorphism $\text{Gal}(\mathbb{K}/\mathbb{k}) \cong \text{Gal}(\mathbb{L}/\mathbb{k})/\text{Gal}(\mathbb{L}/\mathbb{K})$. Then Step 6 will follow from Step 5 if it is shown that any homomorphic image of a solvable group is solvable. Thus let G be a solvable group, and let $\varphi : G \to H$ be an onto homomorphism. Write $G = G_1 \supseteq \cdots \supseteq G_m = \{1\}$ with abelian quotients, and define $H_i = \varphi(G_i)$. Passage to the quotient gives us a homomorphism φ_i carrying G_i onto H_i/H_{i+1}. Since $\varphi(G_{i+1}) \subseteq H_{i+1}$, φ induces a homomorphism $\overline{\varphi}_i$ of G_i/G_{i+1} onto H_i/H_{i+1}. As the image of an abelian group under a homomorphism, H_i/H_{i+1} is abelian. Therefore H is solvable. This completes the proof. □

12. Construction of Regular Polygons

Theorem 9.25 proved the constructibility of regular n-gons when n is the product of a power of 2 and distinct Fermat primes, but it gave little clue how to carry out the construction. In this section we supply enough further detail so that one can actually carry out the construction. It is enough to handle the case that n is a Fermat prime, $n = 2^{2^N} + 1$, and we shall suppose that n is a prime of this form.

Let $\zeta = e^{2\pi i/n}$. The field of interest is $\mathbb{Q}(\zeta)$, with $[\mathbb{Q}(\zeta) : \mathbb{Q}] = n - 1$. The usual basis of $\mathbb{Q}(\zeta)$ over $\mathbb{Q}$ is $\{1, \zeta, \zeta^2, \ldots, \zeta^{n-2}\}$, but we shall use the basis

$$\{\zeta, \zeta^2, \zeta^3, \ldots, \zeta^{n-1}\}$$

instead, in order to identify the Galois group $\mathrm{Gal}(\mathbb{Q}(\zeta)/\mathbb{Q})$ more readily with $\mathbb{F}_n^\times$, where $\mathbb{F}_n = \mathbb{Z}/n\mathbb{Z}$ is the field of n elements. In more detail we associate the additive group of $\mathbb{F}_n$ with the additive group of exponents of the members of the cyclic group $\{1, \zeta, \zeta^2, \zeta^3, \ldots, \zeta^{n-1}\}$, and members of the Galois group correspond to the various multiplications of these exponents by $\mathbb{F}_n^\times = \{1, 2, \ldots, n-1\}$. The group $\mathbb{F}_n^\times$ is known to be cyclic of order $n - 1$, and thus the isomorphic Galois group is cyclic. If a generator σ of the Galois group is to correspond to multiplication by a generator g of $\mathbb{F}_n^\times$, then $\sigma(\zeta^s) = \zeta^{gs}$ for all s. With the prime n of the form $2^{2^N} + 1$, let us note for the sake of completeness why we can always take $g = 3$.

Lemma 9.46. The number 3 is a generator of $\mathbb{F}_n^\times$ when n is prime of the form $2^{2^N} + 1$ with $N > 0$.

REMARKS. We verified this assertion for $n = 17$ in Section 6, and in principle one could verify the lemma in any particular case in the same way. Here is a general argument using the law of quadratic reciprocity, whose full statement and proof will be given in *Advanced Algebra*. For a prime number n that is congruent to 1 modulo 4, quadratic reciprocity implies that 3 is a square modulo n if and only if n is a square modulo 3. Since

$$2^{2^N} - 1 = (2^{2^{N-1}} + 1)(2^{2^{N-2}} + 1) \cdots (2^{2^1} + 1)(2^{2^1} - 1)$$

and $2^{2^1} - 1 = 3$, 3 divides $2^{2^N} - 1$. Thus n is congruent to 2 modulo 3, n is not a square modulo 3, and 3 is not a square modulo n. The nonsquares modulo $n = 2^{2^N} + 1$ are exactly the generators of $\mathbb{F}_n^\times$, and therefore 3 is a generator.

Taking Lemma 9.46 into account, we suppose for the remainder of this section that the generator σ of the Galois group corresponds to multiplication of exponents of ζ by 3. Then $\sigma(\zeta) = \zeta^3$ and $\sigma(\zeta^s) = \zeta^{3s}$. These formulas and $\mathbb{Q}$ linearity tell us explicitly how σ operates on all of $\mathbb{Q}(\zeta)$.

The fixed fields that arise within $\mathbb{Q}(\zeta)$ correspond to subgroups of the group $\mathrm{Gal}(\mathbb{Q}(\zeta)/\mathbb{Q}) \cong \{\sigma^j \mid 0 \leq j < 2^{2^N}\}$, and there is one for each power of 2 from 2^0 to 2^{2^N}. Fix attention on the subgroup H_l of order l, and write $2^{2^N} = kl$, with k and l being powers of 2. A generator of this subgroup is σ^k, and the subgroup is $H_l = \{1, \sigma^k, \sigma^{2k}, \ldots, \sigma^{(l-1)k}\}$. Let $\mathbb{K}_l$ be the fixed field of this subgroup, or equivalently of its generator σ^k; this has dimension k over $\mathbb{Q}$.

We shall determine a basis of $\mathbb{K}_l$ over $\mathbb{Q}$. Since $\sigma(\zeta^s) = \zeta^{3s}$, we have $\sigma^k(\zeta^s) = \zeta^{3^k s}$. For $0 \le r \le k-1$, the k elements

$$\eta_r = \zeta^{3^r} + \zeta^{3^{r+k}} + \zeta^{3^{r+2k}} + \cdots + \zeta^{3^{r+k(l-1)}}$$

are linearly independent over $\mathbb{Q}$ because they involve disjoint sets of basis vectors of $\mathbb{Q}(\zeta)$ as r varies. The computation

$$\begin{aligned}
\sigma^k(\eta_r) &= \sigma^k\left(\zeta^{3^r} + \zeta^{3^{r+k}} + \zeta^{3^{r+2k}} + \cdots + \zeta^{3^{r+k(l-1)}}\right) \\
&= \zeta^{3^{r+k}} + \zeta^{3^{r+2k}} + \zeta^{3^{r+3k}} + \cdots + \zeta^{3^{r+kl}} \\
&= \zeta^{3^r} + \zeta^{3^{r+k}} + \zeta^{3^{r+2k}} + \cdots + \zeta^{3^{r+k(l-1)}} \\
&= \eta_r
\end{aligned}$$

shows that each of these vectors is in $\mathbb{K}_l$. Hence $\{\eta_0, \ldots, \eta_{k-1}\}$ is a basis of $\mathbb{K}_l$ over $\mathbb{Q}$. The elements of this basis are called the **periods** of l terms of the cyclotomic field.

The extreme cases for the periods are $(k, l) = (2^{2^N}, 1)$, for which $0 \le r \le 2^{2^N} - 1$ with $\eta_r = \zeta^{3^r}$, and $(k, l) = (1, 2^{2^N})$, for which $r = 0$ with

$$\eta_0 = \zeta^{3^0} + \zeta^{3^1} + \zeta^{3^2} + \cdots + \zeta^{3^{2^{2^N}-1}} = \zeta + \zeta^2 + \zeta^3 + \cdots + \zeta^{n-1} = -1.$$

Two facts enter into determining how to write ζ in terms of rationals and square roots. The first is that at stage k for $k \ge 2$, the sum of certain pairs of η_r's is an η for stage $k-1$. The second is that the product of two η_r's at stage k is an integer combination of η's from the same stage and that the sum formulas express this combination in terms of η's from earlier stages. The result is that at the kth stage we obtain expressions for the sum and product of two η_r's in terms of η's from earlier stages. Therefore the two η_r's at stage k are the roots of a quadratic equation whose coefficients involve η's from earlier stages. Consequently we can compute the η_r's explicitly by induction on k. To proceed further, we need to know the formula for the product of two η_r's, which is due to Gauss.

To multiply two η_r's, we need to multiply various powers of ζ, and the exponents get added in the process. This addition is not readily compatible with terms like ζ^{3^r} and ζ^{3^s}, and for that reason Gauss introduced new notation. Define

$$\eta^{(t)} = \zeta^t + \zeta^{t3^k} + \zeta^{t3^{2k}} + \cdots + \zeta^{t3^{k(l-1)}} = \sum_{v \bmod l} \zeta^{t3^{kv}}$$

for $0 \le t \le n-1$. Then $\eta^{(0)} = l$, and for $0 < t \le n-1$, $\eta^{(t)}$ is the η_r in which ζ^t occurs. Gauss's product formula is given by

$$\eta^{(s)}\eta^{(t)} = \sum_{u \bmod l} \left(\sum_{v \bmod l} \zeta^{s3^{ku}+t3^{kv}} \right)$$
$$= \sum_{u \bmod l} \left(\sum_{w \bmod l} \zeta^{s3^{ku}+t3^{k(u+w)}} \right) \quad \text{with } v \mapsto u+w$$
$$= \sum_{w \bmod l} \left(\sum_{u \bmod l} \zeta^{(s+t3^{kw})3^{ku}} \right)$$
$$= \sum_{w \bmod l} \eta^{(s+t3^{kw})}.$$

In words, this says that to multiply two η's, we add the η's for the exponents obtained by multiplying the first term of $\eta^{(s)}$ by all the terms of $\eta^{(t)}$.

At this point it is more illuminating to work some examples than to try for a general result.

EXAMPLE 1. $n = 5$, $N = 1$, $2^{2^N} = 4$. The relevant pairs (k, l) to study in sequence are $(k, l) = (1, 4), (2, 2), (4, 1)$, and the case $(k, l) = (1, 4)$ is trivial since the only subscripted η is $\sum_{s=0}^{3} \zeta^{3^s} = -1$.

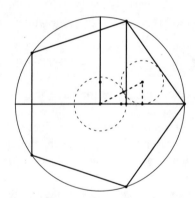

FIGURE 9.3. Construction of a regular pentagon. The circle with center $\left(\frac{1}{2}, \frac{1}{4}\right)$ and radius $\frac{1}{4}$ meets the line from $\left(\frac{1}{2}, \frac{1}{4}\right)$ to the origin at a point at distance $\cos(2\pi/5)$ from the origin.

For $k = 2$, i.e., for the case that there are 2 periods of 2 terms each, we go back to the definition of the η's and find that

$$\eta_0 = \zeta^{3^{0+2\cdot 0}} + \zeta^{3^{0+2\cdot 1}} = \zeta^1 + \zeta^4,$$
$$\eta_1 = \zeta^{3^{1+2\cdot 0}} + \zeta^{3^{1+2\cdot 1}} = \zeta^3 + \zeta^2.$$

12. Construction of Regular Polygons

We form those sums of pairs of η's that yield an η from the previous step. Here there is only one pair, and the sum is given by
$$\eta_0 + \eta_1 = -1.$$
Next we form the elements $\eta^{(t)}$, remembering that for $t > 0$, $\eta^{(t)}$ is the η_r in which ζ^t occurs. Then
$$\eta^{(0)} = 2, \quad \eta^{(1)} = \eta_0, \quad \eta^{(2)} = \eta_1, \quad \eta^{(3)} = \eta_1, \quad \eta^{(4)} = \eta_0.$$
We apply Gauss's product formula to compute the product of the two η's whose sum we have identified. The formula gives
$$\eta_0 \eta_1 = \eta^{(1)} \eta^{(2)} = \eta^{(4)} + \eta^{(3)} = \eta_0 + \eta_1 = -1,$$
the second equality following since the rule for the indices is to extract a power of ζ appearing in $\eta^{(1)}$ and add that index to all the powers of ζ appearing in $\eta^{(2)}$. Since η_0 and η_1 have sum -1 and product -1, they are the roots of the quadratic equation
$$x^2 + x - 1 = 0, \quad \text{namely } \tfrac{1}{2}(-1 \pm \sqrt{5}).$$
Deciding which root is η_0 and which is η_1 involves looking at signs. The two roots of the quadratic equation are of opposite sign because the constant term of the quadratic equation is negative. Since $\eta_0 = \zeta + \zeta^{-1} = e^{2\pi i/5} + e^{-2\pi i/5} = 2\cos(2\pi/5)$ is positive, we obtain
$$\eta_0 = \tfrac{1}{2}(-1 + \sqrt{5}) \quad \text{and} \quad \eta_1 = \tfrac{1}{2}(-1 - \sqrt{5}).$$
The computation can in principle stop here, since knowing $\cos(2\pi/5)$ gives us $\sin(2\pi/5)$ and therefore $e^{2\pi i/5}$. See Figure 9.3. But it is instructive to carry out the algorithm anyway. We are thus to treat $k = 4$. The periods of 1 term are
$$\xi_0 = \zeta, \quad \xi_1 = \zeta^3, \quad \xi_2 = \zeta^4, \quad \xi_3 = \zeta^2.$$
The corresponding objects with superscripts are
$$\xi^{(0)} = 1, \quad \xi^{(1)} = \xi_0, \quad \xi^{(2)} = \xi_3, \quad \xi^{(3)} = \xi_1, \quad \xi^{(4)} = \xi_2.$$
The relevant sums of pairs are
$$\xi_0 + \xi_2 = \eta_0,$$
$$\xi_1 + \xi_3 = \xi_1.$$
We again use Gauss's product formula, and this time we obtain
$$\xi_0 \xi_2 = \xi^{(1)} \xi^{(4)} = \xi^{(5)} = \xi^{(0)} = 1.$$
Hence ξ_0 and ξ_2 are the roots of the quadratic equation
$$y^2 - \eta_0 y + 1 = 0, \quad \text{namely} \quad \frac{\frac{-1+\sqrt{5}}{2} \pm i\sqrt{4 - \left(\frac{-1+\sqrt{5}}{2}\right)^2}}{2}.$$
The root y involving the plus sign is $e^{2\pi i/5}$.

EXAMPLE 2.[11] $n = 17$, $N = 2$, $2^{2^N} = 16$. The relevant pairs (k, l) have $kl = 16$, and the case $(k, l) = (1, 16)$ is trivial since the only subscripted η is $\sum_{s=0}^{15} \zeta^{3^s} = -1$.

For $k = 2$, the 2 periods have 8 terms each, and

$$\eta_0 = \zeta^{3^{0+2\cdot 0}} + \zeta^{3^{0+2\cdot 1}} + \zeta^{3^{0+2\cdot 2}} + \zeta^{3^{0+2\cdot 3}} + \zeta^{3^{0+2\cdot 4}} + \zeta^{3^{0+2\cdot 5}} + \zeta^{3^{0+2\cdot 6}} + \zeta^{3^{0+2\cdot 7}}$$
$$= \zeta^1 + \zeta^9 + \zeta^{13} + \zeta^{15} + \zeta^{16} + \zeta^8 + \zeta^4 + \zeta^2,$$
$$\eta_1 = \zeta^{3^{1+2\cdot 0}} + \zeta^{3^{1+2\cdot 1}} + \zeta^{3^{1+2\cdot 2}} + \zeta^{3^{1+2\cdot 3}} + \zeta^{3^{1+2\cdot 4}} + \zeta^{3^{1+2\cdot 5}} + \zeta^{3^{1+2\cdot 6}} + \zeta^{3^{1+2\cdot 7}}$$
$$= \zeta^3 + \zeta^{10} + \zeta^5 + \zeta^{11} + \zeta^{14} + \zeta^7 + \zeta^{12} + \zeta^6.$$

We form those sums of pairs of η's that yield an η from the previous step. Here there is only one pair, and the sum is given by

$$\eta_0 + \eta_1 = -1.$$

Next we form the elements $\eta^{(t)}$, remembering that for $t > 0$, $\eta^{(t)}$ is the η_r in which ζ^t occurs. Then $\eta^{(0)} = 2$,

$$\eta^{(1)} = \eta^{(9)} = \eta^{(13)} = \eta^{(15)} = \eta^{(16)} = \eta^{(8)} = \eta^{(4)} = \eta^{(2)} = \eta_0,$$
$$\eta^{(3)} = \eta^{(10)} = \eta^{(5)} = \eta^{(11)} = \eta^{(14)} = \eta^{(7)} = \eta^{(12)} = \eta^{(6)} = \eta_1.$$

To compute $\eta_0 \eta_1$ by means of Gauss's product formula, we use $\eta_0 = \eta^{(1)}$ and $\eta_1 = \eta^{(3)}$. Then

$$\eta_0 \eta_1 = \eta^{(1)} \eta^{(3)} = \eta^{(4)} + \eta^{(11)} + \eta^{(6)} + \eta^{(12)} + \eta^{(15)} + \eta^{(8)} + \eta^{(13)} + \eta^{(7)},$$

the indices on the right side being the indices for η_1 plus one. Resubstituting in terms of η_0 and η_1, we obtain

$$\eta_0 \eta_1 = 4\eta_0 + 4\eta_1 = -4.$$

Therefore η_0 and η_1 are the roots of the quadratic equation

$$x^2 + x - 4 = 0, \qquad \text{namely } \tfrac{1}{2}(-1 \pm \sqrt{17}).$$

Deciding which root is η_0 and which is η_1 involves looking at signs. The two roots of the quadratic equation are of opposite sign. Since

$$\eta_0 = (\zeta^1 + \zeta^{-1}) + (\zeta^2 + \zeta^{-2}) + (\zeta^4 + \zeta^{-4}) + (\zeta^8 + \zeta^{-8})$$
$$= 2\big(\cos(2\pi/17) + \cos(4\pi/17) + \cos(8\pi/17) + \cos(16\pi/17)\big)$$
$$> 2(\tfrac{1}{2} + \tfrac{1}{2} + 0 + (-1)) = 0,$$

---

[11]The discussion of this example closely follows that in Van der Waerden, Vol. I, Section 54.

12. Construction of Regular Polygons

η_0 is the positive root, and we have

$$\eta_0 = \tfrac{1}{2}(-1 + \sqrt{17}) \quad \text{and} \quad \eta_1 = \tfrac{1}{2}(-1 - \sqrt{17}).$$

For $k = 4$, the 4 periods have 4 terms each, and

$$\xi_0 = \zeta^{3^0+4\cdot 0} + \zeta^{3^0+4\cdot 1} + \zeta^{3^0+4\cdot 2} + \zeta^{3^0+4\cdot 3} = \zeta^1 + \zeta^{13} + \zeta^{16} + \zeta^4,$$

$$\xi_1 = \zeta^{3^1+4\cdot 0} + \zeta^{3^1+4\cdot 1} + \zeta^{3^1+4\cdot 2} + \zeta^{3^1+4\cdot 3} = \zeta^3 + \zeta^5 + \zeta^{14} + \zeta^{12},$$

$$\xi_2 = \zeta^{3^2+4\cdot 0} + \zeta^{3^2+4\cdot 1} + \zeta^{3^2+4\cdot 2} + \zeta^{3^2+4\cdot 3} = \zeta^9 + \zeta^{15} + \zeta^8 + \zeta^2,$$

$$\xi_3 = \zeta^{3^3+4\cdot 0} + \zeta^{3^3+4\cdot 1} + \zeta^{3^3+4\cdot 2} + \zeta^{3^3+4\cdot 3} = \zeta^{10} + \zeta^{11} + \zeta^7 + \zeta^6.$$

The sums of pairs of these that yield η's are

$$\xi_0 + \xi_2 = \eta_0$$
$$\xi_1 + \xi_3 = \eta_1.$$

We can read off superscripted ξ's from the exponents on the right sides of the formulas for $\xi_0, \ldots, \xi_3$, and the results are

$$\xi^{(1)} = \xi^{(13)} = \xi^{(16)} = \xi^{(4)} = \xi_0,$$
$$\xi^{(3)} = \xi^{(5)} = \xi^{(14)} = \xi^{(12)} = \xi_1,$$
$$\xi^{(9)} = \xi^{(15)} = \xi^{(8)} = \xi^{(2)} = \xi_2,$$
$$\xi^{(10)} = \xi^{(11)} = \xi^{(7)} = \xi^{(6)} = \xi_3.$$

Then the relevant products are

$$\xi_0\xi_2 = \xi^{(1)}\xi^{(9)} = \xi^{(10)} + \xi^{(16)} + \xi^{(9)} + \xi^{(3)} = \xi_3 + \xi_0 + \xi_2 + \xi_1 = -1,$$
$$\xi_1\xi_3 = \xi^{(3)}\xi^{(6)} = \xi^{(13)} + \xi^{(14)} + \xi^{(10)} + \xi^{(9)} = \xi_0 + \xi_1 + \xi_3 + \xi_2 = -1.$$

Thus ξ_0 and ξ_2 are the roots of the quadratic equation

$$y^2 - \eta_0 y - 1 = 0,$$

while ξ_1 and ξ_3 are the roots of the quadratic equation

$$y^2 - \eta_1 y - 1 = 0.$$

Since $\xi_0\xi_2$ and $\xi_1\xi_3$ are negative, these equations each have roots of opposite sign. We observe that $\xi_0 = 2(\cos(2\pi/17) + \cos(8\pi/17)) > 0$ and that $\xi_3 = 2(\cos(14\pi/17) + \cos(12\pi/17)) < 0$, and we conclude that the signs are

$$\xi_0 > 0 \quad \text{and} \quad \xi_2 < 0,$$
$$\xi_1 > 0 \quad \text{and} \quad \xi_3 < 0.$$

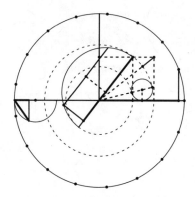

FIGURE 9.4. Construction of a regular 17-gon. The small circle has center $\left(\frac{1}{2}, \frac{1}{8}\right)$ and radius $\frac{1}{8}$. Two circles are drawn tangent to it with center $(0, 0)$; their radii are $\eta_0/4$ and $|\eta_1|/4$. Their x intercepts and height $\frac{1}{2}$ determine the dashed box. The diameter of the large solid semicircle is $\xi_0/2$, and its heavy part is $\lambda_0/2$. The separate semicircle at the left constructs $\sqrt{\xi_1/4}$ from $\xi_1/2$, and the chord in the large semicircle is at distance $\sqrt{\xi_1/4}$ from the diameter.

For $k = 8$, the 8 periods have 2 terms each, and the two with sum ξ_0 are
$$\lambda_0 = \zeta^{3^{0+8\cdot 0}} + \zeta^{3^{0+8\cdot 1}} = \zeta^1 + \zeta^{16},$$
$$\lambda_4 = \zeta^{3^{4+8\cdot 0}} + \zeta^{3^{4+8\cdot 1}} = \zeta^{13} + \zeta^4.$$
Their sum and their product are given by
$$\lambda_0 + \lambda_4 = \xi_0,$$
$$\lambda_0 \lambda_4 = \zeta^{14} + \zeta^5 + \zeta^{12} + \zeta^3 = \xi_1.$$
Thus λ_0 and λ_4 are the roots of the quadratic equation
$$z^2 - \xi_0 z + \xi_1 = 0.$$
Since $\lambda_0 = 2\cos(2\pi/17) > 2\cos(8\pi/17) = \lambda_4$, λ_0 is the larger of the two roots of the equation.

In summary, we have successively defined
$$\eta_0 = \tfrac{1}{2}\left(-1 + \sqrt{17}\right) \quad \text{and} \quad \eta_1 = \tfrac{1}{2}\left(-1 - \sqrt{17}\right),$$
$$\xi_0 = \tfrac{1}{2}\left(\eta_0 + \sqrt{\eta_0^2 + 4}\right) \quad \text{and} \quad \xi_2 = \tfrac{1}{2}\left(\eta_0 - \sqrt{\eta_0^2 + 4}\right),$$
$$\xi_1 = \tfrac{1}{2}\left(\eta_1 + \sqrt{\eta_1^2 + 4}\right) \quad \text{and} \quad \xi_3 = \tfrac{1}{2}\left(\eta_1 - \sqrt{\eta_1^2 + 4}\right),$$
$$\lambda_0 = \tfrac{1}{2}\left(\xi_0 + \sqrt{\xi_0^2 - 4\xi_1}\right).$$

Since $\lambda_0 = 2\cos(2\pi/17)$, these formulas explicitly point to how to construct a regular 17-gon. See Figure 9.4.

13. Solution of Certain Polynomial Equations with Solvable Galois Group

In this section we investigate what specific information can be deduced about a finite Galois extension in characteristic 0 when the Galois group is solvable. The tool is a precursor of modern harmonic analysis[12] known as "Lagrange resolvents." The argument of the previous section could be regarded as an instance of applying the theory of Lagrange resolvents, but Lagrange resolvents give only the simpler formulas of the previous section, not the Gauss product formula.

Proposition 9.47. Let $\mathbb{K}$ be a finite normal extension of a field $\Bbbk$ of characteristic 0, suppose that $\text{Gal}(\mathbb{K}/\Bbbk)$ is cyclic of order n with σ as a generator, and suppose that $X^n - 1$ splits in $\Bbbk$. Fix a generator σ of $\text{Gal}(\mathbb{K}/\Bbbk)$ and a primitive n^{th} root ω of 1 in $\Bbbk$. For $0 \le r < n$, define $\Bbbk$ linear maps $E_r : \mathbb{K} \to \mathbb{K}$ by

$$E_r x = n^{-1} \sum_{k \bmod n} \omega^{-kr} \sigma^k x \qquad \text{for } x \in \mathbb{K}.$$

Then

(a) $E_r E_s$ equals E_s if $r = s$ and equals 0 if $r \not\equiv s \bmod n$, so that the E_r's are commuting projection operators whose images are linearly independent,
(b) $\sum_{r \bmod n} E_r = I$, so that the direct sum of the images of the E_r's is all of $\mathbb{K}$,
(c) $\sigma(x) = \omega^r x$ for all r and for all x in image E_r,
(d) image $E_0 = \Bbbk$.

REMARKS. The integers k and r depend only on their values modulo n, and the summation indices "$k \bmod n$" and "$r \bmod n$" are to be interpreted accordingly. The operators E_r are known classically as **Lagrange resolvents**, apart from the constant n^{-1}. The proposition says that the $\Bbbk$ linear map σ has a basis of eigenvectors, that the eigenvalues are a subset of the powers ω^r, and that each E_r is the projection operator on the eigenspace for the eigenvalue ω^r along the sum of the remaining eigenspaces.

---

[12]Lagrange resolvents give a certain specific Fourier decomposition relative to a cyclic group. Similar formulas apply whenever a cyclic group acts linearly on a vector space over $\Bbbk$ and the relevant roots of 1 lie in $\Bbbk$. For the corresponding decomposition of a vector space over $\mathbb{C}$ when a finite group G acts linearly, see Problems 47–52 at the end of Chapter VII. The decomposition in those problems can be seen to work for any field $\Bbbk$ of characteristic 0 for which the values of all irreducible characters of G lie in $\Bbbk$. The values of the characters are sums of certain roots of 1, and thus it is enough that $\Bbbk$ contain a certain finite set of roots of 1.

PROOF. For x in $\mathbb{K}$, we compute

$$E_r E_s x = n^{-2} \sum_{k \bmod n} \omega^{-kr} \sigma^k \Big(\sum_{l \bmod n} \omega^{-ls} \sigma^l x \Big)$$

$$= n^{-2} \sum_{k \bmod n} \sum_{m \bmod n} \omega^{-kr} \sigma^k \omega^{-ms+ks} \sigma^{m-k} x$$

$$= n^{-2} \sum_{m \bmod n} \Big(\sum_{k \bmod n} \omega^{k(s-r)} \Big) \omega^{-ms} \sigma^m x.$$

The expression in parentheses on the right side is the sum of a finite geometric series. If $s \equiv r \bmod n$, then every term in the sum is 1, and the sum is n. If $s \not\equiv r \bmod n$, then the sum is $\frac{1-\omega^{n(s-r)}}{1-\omega^{s-r}} = 0$. Thus (a) follows.

Next we calculate

$$\sum_{r \bmod n} E_r x = \sum_{r \bmod n} n^{-1} \sum_{k \bmod n} \omega^{-kr} \sigma^k x = \sum_{k \bmod n} n^{-1} \Big(\sum_{r \bmod n} \omega^{-kr} \Big) \sigma^k x.$$

As in the previous paragraph, the sum in parentheses is n if $k = 0$ and it is 0 if $k \not\equiv 0 \bmod n$. Therefore only the $k = 0$ term on the right side contributes, and the right side simplifies to x. This proves (b).

The computation

$$\sigma(E_r x) = n^{-1} \sum_{k \bmod n} \omega^{-kr} \sigma^{k+1} x$$

$$= n^{-1} \sum_{l \bmod n} \omega^{(-l+1)r} \sigma^l x$$

$$= \omega^r n^{-1} \sum_{l \bmod n} \omega^{-lr} \sigma^l x = \omega^r E_r x$$

shows that $\sigma(y) = \omega^r y$ for every y of the form $E_r x$, and these y's are the members of the image of E_r. This proves (c).

Combining (b) and (c), we see that $\sigma(x) = x$ if and only if x is in image E_0. Since $\text{Gal}(\mathbb{K}/\Bbbk)$ is cyclic, the members of $\mathbb{K}$ fixed by σ are the members fixed by the Galois group, and these are the members of $\Bbbk$ by Proposition 9.35d. This proves (d). $\square$

Corollary 9.48. Let $\mathbb{K}$ be a finite normal extension of a field $\Bbbk$ of characteristic 0, suppose that $\text{Gal}(\mathbb{K}/\Bbbk)$ is cyclic of prime order p, and suppose that $X^p - 1$ splits in $\Bbbk$. Then there exist a in $\Bbbk$ and x in $\mathbb{K}$ such that $x^p = a$ and $\mathbb{K} = \Bbbk(x)$.

REMARKS. In other words, a finite normal extension field in characteristic 0 with Galois group cyclic of prime order p is necessarily obtained by adjoining a pth root of some element of the base field, provided that the base field contains all the pth roots of 1. Once the extension field contains one p^{th} root of an element of the base field, it has to contain all p^{th} roots, since the base field by assumption contains a full complement of pth roots of 1.

PROOF. We apply Proposition 9.47 with $n = p$. Since $[\mathbb{K} : \Bbbk] = p > 1$, (d) shows that E_0 is not the identity. By (b), some E_r with $r \neq 0$ is not the 0 operator. Let x be a nonzero element in image E_r. Since the generator σ of the Galois group is a field automorphism, $\sigma(x^p) = \sigma(x)^p = (\omega^r x)^p = \omega^{rp} x^p = x^p$. Since x^p is fixed by the Galois group, x^p lies in $\Bbbk$. Then the element $a = x^p$ has the property that $x^p = a$ and $\mathbb{K} \supseteq \Bbbk(x) \supsetneq \Bbbk$. Since $[\mathbb{K} : \Bbbk]$ is prime, Corollary 9.7 shows that there are no intermediate fields between $\mathbb{K}$ and $\mathbb{K}$. Therefore $\mathbb{K} = \Bbbk(x)$. □

We shall apply Corollary 9.48 to prove the converse statement in Theorem 9.44—that solvability of the Galois group for a polynomial equation in characteristic 0 implies that the solutions of the equation are expressible in terms of radicals and the base field. We begin with a lemma that handles a special case.

Lemma 9.49. Let $\Bbbk$ be a field of characteristic 0, let $n > 0$ be an integer, and let $\mathbb{K}$ be a splitting field for $\prod_{r=1}^{n}(X^r - 1)$ over $\Bbbk$. Then $\mathbb{K}/\Bbbk$ is a Galois extension, the Galois group of $\mathrm{Gal}(\mathbb{K}/\Bbbk)$ is abelian, and $\mathbb{K}$ has a root tower over $\Bbbk$.

PROOF. Being a splitting field in characteristic 0, $\mathbb{K}$ is a finite Galois extension of $\Bbbk$. For $1 \leq r \leq n$, let ω_r be a primitive r^{th} root of 1 in $\mathbb{K}$. The primitive r^{th} roots of 1 are parametrized by the group $(\mathbb{Z}/r\mathbb{Z})^\times$ once some ω_r is specified, the parametrization being $k \mapsto \omega_r^k$. If σ is in $\mathrm{Gal}(\mathbb{K}/\Bbbk)$, then $\sigma(\omega_r) = \omega_r^k$ for some such k. This correspondence respects multiplication in $(\mathbb{Z}/r\mathbb{Z})^\times$ since if $\sigma(\omega_r) = \omega_r^k$ and $\sigma'(\omega_r) = \omega_r^l$, then $\sigma'(\sigma(\omega_r)) = \sigma'(\omega_r^k) = \sigma'(\omega_r)^k = \omega_r^{kl}$. Thus for each r, we have a homomorphism of $\mathrm{Gal}(\mathbb{K}/\Bbbk)$ into the abelian group $(\mathbb{Z}/r\mathbb{Z})^\times$. Putting these homomorphisms together as r varies and using the fact that the ω_r's generate $\mathbb{K}$ over $\Bbbk$, we obtain a one-one homomorphism of $\mathrm{Gal}(\mathbb{K}/\Bbbk)$ into the abelian group $\prod_{r=1}^{n}(\mathbb{Z}/r\mathbb{Z})^\times$. Consequently $\mathrm{Gal}(\mathbb{K}/\Bbbk)$ is isomorphic to a subgroup of an abelian group and is abelian.

It follows from Corollary 9.39 that every extension of intermediate fields is Galois and has abelian Galois group. For $1 \leq r \leq n$, we introduce the intermediate field $\mathbb{K}_r = \Bbbk(\omega_1, \omega_2, \ldots, \omega_r)$. Here $\mathbb{K}_1 = \Bbbk(1) = \Bbbk$. For $1 < r < n$, $\mathbb{K}_r$ is generated as a vector space over $\mathbb{K}_{r-1}$ by $\omega_r, \omega_r^2, \ldots, \omega_r^{r-1}$ since $\sum_{k=0}^{r-1} \omega_r^k = 0$ for $r > 1$, and thus $[\mathbb{K}_r : \mathbb{K}_{r-1}] \leq r - 1$. Since $\mathrm{Gal}(\mathbb{K}_r/\mathbb{K}_{r-1})$ is abelian, it has a composition series whose consecutive quotients are cyclic of prime order, the prime order necessarily being $\leq [\mathbb{K}_r : \mathbb{K}_{r-1}] \leq r - 1$. Applying Galois theory, form the chain of intermediate extensions between $\mathbb{K}_{r-1}$ and $\mathbb{K}_r$. The degree of each extension is some prime p with $p \leq r - 1$, the prime depending on the two fields in the chain. The pth roots of unity are in the smaller of any two consecutive fields because they are in $\mathbb{K}_{r-1}$. By Corollary 9.48, such a degree-p extension between $\mathbb{K}_{r-1}$ and $\mathbb{K}_r$ is generated by the smaller field and the pth root of an element in the smaller field. Since $\mathbb{K}_1 = \Bbbk$, we see inductively that $\mathbb{K}_r$ has a root tower over $\mathbb{K}_{r-1}$ for each r. Since $\mathbb{K} = \mathbb{K}_n$, $\mathbb{K}$ has a root tower over $\Bbbk$. □

PROOF OF SUFFICIENCY IN THEOREM 9.44 THAT $\text{Gal}(\mathbb{K}/\Bbbk)$ BE SOLVABLE. Let $F(X)$ be in $\Bbbk[X]$, and suppose that $\mathbb{K}$ is a splitting field of $F(X)$ over $\Bbbk$. Under the assumption that $\text{Gal}(\mathbb{K}/\Bbbk)$ is solvable, we are to prove that there exists a finite extension $\mathbb{K}'$ of $\mathbb{K}$ having a root tower.

Since $G = \text{Gal}(\mathbb{K}/\Bbbk)$ is solvable, we can find a finite sequence of subgroups of G, each normal in the next larger one, such that the quotient of any consecutive pair is cyclic of prime order. We write

$$G = H_0 \supseteq H_1 \supseteq \cdots \supseteq H_{k-1} \supseteq H_k = \{1\}$$

with H_j/H_{j+1} cyclic of prime order p_j for $0 \leq j < k$. Let

$$\Bbbk = \mathbb{K}_0 \subseteq \mathbb{K}_1 \subseteq \cdots \subseteq \mathbb{K}_{k-1} \subseteq \mathbb{K}_k = \mathbb{K}$$

be the corresponding sequence of intermediate fields given by the Fundamental Theorem of Galois Theory (Theorem 9.38). Here $\mathbb{K}_j = \mathbb{K}^{H_j}$, and $H_j = \text{Gal}(\mathbb{K}/\mathbb{K}_j)$.

According to Corollary 9.39, $\mathbb{K}_{j+1}$ is a normal extension of $\mathbb{K}_j$ if and only if $\text{Gal}(\mathbb{K}/\mathbb{K}_{j+1})$ is a normal subgroup of $\text{Gal}(\mathbb{K}/\mathbb{K}_j)$, and in this case we have a group isomorphism $\text{Gal}(\mathbb{K}/\mathbb{K}_j)/\text{Gal}(\mathbb{K}/\mathbb{K}_{j+1}) \cong \text{Gal}(\mathbb{K}_{j+1}/\mathbb{K}_j)$. Since H_{j+1} is a normal subgroup of H_j with quotient cyclic of order p_j, it follows that $\mathbb{K}_{j+1}/\mathbb{K}_j$ is indeed normal and the Galois group is cyclic of order p_j.

Let us use Theorem 9.22 to regard $\mathbb{K}$ as lying in a fixed algebraic closure $\overline{\mathbb{K}}'$. Let n be the product of all the primes p_j, and let $\mathbb{K}'_0$ be the splitting field over $\Bbbk$ for $\prod_{r=1}^{n}(X^r - 1)$ within $\overline{\mathbb{K}}'$. For $1 \leq j \leq k$, let $\mathbb{K}'_j$ be the subfield of $\overline{\mathbb{K}}'$ generated by $\mathbb{K}_j$ and $\mathbb{K}'_0$. We define $\mathbb{K}' = \mathbb{K}'_k$. Then we have

$$\Bbbk \subseteq \mathbb{K}'_0 \subseteq \mathbb{K}'_1 \subseteq \cdots \subseteq \mathbb{K}'_{k-1} \subseteq \mathbb{K}'_k = \mathbb{K}'.$$

Lemma 9.49 shows that $\mathbb{K}'_0$ has a root tower over $\mathbb{K}'$. To complete the proof, it is enough to show for each $j \geq 0$ that either $\mathbb{K}'_{j+1} = \mathbb{K}'_j$ or else $[\mathbb{K}'_{j+1} : \mathbb{K}'_j] = p_j$ and $\mathbb{K}'_{j+1}$ is generated by $\mathbb{K}'_j$ and the p_j^{th} root of some member of $\mathbb{K}'_j$.

For each $j \geq 0$, suppose that $\mathbb{K}_{j+1} = \mathbb{K}_j(x_j)$. Let $F_j(X)$ be the minimal polynomial of x_j over $\mathbb{K}_j$. Since $\mathbb{K}_{j+1}/\mathbb{K}_j$ is normal, $\mathbb{K}_{j+1}$ is the splitting field of $F_j(X)$ over $\mathbb{K}_j$. Then $\mathbb{K}'_{j+1} = \mathbb{K}'_j(x_j)$ is the splitting field of $F_j(X) \prod_{r=1}^{n}(X^r - 1)$ over $\mathbb{K}'_j$, and consequently $\mathbb{K}'_{j+1}/\mathbb{K}'_j$ is a normal extension. If g is in $\text{Gal}(\mathbb{K}'_{j+1}/\mathbb{K}'_j)$, then g sends x_j into a root of $F_j(X)$ and is determined by this root. The restriction $g|_{\mathbb{K}_{j+1}}$ therefore carries $\mathbb{K}_{j+1}$ into itself and is in $\text{Gal}(\mathbb{K}_{j+1}/\mathbb{K}_j)$. Since g is determined by $g(x_j)$, the group homomorphism $g \mapsto g|_{\mathbb{K}_{j+1}}$ is one-one. The image of this homomorphism must be a subgroup of $\text{Gal}(\mathbb{K}_{j+1}/\mathbb{K}_j)$ and therefore must be trivial or have p_j elements. In the first case, $\mathbb{K}'_{j+1} = \mathbb{K}'_j$, and in the second case, $[\mathbb{K}'_{j+1} : \mathbb{K}'_j] = p_j$. In the latter case, $\mathbb{K}'_j$ contains all p_j of the p_j^{th} roots of 1 since these roots of 1 are in $\mathbb{K}'_0$; by Corollary 9.48, $\mathbb{K}'_{j+1}$ is generated by $\mathbb{K}'_j$ and a p_j^{th} root of some member of $\mathbb{K}'_j$. This completes the proof. □

13. Solution of Certain Polynomial Equations with Solvable Galois Group

We turn now to apply our methods to irreducible cubics over a field $\Bbbk$ of characteristic 0. In effect we shall derive Cardan's formula,[13] which was mentioned at the beginning of Section 11.

The Galois group of a splitting field of a cubic polynomial has to be a subgroup of the symmetric group $\mathfrak{S}_3$, and irreducibility of the cubic implies that the Galois group has to contain a 3-cycle. Therefore the Galois group has to be either $\mathfrak{S}_3$ or the alternating group $\mathfrak{A}_3 \cong C_3$.

Let the cubic be $X^3 + a_2 X^2 + a_1 X + a_0$, the coefficients being in $\Bbbk$. Substituting $X = Z - \frac{1}{3}a_2$ converts the polynomial into

$$(Z - \tfrac{1}{3}a_2)^3 + a_2(Z - \tfrac{1}{3}a_2)^2 + a_1(Z - \tfrac{1}{3}a_2) + a_0$$
$$= Z^3 + (a_1 - \tfrac{1}{3}a_2^2)Z + (a_0 - \tfrac{1}{3}a_1 a_2 + \tfrac{2}{27}a_2^3),$$

and therefore we can assume whenever convenient that the given polynomial has $a_2 = 0$.

Suppose for the moment that the Galois group is $G = \mathfrak{S}_3$. A composition series is

$$G = \mathfrak{S}_3 \supseteq \mathfrak{A}_3 \supseteq \{1\},$$

and we can write the corresponding sequence of fixed fields as

$$\Bbbk \subseteq \mathbb{L} \subseteq \mathbb{K},$$

where $\mathbb{K}$ is the splitting field and $\mathbb{L}$ is $\mathbb{K}^{\mathfrak{A}_3}$. The dimensions satisfy $[\mathbb{L} : \Bbbk] = 2$ and $[\mathbb{K} : \mathbb{L}] = 3$.

Let the roots in $\mathbb{K}$ of the given cubic be r_1, r_2, r_3. Since G is solvable, Theorem 9.44 tells us that the roots are expressible in terms of radicals and members of $\Bbbk$. To derive explicit formulas for the roots, the idea is to use a two-step process with Lagrange resolvents, arguing as in the proof of Corollary 9.48 at each step.

The first step involves passing from $\Bbbk$ to $\mathbb{L}$. The square roots of 1 are already in $\Bbbk$, and $\mathbb{L}$ is to be obtained from $\Bbbk$ by adjoining one of the square roots of some element of $\Bbbk$. In Proposition 9.47 the Galois group $\text{Gal}(\mathbb{L}/\Bbbk)$ is a 2-element quotient group, the sum is over members of the quotient group, and the element x is in $\mathbb{L}$. It is a little more convenient to pull the sum back to one over the 6-element symmetric group, taking ω to be the sign function on $\mathfrak{S}_3$ and taking x to be any element of $\mathbb{K}$. The formulas for the projection operators E_0 and E_1 are then

$$E_0 x = \tfrac{1}{6} \sum_{\sigma \in \mathfrak{S}_3} \sigma(x),$$

$$E_1 x = \tfrac{1}{6} \sum_{\sigma \in \mathfrak{S}_3} (\text{sgn}\,\sigma) \sigma(x),$$

---

[13] We discuss only Cardan's cubic formula, omitting any discussion of the corresponding quartic formula, which often bears Cardan's name and which can be handled with the same techniques. See Van der Waerden, Vol. I, Section 58, for details.

with x in $\mathbb{K}$, and the proof of Corollary 9.48 tells us to adjoin to $\mathbb{k}$ the square root of any element of image E_1, i.e., any element with $\sigma(x) = (\operatorname{sgn} x)x$ for all σ in $\mathfrak{S}_3$.

The only elements of $\mathbb{K}$ for which we have good control of the action of the Galois group, apart from the elements of $\mathbb{k}$, are the elements that are expressed directly in terms of the roots r_1, r_2, r_3 of the polynomial. By renumbering the roots if necessary, we may assume that the roots are permuted by $\mathfrak{S}_3$ according to their subscripts. An example of a polynomial function of r_1, r_2, r_3 that transforms according to the sign of the permutation played a role in Section I.4 in defining the sign of a permutation. It is the **difference product** of the polynomial, namely

$$\prod_{1 \leq i < j \leq 3} (r_j - r_i).$$

This is a square root of the **discriminant** D of the polynomial, which is given by

$$D = \prod_{1 \leq i < j \leq 3} (r_j - r_i)^2.$$

We shall compute D in terms of the coefficients of the cubic shortly. In the meantime, the proof of Corollary 9.48 thus tells us that $\mathbb{L} = \mathbb{k}(\sqrt{D})$. Here $\sqrt{D}$ is given by

$$\sqrt{D} = (r_3 - r_2)(r_3 - r_1)(r_2 - r_1)$$
$$= (r_1 r_2^2 + r_2 r_3^2 + r_3 r_1^2) - (r_1^2 r_2 + r_2^2 r_3 + r_3^2 r_1).$$

The second step is to pass from $\mathbb{L}$ to $\mathbb{K}$. Corollary 9.48 says to expect $\mathbb{K}$ to be obtained by adjoining the cube root of something if the cube roots of 1 are already present in $\mathbb{L}$. The proof of the second half of Theorem 9.44, which follows Corollary 9.48, indicates how we can incorporate the cube roots of 1 into the fields in order to have a root tower. What we can do is to replace $\mathbb{k}$ at the start by a splitting field for $\prod_{1 \leq r \leq 3} (X^r - 1)$. Since ± 1 are already in $\mathbb{k}$, we are to adjoin the nontrivial cube roots of 1, i.e., the roots of $X^2 + X + 1$, if they are not already present. In other words, what we do is replace $\mathbb{k}$ at the start by $\mathbb{k}(\sqrt{-3})$. Changing notation, we assume that $\sqrt{-3}$ lies in $\mathbb{k}$ from the outset.

We can now use Lagrange resolvents. Let σ be the generator (1 2 3) of $\mathfrak{A}_3$, sending r_1 to r_2, r_2 to r_3, and r_3 to r_1. Let $\omega = \frac{1}{2}(-1 + \sqrt{-3})$ be a primitive cube root of 1. Then we have

$$E_0 x = \tfrac{1}{3}(x + \sigma x + \sigma^2 x),$$
$$E_1 x = \tfrac{1}{3}(x + \omega^{-1} \sigma x + \omega^{-2} \sigma^2 x),$$
$$E_2 x = \tfrac{1}{3}(x + \omega^{-2} \sigma x + \omega^{-1} \sigma^2 x).$$

Again we can use any x, but the roots of the cubic are the simplest nontrivial elements for which we know the action of σ. Corollary 9.48 shows that $\mathbb{K} = \mathbb{L}(E_1 x)$ if $E_1 x \neq 0$. Proposition 9.47 says that $(E_1 x)^3$ is fixed by σ, and it therefore lies in $\mathbb{L}$. Hence $\mathbb{K}$ is identified as obtained from $\mathbb{L}$ by adjoining a cube root of the element $(E_1 x)^3$ of $\mathbb{L}$.

Taking $x = r_1$, we have $\sigma x = r_2$ and $\sigma^2 x = r_3$. Also, $\omega^{\pm 1} = \frac{1}{2}(-1 \pm \sqrt{-3})$. Using the formula for $E_1 x$ and substituting for $\sqrt{D}$ and $\omega^{\pm 1}$ then gives

$$(3E_1 r_1)^3 = r_1^3 + r_2^3 + r_3^3 + 6r_1 r_2 r_3$$
$$+ 3\omega^{-1}(r_1^2 r_2 + r_2^2 r_3 + r_3^2 r_1) + 3\omega(r_1 r_2^2 + r_2 r_3^2 + r_3 r_1^2)$$
$$= \sum_i r_i^3 + 6r_1 r_2 r_3 - \tfrac{3}{2} \sum_{i \neq j} r_i^2 r_j + \tfrac{3}{2}\sqrt{-3}\sqrt{D}.$$

To proceed further, we shall want to substitute expressions involving the coefficients of the cubic for the above symmetric expressions in the roots.[14] These expressions will be considerably simplified if we assume that the coefficient of X^2 in the cubic is 0. We know that this assumption involves no loss of generality. Thus we assume for the remainder of this section that the cubic is $X^3 + pX + q$. The relevant formulas relating the roots and the coefficients are

$$r_1 + r_2 + r_3 = 0,$$
$$r_1 r_2 + r_1 r_3 + r_2 r_3 = p,$$
$$r_1 r_2 r_3 = -q.$$

Aiming for the right side of the displayed formula for $(3E_1 r_1)^3$, we have

$$0 = (r_1 + r_2 + r_3)^3 = \sum_i r_i^3 + 3 \sum_{i \neq j} r_i^2 r_j + 6 r_1 r_2 r_3,$$
$$0 = (r_1 + r_2 + r_3)(r_1 r_2 + r_1 r_3 + r_2 r_3) = -\tfrac{9}{2} \sum_{i \neq j} r_i^2 r_j - \tfrac{27}{2} r_1 r_2 r_3,$$
$$-\tfrac{27}{2} q = \tfrac{27}{2} r_1 r_2 r_3.$$

Addition of these three lines and comparison with the expression for $3(E_1 r_1)^3$ yields

$$-\tfrac{27}{2} q = \sum_i r_i^3 - \tfrac{3}{2} \sum_{i \neq j} r_i^2 r_j + 6 r_1 r_2 r_3 = (3E_1 r_1)^3 - \tfrac{3}{2}\sqrt{-3}\sqrt{D}.$$

Consequently
$$(3E_1 r_1)^3 = -\tfrac{27}{2} q + \tfrac{3}{2}\sqrt{-3}\sqrt{D}.$$

---

[14]Problems 36–39 at the end of Chapter VIII assure us that this rewriting is possible. For our derivation this assurance is not logically necessary, since we will be producing explicit formulas.

Similarly
$$(3E_2r_1)^3 = -\tfrac{27}{2}q - \tfrac{3}{2}\sqrt{-3}\sqrt{D}.$$

Since $3E_0r_1 = r_1 + r_2 + r_3 = 0$, we have expressions for E_0r_1, E_1r_1, and E_2r_1, apart from the choices of the cube roots. Proposition 9.47b says that we recover r_1 by addition: $r_1 = E_0r_1 + E_1r_1 + E_2r_1$. Thus we have found a root explicitly as soon as we sort out the ambiguity in the choices of cube roots and determine the value of D in terms of the coefficients p and q.

Theorem 9.50 (Cardan's formula). Let $\mathbb{k}$ be a field of characteristic 0 containing $\sqrt{-3}$, and let $X^3 + pX + q$ be an irreducible cubic in $\mathbb{k}[X]$. For this polynomial the discriminant D is given by
$$D = -4p^3 - 27q^2.$$

The Galois group of a splitting field of the cubic is $\mathfrak{S}_3$ if D is a nonsquare in $\mathbb{k}$ and is $\mathfrak{A}_3$ if D is a square in $\mathbb{k}$. In either case, fix a square root of D, denote it by $\sqrt{D}$, and let $\omega^{\pm 1} = \tfrac{1}{2}(-1 \pm \sqrt{-3})$ be the primitive cube roots of 1. Then it is possible to determine cube roots of the form

$$3E_1r_1 = \sqrt[3]{-\tfrac{27}{2}q + \tfrac{3}{2}\sqrt{-3}\sqrt{D}} \quad \text{and} \quad 3E_2r_1 = \sqrt[3]{-\tfrac{27}{2}q - \tfrac{3}{2}\sqrt{-3}\sqrt{D}}$$

in such a way that their product is $(3E_1r_1)(3E_1r_2) = -3p$, and in this case the three roots of $X^3 + pX + q$ are given by

$$r_1 = E_1r_1 + E_2r_1,$$
$$r_2 = \omega E_1r_1 + \omega^2 E_2r_1,$$
$$r_3 = \omega^2 E_1r_1 + \omega E_2r_1.$$

PROOF. Define $\sigma_k = r_1^k + r_2^k + r_3^k$ for $1 \leq k \leq 4$. By inspection we have

$$\begin{pmatrix} 1 & 1 & 1 \\ r_1 & r_2 & r_3 \\ r_1^2 & r_2^2 & r_3^2 \end{pmatrix} \begin{pmatrix} 1 & r_1 & r_1^2 \\ 1 & r_2 & r_2^2 \\ 1 & r_3 & r_3^2 \end{pmatrix} = \begin{pmatrix} 3 & \sigma_1 & \sigma_2 \\ \sigma_1 & \sigma_2 & \sigma_3 \\ \sigma_2 & \sigma_3 & \sigma_4 \end{pmatrix}.$$

Taking the determinant of both sides and applying Corollary 5.3, we obtain

$$D = \det \begin{pmatrix} 3 & \sigma_1 & \sigma_2 \\ \sigma_1 & \sigma_2 & \sigma_3 \\ \sigma_2 & \sigma_3 & \sigma_4 \end{pmatrix} = 3\sigma_2\sigma_4 - \sigma_2^3 - 3\sigma_3^2.$$

The given cubic shows that $\sigma_1 = r_1 + r_2 + r_3 = 0$. For the other σ_i's, we have

$$\sigma_2 = r_1^2 + r_2^2 + r_3^2 = (r_1 + r_2 + r_3)^2 - 2(r_1 r_2 + r_1 r_3 + r_2 r_3) = -2p,$$

$$\sigma_3 = r_1^3 + r_2^3 + r_3^3 = (r_1 + r_2 + r_3)(r_1^2 + r_2^2 + r_3^2)$$
$$\qquad - (r_1^2 r_2 + r_1^2 r_3 + r_2^2 r_1 + r_2^2 r_3 + r_3^2 r_1 r_3^2 r_2)$$
$$\qquad = -(r_1 + r_2 + r_3)(r_1 r_2 + r_1 r_3 + r_2 r_3) + 3 r_1 r_2 r_3 = -3q,$$

$$\sigma_4 = r_1^4 + r_2^4 + r_3^4 = (r_1^2 + r_2^2 + r_3^2)^2 - 2(r_1^2 r_2^2 + r_1^2 r_3^2 + r_2^2 r_3^2)$$
$$\qquad = (-2p)^2 - 2(r_1 r_2 + r_1 r_3 + r_2 r_3)^2$$
$$\qquad + 4 r_1 r_2 r_3 (r_1 + r_2 + r_3) = (-2p)^2 - 2(p)^2 = 2p^2.$$

Substituting, we obtain $D = -12p^3 + 8p^3 - 27q^2 = -4p^3 - 27q^2$. This proves the formula for D. In particular, it confirms that D lies in $\Bbbk$.

The Galois group of the splitting field of the polynomial must be $\mathfrak{S}_3$ or $\mathfrak{A}_3$. If it is $\mathfrak{S}_3$, then we saw above that $\mathbb{L} = \Bbbk(\sqrt{D})$ and that $[\mathbb{L} : \Bbbk] = 2$. Hence D is a nonsquare in $\Bbbk$. If the Galois group is $\mathfrak{A}_3$, then $(r_3 - r_2)(r_3 - r_1)(r_2 - r_1)$ is fixed by the Galois group and lies in $\Bbbk$. The square of this element is D, and hence D is a square in $\Bbbk$.

With either Galois group the calculations with the cubic extension that precede the statement of the theorem are valid. If r_1 is one of the roots, then we know that

$$r_1 = E_0 r_1 + E_1 r_1 + E_2 r_1 = E_1 r_1 + E_2 r_1,$$

$$(3 E_1 r_1)^3 = -\tfrac{27}{2} q + \tfrac{3}{2}\sqrt{-3}\sqrt{D},$$

$$(3 E_2 r_1)^3 = -\tfrac{27}{2} q - \tfrac{3}{2}\sqrt{-3}\sqrt{D}.$$

The uniqueness of simple extensions (Theorem 9.11) says that we can make any choice of cube root to determine $3 E_1 r_1$. Then

$$(3 E_1 r_1)(3 E_2 r_1) = (r_1 + \omega^{-1} \sigma r_1 + \omega^{-2} \sigma^2 r_1)(r_1 + \omega^{-2} \sigma r_1 + \omega^{-1} \sigma^2 r_1)$$
$$= (r_1 + \omega^{-1} r_2 + \omega r_3)(r_1 + \omega r_2 + \omega^{-1} r_3)$$
$$= (r_1^2 + r_2^2 + r_3^2) + (\omega + \omega^{-1})(r_1 r_2 + r_1 r_3 + r_2 r_3)$$
$$= (r_1^2 + r_2^2 + r_3^2) - (r_1 r_2 + r_1 r_3 + r_2 r_3).$$

The first term on the right side we calculated in the first paragraph of the proof as $\sigma_2 = -2p$, and the second term gives $-p$. Thus $(3 E_1 r_1)(3 E_2 r_1) = -3p$ as asserted. Since σ operates on image E_1 as multiplication by ω and on image E_2 as multiplication by ω^2, the fact that $r_1 = E_1 r_1 + E_2 r_1$ implies that

$$r_2 = \sigma(r_1) = \omega E_1 r_1 + \omega^2 E_2 r_1$$

and

$$r_3 = \sigma^2(r_1) = \omega^2 E_1 r_1 + \omega E_2 r_1.$$

This completes the proof. $\square$

14. Proof That π Is Transcendental

In this section and the next three, we combine Galois theory with some of the ring theory in the second half of Chapter VIII. This combination will allow us to prove some striking theorems, see how Galois groups can be used effectively in practice, and develop some techniques for identifying Galois groups explicitly.

The present section is devoted to the proof of the following theorem.

Theorem 9.51 (Lindemann, 1882). The number π is transcendental over $\mathbb{Q}$.

The argument we give is based on that in a book by L. K. Hua.[15] For purposes of having a precise theorem, π is defined as the least positive real number such that $e^{\pi i} = -1$. In addition to Galois theory in the form of Proposition 9.35, the proof here will make use of a few facts about algebraic integers. Algebraic integers were defined in Section VIII.1 and again in Section VIII.9 (as well as in Section VII.4) as complex numbers that are roots of monic polynomials in $\mathbb{Z}[X]$. The algebraic integers form a ring by Corollary 8.38 (or alternatively by Lemma 7.30), the only algebraic integers in $\mathbb{Q}$ are the members of $\mathbb{Z}$ by Proposition 8.41 (or alternatively by Lemma 7.30), and any algebraic number x has the property that nx is an algebraic integer for some integer $n \neq 0$ by Proposition 8.42.

We begin with a lemma.

Lemma 9.52. Let $f(X)$ in $\mathbb{C}[X]$ be given by $f(X) = \sum_{k=0}^{n} a_k X^k$, and define $F(X)$ to be the sum of the derivatives of $f(X)$:

$$F(X) = \sum_{l=0}^{n} f^{(l)}(X).$$

If $Q(z)$ is defined as $Q(z) = F(0)e^z - F(z)$ for $z \in \mathbb{C}$, then $F(0) = \sum_{k=0}^{n} a_k k!$ and

$$|Q(z)| \leq e^{|z|} \sum_{k=0}^{n} |a_k| |z|^k.$$

PROOF. We calculate directly that

$$F(z) = \sum_{l=0}^{n} \sum_{k=l}^{n} \frac{a_k k!}{(k-l)!} z^{k-l} = \sum_{k=0}^{n} a_k \sum_{l=0}^{k} \frac{k!}{(k-l)!} z^{k-l} = \sum_{k=0}^{n} a_k \sum_{l=0}^{k} \frac{k!}{l!} z^l.$$

---

[15] *Introduction to Number Theory*, pp. 484–488. In the same pages Hua establishes the earlier theorem of Hermite that e is transcendental, using a related but simpler argument.

Evaluation at $z = 0$ gives $F(0) = \sum_{k=0}^{n} a_k k!$. Then

$$|Q(z)| \le \left| \sum_{k=0}^{n} a_k \sum_{l=0}^{\infty} \frac{k!}{l!} z^l - \sum_{k=0}^{n} \sum_{l=0}^{k} \frac{k!}{l!} z^l \right|$$

$$= \left| \sum_{k=0}^{n} a_k \sum_{l=k+1}^{\infty} \frac{k!}{l!} z^l \right|$$

$$\le \sum_{k=0}^{n} |a_k| \sum_{l=k+1}^{\infty} \frac{|z|^l}{(l-k)!} \qquad \text{since } \binom{l}{k}^{-1} \le 1$$

$$= \sum_{k=0}^{n} |a_k| |z|^k \sum_{m=1}^{\infty} \frac{|z|^m}{m!}$$

$$\le e^{|z|} \sum_{k=0}^{n} |a_k| |z|^k. \qquad \square$$

PROOF OF THEOREM 9.51. Arguing by contradiction, suppose that π is algebraic over $\mathbb{Q}$, so that $\alpha = \pi i$ is algebraic over $\mathbb{Q}$ as well. Let $M(X)$ be the minimal polynomial of α over $\mathbb{Q}$, and let $\mathbb{K}$ be the splitting field of $M(X)$ in $\mathbb{C}$. This exists since $\mathbb{C}$ is algebraically closed. We write $\alpha_1, \ldots, \alpha_m$ for the roots of $M(X)$ in $\mathbb{K}$, with $\alpha_1 = \alpha$. These are distinct algebraic numbers, and they are permuted by the Galois group, $G = \text{Gal}(\mathbb{K}/\mathbb{Q})$. What we shall show is that

$$R = \prod_{j=1}^{m} (1 + e^{\alpha_j}) \ne 0.$$

This will be a contradiction since $1 + e^{\alpha_1} = 0$ for $\alpha_1 = i\pi$.

We expand the product defining R, obtaining

$$R = 1 + \sum_{j} e^{\alpha_j} + \sum_{j,k} e^{\alpha_j + \alpha_k} + \cdots,$$

Whenever one of the exponentials has total exponent 0, we lump that term with the constant 1. Otherwise we write the term as e^{β_l}, allowing repetitions among terms e^{β_l}. Thus

$$R = N + e^{\beta_1} + e^{\beta_2} + \cdots + e^{\beta_r},$$

with N an integer ≥ 1, with each $\beta_l \ne 0$, and with $N + r = 2^m$.

Each member of $G = \text{Gal}(\mathbb{K}/\mathbb{Q})$ permutes $\alpha_1, \ldots, \alpha_m$, and it therefore permutes the β_l's that are single α_j's, permutes the β_l's that are the nonzero sums of two α_j's, permutes the β_l's that are the nonzero sums of three α_j's, and so on.

Choose an integer $a > 0$ such that $a\alpha_1, \ldots, a\alpha_m$ are algebraic integers, let p be a prime number large enough to satisfy some conditions to be specified shortly, and define

$$f(X) = \frac{(aX)^{p-1}}{(p-1)!} \prod_{l=1}^{r} (aX - a\beta_l)^p = \sum_{k=0}^{n} a_k X^n.$$

The members σ of G act on $f(X)$ as usual by acting on the coefficients. Each β_l that is the nonzero sum of a certain number of α_j's is sent into another $\beta_{l'}$ of the same kind, and thus σ just permutes the factors of the product defining f, leaving $f(X)$ unchanged. The coefficients of $(p-1)! f(a^{-1}X)$ are algebraic integers in $\mathbb{K}$. Being fixed by G, they are in $\mathbb{Q}$ by Proposition 9.35d, and hence they are in $\mathbb{Z}$. Therefore

$$f(X) = \frac{A_{p-1}a^{p-1}X^{p-1} + A_p a^p X^p + \cdots}{(p-1)!}$$

with $A_{p-1}, A_p, \ldots$ in $\mathbb{Z}$. Since $A_{p-1} = \prod_{l=1}^{r}(-a\beta_l)^p$, we can arrange that p does not divide $A_{p-1}a^{p-1}$ by choosing p greater than a and greater than $\left|\prod_{l=1}^{r}(a\beta_l)\right|$. If we look at the lth factor in the product defining $f(X)$, we see that $(X - \beta_l)^p$ divides $f(X)$ in $\mathbb{K}[X]$. Therefore we have further formulas for $f(X)$, namely

$$f(X) = \frac{\gamma_{p,l}(X - \beta_l)^p + \gamma_{p+1,l}(X - \beta_l)^{p+1} + \cdots}{(p-1)!} \qquad \text{for } 1 \leq l \leq r.$$

As in Lemma 9.52, we define

$$F(X) = \sum_{l=0}^{n} f^{(l)}(X) \qquad \text{and} \qquad Q(z) = F(0)e^z - F(z).$$

Then we have $F(0) = \sum_{k=0}^{n} a_k k!$. For $1 \leq l \leq r$, the definition of $Q(z)$ gives $F(0)e^{\beta_l} = F(\beta_l) + Q(\beta_l)$. Substituting from the definition of R, we obtain

$$F(0)R = F(0)\Big(N + \sum_{l=1}^{r} e^{\beta_l}\Big) = NF(0) + \sum_{l=1}^{r} F(\beta_l) + \sum_{l=1}^{r} Q(\beta_l). \qquad (*)$$

A further condition that we impose on the size of p is that $p > N$. Then the computation

$$NF(0) = N\sum_{k=0}^{n} a_k k! = N(A_{p-1}a^{p-1} + pA_p a^p + p(p+1)A_{p+1}a^{p+1} + \cdots)$$

and the properties of $A_{p-1}, A_p, \ldots$ together imply that $NF(0)$ is an integer and is not divisible by p.

Let us compute $F(\beta_l)$. The derivatives through order $p-1$ of $f(X)$ are 0 at β_l. For the pth derivative we have

$$p\gamma_{p,l} = f^{(p)}(\beta_l) = pA_p a^p + \sum_{j \geq 1} \frac{(p+j)\cdots(j+1)}{(p-1)!} A_{p+j} a^{p+j} \beta_l^j.$$

The coefficient of $A_{p+j}a^{p+j}\beta_l^j$ inside the sum equals

$$\frac{(p+j)\cdots(j+1)j!p}{p(p-1)!j!} = p\binom{p+j}{j},$$

and thus

$$p\gamma_{p,l} = f^{(p)}(\beta_l) = a^p\left(pA_p + \sum_{j\geq 1} p\binom{p+j}{j}A_{p+j}(a\beta_l)^j\right).$$

The higher-order derivatives are computed and simplified similarly. For the $(p+k)^{\text{th}}$ derivative with $k \geq 1$, we find that

$$(p+k)\cdots(p+1)p\gamma_{p+k,l} = f^{(p+k)}(\beta_l)$$
$$= a^{p+k}\big((p+k)\cdots(p+1)pA_{p+k} \qquad (**)$$
$$+ \sum_{j\geq 1}(p+k)\cdots(p+1)p\binom{p+j+k}{j}A_{p+j+k}(a\beta_l)^j\big).$$

Put $C_{p+k} = \sum_{l=1}^r \gamma_{p+k,l}$. Summing the left and right members of $(**)$ over l gives

$$C_{p+k} = a^{p+k}\left(rA_{p+k} + \sum_{j\geq 1}\binom{p+j+k}{j}A_{p+j+k}\sum_{j=1}^l(a\beta_l)^j\right).$$

The sum $\sum_{j=1}^l(a\beta_l)^j$ is an algebraic integer fixed by G, and it is therefore an integer. Consequently each C_{p+k} is an integer. Summing the left and middle members of $(**)$ over k and l gives

$$\sum_{l=1}^r F(\beta_l) = \sum_{k\geq 0}(p+k)\cdots(p+1)pC_{p+k},$$

and this is an integer divisible by p.

Since $NF(0)$ is an integer not divisible by p, $NF(0) + \sum_{l=1}^r F(\beta_l)$ is an integer not divisible by p, and we have

$$\left|NF(0) + \sum_{l=1}^r F(\beta_l)\right| \geq 1.$$

In view of $(*)$, we will have a contradiction to $R = 0$ if we show that

$$\left|\sum_{l=1}^r Q(\beta_l)\right| < 1.$$

An easy argument by induction on m shows that if $\sum_{k=0}^m d_k z^k = \prod_{j=1}^s (z - c_j)$, then $\sum_{k=0}^m |d_k||z|^k \leq \prod_{j=1}^s (|z| + |c_j|)$. Applying this observation to the sum and product defining $f(X)$ and using Lemma 9.52, we see that

$$e^{-|z|}|Q(z)| \leq \sum_{k=0}^n |a_k||z|^k \leq \frac{(a|z|)^{p-1}\prod_{l=1}^r(a|z|+a|\beta_l|)^p}{(p-1)!}.$$

For each fixed z, the right side is the $(p-1)^{\text{st}}$ term of the convergent series for an exponential function at an appropriate point, and hence the right side is less than $r^{-1}e^{-|z|}$ for p sufficiently large, p depending on z. Choosing p large enough to make the right side less than $r^{-1}e^{-|z|}$ for $z = \beta_1, \ldots, \beta_l$ and summing over these z's, we obtain $\left|\sum_{l=1}^{r} Q(\beta_l)\right| < 1$, and we have arrived at the contradiction we anticipated. $\square$

15. Norm and Trace

This is the second of four sections in which we combine Galois theory with some of the ring theory in the second half of Chapter VIII. We shall make use of a little more linear algebra than we have used thus far in this chapter, and we shall conclude the section by completing the proof of Theorem 8.54 concerning extensions of Dedekind domains.

Let $\Bbbk$ be a field, not necessarily of characteristic 0, and let $\mathbb{K}$ be a finite algebraic extension. We take advantage of the fact that $\mathbb{K}$ is a vector space over $\Bbbk$. If a is in $\mathbb{K}$, let us write $M(a)$ for the $\Bbbk$ linear mapping from $\mathbb{K}$ to $\mathbb{K}$ given by multiplication by a. The characteristic polynomial $\det(XI - M(a))$ is called the **field polynomial** of a and is a monic polynomial in $\Bbbk[X]$ of degree $[\mathbb{K} : \Bbbk]$. The **norm** and **trace** of a relative to $\mathbb{K}/\Bbbk$ are defined to be the determinant and trace of the linear mapping $M(a)$. In symbols,

$$N_{\mathbb{K}/\Bbbk}(a) = \det(M(a)),$$
$$\text{Tr}_{\mathbb{K}/\Bbbk}(a) = \text{Tr}(M(a)).$$

Both $N_{\mathbb{K}/\Bbbk}$ and $\text{Tr}_{\mathbb{K}/\Bbbk}$ are functions from $\mathbb{K}$ to $\Bbbk$. If $n = [\mathbb{K} : \Bbbk]$, then $N_{\mathbb{K}/\Bbbk}(a)$ is $(-1)^n$ times the constant term of $\det(XI - M(a))$, and $\text{Tr}_{\mathbb{K}/\Bbbk}(a)$ is minus the coefficient of X^{n-1}. The subscript $\mathbb{K}/\Bbbk$ may be omitted when there is no chance of ambiguity.

EXAMPLE. $\Bbbk = \mathbb{Q}$, $\mathbb{K} = \mathbb{Q}(\sqrt{2})$, $a = \sqrt{2}$. If we use $\Gamma = (1, \sqrt{2})$ as an ordered basis of $\mathbb{K}$ over $\Bbbk$, then the matrix of $M(a)$ relative to Γ is $\left(\dfrac{M(a)}{\Gamma\Gamma}\right) = \begin{pmatrix} 0 & 2 \\ 1 & 0 \end{pmatrix}$. Since characteristic polynomials are independent of the choice of basis, the field polynomial of a can be computed in this basis and is given by

$$\det\left(\frac{XI - M(a)}{\Gamma\Gamma}\right) = \det\begin{pmatrix} X & -2 \\ -1 & X \end{pmatrix} = X^2 - 2.$$

We can read off the norm and trace as $N(a) = -2$ and $\text{Tr}(a) = 0$.

Proposition 9.53. If $\mathbb{K}/\mathbb{k}$ is a finite extension of fields with $n = [\mathbb{K} : \mathbb{k}]$, then norms and traces relative to $\mathbb{K}/\mathbb{k}$ have the following properties:

(a) $N(ab) = N(a)N(b)$,
(b) $N(ca) = c^n N(a)$ for $c \in \mathbb{k}$,
(c) $N(1) = 1$, and consequently $N(c) = c^n$ for $c \in \mathbb{k}$,
(c) $\text{Tr}(a+b) = \text{Tr}(a) + \text{Tr}(b)$,
(d) $\text{Tr}(ca) = c \, \text{Tr}(a)$ for $c \in \mathbb{k}$,
(e) $\text{Tr}(1) = n$, and consequently $\text{Tr}(c) = nc$ for $c \in \mathbb{k}$.

PROOF. Properties (a) and (b) follow from properties of the determinant in combination with the identities $M(ab) = M(a)M(b)$ and $M(ca) = cM(a)$. Properties (c) and (d) follow from properties of the trace in combination with the identities $M(a+b) = M(a) + M(b)$ and $M(ca) = cM(a)$. Since $M(1)$ is the identity, the norm and trace of 1 are 1 and n, respectively. The other conclusions in (c) and (e) are then consequences of this fact in combination with (b) and (d). □

Proposition 9.54. Let $\mathbb{K}/\mathbb{k}$ and $\mathbb{L}/\mathbb{K}$ be finite extensions of fields with $[\mathbb{K} : \mathbb{k}] = n$ and $[\mathbb{L} : \mathbb{K}] = m$, and let a be in $\mathbb{K}$. The element a acts by multiplication on $\mathbb{K}$ and also on $\mathbb{L}$, yielding $\mathbb{k}$ linear maps in each case that will be denoted by $M_{\mathbb{K}/\mathbb{k}}(a)$ and $M_{\mathbb{L}/\mathbb{k}}(a)$. Then in suitable ordered vector-space bases the matrix of $M_{\mathbb{L}/\mathbb{k}}(a)$ is block diagonal, each block being the matrix of $M_{\mathbb{K}/\mathbb{k}}(a)$.

PROOF. We choose the bases as in Theorem 7.6. Thus let $\Gamma = (\omega_1, \omega_2, \dots)$ be an ordered basis of $\mathbb{K}$ over $\mathbb{k}$, and let $\Delta = (\xi_1, \xi_2, \dots)$ be a basis of $\mathbb{L}$ over $\mathbb{K}$. Theorem 7.6 observes that the mn products $\xi_i \omega_j$ form a basis of $\mathbb{L}$ over $\mathbb{k}$, and we make this set into an ordered basis Ω by saying that $(i_1, j_1) < (i_2, j_2)$ if $i_1 < i_2$ or if $i_1 = i_2$ and $j_1 < j_2$. Let $M_{\mathbb{K}/\mathbb{k}}(a)\omega_j = \sum_l c_{lj} \omega_l$. Then

$$M_{\mathbb{L}/\mathbb{k}}(a)\xi_i \omega_j = \Big(\sum_{l=1}^{n} c_{lj}\omega_l\Big)\xi_i = \sum_{k=1}^{m}\sum_{l=1}^{n}(\delta_{ki} c_{lj})\xi_k \omega_l,$$

where δ_{ki} is 1 when $k = i$ and is 0 otherwise. The matrix $\binom{M_{\mathbb{L}/\mathbb{k}}(a)}{\Omega \Omega}$ has $((k,l), (i,j))^{\text{th}}$ entry $\delta_{ki} c_{lj}$, and this is 0 unless the primary indices k and i are equal. Thus the matrix is block diagonal, the entries of the ith diagonal block being c_{lj}. □

Corollary 9.55. Let $\mathbb{K}/\mathbb{k}$ and $\mathbb{L}/\mathbb{K}$ be finite extensions of fields with $[\mathbb{L} : \mathbb{K}] = m$, and let a be in $\mathbb{K}$. Let $M_{\mathbb{K}/\mathbb{k}}(a)$ and $M_{\mathbb{L}/\mathbb{k}}(a)$ denote multiplication by a on $\mathbb{K}$ and on $\mathbb{L}$, and let $F_{\mathbb{K}/\mathbb{k}}(X)$ and $F_{\mathbb{L}/\mathbb{k}}(X)$ be the corresponding field polynomials. Then

$$F_{\mathbb{L}/\mathbb{k}}(X) = \big(F_{\mathbb{K}/\mathbb{k}}(X)\big)^m.$$

Consequently $N_{\mathbb{L}/\mathbb{k}}(a) = (N_{\mathbb{K}/\mathbb{k}}(a))^m$ and $\text{Tr}_{\mathbb{L}/\mathbb{k}}(a) = m \, \text{Tr}_{\mathbb{K}/\mathbb{k}}(a)$.

PROOF. Proposition 9.54 shows that the matrix of $XI - M_{\mathbb{L}/\mathbb{k}}(a)$ may be taken to be block diagonal with each of the m diagonal blocks equal to the matrix of $XI - M_{\mathbb{K}/\mathbb{k}}(a)$. The determinant of $XI - M_{\mathbb{L}/\mathbb{k}}(a)$ is the product of the determinants of the diagonal blocks, and the formula relating the field polynomials is proved.

The formulas for the norms and the traces are consequences of this relationship. In fact, let
$$F_{\mathbb{K}/\mathbb{k}}(X) = X^n + c_{n-1}X^{n-1} + \cdots + c_0$$
and
$$F_{\mathbb{L}/\mathbb{k}}(X) = X^{mn} + d_{mn-1}X^{mn-1} + \cdots + d_0.$$

Comparing coefficients of $F_{\mathbb{L}/\mathbb{k}}(X)$ and $(F_{\mathbb{K}/\mathbb{k}}(X))^m$, we see that $d_{mn-1} = mc_{n-1}$ and $d_0 = c_0^m$. Therefore
$$N_{\mathbb{L}/\mathbb{k}}(a) = (-1)^{mn} d_0 = ((-1)^n c_0)^m = (N_{\mathbb{K}/\mathbb{k}}(a))^m$$
and
$$\mathrm{Tr}_{\mathbb{L}/\mathbb{k}}(a) = -d_{mn-1} = -mc_{n-1} = m\,\mathrm{Tr}_{\mathbb{K}/\mathbb{k}}(a).$$

This completes the proof. $\square$

Corollary 9.56. Let $\mathbb{K}/\mathbb{k}$ be a finite extension of fields, and let a be in $\mathbb{K}$. Then the field polynomial of a relative to $\mathbb{K}/\mathbb{k}$ is a power of the minimal polynomial of a over $\mathbb{k}$, the power being $[\mathbb{K} : \mathbb{k}(a)]$. In the special case $\mathbb{K} = \mathbb{k}(a)$, the minimal polynomial of a coincides with the field polynomial.

REMARKS. In the theory of a single linear transformation as in Chapter V, the minimal polynomial of a linear map divides the characteristic polynomial, by the Cayley–Hamilton Theorem (Theorem 5.9). For a multiplication operator in the context of fields, we get a much more precise result—that the characteristic polynomial is a power of the minimal polynomial.

PROOF. If $F(X)$ is in $\mathbb{k}[X]$, then the operation M of multiplication has
$$M(F(a))b = F(a)b = F(M(a))b \qquad \text{for } b \in \mathbb{K}, \tag{$*$}$$

as we see by first considering monomials and then forming $\mathbb{k}$ linear combinations. The minimal polynomial of a over $\mathbb{k}$ is the unique monic $F(X)$ of lowest degree in $\mathbb{k}[X]$ for which $F(a) = 0$, hence such that $M(F(a)) = 0$. Meanwhile, the minimal polynomial of the linear map $M(a)$ is the unique monic $F(X)$ of lowest degree such that $F(M(a)) = 0$. These two polynomials coincide because of $(*)$.

The degree of the minimal polynomial of $M(a)$ thus equals the degree of the minimal polynomial of a, which is $[\mathbb{k}(a) : \mathbb{k}]$. The Cayley–Hamilton Theorem (Theorem 5.9) shows that the minimal polynomial of $M(a)$ divides the characteristic polynomial of $M(a)$, i.e., the field polynomial of a. When the field $\mathbb{K}$ is $\mathbb{k}(a)$, the minimal polynomial of a and the field polynomial of a have the same degree; since they are monic, they are equal. This proves the second conclusion of the corollary.

For the first conclusion we know from Corollary 9.55 that the field polynomial of a relative to a general $\mathbb{K}$ is the $[\mathbb{K} : \mathbb{k}(a)]^{\text{th}}$ power of the field polynomial of a relative to $\mathbb{k}(a)$. Since we have just seen that the latter polynomial is the minimal polynomial of a, the first conclusion of the corollary follows. $\square$

EXAMPLE, CONTINUED. $\mathbb{k} = \mathbb{Q}, \mathbb{K} = \mathbb{Q}(\sqrt{2}), a = \sqrt{2}$. We have seen that the field polynomial of a is $X^2 - 2$, that the norm and trace are $N(a) = -2$ and $\text{Tr}(a) = 0$, and that the matrix of the multiplication operator $M(a)$ in the ordered basis $\Gamma = (1, \sqrt{2})$ is $\begin{pmatrix} M(a) \\ \Gamma\Gamma \end{pmatrix} = \begin{pmatrix} 0 & 2 \\ 1 & 0 \end{pmatrix}$. The eigenvalues of $\begin{pmatrix} M(a) \\ \Gamma\Gamma \end{pmatrix}$ are $\pm\sqrt{2}$, namely the roots of the field polynomial. These are not in the field $\mathbb{k}$. Indeed, they could not possibly be in the field, or we would have $M(a)x = \lambda x$ for some $x \neq 0$ in $\mathbb{K}$ and some λ in $\mathbb{k}$, and this would mean that $\lambda = a$. Since the roots $\pm\sqrt{2}$ of the field polynomial each have multiplicity 1 and lie in $\mathbb{K}$, the matrix $\begin{pmatrix} M(a) \\ \Gamma\Gamma \end{pmatrix}$ is similar over $\mathbb{K}$ to the diagonal matrix $\begin{pmatrix} \sqrt{2} & 0 \\ 0 & -\sqrt{2} \end{pmatrix}$. Since similar matrices have the same trace and the same norm, we can compute the trace and norm of $M(a)$ from this diagonal matrix, namely by adding or multiplying its diagonal entries. The significance of the diagonal entries is that they are the images of $\sqrt{2}$ under the members of the Galois group $\text{Gal}(\mathbb{K}/\mathbb{k})$. We shall now generalize these considerations. Additional complications arise when $\mathbb{K}/\mathbb{k}$ fails to be separable and normal.[16]

Proposition 9.57. Let $\mathbb{k}$ be a field, let $\mathbb{k}(a)$ be an algebraic extension of $\mathbb{k}$, and suppose that the minimal polynomial $F(X)$ of a over $\mathbb{k}$ is separable. Let $\mathbb{K}$ be a splitting field of $F(X)$, and factor $F(X)$ over $\mathbb{K}$ as

$$F(X) = (X - a_1)(X - a_2) \cdots (X - a_n)$$

with all $a_j \in \mathbb{K}$ and with $a_1 = a$. Then the matrix of the multiplication operator $M(a)_{\mathbb{k}(a)/\mathbb{k}}$ of a on $\mathbb{k}(a)$ is similar over $\mathbb{K}$ to a diagonal matrix with diagonal entries $a_1, \ldots, a_n$. Consequently

$$N_{\mathbb{k}(a)/\mathbb{k}}(a) = \prod_{j=1}^{n} a_j \quad \text{and} \quad \text{Tr}_{\mathbb{k}(a)/\mathbb{k}}(a) = \sum_{j=1}^{n} a_j.$$

---

[16]The above argument used a matrix with entries in $\mathbb{k}$ and considered the entries as in the larger field $\mathbb{K}$. The reader may wonder what the corresponding construction is for the $\mathbb{k}$ linear map $M(a)$. It is *not* to treat $M(a)$ as a $\mathbb{K}$ linear map on $\mathbb{K}$, since then $M(a)$ would have just the one eigenvalue $\sqrt{2}$, which would have multiplicity 1. Instead, it is to use tensor products as in Chapter VI, knowledge of which is not being assumed at present. The idea is to extend scalars, replacing $\mathbb{K}$ by $\mathbb{K} \otimes_{\mathbb{k}} \mathbb{K}$ and replacing $M(a)$ by $M(a) \otimes 1$. The $\mathbb{K}$ linearity occurs in the second member of the tensor product, not the first, and the operator $M(a) \otimes 1$ is the $\mathbb{K}$ linear map with eigenvalues $\pm\sqrt{2}$.

REMARKS. The elements $a_1, \ldots, a_n$ of $\mathbb{K}$, with $a_1 = a$, are called the **conjugates** of a over $\mathbb{k}$. The conjugates of a are the images of a under the Galois group when $\mathbb{k}(a)$ is Galois over $\mathbb{k}$, but they extend outside $\mathbb{k}$ when $\mathbb{k}(a)/\mathbb{k}$ is not normal.

PROOF. Corollary 9.56 shows that $F(X)$ equals the field polynomial of a relative to $\mathbb{k}(a)/\mathbb{k}$, i.e., is the characteristic polynomial of the multiplication operator $M_{\mathbb{k}(a)/\mathbb{k}}(a)$. Let A be the matrix of $M_{\mathbb{k}(a)/\mathbb{k}}(a)$ in some ordered basis of $\mathbb{k}(a)$ over $\mathbb{k}$. If we regard A as a matrix with entries in $\mathbb{K}$, then the characteristic polynomial of A splits in $\mathbb{K}$, and the roots of the characteristic polynomial have multiplicity 1, by separability. Consequently A has a basis of eigenvectors, the eigenvectors being column vectors with entries in $\mathbb{K}$ and the eigenvalues being the members $a_1, \ldots, a_n$ of $\mathbb{K}$. It follows that A is similar over $\mathbb{K}$ to a diagonal matrix with diagonal entries $a_1, \ldots, a_n$. The determinant and trace of this diagonal matrix equal the determinant and trace of A, and therefore the norm and trace of a are the product and sum of the members $a_1, \ldots, a_n$ of $\mathbb{K}$. □

Corollary 9.58. Let $\mathbb{K}$ be a finite Galois extension of the field $\mathbb{k}$, let $G = \text{Gal}(\mathbb{K}/\mathbb{k})$, let $\mathbb{L}$ be an intermediate field with $\mathbb{k} \subseteq \mathbb{L} \subseteq \mathbb{K}$, and let $H = \text{Gal}(\mathbb{K}/\mathbb{L})$ as a subgroup of G. Fix an ordered basis Γ of $\mathbb{L}$ over $\mathbb{k}$. Then the expression "$\sigma(a)$ for $\sigma \in G/H$" is well defined for a in $\mathbb{L}$, and there exists a nonsingular matrix C of size $[\mathbb{L} : \mathbb{k}]$ with entries in $\mathbb{K}$ such that every a in $\mathbb{L}$ has $C^{-1} \left(\begin{smallmatrix} M_{\mathbb{L}/\mathbb{k}}(a) \\ \Gamma\Gamma \end{smallmatrix} \right) C$ diagonal with diagonal entries $\sigma(a)$ for $\sigma \in G/H$. In particular, every member a of $\mathbb{L}$ has norm and trace given by

$$N_{\mathbb{L}/\mathbb{k}}(a) = \prod_{\sigma \in G/H} \sigma(a) \quad \text{and} \quad \text{Tr}_{\mathbb{L}/\mathbb{k}}(a) = \sum_{\sigma \in G/H} \sigma(a).$$

PROOF. Let a be in $\mathbb{L}$, σ be in G, and τ be in H. Then $\tau(a) = a$, and therefore $\sigma\tau(a) = \sigma(a)$. Consequently all members of the coset σH of G/H have the same value on a, and "$\sigma(a)$ for $\sigma \in G/H$" is well defined.

Let $n = [\mathbb{L} : \mathbb{k}] = |G/H|$. Fix an ordered basis Γ of $\mathbb{L}$ over $\mathbb{k}$. For each $a \in \mathbb{L}$, let $A(a)$ be the matrix of the multiplication operator $M(a)_{\mathbb{L}/\mathbb{k}}$ relative to Γ.

The Theorem of the Primitive Element (Theorem 9.34) shows that $\mathbb{L} = \mathbb{k}(x)$ for some x. Proposition 9.57 applies to this element x and to a splitting field within $\mathbb{K}$ for its minimal polynomial, showing that there is a nonsingular matrix C with entries in $\mathbb{K}$ such that $C^{-1}A(x)C$ is a diagonal matrix whose diagonal entries are the n conjugates $x_1, \ldots, x_n$ of x in $\mathbb{K}$, x_1 being x; the diagonal entries are necessarily distinct by separability. For each i with $1 \leq i \leq n$, there exists σ_i in G with $\sigma_i(x) = x_i$ by Theorems 9.11 and 9.23. Since H fixes $\mathbb{L}$, every member of the coset $\sigma_i H$ carries x to x_i. On the other hand, every σ in G must carry x to some conjugate, hence must have $\sigma(x) = \sigma_i(x)$ for some i. Then $\sigma_i^{-1}\sigma$ fixes x

and hence $\mathbb{L}$, and it follows that $\sigma_i^{-1}\sigma$ is in H. Thus σ is in $\sigma_i H$. In other words, the conjugates $x_1, \ldots, x_n$ may be regarded exactly as the images of the n cosets $\sigma_j H$.

In this terminology the diagonal entries of $C^{-1}A(x)C$ are the n elements $\sigma(x)$ for σ in G/H. For each j with $0 \leq j \leq n-1$, we have $A(x^j) = A(x)^j$, and hence $C^{-1}A(x^j)C = C^{-1}A(x)^j C$ is diagonal with diagonal entries $\sigma(x)^j = \sigma(x^j)$ for σ in G/H. Forming $\mathbb{k}$ linear combinations, we see for every polynomial $P(X)$ in $\mathbb{k}[X]$ of degree $\leq n-1$ that $C^{-1}A(P(x))C$ is diagonal with diagonal entries $\sigma(P(x))$. Every element a of $\mathbb{K}$ is of the form $P(x)$ for some such $P(X)$, and the existence of C in the statement of the corollary is proved. The formulas for the norm and trace follow by taking the determinant and trace. $\square$

Corollary 9.59. If $\mathbb{K}$ is a finite separable extension of the field $\mathbb{k}$, then the trace function $\mathrm{Tr}_{\mathbb{K}/\mathbb{k}}$ is not identically 0.

REMARKS. This result is trivial in characteristic 0 because $\mathrm{Tr}_{\mathbb{K}/\mathbb{k}}(1) = [\mathbb{K}:\mathbb{k}]$ is not zero. The result is not so evident in characteristic p, and the assumption of separability is crucial. An example for which separability fails and the trace function is identically 0 has $\mathbb{k} = \mathbb{F}(x)$, where $\mathbb{F}$ is a finite field of characteristic p and x is transcendental, and $\mathbb{K} = \mathbb{k}(x^{1/p})$. The basis elements $1, x^{1/p}, x^{2/p}, \ldots, x^{(p-1)/p}$ all have trace 0, and therefore the trace is identically 0.

PROOF. By the Theorem of the Primitive Element (Theorem 9.34), we can write $\mathbb{K} = \mathbb{k}(a)$ for some $a \neq 0$. Let $\mathbb{K}'$ be a splitting field for the minimal polynomial of a over $\mathbb{k}$. Then $\mathbb{K}'/\mathbb{k}$ is a separable extension by Corollary 9.30 and hence is a finite Galois extension. Proposition 9.57 shows that the matrix of $M_{\mathbb{K}/\mathbb{k}}(a)$ in any ordered basis of $\mathbb{K}$ over $\mathbb{k}$ is similar over $\mathbb{K}'$ to a diagonal matrix with entries $a_1, \ldots, a_n$, where $a_1, \ldots, a_n$ are the conjugates of a with $a_1 = a$. These conjugates are necessarily distinct by separability. For $1 \leq k \leq n$, the matrix of $M_{\mathbb{K}/\mathbb{k}}(a^k)$ is similar via the same matrix over $\mathbb{K}'$ to a diagonal matrix with entries $a_1^k, \ldots, a_n^k$. If $\mathrm{Tr}_{\mathbb{K}/\mathbb{k}}(a^k) = 0$ for $1 \leq k \leq n$, then we obtain the homogeneous system of linear equations

$$a_1 x_1 + a_2 x_2 + \cdots + a_n x_n = 0,$$
$$a_1^2 x_1 + a_2^2 x_2 + \cdots + a_n^2 x_n = 0,$$
$$\vdots$$
$$a_1^n x_1 + a_2^n x_2 + \cdots + a_n^n x_n = 0,$$

with $(x_1, \ldots, x_n) = (1, \ldots, 1)$ as a nonzero solution. The coefficient matrix must therefore have determinant 0. This coefficient matrix, however, is a Vandermonde matrix except that the jth column is multiplied by a_j for each j. Since $a_1, \ldots, a_n$

are distinct, Corollary 5.3 shows that the determinant of the coefficient matrix can be 0 only if $a_1 a_2 \cdots a_n = 0$. Since $a \neq 0$, we have arrived at a contradiction, and we conclude that $\mathrm{Tr}_{\mathbb{K}/\mathbb{k}}(a^k) \neq 0$ for some k. □

With the aid of Corollary 9.59, we can complete the proof of Theorem 8.54 in Section VIII.11. Let us restate the part that still needs proof.

THEOREM 8.54. If R is a Dedekind domain with field of fractions F and if K is a finite separable extension field of F, then the integral closure T of R in K is finitely generated as an R module and consequently is a Dedekind domain.

REMARKS. What needs proof is that T is finitely generated as an R module. It was shown in Section VIII.11 how to deduce as a consequence that T is a Dedekind domain.

PROOF. Since R is Noetherian (being a Dedekind domain), Proposition 8.34 shows that it is enough to exhibit T as an R submodule of a finitely generated R module in K. Let $\{u_1, \ldots, u_n\}$ be a vector-space basis of K over F. Proposition 8.42 shows that we may assume that each u_i is in T.

Define an F linear map from K into its F vector-space dual K' by $y \mapsto \ell_y$, where $\ell_y(x) = \mathrm{Tr}_{K/F}(xy)$ for $x \in K$. This map is one-one by Corollary 9.59, and the equality of dimensions of K and K' over F therefore implies that the map is onto. We can thus view every member of K' as uniquely of the form ℓ_y for some y in K. With this understanding, let $\{\ell_{v_1}, \ldots, \ell_{v_n}\}$ be the dual basis of K' with $\ell_{v_j}(u_i) = \delta_{ij}$ for all i and j. Then we have

$$\mathrm{Tr}_{K/F}(u_i v_j) = \delta_{ij} \qquad \text{for all } i \text{ and } j.$$

Applying Proposition 8.42, choose $c \neq 0$ in R with cv_j in T for all j. We shall complete the proof by showing that

$$T \subseteq Rc^{-1}u_1 + \cdots + Rc^{-1}u_n. \tag{$*$}$$

Before doing so, let us observe that

$$\mathrm{Tr}_{K/F}(t) \quad \text{is in } R \text{ if } t \text{ is in } T. \tag{$**$}$$

In fact, Proposition 9.57 shows that $\mathrm{Tr}_{F(t)/F}(t)$ is the sum of all the conjugates of t, whether or not they are in K. The conjugates have the same minimal polynomial over F that t has, and hence they are integral over R. Their sum $\mathrm{Tr}_{F(t)/F}(t)$ must be integral over R by Corollary 8.38, and it must lie in F. Since R is integrally closed (being a Dedekind domain), $\mathrm{Tr}_{F(t)/F}(t)$ lies in R. This proves $(**)$.

Now we can return to the proof of $(*)$. Let x be given in T. Since T is a ring, cxv_j is in T for each j, and $\mathrm{Tr}_{K/F}(cxv_j)$ is in R by $(**)$. Since $\{u_1, \ldots, u_n\}$ is a basis, we can write $x = \sum_i d_i u_i$ with each d_i in F. Since $\text{Tr}(cxv_j)$ is in R, the computation

$$\text{Tr}(cxv_j) = c\,\text{Tr}_{K/F}(xv_j) = c\sum_{i=1}^n d_i\,\text{Tr}(u_i v_j) = cd_j$$

shows that cd_j is in R. Then the expansion $x = \sum_i (cd_i)c^{-1}u_i$ exhibits x as in $Rc^{-1}u_1 + \cdots + Rc^{-1}u_n$ and completes the proof of $(*)$. $\square$

16. Splitting of Prime Ideals in Extensions

Section VIII.7 was a section of motivation showing the importance for number theory and geometry of passing from factorization of elements to factorization of ideals. The later sections of Chapter VIII set the framework for this study, examining the notions of Noetherian domain, integral closure, and localization and putting them together in the notion of Dedekind domain. Only just now were we able to complete the proof of the fundamental result (Theorem 8.54) for constructing Dedekind domains out of other Dedekind domains. However, that proposition does not complete the task of extending what is in Section VIII.7 to a wider context. Much of Section VIII.7 concerned the relationship between prime ideals in one domain and prime ideals in an extension. In the present section we put that relationship in a wider context, showing how the examples of Section VIII.7 are special cases of the present theory.

In two of the examples in Section VIII.7, we worked with the ring $\mathbb{Z}$ of integers inside its field of fractions $\mathbb{Q}$ and with the ring T of algebraic integers within a quadratic extension $\mathbb{K}$ of $\mathbb{Q}$. In the third example in that section, we worked with the ring $\mathbb{C}[x]$, for transcendental x, inside its field of fractions $\mathbb{C}(x)$ and with a certain integral domain T within a quadratic extension of $\mathbb{C}(x)$. For all three examples we saw a correspondence between prime ideals P in T and prime ideals (p) in $\mathbb{Z}$ or $\mathbb{C}[x]$, and that correspondence was formalized in a more general setting in Propositions 8.43 and 8.53. The objective now is to understand that correspondence a little better.

The notation for this section is as follows: Let R be a Dedekind domain, such as $\mathbb{Z}$ or $\mathbb{C}[x]$, and let F be its field of fractions.[17] Let K be a finite separable extension of F, and let T be the integral closure of R in K. Theorem 8.54, including the part just proved in the previous section, shows that T is a Dedekind domain. We make repeated use of the fact about Dedekind domains that every nonzero prime ideal is maximal.

---

[17] It might seem more natural to assume that R is a principal ideal domain, as it is with $\mathbb{Z}$ and $\mathbb{C}[x]$. But that extra assumption will not help us, and it will often not be satisfied when the present results are used in the proof of the important Theorem 9.64 in the next section.

Proposition 8.43 shows that if P is any nonzero prime ideal of T, then $\mathfrak{p} = R \cap P$ is a nonzero prime ideal of R. In the reverse direction Proposition 8.53 shows that if $\mathfrak{p}$ is any nonzero prime ideal in R, then $\mathfrak{p}T \neq T$, and there exists at least one prime ideal P of T with $\mathfrak{p} = R \cap P$. The unique factorization of ideals in T (given as Theorem 8.55) explains this correspondence better. If $\mathfrak{p}$ is given, then $\mathfrak{p}T$ is a proper ideal, hence is contained in some maximal ideal P. Since "to contain is to divide" (by Theorem 8.55d), such P's (and only such P's) are factors in the decomposition of $\mathfrak{p}T$ as the product of nonzero prime ideals. Accordingly let us write

$$\mathfrak{p}T = \prod_{i=1}^{g} P_i^{e_i},$$

where the P_i are the distinct prime ideals of T containing $\mathfrak{p}T$, or equivalently the distinct prime ideals of T satisfying $R \cap P_i = \mathfrak{p}$. The e_i are positive integers called the **ramification indices**.

For each P_i, we can form the composition $R \subseteq T \to T/P_i$ of inclusion followed by passage to the quotient. Since $\mathfrak{p} \subseteq P_i$, this composition descends to a ring homomorphism $R/\mathfrak{p} \to T/P_i$. The ideal $\mathfrak{p}$ is maximal in R, and the ideal P_i is maximal in T. Thus the mapping $R/\mathfrak{p} \to T/P_i$ is in fact a field map. We regard it as an inclusion. Define

$$f_i = [T/P_i : R/\mathfrak{p}],$$

allowing the dimension for the moment possibly to be $+\infty$. It will follow from Theorem 9.60, however, that f_i is finite. The integer f_i is called the **residue class degree**.

Theorem 9.60. Let R be a Dedekind domain, let F be its field of fractions, let K be a finite separable extension of F with $[K : F] = n$, and let T be the integral closure of R in K. If $\mathfrak{p}$ is a nonzero prime ideal in R and $\mathfrak{p}T = \prod_{i=1}^{g} P_i^{e_i}$ is a decomposition of $\mathfrak{p}T$ as the product of powers of distinct nonzero prime ideals in T, then the ramification indices e_i and residue class degrees $f_i = [T/P_i : R/\mathfrak{p}]$ are related by

$$\sum_{i=1}^{g} e_i f_i = n.$$

REMARKS. Consequently each f_i is finite. The cases of interest for our earlier examples have $R = \mathbb{Z}$ or $R = \mathbb{C}[x]$. When $R = \mathbb{Z}$, each $R/\mathfrak{p}$ is a finite field. However, when $R = \mathbb{K}[x]$ for some field $\mathbb{K}$ of characteristic 0 like $\mathbb{K} = \mathbb{C}$, then each $R/\mathfrak{p}$ is a finite extension of $\mathbb{K}$, hence is an infinite field.[18]

---

[18]When $R = \mathbb{C}[x]$, then $T/P_i = R/\mathfrak{p} \cong \mathbb{C}$ since $\mathbb{C}$ is algebraically closed. The last example of the present section will elaborate.

16. Splitting of Prime Ideals in Extensions

PROOF. Corollary 8.63 gives a ring isomorphism

$$T/(\mathfrak{p}T) \cong T/P_1^{e_1} \times \cdots \times T/P_g^{e_g}. \qquad (*)$$

Recall from the definition of residue class degree that we have a field mapping of $R/\mathfrak{p}$ into each T/P_i. Since $\mathfrak{p} \subseteq P_i^e$ for $1 \leq e \leq e_i$ and since $\mathfrak{p} \subseteq \mathfrak{p}T$, it follows similarly that we have a one-one ring homomorphism of $R/\mathfrak{p}$ into each T/P_i^e with $1 \leq e \leq e_i$ and another one-one ring homomorphism of $R/\mathfrak{p}$ into $T/(\mathfrak{p}T)$. Consequently each T/P_i^e with $1 \leq e \leq e_i$, the product $T/P_1^{e_1} \times \cdots \times T/P_g^{e_g}$, and $T/(\mathfrak{p}T)$ may all be regarded as unital $R/\mathfrak{p}$ modules, i.e., as vector spaces over the field $R/\mathfrak{p}$. Fix i. For $1 \leq e \leq e_i$, let us prove by induction on e that

$$\dim_{R/\mathfrak{p}}(T/P_i^e) = ef_i, \qquad (**)$$

the case $e = 1$ being the base case of the induction. Assume inductively that $(**)$ holds for exponents from 1 to $e - 1$. We know from Corollary 8.60 that P_i^{e-1}/P_i^e is a vector space over the field T/P_i with

$$\dim_{T/P_i}(P_i^{e-1}/P_i^e) = 1. \qquad (\dagger)$$

The First Isomorphism Theorem (as in the remark with Theorem 8.3) gives $T/P_i^{e-1} \cong (T/P_i^e)/(P_i^{e-1}/P_i^e)$ as vector spaces over $R/\mathfrak{p}$, and it follows that

$$\dim_{R/\mathfrak{p}}(T/P_i^e) = \dim_{R/\mathfrak{p}}(T/P_i^{e-1}) + \dim_{R/\mathfrak{p}}(P_i^{e-1}/P_i^e)$$
$$= (e-1)f_i + f_i = ef_i,$$

the next-to-last equality following from $(\dagger)$ and the inductive hypothesis for the cases $e - 1$ and 1. This completes the induction and the proof of $(**)$.

In view of the decomposition $(*)$ and the formula $(**)$ when $e = e_i$, the theorem will follow if it is shown that

$$\dim_{R/\mathfrak{p}}(T/(\mathfrak{p}T)) = n. \qquad (\dagger\dagger)$$

To prove $(\dagger\dagger)$ we localize. Let S be the complement of the prime ideal $\mathfrak{p}$ of R. Corollary 8.48 shows that $S^{-1}R$ is a Dedekind domain, Corollary 8.50 shows that $S^{-1}\mathfrak{p}$ is its unique maximal ideal, and Corollary 8.62 shows that $S^{-1}R$ is a principal ideal domain.

The composition $R \subseteq S^{-1}R \to S^{-1}R/S^{-1}\mathfrak{p}$ descends to a field mapping $R/\mathfrak{p} \to S^{-1}R/S^{-1}\mathfrak{p}$. Let us see that this mapping is onto. If $s_0^{-1}r_0 + S^{-1}\mathfrak{p}$ in $S^{-1}R/S^{-1}\mathfrak{p}$ is given, then s_0 is not in $\mathfrak{p}$, and the maximality of $\mathfrak{p}$ as an ideal in R implies that $(s_0) + \mathfrak{p} = R$. Therefore we can choose r in R and x in $\mathfrak{p}$ with $rs_0 + x = r_0$. Under the mapping $R/\mathfrak{p} \to S^{-1}R/S^{-1}\mathfrak{p}$, the image of $r + \mathfrak{p}$ is

$r + S^{-1}\mathfrak{p} = r + s_0^{-1}x + S^{-1}\mathfrak{p} = s_0^{-1}(rs_0 + x) + S^{-1}\mathfrak{p} = s_0^{-1}r_0 + S^{-1}\mathfrak{p}$. Thus our mapping is onto $S^{-1}R/S^{-1}\mathfrak{p}$, and we have an isomorphism of fields

$$R/\mathfrak{p} \cong S^{-1}R/S^{-1}\mathfrak{p}. \qquad (\ddagger)$$

Similarly the composition $T \subseteq S^{-1}T \to S^{-1}T/(S^{-1}\mathfrak{p}T)$ descends to a homomorphism of rings $T/\mathfrak{p}T \to S^{-1}T/(S^{-1}\mathfrak{p}T)$. Let us show that this map too is one-one onto.

If $t + \mathfrak{p}T$ is in the kernel, then the member t of T is in $S^{-1}\mathfrak{p}T$, and st is in $\mathfrak{p}T$ for some s in S. Hence we have $(s)(t) \subseteq P_1^{e_1} \cdots P_g^{e_g}$, and we can write $(s)(t) = P_1^{e_1} \cdots P_g^{e_g} Q$ for some ideal Q. Factoring the principal ideals (s) and (t) and using the uniqueness of factorization of ideals gives

$$(s) = P_1^{u_1} \cdots P_g^{u_g} Q_1 \quad \text{and} \quad (t) = P_1^{v_1} \cdots P_g^{v_g} Q_2$$

with $Q = Q_1 Q_2$ and with $u_j + v_j = e_j$ for all j. If $u_j > 0$, then we must have $(s) \subseteq P_j$ and $sR \subseteq P_j \cap R = \mathfrak{p}$. This says that s is in $\mathfrak{p}$, in contradiction to the fact that S equals the set-theoretic complement of $\mathfrak{p}$ in R. We conclude that $u_j = 0$ for all j. Therefore $(t) = P_1^{e_1} \cdots P_g^{e_g} Q_2 \subseteq P_1^{e_1} \cdots P_g^{e_g} = \mathfrak{p}T$, and t is in $\mathfrak{p}T$. Consequently the kernel consists of the 0 coset alone.

Let us show that $T/\mathfrak{p}T$ maps onto $S^{-1}T/(S^{-1}\mathfrak{p}T)$. If $s_0^{-1}t_0 + S^{-1}\mathfrak{p}T$ in $S^{-1}T/S^{-1}\mathfrak{p}T$ is given, then s_0 is not in $\mathfrak{p}$, and the maximality of $\mathfrak{p}$ as an ideal in R implies that $(s_0) + \mathfrak{p} = R$. Therefore we can choose r in R and x in $\mathfrak{p}$ with $rs_0 + x = 1$, hence with $rs_0t_0 + xt_0 = t_0$. Under the mapping $T/\mathfrak{p}T \to S^{-1}T/(S^{-1}\mathfrak{p}T)$, the image of $rt_0 + \mathfrak{p}T$ is

$$\begin{aligned}rt_0 + S^{-1}\mathfrak{p}T &= rt_0 + s_0^{-1}xt_0 + S^{-1}\mathfrak{p}T \\ &= s_0^{-1}(rs_0t_0 + xt_0) + S^{-1}\mathfrak{p}T \\ &= s_0^{-1}t_0 + S^{-1}\mathfrak{p}T.\end{aligned}$$

Thus our mapping is onto $S^{-1}T/S^{-1}\mathfrak{p}T$, and we conclude that we have an isomorphism of rings

$$T/\mathfrak{p}T \to S^{-1}T/(S^{-1}\mathfrak{p}T). \qquad (\ddagger\ddagger)$$

Since T is finitely generated as an R module (Theorem 8.54), $S^{-1}T$ is finitely generated as an $S^{-1}R$ module with the same generators. Since $S^{-1}R$ is a principal ideal domain, Theorem 8.25c shows that $S^{-1}T$ is the direct sum of cyclic $S^{-1}R$ modules. Each of these cyclic modules must in fact be isomorphic to $S^{-1}R$ since $S^{-1}T$ has no zero divisors, and therefore $S^{-1}T$ is a free $S^{-1}R$ module of some finite rank m. If $t_1, \ldots, t_m$ are free generators, then we have

$$S^{-1}T = S^{-1}Rt_1 + \cdots + S^{-1}Rt_m. \qquad (\S)$$

Let us see that $\{t_1, \ldots, t_m\}$ is an F vector-space basis of K. Suppose $\sum_j c_j t_j = 0$ with all c_j in F. Proposition 8.42 shows that there is an $r \neq 0$ in R with $rc_1, \ldots, rc_m$ in R. Then $\sum_j (rc_j) t_j = 0$, and the independence of $t_1, \ldots, t_m$ over $S^{-1}R$ implies that $rc_j = 0$ for all j. Thus $c_j = 0$ for all j, and we obtain linear independence over F. If $x \in K$ is given, we can choose $r \neq 0$ in R with rx in T by Proposition 8.42. Since $t_1, \ldots, t_m$ span $S^{-1}T$ over $S^{-1}R$, we can find members $d_1, \ldots, d_m$ of $S^{-1}R$ with $rx = \sum_j d_j t_j$. Then $x = \sum_j r^{-1} d_j t_j$ with each coefficient $r^{-1} d_j$ in F. This proves the spanning. Hence $\{t_1, \ldots, t_m\}$ is an F vector-space basis, and $m = n$.

To complete the proof of (††) and hence the theorem, it is enough, in view of the isomorphisms (‡) and (‡‡), to prove that the cosets $t_j + S^{-1}\mathfrak{p}T$ in $S^{-1}T/(S^{-1}\mathfrak{p}T)$ form a vector-space basis over $S^{-1}R/S^{-1}\mathfrak{p}$. If t is in $S^{-1}T$, then (§) says that $t = \sum c_j t_j$ with c_j in $S^{-1}R$. Hence

$$t + S^{-1}\mathfrak{p}T = \sum (c_j + S^{-1}\mathfrak{p})(t_j + S^{-1}\mathfrak{p}T),$$

and we have spanning. If $\sum_j (c_j + S^{-1}\mathfrak{p})(t_j + S^{-1}\mathfrak{p}T) = 0 + S^{-1}\mathfrak{p}T$, then $\sum_j c_j t_j$ is in $S^{-1}\mathfrak{p}T$. Thus we can write $\sum_j c_j t_j = \sum_i a_i t_i'$ with $a_i \in \mathfrak{p}$ and $t_i' \in S^{-1}T$. Expanding each t_i' according to (§), substituting, and using the uniqueness of the expansion (§), we see for each j that c_j is a sum of products of the a_i's by members of $S^{-1}R$. Therefore each c_j is in $S^{-1}\mathfrak{p}$. This proves the linear independence and establishes (††). □

The case of greatest interest is that K is a finite Galois extension of F. In this case the statement of Theorem 9.60 simplifies and will be given in its simplified form as Theorem 9.62. We begin with a lemma.

Lemma 9.61. Let R be a Dedekind domain, let F be its field of fractions, let K be a finite separable extension of F, and let T be the integral closure of R in K. Suppose that K is Galois over F. If $\mathfrak{p}$ is a nonzero prime ideal in R and $\mathfrak{p}T = \prod_{i=1}^{g} P_i^{e_i}$ is a decomposition of $\mathfrak{p}T$ as the product of nonzero prime ideals in T, then $\text{Gal}(K/F)$ is transitive on the set of ideals $\{P_1, \ldots, P_g\}$.

PROOF. Arguing by contradiction, suppose that P_j is not of the form $\sigma(P_1)$ for some σ in $\text{Gal}(K/F)$. By the Chinese Remainder Theorem we can choose an element t of T with $t \equiv 0 \mod P_j$ and $t \equiv 1 \mod \sigma(P_1)$ for all σ. Every σ in $\text{Gal}(K/F)$ carries t to a member of T since t and $\sigma(t)$ have the same minimal polynomial over F. Corollary 9.58 shows that $N_{K/F}(t) = \prod_{\sigma \in \text{Gal}(K/F)} \sigma(t)$, and consequently $N_{K/F}(t)$ is in $T \cap F = R$. Since the factor t itself is in P_j, $N_{K/F}(t)$ is in P_j. Therefore $N_{K/F}(t)$ is in $R \cap P_j = \mathfrak{p} \subseteq \prod_{i=1}^{g} P_i^{e_i}$. The right side is contained in P_1. Since P_1 is prime, some factor $\sigma_l(t)$ of $N_{K/F}(t)$ is in P_1. Then t is in $\sigma_l^{-1}(P_1)$, in contradiction to the fact that $t \equiv 1 \mod \sigma(P_1)$ for all σ. □

Theorem 9.62. Let R be a Dedekind domain, let F be its field of fractions, let K be a finite separable extension of F with $[K : F] = n$, and let T be the integral closure of R in K. Suppose that K is Galois over F. If $\mathfrak{p}$ is a nonzero prime ideal in R and $\mathfrak{p}T = \prod_{i=1}^{g} P_i^{e_i}$ is a decomposition of $\mathfrak{p}T$ as the product of powers of distinct nonzero prime ideals in T, then the ramification indices have $e_1 = \cdots = e_g$, and the residue class degrees $f_i = [T/P_i : R/\mathfrak{p}]$ have $f_1 = \cdots = f_g$. If e and f denote the common value of the e_i's and of the f_j's, then

$$efg = n.$$

PROOF. For σ in $\text{Gal}(K/F)$, apply σ to the factorization $\mathfrak{p}T = \prod_{i=1}^{g} P_i^{e_i}$, obtaining

$$\mathfrak{p}T = \sigma(P_1)^{e_1} \prod_{i=2}^{g} \sigma(P_i)^{e_i}.$$

Lemma 9.61 shows that $\sigma(P_1)$ can be any P_j, and unique factorization of ideals (Theorem 8.55) therefore implies that $e_1 = e_j$. With the same σ, the fact that σ respects the field operations implies that

$$T/P_1 \cong \sigma(T)/\sigma(P_1) = T/P_j,$$

and thus $f_1 = f_j$. Substituting the values of the e_i's and the f_j's into the formula of Theorem 9.60, we obtain $efg = n$. □

EXAMPLES WITH $n = 2$ CONTINUED FROM SECTION VIII.7.

(1) $R = \mathbb{Z}$ and $T = \mathbb{Z}[\sqrt{-1}\,]$. In this case, $\mathbb{Z}$ and T are both principal ideal domains. We found three possible behaviors[19] for the prime factorization of a principal ideal $(p)T$ in T generated by a prime $p > 0$ in $\mathbb{Z}$:

(a) $(p)T$ is prime in T if $p = 4m + 3$. Here $e = g = 1$; so $f = 2$.
(b) $(p)T = (a+ib)(a-ib)$ with $p = a^2 + b^2$ if $p = 4m+1$. Here $e = 1$ and $g = 2$; so $f = 1$.
(c) $(2)T = (1+i)^2$. Here $e = 2$ and $g = 1$; so $f = 1$.

(2) $R = \mathbb{Z}$ and $T = \mathbb{Z}[\sqrt{-5}\,]$. In this case, T is not a unique factorization domain and is in particular not a principal ideal domain. We gave examples of three possible behaviors for the prime factorization of a principal ideal $(p)T$ in T generated by a prime $p > 0$ in $\mathbb{Z}$:

(a) $(11)T$ is prime in T. Here $e = g = 1$; so $f = 2$.
(b) $(2)T = (2, 1+\sqrt{-5})(2, 1-\sqrt{-5})$. Here $e = 1$ and $g = 2$; so $f = 1$.
(c) $(5)T = (\sqrt{-5}\,)^2$. Here $e = 2$ and $g = 1$; so $f = 1$.

---

[19]The notation here fits with the notation in Theorem 9.62 and is different from the notation in Section VIII.7.

(3) $R = \mathbb{C}[x]$ and $T = \mathbb{C}[x, \sqrt{(x-1)x(x+1)}\,]$. In this case, R is a principal ideal domain, and we saw that T is not a unique factorization domain. We found two possible behaviors for the prime factorization of a principal ideal $(p)T$ in T generated by a prime p in $\mathbb{C}[x]$:

(a) $(x - x_0)T = (x - x_0, y - y_0)(x - x_0, y + y_0)$ if the equal expressions $y_0^2 = (x_0 - 1)x_0(x_0 + 1)$ are not 0. Here $e = 1$ and $g = 2$; so $f = 1$.
(b) $(x - x_0)T = (x - x_0, y)^2$ if x_0 is in $\{-1, 0, +1\}$. Here $e = 2$ and $g = 1$; so $f = 1$.

The third type, with $(x - x_0)T$ prime in T, does not arise. It cannot arise since $f > 1$ would point to a quadratic extension of $\mathbb{C}$, yet $\mathbb{C}$ is algebraically closed.

17. Two Tools for Computing Galois Groups

In Section 8 we mentioned that the effect of the Fundamental Theorem of Galois Theory is to reduce the extremely difficult problem of finding intermediate fields to the less-difficult problem of finding a Galois group. In the intervening sections we have seen some illustrations of the power of this reduction, all in cases in which the Galois group was close at hand.

The problem of finding a Galois group in a particular situation is usually not as easy as in those cases, and it by no means can be considered as solved in general. In this section we combine Galois theory with some of the ring theory in the second half of Chapter VIII in order to develop two tools that sometimes help identify particular Galois groups.

Let us think in terms of a finite Galois extension K of the rationals $\mathbb{Q}$. The field K is the splitting field of some irreducible monic polynomial with rational coefficients, and we can scale this polynomial's indeterminate (in effect by multiplying its roots by some nonzero integer) so that the polynomial is monic and has integer coefficients. Thus let $F(X)$ be a monic irreducible polynomial in $\mathbb{Z}[X]$ of some degree d, and let K be its splitting field over $\mathbb{Q}$. The members of $\mathrm{Gal}(K/\mathbb{Q})$ are determined by their effect on the d roots of $F(X)$, and hence $\mathrm{Gal}(K/\mathbb{Q})$ may be regarded as a subgroup of the symmetric group $\mathfrak{S}_d$. If $r_1, \ldots, r_d$ are the roots of $F(X)$, then the **discriminant** of $F(X)$ is the member of K defined by

$$D = \prod_{1 \le i < j \le d} (r_j - r_i)^2.$$

This was defined in Section 13 in the cases $d = 2$ and $d = 3$, and we computed the value of D in those cases. The discriminant is an integer under our hypotheses, and it is computable even though the roots $r_1, \ldots, r_d$ of $F(X)$ are not at hand. In fact, the proof of Theorem 9.50 indicates that the discriminant D is given by the determinant

$$D = \det \begin{pmatrix} d & a_1 & a_2 & \cdots & a_{d-1} \\ a_1 & a_2 & a_3 & \cdots & a_d \\ a_2 & a_3 & a_4 & \cdots & a_{d+1} \\ & & & \vdots & \\ a_{d-1} & a_d & a_{d+1} & \cdots & a_{2d-2} \end{pmatrix},$$

where $a_j = r_1^j + r_2^j + \cdots + r_d^j$. Problems 36–39 at the end of Chapter VIII show that each of $a_1, \ldots, a_{2d-1}$ can be expressed as a polynomial in the elementary symmetric polynomials in $r_1, \ldots, r_d$, i.e., in the coefficients of $F(X)$, and doing so in a symbolic manipulation program is manageable for any fixed degree.[20]

The first of the two tools that sometimes help in identifying particular Galois groups directly concerns the discriminant: the discriminant is a square if and only if the Galois group is a subgroup of the alternating group. Let us state the result in the context of a general finite Galois extension even though we shall use it only for our Galois extension $K/\mathbb{Q}$.

Proposition 9.63. Let $\mathbb{K}/\mathbb{k}$ be a finite Galois extension, and suppose that $\mathbb{K}$ is the splitting field of a separable polynomial $F(X)$ in $\mathbb{k}[X]$ of degree d. Let D be the discriminant of $F(X)$, and regard $G = \text{Gal}(\mathbb{K}/\mathbb{k})$ as a subgroup of the symmetric group $\mathfrak{S}_d$. Then D is in $\mathbb{k}$, and G is a subgroup of the alternating group $\mathfrak{A}_d$ if and only if D is the square of an element of $\mathbb{k}$.

REMARK. The proof will use Galois theory to show that D is in $\mathbb{k}$, and Problems 36–39 at the end of Chapter VIII do not need to be invoked.

PROOF. Let $r_1, \ldots, r_d$ be the roots of $F(X)$, and put $\Delta = \prod_{i<j}(r_j - r_i)$. Under the identification of G with a subgroup of the permutation group $\mathfrak{S}_d$ on $\{1, \ldots, d\}$, each σ in G has

$$\sigma(\Delta) = \prod_{i<j}(\sigma(r_j) - \sigma(r_i)) = \prod_{i<j}(r_{\sigma(j)} - r_{\sigma(i)}) = (\text{sgn } \sigma)\prod_{i<j}(r_j - r_i) = (\text{sgn } \sigma)\Delta.$$

---

[20] For example, when $d = 3$, let $F(X) = X^3 - c_1 X^2 + c_2 X - c_3$. In Mathematica the following program produces a_1, a_2, a_3, a_4 as output:
```
e1={a1==r1+r2+r3,  r1+r2+r3==c1,  r1 r2+r2 r3+r1 r3==c2,
    r1 r2 r3==c3}
Eliminate[e1,{r1,r2,r3}]
e2={a2==r1^2+r2^2+r3^2,  r1+r2+r3==c1,  r1 r2+r2 r3+r1 r3==c2,
    r1 r2 r3==c3}
Eliminate[e2,{r1,r2,r3}]
e3={a3==r1^3+r2^3+r3^3,  r1+r2+r3==c1,  r1 r2+r2 r3+r1 r3==c2,
    r1 r2 r3==c3}
Eliminate[e3,{r1,r2,r3}]
e4={a4==r1^4+r2^4+r3^4,  r1+r2+r3==c1,  r1 r2+r2 r3+r1 r3==c2,
    r1 r2 r3==c3}
Eliminate[e4,{r1,r2,r3}]
```

17. Two Tools for Computing Galois Groups   529

In particular, the element $D = \Delta^2$ has $\sigma(D) = D$. By Proposition 9.35d, D is in $\Bbbk$.

If some σ in G has $\operatorname{sgn}\sigma = -1$, then σ does not fix Δ, and Δ is not in $\Bbbk$. Since Δ is a square root of D and since any two square roots of an element in a field differ at most by a sign, D is not the square of any element of $\Bbbk$.

Conversely if every σ in G has $\operatorname{sgn}\sigma = +1$, then every σ fixes Δ, and Proposition 9.35d shows that Δ is in $\Bbbk$. Since $D = \Delta^2$, D is the square of the member Δ of $\Bbbk$. □

The second tool is complicated to prove but simple to state. We reduce the polynomial $F(X)$ modulo p for each prime number p and form the associated finite splitting field. The Galois group for a finite extension of finite fields is cyclic by Proposition 9.40, and we thus obtain a cyclic subgroup of $\mathfrak{S}_d$. The second tool is this: if p does not divide the discriminant of $F(X)$, then this cyclic group as a permutation group is a subgroup of $\operatorname{Gal}(K/\mathbb{Q})$ as a permutation group, up to a relabeling of the symbols. In other words, the order *and cycle structure* of a generator of the cyclic group are the same as the order and cycle structure of some element of $\operatorname{Gal}(K/\mathbb{Q})$.

Let us formulate the result precisely. In the setting of Theorem 9.62, fix a prime ideal P of T lying in the factorization of $\mathfrak{p}T$. Each member σ of $G = \operatorname{Gal}(K/F)$ carries T to itself, but not every σ in G carries P to itself. Let G_P be the isotropy subgroup of G at P, i.e., let $G_P = \{\sigma \in G \mid \sigma(P) = P\}$. The subgroup G_P is called the **decomposition group** at P. Each σ in G_P descends to an automorphism of the field T/P that fixes the subfield $R/\mathfrak{p}$, since σ fixes each element of R. Thus σ defines a member $\overline{\sigma}$ of $\overline{G} = \operatorname{Gal}((T/P)/(R/\mathfrak{p}))$ by the formula

$$\overline{\sigma}(\overline{x}) = \overline{\sigma(x)}, \qquad \text{where } \overline{y} = y + P \text{ for } y \in T.$$

It is apparent that $\sigma \mapsto \overline{\sigma}$ is a homomorphism of G into $\overline{G}$. This homomorphism turns out to yield the result stated informally in the previous paragraph. It has the key property given in Theorem 9.64.

Theorem 9.64. Let R be a Dedekind domain, let F be its field of fractions, let K be a finite separable extension of F with $[K : F] = n$, and let T be the integral closure of R in K. Suppose that K is Galois over F. Let $\mathfrak{p}$ be a nonzero prime ideal in R, let $P = P_1$ be a prime factor in a decomposition $\mathfrak{p}T = \prod_{i=1}^{g} P_i^{e_i}$ of $\mathfrak{p}T$ as the product of powers of distinct nonzero prime ideals in T, and suppose that T/P is a Galois extension of $R/\mathfrak{p}$. Let $G = \operatorname{Gal}(K/F)$, $G_P = \{\sigma \in G \mid \sigma(P) = P\}$, and $\overline{G} = \operatorname{Gal}((T/P)/(R/\mathfrak{p}))$. Then the group homomorphism $\sigma \mapsto \overline{\sigma}$ of G_P into $\overline{G}$ carries G_P onto $\overline{G}$.

REMARKS. In our application with $R = \mathbb{Z}$, T/P and $R/\mathfrak{p}$ are finite fields, and Proposition 9.40 shows that T/P is a Galois extension of $R/\mathfrak{p}$ with no further assumptions.

PROOF. Let K^d be the fixed field of G_P within K; Theorem 9.38 shows that $\text{Gal}(K/K^d) = G_P$. Let T^d be the integral closure of R in K^d; this is a Dedekind domain, and T is the integral closure of T^d in K. We are going to apply Theorem 9.62 with R in the theorem replaced[21] by T^d.

Proposition 8.43 shows that $\mathcal{P} = T^d \cap P$ is a nonzero prime ideal of T^d. Since every member of G_P carries P to itself and since G_P is the full Galois group of K over K^d, Lemma 9.61 shows that P is the only nonzero prime ideal of T whose intersection with T^d is $\mathcal{P}$. Therefore $\mathcal{P} T^d = P^{e'}$ for some integer $e' \geq 1$.

As always, we have a field mapping $R/\mathfrak{p} \to T^d/\mathcal{P}$. Let us show that this mapping is onto $T^d/\mathcal{P}$. For any given u in T^d, we are to produce r in R with

$$r \equiv u \bmod \mathcal{P}. \tag{$*$}$$

Each σ in G that is not in G_P has $\sigma^{-1} P \neq P$, and the previous paragraph shows that the nonzero prime ideal $\mathcal{P}_\sigma = T^d \cap \sigma^{-1} P$ of T^d has $\mathcal{P}_\sigma \neq T^d \cap P$. Therefore $\mathcal{P}_\sigma + \mathcal{P} = T^d$, and the Chinese Remainder Theorem (Theorem 8.27) shows that we can find an element v of T^d with

$$v \equiv u \bmod \mathcal{P} \quad \text{and} \quad v \equiv 1 \bmod \mathcal{P}_\sigma$$

for all σ that lie in G but not G_P. The first congruence implies that $v - u$ is in $\mathcal{P} = T^d \cap P \subseteq P$, hence that

$$v \equiv u \bmod P, \tag{$**$}$$

while the second congruence implies that $v - 1$ is in $\mathcal{P}_\sigma = T^d \cap \sigma^{-1} P \subseteq \sigma^{-1} P$, hence that $\sigma(v-1)$ lies in P. Therefore

$$\sigma(v) \equiv 1 \bmod P \quad \text{for all } \sigma \text{ in } G \text{ but not } G_P. \tag{$\dagger$}$$

Put $r = N_{K^d/F}(v)$. Since the splitting field of the minimal polynomial of v over F is contained in K, Corollary 9.58 shows that r is the product of the elements $\sigma(v)$ for σ in G/G_P. Each of these is in T, and hence $N_{K^d/F}(v)$ is in T. Since $N_{K^d/F}(v)$ is also in F, $r = N_{K^d/F}(v)$ is in $T \cap F = R$. If we use $\sigma = 1$ as the representative of the identity coset of G/G_P, then we have

$$r = N_{K^d/F}(v) = v \Big(\prod_{\substack{\text{some } \sigma\text{'s} \\ \text{not in } G_P}} \sigma(v) \Big).$$

---

[21]Consequently it would not have been sufficient to prove Theorem 9.62 when the ring R is a principal ideal domain.

17. Two Tools for Computing Galois Groups   531

The factor of v is congruent to u mod P by $(**)$, and each factor in parentheses is congruent to 1 mod P by $(\dagger)$. Therefore $r \equiv u$ mod P, and $r - u$ is in P. Since $r - u$ is in T^d, $r - u$ is in $T^d \cap P = \mathcal{P}$. This proves $(*)$. Consequently we can identify $\overline{G} = \text{Gal}((T/P)/(R/\mathfrak{p}))$ with $\text{Gal}((T/P)/(T^d/\mathcal{P}))$.

Choose $\bar{x}_1$ in T/P with $T/P = (T^d/\mathcal{P})[\bar{x}_1]$; this choice is possible by the assumed separability of $(T/P)/(R/\mathfrak{p})$. Let x_1 be a member of T with $\bar{x}_1 = x_1 + P$, and let $M(X)$ be the minimal polynomial of x_1 over K^d. Since x_1 is in T, the coefficients of $M(X)$ are in T^d. Let $\overline{M}(X)$ be the corresponding member of $(T^d/\mathcal{P})[X]$, given by the substitution homomorphism that takes T^d to $T^d/\mathcal{P}$ and takes X to X. Since K/K^d is normal, $M(X)$ splits over K. Write $x_1, \ldots, x_n$ for its roots; these are in T.

Let τ be given in $\overline{G}$, and suppose that $\tau(\bar{x}_1) = \bar{x}_j$. Since $M(X)$ is irreducible over K^d, the Galois group $\text{Gal}(K/K^d) = G_P$ is transitive on its roots. Choose σ in G_P with $\sigma(x_1) = x_j$. Then $\overline{\sigma}(\bar{x}_1) = \bar{x}_j$. Since $\overline{\sigma}$ and τ agree on the generator $\bar{x}_1$ of T/P over $T^d/\mathcal{P}$, they agree on T/P. Therefore τ is exhibited as the image of σ under the homomorphism of the theorem, and the proof is complete. □

A first consequence of Theorem 9.64 is that we get interpretations of the integers e, f, and g, and they will be helpful to us. Galois theory gives us $|G| = n$, and Theorem 9.62 says that $efg = n$. The transitivity in Lemma 9.61 says that G acts transitively on the set $\{P_1, \ldots, P_g\}$, and the isotropy subgroup at $P = P_1$ is G_P. Hence $g|G_P| = |G|$, and $|G_P| = n/g = ef$. Galois theory gives us $|\overline{G}| = f$, and the fact that G_P maps onto $\overline{G}$ says that $G_P/\text{kernel} \cong \overline{G}$; therefore $|\text{kernel}| = |G_P|/|\overline{G}| = (ef)/f = e$. We conclude that g is the number of cosets modulo G_P, e is the order of the kernel of the homomorphism in Theorem 9.64, and f is the order of the cyclic group $\overline{G}$.

In the setting of interest for current purposes, we are taking $R = \mathbb{Z}$, $F = \mathbb{Q}$, and K equal to the splitting field of a given monic irreducible polynomial $F(X)$ of degree d in $\mathbb{Z}[X]$. We will be using Theorem 9.64 for various choices of $\mathfrak{p} = (p)$ in $\mathbb{Z}$ to make progress on identifying $\text{Gal}(K/\mathbb{Q})$. In order to identify $\overline{G}$ with the subgroup G_P of G, we need the kernel of the homomorphism of G_P onto $\overline{G}$ to be trivial. From the previous paragraph we know that the condition in question is that $e = 1$. We postpone to *Advanced Algebra* any justification of the assertion that $e = 1$ if p does not divide the discriminant of $F(X)$.

In previous sections we have identified $\text{Gal}(K/\mathbb{Q})$ in some cases when the Galois group is relatively small compared with the degree d of the polynomial. The method now is helpful when the Galois group is relatively large compared with d.

Let us be sure when $e = 1$ that the theorem is telling us not only that G_P is isomorphic to $\overline{G}$ as an abstract group, but also that the cycle structure of the elements of $\overline{G}$ is the same as the cycle structure of the elements of G_P. For this purpose we ignore the proof of the theorem and concentrate only on the statement. Assuming that p does not divide the discriminant, let $\overline{F}(X)$ be the reduction of $F(X)$ modulo p, let $r_1, \ldots, r_d$ be the roots of $F(X)$ in T, and let $\bar{r}_1, \ldots, \bar{r}_d$ be the images of $r_1, \ldots, r_d$ under the quotient homomorphism $T \to T/P$. The elements $\bar{r}_1, \ldots, \bar{r}_d$ are distinct since p does not divide the discriminant of $F(X)$. Any member σ of $G = \text{Gal}(K/\mathbb{Q})$ permutes $r_1, \ldots, r_d$ and is determined by the resulting permutation since K is assumed to be generated by $r_1, \ldots, r_d$. Under the assumption that σ is in G_P, σ descends to an automorphism $\bar{\sigma}$ of T/P. This automorphism $\bar{\sigma}$ acts on the set of elements $\bar{r}_1, \ldots, \bar{r}_d$, permuting them. Since the mapping of the r_j's to the $\bar{r}_j$'s is one-one, the resulting permutation of the subscripts $1, \ldots, d$ is the same.

When p varies, we cannot match the elements $\bar{r}_1, \ldots, \bar{r}_d$ for one value of p with those for another value of p, because we have no direct knowledge of $r_1, \ldots, r_d$. Thus we cannot directly compare the permutation groups $\overline{G}$ that we obtain for different p's. But at least we know their cycle structure.

To apply the theory, we factor $\overline{F}(X)$ quickly with a symbolic manipulation program, and we obtain the Galois group of a splitting field of $\overline{F}(X)$ by inspection, together with the cycle structure of its elements. Specifically an irreducible factor of degree m contributes an m-cycle for the element, and the cycles corresponding to distinct irreducible factors are disjoint. Then we put together the information from various p's and see what elements must be in $\text{Gal}(K/\mathbb{Q})$, up to a relabeling of indices.

EXAMPLE 1. $F(X) = X^5 - X - 1$. The discriminant is $D = 2869 = 19 \cdot 151$. Thus the method may be used with any prime number other than 19 and 151. Here is the factorization for a few primes, together with the cycle structure within $\mathfrak{S}_5$ for a generator of $\overline{G}$:

p	$\overline{F}(X)$	Cycle lengths
2	$(X^2 + X + 1)(X^3 + X + 1)$	2, 3
3	$X^5 + 2X + 2$	5
17	$(X+9)(X+11)(X^3 + 14X^2 + 12X + 6)$	1, 1, 3
23	$(X+9)(X^4 + 14X^3 + 12X^2 + 7X + 5)$	1, 4

For comparison, $p = 19$ gives $\overline{F}(X) = (X+6)^2(X^2 + 7X^2 + 13X + 10)$, but we cannot use this prime since it divides the discriminant. It is enough to use the information from $p = 2$ and $p = 3$. The irreducibility modulo 3 implies irreducibility over $\mathbb{Q}$. From $p = 3$, we obtain a 5-cycle in $\text{Gal}(K/\mathbb{Q})$. From $p = 2$, we obtain the product of a 2-cycle and a 3-cycle, and the cube of this element is a 2-cycle. In the example in Section 11 following the statement of Theorem 9.44, we saw in effect that the only subgroup of $\mathfrak{S}_5$ containing a 5-cycle and a 2-cycle is $\mathfrak{S}_5$ itself. Therefore $\text{Gal}(K/\mathbb{Q}) = \mathfrak{S}_5$.

EXAMPLE 2. $F(X) = X^5 + 10X^3 - 10X^2 + 35X - 18$. The discriminant is $D = 3025000000 = 2^6 5^8 11^2$, a perfect square. Thus the Galois group is a subgroup of the alternating group $\mathfrak{A}_5$. The method using reduction modulo p may be used with any prime other than 2, 5, and 11. Here is the factorization for a few primes, together with the cycle structure within $\mathfrak{S}_5$ for a generator of $\overline{G}$:

p	$\overline{F}(X)$	Cycle lengths
3	$X(X+2)(X^3+X^2+2X+1)$	1, 1, 3
7	$X^5 + 3X^3 + 4X^2 + 3$	5
17	$(X+14)(X^2+5X+14)(X^2+15X+15)$	1, 2, 2

It is enough to use the information from $p = 3$ and $p = 7$. The irreducibility modulo 7 implies irreducibility over $\mathbb{Q}$. From $p = 7$, we obtain a 5-cycle in $\mathrm{Gal}(K/\mathbb{Q})$. From $p = 3$, we obtain a 3-cycle. Any 5-cycle and any 3-cycle together generate all of $\mathfrak{A}_5$. In fact, the generated subgroup must have order divisible by 15, hence must have order 15, 30, or 60. It cannot be of order 15 because every group of order 15 is cyclic and $\mathfrak{A}_5$ has no elements of order 15. It cannot be of order 30 because $\mathfrak{A}_5$ is simple and subgroups of index 2 have to be normal. Hence it is all of $\mathfrak{A}_5$.

EXAMPLE 3. Galois group $\mathfrak{S}_d$. Given $d \geq 4$, let us see how to form an irreducible $F(X)$ for which $\mathrm{Gal}(K/\mathbb{Q})$ is all of $\mathfrak{S}_d$. For any degree d and any prime number ℓ, there exists at least one irreducible monic polynomial of degree d in $\mathbb{F}_\ell[X]$; the reason is that the finite field $\mathbb{F}_{\ell^d}$ is a simple extension of $\mathbb{F}_\ell$ by Corollary 9.19. Let $H_{d,2}(X)$ be such a polynomial of degree d for $\ell = 2$, and let $H_{d-1,3}(X)$ be such a polynomial of degree $d-1$ for $\ell = 3$. Then let p be a prime greater than d, and let $H_{2,p}(X)$ be an irreducible monic polynomial of degree 2 in $\mathbb{F}_p[X]$. We can regard each of $H_{d,2}(X)$, $H_{d-1,3}(X)$, and $H_{2,p}(X)$ as in $\mathbb{Z}[X]$ by reinterpreting their coefficients as integers. Consider the congruences

$$F[X] \equiv H_{d,2}(X) \quad \mathrm{mod}\ (2),$$
$$F[X] \equiv X H_{d-1,3}(X) \quad \mathrm{mod}\ (3),$$
$$F[X] \equiv \Big(\prod_{k=0}^{d-3}(X-k)\Big) H_{2,p}(X) \quad \mathrm{mod}\ (p),$$

in $\mathbb{Z}[X]$. Since the sum of any two of the three ideals (2), (3), and (p) of $\mathbb{Z}[X]$ is $\mathbb{Z}[X]$, the Chinese Remainder Theorem (Theorem 8.27) implies that there exists a simultaneous solution $F[X]$ to these congruences in $\mathbb{Z}[X]$, and we may take $F[X]$ to be monic of degree d. Let K be a splitting field for $F[X]$ over $\mathbb{Q}$. Our method applies to the primes 2, 3, and p since none of the three polynomials has any repeated factors. The result of applying the method is that $\mathrm{Gal}(K/\mathbb{Q})$ contains a d-cycle, a $(d-1)$-cycle, and a 2-cycle. Let us see that the subgroup generated by these three elements is all of $\mathfrak{S}_d$. We may assume that the $(d-1)$-cycle is $(1\ 2\ \cdots\ d{-}1)$. Without loss of generality, the 2-cycle is either $(1\ j)$ with $j < d$ or is $(k\ d)$ with $k < d$. In the first case some power of the d-cycle is a permutation τ with $\tau(1) = d$; if σ denotes the 2-cycle $(1\ j)$, then Lemma 4.41 shows that $\tau\sigma\tau^{-1}$ is the 2-cycle $(d\ \tau(j))$, and this is of the form $(k\ d)$ with $k < d$. Thus we may assume in any event that $\mathrm{Gal}(K/\mathbb{Q})$ contains $(1\ 2\ \cdots\ d{-}1)$ and some 2-cycle $(k\ d)$ with $k < d$. Conjugating $(k\ d)$ by powers of $(1\ 2\ \cdots\ d{-}1)$, we see that $\mathrm{Gal}(K/\mathbb{Q})$ contains *every* 2-cycle $(k\ d)$ with $k < d$. For $1 \le k < d-1$, we then find that $\mathrm{Gal}(K/\mathbb{Q})$ contains

$$(k\ d)(k+1\ d)(k\ d) = (k\ k+1).$$

So $\mathrm{Gal}(K/\mathbb{Q})$ contains $(1\ 2), (2\ 3), \ldots, (d{-}2\ d{-}1)$, and we have already seen that it contains $(d{-}1\ d)$. These $d-1$ transpositions generate the full symmetric group, and therefore $\mathrm{Gal}(K/\mathbb{Q}) = \mathfrak{S}_d$.

18. Problems

1. Take as known that the polynomial $X^3 - 3X + 4$ is irreducible over $\mathbb{Q}$, and let r be a complex root of it. In the field $\mathbb{Q}(r)$, find a multiplicative inverse for $r^2 + r + 1$ and express it in the form $ar^2 + br + c$ with a, b, c in $\mathbb{Q}$.

2. Suppose that R is an integral domain and that F is a subring that is a field, so that R can be considered as a vector space over F. Prove that if $\dim_F R$ is finite, then R is a field.

3. Let $\mathbb{K}$ be a subfield of $\mathbb{C}$ that is not a subfield of $\mathbb{R}$. Prove that $\mathbb{K}$ is topologically dense in $\mathbb{C}$.

4. Let $\mathbb{K} = \mathbb{k}(x)$ be a transcendental extension of the field $\mathbb{k}$, and let y be a member of $\mathbb{K}$ that is not in $\mathbb{k}$. Prove that $\mathbb{k}(x)$ is an algebraic extension of $\mathbb{k}(y)$.

5. What is a necessary and sufficient condition on an integer $N > 0$ for the positive square root of N to be in the subfield $\mathbb{Q}(\sqrt[3]{2})$ of $\mathbb{R}$?

6. The polynomials $F(X) = X^3 + X + 1$ and $G(Y) = Y^3 + Y^2 + 1$ are irreducible over $\mathbb{F}_2$. Let $\mathbb{K}$ be the field $\mathbb{K} = \mathbb{F}_2[X]/(F(X))$, and let $\mathbb{L}$ be the field $\mathbb{L} = \mathbb{F}_2[Y]/(G(Y))$. Since $\mathbb{K}$ and $\mathbb{L}$ are two fields of order 8, they must be isomorphic. Find an explicit isomorphism.

7. Can a field of order 8 have a subfield of order 4? Why or why not?

8. If $\mathbb{K}$ is a finite field, prove that the product of the nonzero elements of $\mathbb{K}$ is -1. (Educational note: When $\mathbb{K}$ is $\mathbb{F}_p$, this result reduces to Wilson's Theorem, given as Problem 8 at the end of Chapter IV.)

9. Suppose that $\mathbb{K}/\mathbb{k}$ is a finite extension of the form $\mathbb{K} = \mathbb{k}(r)$ with $[\mathbb{K}:\mathbb{k}]$ odd. Prove that $\mathbb{K} = \mathbb{k}(r^2)$.

10. Suppose that $\mathbb{K}/\mathbb{k}$ is a finite extension of fields and that $\mathbb{K} = \mathbb{k}[r,s]$. Prove that if $[\mathbb{k}(r):\mathbb{k}]$ is relatively prime to $[\mathbb{k}(s),\mathbb{k}]$, then
   (a) the minimal polynomial of r over $\mathbb{k}$ is irreducible over $\mathbb{k}(s)$,
   (b) $[\mathbb{K}:\mathbb{k}] = [\mathbb{k}(r):\mathbb{k}][\mathbb{k}(s):\mathbb{k}]$.

11. In $\mathbb{C}$, let $\beta = \sqrt[3]{2}$, $\omega = \frac{1}{2}(-1+\sqrt{-3})$, and $\alpha = \omega\beta$.
    (a) Prove for all c in $\mathbb{Q}$ that $\gamma = \beta + c\alpha$ is a root of some sixth-degree polynomial of the form $X^6 + aX^3 + b$.
    (b) Prove that the minimal polynomial of $\beta + \alpha$ over $\mathbb{Q}$ has degree 3.
    (c) Prove that the minimal polynomial of $\beta - \alpha$ over $\mathbb{Q}$ has degree 6.

12. Suppose that $\mathbb{k}$ is a finite field and that $F(X)$ is a member of $\mathbb{k}[X]$ whose derivative is the 0 polynomial. Prove that $F(X)$ is reducible over $\mathbb{k}$.

13. Let $\mathbb{k}$ be a field, let $F(X)$ be a separable polynomial in $\mathbb{k}[X]$, let $\mathbb{K}$ be a splitting field of $F(X)$ over $\mathbb{k}$, and let $r_1, \ldots, r_n$ be the roots of $F(X)$ in $\mathbb{K}$. Regard $\text{Gal}(\mathbb{K}/\mathbb{k})$ as a subgroup of the symmetric group $\mathfrak{S}_n$.
    (a) Prove that $\text{Gal}(\mathbb{K}/\mathbb{k})$ is transitive on $\{r_1, \ldots, r_n\}$ if and only if $F(X)$ is irreducible over $\mathbb{k}$.
    (b) Show that the cyclotomic polynomial $\Phi_8(X)$ is an example with $\mathbb{k} = \mathbb{Q}$ and $n = 4$ for which $\text{Gal}(\mathbb{K}/\mathbb{k})$ is transitive but $\text{Gal}(\mathbb{K}/\mathbb{k})$ contains no 4-cycle.
    (c) Prove that if n is prime and $F(X)$ is irreducible over $\mathbb{k}$, then $\text{Gal}(\mathbb{K}/\mathbb{k})$ contains an n-cycle.

14. Let $a_1, \ldots, a_n$ be relatively prime square-free integers ≥ 2, and define $\mathbb{L}_k = \mathbb{Q}(\sqrt{a_1}, \ldots, \sqrt{a_k})$ for $0 \leq k \leq n$.
    (a) Show for each k that $[\mathbb{L}_k : \mathbb{Q}] = 2^l$ with $0 \leq l \leq k$.
    (b) Suppose for a particular k that $[\mathbb{L}_k : \mathbb{Q}] = 2^k$. Exhibit a vector-space basis of $\mathbb{L}_k$ over $\mathbb{Q}$, and describe the members of $\text{Gal}(\mathbb{L}_k/\mathbb{Q})$ by telling the effect of each member on all basis vectors of $\mathbb{L}_k$ over $\mathbb{Q}$.
    (c) Suppose for a particular $k < n$ that $[\mathbb{L}_k : \mathbb{Q}] = 2^k$. Assume that $\sqrt{a_{k+1}}$ lies in $\mathbb{L}_k$, and let $\sqrt{a_{k+1}}$ be expanded in terms of the basis of (b). Show that application of the members of $\text{Gal}(\mathbb{L}_k/\mathbb{Q})$ leads to a contradiction.
    (d) Deduce that $[\mathbb{L}_n : \mathbb{Q}] = 2^n$.

15. Let p be a prime number, and suppose that a is in $\mathbb{Q}$ and r is a member of $\mathbb{C}$ but not $\mathbb{Q}$ with $r^p = a$. Prove that
    (a) the cyclotomic polynomial $\Phi_p(X)$ is irreducible in $\mathbb{Q}(r)$,
    (b) the splitting field $\mathbb{K}$ of $X^p - a$ over $\mathbb{Q}$ has degree $[\mathbb{K} : \mathbb{Q}] = p(p-1)$,
    (c) the Galois group $\text{Gal}(\mathbb{K}/\mathbb{Q})$ is isomorphic to a semidirect product of the multiplicative group of $\mathbb{F}_p$ and the additive group of $\mathbb{F}_p$, with the action of a member m of the multiplicative group on the members n of the additive group being given by $m(n) = mn$.

16. Let $F(X)$ be a polynomial in $\mathbb{k}[X]$ of degree n, where $\mathbb{k}$ is a field of characteristic 0, and let $\mathbb{K}$ be a splitting field for $F(X)$ over $\mathbb{k}$. Prove that $[\mathbb{K} : \mathbb{k}]$ divides $n!$.

17. Suppose that $\mathbb{k}$ is a field of characteristic 0. Let $\mathbb{K}$ be a quadratic extension $\mathbb{k}(r)$, where $r^2 = a$ is a member of $\mathbb{k}$. Determine all elements of $\mathbb{K}$ whose squares are in $\mathbb{k}$.

18. Let G be a finite group. Show that there exist two finite extensions $\mathbb{k}$ and $\mathbb{K}$ of $\mathbb{Q}$ such that $\mathbb{K}$ is a Galois extension of $\mathbb{k}$ and the Galois group $\mathrm{Gal}(\mathbb{K}/\mathbb{k})$ is isomorphic to G.

19. Let $\mathbb{K}/\mathbb{k}$ be a finite normal extension. For $F(X)$ in $\mathbb{K}[X]$ and σ in $\mathrm{Gal}(\mathbb{K}/\mathbb{k})$, let $F^\sigma(X)$ be the result of the substitution homomorphism $\mathbb{K}[X] \to \mathbb{K}[X]$ carrying X to X and extending the action of σ on $\mathbb{K}$, i.e., let $F^\sigma(X)$ be obtained by applying σ to the coefficients of $F(X)$. Prove that $\prod_{\sigma \in \mathrm{Gal}(\mathbb{K}/\mathbb{k})} F^\sigma(X)$ is in $\mathbb{k}[X]$.

20. Corollary 9.37 concerns a separable algebraic extension $\mathbb{K}/\mathbb{k}$ and a finite subgroup H of $\mathrm{Gal}(\mathbb{K}/\mathbb{k})$, showing that $\mathbb{K}/\mathbb{K}^H$ is a finite Galois extension with $H = \mathrm{Gal}(\mathbb{K}/\mathbb{K}^H)$ and $[\mathbb{K} : \mathbb{K}^H] = |H|$. By going over its proof, obtain the conclusion that if $\{x_1, \ldots, x_n\}$ is the H orbit of x_1 in $\mathbb{K}$, then
    (a) the minimal polynomial of x_1 over $\mathbb{K}^H$ is $\prod_{j=1}^n (X - x_j)$.
    (b) n divides $|H|$.
    (c) $\mathbb{K} = \mathbb{K}^H(x_1)$ if the isotropy subgroup of H at x_1 is trivial.

21. Let $\mathbb{K}$ be the transcendental extension $\mathbb{C}(z)$ of $\mathbb{C}$.
    (a) Prove that any linear fractional transformation $\varphi(z) = \frac{az+b}{cz+d}$ with $ad-bc \neq 0$ in $\mathbb{C}$ extends uniquely to a $\mathbb{C}$ automorphism of $\mathbb{K}$.
    (b) Let H be the 4-element subgroup of $\mathrm{Gal}(\mathbb{K}/\mathbb{C})$ generated by the extensions of $\sigma(z) = -z$ and $\tau(z) = 1/z$. Show that $w = z^2 + z^{-2}$ is invariant under H, and conclude that every member of $\mathbb{C}(w)$ lies in $\mathbb{K}^H$.
    (c) Applying the previous problem to the element $x_1 = z$ of $\mathbb{K}$, show that the minimal polynomial of z over $\mathbb{C}(w)$ has degree 4.
    (d) Conclude that $\mathbb{K}^H = \mathbb{C}(z^2 + z^{-2})$.

22. In characteristic 0, let $\mathbb{L}/\mathbb{K}$ and $\mathbb{K}/\mathbb{k}$ be quadratic extensions.
    (a) Show that there exists an irreducible polynomial $F(X) = X^4 + bX^2 + c$ in $\mathbb{k}[X]$ such that $F(r) = 0$ for some r in $\mathbb{L}$.
    (b) Show that the element r in (a) has $\mathbb{L} = \mathbb{k}(r)$.
    (c) Show that $\mathbb{L}$ is a normal extension of $\mathbb{k}$ with Galois group $C_2 \times C_2$ if and only if c is a square in $\mathbb{k}$ for some polynomial as in (a), if and only if c is a square in $\mathbb{k}$ for every polynomial as in (a).
    (d) Show that $\mathbb{L}$ is a normal extension of $\mathbb{k}$ with Galois group C_4 if and only if $c^{-1}(b^2 - 4c)$ is a square in $\mathbb{k}$ for some polynomial as in (a), if and only if $c^{-1}(b^2 - 4c)$ is a square in $\mathbb{k}$ for every polynomial as in (a).

(e) Give an example of quadratic extensions $\mathbb{L}/\mathbb{K}$ and $\mathbb{K}/\mathbb{k}$ in characteristic 0 such that $\mathbb{L}/\mathbb{k}$ is not normal.

23. Determine Galois groups for splitting fields over $\mathbb{Q}$ for the two polynomials $X^3 - 3X + 1$ and $X^3 + X + 1$.

24. Suppose that $F(X)$ is an irreducible cubic polynomial in $\mathbb{Q}[X]$ whose splitting field $\mathbb{K}$ has $\text{Gal}(\mathbb{K}/\mathbb{Q})$ isomorphic to $\mathfrak{S}_3$. What are the possibilities, up to isomorphism, for the Galois group of a splitting field of $(X^3 - 1)F(X)$ over $\mathbb{Q}$?

25. Let $\mathbb{K}/\mathbb{k}$ be a finite Galois extension whose Galois group is isomorphic to $\mathfrak{S}_3$. Is $\mathbb{K}$ necessarily a splitting field of some irreducible cubic polynomial in $\mathbb{k}[X]$? Why or why not?

26. Is Cardan's cubic formula valid for finding roots of reducible cubics $X^3 + pX + q$ in characteristic 0?

27. Prove that the discriminant of a real cubic with distinct roots is positive if all the roots are real, and is negative if two of the roots are complex.

28. Let $F(X) = X^3 + pX + q$ be irreducible in $\mathbb{Q}[X]$, and suppose that $X - r$ is a factor for some r in $\mathbb{C}$.
    (a) Show that $F(X)$ factors in $\mathbb{Q}(r)[X]$ as $F(X) = (X-r)(X^2+rX+(r^2+p))$.
    (b) We know that $\mathbb{Q}(r)$ is a splitting field for $F(X)$ over $\mathbb{Q}$ if and only if the discriminant $-4p^3 - 27q^2$ is a square in $\mathbb{Q}$. On the other hand, it is evident from the factorization of $F(X)$ that it splits is $\mathbb{Q}(r)$ if and only if the discriminant $r^2 - 4(r^2 + p)$ is a square in $\mathbb{Q}(r)$. Show by a direct calculation that these two conditions are equivalent.

29. Let $\mathbb{K}$ be a splitting field of an irreducible cubic polynomial $F(X)$ in $\mathbb{Q}[X]$. If $\text{Gal}(\mathbb{K}/\mathbb{Q})$ is $\mathfrak{S}_3$, does it follow that $\mathbb{K}$ contains all three cube roots of 1? Why or why not?

30. In characteristic 0, let $\mathbb{K}$ be the splitting field over $\mathbb{k}$ of an irreducible polynomial in $\mathbb{k}[X]$ of degree 5. Assuming that the discriminant of the polynomial is a square in $\mathbb{k}$, what are the possibilities for $\text{Gal}(\mathbb{K}/\mathbb{k})$ up to a relabeling of the indices?

31. Determine the Galois group of a splitting field over $\mathbb{Q}$ for the polynomial $X^5 + 6X^3 - 12X^2 + 5X - 4$. Use of a computer may be helpful for this problem.

32. The proof of Theorem 9.64 introduced a positive integer e' in its second paragraph. Prove that e' equals the integer e_1 in the statement of the theorem.

33. Let R be a Dedekind domain, let F be its field of fractions, let K be a finite separable extension of F, and let L be a finite separable extension of K. Let T be the integral closure of R in K, and let U be the integral closure of R in L. Let $\mathfrak{p}$, P, and Q be nonzero prime ideals in R, T, and U, respectively, and let the

ramification indices and decomposition degrees for the extensions L/K, L/F, and K/F be

$$e(Q|P), e(P|\mathfrak{p}), e(Q|\mathfrak{p}) \quad \text{and} \quad f(Q|P), f(P|\mathfrak{p}), f(Q|\mathfrak{p}).$$

Prove that

$$e(Q|\mathfrak{p}) = e(Q|P)e(P|\mathfrak{p}) \quad \text{and} \quad f(Q|\mathfrak{p}) = f(Q|P)f(P|\mathfrak{p}).$$

Problems 34–40 concern norms and traces.

34. Let m be a square-free integer, and let N and Tr denote the norm and trace from $\mathbb{Q}(\sqrt{m})$ to $\mathbb{Q}$.
    (a) Show that $N(a + b\sqrt{m}) = a^2 - mb^2$ and $\text{Tr}(a + b\sqrt{m}) = 2a$.
    (b) Let T be the ring of algebraic integers in $\mathbb{Q}(\sqrt{m})$. It was shown in Section VIII.9 that T consists of all $a + b\sqrt{m}$ with a, b in $\mathbb{Z}$ if $m \equiv 2 \bmod 4$ or $m \equiv 3 \bmod 4$, and of all $a + b\sqrt{m}$ with a, b in $\mathbb{Z}$ or a, b in $\mathbb{Z} + \frac{1}{2}$ if $m \equiv 1 \bmod 4$. Prove for $a + b\sqrt{m}$ in $\mathbb{Q}(\sqrt{m})$ that $a + b\sqrt{m}$ is in T if and only if $N(a + b\sqrt{m})$ and $\text{Tr}(a + b\sqrt{m})$ are both in $\mathbb{Z}$.
    (c) Assume that $a + b\sqrt{m}$ is in T. Prove that $N(a + b\sqrt{m})$ is in $\mathbb{Z}^\times$ if and only if $a + b\sqrt{m}$ is in $T^\times$.
    (d) For $m = 2$, give an example of a member of $T^\times$ other than ± 1.

35. For the extension $\mathbb{Q}(\sqrt[3]{2})/\mathbb{Q}$, find the value of the norm N and the trace Tr on a general element $a + b\sqrt[3]{2} + c(\sqrt[3]{2})^2$ of $\mathbb{Q}(\sqrt[3]{2})$; here a, b, c are in $\mathbb{Q}$.

36. Let $N(\cdot)$ be the norm relative to the extension $\mathbb{Q}(\zeta)/\mathbb{Q}$, where ζ is a primitive n^{th} root of 1.
    (a) Show that $N(1-\zeta) = \Phi_n(1)$, where $\Phi_n(X)$ is the nth cyclotomic polynomial.
    (b) Using the formula $\prod_{d|n,\, d>1} \Phi_d(X) = X^{n-1} + X^{n-2} + \cdots + 1$, show that $N(1 - \zeta) = \Phi_n(1)$ equals p if n is a power of the positive prime p and equals 1 if n is divisible by more than one positive prime.

37. Let $p > 0$ be a prime in $\mathbb{Z}$ of the form $4n + 1$. It was shown in Problem 31 at the end of Chapter VIII that such a prime is the sum of two squares. This problem gives a shorter proof. Take as known from Section VIII.4 that the ring $\mathbb{Z}[\sqrt{-1}]$ of Gaussian integers is a Euclidean domain, and from Problem 30 at the end of Chapter VIII that $x^2 \equiv -1 \bmod p$ has an integer solution x.
    (a) Write

$$\frac{x \pm \sqrt{-1}}{p} = \frac{1}{p} x \pm \frac{1}{p}\sqrt{-1}.$$

If p were prime in $\mathbb{Z}[\sqrt{-1}]$, then it would follow from the divisibility of $x^2 + 1$ by p that p divides $x + \sqrt{-1}$ or p divides $x - \sqrt{-1}$. Deduce from the displayed equation that neither alternative is viable, and conclude that p cannot be prime in $\mathbb{Z}[\sqrt{-1}]$.

(b) Using the conclusion of (a) to write p as a nontrivial product in $\mathbb{Z}[\sqrt{-1}]$ and applying the norm function, prove that there exist integers a and b such that $p = a^2 + b^2$.

38. Let $p > 0$ be a prime in $\mathbb{Z}$ of the form $8n+1$. Take as known from Problem 13 at the end of Chapter VIII that $\mathbb{Z}[\sqrt{-2}]$ is a Euclidean domain, and from the law of quadratic reciprocity (to be proved in *Advanced Algebra*) that $x^2 \equiv -2 \bmod p$ has an integer solution x. Guided by the argument for the previous problem, prove that there exist integers a and b such that $p = a^2 + 2b^2$.

39. Let $p > 0$ be a prime in $\mathbb{Z}$ of the form $6n + 1$. Take as known from Problem 26 at the end of Chapter VIII that $\mathbb{Z}[\frac{1}{2}(1 + \sqrt{-3})]$ is a Euclidean domain, and from the law of quadratic reciprocity (to be proved in *Advanced Algebra*) that $x^2 \equiv -3 \bmod p$ has an integer solution x. Guided by the argument for the previous problem, prove that there exist integers a and b such that $p = a^2 + 3b^2$.

40. Let $\mathbb{k} \subseteq \mathbb{L} \subseteq \mathbb{L}'$ be fields such that $\mathbb{L}'/\mathbb{k}$ is a finite separable extension. Using Corollary 9.58, prove that the norm and trace satisfy

$$N_{\mathbb{L}'/\mathbb{k}} = N_{\mathbb{L}/\mathbb{k}} \circ N_{\mathbb{L}'/\mathbb{L}} \quad \text{and} \quad \text{Tr}_{\mathbb{L}'/\mathbb{k}} = \text{Tr}_{\mathbb{L}/\mathbb{k}} \circ \text{Tr}_{\mathbb{L}'/\mathbb{L}}.$$

Problems 41–45 make use of the theory of symmetric polynomials, which was introduced in Problems 36–39 at the end of Chapter VIII.

41. Let $\mathbb{k}$ be a field, let $F(X)$ be a polynomial in $\mathbb{k}[X]$, let $\mathbb{K}$ be an extension field in which $F(X)$ splits, and let $r_1, \ldots, r_n$ be the roots of $F(X)$ in $\mathbb{K}$, repeated according to their multiplicities. If $P(X_1, \ldots, X_n)$ is a symmetric polynomial in $\mathbb{k}[X_1, \ldots, X_n]$, prove that $P(r_1, \ldots, r_n)$ is a member of $\mathbb{k}$.

42. Let $\mathbb{k}$ be a field, let $F(X)$ and $G(X)$ be polynomials over $\mathbb{k}$, let $\mathbb{K}$ be an extension field in which $F(X)$ and $G(X)$ both split, and let $r_1, \ldots, r_m$ and $s_1, \ldots, s_n$ be the respective roots of $F(X)$ and $G(X)$ in $\mathbb{K}$, repeated according to their multiplicities. Deduce from the previous problem that the polynomials

$$H_1(X) = \prod_{i=1}^{m} \prod_{j=1}^{n} (X - r_i - s_j) \quad \text{and} \quad H_2(X) = \prod_{i=1}^{m} \prod_{j=1}^{n} (X - r_i s_j)$$

lie in $\mathbb{k}[X]$.

43. (a) Find a nonzero polynomial with rational coefficients having $\sqrt{2} + \sqrt{3}$ as a root. What is the minimal polynomial of $\sqrt{2} + \sqrt{3}$ over $\mathbb{Q}$?
    (b) Find a nonzero polynomial with rational coefficients having $\sqrt{2} + \sqrt[3]{2}$ as a root. What is the minimal polynomial of $\sqrt{2} + \sqrt[3]{2}$ over $\mathbb{Q}$?

44. Let $\mathbb{k}$ be a field of characteristic 0, and let $\mathbb{K} = \mathbb{k}(r_1, \ldots, r_n)$ be the field of fractions of the polynomial ring $\mathbb{k}[r_1, \ldots, r_n]$ in n indeterminates. Show that any σ in the symmetric group $\mathfrak{S}_n$ defines a member of $\text{Gal}(\mathbb{K}/\mathbb{k})$ such that $\sigma(r_j) = r_{\sigma(j)}$ for all σ in $\mathfrak{S}_n$. Then define $F(X)$ to be the polynomial
$$F(X) = (X - r_1) \cdots (X - r_n)$$
in $\mathbb{K}[X]$, and show that
   (a) $F(X)$ is irreducible over the fixed field $\mathbb{K}^{\mathfrak{S}_n}$,
   (b) $\mathbb{K}$ is a splitting field for $F(X)$ over $\mathbb{K}^{\mathfrak{S}_n}$,
   (c) $\mathbb{K}^{\mathfrak{S}_n} = \mathbb{k}(u_1, \ldots, u_n)$, where $u_1, \ldots, u_n$ are given by
$$u_1 = \sum_i r_i, \qquad u_2 = \sum_{i<j} r_i r_j, \qquad \ldots, \qquad u_n = \prod_i r_i,$$
   (d) the Galois group of the splitting field of $F(X)$ over $\mathbb{k}(u_1, \ldots, u_n)$ is $\mathfrak{S}_n$.

45. **(Cubic resolvent)** This problem carries out one step in finding the roots of an arbitrary quartic polynomial. Let $\mathbb{k}$ be a field of characteristic 0, let $\mathbb{K} = \mathbb{k}(p, q, r)$ be the field of fractions of the polynomial ring $\mathbb{k}[p, q, r]$ in n indeterminates, and let $\mathbb{L}$ be a splitting field of the polynomial
$$F(X) = X^4 + pX^2 + qX + r$$
in $\mathbb{K}[X]$. The Galois group $\text{Gal}(\mathbb{L}/\mathbb{K})$ is $\mathfrak{S}_4$ by the previous problem. Let $B_4 = \{(1), (1\ 2)(3\ 4), (1\ 3)(2\ 4), (1\ 4)(2\ 3)\}$. In the composition series $\mathfrak{S}_4 \supseteq \mathfrak{A}_4 \supseteq B_4 \supseteq \{(1), (1\ 2)\} \supseteq \{1\}$, Proposition 9.63 shows that the fixed field of $\mathfrak{A}_4$ is $\mathbb{K}(\sqrt{D})$, where D is the discriminant. To obtain the fixed field of B_4, we adjoin to $\mathbb{K}(\sqrt{D})$ an element of $\mathbb{L}$ invariant under B_4 but not under $\mathfrak{A}_4$. If s_1, s_2, s_3, s_4 denote the roots of $F(X)$ in $\mathbb{L}$, then such an element is $(s_1 + s_2)(s_3 + s_4)$. Its three conjugates under $\mathfrak{A}_4/B_4$ are
$$\theta_1 = (s_1 + s_2)(s_3 + s_4),$$
$$\theta_2 = (s_1 + s_3)(s_2 + s_4),$$
$$\theta_3 = (s_1 + s_4)(s_2 + s_3),$$
which are the three roots of the "cubic resolvent" polynomial
$$\theta^3 - c_1\theta^2 + c_2\theta - c_3,$$
where c_1, c_2, c_3 are the elementary symmetric polynomials in $\theta_1, \theta_2, \theta_3$ given by
$$c_1 = \sum_i \theta_i, \qquad c_2 = \sum_{i<j} \theta_i \theta_j, \qquad c_3 = \prod_i \theta_i.$$
   (a) Show that c_1, c_2, c_3 are symmetric polynomials in s_1, s_2, s_3, s_4, hence are polynomials in the coefficients p, q, r.

(b) Verify that $c_1 = 2p$, $c_2 = p^2 - 4r$, and $c_3 = q^2$.

(c) Show that the discriminant of the cubic resolvent equals the discriminant of the original quartic polynomial.

Problems 46–50 concern Galois groups of splitting fields of quartic polynomials. Take as known that the discriminant of a quartic polynomial $F(X) = X^4 + pX^2 + qX + r$ is given by

$$-4p^3q^2 - 27q^4 + 16p^4r + 144pq^2r - 128p^2r^2 + 256r^3.$$

Let $\mathbb{K}$ be a splitting field for $F(X)$ over $\mathbb{Q}$, and let $G = \mathrm{Gal}(\mathbb{K}/\mathbb{Q})$. Regard G as a subgroup of the symmetric group $\mathfrak{S}_4$.

46. (a) Identify all transitive subgroups of the alternating group $\mathfrak{A}_4$, up to a relabeling of the four indices.

(b) Identify all transitive subgroups of the symmetric group $\mathfrak{S}_4$ other than those in (a), up to a relabeling of the four indices.

47. Suppose $q = 0$.

(a) Show that G is a subgroup of $\mathfrak{A}_4$ if and only if r is a square in $\mathbb{Q}$.

(b) Show by solving $F(X) = 0$ explicitly that $[\mathbb{K} : \mathbb{Q}]$ is a power of 2, and conclude that G has no element of order 3.

(c) Deduce when r is a square that $G = \{(1), (1\ 2)(3\ 4), (1\ 3)(2\ 4), (1\ 4)(2\ 3)\}$ if $F(X)$ is irreducible over $\mathbb{Q}$.

(d) Deduce when r is a nonsquare that G is cyclic of order 4 or is dihedral of order 8 if $F(x)$ is irreducible over $\mathbb{Q}$; in the dihedral case, G is generated by a 4-cycle and the group listed in (c). (Problem 22 shows how to distinguish between the two cases.)

48. For $F(X) = X^4 + X + 1$, show by considering reduction modulo 2 and modulo 3 that $G = \mathfrak{S}_4$.

49. Let $F(X) = X^4 + 8X + 12$.

(a) Compute the discriminant of $F(X)$, and verify that it is a square.

(b) Show that $F(X) \equiv (1 + X)(2 + X + 4X^2 + X^3) \bmod 5$ with the two factors on the right side irreducible in $\mathbb{F}_5$.

(c) Show from (a) and (b) that if $F(X)$ is reducible over $\mathbb{Q}$, then it must have a root that is an integer. Check that there is no such root.

(d) Conclude that $G = \mathfrak{A}_4$.

50. For each transitive group G as in Problem 46, find a polynomial $F(X)$ of degree 4 over $\mathbb{Q}$ whose splitting field $\mathbb{K}$ over $\mathbb{Q}$ has $\mathrm{Gal}(\mathbb{K}/\mathbb{Q})$ isomorphic to G.

Problems 51–56 continue the introduction to error-correcting codes begun in Problems 63–73 at the end of Chapter IV and continued in Problems 25–28 at the end of Chapter VII. The current problems will not make use of the problems in Chapter VII. As in the problems in Chapter IV, we work with the field $\mathbb{F} = \mathbb{Z}/2\mathbb{Z}$, with Hamming space

$\mathbb{F}^n$, and with linear codes C in $\mathbb{F}^n$. The minimal distance of C is denoted by $\delta(C)$. Problem 72 in Chapter IV introduced cyclic redundancy codes, which are determined by a generating polynomial $G(X)$ of some degree g suitably less than n. Such a code C is built from all polynomials $G(X)B(X)$ with $B(X) = 0$ or deg $B(X) \leq n - g - 1$. A given polynomial $c_0 + c_1 X + \cdots$ becomes the n-tuple $(c_0, c_1, \ldots)$ of C; the code C has dimension $n - g$. This set of problems will discuss a special class of cyclic redundancy codes called cyclic codes, and then a special subclass called BCH codes.

51. A linear code C in $\mathbb{F}^n$ is called a **cyclic code** if whenever $(c_0, c_1, \ldots, c_{n-1})$ is in C, then so is $(c_{n-1}, c_0, c_1, \ldots, c_{n-2})$.
    (a) Prove that a linear code C is cyclic if and only if the set of all polynomials $c_0 + c_1 X + \cdots + c_{n-1} X^{n-1}$ corresponding to members $(c_0, c_1, \ldots, c_{n-1})$ of C is an ideal in the ring $\mathbb{F}[X]/(X^n - 1)$. (In this case the members of C will be identified with the set of such polynomials.)
    (b) Prove that if C is cyclic and nonzero, then there exists a unique $G(X)$ in C of lowest possible degree. Moreover, $G(X)$ divides $X^n - 1$ in $\mathbb{F}[X]$, and C consists exactly of the polynomials $G(X)F(X)$ mod $(X^n - 1)$ such that $F(X) = 0$ or deg $F(X) \leq n - \deg G(X) - 1$, and C has dimension $n - \deg G(X)$. (The polynomial $G(X)$ is called the **generating polynomial** of C. A cyclic code C over the field $\mathbb{Z}/2\mathbb{Z}$ having block length n and dimension k is called a **binary cyclic (n, k) code**.)
    (c) Prove that if $G(X)$ has degree $n - k$, then a basis of C consists of the polynomials $G(X), XG(X), X^2 G(X), \ldots, X^{k-1} G(X)$.
    (d) Under the assumption that C is cyclic and nonzero, (b) says that it is possible to write $X^n - 1 = G(X)H(X)$ for some $H(X)$ in $\mathbb{F}[X]$. Prove that a member $B(X)$ of $\mathbb{F}[X]/(X^n - 1)$ lies in C if and only if $H(X)B(X) \equiv 0$ mod $(X^n - 1)$.

52. (a) Show that the row space C of the matrix $\mathcal{G} = \begin{pmatrix} 1 & 0 & 0 & 1 & 0 & 1 & 1 \\ 0 & 1 & 0 & 1 & 1 & 1 & 0 \\ 0 & 0 & 1 & 0 & 1 & 1 & 1 \end{pmatrix}$ is a cyclic $(7, 3)$ code with generating polynomial $G(X) = 1 + X^2 + X^3 + X^4$.
    (b) Show directly from $\mathcal{G}$ that C has minimal distance $\delta = 4$.
    (c) The polynomial $H(X) = 1 + X^2 + X^3$ has the property that $G(X)H(X) = X^7 - 1$ in $\mathbb{F}[X]$. Find a 4-by-7 matrix $\mathcal{H}$ such that the column vectors $v \in \mathbb{F}^7$ that lie in C are exactly the ones with $\mathcal{H}v = 0$.
    (d) The matrix $\mathcal{H}$ in (c) is called the **check matrix** for the code. Describe a procedure for constructing the check matrix when starting from a general binary cyclic (n, k) code whose generating polynomial $G(X)$ is known and whose polynomial $H(X)$ with $G(X)H(X) = X^n - 1$ is known. Prove that the procedure works.

53. Show that $X^n - 1$ is a separable polynomial over $\mathbb{F}$ if n is odd but not if n is even.

54. Let C be a binary cyclic (n, k) code with generating polynomial $G(X)$, and suppose that n is odd. Let $\mathbb{K}$ be a finite extension field of $\mathbb{F}$ in which $X^n - 1$

splits, and let α be a primitive n^{th} root of 1, i.e., a root of $X^n - 1$ in $\mathbb{K}$ such that $\alpha^m \neq 1$ for $0 < m < n$. Suppose that r and s are integers with $0 \leq s < n$ and
$$G(\alpha^r) = G(\alpha^{r+1}) = \cdots = G(\alpha^{r+s}) = 0.$$

(a) Let $P(X) = G(X)F(X)$ with $F(X) \neq 0$ and $\deg F < k$ be an arbitrary nonzero member of C, so that $P(\alpha^r) = P(\alpha^{r+1}) = \cdots = P(\alpha^{r+s}) = 0$. Write $P(X) = c_0 + c_1 X + \cdots + c_{n-1} X^{n-1}$, and use the values of $P(\alpha^j)$ for $r \leq j \leq r+s$ to set up a homogeneous system of $s+1$ linear equations with n unknowns $c_0, \ldots, c_{n-1}$.

(b) Using an argument with Vandermonde determinants, show that every $(s+1)$-by-$(s+1)$ submatrix of the coefficient matrix of the system in (a) is invertible.

(c) Obtain a contradiction from (b) if $s+1$ or fewer of the coefficients of $P(X)$ are nonzero.

(d) Conclude that the minimal distance $\delta(C)$ is $\geq s + 2$.

55. **(BCH codes, or Bose–Chaudhuri–Hocquenghem codes)** Let n be an odd positive integer, let e be a positive integer $< n/2$, let $\mathbb{K}$ be a finite extension field of $\mathbb{F}$ in which $X^n - 1$ splits, and let α be a primitive n^{th} root of 1 in $\mathbb{K}$. For $1 \leq j \leq 2e$, let $F_j(X)$ be the minimal polynomial of α^j over $\mathbb{F}$, and define $G(X) = (1+X) \operatorname{LCM}(F_1(X), \ldots, F_{2e}(X))$. Prove that $G(X)$ divides $X^n - 1$ and that $G(X)$ is the generating polynomial for a cyclic code C in $\mathbb{F}^n$ with minimal distance $\delta(C) \geq 2e + 2$. (Educational note: Therefore C has the built-in capability of correcting at least e errors.)

56. In the setting of the previous problem, let $n = 2^m - 1$ for a positive integer m, and let $\mathbb{K}$ be a field of order 2^m.
    (a) Prove that any irreducible polynomial in $\mathbb{F}[X]$ with a root in $\mathbb{K}$ has order dividing m, and conclude that the order of the generating polynomial $G(X)$ in the previous problem is at most $2em + 1$.
    (b) Prove that there exists a sequence C_r of binary cyclic (n_r, k_r) codes of BCH type such that k_r/n_r tends to 1 and the minimal distance $\delta(C_r)$ tends to infinity. (Educational note: The fraction k_r/n_r tells the fraction of message bits to total bits in each transmitted block. Thus the problem says that there are linear codes capable of correcting as large a number of errors as we please while having as large a percentage of message bits as we please.)

57. Take as known that $F_1(X) = 1 + X + X^4$ is irreducible over $\mathbb{F}$. Let $\mathbb{K}$ be the field $\mathbb{F}[X]/(F_1(X))$ of order 16, and let α be the coset $X + (F_1(X))$ in $\mathbb{K}$.
    (a) Explain why $F_1(X)$ factors as $F_1(X) = (X - \alpha)(X - \alpha^2)(X - \alpha^4)(X - \alpha^8)$ over $\mathbb{K}$.
    (b) Find the minimal polynomial $F_3(X)$ of α^3.
    (c) Show in $\mathbb{F}^{15}$ that the binary cyclic code C with generating polynomial $G(X) = (1+X)F_1(X)F_3(X)$ has $\dim C = 6$ and $\delta(C) \geq 6$.

CHAPTER X

Modules over Noncommutative Rings

Abstract. This chapter contains two sets of tools for working with modules over a ring R with identity. The first set concerns finiteness conditions on modules, and the second set concerns the Hom and tensor product functors.

Sections 1–3 concern finiteness conditions on modules. Section 1 deals with simple and semisimple modules. A simple module over a ring is a nonzero unital module with no proper nonzero submodules, and a semisimple module is a module generated by simple modules. It is proved that semisimple modules are direct sums of simple modules and that any quotient or submodule of a semisimple module is semisimple. Section 2 establishes an analog for modules of the Jordan–Hölder Theorem for groups that was proved in Chapter IV; the theorem says that any two composition series have matching consecutive quotients, apart from the order in which they appear. Section 3 shows that a module has a composition series if and only if it satisfies both the ascending chain condition and the descending chain condition for its submodules.

Sections 4–6 concern the Hom and tensor product functors. Section 4 regards $\operatorname{Hom}_R(M, N)$, where M and N are unital left R modules, as a contravariant functor of the M variable and as a covariant functor of the N variable. The section examines the interaction of these functors with the direct sum and direct product functors, the relationship between Hom and matrices, the role of bimodules, and the use of Hom to change the underlying ring. Section 5 introduces the tensor product $M \otimes_R N$ of a unital right R module M and a unital left R module N, regarding tensor product as a covariant functor of either variable. The section examines the effect of interchanging M and N, the interaction of tensor product with direct sum, an associativity formula for triple tensor products, an associativity formula involving a mixture of Hom and tensor product, and the use of tensor product to change the underlying ring. Section 6 introduces the notions of a complex and an exact sequence in the category of all unital left R modules and in the category of all unital right R modules. It shows the extent to which the Hom and tensor product functors respect exactness for part of a short exact sequence, and it gives examples of how Hom and tensor product may fail to respect exactness completely.

1. Simple and Semisimple Modules

This chapter develops further theory for unital modules over a ring with identity beyond what is in Section VIII.1. Results about modules that take advantage of commutativity of the ring were included in Chapter VIII. In the present chapter the ring may or may not be commutative. We shall be interested in those modules whose structure is especially easy to analyze and in constructions that create new modules from old ones. The chapter consists of tools for working with such modules and their related rings and algebras. There are no major theorems in the chapter, but the material here is essential for the developments in *Advanced Algebra*.

Throughout this chapter, R will denote a ring with identity. We shall work with the category $\mathcal{C}$ of all unital left R modules. Specifically the objects of $\mathcal{C}$ are left unital R modules, and the space of morphisms between two such modules M and N consists of all R homomorphisms from M into N. It is customary to write $\text{Hom}_R(M, N)$ for this set of morphisms.[1] In the special case that R is a field, the notation $\text{Hom}_R(M, N)$ reduces to notation we introduced in Section II.3 for the set of linear maps from one vector space over R to another. For general R, the set $\text{Hom}_R(M, N)$ is an abelian group under addition of the values: $(\varphi_1 + \varphi_2)(m) = \varphi_1(m) + \varphi_2(m)$. Without some further hypothesis on R, $\text{Hom}_R(M, N)$ does not have a natural R module structure.

However, there is some residual action by scalars. Any element z in the **center** Z of R, i.e., any element with $cr = rc$ for all r in R, acts on $\text{Hom}_R(M, N)$. The definition is that $(c\varphi)(m) = \varphi(cm)$. The function $c\varphi$ certainly respects addition, and it respects action by a scalar r in R because $(c\varphi)(rm) = \varphi(crm) = \varphi(rcm) = r\varphi(cm) = r(c\varphi)(m)$; thus $c\varphi$ is in $\text{Hom}_R(M, N)$, and $\text{Hom}_R(M, N)$ becomes a Z module. The center Z automatically contains the multiplicative identity 1 and its integer multiples $\mathbb{Z}1$.

We shall tend to ignore this action by the center except in two special cases. One is that R is commutative, and then $\text{Hom}_R(M, N)$ is an R module. The other is that R is an associative algebra (with identity) over a field F. In this case the action of members of F on the identity of R embeds F into R, and F may thus be identified with a subfield of the center of R. The result is that when R is an associative algebra over a field F, then $\text{Hom}_R(M, N)$ is a vector space over F.

We write $\text{End}_R(M)$ for $\text{Hom}_R(M, M)$. This abelian group has the structure of a ring with identity, multiplication being composition: $(\varphi\psi)(m) = \varphi(\psi(m))$. The distributive laws need to be checked: the formula $(\varphi_1 + \varphi_2)\psi = \varphi_1\psi + \varphi_2\psi$ is immediate from the calculation

$$((\varphi_1 + \varphi_2)\psi)(m) = (\varphi_1 + \varphi_2)(\psi(m))$$
$$= \varphi_1(\psi(m)) + \varphi_2(\psi(m)) = (\varphi_1\psi + \varphi_2\psi)(m),$$

while the formula $\varphi(\psi_1 + \psi_2) = \varphi\psi_1 + \varphi\psi_2$ makes use of the fact that φ respects addition and is proved by the calculation

$$(\varphi(\psi_1 + \psi_2))(m) = \varphi(\psi_1(m) + \psi_2(m))$$
$$= \varphi(\psi_1(m)) + \varphi(\psi_2(m)) = (\varphi\psi_1 + \varphi\psi_2)(m).$$

---

[1] The notation $\text{Hom}(M, N)$ with no subscript is sometimes used for $\text{Hom}_{\mathbb{Z}}(M, N)$, i.e., to denote the group of homomorphisms from one abelian group to another.

If Z is the center of R, then $\operatorname{End}_R(M)$ is a Z module, as well as a ring, and the two structures are compatible; the result is that $\operatorname{End}_R(M)$ is an associative Z algebra in the sense of Example 15 in Section VIII.1. In particular, when R is an associative algebra over a field F, then $\operatorname{End}_R(M)$ is an associative F algebra.

There is usually no need to re-prove for right R modules an analog of each result about left R modules. The reason is that we can make use of the **opposite ring** R^o of R, defined to be the same underlying abelian group but with reversed multiplication: $a \circ b = ba$. Any left R module M then becomes a right R^o module M^o under the definition $mr^o = rm$ for r in R, m in M, and r^o equal to the same set-theoretic member of R^o as r. The theory of unital left R modules for all R thereby yields a theory of unital right R modules for all R.

A unital left R module M is said to be **simple**, or **irreducible**, if $M \neq 0$ and if M has no proper nonzero R submodules. If M is simple, then $M = Rx$ for each $x \neq 0$ in M; conversely if $M \neq 0$ has $M = Rx$ for each $x \neq 0$ in M, then M is simple. Whenever $M = Rx$ for an element x, then M is isomorphic as a unital left R module to R/I, where I is the left ideal $I = \{r \in R \mid rx = 0\}$.

A unital left R module M is said to be **semisimple** if M is generated by simple left R submodules, i.e., if it is the sum of simple left R submodules. In this definition, the sum may be empty (and then $M = 0$), it may be finite, or it may be infinite. Evidently simple implies semisimple for unital left R modules.

We come to examples in a moment. First we prove that the sum of simple left R modules in a semisimple module may always be taken to be a direct sum, i.e., that semisimple modules are **completely reducible**.

Proposition 10.1. If the unital left R module M is semisimple, then M is the direct sum of some family of simple R submodules. In more detail if $\{M_s \mid s \in S\}$ is a family of simple R submodules of the unital left R module M whose sum is M, then there is a subset T of S with the property that

$$M = \bigoplus_{t \in T} M_t.$$

PROOF. Call a subset U of S "independent" if the sum $\sum_{u \in U} M_u$ is direct. This condition means that for every finite subset $\{u_1, \ldots, u_n\}$ of U and every set of elements $m_i \in M_{u_i}$, the equation $m_1 + \cdots + m_n = 0$ implies that each m_i is 0. From this formulation it follows that the union of any increasing chain of independent subsets of S is itself independent. By Zorn's Lemma there is a maximal independent subset T of S. By definition the sum $M_0 = \sum_{t \in T} M_t$ is direct. Consequently it suffices to show that M_0 is all of M. By the hypothesis on S, it is enough to show that each M_s is contained in M_0. For s in T, this conclusion is clear. Thus suppose s is not in T. By the maximality of T, $T \cup \{s\}$ is not independent. Consequently the sum $M_s + M_0$ is not direct, and it follows that $M_s \cap M_0 \neq 0$. But this intersection is an R submodule of M_s. Since M_s is simple, a nonzero R submodule of M_s must be all of M_s. Thus $M_s \cap M_0 = M_s$, and M_s is contained in M_0. □

EXAMPLES OF SEMISIMPLE MODULES.

(1) Let F be a field. Left and right amount to the same thing for modules when the underlying ring is commutative. We know that the unital F modules are just the vector spaces over F. Such a vector space V is a simple F module if and only if it is 1-dimensional, since 1-dimensionality is the necessary and sufficient condition to have $V \neq 0$ be of the form $V = Fx$ for all $x \neq 0$ in V. Any vector space V is the sum of all of its 1-dimensional subspaces, and consequently every unital F module is semisimple. Theorem 2.42 shows that each vector space V has a basis; this theorem is therefore a special case of Proposition 10.1, which says that any semisimple module is the *direct* sum of simple modules.

(2) Let D be a division ring. Division rings were defined in Section IV.4 as rings with identity $1 \neq 0$ such that the nonzero elements form a group under multiplication. Every field is a division ring, and the quaternions form a division ring that is not a field. Let M be a unital left D module, and let $x \neq 0$ be in M. Then the left D module Dx is simple because if $N \subseteq Dx$ is a nonzero D submodule and if y is in N, then we can write $y = dx$ with d in D and see from the formula $d^{-1}y = x$ that x is in N and $N = Dx$. Any unital left D module is the sum of its D submodules Dx for x in M, and therefore every unital left D module is semisimple. From Proposition 10.1 we can conclude that every unital left D module M is the direct sum of simple modules. In other words, M has a basis, just as if D were a field. Consequently it is customary to refer to unital left D modules as **left vector spaces** over D. A notion of (left) dimension, equal to a well-defined nonnegative integer or ∞, will emerge from the discussion in the next section.

(3) Let D be a division ring. Section V.2 introduced the ring of n-by-n matrices over any commutative ring with identity, and Example 4 of rings in Section VIII.1 extended the definition to the case that the ring is noncommutative. Thus let R be the ring $M_n(D)$. Let $M = D^n$ be the abelian group of n-component column vectors with entries in D. Under multiplication of matrices times column vectors, M becomes a unital left R module. Let us prove that M is simple. It is enough to show that $Rm = M$ for every nonzero m in M. Let m' be in M with entries m'_i, and suppose that the i_0^{th} component m_{i_0} of m is $\neq 0$. Then we can multiply on the left of m by the matrix r whose $(i, j)^{\text{th}}$ entry r_{ij} is $m'_i m_{i_0}^{-1}$ if $(i, j) = (i, j_0)$ and is 0 otherwise, and the product is the column vector m'. Thus m' is in Rm, and $Rm = M$ as required. Hence $M = D^n$ is an example of a simple R module.

(4) Again let D be a division ring, and let $R = M_n(D)$. Let us see that the left R module R is semisimple. In fact, if R_j is the additive subgroup of R whose

nonzero entries are all in the jth column, then R_j is a left R submodule of R that is R isomorphic to D^n. Thus we see that $R = R_1 \oplus \cdots \oplus R_n$ as left R modules, and the left R module R is semisimple as a consequence of Example 3.

(5) Let G be a group, and let $\mathbb{C}G$ be the complex group algebra defined in Example 16 in Section VIII.1. Let V be a vector space over $\mathbb{C}$, and let $\Phi : G \to GL(V)$ be a representation of G on V. The universal mapping property of complex group algebras described in that example and pictured in Figure 8.4 shows that the representation Φ of G extends to $\mathbb{C}G$ and makes V into a unital left $\mathbb{C}G$ module. Conversely if the complex vector space V is a unital left $\mathbb{C}G$ module, then we obtain a representation of G by restriction from $\mathbb{C}G$ to G. What needs to be checked here is that each member of G acts by an invertible linear mapping. This is a consequence of the unital property; since 1 acts as 1, the action by g^{-1} inverts the action of g. Thus we have a one-one correspondence of representations of G on complex vector spaces with unital left $\mathbb{C}G$ module structures. Under this correspondence, irreducible representations of G (i.e., nonzero representations having no proper nonzero invariant subspace) correspond to simple $\mathbb{C}G$ modules. Now suppose that G is finite. Readers who have looked at Section VII.4 know from Corollary 7.21 that every finite-dimensional representation of a finite group G on a complex vector space is the direct sum of irreducible representations; the corresponding $\mathbb{C}G$ modules are therefore semisimple. But more is true. If V is *any* $\mathbb{C}G$ module for the finite group G and if x is in V, then $\mathbb{C}Gx$ is a vector subspace spanned by $\{gx \mid g \in G\}$ and consequently is finite-dimensional. Applying what is known from Section VII.4, we can write $\mathbb{C}Gx$ as the direct sum of simple $\mathbb{C}G$ modules. Therefore the sum of all simple $\mathbb{C}G$ modules in V is all of V, and V is semisimple. From Proposition 10.1 we conclude that every unital left $\mathbb{C}G$ module is semisimple if G is a finite group.

The next proposition shows that decompositions of semisimple modules as direct sums of simple modules behave in a fashion analogous to decompositions of vector spaces as direct sums of 1-dimensional vector subspaces. However, the simple modules need not all be isomorphic to one another, as is shown by Example 5. A theory that takes the isomorphism types of simple modules into account appears in Problems 12–20 at the end of the chapter.

Proposition 10.2. Let M be a semisimple left R module, and suppose that $M = \bigoplus_{s \in S} M_s$ is the direct sum of simple R modules M_s. Let N be any R submodule of M. Then

(a) the quotient module M/N is semisimple. In more detail there is a subset T of S with the property that the submodule $M_T = \bigoplus_{t \in T} M_t$ of M maps R isomorphically onto M/N.

(b) N is a direct summand of M. In more detail, $M = N \oplus M_T$, where M_T is as in (a).
(c) N is semisimple. In more detail choose T as in (a), and write T' for the complement of T in S. Then the quotient mapping $M \to M/M_T$ restricts to an R isomorphism of N onto M/M_T, and M/M_T is R isomorphic to $M_{T'}$.

PROOF. Each simple R submodule M_s of M maps to an R submodule $\overline{M}_s$ of M/N. This image either is simple (and then is R isomorphic to M_s) or is zero. We let U be the subset of S for which it is simple. Then M/N is evidently the sum of the simple R submodules $\{\overline{M}_s \mid s \in U\}$. By Proposition 10.1 there is a subset T of U such that
$$M/N = \bigoplus_{t \in T} \overline{M}_t.$$
This proves (a).

For (b), we use the following elementary observation: if N and N' are R submodules of M, then $M = N \oplus N'$ if and only if the quotient map $M \to M/N$ carries N' isomorphically onto the quotient M/N. Taking $N' = M_T$ and applying (a), we obtain (b).

For (c), the same observation when applied first to $M = N \oplus M_T$ and then to $M = M_{T'} \oplus M_T$ shows that the quotient map $M \to M/M_T$ carries N isomorphically onto M/M_T and carries $M_{T'}$ isomorphically onto M/M_T. Therefore $N \cong M/M_T \cong M_{T'}$, and (c) is proved. $\square$

In the context of simple modules, $\mathrm{Hom}_R(M, N)$ has special properties. Readers who have looked at Section VII.4 have seen these special properties in the context of representations of finite groups on complex vector spaces. There they were captured by Schur's Lemma (Proposition 7.18). If we pass from representations on complex vector spaces to $\mathbb{C}G$ modules, following the prescription in Example 5, we obtain a result about $\mathrm{Hom}_{\mathbb{C}G}(M, N)$ when G is a finite group. Lemma 10.3 and Proposition 10.4 generalize this to a result about $\mathrm{Hom}_R(M, N)$ for arbitrary R.

Lemma 10.3. Suppose that E is a simple left R module and that $M = \bigoplus_{a \in A} M_a$ is a direct-sum decomposition of the unital left R module M into arbitrary R submodules, not necessarily simple. Then
$$\mathrm{Hom}_R(E, M) \cong \bigoplus_{a \in A} \mathrm{Hom}_R(E, M_a)$$
as an isomorphism of abelian groups.

REMARKS. The hypothesis that E is simple is critical here. Without it a map into a direct sum might have nonzero projections into infinitely many of the summands, and then it could not be represented as a finite sum of maps into summands. Proposition 10.12 below will point out that the correct identity without a special hypothesis on E is $\operatorname{Hom}_R(E, \prod_{a \in A} M_a) \cong \prod_{a \in A} \operatorname{Hom}_R(E, M_a)$.

PROOF. Suppose φ is in $\operatorname{Hom}_R(E, M)$. Write φ_a for the composition of φ with the projection $M \to M_a$. The map from left to right in the displayed isomorphism is to be $\varphi \mapsto \{\varphi_a\}_{a \in A}$. Suppose for the moment that the image is contained in the direct sum on the right. The mapping is one-one since M is the sum of the M_a's, and it is onto since the mapping is the identity on each subgroup $\operatorname{Hom}_R(E, M_a)$ of $\operatorname{Hom}_R(E, M)$.

Thus we must show for each φ that only finitely many of the maps φ_a are nonzero. Choose e in E with $\varphi(e) \neq 0$, and write

$$\varphi(e) = m_1 + \cdots + m_n \qquad \text{with } m_i \in M_{a_i}.$$

Since E is simple, $E = Re$. Therefore

$$\varphi(E) = R\varphi(e) = R(m_1 + \cdots + m_n) \subseteq Rm_1 + \cdots + Rm_n$$
$$\subseteq M_{a_1} \oplus \cdots \oplus M_{a_n}.$$

Consequently only $\varphi_{a_1}, \ldots, \varphi_{a_n}$ can be nonzero. □

Lemma 10.3 enables us to study maps between semisimple modules in terms of maps between simple modules. The latter are described by the next result.

Proposition 10.4 (Schur's Lemma). Suppose that M and N are simple left R modules.

(a) If M and N are not R isomorphic, then $\operatorname{Hom}_R(M, N) = 0$.
(b) $\operatorname{End}_R(M)$ is a division ring.
(c) (Dixmier) If R is an associative algebra over an algebraically closed field F and if the vector-space dimension of M over F is less than the cardinality of F, then $\operatorname{End}_R(M)$ consists of the F multiples of the identity.

REMARK. In the setting of representations of a finite group G as in Section VII.4, or in the case that G is a finite group and $R = \mathbb{C}G$ in the current setting, any singly generated R module such as M or N is finite-dimensional over $\mathbb{C}$. Part (a) in that case reduces to the statement that the vector space of intertwining operators between two inequivalent irreducible representations is 0. Part (c) in that case says that the space of self-intertwining operators for an irreducible representation consists of the scalar multiples of the identity. For a general R, we get only the weaker conclusion of (b) that $\operatorname{End}_R(M)$ is a division ring. If R is an associative algebra over a field F, we have seen that $\operatorname{End}_R(M)$ is an associative algebra over F, and (c) gives a condition under which we can improve upon (b).

PROOF. Suppose that φ is nonzero in $\text{Hom}_R(M, N)$. Then $\ker \varphi$ is a proper R submodule of M, and we must have $\ker \varphi = 0$ since M is simple. Similarly image φ is a nonzero R submodule of N, and we must have image $\varphi = F$ since N is simple. Therefore φ is an R isomorphism of M onto N. This proves (a) and (b).

For (c) let m be a nonzero element of M. The map $\varphi \mapsto \varphi(m)$ is F linear and one-one from $\text{End}_R(M)$ into M by (b). Thus $\text{End}_R(M)$ as an associative division algebra over F has vector-space dimension at most the vector-space dimension of M, and the latter by hypothesis is strictly less than the cardinality of F. Arguing by contradiction, let us assume that $\text{End}_R(M)$ is not equal to F; say $\text{End}_R(M)$ contains an element φ not in F.

The smallest division subalgebra of $\text{End}_R(M)$ containing F and φ is the field $\overline{F}$ generated by F and φ. Since F is algebraically closed, φ is not a root of any nonzero polynomial with coefficients in F. Thus the substitution homomorphism equal to the identity on F and carrying X to φ is one-one from $F[X]$ into $\overline{F}$. By the universal mapping property of fields of fractions (Proposition 8.6), the substitution homomorphism factors through the field of fractions $F(X)$. Thus we may regard $F(X)$ as a subfield of $\overline{F}$. In the field $F(X)$, the set of elements $\{(X - c)^{-1} \mid c \in F\}$ is linearly independent over F, as we see by assuming a nontrivial linear dependence and clearing fractions, and hence $\dim_F F(X)$ is $\geq$ the cardinality of F. Since $\text{End}_R(M) \supseteq \overline{F} \supseteq F(X)$ under our identification, the dimension of $\text{End}_R(M)$ over F is $\geq$ the cardinality of F. This conclusion contradicts the observation of the previous paragraph that the dimension of $\text{End}_R(M)$ is strictly less than the cardinality of F. So the assumption that $\text{End}_R(M)$ contains an element not in F must be false, and (c) follows. □

2. Composition Series

We continue with R as a ring with identity, and we work with the category of all unital left R modules. In this section we shall say what is meant by a unital left R module of "finite length," and we shall investigate semisimplicity for such modules.

A **finite filtration** of a unital left R module M is a finite descending chain

$$M = M_0 \supseteq M_1 \supseteq \cdots \supseteq M_n = 0$$

of R submodules. We do not insist on this particular indexing, and with the obvious adjustments, we allow also a finite increasing chain to be called a finite filtration. Relative to the displayed inclusions, the modules M_i/M_{i+1} for $0 \leq i \leq n - 1$ are called the **consecutive quotients** of the filtration. The finite filtration is called a **composition series** if the consecutive quotients are all simple

R modules; in particular, they are to be nonzero. The consecutive quotients in this case are called **composition factors**.

We encountered an analogous notion with groups in Section IV.8, but there was a complication in that case. The complication was that each subgroup had to be normal in the next-larger subgroup in order for the consecutive quotients to be groups. The overlap between the current treatment and the earlier treatment occurs for abelian groups, which on the one hand are unital $\mathbb{Z}$ modules and on the other hand are groups whose subgroups are automatically normal.

We are going to obtain analogs for the category of unital left R modules of the group-theoretic results of Zassenhaus, Schreier, and Jordan–Hölder in Section IV.8. The ones here will be a little easier to prove than those in Section IV.8 since we do not have the complication of checking whether subgroups are normal. Let

$$M = M_0 \supseteq M_1 \supseteq \cdots \supseteq M_m = 0$$

and
$$M = N_0 \supseteq N_1 \supseteq \cdots \supseteq N_n = 0$$

be two finite filtrations of M. We say that the second is a **refinement** of the first if there is a one-one function $f : \{0, \ldots, m\} \to \{0, \ldots, n\}$ with $M_i = N_{f(i)}$ for $0 \leq i \leq m$. The two finite filtrations of M are said to be **equivalent** if $m = n$ and if the order of the consecutive quotients $M_0/M_1, M_1/M_2, \ldots, M_{m-1}/M_m$ may be rearranged so that they are respectively isomorphic to $N_0/N_1, N_1/N_2, \ldots, N_{m-1}/N_m$.

Lemma 10.5 (Zassenhaus). Let M_1, M_2, M_1', and M_2' be R submodules of a unital left R module M with $M_1' \subseteq M_1$ and $M_2' \subseteq M_2$. Then
$$((M_1 \cap M_2) + M_1')/((M_1 \cap M_2') + M_1')$$
$$\cong ((M_1 \cap M_2) + M_2')/((M_1' \cap M_2) + M_2').$$

PROOF. By the Second Isomorphism Theorem (Theorem 8.4),
$$(M_1 \cap M_2)/(((M_1 \cap M_2') + M_1') \cap (M_1 \cap M_2))$$
$$\cong ((M_1 \cap M_2) + (M_1 \cap M_2') + M_1')/((M_1 \cap M_2') + M_1')$$
$$= ((M_1 \cap M_2) + M_1')/((M_1 \cap M_2') + M_1').$$

Since we have
$$((M_1 \cap M_2') + M_1') \cap (M_1 \cap M_2) = ((M_1 \cap M_2') + M_1') \cap M_2$$
$$= (M_1 \cap M_2') + (M_1' \cap M_2),$$

we can rewrite the above isomorphism as
$$(M_1 \cap M_2)/((M_1 \cap M_2') + (M_1' \cap M_2))$$
$$\cong ((M_1 \cap M_2) + M_1')/((M_1 \cap M_2') + M_1').$$

The left side of this isomorphism is symmetric under interchange of the indices 1 and 2. Hence so is the right side, and the lemma follows. □

2. Composition Series

Theorem 10.6 (Schreier). Any two finite filtrations of a module M in $\mathcal{C}$ have equivalent refinements.

PROOF. Let the two finite filtrations be
$$M = M_0 \supseteq M_1 \supseteq \cdots \supseteq M_m = 0$$
and
$$M = N_0 \supseteq N_1 \supseteq \cdots \supseteq N_n = 0,$$
and define
$$M_{ij} = (M_i \cap N_j) + M_{i+1} \quad \text{for } 0 \le i \le m-1 \text{ and } 0 \le j \le n,$$
$$N_{ji} = (M_i \cap N_j) + N_{j+1} \quad \text{for } 0 \le i \le m \text{ and } 0 \le j \le n-1.$$
Then
$$M = M_{00} \supseteq M_{01} \supseteq \cdots \supseteq M_{0n}$$
$$\supseteq M_{10} \supseteq M_{11} \supseteq \cdots \supseteq M_{1n} \supseteq \cdots \supseteq M_{m-1,n} = 0$$
and
$$M = N_{00} \supseteq N_{01} \supseteq \cdots \supseteq N_{0m}$$
$$\supseteq N_{10} \supseteq N_{11} \supseteq \cdots \supseteq N_{1m} \supseteq \cdots \supseteq N_{n-1,m} = 0$$
are refinements of the respective given filtrations. The containments $M_{in} \supseteq M_{i+1,0}$ and $N_{jm} \supseteq N_{j+1,0}$ are equalities here, and the only nonzero consecutive quotients are therefore of the form $M_{ij}/M_{i,j+1}$ and $N_{ji}/N_{j,i+1}$. For these we have

$M_{ij}/M_{i,j+1}$
$= ((M_i \cap N_j) + M_{i+1})/((M_i \cap N_{j+1}) + M_{i+1})$ by definition
$\cong ((M_i \cap N_j) + N_{j+1})/((M_{i+1} \cap N_j) + N_{j+1})$ by Lemma 10.5
$= N_{ji}/N_{j,i+1}$ by definition,

and thus the above refinements are equivalent. $\square$

Corollary 10.7 (Jordan–Hölder Theorem). If M is a unital left R module with a composition series, then

(a) any finite filtration of M in which all consecutive quotients are nonzero can be refined to a composition series, and
(b) any two composition series of M are equivalent.

PROOF. We apply Theorem 10.6 to a given filtration and a known composition series. After discarding redundant terms from each refinement (those that lead to 0 as a consecutive quotient), we arrive at a refinement of our given finite filtration that is equivalent to the known composition series. Hence the refinement is a composition series. This proves (a). If we specialize this argument to the case that the given filtration is a composition series, then we obtain (b). $\square$

Corollary 10.7 implies that the composition factors for a given composition series depend only on M, not on the particular composition series. Moreover, if $M' \supseteq M''$ are R submodules of an M with a composition series such that M'/M'' is simple, then M'/M'' is a composition factor of M. This fact follows by eliminating redundant terms from the finite filtration $M \supseteq M' \supseteq M'' \supseteq 0$ and applying Corollary 10.7a to the result.

If a unital left R module M has a composition series, then we say that M has **finite length**. This notion is closed under passage to submodules and quotients. In fact, if
$$M = M_0 \supseteq M_1 \supseteq \cdots \supseteq M_n = 0$$
is a composition series of M and if M' is an R submodule of M, then
$$M' = M_0 \cap M' \supseteq M_1 \cap M' \supseteq \cdots \supseteq M_n \cap M' = 0$$
is a finite filtration of M' in which each consecutive quotient is simple or 0. Discarding redundant terms (which lead to 0 as a consecutive quotient), we obtain a composition series for M'. A similar argument works for M/M'.

Let us see that if the unital left R modules M' and M/M' have finite length, then so does M. In fact, we take a composition series for M/M', pull it back to M, and concatenate it to a composition series for M'. The result is a composition series for M, and the assertion follows. In particular, the direct sum of two unital left R modules of finite length has finite length.

If M has a composition series of the form $M = M_0 \supseteq M_1 \supseteq \cdots \supseteq M_n = 0$, then we say that M has **length** n. If it has no composition series, we say it has infinite length. According to Corollary 10.7, this notion of length is independent of the particular composition series that we use. The argument in the previous paragraph shows that if M' is an R submodule of M, then
$$\text{length}(M) = \text{length}(M') + \text{length}(M/M'),$$
with the finiteness of either side implying the finiteness of the other side. One consequence is that if M' is a length-n submodule of a length-n module M with n finite, then $M' = M$. Another consequence is that if M is a semisimple left R module, then M has a composition series if and only if M is the *finite* direct sum of simple left R modules.

From the last of these observations, we see that if F is a field, then the vector spaces over F that have a composition series are the finite-dimensional vector spaces, and in this case the length of the vector space is its dimension. The structure of finite-dimensional vector spaces is so elementary that the Jordan–Hölder Theorem is of no interest in this case, and it was for that reason that no version of the Jordan–Hölder Theorem for vector spaces appeared earlier in the book.

2. Composition Series

In the case that $R = D$ is a division ring, matters are slightly subtler. We know from Example 2 in Section 1 that every unital left D module is semisimple, and we noted that such D modules are therefore called left vector spaces. Corollary 10.7 shows that the number of summands in any decomposition of a left vector space V as the direct sum of simple D modules is either an integer $n \geq 0$ independent of the decomposition or is infinite, independently of the decomposition. This number, the integer n or ∞, is called the **dimension** of the left vector space V.

We saw one other example of a semisimple left R module. Specifically if D is a division ring, then we saw in Example 4 of Section 1 that $R = M_n(D)$ is semisimple as a left R module. The number of simple summands is n, and hence R has length n. So R has a composition series when considered as a left R module.

There are two other cases in which composition series give something familiar. One is the case that R is the ring $\mathbb{Z}$ of integers. A unital $\mathbb{Z}$ module is an abelian group, and we know that the simple abelian groups are the cyclic groups of prime order. For an abelian group with a composition series, the order of the group is the product of the orders of the consecutive quotients and hence is finite. Consequently an abelian group has a composition series if and only if it is a finite abelian group. Such a group need not be semisimple; the group C_4, for example, is not the direct sum of cyclic groups of prime order.

The other case concerns triangular form, Jordan canonical form, and related decompositions, as explained in Sections V.3 and V.6 and as reinterpreted after Corollary 8.29. Let V be a finite-dimensional vector space over a field $\mathbb{K}$, and let $L : V \to V$ be a linear mapping from V to itself. Put $R = \mathbb{K}[X]$, and make V into a unital R module by the definition $A(X)(v) = A(L)v$ for any $A(X)$ in $\mathbb{K}[X]$ and v in V. The R submodules are the vector subspaces of V that are invariant under L. The finite dimensionality of V forces V to have a composition series as an R module. Let us suppose for a moment that $\mathbb{K}$ is algebraically closed. Proposition 5.6 says that the matrix of L in some ordered basis is upper triangular, and linear combinations of the first k vectors in this basis form an invariant subspace under L of dimension k. These subspaces are nested, and thus we obtain a composition series. Thus obtaining a composition series when $\mathbb{K}$ is algebraically closed is equivalent to obtaining triangular form. The existence of Jordan form is a finer result. The discussion after Corollary 8.29 shows that V is a finite direct sum of R modules $R/(X - c_j)^{k_j}$ with c_j in $\mathbb{K}$ and $k_j > 0$. For each of these, the discussion at the end of Section VIII.6 shows how to refine $R/(X - c_j)^{k_j}$ to a composition series for which there is an R submodule of each possible dimension from 0 to k_j; the finer structure is hidden in the way that each invariant subspace is obtained from the next smaller invariant subspace. If $\mathbb{K}$ is not necessarily algebraically closed, then $(X - c_j)^{k_j}$ is to be replaced by $P_j(X)^{k_j}$ for some prime polynomial $P_j(X)$, and the consecutive quotients for $R/(P_j(X))^{k_j}$ have dimension equal to the degree of $P_j(X)$.

3. Chain Conditions

We continue with R as a ring with identity, and we work with the category of all unital left R modules. Except in special cases we did not address conditions in Section 2 under which a unital left R module M has a composition series. In this section we shall see that a necessary and sufficient condition for M to have a composition series is that it satisfy two "chain conditions," an ascending one and a descending one, that we shall define. We already encountered the ascending chain condition in Proposition 8.30 for the special case that R is a commutative ring with identity, and the proof for general R requires only cosmetic changes.

Proposition 10.8. If R is a ring with identity and M is a unital left R module, then the following conditions on R submodules of M are equivalent:
   (a) (**ascending chain condition**) every strictly ascending chain of R submodules $M_1 \subsetneq M_2 \subsetneq \cdots$ terminates in finitely many steps,
   (b) (**maximum condition**) every nonempty collection of R submodules has a maximal element under inclusion,
   (c) (**finite basis condition**) every R submodule is finitely generated.

PROOF. To see that (a) implies (b), let S be a nonempty collection of R submodules of M. Take M_1 in S. If M_1 is not maximal, choose M_2 in S properly containing M_1. If M_2 is not maximal, choose M_3 in S properly containing M_2. Continue in this way. By (a), this process must terminate, and then we have found a maximal R submodule in S.

To see that (b) implies (c), let N be an R submodule of M, and let S be the collection of all finitely generated R submodules of N. This collection is nonempty since 0 is in it. By (b), S has a maximal element, say N'. If x is in N but x is not in N', then $N' + Rx$ is a finitely generated R submodule of N that properly contains N' and therefore gives a contradiction. We conclude that $N' = N$, and therefore N is finitely generated.

To see that (c) implies (a), let $M_1 \subsetneq M_2 \subsetneq \cdots$ be given, and put $N = \bigcup_{n=1}^{\infty} M_n$. By (c), N is finitely generated. Since the M_n are increasing with n, we can find some M_{n_0} containing all the generators. Then the sequence stops no later than at M_{n_0}. □

The corresponding result for descending chains is as follows.

Proposition 10.9. If R is a ring with identity and M is a unital left R module, then the following conditions on R submodules of M are equivalent:
   (a) (**descending chain condition**) every strictly descending chain of R submodules $M_1 \supsetneq M_2 \supsetneq \cdots$ terminates in finitely many steps,
   (b) (**minimum condition**) every nonempty collection of R submodules has a minimal element under inclusion.

3. Chain Conditions 557

PROOF. To see that (a) implies (b), let S be a nonempty collection of R submodules of M. Take M_1 in S. If M_1 is not minimal, choose M_2 in S properly contained in M_1. If M_2 is not minimal, choose M_3 in S properly contained in M_2. Continue in this way. By (a), this process must terminate, and then we have found a minimal R submodule in S.

To see that (b) implies (a), we observe that the members of any strictly descending chain would be a family without a minimal element. Since (b) says that any nonempty family has a minimal element, there can be no such chain. $\square$

Proposition 10.10. Let R be a ring with identity, let M be a unital left R module, and let N be an R submodule of M. Then

(a) M satisfies the ascending chain condition if and only if N and M/N satisfy the ascending chain condition,
(b) M satisfies the descending chain condition if and only if N and M/N satisfy the descending chain condition.

PROOF. We prove (a), and the proof of (b) is completely similar. Suppose M satisfies the ascending chain condition and hence also the maximum condition by Proposition 10.8. The R submodules of N are in particular R submodules of M and hence satisfy the maximum condition. The R submodules of M/N lift back to R submodules of M containing N, and they too must satisfy the maximum condition. By Proposition 10.8, N and M/N satisfy the ascending chain condition.

Conversely suppose that N and M/N satisfy the ascending chain condition. Let $\{M_l\}$ be an ascending chain of R submodules of M; we are to show that $\{M_l\}$ is constant from some point on. Since N and M/N satisfy the ascending chain condition, we can find an n such that

$$M_{n+k} \cap N = M_n \cap N \quad \text{and} \quad (M_{n+k} + N)/N = (M_n + N)/N$$

for all $k \geq 0$. Combining the Second Isomorphism Theorem (Theorem 8.4) and the first of these identities gives

$$(M_{n+k} + N)/N \cong M_{n+k}/(M_{n+k} \cap N) = M_{n+k}/(M_n \cap N)$$

for all $k \geq 0$. Combining this result and two applications of the second of the identities gives

$$M_{n+k}/(M_n \cap N) = M_n/(M_n \cap N).$$

The First Isomorphism Theorem (Theorem 8.3) shows that

$$\bigl(M_{n+k}/(M_n \cap N)\bigr)/\bigl(M_n/(M_n \cap N)\bigr) \cong M_{n+k}/M_n.$$

Since the left side is the 0 module, the right side is the 0 module. Therefore $M_{n+k} = M_n$ for all $k \geq 0$. $\square$

Proposition 10.11. If R is a ring with identity and M is a unital left R module, then M has a composition series if and only if M satisfies both the ascending chain condition and the descending chain condition.

PROOF. If M has a composition series of length n, then the Jordan–Hölder Theorem (Corollary 10.7a) shows that every finite filtration of M with nonzero consecutive quotients has length $\leq n$, and hence M satisfies both chain conditions.

Conversely suppose that M satisfies both chain conditions. By the maximum condition, choose if possible a maximal proper R submodule N_1 of M, then choose if possible a maximal proper R submodule N_2 of N_1, and so on. If all these choices are possible, we obtain a strictly descending chain $M \supsetneq N_1 \supsetneq N_2 \supsetneq \cdots$, and the consecutive quotients will be simple at each stage. The minimum condition says that we cannot have such a chain, and thus the choice is impossible for the first time at some stage k. That means that some N_k has no proper R submodule, and N_k must be 0. Then $M = N_1 \supsetneq N_2 \supsetneq \cdots \supsetneq N_k = 0$ is a composition series. □

4. Hom and End for Modules

We continue to work with the category $\mathcal{C}$ of unital left R modules, where R is a ring with identity, not necessarily commutative. Our interest in this section is with $\text{Hom}_R(M, N)$ and $\text{End}_R(M)$, where M and N are modules in $\mathcal{C}$. Recall from Section 1 that $\text{Hom}_R(M, N)$ is a unital Z module, where Z is the center of R, and that $\text{End}_R(M)$ is a Z algebra, the multiplication being composition. We shall tend to ignore Z except when R is commutative or R is an associative algebra over a field. However, Z will implicitly play a role in the context of bimodules, which we introduce near the end of this section.

In this section we shall be interested in interactions of $\text{Hom}_R(M, N)$ and $\text{End}_R(M)$ within the category $\mathcal{C}$, in identities that they satisfy, in the naturality of such identities, and in the use of $\text{Hom}_R(M, N)$ in "change of rings," also known as "extension of scalars." The next section will carry out a similar investigation for a notion of tensor product that generalizes the tensor products in Chapter VI, and we shall obtain in addition one important formula involving Hom and tensor products at the same time. Finally in Section VI we shall examine the effect of Hom and tensor product on "exact sequences."

The first observation is that Hom_R is a functor, either a functor of one variable with the other variable held fixed or, less satisfactorily, a functor of two variables. To be precise, let $\mathcal{D}$ be the category of all abelian groups. For fixed M in $\text{Obj}(\mathcal{C})$, we define

$$F(N) = \text{Hom}_R(M, N).$$

4. Hom and End for Modules

If φ is in $\operatorname{Hom}_R(N, N')$, we define $F(\varphi)$ in $\operatorname{Hom}_{\mathbb{Z}}\big(\operatorname{Hom}_R(M,N), \operatorname{Hom}_R(M,N')\big)$ by the formula

$$F(\varphi)(\tau) = \varphi\tau \qquad \text{for } \tau \in \operatorname{Hom}_R(M, N),$$

where $\varphi\tau$ denotes the composition of τ followed by φ. In other words, $F(\varphi)$ is given by *post*multiplication by φ. By inspection we see that $F(1_N)$ is the identity from $\operatorname{Hom}_R(M, N)$ to itself if 1_N is the identity on N and that $F(\varphi'\varphi) = F(\varphi')F(\varphi)$ if φ' is in $\operatorname{Hom}_R(N', N'')$; the latter formula comes down to the associativity formula $(\varphi'\varphi)\tau = \varphi'(\varphi\tau)$ for functions under composition. Therefore F is a covariant functor from the category $\mathcal{C}$ to the category $\mathcal{D}$. We write $\operatorname{Hom}(1, \varphi)$ for $F(\varphi)$, so that $\operatorname{Hom}(1, \varphi)(\tau) = \varphi\tau$.

Similarly for fixed N in $\operatorname{Obj}(\mathcal{C})$, we define

$$G(M) = \operatorname{Hom}_R(M, N).$$

On morphisms, G is given by *pre*multiplication. Specifically for a morphism ψ in $\operatorname{Hom}_R(M, M')$, we define $G(\psi)$ in $\operatorname{Hom}_{\mathbb{Z}}\big(\operatorname{Hom}_R(M', N), \operatorname{Hom}_R(M, N)\big)$ by the formula

$$G(\psi)(\tau) = \tau\psi \qquad \text{for } \tau \in \operatorname{Hom}_R(M', N).$$

We readily check that G is a contravariant functor from $\mathcal{C}$ to $\mathcal{D}$. We write $\operatorname{Hom}(\psi, 1)$ for $G(\psi)$, so that $\operatorname{Hom}(\psi, 1)(\tau) = \tau\psi$.

To create a single functor H from F and G, we can try to define a functor H from $\mathcal{C}^2$ to $\mathcal{D}$ by $H(M, N) = \operatorname{Hom}_R(M, N)$. If $\varphi \in \operatorname{Hom}_R(N, N')$ and $\psi \in \operatorname{Hom}_R(M, M')$ are given, we can try the formula $H(\psi, \varphi)(\tau) = \varphi\tau\psi$ as a definition for τ in $\operatorname{Hom}_R(M', N)$. The trouble is that H is mixed as contravariant in the first variable and covariant in the second variable. To get H to be covariant, we can use the same formulas but regard H as defined on $\mathcal{C}^{\operatorname{opp}} \times \mathcal{C}$, where $\mathcal{C}^{\operatorname{opp}}$ is the opposite category of $\mathcal{C}$, as defined in Problems 78–80 at the end of Chapter IV. But this is getting to be a complicated structure for describing something simple, and we shall simply avoid this construction altogether,[2] working with F or G as circumstances dictate.

Even though we shall not work with H as a functor, it is convenient to combine $\operatorname{Hom}(1, \varphi)$ and $\operatorname{Hom}(\psi, 1)$ into a single definition of $\operatorname{Hom}(\psi, \varphi)$ as $\operatorname{Hom}(\psi, \varphi)(\tau) = \varphi\tau\psi$. In particular, $\operatorname{Hom}(1, \varphi)$ and $\operatorname{Hom}(\psi, 1)$ commute with each other; the commutativity follows from the associative law

$$\operatorname{Hom}(\psi, 1) \circ \operatorname{Hom}(1, \varphi)(\tau) = (\varphi\tau)\psi = \varphi(\tau\psi) = \operatorname{Hom}(1, \varphi) \circ \operatorname{Hom}(\psi, 1)(\tau).$$

---

[2]In category theory one sometimes proceeds in another way, defining a "bifunctor" to be a functor-like thing depending on two variables, covariant or contravariant in each but maybe not the same in each, and satisfying an appropriate commutativity property for the two variables.

Now let us turn to three identities involving Hom_R and to their ramifications. Each identity will assert some isomorphism involving Hom, and we consider each side of the identity as the value of a functor. We shall be interested in knowing that the isomorphism is natural in each case, the notion of naturality having been defined in Section VI.6. The naturality need be proved in just one direction in each case, since the inverse of an isomorphism that is natural is an isomorphism that is natural.

The first two identities concern the interaction of Hom_R with direct products and direct sums. Direct products and direct sums of unital left R modules were defined in Examples 7 and 8 of modules in Section VIII.1, and they were seen to be the product and coproduct functors for the category $\mathcal{C}$. If S is a nonempty set, then the direct product $\prod_{s \in S} M_s$ of a family of unital left R modules $\{M_s \mid s \in S\}$ is the module whose underlying set is the Cartesian product of the sets M_s and whose operations are defined coordinate by coordinate. The direct sum $\bigoplus_{s \in S} M_s$ is the R submodule of elements of $\prod_{s \in S} M_s$ that are nonzero in only finitely many coordinates.

Proposition 10.12. Let S be a nonempty set, let M_s and N_s be unital left R modules for each $s \in S$, and let M and N be unital left R modules. Then there are isomorphisms of abelian groups

(a) $\text{Hom}_R \left(\bigoplus_{s \in S} M_s, N \right) \cong \prod_{s \in S} \text{Hom}_R(M_s, N)$,
(b) $\text{Hom}_R \left(M, \prod_{s \in S} N_s \right) \cong \prod_{s \in S} \text{Hom}_R(M, N_s)$.

Moreover, the isomorphism in (a) is natural in the variable $\{M_s\}_{s \in S}$ and in the variable N, and the isomorphism in (b) is natural in the variable M and in the variable $\{N_s\}_{s \in S}$.

REMARKS. In each instance the assertion of naturality is that some square diagram is commutative, as illustrated in Figure 6.3. For example, if the mapping from left to right in the isomorphism (a) is denoted for fixed N by $\Phi_{\{M_s\}_{s \in S}}$ and if a system of R homomorphisms $\varphi_s : M_s \to M'_s$ is given, then one assertion of naturality for (a) is that $\Phi_{\{M'_s\}_{s \in S}} \circ \{\text{Hom}(\oplus \varphi_s, 1)\} = \{\text{Hom}(\oplus \varphi_s, 1)\} \circ \Phi_{\{M_s\}_{s \in S}}$. The other says for fixed $\{M_s\}_{s \in S}$ and for an R homomorphism $\psi : N \to N'$ that $\Phi_{N'} \circ \text{Hom}(1, \psi) = \text{Hom}(1, \psi) \circ \Phi_N$ if the isomorphism (a) is denoted for fixed $\bigoplus M_s$ by Φ_N and if $\psi : N \to N'$ is an R homomorphism. Two corresponding assertions are made about (b). To simplify the notation, we shall usually drop the subscripts from Φ.

PROOF. For (a), let $e_s : M_s \to \bigoplus_t M_t$ be the sth inclusion, and let $p_s : \bigoplus_t M_t \to M_s$ be the sth projection; the latter is defined as the restriction of the projection associated with the direct product. The map from left to right in (a) is given by $\Phi(\sigma) = \{\sigma \circ e_s\}_{s \in S}$ for σ in $\text{Hom}_R \left(\bigoplus_s M_s, N \right)$, and the expected formula for the inverse is $\Phi'(\{\tau_s\}_{s \in S}) = \sum_s (\tau_s \circ p_s)$. Then we have

$$\Phi'(\Phi(\sigma)) = \Phi'(\{\sigma \circ e_s\}_s) = \sum_s (\sigma \circ e_s \circ p_s) = \sigma$$

and
$$\Phi(\Phi'(\{\tau_s\}_s)) = \Phi\left(\sum_s (\tau_s \circ p_s)\right) = \left\{\left(\sum_s (\tau_s \circ p_s)\right) \circ e_t\right\}_t$$
$$= \{\tau_s \circ p_s \circ e_s\}_s = \{\tau_s\}_s.$$

Hence Φ is an isomorphism with inverse Φ'.

Next let the system of R homomorphisms $\varphi_s : M'_s \to M_s$ be given, let $e'_s : M'_s \to \bigoplus_t M'_t$ be the sth inclusion, and fix N. For σ in $\text{Hom}_R\left(\bigoplus_s M_s, N\right)$, we have

$$\{\text{Hom}(\oplus \varphi_s, 1)\}_s(\Phi(\sigma)) = \{\text{Hom}(\oplus \varphi_s, 1)\}_s(\{\sigma \circ e_s\}_s) = \{\sigma \circ e_s\}_s \circ \{\varphi_s\}_s$$
$$= \{\sigma \circ e_s \circ \varphi_s\}_s = \{\sigma \circ \varphi_s \circ e'_s\}_s = \{\sigma \circ \{\varphi_s\}_s \circ e'_t\}_t$$
$$= \Phi(\sigma \circ \{\varphi_s\}_s) = \Phi(\{\text{Hom}(\oplus \varphi_s, 1)\}_s(\sigma)).$$

This proves naturality in the variable $\{M_s\}_s$. If an R homomorphism $\varphi : N \to N'$ is given and if σ is in $\text{Hom}_R\left(\bigoplus_s M_s, N\right)$, then

$$\Phi(\text{Hom}(1, \varphi)(\sigma)) = \Phi(\varphi \circ \sigma) = \{\varphi \circ \sigma \circ e_s\}_s$$
$$= \text{Hom}(1, \varphi)(\{\sigma \circ e_s\}_s) = \text{Hom}(1, \varphi)(\Phi(\sigma)).$$

This proves naturality in the variable N.

For (b), let $p_s : \prod N_t \to N_s$ be the sth projection. The map from left to right in (b) is given by $\Phi(\sigma) = \{p_s \circ \sigma\}_s$ for σ in $\text{Hom}_R\left(M, \prod_s N_s\right)$, and the inverse is given by $\Phi'(\{\tau_s\}_s) = \tau$, where $\tau(m) = \{\tau_s(m)\}_s$. The proof of naturality is similar to the corresponding proof in (a) and is omitted. □

One ramification of Proposition 10.12 is the correspondence of "linear" maps to matrices when the ring R of scalars is noncommutative. If R is a field and V is an n-dimensional vector space over R, then we know that $\text{End}_R(V)$ is isomorphic as an R algebra to the space $M_n(R)$ of n-by-n matrices over R, the isomorphism being fixed once we choose an ordered basis of V. Things are more subtle when R is noncommutative.

Corollary 10.13. Let V be a unital left R module, and let S be the ring $S = \text{End}_R(V)$. For integers $m \geq 1$ and $n \geq 1$, there is a canonical isomorphism of abelian groups
$$\text{Hom}_R(V^n, V^m) \cong M_{mn}(S)$$
such that composition of R homomorphisms, given as a mapping
$$\text{Hom}_R(V^n, V^m) \times \text{Hom}_R(V^p, V^n) \longrightarrow \text{Hom}_R(V^p, V^m),$$

corresponds to matrix multiplication

$$M_{mn}(S) \times M_{np}(S) \longrightarrow M_{mp}(S).$$

In particular, in the special case that $m = n$, this canonical isomorphism becomes an isomorphism of rings

$$\operatorname{End}_R(V^n) \cong M_n(S).$$

REMARKS. For $V = R$, this isomorphism takes the form

$$\operatorname{End}_R(R^n) \cong M_n(\operatorname{End}_R(R))$$

and looks like something familiar from the case that R is a field. If $\operatorname{End}_R(R)$ were to be isomorphic as a ring to R, then the correspondence would be exactly what we might expect between R linear mappings from a free R module of rank n into itself, with n-by-n matrices with entries in R. However, $\operatorname{End}_R(R)$ is not ordinarily isomorphic to R, and the correspondence is something different and unexpected. We shall sort out these matters in Proposition 10.14 and Corollary 10.15.

PROOF. Let $e_j : V \to V^n = \bigoplus_{k=1}^n V = \prod_{k=1}^n V$ be the jth inclusion for whatever n is under discussion, and let $p_i : V^m \to V$ be the ith projection for whatever m is under discussion. For f in $\operatorname{Hom}_R(V^n, V^m)$, define $f_{ij} = p_i f e_j$. Then f_{ij} is R linear from V into V, hence is in $S = \operatorname{End}_R(V)$. If also g is in $\operatorname{Hom}_R(V^p, V^n)$, so that $f \circ g$ is in $\operatorname{Hom}_R(V^p, V^m)$, then the formula $\sum_{k=1}^n e_k p_k = 1$ on V^n gives

$$(f \circ g)_{ij} = p_i f g e_j = \sum_{k=1}^n p_i f e_k p_k g e_j = \sum_{k=1}^n f_{ik} g_{kj}.$$

Thus $f \circ g$ corresponds to the matrix product $[f_{ij}][g_{ij}]$, and the mapping is a ring homomorphism. Since

$$\sum_{i,j} e_i f_{ij} p_j = \sum_{i,j} e_i p_i f e_j p_j = \left(\sum_i e_i p_i\right) f \left(\sum_j e_j p_j\right) = 1 f 1 = f,$$

the mapping is one-one. If an arbitrary member $[u_{ij}]$ of $M_{mn}(S)$ is given, then we can define $f = \sum_{k,l} e_k u_{kl} p_l$, obtain $f_{ij} = p_i f e_j = \sum_{k,l} p_i e_k u_{kl} p_l e_j = p_i e_i u_{ij} p_j e_j = u_{ij}$, and conclude that the mapping is onto. □

Proposition 10.14. The mapping $\varphi \mapsto \varphi(1)$ is a ring isomorphism $\operatorname{End}_R(R) \cong R^o$ of $\operatorname{End}_R(R)$ onto the opposite ring R^o of R.

PROOF. The mapping $\varphi \mapsto \varphi(1)$ certainly respects addition. If φ maps to $\varphi(1)$ and τ maps to $\tau(1)$, then $\varphi\tau$ maps to $(\varphi\tau)(1) = \varphi(\tau(1)) = \varphi(\tau(1)1) = \tau(1)\varphi(1)$ since φ respects left multiplication by the element $\tau(1)$ of R. The order of multiplication is therefore reversed, and $\varphi \mapsto \varphi(1)$ is a ring homomorphism of $\mathrm{End}_R(R)$ into R^o.

If r is given in R^o, define $\varphi_r(s) = sr$ for s in R. Then φ_r respects addition, and it respects left multiplication by R because $\varphi_r(r's) = r'sr = r'\varphi_r(s)$. Therefore φ_r is a member of $\mathrm{End}_R(R)$ such that $\varphi_r(1) = r$, and $\varphi \mapsto \varphi(1)$ is onto R^o.

If φ in $\mathrm{End}_R(R)$ has $\varphi(1) = 0$, then the R linearity of φ implies that $\varphi(r) = \varphi(r1) = r\varphi(1) = r0 = 0$, so that $\varphi = 0$. Consequently the map $\varphi \mapsto \varphi(1)$ is one-one. $\square$

Corollary 10.15. For any integer $n \geq 1$, $\mathrm{End}_R(R^n)$ is ring isomorphic to $M_n(R^o)$.

REMARKS. Now we can complete the remarks with Corollary 10.13: the case in which R is commutative might lead us to believe that $\mathrm{End}_R(R^n)$ is isomorphic to $M_n(R)$, but the correct isomorphism is with $M_n(R^o)$ instead.

PROOF. Corollary 10.13 shows that $\mathrm{End}_R(R^n)$ is isomorphic to $M_n(\mathrm{End}_R(R))$, and Proposition 10.14 shows that the latter ring is isomorphic to $M_n(R^o)$. $\square$

The third identity involving Hom_R concerns $\mathrm{Hom}_R(R, M)$, where M is a unital left R module. Ordinarily $\mathrm{Hom}_R(N, M)$, when N and M are two unital left R modules, is not an R module, but in the case that $N = R$, it is. The definition of the scalar multiplication by $r \in R$ is $(r\varphi)(r') = \varphi(r'r)$ for $r' \in R$ and $\varphi \in \mathrm{Hom}_R(R, M)$. To see that $r\varphi$ is in $\mathrm{Hom}_R(R, M)$, we let s be in R and compute that $(r\varphi)(sr') = \varphi((sr')r) = \varphi(s(r'r)) = s(\varphi(r'r)) = s((r\varphi)(r'))$, as required. To see that $(sr)\varphi = s(r\varphi)$, we compute that $((sr)\varphi)(r') = \varphi(r'(sr)) = \varphi((r's)r) = (r\varphi)(r's) = (s(r\varphi))(r')$. Proposition 10.16 identifies $\mathrm{Hom}_R(R, M)$ as an R module.

Proposition 10.16. For any unital left R module M, there is a canonical R isomorphism
$$\mathrm{Hom}_R(R, M) \cong M,$$
and this isomorphism is natural in the variable M.

PROOF. The map Φ from left to right is given by $\Phi(\sigma) = \sigma(1)$, and the inverse will be seen to be given by $\Phi'(m) = \tau_m$ with $\tau_m(r) = rm$. The computation $\Phi(r\sigma) = (r\sigma)(1) = \sigma(1r) = \sigma(r1) = r(\sigma(1)) = r(\Phi(\sigma))$ shows that Φ is an R homomorphism, and the computation $\tau_m(sr) = (sr)m = s(rm) = s(\tau_m(r))$ shows that τ_m is in $\mathrm{Hom}_R(R, M)$.

To see that Φ is an isomorphism with inverse Φ', we observe that $\Phi'\Phi$ carries $\mathrm{Hom}_R(R, M)$ into itself and has $(\Phi'\Phi)(\sigma) = \Phi'(\sigma(1)) = \tau_{\sigma(1)}$, where $\tau_{\sigma(1)}(r) = r\sigma(1) = \sigma(r)$; thus $(\Phi'\Phi)(\sigma) = \sigma$, and $\Phi'\Phi$ is the identity. Also, $(\Phi\Phi')(m) =$

$\Phi(\tau_m) = \tau_m(1) = 1m = m$, and $\Phi\Phi'$ is the identity.

For the naturality let $\varphi : M \to M'$ be an R homomorphism. Then we have $\Phi(\text{Hom}(1, \varphi)(\sigma)) = \Phi(\varphi\sigma) = \varphi\sigma(1) = \varphi(\Phi(\sigma))$, and naturality is proved. $\square$

A relevant observation about the construction whose result is identified in Proposition 10.16 is that we could get by with something more general than R in the first variable of Hom_R. In fact, the construction would have worked for $\text{Hom}_R(P, M)$ for any unital (R, R) "bimodule" P, i.e., any abelian group P that is a unital left R module and unital right R module in such a way that the two actions commute: $(rp)r' = r(pr')$. More generally let S be a second ring with identity. We say that P is a unital (R, S) **bimodule** if P is simultaneously a unital left R module and a unital right S module in such a way that $(rp)s = r(ps)$ for $r \in R, s \in S$, and $p \in P$. The following proposition shows that P allows us to construct a unital left S module out of any unital left R module M.

Proposition 10.17. If R and S are two rings with identity, if P is a unital (R, S) bimodule, and if M is any unital left R module, then the abelian group $\text{Hom}_R(P, M)$ becomes a unital left S module under the definition $(s\varphi)(p) = \varphi(ps)$ for $s \in S, \varphi \in \text{Hom}_R(P, M)$, and $p \in P$.

PROOF. To see that $s\varphi$ is an R homomorphism, we compute that $(s\varphi)(rp) = \varphi((rp)s) = \varphi(r(ps)) = r(\varphi(ps)) = r((s\varphi)(p))$. It is clear that 1 acts as 1, and the distributive laws are routine. What needs checking is the formula $(ss')\varphi = s(s'\varphi)$ for s and s' in S and φ in $\text{Hom}_R(P, M)$. We compute that $((ss')\varphi)(p) = \varphi(p(ss')) = \varphi((ps)s') = (s'\varphi)(ps) = s((s'\varphi))(p)$, and the result follows. $\square$

An example of a unital (R, S) bimodule P is a ring S with identity such that R is a subring of S with the same identity. Then we can take $P = S$, with the result that R acts on the left, S acts on the right, and the two actions commute by the associative law for multiplication in S. In this situation the passage from R to $\text{Hom}_R(S, M)$ is called a **change of rings**, or **extension of scalars**, for M.

In the special case that the rings are fields and the modules are vector spaces, we saw a different kind of change of rings in Section VI.6. What we saw there is that if $\mathbb{K} \subseteq \mathbb{L}$ is an inclusion of fields and if E is a vector space over $\mathbb{K}$, then $E^{\mathbb{L}} = E \otimes_{\mathbb{K}} \mathbb{L}$ has a canonical scalar multiplication by members of $\mathbb{L}$ under the definition that multiplication by $c \in \mathbb{L}$ is the linear mapping $1 \otimes (l \mapsto cl)$. In the next section we shall see that this change of rings by means of tensor products for vector spaces generalizes to give a second construction of a change of rings for modules over a ring with identity.

5. Tensor Product for Modules

In this section, R is still a ring with identity, and others rings will play a role as well. We are going to generalize the discussion of tensor products of Section VI.6, extending the notion from the tensor product of two vector spaces over a field to the tensor product of a unital right R module and a unital left R module. The tensor product will ordinarily not have the structure of an R module; it will be just an abelian group. Additional structure on the tensor product will come from a bimodule structure on one or both of the given R modules. For example it will be seen that the tensor product, in the current sense, of two vector spaces over a field F is a vector space over F because both vector spaces can be regarded as unital bimodules over F. We return to this detail after giving the definition and the theorem. Later in this section we shall obtain two fundamental associativity formulas, one for triple tensor products and one involving tensor product and Hom together.

Let M be a unital *right* R module, and let N be a unital *left* R module. An R **bilinear function** from $M \times N$ into an abelian group is a function b such that

$$b(m_1 + m_2, n) = b(m_1, n) + b(m_2, n) \quad \text{for all } m_1 \in M, m_2 \in M, n \in N,$$
$$b(m, n_1 + n_2) = b(m, n_1) + b(m, n_2) \quad \text{for all } m \in M, n_1 \in N, n_2 \in N,$$
$$b(mr, n) = b(m, rn) \quad \text{for all } m \in M, n \in N, r \in R.$$

The first two conditions are summarized by saying that b is **additive** in each variable. A **tensor product** of M and N over R is a pair (V, ι) consisting of an abelian group V and an R bilinear map $\iota : M \times N \to V$ having the following universal mapping property: whenever b is an R bilinear function from $M \times N$ into an abelian group A, then there exists a unique abelian-group homomorphism $L : V \to A$ such that the diagram in Figure 10.1 commutes, i.e., such that $L\iota = b$ holds in the diagram. When ι is understood, one frequently refers to V itself as the tensor product. The abelian-group homomorphism $L : V \to A$ is called the **additive extension** of b to the tensor product.[3] Theorem 10.18 below will address existence and essential uniqueness of the tensor product. Because of the essential uniqueness, it is customary to denote a tensor product by $M \otimes_R N$, and Figure 10.1 incorporates this notation.[4] The image $\iota(m, n)$ of the member (m, n) of $M \times N$ under ι is denoted by $m \otimes n$.

---

[3] *Warning*. The name "additive extension" is in analogy with the situation for the tensor product of vector spaces over a field, in which the extension is linear and really is an extension. Example 2 below will show that the tensor product of nonzero modules can be 0, and hence we do not always get something for general R that we can regard intuitively as an extension.

[4] Sometimes the notation $M \otimes_R N$ refers to the constructed abelian group in the proof of Theorem 10.18, and sometimes it refers to any abelian group as in the definition of tensor product.

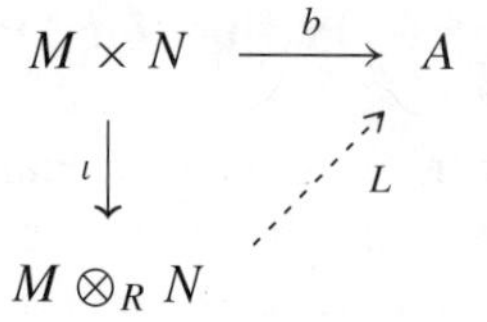

$$M \times N \xrightarrow{b} A$$
$$\iota \downarrow \quad \nearrow L$$
$$M \otimes_R N$$

FIGURE 10.1. Universal mapping property of a tensor product of a right R module M and a left R module N.

Theorem 10.18. Let R be a ring with identity. If M is a unital right R module and N is a unital left R module, then there exists a tensor product $(M \otimes_R N, \iota)$ of M and N over R, and it is unique in the following sense: if (V_1, ι_1) and (V_2, ι_2) are two tensor products, then there exists a unique abelian-group homomorphism $\Phi : V_1 \to V_2$ such that $\Phi \circ \iota_1 = \iota_2$, and it is an isomorphism. Any tensor product is generated as an abelian group by the image of $M \times N$ in it. Moreover, tensor product is a covariant functor from the category of pairs consisting of a unital right R module and a unital left R module to the category of abelian groups under the following definition: if $\varphi : M \to M'$ is a homomorphism of unital right R modules and $\psi : N \to N'$ is a homomorphism of unital left R modules, then there exists a unique homomorphism of abelian groups $\varphi \otimes \psi : M \otimes_R N \to M' \otimes_R N'$ such that $(\varphi \otimes \psi)(m \otimes n) = \varphi(m) \otimes \psi(n)$ for all $m \in M$ and $n \in N$.

PROOF. Form the free abelian group G with a $\mathbb{Z}$ basis parametrized by the elements of $M \times N$. We write $e(m, n)$ for the basis element in G corresponding to the element (m, n) of $M \times N$, and we regard e as a one-one function from $M \times N$ onto the $\mathbb{Z}$ basis of G. Let H be the subgroup of G generated by all elements of any of the forms

$$e(m_1 + m_2, n) - e(m_1, n) - e(m_2, n),$$
$$e(m, n_1 + n_2) - e(m, n_1) - e(m, n_2), \quad (*)$$
$$e(mr, n) - e(m, rn),$$

where the elements m, m_1, m_2 are in M, the elements n, n_1, n_2 are in N, and the scalar r is in R. We define $M \otimes_R N$ to be the quotient group G/H, $q : G \to G/H$ to be the quotient homomorphism, and ι to be the function $(m, n) \mapsto e(m, n) + H$ from $M \times N$ into G/H. The function ι is therefore given by $\iota = q \circ e$.

Let us prove that $(M \otimes_R N, \iota)$ is a tensor product of M and N over R. Each of the elements in $(*)$ lies in H and hence is mapped by q into the 0 coset of G/H. Since q is a homomorphism and since $\iota = q \circ e$, we obtain

$$\iota(m_1 + m_2, n) = \iota(m_1, n) + \iota(m_2, n)$$

5. Tensor Product for Modules

from the first relation in $(*)$ and similar equalities from the other two relations. Therefore $\iota : M \times N \to M \otimes_R N$ is an R bilinear function.

Now let $b : M \times N \to A$ be an R bilinear function from $M \times N$ into an abelian group A. The universal mapping property in Figure 8.2 for free abelian groups shows that there exists a unique group homomorphism $\widetilde{L} : G \to A$ such that $\widetilde{L}(e(m, n)) = b(m, n)$ for all (m, n) in $M \times N$. For the first expression in $(*)$, we have

$$\widetilde{L}\big(e(m_1 + m_2, n) - e(m_1, n) - e(m_2, n)\big)$$
$$= \widetilde{L}(e(m_1 + m_2, n)) - \widetilde{L}(e(m_1, n)) - \widetilde{L}(e(m_2, n))$$
$$= b(m_1 + m_2, n) - b(m_1, n) - b(m_2, n).$$

The right side is 0 since b is R bilinear, and a similar conclusion applies to the other two expressions in $(*)$. Therefore each member of $(*)$ lies in the kernel of $\widetilde{L}$, and the generated subgroup H lies in the kernel of $\widetilde{L}$. Consequently $\widetilde{L}$ descends to a group homomorphism $L : G/H \to A$, i.e., there exists L with $\widetilde{L} = L \circ q$. On any element (m, n) in $M \times N$, we then have $L \circ \iota = L \circ q \circ e = \widetilde{L} \circ e = b$. This proves the existence asserted by the universal mapping property for a tensor product over R. For the asserted uniqueness, the formula $L \circ \iota = b$ shows that L is determined uniquely by b on $\iota(M \times N)$. It is immediate from the definition of $M \otimes_R N$ that $\iota(M \times N)$ generates $M \otimes_R N$, and thus L is determined uniquely on all of $M \otimes_R N$.

Therefore $(M \otimes_R N, \iota)$ is a tensor product. Problems 18–22 at the end of Chapter VI show that the uniqueness up to the asserted isomorphism follows from general category theory.

We are left with defining $\varphi \otimes \psi$ when $\varphi : M \to M'$ and $\psi : N \to N'$ are given, and to showing that this definition makes tensor product into a covariant functor. Define $b : M \times N \to M' \otimes_R N'$ by $b(m, n) = \varphi(m) \otimes \psi(n)$. Then b is R bilinear into an abelian group, the property $b(mr, n) = b(m, rn)$ being verified by the calculation

$$b(mr, n) = \varphi(mr) \otimes \psi(n) = \varphi(m)r \otimes \psi(n)$$
$$= \varphi(m) \otimes r\psi(n) = \varphi(m) \otimes \psi(rn) = b(m, rn).$$

The additive extension of b to $M \otimes_R N$ is taken to be $\varphi \otimes \psi$. The formula is therefore $(\varphi \otimes \psi)(m \otimes n) = \varphi(m) \otimes \psi(n)$. If we are given also $\varphi' : M' \to M''$ and $\psi' : N' \to N''$, then

$$(\varphi' \otimes \psi')(\varphi \otimes \psi)(m \otimes n) = (\varphi' \otimes \psi')(\varphi(m) \otimes \psi(n)) = \varphi'\varphi(m) \otimes \psi'\psi(n)$$
$$= (\varphi'\varphi \otimes \psi'\psi)(m \otimes n).$$

Since the elements $m \otimes n$ generate $M \otimes_R N$, we obtain $(\varphi' \otimes \psi')(\varphi \otimes \psi) = \varphi'\varphi \otimes \psi'\psi$. Similarly we check that $1_M \otimes 1_N = 1_{M \otimes N}$. Therefore tensor product is a covariant functor. □

As in the last part of the above proof, the general procedure for constructing an abelian-group homomorphism $L : M \otimes_R N \to A$ is somehow to define an R bilinear function $b : M \times N \to A$ and to take the additive extension from Theorem 10.18 as the desired homomorphism. Once one has observed that the expression $b(m, n)$ is of a form that makes it R bilinear, then the homomorphism L is defined and is uniquely determined by its values on elements $m \otimes n$, according to the theorem.

In practice, M or N often has some additional structure, and that structure may be reflected in some additional property of the tensor product. The corollary below addresses some situations of this kind.

Corollary 10.19. Let R, S, and T be rings with identity, and suppose that M is a unital right R module and N is a unital left R module. Under the additional hypothesis that

(a) M is a unital (S, R) bimodule, then $M \otimes_R N$ is a unital left S module in a unique way such that $s(m \otimes n) = sm \otimes n$ for all $m \in M$, $n \in N$, and $s \in S$,

(b) N is a unital (R, T) bimodule, then $M \otimes_R N$ is a unital right T module in a unique way such that $(m \otimes n)t = m \otimes nt$ for all $m \in M$, $n \in N$, and $t \in T$,

(c) M is a unital (S, R) bimodule and N is a unital (R, T) bimodule, then $M \otimes_R N$ is a unital (R, T) bimodule under the left R module structure in (a) and the right T module structure in (b).

PROOF. In (a), let left multiplication by $s \in S$ within M be given by $\varphi_s : M \to M$ with $\varphi_s(m) = sm$. Then multiplication by s in S within $M \otimes_R N$ is given by $\varphi_s \otimes 1$. The covariant-functor property makes $\varphi_s \varphi_{s'} = \varphi_{ss'}$ and $\varphi_1 = 1$, and the distributive properties follow from the definitions and the fact that each φ_s is a homomorphism of the additive group M. This proves (a), and (b) is proved similarly. For (c), if left multiplication by $s \in S$ within M is given by φ_s and if right multiplication by $t \in T$ within N is given by ψ_t, then the commutativity of the operations on $M \otimes_R N$ follows from the fact that the additive homomorphisms $\varphi_s \otimes 1$ and $1 \otimes \psi_t$ commute with each other. □

EXAMPLES.

(1) $R \otimes_R M \cong M$ as an isomorphism of left R modules whenever M is a left R module. Here we regard R as a unital (R, R) bimodule, so that $R \otimes_R M \cong M$ has the structure of a unital left R module by Corollary 10.19a. The mapping of left to right is the additive extension Φ of the R bilinear function $b(r, m) = rm$, satisfying $\Phi(r \otimes m) = rm$. It respects the left action by R. The two-sided inverse Φ' to Φ is given by $\Phi'(m) = 1 \otimes m$. Then $\Phi' \circ \Phi$ is the identity since

$\Phi'(\Phi(r \otimes m)) = \Phi'(rm) = 1 \otimes rm = r \otimes m$, and $\Phi \otimes \Phi'$ is the identity since $\Phi(\Phi'(m)) = \Phi(1 \otimes m) = 1m = m$. The R isomorphism $R \otimes_R M \cong M$ is natural in M. In fact, if $\varphi : M \to M'$ is given, then

$$\varphi(\Phi(r \otimes m)) = \varphi(rm) = r\varphi(m)$$
$$= \Phi(r \otimes \varphi(m)) = \Phi((1 \otimes \varphi)(r \otimes m)).$$

(2) $\mathbb{R} = \mathbb{Z}$. In this case, $M \otimes_{\mathbb{Z}} N$ is the **tensor product of abelian groups**. Let us consider what abelian group we obtain when M and N are both finitely generated. Proposition 10.21 below shows that direct sums pull out of any tensor product, and hence it is enough to treat the tensor product of two cyclic groups. For $\mathbb{Z} \otimes_{\mathbb{Z}} A$, we get A by Example 1, and Proposition 10.20 below shows that $A \otimes_{\mathbb{Z}} \mathbb{Z}$ gives the same thing. Problem 3 at the end of the chapter identifies the tensor product of two arbitrary finite cyclic groups $(\mathbb{Z}/k\mathbb{Z}) \otimes_{\mathbb{Z}} (\mathbb{Z}/l\mathbb{Z})$. For now, let us verify in the special case that $\mathrm{GCD}(k, l) = 1$ that $(\mathbb{Z}/k\mathbb{Z}) \otimes_{\mathbb{Z}} (\mathbb{Z}/l\mathbb{Z}) = 0$. This tensor product is a unital $\mathbb{Z}$ module, being an abelian group, and Corollary 10.19a shows that the action by $\mathbb{Z}$ is given by $c(a \otimes b) = ca \otimes b$ for any integer c. Then we have $0 = (k1) \otimes 1 = k(1 \otimes 1)$ and $0 = 1 \otimes (l1) = (1l) \otimes 1 = (l1) \otimes 1 = l(1 \otimes 1)$. Choosing integers x and y such that $xk + yl = 1$, we see that $1 \otimes 1 = x(k(1 \otimes 1)) + y(l(1 \otimes 1)) = 0 + 0 = 0$. The tensor product is generated by $1 \otimes 1$, and thus the tensor product is 0.

(3) R equal to a commutative ring with identity. Then M is an (R, R) bimodule, since any unital left module for a commutative ring is a right module under the definition $mr = rm$ and vice versa. Corollary 10.19 shows therefore that $M \otimes_R N$ is a unital R module. The special case that R is a field was treated in Section VI.6.

(4) M equal to a ring S with R as a subring with the same identity. Then we can regard S as a unital (S, R) bimodule, and Corollary 10.19a shows that $S \otimes_R M$ is a unital left S module. The passage from M to $S \otimes_R M$ is a second kind of **change of rings**, or **extension of scalars**, the first kind being the passage from M to $\mathrm{Hom}_R(S, M)$ as in the previous section. Complexification of a real vector space V as $V \otimes_{\mathbb{R}} \mathbb{C}$ is an instance of this change of rings by means of tensor products. (Here we are taking into account the isomorphism $V \otimes_{\mathbb{R}} \mathbb{C} \cong \mathbb{C} \otimes_{\mathbb{R}} V$ given in Proposition 10.20 below.)

(5) M and N equal to associative R algebras with identity over a commutative ring R with identity. Proposition 10.24 below shows that $M \otimes_R N$ is another associative algebra with identity over R, with a multiplication such that

$$(m_1 \otimes n_1)(m_2 \otimes n_2) = m_1 m_2 \otimes n_1 n_2.$$

In this case the additional structure on the tensor product is not a consequence of Corollary 10.19, and additional argument is necessary.

The rest of this section will be devoted to establishing some identities for tensor product, together with their naturality, and to proving that the tensor product over R of two R algebras, for a commutative ring R with identity, is again an R algebra. Each identity involves setting up a homomorphism involving one or more tensor products, and it is necessary to prove in each case that the homomorphism is an isomorphism. For this purpose it is often inconvenient to prove directly that the homomorphism has 0 kernel and is onto. In such cases one constructs what ought to be the inverse homomorphism and proves that it is indeed a two-sided inverse.

Proposition 10.20. Let R be a ring with identity, let M be a unital right R module, and let N be a unital left R module. Let R^o be the opposite ring of R, let M^o be M regarded as a left R^o module, and let N^o be N regarded as a right R^o module. Then

$$M \otimes_R N \cong N^o \otimes_{R^o} M^o$$

under the unique homomorphism of abelian groups carrying $m \otimes n$ in $M \otimes_R N$ into $n \otimes m$ in $N^o \otimes_{R^o} M^o$. The isomorphism is natural in the variables M and N.

REMARK. To make the proof below a little clearer, we shall distinguish between elements of M and M^o, writing m in the first case and m^o in the second case, even though $m^o = m$ under our definitions. A similar notational convention will be in force for N.

PROOF. The map $(m, n) \mapsto n^o \otimes m^o$ is additive in each variable and carries (m, rn) to $(rn)^o \otimes m^o = n^o r^o \otimes m^o = n^o \otimes r^o m^o = n^o \otimes (mr)^o$. This expression is the image also of (mr, n), and hence $(m, n) \mapsto n^o \otimes m^o$ is R bilinear and has an additive extension Φ to $M \otimes_R N$. Arguing similarly, we readily construct a homomorphism $\Phi' : N^o \otimes_{R^o} M^o \to M \otimes_R N$. It is immediate that Φ' is a two-sided inverse to Φ, and the isomorphism follows. For the naturality in M, suppose that $\varphi : M \to M'$ is an R homomorphism. Write φ^o for the homomorphism with $\varphi^o(m^o) = (\varphi(m))^o$. Then $(1 \otimes \varphi^o)(\Phi(m \otimes n)) = (1 \otimes \varphi^o)(n^o \otimes m^o) = n^o \otimes \varphi^o(m^o) = n^o \otimes (\varphi(m))^o = \Phi(\varphi(m) \otimes n) = \Phi((\varphi \otimes 1)(m \otimes n))$. This proves the naturality in the M variable, and naturality in the N variable is proved similarly. $\square$

Proposition 10.21. Let R be a ring with identity, let S be a nonempty set, let M_s be a unital right R module for each $s \in S$, and let N be a unital left R module. Then

$$\Big(\bigoplus_{s \in S} M_s\Big) \otimes_R N \cong \bigoplus_{s \in S} (M_s \otimes_R N)$$

as abelian groups, and the isomorphism is natural in the tuple $(\{M_s\}_{s \in S}, N)$.

5. Tensor Product for Modules

REMARKS. A similar conclusion holds if the direct sum occurs in the second member of the tensor product, as a consequence of Proposition 10.20. The naturality carries with it some additional conclusions. For example, if each M_s is a unital (T, R) bimodule for a ring T with identity, then the displayed isomorphism is an isomorphism of left T modules.

PROOF. The map $(\{m_s\}_s, n) \mapsto \{m_s \otimes n\}_s$ is R bilinear from $\left(\bigoplus_{s \in S} M_s\right) \times N$ into $\bigoplus_{s \in S} (M_s \otimes_R N)$, and its additive extension Φ is the homomorphism from left to right in the displayed isomorphism. It has $\Phi(\{m_s\}_s \otimes n) = \{m_s \otimes n\}_s$. To construct the inverse, let $i_s : M_s \to \bigoplus_{t \in S} M_t$ be the sth inclusion. Then $(m_s, n) \mapsto i_s(m_s) \otimes n$ is R bilinear into $\left(\bigoplus_{s \in S} M_s\right) \otimes_R N$ and has an additive extension carrying $m_s \otimes n$ to $i_s(m_s) \otimes n$ in $\left(\bigoplus_{s \in S} M_s\right) \otimes_R N$. The universal mapping property of direct sums of abelian groups then gives us a corresponding abelian-group homomorphism $\Phi' : \bigoplus_{s \in S} (M_s \otimes_R N) \to \left(\bigoplus_{s \in S} M_s\right) \otimes_R N$. It has $\Phi'(\{m_s \otimes n\}_s) = \{m_s\}_s \otimes n$. It is immediate that $\Phi' \circ \Phi$ fixes each $\{m_s\}_s \otimes n$ and hence is the identity, and that $\Phi \circ \Phi'$ fixes each $\{m_s \otimes n\}_s$ and hence is the identity.

For the naturality let $\varphi_s : M_s \to M'_s$ be an R homomorphism of right R modules, and let $\psi : N \to N'$ be an R homomorphism of left R modules. Then

$$\Phi\big((\{\varphi_s\}_s \otimes \psi)(\{m_s\}_s \otimes n)\big) = \Phi\big(\{\varphi_s(m_s)\}_s \otimes \psi(n)\big) = \{\varphi_s(m_s) \otimes \psi(n)\}_s$$
$$= \{\varphi_s \otimes \psi\}_s(\{m_s \otimes n\}) = \{\varphi_s \otimes \psi\}_s(\Phi(\{m_s\} \otimes n)),$$

and naturality is proved. $\square$

Proposition 10.22. Let R and S be rings with identity, let M be a unital right R module, let N be a unital (R, S) bimodule, and let P be a unital left S module. Then

$$(M \otimes_R N) \otimes_S P \cong M \otimes_R (N \otimes_S P)$$

under the unique homomorphism Φ of abelian groups such that $\Phi((m \otimes n) \otimes p) = m \otimes (n \otimes p)$. The isomorphism is natural in the triple (M, N, P).

REMARKS. As with Proposition 10.21, the naturality carries with it some additional conclusions. For example, if T is a ring with identity and M is actually a unital (T, R) bimodule, then the isomorphism is one of left T modules.

PROOF. For fixed p, the map $(m, n, p) \mapsto m \otimes (n \otimes p)$ is R bilinear. In fact, the map is certainly additive in m and in n. For the transformation law with an element r of R, the calculation is $(mr, n, p) \mapsto mr \otimes (n \otimes p) = m \otimes r(n \otimes p) = m \otimes (rn \otimes p)$, and this is the image of (m, rn, p).

Thus for each fixed p, we have a unique well-defined extension, additive in m and n, carrying $(m \otimes n, p)$ to $m \otimes (n \otimes p)$. Using the uniqueness, we see that this extended map is additive in the variables $m \otimes n$ and p. Also, if s is in S, then $((m \otimes n)s, p) = (m \otimes ns, p)$ maps to $m \otimes (ns \otimes p) = m \otimes (n \otimes sp)$, which is the image of $(m \otimes n, sp)$, and therefore $(m \otimes n, p) \mapsto m \otimes (n \otimes p)$ is S bilinear. Consequently there exists a homomorphism Φ of abelian groups as in the statement of the proposition.

A similar argument produces a homomorphism Φ' of abelian groups carrying the right member of the display to the left member such that $\Phi'(m \otimes (n \otimes p)) = (m \otimes n) \otimes p$. On the generating elements, we see that $\Phi' \circ \Phi$ and $\Phi \circ \Phi'$ are the identity. This proves the isomorphism.

For the naturality, let $\varphi : M \to M'$, $\psi : N \to N'$, and $\tau : P \to P'$ be maps respecting the appropriate module structure in each case. Then

$$\Phi\big(((\varphi \otimes \psi) \otimes \tau)((m \otimes n) \otimes p)\big) = \Phi\big((\varphi \otimes \psi)(m \otimes n) \otimes \tau(p)\big)$$
$$= \Phi\big((\varphi(m) \otimes \psi(n)) \otimes \tau(p)\big) = \varphi(m) \otimes (\psi(n) \otimes \tau(p))$$
$$= (\varphi \otimes (\psi \otimes \tau))(m \otimes (n \otimes p)) = (\varphi \otimes (\psi \otimes \tau))(\Phi((m \otimes n) \otimes p)),$$

and naturality is proved. □

Proposition 10.23. Let R and S be rings with identity, let M be a unital left R module, let N be a unital (S, R) bimodule, and let P be a unital left S module. Then

$$\operatorname{Hom}_S(N \otimes_R M, P) \cong \operatorname{Hom}_R(M, \operatorname{Hom}_S(N, P))$$

under the homomorphism Φ of abelian groups defined by $\Phi(\varphi)(m)(n) = \varphi(n \otimes m)$ for $m \in M$, $n \in N$, and $\varphi \in \operatorname{Hom}_S(M \otimes_R N, P)$. The isomorphism is natural in the variables (N, M) and P.

REMARKS. In the displayed isomorphism, $N \otimes_R M$ on the left side is automatically a left S module, and hence $\operatorname{Hom}_S(N \otimes_R M, P)$ is a well-defined abelian group. For the right side, Proposition 10.17 shows that $\operatorname{Hom}_S(N, P)$ is a left R module under the definition $(r\tau)(n) = \tau(nr)$; consequently $\operatorname{Hom}_R(M, \operatorname{Hom}_S(N, P))$ is a well-defined abelian group. The naturality in the conclusion allows one to conclude, for example, that if M is in fact a unital (R, T) bimodule for a ring T with identity, then the displayed isomorphism is an isomorphism of left T modules.

PROOF. The homomorphism Φ is well defined. We construct its inverse. If ψ is in $\operatorname{Hom}_R(M, \operatorname{Hom}_S(N, P))$, then the map $(n, m) \mapsto \psi(m)(n)$ sends (nr, m) to $\psi(m)(nr) = (r(\psi(m))(n) = (\psi(rm))(n)$, and this is the image of (n, rm). Hence $(n, m) \mapsto \psi(m)(n)$ is R bilinear and yields a map of $N \otimes_R M$ into P such that $n \otimes m$ maps to $\psi(m)(n)$. The latter map is an S homomorphism since $sn \otimes m$ maps to $\psi(m)(sn) = s(\psi(m)(n))$, which is s applied to the image of $n \otimes m$. We define $\Phi'(\psi)$ to be the map defined on $N \otimes_R M$ with $\Phi'(\psi)(n \otimes m) = \psi(m)(n)$.

Then $\Phi'(\Phi(\varphi))(n \otimes m) = \Phi(\varphi)(m)(n) = \varphi(n \otimes m)$ shows that $\Phi' \circ \Phi$ is the identity, and $\Phi(\Phi'(\psi))(m)(n) = \Phi'(\psi)(n \otimes m) = \psi(m)(n)$ shows that $\Phi \circ \Phi'$ is the identity. Hence Φ is an isomorphism of abelian groups.

For naturality in (N, M), let $\sigma : N' \to N$ and $\tau : M' \to M$ be given. Then

$$\Phi(\text{Hom}(\sigma \otimes \tau, 1)\varphi)(m')(n') = (\text{Hom}(\sigma \otimes \tau, 1)(\varphi))(n' \otimes m')$$
$$= \varphi(\sigma \otimes \tau)(n' \otimes m') = \varphi(\sigma(n') \otimes \tau(m')) = \Phi(\varphi)(\tau(m'))(\sigma(n'))$$
$$= \text{Hom}(\tau, \text{Hom}(\sigma, 1))(\Phi(\varphi))(m')(n'),$$

and naturality is proved in (N, M). For naturality in P, let $\sigma : P \to P'$ be given. Then

$$\Phi(\text{Hom}(1, \sigma)\varphi)(m)(n) = (\text{Hom}(1, \sigma)\varphi)(n \otimes m) = \sigma\varphi(n \otimes m)$$
$$= \sigma\big((\Phi(\varphi))(m)(n)\big) = \text{Hom}(1, \text{Hom}(1, \sigma))(\Phi(\varphi))(m)(n),$$

and naturality is proved in P. □

Proposition 10.24. Let R be a commutative ring with identity, and let M and N be associative R algebras with identity. Then $M \otimes_R N$ is an associative R algebra with identity under the unique multiplication law satisfying

$$(m \otimes n)(m' \otimes n') = mm' \otimes nn'.$$

PROOF. What we know from Example 3 is that $M \otimes_R N$ is a unital R module. We need to define the associative-algebra multiplication in $M \otimes_R N$ and check that it satisfies the required properties.

Let $\mu(m)$ and $\nu(n)$ be the left multiplication operators in M and N defined by $\mu(m)(m') = mm'$ and $\nu(n)(n') = nn'$. The fact that R is central in M means that $\mu(m)(rm') = mrm' = rmm' = r\mu(m)(m')$ and hence that the mapping $\mu(m) : M \to M$ is a homomorphism of R modules. Similarly $\nu(n) : N \to N$ is a homomorphism of R modules. Therefore $\mu(m) \otimes \nu(n)$ is a well-defined homomorphism of abelian groups for each (m, n) in $M \times N$, and $b(m, n) = \mu(m) \otimes \nu(n)$ is a well-defined map of $M \times N$ into the abelian group $\text{End}_{\mathbb{Z}}(M \otimes_R N)$. The map b is certainly additive in the M variable and in the N variable. If r is in R, then $b(mr, n) = \mu(mr) \otimes \nu(n)$. Since

$$(\mu(mr) \otimes \nu(n))(m' \otimes n') = mrm' \otimes nn' = mm'r \otimes nn'$$
$$= mm' \otimes rnn' = (\mu(m) \otimes \nu(rn))(m \otimes n'),$$

we see that $b(mr, n) = b(m, rn)$. Thus b is R bilinear and extends to a homomorphism $L : M \otimes_R N \to \text{End}_{\mathbb{Z}}(M \otimes_R N)$ of abelian groups.

For x and y in $M \otimes_R N$, we define a product by $xy = L(x)(y)$. Since $L(x)$ is in $\operatorname{End}_{\mathbb{Z}}(M \otimes_R N)$, we have $x(y_1 + y_2) = xy_1 + xy_2$. Since L is a homomorphism, $L(x_1 + x_2) = L(x_1) + L(x_2)$, and therefore $(x_1 + x_2)y = x_1 y + x_2 y$. The element $1_M \otimes 1_N$, where 1_M and 1_N are the respective identities of M and N, is a two-sided identity for $M \otimes_R N$. Since $M \otimes_R N$ is a two-sided unital R module, we have $rx = xr$, and thus $R(1_M \otimes 1_N)$ lies in the center of $M \otimes_R N$. Therefore the product operation is R linear in each variable.

Suppose that $x = m \otimes n$ and $y = m' \otimes n'$. Then we have

$$xy = L(x)(y) = L(m \otimes n)(m' \otimes n') = b(m,n)(m' \otimes n')$$
$$= (\mu(m) \otimes v(n))(m' \otimes n') = mm' \otimes nn'$$

as asserted in the statement of the proposition. Consequently

$$(m \otimes n)\big((m' \otimes n')(m'' \otimes n'')\big) = (m \otimes n)(m'm'' \otimes n'n'') = m(m'm'') \otimes n(n'n'')$$
$$= (mm')m'' \otimes (nn')n'' = (mm' \otimes nn')(m'' \otimes n'')$$
$$= \big((m \otimes n)(m' \otimes n')\big)(m'' \otimes n'').$$

This proves associativity of multiplication on elements of the form $m \otimes n$. Since these elements generate the tensor product as an abelian group and since the distributive laws hold, associativity holds in general. □

6. Exact Sequences

Consider a diagram of abelian groups and group homomorphisms of the form

$$\cdots \xrightarrow{\varphi_{n-1}} M_{n-1} \xrightarrow{\varphi_n} M_n \xrightarrow{\varphi_{n+1}} M_{n+1} \xrightarrow{\varphi_{n+2}} \cdots,$$

where M_{n-1}, M_n, M_{n+1}, etc., are abelian groups and $\varphi_{n-1}, \varphi_n, \varphi_{n+1}, \varphi_{n+2}$, etc., are homomorphisms. The diagram can be finite or infinite, and the particular kind of indexing is not important. The sequence in question is called a **complex** if all consecutive compositions are 0, i.e., if $\varphi_{k+1}\varphi_k = 0$ for all k. This condition is equivalent to having $\operatorname{image}(\varphi_k) \subseteq \ker(\varphi_{k+1})$ and is the backdrop for the traditional definitions of homology and cohomology groups, which are the various quotients $\ker(\varphi_{k+1})/\operatorname{image}(\varphi_k)$.

EXAMPLES OF COMPLEXES.

(1) The simplicial homology of a simplicial complex. For this situation the indexing is reversed (say by replacing n by $-n$), so that the homomorphisms lower the index. Each group M_n is a group whose elements are called "chains," and the homomorphisms are called "boundary maps." The chains in the kernel of one of the homomorphisms are said to be "closed," and those in the image

of a homomorphism are said to be "exact." The quotient of the two, taking into account the reversal of the indexing, is the system of simplicial homology groups of the simplicial complex.

(2) The de Rham cohomology of a smooth manifold. For this situation the indexing goes upward as indicated, the group M_n is the vector space of smooth differential forms of degree n, the homomorphisms are the restrictions to these spaces of the linear de Rham operator d, $\ker(\varphi_{n+1})$ is the vector subspace of "closed" forms, $\operatorname{image}(\varphi_n)$ is the vector subspace of "exact" forms, and the quotient $\ker(\varphi_{n+1})/\operatorname{image}(\varphi_n)$ is the nth de Rham cohomology space of the manifold.

(3) Cohomology of groups. This was defined in Section VII.6, knowledge of which is not assumed in the present chapter. The result that shows that the appropriate sequence is a complex is Proposition 7.39, for which we gave a direct but complicated combinatorial proof.

The above sequence is said to be **exact** at M_n if $\ker(\varphi_{n+1}) = \operatorname{image}(\varphi_n)$. It is said to be an **exact sequence** if it is exact at every group in the sequence. The condition of exactness may be viewed as having two parts to it. One is the inclusion $\operatorname{image}(\varphi_n) \subseteq \ker(\varphi_{n+1})$ that enters the definition of complex. Since this condition says that $\varphi_{n+1}\varphi_n = 0$, it is often easy to check. The other condition is that $\ker(\varphi_{n+1}) \subseteq \operatorname{image}(\varphi_n)$, a condition that often is more difficult to check.

The extent to which a complex fails to be exact plays a fundamental role in the subject of homological algebra. This is a subject that for the most part is left to *Advanced Algebra*, that puts the examples above into a wider context, and that develops techniques for working with homology and cohomology. In this section we shall give the barest hint of an introduction to the subject by discussing some of the effects of the Hom functor and the tensor product functor on exact sequences.

Let us establish a setting for applying a functor F to an exact sequence or more general complex. For current purposes we have in mind that F is Hom in one of its two variables or is tensor product in one of its two variables. First we need to have two categories available so that F carries the one category to the other. These categories will have to satisfy some properties, but we shall not attempt to list such properties at this time.[5] Let us be content with some familiar examples of categories whose objects are abelian groups with additional structure and whose morphisms are group homomorphisms with additional structure. Specifically let R be a ring with identity, let $\mathcal{C}_R$ be the category of all unital left R modules, and let $\mathcal{D}_R$ be the category of all unital right R modules. We suppose that our functor F carries some $\mathcal{C}_R$ or $\mathcal{D}_R$ to another such category, possibly for a different ring.

---

[5]The appropriate notion is that of an "abelian category," which is defined in *Advanced Algebra*.

The functor F can be covariant or contravariant. We require also of F that it be an **additive functor**, i.e., that $F(\varphi_1 + \varphi_2) = F(\varphi_1) + F(\varphi_2)$ for any maps φ_1 and φ_2 that lie in the same Hom group.

With the additional structure in place, we can now introduce the notions of **complex** and **exact sequence** for the domain and range categories of F, not just for the category of abelian groups. In this case the abelian groups in the sequence are to be objects in the category, and the group homomorphisms in the sequence are to be morphisms in the category; otherwise the definitions are unchanged. The condition that F be additive implies that F carries any 0 map to a 0 map, and that property will be key for us. In fact, we can apply F to any complex in the domain category (by applying it to each object and morphism in the sequence); after F is applied, the arrows point the same way if F is covariant, and they point the opposite way if F is contravariant. If F is covariant, it sends any consecutive composition $0 = \varphi_{k+1}\varphi_k$ to $0 = F(0) = F(\varphi_{k+1}\varphi_k) = F(\varphi_{k+1})F(\varphi_k)$; therefore the consecutive composition of F of the maps is 0, and we obtain a complex. If F is contravariant, we have $0 = F(0) = F(\varphi_{k+1}\varphi_k) = F(\varphi_k)F(\varphi_{k+1})$; the consecutive composition of F of the maps is still 0, and we still obtain a complex. Thus the additive functor F sends any complex to a complex.

However, not all additive functors invariably send exact sequences to exact sequences, as we shall see with Hom and tensor product in the category $C_\mathbb{Z}$. Yet they each preserve some features of certain exact sequences, even when $\mathbb{Z}$ is replaced by a general ring with identity. To be precise we introduce the following definition.

A **short exact sequence** in our category is an exact sequence of the form

$$0 \longrightarrow M \xrightarrow{\varphi} N \xrightarrow{\psi} P \longrightarrow 0.$$

Exactness of this sequence incorporates three conditions:

(i) φ is one-one,
(ii) $\ker \psi = \operatorname{image} \varphi$,
(iii) ψ is onto.

In fact, the three conditions are precisely the conditions of exactness at M, N, and P, respectively, since the maps at either end are 0 maps. If we think of φ as an inclusion map, then the short exact sequence corresponds to the isomorphism $N/M \cong P$ obtained because ψ factors through to the quotient N/M.

Proposition 10.25. Let R be a ring with identity, let

$$0 \longrightarrow M \xrightarrow{\varphi} N \xrightarrow{\psi} P \longrightarrow 0$$

be a short exact sequence in the category C_R, let E be a module in C_R, and let E' be a module in $\mathcal{D}_R$. Then the following sequences in $C_\mathbb{Z}$ are exact:

6. Exact Sequences

$$E' \otimes_R M \xrightarrow{1 \otimes \varphi} E' \otimes_R N \xrightarrow{1 \otimes \psi} E' \otimes_R P \longrightarrow 0,$$

$$0 \longrightarrow \operatorname{Hom}_R(E, M) \xrightarrow{\operatorname{Hom}(1,\varphi)} \operatorname{Hom}_R(E, N) \xrightarrow{\operatorname{Hom}(1,\psi)} \operatorname{Hom}_R(E, P),$$

$$\operatorname{Hom}_R(M, E) \xleftarrow{\operatorname{Hom}(\varphi,1)} \operatorname{Hom}_R(N, E) \xleftarrow{\operatorname{Hom}(\psi,1)} \operatorname{Hom}_R(P, E) \longleftarrow 0.$$

REMARKS. Similarly tensor product in the first variable, which carries $\mathcal{D}_R$ to $\mathcal{C}_{\mathbb{Z}}$, retains the same exactness as in the first of these three sequences. In each case when we specialize to $R = \mathbb{Z}$, there are examples to show that exactness fails if we try to include the expected remaining 0 in the above three sequences. We give such examples after the proof of the proposition.

PROOF. For the first sequence in $\mathcal{C}_{\mathbb{Z}}$, we are to show that $1 \otimes \psi$ is onto $E' \otimes_R P$ and that every member of the kernel of $1 \otimes \psi$ is in the image of $1 \otimes \varphi$. (Recall that $\ker(1 \otimes \psi) \supseteq \operatorname{image}(1 \otimes \varphi)$ since the sequence is a automatically a complex.)

Thus let $p \in P$ be given. Since $\psi : N \to P$ is onto, choose $n \in N$ with $\psi(n) = p$. Then $(1 \otimes \psi)(e \otimes n) = e \otimes p$. The elements $e \otimes p$ generate $E' \otimes_R P$ as an abelian group, and hence $1 \otimes \psi$ is onto $E' \otimes_R P$.

To show that $\ker(1 \otimes \psi) \subseteq \operatorname{image}(1 \otimes \varphi)$, we observe from the exactness of the given sequence at N that $E \otimes_R \ker \psi = E \otimes_R \operatorname{image} \varphi$ is generated by all elements $e \otimes \varphi(m)$, hence by all elements $(1 \otimes \varphi)(e \otimes m)$. Therefore $E \otimes_R \operatorname{image} \varphi = \operatorname{image}(1 \otimes \varphi)$, and it is enough to prove that

$$\ker(1 \otimes \psi) \subseteq E \otimes_R \ker \psi. \tag{$*$}$$

To prove ($*$), we use the fact that ψ is onto P. Define $W = E \otimes_R \ker \psi$ as a subgroup of $E \otimes_R N$, and let $q : E \otimes_R N \to (E \otimes_R N)/W$ be the quotient homomorphism. Define $b : E \times P \to (E \otimes_R N)/W$ by

$$b(e, p) = (e \otimes n) + W, \qquad \text{where } n \text{ is chosen such that } \psi(n) = p.$$

The expression $b(e, p)$ does not depend on the choice of the element n having $\psi(n) = p$ since another choice n' will differ from n by a member of $\ker \psi$ and will affect the definition only by a member of W. The function b is certainly additive in each variable, and it evidently has $b(er, p) = b(e, rp)$ for $r \in R$ as well. Thus b is R bilinear. Let $L : E \otimes_R P \to (E \otimes_R N)/W$ be the additive extension. From $b(e, \psi(n)) = (e \otimes n) + W$, we see that $L(e \otimes \psi(n)) = (e \otimes n) + W$, hence that $L \circ (1 \otimes \psi) = q$. This formula shows that $\ker(1 \otimes \psi) \subseteq \ker q = W$, and this is the inclusion ($*$).

For the second sequence in $\mathcal{C}_{\mathbb{Z}}$, we are to show that $\operatorname{Hom}(1, \varphi)$ is one-one and that every member of the kernel of $\operatorname{Hom}(1, \psi)$ is in the image of $\operatorname{Hom}(1, \varphi)$. If

σ is in $\mathrm{Hom}_R(E, M)$ with $\mathrm{Hom}(1, \varphi)(\sigma) = 0$, then $\varphi(\sigma(e)) = 0$ for all $e \in E$. Since φ is one-one, $\sigma(e) = 0$ for all e, and $\sigma = 0$.

If τ in $\mathrm{Hom}_R(E, N)$ is in the kernel of $\mathrm{Hom}(1, \psi)$, so that $\psi(\tau(e)) = 0$ for all $e \in E$, then $\tau(e) = \varphi(m)$ for some $m \in M$ depending on e, by exactness of the given sequence at N; the element m is unique because φ is one-one. Define τ' in $\mathrm{Hom}_R(E, M)$ by $\tau'(e) = $ this m; the uniqueness of m for each e ensures that τ' is in $\mathrm{Hom}_R(E, M)$. Then we have $\tau(e) = \varphi(m) = \varphi(\tau'(e))$, and we conclude that $\tau = \mathrm{Hom}(1, \varphi)(\tau')$.

For the third sequence in $\mathcal{C}_\mathbb{Z}$, we are to show that $\mathrm{Hom}(\psi, 1)$ is one-one and that every member of the kernel of $\mathrm{Hom}(\varphi, 1)$ is in the image of $\mathrm{Hom}(\psi, 1)$. If σ is in $\mathrm{Hom}_R(P, E)$ with $\mathrm{Hom}(\psi, 1)(\sigma) = 0$, then $\sigma(\psi(n)) = 0$ for all n in N. Since ψ carries N onto P, $\sigma = 0$.

If τ in $\mathrm{Hom}_R(N, E)$ is in the kernel of $\mathrm{Hom}(\varphi, 1)$, then $\mathrm{Hom}(\varphi, 1)(\tau) = 0$. So $\tau(\varphi(m)) = 0$ for all $m \in M$. Thus τ vanishes on image $\varphi = \ker \psi$, and τ descends to an R homomorphism $\overline{\tau} : N/\ker \psi \to E$. That is, τ is of the form $\tau = \overline{\tau}\psi = \mathrm{Hom}(\psi, 1)(\overline{\tau})$. $\square$

EXAMPLES OF FAILURE OF EXACTNESS IN $\mathcal{C}_\mathbb{Z}$. We start from the exact sequence

$$0 \longrightarrow \mathbb{Z} \xrightarrow{\varphi} \mathbb{Z} \xrightarrow{\psi} \mathbb{Z}/2\mathbb{Z} \longrightarrow 0,$$

where φ is multiplication by 2 and ψ is the usual quotient homomorphism.

(1) We apply $\mathbb{Z}/2\mathbb{Z} \otimes_\mathbb{Z} (\cdot)$ to the given exact sequence, and the claim is that $1 \otimes \varphi : (\mathbb{Z}/2\mathbb{Z} \otimes_\mathbb{Z} \mathbb{Z}) \to (\mathbb{Z}/2\mathbb{Z} \otimes_\mathbb{Z} \mathbb{Z})$ is not one-one. In fact, $\mathbb{Z}/2\mathbb{Z} \otimes_\mathbb{Z} \mathbb{Z} \cong \mathbb{Z}/2\mathbb{Z}$, and $1 \otimes \varphi$ acts as multiplication by 2 under the isomorphism, hence is the 0 map and is not one-one.

(2) We apply $\mathrm{Hom}_\mathbb{Z}(\mathbb{Z}/2\mathbb{Z}, \cdot)$ to the given exact sequence, and the claim is that $\mathrm{Hom}(1, \psi) : \mathrm{Hom}_\mathbb{Z}(\mathbb{Z}/2\mathbb{Z}, \mathbb{Z}) \to \mathrm{Hom}_\mathbb{Z}(\mathbb{Z}/2\mathbb{Z}, \mathbb{Z}/2\mathbb{Z})$ is not onto. In fact, $\mathrm{Hom}_\mathbb{Z}(\mathbb{Z}/2\mathbb{Z}, \mathbb{Z}) = 0$, and the identity map in $\mathrm{Hom}_\mathbb{Z}(\mathbb{Z}/2\mathbb{Z}, \mathbb{Z}/2\mathbb{Z})$ is nonzero; therefore $\mathrm{Hom}(1, \psi)$ cannot be onto.

(3) We apply $\mathrm{Hom}_\mathbb{Z}(\cdot, \mathbb{Z}/2\mathbb{Z}))$ to the given exact sequence, and the claim is that $\mathrm{Hom}(\varphi, 1) : \mathrm{Hom}_\mathbb{Z}(\mathbb{Z}, \mathbb{Z}/2\mathbb{Z}) \to \mathrm{Hom}_\mathbb{Z}(\mathbb{Z}, \mathbb{Z}/2\mathbb{Z})$ is not onto. In fact, $\mathrm{Hom}(\varphi, 1)$ is premultiplication by 2 and carries any σ in $\mathrm{Hom}_\mathbb{Z}(\mathbb{Z}, \mathbb{Z}/2\mathbb{Z})$ to the homomorphism $k \mapsto \sigma(2k) = 2\sigma(k) = 0$. Since the usual quotient homomorphism $\mathbb{Z} \to \mathbb{Z}/2\mathbb{Z}$ is a nonzero member of $\mathrm{Hom}_\mathbb{Z}(\mathbb{Z}, \mathbb{Z}/2\mathbb{Z})$, $\mathrm{Hom}(\varphi, 1)$ is not onto $\mathrm{Hom}_\mathbb{Z}(\mathbb{Z}, \mathbb{Z}/2\mathbb{Z})$.

7. Problems

1. Suppose that the commutative ring R is an integral domain. As usual, the R submodules of R are the ideals. Prove that the ideals satisfy the descending chain condition if and only if R is a field.

2. Let $\mathbb{F} = \mathbb{F}_2$ be a field with two elements.
   (a) Give an example of a representation of the cyclic group C_2 on $\mathbb{F}^2$ with the property that there is a 1-dimensional invariant subspace U but there is no invariant subspace V with $\mathbb{F}^2 = U \oplus V$.
   (b) How can one conclude from (a) that the group algebra $R = \mathbb{F} C_2$ has a unital left R module of finite length that is not semisimple? (Educational note: Compare this conclusion with Example 5 in Section 1, which shows that every unital left $\mathbb{C}G$ module is semisimple if G is a finite group.)

3. Let G be the abelian group $(\mathbb{Z}/k\mathbb{Z}) \otimes_\mathbb{Z} (\mathbb{Z}/l\mathbb{Z})$, where k and l are nonzero integers.
   (a) Prove that G is generated by the element $1 \otimes 1$.
   (b) Prove that if k divides l, then $(\mathbb{Z}/k\mathbb{Z}) \otimes_\mathbb{Z} (\mathbb{Z}/l\mathbb{Z}) \cong (\mathbb{Z}/k\mathbb{Z}) \otimes_\mathbb{Z} (\mathbb{Z}/k\mathbb{Z})$.
   (c) Using multiplication as a $\mathbb{Z}$ bilinear form on $(\mathbb{Z}/k\mathbb{Z}) \times (\mathbb{Z}/k\mathbb{Z})$, prove that $(\mathbb{Z}/k\mathbb{Z}) \otimes_\mathbb{Z} (\mathbb{Z}/k\mathbb{Z})$ has at least $|k|$ elements.
   (d) Conclude that $(\mathbb{Z}/k\mathbb{Z}) \otimes_\mathbb{Z} (\mathbb{Z}/l\mathbb{Z}) \cong \mathbb{Z}/d\mathbb{Z}$, where $d = \mathrm{GCD}(k, l)$.

4. **(Fitting's Lemma)** Let R be a ring with identity, let M be a unital left R module, and suppose that M has a composition series. Let φ be a member of $\mathrm{End}_R(M)$.
   (a) Prove for the composition powers φ^n of φ that there exists an integer N such that $\ker \varphi^n = \ker \varphi^{n+1}$ and $\mathrm{image}\, \varphi^n = \mathrm{image}\, \varphi^{n+1}$ for all $n \geq N$.
   (b) Let $\mathcal{K}$ and $\mathcal{I}$ be the respective R submodules of M obtained for $n \geq N$ in (a). Prove that $\mathcal{K} \cap \mathcal{I} = 0$.
   (c) For x in M, show that there is some y in image φ^N with $\varphi^N(x) = \varphi^N(y)$.
   (d) Deduce from (c) that $M = \mathcal{K} + \mathcal{I}$, and conclude from (b) that $M = \mathcal{K} \oplus \mathcal{I}$.
   (e) Prove that φ carries $\mathcal{I}$ one-one onto $\mathcal{I}$ and that $(\varphi|_\mathcal{K})^n = 0$ for some n.

5. Let R be a ring with identity, and let
$$0 \longrightarrow M \xrightarrow{\varphi} N \xrightarrow{\psi} P \longrightarrow 0$$
be an exact sequence of unital left R modules. Prove that the following conditions are equivalent:
   (i) N is a direct sum $N' \oplus \ker \psi$ of R submodules for some N',
   (ii) there exists an R homomorphism $\sigma : P \to N$ such that $\psi\sigma = 1_P$,
   (iii) there exists an R homomorphism $\tau : N \to M$ such that $\tau\varphi = 1_M$.

   (Educational note: In this case one says that the exact sequence is **split**.)

6. (a) If R is the ring of quaternions, prove that $\text{End}_R(R)$ is isomorphic to R as a ring.
   (b) Give an example of a noncommutative ring with identity for which $\text{End}_R(R)$ is not isomorphic to R, and explain why it is not isomorphic.

7. Let R be a ring with identity, and let M be a unital left R module. Prove that M has a unique maximal semisimple R submodule N. (Educational note: The R submodule N is called the **socle** of M.)

8. Let $\mathbb{F} \subseteq \mathbb{K}$ be an inclusion of fields, and let A be an associative algebra with identity over $\mathbb{F}$. Proposition 10.24 makes $A \otimes_{\mathbb{F}} \mathbb{K}$ into an associative algebra over $\mathbb{F}$ with a multiplication such that $(a_1 \otimes k_1)(a_2 \otimes k_2) = a_1 a_2 \otimes k_1 k_2$. Show that $A \otimes_{\mathbb{F}} \mathbb{K}$ is in fact an associative algebra over $\mathbb{K}$ with scalar multiplication by k in $\mathbb{K}$ equal to left multiplication by $1 \otimes k$.

9. A Lie algebra $\mathfrak{g}$ over a field $\mathbb{K}$ is defined, according to Problems 31–35 at the end of Chapter VI, to be a nonassociative algebra over $\mathbb{K}$ with a multiplication written $[x, y]$ that is alternating as a function of the pair (x, y) and satisfies $[x, [y, z]] + [y, [z, x]] + [z, [x, y]] = 0$ for all x, y, z in $\mathfrak{g}$. If $\mathbb{L}$ is a field containing $\mathbb{K}$, prove that $\mathfrak{g}^{\mathbb{L}} = \mathfrak{g} \otimes_{\mathbb{K}} \mathbb{L}$ becomes a Lie algebra over $\mathbb{L}$ in a unique way such that its multiplication satisfies $[x \otimes c, y \otimes d] = [x, y] \otimes cd$ for x, y in $\mathfrak{g}$ and c, d in $\mathbb{L}$.

10. Let R be a ring with identity, let A be a unital right R module, and let B be a unital left R module. Since $\mathbb{Z} \subseteq R$, A and B can be considered also as $\mathbb{Z}$ modules. Form a version of $A \otimes_R B$ with associated R bilinear map $b_1 : A \times B \to A \otimes_R B$, and form a version of $A \otimes_{\mathbb{Z}} B$ with associated $\mathbb{Z}$ bilinear map $b_2 : A \times B \to A \otimes_{\mathbb{Z}} B$. Let H be the subgroup of $A \otimes_{\mathbb{Z}} B$ generated by all elements $b_2(ar, b) - b_2(a, rb)$ with $a \in A$, $b \in B$, $r \in R$, and let $q : A \otimes_{\mathbb{Z}} B \to (A \otimes_{\mathbb{Z}} B)/H$ be the quotient homomorphism. Prove that there is an abelian group isomorphism $\Phi : (A \otimes_{\mathbb{Z}} B)/H \to A \otimes_R B$ such that $\Phi(q(b_2(a, b))) = b_1(a, b)$ for all $a \in A$ and $b \in B$.

11. Let R be a commutative ring with identity, and let $\mathcal{C}$ be the category of all commutative associative R algebras with identity. Prove that if A_1 and A_2 are in $\text{Obj}(\mathcal{C})$, then $(A_1 \otimes_R A_2, \{i_1, i_2\})$ is a coproduct, where $i_1 : A_1 \to A_1 \otimes_R A_2$ is given by $i_1(a_1) = a_1 \otimes 1$ and $i_2 : A_2 \to A_1 \otimes_R A_2$ is given by $i_2(a_2) = 1 \otimes a_2$.

Problems 12–20 partition simple left R modules into isomorphism types, where R is a ring with identity. For each simple left R module E and each unital left R module M, one forms the sum M_E of all simple R submodules that are isomorphic to E and calls it an **isotypic R submodule** of M. The problems introduce a calculus for working with the members of $\text{End}_R(M_E)$ in terms of right vector spaces over a certain division ring. They show that if M is semisimple, then M is the direct sum of all its isotypic R submodules, each of these is mapped to itself by every member of $\text{End}_R(M)$, and consequently one can understand $\text{End}_R(M)$ in terms of right vector spaces over certain division rings. These problems generalize and extend Problems

47–52 at the end of Chapter VII, which in effect deal with what happens for the ring $\mathbb{C}G$ when G is a finite group; however, the material of Chapter VII is not prerequisite for these problems. The following notation is in force: M is any unital left R module, E is a simple left R module, $D_E = \operatorname{Hom}_R(E, E)$ is the ring known from Proposition 10.4b to be a division ring,

$$M_E = (\text{sum of all } R \text{ submodules of } M \text{ that are } R \text{ isomorphic to } E),$$

and $$M^E = \operatorname{Hom}_R(E, M).$$

Unital right D_E modules are right vector spaces over D_E. In Problems 18–20, $\mathcal{E}$ denotes a set of representatives of all R isomorphism classes of simple left R modules.

12. Prove that
    (a) M_E is a direct sum of simple R modules that are R isomorphic to E,
    (b) the image of every mapping in M^E belongs to M_E,
    (c) redefinition of the range from M to M_E defines an isomorphism $M^E \cong \operatorname{Hom}_R(E, M_E)$ of abelian groups.

13. Prove that
    (a) M^E is a unital right D_E module under composition of R homomorphisms,
    (b) E is a unital left D_E module under the operation of the members of D_E,
    (c) the left R module action and the left D_E module action on E commute with each other.

14. Show that $M^E \otimes_{D_E} E$ is a unital left R module in such a way that $r(m \otimes e) = m \otimes re$.

15. Prove that there is a well-defined R homomorphism $\Phi : M^E \otimes_{D_E} E \to M$ such that $\Phi(\psi \otimes e) = \psi(e)$ and such that Φ is an R isomorphism onto M_E.

16. Prove that the left R submodules N of M_E are in one-one correspondence with the right D_E vector subspaces W of M^E by the maps

$$N \mapsto \operatorname{Hom}_R(E, N) \subseteq \operatorname{Hom}_R(E, M) = M^E \quad \text{if } N \subseteq M_E$$

and $$W \mapsto W \otimes_{D_E} E \subseteq M^E \otimes_{D_E} E \cong M_E \quad \text{if } W \subseteq M^E.$$

17. Prove for any unital left R module N that there is a canonical R isomorphism

$$\operatorname{Hom}_R(M_E, N_E) \cong \operatorname{Hom}_{D_E}(M^E, N^E)$$

of abelian groups defined as follows. Suppose φ is in $\operatorname{Hom}_R(M_E, N_E)$. Composition with φ carries $\operatorname{Hom}_R(E, M)$ into $\operatorname{Hom}_R(E, N)$; this map respects the right action of D_E and hence induces a map

$$\varphi^E \in \operatorname{Hom}_{D_E}(M^E, N^E).$$

The isomorphism is given in terms of the isomorphisms Φ_M for M and Φ_N for N in Problem 15 by

$$\varphi(\Phi_M(\psi \otimes e)) = \Phi_N(\varphi^E(\psi) \otimes e) \qquad \text{for } \psi \in M^E.$$

18. If M is semisimple, prove that

$$M = \bigoplus_{E \in \mathcal{E}} M_E \cong \bigoplus_{E \in \mathcal{E}} (M^E \otimes_{D_E} E).$$

19. Still with M semisimple, prove that the left R submodules of M are in one-one correspondence with families $\{W^E \mid E \in \mathcal{E}\}$ of right D_E vector subspaces of M^E.

20. Suppose that M and N are two semisimple left R modules. Prove that there is a canonical isomorphism of abelian groups

$$\operatorname{Hom}_R(M, N) \cong \prod_{E \in \mathcal{E}} \operatorname{Hom}_{D_E}(M^E, N^E).$$

More precisely prove that an R module map from M to N is specified by giving, for a representative E of each class of simple left R modules, an arbitrary right vector-space map from M^E to N^E.

APPENDIX

Abstract. This appendix treats some topics that are likely to be well known by some readers and less known by others. Most of it already comes into play by Chapter II. Section A1 deals with set theory and with functions: it discusses the role of formal set theory, it works in a simplified framework that avoids too much formalism and the standard pitfalls, it establishes notation, and it mentions some formulas. Some emphasis is put on distinguishing the image and the range of a function, since this distinction is important in algebra and algebraic topology.

Section A2 defines equivalence relations and establishes the basic fact that they lead to a partitioning of the underlying set into equivalence classes.

Section A3 reviews the construction of rational numbers from the integers, and real numbers from the rational numbers. From there it concentrates on the solvability within the real numbers of certain polynomial equations.

Section A4 is a quick review of complex numbers, real and imaginary parts, complex conjugation, and absolute value.

Sections A5 and A6 return to set theory. Section A5 defines partial orderings and includes Zorn's Lemma, which is a powerful version of the Axiom of Choice, while Section A6 concerns cardinality.

A1. Sets and Functions

Algebra typically makes use of an informal notion of set theory and notation for it in which sets are described by properties of their elements and by operations on sets. This informal set theory, if allowed to be too informal, runs into certain paradoxes, such as **Russell's paradox:** "If S is the set of all sets that do not contain themselves as elements, is S a member of S or is it not?" The conclusion of Russell's paradox is that the "set" of all sets that do not contain themselves as elements is not in fact a set.

Mathematicians' experience is that such pitfalls can be avoided completely by working within some formal axiom system for sets, of which there are several that are well established. A basic one is "Zermelo–Fraenkel set theory," and the remarks in this section refer specifically to it but refer to the others at least to some extent.[1]

The standard logical paradoxes are avoided by having sets, elements (or "entities"), and a membership relation $\in$ such that $a \in S$ is a meaningful statement,

---

[1] Mathematicians have no proof that this technique avoids problems completely. Such a proof would be a proof of the consistency of a version of mathematics in which one can construct the integers, and it is known that this much of mathematics cannot be proved to be consistent unless it is in fact inconsistent.

true or false, if and only if a is an element and S is a set. The terms **set**, **element**, and $\in$ are taken to be primitive terms of the theory that are in effect defined by a system of axioms. The axioms ensure the existence of many sets, including infinite sets, and operations on sets that lead to other sets. To make full use of this axiom system, one has to regard it as occurring in the framework of certain rules of logic that tell the forms of basic statements (namely, $a = b$, $a \in S$, and "S is a set"), the connectives for creating complicated statements from simple ones ("or," "and," "not," and "if ... then"), and the way that quantifiers work ("there exists" and "for all").

Working rigorously with such a system would likely make the development of mathematics unwieldy, and it might well obscure important patterns and directions. In practice, therefore, one compromises between using a formal axiom system and working totally informally; let us say that one works "informally but carefully." The logical problems are avoided not by rigid use of an axiom system, but by taking care that sets do not become too "large": one limits the sets that one uses to those obtained from other sets by set-theoretic operations and by passage to subsets.[2]

A feature of the axiom system lying behind working informally but carefully is that it does not preclude the existence of additional sets beyond those forced to exist by the axioms. Thus, for example, in the subject of coin-tossing within probability, it is normal to work with the set of possible outcomes as $S = \{$heads, tails$\}$ even though it is not immediately apparent that requiring this S to be a set does not introduce some contradiction.

It is worth emphasizing that the points of the theory at which one takes particular care vary somewhat from subject to subject within mathematics. For example it is sometimes of interest in calculus of several variables to distinguish between the range of a function and its image in a way that will be mentioned below, but it is usually not too important. In homological algebra, however, the distinction is extremely important, and the subject loses a great deal of its impact if one blurs the notions of range and image.

Some references for set theory that are appropriate for reading *once* are Halmos's *Naive Set Theory*, Hayden–Kennison's *Zermelo–Fraenkel Set Theory*, and Chapter 0 and the appendix of Kelley's *General Topology*. The Kelley book is one that uses the word "class" as a primitive term more general than "set"; it develops von Neumann set theory.

All that being said, let us now introduce the familiar terms, constructions, and notation that one associates with set theory. To cut down on repetition, one

---

[2]Not every set so obtained is to be regarded as "constructed." The Axiom of Choice, which we come to shortly, is an existence statement for elements in products of sets, and the result of applying the axiom is a set that can hardly be viewed as "constructed."

allows some alternative words for "set," such as **family** and **collection**. The word "class" is used by some authors as a synonym for "set," but the word **class** is used in some set-theory axiom systems to refer to a more general notion than "set," and it will be useful to preserve this possibility. Thus a class can be a set, but we allow ourselves to speak, for example, of the class of all groups even though this class is too large to be a set. Alternative terms for "element" are **member** and **point**; we shall not use the term "entity." Instead of writing $\in$ systematically, we allow ourselves to write "in." Generally, we do not use $\in$ in sentences of text as an abbreviation for an expression like "is in" that contains a verb.

If A and B are two sets, some familiar operations on them are the **union** $A \cup B$, the **intersection** $A \cap B$, and the **difference** $A - B$, all defined in the usual way in terms of the elements they contain. Notation for the difference of sets varies from author to author; some other authors write $A \setminus B$ or $A \sim B$ for difference, but this book uses $A - B$. If one is thinking of A as a universe, one may abbreviate $A - B$ as B^c, the **complement** of B in A. The empty set $\emptyset$ is a set, and so is the set of all subsets of a set A, which is sometimes denoted by 2^A. Inclusion of a subset A in a set B is written $A \subseteq B$ or $B \supseteq A$; then B is a superset of A. Inclusion that does not permit equality is denoted by $A \subsetneq B$ or $B \supsetneq A$; in this case one says that A is a **proper subset** of B or that A is **properly contained** in B.

If A is a set, the **singleton** $\{A\}$ is a set with just the one member A. Another operation is **unordered pair**, whose formal definition is $\{A, B\} = \{A\} \cup \{B\}$ and whose informal meaning is a set of two elements in which we cannot distinguish either element over the other. Still another operation is **ordered pair**, whose formal definition is $(A, B) = \{\{A\}, \{A, B\}\}$. It is customary to think of an ordered pair as a set with two elements in which one of the elements can be distinguished as coming first.[3]

Let A and B be two sets. The set of all ordered pairs of an element of A and an element of B is a set denoted by $A \times B$; it is called the **product** of A and B or the **Cartesian product**. A **relation** between a set A and a set B is a subset of $A \times B$. Functions, which are to be defined in a moment, provide examples. Two examples of relations that are usually not functions are "equivalence relations," which are discussed in Section A2, and "partial orderings," which are discussed in Section A5.

If A and B are sets, a relation f between A and B is said to be a **function**, written $f : A \to B$, if for each $x \in A$, there is exactly one $y \in B$ such that (x, y) is in f. If (x, y) is in f, we write $f(x) = y$. In this informal but careful definition of function, the function consists of more than just a set of ordered

---

[3]Unfortunately a "sequence" gets denoted by $\{x_1, x_2, \ldots\}$ or $\{x_n\}_{n=1}^{\infty}$. If its notation were really consistent with the above definitions, we might infer, inaccurately, that the order of the terms of the sequence does not matter. The notation for unordered pairs, ordered pairs, and sequences is, however, traditional, and it will not be changed here.

pairs; it consists of the set of ordered pairs regarded as a subset of $A \times B$. This careful definition makes it meaningful to say that the set A is the **domain**, the set B is the **range**, and the subset of $y \in B$ such that $y = f(x)$ for some $x \in A$ is the **image** of f. The image is also denoted by $f(A)$. Sometimes a function f is described in terms of what happens to typical elements, and then the notation is $x \mapsto f(x)$ or $x \mapsto y$, possibly with y given by some formula or by some description in words about how it is obtained from x. Sometimes a function f is written as $f(\cdot)$, with a dot indicating the placement of the variable; this notation is especially helpful in working with restrictions, which we come to in a moment, and with functions of two variables when one of the variables is held fixed. This notation is useful also for functions that involve unusual symbols, such as the absolute value function $x \mapsto |x|$, which in this notation becomes $|\cdot|$. The word **map** or **mapping** is used for "function" and for the operation of a function, especially when a geometric setting for the function is of importance.

Often mathematicians are not so careful with the definition of function. Depending on the degree of informality that is allowed, one may occasionally refer to a function as $f(x)$ when it should be called f or $x \mapsto f(x)$. If any confusion is possible, it is wise to use the more rigorous notation. Another habit of informality is to regard a function $f : A \to B$ as simply a set of ordered pairs. Thus two functions $f_1 : A \to B$ and $f_2 : A \to C$ become the same if $f_1(a) = f_2(a)$ for all a in A. With the less-careful definition, the notion of the range of a function is not really well defined. The less-careful definition can lead to trouble in algebra and topology, but it does not often lead to trouble in analysis until one gets to a level where algebra and analysis merge somewhat. One place where it comes into play in algebra is in the notion of an exact sequence of three abelian groups $A \xrightarrow{\varphi} B \xrightarrow{\psi} C$, which is defined as a system of three abelian groups and homomorphisms as indicated such that the kernel of ψ equals the image of φ. In this definition one is not free to adjust B to be the image of φ since that adjustment will affect the kernel of ψ as well.

The set of all functions from a set A to a set B is a set. It is sometimes denoted by B^A. The special case 2^A that arises with subsets comes by regarding 2 as a set $\{1, 2\}$ and identifying a function f from A into $\{1, 2\}$ with the subset of all elements x of A for which $f(x) = 1$.

If a subset B of a set A may be described by some distinguishing property P of its elements, we may write this relationship as $B = \{x \in A \mid P\}$. For example the function f in the previous paragraph is identified with the subset $\{x \in A \mid f(x) = 1\}$. Another example is the image of a general function $f : A \to B$, namely $f(A) = \{y \in B \mid y = f(x) \text{ for some } x \in A\}$. Still more generally along these lines, if E is any subset of A, then $f(E)$ denotes the set $\{y \in B \mid y = f(x) \text{ for some } x \in E\}$. Some authors use a colon or semicolon or comma instead of a vertical line in this notation.

This book frequently uses sets denoted by expressions like $\bigcup_{x \in S} A_x$, an indexed union, where S is a set that is usually nonempty. If S is the set $\{1, 2\}$, this reduces to $A_1 \cup A_2$. In the general case it is understood that we have an unnamed function, say f, given by $x \mapsto A_x$, having domain S and range the set of all subsets of an unnamed set T, and $\bigcup_{x \in S} A_x$ is the set of all $y \in T$ such that y is in A_x for some $x \in S$. When S is understood, we may write $\bigcup_x A_x$ instead of $\bigcup_{x \in S} A_x$. Indexed intersections $\bigcap_{x \in S} A_x$ are defined similarly, and this time it is essential to disallow S empty because otherwise the intersection cannot be a set in any useful set theory.

There is also an indexed **Cartesian product** $\bigtimes_{x \in S} A_x$ that specializes in the case that $S = \{1, 2\}$ to $A_1 \times A_2$. Usually S is assumed nonempty. This Cartesian product is the set of all functions f from S into $\bigcup_{x \in S} A_x$ such that $f(x)$ is in A_x for all $x \in S$. In the special case that S is $\{1, \ldots, n\}$, the Cartesian product is the set of ordered n-tuples from n sets $A_1, \ldots, A_n$ and may be denoted by $A_1 \times \cdots \times A_n$; its members may be denoted by $(a_1, \ldots, a_n)$ with $a_j \in A_j$ for $1 \leq j \leq n$. When the factors of a Cartesian product have some additional algebraic structure, the notation for the Cartesian product is often altered; for example the Cartesian product of groups A_x is denoted by $\prod_{x \in S} A_x$.

It is completely normal in algebra, and it is the practice in this book, to take the following axiom as part of one's set theory; the axiom is customarily used without specific mention.

Axiom of Choice. The Cartesian product of nonempty sets is nonempty.

If the index set is finite, then the Axiom of Choice reduces to a theorem of set theory. The axiom is often used quite innocently with a countably infinite index set. For example a theorem of analysis asserts that any bounded sequence $\{a_n\}$ of real numbers has a subsequence converging to $\limsup a_n$, and the proof constructs one member of the sequence at a time. When the proof is written in such a way that these members have some flexibility in their definitions, the Axiom of Choice is usually being invoked. The proof can be rewritten so that the members of the subsequence have specific definitions, such as "the term a_n such that n is the smallest integer satisfying such-and-such properties." In this case the axiom is not being invoked. In fact, one can often rewrite proofs involving a countably infinite choice so that they involve specific definitions and therefore avoid invoking the axiom, but there is no point in undertaking this rewriting. In algebra the axiom is often invoked in situations in which the index set is uncountable; selection of a representative from each of uncountably many equivalence classes is such a choice if all equivalence classes have more than one element.

From the Axiom of Choice, one can deduce a powerful tool known as Zorn's Lemma, whose use it is customary to acknowledge. Zorn's Lemma appears in Section A5.

If $f : A \to B$ is a function and B is a subset of B', then f can be regarded as a function with range B' in a natural way. Namely, the set of ordered pairs is unchanged but is to be regarded as a subset of $A \times B'$ rather than $A \times B$.

Let $f : A \to B$ and $g : B \to C$ be two functions such that the range of f equals the domain of g. The **composition** $g \circ f : A \to C$, written sometimes as $gf : A \to C$, is the function with $(g \circ f)(x) = g(f(x))$ for all x. Because of the construction in the previous paragraph, it is meaningful to define the composition more generally when the range of f is merely a subset of the domain of g.

A function $f : A \to B$ is said to be **one-one** if $f(x_1) \neq f(x_2)$ whenever x_1 and x_2 are distinct members of A. The function is said to be **onto**, or often "onto B," if its image equals its range. The terminology "onto B" avoids confusion: it specifies the image and thereby guards against the use of the less careful definition of function mentioned above. A mathematical audience often contains some people who use the more careful definition of function and some people who use the less careful definition. For the latter kind of person, a function is always onto something, namely its image, and a statement that a particular function is onto might be regarded as a tautology. A function from one set to another is said to put the sets in **one-one correspondence** if the function is one-one and onto.

When a function $f : A \to B$ is one-one and is onto B, there exists a function $g : B \to A$ such that $g \circ f$ is the identity function on A and $f \circ g$ is the identity function on B. The function g is unique, and it is defined by the condition, for $y \in B$, that $g(y)$ is the unique $x \in A$ with $f(x) = y$. The function g is called the **inverse function** of f and is often denoted by f^{-1}.

Conversely if $f : A \to B$ has an inverse function, then f is one-one and is onto B. The reason is that a composition $g \circ f$ can be one-one only if f is one-one, and in addition, that a composition $f \circ g$ can be onto the range of f only if f is onto its range.

If $f : A \to B$ is a function and E is a subset of A, the **restriction** of f to E, denoted by $f\big|_E$, is the function $f : E \to B$ consisting of all ordered pairs $(x, f(x))$ with $x \in E$, this set being regarded as a subset of $E \times B$, not of $A \times B$. One especially common example of a restriction is restriction to one of the variables of a function of two variables, and then the idea of using a dot in place of a variable can be helpful notationally. Thus the function of two variables might be indicated by f or $(x, y) \mapsto f(x, y)$, and the restriction to the first variable, for fixed value of the second variable, would be $f(\,\cdot\,, y)$ or $x \mapsto f(x, y)$.

We conclude this section with a discussion of direct and inverse images of sets under functions. If $f : A \to B$ is a function and E is a subset of A, we have defined $f(E) = \{y \in B \mid y = f(x) \text{ for some } x \in E\}$. This is the same as the image of $f\big|_E$ and is frequently called the image or **direct image** of E under f. The notion of direct image does not behave well with respect to some set-theoretic operations: it respects unions but not intersections. In the case of unions, we have

$$f\left(\bigcup_{s\in S} E_s\right) = \bigcup_{s\in S} f(E_s);$$

the inclusion $\supseteq$ follows since $f\left(\bigcup_{s\in S} E_s\right) \supseteq f(E_s)$ for each s, and the inclusion $\subseteq$ follows because any member of the left side is f of a member of some E_s. In the case of intersections, the question $f(E\cap F) \stackrel{?}{=} f(E) \cap f(F)$ can easily have a negative answer, the correct general statement being $f(E\cap F) \subseteq f(E) \cap f(F)$. An example with equality failing occurs when $A = \{1, 2, 3\}$, $B = \{1, 2\}$, $f(1) = f(3) = 1$, $f(2) = 2$, $E = \{1, 2\}$ and $F = \{2, 3\}$ because $f(E \cap F) = \{2\}$ and $f(E) \cap f(F) = \{1, 2\}$.

If $f : A \to B$ is a function and E is a subset of B, the **inverse image** of E under f is the set $f^{-1}(E) = \{x \in A \mid f(x) \in E\}$. This is well defined even if f does not have an inverse function. (If f does have an inverse function f^{-1}, then the inverse image of E under f coincides with the direct image of E under f^{-1}.)

Unlike direct images, inverse images behave well under set-theoretic operations. If $f : A \to B$ is a function and $\{E_s \mid s \in S\}$ is a set of subsets of B, then

$$f^{-1}\left(\bigcap_{s\in S} E_s\right) = \bigcap_{s\in S} f^{-1}(E_s),$$

$$f^{-1}\left(\bigcup_{s\in S} E_s\right) = \bigcup_{s\in S} f^{-1}(E_s),$$

$$f^{-1}(E_s^c) = (f^{-1}(E_s))^c.$$

In the third of these identities, the complement on the left side is taken within B, and the complement on the right side is taken within A. To prove the first identity, we observe that $f^{-1}\left(\bigcap_{s\in S} E_s\right) \subseteq f^{-1}(E_s)$ for each $s \in S$ and hence $f^{-1}\left(\bigcap_{s\in S} E_s\right) \subseteq \bigcap_{s\in S} f^{-1}(E_s)$. For the reverse inclusion, if x is in $\bigcap_{s\in S} f^{-1}(E_s)$, then x is in $f^{-1}(E_s)$ for each s and thus $f(x)$ is in E_s for each s. Hence $f(x)$ is in $\bigcap_{s\in S} E_s$, and x is in $f^{-1}\left(\bigcap_{s\in S} E_s\right)$. This proves the reverse inclusion. The second and third identities are proved similarly.

A2. Equivalence Relations

An **equivalence relation** on a set S is a relation between S and itself, i.e., is a subset of $S \times S$, satisfying three defining properties. We use notation like $a \simeq b$, written "a is equivalent to b," to mean that the ordered pair (a, b) is a member of the relation, and we say that "$\simeq$" is the equivalence relation. The three defining properties are

(i) $a \simeq a$ for all a in S, i.e., $\simeq$ is **reflexive**,
(ii) $a \simeq b$ implies $b \simeq a$ if a and b are in S, i.e., $\simeq$ is **symmetric**.
(iii) $a \simeq b$ and $b \simeq c$ together imply $a \simeq c$ if a, b, and c are in S, i.e., $\simeq$ is **transitive**.

An example occurs with S equal to the set $\mathbb{Z}$ of integers with $a \simeq b$ meaning that the difference $a - b$ is even. The properties hold because (i) 0 is even, (ii) the negative of an even integer is even, and (iii) the sum of two even integers is even.

There is one fundamental result about abstract equivalence relations. The **equivalence class** of a, written $[a]$ for now, is the set of all members b of S such that $a \simeq b$.

Proposition. If $\simeq$ is an equivalence relation on a set S, then any two equivalence classes are disjoint or equal, and S is the union of all the equivalence classes.

PROOF. Let $[a]$ and $[b]$ be the equivalence classes of members a and b of S. If $[a] \cap [b] \neq \varnothing$, choose c in the intersection. Then $a \simeq c$ and $b \simeq c$. By (ii), $c \simeq b$, and then by (iii), $a \simeq b$. If d is any member of $[b]$, then $b \simeq d$. From (iii), $a \simeq b$ and $b \simeq d$ together imply $a \simeq d$. Thus $[b] \subseteq [a]$. Reversing the roles of a and b, we see that $[a] \subseteq [b]$ also, whence $[a] = [b]$. This proves the first conclusion. The second conclusion follows from (i), which ensures that a is in $[a]$, hence that every member of S lies in some equivalence class. $\square$

EXAMPLE. With the equivalence relation on $\mathbb{Z}$ that $a \simeq b$ if $a - b$ is even, there are two equivalence classes—the subset of even integers and the subset of odd integers.

The first two examples of equivalence relations in this book arise in Section II.3. The first example, which is captured in the definition of square matrices that are "similar," yields equivalence classes exactly as above. A square matrix A is similar to a square matrix B if there is a matrix C with $B = C^{-1}AC$. The text does not mention in Chapter II that similarity is an equivalence relation, but it is routine to check that it is reflexive, symmetric, and transitive. The second example is a relation "is isomorphic to" and implicitly is defined on the class of all vector spaces. This class is not a set, and Section A1 of this appendix suggested avoiding using classes that are not sets in order to avoid the logical paradoxes mentioned at the beginning of the appendix. There is not much problem with using general classes in this particular situation, but there is a simple approach in this situation for eliminating classes that are not sets and thereby following the suggestion of Section A1 without making an exception. The approach is to work with any subclass of vector spaces that is a set. The equivalence relation is well defined on the set of vector spaces in question, and the proposition yields equivalence classes within that set. This set can be an arbitrary subclass of the class of all vector spaces that happens to be a set, and the practical effect is the same as if the equivalence relation had been defined on the class of all vector spaces.

A3. Real Numbers

Real numbers are taken as known, as are the rational numbers from which they are constructed. It will be useful, however, to review the constructions of both these number systems so as to be able to discuss the solvability of polynomial equations better.

We take the set $\mathbb{Z}$ of integers as given, along with its ordering and its operations of addition, subtraction, and multiplication. The set $\mathbb{Q}$ of rational numbers is constructed rigorously from $\mathbb{Z}$ as follows. We start from the set of ordered pairs (a, b) of integers such that $b \neq 0$. The idea is that (a, b) is to correspond to a/b and that we want (na, nb) to correspond to the same a/b if n is any nonzero integer. Thus we say that two such pairs have $(a, b) \sim (c, d)$ if $ad = bc$. This relation is evidently reflexive and symmetric, and it will be an equivalence relation if it is transitive. If $(a, b) \sim (c, d)$ and $(c, d) \sim (e, f)$, then $ad = bc$ and $cf = de$. So $adf = bcf = bde$. Since $d \neq 0$, $af = be$ and $\sim$ is transitive.

From Section A2 the set of such pairs is partitioned into equivalence classes by means of $\sim$. Each equivalence class is called a **rational number**. To define the arithmetic operations on rational numbers, we first define operations on pairs, and then we check that the operations respect the partitioning into classes. For addition, the definition is $(a, b) + (c, d) = (ad + bc, bd)$. What needs checking is that if $(a, b) \sim (a', b')$ and $(c, d) \sim (c', d')$, then $(ad + bc, bd) \sim (a'd' + b'c', b'd')$. This is a routine matter: $(ad+bc)(b'd') = ab'dd' + bb'cd' = a'bdd' + bb'c'd = (a'd' + b'c')bd$, and thus addition of rational numbers is well defined. The operations on pairs for negative, multiplication, and reciprocal are $-(a, b) = (-a, b)$, $(a, b)(c, d) = (ac, bd)$, and $(a, b)^{-1} = (b, a)$, and we readily check that these define corresponding operations on rational numbers. Finally one derives the familiar associative, commutative, and distributive laws for these operations on $\mathbb{Q}$.

The above construction is repeated, with more details, in the more general construction of "fields of fractions" in Chapter VIII.

Inequalities on rational numbers are defined from inequalities on integers, taking into account that an inequality between integers is preserved when multiplied by a positive integer. Each rational number has a representative pair (a, b) with $b > 0$ because any pair can always be replaced by the pair of negatives. Thus let (a, b) and (c, d) be given with $b > 0$ and $d > 0$. We say that $(a, b) \leq (c, d)$ if $ad \leq bc$. One readily checks that this ordering respects equivalence classes and leads to the usual properties of the ordering on $\mathbb{Q}$. The positive rationals are those greater than 0, and the negative rationals are those less than 0.

The formal definition is that a **real number** is a **cut** of rational numbers, i.e., a subset of rational numbers that is neither $\mathbb{Q}$ nor the empty set, has no largest element, and contains all rational numbers less than any rational that it contains. The set of cuts, i.e., the set of real numbers, is denoted by $\mathbb{R}$. The idea of the construction is as follows: Each rational number q determines a cut q^*, namely the set of all rationals less than q. Under the identification of $\mathbb{Q}$ with a subset of $\mathbb{R}$, the cut defining a real number consists of all rational numbers less than the given real number.

The set of cuts gets a natural ordering, given by inclusion. In place of $\subseteq$, we write $\leq$. For any two cuts r and s, we have $r \leq s$ or $s \leq r$, and if both occur, then $r = s$. We can then define $<, \geq$, and $>$ in the expected way. The positive cuts r are those with $0^* < r$, and the negative cuts are those with $r < 0^*$.

Once cuts and their ordering are in place, one can go about defining the usual operations of arithmetic and proving that $\mathbb{R}$ with these operations satisfies the familiar associative, commutative, and distributive laws, and that these interact with inequalities in the usual ways. The definitions of addition and subtraction are easy: the sum or difference of two cuts is simply the set of sums or differences of the rationals from the respective cuts. For multiplication and reciprocals one has to take signs into account. For example the product of two positive cuts consists of all products of positive rationals from the two cuts, as well as 0 and all negative rationals. After these definitions and the proofs of the usual arithmetic operations are complete, it is customary to write 0 and 1 in place of 0^* and 1^*.

This much allows us to define n^{th} roots. The following proposition gives the precise details.

Proposition. If r is a positive real number and n is a positive integer, then there exists a unique positive real number s such that $s^n = r$.

REMARK. In the terminology and notation introduced in Section I.3, the polynomial $X^n - r$ in $\mathbb{R}[X]$ has a unique positive root if r is positive in $\mathbb{R}$.

SKETCH OF PROOF. Let s consist of all positive rationals q such that $q^n < r$, together with all rationals ≤ 0. One checks that s is a cut and that $s^n = r$. This proves existence. For uniqueness any positive cut s' with $(s')^n = r$ must contain exactly the same rationals and hence must equal s. $\square$

To make efficient use of cuts in connection with arithmetic and algebra, one needs to develop a certain amount of real-variable theory. This theory will not be developed in any detail here; let us be content with a sketch, giving a proof of the one specific result that we shall need.[4]

The first step in the process is to observe that any nonempty subset of reals with an upper bound has a least upper bound (the **supremum**, written as sup). This is proved by taking the union of the cuts for each of the given real numbers and showing that the result is a cut. Similarly any nonempty subset of reals with a lower bound has a greatest lower bound (the **infimum**, written as inf). This property follows by applying the least-upper-bound property to the negatives of the given reals and then taking the negative of the resulting least upper bound.

Meanwhile, we can introduce sequences of real numbers and convergence of sequences in the usual way. In terms of convergence, the key property of sequences of real numbers is given by the Bolzano–Weierstrass Theorem: any bounded sequence has a convergent subsequence. In fact, if the given bounded sequence is $\{s_n\}$, it can be shown that there is a subsequence convergent to the greatest lower bound over m of the least upper bound for $k \geq m$ of the numbers s_k.

Next one introduces continuity of functions in the usual way. The Bolzano–Weierstrass Theorem may readily be used to prove that any continuous real-valued function on a closed bounded interval takes on its maximum and minimum values. With a little more effort the Bolzano–Weierstrass Theorem may be used also to show that any continuous real-valued function on a closed bounded interval is uniformly continuous. That brings us to the theorem that we shall use in developing basic algebra.

Theorem (Intermediate Value Theorem). Let $a < b$ be real numbers, and let $f : [a, b] \to \mathbb{R}$ be continuous. Then f, in the interval $[a, b]$, takes on all values between $f(a)$ and $f(b)$.

PROOF. Let $f(a) = \alpha$ and $f(b) = \beta$, and let γ be between α and β. We may assume that γ is in fact strictly between α and β. Possibly by replacing f by $-f$, we may assume that also $\alpha < \beta$. Let

$$A = \{x \in [a, b] \mid f(x) \leq \gamma\} \quad \text{and} \quad B = \{x \in [a, b] \mid f(x) \geq \gamma\}.$$

These sets are nonempty since a is in A and b is in B, and f is bounded since any continuous function on a closed bounded interval takes on finite maximum and minimum values. Thus the numbers $\gamma_1 = \sup\{f(x) \mid x \in A\}$ and $\gamma_2 = \inf\{f(x) \mid x \in B\}$ are well defined and have $\gamma_1 \leq \gamma \leq \gamma_2$.

If $\gamma_1 = \gamma$, then we can find a sequence $\{x_n\}$ in A such that $f(x_n)$ converges to γ. Using the Bolzano–Weierstrass Theorem, we can find a convergent subsequence

---

[4]Details of the omitted steps may be found, for example, in Section I.1 of the author's book *Basic Real Analysis*.

$\{x_{n_k}\}$ of $\{x_n\}$, say with limit x_0. By continuity of f, $\{f(x_{n_k})\}$ converges to $f(x_0)$. Then $f(x_0) = \gamma_1 = \gamma$, and we are done. Arguing by contradiction, we may therefore assume that $\gamma_1 < \gamma$. Similarly we may assume that $\gamma < \gamma_2$, but we do not need to do so.

Let $\epsilon = \gamma_2 - \gamma_1$, and choose, since the continuous function f is necessarily uniformly continuous, $\delta > 0$ such that $|x_1 - x_2| < \delta$ implies $|f(x_1) - f(x_2)| < \epsilon$ whenever x_1 and x_2 both lie in $[a, b]$. Then choose an integer n such that $2^{-n}(b-a) < \delta$, and consider the value of f at the points $p_k = a + k2^{-n}(b-a)$ for $0 \le k \le 2^n$. Since $p_{k+1} - p_k = 2^{-n}(b-a) < \delta$, we have $|f(p_{k+1}) - f(p_k)| < \epsilon = \gamma_2 - \gamma_1$. Consequently if $f(p_k) \le \gamma_1$, then

$$f(p_{k+1}) \le f(p_k) + |f(p_{k+1}) - f(p_k)| < \gamma_1 + (\gamma_2 - \gamma_1) = \gamma_2,$$

and hence $f(p_{k+1}) \le \gamma_1$. Now $f(p_0) = f(a) = \alpha \le \gamma_1$. Thus induction shows that $f(p_k) \le \gamma_1$ for all $k \le 2^n$. However, for $k = 2^n$, we have $p_{2^n} = b$. Hence $f(b) = \beta \ge \gamma > \gamma_1$, and we have arrived at a contradiction. □

A4. Complex Numbers

Complex numbers are taken as known, and this section reviews their notation and basic properties.

Briefly, the system $\mathbb{C}$ of complex numbers is a two-dimensional vector space over $\mathbb{R}$ with a distinguished basis $\{1, i\}$ and a multiplication defined initially by $11 = 1$, $1i = i1 = i$, and $ii = -1$. Elements may then be written as $a + bi$ or $a + ib$ with a and b in $\mathbb{R}$; here a is an abbreviation for $a1$. The multiplication is extended to all of $\mathbb{C}$ so that the distributive laws hold, i.e., so that $(a+bi)(c+di)$ can be expanded in the expected way. The multiplication is associative and commutative, the element 1 acts as a multiplicative identity, and every nonzero element has a multiplicative inverse: $(a + bi)\left(\frac{a}{a^2+b^2} - i\frac{b}{a^2+b^2}\right) = 1$.

Complex conjugation is indicated by a bar: the conjugate of $a + bi$ is $a - bi$ if a and b are real, and we write $\overline{a + bi} = a - bi$. Then we have $\overline{z + w} = \bar{z} + \bar{w}$, $\overline{rz} = r\bar{z}$ if r is real, and $\overline{zw} = \bar{z}\bar{w}$.

The **real** and **imaginary parts** of $z = a + bi$ are $\operatorname{Re} z = a$ and $\operatorname{Im} z = b$. These may be computed as $\operatorname{Re} z = \frac{1}{2}(z + \bar{z})$ and $\operatorname{Im} z = -\frac{i}{2}(z - \bar{z})$.

The **absolute value** function of $z = a + bi$ is given by $|z| = \sqrt{a^2 + b^2}$, and this satisfies $|z|^2 = z\bar{z}$. It has the simple properties that $|\bar{z}| = |z|$, $|\operatorname{Re} z| \le |z|$, and $|\operatorname{Im} z| \le |z|$. In addition, it satisfies

$$|zw| = |z||w|$$

because
$$|zw|^2 = zw\overline{zw} = zw\bar{z}\bar{w} = z\bar{z}w\bar{w} = |z|^2|w|^2,$$

and it satisfies the **triangle inequality**

$$|z+w| \leq |z|+|w|$$

because
$$|z+w|^2 = (z+w)\overline{(z+w)} = z\bar{z} + z\bar{w} + w\bar{z} + w\bar{w}$$
$$= |z|^2 + 2\operatorname{Re}(z\bar{w}) + |w|^2 \leq |z|^2 + 2|z\bar{w}| + |w|^2$$
$$= |z|^2 + 2|z||w| + |w|^2 = (|z|+|w|)^2.$$

A5. Partial Orderings and Zorn's Lemma

A **partial ordering** on a set S is a relation between S and itself, i.e., a subset of $S \times S$, satisfying two properties. We define the expression $a \leq b$ to mean that the ordered pair (a, b) is a member of the relation, and we say that "$\leq$" is the partial ordering. The properties are

(i) $a \leq a$ for all a in S, i.e., $\leq$ is **reflexive**,
(ii) $a \leq b$ and $b \leq c$ together imply $a \leq c$ whenever a, b, and c are in S, i.e., $\leq$ is **transitive**.

An example of such an S is any set of subsets of a set X, with $\leq$ taken to be inclusion $\subseteq$. This particular partial ordering has a third property of interest, namely

(iii) $a \leq b$ and $b \leq a$ with a and b in S imply $a = b$.

However, the validity of (iii) has no bearing on Zorn's Lemma below. A partial ordering is said to be a **total ordering** or **simple ordering** if (iii) holds and also

(iv) any a and b in S have $a \leq b$ or $b \leq a$ or both.

For the sake of a result to be proved at the end of the section, let us interpolate one further definition: a totally ordered set is said to be **well ordered** if every nonempty subset has a least element, i.e., if each nonempty subset contains an element a such that $a \leq b$ for all b in the subset.

A **chain** in a partially ordered set S is a totally ordered subset. An **upper bound** for a chain T is an element u in S such that $c \leq u$ for all c in T. A **maximal element** in S is an element m such that whenever $m \leq a$ for some a in S, then $a \leq m$. (If (iii) holds, we can conclude in this case that $m = a$.)

Zorn's Lemma. If S is a nonempty partially ordered set in which every chain has an upper bound, then S has a maximal element.

REMARKS. Zorn's Lemma will be proved below using the Axiom of Choice, which was stated in Section A1. It is an easy exercise to see, conversely, that Zorn's Lemma implies the Axiom of Choice. It is customary with many mathematical writers to mention Zorn's Lemma each time it is invoked, even though most writers nowadays do not ordinarily acknowledge uses of the Axiom of Choice. Before coming to the proof, we give an example of how Zorn's Lemma is used. This example uses vector spaces and is expanded upon in Section II.9.

EXAMPLE. Zorn's Lemma gives a quick proof that any real vector space V has a basis. In fact, let S be the set of all linearly independent subsets of V, and order S by inclusion upward as in the example above of a partial ordering. The set S is nonempty because $\varnothing$ is a linearly independent subset of V. Let T be a chain in S, and let u be the union of the members of T. If t is in T, we certainly have $t \subseteq u$. Let us see that u is linearly independent. For u to be dependent would mean that there are vectors $x_1, \ldots, x_n$ in u with $r_1 x_1 + \cdots + r_n x_n = 0$ for some system of real numbers not all 0. Let x_j be in the member t_j of the chain T. Since $t_1 \subseteq t_2$ or $t_2 \subseteq t_1$, x_1 and x_2 are both in t_1 or both in t_2. To keep the notation neutral, say they are both in t_2'. Since $t_2' \subseteq t_3$ or $t_3 \subseteq t_2'$, all of x_1, x_2, x_3 are in t_2' or they are all in t_3. Say they are both in t_3'. Continuing in this way, we arrive at one of the sets $t_1, \ldots, t_n$, say t_n', such that all of $x_1, \ldots, x_n$ are all in t_n'. The members of t_n' are linearly independent by assumption, and we obtain the contradiction $r_1 = \cdots = r_n = 0$. We conclude that the chain T has an upper bound in S. By Zorn's Lemma, S has a maximal element, say m. If m is not a basis, it fails to span. If a vector x is not in its span, it is routine to see that $m \cup \{x\}$ is linearly independent and properly contains m, in contradiction to the maximality of m. We conclude that m is a basis.

We now begin the proof of Zorn's Lemma. If T is a chain in a partially ordered set S, then an upper bound u_0 for T is a **least upper bound** for T if $u_0 \leq u$ for all upper bounds of T. If (iii) holds in S, then there can be at most one least upper bound for T. In fact, if u_0 and u_0' are least upper bounds, then $u_0 \leq u_0'$ since u_0 is a least upper bound, and $u_0' \leq u_0$ since u_0' is a least upper bound; by (iii), $u_0 = u_0'$. The proof follows that in Dunford–Schwartz's *Linear Operators I*.

Lemma. Let X be a nonempty partially ordered set such that (iii) holds, and write $\leq$ for the partial ordering. Suppose that X has the additional property that each nonempty chain in X has a least upper bound in X. If $f : X \to X$ is a function such that $x \leq f(x)$ for all x in X, then there exists an x_0 in X with $f(x_0) = x_0$.

PROOF. A nonempty subset E of X will be called *admissible* for purposes of this proof if $f(E) \subseteq E$ and if the least upper bound of each nonempty chain in E, which exists in X by assumption, actually lies in E. By assumption, X is an admissible subset of X. If x is in X, then the intersection of admissible subsets of X containing x is admissible. Thus the intersection A_x of all admissible subsets containing x is an admissible subset containing x. The set of all y in X with $x \leq y$ is an admissible subset of X containing x, and it follows that $x \leq y$ for all y in A_x.

By hypothesis, X is nonempty. Fix an element a in X, and let $A = A_a$. The main step is to prove that A is a chain. Once that is established, we argue as follows: Since A is a chain, its least upper bound x_0 lies in X, and since A is an admissible subset, x_0 lies in A. By admissibility, $f(A) \subseteq A$. Hence $f(x_0)$ is in A. Since x_0 is an upper bound of A, $f(x_0) \leq x_0$. On the other hand, $x_0 \leq f(x_0)$ by the assumed property of f. Therefore $f(x_0) = x_0$ by (iii).

To prove that A is a chain, consider the subset C of members x of A with the property that there is a nonempty chain C_x in A containing a and x such that

- $a \leq y \leq x$ for all y in C_x,
- $f(C_x - \{x\}) \subseteq C_x$, and
- the least upper bound of any nonempty subchain of C_x is in C_x.

The element a is in C because we can take $C_a = \{a\}$. If x is in C, so that C_x exists, let us use the bulleted properties to see that

$$A = A_x \cup C_x. \qquad (*)$$

We have $A \supseteq C_x$ by definition; also $A \cap A_x$ is an admissible set containing x and hence containing A_x, and thus $A \supseteq A_x$. Therefore $A \supseteq A_x \cup C_x$. For the reverse inclusion it is enough to prove that $A_x \cup C_x$ is an admissible subset of X containing a. The element a is in C_x and thus is in $A_x \cup C_x$. For the admissibility we have to show that $f(A_x \cup C_x) \subseteq A_x \cup C_x$ and that the least upper bound of any nonempty chain in $A_x \cup C_x$ lies in $A_x \cup C_x$. Since x lies in A_x, $A_x \cup C_x = A_x \cup (C_x - \{x\})$ and $f(A_x \cup C_x) = f(A_x) \cup f(C_x - \{x\}) \subseteq A_x \cup C_x$, the inclusion following from the admissibility of A_x and the second bulleted property of C_x.

To complete the proof of $(*)$, take a nonempty chain in $A_x \cup C_x$, and let u be its least upper bound in X; it is enough to show that u is in $A_x \cup C_x$. The element u is necessarily in A since A is admissible. Observe that

$$y \leq x \quad \text{and} \quad x \leq z \qquad \text{whenever } y \text{ is in } C_x \text{ and } z \text{ is in } A_x. \qquad (**)$$

If the chain has at least one member in A_x, then $(**)$ implies that $x \leq u$, and hence the set of members of the chain that lie in A_x forms a nonempty chain in A_x with least upper bound u. Since A_x is admissible, u is in A_x. Otherwise the chain has all its members in C_x, and then u is in C_x by the third bulleted property of C_x.

This completes the proof of $(*)$. Although we do not need the fact, let us observe that combining $(**)$ and $(*)$ yields $A_x \cap C_x = \{x\}$ whenever C_x exists. Thus $C_x = (A - A_x) \cup \{x\}$ if C_x exists. In particular the defining properties of C_x determine C_x completely.

Recalling that C is the subset of members of A such that C_x exists, we shall show that C is an admissible set containing a. If we can do so, then it follows that $C \supseteq A_a = A$ and hence $C = A$. This fact, in combination with $(*)$ and $(**)$, proves that A is a chain: if x and y are in A, $(*)$ shows that y is in A_x or C_x, and $(**)$ shows that $x \leq y$ in the first case and $y \leq x$ in the second case.

Thus the proof will be complete if we show that C is admissible and contains a. We already observed that it contains a. We need to see that $f(C) \subseteq C$ and that the least upper bound of any nonempty chain of C lies in C. For the first of these conclusions, let us see that if x is in C, then $C_{f(x)}$ exists and can be taken to be $C_x \cup \{f(x)\}$. Property $(*)$ proves that $C_x \cup \{f(x)\}$ satisfies the first bulleted property of $C_{f(x)}$, and the second bulleted property follows from that same property of C_x and the fact that $x \leq f(x)$. Any nonempty chain in $C_x \cup \{f(x)\}$ either lies in C_x and has its least upper bound in C_x or else contains $f(x)$ as a member and then has $f(x)$ as its least upper bound. Thus $C_x \cup \{f(x)\}$ satisfies the third bulleted property of $C_{f(x)}$ and can be taken to be $C_{f(x)}$. Hence $f(x)$ is in C, and $f(C) \subseteq C$.

Finally take any nonempty chain $\{x_\alpha\}$ in C, and let u be its least upper bound, which is necessarily in A. Form the set $\left(\bigcup_\alpha C_{x_\alpha}\right) \cup \{u\}$. It is immediate that this set satisfies the three bulleted properties of C_u and therefore can be taken to be C_u. Hence u is in C, and C contains the least upper bound of any nonempty chain in C. Then C is indeed an admissible set containing a. □

PROOF OF ZORN'S LEMMA. Let S be a partially ordered set, with partial ordering $\leq$, in which every chain has an upper bound. Let X be the partially ordered system, ordered by inclusion upward $\subseteq$, of nonempty chains[5] in S. The partially ordered system X, being given by ordinary inclusion, satisfies property (iii). A nonempty chain C in X is a nested system of chains c_α of S, and $\bigcup_\alpha c_\alpha$ is a chain in S that is a least upper bound for C. The lemma is therefore applicable to any function $f : X \to X$ such that $c \subseteq f(c)$ for all c in X. We use the lemma to produce a maximal chain in X.

Arguing by contradiction, suppose that no chain within S is maximal under inclusion. For each nonempty chain c within S, let $f(c)$ be a chain with $c \subseteq f(c)$ and $c \neq f(c)$. (This choice of $f(c)$ for each c is where we use the Axiom of Choice.) The result is a function $f : X \to X$ of the required kind, the lemma says that $f(c) = c$ for some c in X, and we arrive at a contradiction. We conclude that there is some maximal chain c_0 within S.

By assumption in Zorn's Lemma, every nonempty chain within S has an upper bound. Let u_0 be an upper bound for the maximal chain c_0. If u is a member of S

---

[5]Here a chain is simply a certain kind of subset of S, and no element of S can occur more than once in it even if (iii) fails for the partial ordering. Thus if $S = \{x, y\}$ with $x \leq y$ and $y \leq x$, then $\{x, y\}$ is in X and in fact is maximal in X.

with $u_0 \leq u$, then $c_0 \cup \{u\}$ is a chain and maximality implies that $c_0 \cup \{u\} = c_0$. Therefore u is in c_0, and $u \leq u_0$. This is the condition that u_0 is a maximal element of S. □

Corollary (Zermelo's Well-Ordering Theorem). Every set has a well ordering.

PROOF. Let S be a set, and let $\mathcal{E}$ be the family of all pairs $(E, \leq_E)$ such that E is a subset of S and $\leq_E$ is a well ordering of E. The family $\mathcal{E}$ is nonempty since $(\emptyset, \emptyset)$ is a member of it. We partially order $\mathcal{E}$ by a notion of "inclusion as an initial segment," saying that $(E, \leq_E) \leq (F, \leq_F)$ if

(i) $E \subseteq F$,
(ii) a and b in E with $a \leq_E b$ implies $a \leq_F b$,
(iii) a in E and b in F but not E together imply $a \leq_F b$.

In preparation for applying Zorn's Lemma, let $\mathcal{C} = \{(E_\alpha, \leq_\alpha)\}$ be a chain in $\mathcal{E}$, with the α's running through some set I. Define $E_0 = \bigcup_\alpha E_\alpha$ and define $\leq_0$ as follows: If e_1 and e_2 are in E_0, let e_1 be in E_{α_1} with α_1 in I, and let e_2 be in E_{α_2} with α_2 in I. Since $\mathcal{C}$ is a chain, we may assume without loss of generality that $(E_{\alpha_1}, \leq_{\alpha_1}) \leq (E_{\alpha_2}, \leq_{\alpha_2})$, so that $E_{\alpha_1} \subseteq E_{\alpha_2}$ in particular. Then e_1 and e_2 are both in E_{α_2} and we define $e_1 \leq_0 e_2$ if $e_1 \leq_{\alpha_2} e_2$, and $e_2 \leq_0 e_1$ if $e_2 \leq_{\alpha_2} e_1$. Because of (i) and (ii) above, the result is well defined independently of the choice of α_1 and α_2. Similar reasoning shows that $\leq_0$ is a total ordering of E_0. If we can prove that $\leq_0$ is a well ordering, then $(E_0, \leq_0)$ is evidently an upper bound in $\mathcal{E}$ for the chain $\mathcal{C}$, and Zorn's Lemma is applicable.

Now suppose that F is a nonempty subset of E_0. Pick an element of F, and let E_{α_0} be a set in the chain that contains it. Since $(E_{\alpha_0}, \leq_{\alpha_0})$ is well ordered and $F \cap E_{\alpha_0}$ is nonempty, $F \cap E_{\alpha_0}$ contains a least element f_0 relative to $\leq_{\alpha_0}$. We show that $f_0 \leq_0 f$ for all f in F. In fact, if f is given, there are two possibilities. One is that f is in E_{α_0}; in this case, the consistency of $\leq_0$ with $\leq_{\alpha_0}$ forces $f_0 \leq_0 f$. The other is that f is not in E_{α_0} but is in some E_{α_1}. Since $\mathcal{C}$ is a chain and $E_{\alpha_1} \subseteq E_{\alpha_0}$ fails, we must have $(E_{\alpha_0}, \leq_{\alpha_0}) \leq (E_{\alpha_1}, \leq_{\alpha_1})$. Then f is in E_{α_1} but not E_{α_0}, and property (iii) above says that $f_0 \leq_{\alpha_1} f$. By the consistency of the orderings, $f_0 \leq_0 f$. Hence f_0 is a least element in F, and E_0 is well ordered.

Application of Zorn's Lemma produces a maximal element $(E, \leq_E)$ of $\mathcal{E}$. If E were a proper subset of S, we could adjoin to E a member s of S not in E and define every element e of E to be $\leq s$. The result would contradict maximality. Therefore $E = S$, and S has been well ordered. □

A6. Cardinality

Two sets A and B are said to have the same **cardinality**, written card $A =$ card B, if there exists a one-one function from A onto B. On any set $\mathcal{A}$ of sets, "having the same cardinality" is plainly an equivalence relation and therefore partitions $\mathcal{A}$ into disjoint equivalence classes, the sets in each class having the same cardinality. The question of what constitutes cardinality (or a "cardinal number") in its own right is one that is addressed in set theory but that we do not need to address carefully here; the idea is that each equivalence class under "having the same cardinality" has a distinguished representative, and the **cardinal number** is defined to be that representative. We write card A for the cardinal number of a set A.

Having addressed equality, we now introduce a partial ordering, saying that card $A \leq$ card B if there is a one-one function from A into B. The first result below is that card $A \leq$ card B and card $B \leq$ card A together imply card $A =$ card B.

Proposition (Schroeder–Bernstein Theorem). If A and B are sets such that there exist one-one functions $f : A \to B$ and $g : B \to A$, then A and B have the same cardinality.

PROOF. Define the function g^{-1} : image $g \to A$ by $g^{-1}(g(a)) = a$; this definition makes sense since g is one-one. Write $(g \circ f)^{(n)}$ for the composition of $g \circ f$ with itself n times, and define $(f \circ g)^{(n)}$ similarly. Define subsets A_n and A'_n of A and subsets B_n and B'_n for $n \geq 0$ by

$$A_n = \text{image}((g \circ f)^{(n)}) - \text{image}((g \circ f)^{(n)} \circ g),$$
$$A'_n = \text{image}((g \circ f)^{(n)} \circ g) - \text{image}((g \circ f)^{(n+1)}),$$
$$B_n = \text{image}((f \circ g)^{(n)}) - \text{image}((f \circ g)^{(n)} \circ f),$$
$$B'_n = \text{image}((f \circ g)^{(n)} \circ f) - \text{image}((f \circ g)^{(n+1)}),$$

and let

$$A_\infty = \bigcap_{n=0}^{\infty} \text{image}((g \circ f)^{(n)}) \quad \text{and} \quad B_\infty = \bigcap_{n=0}^{\infty} \text{image}((f \circ g)^{(n)}).$$

Then we have

$$A = A_\infty \cup \bigcup_{n=0}^{\infty} A_n \cup \bigcup_{n=0}^{\infty} A'_n \quad \text{and} \quad B = B_\infty \cup \bigcup_{n=0}^{\infty} B_n \cup \bigcup_{n=0}^{\infty} B'_n,$$

with both unions disjoint.

Let us prove that f carries A_n one-one onto B'_n. If a is in A_n, then $a = (g \circ f)^{(n)}(x)$ for some $x \in A$ and a is not of the form $(g \circ f)^{(n)}(g(y))$ with $y \in B$. Applying f, we obtain $f(a) = (f \circ ((g \circ f)^{(n)})(x) = (f \circ g)^{(n)}(f(x))$, so that $f(a)$ is in the image of $((f \circ g)^{(n)} \circ f)$. Meanwhile, if $f(a)$ is in the image of $(f \circ g)^{(n+1)}$, then $f(a) = (f \circ g)^{(n+1)}(y) = f((g \circ f)^{(n)}(g(y)))$ for some $y \in B$. Since f is one-one, we can cancel the f on the outside and obtain $a = (g \circ f)^{(n)}(g(y))$, in contradiction to the fact that a is in A_n. Thus f carries A_n into B'_n, and it is certainly one-one. To see that $f(A_n)$ contains all of B'_n, let

$b \in B'_n$ be given. Then $b = (f \circ g)^{(n)}(f(x))$ for some $x \in A$ and b is not of the form $(f \circ g)^{(n+1)}(y)$ with $y \in B$. Hence $b = f((g \circ f)^{(n)}(x))$, i.e., $b = f(a)$ with $a = (g \circ f)^{(n)}(x)$. If this element a were in the image of $(g \circ f)^{(n)} \circ g$, we could write $a = (g \circ f)^{(n)}(g(y))$ for some $y \in B$, and then we would have $b = f(a) = f((g \circ f)^{(n)}(g(y))) = (f \circ g)^{(n+1)}(y)$, contradiction. Thus a is in A_n, and f carries A_n one-one onto B'_n.

Similarly g carries B_n one-one onto A'_n. Since A'_n is in the image of g, we can apply g^{-1} to it and see that g^{-1} carries A'_n one-one onto B_n.

The same kind of reasoning as above shows that f carries A_∞ one-one onto B_∞. In summary, f carries each A_n one-one onto B'_n and carries A_∞ one-one onto B_∞, while g^{-1} carries each A'_n one-one onto B_n. Then the function

$$h = \begin{cases} f & \text{on } A_\infty \text{ and each } A_n, \\ g^{-1} & \text{on each } A'_n, \end{cases}$$

carries A one-one onto B. □

Next we show that any two sets A and B have comparable cardinalities in the sense that either card $A \leq$ card B or card $B \leq$ card A.

Proposition. If A and B are two sets, then either there is a one-one function from A into B or there is a one-one function from B into A.

PROOF. Consider the set S of all one-one functions $f : E \to B$ with $E \subseteq A$, the empty function with $E = \emptyset$ being one such. Each such function is a certain subset of $A \times B$. If we order S by inclusion upward, then the union of the members of any chain is an upper bound for the chain. By Zorn's Lemma let $G : E_0 \to B$ be a maximal one-one function of this kind, and let F_0 be the image of G. If $E_0 = A$, then G is a one-one function from A into B. If $F_0 = B$, then G^{-1} is a one-one function from B into A. If neither of these things happens, then there exist $x_0 \in A - E_0$ and y_0 in $B - F_0$, and the function $\widetilde{G}$ equal to G on E_0 and having $\widetilde{G}(x_0) = y_0$ extends G and is still one-one; thus it contradicts the maximality of G. □

Corollary. If E is an infinite set, then E has a countably infinite subset.

PROOF. The proposition shows that either there is a one-one function from the set of positive integers into E, in which case we are done, or there is a one-one function from E into the set of positive integers. In the latter case the image cannot be finite since E is assumed infinite. Then the image must be an infinite subset of the positive integers. This set can be enumerated and is therefore countably infinite. Thus E is countably infinite. □

Cantor's proof that there exist uncountable sets, done with a diagonal argument, in fact showed how to start from any set A and construct a set with strictly larger cardinality.

Proposition (Cantor). If A is a set and 2^A denotes the set of all subsets of A, then card 2^A is strictly larger than card A.

PROOF. The map $A \mapsto \{A\}$ is a one-one function from A into 2^A. If we are given a one-one function $F : A \to 2^A$, let E be the set of all x in A such that x is not in $F(x)$. If we define $E = F(x_0)$, then $x_0 \in E$ implies $x_0 \notin F(x_0) = E$, while $x_0 \notin E$ implies $x \in F(x_0) = E$. We have a contradiction in any case, and hence E cannot be of the form $F(x_0)$. We conclude that F cannot be onto 2^A. $\square$

Proposition. If E is an infinite set, then E is the disjoint union of sets that are each countably infinite.

PROOF. Let S be the set of all disjoint unions of countably infinite subsets of E. If $A = \bigcup_\alpha U_\alpha$ and $B = \bigcup_\beta V_\beta$ are members of S, say that $A \leq B$ if each A_α is some B_β. The result is a partial ordering on S. If $\mathcal{U}$ is a chain in S, then the collection C of all countably infinite sets that are U_α's in some member of $\mathcal{U}$ is a collection of countably infinite subsets of E that contains each member of $\mathcal{U}$. If U_α and U_β are distinct members of C, then U_α and U_β must both be in some member of $\mathcal{U}$ and hence must be disjoint. Thus C is an upper bound for $\mathcal{U}$. Also, the empty union is a member of S. By Zorn's Lemma, S has a maximal element M. Let F be the union of the members of M. If $E - F$ were to be infinite, then the corollary above would show that $E - F$ has a countably infinite subset Z, and $M \cup \{Z\}$ would contradict the maximality of M. Thus $E - F$ is finite. Since E is infinite, the corollary shows that E contains at least one countably infinite subset. Thus M has some member T. The set $T' = T \cup (E - F)$ is countably infinite, and $(M - \{T\}) \cup T'$ is the required decomposition of E as the disjoint union of countably infinite sets. $\square$

Corollary. Let S and E be nonempty sets with S infinite, and suppose that to each element s of S is associated a countable subset E_x of E in such a way that $E = \bigcup_{s \in S} E_s$. Then card $E \leq$ card S.

PROOF. The proposition allows us to write S as the disjoint union of countably infinite sets. If U is one of these sets, then $E_U = \bigcup_{s \in U} E_s$ is countable, being the countable union of countable sets. Therefore there exists a function from U onto E_U. The union of these functions, as U varies, yields a function from S onto $\bigcup E_U = E$, and consequently card $E \leq$ card S. $\square$

Addition is well defined for cardinals: the **sum of two cardinal numbers** is defined to be the cardinality of the disjoint union of the two sets in question. If at least one of the two cardinals is infinite, the sum equals the larger of the two, as an immediate consequence of the above corollary.

HINTS FOR SOLUTIONS OF PROBLEMS

Chapter I

1. 582.

2. The Euclidean algorithm gives $11 = 1 \cdot 7 + 4$, $7 = 1 \cdot 4 + 3$, $4 = 1 \cdot 3 + 1$, $3 = 3 \cdot 1 + 0$. So the GCD is 1. Reversing the steps gives $1 = 4 - 1 \cdot 3 = (11 - 1 \cdot 7) - 1 \cdot (7 - 1 \cdot 4) = (11 - 1 \cdot 7) - 1 \cdot (7 - 1 \cdot (11 - 1 \cdot 7)) = 2 \cdot 11 - 3 \cdot 7$. So $(x, y) = (2, -3)$ is a solution in (a). For (b), the difference of any two solutions solves $11x + 7y = 0$, and the solutions of this are of the form $(x, y) = n(7, -11)$.

3. Let $d_n = \text{GCD}(a_1, \ldots, a_n)$. The sequence d_n is a monotone decreasing sequence of positive integers, and it must eventually be constant. This eventual constant value is d, and thus $d_n = d$ for suitably large n.

4. These n's divide $x + y - 2$ and the sum of $2x - 3y - 3$ and -2 times the $x + y - 2$, hence $x + y - 2$ and $-5y + 1$. A necessary and sufficient condition for $-5y + 1 = na$ to be solvable for the pair (a, y) is that $\text{GCD}(5, n) = 1$ by Proposition 1.2c. Let us see that the answer to the problem is $\text{GCD}(5, n) = 1$.

The n's we seek must further divide $5(x+y-2) = 5x+5y-10$ and $-5y+1$, hence also the sum $5x-9$, as well as $-5y+1$. If $\text{GCD}(5, n) = 1$, then $5x-9 = nb$ is solvable for (b, x). With our solutions (a, y) and (b, x), we have $5x + 5y - 10 = n(b - a)$. Since 5 divides the left side and $\text{GCD}(5, n) = 1$, 5 divides $b - a$. Write $b - a = 5c$. Then $x + y - 2 = nc$ and $-5y + 1 = na$, and we obtain $2x - 3y - 3 = n(2c + a)$.

5. $Q(x) = (X - 1)P(X) + (X^3 + X^2 + X + 1)$, $P(X) = X(X^3 + X^2 + X + 1) + (X^2+1)$, $X^3+X^2+X+1 = (X+1)(X^2+1)+0$. Hence the GCD is $D(X) = X^2+1$. For (b), we retrace the steps, letting $R(X) = X^3 + X^2 + X + 1$. We have $D(X) = P(X) - XR(X) = P(X) - X(Q(X)-(X-1)P(X)) = (X^2-X+1)P(X) - XQ(X)$. Thus $A(X) = X^2 - X + 1$ and $B(X) = -X$.

6. The computation via the Euclidean algorithm, done within $\mathbb{C}[X]$, retains real numbers as coefficients throughout. By Proposition 1.15a one GCD has real coefficients. By Proposition 1.15c any GCD is a complex multiple of this polynomial with real coefficients.

7. In (a), we may assume, without loss of generality, that P has leading coefficient 1, so that $P(X) = X^n + a_{n-1}X^{n-1} + \cdots + a_0 = \prod_j (X - z_j)^{m_j}$. Define $Q(X) = \prod_j (X - \bar{z}_j)^{m_j}$. Then $Q(\bar{z}) = \prod_j (\bar{z} - \bar{z}_j)^{m_j} = \overline{\prod_j (z - z_j)} = \overline{P(z)}$. Replacing $\bar{z}$ by z gives $Q(z) = \overline{P(\bar{z})} = z^n + \overline{a_{n-1}}z^{n-1} + \cdots + \overline{a_0} = z^n + \overline{a_{n-1}}z^{n-1} + \cdots + \overline{a_0}$.

603

Since P has real coefficients, $Q(z) = P(z)$ for all z. Then $Q - P$ has every z as a root and in particular has more than n roots. Hence it must be the 0 polynomial. So $\prod_j (X - \bar{z}_j)^{m_j} = \prod_j (X - z_j)^{m_j}$, and the result follows from unique factorization (Theorem 1.17).

In (b), the result of (a) shows that we may factor any real polynomial in $\mathbb{C}[X]$ with leading coefficient 1 in the form $\prod_{x_j \text{ real}} (X - x_j)^{m_j} \prod_{z_j \text{ nonreal}} ((X - z_j)^{n_j} (X - \bar{z}_j)^{n_j})$. The right side equals $\prod_{x_j \text{ real}} (X - x_j)^{m_j} \prod_{z_j \text{ nonreal}} (X^2 - (z_j + \bar{z}_j)X + z_j \bar{z}_j)^{n_j}$. Every factor on the right side is in $\mathbb{R}[X]$, and the only way that the polynomial can be prime in $\mathbb{R}[X]$ is if only one factor is present. Thus the polynomial has degree at most 2.

8. For (a), let deg $A = d$ and form the equation $A(p/q) = 0$. Multiply through by q^d in order to clear fractions. Every term in the equation except the leading term has q as a factor, and thus q divides the leading term p^d. Since GCD$(p, q) = 1$, no prime can divide q. Thus $q = \pm 1$, and $n = p/q$ is an integer. Forming the equation $A(n) = 0$, we see that n is a factor of each term except possibly the constant term a_0. Thus n divides a_0.

For (b), we apply (a) to both polynomials. The only possible rational roots of $X^2 - 2$ are ± 1 and ± 2, while the only possible rational roots of $X^3 + X^2 + 1$ are ± 1. Checking directly, we see that none of these possibilities is actually a root. By the Factor Theorem, neither $X^2 - 2$ nor $X^3 + X^2 + 1$ has a first-degree factor in $\mathbb{Q}[X]$. If a polynomial of degree ≤ 3 has a nontrivial factorization, then it has a first-degree factor. We conclude that $X^2 - 2$ and $X^3 + X^2 + 1$ are prime.

9. Computation gives GCD(8645, 10465) = 455. Therefore 8645/10465 equals 19/23 in lowest terms.

10. Apart from the identity, the cycle structures are those of (1 2) with 6 representatives, (1 2 3) with 8 representatives, (1 2 3 4) with 6 representatives, and (1 2)(3 4) with 3 representatives. This checks, since there are $4! = 24$ permutations in all.

11. Check that the function $\sigma \mapsto \sigma(1\ 2)$ is one-one from the set of permutations of sign $+1$ onto the set of permutations of sign -1.

12. (a) $x_3 \begin{pmatrix} 1 \\ -2 \\ 1 \end{pmatrix}$. (b) None. (c) $\begin{pmatrix} -11/3 \\ 10/3 \\ 0 \end{pmatrix} + x_3 \begin{pmatrix} 1 \\ -2 \\ 1 \end{pmatrix}$.

13. By the definition of "step," an interchange of two rows (type (i)) takes n steps, and a multiplication of a row by a nonzero scalar (type (ii)) takes n steps. Also, replacement of a row by the sum of it and a multiple of another row (type (iii)) takes $2n$ steps. We proceed through the row-reduction algorithm column by column. For each of the n columns, we do possibly one operation of type (i) and then possibly an operation of type (ii). This much requires $\leq 2n$ steps. Then we do at most $n - 1$ operations of type (iii), requiring $\leq 2n(n - 1)$ steps. Thus a single column is handled in $\leq 2n(n - 1) + 2n = n^2$ steps, and the entire row reduction requires $\leq 2n^3$ steps.

14. $A + B = \begin{pmatrix} -2 & 11 \\ 3 & 8 \end{pmatrix}$, and $AB = \begin{pmatrix} -11 & 25 \\ -21 & 47 \end{pmatrix}$.

15. We induct on n, the result being clear for $n = 1$. Taking into account the fact that B commutes with A, we have $(A + B)^n = (A + B)(A + B)^{n-1} =$

$(A + B) \sum_{k=0}^{n-1} \binom{n-1}{k} A^{n-1-k} B^k = \sum_{k=0}^{n-1} \binom{n-1}{k} A^{n-k} B^k + \sum_{k=0}^{n-1} \binom{n-1}{k} A^{n-1-k} B^{k+1} = \sum_{k=0}^{n-1} \binom{n-1}{k} A^{n-k} B^k + \sum_{k=1}^{n} \binom{n-1}{k-1} A^{n-k} B^k = A^n + \sum_{k=1}^{n-1} [\binom{n-1}{k} + \binom{n-1}{k-1}] A^{n-k} B^k + B^n.$
In turn, the right side equals $\sum_{k=0}^{n} \binom{n}{k} A^{n-k} B^k$ by the Pascal-triangle identity for binomial coefficients.

16. Write $\begin{pmatrix} 1 & 1 & 0 \\ 0 & 1 & 1 \\ 0 & 0 & 1 \end{pmatrix}$ as $I + B$, where $B = \begin{pmatrix} 0 & 1 & 0 \\ 0 & 0 & 1 \\ 0 & 0 & 0 \end{pmatrix}$, and apply Problem 15. Since $B^2 = \begin{pmatrix} 0 & 0 & 1 \\ 0 & 0 & 0 \\ 0 & 0 & 0 \end{pmatrix}$ and $B^3 = 0$, we obtain $(I + B)^n = I + nB + \frac{1}{2}n^2 B^2 = \begin{pmatrix} 1 & n & \frac{1}{2}n^2 \\ 0 & 1 & n \\ 0 & 0 & 1 \end{pmatrix}$.

17. $(AD)_{ij} = A_{ij} d_j$ and $(DA)_{ij} = d_i A_{ij}$. Thus $AD = DA$ if and only if $d_i = d_j$ for all (i, j) for which $A_{ij} \neq 0$.

18. $E_{kl} E_{pq} = \delta_{lp} E_{kq}$.

19. Check that $\begin{pmatrix} a & b \\ c & d \end{pmatrix}$ times the asserted inverse is the identity. Then the matrix actually is the inverse. Apply the inverse to $\begin{pmatrix} a & b \\ c & d \end{pmatrix} \begin{pmatrix} x \\ y \end{pmatrix} = \begin{pmatrix} p \\ q \end{pmatrix}$ to obtain the value for $\begin{pmatrix} x \\ y \end{pmatrix}$.

20. (a) No inverse. (b) $A^{-1} = \begin{pmatrix} -2/3 & -4/3 & 1 \\ -2/3 & 11/3 & -2 \\ 1 & -2 & 1 \end{pmatrix}$. (c) $A^{-1} = \begin{pmatrix} 1 & -1 & 0 \\ -2 & 3 & -1 \\ 2 & -5 & 4 \end{pmatrix}$.

21. No. If the algorithm is followed, then the row of 0's persists throughout the row reduction, at worst moving to a different row at various stages.

22. If $C = (AB)^{-1}$, then $ABC = I$ shows that BC is the inverse of A and $CAB = I$ shows that CA is the inverse of B.

23. $(I + A)(I - A + A^2 - A^3 + \cdots + (-A)^{k-1}) = I - (-A)^k = I$ shows that $I - A + A^2 - A^3 + \cdots + (-A)^{k-1}$ is an inverse.

24. Let S be the set of positive integers, and let $f(n) = n + 1$. Take $g(n)$ to be $n - 1$ for $n > 1$ and $g(1) = 1$. Then $g \circ f$ is the identity. But f is not onto S, and g is not one-one.

25. Take $A = (1\ 0)$ and $B = \begin{pmatrix} 1 \\ 0 \end{pmatrix}$. Then $BA = \begin{pmatrix} 1 & 0 \\ 0 & 0 \end{pmatrix}$. More generally, if $A = (a\ b)$ and $B = \begin{pmatrix} c \\ d \end{pmatrix}$, then $BA = \begin{pmatrix} ca & cb \\ da & db \end{pmatrix}$. If the upper right entry is 0, then $c = 0$ or $b = 0$. But then one of the two diagonal entries must be 0, and hence BA cannot be the identity.

26. The set of common multiples is a nonempty set of positive integers because ab is in it. Therefore it has a least element.

27. This is a restatement of Corollary 1.7.

28. Let a and b have prime factorizations $a = p_1^{k_1} \cdots p_r^{k_r}$ and $b = p_1^{l_1} \cdots p_r^{l_r}$. Problem 27 shows that any positive common multiple N of a and b is of the form $p_1^{m_1} \cdots p_r^{m_r} q_1^{n_1} \cdots q_s^{n_s}$ with $m_j \geq k_j, m_j \geq l_j$, and $n_j \geq 0$, and certainly any positive integer of this form is a common multiple. The inequalities for m_j are equivalent with the condition $m_j \geq \max(k_j, l_j)$. The smallest positive integer of this kind has

$m_j = \max(k_j, l_j)$ and $n_j = 0$. This proves (a). In combination with the form of N, the formula for LCM(a, b) proves (b). Conclusion (c) follows from Corollary 1.8 and the identity $k_j + l_j = \min(k_j, l_j) + \max(k_j, l_j)$.

29. If $a_j = p_1^{k_{1,j}} \cdots p_r^{k_{r,j}}$ is a prime factorization of a_j, then LCM$(a_1, \ldots, a_t) = p_1^{\max_{1 \le j \le r}\{k_{1,j}\}} \cdots p_r^{\max_{1 \le j \le r}\{k_{r,j}\}}$, just as with Corollary 1.11.

Chapter II

1. The methods at the end of Section 2 lead to the basis $\{(\frac{2}{3}, 1, 0), (-\frac{5}{3}, 0, 1)\}$ for (a) and to the basis $\{(1, -\frac{1}{2}, 2)\}$ for (b).

2. For $0 \le k < n$, the two recursive formulas and one application of associativity give $v_{(k+1)} + v^{(k+2)} = (v_{(k)} + v_{k+1}) + v^{(k+2)} = v_{(k)} + (v_{k+1} + v^{(k+2)}) = v_{(k)} + v^{(k+1)}$, and (a) follows.

For (b), we proceed by induction on n, the cases $n \le 3$ being handled by associativity. Suppose that the result holds for sums of fewer than n vectors, with $n \ge 4$. In a sum of n vectors, there is some outer plus sign, and the inductive hypothesis means that the sum is of the form $(v_1 + \cdots + v_k) + (v_{k+1} + \cdots + v_n)$, the expressions $v_1 + \cdots + v_k$ and $v_{k+1} + \cdots + v_n$ being unambiguous. The inductive hypothesis means that we have $v_1 + \cdots + v_k = v_{(k)}$ and $v_{k+1} + \cdots + v_n = v^{(k+1)}$, and hence the expression we are studying is of the form $v_{(k)} + v^{(k+1)}$. Part (a) shows that this is independent of k, and hence (b) follows.

3. From Section I.4, σ is a product of transpositions, and hence it is enough to prove the result for a transposition. When $r + 1 < s$, iteration of the identity $(r \ s) = (r \ r+1)(r+1 \ s)(r \ r+1)$ shows that any transposition is a product of transpositions of the form $(r \ r+1)$, and hence it is enough to prove the formula for $\sigma = (r \ r+1)$. This case is just the commutative law, and the result follows.

4. (a) $\{(1 \ 2 \ -1), (0 \ 0 \ 1)\}$; (b) $\left\{\begin{pmatrix}1\\2\\0\end{pmatrix}, \begin{pmatrix}0\\1\\-1\end{pmatrix}\right\}$; (c) 2.

5. If R is a reduced row-echelon form of A, then we know that $R = EA$, where E is a product of invertible elementary matrices. Since A has rank one, R has a single nonzero row r and is of the form $e_1 r$, where e_1 is the first standard basis vector. Then $A = E^{-1} R = (E^{-1} e_1) r$, and we can take $c = E^{-1} e_1$.

6. In (a), let $u_1, \ldots, u_s$ be the rows of R having at least one of the first r entries nonzero, and let $u_{s+1}, \ldots, u_m$ be the other rows. For each i with $1 \le i \le s$, the first nonzero entry of u_i corresponds to a corner variable and occurs in the $j(i)^{\text{th}}$ position with $j(i) \le r$. The most general member of the row space of A is of the form $c_1 u_1 + \cdots + c_m u_m$, and the $j(i)^{\text{th}}$ entry of this is c_i. For this row vector to be in the indicated span, we must have $c_i = 0$ for $i \le s$.

In (b), let R' be a second reduced row-echelon form, and let its nonzero rows be $v_1, \ldots, v_m$. From part (a), it follows that the linear span of $u_{s+1}, \ldots, u_m$ equals the linear span of $v_{s+1}, \ldots, v_m$ for each s. Moreover, the value of each $j(i)$ has to be the same for u_i as for v_i. Inducting downward, we prove that $u_i = v_i$ for each i. For $i = m$, this follows since the first nonzero entry is 1 for both u_m and v_m. Assuming the result for $s + 1$, we write $v_s = c_s u_s + c_{s+1} u_{s+1} + \cdots + c_m u_m$. We have $c_s = 1$ since the first nonzero entry of u_s and v_s is 1, and we have $c_{s+i} = 0$ for $i > 0$ since the $j(s+i)^{\text{th}}$ entry of this equality of row vectors is $0 = c_{s+i}$. Thus $v_s = u_s$, and the induction is complete.

7. Let $E = \{x_1, \ldots, x_N\}$, and let $f_1, \ldots, f_n$ be a basis of U. Form the matrix
$A = \begin{pmatrix} f_1(x_1) & \cdots & f_1(x_N) \\ \vdots & \ddots & \vdots \\ f_n(x_1) & \cdots & f_n(x_N) \end{pmatrix}$. By assumption, A has row rank n. Therefore it has column rank n, and there exist n linearly independent columns, say columns $j_1, \ldots, j_n$. Then $D = \{x_{j_1}, \ldots, x_{j_n}\}$.

8. Let the listed basis be Γ, and let Σ be the standard basis. Then $\begin{pmatrix} I \\ \Sigma\Gamma \end{pmatrix} = \begin{pmatrix} 3 & -4 \\ -2 & 3 \end{pmatrix}$, the inverse matrix is $\begin{pmatrix} I \\ \Gamma\Sigma \end{pmatrix} = \begin{pmatrix} 3 & 4 \\ 2 & 3 \end{pmatrix}$, and $\begin{pmatrix} L \\ \Gamma\Gamma \end{pmatrix}$ is the product $\begin{pmatrix} 3 & 4 \\ 2 & 3 \end{pmatrix} \begin{pmatrix} -6 & -12 \\ 6 & 11 \end{pmatrix} \begin{pmatrix} 3 & -4 \\ -2 & 3 \end{pmatrix} = \begin{pmatrix} 2 & 0 \\ 0 & 3 \end{pmatrix}$.

9. One could compute the matrix of $I - D^2$ in an explicit basis, but an easier way is to observe that $D^3 = 0$ and hence $(I - D^2)(I + D^2) = I - D^4 = I$.

10. Since image$(AB) \subseteq$ image A, we have rank$(AB) = \dim$ image$(AB) \leq \dim$ image $A = $ rank A. Similarly rank$((AB)^t) = $ rank$(B^t A^t) \leq $ rank B^t. Since a matrix and its transpose have the same rank (by the equality of row rank and column rank), rank$(AB) \leq $ rank B.

11. Since A has n columns, rank $A \leq n$. Applying Problem 10 gives rank$(AB) \leq $ rank $A \leq n$. Since $n < k = $ rank I, we cannot have $AB = I$.

12. Take $A = \begin{pmatrix} 1 & 0 \\ 0 & 0 \end{pmatrix}$ and $B = \begin{pmatrix} 0 & 1 \\ 0 & 0 \end{pmatrix}$. Then $AB = B$ has rank 1 while $BA = 0$ has rank 0.

13. $\{\cosh t, \sinh t\}$.

14. Let $\{v_n \mid n \in \mathbb{Z}\}$ be a countably infinite basis. For each subset S of $\mathbb{Z}$, define v'_S to be the member of V' such that $v'_S(v_n)$ is 1 if n is in S and is 0 if not. Choose by Theorem 2.42 a subset of $\{v'_S\}$ that is a basis for the linear span of all v'_S. Arguing by contradiction, assume that this basis is countable. Number the S's in question as $S_1, S_2, \ldots$. Any v'_S then has a unique expansion as $v'_S = c_1 v'_{S_1} + \cdots + c_k v'_{S_k}$ for some k. Fix k, and let v'_S be expandable for this k. Let $E \subseteq \{1, \ldots, k\}$. Let m and n be such that v_m and v_n are in S_j for j in E and are not in S_j for j in $\{1, \ldots, k\} - E$. Then $v'_{S_j}(v_m) = v'_{S_j}(v_n)$ for $j = 1, \ldots, k$, and hence $v'_S(v_m) = v'_S(v_n)$. Thus with k fixed, the number of S's for which v'_S is expandable is at most 2^k. In particular, it is finite. Taking the union over k, we find that there are only countably many v'_S in the linear span of $v'_{S_1}, v'_{S_2}, \ldots$. But there are uncountably many subsets S of $\mathbb{Z}$, and we

have thus arrived at a contradiction. We conclude that our subset of all v'_S that is a basis for the linear span must have been uncountable.

15. For (a), take L, M, and N to be the three 1-dimensional subspaces of $\mathbb{R}^2$ shown in Figure 2.1. Then $L \cap (M + N) = L$ while $(L \cap M) + (L \cap N) = 0$.

For (b), we always have $\supseteq$ since $L \cap (M+N) \supseteq L \cap M$ and $L \cap (M+N) \supseteq L \cap N$.

For (c), if $l = m + n$ is in $L \cap (M + N)$, then $L \supseteq M$ implies that $n = l - m$ is in L. So $l = m + n$ has $m \in L \cap M$ and $n \in L \cap N$.

16. Take M, N_1, and N_2 to be the three 1-dimensional subspaces of $V = \mathbb{R}^2$ shown in Figure 2.1. Then $M \oplus N_1 = M \oplus N_2 = \mathbb{R}^2$, but $N_1 \neq N_2$.

17. (b) only.

18. In $V_1 \oplus \cdots \oplus V_n$, let p_j pick off the jth coordinate, and let i_j carry v_j to $(0, \ldots, 0, v_j, 0, \ldots, 0)$. Then $p_r i_s$ is I on V_s if $r = s$ and is 0 on V_s if $r \neq s$. Also, $\sum_{k=1}^{n} i_k p_k = I$ on $V_1 \oplus \cdots \oplus V_n$.

19. Corollary 2.15 shows that $\dim \ker T + \dim \operatorname{image} T = n$. Since $\ker T$ and $\operatorname{image} T$ have 0 intersection, the union of bases of $\ker T$ and $\operatorname{image} T$ is a linearly independent set of n vectors in $\mathbb{R}^n$. This set must be a basis of $\mathbb{R}^n$, and hence $\mathbb{R}^n = \ker T \oplus \operatorname{image} T$. This proves (a).

For (b), let $T^2 = T$ and suppose that v is in $\ker T \cap \operatorname{image} T$. Since v is in $\operatorname{image} T$, we have $v = T(w)$ for some w. Then $v = T(w) = T^2(w) = T(T(w)) = T(v)$, and the right side is 0 since v is in $\ker T$. Consequently $\ker T \cap \operatorname{image} T = 0$.

20. Define $L : V'_1 \oplus V'_2 \to (V_1 \oplus V_2)'$ by $L(\mu_1, \mu_2)(v_1, v_2) = \mu_1(v_1) + \mu_2(v_2)$.

21. Proposition 2.25 shows that $y \mapsto z$ is onto the subset of z's in V' such that $M \subseteq \ker z$, i.e., is onto $\operatorname{Ann} M$. Since q is onto V/M, $y \mapsto z$ is one-one.

22. The kernel of q is M, and thus the kernel of $q\big|_N$ is $M \cap N$. So $q\big|_N$ is one-one if and only if $M \cap N = 0$.

If $M + N = V$, then any $v \in V$ is of the form $m+n$; so v has $v + M = m+n+M = n + M = q(n)$, and q carries N onto V/M. Conversely if q carries N onto V/M, let $v \in V$ be given, and choose n with $q(n) = v + M$. Then $n + M = v + M$, and hence $v - n$ is in M. This says that $V = M + N$.

Consequently $q\big|_N : N \to V/M$ is an isomorphism if and only if $M \cap N = 0$ and $M + N = V$, and we know from Proposition 2.30 that this pair of conditions is equivalent to the single condition $V = M \oplus N$.

23. If A^{-1} has integer entries, then $\det A$ and $\det A^{-1}$ are integers that are reciprocals, and we conclude that $\det A = \pm 1$. If $\det A = \pm 1$, then Cramer's rule shows that A^{-1} has integer entries.

24. When $r = \operatorname{rank} A$, there exist r linearly independent rows. Say that these are the ones numbered $i_1, \ldots, i_r$. Let A_1 be the r-by-n matrix obtained by deleting the remaining rows. Since A_1 has rank r, it has r linearly independent columns. Say that these are the ones numbered $j_1, \ldots, j_r$. Let A_2 be the r-by-r matrix obtained by deleting the remaining columns. Then A_2 is a square matrix of rank r, is therefore

invertible, and must have nonzero determinant. In the reverse direction if some s-by-s submatrix has nonzero determinant, then the rows of the submatrix are linearly independent, and certainly the corresponding rows of A are linearly independent. Thus $s \leq \operatorname{rank} A$.

25. Let the expression in question be $f(t) = \sum_{i=1}^{n} a_i e^{c_i t}$. Put $r_i = e^{c_i}$. The numbers r_i are distinct. The fact that $f(0) = f(1) = \cdots = f(n-1) = 0$ says that the product of the Vandermonde matrix formed from $r_1, \ldots, r_n$ times the column vector $(a_1, \ldots, a_n)$ is the 0 vector. Since the Vandermonde matrix is invertible, it follows that $(a_1, \ldots, a_n)$ is the 0 vector.

26. The characteristic polynomial is $\lambda^2 - 5\lambda + 6 = (\lambda - 2)(\lambda - 3)$. The eigenvectors for $\lambda = 2$ are all nonzero multiples of $\begin{pmatrix} 1 \\ 2 \end{pmatrix}$, and the eigenvectors for $\lambda = 3$ are all nonzero multiples of $\begin{pmatrix} 1 \\ 3 \end{pmatrix}$.

27. $\sum_i (C^{-1} A C)_{ii} = \sum_i \sum_{j,k} (C^{-1})_{ij} A_{jk} C_{ki} = \sum_{j,k} A_{jk} \sum_i C_{ki} (C^{-1})_{ij} = \sum_{j,k} A_{jk} \delta_{kj} = \sum_j A_{jj}$.

28. For $n = 2$, direct computation gives $\lambda^2 - a_1 \lambda - a_0$. Similarly we obtain $\lambda^3 - a_2 \lambda^2 - a_1 \lambda - a_0$ when $n = 3$. We are thus led to the guess in the general case that the determinant is $\lambda^n - a_{n-1} \lambda^{n-1} - \cdots - a_1 \lambda - a_0$. This is proved by induction, using expansion in cofactors about the first column. The term from the $(1, 1)$ entry, by the inductive hypothesis, is $\lambda(\lambda^{n-1} - a_{n-1} \lambda^{n-2} - \cdots - a_1)$, and the term from the $(1, n)$ entry is $(-1)^{n+1}(-a_0) \det B$, where B is a lower triangular matrix of size $n - 1$ with -1 in every diagonal entry. Then $\det B = (-1)^{n-1}$, and substitution completes the induction.

29. In (a), we have $\det(\lambda I - AB) = \det(A(\lambda A^{-1} - B)) = \det A \det(\lambda A^{-1} - B) = \det(\lambda A^{-1} - B) \det A = \det((\lambda A^{-1} - B) A) = \det(\lambda I - BA)$.

For (b), we know from the fact that the characteristic polynomial of A is a polynomial that there are only finitely many ϵ for which $A + \epsilon I$ fails to be invertible. Thus there is some $\epsilon_0 > 0$ such that $A + \epsilon I$ is invertible when $0 < \epsilon < \epsilon_0$. By (a), these ϵ's have $\det(\lambda I - (A + \epsilon I)B) = \det(\lambda I - B(A + \epsilon I))$. Since det is a polynomial in the entries of the matrix it is applied to, $\det(\lambda I - C)$ is a continuous function of the entries of C. Taking $C = (A + \epsilon I)B$ and then $C = B(A + \epsilon I)$, and letting ϵ tend to 0, we obtain $\det(\lambda I - AB) = \det(\lambda I - BA)$.

30. In $\mathbb{R}^1$, let the nth spanning set consist of $\{(r) \mid 0 < r < 1/n\}$. These each span $\mathbb{R}^1$, but their intersection is empty and the empty set does not span $\mathbb{R}^1$.

31. One can appeal to Problem 14 and take into account that $\dim V'' \geq \dim V'$, but the following argument is a little simpler: Let $\{v_n\}$ be a countably infinite basis of V, let U_0 be the vector subspace of all $v' \in V'$ such that $v'(v_n)$ is 0 for all n sufficiently large, and define v'_∞ in V' by $v'_\infty(v_n) = 1$ for all n. The version of Proposition 2.20 in Section 9 shows that there exists a member v'' of V'' such that $v''(U_0) = 0$ and $v''(v'_\infty) = 1$. Arguing by contradiction, suppose that some $v \in V$ has $v'' = \iota(v)$. Then $v''(v') = \iota(v)(v') = v'(v)$ for every v' in V'. For v' in U_0, this equality gives

$0 = v''(v') = v'(v)$. An example of a member of U_0 is v'_k defined by $v'_k(v_n) = \delta_{kn}$, and we must have $v'_k(v) = 0$ for all k. Expanding v in terms of the basis $\{v_n\}$, we see that all the coefficients of v are 0. Thus $v = 0$ and $v'' = \iota(v) = 0$. But $v''(v'_\infty) = 1$, and we have a contradiction.

32. $\operatorname{Ann}(M + N) \subseteq \operatorname{Ann} M$, and $\operatorname{Ann}(M + N) \subseteq \operatorname{Ann} N$; thus $\operatorname{Ann}(M + N) \subseteq \operatorname{Ann} M \cap \operatorname{Ann} N$. If v' is 0 on M and is 0 on N, then it is 0 on $M + N$. Hence $\operatorname{Ann}(M + N) \supseteq \operatorname{Ann} M \cap \operatorname{Ann} N$.

33. $\operatorname{Ann}(M \cap N) \supseteq \operatorname{Ann} M$, and $\operatorname{Ann}(M \cap N) \supseteq \operatorname{Ann} N$; thus $\operatorname{Ann}(M \cap N) \supseteq \operatorname{Ann} M + \operatorname{Ann} N$. Let $\{u_\alpha\}$ be a basis of $M \cap N$, let v_β be vectors added to $\{u_\alpha\}$ to obtain a basis of M, and let w_γ be vectors added to $\{u_\alpha\}$ to obtain a basis of N. Then $\{u_\alpha\} \cup \{v_\beta\} \cup \{w_\gamma\}$ is a basis of $M + N$. Let x_δ be vectors added to this to obtain a basis of V. If v' is given in $\operatorname{Ann}(M \cap N)$, define v'_1 to be v' on all the basis vectors but the v_β, where it is to be 0, and define $v'_2 = v' - v'_1$. Then $v' = v'_1 + v'_2$ with $v'_1 \in \operatorname{Ann} M$ and $v'_2 \in \operatorname{Ann} N$. So $\operatorname{Ann}(M \cap N) \subseteq \operatorname{Ann} M + \operatorname{Ann} N$.

34. Let v be in M, and let v' be in $\operatorname{Ann} M$. Then $\iota(v)(v') = v'(v) = 0$. This proves (a).

For (b), Propositions 2.19 and 2.20a give $\dim \operatorname{Ann} M = \dim V' - \dim M$ and $\dim \operatorname{Ann}(\operatorname{Ann} M) = \dim V'' - \dim \operatorname{Ann} M = \dim V'' - (\dim V' - \dim M) = \dim M = \dim \iota(M)$. This equality in the presence of the inclusion $\iota(M) \subseteq \operatorname{Ann}(\operatorname{Ann} M)$ implies $\iota(M) = \operatorname{Ann}(\operatorname{Ann} M)$ by Corollary 2.4.

For (c), let V be as in Problem 31, and put $M = V$. Then $\operatorname{Ann}(M) = 0$ and $\operatorname{Ann}(\operatorname{Ann} M) = V'' \neq \iota(V)$.

35. Parts (a) and (b) follow by writing out individual entries of the products as appropriate sums.

36. If A or D is not invertible, then suitable row operations on the matrix on the left side exhibit the matrix on the left as not invertible, and hence both sides are 0. Thus we may assume that A^{-1} and D^{-1} exist. Problem 35c allows us to decompose the given matrix as $\begin{pmatrix} A & B \\ 0 & D \end{pmatrix} = \begin{pmatrix} A & 0 \\ 0 & I \end{pmatrix} \begin{pmatrix} I & 0 \\ 0 & D \end{pmatrix} \begin{pmatrix} I & A^{-1}B \\ 0 & I \end{pmatrix}$. The determinant of the product is the product of the determinants. Using the defining formula for det, we see that the first two determinants from the right side are $\det A$ and $\det D$. The third determinant is 1 since the matrix is triangular with 1's on the diagonal.

37. In effect, we do row reduction with blocks, taking advantage of Problem 35c. We have $\begin{pmatrix} A & B \\ C & D \end{pmatrix} = \begin{pmatrix} A & 0 \\ 0 & I \end{pmatrix} \begin{pmatrix} I & A^{-1}B \\ C & D \end{pmatrix} = \begin{pmatrix} A & 0 \\ 0 & I \end{pmatrix} \begin{pmatrix} I & 0 \\ C & I \end{pmatrix} \begin{pmatrix} I & A^{-1}B \\ 0 & D - CA^{-1}B \end{pmatrix}$. Taking the determinant of both sides and using Problem 36, we obtain $\det \begin{pmatrix} A & B \\ C & D \end{pmatrix} = (\det A) \det(D - CA^{-1}B) = \det(AD - ACA^{-1}B)$, and this equals $\det(AD - CB)$ since $AC = CA$.

38. The matrices $\begin{pmatrix} A \\ 0 \end{pmatrix}$ and $\begin{pmatrix} B & 0 \end{pmatrix}$ are of size n-by-n, and their products in the two orders are $\begin{pmatrix} AB & 0 \\ 0 & 0 \end{pmatrix}$ and BA. Problem 29 shows that $\det \left(\lambda I_n - \begin{pmatrix} AB & 0 \\ 0 & 0 \end{pmatrix}\right) = \det(\lambda I_n - BA)$. The left side equals $\lambda^{n-k} \det(\lambda I_k - AB)$, and the result follows.

39. Substitute the definitions of the determinants of $A(S)$ and $\widehat{A}(S)$ into the right side, sort out the signs, and verify that the result is the defining expression for det A.

40. Expansion in cofactors about the last row gives $\det A_n = (A_n)_{nn} \det \widehat{(A_n)}_{nn} - (A_n)_{n-1,n} \det \widehat{(A_n)}_{n-1,n} = 2 \det A_{n-1} + \det B$, where B in block form is the square matrix of size $n-1$ given by $B = \begin{pmatrix} A_{n-2} & 0 \\ 0 & -1 \end{pmatrix}$. Expansion by cofactors of det B about the last row shows that $\det B = -\det A_{n-2}$, and the stated formula results.

41. Inspection gives $\det A_1 = 2$ and $\det A_2 = 3$. The function f with $f(n) = \det A_n - (n+1)$ thus has $f(1) = f(2) = 0$ and $f(n) = 2f(n-1) - f(n-2)$ for $n \geq 3$, and it must be 0 for all $n \geq 1$.

42. The only changes in (a) are notational. For (b), we compute $\det C_2 = \det C_3 = 2$, and the formula $\det C_n = 2$ follows as in Problem 41.

43. For (b), we interchange the first two rows and then interchange the first two columns. The determinant does not change.

44. For (b), we interchange the third and fourth rows and then interchange the third and fourth columns. For (c) we change the list of rows and columns of A_n from $1, 2, 3, 4, 5$ to $3, 5, 4, 2, 1$.

Chapter III

1. Since $\operatorname{Tr} B^* A = \sum_{i,j} A_{ij} \overline{B_{ij}}$, the inner product is the usual inner product on the n^2 entries. Then (a) and (b) are immediate. For (c), (a) gives the result of Parseval's equality relative to the orthonormal basis in (b).
For (d), let U be the unitary matrix with columns $u_1, \ldots, u_n$, i.e., the matrix $\begin{pmatrix} I \\ \Sigma \Gamma \end{pmatrix}$, where $\Gamma = (u_1, \ldots, u_n)$ and Σ is the standard ordered basis. Then $\|A\|_{\mathrm{HS}}^2 = \operatorname{Tr}(A^*A) = \operatorname{Tr}(U^{-1}A^*AU) = \sum_{i,j} |(AU)_{ij}|^2 = \sum_j \|Au_j\|^2$, and this equals $\sum_{i,j} |v_i^* Au_j|^2$ by Parseval's equality.
In (e), $W^\perp$ consists of all matrices that are 0 along the diagonal. It has dimension $n^2 - n$.

2. The system has unknowns $c_0, c_1, \ldots, c_n$, where $p_n(x) = c_0 + c_1 x + \cdots + c_n x^n$, and the kth equation, for $0 \leq k \leq n$, comes from the equality for $f(x) = x^k$, namely $2^{-k} = \sum_{j=0}^n (j+k+1)^{-1} c_j$.

3. $(LM)^* = M^*L^* = ML$ is equal to LM if and only if $LM = ML$.

4. A vector u is in $\ker L$ if and only if $(L(u), v) = 0$ for all v, if and only if $(u, L^*(v)) = 0$ for all v, if and only if u is in $(\operatorname{image} L^*)^\perp$.

5. There are none. The characteristic polynomial has no real roots, but all roots must be real if A is Hermitian.

6. The map $v_1 \mapsto (L(v_1), v_2)_2$ is a linear functional on V_1 and hence is given by the inner product with a unique member u_1 of V_1, i.e., $((L(v_1), v_2)_2 = (v_1, u_1)_1$, and

we define this element u_1 to be $L^*(v_2)$. We readily check that L^* is linear, and (a) is then proved. The proof of (b) proceeds in the same way as in the case that $V_1 = V_2$.

7. In (a), if v is in $S^\perp \cap T^\perp$, then v is in $V^\perp = 0$. Thus $S^\perp + T^\perp$ is a direct sum. We have dim $V =$ dim $S +$ dim $T =$ (dim $V -$ dim $S^\perp) +$ (dim $V -$ dim $T^\perp) =$ 2 dim $V -$ dim $S^\perp -$ dim $T^\perp$. Therefore dim $V =$ dim$(S^\perp + T^\perp)$. The inclusion plus the equality of the finite dimensions forces $V = S^\perp + T^\perp$.

In (b), let λ be 0 or 1. Then $E^*u = \lambda u$ if and only if $(E^*u, v) = \lambda(u, v)$ for all v, if and only if $(u, Ev) = \lambda(u, v)$ for all v. When $\lambda = 1$, this says that $E^*u = u$ if and only if $u \perp (I - E)v$ for all v, hence if and only if $u \perp T$, hence if and only if u is in $T^\perp$. When $\lambda = 0$, it says that $E^*u = 0$ if and only if $u \perp Ev$ for all v, hence if and only if $u \perp S$, hence if and only if u is in $S^\perp$.

8. The formulas of the Gram–Schmidt orthogonalization process have $v_j = c_j(ge_j) + \sum_{i<j} a_{ij} v_i$ with $c_j > 0$. Therefore $ge_j = c_j^{-1} v_j + \sum_{i<j} b_{ij} v_i$, and

$$(k^{-1}g)_{ij} = \sum_l (k^{-1})_{il} g_{lj} = \sum_l (k^{-1})_{il} (ge_j)_l$$
$$= c_j^{-1} \sum_l (k^{-1})_{il} (v_j)_l + \sum_l \sum_{m<j} (k^{-1})_{il} b_{mj} (v_m)_l$$
$$= c_j^{-1} (k^{-1} v_j)_i + \sum_{m<j} b_{mj} (k^{-1} v_m)_i = c_j^{-1} \delta_{ij} + \sum_{m<j} b_{mj} \delta_{im}.$$

If $i = j$, the right side is c_j^{-1} and is positive. If $i > j$, then every term on the right side is 0. Thus $k^{-1}g$ is upper triangular with positive diagonal entries. Since k carries the standard orthonormal basis to the orthonormal basis $\{v_1, \ldots, v_n\}$, k is unitary.

9. For (a), the Spectral Theorem and Corollary 3.22 show that A is similar to a diagonal matrix with positive diagonal entries. Thus det $A > 0$. In (b), we specialize the inequality $\bar{x}^t Ax > 0$ to x's that are 0 except in the entries numbered $i_1, \ldots, i_n$, and we find that the submatrix is positive definite. Then the result follows from Corollary 3.22.

10. Take $g = \sqrt{A}$ in Problem 8 and obtain $\sqrt{A} = kt$ with k unitary and t upper triangular with positive diagonal entries. Then $A = (\sqrt{A})^*(\sqrt{A}) = (kt)^*(kt) = t^*t$.

11. The roots of the characteristic polynomial are $\frac{1}{2}(a + d + s)$ and $\frac{1}{2}(a + d - s)$, where $s = \sqrt{(a-d)^2 + 4|b|^2}$. Let $r = \frac{1}{2}(-a + d + s)$. Then $D = \begin{pmatrix} \frac{1}{2}(a+d+s) & 0 \\ 0 & \frac{1}{2}(a+d-s) \end{pmatrix}$ and $U = (b^2 + r^2)^{-1/2} \begin{pmatrix} b & -r \\ r & \bar{b} \end{pmatrix}$.

12. In (a), the conditions $ad - |b|^2 > 0$ and $a + d > 0$ together are necessary and sufficient. In (b), let $\sqrt{D} = \begin{pmatrix} \sqrt{\frac{1}{2}(a+d+s)} & 0 \\ 0 & \sqrt{\frac{1}{2}(a+d-s)} \end{pmatrix}$, and let U be as in the previous problem. Then the positive definite square root of A is $U\sqrt{D}U^{-1}$.

13. The Spectral Theorem shows that A has a basis of eigenvectors, each with a real eigenvalue. If v is an eigenvector with eigenvalue λ, then $v^t Av = 0$ says that $\lambda \|v\|^2 = 0$. So every eigenvalue is 0, and A, being similar to a diagonal matrix, has to be 0.

14. Choosing a basis of eigenvectors, we may solve the corresponding problem for diagonal matrices. Thus let A be a diagonal matrix, and assume, without loss of generality, that $A_{11} = \cdots = A_{kk} = 1$ and $A_{jj} \neq 1$ for $j > k$. Then the given equation $(I - A)^2 y = (I - A)x$ says that $(1 - A_{jj})^2 y_j = (1 - A_{jj})x_j$ for all j. Thus define y_j to be 0 if $j \leq k$, and choose $y_j = (1 - A_{jj})^{-1} x_j$ for $j > k$.

15. $LL^* = (UP)(UP)^* = (PU)(PU)^* = PUU^*P = P^2 = PU^*UP = (UP)^*(UP) = L^*L$.

16. The family has a basis of simultaneous eigenvectors, and the matrices are all diagonal in this basis. So the answer is the dimension of the vector space of diagonal matrices, namely n.

17. In (a), $c' G(v_1, \ldots, v_n) \bar{c} = \sum_{i,j} c_i (v_i, v_j) \bar{c}_j = \left(\sum_i c_i v_i, \sum_j c_j v_j \right) = \|c_1 v_1 + \cdots + c_n v_n\|^2$. Thus Corollary 3.22 shows that $G(v_1, \ldots, v_n)$ is positive semidefinite. Moreover, $\|c_1 v_1 + \cdots + c_n v_n\|^2 = 0$ for some $c \neq 0$ if and only if $v_1, \ldots, v_n$ are linearly dependent. Thus $G(v_1, \ldots, v_n)$ is definite if and only if $v_1, \ldots, v_n$ are linearly independent. We know that a positive semidefinite matrix is definite if and only if it is invertible, and thus $\det G(v_1, \ldots, v_n) > 0$ if and only if $v_1, \ldots, v_n$ are linearly independent; this proves (b). In (c), equality holds in the Schwarz inequality if and only if the two vectors are linearly dependent, i.e., if and only if one of them is a multiple of the other.

18. This is immediate by induction.

19. For (a), the left side is $D^2(X^{n+1}) = (n+1) D(X^n X')$. Comparing with the expected right side, we see that we are to show that

$$nD(X^n X') \stackrel{?}{=} (2n+1) n X'' X^n + 4n^2 X^{n-1}.$$

The left side equals nX^{n-1} times $n(X')^2 + XX''$, while the right side equals nX^{n-1} times $(2n+1)X''X + 4n$. Since

$$n(X')^2 + XX'' = 4nx^2 + 2x^2 - 2$$
$$= (4n+2)x^2 - (4n+2) + 4n = (2n+1)X''X + 4n,$$

(a) is proved.

For (b), the Leibniz rule gives $D^n(X'Y) = X'D^n Y + nX''D^{n-1}Y$ for any Y. Meanwhile, application of D^{n-1} to (a) yields

$$D^{n+1}(X^{n+1}) = (2n+1)D^n(X'X^n) - n(2n+1)X''D^{n-1}(X^n) - 4n^2 D^{n-1}(X^{n-1}).$$

Substituting with $Y = (2n+1)X^n$, we obtain (b). The recursion in conclusion (c) follows immediately by multiplying by $(2^{n+1} n!)^{-1}$.

For (d), conclusion (c) and the definition of P_n show that $Q_n = P_n - R_n$ satisfies $Q_0 = Q_1 = 0$ and $(n+1)Q_{n+1}(x) - (2n+1)x Q_n(x) - n Q_{n-1}(x)$. Thus $Q_n(x) = 0$ for every n by induction.

20–21. Write $X = x^2 - 1$. Since $X^n = (x-1)^n f(x)$, the function X^n has all derivatives through order $n-1$ equal to 0 at $x=1$. The same conclusion applies also at $x=-1$. If $m \leq n$, integration by parts gives

$$\int_{-1}^{1} D^m(X^m) D^n(X^n)\, dx = [D^m(X^m) D^{n-1}(X^n)]_{-1}^{1} - \int_{-1}^{1} D^{m+1}(X^m) D^{n-1}(X^n)\, dx$$
$$= -\int_{-1}^{1} D^{m+1}(X^m) D^{n-1}(X^n)\, dx$$
$$= \cdots = (-1)^k \int_{-1}^{1} D^{m+k}(X^m) D^{n-k}(X^n)\, dx$$

for $k \leq n$. If $m < n$, then taking $k = m+1$ gives 0 on the right side because $D^{2m+1}(X^m) = 0$. If $m = n$, then taking $k = n$ gives $(-1)^n \int_{-1}^{1} X^n D^{2n}(X^n)\, dx = (-1)^n (2n)! \int_{-1}^{1} X^n\, dx$ on the right side. Therefore

$$\langle D^n(X^n), D^n(X^n) \rangle = (-1)^n (2n)! (-1)^n \frac{2(2^n n!)^2}{(2n+1)!} = \frac{2(2^n n!)^2}{2n+1},$$

and $\langle P_n, P_n \rangle = \frac{2}{2n+1}$.

22. The expansion for (a) is

$$D^{n+1}[(D(X^n))X]$$
$$= D^{n+1}(D(X^n))X + (n+1)D^n(D(X^n))X' + \tfrac{1}{2}n(n+1)D^{n-1}(D(X^n))X''$$
$$= XD^2(D^n(X^n)) + (n+1)X'D(D^n(X^n)) + \tfrac{1}{2}n(n+1)X''D^n(X^n),$$

and the expansion for (b) is

$$D^{n+1}[(D(X^n))X] = D^{n+1}(nX^n X') = D^{n+1}(nX^n)X' + (n+1)D^n(nX^n)X''$$
$$= nD(D^n(X^n))X' + n(n+1)D^n(X^n)X''.$$

Thus, for (c), we get $(x^2-1)D^2(P_n(x)) + (n+1)2xD(P_n(x)) + \tfrac{1}{2}n(n+1)2P_n(x) = nD(P_n(x))2x + n(n+1)P_n(x)2$. This simplifies to

$$(x^2-1)P_n'' + 2(n+1)xP_n' + n(n+1)P_n = 2nxP_n' + 2n(n+1)P_n$$

and then to $(1-x^2)P_n'' - 2xP_n' + n(n+1)P_n = 0$.

24. In Problems 24–28, there is no difficulty with addition, and we have to check something only about scalar multiplication. For Problem 24, we need to check in $\overline{V}$ that $(ab)v = a(bv)$, $1v = v$, $a(u+v) = au + av$, and $(a+b)v = av + bv$. These are satisfied in $\overline{V}$ because the identities $\overline{(ab)}v = \bar{a}(\bar{b}v)$, $1v = v$, $\bar{a}(u+v) = \bar{a}u + \bar{a}v$, and $\overline{(a+b)}v = \bar{a}v + \bar{b}v$ hold in V.

25. We are to see that $\overline{L}$ respects scalar multiplication, and the argument is that $\overline{L}(cv) = L(\bar{c}v) = \bar{c}L(v) = c\overline{L}(v)$.

26. We have $(au, bv)_{\overline{V}} = (\overline{b}v, \overline{a}u)_V = a\overline{b}(v, u)_V = a\overline{b}(u, v)_{\overline{V}}$, as required.

27. Let ℓ in V' correspond to v in V, so that $\ell(u) = (u, v)_V = (v, u)_{\overline{V}}$. Then ℓ in V' corresponds to v in $\overline{V}$, while $(c\ell)(u) = c(v, u)_{\overline{V}} = (cv, u)_{\overline{V}}$ shows that $c\ell$ corresponds to cv in $\overline{V}$.

28. Let ℓ in V' correspond to v in $\overline{V}$. Then $L^t(\ell)(u) = \ell(L(u)) = (v, L(u))_V = (v, \overline{L}(u))_{\overline{V}} = ((\overline{L})^*(v), u)_{\overline{V}}$, and this says that $L^t(\ell)$ corresponds to $(\overline{L})^*(v)$, i.e., L^t corresponds to $(\overline{L})^*$.

29. In (a), it is enough to check the result for p and q equal to monomials, and (b) is a direct calculation. In (c), let $p(x) = \sum c_{k_1,\ldots,k_n} x_1^{k_1} \cdots x_n^{k_n}$. The bilinearity and (b) show that $\langle p, p \rangle = \sum (c_{k_1,\ldots,k_n})^2 \langle x_1^{k_1} \cdots x_n^{k_n}, x_1^{k_1} \cdots x_n^{k_n} \rangle$, and this is positive unless all the coefficients are 0.

30. The polynomial p is in H_N if and only if $\partial(|x|^2)p = 0$, if and only if $\langle \partial(|x|^2)p, q \rangle = 0$ for all q in V_{N-2}, if and only if $\partial(q)(\partial(|x|^2)p) = 0$ for all q in V_{N-2}, if and only if $\partial(|x|^2 q)p = 0$ for all q in V_{N-2}, if and only if $\langle p, |x|^2 q \rangle = 0$ for all q in V_{N-2}, if and only if p is in $(|x|^2 V_{N-2})^\perp$.

31. Problem 30 gives $V_N = H_N \oplus |x|^2 V_{N-2}$, and we iterate this decomposition.

32. A basis of $|x|^2 V_2$ is $\{|x|^2 x_1^2, |x|^2 x_1 x_2, |x|^2 x_2^2\}$. Apply the Gram–Schmidt orthogonalization procedure to obtain an orthonormal basis $\{|x|^2 u_1, |x|^2 u_2, |x|^2 u_3\}$, and write $x_1^4 + y_1^4 = h_4 + \sum_{j=1}^{3} (x_1^4 + y_1^4, |x|^2 u_j) |x|^2 u_j$. Then h_4 is harmonic by Problem 30. A basis of $|x|^2 V_0$ is $|x|^2$, and hence an orthonormal basis consists of the single vector $w = \||x|^2\|^{-1} |x|^2$. Write $u_j = h_{2,j} + (u_j, w)w$ for each j, and substitute. Each $h_{2,j}$ is harmonic. Then we have

$$x_1^4 + y_1^4 = h_4 + \sum_{j=1}^{3} (x_1^4 + y_1^4, |x|^2 u_j) |x|^2 (h_{2,j} + (u_j, w)w)$$
$$= h_4 + |x|^2 \sum_{j=1}^{3} (x_1^4 + y_1^4, |x|^2 u_j) h_{2,j}$$
$$+ |x|^4 \sum_{j=1}^{3} (x_1^4 + y_1^4, |x|^2 u_j)(u_j, w) \||x|^2\|^{-1}$$

with h_4 in H_4, each $h_{2,j}$ in H_2, and the last sum in H_0.

33. Let P be the positive semidefinite square root of B. Then $AB = APP$, and hence $\det(\lambda I - AB) = \det(\lambda I - PAP)$. Consequently AB has the same eigenvalues as PAP. The latter is positive semidefinite since $(PAPv, v) = (A(Pv), Pv) \geq 0$. Therefore all the eigenvalues of AB are ≥ 0.

34. Since $(P^{-1}ABCP^{-1}v, v) = (ABC(P^{-1}v), P^{-1}v)$, ABC is positive semidefinite if and only if $P^{-1}ABCP^{-1}$ is positive semidefinite, if and only if $P^{-1}ABCP^{-1}$ has all eigenvalues ≥ 0. But $P^{-1}ABCP^{-1}$ has the same eigenvalues as $ABCP^{-1}P^{-1} = AB$, which has all eigenvalues ≥ 0 by the previous problem.

Chapter IV

1. If $a^2 = b^2 = (ab)^2 = 1$, then $a^{-1} = a$, $b^{-1} = b$, and $(ab)^{-1} = ab$. So $ab = (ab)^{-1} = b^{-1}a^{-1} = ba$.

2. Number the vertices counterclockwise as 1, 2, 3, 4. The motions in D_4 are then given by permutations as 1, (1 2)(3 4), (1 4)(2 3), (1 3), (2 4), (1 2 3 4), (1 3)(2 4), (1 4 3 2).

3. Choose integers x and y with $xl + y|G| = 1$. Then $a = a^{xl+y|G|} = (a^l)^x(a^{|G|})^y = (a^l)^x$ since $a^{|G|} = 1$, and this is a power of an element of H.

4. Define $\varphi : G \to G'$ by $\varphi(a) = a$. Then $\varphi(a) \circ \varphi(b) = a \circ b = ba = \varphi(ba) = \varphi(a \circ b)$. From this equality it follows that G' is a group and that φ is an isomorphism.

5. For $n > 0$, $(ab)^n = abab \cdots ab = a^n b^n$; also $(ab)^{-n} = ((ab)^{-1})^n = (b^{-1}a^{-1})^n = (a^{-1}b^{-1})^n = (a^{-1})^n(b^{-1})^n = a^{-n}b^{-n}$. In $\mathfrak{S}_3$, $a \mapsto a^2$ is not a homomorphism since four elements are sent to 1 and since 4 does not divide $|\mathfrak{S}_3| = 6$.

6. Define $\varphi : H \times K \to HK$ by $\varphi(h, k) = hk$. What needs proof is that members of H commute with members of K. If h is in H and k is in K, then $(hkh^{-1})h = hk = k(k^{-1}hk)$. Since H and K are normal, hkh^{-1} is in K and $k^{-1}hk$ is in H. Then $k^{-1}(hkh^{-1}) = (k^{-1}hk)h^{-1}$ and $H \cap K = \{1\}$ together imply $k^{-1}(hkh^{-1}) = 1 = (k^{-1}hk)h^{-1} = 1$. From the first of these, $k = hkh^{-1}$. Therefore $hk = kh$.

7. Since GCD(1234, 8191) = 1, there exist x and y with $1234x + 8191y = 1$, and x and y can be found explicitly by the Euclidean algorithm of Section I.1. For this x, $1234x \equiv 1 \mod 8191$.

8. The members $1, 2, \ldots, p - 1$ of $\mathbb{F}_p$ are roots of $X^{p-1} - 1 = 0$. By iterated use of the Factor Theorem, $X^{p-1} - 1 = (X - 1)(X - 2) \cdots (X - (p - 1))Q(X)$, and $Q(X)$ must have degree 0. Checking the coefficient of X^{p-1} on both sides shows that $Q(X) = 1$. Evaluating at $X = 0$ gives $-1 = (-1)(-2) \cdots (-(p - 1)) \mod p$. Since p is odd, this equation reads $(p - 1)! \equiv -1 \mod p$.

9. Corollary 4.39 shows that such a group has to be abelian, and Theorem 4.56 shows that it is the direct sum of cyclic groups. Thus it must be C_{p^2} or $C_p \times C_p$, up to isomorphism.

10. If $y = axa^{-1}$, then $y^n = ax^na^{-1}$. This proves (a). Also, $ba = a^{-1}(ab)a$ shows that ba and ab are conjugate. This proves (b).

11. There are four classes: $C_1 = \{1\}$, $C_2 = \{(1\ 2)(3\ 4), (1\ 3)(2\ 4), (1\ 4)(2\ 3)\}$, $C_3 = \{(1\ 2\ 3), (3\ 4\ 1), (2\ 1\ 4), (4\ 3\ 2)\}$, $C_4 = \{(1\ 3\ 2), (3\ 1\ 4), (2\ 4\ 1), (4\ 2\ 3)\}$. The centralizer of the first element of each class is $\mathfrak{A}_4$ for C_1, $C_1 \cup C_2$ for C_2, $\{(1\ 2\ 3), (1\ 3\ 2)\}$ for C_3 and C_4. Since $\mathfrak{A}_4$ has no element of order 6, it has no subgroup C_6. In a subgroup $\mathfrak{S}_3$, an element of order 3 is conjugate to its square, but no element of order 3 in $\mathfrak{A}_4$ is conjugate to its square.

12. A subgroup of order 30 would have index 2 and would thus be normal, in contradiction to Theorem 4.47.

13. This is a special case of Proposition 4.36.

14. Since H is normal, G acts on H by conjugation. The number of elements in an orbit has to be a divisor of $|G|$, and the smallest divisor of $|G|$ apart from 1 is p, by hypothesis. Since $\{1\}$ is one orbit and there are only $p-1$ other elements in H, each orbit must contain one element. Therefore $ghg^{-1} = h$ for each $g \in G$ and $h \in H$, and each h is in Z_G.

15. Certainly the inner automorphisms are closed under composition and inversion and therefore form a subgroup. If φ is an automorphism and ψ is the inner automorphism $\psi(x) = axa^{-1}$, then $\varphi \circ \psi \circ \varphi^{-1}(x) = \varphi(a\varphi^{-1}(x)a^{-1}) = \varphi(a)x\varphi(a)^{-1}$ shows that $\varphi \circ \psi \circ \varphi^{-1}$ is inner. Hence the subgroup of inner automorphisms is normal. Define a mapping Φ of G into the inner automorphisms by $\Phi(a) = \{x \mapsto axa^{-1}\}$. Then $\Phi(ab) = \Phi(a)\Phi(b)$, and hence Φ is a homomorphism. Certainly Φ is onto the inner automorphisms, and its kernel consists of all elements $a \in G$ with $axa^{-1} = x$ for all x, hence consists of all a in Z_G. Thus Φ exhibits G/Z_G as isomorphic to the group of inner automorphisms.

16. Part (a) is proved in the same way as Lemma 4.45. For (b), choose $m = 8$; then Aut C_m is $C_2 \times C_2$.

17. In (a), each C_k is a conjugacy class, by Proposition 4.42, and it is evident that the C_k's are the only conjugacy classes whose members have order 2. If x and y are in $\mathfrak{S}_n$, then $\tau(xyx^{-1}) = \tau(x)\tau(y)\tau(x)^{-1}$ shows that τ carries any conjugate of y to a conjugate of $\tau(y)$. Therefore conjugacy classes map to conjugacy classes under τ, and $\tau(C_1)$ has to be some C_k.

In (b), the number of ways of selecting $2k$ elements from n is $\binom{n}{2k}$. For each of these, the number of ways of selecting k unordered pairs of elements from $2k$ elements is the multinomial coefficient $\binom{2k}{2,\ldots,2} = \frac{(2k)!}{2^k}$. Although the individual pairs are unordered, this enumeration counts one for each different ordering of the k pairs. There are $k!$ orderings, and hence the multinomial coefficient must be divided by $k!$ to discount the enumeration of the pairs. Thus $|C_k|$ is the product of the integer $\binom{n}{2k}$ and the integer $\frac{(2k)!}{2^k k!}$.

In (c), we saw in (b) that $N_k = \frac{(2k)!}{2^k k!}$ is always an integer. Let us bound it below. Canceling every even factor of the numerator by a factor of $k!$ and a factor of 2^k, we see that $N_k = (2k-1)(2k-3)(2k-5)\cdots(3)(1)$. Thus $N_k \geq 2k-1$ with equality only if $2k-1 = 1$, in which case $k = 1$. Also, $N_k \geq (2k-1)(2k-3)$ with equality holding for a value of $k > 1$ only if $2k-3 = 1$, in which case $k = 2$.

Now let us compare $|C_k|$ and $|C_1|$. We have $N_1 = 1$. Also, $|C_k| = \binom{n}{2k}\frac{(2k)!}{2^k k!} = N_k \binom{n}{2k}$ and $|C_1| = N_1 \binom{n}{2} = \binom{n}{2}$. The easy comparison is that $|C_k| \geq \binom{n}{2k}$ and this is $> \binom{n}{2} = |C_1|$ unless $k = 1$ or $|n - 2k| \leq 2$. Thus $|C_k| > |C_1|$ unless k equals 1 or $\frac{1}{2}n$ or $\frac{1}{2}(n-1)$ or $\frac{1}{2}(n-2)$. We can discard $k = \frac{1}{2}(n-2)$ because in this case $|C_k| = N_k \binom{n}{2} > N_1 \binom{n}{2} = |C_1|$ except when $k = 1$.

Consider $k = \frac{1}{2}(n-1)$ with $k > 1$. Then $|C_1| = \frac{1}{2}n(n-1) = nk$ and $|C_k| = N_k \binom{n}{n-1} = nN_k$. From above, the latter is $> n(2k-1) \geq nk = |C_1|$.

Finally consider $k = \frac{1}{2}n$ with $k > 1$. Then $|C_1| = \frac{1}{2}n(n-1) = (n-1)k$ and $|C_k| = N_k\binom{n}{n} = N_k$. From above, the latter for $k > 1$ is $\geq (2k-1)(2k-3) = (n-1)(n-3)$, and this is $> (n-1)k = |C_1|$ unless $k \geq n-3$. When $k \geq n-3$, we obtain $\frac{1}{2}n \geq n-3$ and $n \leq 6$. Since $k = \frac{1}{2}n$, n has to be even with $n \leq 6$. The case $n = 6$ (with $k = 3$) we are allowing, and the case $n = 4$ with $k = 2$ has $|C_2| = 3 \neq 6 = |C_1|$. Thus the only exceptions have $k = 1$ or $n = 6$.

18. In the composition series given for $\mathfrak{S}_4$ in Section 7, take G to be $\mathfrak{A}_4$, N to be the 4-element subgroup in the series, and M to be the 2-element subgroup.

19. If $\mathrm{GCD}(r,s) = 1$, define a homomorphism $\varphi : \mathbb{Z} \to (\mathbb{Z}/r\mathbb{Z}) \times (\mathbb{Z}/s\mathbb{Z})$ by $\varphi(n) = (n \bmod r, n \bmod s)$. This is 0 for $n = rs$. Thus it descends to a homomorphism $\overline{\varphi} : \mathbb{Z}/rs\mathbb{Z} \to (\mathbb{Z}/r\mathbb{Z}) \times (\mathbb{Z}/s\mathbb{Z})$. The kernel of φ consists of all integers n divisible by r and s. Since r and s are relatively prime, such integers are divisible by rs. Thus $\ker \varphi = rs\mathbb{Z}$, and $\overline{\varphi}$ is one-one. Since the domain and range have the same number of elements, φ is onto.

Conversely if $\mathrm{GCD}(r,s) \neq 1$, then some prime p divides both r and s. The number of elements in C_{rs} of order p is then $p-1$, while the number of elements in $C_r \times C_s$ of order p is $p(p-1)+(p-1) = p^2-1$. So C_{rs} cannot be isomorphic to $C_r \times C_s$.

20. Three, namely C_{27}, $C_9 \times C_3$, and $C_3 \times C_3 \times C_3$.

21. The matrix relating the bases is $C = \begin{pmatrix} 3 & 2 & 5 \\ 0 & 1 & 3 \\ 0 & 1 & 5 \end{pmatrix}$. A row interchange and a column interchange move the entry 1 in the center to the upper left and give $\begin{pmatrix} 1 & 0 & 3 \\ 2 & 3 & 5 \\ 1 & 0 & 5 \end{pmatrix}$. Two row operations and one column operation eliminate the other entries in the first column and first row, yielding $\begin{pmatrix} 1 & 0 & 0 \\ 0 & 3 & -1 \\ 0 & 0 & 2 \end{pmatrix}$. The remaining steps pass from there to

$$\begin{pmatrix} 1 & 0 & 0 \\ 0 & -1 & 3 \\ 0 & 2 & 0 \end{pmatrix} \mapsto \begin{pmatrix} 1 & 0 & 0 \\ 0 & 1 & -3 \\ 0 & 2 & 0 \end{pmatrix} \mapsto \begin{pmatrix} 1 & 0 & 0 \\ 0 & 1 & -3 \\ 0 & 0 & 6 \end{pmatrix} \mapsto \begin{pmatrix} 1 & 0 & 0 \\ 0 & 1 & 0 \\ 0 & 0 & 6 \end{pmatrix}.$$

Hence $H = \mathbb{Z} \oplus \mathbb{Z} \oplus 6\mathbb{Z}$, and $G/H \cong C_6$.

22. Let the four generators for G be x_1, x_2, x_3, x_4, and let the four generators for H be y_1, y_2, y_3, y_4. Since each is linearly independent over $\mathbb{Q}$, it is linearly independent over $\mathbb{Z}$. The matrix of the y_i's in terms of the x_j's is $C = \begin{pmatrix} 2 & -1 & 0 & 0 \\ -1 & 2 & -1 & -1 \\ 0 & -1 & 0 & 2 \\ 0 & -1 & 2 & 0 \end{pmatrix}$. The reduction procedure on this leads to $\begin{pmatrix} 1 & 0 & 0 & 0 \\ 0 & 1 & 0 & 0 \\ 0 & 0 & 2 & 0 \\ 0 & 0 & 0 & 2 \end{pmatrix}$. Hence $G/H \cong C_2 \times C_2$.

23. Each step of row reduction or column reduction preserves the rank of the matrix as a member of $M_{nn}(\mathbb{Q})$ since row rank equals column rank. Following through the steps of the procedure, we may assume that the matrix is diagonal with diagonal entries $D_{11}, \ldots, D_{nn}$ with $D_{jj} \neq 0$ exactly for $1 \leq j \leq r$. Then $H = \bigoplus_{j=1}^{r} D_{jj}\mathbb{Z}$, and we can read off that H has rank r and the $\mathbb{Q}$ rank of the matrix is r.

Chapter IV   619

24. Let G be an abelian group, and let $\widetilde{G} = \bigoplus_{g \in G} \mathbb{Z}$. For each g, form the homomorphism $\varphi_g : \mathbb{Z} \to G$ given in additive notation by $\varphi_g(n) = ng$. Then the universal mapping property of direct sums gives the desired homomorphism of the free abelian group $\widetilde{G}$ onto G.

25. For (a), right translation by any element of $H \cap K$ sends xH to itself and yK to itself, hence sends $xH \cap yK$ to itself. Therefore $xH \cap yK$ is a union of left cosets of $H \cap K$. We are to see that at most one left coset is involved. Thus suppose we have two elements g_1 and g_2 in $xH \cap yK$. Write $g_1 = xh_1 = yk_1$ and $g_2 = xh_2 = yk_2$. Then $g_2^{-1}g_1 = h_2^{-1}h_1 = k_2^{-1}k_1$, and $g_2^{-1}g_1$ is exhibited as in $H \cap K$. So g_1 is in $g_2(H \cap K)$.

For (b), if the sets $x_1 H, \ldots, x_m H$ exhaust G and the sets $y_1 K, \ldots, y_n K$ exhaust G, then G is the union of the mn sets $x_i H \cap y_j K$. By (a), G is exhibited as the union of $\leq mn$ left cosets of $H \cap K$.

26. Returning to Problem 23, we see that $H = \bigoplus_{j=1}^{n} D_{jj}\mathbb{Z}$ with each $D_{jj} \neq 0$. Then the index of H in G is $\prod_{j=1}^{n} D_{jj}$.

27. In (a), take $H_2 = \{(1), (1\ 2)(3\ 4), (1\ 3)(2\ 4), (1\ 4)(2\ 3), (1\ 3), (2\ 4), (1\ 2\ 3\ 4), (1\ 4\ 3\ 2)\}$. The number of such subgroups is $2k + 1$ and divides 3. Since H_2 is not normal, the number is > 1. Therefore it is 3.

In (b), take $H_3 = \{(1), (1\ 2\ 3), (1\ 3\ 2)\}$. The number of such subgroups is $3k + 1$ and divides 8. Since H_3 is not normal, the number is 4.

28. Disproof: In $\mathfrak{S}_3$, take $H = \{(1), (1\ 2)\}$. Then $N(H) = H$, and this is not normal.

29. Since $168/7 = 24$, the number of Sylow 7-subgroups is $7k + 1$ and divides 24. The group G is assumed simple, and so $k \neq 0$. Then k must be 1, and there are 8 distinct Sylow 7-subgroups. Any two of these intersect only in the identity, and each contains 6 elements of order 7. Hence there are 48 elements of order 7.

30. The number of Sylow q-subgroups is $qk + 1$ and divides p, hence must be 1. So S_q is normal, and the set $S_p S_q$ of products is a subgroup. An argument in the proof of Proposition 4.60 shows that each element of G is uniquely a product of a member of S_p and a member of S_q, and hence G is a semidirect product.

31. Let Γ be the set of subgroups conjugate to H, and form the action $G \times \Gamma \to \Gamma$ by conjugation. The isotropy subgroup at H is $N(H)$, which must have index 1 or index p in G. If it has index 1, then H is normal, and $|\Gamma| = 1$. Otherwise it has index p. Then $N(H) = H$, the orbit of H has $|G|/|H| = p$ elements, and $|\Gamma| = p$.

32. In (a), the subgroup H is a Sylow 2-subgroup, and the number of its conjugates must then be $2k+1$ and divide $24/8 = 3$. Since H is assumed not normal, the number of conjugates has to be 3.

In (b), call the conjugates H, H', and H''. Each member g of G acts on the set $\{H, H', H''\}$ by conjugation of the subgroups, sending H to gHg^{-1}, H' to $gH'g^{-1}$, and H'' to $gH''g^{-1}$. The result is that we obtain a function Φ from G to the permutation group $\mathfrak{S}_3$ on $\{H, H', H''\}$. This function Φ is a group homomorphism.

In (c), the subgroup ker Φ is normal, and it is enough to show that this subgroup is neither $\{1\}$ nor G. The image of Φ is not the identity subgroup since some member g of G has $gHg^{-1} = H'$; thus ker $\Phi \neq G$. Since $24/|\ker \Phi| = |G|/|\ker \Phi| = |\text{image }\Phi| \leq 6$, we have $6|\ker \Phi| \geq 24$ and $|\ker \Phi| \geq 4$; thus ker $\Phi \neq \{1\}$.

33. Let H be a Sylow 3-subgroup, of order 9. If H is normal, then G/H is a subgroup of order 4, necessarily either $C_2 \times C_2$ or C_4. Both of these groups of order 4 are isomorphic to subgroups of $\mathfrak{S}_4$, and thus there is a nontrivial homomorphism of G onto a subgroup of order 4 in $\mathfrak{S}_4$.

If H is not normal, then the number of conjugates of H is $3k + 1$ and divides 4. Then the number of conjugates must be 4. Arguing as in the previous problem we obtain a homomorphism of G into $\mathfrak{S}_4$ by having each element of g map to the corresponding permutation of the conjugates of H. This homomorphism is nontrivial since H can be moved to any of its conjugates by some element of G and since the number of such conjugates is > 1.

34. Let K be a Sylow q-subgroup. The number of conjugates of K is of the form $kq + 1$ and divides $2p$. If $k = 0$, then K is normal. This conclusion disposes of (a) and the first statement of (b) for this case. We come back to the remainder of (b) for this case in a moment.

If $k > 1$, then $kq + 1 \leq 2p$ is impossible since $p < q$. Thus the only other possibility besides $k = 0$ is $k = 1$. Then $q + 1$ divides $2p$. So $q + 1$ equals $1, 2, p$, or $2p$. Since $q > p$, the only possibility is $q + 1 = 2p$. This completes the argument for (a).

For the rest we may assume that $q + 1 = 2p$. If either of H or K is normal, then an argument in the proof of Proposition 4.60 shows that HK is a subgroup with pq elements. Since $2p = q + 1$, p divides $q + 1$. If p also divides $q - 1$, then p divides the difference, which is 2, and we obtain a contradiction. So p does not divide $q - 1$, and Proposition 4.60 shows that HK is abelian, hence cyclic.

Thus we are reduced to the situation that $q + 1 = 2p$ and K is not normal; we are to prove that H is normal. We have seen in this case that the number of conjugates of K is $q + 1$, and hence the number of elements of order q is $(q+1)(q-1) = 2p(q-1) = 2pq - 2p$. The number of conjugates of H is of the form $lp + 1$ and divides $2q$. If $l = 0$, then H is normal, and we are done. If $l \geq 1$, then the number of elements of order p is $(lp+1)(p-1) \geq (p+1)(p-1) = p^2 - 1$. Thus the total number of elements of order $1, p$, or q is $\geq 1 + (p^2 - 1) + (2pq - 2p) = 2pq + (p-1)^2 - 1 \geq 2pq + 2^2 - 1 > 2pq$, and we have obtained a contradiction.

35. Certainly ψ is one-one and onto. For (h, k) and (h', k') in $H \times_{\varphi_2} K$, we have

$$\psi((h,k)(h',k')) = \psi(hh', ((\varphi_2)_{h'^{-1}}(k))k') = (\varphi(hh'), ((\varphi_2)_{h'^{-1}}(k))k')$$

and

$$\psi(h,k)\psi(h',k') = (\varphi(h),k)(\varphi(h'),k') = (\varphi(hh'), ((\varphi_1)_{\varphi(h')^{-1}}(k))k').$$

The right sides are equal because $(\varphi_2)_{h'^{-1}} = (\varphi_1 \circ \varphi)_{h'^{-1}} = (\varphi_1)_{\varphi(h'^{-1})} = (\varphi_1)_{\varphi(h')^{-1}}$.

36. Again ψ is visibly one-one and onto. The formula for φ_2 in terms of φ_1 is given more concretely as $(\varphi_2)_h(k) = a\big((\varphi_1)_h(a^{-1}(k))\big)$. For (h, k) and (h', k') in $H \times_{\varphi_1} K$, we then have

$$\psi((h,k)(h',k')) = \psi(hh', ((\varphi_1)_{h'^{-1}}(k))k')$$
$$= \big(hh', a(((\varphi_1)_{h'^{-1}}(k))k')\big) = \big(hh', a((\varphi_1)_{h'^{-1}}(k))a(k')\big)$$

and

$$\psi(h,k)\psi(h',k') = (h, a(k))(h', a(k')) = \big(hh', ((\varphi_2)_{h'^{-1}}(a(k)))a(k')\big)$$
$$= \big(hh', (a((\varphi_1)_{h'^{-1}}(a^{-1}(a(k)))))a(k')\big).$$

The right sides are equal because $a^{-1}(a(k)) = k$.

37. An action of C_p on C_q is a homomorphism of C_p into $\operatorname{Aut} C_q \cong C_{q-1}$. If a is a generator of C_p and b is a generator of C_{q-1}, we may assume that $a \mapsto b^k$ for some k. Since the action is nontrivial, $0 < k < q - 1$. Then $1 = a^p$ maps to b^{kp}, and therefore b^{kp} must be 1. This means that kp must be a multiple of $q - 1$. So $kp = r(q - 1)$. Since $0 < k < q - 1$, we see that $p > r$. Therefore p does not divide r and must divide $q - 1$.

38. Put $n = (q - 1)/p$. Let a be a generator of C_p, and let b be a generator of $\operatorname{Aut} C_q \cong C_{q-1}$. For reference, take $\tau(a) = b^n$. This defines a nontrivial homomorphism of C_p into C_{q-1}. Any other one is of the form $\tau_1(a) = b^{k_1}$ with $0 \le k_1 < q - 1$. As in the previous problem, we know that $k_1 p = r(q - 1)$. Hence $k_1 = nr$ for some r with $1 \le r \le p - 1$. The mapping $\varphi(a^s) = a^{rs}$ is then an automorphism of C_p, and $\tau_1(a) = b^{k_1} = b^{nr} = \tau(a^r) = (\tau \circ \varphi)(a)$. So $\tau_1 = \tau \circ \varphi$. Problem 35 applies and yields the desired isomorphism.

40. For (a), $D_4 \supseteq C_4 \supseteq C_2 \supseteq \{1\}$, where C_4 is the subgroup of rotations. For (b), $H_8 \supseteq C_4 \supseteq C_2 \supseteq \{1\}$, where C_4 is the subgroup $\{\pm 1, \pm i\}$.

41. For (a), the trivial subgroup, the whole group, and all subgroups of index 2 are automatically normal. The only other possibility is order 2. Since -1 is the only element of order 2, the only subgroup of order 2 is $\{\pm 1\}$. This is the center of H_8 and hence is normal.

For (b), the five conjugacy classes are $\{\pm \mathbf{i}\}, \{\pm \mathbf{j}\}, \{\pm \mathbf{k}\}, \{-1\}$, and $\{1\}$.

For (c), Problem 15 shows that the inner automorphisms form a normal subgroup isomorphic to the quotient of H_8 by its center. The center is $\{\pm 1\}$, and thus the inner automorphisms form a subgroup of the group of all automorphisms isomorphic to $C_2 \times C_2$. The nontrivial inner automorphisms multiply two of $\mathbf{i}, \mathbf{j}, \mathbf{k}$ by -1 and fix the third one. In addition, the cyclic map $\mathbf{i} \mapsto \mathbf{j} \mapsto \mathbf{k} \mapsto \mathbf{i}$ is an automorphism and gives an automorphism of order 3. So is its square. One more automorphism fixes $\mathbf{i}$ and has $\mathbf{j} \mapsto \mathbf{k} \mapsto -\mathbf{j} \mapsto -\mathbf{k}$. Consequently the group of automorphisms G acts transitively on the set of six elements of order 4, and $|G| = 6|H|$, where H is the subgroup fixing **i**. With **i** fixed, an automorphism can carry **j** to any of $\pm$**j** and $\pm$**k**. Thus $|H| = 4|K|$, where K is the subgroup fixing **i** and **j**. Since **i** and **j** generate H_8, K is trivial. Hence $|\operatorname{Aut} H_8| = 24$.

42. The only possible orders are the divisors of 8. If it were to have an element of order 8, it would be cyclic, hence abelian. If all elements other than the identity were to have order 2, it would be abelian by Problem 1. Hence it must have an element of order 4.

43. Let C_2 be the subgroup generated by the element of order 2. Proposition 4.44 shows that G is a semidirect product $C_2 \times_\tau K$, and τ has to be nontrivial for G to be nonabelian. By Problem 16a, there is only one possibility for τ. Since D_4 is one such semidirect product, G must be isomorphic to D_4.

44. Let the elements of K be the powers of **i**. By assumption every element outside K has order 4. Thus $\mathbf{i}^2$ is the only element of order 2. Its conjugacy class therefore contains no other element, and it is central. Let us write -1 for this element. No element other than ± 1 can be central since if the center has order 4, then it commutes with any other element and together they generate an abelian G. So $Z_G = \{\pm 1\}$. Next let **j** be an element of order 4 not in K. Define $\mathbf{k} = \mathbf{ij}$. We know that $\mathbf{j}^2 = \mathbf{k}^2 = -1$, and thus the 8 elements are $\pm 1, \pm \mathbf{i}, \pm \mathbf{j}, \pm \mathbf{k}$. From $\mathbf{k} = \mathbf{ij}$, we obtain $\mathbf{kj} = (\mathbf{i})(-1) = -\mathbf{i}$ and similarly $\mathbf{ik} = -\mathbf{j}$. Finally we know that **i** and **j** do not commute (since G would otherwise be abelian) and that neither **ij** nor **ji** is a power of **i** or **j**. Thus **ji** has to be $\pm \mathbf{k}$ and cannot be **k**. So $\mathbf{ji} = -\mathbf{k}$, and we then obtain $\mathbf{jk} = \mathbf{i}$ and $\mathbf{ki} = \mathbf{j}$. Thus the multiplication table in G matches that in H_8, and we have an isomorphism.

46. Suppose $K \cong C_4$. If H acts nontrivially on K, then there is a nontrivial homomorphism of $H \cong C_3$ into $\operatorname{Aut} K \cong \operatorname{Aut} C_4 \cong C_2$. Since C_2 has no element of order 3, this is impossible.

If $K \cong C_2 \times C_2$, then $\operatorname{Aut} K \cong \mathfrak{S}_3$, the automorphisms being the permutations of the set $\{(1, 0), (0, 1), (1, 1)\}$. Thus there are two nontrivial homomorphisms of C_3 into $\operatorname{Aut} K$. Since the elements of order 3 in $\mathfrak{S}_3$ are conjugate in $\mathfrak{S}_3$, Problem 36 applies and shows that the two resulting semidirect products are isomorphic. The group $\mathfrak{A}_4$ meets the conditions of this problem, and hence the given G must be isomorphic to $\mathfrak{A}_4$.

47. Certainly one of those conditions holds, and G is abelian if (i) holds. If (ii) holds, then τ has order 2, and τ is determined by its kernel. Let us rewrite the group K as $C_2 \times C_2$ with the second factor as the kernel of τ, so that τ factors through to a homomorphism of the first factor. Then $(C_2 \times C_2) \times_\tau C_3 \cong C_2 \times_{\overline{\tau}} (C_2 \times C_3) \cong C_2 \times_{\overline{\tau}} C_6 \cong D_6$. If (iii) holds, we have a nonnormal subgroup of order 4 in G, and this does not happen in $\mathfrak{A}_4$ or D_6.

48. If (iii) holds, the homomorphism $C_4 \to \operatorname{Aut} C_3$ has to be nontrivial and is then uniquely determined since $\operatorname{Aut} C_3 \cong C_2$. This proves the uniqueness of the group up to isomorphism. The group has 1 element of order 1, 3 elements of order 2, 2 elements of order 3, and 6 elements of order 4.

Chapter IV   623

49. Let H be a Sylow q-subgroup, and let K be a Sylow p-subgroup. The number of conjugates of H is of the form $qk + 1$ and divides p^2. Since p is prime, $qk + 1$ must be 1, p, or p^2. If H is not normal, then $k > 0$ and we cannot have $qk + 1 = p$ since $p < q$; therefore $qk + 1 = p^2$. In this case the number of elements of order q is $(qk + 1)(q - 1) = p^2(q - 1) = p^2 q - p^2$, and a Sylow p-subgroup then accounts for all the remaining elements. Consequently H not normal implies K normal.

Now let us analyze what k must be when $qk = p^2 - 1$. Since q is prime, q divides $p + 1$ or q divides $p - 1$. But the condition q divides $p - 1$ is impossible since $p < q$, and thus q divides $p + 1$. Since $2q > p + q > p + 1$, we must in fact have $q = p + 1$. Since all primes but 2 are odd, this says that $p = 2$ and $q = 3$. We conclude that either $p^2 q = 12$ or else the condition $qk = p^2 - 1$ is impossible; when $qk = p^2 - 1$ fails, we have seen that H is normal.

50. We form three distinct semidirect products, two with Sylow p-subgroup C_{p^2} and one with Sylow p-subgroup $C_p \times C_p$. For each a Sylow q-subgroup C_q is to be normal. We know from Problem 16a and Corollary 4.27 that the group of automorphisms of the cyclic group C_q is isomorphic with C_{q-1}. We obtain one homomorphism $C_{p^2} \to C_{q-1}$ by mapping a generator of C_{p^2} to an element in C_{q-1} of order p^2 and a second homomorphism by mapping a generator of C_{p^2} to an element in C_{q-1} of order p. The third semidirect product comes by having the first factor C_p of $C_p \times C_p$ act trivially on C_q and having the second factor act with a generator of C_p mapping to an element of order p in C_{q-1}.

51. The second and third groups constructed in the previous problem make sense when p divides $q - 1$.

52. If p does not divide $q - 1$, then $p^2 q \neq 12$. Problem 49 then shows that a Sylow q-subgroup is normal. Hence the group has to be a semidirect product. The action of a Sylow p-subgroup on C_q corresponds to a homomorphism of C_{p^2} or $C_p \times C_p$ into C_{q-1}, and the condition that p not divide $q - 1$ means that C_{p^2} or $C_p \times C_p$ must map to the identity. Therefore the group is abelian.

53. In (a) and (b), the automorphism group of $\mathbb{Z}/9\mathbb{Z}$ is given by multiplication by the members of $(\mathbb{Z}/9\mathbb{Z})^\times = \{1, 2, 4, 5, 7, 8\}$. The element 4 has square 7 and cube 1 modulo 9, and hence the multiplications by 1, 4, 7 yield a group of automorphisms of order 3 of C_9. Hence C_3 has a nontrivial action by automorphisms on C_9, and there exists a nonabelian semidirect product of C_3 and C_9 with C_9 normal.

In (c), let a be a generator of C_9, let b be a generator of C_3, and let τ_b be the automorphism $a \mapsto a^7$. Then $\tau_{b^{-1}}$ is the automorphism $a^n \mapsto a^{4n}$, and $\tau_{b^{-p}}(a^n) = a^{4^p n}$. Proposition 4.43 says that $(b^m, a^n)(b^p, a^q) = (b^{m+p}, (\tau_{b^{-p}}(a^n))a^q)$, and the right side equals $(b^{m+p}, a^{4^p n + q})$. Taking $m = -1$, $n = 1$, $p = 1$, and $q = 0$, we obtain $(b^{-1}, a)(b, 1) = (1, a^4)$. Abbreviating $(1, a)$ as a and $(b, 1)$ as b, we obtain $a^9 = b^3 = b^{-1} a b a^{-4} = 1$.

54. In such a group the subgroup H is normal by Proposition 4.36, and thus the group of order 27 is a semidirect product of C_3 and C_9 with C_9 normal. A nonabelian such semidirect product must have a generator of C_3 mapping into an automorphism

of order 3 of C_9. There are two possibilities, and Problem 35 shows that they lead to isomorphic semidirect products.

55. $|GL(2, \mathbb{F})| = (q^2 - 1)(q^2 - q)$ and $|SL(2, \mathbb{F})| = (q - 1)^{-1}|GL(2, \mathbb{F})|$ because $|GL(2, \mathbb{F})| = |\ker \det| \, |\, \text{image} \det|$. This handles (a) and (b). For (c), the scalar matrices of determinant 1 are those for which the scalar has square 1. Since the characteristic is not 2, both ± 1 qualify. Since $\mathbb{F}$ is a field, the polynomial $X^2 - 1$ can have only two roots. So we factor by a group of order 2, and the number of elements is cut in half. For (d), the order in general is $\frac{(q^2-1)(q^2-q)}{2(q-1)} = \frac{1}{2}(q-1)q(q+1)$. Then $|PSL(2, \mathbb{F}_7)| = 168$.

56. Regard G as a group of invertible linear mappings that is to be written in the standard basis Σ. Let $\Gamma = (u, v)$. If $A = \begin{pmatrix} M \\ \Gamma\Gamma \end{pmatrix}$, then $A = \begin{pmatrix} 0 & -1 \\ 1 & c \end{pmatrix}$, the upper right entry being -1 because $\det M = 1$. Then $\begin{pmatrix} M \\ \Sigma\Sigma \end{pmatrix} = \begin{pmatrix} M \\ \Sigma\Gamma \end{pmatrix} A \begin{pmatrix} M \\ \Sigma\Gamma \end{pmatrix}^{-1}$. Products AB go into products of such expressions, and conjugates hAh^{-1} by matrices of determinant 1 go into expressions

$$\left(\begin{pmatrix} M \\ \Sigma\Gamma \end{pmatrix} h \begin{pmatrix} M \\ \Sigma\Gamma \end{pmatrix}^{-1}\right)\left(\begin{pmatrix} M \\ \Sigma\Gamma \end{pmatrix} A \begin{pmatrix} M \\ \Sigma\Gamma \end{pmatrix}^{-1}\right)\left(\begin{pmatrix} M \\ \Sigma\Gamma \end{pmatrix} h \begin{pmatrix} M \\ \Sigma\Gamma \end{pmatrix}^{-1}\right)^{-1}$$

that are conjugates of such expressions. Thus if A and such expressions generate $SL(2, \mathbb{F})$, then the conjugates generate the conjugates, again giving $SL(2, \mathbb{F})$.

57. In (a), $B^{-1}A^{-1}BA$ is the product of the conjugate $B^{-1}A^{-1}B$ of the inverse of A by A itself and hence is in G. Direct computation shows that the matrix in question is $\begin{pmatrix} a^{-2} & c(a^{-2}-1) \\ 0 & a^2 \end{pmatrix}$. In (b), the diagonal entries are equal if and only if $a^{-2} = a^2$, hence if and only if $a^4 = 1$. In (c), the result of (b) shows that there are at most 4 choices of a to avoid. We must also avoid $a = 0$. Thus if the field has more than 5 elements, a can be chosen nonzero so that $a^4 \neq 1$.

58. As in Problem 57a, the conditions that C is in G and $\det D = 1$ imply that $CDC^{-1}D^{-1}$ is in G. The product in question is $\begin{pmatrix} 1 & x^2-1 \\ 0 & 1 \end{pmatrix}$. Since $x \neq \pm 1, \lambda = x^2 - 1$ is not 0.

59. Let Λ be the set of λ such that $E(\lambda) = \begin{pmatrix} 1 & \lambda \\ 0 & 1 \end{pmatrix}$ is in G. Since $E(\lambda + \lambda') = E(\lambda)E(\lambda')$ and $E(\lambda)^{-1} = E(-\lambda)$, Λ is closed under addition and negation. Since $\begin{pmatrix} \alpha & 0 \\ 0 & \alpha^{-1} \end{pmatrix} E(\lambda) \begin{pmatrix} \alpha & 0 \\ 0 & \alpha^{-1} \end{pmatrix}^{-1} = E(\alpha^2 \lambda)$, Λ is closed under multiplication by squares of nonzero elements.

60. The previous problems produce some $\lambda_0 \neq 0$ in Λ, and $-\lambda_0$ is in Λ since Λ is closed under negatives. If $x \neq \pm 1$, then $\frac{1}{4}(x+1)^2$ and $\frac{1}{4}(x-1)$ are nonzero squares, and hence $\frac{1}{4}(x+1)^2\lambda_0$ and $\frac{1}{4}(x-1)\lambda_0$ are in λ. Subtracting, we see that $x\lambda_0$ is in Λ. Thus all multiples of λ_0 except possibly for those by $0, +1, -1$ are in Λ. However, we have seen separately that $0, \lambda_0, -\lambda_0$ are in Λ. Hence $\Lambda = \mathbb{F}$.

61. The conjugacy follows from $\begin{pmatrix} 0 & 1 \\ -1 & 0 \end{pmatrix} \begin{pmatrix} 1 & \lambda \\ 0 & 1 \end{pmatrix} \begin{pmatrix} 0 & 1 \\ -1 & 0 \end{pmatrix}^{-1} = \begin{pmatrix} 1 & 0 \\ -\lambda & 1 \end{pmatrix}$. Next we have $\begin{pmatrix} 1 & a \\ 0 & 1 \end{pmatrix} \begin{pmatrix} 1 & 0 \\ b & 1 \end{pmatrix} \begin{pmatrix} 1 & c \\ 0 & 1 \end{pmatrix} = \begin{pmatrix} 1+ab & c+cab+a \\ b & bc+1 \end{pmatrix}$, and it follows that every member of SL(2, $\mathbb{F}$) with lower left entry nonzero is in G. Conjugating by $\begin{pmatrix} 0 & 1 \\ -1 & 0 \end{pmatrix}$, we obtain the same conclusion when the upper right entry is nonzero. Finally $\begin{pmatrix} 1 & 1 \\ 0 & 1 \end{pmatrix} \begin{pmatrix} \alpha & 0 \\ 0 & \alpha^{-1} \end{pmatrix} = \begin{pmatrix} \alpha & \alpha^{-1} \\ 0 & \alpha^{-1} \end{pmatrix}$ says $\begin{pmatrix} \alpha & 0 \\ 0 & \alpha^{-1} \end{pmatrix} = \begin{pmatrix} 1 & -1 \\ 0 & 1 \end{pmatrix} \begin{pmatrix} \alpha & \alpha^{-1} \\ 0 & \alpha^{-1} \end{pmatrix}$ and shows that every matrix $\begin{pmatrix} \alpha & 0 \\ 0 & \alpha^{-1} \end{pmatrix}$ is in G. Hence $G = \text{SL}(2, \mathbb{F})$.

62. Let $\varphi : \text{SL}(2, \mathbb{F}) \to \text{PSL}(2, \mathbb{F})$ be the quotient homomorphism. If H is a normal subgroup $\neq \{1\}$ in PSL(2, $\mathbb{F}$), then $\varphi^{-1}(H)$ is a normal subgroup of SL(2, $\mathbb{F}$) containing an element not in the center. By Problem 61, $\varphi^{-1}(H) = \text{SL}(2, \mathbb{F})$. Therefore $H = \varphi(\varphi^{-1}(H)) = \varphi(\text{SL}(2, \mathbb{F})) = \text{PSL}(2, \mathbb{F})$.

63. If a differs from c in a set A of k places and if b differs from c in a set B of l places, then a differs from b at most in the places of $A \cup B$, hence in at most $k + l$ places. Therefore $d(a, b) \leq d(a, c) + d(c, b)$.

If $d(w, a) \leq (D-1)/2$ and $d(w, b) \leq (D-1)/2$ with a and b distinct in C, then it follows that $d(a, b) \leq (D-1)$ and hence that $\delta(C) = \min_{x \neq y \text{ in } C} d(x, y) \leq d(a, b) \leq (D-1) < D$.

64. Since C is linear, 0 is in C. Then $\delta(C) \leq d(0, c)$ for every c in C, and we obtain $\delta(C) \leq \min_{c \in C} d(0, c)$. On the other hand, we certainly have $d(a, b) = d(0, a - b)$ for all a, b in $\mathbb{F}^n$. If a and b are in C, then the linearity of C forces $a - b$ to be in C, and hence $d(a, b) = d(0, a - b) \geq \min_{c \in C} d(0, c)$. Taking the minimum over all a and b, we obtain $\delta(C) \geq \min_{c \in C} d(0, c)$. Hence equality holds.

65. $n + 1$ and 0, 1 and n, n and 1, 2 and $n - 1$.

66. In (a), a basis vector c is 1 in one of the entries corresponding to the corner variables, and it is 0 in the other entries corresponding to corner variables. At worst it could be 1 in every entry corresponding to an independent variable. The number of independent variables is n minus the rank, i.e., n minus $\dim C$. Thus $\text{wt}(c) \leq 1 + n - \dim C$. Since $\delta(C) \leq \text{wt}(c)$, $\dim C + \delta(C) \leq n + 1$.

For (b), one can take the parity-check code.

For (c), the alternative would be $\dim C + \delta(C) = n+1$. Then $\dim C + \text{wt}(c) \geq n+1$ for every c in C. Consequently every basis vector of C must have a 1 in every position corresponding to an independent variable. Since $\dim C \geq 2$, there are at least two such basis vectors. Their sum gets a contribution of 2 to its weight from the corner variables and can have a 0 in at most 1 position corresponding to an independent variable. But their sum is 0 in every position corresponding to an independent variable. Hence there is at most one such position, and we conclude that $n - \dim C = 1$, in contradiction to the hypothesis $\dim C \leq n - 2$.

67. A direct check of all seven nonzero elements of C shows that each has weight 3. Therefore $\delta(C) = 3$.

68. In (a), the basis vectors each have one 1 in positions 3, 5, 6, 7, and at least two of the parity bits in positions 1, 2, 4 are 1 since none of 3, 5, 6, 7 is a power of 2. Any sum of two distinct basis vectors has two 1's in positions 3, 5, 6, 7, and the parity bits cannot all be 0 since the parity bits for each of the basis vectors identify the basis vector and since the two basis vectors in question are distinct. Finally the sum of three or more basis vectors has 1 in three or more positions 3, 5, 6, 7 and hence has weight ≥ 3. Thus all code words have weight ≥ 3, and therefore $\delta(C_7) \geq 3$. Since the first basis vector has weight 3, $\delta(C_7) = 3$.

In (b), each word in C_8 is a word of C_7 plus a parity bit. The part from C_7 has weight ≥ 3, by (a), and the parity bit means that the weight has to be even. Thus the weight of every word in C_8 is ≥ 4.

In (c) for C_{2^r-1}, we distinguish between the r bits whose indices are a power of two and the other $2^r - 1 - r$ bits. The first are the check bits, and the others are the message bits. The message bits are allowed to be arbitrary, and the check bits will depend on them. Thus dim $C_8 = 2^r - r - 1$. For a given pattern of message bits, the check bit in position 2^j counts, modulo 2, the number of 1's in message bits that occur is positions requiring 2^j in their binary expansions. Then C_{2^r} is obtained by adjoining a parity bit to each word of C_{2^r-1}.

The first conclusion of (d) was proved in the course of answering (c), and the other two conclusions follow by the same argument that was given for $r = 3$ in (a) and (b).

69. In (a), the dimension of the null space of H is the number of columns minus the rank, hence is $7 - 3 = 4$. Since C_7 lies in the null space and dim $C_7 = 4$, the null space equals C_7.

In (b), let c be in C_7. If e_i denotes the usual i^{th} basis vector, then $H(c + e_i) = Hc + He_i = He_i$, and this is the ith column of H.

70. Take a basis of C, write it as the rows of a matrix, row reduce the matrix, and permute the variables so that all the corner variables precede all the independent variables. The resulting matrix in block form is $(I \ A)$ for some matrix A with dim C rows and $n - \dim C$ columns. Since each basis vector has weight ≥ 3, each row of A has at least two 1's. Since each sum of two distinct basis vectors has weight ≥ 3, the sum of two distinct rows of A cannot be 0. Thus the rows of A must be distinct.

Arguing by contradiction, suppose that dim $C > n - r$, so that A has $\leq r - 1$ columns. The number of possible rows in A with at least two 1's is then $\leq 2^{r-1} - 1 - (r - 1) = 2^{r-1} - r$. Hence $n - r < \dim C \leq 2^{r-1} - r$, and $n < 2^{r-1}$, contradiction.

71. For (a), the answers are X^n, $(X + Y)^n$, $X^n + Y^n$, $\frac{1}{2}((X + Y)^n + \frac{1}{2}(X - Y)^n)$, $X^6 + 7X^3Y^3$, $X^7 + 7X^4Y^3 + 7X^3Y^4 + Y^7$, and $X^8 + 14X^4Y^4 + Y^8$. The last three are by a direct count of the number of code words of each weight.

In (b), the 0 word is the unique code word of weight 0, and it is present in every linear code.

In (c), the expression $X^{n-\text{wt}(c)} Y^{\text{wt}(c)}$ makes a contribution of 0 to the coefficient $N_k(C)$ of $X^{n-k}Y^k$ if wt$(c) \neq k$ and makes a contribution of 1 to the coefficient if wt$(c) = k$. Summing on c yields $\sum_{k=0}^{n} N_k(C) X^{n-k} Y^k = \sum_{c \in C} X^{n-\text{wt}(c)} Y^{\text{wt}(c)}$.

72. The equality $(1+X+X^2+X^4)(1+X+X^3) = 1+X^7$ produces a member of C with weight 2. Therefore $\delta(C) \leq 2$. On the other hand, the product of $1+X+X^2+X^4$ with a polynomial can never be a monomial, and therefore no code word has weight 1. Thus $\delta(C) > 1$.

73. In essence we use the method suggested by the solution to Problem 70, except that we put coefficients corresponding to low degrees on the left and we row reduce the matrix into the form $(A\ I)$. Let $8 \leq n \leq 19$. Form the images of as many of the following polynomials as have degree $\leq n$:

$$1,\ X,\ X^2,\ X^3,\ X^4,\ X^5,\ X^6+1,\ X^7+X+1,\ X^k(X^8+X^2+X+1) \text{ for } k \geq 0.$$

The list stops with $k = n - 16$. Assemble the coefficients of the image polynomials as the rows of a matrix as in Problem 70. The images form a basis of C. They all have weight 4, and thus every member of C has even weight. Since the image of 1 has weight 4, $\delta(C)$ must be 2 or 4.

Imagine doing a row reduction as in the solution of Problem 70. We want to rule out $\delta(C) = 2$, and it is enough to show that the basis vectors and all sums of two distinct basis vectors have weight > 2. To handle the basis vectors, it is enough to show that the A part of the reduced matrix $(A\ I)$ never has just one 1 in a row. To handle the sums of two distinct basis vectors, it is enough to show that the sum of two rows of A is never 0, i.e., that the rows of A are distinct.

The matrix A will have 8 columns, corresponding to powers X^l with $l \leq 7$. The rows of $(A\ I)$ are thus to correspond to polynomials of the form $X^m +$ "lower," where each expression "lower" has degree at most 7 and m takes on the values $8, 9, \ldots, n$. The polynomials whose images correspond to the rows of the reduced matrix are

$$1,\ X,\ \ldots,\ X^5,\ X^6+1,\ X^7+X+1,$$
$$X^8+X^2+X+1,\ X(X^8+X^2+X+1),\ \ldots,\ X^3(X^8+X^2+X+1),$$

and the left part A of the reduced matrix is

$$A = \begin{pmatrix} 1&1&1&0&0&0&0&0 \\ 0&1&1&1&0&0&0&0 \\ 0&0&1&1&1&0&0&0 \\ 0&0&0&1&1&1&0&0 \\ 0&0&0&0&1&1&1&0 \\ 0&0&0&0&0&1&1&1 \\ 1&1&1&0&0&0&1&1 \\ 1&0&0&1&0&0&0&1 \\ 1&0&1&0&1&0&0&0 \\ 0&1&0&1&0&1&0&0 \\ 0&0&1&0&1&0&1&0 \\ 0&0&0&1&0&1&0&1 \end{pmatrix}.$$

No row of A is 0, and no two distinct rows are equal. This completes the proof.

74. Suppose that $\{X_s\}_{s \in S}$ is an object in $\mathcal{C}^S$ and that $f_s : X_s \to A$ for each s is a function, A being a particular set. The disjoint union of the X_s's consists of all

ordered pairs (x_s, s) with $s \in S$ and $x_s \in X_s$, and we define $i_s(x_s) = (x_s, s)$. To define a function f from the disjoint union of the X_s's into A such that $fi_s = f_s$ for all s, we let $f(x_s, s) = f_s(x_s)$. Then $fi_s(x_s) = f(x_s, s) = f_s(x_s)$. Thus f exists. On the other hand, the condition that $fi_s = f_s$ forces $f(x_s, s)$ to be $f_s(x_s)$, and hence the f in the universal mapping property is unique, as it is required to be.

76. Peeking ahead to Problem 80, we take the category to be $\mathcal{C}^{\text{opp}}$, where $\mathcal{C}$ is the category defined in Section 11 after Example 4 of products. The category $\mathcal{C}$ has no product functor when S has two elements.

77. The existence of the identity and associativity are part of the definition. The existence of inverses is given in the hypothesis. The answer to the question is "yes"; if a group G is given, define a category with one object, namely the set G, define $\text{Morph}(G, G)$ to be the set G, and let the law of composition be the group law.

78. To see that $\circ^{\text{opp}}$ is well defined, let f be in $\text{Morph}_{\mathcal{C}^{\text{opp}}}(A, B)$, and let g be in $\text{Morph}_{\mathcal{C}^{\text{opp}}}(B, C)$. The definition is $g \circ^{\text{opp}} f = f \circ g$, and this is meaningful since g is in $\text{Morph}_{\mathcal{C}}(C, B)$ and f is in $\text{Morph}_{\mathcal{C}}(B, A)$. The associativity and the existence of the identity are straightforward to check. It is clear from the definition that $(\mathcal{C}^{\text{opp}})^{\text{opp}} = \mathcal{C}$.

In a diagram the vertices stay where they are, and so do the morphisms, since the objects and the sets of morphisms do not change. However, the direction of each arrow is reversed since "domain" and "range" are interchanged in passing from $\mathcal{C}$ to $\mathcal{C}^{\text{opp}}$. Thus diagrams map to diagrams with the arrows reversed.

Compositions correspond because of the definition of $\circ^{\text{opp}}$, and it follows that commutative diagrams map to commutative diagrams.

79. Let A and B be sets such that A has three elements and B has one element. The number of functions from A to B is then one, and the number of functions from B to A is three. Since $\text{Morph}_{\mathcal{C}^{\text{opp}}}(A, B) = \text{Morph}_{\mathcal{C}}(B, A)$, $\text{Morph}_{\mathcal{C}^{\text{opp}}}(A, B)$ has three elements and cannot be accounted for by functions from A to B.

80. For (a), if $(X, \{p_s\}_{s \in S})$ is a product of $\{X_s\}_{s \in S}$, we set up the diagram of the universal mapping property of the product. Passing to $\mathcal{C}^{\text{opp}}$ and using Problem 78, we obtain the same diagram in $\mathcal{C}^{\text{opp}}$ but with the arrows reversed. Then it follows that $(X, \{p_s\}_{s \in S})$, when interpreted in $\mathcal{C}^{\text{opp}}$, satisfies the condition of being a coproduct. The other half proceeds in the same way.

For (b), we start with two coproducts in $\mathcal{C}$ and pass to $\mathcal{C}^{\text{opp}}$, where they become products, according to (a). Proposition 4.63 shows that the two products are canonically isomorphic in $\mathcal{C}^{\text{opp}}$. This isomorphism, when reinterpreted in $\mathcal{C}$, is still an isomorphism, and the result is that the two coproducts in $\mathcal{C}$ are canonically isomorphic.

Chapter V

1. For (a), we have $((g_1, h_1)((g_2, h_2)x)) = (g_1, h_1)(g_2 x h_2^{-1}) = g_1 g_2 x h_2^{-1} h_1^{-1} = (g_1 g_2) x (h_1 h_2)^{-1} = (g_1 g_2, h_1 h_2)x$ and $(1, 1)x = 1x1^{-1} = x$.

For (b), left multiplications by $GL(m, \mathbb{C})$ preserve the row space, hence the rank, and right multiplications by $GL(n, \mathbb{C})$ preserve the column space, hence the rank. Hence all members of an orbit have the same rank.

Row operations, which correspond to left multiplications by elementary matrices, can be used to bring the matrix into reduced row-echelon form, and then column operations, which correspond to right multiplications by elementary matrices, can be used to bring the result into reduced column-echelon form. If $r = \min(m, n)$, then the resulting matrix is 1 in entries $(1, 1), (2, 2), \ldots, (l, l)$ for some $l \leq r$ and 0 elsewhere. This has rank l and answers (c) and the remainder of (b).

2. If A has minimal polynomial $\lambda^k + c_{k-1}\lambda^{k-1} + \cdots + c_1\lambda + c_0$, with $c_0 \neq 0$, then $I = A(-c_0^{-1}(A^{k-1} + c_{k-1}A^{k-2} + \cdots + c_1 I))$, and A is invertible. Conversely if $c_0 = 0$, then λ is a factor of the minimal polynomial and must be a factor of the characteristic polynomial, by Corollary 5.10. Then 0 is an eigenvalue, and the null space is nonzero. Hence A is not invertible.

3. Proposition 5.12 shows that $l_j \geq \max(r_j, s_j)$. For u in U, we know that $P_1(L)^{r_1} \cdots P_k(L)^{r_k}(u) = 0$. For w in W, we know that $P_1(L)^{s_1} \cdots P_k(L)^{s_k}(w) = 0$. Thus any v in U or W has $P_1(L)^{\max(r_1,s_1)} \cdots P_k(L)^{\max(r_k,s_k)}(v) = 0$. Forming sums, we see that $P_1(L)^{\max(r_1,s_1)} \cdots P_k(L)^{\max(r_k,s_k)}(v) = 0$ for all v in V. Thus the minimal polynomial divides $P_1(\lambda)^{\max(r_1,s_1)} \cdots P_k(\lambda)^{\max(r_k,s_k)}$, and we must have $l_j \leq \max(r_j, s_j)$.

4. For any monomial $P(\lambda) = \lambda^j$, the monomial $Q(\lambda) = \lambda P(\lambda) = \lambda^{j+1}$ has $Q(BA) = BA(BA)^j = B(AB)^j A = BP(AB)A$. Taking suitable linear combinations of this result as j varies, we obtain (a).

For (b), let $M_{AB}(\lambda)$ and $M_{BA}(\lambda)$ be the minimal polynomials of AB and BA. Part (a) implies that $M_{BA}(\lambda)$ divides $\lambda M_{AB}(\lambda)$. Reversing the roles of A and B, we see that $M_{AB}(\lambda)$ divides $\lambda M_{BA}(\lambda)$. By unique factorization all the prime powers in the prime factorizations of $M_{AB}(\lambda)$ and $M_{BA}(\lambda)$ are the same except for the power of λ. The powers of λ in the factorizations of $M_{AB}(\lambda)$ and $M_{BA}(\lambda)$ differ at most by 1.

5. Theorem 5.14 allows us to write $\mathbb{K}^n = U_1 \oplus \cdots \oplus U_k$ and $\mathbb{K}^n = W_1 \oplus \cdots \oplus W_l$, where the U_j are the eigenspaces for the distinct eigenvalues of D and the W_j are the eigenspaces for the distinct eigenvalues of D'. These decompositions are the primary decompositions as in Theorem 5.19, and (e) of that theorem shows that $W_j = (W_j \cap U_1) \oplus \cdots \oplus (W_j \cap U_k)$ for $1 \leq j \leq l$. Summing on j, we see that $\mathbb{K}^n$ is the direct sum of all $U_i \cap W_j$. Each of D and D' is scalar on $U_i \cap W_j$, and (a) follows by translating this result into a statement about matrices.

The matrices $N = \begin{pmatrix} 0 & 1 \\ 0 & 0 \end{pmatrix}$ and $N' = \begin{pmatrix} 0 & 2 \\ 0 & 0 \end{pmatrix}$ commute, and both have N uniquely as Jordan form. If C were to exist with $C^{-1}NC$ and $C^{-1}N'C$ both in Jordan form, we would have $C^{-1}NC = C^{-1}N'C$ and $N = N'$, contradiction. This answers (b).

6. If E is the projection of V on U along W, then each member of U is an eigenvector with eigenvalue 1, and each member of W is an eigenvector with eigenvalue 0. The union of bases of U and W is then a basis of eigenvectors for E, and (a) follows

from Theorem 5.14. In view of Proposition 5.15, two projections are given by similar matrices if and only if they have the same rank.

7. For (a), $EF = F$ implies image $F \subseteq$ image E, which implies $EF = F$. Reversing the roles of E and F, we see that $FE = E$ if and only if image $E \subseteq$ image F.

For (b), $EF = E$ implies $\ker F \subseteq \ker E$, while $FE = F$ implies $\ker E \subseteq \ker F$. So $EF = E$ and $FE = F$ implies $\ker E = \ker F$. Conversely if $\ker F \subseteq \ker E$, then $EF = E$ on $\ker F$ and $EF = E$ on image F; so $EF = E$. Reversing the roles of E and F, we see that $\ker E \subseteq \ker F$ implies $FE = F$.

8. If $EF = FE$, then $(EF)^2 = EFEF = E(FE)F = E(EF)F = E^2F^2 = EF$. So EF is a projection. This proves (a).

For (b), let $E = \begin{pmatrix} 1 & 0 \\ 0 & 0 \end{pmatrix}$ and $F = \begin{pmatrix} 1 & 1 \\ 0 & 0 \end{pmatrix}$. Each is a projection, and $EF = F$, so that EF is a projection. However, $FE = E$. Since $E \neq F$, $EF \neq FE$.

9. If E is a projection, then $U = 2E - I$ has $U^2 = 4E^2 - 4E + I = 4E - 4E + I = I$; so U is an involution. If U is an involution, then $E = \frac{1}{2}(U + I)$ has $E^2 = \frac{1}{4}(U^2 + 2U + I) = \frac{1}{4}(I + 2U + I) = \frac{1}{2}(U + I) = E$. So E is a projection. The two formulas $U = 2E - I$ and $E = \frac{1}{2}(U + I)$ are inverse to each other.

10. Apply Theorem 5.19, and take U to be the primary subspace for the prime polynomial λ and W to be the sum of the remaining primary subspaces. Then (i), (ii), and (iii) are immediate from the theorem. For (iv), let U_j be the primary subspace for some other prime polynomial $P(\lambda)$. The theorem shows that $L\big|_{U_j}$ has a power of $P(\lambda)$ as minimal polynomial. Since λ does not divide $P(\lambda)$, Problem 2 shows that $L\big|_{U_j}$ is invertible. Hence $L\big|_{U_j}$ is invertible on the direct sum of the U_j's other than the one for the polynomial λ.

11. Let $V = U_1 \oplus \cdots \oplus U_k$ be the primary decomposition, with U_1 corresponding to the prime λ. By (ii) and Theorem 5.19e, $U = (U_1 \cap U) \oplus \cdots \oplus (U_k \cap U)$ and similarly for W. Then $U_j \cap U = 0$ for $j \geq 2$ by (iii), and hence $U \subseteq U_1$. By (iv), $U_1 \cap W = 0$, so that $W \subseteq U_2 \oplus \cdots \oplus U_k$. By (i), $U = U_1$ and $W = U_2 \oplus \cdots \oplus U_k$.

12. Part (a) is immediate, and a basis for (b) consists of the union of bases for the individual U_j's. Part (f) is evident.

Since D is a linear combination of the E_j's and each E_j is a polynomial in L, D is a polynomial in L. Hence N is a polynomial in L. This proves (d) and (e).

If $Q_j(\lambda)$ is the polynomial $(\lambda - \lambda_0)^{l_j}$, then $N^{l_j} = (L - D)^{l_j} = Q_j(L)$ on U_j, and Theorem 5.19f shows that $Q_j(L)$ is 0 on U_j. Therefore a power of N is 0 on each U_j, and N is nilpotent. This proves (c). Part (g) now follows.

13. Each eigenvector of D must lie in some U_j by Theorem 5.19e. If V_i is the eigenspace of D with eigenvalue c_i, it follows that $V_i \subseteq U_{j(i)}$ for some $j = j(i)$. Thus each U_j is the sum of full eigenspaces of D. Property (d) forces N to carry V_i into itself. By (c), $(L - D)^n$ is 0 on V_i for $n = \dim V$; hence $(L - c_i I)^n$ is 0 on V_i. Since $V_i \subseteq U_j$, $(L - \lambda_j I)^n$ is 0 on U_j. Application of Problem 10 to $L - c_i I$

shows that $L - \lambda_j I$ is nonsingular on V_i if $c_i \neq \lambda_j$, in contradiction to the fact that $(L - \lambda_j I)^n$ is 0 on U_j, and therefore $c_i = \lambda_j$. The conclusion is that $V_i = U_{j(i)}$, and the desired uniqueness follows.

14. Since $-A$ is nilpotent, Lemma 5.22 shows that $\det(\lambda I - (-A)) = \lambda^n$, where n is the size of the matrix. Substitution of $\lambda = 1$ gives the desired result.

15. The characteristic polynomial is $\lambda^2 - 2\lambda + 1 = (\lambda - 1)^2$. Since $A - I \neq 0$, the minimal polynomial is $(\lambda - 1)^2$ rather than $\lambda - 1$. Thus the Jordan form is $J = \begin{pmatrix} 1 & 1 \\ 0 & 1 \end{pmatrix}$. Solving shows that $\ker(A - I)$ consists of the multiples of $\begin{pmatrix} 3/2 \\ 1 \end{pmatrix}$. Use $\begin{pmatrix} 3 \\ 2 \end{pmatrix}$ as the first column of C, and solve $(A - I)X = \begin{pmatrix} 3 \\ 2 \end{pmatrix}$ to get $X = \begin{pmatrix} 1 \\ 1 \end{pmatrix}$ as one answer for the second column. Then $C = \begin{pmatrix} 3 & 1 \\ 2 & 1 \end{pmatrix}$, $C^{-1} = \begin{pmatrix} 1 & -1 \\ -2 & 3 \end{pmatrix}$, and one readily checks that $C^{-1}AC = J$.

16. The characteristic polynomial is $P(\lambda) = \det(\lambda I - A) = \lambda^3$. Thus A is nilpotent, and in fact $A^2 = 0$. Then $J = \begin{pmatrix} 0 & 1 & 0 \\ 0 & 0 & 0 \\ 0 & 0 & 0 \end{pmatrix}$, and the computation proceeds as in Example 1 in Section 7, yielding $C = \begin{pmatrix} 4 & 1 & -1 \\ -8 & 0 & 4 \\ 8 & 0 & 0 \end{pmatrix}$ and $C^{-1} = \begin{pmatrix} 0 & 0 & \frac{1}{8} \\ 1 & \frac{1}{4} & -\frac{1}{4} \\ 0 & \frac{1}{4} & \frac{1}{4} \end{pmatrix}$.

17. The characteristic polynomial is $(\lambda - 2)^6(\lambda - 3)$ by inspection. Thus there is a primary subspace for $\lambda - 2$ with dimension 6 and a primary subspace for $\lambda - 3$ with dimension 1. For the Jordan form let $K_j = \ker(A - 2I)^j$. By raising $A - 2I$ to powers and row reducing, we see that $\dim K_3 = 6$, $\dim K_2 = 5$, and $\dim K_1 = 3$. We do not have to proceed beyond K_3 since we have reached the full dimension 6 of the primary subspace for $\lambda - 2$. Therefore the number of Jordan blocks for $\lambda - 2$ of size ≥ 3 is $6 - 5 = 1$, of size ≥ 2 is $5 - 3 = 2$, and of size ≥ 1 is 3. Hence there is one block of each size 1, 2, and 3, and

$$J = \begin{pmatrix} 2 & 1 & 0 & & & & \\ & 2 & 1 & & & & \\ & & 2 & & & & \\ & & & 2 & 1 & & \\ & & & & 2 & & \\ & & & & & 2 & \\ & & & & & & 3 \end{pmatrix}.$$

Solving $(A - 3I)X = 0$, we find that the eigenvectors for eigenvalue 3 are the multiples of $(5, 2, 2, 3, 2, 1, 1)$. Thus this vector can be taken to be the last column of C.

The next step is to express K_1, K_2, and K_3 explicitly in terms of parameters by using the standard solution procedure for systems of homogeneous linear equations.

The result is that

$$K_1 = \left\{ \begin{pmatrix} x_1 \\ x_2 \\ x_3 \\ 0 \\ 0 \\ 0 \\ 0 \end{pmatrix} \right\}, \quad K_2 = \left\{ \begin{pmatrix} x_1 \\ x_2 \\ x_3 \\ x_4 \\ x_5 \\ 0 \\ 0 \end{pmatrix} \right\}, \quad K_3 = \left\{ \begin{pmatrix} x_1 \\ x_2 \\ x_3 \\ x_4 \\ x_5 \\ x_6 \\ 0 \end{pmatrix} \right\}.$$

Following the method of Example 1 in Section 7, we choose W_2 such that $K_3 = K_2 \oplus W_2$, and then we form $U_1 = (A - 2I)(W_2)$:

$$W_2 = \left\{ \begin{pmatrix} 0 \\ 0 \\ 0 \\ 0 \\ 0 \\ x_6 \\ 0 \end{pmatrix} \right\} \quad \text{and} \quad U_1 = \left\{ \begin{pmatrix} 0 \\ x_6 \\ 0 \\ x_6 \\ x_6 \\ 0 \\ 0 \end{pmatrix} \right\}.$$

We choose W_1 such that $K_2 = K_1 \oplus U_1 \oplus W_1$, and we form $U_0 = (A-2I)(U_1+W_1)$:

$$W_1 = \left\{ \begin{pmatrix} 0 \\ 0 \\ 0 \\ x_4 \\ 0 \\ 0 \\ 0 \end{pmatrix} \right\} \quad \text{and} \quad U_0 = \left\{ \begin{pmatrix} x_4+2x_6 \\ 0 \\ x_6 \\ 0 \\ 0 \\ 0 \\ 0 \end{pmatrix} \right\}.$$

Finally we choose W_0 such that $K_1 = K_0 \oplus U_0 \oplus W_0$. Here $K_0 = 0$, and we can take $W_0 = \{(0, x_2, 0, 0, 0, 0, 0)\}$.

To form C we take a basis of each W_j, apply powers of $A-2I$ in turn to its members, and line up the resulting columns, along with the eigenvector for eigenvalue 3, as C:

$$C = \begin{pmatrix} 2 & 0 & 0 & 1 & 0 & 0 & 5 \\ 0 & 1 & 0 & 0 & 0 & 1 & 2 \\ 1 & 0 & 0 & 0 & 0 & 0 & 2 \\ 0 & 1 & 0 & 0 & 1 & 0 & 3 \\ 0 & 1 & 0 & 0 & 0 & 0 & 2 \\ 0 & 0 & 1 & 0 & 0 & 0 & 1 \\ 0 & 0 & 0 & 0 & 0 & 0 & 1 \end{pmatrix}.$$

18. In (a), if every prime-power factor of the minimal polynomial is of degree 1, then the matrix is similar to a diagonal matrix, and the multiplicities of the eigenvalues can be seen from the characteristic polynomial. If the minimal polynomial is $(\lambda - c)^2$, then the matrix has to be similar to $\begin{pmatrix} c & 1 & 0 \\ 0 & c & 0 \\ 0 & 0 & c \end{pmatrix}$. If the minimal polynomial instead is $(\lambda - c)^2(\lambda - d)$, then the matrix has to be similar to $\begin{pmatrix} c & 1 & 0 \\ 0 & c & 0 \\ 0 & 0 & d \end{pmatrix}$. If the minimal

Chapter V   633

polynomial is $(\lambda - c)^3$, then the matrix has to be similar to $\begin{pmatrix} c & 1 & c \\ 0 & c & 1 \\ 0 & 0 & c \end{pmatrix}$. There are no other possibilities.

For (b), $\begin{pmatrix} 0 & 1 & 0 & 0 \\ 0 & 0 & 0 & 0 \\ 0 & 0 & 0 & 1 \\ 0 & 0 & 0 & 0 \end{pmatrix}$ and $\begin{pmatrix} 0 & 1 & 0 & 0 \\ 0 & 0 & 0 & 0 \\ 0 & 0 & 0 & 0 \\ 0 & 0 & 0 & 0 \end{pmatrix}$ both have minimal polynomial λ^2 and characteristic polynomial λ^4, but they are not similar because their ranks are unequal.

19. If the diagonal entries are c and N denotes the strictly upper-triangular part, then $J^k = (cI + N)^k = \sum_{j=0}^{k} \binom{k}{j} c^{k-j} N^j$. The term from $j = 1$ is not canceled by any other term, and hence J^k is not diagonal.

20. Choose J in Jordan form and C invertible with $J = C^{-1}AC$. Then $J^n = CA^nC^{-1} = CC^{-1} = I$. By Problem 19, every Jordan block in J is of size 1-by-1. Thus A is similar to a diagonal matrix D, and each diagonal entry of D must be an nth root of unity. Any n-tuple of nth roots of unity can form the diagonal entries, and the corresponding matrices are similar if and only if one is a permutation of the other.

21. The minimal polynomial has to divide $\lambda(\lambda^2 - 1) = \lambda(\lambda + 1)(\lambda - 1)$. Hence there is a basis of eigenvectors, the allowable eigenvalues being 1, -1, and 0. A similarity class is therefore given by an unordered triple of elements from the set $\{1, -1, 0\}$. There are three possibilities for a single eigenvalue, six possibilities for one eigenvalue of multiplicity 2 and one of multiplicity 1, and one possibility with all three eigenvalues present. So the answer is ten.

22. If $A^2 = N$ and $N^n = 0$, then $A^{2n} = 0$. So A is nilpotent and $A^n = 0$. Since $N^{n-1} \neq 0$, $A^{2n-2} \neq 0$. Therefore $n > 2n - 2$, and $n = 1$.

23. If J is of size n, then the matrix C with $C_{i,n+1-i} = 1$ for $1 \leq i \leq n$ and $C_{ij} = 0$ otherwise has $C^{-1}JC = J^t$.

24. Choose C with $C^{-1}AC = J$ in Jordan form. Problem 23 shows that there is a block-diagonal matrix B with $B^{-1}JB = J^t$. Then $B^{-1}C^{-1}ACB = J^t$ and $C^t A^t (C^{-1})^t = J^t$. So $B^{-1}C^{-1}ACB = C^t A^t (C^{-1})^t$, and the result follows.

25. The matrices A and B have $A^2 = B^2 = 0$ and hence are nilpotent. Since each of A and B has rank 2, $\dim \ker A = \dim \ker B = 2$. The numbers $\dim \ker A^k$ and $\dim \ker B^k$ being equal for all k, the two matrices have the same Jordan form and are therefore similar.

26. If $M(\lambda)$ is the minimal polynomial of L, then $M(L)v = 0$. Hence $M(\lambda)$ is in $\mathcal{I}_v$. Then Proposition 5.8 shows that $M_v(\lambda)$ exists.

27. The polynomial $M_v(\lambda)$ has to divide the minimal polynomial of $L|_{\mathcal{P}(v)}$, and the latter has degree $\leq \dim \mathcal{P}(v)$. Hence $\deg M_v(\lambda) \leq \dim \mathcal{P}(v)$. If $v, L(v), \ldots, L^{\deg M_v - 1}(v)$ are linearly dependent, then there is a nonzero polynomial $Q(\lambda)$ of degree $\leq \deg M_v - 1$ with $Q(L)(v) = 0$, and that fact contradicts the minimality of the degree of $M_v(\lambda)$. Hence they are independent, and $\deg M_v(\lambda) \geq \dim \mathcal{P}(v)$. Thus equality holds, and the linearly independent set is a basis. This proves (a) and (b).

Since $M_v(\lambda)$ divides the minimal polynomial of $L\big|_{\mathcal{P}(v)}$, which divides the characteristic polynomial of $L\big|_{\mathcal{P}(v)}$, and since the end polynomials have degree $\dim \mathcal{P}(v)$, these three polynomials are all equal. This proves (c).

28. Use the ordered basis $(L^{d-1}(v), L^{d-2}(v), \ldots, L(v), v)$.

29. Since $P(\lambda)$ is prime and does not divide $Q(\lambda)$, there exist polynomials $A(\lambda)$ and $B(\lambda)$ with $A(\lambda)P(\lambda) + B(\lambda)Q(\lambda) = 1$. Using the substitution that sends λ to L and applying both sides to v, we obtain $B(L)Q(L)(v) = v$. Hence $\mathcal{P}(Q(L)(v)) \supseteq \mathcal{P}(v)$. Since the reverse inclusion is clear, the result follows.

30. In (a), the base case of the induction is that $\dim V = \deg P(\lambda)$, and then the result follows from Problem 27. For the inductive step, the same problem shows that there must be a nontrivial invariant subspace U. Proposition 5.12 shows that the minimal polynomial for U and V/U is $P(\lambda)$, and induction shows that the characteristic polynomial for U and V/U is a power of $P(\lambda)$. Proposition 5.11 then shows that the characteristic polynomial for V is a power of $P(\lambda)$.

For (b), we induct on l, using (a) to handle the case $l = 1$. For general l, form the invariant subspace $U = \ker P(\lambda)^{l-1}$, for which the minimal polynomial is some $P(\lambda)^r$ with $r < l$. The minimal polynomial of V/U is certainly $P(\lambda)$. By induction, U and V/U have characteristic polynomials equal to powers of $P(\lambda)$, and Proposition 5.11 shows that the same thing is true for V.

In (c), (b) says that the characteristic polynomial is of the form $P(\lambda)^r$ for some r. Then the degree of the characteristic polynomial is rd, where $d = \deg P(\lambda)$.

31. Part (a) follows from Theorem 5.19 and Problem 30b.

For (b), Proposition 5.11 shows that it is enough to prove the result for the restrictions of D and L to U_j. The characteristic polynomial of $D\big|_{U_j}$ is $(\lambda - \lambda_j)^{\dim U_j}$ since $D\big|_{U_j}$ is scalar. Theorem 5.19f and (a) in the present problem show that the characteristic polynomial of $L\big|_{U_j}$ is a power of $\lambda - \lambda_j$. The power must be $\dim U_j$ since the degree of the characteristic polynomial is the dimension of the space. Thus the restrictions to U_j of D and L have the same characteristic polynomial.

32–34. These are proved word-for-word in the same way as Lemmas 5.23 through 5.25 except that n is to be replaced by l and N is to be replaced by $P(L)$.

35. If $Q(\lambda)$ is in $\mathbb{K}[\lambda]$, we successively apply the division algorithm to write

$$Q = A_0 P + B_0 \quad \text{with } \deg B_0 < \deg P,$$
$$A_0 = A_1 P + B_1 \quad \text{with } \deg B_1 < \deg P,$$
$$A_1 = A_2 P + B_2 \quad \text{with } \deg B_2 < \deg P,$$

etc., and then we substitute and find that

$$Q = A_0 P + B_0 = A_1 P^2 + B_1 P + B_0 = A_2 P^3 + B_2 P^2 + B_1 P + B_0$$
$$= \cdots = A_j P^{j+1} + B_j P^j + \cdots + B_2 P^2 + B_1 P + B_0$$

with each B_i equal to 0 or of degree $<\deg P$. The fact that $W_j \subseteq K_{j+1}$ implies that $P^{j+1}(L)(v) = 0$. Consequently

$$\mathcal{P}(v) = \{(B_j P^j + \cdots + B_1 P + B_0)(L)(v) \mid B_i = 0 \text{ or } \deg B_i < d \text{ for } 0 \leq i \leq j\},$$

and the given set spans $\mathcal{P}(v)$.

For the linear independence suppose that some such expression is 0 with not all $B_i(\lambda)$ equal to 0. Fix i as small as possible with $B_i(\lambda) \neq 0$. Since $P(L)^{j+1}(v) = 0$, $B_r(L)P(L)^r(v)$ is annihilated by $P(L)^{j-i}$ if $r > i$. Application of $P(L)^{j-i}$ to the dependence relation yields

$$P(L)^{j-i}(B_j(L)P(L)^j(v) + \cdots + B_{i+1}(L)P(L)^{i+1} + B_i(L)P(L)^i)(v) = 0$$

and therefore also $B_i(L)P(L)^j(v) = 0$. Since $\deg B_i < \deg P$, Problem 29 shows that $P(L)^j(v) = 0$. Therefore v is in K_j. Since $W_j \cap K_j$, we conclude $v = 0$, contradiction.

36. We show at the same time that it is possible to arrange for each U_j and W_j to be such that $K_j + U_j$ and $K_j + W_j$ are invariant under L. We proceed by induction downward on j. The construction begins with $U_{l-1} = 0$ and W_{l-1} chosen such that $K_l = K_{l-1} \oplus W_{l-1}$. Then we have $L(W_{l-1}) \subseteq W_{l-1} + K_{l-1}$ and $L(U_{l-1}) \subseteq U_{l-1} + K_{l-1}$. Select some $v_1^{(l-1)} \neq 0$ in W_{l-1}. If there is a polynomial $B(\lambda) \neq 0$ with $\deg B < \deg P$ such that $B(L)(v_1^{(l-1)})$ is in K_{l-1}, then it follows from Problem 29 and the invariance of K_{l-1} under L that $v_1^{(l-1)}$ is in K_{l-1}, contradiction. So there is no such polynomial, and the vectors $v_1^{(l-1)}$, $L(v_1^{(l-1)}), \ldots, L^{d-1}(v_1^{(l-1)})$ are linearly independent with span $T_1^{(l-1)}$ such that $K_{l-1} + T_1^{(l-1)}$ is a direct sum.

If $K_{l-1} + T_1^{(l-1)} \neq K_l$, then we form $v_2^{(l-1)}$ and $T_2^{(l-1)}$ in the same way. If there is a polynomial $B(\lambda) \neq 0$ with $\deg B < \deg P$ such that $B(L)(v_2^{(l-1)})$ is in $K_{l-1} + T_1^{(l-1)}$, then Problem 29 shows that $v_2^{(l-1)}$ is in $K_{l-1} + T_1^{(l-1)}$, contradiction. We conclude that $K_{l-1} + T_1^{(l-1)} + T_2^{(l-1)}$ is a direct sum. Continuing in this way, we obtain enough linearly independent vectors to have a basis for a complement $W_{l-1} = T_1^{(l-1)} + T_2^{(l-1)} + \cdots$ to K_{l-1}.

Now suppose inductively in the construction of U_j and W_j that $j \leq l - 2$ and that $U_{j+1} + K_{j+1}$ and $W_{j+1} + K_{j+1}$ are invariant under L. We define $U_j = P(L)(U_{j+1} \oplus W_{j+1})$, and the assumed invariance implies that $U_j + K_j$ is invariant under L. We now construct W_j in the same way that we constructed W_{l-1}, insisting that $(U_j + K_j) \cap W_j = 0$. If we choose $v_1^{(j)}$ in K_{j+1} but not $U_j + K_j$, then the invariance of $U_j + K_j$ under L implies that the vectors $v_1^{(j)}, L(v_1^{(j)}), \ldots, L^{d-1}(v_1^{(j)})$ are linearly independent and their linear span $T_1^{(j)}$ is such that $U_j + K_j + T_1^{(j)}$ is a direct sum. Continuing in this way, we obtain the required basis of a complement W_j to $K_j \oplus U_j$.

37. Problem 36 arranges that the vectors $L^r(v_{ij}^{(j)})$ for $0 \leq r \leq d-1$ and all i_j form a basis of W_j. We show by induction downward for $j \leq l-1$ that the vectors $L^r P(L)^k(v_{ij+k}^{(j+k)})$ for $0 \leq r \leq d-1, k > 0$, and all i_{j+k} form a basis of U_j. This holds for $j = l-1$ since $U_{l-1} = 0$. If it is true for $j+1$, then $U_{j+1} \oplus W_{j+1}$ has a basis consisting of all $L^r P(L)^k(v_{ij+1+k}^{(j+1+k)})$ for $0 \leq r \leq d-1, k \geq 0$, and all i_{j+1+k}. Since Problem 33 shows that $P(L)$ is one-one from $U_{j+1} \oplus W_{j+1}$ onto U_j, U_j has a basis consisting of all $L^r P(L)^{k+1}(v_{ij+1+k}^{(j+1+k)})$ for $0 \leq r \leq d-1, k \geq 0$, and all i_{j+1+k}, i.e., all $L^r P(L)^k(v_{ij+k}^{(j+k)})$ for $0 \leq r \leq d-1, k > 0$, and all i_{j+k}. This completes the induction.

38. Problem 35 gives a basis for the cyclic subspace generated by $v_{ij}^{(j)}$, Problem 37 shows that the members within $U_i \oplus W_i$ of the union of these bases, as j and i_j vary, form a basis of $U_i \oplus W_i$, and Problem 34 allows us to conclude that as i varies, we obtain a basis of V.

39. Because of the linear independence proved in Problem 38, the left side of the formula in question equals the number of vectors $v_{i_k}^{(k)}$ in any W_k with $k \geq j$, which equals $\sum_{k \geq j}(\dim W_k)/d$. Iterated application of Problem 33 gives

$$\dim K_{j+1} - \dim K_j = \dim U_j + \dim W_j = \dim U_{j+1} + \dim W_{j+1} + \dim W_j$$
$$= \cdots = \sum_{k \geq j} \dim W_k,$$

and the result follows.

40. The minimal polynomial for any cyclic subspace must divide the minimal polynomial for V and hence must be a power of $P(\lambda)$. Problem 28 shows that the restrictions of L to any two cyclic subspaces with the same minimal polynomial are isomorphic. Hence the decomposition into cyclic subspaces will be unique up to isomorphism as soon as it is proved that the number of cyclic direct summands with minimal polynomial of the form $P(\lambda)^k$ with $k \geq j+1$ equals $(\dim K_{j+1} - \dim K_j)/d$.

Suppose that V is the direct sum of cyclic subspaces C_i, with v_i as the generator of C_i. Since each C_i is invariant under L, each K_r is the direct sum of the subspaces $K_r \cap C_i$. Thus

$$\dim K_{j+1} - \dim K_j = \sum_i \Big(\dim(K_{j+1} \cap C_i) - \dim(K_j \cap C_i)\Big).$$

If $P(\lambda)^k$ is the minimal polynomial of C_i, it is enough to show that the right side of this displayed formula equals d if $k \geq j+1$ and equals 0 if $k \leq j$. By Problem 35, C_i has a basis consisting of all vectors $L^r P(L)^s(v_i)$ with $0 \leq r \leq d-1$ and $0 \leq s \leq k-1$. The nonzero vectors among the $L^r P(L)^{s+j+1}(v_i)$ are still linearly independent; these are the ones with $s+j+1 < k$, i.e., $s < k-j-1$. The vectors $L^r P(L)^s(v_i)$ that are not sent to 0 by $P(L)^{j+1}$ are a basis of $K_{j+1} \cap C_i$. These are the ones with $s \geq k - j - 1$. This is the full basis of C_i if $j + 1 > k$, and there are $d(j + 1)$ such vectors if $j + 1 \leq k$. Thus

$$\dim K_{j+1} \cap C_i = \begin{cases} dk & \text{if } j + 1 > k, \\ d(j+1) & \text{if } j + 1 \leq k. \end{cases}$$

Similarly

$$\dim K_j \cap C_i = \begin{cases} dk & \text{if } j > k, \\ dj & \text{if } j \leq k. \end{cases}$$

Subtracting and taking the cases into account, we see that

$$\dim(K_{j+1} \cap C_i) - \dim(K_j \cap C_i) = \begin{cases} d & \text{if } j + 1 \leq k, \\ 0 & \text{otherwise.} \end{cases}$$

41. (a) $\begin{pmatrix} \cos t & \sin t \\ -\sin t & \cos t \end{pmatrix}$, (b) $\begin{pmatrix} \cosh t & \sinh t \\ \sinh t & \cosh t \end{pmatrix}$, (c) the diagonal matrix with diagonal entries $e^{d_1}, \ldots, e^{d_n}$.

42. Suppose that J has diagonal entry c. Let N be the strictly upper-triangular part of J. Then $e^{tJ} = e^{tcI + tN} = e^{tc}e^{tN}$. Here $e^{tN} = I + tN + \frac{1}{2!}t^2 N^2 + \cdots + \frac{1}{(n-1)!}t^{n-1}N^{n-1}$ since $N^n = 0$. The powers of N were observed to have the diagonal of 1's move one step at a time up and to the right.

43. $\frac{d}{dt}(e^{tA}v) = (Ae^{tA})v = A(e^{tA}v)$.

44. Suppose that $y(t)$ is a solution. The product rule for derivatives is valid in this situation by the usual derivation. Hence $\frac{d}{dt}(e^{-tA}y(t)) = \frac{d}{dt}(e^{-tA})y(t) + e^{-tA}y'(t) = -e^{-tA}Ay(t) + e^{-tA}y'(t) = e^{-tA}(-Ay(t) + y'(t))$. The right side is 0 since $y(t)$ solves the differential equation. Since $\frac{d}{dt}(e^{-tA}y(t)) = 0$, each component of $e^{-tA}y(t)$ is constant. Thus for a suitable vector v of complex constants, $e^{-tA}y(t) = v$, and the conclusion is that $y(t) = e^{tA}v$.

45. The first formula follows by making a term-by-term calculation with the defining series. Multiplication of C has to be interchanged with the infinite sum, and similarly for C^{-1}, but these operations are simply the operations of taking certain linear combinations of limits.

Suppose that $z(t)$ satisfies $\frac{d}{dt}z(t) = (C^{-1}AC)z(t)$ and $z(0) = u$. Multiplying by C gives $\frac{d}{dt}Cz(t) = ACz(t)$. Thus $y(t) = Cz(t)$ satisfies $\frac{d}{dt}y(t) = Ay(t)$ and $y(0) = Cz(0) = Cu$. We can invert the correspondence by using C^{-1}.

46. Example 3 in Section 7 says that $C^{-1}AC = J$ holds for $J = \begin{pmatrix} 3 & 1 & 0 \\ 0 & 3 & 0 \\ 0 & 0 & 2 \end{pmatrix}$ and $C = \begin{pmatrix} -1 & 1 & 0 \\ -1 & 0 & 0 \\ -1 & 0 & 1 \end{pmatrix}$. Define $u = C^{-1}\begin{pmatrix} 1 \\ 2 \\ 3 \end{pmatrix} = \begin{pmatrix} 0 & -1 & 0 \\ 1 & -1 & 0 \\ 0 & -1 & 1 \end{pmatrix}\begin{pmatrix} 1 \\ 2 \\ 3 \end{pmatrix} = \begin{pmatrix} -2 \\ -1 \\ 1 \end{pmatrix}$. Problems 42–43 show that the unique solution of $\frac{d}{dt}z(t) = Jz(t)$ with $z(0) = u$ is $z(t) = e^{tJ}u$.

Problem 45 shows that the unique solution to $\frac{d}{dt}y(t) = Ay(t)$ with $y(0) = Cu = \begin{pmatrix} 1 \\ 2 \\ 3 \end{pmatrix}$
is $y(t) = Cz(t) = Ce^{tJ}u$. By Problem 42, this is

$$y(t) = \begin{pmatrix} -1 & 1 & 0 \\ -1 & 0 & 0 \\ -1 & 0 & 1 \end{pmatrix} \begin{pmatrix} e^{3t} & 0 & 0 \\ 0 & e^{3t} & 0 \\ 0 & 0 & e^{2t} \end{pmatrix} \begin{pmatrix} 1 & t & 0 \\ 0 & 1 & 0 \\ 0 & 0 & 1 \end{pmatrix} \begin{pmatrix} -2 \\ -1 \\ 1 \end{pmatrix} = \begin{pmatrix} -1 & 1 & 0 \\ -1 & 0 & 0 \\ -1 & 0 & 1 \end{pmatrix} \begin{pmatrix} e^{3t} & te^{3t} & 0 \\ 0 & e^{3t} & 0 \\ 0 & 0 & e^{2t} \end{pmatrix} \begin{pmatrix} -2 \\ -1 \\ 1 \end{pmatrix}$$

$$= \begin{pmatrix} -e^{3t} & -te^{3t}+e^{3t} & 0 \\ -e^{3t} & -te^{3t} & 0 \\ -e^{3t} & -te^{3t} & e^{2t} \end{pmatrix} \begin{pmatrix} -2 \\ -1 \\ 1 \end{pmatrix} = \begin{pmatrix} e^{3t}+te^{3t} \\ 2e^{3t}+te^{3t} \\ 2e^{3t}+te^{3t}+e^{2t} \end{pmatrix}.$$

Chapter VI

1. In (a), the linear function $\varphi : V \to V'$ given by $\varphi(v) = \langle v, \cdot \rangle$ has kernel equal to the left radical of the bilinear form, hence 0. Therefore φ is one-one, and dim image φ = dim V = dim V'. Since dim $V' < \infty$, φ is onto V'. In (b), $v \mapsto (v, \cdot)$ is a linear functional and by (a) is of the form $(v, u) = \langle w, u \rangle$ for some unique w depending on v. Set $w = L(v)$. The uniqueness shows that $L(v_1 + v_2) = L(v_1) + L(v_1)$ and $L(cv) = cL(v)$. Hence L is linear.

2. Since $M^t AM$ would have to be nonsingular, the only possibility would be $M^t AM$ equal to the identity. Writing M^{-1} as $\begin{pmatrix} a & b \\ c & d \end{pmatrix}$, we obtain the conditions $a+c = b+d = 0$ and $ab + cd = 1$. A check of cases shows that these have no solution.

3. Take $M = \begin{pmatrix} -1 & 1 \\ 1 & 1 \end{pmatrix}$.

5. Define $(a + bi)w = aw + bJ(w)$ for a and b real. The crucial property to show in order to obtain a complex vector space is that $((a + bi)(c + di))(w) = (a + bi)((c + di)w)$; expansion of both sides shows that both sides are equal to $(ac - bd)w + (bc + ad)J(w)$ since $J^2 = -I$. Thus $W = V_{\mathbb{R}}$ for a suitable V.

Next define $(v, w) = \langle J(v), w \rangle + i \langle v, w \rangle$. This is bilinear over $\mathbb{R}$. It is complex linear in the first variable because $(J(v), w) = \langle J^2(v), w \rangle + i \langle J(v), w \rangle = -\langle v, w \rangle + i \langle J(v), w \rangle = i(v, w)$. It is Hermitian because $\overline{(w, v)} = \langle J(w), v \rangle - i \langle v, w \rangle = \langle J^2(w), J(v) \rangle - i \langle w, v \rangle = -\langle w, J(v) \rangle - i \langle w, v \rangle = \langle J(v), w \rangle + i \langle v, w \rangle = (v, w)$.

6. For (a), U isotropic implies $U^\perp \supseteq U$. If v is a vector in $U^\perp$ but not U, then $U \oplus \mathbb{K}v$ is isotropic. Maximality thus implies that $U^\perp = U$. Proposition 6.3 says that dim V = dim U + dim $U^\perp$, and we conclude that dim V = 2 dim U. So dim $U = n$.

The proof of (b) goes by induction on the dimension, the base case being dimension 2, where there is no problem. Assuming the result for spaces of dimension less than dim V, let S_1 be maximal isotropic in V, so that dim $S_1 = \frac{1}{2}$ dim V by (a). Fix a basis $\{v_1, \ldots, v_n\}$ of S_1. Choose u_1 with $\langle v_1, u_1 \rangle = 1$; this exists by nondegeneracy. Put $U = \mathbb{K}v_1 \oplus \mathbb{K}u_1$. Then $\langle \cdot, \cdot \rangle|_{U \times U}$ is evidently nondegenerate, and Corollary 6.4 shows that $V = U \oplus U^\perp$. Certainly $S_1 \cap U^\perp$ is an isotropic subspace of $U^\perp$. It contains the $n - 1$ linearly independent elements $v_j - \langle v_j, u_1 \rangle v_1$ for $2 \leq j \leq n$ and hence has dimension $\geq n-1$. Therefore it is maximal isotropic. By induction, there is a maximal isotropic subspace T of $U^\perp$ with $(S_1 \cap U^\perp) \cap T = 0$. Put $S_2 = T \oplus \mathbb{K}u_1$. Since $\langle u_1, U^\perp \rangle = 0, \langle u_1, T \rangle = 0$. Therefore S_2 is isotropic, hence maximal isotropic in V. Suppose that the element $t + cu_1$ of S_2 lies in S_1. From $\langle v_1, t + cu_1 \rangle = 0$, $v_1 \in U, t \in U^\perp$, and $\langle v_1, u_1 \rangle = 1$, we obtain $c = 0$. Then $t + cv_1$ lies in $(S_1 \cap U^\perp) \cap T$, which is 0. We conclude that $S_1 \cap S_2 = 0$.

For (c), if $\langle \cdot, s_2 \rangle$ is the 0 function on S_1, then the fact that S_1 is maximal isotropic implies that $s_2 = 0$. Therefore the mapping $s_2 \mapsto \langle \cdot, s_2 \rangle |_{S_1}$ is one-one. A count of dimensions shows that it is onto S_1'.

In (d), choose any basis $\{p_1, \ldots, p_n\}$ of S_1, and let $\{q_1, \ldots, q_n\}$ be the dual basis of S_1', which has been identified with S_2 by (c).

7. In (a), first suppose that $h : \bigoplus_s U_s \to V$ is given. Then hi_s is in $\text{Hom}_\mathbb{K}(U_s, V)$, and the map from left to right may be taken to be $h \mapsto \{hi_s\}_{s \in S}$. Next suppose that $h_s : U_s \to V$ is given for each s. Then the universal mapping property of $\bigoplus_s U_s$ supplies $h : \bigoplus_s U_s \to V$ with $hi_s = h_s$ for all s. The map from right to left may be taken as $\{h_s\}_{s \in S} \mapsto h$. These two maps invert each other.

In (b), first suppose that $h_s : U \to V_s$ is given for each s. Then the universal mapping property of the direct product produces $h : U \to \prod_s V_s$. The map from right to left may be taken as $\{h_s\}_{s \in S} \mapsto h$. Next suppose that $h : U \to \prod_s V_s$ is given. Then $p_s h$ is in $\text{Hom}_\mathbb{K}(U, V_s)$ for each $s \in S$. Consequently the S-tuple $\{h_s\}_{s \in S}$ is in $\prod_s \text{Hom}_\mathbb{K}(U, V_s)$. Then the map from left to right can be taken as $h \mapsto \{p_s h\}_{s \in S}$. These two maps invert each other.

For (c), we treat (a) and (b) separately. In the case of (a), take S countably infinite with each $U_s = \mathbb{K}$ and with $V = \mathbb{K}$. Then $\text{Hom}_\mathbb{K}(\bigoplus_{s \in S} U_s, V)$ has uncountable dimension and $\bigoplus_{s \in S} \text{Hom}_\mathbb{K}(U_s, V)$ has countable dimension.

In the case of (b), take S to be countably infinite with each $V_s = \mathbb{K}$ and with $U = \bigoplus_{s \in S} V_s$. Each member of $\text{Hom}_\mathbb{K}(U, V_{s_0})$ has its values in V_{s_0}, and hence each member of $\bigoplus_s \text{Hom}_\mathbb{K}(U, V_s)$ has its values in finitely many V_s. On the other hand, the identity function from U into $\bigoplus_s V_s$ is in $\text{Hom}_\mathbb{K}(U, \bigoplus V_s)$ and takes values in all V_s's.

8. For (a), we have $g_1(g_2(x)) = g_1(g_2 x g_2^t) = g_1 g_2 x g_2^t g_1^t = (g_1 g_2) x (g_1 g_2)^t = (g_1 g_2)(x)$. If x is alternating, then $(g x g^t)^t = g x^t g^t = -g x g^t$, and $(g x g^t)_{ii} = \sum_{j,k} g_{ij} x_{jk} g_{ik} = \sum_{j<k} g_{ij} x_{jk} g_{ik} + \sum_{j>k} g_{ij} x_{jk} g_{ik} = \sum_{j<k} g_{ij} (x_{jk} - x_{jk}) g_{ik} = 0$; hence $g x g^t$ is alternating. If x is symmetric, then $(g x g^t)^t = g x^t g^t = g x g^t$, and $g x g^t$ is symmetric.

For (b), certainly x and $g x g^t$ have the same rank if g is nonsingular. Theorem 6.7 shows that an alternating matrix x can be transformed by some nonsingular g to a matrix $g x g^t$ that is block diagonal with k blocks of the form $\begin{pmatrix} 0 & 1 \\ -1 & 0 \end{pmatrix}$, where $2k$ is the rank, followed by 0's down the diagonal. This proves that any two alternating matrices of the same rank lie in the same orbit. It also gives an example of a matrix in each orbit.

For (c), certainly x and $g x g^t$ have the same rank if g is nonsingular. The Principal

Axis Theorem (Theorem 6.5) shows that any symmetric matrix over $\mathbb{C}$ can be transformed by some nonsingular g to a matrix gxg^t that is diagonal, say with diagonal entries $d_1, \ldots, d_n$. We may assume that $d_1, \ldots, d_k$ are nonzero and the others are 0. Taking h to be the diagonal matrix with diagonal entries $(d_1^{-1/2}, \ldots, d_k^{-1/2}, 0, \ldots, 0)$ and forming $h(gxg^t)h^t$, we obtain a diagonal matrix in the same orbit whose first k diagonal entries are 1 and whose other diagonal entries are 0. As k varies, these matrices have different ranks and hence lie in different orbits. They provide examples of matrices in each orbit.

9. In (a), the formula is $T_{UV}\left(\sum_i(u_i' \otimes v_i)\right)(u) = \sum_i u_i'(u)v_i$, and we may assume that $\{v_i\}$ is linearly independent. If this is 0 for all u, then the linear independence of the v_i's implies that $u_i'(u) = 0$ for all i and all u. Then all u_i' are 0, and hence $\sum_i(u_i' \otimes v_i) = 0$. Thus T_{UV} is one-one.

In (b), Problem 7a shows that it is enough to handle $U = \mathbb{K}$. Thus we are to show that $\mathbb{K}' \otimes_{\mathbb{K}} V$ maps onto $\mathrm{Hom}_{\mathbb{K}}(\mathbb{K}, V) \cong V$. One member of $\mathbb{K}'$ is the identity function $1'$ on $\mathbb{K}$, and $1' \otimes V$ certainly maps onto V.

For (c), if $U = V$ and if $\dim U$ is infinite, every member of the image of T_{UU} has finite rank, but $\mathrm{Hom}_{\mathbb{K}}(U, U)$ contains the identity function, which has infinite rank.

In (d), let $L : U_1 \to U$ and $M : V \to V_1$ be given, so that $F(L, M)$ carrying $(U' \otimes_{\mathbb{K}} V)$ to $(U_1' \otimes_{\mathbb{K}} V_1)$ is given by $F(L, M)(u' \otimes v) = L^t(u') \otimes M(v)$ and $G(L, M)$ carrying $\mathrm{Hom}_{\mathbb{K}}(U, V)$ to $\mathrm{Hom}_{\mathbb{K}}(U_1, V_1)$ has $(G(L, M)(\varphi))(u_1) = M(\varphi(L(u_1)))$ Then

$$T_{U_1 V_1} F(L, M)(u' \otimes v)(u_1) = T_{U_1 V_1}\bigl(L^t(u') \otimes M(v)\bigr)(u_1)$$
$$= L^t(u')(u_1)M(v) = u'(L(u_1))M(v),$$
$$G(L, M)T_{UV}(u' \otimes v)(u_1) = M((T_{UV}(u' \otimes v))(L(u_1)))$$
$$= M(u'(L(u_1))v) = u'(L(u_1))M(v).$$

The right sides are equal, and hence $\{T_{UV}\}$ is a natural transformation.

In (e), the answer is no because the maps T_{UV} need not be isomorphisms, according to (c).

10. To see that $\Psi(E)$ is a vector space, one has to verify that $(l + l')\varphi = l\varphi + l'\varphi$, $l(\varphi + \varphi') = l\varphi + l\varphi'$, and $(ll')\varphi = l(l'\varphi)$, and these are all routine. If μ is in $\mathrm{Hom}_{\mathbb{K}}(E, F)$, then $\Psi(\mu) : \mathrm{Hom}_{\mathbb{K}}(\mathbb{L}, E) \to \mathrm{Hom}_{\mathbb{K}}(\mathbb{L}, F)$ has to be given by left-by-μ, and the key step is to show that $\Psi(\mu)$ is $\mathbb{L}$ linear, not merely $\mathbb{K}$ linear. For φ in $\mathrm{Hom}_{\mathbb{K}}(\mathbb{L}, E)$ and l, l' in $\mathbb{L}$, we have $(\Psi(\mu)(l\varphi))(l') = \mu((l\varphi)(l')) = \mu(\varphi(ll')) = (\Psi(\mu)\varphi)(ll') = (l(\Psi(\mu)\varphi))(l')$. Hence $\Psi(\mu)(l\varphi) = l(\Psi(\mu)\varphi)$ as required. It is routine to check that $\Psi(1) = 1$ and that $\mu \to \Psi(\mu)$ respects compositions, and hence Ψ is a functor.

11. Let $\Gamma = (v_1, \ldots, v_n)$ be an ordered basis of E, $\Delta = (w_1, \ldots, w_m)$ be an ordered basis of F, and $A = [A_{ij}]$ be the matrix of L in these ordered bases. Put $\Gamma_{\mathbb{R}} = (v_1, iv_1, \ldots, v_n, iv_n)$ and $\Delta_{\mathbb{R}} = (w_1, iw_1, \ldots, w_m, iw_m)$. Then the matrix of $L_{\mathbb{R}}$ in these ordered bases is obtained by replacing A_{ij} by the 2-by-2 block $\begin{pmatrix} \mathrm{Re}\,A_{ij} & -\mathrm{Im}\,A_{ij} \\ \mathrm{Im}\,A_{ij} & \mathrm{Re}\,A_{ij} \end{pmatrix}$.

12. Let $\Gamma_1 = (u_1, \ldots, u_m)$ and $\Delta_1 = (v_1, \ldots, v_n)$, and put

$$\Omega_1 = (u_1 \otimes v_1, u_1 \otimes v_2, \ldots, u_1 \otimes v_n, u_2 \otimes v_1, \ldots, u_2 \otimes v_n, \ldots, u_m \otimes v_n).$$

Form Ω_2 from the ordered bases Γ_2 and Δ_2 similarly. Members of Ω_1 are indexed by pairs (i, j) with $1 \leq i \leq m$ and $1 \leq j \leq n$, and members of Ω_2 are indexed similarly by pairs (r, s). Then $C_{(r,s),(i,j)} = A_{ri} B_{sj}$.

13. Define F to be the vector space $\mathbb{K}U \oplus \mathbb{K}V$, and let l be the linear map $l : F \to T(E)$ given by $l(U) = Y$ and $l(V) = X^2 + XY + Y^2$. Let L be the extension of l to an algebra homomorphism $L : T(F) \to T(E)$ with $L(1) = 1$. The subalgebra in question is the image of L, and the affirmative answer to the question comes by showing that L is one-one. It is enough to show that the basis elements consisting of all iterated products $U^{i_1} \otimes V^{j_1} \otimes U^{i_2} \otimes \cdots \otimes V^{j_n}$ are carried by L to linearly independent elements. The image of this element is homogeneous of degree $\sum_{k=1}^{n}(i_k + 2j_k)$, and it is enough to consider only those images with the same homogeneity, i.e., with $\sum_{k=1}^{n}(i_k + 2j_k)$ constant. A failure of linear independence would mean that among these, the ones with the highest total power of X, namely with $\sum_{k=1}^{n} 2j_k$ maximal, must cancel together. These terms are monomials with $\sum i_k$ factors of Y and $\sum j_k$ factors of X^2, and all such monomials, being also monomials in X and Y, are linearly independent.

14. Let $\iota_E : E \to S(E)$ be the one-one linear map that embeds E as $S^1(E) \subseteq S(E)$, and define ι_F similarly. The composition $\iota_F \varphi$ is a linear map of E into the commutative associative algebra $S(F)$, and Proposition 6.23b yields a homomorphism $\Phi : S(E) \to S(F)$ of algebras with identity such that $\iota_F \varphi = \Phi \iota_E$. We take Φ as $S(\varphi)$, and this addresses (a). Part (c) is part of the construction of $S(\varphi)$. For (b), it is plain that $S(1_E) = 1_{S(E)}$. For compositions, suppose that $\psi : F \to G$ is linear and that $S(\psi)$ is formed similarly. Proposition 6.23b says that $S(\psi \varphi)$ is the unique homomorphism of $S(E)$ into $S(G)$ carrying 1 into 1 and satisfying $\iota_G \psi \varphi = S(\psi \varphi) \iota_E$. On the other hand, $S(\psi)S(\varphi)$ is another homomorphism of $S(E)$ into $S(G)$ carrying 1 into 1, and it satisfies $\iota_G(\psi \varphi) = (\iota_G \psi)\varphi = (S(\psi)\iota_F)\varphi = S(\psi)(\iota_F \varphi) = S(\psi)(S(\varphi)\iota_E) = (S(\psi)S(\varphi))\iota_E$. Therefore $S(\psi \varphi) = S(\psi)S(\varphi)$ by uniqueness, and S is a functor.

15. The homomorphism $\widetilde{\Phi}$ carries each $T^n(E)$ into itself. Since $\widetilde{\Phi}$ carries commutators into commutators, $\widetilde{\Phi}(I) \subseteq I$. Thus $\widetilde{\Phi}(T^n(E) \cap I) \subseteq T^n(E) \cap I$. Also, $\widetilde{\Phi}$ commutes with the symmetrizer operator and hence carries $\widetilde{S}^n(E)$ into itself. We are given the equation $q\widetilde{\Phi}(x) = \Phi q(x)$ on all of $T^n(E)$. Since $\widetilde{\Phi}$ carries $\widetilde{S}^n(E)$ into itself, we can interpret this as saying that $\widetilde{\Phi}|_{\widetilde{S}^n(E)}$ is well defined, and then all the assertions in the problem have been addressed.

16. Fix an ordered basis and check the result directly for L's that correspond to elementary matrices. The determinant and the scalar effect on $\bigwedge^{\dim E}(E)$ both multiply under composition, and the result follows.

17. Part (a) is a consequence of uniqueness. The formula for (b) is $\Phi(g)P(v) = \Phi(g^{-1}v)$ for v in $\mathbb{K}^n$.

18. For (a), take $\mathcal{A}$ to be the category of commutative associative algebras over $\mathbb{K}$ with identity, $\mathcal{V}$ to be the category of vector spaces over $\mathbb{K}$, and $\mathcal{F} : \mathcal{A} \to \mathcal{V}$ to be the forgetful functor that takes an algebra and retains only the vector-space structure. If a vector space E is given, then (S, ι) is taken to be $(S(E), \iota_E)$, where $S(E)$ is the symmetric algebra of E and $\iota_E : E \to \mathcal{F}(S(E))$ is the identification of E with the first-order symmetric tensors.

For (b), take $\mathcal{V}$ again to be the category of vector spaces over $\mathbb{K}$. Define $\mathcal{A}$ to be the category whose objects are pairs (A, F) in which A is an associative algebra over $\mathbb{K}$ with identity and F is a vector subspace of A such that every element f of F has $f^2 = 0$ and whose morphisms $\varphi \in \text{Morph}((A, F), (A_1, F_1))$ are algebra homomorphisms $\varphi : A \to A'$ such that $\varphi(F) \subseteq F_1$. The functor $\mathcal{F} : \mathcal{A} \to \mathcal{V}$ is to take the pair (A, F) to F and is to take the morphism φ to $\varphi|_F : F \to F_1$. If a vector space E is given, we take (S, ι) to be $((\bigwedge E, \bigwedge^1 E), \iota_E)$, where $\iota_E : E \to \bigwedge^1 E = \mathcal{F}(\bigwedge E, \bigwedge^1(E))$ is the identification of E with the first-order alternating tensors.

For (c), let the nonempty index set be J. Take $\mathcal{V} = \mathcal{C}^J$ and $\mathcal{A} = \mathcal{C}$. The functor $\mathcal{F} : \mathcal{C} \to \mathcal{C}^J$ is the "diagonal functor" taking an object A to the J-tuple whose j^{th} coordinate is A for every j; this functor takes any morphism $\varphi \in \text{Morph}_\mathcal{C}(A, A')$ to the J-tuple whose j^{th} coordinate is φ for every j. The given E is to be a J-tuple of objects $\{X_j\}_{j \in J}$, S is to be the coproduct $\coprod_{j \in J} X_j$, and $\iota : \{X_j\}_{j \in J} \to \mathcal{F}(S)$ is to be the given J-tuple $\{i_j\}_{j \in J}$ of morphisms of X_j into X.

19. Let L be the unique member of $\text{Morph}_\mathcal{A}(S, S')$ given as corresponding to ι' in $\text{Morph}_\mathcal{V}(E, \mathcal{F}(S'))$, i.e., satisfying $\mathcal{F}(L)\iota = \iota'$. Similarly let L' be the unique member of $\text{Morph}_\mathcal{A}(S, S')$ corresponding to ι in $\text{Morph}_\mathcal{V}(E, \mathcal{F}(S))$, i.e., satisfying $\mathcal{F}(L')\iota' = \iota$. Then $L'L$ and 1_S are in $\text{Morph}_\mathcal{A}(S, S)$ and have $\mathcal{F}(1_S)\iota = 1_{\mathcal{F}(S)}\iota = \iota$ and $\mathcal{F}(L'L)\iota = (\mathcal{F}(L')\mathcal{F}(L))\iota = \mathcal{F}(L')(\mathcal{F}(L)\iota) = \mathcal{F}(L')\iota' = \iota$. By uniqueness, $1_S = L'L$. Similarly $LL' = 1_{S'}$.

20. By definition, T_A satisfies $T_A(L) = \mathcal{F}(L)\iota$ for $L \in \text{Morph}_\mathcal{A}(S, A)$. For φ in $\text{Morph}_\mathcal{A}(A, A')$, we are to show that $G(\varphi)(T_A(L)) = T_{A'}(F(\varphi)(L))$. Substitution from the definitions gives $G(\varphi)(T_A(L)) = \mathcal{F}(\varphi)\mathcal{F}(L)\iota = \mathcal{F}(\varphi L)\iota$ and $T_{A'}(F(\varphi)(L)) = T_{A'}(\varphi L) = \mathcal{F}(\varphi L)\iota$. These are equal, and hence $\{T_A\}$ is a natural transformation. Since each T_A is one-one onto by hypothesis, the system $\{T_A\}$ is a natural isomorphism.

21. The previous problem shows that F is naturally isomorphic to G and that F' is naturally isomorphic to G. Hence F is naturally isomorphic to F'. The hypotheses of Proposition 6.16 are satisfied, and the conclusion is that the object S is isomorphic in $\mathcal{A}$ to the object S' by a specific isomorphism described in the proposition.

22. Let E and F be in $\text{Obj}(\mathcal{V})$, and let φ be in $\text{Morph}_\mathcal{V}(E, F)$. Then $\iota_F \varphi$ is in $\text{Morph}_\mathcal{V}(E, \mathcal{F}(S(F)))$, and the universal mapping property of $(S(E), \iota_E)$ produces a unique Φ in $\text{Morph}_\mathcal{A}(S(E), S(F))$ such that $\mathcal{F}(\Phi)\iota_E = \iota_F \varphi$. We define $S(\varphi) = \Phi$. There is no difficulty in checking that $S(1_E) = 1_{S(E)}$. Let us check that if we are given also ψ in $\text{Morph}_\mathcal{V}(F, G)$, then $S(\psi)S(\varphi) = S(\psi\varphi)$. We know that $S(\psi\varphi)$ is

the unique member of $\text{Morph}_A(S(E), S(G))$ satisfying $\iota_G \psi \varphi = \mathcal{F}(S(\psi\varphi))\iota_E$. On the other hand, $S(\psi)S(\varphi)$ is another member of $\text{Morph}_A(S(E), S(G))$, and it satisfies $\iota_G(\psi\varphi) = (\iota_G\psi)\varphi = (\mathcal{F}(S(\psi))\iota_F)\varphi = \mathcal{F}(S(\psi))(\iota_F\varphi) = \mathcal{F}(S(\psi))(\mathcal{F}(S(\varphi))\iota_E) = (\mathcal{F}(S(\psi))\mathcal{F}(S(\varphi)))\iota_E = \mathcal{F}(S(\psi)S(\varphi))\iota_E$. Therefore $S(\psi\varphi) = S(\psi)S(\varphi)$ by uniqueness, and S is a functor.

23. $\text{Pfaff}(J) = 1$ because the only nonzero term comes from $\tau = 1$.

24. The terms in which σ contains a 1-cycle are each 0 because the diagonal entries of X are 0. The remaining terms in which σ contains some cycle of odd length will be grouped in disjoint pairs that add to 0. If such a σ is given, choose the smallest label $1, \ldots, 2n$ that is moved by a cycle of odd length within σ, and let τ be that cycle. Let σ' be the product of τ^{-1} and the remaining cycles of σ. The resulting unordered pairs $\{\sigma, \sigma'\}$ are disjoint. For the indices i moved by τ, $x_{i,\sigma(i)} = x_{i,\tau(i)}$ while $x_{i,\sigma'(i)} = x_{i,\tau^{-1}(i)} = -x_{\tau^{-1}(i),i}$. Then $\prod_{\tau(i)\neq i} x_{i,\sigma(i)} = \prod_{\tau(i)\neq i} x_{i,\tau(i)}$ and we obtain $\prod_{\tau(i)\neq i} x_{i,\sigma'(i)} = \prod_{\tau(i)\neq i} x_{i,\tau^{-1}(i)} = (-1)^{\text{length }\tau} \prod_{\tau(i)\neq i} x_{\tau^{-1}(i),i} = (-1)^{\text{length }\tau} \prod_{\tau(i)\neq i} x_{i,\tau(i)} = (-1)^{\text{length }\tau} \prod_{\tau(i)\neq i} x_{i,\sigma(i)} = -\prod_{\tau(i)\neq i} x_{i,\sigma(i)}$. If $\tau(i) = i$, then $x_{i,\sigma(i)} = x_{i,\sigma'(i)}$. Thus $\prod_i x_{i,\sigma(i)} = -\prod_i x_{i,\sigma'(i)}$. Since $\text{sgn}\,\sigma = \text{sgn}\,\sigma'$, the terms for σ and σ' sum to 0.

25. If σ is good, let A_0 consist of the smallest index in each cycle of σ, let A be the union of all $\sigma^{2k}(A_0)$ for $k \geq 0$, and let B be the union of all $\sigma^{2k+1}(A_0)$ for all $k \geq 0$. Certainly $A \cup B = \{1, \ldots, 2n\}$, $\sigma(A) = B$, and $\sigma(B) = A$. We have to prove that $A \cap B = \emptyset$. If the intersection is nonempty, we have $\sigma^{2k}(a_0) = \sigma^{2l+1}(a_0')$ for some a_0 and a_0' in A_0. Possibly by increasing l by an even multiple of the order of σ, we may assume that $l \geq k$. Then $\sigma^{2(l-k)+1} a_0' = a_0$. This says that a_0' and a_0 lie in the same cycle. Being least indices in cycles, they must be equal. Then some odd power of σ fixes a_0, and the cycle of σ whose least element is a_0 must have odd length, contradiction.

The definitions of A and B in terms of A_0 are forced by the conditions in the statement of the problem, and therefore A and B are unique.

26. Since $A \cup B = \{1, \ldots, 2n\}$ and $A \cap B = \emptyset$, we have $y(\sigma)z(\sigma) = \prod_{l=1}^{2n} x_{l,\sigma(l)}$. The definitions of τ and τ' make $y(\sigma) = s(\tau) \prod_{k=1}^{n} x_{\tau(2k-1),\tau(2k)}$ and $z(\sigma) = s'(\tau') \prod_{k=1}^{n} x_{\tau'(2k-1),\tau'(2k)}$. The construction has made the integers $\tau(2k-1)$ increasing and has made the inequalities $\tau(2k-1) < \tau(2k)$ hold, and similarly for τ'. This proves the desired equality, apart from signs.

27. The previous problem shows that $(\text{sgn}\,\sigma) \prod_{l=1}^{2n} x_{l,\sigma(l)}$ equals

$$(\text{sgn}\,\sigma) s(\tau) s'(\tau') \prod_{k=1}^{n} x_{\tau(2k-1),\tau(2k)} \prod_{k=1}^{n} x_{\tau'(2k-1),\tau'(2k)}.$$

Thus we want to see that

$$(\text{sgn}\,\sigma) s(\tau) s'(\tau') = (\text{sgn}\,\tau)(\text{sgn}\,\tau'). \qquad (*)$$

In proving (∗), we retain the step in which factors x_{ij} of $y(\sigma)$ and $z(\sigma)$ are replaced by x_{ji} with a minus sign if $j < i$, but we may disregard the step in which the factors are then rearranged so that τ and τ' can be defined. In fact, this rearranging does not affect the signs of τ and τ'. The reason is that if ρ is in $\mathfrak{S}_n$ and if $\widetilde{\rho}$ in $\mathfrak{S}_{2n}$ is defined by $\widetilde{\rho}(2k-1) = 2\rho(k)-1$ and $\widetilde{\rho}(2k) = 2\rho(k)$, then $\operatorname{sgn}\widetilde{\rho} = +1$; it is enough to check this fact when ρ is a consecutive transposition, and in this case $\widetilde{\rho}$ is the product of two transpositions and is even.

Turning to (∗), we first consider the case in which σ, when written as a disjoint product of cycles, takes the integers $1, \ldots, 2n$ in order. In this case we compute directly that $\tau = 1$, that $s(\tau)$ involves no sign changes, and that τ' is the product of cycles of odd length, with an individual cycle of τ' permuting cyclically all but the last member of a cycle of σ. Thus τ' is even. In the adjustment of factors of $z(\sigma)$, one minus sign is introduced because of each cycle in σ and comes from the last and first indices in the cycle. Thus $s'(\tau')$ is $(-1)^p$, where p is the number of cycles in σ, and this is also the value of $\operatorname{sgn}\sigma$. Hence (∗) holds for this σ.

A general σ is conjugate in $\mathfrak{S}_{2n}$ to the one in the previous paragraph. Thus it is enough to show that if (∗) holds for σ, then it holds for $\sigma' = (a \ a+1)\sigma(a \ a+1)$. First suppose that $\sigma(a) \neq a+1$ and $\sigma(a+1) \neq a$. Then a factor of $y(\sigma)$ gets replaced with a minus sign for σ if and only if it gets replaced for σ', and similarly for $z(\sigma)$. Hence $s(\tau)$ and $s'(\tau')$ are unchanged in passing from σ to σ'. The effect on τ and τ', in view of the observation immediately after (∗), is to multiply each on the left by $(a \ a+1)$. Thus $\operatorname{sgn}\tau$ and $\operatorname{sgn}\tau'$ are each reversed. Since $\operatorname{sgn}\sigma = \operatorname{sgn}\sigma'$, (∗) remains valid for σ'.

Now suppose that $\sigma(a) = a+1$. We may assume that $\sigma(a+1) \neq a$ since otherwise $\sigma' = \sigma$. To fix the ideas, first suppose that a is in A. Then one factor in $y(\sigma)$ is $x_{a,a+1}$, and the corresponding factor of $y(\sigma')$ is $x_{a+1,a}$. As a result τ is unchanged under the passage from σ to σ', but the number of minus signs contributing to $s(\tau)$ is increased by 1 and $s(\tau)$ is therefore reversed. Meanwhile, τ' is left multiplied by $(a \ a+1)$, and $s'(\tau')$ is unchanged. Thus (∗) remains valid for σ'. If a instead is in B, then the roles of τ and τ' are reversed in the above argument, but the conclusion about (∗) is not affected. Finally suppose that $\sigma(a+1) = a$ and $\sigma(a) \neq a+1$. Then the argument is the same except that the number of signs contributing to $s(\tau)$ or $s'(\tau')$ is decreased by 1. In any event, (∗) remains valid for σ'.

28. What is needed is an inverse construction that passes from the pair (τ, τ') to σ. Define $\omega \in \mathfrak{S}_{2n}$ to be the commuting product of the n transpositions $(2k-1 \ 2k)$ for $1 \leq k \leq n$.

Assuming for the moment that we know that some index a is to be in A, we see from the definitions above that $b = \sigma(a)$ is to be given by $b = \tau(\omega(\tau^{-1}(a)))$ and b is to be in B. If, on the other hand, we know that some index b is to be in B, then $\sigma(b)$ is to be given by $\tau'(\omega(\tau'^{-1}(b)))$ and is to be in A. Thus the cycle within σ to which a belongs has to be given by applying alternately $\tau\omega\tau^{-1}$ and then $\tau'\omega\tau'^{-1}$.

The critical fact is that this cycle is necessarily even. In the contrary case we would have $\tau\omega\tau^{-1}(\tau'\omega\tau'^{-1}\tau\omega\tau^{-1})^k(a) = a$ for some k. If $k = 2l$, then this equality gives $(\tau\omega\tau^{-1}\tau'\omega\tau'^{-1})^l(\tau\omega\tau^{-1})(\tau'\omega\tau'^{-1}\tau\omega\tau^{-1})^l(a) = a$, which we can rewrite as $(\tau\omega\tau^{-1})(\tau'\omega\tau'^{-1}\tau\omega\tau^{-1})^l(a) = (\tau'\omega\tau'^{-1}\tau\omega\tau^{-1})^l(a)$; this equation is contradictory since $\tau\omega\tau^{-1}$ is a permutation that moves every index. If $k = 2l+1$, then this equality gives $(\tau\omega\tau^{-1})(\tau'\omega\tau'^{-1}\tau\omega\tau^{-1})^l(\tau'\omega\tau'^{-1})(\tau\omega\tau^{-1}\tau'\omega\tau'^{-1})^l(\tau\omega\tau^{-1})(a) = a$ and hence $(\tau'\omega\tau'^{-1})(\tau\omega\tau^{-1}\tau'\omega\tau'^{-1})^l(\tau\omega\tau^{-1})(a) = (\tau\omega\tau^{-1}\tau'\omega\tau'^{-1})^l(\tau\omega\tau^{-1})(a)$; this equation is contradictory since $\tau\omega\tau^{-1}$ is a permutation that moves every index.

What we know is that the smallest index in each cycle is to be in A. Thus we can use this process to construct σ from (τ, τ'), one cycle at a time. For the first cycle the index 1 is to be in A; for the next cycle the smallest remaining index is to be in A, and so on. We have seen that the constructed σ will be the product of even cycles, and we can define A as the union of the images of the even powers of σ on the least indices of each cycle, with B as the complement. In this way we have formed σ and its disjoint decomposition $\{1, \ldots, 2n\} = A \cup B$, and it is apparent that τ and τ' are indeed the permutations formed in the usual passage from σ to (τ, τ') via (A, B).

29. It is enough to prove that $\varphi|_{V_n} : V_n \to V_n^{\#}$ is an isomorphism for every n. We establish this property by induction on n, the trivial case for the induction being $n = -1$. Suppose that

$$\varphi|_{V_{n-1}} : V_{n-1} \to V_{n-1}^{\#} \quad \text{is an isomorphism.} \tag{$*$}$$

By assumption

$$\operatorname{gr}^n \varphi : (V_n/V_{n-1}) \to (V_n^{\#}/V_{n-1}^{\#}) \quad \text{is an isomorphism.} \tag{$**$}$$

If v is in $\ker(\varphi|_{V_n})$, then $(\operatorname{gr}^n \varphi)(v+V_{n-1}) = 0+V_{n-1}^{\#}$, and $(**)$ shows that v is in V_{n-1}. By $(*)$, $v = 0$. Thus $\varphi|_{V_n}$ is one-one. Next suppose that $v^{\#}$ is in $V_n^{\#}$. By $(**)$ there exists v_n in V_n such that $(\operatorname{gr}^n \varphi)(v_n + V_{n-1}) = v^{\#} + V_{n-1}^{\#}$. Write $\varphi(v_n) = v^{\#} + v_{n-1}^{\#}$ with $v_{n-1}^{\#}$ in $V_{n-1}^{\#}$. By $(*)$ there exists v_{n-1} in V_{n-1} with $\varphi(v_{n-1}) = v_{n-1}^{\#}$. Then $\varphi(v_n - v_{n-1}) = v^{\#}$, and thus $\varphi|_{V_n}$ is onto. This completes the induction.

30. We define a product $(A_m/A_{m-1}) \times (A_n/A_{n-1}) \to A_{m+n}/A_{m+n-1}$ by

$$(a_m + A_{m-1})(a_n + A_{n-1}) = a_m a_n + A_{m+n-1}.$$

This is well defined since $a_m A_{n-1}$, $A_{m-1} a_n$, and $A_{m-1} A_{n-1}$ are all contained in A_{m+n-1}. It is clear that this multiplication is distributive and associative as far as it is defined. We extend the definition of multiplication to all of $\operatorname{gr} A$ by taking sums of products of homogeneous elements, and the result is an associative algebra. The identity is the element $1 + A_{-1}$ of A_0/A_{-1}.

31. $[x, x] = xx - xx = 0$, and also $[x, [y, z]] + [y, [z, x]] + [z, [x, y]] = (xyz - xzy - yzx + zyx) + (yzx - yxz - zxy + xzy) + (zxy - zyx - xyz + yxz) = 0$.

32. In (a), let x and y be in $\mathfrak{g}$. Then we have

$$[x, y]^t A + A[x, y] = (xy - yx)^t A + A(xy - yx)$$
$$= y^t x^t A - x^t y^t A + Axy - Ayx$$
$$= y^t(x^t A + Ax) - x^t(y^t A + Ay) + (x^t A + Ax)y - (y^t A + Ay)x = 0.$$

Part (b) is the special case $A = I$.

33. Uniqueness follows from the fact that 1 and $\iota(\mathfrak{g})$ generate $U(\mathfrak{g})$. For existence let $\widetilde{L} : T(\mathfrak{g}) \to A$ be the extension given by the universal mapping property of $T(\mathfrak{g})$ in Proposition 6.22. To obtain L, we are to show that $\widetilde{L}$ annihilates the ideal I''. It is enough to consider $\widetilde{L}$ on a typical generator of I'', where we have

$$\widetilde{L}(\iota X \otimes \iota Y - \iota Y \otimes \iota X - \iota[X, Y]) = \widetilde{L}(\iota X)\widetilde{L}(\iota Y) - \widetilde{L}(\iota Y)\widetilde{L}(\iota X) - \widetilde{L}(\iota[X, Y])$$
$$= l(X)l(Y) - l(Y)l(X) - l[X, Y]$$
$$= 0.$$

34. First one proves the following: if $Z_1, \ldots, Z_p$ are in $\mathfrak{g}$ and σ is a permutation of $\{1, \ldots, p\}$; then $(\iota Z_1) \cdots (\iota Z_p) - (\iota Z_{\sigma(1)}) \cdots (\iota Z_{\sigma(p)})$ is in $U_{p-1}(\mathfrak{g})$. In fact, it is enough to prove this statement when σ is the transposition of j with $j + 1$. In this case the statement follows from the identity $(\iota Z_j)(\iota Z_{j+1}) - (\iota Z_{j+1})(\iota Z_j) = \iota[Z_j, Z_{j+1}]$ by multiplying through on the left by $(\iota Z_1) \cdots (\iota Z_{j-1})$ and on the right by $(\iota Z_{j+2}) \cdots (\iota Z_p)$.

For the assertion in the problem, if we use *all* monomials with $\sum_m j_m \leq p$, we certainly have a spanning set, since the obvious preimages in $T(\mathfrak{g})$ span $\bigoplus_{k \leq p} T_k(\mathfrak{g})$. The result of the previous paragraph then implies inductively that the monomials with monotone increasing indices suffice.

35. We shall construct the map in the opposite direction without using the Poincaré–Birkhoff–Witt Theorem, appeal to the theorem to show that we have an isomorphism, and then compute what the map is in terms of a basis. Let $T_n(\mathfrak{g}) = \bigoplus_{k=0}^{n} T^k(\mathfrak{g})$ be the nth member of the usual filtration of $T(\mathfrak{g})$. Define $U_n(\mathfrak{g})$ to be the image in $U(\mathfrak{g})$ of $T_n(\mathfrak{g})$ under the passage $T(\mathfrak{g}) \to T(\mathfrak{g})/I''$. Form the composition

$$T_n(\mathfrak{g}) \to (T_n(\mathfrak{g}) + I'')/I'' = U_n(\mathfrak{g}) \to U_n(\mathfrak{g})/U_{n-1}(\mathfrak{g}).$$

This composition is onto and carries $T_{n-1}(\mathfrak{g})$ to 0. Since $T^n(\mathfrak{g})$ is a vector-space complement to $T_{n-1}(\mathfrak{g})$ in $T_n(\mathfrak{g})$, we obtain an onto linear map $T^n(\mathfrak{g}) \to U_n(\mathfrak{g})/U_{n-1}(\mathfrak{g})$. Taking the direct sum over n gives an onto linear map

$$\widetilde{\psi} : T(\mathfrak{g}) \to \operatorname{gr} U(\mathfrak{g})$$

that respects the grading.

Let I be the two-sided ideal in $T(\mathfrak{g})$ such that $S(\mathfrak{g}) = T(\mathfrak{g})/I$. It is generated by all $X \otimes Y - Y \otimes X$ with X and Y in $T^1(\mathfrak{g})$. Let us show that the linear map $\widetilde{\psi} : T(\mathfrak{g}) \to \operatorname{gr} U(\mathfrak{g})$ respects multiplication and annihilates the defining ideal I for $S(\mathfrak{g})$; then we can conclude that ψ descends to an algebra homomorphism

$$\psi : S(\mathfrak{g}) \to \operatorname{gr} U(\mathfrak{g})$$

that respects the grading.

To do so, let x be in $T^r(\mathfrak{g})$ and let y be in $T^s(\mathfrak{g})$. Then $x + I''$ is in $U_r(\mathfrak{g})$, and we may regard $\widetilde{\psi}(x)$ as the coset $x + T_{r-1}(\mathfrak{g}) + I''$ in $U_r(\mathfrak{g})/U_{r-1}(\mathfrak{g})$, with 0 in all other coordinates of $\operatorname{gr} U(\mathfrak{g})$ since x is homogeneous. Arguing in a similar fashion with y and xy, we obtain

$$\widetilde{\psi}(x) = x + T_{r-1}(\mathfrak{g}) + I'', \qquad \widetilde{\psi}(y) = y + T_{s-1}(\mathfrak{g}) + I'',$$
$$\text{and} \qquad \widetilde{\psi}(xy) = xy + T_{r+s-1}(\mathfrak{g}) + I''.$$

Since I'' is an ideal, $\widetilde{\psi}(x)\widetilde{\psi}(y) = \widetilde{\psi}(xy)$. General members x and y of $T(\mathfrak{g})$ are sums of homogeneous elements, and hence $\widetilde{\psi}$ respects multiplication.

Consequently ker $\widetilde{\psi}$ is a two-sided ideal. To show that ker $\widetilde{\psi} \supseteq I$, it is enough to show that ker $\widetilde{\psi}$ contains all generators $X \otimes Y - Y \otimes X$. We have

$$\widetilde{\psi}(X \otimes Y - Y \otimes X) = X \otimes Y - Y \otimes X + T_1(\mathfrak{g}) + I''$$
$$= [X, Y] + T_1(\mathfrak{g}) + I''$$
$$= T_1(\mathfrak{g}) + I'',$$

and thus $\widetilde{\psi}$ maps the generator to 0. Hence $\widetilde{\psi}$ descends to a homomorphism ψ as asserted.

Finally we show that this homomorphism is an isomorphism. Let $\{X_i\}$ be an ordered basis of $\mathfrak{g}$. We know that the monomials $X_{i_1}^{j_1} \cdots X_{i_k}^{j_k}$ in $S(\mathfrak{g})$ with $i_1 < \cdots < i_k$ and with $\sum_m j_m = n$ form a basis of $S^n(\mathfrak{g})$. Let us follow the effect of ψ on such a monomial. A preimage of this monomial in $T^n(\mathfrak{g})$ is the element

$$X_{i_1} \otimes \cdots \otimes X_{i_1} \otimes \cdots \otimes X_{i_k} \otimes \cdots \otimes X_{i_k},$$

in which there are j_m factors of X_{i_m} for $1 \leq m \leq k$. This element maps to the monomial in $U_n(\mathfrak{g})$ that we have denoted by $X_{i_1}^{j_1} \cdots X_{i_k}^{j_k}$, and then we pass to the quotient $U_n(\mathfrak{g})/U_{n-1}(\mathfrak{g})$. The Poincaré–Birkhoff–Witt Theorem shows that such monomials modulo $U_{n-1}(\mathfrak{g})$ form a basis of $U_n(\mathfrak{g})/U_{n-1}(\mathfrak{g})$. Consequently ψ is an isomorphism.

36. This is quite similar to Problem 33.

37. This is similar to Problem 34.

38. What is needed here is a description of a triple product of generators in terms of permuting indices and replacing repeated pairs of indices by a scalar; the description does not depend on the way that the parentheses are inserted in a triple product, and then associativity follows. The details are omitted.

39. Using the universal mapping property of Problem 36, construct an algebra homomorphism $L : \text{Cliff}(E, \langle \cdot, \cdot \rangle) \to C$ carrying 1 into 1 and extending the mapping $e_i \mapsto e_i$. Since the e_i's and 1 generate C, L is onto C. Problem 37 shows that $\dim \text{Cliff}(E, \langle \cdot, \cdot \rangle) \leq 2^n$, and we know that $\dim C = 2^n$. Since L is onto, L must be one-one, as well as onto.

40. This is similar to Problem 35. The substitute for the Poincaré–Birkhoff–Witt Theorem is the fact established by Problem 39 that the spanning set of 2^n elements in Problem 37 is actually a basis.

41. The matrix that corresponds to X_0 has $r = -2$.

42. To see that $\widetilde{\iota}$ has the asserted properties, form the quotient map $T(H(V)) \to T(V)$ by factoring out the two-sided ideal generated by $X_0 - 1$. The composition $T(H(V)) \to W(V)$ is obtained by factoring out the two-sided ideal generated by $X_0 - 1$ and all $u \otimes v - v \otimes u - \langle u, v \rangle 1$, hence by all $u \otimes v - v \otimes u - \langle u, v \rangle X_0$ and by $X_0 - 1$. Thus $T(H(V)) \to W(V)$ factors into the standard quotient map $T(H(V)) \to U(H(V))$ followed by the quotient map of $U(H(V))$ by the ideal generated by $X_0 - 1$. By uniqueness in the universal mapping property for universal enveloping algebras, $\widetilde{\iota}$ is given by factoring out by $X_0 - 1$.

43. Let P be the extension of φ to an associative algebra homomorphism of $U(H(V))$ into A. Then $P(X_0) = 1$ since $\varphi(X_0) = 1$. The previous problem shows that P descends to $W(V)$, i.e., that there exists $\widetilde{\varphi}$ with $P = \widetilde{\varphi} \circ \widetilde{\iota}$. Restriction to V gives $\varphi = \widetilde{\varphi} \circ \iota$.

44. This is immediate from Problem 42 and the spanning in Problem 34.

46. The linear combination $L_j = \varphi(p_j) + 2\pi \varphi(q_j)$ of the two given linear mappings $\varphi(p_j) = \partial/\partial x_j$ and $\varphi(q_j) = m_j$ replaces $P(x)$ in $e^{-\pi |x|^2} P(x)$ by $\partial P/\partial x_j$. Take a nonzero $e^{-\pi |x|^2} P(x)$ in an invariant subspace U, let $x_1^{k_1} \cdots x_n^{k_n}$ be a monomial of maximal total degree in $P(x)$, and apply $L_1^{k_1} \cdots L_n^{k_n}$ to $e^{-\pi |x|^2} P(x)$ to see that $e^{-\pi |x|^2}$ is in U. Then apply products of powers of the various m_j's to this to see that all of V is contained in U.

47. Let $r_i = p_i + 2\pi q_i$, so that $\varphi(r_i)(Pe^{-\pi |x|^2}) = (\partial P/\partial x_i)e^{-\pi |x|^2}$. It is enough to prove that no nontrivial linear combination of the members of the spanning set $q_1^{k_1} \cdots q_n^{k_n} r_1^{l_1} \cdots r_n^{l_n}$ maps to 0 under $\widetilde{\varphi}$. Let a linear combination of such terms map to 0 under $\widetilde{\varphi}$. Among all the terms that occur in the linear combination with nonzero coefficient, let $(L_1, \ldots, L_n)$ be the largest tuple of exponents $(l_1, \ldots, l_n)$ that occurs; here "largest" refers to the lexicographic ordering taking l_1 first, then l_2, and so on. Put $P(x_1, \ldots, x_n) = x_1^{L_1} \cdots x_n^{L_n}$. If $(l_1, \ldots, l_n) < (L_1, \ldots, L_n)$ lexicographically, then $\widetilde{\varphi}(r_1^{l_1} \cdots r_n^{l_n})(Pe^{-\pi |x|^2}) = 0$.

Thus $\widetilde{\varphi}(q_1^{k_1}\cdots q_n^{k_n}r_1^{l_1}\cdots r_n^{l_n})(Pe^{-\pi|x|^2})$ is 0 if $(l_1,\ldots,l_n) < (L_1,\ldots,L_n)$ lexicographically and equals $x_1^{k_1}\cdots x_n^{k_n}L_1!\cdots L_n!e^{-\pi|x|^2}$ if $(l_1,\ldots,l_n) = (L_1,\ldots,L_n)$. The linear independence follows immediately.

48. This is similar to Problems 35 and 40. The key fact needed is the linear independence established in the previous problem.

52. In (a), for $[a,b,c]$ to be alternating means that $[a,a,c] = [a,b,a] = [b,a,a] = 0$. These say that $(aa)c - a(ac) = (ab)a - a(ba) = (ba)a - b(aa) = 0$. For (b), $[a,a,c] = [b,a,a] = 0$ and the 3-linearity together imply that $[a,b,a] = [a,b,a]+[b,b,a] = [a+b,b,a] = [a+b,b,a]+[a+b,a,a] = [a+b,a+b,a] = 0$.

53. For (a), $(1,0)(c,d) = (c,d)$ and $(a,b)(1,0) = (a,b)$ directly from the definition. Also, the definition $(a,b)^* = (a^*,-b)$ makes $(1,0)^* = (1,0)$, $(a,b)^{**} = (a^*;-b)^* = (a^{**},b) = (a,b)$, and $(c,d)^*(a,b)^* = (c^*,-d)(a^*,-b) = (c^*a^* - bd^*, -c^{**}b - a^*d) = ((c^*a^* - bd^*)^*, a^*d + cb)^* = (ac - db^*, a^*d + cb)^* = ((a,b)(c,d))^*$.

For (b), (c), and (d), we observe that

$$((a,b)(c,d))(e,f) = (ac\cdot e - db^*\cdot e - f\cdot d^*a + f\cdot b^*c^*, c^*a^*\cdot f - bd^*\cdot f + e\cdot a^*d + e\cdot cb)$$

and

$$(a,b)((c,d)(e,f)) = (a\cdot ce - a\cdot fd^* - c^*f\cdot b^* - ed\cdot b^*, a^*\cdot c^*f + a^*\cdot ed + ce\cdot b - fd^*\cdot b),$$

and the results are immediate.

In (e), (i) is the usual construction, and (ii) has $\mathbf{1} = (1,0), \mathbf{i} = (i,0), \mathbf{j} = (0,1)$, and $\mathbf{k} = (0,-i)$, with the identity of $\mathbb{H}$ written now as $\mathbf{1}$.

54. For (a), $(a,b)^* + (a,b) = (a^*,-b) + (a,b) = (a^*+a, 0)$, which is a real multiple of $(1,0)$. Also, $(a,b)(a,b)^* = (a,b)(a^*,-b) = (aa^* + bb^*, a^*(-b) + a^*b) = (aa^* + bb^*, 0)$, and this is a positive multiple of $(1,0)$ since aa^* and bb^* are ≥ 0 and at least one of them is positive. A similar argument applies to $(a,b)^*(a,b)$.

In (b), certainly (a,b) is bilinear over $\mathbb{R}$, the expression for (a,b) is manifestly symmetric, and we know that $(a,a) = aa^*$ is ≥ 0 with equality only for $a = 0$.

In (c), we are to prove that $(xx)y = x(xy)$ and $(yx)x = y(xx)$ in B. It is enough to prove the first identity since application of $*$ to it gives the second identity. We use $(c,d) = (a,b)$ and substitute into the displayed formulas above for Problem 53. We find that $((a,b)(a,b))(e,f)$ equals

$$(aa\cdot e - bb^*\cdot e - f\cdot b^*a - f\cdot b^*a^*, \quad a^*a^*\cdot f - bb^*\cdot f + e\cdot a^*b + e\cdot ab)$$

and that $(a,b)((a,b)(e,f))$ equals

$$(a\cdot ae - a\cdot fb^* - a^*f\cdot b^* - eb\cdot b^*, \quad a^*\cdot a^*f + a^*\cdot eb + ae\cdot b - fb^*\cdot b).$$

Taking into account the associativity of A, we see that it is enough to show that $(bb^*)e = e(bb^*)$, $fb^*(a+a^*) = (a+a^*)fb^*$, $(bb^*)f = f(bb^*)$, and $e(a+a^*) = (a+a^*)e$. These all follow from the fact that A is nicely normed.

55. Part (a) follows from (a) and (c) of the previous problem.

In (b), we have $(xx^*)y = (x(c1-x))y = cxy - (xx)y = cxy - x(xy) = x(cy - xy) = x((c1-x)y) = x(x^*y)$. The equality $x(yy^*) = (xy)y^*$ follows by applying $*$ and renaming the variables.

In (c), use of (b) and the definitions of the norm and $*$ gives $\|ab\|^2 a = ((ab)(ab)^*)a = (ab)((ab)^*a) = (ab)((b^*a^*)a) = (ab)(b^*(a^*a)) = \|a\|^2((ab)b^*) = \|a\|^2 a(bb^*) = \|a\|^2 \|b\|^2 a$.

For (d), the norm equality of (c) implies that the $\mathbb{R}$ linear maps left-by-a and right-by-a are one-one, and the finite dimensionality of $\mathbb{O}$ allows us to conclude that they are onto. Hence they are invertible.

For (e), use of (b) gives $a(\|a\|^{-2}a^*b) = \|a\|^{-2}a(a^*b) = \|a\|^{-2}(aa^*)b = \|a\|^{-2}\|a\|^2 b = b$. This proves the result for left multiplication, and the argument for right multiplication is similar.

For (f), the table is as follows, with each entry representing the product of the element at the left (the row index) by the element at the top (the column index):

$(1,0)$	$(\mathbf{i},0)$	$(\mathbf{j},0)$	$(\mathbf{k},0)$	$(0,1)$	$(0,\mathbf{i})$	$(0,\mathbf{j})$	$(0,\mathbf{k})$
$(\mathbf{i},0)$	$-(1,0)$	$(\mathbf{k},0)$	$-(\mathbf{j},0)$	$-(0,\mathbf{i})$	$(0,1)$	$-(0,\mathbf{k})$	$(0,\mathbf{j})$
$(\mathbf{j},0)$	$-(\mathbf{k},0)$	$-(1,0)$	$(\mathbf{i},0)$	$-(0,\mathbf{j})$	$(0,\mathbf{k})$	$(0,1)$	$-(0,\mathbf{i})$
$(\mathbf{k},0)$	$(\mathbf{j},0)$	$-(\mathbf{i},0)$	$-(1,0)$	$-(0,\mathbf{k})$	$-(0,\mathbf{j})$	$(0,\mathbf{i})$	$(0,1)$
$(0,1)$	$(0,\mathbf{i})$	$(0,\mathbf{j})$	$(0,\mathbf{k})$	$-(0,1)$	$-(0,\mathbf{i})$	$-(0,\mathbf{j})$	$-(0,\mathbf{k})$
$(0,\mathbf{i})$	$-(1,0)$	$-(\mathbf{k},0)$	$(\mathbf{j},0)$	$(0,\mathbf{i})$	$-(0,1)$	$-(0,\mathbf{k})$	$(0,\mathbf{j})$
$(0,\mathbf{j})$	$(\mathbf{k},0)$	$-(1,0)$	$-(\mathbf{i},0)$	$(0,\mathbf{j})$	$(0,\mathbf{k})$	$-(0,1)$	$-(0,\mathbf{i})$
$(0,\mathbf{k})$	$-(\mathbf{j},0)$	$(\mathbf{i},0)$	$-(1,0)$	$(0,\mathbf{k})$	$-(0,\mathbf{j})$	$(0,\mathbf{i})$	$-(0,1)$

56. Although B is nicely normed, the steps of (b) in Problem 55 are not justified for it because we cannot conclude that B is alternative. Since the argument for (b) breaks down, so do the arguments for (c) and (d).

Chapter VII

1. The only integer < 60 that is not the product of powers of at most two primes is 30. Thus Burnside's Theorem assures us that the only possible order less than 60 for a nonabelian simple group is 30. The integer 30 is of the form $2pq$ with $p = 3$ and $q = 5$, and $q + 1 = 2p$. Part (b) of Problem 34 at the end of Chapter IV is applicable and shows that the group has a subgroup of index 2; subgroups of index 2 are always normal.

2. For (a) and (b), $(xyx^{-1}y^{-1})^{-1} = yxy^{-1}x^{-1}$ is a commutator, and so is $a(xyx^{-1}y^{-1})a^{-1} = (axa^{-1})(aya^{-1})(axa^{-1})^{-1}(aya^{-1})^{-1}$.

3. Let H be generated by a and b, and let K be generated by bab^2 and bab^3. Certainly $K \subseteq H$. Since bab^2 and bab^3 are in K, so is $(bab^2)^{-1}(bab^3) = b$ and then so is $(b^{-1})(bab^2)(b^{-2}) = a$. Hence $H \subseteq K$.

Chapter VII   651

4. If H is characteristic, then in particular every inner automorphism $x \to gxg^{-1}$ carries H to itself, and H is normal. If $\varphi : G \to G$ is an automorphism and z is in Z_G, then the equality $\varphi(z)\varphi(g) = \varphi(zg) = \varphi(gz) = \varphi(g)\varphi(z)$ and the fact that φ is onto G show that $\varphi(z)$ is in Z_G. If $\psi : G \to G$ is an automorphism, then $\psi(xyx^{-1}y^{-1}) = \psi(x)\psi(y)(\psi(x))^{-1}(\psi(y))^{-1}$ shows that ψ carries commutators to commutators; hence ψ carries the generated subgroup G' to itself.

5. H_8, Z_{H_8}, and $\{1\}$ are characteristic. But the subgroups of order 4 are not, because, for example, there exists an automorphism of H_8 carrying $\mathbf{i}$ to $\mathbf{j}$.

6. Yes. The proof of Proposition 7.7, which takes $S = G$, gives a finite presentation.

7. In (a), $\begin{pmatrix} \sqrt{2} & 0 \\ 0 & \frac{1}{\sqrt{2}} \end{pmatrix} \begin{pmatrix} 1 & t \\ 0 & 1 \end{pmatrix} \begin{pmatrix} \sqrt{2} & 0 \\ 0 & \frac{1}{\sqrt{2}} \end{pmatrix}^{-1} \begin{pmatrix} 1 & t \\ 0 & 1 \end{pmatrix}^{-1} = \begin{pmatrix} 1 & t \\ 0 & 1 \end{pmatrix}$.

In (b), we have also $\begin{pmatrix} 0 & 1 \\ -1 & 0 \end{pmatrix} \begin{pmatrix} e^s & 0 \\ 0 & e^{-s} \end{pmatrix} \begin{pmatrix} 0 & 1 \\ 1 & 0 \end{pmatrix}^{-1} \begin{pmatrix} e^s & 0 \\ 0 & e^{-s} \end{pmatrix}^{-1} = \begin{pmatrix} e^{-2s} & 0 \\ 0 & e^{2s} \end{pmatrix}$ and $\begin{pmatrix} \frac{1}{\sqrt{2}} & 0 \\ 0 & \sqrt{2} \end{pmatrix} \begin{pmatrix} 1 & 0 \\ r & 1 \end{pmatrix} \begin{pmatrix} \frac{1}{\sqrt{2}} & 0 \\ 0 & \sqrt{2} \end{pmatrix}^{-1} \begin{pmatrix} 1 & 0 \\ r & 1 \end{pmatrix}^{-1} = \begin{pmatrix} 1 & 0 \\ r & 1 \end{pmatrix}$. Thus $\begin{pmatrix} 1 & 0 \\ r & 1 \end{pmatrix}$, $\begin{pmatrix} 1 & t \\ 0 & 1 \end{pmatrix}$, and $\begin{pmatrix} a & 0 \\ 0 & a^{-1} \end{pmatrix}$ are in G' for $a > 0$. Since $\begin{pmatrix} a & b \\ c & d \end{pmatrix} = \begin{pmatrix} 1 & 0 \\ c/a & 1 \end{pmatrix} \begin{pmatrix} a & 0 \\ 0 & a^{-1} \end{pmatrix} \begin{pmatrix} 1 & b/a \\ 0 & 1 \end{pmatrix}$, the matrix $\begin{pmatrix} a & b \\ c & d \end{pmatrix}$ is in G' if $a > 0$. If $a < 0$, we have $\begin{pmatrix} a & b \\ c & d \end{pmatrix} \begin{pmatrix} 1 & 0 \\ r & 1 \end{pmatrix} = \begin{pmatrix} a+br & b \\ c+dr & d \end{pmatrix}$; if $b \neq 0$, then $a + br > 0$ for suitable r and therefore the equality $\begin{pmatrix} a & b \\ c & d \end{pmatrix} = \begin{pmatrix} a+br & b \\ c+dr & d \end{pmatrix} \begin{pmatrix} 1 & 0 \\ -r & 1 \end{pmatrix}$ exhibits $\begin{pmatrix} a & b \\ c & d \end{pmatrix}$ as in G'. Similarly if $c \neq 0$, then $a + cr > 0$ for suitable r and hence $\begin{pmatrix} 1 & r \\ 0 & 1 \end{pmatrix} \begin{pmatrix} a & b \\ c & d \end{pmatrix} = \begin{pmatrix} a+cr & b+dr \\ c & d \end{pmatrix}$ exhibits $\begin{pmatrix} a & b \\ c & d \end{pmatrix}$ as in G'. Thus all members of G are in G' except possibly for $\begin{pmatrix} a & 0 \\ 0 & a^{-1} \end{pmatrix}$ with $a < 0$. So it is enough to prove that $\begin{pmatrix} -1 & 0 \\ 0 & -1 \end{pmatrix}$ is in G'. This follows since $\begin{pmatrix} 0 & 1 \\ -1 & 0 \end{pmatrix}$ has been shown to be in G' and has square equal to $\begin{pmatrix} -1 & 0 \\ 0 & -1 \end{pmatrix}$.

In (c), suppose that $(xyx^{-1})y^{-1} = \begin{pmatrix} -1 & 0 \\ 0 & -1 \end{pmatrix}$. Then $xyx^{-1} = -y$. Taking the trace of both sides and using the fact that $\text{Tr}\, xyx^{-1} = \text{Tr}\, y$, we see that $\text{Tr}\, y = -\text{Tr}\, y$ and $\text{Tr}\, y = 0$. Put $x = \begin{pmatrix} r & s \\ t & u \end{pmatrix}$ and $y = \begin{pmatrix} a & b \\ c & -a \end{pmatrix}$, and substitute into the equality $xy = -yx$. The entry-by-entry equations are $ra + sc = -ra - tb$, $rb = -ub$, $uc = -rc$, and $tb - ua = -sc + ua$. The first and fourth equations together say that $2ra = -tb - sc = -2ua$. Thus we have $(r + u)a = 0$, $(r + u)b = 0$, and $(r + u)c = 0$. Since at least one of a, b, c is nonzero, $r + u = 0$ and $x = \begin{pmatrix} r & s \\ t & -r \end{pmatrix}$. Writing out the equality $xy = -yx$, we obtain the necessary and sufficient condition

$$2ra = -sc - tb. \qquad (*)$$

The determinant conditions are $-r^2 - st = 1$ and $-a^2 - bc = 1$. Multiplying $(*)$ by sc and substituting $st = -1 - r^2$ and $bc = -1 - a^2$, we obtain $2rsac =$

$-s^2c^2 - (-1 - r^2)(-1 - a^2)$ and then $0 = -s^2c^2 - 2rsac - 1 - a^2 - r^2 - r^2a^2 = -(ra + sc)^2 - 1 - a^2 - r^2$, contradiction. Thus $\begin{pmatrix} -1 & 0 \\ 0 & -1 \end{pmatrix}$ is not a commutator.

8. By Proposition 7.8 the constructed group is a quotient of the group given by generators and relations. We actually have an isomorphism if each element of the group given by generators and relations is of the form $b^p a^q$ with $0 \le p \le 2$ and $0 \le q \le 8$ because the group given by generators and relations then has order ≤ 27. Right multiplication by a carries this set to itself. Right multiplication by b has $b^p a^q b = b^p b (b^{-1} a^q b) = b^{p+1} (b^{-1} a b)^q = b^p (a^4)^q = b^p a^{4q}$, and this equals a suitable element $b^{p'} a^{q'}$ with $0 \le p' \le 2$ and $0 \le q' \le 8$. Hence the group defined by generators and relations has at most 27 elements, and we have the desired isomorphism.

9. Let F_n be free on $x_1, y_1, \ldots, x_n, y_n$, let $\varphi : F_n \to F_n/F_n'$ be the homomorphism of Corollary 7.5, and let $\Psi : F_n \to G_n$ be the given quotient homomorphism. Then $\ker \varphi \subseteq \ker \Psi$, and Proposition 4.11 shows that there exists a group homomorphism $\psi : G_n \to F_n/F_n'$ such that $\psi \circ \Psi = \varphi$. Since F_n/F_n' is abelian, ψ factors as $\bar{\psi} \circ q$, where $q : G_n \to G_n/G_n'$ is the quotient and $\bar{\psi} : G_n/G_n' \to F_n/F_n'$ is a homomorphism. Thus $\bar{\psi} \circ q \circ \Psi = \varphi$. Since φ is onto, $\bar{\psi}$ is onto; thus the image of $\bar{\psi}$ is isomorphic with F_n/F_n', which is free abelian of rank $2n$. The group G_n/G_n' is abelian and has a generating set of $2n$ generators, thus is a homomorphic image $\xi : A_n \to G_n/G_n'$, where A_n is free abelian with $2n$ generators. The composition $\bar{\psi} \circ \xi$ is a homomorphism from a free abelian group of rank $2n$ onto a free abelian group of rank $2n$. Taking into account the proof of Theorem 4.46, we see that $\bar{\psi} \circ \xi$ is one-one. Since ξ is onto G_n/G_n', $\bar{\psi}$ is one-one. Therefore G_n/G_n' is free abelian of rank $2n$.

10. Let F be a free group of rank n, let $q : F \to F/F'$ be the quotient homomorphism, let $x_1, \ldots, x_k$ with $k < n$ be generators of F, let $\widetilde{F} = F(\{x_1, \ldots, x_k\})$, and let $\Phi : \widetilde{F} \to F$ be the quotient homomorphism. The composition $q \circ \Phi$ is a homomorphism of $\widetilde{F}$ onto the abelian group F/F', and it factors through to a homomorphism of $\widetilde{F}/\widetilde{F}'$ onto F/F'. Here the domain is abelian with k generators, and the image is free abelian with n generators, and there can be no such homomorphism.

11. For (a), we can use 1 and a. For (b), the proof of Theorem 7.10 says that we are to multiply each of these by a, b, c on the right and take the H part of the result. The H parts that are not 1 form a free basis. We have $1a = a$ and $1a\rho(a)^{-1} = 1, 1b = ba^{-1}a$ and $1b\rho(b)^{-1} = ba^{-1}, 1c = c = ca^{-1}a$ and $1c\rho(c)^{-1} = ca^{-1}, aa = a^2 1$ and $aa\rho(a^2)^{-1} = a^2, ab = ab 1$ and $ab\rho(ab)^{-1} = ab$, and $ac = ac 1$ and $ac\rho(ac)^{-1} = ac$. Thus a free basis of the generated subgroup is $\{ba^{-1}, ca^{-1}, a^2, ab, ac\}$.

12. The thing to prove, by induction on n, is that if $a_1 a_2 \cdots a_n$ is a reduced word in variables $u_0, u_1, u_2, \ldots$ and their inverses, and if we then substitute $x^k y x^{-k}$ for u_k and reduce in terms of x, y, then the reduced form involves a total of n factors of y or y^{-1}, the factor to the left of the first y or y^{-1} is x^p if $a_1 = u_p^{\pm 1}$, and the factor to the right of the last y or y^{-1} is x^{-q} if $a_n = u_q^{\pm 1}$.

13. The remarks with Proposition 7.15 show that the reduced words in $C_2 * C_2$ are all words whose terms are alternately x and y. Let H be a normal subgroup $\neq \{1\}$. Then H contains a conjugate of a nontrivial such word. Form the shortest such word $\neq 1$ in H. If the word begins and ends with x and has length > 1, we can conjugate by x and reduce the length by 2; similarly if it begins and ends with y and has length > 1, we can conjugate by y and reduce the length by 2. We conclude that the word has length 1. Then H contains x or y and is a quotient of either $\langle y;\ y^2 \rangle$ or $\langle x;\ x^2 \rangle$, which give C_2 and $\{1\}$.

Thus we may assume that a shortest nontrivial reduced word in H is a product $xy \cdots xy$ with $2n$ factors or a product $yx \cdots yx$ with $2n$ factors. Then G/H is a quotient of $\langle a, b;\ a^2, b^2, (ab)^n \rangle$, and we saw in an example in Section 2 that this group is D_n. We readily check that all quotients of D_n are of the form $\{1\}$, C_2, $C_2 \times C_2$, and D_m for certain values of $m \geq 3$.

14. Argument #1: When the irreducible representations are all 1-dimensional, Corollary 7.25 shows that the number of irreducible representations must be $|G|$, and Corollary 7.28 shows that the number of conjugacy classes must be $|G|$. Therefore each conjugacy class contains just one element, and G is abelian.

Argument #2: Theorem 7.24 shows that the irreducible representations separate points in G in the sense that for any pair x, y in the group, there is some irreducible R with $R(x) \neq R(y)$. When the irreducible representations are all 1-dimensional, the multiplicative characters separate points. Since every multiplicative character is trivial on the commutator subgroup, the commutator subgroup must be $\{1\}$. Then every pair x, y has $xyx^{-1}y^{-1} = 1$ and $xy = yx$.

15. This is immediate from Lemma 7.11.

16. For (a), every cochain f has the property that $mf = 0$. Hence the same thing is true of cocycles and of cohomology elements.

For (b), the cocycle condition for f says that

$$(-1)^n f(g_1, \ldots, g_n) = g_1(f(g_2, \ldots, g_{n+1}))$$
$$+ \sum_{i=1}^{n-1} (-1)^i f(g_1, \ldots, g_{i-1}, g_i g_{i+1}, g_{i+2}, \ldots, g_{n+1})$$
$$+ (-1)^n f(g_1, \ldots, g_{n-1}, g_n g_{n+1}).$$

Summing over g_{n+1} in G gives

$$(-1)^n |G| f(g_1, \ldots, g_n) = g_1(F(g_2, \ldots, g_n))$$
$$+ \sum_{i=1}^{n-1} (-1)^i F(g_1, \ldots, g_{i-1}, g_i g_{i+1}, g_{i+2}, \ldots, g_n)$$
$$+ (-1)^n F(g_1, \ldots, g_{n-1}).$$

The right side we recognize as $(\delta_{n-1} F)(g_1, \ldots, g_n)$, which is the value of a coboundary at $(g_1, \ldots, g_n)$. Therefore $|G| f$ is a coboundary and becomes the 0 element in

$H^2(G, N)$. Thus f, when regarded as an element of $H^2(G, N)$ has order dividing $|G|$.

17. The two parts of the previous problem show that every element of $H^2(G, N)$ is of finite order dividing both $|G|$ and $|G/N|$. Since $\mathrm{GCD}(|G|, |G/N|) = 1$, every element of $H^2(G, N)$ has order 1. Thus $H^2(G, N) = 0$, and the only extension is the semidirect product.

18. The only automorphism of C_2 is the trivial automorphism, and therefore τ is trivial. The two possibilities for G are $C_2 \times C_2$ and C_4. With $G = C_2 \times C_2$, the group E can be $C_2 \times C_2 \times C_2$ or H_8, and with $G = C_4$, E can be $C_2 \times C_4$ or C_8. For the cases $E = C_2 \times C_2 \times C_2$ and $E = C_2 \times C_4$, the extension is the direct product, and no further discussion is necessary. For the cases $E = H_8$ and $E = C_8$, the embedding of $N = C_2$ is unique, and we therefore get only one extension in each case. Thus there are exactly two inequivalent extensions for each choice of G.

19. If N embeds as a summand C_2, then the quotient E/N has one fewer summand C_2, is still the countable direct sum of copies of C_2 and C_4, and is therefore isomorphic to E. If N embeds as a 2-element subgroup of a summand C_4, then the quotient E/N has one fewer summand C_4 and one more summand C_2, is still the countable direct sum of copies of C_2 and C_4, and is therefore isomorphic to E.

The action τ has to be trivial because C_2 has only the trivial automorphism.

If an equivalence Φ of extensions were to exist, it would have to satisfy $\Phi i_1(x) = i_2(x)$ for the nontrivial element x of $N = C_2$. But $i_1(x)$ is an element of order 2 that is not the square of an element of order 4, while $i_2(x)$ is an element of order 2 that is the square of an element of order 4. Since Φ is an isomorphism, it has to carry nonsquares to nonsquares, and we cannot have $\Phi i_1(x) = i_2(x)$.

20. Let us write i_1 and i_2 for the inclusions of N into E_1 and E_2. For $(i_1(x), 1)$ to be in Q, $i_1(x)$ must be 1; hence x must be 1. Thus $x \mapsto (i_1(x), 1)Q$ is one-one. The image of φ is the same as the image of φ_1, which is G. Suppose that (e_1, e_2) is in $(E_1, E_2) \cap Q$. Then $\varphi_1(e_1) = \varphi_2(e_2)$ and $(e_1, e_2) = (i_1(x), i_2(x)^{-1})$ for some $x \in N$. Then $\varphi(e_1, e_2) = \varphi_1(i_1(x)) = 1$, and φ descends to the quotient.

If $(e_1, e_2)Q$ is in the kernel of the descended φ, then (e_1, e_2) is in the kernel of the original φ, and e_1 is in the kernel of φ_1. Therefore $e_1 = i_1(x)$ for some $x \in N$. Since $\varphi_2(e_2) = \varphi_1(e_1)$, e_2 is in the kernel of φ_2 and $e_2 = i_2(y)$ for some $y \in N$. The element $(i_1(y), i_2(y)^{-1})$ is in Q, and we therefore have $(i_1(x), i_2(y))Q = (i_1(x), i_2(y))(i_1(y), i_2(y)^{-1})Q = (i_1(xy), 1)Q$. Thus $(i_1(x), i_2(y))Q$ is exhibited as in the image of the embedded copy of N.

21. Since Q is normal, we have $(\bar{u}, \widetilde{u})(\bar{v}, \widetilde{v})Q = (a(u, v)\overline{uv}, b(u, v)\widetilde{uv})Q = (b(u, v), b(u, v)^{-1})(a(u, v)\overline{uv}, b(u, v)\widetilde{uv})Q = (b(u, v)a(u, v)\overline{uv}, 1\widetilde{uv})Q = (b(u, v)a(u, v), 1)(\overline{uv}, \widetilde{uv})Q)$. Thus the cocycle for $(E_1, E_2)Q$ is $\{b(u, v)a(u, v)\} = \{a(u, v)b(u, v)\}$.

22. Let $\Phi_1 : E_1 \to E_1'$ and $\Phi_2 : E_2 \to E_2'$ be isomorphisms exhibiting the equivalences of the extensions. Define $\Phi(e_1, e_2) = (\Phi(e_1), \Phi(e_2))Q'$, and check that this descends to the required isomorphism $\Phi : (E_1, E_2)/Q \to (E_1', E_2')/Q'$.

23. $\widehat{f}(\chi) = \sum_{t \in G} f(t)\overline{\chi(t)} = \sum_{\dot{t} \in G/H} \sum_{h \in H} f(t+h)\overline{\dot{\chi}(\dot{t})} = \sum_{\dot{t} \in G/H} F(\dot{t})\overline{\dot{\chi}(\dot{t})}$
$= \widehat{F}(\dot{\chi})$.

24. Fourier inversion and Problem 23 give $F(x) = |G/H|^{-1} \sum_{\dot{\chi} \in \widehat{G/H}} \widehat{F}(\dot{\chi})\dot{\chi}(x)$
$= |G/H|^{-1} \sum_{\dot{\chi} \in \widehat{G/H}} \widehat{f}(\chi)\dot{\chi}(x)$. Pulling back $\dot{\chi}$ to the member χ of $\widehat{G}$ with $\chi|_H = 1$ and substituting the definition of F, we obtain the desired result.

25. For (a), if $C = 0$, then all $a \in \mathbb{F}^n$ have $(a, 0) = 0$, and hence $C^\perp = \mathbb{F}^n$. For (b), the repetition code has $C = \{0, (1, \ldots, 1)\}$. The members a of $\mathbb{F}^n$ with $(a, (1, \ldots, 1)) = 0$ are the members of even weight, hence the members of the parity-check code. For (c), it is enough to check that $(a, c) = 0$ for each pair of members a, c of a basis of C, and this one can do by hand.

For (d), Proposition 6.3 shows that $n = \dim C + \dim C^\perp$. Since $C = C^\perp$, $\dim C = n/2$.

For (e), every member c of C is in $C^\perp$ and must in particular have $(c, c) = 0$. Therefore c has even weight.

For (f), let c and c' be in C, and write cc' for the entry-by-entry product (logical "and"). Then $\text{wt}(c + c') = \text{wt}(c) + \text{wt}(c') - 2\text{wt}(cc')$, and hence $\frac{1}{2}\text{wt}(c + c') = \frac{1}{2}\text{wt}(c) + \frac{1}{2}\text{wt}(c') - \text{wt}(cc')$. Considering this equality modulo 2 shows that it is enough to prove that $C \subseteq C^\perp$ implies that $\text{wt}(cc')$ is even whenever c and c' are in C. Modulo 2, we have $\text{wt}(cc') \equiv (c, c')$, and $(c, c') = 0$ since $C \subseteq C^\perp$.

26. In (a), every element of $\mathbb{F}^n$ has order at most 2, and thus χ takes only the values ± 1. Define $(a_\chi)_i$ to be 0 if $\chi(e_i) = +1$ and to be 1 if $\chi(e_i) = -1$. Then $\chi(e_i) = (-1)^{(a, e_i)}$ for each i. The two sides extend uniquely as homomorphisms of $\mathbb{F}^n$ to $\{\pm 1\}$, and it follows that $\chi(c) = (-1)^{(a,c)}$ for all $c \in \mathbb{F}^n$. The remainder of (a) is routine.

In (b), let χ correspond to a. Then $\widehat{f}(a) = \widehat{f}(\chi) = \sum_{c \in \mathbb{F}^n} f(c)\overline{\chi(c)} = \sum_{c \in \mathbb{F}^n} f(c)(-1)^{(a,c)}$.

In (c), we have

$$\prod_i \widehat{f_i}(a_i) = \prod_i \sum_{c_i \in \mathbb{F}} f_i(c_i)(-1)^{a_i c_i}$$
$$= \sum_{c_1 \in \mathbb{F}} f_1(c_1)(-1)^{a_1 c_1} \cdots \sum_{c_n \in \mathbb{F}} f_n(c_n)(-1)^{a_n c_n}$$
$$= \sum_{c \in \mathbb{F}^n} f_1(c_1)(-1)^{a_1 c_1} \cdots f_n(c_n)(-1)^{a_n c_n} = \sum_{c \in \mathbb{F}^n} f(c)(-1)^{(a,c)} = \widehat{f}(a).$$

27. In (a), $\widehat{f_0}(0) = \sum_{c_0 \in \mathbb{F}} f_0(c_0)(-1)^{0 c_0} = f_0(0)(+1) + f_0(1)(+1) = x + y$ and $\widehat{f_0}(1) = \sum_{c_0 \in \mathbb{F}} f_0(c_0)(-1)^{1 c_0} = f_0(0)(+1) + f_0(1)(-1) = x - y$.
In (b), Problem 26c gives

$$\widehat{f}(a) = \prod_{i=1}^n \widehat{f_0}(a_i) = \Big(\prod_{i \text{ with } a_i = 0}(x + y)\Big)\Big(\prod_{i \text{ with } a_i = 1}(x - y)\Big)$$
$$= (x + y)^{n - \text{wt}(a)}(x - y)^{\text{wt}(a)}.$$

28. In (a), the members of $\widehat{G/H}$ lift exactly to the members ω of $\widehat{G}$ with $\omega\big|_H = 1$. Under the mapping of Problem 26a, any member χ of $\widehat{G}$ yields a unique member a_χ of $\mathbb{F}^n$ with $\chi(c) = (-1)^{(a_\chi, c)}$ for all $c \in \mathbb{F}^n$. If a_χ is in $C^\perp$, then this formula gives $\chi(c) = 1$, i.e., $\chi\big|_H = 1$. If a_χ is not in $C^\perp$, then $\chi(c_0) \neq 1$ for some $c_0 \in C$, i.e., $\chi\big|_H \neq 1$.

In (b), we apply the special case of Problem 24 mentioned in the educational note. Then the result is immediate, in view of (a).

In (c), we let $f(c) = x^{n-\text{wt}(c)} y^{\text{wt}(c)}$. Problem 27b says that $\widehat{f}(a) = (x+y)^{n-\text{wt}(a)}(x-y)^{\text{wt}(a)}$. Substituting into the formula of the previous part gives $\sum_{c \in C} x^{n-\text{wt}(c)} y^{\text{wt}(c)} = |C^\perp|^{-1} \sum_{a \in C^\perp} (x+y)^{n-\text{wt}(a)}(x-y)^{\text{wt}(a)}$, and this says that $W_C(x, y) = |C^\perp|^{-1} W_{C^\perp}(x+y, x-y)$.

In (e), parts (d) and (e) of Problem 25 show that the only monomials $X^k Y^l$ in $W_C(X, Y)$ with nonzero coefficients are those with k and l even. Therefore $W_C(X, Y)$ is invariant under the transformations $X \mapsto -X$ and $Y \mapsto -Y$. The MacWilliams identity shows that $W_C(X, Y)$, apart from a constant, is the same polynomial in $X + Y$ and $X - Y$. Therefore $W_C(X, Y)$ is invariant also under $(X + Y) \mapsto -(X + Y)$ and under $(X - Y) \mapsto -(X - Y)$. Thus $W_C(X, Y)$ is invariant under the group of symmetries of a regular octagon centered at 0 with one of its sides centered at $(1, 0)$. This symmetry group is D_8.

29. The characters of G are the ones with $\chi_n(1) = \zeta_m^n$ for $0 \leq n < m$. Such a character is trivial on H if and only if $\chi_n(q) = 1$, i.e., if and only if $\zeta_m^{nq} = 1$; this means that nq is a multiple of m, hence that n is a multiple of p.

The element 1 of H is the element q of G. Thus the question about the identification of the descended characters asks the value of $\chi_n(1)$ when n is a multiple jp of p. The value is $\chi_n(1) = \zeta_m^n = \zeta_{pq}^{jp} = \zeta_q^j$.

If we have computed F on G/H and want to compute $\widehat{F}$ from the definition of Fourier coefficients, we have to multiply each of the q values of F by the values of each of the q characters of G/H and then add. The number of multiplications is q^2. The actual computation of F from f involves p additions for each of the q values of i, hence pq additions.

30. $\widehat{f}(\zeta_m^{jp+k}) = \sum_{i=0}^{m-1} f(i) \zeta_m^{-(jp+k)i} = \sum_{i=0}^{m-1} (f(i) \zeta_m^{-ki}) \zeta_m^{-jp}$. The variant of f for the number k is then $i \mapsto f(i) \zeta_m^{-ki}$. Handling each value of k involves $m = pq$ steps to compute the variant of f and then the $q^2 + pq$ steps of Problem 29. Thus we have $q^2 + 2pq$ steps for each k, which we regard as of order $q^2 + pq$. This means $p(q^2 + pq)$ steps when all k's are counted, hence $pq(p+q)$ steps.

32. By inspection, $(\ell_{v_1}, \ell_{v_2})_{V'} = (v_1, v_2)_{\overline{V}}$ has the properties of an inner product. The definition is set up so that the linear mapping $\ell_v \mapsto v$ of V' into $\overline{V}$ preserves inner products.

33. The contragredient has $(R^c(x)\ell_v)(v') = \ell_v(R(x^{-1})v') = (R(x^{-1})v', v)_V = (v', R(x)v)_V = \ell_{R(x)v}(v')$. Hence $R^c(x)\ell_v = \ell_{R(x)v}$, and $(R^c(x)\ell_v, R^c(x)\ell'_v)_{V'} = (R(x)v, R(x)v')_{\overline{V}} = \overline{(R(x)v', R(x)v)_V} = \overline{(v', v)_V} = (v, v')_{\overline{V}} = (\ell_v, \ell_{v'})_{V'}$.

34. If $\{v_j\}$ is an orthonormal basis of V, then $\{\ell_{v_j}\}$ is an orthonormal basis of V' by Problem 32, and $(R^c(x)\ell_{v_j}, \ell_{v_j})_{V'} = (\ell_{R(x)v_j}, \ell_{v_j})_{V'} = (v_j, R(x)v_j)_V = \overline{(R(x)v_j, v_j)_V}$. Summing on j gives the desired equality of group characters.

35. In view of Problem 34 a necessary condition on a 1-dimensional representation for it to be equivalent to its contragredient is that it be real-valued. Hence the two nontrivial multiplicative characters of C_3 are not equivalent to their contragredients.

36. Following the notation in the discussion before Theorem 7.23, let $\rho_{ij}(x) = (R(x)u_j, u_i)$, let l be the left-regular representation, and let $\ell_v(u) = (u, v)_V$ be as above. Consider, for fixed j_0, the image of $R^c(g)\ell_{u_i}$ under the linear extension to V' of the map $E'(\ell_{u_k})(x) = (R(x)u_{j_0}, u_k)_V$. This is $E'(\ell_{\sum_k \bar{c}_k u_k})(x) = E'\left(\sum_k \bar{c}_k \ell_{u_k}\right)(x) = \sum_k \bar{c}_k E'(\ell_{u_k})(x) = \sum_k \bar{c}_k (R(x)u_{j_0}, u_k)_V = (R(x)u_{j_0}, \sum_k c_k u_k)_V$, and hence $E'(\ell_v)(x) = (R(x)u_{j_0}, v)_V$. Then the image of interest is

$$E'(R^c(g)\ell_{u_i})(x) = E'(\ell_{R(g)u_i})(x) = (R(x)u_{j_0}, R(g)u_i)_V$$
$$= (R(g^{-1}x)u_{j_0}, u_i)_V = (l(g)\rho_{ij_0})(x).$$

Therefore l carries a column of matrix coefficients to itself and is equivalent on such a column to R^c.

37. Let $x = \begin{pmatrix} 0 & -1 \\ 1 & 0 \end{pmatrix}$ and $y = \begin{pmatrix} 0 & 1 \\ -1 & -1 \end{pmatrix}$, and let Γ be the subgroup generated by x and y. Observe that $-I = x^2$, $y^{-1} = \begin{pmatrix} -1 & -1 \\ 1 & 0 \end{pmatrix}$, and $yx = \begin{pmatrix} 1 & 0 \\ -1 & 1 \end{pmatrix}$ are in Γ. Arguing by contradiction, suppose that $\Gamma \neq \mathrm{SL}(2, \mathbb{Z})$. Choose a matrix $z = \begin{pmatrix} a & b \\ c & d \end{pmatrix}$ in $\mathrm{SL}(2, \mathbb{Z})$ but not Γ such that $\max(|a|, |b|)$ is as small as possible. If $ab = 0$, then one of $|a|$ and $|b|$ is 1 and the other is 0 because the matrix has determinant 1. If $|a| = 0$, then zy^{-1} has top row $(\pm 1 \ 0)$; so in either event we see that some member of $\mathrm{SL}(2, \mathbb{Z})$ outside Γ is of the form $\pm \begin{pmatrix} 1 & 0 \\ t & 1 \end{pmatrix}$. Since $x^2 = -I$ is in Γ and $yx = \begin{pmatrix} 1 & 0 \\ -1 & 1 \end{pmatrix}$ is in Γ, this is a contradiction.

Thus the matrix z cannot have $ab = 0$. Suppose that $ab > 0$. Then zy has top row $(-b \ a - b)$, and zy^{-1} has top row $(-a + b \ -a)$. The minimality of $\max(|a|, |b|)$ for z says that

$$\max(|a|, |b|) \leq \max(|-b|, |a - b|) \quad \text{and} \quad \max(|a|, |b|) \leq \max(|-a + b|, |-a|).$$

Now $|a - b| < \max(|a|, |b|)$ since $ab > 0$, and the only way that we can have the above inequalities is if $a = b$. In this case, zy is a member of $\mathrm{SL}(2, \mathbb{Z})$ outside Γ whose top-row entries have product 0, and we have seen that this is a contradiction.

Thus we must have $ab < 0$. Then zx has top row $(b \ -a)$. The product of these entries is positive and the maximum of their absolute values is the same as that for z. So we are reduced to the situation in the previous paragraph, which we saw leads to a contradiction. We conclude that $\Gamma = \mathrm{SL}(2, \mathbb{Z})$.

38. In PSL$(2, \mathbb{Z})$, we have $x^2 = y^3 = 1$, and Problem 37 shows that x and y generate PSL$(2, \mathbb{Z})$. Proposition 7.8 therefore produces a homomorphism carrying $\langle X, Y; X^2, Y^3 \rangle$ onto PSL$(2, \mathbb{Z})$. Proposition 7.16 shows that $C_2 * C_3 \cong \langle X, Y; X^2, Y^3 \rangle$, and the composition of these two maps yields the desired homomorphism Φ.

39. Let us drop the "mod $\pm I$" in order to simplify the notation. In (a), $yx = \begin{pmatrix} 0 & 1 \\ -1 & -1 \end{pmatrix} \begin{pmatrix} 0 & -1 \\ 1 & 0 \end{pmatrix} = \begin{pmatrix} 1 & 0 \\ -1 & 1 \end{pmatrix}$ and $y^{-1}x = \begin{pmatrix} -1 & -1 \\ 1 & 0 \end{pmatrix} \begin{pmatrix} 0 & -1 \\ 1 & 0 \end{pmatrix} = \begin{pmatrix} -1 & 1 \\ 0 & -1 \end{pmatrix}$. Then $zyx = \begin{pmatrix} a-b & b \\ c-d & d \end{pmatrix}$, and $\mu(zyx) = \max(|a-b|, |b|)$. If $ab \leq 0$, then $|a-b| \geq |a|$ and hence $\mu(zyx) \geq \mu(z)$. Similarly $zy^{-1}x = \begin{pmatrix} -a & a-b \\ -c & c-d \end{pmatrix}$, and $\mu(zy^{-1}x) = \max(|a|, |a-b|)$. If $ab \leq 0$, then $|a-b| \geq |b|$ and hence $\mu(zy^{-1}x) \geq \mu(z)$. The arguments with ν are similar.

In (b), we have $zx = \begin{pmatrix} b & -a \\ d & -c \end{pmatrix}$. Then $\mu(zx) = \max(|b|, |a|) = \mu(z)$ and $\nu(zx) = \max(|d|, |c|) = \nu(z)$.

In (c), the entries of z are limited to ± 1 and 0. We may take the first nonzero entry in the first column to be $+1$ by adjusting by $-I$ if necessary. Then the possibilities with determinant 1 are $\begin{pmatrix} 1 & 0 \\ 0 & 1 \end{pmatrix}, \begin{pmatrix} 1 & 1 \\ 0 & 1 \end{pmatrix}, \begin{pmatrix} 1 & -1 \\ 0 & 1 \end{pmatrix}, \begin{pmatrix} 1 & 0 \\ 1 & 1 \end{pmatrix}, \begin{pmatrix} 1 & -1 \\ 1 & 0 \end{pmatrix}, \begin{pmatrix} 1 & 0 \\ -1 & 1 \end{pmatrix}, \begin{pmatrix} 1 & 1 \\ -1 & 0 \end{pmatrix}, \begin{pmatrix} 0 & -1 \\ 1 & 0 \end{pmatrix}, \begin{pmatrix} 0 & -1 \\ 1 & 1 \end{pmatrix}$, and $\begin{pmatrix} 0 & -1 \\ 1 & -1 \end{pmatrix}$.

In (d), let us prove by induction on n that if $Z = a_1 \cdots a_n$ is reduced and ends in X, then $\Phi(Z) = \begin{pmatrix} a & b \\ c & d \end{pmatrix}$ has $ab \leq 0$. The base cases of the induction are $n = 1$ and $n = 2$, where we have $Z = X$, $Z = YX$, and $Z = Y^{-1}X$; since $\Phi(Z)$ is $\begin{pmatrix} 0 & -1 \\ 1 & 0 \end{pmatrix}, \begin{pmatrix} 1 & 0 \\ -1 & 1 \end{pmatrix}$, and $\begin{pmatrix} -1 & 1 \\ 0 & -1 \end{pmatrix}$ in the three cases, we have $ab \leq 0$ for each. For the inductive step we pass from Z, which ends in X, to anything obtained by adjoining factors at the right in such a way that the new word is still reduced and has X at the right end. This means that Z is replaced by ZYX or by $ZY^{-1}X$. Suppose that $\Phi(Z) = \begin{pmatrix} a & b \\ c & d \end{pmatrix}$. We are assuming that $ab \leq 0$. According to the calculation in the solution of (a), the entries in the first row of $\Phi(ZYX)$ are $a - b$ and b, with product $(a - b)b = ab - b^2 \leq ab \leq 0$, and the entries in the first row of $\Phi(ZY^{-1}X)$ are $-a$ and $a - b$, with product $-a(a - b) = -a^2 + ab \leq ab \leq 0$. Thus the induction goes forward, and our assertion follows.

Now we can prove by induction that

$$\mu(\Phi(a_1 \cdots a_n)) \geq \mu(\Phi(a_1 \cdots a_{n-1})) \tag{$*$}$$

if $Z = a_1 \cdots a_n = Z'a_n$ is reduced. The result is trivial for $n = 1$, and we let $n \geq 2$ be given and assume the inequality for words of length $< n$. Let a word of length $n \geq 2$ be given. If $a_n = X$, then $(*)$ is immediate from (b). If $a_n \neq X$, then $a_{n-1} = X$ and a_n is Y or Y^{-1}. Also, ZX is a reduced word. From the previous paragraph we know that the product of the entries in the first row of $\mu(\Phi(Z'))$ is ≤ 0. Applying (b) and then (a), we obtain $\mu(\Phi(Z)) = \mu(\Phi(ZX)) = \mu(\Phi(Z'a_n X)) \geq \mu(\Phi(Z'))$, and this proves (∗). Similar arguments apply to ν.

For (e), we are to prove that if W is a nonempty reduced word, then $\Phi(W)$ is not the identity of PSL(2, $\mathbb{Z}$). Assuming the contrary, we may assume without loss of generality that W is as short as possible with this property. If $W = a_1 \cdots a_n$, and $\Phi(W)$ is the identity, then $\mu(\Phi(W)) = \mu(I) = 1$ and similarly $\nu(\Phi(W)) = 1$. By (d), we must have $\mu(\Phi(a_1 \cdots a_k)) = \nu(\Phi(a_1 \cdots a_k)) = 1$ for $1 \leq k \leq n$. Then, for each k with $1 \leq k < n$, $\Phi(a_1 \cdots a_k)$ lies in the set of 10 matrices in (c) but is not the identity. The 10 matrices in (c) are obtained by applying Φ to the elements 1, XY, $Y^{-1}X$, XY^{-1}, XYX, YX, Y^{-1}, X, Y, and $XY^{-1}X$. The remaining words W of length 3 are YXY, YXY^{-1}, $Y^{-1}XY$, $Y^{-1}XY^{-1}$, and the ones of length 4 are $XYXY$, $XYXY^{-1}$, $XY^{-1}XY$, $XY^{-1}XY^{-1}$, $YXYX$, $YXY^{-1}X$, $Y^{-1}XYX$, $Y^{-1}XY^{-1}X$. We compute Φ directly on these 12 reduced words and obtain $\begin{pmatrix} 0 & 1 \\ -1 & -2 \end{pmatrix}$, $\begin{pmatrix} -1 & -1 \\ 2 & 1 \end{pmatrix}$, $\begin{pmatrix} -1 & -2 \\ 1 & 1 \end{pmatrix}$, $\begin{pmatrix} 2 & 1 \\ -1 & 0 \end{pmatrix}$, $\begin{pmatrix} 1 & 2 \\ 0 & 1 \end{pmatrix}$, $\begin{pmatrix} 2 & 1 \\ 1 & 1 \end{pmatrix}$, $\begin{pmatrix} 1 & 1 \\ 1 & 2 \end{pmatrix}$, $\begin{pmatrix} 1 & 0 \\ 2 & 1 \end{pmatrix}$, $\begin{pmatrix} 1 & 0 \\ -2 & 1 \end{pmatrix}$, $\begin{pmatrix} 1 & -1 \\ -1 & 2 \end{pmatrix}$, $\begin{pmatrix} 2 & -1 \\ -1 & 1 \end{pmatrix}$, $\begin{pmatrix} 1 & -2 \\ 0 & 1 \end{pmatrix}$. Consequently $\Phi(W)$ is not the identity for W of positive length ≤ 4. The inequality of (d) shows that $\mu(\Phi(W)) \geq 2$ if W has length > 4, and therefore $\Phi(W)$ is the identity only if W is the empty word.

40. The definition of $\widetilde{\sigma}_m$ is $\widetilde{\sigma}_m \begin{pmatrix} a & b \\ c & d \end{pmatrix} = \begin{pmatrix} a+m\mathbb{Z} & b+m\mathbb{Z} \\ c+m\mathbb{Z} & d+m\mathbb{Z} \end{pmatrix}$. We readily check that $\widetilde{\sigma}_m$ respects multiplication and hence is a homomorphism into some group of matrices. Since $(a+m\mathbb{Z})(d+m\mathbb{Z}) - (b+m\mathbb{Z})(c+m\mathbb{Z}) = (ad-bc) + m\mathbb{Z} = 1 + m\mathbb{Z}$, the image group is contained in SL(2, $\mathbb{Z}/m\mathbb{Z}$). The kernel is the set of matrices $\begin{pmatrix} a & b \\ c & d \end{pmatrix}$ in SL(2, $\mathbb{Z}$) with $a + m\mathbb{Z} = 1 + m\mathbb{Z}$, $b + m\mathbb{Z} = 0 + m\mathbb{Z}$, $c + m\mathbb{Z} = 0 + m\mathbb{Z}$, $d + m\mathbb{Z} = 1 + m\mathbb{Z}$, and these are exactly the matrices M in SL(2, $\mathbb{Z}$) with every entry of $M - I$ divisible by m. Therefore $\ker \widetilde{\sigma}_m = \Gamma(m)$. This proves (a).

In (b), let $\gamma = \text{GCD}(\alpha, m)$, so that $\alpha\gamma^{-1}$ and $m\gamma^{-1}$ are relatively prime. Applying Dirichlet's theorem on primes in arithmetic progressions, take $p > |\beta|$ to be a prime of the form $p = \alpha\gamma^{-1} + rm\gamma^{-1}$ for some r. Then $\alpha + rm = p\gamma$, and $\text{GCD}(\alpha + rm, \beta) = \text{GCD}(p\gamma, \beta) = \text{GCD}(\gamma, \beta) = \text{GCD}(\text{GCD}(\alpha, m), \beta) = \text{GCD}(\alpha, \beta, m) = 1$.

For (c), corresponding to any member of SL(2, $\mathbb{Z}/m\mathbb{Z}$) is a matrix $\begin{pmatrix} a & b \\ c & d \end{pmatrix}$ with integer entries with $ad - bc \equiv 1 \bmod m$. If p is a prime dividing $a - b$ and $c - d$, then $ad - bc \equiv bd - bd \equiv 0 \bmod p$, and hence p does not divide m. Therefore $\text{GCD}(a - b, c - d, m) = 1$. Applying (b), we obtain an integer r such that $\text{GCD}(a + rm - b, c - d) = 1$. Let us then work instead with $\begin{pmatrix} a+rm & b \\ c & d \end{pmatrix}$. Adjusting notation to call this matrix $\begin{pmatrix} a & b \\ c & d \end{pmatrix}$, we may assume that $\text{GCD}(a-b, c-d) = 1$. Since m divides $ad - bc - 1$, there exist integers C and A with

$$(a - b)C + (d - c)A = \tfrac{1-(ad-bc)}{m}.$$

Then $\det \begin{pmatrix} a+mA & b+mA \\ c+mC & d+mC \end{pmatrix}$ is equal to

$$(ad - bc) + (d - c)mA + (a - b)mC = (ad - bc) + m\left(\tfrac{1-(ad-bc)}{m}\right) = 1,$$

and $\begin{pmatrix} a+mA & b+mA \\ c+mC & d+mC \end{pmatrix}$ is a member of SL(2, $\mathbb{Z}$) whose image under $\tilde{\sigma}_m$ is the given matrix in SL(2, $\mathbb{Z}/m\mathbb{Z}$).

41. For the remainder of the problems in this set, it will be convenient to regard the isomorphism $C_2 * C_3 \cong \langle X, Y; \ X^2, Y^3 \rangle$ of Proposition 7.16 as an equality: $C_2 * C_3 = \langle X, Y; \ X^2, Y^3 \rangle$.

In (a), Φ_m is well defined as a consequence of the second conclusion of Proposition 7.8.

In (b), it is immediate from Proposition 7.8 that the kernel of Φ_m is the smallest normal subgroup of $C_2 * C_3$ containing the element $(XY)^m$. Under the isomorphism $\Phi : C_2 * C_3 \to \mathrm{PSL}(2, \mathbb{Z})$, we have $\Phi((XY)^m) = (xy)^m \mod \pm I$. Since the smallest normal subgroup H_m of PSL(2, $\mathbb{Z}$) containing $(xy)^m \mod \pm I = \Phi((XY)^m)$ is Φ of the smallest normal subgroup of $C_2 * C_3$ containing $(XY)^m$, we have $H_m = \Phi(\ker \Phi_m)$.

In (c), if passage to the quotient is denoted by q_m, Proposition 4.11 shows that the point needing verification is that the scalar matrices in SL(2, $\mathbb{Z}$) lie in the kernel of $q_m \circ \tilde{\sigma}_m$, and this follows since $\begin{pmatrix} -1 & 0 \\ 0 & -1 \end{pmatrix}$ maps under $\tilde{\sigma}_m$ to the matrix with entries taken modulo m and then maps to the identity under q_m.

In (d), K_m is a normal subgroup of PSL(2, $\mathbb{Z}$), and it is thus enough to show that the element $(xy)^m \mod \pm I$ of H_m is in K_m. Since $\begin{pmatrix} 0 & -1 \\ 1 & 0 \end{pmatrix} \begin{pmatrix} 0 & 1 \\ -1 & -1 \end{pmatrix} = \begin{pmatrix} 1 & 1 \\ 0 & 1 \end{pmatrix}$ and since the mth power of this matrix is in $\Gamma(m)$, $(xy)^m \mod \pm I$ is indeed in K_m.

For (e), part (d) shows that $\begin{pmatrix} 1 & m \\ 0 & 1 \end{pmatrix} \mod \pm I$ is in K_m, and its tth power $\begin{pmatrix} 1 & tm \\ 0 & 1 \end{pmatrix} \mod \pm I$, for t an integer, has to be in K_m. Then $\begin{pmatrix} 1 & 0 \\ -tm & 1 \end{pmatrix} = x \begin{pmatrix} 1 & tm \\ 0 & 1 \end{pmatrix} x^{-1} \mod \pm I$ is in K_m since K_m is normal, and so are $\begin{pmatrix} 1+tm & tm \\ -tm & 1-tm \end{pmatrix} = y^{-1} \begin{pmatrix} 1 & tm \\ 0 & 1 \end{pmatrix} y$ and $\begin{pmatrix} 1-tm & tm \\ -tm & 1+tm \end{pmatrix} = xy^{-1} \begin{pmatrix} 1 & tm \\ 0 & 1 \end{pmatrix} yx^{-1}$, for the same reason.

42. Let x and y be the listed images in the stated permutation groups of X and Y. The homomorphisms in this problem come from Proposition 7.8 since in each case $x^2 = 1$, $y^2 = 1$, and $(xy)^m$ can be verified to be 1. What needs to be verified in each case is that x and y generate the stated permutation group.

In (a), the image group has a subgroup of order 2 and a subgroup of order 3 and hence must be the whole 6-element $\mathfrak{S}_3$.

In (b), Lemma 4.41 shows that $(1\ 2\ 3)(1\ 2)(3\ 4)(1\ 2\ 3)^{-1} = (2\ 3)(1\ 4)$, and hence the image group has a subgroup of 4 even permutations and a subgroup of 3 even permutations, therefore must be all of $\mathfrak{A}_4$.

In (c), we have $(1\ 2)(2\ 3\ 4) = (1\ 2\ 3\ 4)$. Thus the image group contains $(1\ 2\ 3\ 4)^2 = (1\ 3)(2\ 4)$, $(2\ 3\ 4)(1\ 3)(2\ 4)(2\ 3\ 4)^{-1} = (1\ 4)(2\ 3)$, and $(2\ 3\ 4)(1\ 2)(2\ 3\ 4)^{-1} = (1\ 3)$, hence a subgroup of order 8 and a subgroup of order 3. Therefore it is all of $\mathfrak{S}_4$.

In (d), we have $(1\ 2)(3\ 4)(1\ 3\ 5) = (1\ 4\ 3\ 5\ 2)$. Thus the image group contains a subgroup of order 5, a subgroup of order 3, and a subgroup of order 2, all contained in $\mathfrak{A}_5$. The image group is not of order 30 because $\mathfrak{A}_5$ has no nontrivial normal subgroups, and hence it must be all of $\mathfrak{A}_5$.

43. As with Problem 39, let us drop the "mod $\pm I$" in order to simplify the notation. In (a), we can take $g_1 = \begin{pmatrix} 1 & 0 \\ 0 & 1 \end{pmatrix}$, $g_2 = \begin{pmatrix} 1 & -1 \\ 0 & 1 \end{pmatrix}$, $g_3 = \begin{pmatrix} 0 & 1 \\ -1 & -1 \end{pmatrix}$, $g_4 = \begin{pmatrix} 0 & 1 \\ -1 & 0 \end{pmatrix}$, $g_5 = \begin{pmatrix} -1 & -1 \\ 1 & 0 \end{pmatrix}$, $g_6 = \begin{pmatrix} 1 & 0 \\ -1 & 1 \end{pmatrix}$.

For (b), first we compute the six values of $g_i b_1$ as $g_1 b_1 = \begin{pmatrix} 0 & 1 \\ -1 & 0 \end{pmatrix}$, $g_2 b_1 = \begin{pmatrix} 1 & 1 \\ -1 & 0 \end{pmatrix}$, $g_3 b_1 = \begin{pmatrix} -1 & 0 \\ 1 & -1 \end{pmatrix}$, $g_4 b_1 = \begin{pmatrix} -1 & 0 \\ 0 & -1 \end{pmatrix}$, $g_5 b_1 = \begin{pmatrix} 1 & -1 \\ 0 & 1 \end{pmatrix}$, $g_6 b_1 = \begin{pmatrix} 0 & 1 \\ -1 & -1 \end{pmatrix}$, and then we compute the six values of $g_i b_2$ as $g_1 b_2 = \begin{pmatrix} 0 & 1 \\ -1 & -1 \end{pmatrix}$, $g_2 b_2 = \begin{pmatrix} 1 & 2 \\ -1 & -1 \end{pmatrix}$, $g_3 b_2 = \begin{pmatrix} -1 & -1 \\ 1 & 0 \end{pmatrix}$, $g_4 b_2 = \begin{pmatrix} -1 & -1 \\ 0 & -1 \end{pmatrix}$, $g_5 b_2 = \begin{pmatrix} 1 & 0 \\ 0 & 1 \end{pmatrix}$, $g_6 b_2 = \begin{pmatrix} 0 & 1 \\ -1 & -2 \end{pmatrix}$. Next we locate each of these products in a coset, writing them with some g_i on the right. We find that, up to mod $\pm I$, the results are $g_1 b_1 = g_4$, $g_2 b_1 = g_5$, $g_3 b_1 = g_6$, $g_4 b_1 = g_1$, $g_5 b_1 = g_2$, $g_6 b_1 = g_3$, $g_1 b_2 = g_3$, $g_2 b_2 = \begin{pmatrix} 3 & 2 \\ -2 & -1 \end{pmatrix} g_6$, $g_3 b_2 = g_5$, $g_4 b_2 = \begin{pmatrix} 1 & 2 \\ 0 & 1 \end{pmatrix} g_2$, $g_5 b_2 = g_1$, $g_6 b_2 = \begin{pmatrix} 1 & 0 \\ -2 & 1 \end{pmatrix} g_4$. The conclusion is that generators of K_2 are the three matrices $\begin{pmatrix} 3 & 2 \\ -2 & -1 \end{pmatrix}$, $\begin{pmatrix} 1 & 2 \\ 0 & 1 \end{pmatrix}$, $\begin{pmatrix} 1 & 0 \\ -2 & 1 \end{pmatrix}$.

For (c), the second and third of the generators in (b) are in H_2 by Problem 41e. The equality $\begin{pmatrix} 3 & 2 \\ -2 & -1 \end{pmatrix} = -\begin{pmatrix} 1 & -2 \\ 0 & 1 \end{pmatrix} \begin{pmatrix} 1 & 0 \\ 2 & 1 \end{pmatrix}$ exhibits the first of the generators as in H_2. Hence all the generators are in H_2 and $K_2 \subseteq H_2$. Therefore $K_2 = H_2$.

For (d) with $m = 3$, we can take the 12 coset representatives to be $g_1 = \begin{pmatrix} 1 & 0 \\ 0 & 1 \end{pmatrix}$, $g_2 = \begin{pmatrix} 1 & -1 \\ 0 & 1 \end{pmatrix}$, $g_3 = \begin{pmatrix} 1 & 1 \\ 0 & 1 \end{pmatrix}$, $g_4 = \begin{pmatrix} 0 & 1 \\ -1 & 0 \end{pmatrix}$, $g_5 = \begin{pmatrix} 0 & 1 \\ -1 & 1 \end{pmatrix}$, $g_6 = \begin{pmatrix} 0 & 1 \\ -1 & -1 \end{pmatrix}$, $g_7 = \begin{pmatrix} 1 & -1 \\ 1 & 0 \end{pmatrix}$, $g_8 = \begin{pmatrix} 1 & 1 \\ 1 & 2 \end{pmatrix}$, $g_9 = \begin{pmatrix} 1 & 1 \\ -1 & 0 \end{pmatrix}$, $g_{10} = \begin{pmatrix} 1 & -1 \\ -1 & 2 \end{pmatrix}$, $g_{11} = \begin{pmatrix} 1 & 0 \\ 1 & 1 \end{pmatrix}$, $g_{12} = \begin{pmatrix} 1 & 0 \\ -1 & 1 \end{pmatrix}$. Then we compute that $g_1 b_1 = \begin{pmatrix} 0 & 1 \\ -1 & 0 \end{pmatrix} = g_4$, $g_2 b_1 = \begin{pmatrix} 1 & 1 \\ -1 & 0 \end{pmatrix} = g_9$, $g_3 b_1 = \begin{pmatrix} -1 & 1 \\ -1 & 0 \end{pmatrix} = g_7$, $g_4 b_1 = \begin{pmatrix} -1 & 0 \\ 0 & -1 \end{pmatrix} = g_1$, $g_5 b_1 = \begin{pmatrix} -1 & 0 \\ -1 & -1 \end{pmatrix} = g_{11}$, $g_6 b_1 = \begin{pmatrix} -1 & 0 \\ 1 & -1 \end{pmatrix} = g_{12}$, $g_7 b_1 = \begin{pmatrix} 1 & 1 \\ 0 & 1 \end{pmatrix} = g_3$, $g_8 b_1 = \begin{pmatrix} -1 & 1 \\ -2 & 1 \end{pmatrix} = \begin{pmatrix} -1 & 0 \\ -3 & -1 \end{pmatrix} g_{10}$, $g_9 b_1 = \begin{pmatrix} -1 & 1 \\ 0 & -1 \end{pmatrix} = g_2$, $g_{10} b_1 = \begin{pmatrix} 1 & 1 \\ -2 & -1 \end{pmatrix} = \begin{pmatrix} 1 & 0 \\ -3 & 1 \end{pmatrix} g_8$, $g_{11} b_1 = \begin{pmatrix} 0 & 1 \\ -1 & 1 \end{pmatrix} = g_5$, $g_{12} b_1 = \begin{pmatrix} 0 & 1 \\ -1 & -1 \end{pmatrix} = g_6$.

Also, $g_1 b_2 = \begin{pmatrix} 0 & 1 \\ -1 & -1 \end{pmatrix} = g_6$, $g_2 b_2 = \begin{pmatrix} 1 & 2 \\ -1 & -1 \end{pmatrix} = \begin{pmatrix} 4 & 3 \\ -3 & -2 \end{pmatrix} g_{10}$, $g_3 b_2 = \begin{pmatrix} -1 & 0 \\ -1 & -1 \end{pmatrix} = g_{11}$, $g_4 b_2 = \begin{pmatrix} -1 & -1 \\ 0 & -1 \end{pmatrix} = g_3$, $g_5 b_2 = \begin{pmatrix} -1 & -1 \\ -1 & -2 \end{pmatrix} = g_8$, $g_6 b_2 = \begin{pmatrix} -1 & -1 \\ 1 & 0 \end{pmatrix} = g_9$, $g_7 b_2 = \begin{pmatrix} 1 & 2 \\ 0 & 1 \end{pmatrix} = \begin{pmatrix} 1 & 3 \\ 0 & 1 \end{pmatrix} g_2$, $g_8 b_2 = \begin{pmatrix} -1 & 0 \\ -2 & -1 \end{pmatrix} = \begin{pmatrix} -1 & 0 \\ -3 & -1 \end{pmatrix} g_{12}$, $g_9 b_2 = \begin{pmatrix} -1 & 0 \\ 0 & -1 \end{pmatrix} = g_1$, $g_{10} b_2 = \begin{pmatrix} 1 & 2 \\ -2 & -3 \end{pmatrix} = \begin{pmatrix} -2 & 3 \\ 3 & -5 \end{pmatrix} g_7$, $g_{11} b_2 = \begin{pmatrix} 0 & 1 \\ -1 & 0 \end{pmatrix} = g_4$, $g_{12} b_2 = \begin{pmatrix} 0 & 1 \\ -1 & -2 \end{pmatrix} = \begin{pmatrix} 1 & 0 \\ -3 & 1 \end{pmatrix} g_5$.

Thus generators of K_3 are $\begin{pmatrix} -1 & 0 \\ -3 & -1 \end{pmatrix}$, $\begin{pmatrix} 1 & 0 \\ -3 & 1 \end{pmatrix}$, $\begin{pmatrix} 4 & 3 \\ -3 & -2 \end{pmatrix}$, $\begin{pmatrix} 1 & 3 \\ 0 & 1 \end{pmatrix}$, $\begin{pmatrix} -1 & 0 \\ -3 & -1 \end{pmatrix}$, $\begin{pmatrix} -2 & 3 \\ 3 & -5 \end{pmatrix}$,

$\begin{pmatrix} 1 & 0 \\ -3 & 1 \end{pmatrix}$. All but $\begin{pmatrix} 4 & 3 \\ -3 & -2 \end{pmatrix}$ and $\begin{pmatrix} -2 & 3 \\ 3 & -5 \end{pmatrix}$ are certainly in H_3. The expressions $\begin{pmatrix} 4 & 3 \\ -3 & -2 \end{pmatrix}$ = $\begin{pmatrix} 1+3 & 3 \\ -3 & 1-3 \end{pmatrix}$ and $\begin{pmatrix} -2 & 3 \\ 3 & -5 \end{pmatrix}$ = $\begin{pmatrix} 1-3 & -3 \\ 3 & 1+3 \end{pmatrix} \begin{pmatrix} 1 & -3 \\ 0 & 1 \end{pmatrix}$ show that these two generators are in H_3. Therefore $K_3 = H_3$.

44. Problem 41 produces a homomorphism σ_m of G_m onto $\mathrm{PSL}(2, \mathbb{Z}/m\mathbb{Z})$ with kernel isomorphic to K_m/H_m. The given fact $H_m = K_m$ for $2 \leq m \leq 5$ implies that σ_m is an isomorphism for these values of m. This proves the first isomorphism in each part. Problem 42 gives us homomorphisms of G_m for these m's onto the third group listed in each part. Composition with σ_m^{-1} then gives a homomorphism of $\mathrm{PSL}(2, \mathbb{Z}/m\mathbb{Z})$ onto the third group. In each case the statement of Problem 43 gives the number of elements in $\mathrm{PSL}(2, \mathbb{Z}/m\mathbb{Z})$, and this matches the number of elements in the third group. It follows that these homomorphisms are isomorphisms.

45. For (a), linearity gives $R_\theta T_{(a,b)} R_\theta^{-1}(x, y) = R_\theta(R_\theta^{-1}(x, y) + (a, b)) = R_\theta R_\theta^{-1}(x, y) + R_\theta(a, b) = (x, y) + R_\theta(a, b) = T_{R_\theta(a,b)}(x, y)$.

For (b), the result of (a) says that we get a semidirect product. Let us show that the two sets—the elements of the semidirect product and the union of the translations and rotations—coincide. In one direction a rotation about (x_0, y_0) is of the form $(x, y) \mapsto R_\theta(x - x_0, y - y_0) + (x_0, y_0) = R_\theta(x, y) + (a, b) = T_{(a,b)} R_\theta(x, y)$, where $(a, b) = -R_\theta(x_0, y_0) + (x_0, y_0)$. Hence it is in the semidirect product. In the reverse direction suppose that $T_{(a,b)} R_\theta$ is in the semidirect product and is not a translation. Then θ is not a multiple of 2π, and we can put $(x_0, y_0) = (1 - R_\theta)^{-1}(a, b)$. Then we have $T_{(a,b)} R_\theta(x, y) = R_\theta(x, y) + (a, b) = R_\theta(x - x_0, y - y_0) + R_\theta(x_0, y_0) + (a, b) = R_\theta(x - x_0, y - y_0) + R_\theta(1 - R_\theta)^{-1}(a, b) + (a, b) = R_\theta(x - x_0, y - y_0) - (1 - R_\theta)(1 - R_\theta)^{-1}(a, b) + (a, b) + (1 - R_\theta)^{-1}(a, b) = R_\theta(x - x_0, y - y_0) + (x_0, y_0)$. Hence $T_{(a,b)} R_\theta$ is a rotation about (x_0, y_0).

46. In (a), we need to show only that $r_c = r_a r_b$. In (b), we need to show that $r_b r_a r_b r_a r_b$ is a translation but not the identity. Then it follows from (b) that the group G generated by r_a and r_b is infinite. Since (a) and Proposition 7.8 yield a homomorphism of $G_6 = \langle X, Y; X^2, Y^3, (XY)^6 \rangle$ onto the infinite group G, it follows that G_6 is infinite. Since $\mathrm{PSL}(\mathbb{Z}/6\mathbb{Z})$ is finite, (c) follows.

To establish the two facts that need checking, we may, without loss of generality, take T to be the triangle with vertices $a = (0, 0)$, $b = (0, -1)$, and $c = (\sqrt{3}, 0)$. The formulas for r_a, r_b, and r_c are $r_a(x, y) = (-x, -y)$,

$$r_b(x, y) = (x \cos \tfrac{2\pi}{3} - (y+1) \sin \tfrac{2\pi}{3},\ x \sin \tfrac{2\pi}{3} + (y+1) \cos \tfrac{2\pi}{3} - 1)$$
$$= \bigl(-x/2 - y\sqrt{3}/2 - \sqrt{3}/2,\ x\sqrt{3}/2 - y/2 - 1/2 - 1\bigr),$$

and

$$r_c(x, y) = ((x - \sqrt{3}) \cos \tfrac{\pi}{3} + y \sin \tfrac{\pi}{3} + \sqrt{3},\ -(x - \sqrt{3}) \sin \tfrac{\pi}{3} + y \cos \tfrac{\pi}{3})$$
$$= \bigl((x - \sqrt{3})/2 + y\sqrt{3}/2 + \sqrt{3},\ -(x - \sqrt{3})\sqrt{3}/2 + y/2\bigr).$$

Then $r_a r_b(x, y) = -r_b(x, y) = r_c(x, y)$ by inspection.

To verify that $r_b r_a r_b r_a r_b$ is a translation, we write $r_b r_a r_b r_a r_b(x, y) = r_b r_c^2(x, y)$. The formula above for r_c gives

$$r_c^2(x, y) = ((x-\sqrt{3})\cos\tfrac{2\pi}{3} + y\sin\tfrac{2\pi}{3} + \sqrt{3}, -(x-\sqrt{3})\sin\tfrac{2\pi}{3} + y\cos\tfrac{2\pi}{3})$$
$$= (-(x-\sqrt{3})/2 + y\sqrt{3}/2 + \sqrt{3}, -(x-\sqrt{3})\sqrt{3}/2 - y/2).$$

Then the first coordinate of $r_b r_c^2(x, y)$ is $-\tfrac{1}{2}(-(x-\sqrt{3})/2 + y\sqrt{3}/2 + \sqrt{3}) + ((x-\sqrt{3})\sqrt{3}/2 + y/2)\sqrt{3}/2 - \sqrt{3}/2 = x - 2\sqrt{3}$, while the second coordinate is $(-(x-\sqrt{3})/2 + y\sqrt{3}/2 + \sqrt{3})\sqrt{3}/2 + ((x-\sqrt{3})\sqrt{3}/2 + y/2)/2 - 3/2 = y$. So $r_b r_c^2(x, y) = (x - 2\sqrt{3}, y)$ is a translation.

47. We may suppose that the representations are unitary. Let $\{v_{1,i}\}$ and $\{v_{2,j}\}$ be orthonormal bases of V_1 and V_2. Then

$$(\chi_{R_1} * \chi_{R_2})(x) = \sum_y \chi_{R_1}(xy^{-1})\chi_{R_2}(y)$$
$$= \sum_{y,i,j} (R_1(xy^{-1})v_{1,i}, v_{1,i})(R_2(y)v_{2,j}, v_{2,j})$$
$$= \sum_{y,i,j,k} (R_1(x)(R_1(y^{-1})v_{1,i}, v_{1,k})v_{1,k}, v_{1,i})(R_2(y)v_{2,j}, v_{2,j})$$
$$= \sum_{i,j,k} (R_1(x)v_{1,k}, v_{1,i}) \sum_y \overline{(R_1(y)v_{1,k}, v_{1,i})}(R_2(y)v_{2,j}, v_{2,j})$$

For (a), the inside sum is 0, and the argument is complete. For (b), let $R_1 = R_2$ and $v_{2,j} = v_{1,j}$. Then the right side of the display continues as

$$= \sum_{i,j,k} (R_1(x)v_{1,k}, v_{1,i})|G|d_{R_1}^{-1}(v_{1,j}, v_{1,k})\overline{(v_{1,j}, v_{1,i})}$$
$$= |G|d_{R_1}^{-1} \sum_{i,j,k}(R_1(x)v_{1,k}, v_{1,i})\delta_{jk}\delta_{ji}$$
$$= |G|d_{R_1}^{-1} \sum_i (R_1(x)v_{1,i}, v_{1,i}) = |G|d_{R_1}^{-1}\chi_{R_1}(x).$$

48. We have $E_\alpha E_\beta = |G|^{-2} d_\alpha d_\beta R(\overline{\chi_\alpha})R(\overline{\chi_\beta}) = |G|^{-2} d_\alpha d_\beta R(\overline{\chi_\alpha * \chi_\beta})$. Problem 47a shows that this is 0 if R_α and R_β are inequivalent; this proves (b). Problem 47b shows that the computation with $R_\alpha = R_\beta$ continues as $= |G|^{-1} d_\alpha R(\overline{\chi_\alpha}) = E_\alpha$; this proves (a).

49. Let S be the set of all finite-dimensional irreducible invariant subspaces V_s of V. Call a subset T of S "independent" if the sum $\sum_{t \in T} V_t$ is direct. This condition means that for every finite subset $\{t_1, \ldots, t_n\}$ of T and every set of elements $v_i \in V_{t_i}$, the equation

$$v_1 + \cdots + v_n = 0$$

implies that each v_i is 0. From this formulation it follows that the union of any increasing chain of independent subsets of S is itself independent. By Zorn's Lemma there is a maximal independent subset T_0 of S. By definition the sum $V_0 = \sum_{t \in T_0} V_t$ is direct. Consequently the problem is to show that V_0 is all of V. Since every member of V lies in a finite direct sum of finite-dimensional irreducible invariant subspaces of V, it suffices to show that each V_s is contained in V_0. If s is in T_0, this conclusion is obvious. Thus suppose s is not in T_0. By the maximality of T_0, $T_0 \cup \{s\}$ is not independent. Consequently the sum $V_0 + V_s$ is not direct, and it follows that $V_0 \cap V_s \neq 0$. But this intersection is an invariant subspace of V_s. Since V_s is irreducible, a nonzero invariant subspace must be all of V_s. Thus V_s is contained in V_0, as we wished to show.

50. Let us impose an inner product on V_0 that makes $R\big|_{V_0}$ unitary. Let $\{v_1, \ldots, v_n\}$ be an orthonormal basis of V_0. If we write $R(x)v_j = \sum_{i=1}^n R_{ij}(x)v_i$, then $R_{ij}(x) = (R(x)v_j, v_i)$. Consequently the character χ_α of $R\big|_{V_0}$ is given by $\chi_\alpha(x) = \sum_i R_{ii}(x)$. Then we have

$$E_\alpha v_j = |G|^{-1} d_\alpha \sum_{x \in G} \overline{\chi_\alpha(x)} R(x)v_j = |G|^{-1} d_\alpha \sum_{x \in G} \sum_{i,k} \overline{R_{kk}(x)} R_{ij}(x) v_i = v_j,$$

and E_α is the identity on V_0.

51. Problem 49 allows us to write V as the direct sum of possibly infinitely many finite-dimensional irreducible invariant subspaces $V = \bigoplus_\gamma V_\gamma$. If any v in V is given, we can write $v = \sum_\gamma v_\gamma$ with only finitely many terms nonzero. Applying E_α and using Problem 50, we see that $E_\alpha v$ is the sum of those v_γ such that $R\big|_{V_\gamma}$ is equivalent to R_α. Thus each nonzero v_γ has the property that $E_\alpha v_\gamma = v_\gamma$ for some α.

On the other hand, this equality cannot hold for two distinct α's. In fact, if R_α and R_β are inequivalent and we have $E_\alpha v_\gamma = v_\gamma$ and $E_\beta v_\gamma = v_\gamma$, then application of E_α to the second equality gives $E_\alpha E_\beta v_\gamma = E_\alpha v_\gamma = v_\gamma$. But $E_\alpha E_\beta = 0$ by Problem 48b, and hence $v_\gamma = 0$.

The conclusion is that for each nonzero v_γ, there is one and only one E_α such that $E_\alpha v_\gamma \neq 0$, and that α has $E_\alpha v_\gamma = v_\gamma$. Applying $\sum_\alpha E_\alpha$ to $v = \sum_\gamma v_\gamma$, we obtain $\sum_\alpha E_\alpha v = \sum_{\gamma,\alpha} E_\alpha v_\gamma = \sum_\gamma v_\gamma = v$. Thus $\sum_\alpha E_\alpha = I$. Problem 50 shows that E_α is the identity on any finite sum of vectors lying in finite-dimensional irreducible invariant subspaces equivalent to R_α. The direct-sum decomposition just proved shows that E_α is 0 on any vector in the direct sum of the images of the other E_β's. Thus the image of E_α is as asserted.

52. For α as given and for any v in V, we have $E_\alpha v = |G|^{-1} \sum_{x \in G} \overline{\omega(x)} R(x)v$. The members of the image of E_α are exactly the vectors v for which $E_\alpha v = v$, hence exactly the vectors v for which $|G|^{-1} \sum_{x \in G} \overline{\omega(x)} R(x)v = v$. Applying $R(y)$ to both sides gives $R(y)v = |G|^{-1} \sum_{x \in G} \overline{\omega(x)} R(yx)v = |G|^{-1} \sum_{x \in G} \overline{\omega(y^{-1}x)} R(x)v = \overline{\omega(y^{-1})} |G|^{-1} \sum_{x \in G} \overline{\omega(x)} R(x)v = \overline{\omega(y^{-1})} v = \omega(y)v.$

Chapter VIII

1. In (a), φ fixes 1 and must therefore fix the subfield generated by 1; this is $\mathbb{Q}$. For (b), $\varphi(a^2) = \varphi(a)^2$. For (c), if $a \leq b$, then $b - a = c^2$ for some c. Hence $\varphi(b) - \varphi(a) = \varphi(c)^2$, and $\varphi(a) \leq \varphi(b)$. For (d), let r be any real, let $\epsilon > 0$ be given, and choose rationals q_1 and q_2 with $q_1 \leq r \leq q_2$ and $q_2 - q_1 < \epsilon$. Then $q_1 = \varphi(q_1) \leq \varphi(r) \leq \varphi(q_2) = q_2$ by (a) and (c). Hence $|\varphi(r) - r| < \epsilon$. Since ϵ is arbitrary, $\varphi(r) = r$.

2. $(1+r)^{-1} = 1 - r + r^2 - r^3 + \cdots \pm r^{n-1}$ if $r^n = 0$.

3. This follows from the universal mapping property of the field of fractions.

4. Suppose that X divides $A(X)B(X)$, i.e., $A(X)B(X) = XC(X)$. If a_0 and b_0 are the constant terms of $A(X)$ and $B(X)$, we then have $a_0 b_0 = 0$. If $a_0 = 0$, then X divides $A(X)$; if $b_0 = 0$, then X divides $B(X)$. Hence X is prime.

5. In (a), take (X) as the ideal. It is prime by Problem 4. Suppose that a is a member of R with no inverse in R. then (X) is not maximal since (a, X) strictly contains it and does not contain 1. For (b), we can use (a, X).

6. In (a), I_{x_0} is certainly an ideal. Suppose J is an ideal with $I_{x_0} \subsetneq J$. Choose f in J that is not in I_{x_0}. The function $x - x_0$ is in I_{x_0}. Therefore $g = f^2 + (x - x_0)^2$ is in J. This function is everywhere > 0, and consequently $1/g$ is in R. Hence $1 = (1/g)g$ is in J, and J cannot be proper. So I_{x_0} is maximal.

Part (b) uses the Heine–Borel Theorem. For each point p in $[0, 1]$, choose a function f_p in I with $f_p(p) \neq 0$. By continuity, f_p is nonvanishing on some open set N_p containing p. As p varies, these open sets N_p cover $[0, 1]$. The Heine–Borel Theorem produces finitely many $N_{p_1}, \ldots, N_{p_k}$ that cover $[0, 1]$. Then f_{p_j} is nonvanishing on N_{p_j}. If x is a member of $[0, 1]$, then x is in some N_{p_j}, and f_{p_j} does not vanish at x. Thus the functions $f_{p_1}, \ldots, f_{p_k}$ have no common zero.

For (c), suppose that the maximal ideal I is not some I_{x_0}. Using (b), we form the function $g = f_{p_1}^2 + \cdots + f_{p_k}^2$. This is in I and is everywhere positive. The function $1/g$ is therefore in R, and $1 = (1/g)g$ is in I. Hence $I = R$, in contradiction to the fact that I is proper.

7. In (a), I_∞ is an ideal, and it is properly contained in the proper ideal of all members of R vanishing at $-\infty$. Part (b) follows from Proposition 8.8. The reason for (c) is that for each x_0 in $\mathbb{R}$, there is a member of R that is nonzero at x_0 and vanishes at infinity; this function has to be in I, and thus I cannot equal I_{x_0}.

8. For (a), let $a + bi$ be a nonzero member of I. Then $(a + b\sqrt{-5})(a - b\sqrt{-5}) = a^2 + 5b^2$ is a positive integer in I.

For (b), I is an additive subgroup of $\mathbb{Z} + \mathbb{Z}\sqrt{-5}$, which is free abelian of rank 2. Therefore I is free abelian of rank 1 or 2. We can rule out rank 1 because I contains a nonzero integer and also the product of that integer and $\sqrt{-5}$.

For (c), a $\mathbb{Z}$ basis of I consists of $x_1 = a_1 + b_1\sqrt{-5}$ and $x_2 = a_2 + b_2\sqrt{-5}$. Put $y_1 = rx_1 + sx_2 = (ra_1 + sa_2) + (rb_1 + sb_2)\sqrt{-5}$ and $y_2 = tx_1 + ux_2$, and aim to have y_1, y_2 form a $\mathbb{Z}$ basis with y_1 not involving $\sqrt{-5}$. We thus want $rb_1 + sb_2 = 0$, and the most economical way of achieving this equality is to put $d = \text{GCD}(b_1, b_2)$ and to take $r = b_2 d^{-1}$ and $s = -b_1 d^{-1}$. Then $\text{GCD}(r, s) = 1$, and we can choose t and u with $ru - st = 1$. With these choices we have $\begin{pmatrix} y_1 \\ y_2 \end{pmatrix} = \begin{pmatrix} r & s \\ t & u \end{pmatrix}\begin{pmatrix} x_1 \\ x_2 \end{pmatrix}$. Since $\det\begin{pmatrix} r & s \\ t & u \end{pmatrix} = 1$, this change is invertible. In other words, y_1 and y_2 form a $\mathbb{Z}$ basis in which y_1 is some nonzero integer n. We may assume that $n > 0$. Let m be the smallest positive integer in I. Then n must be a multiple of m by an application of the division algorithm. Since y_1 and y_2 form a $\mathbb{Z}$ basis of I, we see that n equals m.

9. It is straightforward to see that P is an ideal and that $xy \in P$ implies $x \in P$ or $y \in P$. The ideal P is proper since the presence of 1 in $\varphi^{-1}(P')$ would mean that $\varphi(1) = 1$ is in P'. But P' is proper, and thus 1 is not in P'.

10. (a) $\{(r, 0) \mid r \in \mathbb{R}\}$ and $\{(0, r) \mid r \in \mathbb{R}\}$.
(b) (X).
(c) $(X - 1)$ and $(X - 2)$.
(d) (0).

11. For (a), $\mathbb{Q}[X]/I$ is a field and hence is a unique factorization domain. For (b), one can give a counterexample. The ring $\mathbb{Z}[\sqrt{-5}]$ is an integral domain and is the quotient of $\mathbb{Z}[X]$ by the ideal $(X^2 + 5)$; therefore $I = (X^2 + 5)$ is prime. On the other hand, $\mathbb{Z}[\sqrt{-5}]$ is not a unique factorization domain.

12. For (a), choose x and y with $xd + yc = 1$. Dividing by n gives $xc^{-1} + yd^{-1} = n^{-1}$. Then (a) follows by multiplying through by m. Part (b) uses an induction. Group n as $(p_1^{k_1} \cdots p_{r-1}^{k_{r-1}}) p_r^{k_r}$ and apply (a) to write $mn^{-1} = a(p_1^{k_1} \cdots p_{r-1}^{k_{r-1}})^{-1} + b p_r^{-k_r}$. Repeat the process with $a(p_1^{k_1} \cdots p_{r-1}^{k_{r-1}})^{-1}$, and continue.

13. For (a), proceed as in the argument in Section 4 until near the end, obtaining x and y just as in that construction. Then $\delta(x + y\sqrt{-2}) = x^2 + 2y^2 \leq \frac{1}{4} + 2 \cdot \frac{1}{4} = \frac{3}{4}$. Then we have $\delta(r + s\sqrt{-2}) < \delta(c + d\sqrt{-2})$, and the argument goes through.
For (b), we would get $\delta(x + y\sqrt{-3}) = x^2 + 3y^2 \leq \frac{1}{4} + 3 \cdot \frac{1}{4} = 1$, and then the step $\delta(r + s\sqrt{-3}) < \delta(c + d\sqrt{-3})$ fails.

14. The map extends to an R module homomorphism by the universal mapping property of RG, and it is one-one onto by inspection. To check that it respects multiplication, it is enough to show that the product $g_1 g_2$ in RG maps to $f_{g_1} * f_{g_2}$, i.e., that $f_{g_1} * f_{g_2} = f_{g_1 g_2}$. The computation is $(f_{g_1} * f_{g_2})(x) = \sum_{y \in G} f_{g_1}(xy^{-1}) f_{g_2}(y) = f_{g_1}(xg_2^{-1})$, and this is 1 if and only if $xg_2^{-1} = g_1$, i.e., if and only if $x = g_1 g_2$. For other values of x, it is 0. Therefore $(f_{g_1} * f_{g_2})(x) = f_{g_1 g_2}(x)$ for all x.

15. Let the monic polynomial in question be $P(X)$. We prove by induction on m that any polynomial $A(X)$ in I of degree m is a multiple of $P(X)$. The base case of the induction is all polynomials of degree $< n$ in I; only 0 fits this description. Assume the result for all degrees $< m$, and let $A(X)$ be any polynomial in I, say with leading term $a_m X^m$, $a_m \neq 0$. Then $a_m X^{m-n} P(X)$ is in I, and so is $B(X) = A(X) - a_m X^{m-n} P(X)$. The coefficient of X^m in $B(X)$ is 0, and hence $B(X) = 0$

or else $\deg B(X) < m$. If $B(X) = 0$, then $A(X) = a_m X^{m-n} P(X)$, and $A(X)$ is a multiple of $P(X)$. If $\deg B(X) < m$, then induction gives $B(X) = C(X)P(X)$, and therefore $A(X) = (a_m X^{m-n} + C(X))P(X)$. So again $A(X)$ is a multiple of $P(X)$.

16. Let $p_1, \ldots, p_n$ be n distinct positive primes in $\mathbb{Z}$, put $q_k = p_1 \cdots p_k$ for $0 \le k \le n$, and take $I_{n+1} = (q_n, q_{n-1}X, q_{n-2}X^2, \ldots, q_0 X^n)$. This can be written with $n+1$ generators but not with fewer than that.

17. In (a), certainly $\ker \varphi \supseteq (y^2 - x^3)$. In the reverse direction, suppose that $\sum_{n=0}^{N} P_n(x)y^n$ is in $\ker \varphi$. Since $y^2 \equiv x^3 \bmod (y^2 - x^3)$, we can reduce this element of $\ker \varphi$ to the form $Q_0(x) + Q_1(x)y$. Substituting with t gives $Q_0(t^2) + Q_1(t^2)t^3 = 0$. The first term involves only even powers of t, and the second term involves only odd powers. Thus each is 0 separately. We are thus to determine what members $Q_0(x)$ and $Q_1(x)y$ of $\mathbb{K}[x, y]$ are in $\ker \varphi$. For $Q_0(t^2)$ to be 0, every coefficient of Q_0 must be 0. For $Q_1(t^2)t^3$ to be 0, every coefficient of Q_1 must be 0. Therefore only 0 is of the stated form, and every member of $\ker \varphi$ lies in $(y^2 - x^3)$.

For (b), image φ contains t^2, t^3, and every power t^n such that $n = 2a + 3b$ with a and b nonnegative integers. It follows that image φ consists of all linear combinations of powers t^n for $n \ge 2$.

19. Write $A(X) = B(X)Q(X)$ in $F[X]$, and let $A(X) = c(A)(c(A)^{-1}A(X))$, $B(X) = c(B)(c(B)^{-1}B(X))$, and $Q(X) = c(Q)(c(Q)^{-1}Q(X))$ be the decompositions of Proposition 8.19. Then we have

$$c(A)(c(A)^{-1}A(X)) = c(B)c(Q)\big((c(B)^{-1}B(X))(c(Q)^{-1}Q(X))\big).$$

By Gauss's Lemma and the uniqueness in Proposition 8.19, we obtain $c(A)^{-1}A(X) = (c(B)^{-1}B(X))(c(Q)^{-1}Q(X))$, apart from unit factors. Therefore the member $B_0(X) = c(B)^{-1}B(X)$ of $R[X]$ is exhibited as dividing $A_0(X) = c(A)^{-1}A(X)$ with a quotient $c(Q)^{-1}Q(X)$ in $R[X]$.

20. Let R be a finite integral domain, and let $a \ne 0$. Multiplication by a is one-one since R is an integral domain, and it must be onto R by the finiteness. Therefore there is some b with $ab = 1$, and we have produced an inverse for a.

21. Let $R' = R/(p)$. Suppose that $A(X) = B(X)C(X)$ nontrivially in $R[X]$ with $B(X) = b_k X^k + \cdots + b_0$, $C(X) = b_l X^l + \cdots + c_0$, and $k + l = N$. Since p divides a_0 but p^2 does not, p divides exactly one of b_0 and c_0, say the former. In $R'[X]$, we have $A(X) \equiv a_N X^N$, $C(X) \equiv c_l X^l + \cdots + c_0$, and $A(X) = B(X)C(X)$. Now X is prime in $R'[X]$ by Problem 4, and X^N divides $B(X)C(X)$ in $R'[X]$. Using the defining property of a prime, one power at a time, we find that X^N divides $B(X)$. Since $\deg B < N$, we must have $B(X) \equiv 0$ in $R'[X]$. Thus p divides b_k in R, and p divides a_N, contradiction.

22. In (a), we regard $WZ - XY$ as a first-degree polynomial in W, with Z being a prime in the ring of coefficients. A nontrivial factorization of $WZ - XY$ must be of the form $A(X, Y, Z)\big(B(X, Y, Z)W + C(X, Y, Z)\big)$ with $Z = A(X, Y, Z)B(X, Y, Z)$. Since Z is prime, one of these factors must be a unit, hence a scalar. If $A(X, Y, Z)$

is a scalar, then the factorization of $WZ - XY$ is trivial. Otherwise we may assume that the factorization is $WZ - XY = Z(W + C(X, Y, Z))$. Then Z divides XY, and we arrive at a contradiction since Z does not appear in XY.

In (b), we expand in cofactors about the top row. Using induction, we see that we can regard the determinant $\det[X_{ij}]$ as a first-degree polynomial in X_{11} with an irreducible coefficient $P(X_{22}, X_{23}, \ldots, X_{nn})$. A nontrivial factorization must be of the form $\det[X_{ij}] = PX_{11} + Q = A(BX_{11}+C)$, where Q, A, B, C are polynomials in the remaining indeterminates. Then $AB = P$ and P irreducible implies that A or B is a unit, hence a scalar. If A is a scalar, our factorization of $\det[X_{ij}]$ is trivial. Otherwise we may assume that the factorization is $\det[X_{ij}] = PX_{11} + Q = P(X_{11}+C)$. Then P must divide Q. Taking the degrees of homogeneity into account, we see that Q must be the product of P and a homogeneous polynomial of degree 1. Every term of P is of the form $\prod_{i=2}^n X_{2,\sigma(2)}$ for some permutation σ of $\{2, \ldots, n\}$, and thus such a factor must appear in every term of Q. However, the only terms of $\det[X_{ij}]$ that contain a factor $\prod_{i=2}^n X_{2,\sigma(2)}$ also contain the factor X_{11}, and this factor is absent in Q. Thus the assumed reducibility has led to a contradiction.

23. The ideal of $\mathbb{Z}[X]$ generated by $A(X)$ and $B(X)$ consists of all polynomials $A(X)C(X) + B(X)D(X)$ with $C(X)$ and $D(X)$ in $\mathbb{Z}[X]$. If such an expression equals some integer n, then a GCD within $\mathbb{Q}[X]$ of $A(X)$ and $B(X)$ divides $A(X)$ and $B(X)$ and hence must divide n. It is therefore of degree 0 and is a unit in $\mathbb{Q}[X]$. Thus $A(X)$ and $B(X)$ are relatively prime in $\mathbb{Q}[X]$.

Conversely if $A(X)$ and $B(X)$ are members of $\mathbb{Z}[X]$ that are relatively prime in $\mathbb{Q}[X]$, we can find $P(X)$ and $Q(X)$ in $\mathbb{Q}[X]$ with $A(X)P(X) + B(X)Q(X) = 1$. Multiplying by a common denominator of the coefficients of $P(X)$ and $Q(X)$, we obtain a relation $A(X)C(X) + B(X)D(X) = n$ with all polynomials in $\mathbb{Z}[X]$. Thus n is in the ideal of $\mathbb{Z}[X]$ generated by $A(X)$ and $B(X)$.

24. We are given $\begin{pmatrix} 1+i & 2-i \\ 3 & 5i \end{pmatrix} \begin{pmatrix} u_1 \\ u_2 \end{pmatrix} = \begin{pmatrix} 0 \\ 0 \end{pmatrix}$ with coefficient matrix $C = \begin{pmatrix} 1+i & 2-i \\ 3 & 5i \end{pmatrix}$. Left multiplication on C by a matrix with determinant a unit does not change the total set of conditions on $\begin{pmatrix} u_1 \\ u_2 \end{pmatrix}$, and right multiplication by such a matrix changes the generators but not the module they generate. In the first column of C, we observe that $\gcd(1+i, 3) = 1$ because $1+i$ divides 2 and $\gcd(2, 3) = 1$. Then we have $-(1-i)(1+i) + 1 \cdot 3 = 1$, and we are led to the matrix $A = \begin{pmatrix} -(1-i) & 1 \\ -3 & 1+i \end{pmatrix}$, which has determinant 1. We can thus replace C by $AC = \begin{pmatrix} 1 & -1+8i \\ 0 & -11+8i \end{pmatrix}$. An invertible column operation replaces the upper right entry by 0. Thus we are led to the diagonal matrix $\begin{pmatrix} 1 & 0 \\ 0 & -11+8i \end{pmatrix}$. In other words, we may assume that the $\mathbb{Z}[i]$ module was given to us with generators t_1, t_2 satisfying $t_1 = 0$ and $(-11+8i)t_2 = 0$. Therefore the given $\mathbb{Z}[i]$ module is cyclic and is $\mathbb{Z}[i]$ isomorphic to $\mathbb{Z}[i]/(-11+8i)$.

25. In (a), $\delta(z) = z\bar{z}$. Then $\delta(zw) = zw\bar{z}\bar{w} = z\bar{z}w\bar{w} = \delta(z)\delta(w)$.

In (b), we start with two nonzero members α and β of R. We are to find γ and ρ in R with $\alpha = \beta\gamma + \rho$ and $\delta(\rho) < \delta(\beta)$. It is the same to find γ and ρ with

$\alpha/\beta = \gamma + \rho/\beta$ and $\delta(\rho/\beta) < 1$. Apply the hypothesis with $z = \alpha/\beta$, and let γ be the element r such that $\delta(z - r) < 1$. Then ρ may be defined as $\beta(z - r)$, and all the conditions are satisfied.

26. Given $z = x + y\sqrt{-m}$, define $r = a + \frac{1}{2}b(1 + \sqrt{-m})$ in $\mathbb{Z}[\frac{1}{2}(1 + \sqrt{-m})]$ by choosing b to be an integer with $|2y - b| \leq \frac{1}{2}$ and then choosing a to be an integer with $|x - a - \frac{1}{2}b| \leq \frac{1}{2}$. Since $|y - \frac{1}{2}b| \leq \frac{1}{4}$, we then have

$$\delta(z - r) = (x - a - \tfrac{1}{2}b)^2 + m(y - \tfrac{1}{2}b)^2 \leq \tfrac{1}{4} + m\tfrac{1}{16} \leq \tfrac{1}{4} + \tfrac{11}{16} < 1.$$

27. In (a), complex conjugation is an automorphism of $\mathbb{Z}[i]$ and must therefore carry primes to primes.

In (b), we know that $(a + bi)(a - bi)$ is the integer $N(a + bi)$. Suppose that $N(a + bi) = mn$ nontrivially with $\mathrm{GCD}(m, n) = 1$. Since $a + bi$ is prime, it divides one of m and n. Say that $m = (a + bi)(c + di)$. Then $m^2 = N(m) = N(a + bi)N(c + di) = mnN(c + di)$. Any prime number dividing n must divide the left side m^2, and hence there can be no such prime. We conclude that $N(a + bi)$ does not have nontrivial relatively prime divisors. Hence it is a power of some prime number p.

In (c), let $N(a + bi) = p^k$. The left side is the product of two primes of $\mathbb{Z}[i]$. If p is the product of l primes of $\mathbb{Z}[i]$, then p^k is the product of kl primes. Then we must have $kl = 2$, and k must divide 2.

In (d), suppose $N(a + bi) = p^2$, so that $k = 2$ in (c). Then $l = 1$, and p is prime in $\mathbb{Z}[i]$.

28. The equation $N(a + bi) = p$ says that $a^2 + b^2 = p$. The right side is $\equiv 3 \bmod 4$, but 3 is not the sum of two squares modulo 4. Hence $N(a + bi) = p$ is impossible when $p \equiv 3 \bmod 4$. Problem 27c then forces $N(a + bi) = p^2$, and Problem 27d says that p is prime in $\mathbb{Z}[i]$.

29. If $N(a + bi) = 2$, then $|a| = |b| = 1$, and we obtain $1 + i$ and its associates. If $N(a + bi) = 4$, then $a = \pm 2$ with $b = 0$ or else $a = 0$ with $b = \pm 2$; in these cases $a + bi$ is an associate of 2, which is $(1 + i)(1 - i)$ and is not prime in $\mathbb{Z}[i]$.

30. The multiplicative group of $\mathbb{F}_p$ is cyclic of order $p - 1$. If p is of the form $4n + 1$, then $\mathbb{F}_p^\times$ has order $4n$. The nth power of a generator then has to be an integer whose square is $\equiv -1 \bmod p$.

31. For (a), we obtain φ_1 by mapping $\mathbb{Z}[X]$ to $\mathbb{F}_p[X]$ with a substitution homomorphism and following this with a passage to the quotient. Similarly φ_2 is obtained from the substitution homomorphism $\mathbb{Z}[X] \to \mathbb{Z}[i]$ followed by the passage to the quotient.

For (b), the kernel of φ_1 consists of all polynomials that are multiples of $X^2 + 1$ when their coefficients are taken modulo p. This is $p\mathbb{Z}[X] + (X^2 + 1)\mathbb{Z}[X] = (p, X^2 + 1)$. The kernel of φ_2 consists of all polynomials with the property that when taken modulo $X^2 + 1$, they are multiples of p. This too is the ideal $(p, X^2 + 1)$.

For (c), Problem 30 shows that the polynomial X^2+1 factors nontrivially in $\mathbb{F}_p[X]$. Therefore X^2+1 is not prime, the ideal (X^2+1) is not prime, and $\mathbb{F}_p[X]/(X^2+1)$ is not an integral domain. By (b), $\mathbb{Z}[i]/(p)$ is not an integral domain, and the ideal (p) is not prime. Hence p is not prime in $\mathbb{Z}[i]$. By (c) and (d) in Problem 27, p is of the form $N(a+bi)$ for some prime $a+bi$ in $\mathbb{Z}[i]$.

For (d), if we have $p = N(a+bi) = N(a'+b'i)$, we obtain two prime factorizations of p in $\mathbb{Z}[i]$ as $p = (a+bi)(a-bi) = (a'+b'i)(a'-b'i)$, and unique factorization in $\mathbb{Z}[i]$ implies that $a'+b'i$ is an associate of $a+bi$ or $a-bi$.

32. For (a), multiply C on the left by the matrix A that is the identity except in the first column, where the ith entry is C_{ii}.

For (b) and (c), the step of row reduction leads to a first column that is 0 in all entries but the first, where it is $\text{GCD}(C_{11}, \ldots, C_{nn})$. In other words, the new entry in position $(1, 1)$ divides all entries in the new C. Therefore one step of column reduction leaves the entry unchanged in position $(1, 1)$, leaves the remainder of the first column equal to 0, and makes the remainder of the first row equal to 0. What is left in the rows and columns other than the first is a matrix whose entries are all divisible by $\text{GCD}(C_{11}, \ldots, C_{nn})$. Hence we can induct on the size.

33. Changing notation slightly from Lemma 8.26, write $AE = DB$ with $\det A$ and $\det B$ in $R^\times$. Over the field of fractions of R, the m-by-n matrices E and D must have the same rank since A and B are invertible, and consequently D and E have the same number of nonzero diagonal entries. Thus for some l with $0 \leq l \leq k$, we are given that D_{jj} divides $D_{j+1, j+1}$ and E_{jj} divides $E_{j+1, j+1}$ whenever $1 \leq j < l$. Fix i with $1 \leq i \leq l$, and consider all possible i-by-i determinants that can be formed using the first i rows of A and one of the $\binom{n}{i}$ sets of i columns. Since $\det A$ is in $R^\times$, it follows from the expansion-by-cofactors formula that these determinants have GCD equal to 1. Each corresponding determinant for DB equals $D_{11} \cdots D_{ii}$ times such a determinant, and hence the GCD for DB is $D_{11} \cdots D_{ii}$.

Meanwhile, the GCD of the determinants for A is also 1, and, because of the divisibility property of the diagonal entries of E, $E_{11} \cdots E_{ii}$ divides each of the determinants for AE. Hence $E_{11} \cdots E_{ii}$ divides the GCD of the determinants for AE, which equals the GCD of the determinants for DB, which equals $D_{11} \cdots D_{ii}$. Thus $E_{11} \cdots E_{ii}$ divides $D_{11} \cdots D_{ii}$.

Arguing similarly with the determinants formed from the first i columns of A, AE, B, and BD, we see that $D_{11} \cdots D_{ii}$ divides $E_{11} \cdots E_{ii}$. Therefore $D_{11} \cdots D_{ii}$ and $E_{11} \cdots E_{ii}$ are associates for $1 \leq i \leq l$. Since none of the factors in question is 0, we see that each of the first l diagonal entries of D is an associate of the corresponding diagonal entry of E. This proves the desired uniqueness.

34. We have seen in this setting that the decomposition of V as a direct sum of cyclic $\mathbb{K}[X]$ modules means a decomposition of V as a direct sum of vector subspaces, each of which is invariant under L. Also, if V_0 is one of these vector subspaces, the cyclic nature of the module means that there is some vector v_0 in V_0 such that $\mathbb{K}[X]v_0 = V_0$, and the diagonal entry of the matrix D in the proof of Theorem 8.25 is a polynomial

$M[X]$ such that $V_0 \cong K[X]/(M(X))$ as a $K[X]$ module. Referring to Problems 26–31 of Chapter V, we see that v_0 is a cyclic vector for the cyclic subspace V_0, and $M[X]$ is the minimal polynomial of L on this subspace.

The divisibility property of the minimal polynomials and also the uniqueness assertion now follow from what has been proved in Problems 32–33. We know from Problem 28 in Chapter V that the data of a cyclic subspace and the minimal polynomial yield a particular matrix for the linear mapping and hence determine the linear mapping on that subspace up to similarity. Consequently the uniqueness statement that has just been observed says that L is determined up to similarity by the integer r and the sequence of minimal polynomials.

35. Let A and B be members of $M_n(\mathbb{K})$. Form the data for each from the rational canonical form in Problem 34. Now consider everything as involving vector spaces over the larger field $\mathbb{L}$. We are given that the two matrices are similar over $\mathbb{L}$, i.e., are conjugate via $\text{GL}(n, \mathbb{L})$. Problem 34 shows that the respective decompositions have the same data. The two matrices still have the same data when we again consider the field to be $\mathbb{K}$. Hence they are similar over $\mathbb{K}$, i.e., are conjugate via $\text{GL}(n, \mathbb{K})$.

36. The fact that the homomorphisms are isomorphisms follows from the composition rule.

38. In (b), we can write any member of $F[X_1, \ldots, X_n, X]$ as

$$A_n(X_1, \ldots, X_n)X^n + \cdots + A_1(X_1, \ldots, X_n)X + A_0(X_1, \ldots, X_n),$$

and σ^{**} acts by having σ^* act on each coefficient. Invariance under all σ^{**}'s therefore means that each coefficient is invariant under all σ^*'s and hence is a symmetric polynomial.

39. In (a), if, for example $i < j$ and $k_i < k_j$, then the monomial $aX_1^{k_1} \cdots X_n^{k_n}$ is increased in the ordering by replacing the factors $X_i^{k_i} X_j^{k_j}$ by $X_i^{k_j} X_j^{k_i}$.

For (b), we need only take the largest monomial in each E_i, raise it to the c_i power, and multiply the results.

For (c), let the largest monomial in A be $aX_1^{k_1} \cdots X_n^{k_n}$. To define M, choose $r = a$ and define $c_j = k_j - k_{j+1}$ for $1 \leq j < n$ and $c_n = k_n$.

For (d), the construction in (c) yields 0 coefficient for $X_1^{k_1} \cdots X_n^{k_n}$, and $A - rM$ has no larger monomials. So if $A - rM = 0$, the largest monomial is below that monomial $X_1^{k_1} \cdots X_n^{k_n}$.

For (e), iteration of the construction in (c) and (d) shows that any homogeneous symmetric polynomial equals a homogeneous polynomial in the elementary symmetric polynomials. Problem 37 shows that any symmetric polynomial is a linear combination of homogeneous symmetric polynomials, and hence every symmetric polynomial is a polynomial in the elementary symmetric polynomials.

40. Suppose that z_0 and w_0 in $\mathbb{C}^m$ have $P(z_0) \neq 0$ and $P(w_0) \neq 0$. As a function of $t \in \mathbb{C}$, $P(z_0 + t(w_0 - z_0))$ is a polynomial function nonvanishing at $t = 0$ and

$t = 1$. The subset of $t \in \mathbb{C}$ where it vanishes is finite, and its complement in $\mathbb{C}$ is necessarily pathwise connected and therefore connected. Thus z_0 and w_0 lie in a connected subset of $\mathbb{C}^m$ where P is nonvanishing. Taking the union of these connected sets with z_0 fixed and w_0 varying, we see that the set of $w_0 \in \mathbb{C}^m$ where $P(w_0) \neq 0$ is connected.

41. For (a), two applications of the formula relating Pfaffians and determinants gives us $\text{Pfaff}(A^t X A)^2 = \det(A^t X A) = (\det A)^2 \det X = (\det A)^2 \text{Pfaff}(X)^2$. Taking the square root gives the desired result.

For (b), we fix X with $\text{Pfaff}(X) \neq 0$ and allow A to vary. On the set where $\det A \neq 0$, the function $A \mapsto \text{Pfaff}(A^t X A)/\det A$ is a continuous function with image in the two-point set $\{\pm \text{Pfaff}(X)\}$, by (a). The domain of the function is connected by Problem 40, and therefore the image has to be connected. Hence the function has to be constant. Checking the value of the function at $A = I$, we see that the function has to be constantly equal to $\text{Pfaff}(X)$.

42. Form the ring $S = \mathbb{Z}[\{A_{ij}\}, \{X_{ij}\}]$. We can then regard $\text{Pfaff}(A^t X A)$ and $(\det A)\text{Pfaff}(X)$ as two polynomials with entries in S. If we fix arbitrary elements $a_{ij} \in \mathbb{Z}$ for all i and j and also $x_{ij} \in \mathbb{Z}$ for $i < j$, then Proposition 4.30 gives us a unique substitution homomorphism $\Psi \to \mathbb{Z}$ such that $\Psi(1) = 1$, $\Psi(A_{ij}) = a_{ij}$, and $\Psi(X_{ij}) = x_{ij}$. Assemble the a_{ij} and x_{ij} into matrices $a = [a_{ij}]$ and $x = [x_{ij}]$ with x alternating. Problem 41b shows that the identity in question holds when the entries are in $\mathbb{C}$, and in particular it holds when the entries are in $\mathbb{Z}$. Therefore $\text{Pfaff}(a^t x a) = (\det a)\text{Pfaff}(x)$. Since $\mathbb{Z}$ is an integral domain and since a and x are arbitrary with x alternating, Corollary 4.32 allows us to conclude that $\text{Pfaff}(A^t X A) = (\det A)\text{Pfaff}(X)$ as an equality in S.

To pass from S to $\mathbb{K}$, let $1_{\mathbb{K}}$ be the identity of $\mathbb{K}$, and let $\varphi_1 : \mathbb{Z} \to \mathbb{K}$ be the unique homomorphism of rings such that $\varphi_1(1) = 1_{\mathbb{K}}$. If we fix arbitrary elements a_{ij} of $\mathbb{K}$ for all i and j, as well as arbitrary elements x_{ij} of $\mathbb{K}$ for $i < j$, then Proposition 4.30 gives us a unique substitution homomorphism $\Phi : S \to \mathbb{K}$ such that $\Phi(1) = \varphi_1(1) = 1_{\mathbb{K}}$, $\Phi(A_{ij}) = a_{ij}$ for all i and j, and $\Phi(X_{ij}) = x_{ij}$ whenever $i < j$. Applying Φ to our identity in S, we obtain $\text{Pfaff}(a^t x a) = (\det a)\text{Pfaff}(x)$ as an equality in $\mathbb{K}$.

43. From Problem 42 and the hypothesis on g, we have $1 = \text{Pfaff}(J) = \text{Pfaff}(g^t J g) = (\det g)\text{Pfaff}(J) = \det g$. Hence $\det g = 1$.

45. For (a), if $\varphi : R \to R/P^k$ is the quotient homomorphism, then φ^{-1} of any ideal of R/P^k is an ideal I of R containing P^k. If Q is a prime ideal dividing I, then Q divides P^k, and it follows that $Q = P$. Thus the only possibilities for I are the powers P^i of P, necessarily stopping with $i = k$.

For (b), we know that π^i lies in P^i but not P^{i+1}. For $1 \leq i \leq k-1$, it follows that the principal ideal $(\pi^i + P^k)/P^k$ is contained in the ideal P^i/P^k but not in P^{i+1}/P^k. Since the ideals P^j/P^k for $j \leq k$ are nested and there are no other ideals in R/P^k, we must have $(\pi^i + P^k)/P^k = P^i/P^k$. Thus P^i/P^k is principal.

46. Corollary 8.63 and Problem 44 together show that every ideal of R/I is principal if it can be shown that every ideal of R/P^k is principal when P is a nonzero

prime ideal. The two parts of Problem 45 together show that every ideal of R/P^k is principal.

47. We may assume that $(a) \subsetneq I$ since otherwise the result follows with $b = 0$. Since $a \neq 0$, the ideal $I/(a)$ in $R/(a)$ is a principal ideal by Problem 46c. If b_0 is a generator of this ideal, then $(R/(a))b_0 = I/(a)$. Since b_0 is in $I/(a)$, we can write it as $b_0 = b + (a)$ for some b in I. Every member of $I/(a)$ is then of the form $(r + (a))(b + (a)) = rb + (a)$, and we conclude that every member of I is of the form $rb + sa$ with r and s in R.

48. Any R submodule of R is an ideal.

49. Write $M = Rx_1 + \cdots + Rx_n$ with $x_1, \ldots, x_n$ in F. Each x_i is of the form $r_i s_i^{-1}$ with r_i and s_i in R and with $s_i \neq 0$. Then aM lies in R for $a = \prod_{i=1}^n s_i$. So aM is an ideal in R, by Problem 48. If N is a second fractional ideal, choose $b \neq 0$ such that bN is an ideal in R. Then $(aM)(bN)$ is an ideal in R, and the formula $MN = (ab)^{-1}(aM)(bN)$ shows that MN is a fractional ideal.

50. Since I is a finitely generated R module, we can write $I = Ra_1 + \cdots + Ra_n$ with all a_i in R. The condition for $x \in F$ to be in I^{-1} is that $xI \subseteq R$, and it is necessary and sufficient that xa_i be in R for all i. Thus it is necessary that x be in $(a_1 \cdots a_n)^{-1} R$. Consequently I^{-1} is an R submodule of the singly generated R module $(a_1 \cdots a_n)^{-1} R$. Since R is Noetherian, I^{-1} is finitely generated.

51. If I is maximal among the nonzero ideals of R for which there is no fractional ideal M of F with $IM = R$, then Lemma 8.58 shows that I is not prime. Choose a nonzero prime ideal P with $I \subsetneq P$. Then Lemma 8.58 and the definitions give $I \subseteq IP^{-1} \subseteq II^{-1} \subseteq R$. We cannot have $I = IP^{-1}$ since otherwise $IP = (IP^{-1})P = I(P^{-1}P) = I$ and Proposition 8.52 gives $I = 0$. By maximality of I, we can find some fractional ideal N with $(IP^{-1})N = R$. Then $I(P^{-1}N) = R$, and we can take $M = P^{-1}N$, by Problem 49.

52. Every member x of M has $xI \subseteq R$, and thus $M \subseteq I^{-1}$. On the other hand, if x is in I^{-1}, then $xI \subseteq R$, $x = xIM \subseteq RM = M$, and x is in M.

53. If M is a fractional ideal, then Problem 49 produces $c \neq 0$ in F with $cM \subseteq R$, and Problem 48 shows that cM is an ideal of R. Using Problem 52, we can write $M = (c)^{-1}(c)M = (c)^{-1}(cM)$. This proves that $M = IJ^{-1}$ for ideals I and J. Then (a) follows from Theorem 8.55 and Problem 52, and (b) follows from Problem 52.

Chapter IX

1. The equation for r gives $r^3 = 3r - 4$ and $r^4 = 3r^2 - 4r$. Therefore the inverse has $1 = (r^2 + r + 1)(ar^2 + br + c) = ar^4 + (a+b)r^3 + (a+b+c)r^2 + (b+c)r + c =$

$r^2(4a+b+c)+r(-a+4b+c)+1(-4a-4b+c)$, and we are led to the system of linear equations

$$4a+b+c=0,$$
$$-a+4b+c=0,$$
$$-4a-4b+c=1.$$

Then $(a,b,c)=(-\frac{3}{49},-\frac{5}{49},\frac{17}{49})$, and $(r^2+r+1)^{-1}=-\frac{3}{49}r^2-\frac{5}{49}r+\frac{17}{49}$.

2. Multiplication by a nonzero r is a one-one F linear mapping from the F vector space R onto itself. Since $\dim_F R < \infty$, this linear mapping must be onto. The element s such that $rs=1$ is a multiplicative inverse of r.

3. Let z_0 be a nonreal element of $\mathbb{K}$. Then the closure of the $\mathbb{Q}$ vector space $\mathbb{Q}+\mathbb{Q}z_0$ contains $\mathbb{R}+\mathbb{R}z_0=\mathbb{C}$.

4. If $y=F(x)/G(x)$, then $G(x)y=F(x)$. Arranging the terms as powers of x with coefficients of the form $ay+b$ with a and b in $\mathbb{k}$, we see that x is a root of a polynomial in one indeterminate over $\mathbb{k}(y)$. Therefore x is algebraic over $\mathbb{k}(y)$.

5. The condition is that N be the square of an integer. For any other N, X^2-N is irreducible over $\mathbb{Q}$, and $[\mathbb{Q}(\sqrt{N}):\mathbb{Q}]=2$. Since 2 does not divide 3, $\mathbb{Q}(\sqrt{N})$ cannot be a subfield of $\mathbb{Q}(\sqrt[3]{2})$.

6. $X \mapsto Y+1$.

7. No, since 8 is not a power of 4. See Corollary 9.19.

8. Let g be a generator of the cyclic group $\mathbb{K}^\times$, and let q be the order of $\mathbb{K}$. Then

$$g \cdot g^2 \cdot g^3 \cdots g^{q-1} = g^{1+2+3+\cdots+(q-1)} = g^{\frac{1}{2}q(q-1)}.$$

If q is even, then this is $(g^{q-1})^{q/2}=1^{q/2}=1=-1$. If q is odd, it is $(g^{\frac{1}{2}(q-1)})^q=(-1)^q=-1$.

9. Let $F(X)=X^n+c_{n-1}X^{n-1}+\cdots+c_0$ be the minimal polynomial of r. We are given that n is odd. Write the equation $F(r)=0$ as

$$r(r^{n-1}+c_{n-2}r^{n-3}+\cdots+c_1)=-c_{n-1}r^{n-1}-c_{n-3}r^{n-3}-\cdots-c_0.$$

Then r is expressed as an element of $\mathbb{k}(r^2)$ unless $r^{n-1}+c_{n-2}r^{n-3}+\cdots+c_1=0$. But this expression cannot be 0 because this polynomial has degree $n-1$ and the minimal polynomial for r has degree n.

10. Let $d_r=[\mathbb{k}(r):\mathbb{k}]$ and $d_s=[\mathbb{k}(s):\mathbb{k}]$. Since $\mathbb{K}$ contains $\mathbb{k}(r)$ and $\mathbb{k}(s)$, we see that d_r and d_s divide $[\mathbb{K}:\mathbb{k}]$. Since $\gcd(d_r,d_s)=1$, $d_r d_s$ divides $[\mathbb{K}:\mathbb{k}]$. The minimal polynomial $M(X)$ of r over $\mathbb{k}$ is a polynomial over $\mathbb{k}(s)$ such that $M(r)=0$. Thus the minimal polynomial $N(X)$ of r over $\mathbb{k}(s)$ divides $M(X)$. If c is the degree of $N(X)$, we then have $c \leq d_r$. Since $d_r d_s$ divides $[\mathbb{K}:\mathbb{k}]$, we obtain

$$d_r d_s \leq [\mathbb{K}:\mathbb{k}] = [\mathbb{k}(r,s):\mathbb{k}] = [\mathbb{k}(s)(r):\mathbb{k}] = c[\mathbb{k}(s):\mathbb{k}] = cd_s \leq d_r d_s.$$

Equality must hold throughout. Equality at the right end says that $c=d_r$, and this proves (a). Equality at the left end says that $d_r d_s = [\mathbb{K}:\mathbb{k}]$, and this proves (b).

11. In (a), we have $\gamma = \beta + c\alpha = \beta(1 + c\omega)$. Here $r = 1 + c\omega$ lies in $\mathbb{Q}(\sqrt{-3})$, and so does r^3. Therefore r^3 is a root of a quadratic polynomial $Y^2 + pY + q$. Then $\gamma^6 + a\gamma^3 + b = r^6\beta^6 + ar^3\beta^3 + b = 4r^6 + 2ar^3 + b = 4(r^6 + \frac{1}{2}ar^3 + \frac{1}{4}b)$, and the right side is 0 if a and b are chosen such that $p = \frac{1}{2}a$ and $q = \frac{1}{4}b$.

In (b), $\gamma = \beta + \alpha = \beta(1 + \omega)$, and $\gamma^3 = \beta^3(\frac{1}{2}(1 + \sqrt{-3}))^3 = 2(-1) = -2$. Then γ satisfies $\gamma^3 + 2 = 0$, and this is irreducible since -2 is not a cube in $\mathbb{Q}$.

In (c), the field $\mathbb{Q}(\gamma)$ contains $\gamma^3 = \beta^3(\frac{1}{2}(3 - \sqrt{-3}))^3 = \frac{1}{4}(3 - \sqrt{-3})^3 = \frac{1}{4}(27 - 9\sqrt{-3} + 3 \cdot 3(-3) - (-3)\sqrt{-3}) = -\frac{3}{2}\sqrt{-3}$. Thus $\mathbb{Q}(\sqrt{-3})$ is a subfield of $\mathbb{Q}(\gamma)$, and 2 divides $[\mathbb{Q}(\gamma) : \mathbb{Q}]$. Since $\mathbb{Q}(\sqrt{-3})$ is a subfield, $\beta = \gamma(1-\omega)^{-1}$ lies in $\mathbb{Q}(\gamma)$. Thus $\mathbb{Q}(\sqrt[3]{2})$ is a subfield of $\mathbb{Q}(\gamma)$, and 3 divides $[\mathbb{Q}(\gamma) : \mathbb{Q}]$. Consequently 6 divides $[\mathbb{Q}(\gamma) : \mathbb{Q}]$, and the minimal polynomial of γ has degree ≥ 6. By (a), it has degree exactly 6.

12. Let the characteristic be p. If $F(X)$ has $F'(X) = 0$, then all the exponents of X appearing in $F(X)$ are multiples of p. Let $F(X) = a_n X^{np} + a_{n-1} X^{(n-1)p} + \cdots + a_1 X^p + a_0$. Since the Frobenius map is onto in the case of a finite field, we can choose members $c_n, \ldots, c_0$ of $\mathbb{k}$ such that $c_n^p = a_n, c_{n-1}^p = a_{n-1}, \ldots, c_0^p = a_0$. Put $G(X) = c_n X^n + c_{n-1} X^{n-1} + \cdots + c_0$. Then $F(X) = G(X)^p$, and $F(X)$ is reducible.

13. In (a), if $F(X) = G(X)H(X)$ is reducible and r_1 is a root of $G(X)$, then $\sigma(r_1)$ is a root of $G(X)$ for any $\sigma \in \text{Gal}(\mathbb{K}/\mathbb{k})$. Consequently the orbit of r_1 under $\text{Gal}(\mathbb{K}/\mathbb{k})$ is a proper subset of the set of roots of $F(X)$. Conversely if $F(X)$ is irreducible and r_j is given, then the uniqueness of simple extensions gives us a $\mathbb{k}$ isomorphism of $\mathbb{k}(r_1)$ onto $\mathbb{k}(r_j)$. Theorem 9.13' shows that this isomorphism extends to a $\mathbb{k}$ automorphism of $\mathbb{K}$, and hence $\text{Gal}(\mathbb{K}/\mathbb{k})$ is transitive on the set of roots of $F(X)$.

In (b), the transitivity follows from (a) and the irreducibility of $\Phi_8(X)$ over $\mathbb{Q}$. Let $\zeta = e^{2\pi i/8}$. The roots of $\Phi_8(X) = X^4 + 1$ are $\zeta, \zeta^3, \zeta^5, \zeta^7$. So if σ is in $\text{Gal}(\mathbb{K}/\mathbb{Q})$, then $\sigma(\zeta) = \zeta^k$ with k odd. Then $\sigma^2(\zeta) = \sigma(\zeta^k) = \sigma(\zeta)^k = (\zeta^k)^k = \zeta^{k^2}$. Since the square of any odd integer is congruent to 1 modulo 8, $\sigma^2(\zeta) = \zeta$. Thus each σ has $\sigma^2 = 1$, and $\text{Gal}(\mathbb{K}/\mathbb{Q})$ cannot contain a 4-cycle.

In (c), the irreducibility of $F(X)$ implies that $F(X)$ is the minimal polynomial of r_1. Hence $[\mathbb{k}(r_1) : \mathbb{k}] = n$. Since $\mathbb{k}(r_1) \subseteq \mathbb{K}$, $[\mathbb{k}(r_1) : \mathbb{k}]$ must divide $[\mathbb{K} : \mathbb{k}]$, and n divides $[\mathbb{K} : \mathbb{k}]$. Therefore n divides the equal integer $\text{Gal}(\mathbb{K}/\mathbb{k})$. If n is prime, then the fact that n divides the order of $\text{Gal}(\mathbb{K}/\mathbb{k})$ implies that $\text{Gal}(\mathbb{K}/\mathbb{k})$ contains an element of order n, by Sylow's Theorems. The only elements of order n in $\mathfrak{S}_n$ are the n-cycles, and hence $\text{Gal}(\mathbb{K}/\mathbb{k})$ contains at least one n-cycle.

14. In (a), we have $\mathbb{L}_{k+1} = \mathbb{L}_k(\sqrt{a_{k+1}})$, and hence $[\mathbb{L}_{k+1} : \mathbb{L}_k]$ equals 1 or 2. By induction, $[\mathbb{L}_k : \mathbb{Q}]$ is a power of 2, and the power is at most the number of steps in the induction, namely k.

In (b), associate to each subset S of $\{1, \ldots, k\}$ the element $v_S = \prod_{j \in S} \sqrt{a_j}$ in $\mathbb{L}_k$. The product of any two such elements is an integer multiple of a third such element, and hence the elements v_S span $\mathbb{L}_k$ linearly over $\mathbb{Q}$. Since there are 2^k such elements, they form a vector-space basis. The extension $\mathbb{L}_k/\mathbb{Q}$ is separable, being in characteristic 0, and it is normal as the splitting field of $\prod_{j=1}^{k}(X^2 - a_j)$. So it is a finite Galois extension. Any member σ of $\text{Gal}(\mathbb{L}_k/\mathbb{Q})$ must permute the roots of each $X^2 - a_j$ and hence must send $\sqrt{a_j}$ to $\pm\sqrt{a_j}$. On the other hand, σ is determined by its effect on each $\sqrt{a_j}$. Since $\text{Gal}(\mathbb{L}_k/\mathbb{Q})$ has order 2^k, there exists for each subset S of $\{1, \ldots, k\}$ one and only one σ such that $\sigma(\sqrt{a_j}) = -\sqrt{a_j}$ for $j \in S$ and $\sigma(\sqrt{a_j}) = +\sqrt{a_j}$ for $j \notin S$. The group $\text{Gal}(\mathbb{L}_k/\mathbb{Q})$ consists exactly of these elements.

In (c), let σ_j be the member of $\text{Gal}(\mathbb{L}_k/\mathbb{Q})$ with $\sigma_j(\sqrt{a_i}) = -\sqrt{a_i}$ for $i = j$ and $\sigma_j(\sqrt{a_i}) = +\sqrt{a_i}$ for $i \neq j$. Then $\sigma_j(v_S) = -v_S$ if j is in S, and $\sigma_j(v_S) = +v_S$ if j is not in S.

Arguing by contradiction, let $\sqrt{a_{k+1}} = \sum c_S v_S$ with each c_S in $\mathbb{Q}$. If $\sigma_j(\sqrt{a_{k+1}}) = \sqrt{a_{k+1}}$, then we have

$$\sum_{\text{all } S} c_S v_S = \sqrt{a_{k+1}} = \sigma_j(\sqrt{a_{k+1}}) = \sum_{\text{all } S} c_S \sigma_j(v_S) = -\sum_{S \text{ with } j \in S} c_S v_S + \sum_{S \text{ with } j \notin S} c_S v_S,$$

and it follows that $c_S = 0$ whenever j is in S. On the other hand, if $\sigma_j(\sqrt{a_{k+1}}) = -\sqrt{a_{k+1}}$, then we have

$$\sum_{\text{all } S} c_S v_S = \sqrt{a_{k+1}} = -\sigma_j(\sqrt{a_{k+1}}) = -\sum_{\text{all } S} c_S \sigma_j(v_S) = \sum_{S \text{ with } j \in S} c_S v_S - \sum_{S \text{ with } j \notin S} c_S v_S,$$

and it follows that $c_S = 0$ whenever j is not in S.

Define $S_0 = \{j \mid \sigma_j(\sqrt{a_{k+1}}) = -\sqrt{a_{k+1}}\}$. From the above it follows that $c_S = 0$ whenever some member of S_0 is not in S, and that $c_S = 0$ whenever some member of the complement of S_0 is in S. In other words, $c_S = 0$ except for c_{S_0}. We conclude that $\sqrt{a_{k+1}} = c_{S_0} v_{S_0} = c_{S_0} \sqrt{\prod_{j \in S_0} a_j}$ and hence that $a_{k+1} = c_{S_0}^2 \prod_{j \in S_0} a_j$. This contradicts the hypothesis that $\{a_1, \ldots, a_n\}$ are relatively prime and square free. Hence $\sqrt{a_{k+1}}$ does not lie in $\mathbb{L}_k$. This proves (c), and we obtain $[\mathbb{L}_{k+1} : \mathbb{L}_k] = 2$. By induction we see that $[\mathbb{L} : \mathbb{Q}] = 2^n$. This proves (d).

15. In (a) and (b), let ζ be a primitive p^{th} root of 1. Then $[\mathbb{Q}(\zeta) : \mathbb{Q}] = p - 1$ is relatively prime to $[\mathbb{Q}(r) : \mathbb{Q}] = p$. Problem 10a shows that $\Phi_p(X)$ is irreducible in $\mathbb{Q}(r)$. Since ζ and r generate $\mathbb{K}$, Problem 10b shows that $[\mathbb{K} : \mathbb{Q}] = [\mathbb{Q}(r) : \mathbb{Q}][\mathbb{Q}(\zeta) : \mathbb{Q}] = p(p-1)$.

In (c), the Galois correspondence between intermediate fields and subgroups of $G = \text{Gal}(\mathbb{K}/\mathbb{Q})$ associates $\mathbb{Q}(\zeta)$ to the subgroup $N = \text{Gal}(\mathbb{K}/\mathbb{Q}(\zeta))$, and it associates $\mathbb{Q}(r)$ to the subgroup $H = \text{Gal}(\mathbb{K}/\mathbb{Q}(r))$. Since $\mathbb{Q}(\zeta)/\mathbb{Q}$ is a normal extension, N is a normal subgroup of G. Any member of $H \cap N$ fixes r and ζ, hence fixes all of $\mathbb{K}$; thus $H \cap N = \{1\}$. The order of N is $[\mathbb{K} : \mathbb{Q}(\zeta)] = p$, and the order of H is $[\mathbb{K} : \mathbb{Q}(r)] = p - 1$. Therefore $|G| = |H||N|$, and G is a semidirect product with N normal.

Proposition 4.44 says that the action of an internal semidirect product is given by $\tau_h(n) = hnh^{-1}$. Let us identify τ_h. Let $h \in H = \text{Gal}(\mathbb{K}/\mathbb{Q}(r))$ have $h(r) = r$ and

$h(\zeta) = \zeta^k$, and let n in $N = \text{Gal}(\mathbb{K}/\mathbb{Q}(\zeta))$ have $n(\zeta) = \zeta$ and $n(r) = r\zeta^l$. Then $hnh^{-1}(r) = hn(r) = h(r\zeta^l) = r\zeta^{kl}$, and $hnh^{-1}(\zeta) = hn(\zeta^k) = h(\zeta^k) = \zeta$. So if n sends r to $r\zeta^l$ and $h(\zeta) = \zeta^k$, then hnh^{-1} is the member of N sending r to ζ^{kl}.

This n is the member of N corresponding to $l \in \mathbb{F}_p$, and this h is the member of H corresponding to $k \in \mathbb{F}_p^\times$. We have just shown that hnh^{-1} is the member of N corresponding to $kl \in \mathbb{F}_p$. Hence the action corresponds to multiplication of $\mathbb{F}_p^\times$ on additive $\mathbb{F}_p$.

16. $[\mathbb{K} : \Bbbk] = \text{Gal}(\mathbb{K}/\Bbbk)$, and $\text{Gal}(\mathbb{K}/\Bbbk)$ is a subgroup of $\mathfrak{S}_n$. Being a subgroup, its order divides the order of $\mathfrak{S}_n$, which is $n!$.

17. The most general element of $\mathbb{K}$ is of the form $x + yr$ with x and y in $\Bbbk$, and its square is $(x^2 + y^2 r^2) + 2xyr$. This is in $\Bbbk$ if and only if $xy = 0$, i.e., if and only if $x + yr$ is in $\Bbbk$ or in $r\Bbbk$.

18. The finite group G may be regarded as a subgroup of the symmetric group $\mathfrak{S}_n$ for $n = |G|$. It was shown in Example 3 of Section 17 that there exists a finite Galois extension $\mathbb{K}$ of $\mathbb{Q}$ with Galois group $\mathfrak{S}_n$. Let $\Bbbk$ be the fixed field of G within $\mathfrak{S}_n$. Then $\text{Gal}(\mathbb{K}/\Bbbk) = G$.

19. The polynomial in question in fixed by every element of the Galois group. Hence its coefficients are in the subfield of $\mathbb{K}$ fixed by all elements of $\text{Gal}(\mathbb{K}/\Bbbk)$. This is $\Bbbk$.

20. For (a), define $F(X) = \prod_{j=1}^n (X - x_j)$. For φ in H, we have $F^\varphi(X) = \prod_{j=1}^n (X - \varphi(x_j)) = \prod_{j=1}^n (X - x_j) = F(X)$. Thus $F(X)$ is in $\mathbb{K}^H[X]$. Let $M(X)$ be the minimal polynomial of x_1 over $\mathbb{K}^H$. Since $F(x_1) = 0$, $M(X)$ divides $F(X)$. On the other hand, the equalities $M^\varphi(X) = M(X)$ and $M(x_1) = 0$ imply that $M(x_j) = 0$ for each j. Thus $M(X)$ has degree at least n, and we conclude that $F(X) = M(X)$.

In (b), n is the number of elements in an orbit of H and hence divides $|H|$.

In (c), when the isotropy subgroup of H at x_1 is trivial, $n = |H|$. Therefore $[\mathbb{K}^H(x_1) : \mathbb{K}^H] = n = |H| = [\mathbb{K} : \mathbb{K}^H]$, the last equality following from Corollary 9.37. Since $\mathbb{K}^H(x_1) \subseteq \mathbb{K}$, it follows that $\mathbb{K}^H(x_1) = \mathbb{K}$.

21. For (a), let $\varphi(z) = \frac{az+b}{cz+d}$ with $ad - bc \neq 0$. Then we have a substitution homomorphism of $\mathbb{C}[X]$ into $\mathbb{C}(z)$ fixing $\mathbb{C}$ and sending X into z. Since the range is a field, this factors through the field of fractions of $\mathbb{C}[X]$ to give a field mapping $\mathbb{C}(X) \to \mathbb{C}(z)$. We can regard the result as a map of $\mathbb{C}(z)$ into itself, and we write the map of $\mathbb{C}(z)$ into itself as $\Phi_{\varphi^{-1}}$. The formula is $\Phi_{\varphi^{-1}}(r) = r \circ \varphi$ for $r = r(z)$ in $\mathbb{C}(z)$. Then $\Phi_{\psi\varphi}(r) = r \circ (\psi\varphi)^{-1} = (r \circ \varphi^{-1}) \circ \psi^{-1} = \Phi_\psi(r \circ \varphi^{-1}) = \Phi_\varphi(\Phi_\psi(r))$, and hence $\Phi_{\psi\varphi} = \Phi_\psi \circ \Phi_\varphi$. From this it follows that $\Phi_{\varphi^{-1}}$ is a two-sided inverse of Φ_φ. Hence Φ_φ is an automorphism.

For (b), $\Phi_\sigma(w(z)) = w(\sigma^{-1}(z)) = (-z)^2 + (-z)^{-2} = z^2 + z^{-2} = w(z)$, and $\Phi_\tau(w(z)) = w(\tau^{-1}(z)) = (1/z)^2 + (1/z)^{-2} = z^2 + z^{-2} = w(z)$. Since $\Phi_{\varphi\psi} = \Phi_\varphi \Phi_\psi$ by (a), it follows that every element of H fixes w. Since each Φ_φ is a field automorphism, $\mathbb{C}(w)$ lies in $\mathbb{K}^H$.

For (c), we know from (b) that $\mathbb{C}(w) \subseteq \mathbb{K}^H$. The orbit of z under H has 4 elements, and Problem 20a shows that the minimal polynomial of z over $\mathbb{K}^H$ has degree 4 and is equal to

$$F(X) = (X-z)(X+z)(X-z^{-1})(X+z^{-1}) = (X^2-z^2)(X^2-z^{-2}) = X^4 - w(z)X^2 + 1.$$

The polynomial $F(X)$ is irreducible over $\mathbb{K}^H$, and its formula shows that its coefficients are in the smaller field $\mathbb{C}(w)$. Hence it is irreducible over $\mathbb{C}(w)$ and is the minimal polynomial of z over $\mathbb{C}(w)$.

For (d), (c) shows that $[\mathbb{K}^H(z) : \mathbb{C}(w)] = 4$. Problem 20c shows that $\mathbb{K} = \mathbb{K}^H(z)$, and hence $[\mathbb{K} : \mathbb{C}(w)] = 4$. Since $[\mathbb{K} : \mathbb{C}(w)] = [\mathbb{K} : \mathbb{K}^H][\mathbb{K}^H : \mathbb{C}(w)]$ and since $[\mathbb{K} : \mathbb{K}^H] = 4$ by Corollary 9.37, $\mathbb{K}^H = \mathbb{C}(w)$.

22. For (a), let $\mathbb{L} = \mathbb{K}(\sqrt{u})$ and $\mathbb{K} = \mathbb{k}(\sqrt{v})$. The minimal polynomial of $\sqrt{u}$ over $\mathbb{K}$ is $X^2 - u$, and this must divide the minimal polynomial of $\sqrt{u}$ over $\mathbb{k}$. The degree of the latter polynomial equals $[\mathbb{k}(\sqrt{u}) : \mathbb{k}]$, which must divide 4. Hence it must be 2 or 4. If it is 2, then $X^2 - u$ lies in $\mathbb{k}[X]$, and u is in $\mathbb{k}$. We return to this case in a moment. Suppose that the minimal polynomial of $\sqrt{u}$ over $\mathbb{k}$ has degree 4. Let us write $u = r + s\sqrt{v}$ for some r and s in $\mathbb{k}$. Then $\sqrt{u}$ is a root of $(X^2 - r - s\sqrt{v})(X^2 - r + s\sqrt{v}) = (X^2 - r)^2 - s^2 v = X^4 - 2rX^2 + (r^2 - s^2 v)$, which is a quartic polynomial in $\mathbb{k}[X]$. Since the minimal polynomial over $\mathbb{k}$ has degree 4, this is the minimal polynomial and is irreducible. Thus (a) holds with $r = \sqrt{u}$.

The remaining case is that u is in $\mathbb{k}$ but $\sqrt{u}$ is not in $\mathbb{k}$. Consider $\pm\sqrt{u} \pm \sqrt{v}$. None of these is in $\mathbb{k}$. The computation

$$(X + \sqrt{u} + \sqrt{v}))(X + \sqrt{u} - \sqrt{v})(X - \sqrt{u} + \sqrt{v})(X - \sqrt{u} - \sqrt{v})$$
$$= ((X + \sqrt{u})^2 - v)((X - \sqrt{u})^2 - v)$$
$$= (X^2 + u - v + 2X\sqrt{u})(X^2 + u - v - 2X\sqrt{u})$$
$$= (X^2 + u - v)^2 - 4uX^2 = X^4 + 2uX^2 - 2vX^2 + (u-v)^2 - 4uX^2$$
$$= X^4 - 2(u+v)X^2 + (u-v)^2 = X^4 + bX^2 + c$$

shows that these are all roots of a quartic polynomial in $\mathbb{k}[X]$ of the correct kind, and the question concerns its irreducibility over $\mathbb{k}$. As in the previous paragraph, reducibility implies that it is the product of two irreducible quadratic members of $\mathbb{k}[X]$. Then the product of two of the first-order factors is in $\mathbb{k}[X]$, and the sum of those two roots must be in $\mathbb{k}$. The six possible sums of pairs of roots are $\pm\sqrt{u}$, $\pm\sqrt{v}$, and 0 twice. Since $\sqrt{u}$ and $\sqrt{v}$ are not in $\mathbb{k}$, the irreducible quadratic must be $X^2 - (\sqrt{u} + \sqrt{v})^2$ or $X^2 - (\sqrt{u} - \sqrt{v})^2$. However, the fact that $\sqrt{u}$ is not in $\mathbb{K} = \mathbb{k}(\sqrt{v})$ implies that neither of $\sqrt{u} \pm \sqrt{v}$ is in $\mathbb{k}$. Thus the quartic polynomial is indeed irreducible. This completes (a).

In (b), we have $4 = [\mathbb{L} : \mathbb{k}] = [\mathbb{L} : \mathbb{k}(r)][\mathbb{k}(r) : \mathbb{k}] = 4[\mathbb{L} : \mathbb{k}(r)]$. Thus $[\mathbb{L} : \mathbb{k}(r)] = 1$, and $\mathbb{L} = \mathbb{K}(r)$.

In (c), suppose that $c = t^2$ for the given $F(X)$. Find members u and v of $\Bbbk$ with $-2(u+v) = b$ and $u - v = t$. Then the displayed computation in (a) shows that $\pm\sqrt{u} \pm \sqrt{v}$ are the roots of $X^4 - 2(u+v)X^2 + (u-v)^2 = X^4 + bX^2 + c$. The given root r must be one of these. Say that $r = \sqrt{u} + \sqrt{v}$ without loss of generality. Since $[\Bbbk(r) : \Bbbk] = 4$ and $[\mathbb{L} : \Bbbk] = 4$ and $\Bbbk(r) \subseteq \mathbb{L}$, we have $\mathbb{L} = \Bbbk(r)$. On the other hand, $\Bbbk(r) \subseteq \Bbbk(\sqrt{u}, \sqrt{v})$, and $[\Bbbk(\sqrt{u}, \sqrt{v}) : \Bbbk] = [\Bbbk(\sqrt{u}, \sqrt{v}) : \Bbbk(\sqrt{u})][\Bbbk(\sqrt{u}) : \Bbbk] \leq 2 \cdot 2 = 4$. Hence $\Bbbk(\sqrt{u}, \sqrt{v}) = \Bbbk(r) = \mathbb{L}$. Then all four roots $\pm\sqrt{u} \pm \sqrt{v}$ of $F(X)$ lie in $\mathbb{L}$, $\mathbb{L}$ is the splitting field of $F(X)$ over $\Bbbk$, and $\mathbb{L}/\Bbbk$ is normal. The Galois group is generated by one element that sends $\sqrt{u}$ to $-\sqrt{u}$ and fixes $\sqrt{v}$, and by a second element that fixes $\sqrt{u}$ and sends $\sqrt{v}$ to $-\sqrt{v}$. Hence it is $C_2 \times C_2$.

Conversely suppose that $\mathbb{L}/\Bbbk$ is normal with Galois group $G = \text{Gal}(\mathbb{L}/\Bbbk) = C_2 \times C_2$. Let an irreducible polynomial $X^4 + bX^2 + c$ in $\Bbbk[X]$ with a root r in $\mathbb{L}$ be given. Since $\mathbb{L}/\Bbbk$ is normal, $X^4 + bX^2 + c$ splits in $\mathbb{L}$. Let the four roots be $\pm r$ and $\pm s$. The square u of any of these roots satisfies $u^2 + bu + c = 0$ and therefore lies in a quadratic extension within $\mathbb{K}$, the same quadratic extension for each root. Let us define $\mathbb{K}$ to be this extension. Then $\mathbb{K} = \Bbbk(\sqrt{b^2 - 4c})$. Because of the structure of G, there exists exactly one element σ in G whose fixed field is $\mathbb{K}$. The minimal polynomial of $\pm r$ over $\mathbb{K}$ is $X^2 + \frac{1}{2}b \pm \frac{1}{2}\sqrt{b^2 - 4c}$ for one of the two choices of sign, and the minimal polynomial of $\pm s$ over $\mathbb{K}$ is the one for the other choice of sign. The element σ must then permute the roots of each of these polynomials, and it follows that $\sigma(r) = \pm r$ and $\sigma(s) = \pm s$. Since neither r nor s is in $\mathbb{K}$, we must in fact have $\sigma(r) = -r$ and $\sigma(s) = -s$. Therefore $\sigma(rs) = rs$. One of the other two nontrivial members τ of G has $\tau(r) = s$. Since $\tau^2 = \tau$, we have $\tau(s) = r$. Thus $\tau(rs) = rs$, and we see that every member of G fixes rs. Consequently rs is in $\Bbbk$. Since rs is equal for some choice of signs to

$$\pm\sqrt{-\tfrac{1}{2}b + \tfrac{1}{2}\sqrt{b^2 - 4c}}\sqrt{-\tfrac{1}{2}b - \tfrac{1}{2}\sqrt{b^2 - 4c}} = \pm\sqrt{\tfrac{1}{4}b^2 - \tfrac{1}{4}(b^2 - 4c)} = \pm\sqrt{c},$$

$\sqrt{c}$ is in $\Bbbk$. In other words, c is the square of a member of $\Bbbk$, as asserted.

In (d), suppose that $c^{-1}(b^2 - 4c)$ for the given $F(X)$ is a square in $\Bbbk$. Arguing with r^2 as in (c), we see that $\mathbb{K} = \Bbbk(\sqrt{b^2 - 4c})$. Making the same computation as in the display just above, we see that $rs = \sqrt{c}$. Since $c^{-1}(b^2 - 4c)$ is a square in $\Bbbk$, $\sqrt{c}$ lies in $\mathbb{K}$. One of the roots, say r, lies in $\mathbb{L}$, and the product $rs = \sqrt{c}$ lies in $\mathbb{K}$, hence in $\mathbb{L}$. We conclude that $\pm r$ and $\pm s$ all lie in $\mathbb{L}$. In other words, $\mathbb{L}$ is the splitting field of $F(X)$ over $\Bbbk$ and is normal. Thus $\mathbb{L}/\Bbbk$ is normal. The Galois group must be either $C_2 \times C_2$ or C_4. If it is $C_2 \times C_2$, then (c) shows that $\sqrt{c}$ lies in $\Bbbk$. Under our assumption that $c^{-1}(b^2 - 4c)$ is a square, $\sqrt{b^2 - 4c}$ lies in $\Bbbk$. Consequently $F(X)$ is reducible, contradiction. We conclude that the Galois group is C_4.

Conversely suppose that $\mathbb{L}/\Bbbk$ is normal with Galois group $G = \text{Gal}(\mathbb{L}/\Bbbk) = C_4$. Let an irreducible polynomial $X^4 + bX^2 + c$ in $\Bbbk[X]$ with a root r in $\mathbb{L}$ be given. Arguing with r^2, we see that r^2 lies in $\mathbb{K} = \Bbbk(\sqrt{b^2 - 4c})$. Since $\mathbb{L}$ is generated by $\Bbbk$ and r, a generator of G cannot send r into $\pm r$. On the other hand, some element of G has to send r into $-r$ since $-r$ is a root of the given polynomial. Therefore $\sigma^2(r) = -r$. Then we have $\sigma(r\sigma(r)) = \sigma(r)\sigma^2(r) = -r\sigma(r)$, and we see that $\sigma^2(r\sigma(r)) = r\sigma(r)$. Consequently $r\sigma(r)$ lies in $\mathbb{K}$. Computing as in (c), we find that $\sqrt{c}$ lies in $\mathbb{K}$. This member of $\mathbb{K}$ has its square in $\mathbb{k}$, and Problem 17 shows that $\sqrt{c}$ lies in $\mathbb{k}$ or in the set of products $\mathbb{k}\sqrt{b^2 - 4c}$. By (c), $\sqrt{c}$ cannot lie in $\mathbb{k}$, and therefore $\sqrt{c} = d\sqrt{b^2 - 4c}$ for some d in $\mathbb{k}$. Hence $c^{-1}(b^2 - 4c) = d^{-2}$ for an element d of $\mathbb{k}$.

For (e), one can take $\mathbb{L} = \mathbb{K}(\sqrt[4]{2})$ and $\mathbb{K} = \mathbb{Q}(\sqrt{2})$. We can easily see directly that $\mathbb{L}$ is not normal. But let us use (c) and (d). The minimal polynomial $F(X)$ in question is $X^4 + 2$, with $b = 0$ and $c = 2$. The conditions in (c) and (d) say that $\mathbb{L}/\mathbb{k}$ is normal if and only if either 2 is a square in $\mathbb{Q}$ or -1 is a square in $\mathbb{Q}$. Neither condition is satisfied, and hence $\mathbb{L}/\mathbb{k}$ is not normal.

23. A cubic will be irreducible if it is divisible by no degree-one factor over $\mathbb{Q}$, hence if it has no root in $\mathbb{Q}$. Since these cubics are monic and are in $\mathbb{Z}[X]$, they will be irreducible if they have no integer root. An integer root must divide the constant term, and we check that neither of ± 1 is a root in either case. Hence both cubics are irreducible. By Problem 13 the Galois group in each case is a transitive subgroup of $\mathfrak{S}_3$, hence is $\mathfrak{S}_3$ or $\mathfrak{A}_3$. The discriminant $-4p^3 - 27q^2$ is 81 in the first case and -31 in the second case; this is a square in the first case but not in the second case. Thus $X^3 - 3X + 1$ has Galois group $\mathfrak{A}_3$, and $X^3 + X + 1$ has Galois group $\mathfrak{S}_3$.

24. The extension field is either $\mathbb{K}$ itself, in which case the Galois group remains $\mathfrak{S}_3$, or it is $\mathbb{L} = \mathbb{K}[\sqrt{-3}]$. Since $\mathbb{K}/\mathbb{Q}$ is normal, $\text{Gal}(\mathbb{L}/\mathbb{K})$ is a normal subgroup of $\text{Gal}(\mathbb{L}/\mathbb{Q})$ of order 2 with quotient isomorphic to $\text{Gal}(\mathbb{K}/\mathbb{Q}) = \mathfrak{S}_3$. The groups of order 12 are classified in Problems 45–48 at the end of Chapter IV. Two such groups are abelian, one is $\mathfrak{A}_4$, and one is $D_6 \cong C_2 \times \mathfrak{S}_3$.

Write a general element of $\mathbb{L}$ as $a + b\sqrt{-3}$. Define $\tau(a + b\sqrt{-3}) = a - b\sqrt{-3}$. This is the nontrivial member of the 2-element group $\text{Gal}(\mathbb{L}/\mathbb{K})$. If σ is in $\text{Gal}(\mathbb{K}/\mathbb{Q})$, then σ extends to a member $\overline{\sigma}$ of $\text{Gal}(\mathbb{L}/\mathbb{Q})$ by the definition $\overline{\sigma}(a + b\sqrt{-3}) = \sigma(a) + \sigma(b)\sqrt{-3}$. In fact, $\overline{\sigma}$ respects addition. To see that it respects multiplication, we compute

$$\begin{aligned}
\overline{\sigma}(a + b\sqrt{-3})\overline{\sigma}(c + d\sqrt{-3}) &= \big(\sigma(a) + \sigma(b)\sqrt{-3}\big)\big(\sigma(c) + \sigma(d)\sqrt{-3}\big) \\
&= \big(\sigma(a)\sigma(c) - 3\sigma(b)\sigma(d)\big) + \big(\sigma(b)\sigma(c) + \sigma(a)\sigma(d)\big)\sqrt{-3} \\
&= \sigma(ac - 3bd) + \sigma(bc + ad)\sqrt{-3} \\
&= \overline{\sigma}\big((ac - 3bd) + (bc + ad)\sqrt{-3}\big) \\
&= \overline{\sigma}\big((a + b\sqrt{-3})(c + d\sqrt{-3})\big).
\end{aligned}$$

It follows that $\text{Gal}(\mathbb{L}/\mathbb{Q})$ is the direct product $C_2 \times \mathfrak{S}_3$, the subgroup C_2 being $\text{Gal}(\mathbb{L}/\mathbb{K})$.

25. Yes. Let $\mathbb{L}$ be the intermediate field corresponding to the subgroup $\{(1), (1\ 2)\}$. Since the subgroup is not normal, $\mathbb{L}/\mathbb{k}$ is not normal. Let r be any element of $\mathbb{L}$ not

Chapter IX   681

in $\mathbb{k}$. Then the minimal polynomial of r over $\mathbb{k}$ has degree 3, and it does not split in $\mathbb{L}$ since $\mathbb{L}/\mathbb{k}$ is not normal. Its splitting field has to be something between $\mathbb{L}$ and $\mathbb{K}$, and the only choice is $\mathbb{K}$.

26. Yes, substitute and check it.

28. In (a), direct expansion of the right side gives $(X-r)(X^2+rX+(r^2+p)) = X^3 + pX - r^3 - pr$. Since $-r^3 - pr = q$, the assertion follows.

For (b), let us check that $r^2(-4p^3 - 27q^2) = (-3r^2 - 4p)(3q + 2pr)^2$, from which the assertion follows. In fact, the right side equals

$$-(3r^2 + 4p)(9q^2 + 12pqr + 4p^2r^2)$$
$$= -(36pq^2 + 48p^2qr + 27q^2r^2 + 16p^3r^2 + 36pqr^3 + 12p^2r^4)$$
$$= -r^2(4p^3 + 27q^2) - 12p^3r^2 - 36pq^2 - 48p^2qr - 36pqr^3 - 12p^2r^4$$
$$= -r^2(4p^3 + 27q^2) - 12p^3r^2 - 36pq^2 - 48p^2qr$$
$$\quad - 36pq(-pr - q) - 12p^2r(-pr - q)$$
$$= -r^2(4p^3 + 27q^2).$$

29. No. For example, $F(X)$ could have three real roots, and then $\mathbb{K}$ would be a subfield of $\mathbb{R}$. A concrete example is $X^3 - 12X + 1$, which is < 0 at -4, is > 0 at 0, is < 0 at 1, and is > 0 at 4; the Intermediate Value Theorem shows that $F(X)$ has three real roots.

30. The group in question is a subgroup of $\mathfrak{S}_5$. It is transitive because of the irreducibility, and it is a subgroup of $\mathfrak{A}_5$ since the discriminant is a square. Problem 13c shows that it contains a 5-cycle. The other cycle structures in $\mathfrak{A}_5$ are the 3-cycles and the pairs of 2-cycles. If a 3-cycle is present, then the group is all of $\mathfrak{A}_5$ because 15 divides its order, all groups of order 15 are cyclic, and $\mathfrak{A}_5$ contains no subgroup of order 30, being simple.

Suppose there are no 3-cycles. A Sylow 2-subgroup may be taken to be a subgroup of $H = \{(1), (1\ 2)(3\ 4), (1\ 3)(2\ 4), (1\ 4)(2\ 3)\}$, and it acts on the group of powers of a 5-cycle. The only nontrivial action of a 2-element group on a 5-element group carries elements to their inverses. Since no nontrivial element of H commutes with a 5-cycle (because $\mathfrak{S}_5$ has no elements of order 10), the Sylow 2-subgroup contains at most two elements. If it is trivial, then the group in question is of order 5, consisting of the powers of a 5-cycle. If the Sylow 2-subgroup has 2 elements, we obtain a semidirect product of a 2-element group with the powers of the 5-cycle, and the result has to be isomorphic to the dihedral group D_5.

Thus the only possibilities are C_5, D_5, and $\mathfrak{A}_5$.

31. Computation shows that the discriminant is $2^{12}7^219^2$, which is a square. By Proposition 9.63 the Galois group is a subgroup of $\mathfrak{A}_5$. Modulo 3, the given polynomial is $2 + 2x + x^5$ and is irreducible. By Theorem 9.64 the Galois group contains a 5-cycle. The given polynomial factors as $(7 + x)(7 + 10x + 7x^2 + x^3)$

modulo 11, and Theorem 9.64 shows that the Galois group contains a 3-cycle. The 5-cycle and 3-cycle generate all of $\mathfrak{A}_5$, and thus the Galois group is $\mathfrak{A}_5$.

32. Write e and f for e_1 and f_1. The proof of Theorem 9.64 showed that $f' = f$. Then $e'f' = |G_P| = |G|/g = efg/g = ef = ef'$, and $e' = e$.

33. If $\mathfrak{p}T = \prod_i P_i^{e(P_i|\mathfrak{p})}$ and $P_iU = \prod_j Q_{ij}^{e(Q_{ij}|P_i)}$, then $\mathfrak{p}U = \prod_i (P_i^{e(P_i|\mathfrak{p})}U) = \prod_i (P_iU)^{e(P_i|\mathfrak{p})} = \prod_i (\prod_j Q_{ij}^{e(Q_{ij}|P_i)})^{e(P_i|\mathfrak{p})}$. Hence $e(P_i|\mathfrak{p}) = e(Q_{ij}|P_i)e(P_i|\mathfrak{p})$. The formula for the f's follows from Corollary 9.7.

34. Corollary 9.58 shows that the norm and the trace are the product and sum of $a + b\sqrt{m}$ and $a - b\sqrt{m}$. Hence they are $a^2 - b^2m$ and $2a$. This proves (a).

In (b), the minimal polynomial of $r = a + b\sqrt{m}$ has degree 2 if $b \neq 0$, and this is the same as the degree of the field polynomial. Hence the two polynomials are equal, and the minimal polynomial is $X^2 - (\text{Tr } r)X + N(r)$. An algebraic integer is an algebraic element whose minimal polynomial over $\mathbb{Q}$ has integer coefficients, and (b) follows.

In (c), if $r = a + b\sqrt{m}$ is a unit with inverse s, then $N(r)N(s) = N(rs) = N(1) = 1$ shows that $N(r)$ is a unit with inverse $N(s)$. Conversely if r is in T with $N(r) = \pm 1$, then $r(a - b\sqrt{m}) = \pm 1$, and $\pm(a - b\sqrt{m})$ is an inverse element in T.

For (d), $\sqrt{2} - 1$ is a unit in the algebraic integers of $\mathbb{Q}[\sqrt{2}]$. Its inverse is $\sqrt{2} + 1$.

35. With respect to the ordered basis $(1, \sqrt[3]{2}, (\sqrt[3]{2})^2)$, the matrix of multiplication by $a + b\sqrt[3]{2} + c(\sqrt[3]{2})^2$ is

$$\begin{pmatrix} a & 2c & 2b \\ b & a & 2c \\ c & b & a \end{pmatrix}.$$

The trace and norm are the trace and determinant of this matrix, namely $3a$ and $a^3 + 2b^3 + 4c^3 - 6abc$.

36. In (a), if ξ is any number algebraic over $\mathbb{Q}$ of degree r, then the norm relative to $\mathbb{Q}(r)/\mathbb{Q}$ of ξ is $(-1)^r M(0)$, where $M(X)$ is the minimal polynomial of ξ over $\mathbb{Q}$. Since $M(1 - (1 - \xi)) = 0$, the minimal polynomial of $1 - \xi$ is the polynomial $M(1 - X)$ adjusted so as to be monic. That is, it is $P(X) = (-1)^r M(1 - X)$. Hence the norm of $1 - \xi$ is $(-1)^r P(0) = (-1)^{2r} M(1) = M(1)$. In the case of the given ζ, the minimal polynomial of ζ is $\Phi_n(X)$, and therefore the norm of $1 - \zeta$ is $\Phi_n(1)$.

For (b), division of both sides of the identity $\prod_{d|n} \Phi_d(X) = X^n - 1$ by $X - 1$ gives $\prod_{d|n,\, d>1} \Phi_d(X) = X^{n-1} + X^{n-2} + \cdots + 1$. Therefore $\prod_{d|n,\, d>1} \Phi_d(1) = n$.

If n is a prime power, say with $n = p^k$, let us see by induction on k that $\Phi_n(1) = p$. The base case of the induction is $k = 1$, and the result of the previous paragraph applies. Assuming that $\Phi_n(1) = p$ for $n = p^k$, we have $p^{k+1} = \prod_{l=1}^{k+1} \Phi_{p^l}(1) = \Phi_{p^{k+1}}(1) \prod_{l=1}^{k} p$. Therefore $\Phi_{p^{k+1}}(1) = p$, and the induction is complete.

Inducting on n, let us now show that $\Phi_n(1) = 1$ if n is divisible by more than one positive prime. The base case of the induction is $n = 2$. Assume that $n = p_1^{k_1} \cdots p_r^{k_r}$

and that the result is known for integers less than n. We may assume that n is divisible by at least two positive primes. Then

$$n = \prod_{d|n,\, d>1} \Phi_d(1) = \prod_{s=1}^{r}\left(\prod_{l=1}^{k_s} \Phi_{p_s^l}(1)\right) \prod_{\text{other } d} \Phi_d(1),$$

where the "other d" are the divisors of n that are divisible by at least two primes. These include n itself. So one of the corresponding factors is $\Phi_n(1)$, and the others are 1 by the inductive hypothesis. The factor in parentheses is $p_l^{k_l}$ by the result of the previous paragraph, and the product of the factors in parentheses is n. Therefore $\Phi_n(1) = 1$, and the induction is complete.

37. For (a), the imaginary part of $p^{-1}x \pm p^{-1}\sqrt{-1}$ is not an integer, and therefore $p^{-1}(x \pm \sqrt{-1})$ is not a Gaussian integer. Consequently p does not divide either of $x \pm \sqrt{-1}$. Since p divides $x^2 + 1$ in $\mathbb{Z}$ and hence in $\mathbb{Z}[\sqrt{-1}]$, p is not prime in $\mathbb{Z}[\sqrt{-1}]$.

For (b), it follows since p is not prime that $p = \alpha\beta$ nontrivially in $\mathbb{Z}[\sqrt{-1}]$. Then $p^2 = N(p) = N(\alpha)N(\beta)$. Problem 34c shows that nontrivial factorization implies that $N(\alpha)$ and $N(\beta)$ are not units. Thus they are both p. If $\alpha = a + b\sqrt{-1}$, then the equation $p = N(\alpha)$ says that $p = a^2 + b^2$.

38. Let N be the norm function in $\mathbb{Q}(\sqrt{-2})$. Since p divides $x^2 + 2 = (x + \sqrt{-2})(x - \sqrt{-2})$ and since neither of $p^{-1}(x \pm \sqrt{-2})$ is of the form $a + b\sqrt{-2}$ with a and b in $\mathbb{Z}$, p is not prime in $\mathbb{Z}[\sqrt{-2}]$. Write $p = \alpha\beta$ nontrivially. Then $p^2 = N(p) = N(\alpha)N(\beta)$ and $N(\alpha) = N(\beta) = p$. If $\alpha = a + b\sqrt{-2}$, then $p = N(\alpha)$ says that $p = a^2 + 2b^2$.

39. This is similar to Problem 38 except that the members of the ring are of the form $a + b\sqrt{-3}$ with a, b in $\mathbb{Z}$ or a, b in $\mathbb{Z} + \frac{1}{2}$. Thus $p = N(\alpha)$ says that $p = a^2 + 3b^2$ either with a, b in $\mathbb{Z}$ or with a, b in $\mathbb{Z} + \frac{1}{2}$. In the latter case, let $\omega^{\pm 1} = \frac{1}{2}(-1 - \sqrt{-3})$. These have $N(\omega^{\pm 1}) = 1$. Therefore

$$p = N(\alpha) = N(\alpha\omega^{\pm 1}) = N\big((a + b\sqrt{-3})(-\tfrac{1}{2} \pm \tfrac{1}{2}\sqrt{-3})\big)$$
$$= N\big(\tfrac{1}{2}(-a \mp 3b) + \tfrac{1}{2}(\pm a - b)\sqrt{-3}\big)$$
$$= \big(\tfrac{1}{2}(-a \mp 3b)\big)^2 + 3\big(\tfrac{1}{2}(\pm a - b)\sqrt{-3}\big)^2.$$

Since a, b are in $\mathbb{Z} + \frac{1}{2}$, one of $a + b$ and $a - b$ is even, and the other is odd, the sum $2a$ being odd. If $a + b$ is even, then $a - 3b$ is even since their difference $4b$ is even, and vice versa. Hence one of the two choices of sign exhibits p as $c^2 + 3d^2$ with c, d in $\mathbb{Z}$.

40. Write $\mathbb{L}' = \mathbb{k}(x)$ by the Theorem of the Primitive Element, and let $\mathbb{K}$ be a splitting field of the minimal polynomial of x over $\mathbb{k}$. Then $\mathbb{K}$ is a finite Galois extension of $\mathbb{k}$ by Corollary 9.30, and we have $\mathbb{k} \subseteq \mathbb{L} \subseteq \mathbb{L}' \subseteq \mathbb{K}$. For a in $\mathbb{L}'$ and b in $\mathbb{L}$, Corollary 9.58 says that $N_{\mathbb{L}'/\mathbb{k}}(a) = \prod_{\sigma \in G/H'} \sigma(a)$, $N_{\mathbb{L}/\mathbb{k}}(b) = \prod_{\sigma \in G/H'} \sigma(b)$,

and $N_{\mathbb{L}'/\mathbb{L}}(a) = \prod_{\tau \in H/H'} \tau(a)$. Hence $N_{\mathbb{L}/\Bbbk}(N_{\mathbb{L}'/\mathbb{L}}(a)) = \prod_{\sigma \in G/H} \sigma(N_{\mathbb{L}'/\mathbb{L}}(a)) = \prod_{\sigma \in G/H} \sigma\left(\prod_{\tau \in H/H'} \tau(a)\right) = \prod_{\sigma \in G/H} \prod_{\tau \in H/H'} \sigma\tau(a) = \prod_{\sigma \in G/H'} \sigma(a) = N_{\mathbb{L}'/\Bbbk}(a)$.
The formula for traces follows similarly by replacing the products by sums in the above computation.

41. Since P is symmetric, $P(X_{\sigma(1)}, \ldots, X_{\sigma(n)}) = P(X_1, \ldots, X_n)$ for every permutation σ. Therefore $P(r_{\sigma(1)}, \ldots, r_{\sigma(n)}) = P(r_1, \ldots, r_n)$ for every σ. Problem 39e at the end of Chapter VIII implies that $P(r_1, \ldots, r_n) = Q(s_1, \ldots, s_n)$ for a polynomial $Q(X_1, \ldots, X_n)$ in $\Bbbk[X_1, \ldots, X_n]$, where $s_1, \ldots, s_n$ are the elementary symmetric polynomials in $r_1, \ldots, r_n$. The elements $s_1, \ldots, s_n$ are the coefficients of $F(X)$, up to sign, and hence are in $\Bbbk$. Therefore $P(r_1, \ldots, r_n) = Q(s_1, \ldots, s_n)$ is in $\Bbbk$.

42. Inspection of the formula gives $H_1(X) = \prod_{i=1}^m G(X - r_i)$. For each i, we can expand $G(X - r_i)$ in powers of X as

$$G(X - r_i) = X^n + b_{n-1}(r_i)X^{n-1} + \cdots + b_1(r_i)X + b_0(r_i),$$

and each of $b_{n-1}, \ldots, b_0$ is a member of $\Bbbk[X]$. When we multiply these for $1 \leq i \leq m$, each power of X in the product has a coefficient that is unchanged if we permute $r_1, \ldots, r_n$. Problem 41 says that the coefficient of each power of X is therefore in $\Bbbk$. Thus $H_1(X)$ is in $\Bbbk[X]$. A similar argument shows that $H_2(X)$ is in $\Bbbk[X]$.

43. For (a), we use $F(X) = X^2 - 2$ and $G(X) = X^2 - 3$ in the previous problem. Then $\sqrt{2} + \sqrt{3}$ is a root of

$$(X - (\sqrt{2} + \sqrt{3}))(X - (\sqrt{2} - \sqrt{3}))(X - (-\sqrt{2} + \sqrt{3}))(X - (-\sqrt{2} - \sqrt{3})),$$

which must have coefficients in $\mathbb{Q}$.

44. Proposition 4.40 extends the action by an element σ in $\mathfrak{S}_n$ uniquely from the set $\{r_1, \ldots, r_n\}$ to $\Bbbk[r_1, \ldots, r_n]$ fixing $\Bbbk$. The extended σ is a one-one homomorphism of $\Bbbk[r_1, \ldots, r_n]$ into itself, hence into $\Bbbk(r_1, \ldots, r_n)$. It extends uniquely to a field mapping of $\Bbbk(r_1, \ldots, r_n)$ into itself by Proposition 8.6. The homomorphism corresponding to a composition is the composition of the homomorphisms, and consequently the homomorphism corresponding to σ^{-1} is a two-sided inverse of the homomorphism corresponding to σ. Thus the extension of σ is an automorphism, as required.

Conclusion (a) is immediate from Problem 20a. For (b), since $\mathbb{K}$ is generated by $\Bbbk$ and $r_1, \ldots, r_n$, $\mathbb{K}$ is certainly generated by $\mathbb{K}^{\mathfrak{S}_n}$ and $r_1, \ldots, r_n$. We have arranged that $F(X)$ splits over $\mathbb{K}$, and hence $\mathbb{K}$ is the splitting field. Conclusion (d) follows from Corollary 9.37 once (c) is proved. Thus we are to prove (c).

The argument for (c) is similar to that in Problem 21. Since $F(X)$ is in $\mathbb{K}^{\mathfrak{S}_n}$, its coefficients are in $\mathbb{K}^{\mathfrak{S}_n}$. Thus $\Bbbk(u_1, \ldots, u_n) \subseteq \mathbb{K}^{\mathfrak{S}_n}$. Consequently Corollary 9.37 gives $n! = [\mathbb{K} : \mathbb{K}^{\mathfrak{S}_n}] \leq [\mathbb{K} : \Bbbk(u_1, \ldots, u_n)]$. Problem 16 shows that the right side divides $n!$. Therefore equality holds throughout, and we see that $[\mathbb{K} : \mathbb{K}^{\mathfrak{S}_n}] = [\mathbb{K} : \Bbbk(u_1, \ldots, u_n)]$. Since $\Bbbk(u_1, \ldots, u_n) \subseteq \mathbb{K}^{\mathfrak{S}_n}$, we must have $\Bbbk(u_1, \ldots, u_n) = \mathbb{K}^{\mathfrak{S}_n}$.

45. For (a), we have

$$c_1 = \sum_i \theta_i = 2 \sum_{i<j} s_i s_j = 2p,$$

$$c_2 = \sum_{i<j} \theta_i \theta_j = \sum_{i<j} s_i^2 s_j^2 + 3 \sum_{i<j<k} s_i s_j s_k (s_i + s_j + s_k) + 6 s_1 s_2 s_3 s_4,$$

$$c_3 = \theta_1 \theta_2 \theta_3 = \sum_{\substack{i,j,k \\ \text{unequal}}} s_i^3 s_j^2 s_k + 2 s_1 s_2 s_3 s_4 \left(\sum_i s_i^2 \right)$$

$$+ 2 \sum_{i<j<k} s_i^2 s_j^2 s_k^2 + 4 s_1 s_2 s_3 s_4 \left(\sum_{i<j} s_i s_j \right).$$

Part (b) is a calculation with symmetric polynomials and is omitted. For (c), we have

$$\theta_1 - \theta_2 = -(s_1 - s_4)(s_2 - s_3),$$
$$\theta_1 - \theta_3 = -(s_1 - s_3)(s_2 - s_4),$$
$$\theta_2 - \theta_3 = -(s_1 - s_2)(s_3 - s_4).$$

The square of the product of the left sides is the discriminant of the cubic resolvent, and the square of the product of the right sides is the discriminant of the given quartic.

46. In (a), the subgroups in question are

$$H = \{(1), (1\ 2)(3\ 4), (1\ 3)(2\ 4), (1\ 4)(2\ 3)\}$$

and $\mathfrak{A}_4$. In (b), one considers the possibilities for a Sylow 2-subgroup and is led to conclude that the only possibilities for the subgroups in question are the powers of a 4-cycle, the dihedral group (generated by H and $(1\ 2\ 3\ 4)$), and $\mathfrak{S}_4$. (The group H and any 2-cycle generate $\mathfrak{S}_4$, and thus the dihedral group cannot be generated by H and a 2-cycle.)

47. In (a), the discriminant reduces when $q = 0$ to $16p^4 r - 128 p^2 r^2 + 256 r^3 = 16 r (p^4 - 8 p^2 r + 16 r^2) = 16 r (p^2 - 4r)^2$. This is 0 if $r = 0$ or $r = p^2/4$. If it is nonzero, it is a square if and only if r is a square. Hence in all cases it is a square if and only if r is a square.

In (b), let $Y = X^2$. The equation is $Y^2 + pY + r = 0$, which can be solved with a square root. For each of the two solutions, we can then solve for X with a square root. Hence all the roots lie in an extension obtained by adjoining at most three square roots. Thus $[\mathbb{K} : \mathbb{Q}]$ divides 8, and $|G|$ divides 8. Consequently G cannot have any element of order 3.

In (c), the irreducibility shows that the possibilities for G are as in Problem 46. Since r is a square, the discriminant is a square, by (a). Proposition 9.63 shows that the possibilities are as in Problem 46a. Part (b) rules out $\mathfrak{A}_4$, and then (c) follows.

In (d), r nonsquare and $F(X)$ irreducible implies that G is a transitive subgroup of $\mathfrak{S}_4$ but not a subgroup of $\mathfrak{A}_4$, by (a). Problem 46b shows that G is $\mathfrak{S}_4$, or the powers of a 4-cycle, or the dihedral group D_4. By (b), there is no element of order 3, and $\mathfrak{S}_4$ is therefore ruled out.

48. The polynomial remains irreducible when reduced modulo 2, and a prime factorization modulo 3 is $(X+2)(X^3+X^2+X+2)$. Thus G is a transitive subgroup of $\mathfrak{S}_4$ containing a 3-cycle. The discriminant is 257, not square. By Problem 46b, $G = \mathfrak{S}_4$.

49. Part (a) is just a computation; the answer is $2^{12}3^4$. The factorization in (b) is routine to check, and the only issue is the irreducibility of the cubic factor. For a cubic polynomial, irreducibility follows if the polynomial has no root in the field. Thus we need only verify that none of 0, 1, 2, 3, 4 is a root modulo 5.

For (c), the conclusion of (b) shows that the only possible reducibility over $\mathbb{Q}$ is into a degree-one factor and a cubic factor. For $X^4 + 8X + 12$ to have a degree-one factor, it must have a rational root, and this root must be an integer dividing 12. Let r be an integer dividing 12. If r is even, then $r^4 + 8r$ is divisible by 16, but 12 is not; so an even r cannot be a root. We are left with ± 1 and ± 3 as the possibilities, and we check that none of these is a root.

In (d), $F(X)$ is irreducible, and G is transitive. It is a subgroup of $\mathfrak{A}_4$ since the discriminant is a square. By (b) and Theorem 9.64, G contains a 3-cycle. Problem 46a shows therefore that $G = \mathfrak{A}_4$.

50. We saw in Problem 49 that $G = \mathfrak{A}_4$ for $X^4 + 8X + 12$, in Problem 47c that $G = \{(1), (1\ 2)(3\ 4), (1\ 3)(2\ 4), (1\ 4)(2\ 3)\}$ for $X^4 + 1$, in Problem 48 that $G = \mathfrak{S}_4$ for $X^4 + X + 1$, and via Eisenstein's criterion that $G = C_4$ for $\Phi_5(X)$. Since $X^4 - 2$ does not split in $\mathbb{Q}(\sqrt[4]{2})$, the Galois group in this case cannot be of order 4, and Problem 47d shows that G must be D_4 in this case.

51. For (a), let C correspond to a set of polynomials I of degree at most $n-1$. If C is cyclic, then I is at least a vector space over $\mathbb{F}$. If $F(X) = c_0 + c_1 X + \cdots + c_{n-1}X^{n-1}$ is in I, then $XF(X) = c_0 X + c_1 X^2 + \cdots + c_{n-1}X^n$ is congruent modulo $(X^n - 1)$ to $c_{n-1} + c_0 X + \cdots + c_{n-2}X^{n-1}$, which is in I since C is cyclic. Hence I is closed under multiplication by X mod $(X^n - 1)$ and hence under arbitrary multiplications modulo $(X^n - 1)$. Therefore I is an ideal in $\mathbb{F}[X]/(X^n - 1)$.

Conversely if I is an ideal in $\mathbb{F}[X]/(X^n - 1)$, then it is a vector space and is closed under multiplication by X mod $(X^n - 1)$ in $\mathbb{F}[X]/(X^n - 1)$. If $F(X) = c_0 + c_1 X + \cdots + c_{n-1}X^{n-1}$ is in I, then $XF(X) = c_0 X + c_1 X^2 + \cdots + c_{n-1}X^n$ mod I has to be in I, and the corresponding member of C is $(c_{n-1}, c_0, c_1, \ldots, c_{n-2})$. Hence C is cyclic.

For the remaining parts, we identify the cyclic code C with the corresponding ideal I in $\mathbb{F}[X]/(X^n - 1)$. In (b), let the lowest degree of a member of I be $n-k$, and let $G(X)$ be a member of I of this degree. If there is a second member of this same degree, then their difference has lower degree since both polynomials are monic, and the difference must be in I, contradiction. Thus $G(X)$ is uniquely defined. Regard $G(X)$ as a member of $\mathbb{F}[X]$ of degree $n-k$, and let $M(X) = \text{GCD}(G(X), X^n - 1)$. Then we can choose $A(X)$ and $B(X)$ in $\mathbb{F}[X]$ with $A(X)(X^n - 1) + B(X)G(X) = M(X)$. Passing to $\mathbb{F}[X]/(X^n - 1)$, we have $B(X)G(X) \equiv M(X)$ mod $(X^n - 1)$. Therefore $M(X)$ is in the ideal I. Since the degree of $M(X)$ is at most deg $G(X)$ and since

$G(X)$ has the minimum degree among the nonzero members of I, either $M(X) = 0$ or $M(X) = G(X)$. The conclusion $M(X) = 0$ is ruled out since $M(X)$ is a greatest common divisor of nonzero polynomials, and thus $M(X) = G(X)$. Therefore $G(X)$ divides $X^n - 1$.

Let $\widetilde{I}$ be the inverse image of I in $\mathbb{F}[X]$. This is an ideal, it contains $G(X)$, and it contains no nonzero element of degree $< \deg G(X)$. Since $\widetilde{I}$ has to be principal, $\widetilde{I} = (G(X))$. In other words, $\widetilde{I}$ consists of all products of $G(X)$ by a member of $\mathbb{F}[X]$. If $F(X)G(X)$ is such a product, then the division algorithm gives $F(X)G(X) = B(X)(X^n - 1) + R(X)$ with $R(X) = 0$ or $\deg R < n$. Since $G(X)$ divides $X^n - 1$, $G(X)$ divides $R(X)$. Therefore every member of $\widetilde{I}$ is congruent modulo $X^n - 1$ to a product $G(X)S(X)$ that is 0 or has degree $< n$. Then (c) is clear.

For (d), (b) showed that $G(X)$ divides $X^n - 1$ in $\mathbb{F}[X]$. Write $X^n - 1 = G(X)H(X)$. If $B(X)$ in $\mathbb{F}[X]/(X^n - 1)$ corresponds to a member of C, then (b) shows that $B(X) = F(X)G(X)$ for some $F(X)$ in $\mathbb{F}[X]$. Multiplying by $H(X)$ gives $B(X)H(X) = F(X)G(X)H(X) = F(X)(X^n - 1)$. Hence $B(X)H(X) \equiv 0 \mod (X^n - 1)$. Conversely if $B(X)H(X) = A(X)(X^n - 1)$, then $B(X)H(X) = A(X)G(X)H(X)$, and $B(X) = A(X)G(X)$.

52. In (a), if r_1, r_2, r_3 denote the rows and if $v_1 = r_1 + r_3$, $v_2 = r_2$, and $v_3 = r_3$, then v_1, v_2, v_3 form a basis for the row space, and they cycle into one another when the columns are shifted in cyclic fashion. Consequently the code is cyclic. Part (b) involves looking at the 7 nonzero members of the space, and one can just do that directly.

In (c), one such matrix is

$$\mathcal{H} = \begin{pmatrix} 0 & 0 & 0 & 1 & 1 & 0 & 1 \\ 0 & 0 & 1 & 1 & 0 & 1 & 0 \\ 0 & 1 & 1 & 0 & 1 & 0 & 0 \\ 1 & 1 & 0 & 1 & 0 & 0 & 0 \end{pmatrix}.$$

A little check shows that the matrix product $\mathcal{H}\mathcal{G}^t$ is the 4-by-3 zero matrix, and hence $\mathcal{H}v = 0$ for each v in C. Thus C is contained in the null space of $\mathcal{H}$. The rank of $\mathcal{H}$ is 4 since the rows are certainly linearly independent. Since the sum of the rank and the dimension of the null space is the number of columns, namely 7, the dimension of the null space is 3. Therefore the null space is C and is no larger.

For (d), the general matrix $\mathcal{H}$ is to have n columns and $n - k$ rows. The entries of the top row are the coefficients of $H(X)$ with the constant term at the right, the coefficient of X in the next-to-last position, and so on. In each successive row these coefficients are shifted one position to the left.

Let $G(X) = g_0 + g_1 X + \cdots + g_{n-k} X^{n-k}$ and $H(X) = h_0 + h_1 X + \cdots + X^k$. We know that $\{0, XG(X), X^2 G(X), \ldots, X^{k-1} G(X)\}$ is a basis of C. In terms of members of $\mathbb{F}^n$ the lth such vector has the entries $g_0, g_1, \ldots, g_{n-k}$ beginning in the lth position. The $(1, j)^{\text{th}}$ entry of $\mathcal{H}$ is h_{n-j} with 0's elsewhere in the row, and the $(i, j)^{\text{th}}$ entry is $h_{n-j-i+1}$ with 0's elsewhere in the row. The product of the ith row of $\mathcal{H}$ and the lth basis vector of C is $\sum_{j=n-k-i+1}^{n-i+1} h_{n-j-i+1} g_{j-l}$, which is the coefficient of $X^{n-i+1-l}$ in $G(X)H(X)$. Here $1 \leq i \leq n-k$ and $1 \leq l \leq k$, so that $2 \leq i+l \leq n$. Thus the power of X in question varies from 1 to $n-1$. Since $G(X)H(X) = X^n - 1$, the coefficient is 0. Thus C lies in the null space of $\mathcal{H}$. The same argument with rank as in the previous paragraph shows that C is exactly the null space.

53. Since $X^n - 1$ has derivative nX^{n-1}, we have $\text{GCD}(X^n - 1, nX^{n-1}) = 1$ when n is odd. Lemma 9.26 then shows that $X^n - 1$ is separable. If n is even, write $n = 2k$. Then $X^n - 1 = (X^k - 1)^2$ in characteristic 2 by Lemma 9.18, and hence every root has multiplicity at least 2.

54. In (a), we have $0 = P(\alpha^j) = c_0 + c_1\alpha^j + c_2\alpha^{2j} + \cdots + c_{n-1}\alpha^{(n-1)j}$ for $r \leq j \leq r+s$, and therefore the column vector $(c_0, c_1, \ldots, c_{n-1})$ satisfies

$$\begin{pmatrix} \alpha^r & \alpha^{2r} & \cdots & \alpha^{(n-1)r} \\ \alpha^{r+1} & \alpha^{2(r+1)} & \cdots & \alpha^{(n-1)(r+1)} \\ & \vdots & & \\ \alpha^{r+s} & \alpha^{2(r+s)} & \cdots & \alpha^{(n-1)(r+s)} \end{pmatrix} \begin{pmatrix} c_0 \\ c_1 \\ c_2 \\ \vdots \\ c_{n-1} \end{pmatrix} = \begin{pmatrix} 0 \\ 0 \\ 0 \\ \vdots \\ 0 \end{pmatrix}.$$

In (b), since $s + 1 \leq n$, the number $s+1$ of rows is $\leq$ the number n of columns. Any square submatrix of size $s+1$ is a Vandermonde matrix after factoring a power of α from each column and transposing, and the determinant of the square submatrix is therefore the product of a power of α and the differences $\alpha^{r+j} - \alpha^{r+i}$ with $j > i$. Since α is nonzero and since two powers of α can be equal only when the exponents differ by a multiple of n, the determinant of the square submatrix is nonzero.

In (c), suppose that $s+1$ or fewer of the coefficients $c_0, c_1, \ldots, c_{n-1}$ are nonzero. Choose $s+1$ of them, say c_{i_j} for $1 \leq j \leq s+1$, such that the remaining ones are 0. If we discard the others from the matrix equation in (a) and discard the corresponding columns of the coefficient matrix, then the matrix equation is still valid since we have discarded only 0's from the given equations. The resulting system is square with an invertible coefficient matrix, and hence the unique solution has $c_{i_j} = 0$ for all j. But then $P(X) = 0$, in contradiction to the assumption that $F(X) \neq 0$.

In (d), if some nonzero member $P(X)$ of C has weight less than $s+2$, then (c) leads to a contradiction. Hence every nonzero weight is $\geq s+2$, and $\delta(C) \geq s+2$.

55. Since α is a root of $X^n - 1$, so is every α^j. Since F_j is the minimal polynomial of α^j, F_j divides $X^n - 1$. Also, $1 + X = X - 1$ divides X^{n-1}, and no F_j equals $X - 1$, since $\alpha^j \neq 1$ for $1 \leq j \leq 2e$ when $2e < n$. Therefore $G(X)$ divides $X^n - 1$. Applying Problem 54 with $r = 0$ and $s = 2e$, we see that the code C generated by $G(X)$ has $\delta(C) \geq s + 2 = 2e + 2$.

56. In (a), if an irreducible polynomial $F(X)$ of degree d has a root β in $\mathbb{K}$, then $\mathbb{K} \supseteq \mathbb{F}(\beta) \supseteq \mathbb{F}$, and $[\mathbb{F}(\beta) : \mathbb{F}] = d$ must divide $[\mathbb{K} : \mathbb{F}] = m$. In the previous problem it follows that each $F_j(X)$ has degree dividing m, hence degree $\leq m$. The worst case for the degree of $G(X)$ is that the LCM equals the product, and then the degree of $G(X)$ is the sum of 1 (from $1 + X$) and the sum of the degrees of the $F_j(X)$'s. Hence $\deg G \leq 2em + 1$ in all cases.

In (b), let $n_r = 2^r - 1$, and let $\mathbb{K}$ be a field with 2^r elements. Theorem 9.14 shows that $\mathbb{K}$ is a splitting field for $X^{2^r} - X$ over $\mathbb{F}$. Hence it is a splitting field for $X^{n_r} - 1$ over $\mathbb{F}$. Let $e = r$, so that $e < n_r/2$ as soon as $r \geq 3$. Using this e in the previous problem, we obtain a cyclic code C_r in $\mathbb{F}^{n_r}$ with $\delta(C_r) \geq 2r + 2$. According to (a), the generating polynomial $G_r(X)$ has degree at most $2er + 1 = 2r^2 + 1$. Therefore $k_r = \dim C_r = n_r - \deg G_r \geq n_r - 2r^2 - 1 = 2^r - 2r^2 - 2$. Then k_r/n_r tends to 1, and $\delta(C_r)$ tends to infinity, as required.

57. In (a), the polynomial $F_1(X)$ splits over $\mathbb{K}$ because every finite extension of a finite field is Galois. The Galois group $\text{Gal}(\mathbb{K}/\mathbb{F})$ consists of the powers of the Frobenius isomorphism $x \mapsto x^2$, by Proposition 9.40, and is transitive on the roots of $F_1(X)$, by Problem 13a. Hence all the roots are of the form α^{2^k}, and all these elements are roots. Taking $k = 0, 1, 2, 3$, we get distinct roots, which is necessary since $\mathbb{K}/\mathbb{F}$ is separable.

For (b), we start from $1 + \alpha + \alpha^4 = 0$ and compute the powers of α in terms of $1, \alpha, \alpha^2, \alpha^3$. The interest is in only the powers $\alpha^0, \alpha^3, \alpha^6, \alpha^9, \alpha^{12}$, but some of the intermediate powers help in the computation. We have

$$\alpha^3 = \alpha^3,$$
$$\alpha^4 = 1 + \alpha,$$
$$\alpha^5 = \alpha + \alpha^2,$$
$$\alpha^6 = \alpha^2 + \alpha^3,$$
$$\alpha^9 = \alpha^3 \alpha^6 = \alpha^3(\alpha^2 + \alpha^3) = \alpha^5 + \alpha^6 = \alpha + \alpha^3,$$
$$\alpha^{12} = (\alpha^2 + \alpha^3)^2 = \alpha^4 + \alpha^6 = 1 + \alpha + \alpha^2 + \alpha^3.$$

Then we form the equation $a + b\alpha^3 + c\alpha^6 + d\alpha^9 + \alpha^{12} = 0$, substitute from above, and equate coefficients. The result is a homogeneous system of four linear equations with five unknowns in $\mathbb{F}$. Solving, we find that the space of solutions is 1-dimensional with $a = b = c = d = e$. Therefore the minimal polynomial of α^3 has degree 4 and is $1 + \alpha + \alpha^2 + \alpha^3 + \alpha^4$.

In (c), we apply Problem 55 with $n = 15$ and $e = 2$. Part (a) shows that $F_1 = F_2 = F_4$, and part (b) computed F_3 as something else of degree 4. Therefore $G(X) = (1 + X)\text{LCM}(F_1, F_2, F_3, F_4) = (1 + X)\text{LCM}(F_1 F_3) = (1 + X)F_1(X)F_3(X)$, which has degree 9. Then $\dim C = 15 - 9 = 6$, and Problem 55 gives $\delta(C) \geq 2e + 2 = 6$.

Chapter X

1. If R is a field, then the only ideals are 0 and R, and they certainly satisfy the descending chain condition. Conversely if the ideals satisfy the descending chain condition, then there is a minimal nonzero ideal I. Fix $m \neq 0$ in I. For any nonzero

element $a \in I$, $Ra = I$ since I is a simple module. If $x \neq 0$ is in R, we apply this observation to xm, which is nonzero since R is an integral domain. Since $Rxm = I$, there exists y in R with $yxm = m$. Then $(1 - yx)m = 0$. Since R is an integral domain and $m \neq 0$, we obtain $1 - yx = 0$. Therefore $y = x^{-1}$.

2. In (a), let $C_2 = \{\pm 1\}$. Define $r(1) = \begin{pmatrix} 1 & 0 \\ 0 & 1 \end{pmatrix}$ and $r(-1) = \begin{pmatrix} 1 & 1 \\ 0 & 1 \end{pmatrix}$. Then r is a representation since $1 + 1 = 0$ in $\mathbb{F}$. The subspace $U = \mathbb{F}\begin{pmatrix} 1 \\ 0 \end{pmatrix}$ is invariant. If there were a complementary invariant subspace, there would be an eigenvector of $r(-1)$ not in U. However, the roots of the characteristic polynomial are both 1, and a second eigenvector would mean that $r(-1)$ is the identity, which it is not. For (b), the representation in (a) makes $\mathbb{F}^2$ into a unital left R module, the R submodules being the invariant subspaces. There is no complementary R submodule to U, and hence $\mathbb{F}^2$ is not semisimple as an R module.

3. If $\{a_s\}$ is a set of generators of M as a right R module and $\{b_t\}$ is a set of generators of N as a left R module, then $\{a_s \otimes b_t\}$ is a set of generators of $M \otimes_R N$ as an abelian group. Then (a) follows from this fact and the fact that 1 generates both $\mathbb{Z}/k\mathbb{Z}$ and $\mathbb{Z}/l\mathbb{Z}$.

In (b), if $l = dk$ for some d and if b has $b = qk + r$ with $0 \leq r < |k|$, then $a1 \otimes b1 = aqk(1 \otimes 1) + (a1 \otimes r1) = aq(k1 \otimes 1) + (a1 \otimes r1) = a1 \otimes r1$, and it follows that the map $a1 \otimes b1 \mapsto a1 \otimes (b \bmod k)1$ is a well-defined group isomorphism of $(\mathbb{Z}/k\mathbb{Z}) \otimes_\mathbb{Z} (\mathbb{Z}/l\mathbb{Z})$ onto $(\mathbb{Z}/k\mathbb{Z}) \otimes_\mathbb{Z} (\mathbb{Z}/k\mathbb{Z})$.

In (c), let $b(x1, y1) = xy \bmod k$ for $x, y \in \mathbb{Z}/k\mathbb{Z}$. This is $\mathbb{Z}$ bilinear from $\mathbb{Z}/k\mathbb{Z} \times \mathbb{Z}/k\mathbb{Z}$ into $\mathbb{Z}/k\mathbb{Z}$ and extends to a group homomorphism $L : \mathbb{Z}/k\mathbb{Z} \otimes_\mathbb{Z} \mathbb{Z}/k\mathbb{Z} \to \mathbb{Z}/k\mathbb{Z}$ with $L(x1 \otimes y1) = xy \bmod k$. In particular, $L(1 \otimes 1) = 1 \bmod k$. Therefore k divides the order of $1 \otimes 1$, and $\mathbb{Z}/k\mathbb{Z} \times \mathbb{Z}/k\mathbb{Z}$ has at least $|k|$ elements.

In (d), we have $0 = k1 \otimes 1 = k(1 \otimes 1)$ and $0 = 1 \otimes l1 = l(1 \otimes 1)$. If $xk + yl = d$, then $d(1 \otimes 1) = x(k(1 \otimes 1)) + y(l(1 \otimes 1)) = 0$. Hence $1 \otimes 1$ has order dividing d. By (c), $1 \otimes 1$ has order at least $|d|$. The result follows.

4. In (a), each $\ker \varphi^n$ is an R submodule of M, and these R submodules form an ascending chain. Hence they are the same from some point on. Similarly each image φ^n is an R submodule of M, and these form a descending chain. Hence they are the same from some point on.

In (b), if x is in $\mathcal{K} \cap \mathcal{I}$, then $\varphi^N x = 0$ and $x = \varphi^N y$ for some y. Then $0 = \varphi^N x = \varphi^{2N} y$. Since y is in $\ker \varphi^{2N} = \ker \varphi^N$, we obtain $0 = \varphi^N y = x$, and $x = 0$.

In (c), if x is in M, then $\varphi^N x$ is in image $\varphi^N =$ image φ^{2N}. Hence $\varphi^N x = \varphi^{2N} z = \varphi^N(\varphi^N z)$ for some $z \in M$, and $\varphi^N x = \varphi^N y$ with $y = \varphi^N z$.

For (d), if x is in M, let y be as in (c), and write $x = (x - y) + y$. Then $\varphi^N(x - y) = \varphi^N x - \varphi^N y = 0$ and $y = \varphi^N z$ show that $x - y$ is in $\mathcal{K}$ and y is in $\mathcal{I}$. Thus $M = \mathcal{K} + \mathcal{I}$. Since $\mathcal{K} \cap \mathcal{I} = 0$ by (b), $M = \mathcal{K} \oplus \mathcal{I}$.

In (e), we know that $\varphi(\text{image } \varphi^n) = \text{image } \varphi^{n+1}$ for all n. Taking $n > N$, we see that $\varphi(\mathcal{I}) = \mathcal{I}$. From (b), $\ker(\varphi|_\mathcal{I}) \subseteq \mathcal{K} \cap \mathcal{I} = 0$. Therefore φ is one-one from $\mathcal{I}$

onto itself. In addition, $\varphi(\ker \varphi^n) \subseteq \ker \varphi^{n-1}$ for all n. Taking $n > N$ shows that $\varphi(\mathcal{K}) \subseteq \mathcal{K}$. For x in $\mathcal{K}$, we have $\varphi^N x = 0$. Therefore $(\varphi|_{\mathcal{K}})^N = 0$.

5. If (i) holds, then $\psi|_{N'}$ is one-one from N' onto P. Let σ be its inverse. Then $\sigma : P \to N'$ is one-one with $\psi\sigma = 1_P$. So (ii) holds.

If (ii) holds, then any n in N has the property that $n - \sigma\psi(n)$ has $\psi(n - \sigma\psi(n)) = \psi(n) - 1_P\psi(n) = 0$ and is therefore in image φ. Write $n - \sigma\psi(n) = \varphi(m)$ for some m depending on n; m is unique since φ is one-one. If $\tau : N \to M$ is defined by $\tau(n) = m$, then τ is an R homomorphism by the uniqueness of m. Consider $\tau(\varphi(m))$ for m in M. The element $n = \varphi(m)$ has $n - \sigma\psi(n) = \varphi(m) - \sigma\psi\varphi(m) = \varphi(m) - \sigma(0) = \varphi(m)$, and the definition of τ says that $\tau(\varphi(m)) = m$. Hence $\tau\varphi = 1_M$, and (iii) holds.

If (iii) holds, then $N' = \ker \tau$ is an R submodule of N. If n is in $N' \cap$ image φ, then $n = \varphi(m)$ for some $m \in M$ and also $0 = \tau(n) = \tau\varphi(m) = 1_M(m) = m$. So $n = 0$, and $N' \cap$ image $\varphi = 0$. If $n \in N$ is given, write $n = (n - \varphi\tau(n)) + \varphi\tau(n)$. Then $\varphi\tau(n)$ is certainly in image φ, and $\tau(n - \varphi\tau(n)) = \tau(n) - 1_M\tau(n) = 0$ shows that $n - \varphi\tau(n)$ is in N'. Therefore $N = N' \oplus$ image φ. Since image $\varphi = \ker \psi$, we see that $N = N' \oplus \ker \psi$ and that (i) holds.

6. For (a), the conjugation mapping C on R, carrying 1 to itself and carrying $\mathbf{i}, \mathbf{j}$, and $\mathbf{k}$ to their negatives, respects addition and satisfies $C(xy) = C(y)C(x)$. Hence it exhibits R and R^o as isomorphic. Then the result follows from Proposition 10.14.

For (b), again by Proposition 10.14, we need a noncommutative ring R with identity such that R is not isomorphic to R^o. Let $\mathbb{F}$ be a field with two elements, and let R be the 8-element ring consisting of all matrices $\begin{pmatrix} a & b \\ 0 & c \end{pmatrix}$ with a, b, c in $\mathbb{F}$. Define x to be the matrix with $a = 1$ and $b = c = 0$, and define y to be the matrix with $b = 1$ and $a = c = 0$. Computation shows that $x^2 = x$, $y^2 = 0$, $xy = y$, and $yx = 0$. A ring isomorphism of R with R^o is the same as an additive isomorphism that reverses the order of multiplication, and we call this an "antiautomorphism" of R. Suppose that an antiautomorphism φ of R exists. We must have $\varphi(1) = 1$. Suppose that $\varphi(x) = u$ and $\varphi(y) = v$. Then $u = \varphi(x) = \varphi(x^2) = \varphi(x)^2 = u^2$ and $0 = \varphi(y^2) = \varphi(y)^2 = v^2$. Expanding u and v in terms of the basis $\{1, x, y\}$ and computing, we find that $u = k1 + lx$ and $v = my$ with k, l, m in $\mathbb{F}$. Since φ reverses the order of multiplication, we have $uv = \varphi(x)\varphi(y) = \varphi(yx) = \varphi(0) = 0$. Thus $0 = (k1 + lx)(my) = km1 + lmxy = (km)1 + (lm)y$, and $km = lm = 0$. Therefore either $m = 0$ or $k = l = 0$. In the first case, $\varphi(y) = v = my = 0$; in the second case $\varphi(x) = u = k1 + lx = 0$. In either case, φ fails to be one-one. We conclude that no antiautomorphism φ of R exists.

7. Take the sum of all simple R submodules of M.

8. Example 4 in Section 5 shows that $A \otimes_\mathbb{F} \mathbb{K}$ is a vector space over $\mathbb{K}$ in such a way that $k_0(a \otimes k) = a \otimes k_0 k$. It is therefore enough to show that the multiplication is $\mathbb{K}$ linear in each variable of the product. Additivity is known, and it is enough to check that $k_0\big((a_1 \otimes k_1)(a_2 \otimes k_2)\big) = \big(k_0(a_1 \otimes k_1)\big)(a_2 \otimes k_2) = (a_1 \otimes k_1)\big(k_0(a_2 \otimes k_2)\big)$.

Since scalar multiplication by k_0 equals left multiplication by $1 \otimes k_0$, the left equality is immediate from associativity of multiplication, and the right equality follows from associativity and from the formula $(a_1 \otimes k_1)(1 \otimes k_0) = a_1 \otimes k_1 k_0 = a_1 \otimes k_0 k_1 = (1 \otimes k_0)(a_1 \otimes k_1)$.

9. Define $\mu(x)(y) = [x, y]$ for x and y in $\mathfrak{g}$, and let $\nu(c)(d) = cd$ for c and d in $\mathbb{L}$. Then $\mu(x) : \mathfrak{g} \to \mathfrak{g}$ and $\nu(c) : \mathbb{L} \to \mathbb{L}$ are $\mathbb{K}$ linear. Therefore $b(x, c) = \mu(x) \otimes \nu(c)$ is $\mathbb{K}$ bilinear from $\mathfrak{g} \times \mathbb{L}$ into the $\mathbb{K}$ vector space $\mathrm{End}_\mathbb{K}(\mathfrak{g} \otimes_\mathbb{K} \mathbb{L})$, and it extends to a $\mathbb{K}$ linear mapping $L : \mathfrak{g} \otimes_\mathbb{K} \mathbb{L} \to \mathrm{End}_\mathbb{K}(\mathfrak{g} \otimes_\mathbb{K} \mathbb{L})$. Define $[X, Y] = L(X)(Y)$.

With the Lie algebra multiplication now well defined in $\mathfrak{g} \otimes_\mathbb{K} \mathbb{L}$, one readily checks the two required properties. Therefore $\mathfrak{g} \otimes_\mathbb{K} \mathbb{L}$ is a Lie algebra over $\mathbb{K}$ satisfying the two required identities.

Meanwhile, we know that $\mathfrak{g} \otimes_\mathbb{K} \mathbb{L}$ is a vector space over $\mathbb{L}$ because of a change of rings. To complete the proof, we need to show that the multiplication is $\mathbb{L}$ linear, not just $\mathbb{K}$ linear. It is enough to check $\mathbb{L}$ linearity in the second variable because of the alternating property. Let s be in $\mathbb{L}$, and let $x \otimes c$ and $y \otimes d$ be elements of $\mathfrak{g} \otimes_\mathbb{K} \mathbb{L}$. Then we have $[x \otimes c, s(y \otimes d)] = [x \otimes c, y \otimes sd] = [x, y] \otimes csd = s([x, y] \otimes cd) = s[x \otimes c, y \otimes d]$. Forming $\mathbb{K}$ linear combinations, we obtain the desired $\mathbb{L}$ linearity in the second variable of the Lie algebra product.

10. This problem will follow from the uniqueness of the tensor product as given in Theorem 10.18 if it is shown that $((A \otimes_\mathbb{Z} B)/H, qb_2)$ is a tensor product of A and B over R. Thus let $\beta : A \times B \to G$ be an R bilinear function from $A \times B$ into an abelian group G. Since β is automatically $\mathbb{Z}$ bilinear, there exists a group homomorphism $\varphi : A \otimes_\mathbb{Z} B \to G$ such that $\varphi(a \otimes b) = \beta(a, b)$ for all $a \in A$ and $b \in B$. Then $\varphi(ar \otimes b - a \otimes rb) = \varphi(ar \otimes b) - \varphi(a \otimes rb) = \beta(ar, b) - \beta(a, rb)$. The right side is in H, and hence φ descends to a group homomorphism $\overline{\varphi} : (A \otimes_\mathbb{Z} B)/H \to G$ such that $\overline{\varphi} q = \varphi$. Then $\beta(a, b) = \varphi(a \otimes b) = \overline{\varphi} q b_2(a, b)$ shows that $\overline{\varphi}(qb_2) = \beta$. Thus $\overline{\varphi}$ is the required additive extension of β. For uniqueness, suppose $\overline{\varphi}'$ is a second additive extension of β. Then $\overline{\varphi}' q b_2(a, b) = \overline{\varphi} q b_2(a, b)$ for all $a \in A$ and $b \in B$, and hence $\overline{\varphi}' q(a \otimes b) = \overline{\varphi} q(a \otimes b)$. The elements $a \otimes b$ generate $A \otimes_\mathbb{Z} B$, and hence $\overline{\varphi}' q = \overline{\varphi} q$ on $A \otimes_\mathbb{Z} B$. Since q maps onto $(A \otimes_\mathbb{Z} B)/H$, $\overline{\varphi}' = \overline{\varphi}$ on $(A \otimes_\mathbb{Z} B)/H$.

11. We are to show that if C is a commutative associative R algebra with identity and if $\varphi_1 : A_1 \to C$ and $\varphi_2 : A_2 \to C$ are homomorphisms of commutative associative R algebras with identity, then there exists a unique homomorphism $\varphi : A_1 \otimes_R A_2 \to C$ of R algebras with identity such that $\varphi i_1 = \varphi_1$ and $\varphi i_2 = \varphi_2$. Define $b(a_1, a_2) = \varphi_1(a_1)\varphi_2(a_2)$. This is R bilinear into C because $b(a_1 r, a_2) = \varphi_1(a_1 r)\varphi_2(a_2) = \varphi_1(a_1) r \varphi_2(a_2) = \varphi_1(a_1)\varphi_2(ra_2) = b(a_2, ra_2)$, and hence there exists a unique homomorphism $\varphi : A_1 \otimes_R A_2 \to C$ of abelian groups such that $\varphi(a_1 \otimes a_2) = b(a_1, a_2) = \varphi_1(a_1)\varphi_2(a_2)$. Then $\varphi i_1(a_1) = \varphi(a_1 \otimes 1) = \varphi_1(a_1)\varphi_2(1) = \varphi_1(a_1)1 = \varphi_1$, and $\varphi i_1 = \varphi_1$. Similarly $\varphi i_2 = \varphi_2$. To complete the proof, it is enough to show that the homomorphism φ of abelian groups is a homomorphism of R algebras. The fact that φ is a homomorphism of R modules is immediate from Corollary 10.19. Also, $\varphi(1 \otimes 1) = \varphi_1(1)\varphi_2(1) = 1$ shows that φ carries identity to identity. Finally

the computation $\varphi\big((a_1 \otimes a_2)(a'_1 \otimes a'_2)\big) = \varphi(a_1 a'_1 \otimes a_2 a'_2) = \varphi_1(a_1 a'_1)\varphi_2(a_2 a'_2) = \varphi_1(a_1)\varphi_1(a'_1)\varphi_2(a_2)\varphi_2(a'_2) = \varphi_1(a_1)\varphi_2(a_2)\varphi_1(a'_1)\varphi_2(a'_2) = \varphi(a_1 \otimes a_2)\varphi(a'_1 \otimes a'_2)$ shows that φ respects multiplication on a set of additive generators of $A_1 \otimes_R A_2$.

12. Part (a) is immediate from Proposition 10.1. If ψ is a nonzero map in M^E, then $\psi(E)$ is a submodule of M isomorphic to E. Hence $\psi(E) \subseteq M_E$ by construction, and (b) follows. Part (c) is immediate from (b).

13. With $d \in D_E = \operatorname{Hom}_R(E, E)$, we can form $\psi d = \psi \circ d$ if ψ is in $\operatorname{Hom}_R(E, M_E)$, and we can form $de = d(e)$ if e is in E. These definitions give the required unital D_E module structures for (a) and (b). The members of $D_E = \operatorname{Hom}_R(E, E)$ commute with the left R action on E by definition, and this is (c).

14. In view of (c) in the previous problem, the left action of R on E can be regarded as a right R^o action on E in such a way that it commutes with the left D_E action on E. In other words, E is a unital (D_E, R^o) bimodule. Corollary 10.19b shows that $M^E \otimes_{D_E} E$ becomes a unital right R^o module, hence a unital left R module.

15. Define a map $b : M^E \times E \to M$, additive in each variable, by $b(\psi, e) = \psi(e)$. For d in D_E, this has $b(\psi \circ d, e) = (\psi \circ d)(e) = \psi(d(e)) = b(\psi, d(e))$. Hence b is D_E bilinear and has an additive extension $\Phi : M^E \otimes_{D_E} E \to M$ with $\Phi(\psi \otimes e) = \psi(e)$.

The map Φ is R linear since $\Phi(r(\psi \otimes e)) = \Phi(\psi \otimes re) = \psi(re) = r(\psi(e)) = r(\Phi(\psi \otimes e))$. Since ψ is in M^E, $\psi(e)$ is in M_E; thus Φ has image in M_E.

To see that Φ is onto M_E, write $M_E = \bigoplus_{s \in T} M_s$ with each M_s simple, and fix an isomorphism $\alpha_s \in \operatorname{Hom}_R(E, M_s)$ for each $s \in T$. For any element $m \in M_E$, we can find a finite subset T' of T such that $m = \sum_{s \in T'} m_s$ with $m_s \in M_s$. If we let $e_s = \alpha_s^{-1}(m_s)$, then $\Phi\big(\sum_{s \in T'} \alpha_s \otimes e_s\big) = m$. Thus Φ maps onto M_E.

To see that Φ is one-one, we observe from Problem 12 and Lemma 10.3 that

$$M^E = \operatorname{Hom}_R(E, M) = \operatorname{Hom}_R(E, M_E) = \operatorname{Hom}_R\Big(E, \bigoplus_{s \in T} M_s\Big) = \bigoplus_{s \in T} \operatorname{Hom}_R(E, M_s).$$

Each summand on the right side is isomorphic to D_E. That is, the collection of isomorphisms $\{\alpha_s\}_{s \in T}$ from the previous paragraph is a basis of M^E as a right D_E vector space. Consequently every element of $M^E \otimes_{D_E} E$ may be written as a finite sum $\sum \alpha_s \otimes e_s$ with $e_s \in E$. The image of the element $\sum \alpha_s \otimes e_s$ is $\sum \alpha_s(e_s)$. If this is 0, then each $\alpha_s(e_s)$ is 0 because of the independence of the M_s's. Since α_s is an isomorphism, it follows that $e_s = 0$ for each s. Therefore $\sum \alpha_s \otimes e_s = 0$. Thus Φ is one-one.

16. The composition in one order is

$$N \mapsto \operatorname{Hom}_R(E, N) \mapsto \operatorname{Hom}_R(E, N) \otimes_{D_E} E. \qquad (*)$$

For $N = M_E$, the map Φ, when applied to the composition, recovers M_E, since Problem 15 says that Φ is onto. For general N, we can write $M_E = N \oplus N'$. When we apply Φ to $(*)$ for N and N' separately, we recover R submodules of N and N', respectively. To have a match for all of M_E, we must recover all of N and N'.

The composition in the other order is

$$W \mapsto W \otimes_{D_E} E \mapsto \operatorname{Hom}_R(E, W \otimes_{D_E} E). \tag{$**$}$$

For $W = M^E$, the image corresponds under the map $\operatorname{Hom}(1, \Phi)$ to $\operatorname{Hom}_R(E, M_E) = M^E$. For general W, we can write $M^E = W \oplus W'$. When we apply $\operatorname{Hom}(1, \Phi)$ to $(**)$ for W, we get an R submodule of M^E that contains W. In fact, for any $w \in W$, $\operatorname{Hom}_R(E, E \otimes_{D_E} E)$ contains the map $e \mapsto w \otimes e$. Composing with Φ gives $e \mapsto w(e)$. Thus the members of W are in the image. Similarly the members of W' are in the image for W'. The direct sum of the images must be M^E, and thus the images must be exactly W and W'.

17. The computation

$$\varphi(\Phi_M(\psi \otimes e)) = \varphi(\psi(e)) = (\varphi \circ \psi)(e)) = \Phi_N((\varphi \circ \psi) \otimes e) = \Phi_N(\varphi^E(\psi) \otimes e)$$

proves the formula in the last line of the statement of the problem. For the inverse, suppose we are given a map $\tau \in \operatorname{Hom}_{D_E}(M^E, N^E)$. Then τ induces an R linear map

$$\tau'_E : M^E \otimes_{D_E} E \to N^E \otimes_{D_E} E$$

defined by

$$\tau'_E(\psi \otimes e) = \tau(\psi) \otimes e.$$

Composition with the isomorphism of Problem 15 gives an R homomorphism

$$\tau_E = \Phi_N \circ \tau'_E \circ \Phi_M^{-1} : M_E \to N_E.$$

We show that $\varphi \mapsto \varphi^E$ and $\tau \mapsto \tau_E$ are inverses. If a map φ in $\operatorname{Hom}_R(M_E, N_E)$ is given, we are to calculate $(\varphi^E)_E \in \operatorname{Hom}_R(M_E, N_E)$. It is enough to find the effect of $(\varphi^E)_E$ on elements $\Phi_M(\psi \otimes e)$ with $\psi \in M^E$ and $e \in E$. For such an element,

$$(\varphi^E)_E(\Phi_M(\psi \otimes e)) = \Phi_N((\varphi^E)'(\psi \otimes e)) = \Phi_N(\varphi^E(\psi) \otimes e)$$
$$= \varphi^E(\psi)(e) = \varphi(\psi(e)) = \varphi(\Phi_M(\psi \otimes e)).$$

Thus $(\varphi^E)_E = \varphi$. Similarly for $\tau \in \operatorname{Hom}_{D_E}(M^E, N^E)$, we find that $(\tau_E)^E = \tau$. Thus $\varphi \mapsto \varphi^E$ and $\tau \mapsto \tau_E$ are inverses.

18. Let us write $M = \bigoplus_{s \in S} M_s$ with each M_s semisimple. Each M_s is contained in some M_E, and hence $M = \sum_{E \in \mathcal{E}} M_E$. Let us see that the sum is direct. If M_E has nonzero intersection with $M_{E_1} + \cdots + M_{E_n}$, where $E_1, \ldots, E_n$ are simple R modules with no two isomorphic, then there is a nonzero R linear map from E

into $M_{E_1} + \cdots + M_{E_n}$. We can write each M_{E_j} as a sum of simple R submodules isomorphic to E_j, and Proposition 10.1 shows that

$$M_{E_1} + \cdots + M_{E_n} = \bigoplus_{s \in T} M'_s$$

with each M'_s isomorphic to one of $E_1, \ldots, E_n$. If all of $E_1, \ldots, E_n$ are nonisomorphic with E, then Lemma 10.3 and Proposition 10.4a show that

$$\text{Hom}_R(E, M_{E_1} + \cdots + M_{E_n}) = 0,$$

contradiction. We conclude that the sum $M = \sum_{E \in \mathcal{E}} M_E$ is direct. This proves the equality at the left in the displayed formula of the problem, and the isomorphism on the right in that display follows from Problem 15.

19. If N is a left R submodule of M, then $N_E \subseteq M_E$ for every E. Conversely the previous problem shows that a system of N_E's defines an R submodule N. Thus this problem is a restatement of Problem 16.

20. We have

$$\text{Hom}_R(M, N) \cong \prod_{E \in \mathcal{E}} \text{Hom}_R(M_E, N) = \prod_{E \in \mathcal{E}} \text{Hom}_R(M_E, N_E),$$

and the rest follows from Problem 17.

SELECTED REFERENCES

Artin, E., *Geometric Algebra*, Interscience Publishers, Inc., New York, 1957; reprinted, John Wiley & Sons, Inc., New York, 1988.

Artin, M., *Algebra*, Prentice–Hall, Englewood Cliffs, NJ, 1991.

Baez, J. C., The octonions, *Bull. Amer. Math. Soc.* 39 (2002), 145–205.

Berlekamp, E. R., *Algebraic Coding Theory*, McGraw–Hill Book Company, 1968.

Berlekamp, E. R. (ed.), *Key Papers in the Development of Coding Theory*, IEEE Press Selected Reprint Series, IEEE Press [Institute of Electrical and Electronics Engineers, Inc.], New York, 1974.

Brown, K. S., *Cohmology of Groups*, Springer-Verlag, New York, 1982; reprinted with corrections, 1994.

Dunford, N., and J. T. Schwartz, *Linear Operators*, Part I, Interscience Publishers, Inc., New York, 1958; reprinted, John Wiley & Sons, Inc., New York, 1988.

Elkies, N. D., Lattices, linear codes, and invariants II, *Notices of the American Mathematical Society* 47 (2000), 1382–1391.

Farb, B., and R. K. Dennis, *Noncommutative Algebra*, Springer-Verlag, New York, 1993.

Hall, M., *The Theory of Groups*, The Macmillan Company, New York, 1959; reprinted, Chelsea Publishing Company, New York, 1976.

Halmos, P. R., *Naive Set Theory*, D. Van Nostrand Company, Inc., Princeton, 1960; reprinted, Springer-Verlag, New York, 1974.

Hasse, H., *Number Theory*, English translation of the original German, Springer-Verlag, Berlin, 1980; reprinted, 2002.

Hayden, S., and J. F. Kennison, *Zermelo–Fraenkel Set Theory*, Charles E. Merrill Publishing Company, Columbus, 1968.

Hecke, E., *Lectures on the Theory of Algebraic Numbers*, English translation of the original German, Springer-Verlag, New York, 1981.

Hermite, C., Sur quelques approximations algébriques, *J. Reine Angew. Math.* 76 (1873), 342–344.

Hilton, P. J., and U. Stammbach, P., *A Course in Homological Algebra*, Springer-Verlag, New York, 1971; second edition, 1997.

Hoffman, K., and R. Kunze, *Linear Algebra*, Prentice–Hall, Englewoord Cliffs, NJ, 1961; second edition, 1971.

Hua, L.-K., *Introduction to Number Theory*, English translation of the original Chinese, Springer-Verlag, Berlin, 1982.

Ireland, K., and M. Rosen *A Classical Introduction to Modern Number Theory*, Springer-Verlag, New York, 1982; second edition, 1990.

Jacobson, N., *Basic Algebra*, Volume I, W. H. Freeman and Company, San Francisco, 1974; second edition, New York, 1985. Volume II, W. H. Freeman and Company, San Francisco, 1980; second edition, New York, 1989.

Jacobson, N., *Lectures in Abstract Algebra*, Volume I, D. Van Nostrand Company, Inc., Princeton, 1951; reprinted, Springer-Verlag, New York, 1975. Volume II, D. Van Nostrand Company, Inc., Princeton, 1953; reprinted, Springer-Verlag, New York, 1975. Volume III, D. Van Nostrand Company, Inc., Princeton, 1964; reprinted with corrections, Springer-Verlag, New York, 1975.

Kelley, J. L., *General Topology*, D. Van Nostrand Company, Inc., Princeton, 1955; reprinted, Springer-Verlag, New York, 1975.

Knapp, A. W., *Basic Real Analysis*, Birkhäuser, Boston, 2005.

Lam, T. Y., *A First Course in Noncommutative Rings*, Springer-Verlag, New York, 1991; second edition, 2001.

Lang, S., *Algebra*, Addison-Wesley, Reading, MA, 1965; second edition 1984; revised third edition, Springer, New York, 2002.

Lang, S., *Algebraic Number Theory*, Springer-Verlag, New York, 1986; second edition, Springer-Verlag, New York, 1994.

Lindemann, F., Über die Zahl π, *Math. Annalen* 20 (1882), 213–225.

Mac Lane, S., *Categories for the Working Mathematician*, Springer, New York, 1971; second edition, 1998

Morgan, S. P., Richard Wesley Hamming (1915–1998), *Notices of the American Mathematical Society* 45 (1998), 972–977.

Pollard, H., *The Theory of Algebraic Numbers*, Carus Monographs, Mathematical Association of America, 1950.

Rostermundt, R., BCH Codes, http://www-math.cudenver.edu/~rrosterm/crypt_proj/crypt_proj.html, 2002.

Rotman, J., *Galois Theory*, Springer-Verlag, New York, 1990; second edition, 1998.

Sah, C.-H., *Abstract Algebra*, Academic Press, New York, 1967.

St. Andrews, School of Mathematics and Statistics, University of St. Andrews, Scotland, *MacTutor History of Mathematics Archive, Biographies of Mathematicians*, updated as of 2006, http://www-groups.dcs.st-and.ac.uk for background, http://www-history.mcs.st-andrews.ac.uk/history/index.html for official entry point, http://www-groups.dcs.st-and.ac.uk/~history/Mathematicians for direct access to list of mathematicians.

Van der Waerden, B. L., *Modern Algebra*, English translation of the original German, Volume I, Frederick Ungar Publishing Company, New York, 1949; multiple later translated editions. Volume II, Frederick Ungar Publishing Company, New York, 1950; multiple later translated editions.

Zariski, O., and P. Samuel, *Commutative Algebra*, Volume I, D. Van Nostrand Company, Inc., Princeton, 1958; reprinted, Springer-Verlag, New York, 1975.

INDEX OF NOTATION

This list indexes recurring symbols introduced in Chapters I through X (pages 1–582). For other recurring symbols, including set-theoretic notation introduced in the appendix (pages 583–602), see the list of Standard Notation on page xviii.

In the list below, each piece of notation is regarded as having a key symbol. The first group consists of those items for which the key symbol is a fixed Latin letter, and the items are arranged roughly alphabetically by that key symbol. The next group consists of those items for which the key symbol is a Greek letter. The final group consists of those items for which the key symbol is a variable or a nonletter, and these are arranged by type.

$\mathfrak{A}_n$, 120
$a(u, v)$, 345
A^{adj}, 72
$\text{Ann}(U)$, 52
$\text{Aut } H$, 166
$B^n(G, N)$, 353
$E^{\mathbb{C}}$, 271
C_m, 125
$C(G, \mathbb{C})$, 327
$C(G, R)$, 378
$c(A)$, 392
$C^n(G, N)$, 353
$C\ell(x)$, 164
$\text{Cliff}(E, \langle \cdot \cdot \rangle)$, 299
D, 506, 527
D_n, 121
deg, 149, 154
det, 67, 213
dim V, 37
$\text{End}_{\mathbb{K}}(V)$, 369
$\text{End}_R(M)$, 545
$e_1, \ldots, e_n$, 36
e_i, f_i, g, 522
$\mathbb{F}$, 9, 34, 157

$\mathbb{F}_4$, 142
$\mathbb{F}_p$, 141, 147
$\mathbb{F}_q$, 457
$F(S)$, 304, 374
$\mathcal{F}(S)$, 158
$\text{Gal}(\mathbb{K}/\mathbb{k})$, 469
GCD, 2, 391
$\text{GL}(V)$, 121
$\text{GL}(n, \mathbb{F})$, 121
$\mathbb{H}$, 127
H_8, 127
$H(V)$, 299
$H^n(G, N)$, 353
$\text{Hom}(\psi, \varphi)$, 559
$\text{Hom}_{\mathbb{F}}(U, V)$, 43, 44
$\text{Hom}_{\mathbb{K}}(U, V)$, 263
$\text{Hom}_R(M, N)$, 545
i, j, k, 127
$\mathbb{K}, \mathbb{k}, \mathbb{K}/\mathbb{k}$, 449
$\ker L$, 46
$\ker \varphi$, 130
l, 329
LCM, 32
lrad, 247

$M_n(R)$, 213
$M_{kn}(\mathbb{F})$, 25
$M_{mn}(R)$, 373
Morph(A, B), 188
$N_{\mathbb{K}/\mathbb{k}}(a)$, 514
$N(H)$, 186
$\mathbb{O}$, 301
$\mathcal{O}$, 340
M^o, R^o, 546
$O(V)$, $O(n)$, 121
Obj$(\mathcal{C})$, 188
$\mathcal{C}^{\text{opp}}$, 190, 208
Pfaff(X), 296, 445
PSL$(2, \mathbb{Z})$, 363
PSL$(2, \mathbb{Z}/m\mathbb{Z})$, 363
PSL$(n, \mathbb{F})$, 203
$\mathbb{Q}[\theta]$, 121, 142
r, 329
rrad, 247
$\mathfrak{S}_n$, 120
$S(E)$, 281
$S^n(E)$, 281
$S' = S \cup S^{-1}$, 304
SO(V), SO(n), 121
SU(V), SU(n), 121
sgn, 17
span$\{v_\alpha\}$, 35
A^t, 41
L^t, 53
$T(E)$, 278
$T^n(E)$, 278
Tr A, 74
Tr$_{\mathbb{K}/\mathbb{k}}(a)$, 514
$U(\mathfrak{g})$, 298
$U(V)$, $U(n)$, 121
$W(S')$, 304
$W(V)$, 299
wt(c), 204
Z_G, 163
$Z_G(x)$, 164
$\mathbb{Z}/m\mathbb{Z}$, 119

$\mathbb{Z}/(m)$, 119
$\mathbb{Z}[\sqrt{-1}]$, 389
$Z^n(G, N)$, 353
$\mathbb{Z}G$, 370

Greek
Γ, Δ, 44
$\binom{u}{\Gamma}$, 45
$\binom{L}{\Delta\Gamma}$, 45
δ_n, 353
$\delta(C)$, 205
$\iota : V \to V''$, 54
Σ, 48
χ_R, 336
φ, 7
φ_x, 450
$\Phi_n(X)$, 484

Operations on sets given by subscripts and superscripts
V', 50
G', 310
$M^\perp$, 95
$U^\perp$, 248
L^*, 99
A^*, 100
$\overline{V}$, 114
$\widehat{G}$, 326
$\mathbb{C}^\times$, $\mathbb{Q}^\times$, $\mathbb{R}^\times$, $\mathbb{Z}^\times$, 119
$(\mathbb{Z}/m\mathbb{Z})^\times$, 141
$R^\times$, 148
P^{-1}, 436

Specific functions
$(\cdot, \cdot)$, 89
$\|\cdot\|$, 90
$[\cdot, \cdot]$, 298
$\langle \cdot, \cdot \rangle$, 246
$[\mathbb{K} : \mathbb{k}]$, 452

Index of Notation

Isolated symbols
$\cong$, 48, 118, 143
$\equiv$, 119
1, 117
$\{1\}$, 117
1_A, 189

Operations on sets and classes
G/H, $G\backslash H$, 129
gH, Hg, 128
$G_1 \times G_2$, 125
$G \times_\tau H$, 167
$G_1 * G_2$, 321
G_p, 162
$G = \langle S; R \rangle$, 311
RG, 377
$\mathbb{F}[X]$, 9
$R[X]$, 148
$R[X_1, \ldots, X_n]$, 154
$\Bbbk[x_1, \ldots, x_n]$, 450
$\Bbbk(x_1, \ldots, x_n)$, 450
$\mathbb{K}(X)$, 381
$\mathcal{C}^S$, 194
V/U, 55
M/N, 375
$I + J$, 402
IJ, 402, 432
$U \oplus V$, 59

$E \otimes_\mathbb{K} F$, 262
$e \otimes f$, 262
$M \otimes_R N$, 565
$m \otimes n$, 565
$\varphi \otimes \psi$, 565
$\bigwedge(E)$, 288
$\bigwedge^n(E)$, 288
$E^\mathbb{L}$, 272
$\bigoplus_{s \in S}$, 62, 137, 373
$\prod_{s \in S}$, 62, 135, 196, 373
$\coprod_{s \in S}$, 198
$\ast_{s \in S}$, 320
$\mathbb{K}^H$, 469
$S^{-1}R$, 425
R_S, 425
R_P, 426
G_P, 529

Miscellaneous
$\begin{pmatrix} 1\,2\,3\,4\,5 \\ 4\,3\,5\,1\,2 \end{pmatrix}$, permutation, 15
(5 2 3), cycle, 16
$f_1 * f_2$, convolution, 336
(a), principal ideal, 387
$(a_1, \ldots, a_n)$, ideal, 387

INDEX

Abel, 489
abelian group, 118
   direct sum for, 137, 138
   finitely generated, 174
   free, 174
   tensor product for, 569
absolute value, 594
addition in abelian group, 118
addition in ring, 140
addition in vector space, 34
addition of cardinal numbers, 602
addition of matrices, 25
additive extension, 565
additive functor, 576
additive in a variable, 565
adjoin, 450
adjoint, 99, 100
   classical, 72
algebra, 277
   alternative, 301
   associative, 277, 369
   associative R, 377
   Clifford, 299
   division, 370
   exterior, 288
   filtered associative, 298
   graded associative, 298
   group, 377, 441
   Heisenberg Lie, 299
   Jordan, 300
   Lie, 278, 298
   polynomial, 286
   symmetric, 281
   tensor, 279
   tensor product for, 573
   universal enveloping, 298
   Weyl, 299
algebraic closure, 461
   existence, 461

   uniqueness, 461
algebraic curve, 408
algebraic element, 450
algebraic extension, 452
   finite, 452
   simple, 453
algebraic integer, 339, 408, 417, 510
algebraic number, 122, 384, 453, 460, 510
algebraic number field, 122, 370, 384, 453
algebraically closed, 460
algebraically closed field, 210
alternating, 67
alternating bilinear form, 250
alternating group, 120, 170
alternating matrix, 254
alternative algebra, 301
annihilator, 52
antisymmetrized tensor, 291
antisymmetrizer, 291
ascending chain condition, 414, 556
associate, 390
associated graded map, 297
associated graded vector space, 297
associated primitive polynomial, 393
associative algebra, 277, 369
   filtered, 298
   graded, 298
   tensor product for, 573
associative law, 25, 34, 82, 117, 140
associative R algebra, 377
associativity formula, 571, 572
associator, 301
automorphism, 449
   of group, 166
   inner, 199
   of number field, 123
Axiom of Choice, 587

Baer multiplication, 352, 358

basis, 36, 174
  dual, 51
  free, 309
  standard, 36
  standard ordered, 48
  Weyl, 293
BCH code, 543
Bessel's inequality, 93
bilinear, 89
bilinear form, 246
  alternating, 250
  invariant, 257
  nondegenerate, 248
  skew-symmetric, 250
  symmetric, 250
bilinear function, 260, 565
bilinear map, 260
bilinear mapping, 260
bimodule, 564
block, 230
Bolzano–Weierstrass Theorem, 593
boundary map, 574
Burnside's Theorem, 342

cancellation law, 117
canonical form, 210
  Jordan, 229, 406
  rational, 243, 443, 444
  of rectangular matrix, 239
canonical-form problem, 212
canonical map into double dual, 54
Cantor, 602
Cardan's formula, 488, 505, 508
Cardano, 488
cardinal number, 600
cardinality, 599
Cartan matrix, 86
Cartesian product, 585
  indexed, 587
category, 53, 134, 188
  opposite, 190, 208
Cayley number, 301
Cayley's Theorem, 124
Cayley–Dickson construction, 301
Cayley–Hamilton Theorem, 219
center, 369, 377, 545
  of group, 163
centralizer of element, 164

chain, 574, 595
chain condition
  ascending, 414, 556
  descending, 556
change of rings, 564, 569
character, 336
  multiplicative, 326
characteristic of a field, 147
characteristic polynomial, 73, 216
characteristic subgroup, 357
check matrix, 542
Chinese Remainder Theorem, 6, 402
class, 584, 585
  equivalence, 590
class equation of group, 185–186
class function, 337
classical adjoint, 72
Clifford algebra, 299
closed, 574
closed form, 575
coboundary, 353
coboundary map, 353
cochain, 353
cocycle, 353
code, 205
  BCH, 543
  cyclic, 542
  cyclic redundancy, 207
  dual, 360
  error-correcting, 204, 360, 541
  Hamming, 206
  linear, 205
  parity-check, 205
  repetition, 205
  self-dual linear, 360
coefficient, 9, 148
  Fourier, 327, 359
  leading, 149
  matrix, 333
cofactor, 70, 215
cohomology group, 353
cohomology of groups, 352, 575
collection, 585
column space, 38
column vector, 25
  in an ordered basis, 45
common multiple, 32
commutative diagram, 193

commutative law, 25, 34, 82, 118
commutative ring, 140
commutator, 357
commutator subgroup, 310
complement, 585
completely reducible, 546
complex, 574, 576
complex conjugate of vector space, 114
complex conjugation, 594
complex number, 594
complexification, 271
composition, 588
composition factor, 172, 552
composition series, 171, 551
congruent modulo, 119
conjugacy class, 164
conjugate, 164
conjugate linear, 89
conjugates of an element, 518
conjugation, complex, 594
consecutive quotient, 171, 551
constant polynomial, 10, 149, 154
constructible coordinates, 465
    field of, 466
constructible regular polygon, 468, 483, 493
contraction of ideal, 428
contragredient, 53
    matrix of, 53
contragredient representation, 362
contravariant functor, 192
convolution, 336, 369, 378
coproduct functor, 198, 373, 580
    in a category, 196
corner variable, 21
correspondence, one-one, 588
coset, 128
countable, xviii
counting formula, 163
covariant functor, 191
Cramer's rule, 24, 72, 215
CRC-8, 207
crossed homomorphism, 354
cubic polynomial, 537
cubic resolvent, 540
cut, 592
cycle, 15
cycle structure, 165
cycles, disjoint, 16

cyclic code, 542
cyclic group, 124
cyclic R module, 398
cyclic redundancy code, 207
cyclic subspace, 241
cyclic vector, 242
cyclotomic field, 485, 495
cyclotomic polynomial, 396, 484, 535

dal Ferro, 488
de Rham cohomology, 575
decomposition group, 529
Dedekind domain, 413, 434, 446, 520
degree, 10, 149, 154, 452
dependent, integrally, 417
derivative of polynomial, 457
descend to, 57, 132, 146, 375
descending chain condition, 556
determinant, 65, 213
    Gram, 113
    of linear map, 66
    of matrix, 66
    properties of, 68, 214
    of square matrix, 67
    Vandermonde, 71, 215
diagonal entry, 24, 178, 443
diagonal matrix, 24, 443
diagram, 193
    commutative, 193
    square, 193
difference, 585
difference product, 506
differential equations, system, 244
differential form, 575
differentiation, 457
dihedral group, 120, 168, 313
dimension, 555
    of vector space, 37, 78
direct image, 588
direct product
    of groups, 125, 126, 135, 136
    of R modules, 373
    of rings, 371
    of vector spaces, 62, 63
direct sum
    of abelian groups, 137, 138
    of R modules, 373
    of vector spaces, 59, 60, 61, 62, 64

Dirichlet's theorem on primes in arithmetic
    progressions, 327, 364
discriminant, 506, 527, 528
disjoint cycles, 16
disjoint union, 197
distributive law, 26, 34, 140
divide, 1, 10, 385, 435
division algebra, 370
division algorithm, 2, 11
division ring, 143, 370
divisor, 1
    elementary, 177, 443
    greatest common, 2, 8, 12, 390
    zero, 143
Dixmier, 550
domain, 586
    Dedekind, 413, 434, 446, 520
    Euclidean, 389, 441, 442
    integral, 143
    principal ideal, 387, 439
    unique factorization, 386
dot product, 89
double a cube, 465, 467
double dual, 54
dual
    double, 54
    of vector space, 50
dual basis, 51
dual code, 360
dual space, 50
duality in category theory, 208

eigenspace of linear function, 76
eigenspace of matrix, 73
eigenvalue of linear function, 76
eigenvalue of matrix, 73
eigenvector of linear function, 76
eigenvector of matrix, 73
Eisenstein's irreducibility criterion, 395
element, 583, 584
elementary divisor, 177, 443
elementary matrix, 28
elementary row operation, 20
elementary symmetric polynomial, 445
entity, 583
entry, 20, 24
    diagonal, 24, 178, 443
enveloping algebra, universal, 298

equality of matrices, 24
equation, linear, 23
equivalence class, 590
equivalence relation, 589
equivalent factor set, 349
equivalent finite filtrations, 552
equivalent group extensions, 349
equivalent normal series, 172
equivalent words, 304
equivariant mapping, 189
error-correcting code, 204, 360, 541
Euclidean algorithm, 2, 13
Euclidean domain, 389, 441, 442
Euler φ function, 7
evaluate, 10
evaluation, 150, 156
even permutation, 120
exact, 575
exact form, 575
exact sequence, 575, 576
    short, 576
    split, 579
expansion
    in cofactors, 70, 215
    homogeneous-polynomial, 154
    monomial, 154
expressible in terms of $\Bbbk$ and radicals, 490
extension
    additive, 565
    algebraic, 452
    field, 449
    finite, 452
    finite algebraic, 452
    finite Galois, 479
    group, 345
    of ideal, 428
    linear, 44, 261
    normal, 476
    of scalars, 272, 564, 569
    separable, 472
    simple algebraic, 453
exterior algebra, 288
external direct product
    of groups, 125, 135
    of R modules, 373
external direct sum
    of abelian groups, 137
    of R modules, 373

of vector spaces, 59, 61
external semidirect product of groups, 167

factor, 1, 10, 135, 385, 435
factor group, 131
factor ring, 145
factor set, 345
Factor Theorem, 11
factor through, 57, 132, 146, 375
factorization, 1, 10
   nontrivial, 2, 10
   prime, 5
   unique, 5, 14
family, 585
fast Fourier transform, 327, 361
Fermat number, 468
Fermat prime, 468
Fermat's Little Theorem, 141
Ferrari, 488
field, 141
   algebraically closed, 210, 460
   characteristic of, 147
   of constructible coordinates, 466
   cyclotomic, 485, 495
   extension, 449
   finite, 142, 152, 158, 370, 457, 483
   fixed, 469
   of fractions, 380, 591
   Galois, 457
   number, 122, 142, 370, 384, 453
   obtained by adjoining, 450
   prime, 147
   quadratic number, 418, 538
   splitting, 454
field isomorphism, 449
field map, 449
field mapping, 449
field polynomial, 514
filtered associative algebra, 298
filtered vector space, 297
filtration finite, 551
finite algebraic extension, 452
finite basis condition, 414, 556
finite extension, 452
finite field, 142, 152, 158, 370, 457, 483
finite filtration, 551
finite Galois extension, 479
finite length, 554

finite linear combination, 35
finite order, 129
finite rank, 176
finite rank of free R module, 398
finite support, 378
finite-dimensional vector space, 37
finitely generated abelian group, 174
   fundamental theorem for, 177
finitely generated group, 312
finitely generated R module, 397
finitely presented group, 312
First Isomorphism Theorem, 57, 132, 376
Fitting's Lemma, 579
fixed field, 469
forgetful functor, 191
form, 260
   bilinear, *see* bilinear form
   Hermitian, 255
   sesquilinear, *see* sesquilinear form
   skew-Hermitian, 255
Fourier coefficient, 327, 359
Fourier inversion formula
   for class functions, 338
   for finite abelian group, 327
   for finite group, 335
Fourier inversion problem, 327
Fourier series, 327
fractional ideal, 446
   unique factorization of, 447
fractions
   field of, 380, 591
   partial, 441
free abelian group, 174
free basis, 309
free group, 305
   rank, 311
free product, 197, 320
free R module, 374
free subset, 309
Frobenius map, 458
function, 585
   bilinear, 260
   class, 337
   k-linear, 260
   k-multilinear, 260
   linear, 42, 44
   multilinear, 260
   polynomial, 152, 157

functional
  linear, 50
  multilinear, 66
functor, 53, 134
  additive, 576
  contravariant, 192
  coproduct, 198, 373, 580
  covariant, 191
  forgetful, 191
  product, 196, 373
Fundamental Theorem
  of Algebra, 14, 460, 486
  of Arithmetic, 5
  of Finitely Generated Abelian Groups, 177
  of Finitely Generated Modules, 399
  of Galois Theory, 342, 479

Galois, 489
Galois extension, finite, 479
Galois field, 457
Galois group, 469
Galois theory, 122, 479
Gauss, 468, 483, 495
Gauss's Lemma, 392
Gaussian integer, 389, 442
general linear group, 121
generated by, 124
generated submodule, 374–375
generating polynomial, 207, 542
generator, 124, 174, 397
  monic, 242
generators, 311
graded associative algebra, 298
graded vector space, 297
Gram determinant, 113
Gram matrix, 113
Gram–Schmidt orthogonalization process, 94
greatest common divisor, 2, 8, 12, 390
greatest lower bound, 593
group, 117
  abelian, 118
  alternating, 120, 170
  automorphism of, 166
  center of, 163
  cohomology, 353
  cyclic, 124
  decomposition, 529
  dihedral, 120, 168, 313
  direct product for, 125, 126, 135, 136
  finitely generated, 312
  finitely presented, 312
  free, 305
  free abelian, 174
  free product for, 320
  Galois, 469
  general linear, 121
  homomorphism of, 130
  icosahedral, 365
  octahedral, 365
  order of, 128
  orthogonal, 121
  quaternion, 127
  quotient of, 131
  rotation, 121
  semidirect product for, 167, 168
  simple, 170
  solvable, 488
  special linear, 121
  special unitary, 121
  symmetric, 120
  tetrahedral, 365
  trivial, 117
  unitary, 121
  of units, 142
group action, 123, 158
  transitive, 162
  trivial, 160
group algebra, 377, 441
group extension, 345
group ring, integral, 370

Hamming code, 206
Hamming distance, 204–205
Hamming space, 204
harmonic analysis, 501
harmonic polynomial, 115
Heisenberg Lie algebra, 299
heptadecagon, 498
Hermite, 510
Hermitian, 100
Hermitian form, 255
Hermitian matrix, 256
Hermitian sesquilinear form, 255
Hermitian symmetric, 89
Hilbert Basis Theorem, 413, 415
Hilbert–Schmidt norm, 111

homogeneous element, 278
homogeneous ideal, 281
homogeneous polynomial, 115, 154
homogeneous system, 23
homogeneous-polynomial expansion, 154
homomorphism
   crossed, 354
   of groups, 130
   of R modules, 372
   of rings, 143
   substitution, 150, 155

icosahedral group, 365
ideal, 144
   contraction of, 428
   extension of, 428
   fractional, 446
   left, 375
   maximal, 382
   prime, 381
   principal, 387
   right, 375
   two-sided, 144
   unique factorization of, 435
identity element, 117
identity in a ring, 141
identity matrix, 27
identity morphism, 189
image, 586
   direct, 588
   of homomorphism, 130
   inverse, 589
imaginary part, 594
independent variable, 21
indeterminate, 9, 148, 153
index of subgroup, 163
indexed Cartesian product, 587
indexed intersection, 587
indexed union, 587
infimum, 593
infinite order, 129
infinite-dimensional vector space, 78
inhomogeneous system, 23
injection, 59, 62
inner automorphism, 199
inner product, 89
inner-product space, 89
integer
   algebraic, 339, 408, 417, 510
   Gaussian, 389, 442
integers modulo, 119
integral, 417
integral closure, 413, 418
integral domain, 143
integral group ring, 370
integrally closed, 422
integrally dependent, 417
Intermediate Value Theorem, 593
internal direct product
   of groups, 126, 136
   of R modules, 373
   of vector spaces, 63
internal direct sum
   of abelian groups, 139
   of R modules, 374
   of vector spaces, 60, 61, 64
internal semidirect product of groups, 168
intersection, 585
   indexed, 587
intertwining operator, 330
invariant
   of group action, 354
   leave a bilinear form, 257
invariant subspace, 73, 330
invariant vector subspace, 216
inverse, 190
   multiplicative, 142
inverse element, 117
inverse function, 588
inverse image, 589
inverse matrix, 27
invertible matrix, 27
involution, 240
irreducible element, 385
irreducible left R module, 546
irreducible representation, 330
isometry, 158
isomorphic, 48, 118, 143, 191, 349, 375
isomorphism, 48, 118, 143, 190, 375, 449
   natural, 265
isotropic subspace, 293
isotropy subgroup, 162
isotypic submodule, 580
Iwasawa decomposition, 112

Jacobi identity, 298

710   Index

Jordan algebra, 300
Jordan block, 229, 406
Jordan canonical form, 229, 406
Jordan form, 229
Jordan normal form, 229
Jordan–Hölder Theorem, 174, 553

$\Bbbk$ isomorphism, 449
k-linear, 66
k-linear function, 260
k-linear map, 260
k-linear mapping, 260
k-multilinear function, 260
k-multilinear map, 260
k-multilinear mapping, 260
kernel
   of homomorphism, 130
   of linear map, 46
Kronecker delta, xviii, 27
Kronecker product, 294

Lagrange resolvents, 501
Lagrange's Theorem, 129
law of composition, 188
law of cosines, 90
law of quadratic reciprocity, 494, 539
leading coefficient, 149
leading term, 149
least common multiple, 32
least upper bound, 593, 596
leave a bilinear form invariant, 257
left coset, 128
left ideal, 375
left R module, 371
left radical, 247
left regular representation, 329, 335, 362
left vector space, 547
left-coset space, 129
Legendre polynomial, 113
length of module, 554
length of word, 304
letter, 120
Lie algebra, 278, 298
   Heisenberg, 299
Lie bracket, 298
Lindemann, 510
linear, 42, 44
linear code, 205

   self-dual, 360
linear combination, 35
linear equation, 23
linear extension, 44, 261
linear fractional transformation, 159
linear function, *see* linear map
linear functional, 50
linear map, 42, 44
   determinant of, 66
   eigenspace of, 76
   eigenvalue of, 76
   eigenvector of, 76
   kernel of, 46
   normal, 109
   orthogonal, 102
   positive definite, 106
   positive semidefinite, 106
   unitary, 102
linear mapping, *see* linear map
linear operator, 42
linear transformation, *see* linear map
linearly independent set, 36, 174
local ring, 430
localization, 413
   of R at the prime P, 427
   of R with respect to S, 425
lower bound, 593

MacWilliams identity, 361
map, 586
   bilinear, 260
   coboundary, 353
   field, 449
   k-linear, 260
   k-multilinear, 260
   linear, 42, 44
   multilinear, 260
mapping, *see* map
matrix, 24
   addition for, 25
   alternating, 254
   Cartan, 86
   check, 542
   coefficient, 333
   column space of, 38
   determinant of, 66, 67
   diagonal, 24, 443
   eigenspace of, 73

eigenvalue of, 73
eigenvector of, 73
elementary, 28
equality for, 24
Gram, 113
Hermitian, 256
identity, 27
inverse, 27
invertible, 27
of a linear map in an ordered basis, 45
multiplication for, 26
nilpotent, 230
nonsingular, 215
null space of, 38
orthogonal, 102
positive definite, 106
positive semidefinite, 106
rank of, 41
row space of, 38
scalar multiplication for, 25
singular, 215
skew-symmetric, 254
square, 24
symmetric, 250
symplectic, 446
trace of, 74
transpose of, 41
unitary, 102
Vandermonde, 71, 215
zero, 25
matrix representation, 329
matrix ring, 368
maximal element, 595
maximal ideal, 382
maximum condition, 414, 556
member, 585
minimal distance, 205
minimal polynomial, 219, 221, 451
minimum condition, 556
module
   cyclic, 398
   direct product for, 373
   direct sum for, 373
   of finite rank, 398
   finitely generated, 397
   free R, 374
   homomorphism of, 372
   irreducible, 546

   left R, 371
   quotient, 375
   rank of, 399
   right R, 372
   semisimple, 546
   simple, 546
   tensor product for, 565
modulo, 119
monic generator, 242
monic polynomial, 149
monomial, 154
monomial expansion, 154
morphism, 188
   identity, 189
multilinear form, symmetric, 280
multilinear function, 260
multilinear functional, 66
multilinear map, 260
multilinear mapping, 260
multiple, 1, 10
   least common, 32
multiplication
   in an algebra, 277
   Baer, 352, 358
   in a group, 117
   of matrices, 26
   in a ring, 140
multiplicative character, 326
multiplicative inverse, 142
multiplicative system, 425
multiplicity of a root, 14

n-fold tensor product, 277
Nakayama's Lemma, 432
natural isomorphism, 265
natural transformation, 265
negative, xviii, 118
nicely normed, 302
Nielsen–Schreier Theorem, 315
nilpotent element, 440
nilpotent matrix, 230
Noetherian ring, 415
nondegenerate bilinear form, 248
nonsingular matrix, 215
nontrivial factorization, 2, 10
norm, 90, 514, 539
   Hilbert–Schmidt, 111
normal extension, 476

normal linear map, 109
normal series
  equivalent, 172
  of groups, 171
normal subgroup, 130
normalizer of subgroup, 186
null space, 38
Nullstellensatz, 409
number
  algebraic, 122, 384, 453, 460, 510
  complex, 594
  rational, 591
  real, 592
number field, 122, 142, 370, 384, 453
  automorphism of, 123
  quadratic, 418, 538

object, 188
octahedral group, 365
octonion, 301
odd permutation, 120
one-one, 588
onto, 588
operation, elementary row, 20
operator
  intertwining, 330
  linear, 42
  projection, 224
opposite category, 190, 208
opposite ring, 546
orbit, 162
order
  finite, 129
  of group, 128
  infinite, 129
ordered pair, 585
ordering
  partial, 595
  simple, 283, 595
  total, 595
  well, 595
ordinary differential equations, system, 244
orthogonal complement, 96
orthogonal group, 121, 259
orthogonal linear map, 102
orthogonal matrix, 102
orthogonal projection, 96
orthogonal set, 92

orthogonal vectors, 92
orthonormal basis, 92
orthonormal set, 92

pair
  ordered, 585
  unordered, 585
parallelogram law, 90
parity-check code, 205
Parseval's equality, 97
partial fractions, 441
partial ordering, 595
pentagon, 496
period of cyclotomic field, 495
permanence of identities, 213
permutation, 15, 120
  even, 120
  odd, 120
Pfaffian, 296, 445
Plancherel formula, 335
Poincaré–Birkhoff–Witt Theorem, 298
point, 585
Poisson summation formula, 359
polar decomposition, 110
polarization, 91
polynomial, 9, 148, 153
  associated primitive, 393
  characteristic, 73, 216
  constant, 10, 149, 154
  cubic, 537
  cyclotomic, 396, 484, 535
  elementary symmetric, 445
  field, 514
  generating, 207, 542
  harmonic, 115
  homogeneous, 115, 154
  Legendre, 113
  minimal, 219, 221, 451
  monic, 149
  primitive, 391
  quartic, 536, 541
  separable, 471
  split, 454
  symmetric, 444, 539
  weight enumerator, 207
  zero, 10, 149
polynomial algebra, 286
polynomial function, 152, 157

Index

polynomial ring, 368
positive, xviii
positive definite linear map, 106
positive definite matrix, 106
positive semidefinite linear map, 106
positive semidefinite matrix, 106
power, 124
presentation, 311
primary block, 230
primary decomposition, 227
Primary Decomposition Theorem, 227
primary subspace, 227
prime, 2, 10
   relatively, 6
prime element, 386
prime factorization, 5
prime field, 147
prime ideal, 381
primitive element, 475
primitive polynomial, 391
   associated, 393
primitive root, 484
Principal Axis Theorem, 251
principal ideal, 387
principal ideal domain, 387, 439
product
   in an algebra, 277
   Cartesian, 585
   in a category, 194
   difference, 506
   dot, 89
   free, 197, 320
   in a group, 117
   indexed Cartesian, 587
   inner, 89
   Kronecker, 294
   of matrices, 26
   n-fold tensor, 277
   of permutations, 15
   set-theoretic, 585
   tensor, 260
   triple tensor, 274
   vector, 278
product functor, 196, 373
projection, 59, 62, 224
   orthogonal, 96
Projection Theorem, 95
proper subset, 585

properly contained, 585
pure tensor, 262
Pythagorean Theorem, 90

quadratic number field, 418, 538
quadratic reciprocity, 494, 539
quartic polynomial, 536, 541
quaternion, 127
quaternion group, 127
quotient group, 131
quotient homomorphism, 131, 145
quotient map, 55
quotient module, 375
quotient ring, 145, 371
quotient space, 55, 129

R homomorphism, 372
R module, 372
R submodule, 374
radical, 247, 250, 254, 490
ramification index, 522, 538
range, 586
rank
   of free abelian group, 176
   of free group, 311
   of free R module, 399
   of matrix, 41
rational canonical form, 243, 443, 444
rational number, 591
real number, 592
real part, 594
reduced row-echelon form, 20
reduced word, 322
reducible element, 386
refinement, 172, 552
reflexive, 590, 595
regular pentagon, 496
regular polygon, 468, 483, 493
regular representation, 329, 334, 335, 362
regular 17-gon, 498
relation, 585
   equivalence, 589
   function as, 585
   partial ordering as, 595
relations, 311
relatively prime, 6
repetition code, 205
representation, 160

contragredient, 362
irreducible, 330
left regular, 329, 335, 362
matrix, 329
right regular, 329, 334, 335
unitary, 329
residue class degree, 522, 538
restriction, 588
of scalars, 274
Riemann sphere, 159
Riesz Representation Theorem, 98
right coset, 128
right ideal, 375
right R module, 372
right radical, 247
right regular representation, 329, 334, 335
rigid motion, 158
ring, 140
commutative, 140
direct product for, 371
division, 143, 370
group, 370
homomorphism of, 143
with identity, 141
local, 430
matrix, 368
Noetherian, 415
opposite, 546
polynomial, 368
quotient of, 145
zero, 141
Rodrigues's formula, 113
root, 10, 151
multiplicity of, 14
primitive, 484
root tower, 490
rotation, 43
rotation group, 121
row-echelon form, 20
row operation, elementary, 20
row reduction, 21
row space, 38
row vector, 25
Russell's paradox, 583

S-tuple, 194
scalar, 9, 19, 34, 88, 209
scalar multiplication

of matrices, 25
in vector space, 34
scalars
extension of, 272, 564, 569
restriction of, 274
Schreier, 173, 345, 553
Schreier set, 316
Schroeder–Bernstein Theorem, 600
Schur orthogonality, 332
Schur's Lemma, 330, 550
Schwarz inequality, 91
Second Isomorphism Theorem, 58, 134, 376
self-adjoint, 100
self-dual linear code, 360
semidirect product of groups, 167, 168
semisimple left R module, 546
separable element, 472
separable extension, 472
separable polynomial, 471
sesquilinear, 89
sesquilinear form, 255
Hermitian, 255
skew-Hermitian, 255
set, 583, 584
set theory
von Neumann, 584
Zermelo–Fraenkel, 583
set-theoretic product, 585
short exact sequence, 576
sign of permutation, 17
signature, 252, 257
significant factor, 318
similar matrices, 48, 211
simple algebraic extension, 453
existence, 453
uniqueness, 454
simple group, 170
simple left R module, 546
simple ordering, 283, 595
simplicial complex, 574
simplicial homology, 574
simply transitive group action, 162
singleton, 585
singular matrix, 215
size, 24
skew-Hermitian form, 255
skew-Hermitian sesquilinear form, 255
skew-symmetric bilinear form, 250

skew-symmetric matrix, 254
socle, 580
solvable group, 488
span, 35, 36
spanning set, 36
special linear group, 121
special unitary group, 121
Spectral Theorem, 104
split exact sequence, 579
split polynomial, 454
splitting field, 454
   existence, 454
   uniqueness, 455
square a circle, 465, 468
square diagram, 193
square matrix, 24
stable subspace, 73
standard basis, 36
standard ordered basis, 48
Steinitz, 461
straightedge and compass, 464
subcategory, 189
subfield, 143
subgroup, 118
   characteristic, 357
   commutator, 310
   index of, 163
   isotropy, 162
   normal, 130
   normalizer of, 186
submodule, 374
   generated, 374–375
   isotypic, 580
subring, 143
subset, 585
subspace, 35
   cyclic, 241
   invariant, 73, 330
   isotropic, 293
   primary, 227
   stable, 73
substitution homomorphism, 150, 155
sum of two cardinal numbers, 602
sum of vector subspaces, 58
superset, 585
support finite, 378
supremum, 593
Sylow p-subgroup, 183

Sylow Theorems, 183
Sylvester's Law, 252, 257
symmetric, 89, 100, 590
   Hermitian, 89
symmetric algebra, 281
symmetric bilinear form, 250
symmetric group, 120
symmetric matrix, 250
symmetric multilinear form, 280
symmetric polynomial, 444, 539
   elementary, 445
symmetrized tensor, 287
symmetrizer, 287
symplectic group, 259
symplectic matrix, 446
system of linear equations, 23
system of ordinary differential equations, 244

Tartaglia, 488
tensor algebra, 279
tensor product, 260
   of abelian groups, 569
   of modules, 565
   n-fold, 277
   of R algebras, 573
   triple, 274
tetrahedral group, 365
Theorem of the Primitive Element, 122, 453, 475, 519
total ordering, 595
trace, 514, 539
   of matrix, 74
transcendental element, 450
transcendental π, 468, 510
transformation
   linear, 42, 44
   linear fractional, 159
   natural, 265
transitive, 590, 595
transitive group action, 162
transpose of matrix, 41
transposition, 16
triangle inequality, 595
triangular form, 216
triple tensor product, 274
trisect an angle, 465, 467
trivial group, 117
trivial group action, 160

tuple, 194
two-sided ideal, 144

UFD1, 386, 416
UFD2, 386
union, 585
   disjoint, 197
   indexed, 587
unique factorization, 5, 14
   of fractional ideal, 447
   of ideal, 435
unique factorization domain, 386
unit, 1, 10
   in a ring, 142
unit vector, 92
unital, 372
unitary group, 121
unitary linear map, 102
unitary matrix, 102
unitary matrix representation, 329
unitary representation, 329
universal enveloping algebra, 298
universal mapping property
   abstract, 198, 295
   of Clifford algebra, 299
   of coproduct in a category, 196
   of direct product of groups, 135, 136
   of direct product of vector spaces, 64
   of direct sum of abelian groups, 137–139
   of direct sum of vector spaces, 60, 64
   of exterior algebra, 289
   of field of fractions, 380
   of free group, 305
   of free R module, 374
   of group algebra, 378
   of integral group ring, 371
   of localization, 428
   of product in a category, 194
   of ring of polynomials, 149, 155–156
   of $S^n(E)$, 282
   of symmetric algebra, 282
   of tensor algebra, 279
   of tensor product of modules, 566
   of tensor product of vector spaces, 260–261
   of universal enveloping algebra, 298
   of $\bigwedge^n(E)$, 289
   of Weyl algebra, 300
unknown, 19

unordered pair, 585
upper bound, 593, 595

Van Kampen Theorem, 320
Vandermonde determinant, 71, 215
Vandermonde matrix, 71, 215
variable, 19
   corner, 21
   independent, 21
vector, 34
   addition for, 34
   column, 25
   cyclic, 242
   row, 25
   scalar multiplication for, 34
   unit, 92
vector product, 278
vector space, 34, 157
   associated graded, 297
   basis of, 36
   complex conjugate of, 114
   dimension of, 37, 78
   direct product for, 62, 63
   direct sum for, 59, 60, 61, 62, 64
   dual of, 50
   filtered, 297
   finite-dimensional, 37
   graded, 297
   infinite-dimensional, 78
   left, 547
   quotient of, 55
vector subspace, 35
   invariant, 216
vector subspaces, sum of, 58
von Neumann set theory, 584

weight, 204
weight enumerator polynomial, 207
well ordering, 595
Wentzel, 468
Weyl algebra, 299
Weyl basis, 293
Wilson's Theorem, 199, 534
word, 304
word problem, 307
   for finitely presented groups, 313
   for free groups, 307
   for free products, 322

Zassenhaus, 172, 552
Zermelo's Well-Ordering Theorem, 462, 599
Zermelo–Fraenkel set theory, 583
zero divisor, 143
zero matrix, 25
zero polynomial, 10, 149
zero ring, 141
Zorn's Lemma, 78, 382, 461, 595